# Elementary Linear Algebra

Fifth Edition

Stanley I. Grossman
University of Montana and University College London

**Harcourt College Publishers**
Fort Worth Philadelphia San Diego New York Orlando Austin
San Antonio Toronto Montreal London Sydney Tokyo

Text Typeface: Times Roman
Compositor: York Graphic Services, Inc.
Acquisitions Editor: Jay Ricci
Developmental Editor: Marc Sherman
Managing Editor: Carol Field
Production Management: York Production Services
Manager of Art and Design: Carol Bleistine
Senior Art Director: Christine Schueler
Text Designer: York Production Services
Cover Designer: Louis Fuiano/Fuiano Art & Design
Director of EDP: Tim Frelick
Production Manager: Carol Florence
Marketing Manager: Monica Wilson
Cover art: Electronic Illustration by Louis Fuiano/Fuiano Art & Design



Printed in the United States of America

ELEMENTARY LINEAR ALGEBRA, 5th ed.

0-03-097354-6

Library of Congress Catalog Card Number: 93-40951

0 1 2 3 4 5 6 7 8 9 039 15 14 13 12 11 10 9

This book was printed on paper made from waste paper, containing 10% post-consumer waste and 40% pre-consumer waste, measured as a percentage of total fiber weight content.

*To Kerstin, Aaron, and Erik*

# Preface

As recently as thirty years ago, the study of linear algebra was largely confined to mathematics and physics majors and to those who needed a knowledge of matrix theory to work in technical areas such as multivariate statistics. Linear algebra is now studied by students in many disciplines due to the invention of high-speed computers and a general increase in the application of mathematics in traditionally nontechnical areas.

## Prerequisites

In writing this text I have had two goals in mind. I have tried to make a large number of linear algebra topics accessible to a wide variety of students who need only a good knowledge of high school algebra. Because many students will have had a year of calculus, I have also included several examples and exercises involving topics from calculus. These are indicated by the symbol [Calculus]. One optional section (Section 6.7) requires calculus, but otherwise *calculus is not a prerequisite.*

## Applications

My second goal was to convince students of the importance of linear algebra in their fields of study. Thus, examples and exercises are drawn from a variety of disciplines. Some of these examples are short, like the application of matrix multiplication to the spread of a contagious disease (page 67). Others are quite a bit longer. These include the Leontief Input-Output Model (pages 18–20 and 108–111), Graph Theory (Section 1.12), Least Squares Approximation (Section 4.10), and a Model of Population Growth (Section 6.2). The booklet *Applications Supplement to Elementary Linear Algebra* contains interesting applications of linear algebra to linear programming, Markov chains, and game theory.

In addition, a large number of relevant, interesting applications can be found in the MATLAB® sections, which are new to this edition. A complete description of the use of MATLAB in this book can be found on pages viii and xviii.

## Theory

For many students linear algebra is their first real *mathematics* course. In this course students are required not only to carry out mathematical computations but also to create proofs. In this book I have tried to strike a balance between technique and theory. All the important techniques are described in great detail with many examples illustrating their use. At the same time, all the theorems that can be proved using the results given in the text are proved in the book. The more difficult proofs are placed at the ends of sections or in separate sections, but they *are* given. The result is a book that will give students both the algebraic skills necessary to solve linear algebra problems that arise in their areas of study and a greater appreciation for the beauty of mathematics.

## FEATURES

Many of the features of the fourth edition have been retained in the fifth edition and are described below. New features are listed on page viii.

### Examples

Students learn mathematics by seeing clearly worked out examples. The fifth edition contains 342 examples and there are 80 more in the *Applications Supplement.* Each example includes all the algebraic steps needed to complete the solution. In many instances explanations are highlighted in color in order to make the steps easier to follow. In addition, the examples are titled so students can grasp more easily the essential concept each example illustrates.

### Exercises

The text contains approximately 2400 exercises (with an additional 455 in the *Applications Supplement*). As in all mathematics books, these are the most important learning tool in the text. Problems are graded in order of increasing difficulty, and there is a balance between technique and proof. The more difficult problems are marked with * and a few exceptionally difficult ones with **. These are supplemented by review exercises at the end of each chapter. Answers to almost *all* odd-numbered problems, including those requiring proofs, are given at the back of the book. Of the 2400 exercises, about 300 are in new, special parts of the problem sections that are labelled "Calculator Box" and "MATLAB." We will say more about these features later.

### The Summing Up Theorem

A major feature of the text is the frequent appearance of the Summing Up Theorem, which ties together seemingly disparate topics in the study of matrices and linear transformations. The theorem is first encountered in Section 1.2 (p. 5). Successively more complete versions of this theorem are found in Section 1.8 (p. 111), Section 1.10 (p. 132), Section 2.4 (p. 216), Section 4.5 (p. 325), Section 4.7 (p. 360), Section 5.4 (p. 515), and Section 6.1 (p. 546).

## Self-Quiz Problems

Almost every problem set begins with multiple-choice and true-false questions that require little or no computation. Answers to these problems appear at the bottom of the last page on which the problems occur (not at the back of the book). These problems are designed to test whether the student understands the basic ideas in the section, and they should be done before tackling the more standard problems that follow. The Self-Quiz Problems are new to this edition.

## Calculator Boxes

Many linear algebra students have access to graphing calculators which perform vector and matrix operations. Consequently, at the end of a number of problem sets—especially in the earlier, more computational parts of the book—I have added "Calculator Boxes" to help students use their calculators in this course.

Each box begins with a detailed description of how to use the TI-85 and, where possible, the Casio fx-7700 GB to solve problems in the preceding section. These descriptions are usually followed by additional problems with "messier" numbers that can be solved easily on a calculator. This feature will be useful to students who own an appropriate calculator. However, it should be stressed that *ownership of a graphing calculator is not required in order to use this book effectively.* The Calculator Boxes are an *optional* feature to be used as the instructor sees fit.

## Chapter Summaries

At the end of each chapter, a detailed review of the important results of that chapter appears. Page references are included.

## Geometry

Some of the important ideas in linear algebra are better understood by observing how they can be interpreted geometrically. I have pointed out geometric interpretations of important concepts in several places in this edition. These include

- The geometry of a system of three equations in three unknowns (p. 20)
- The geometric interpretation of the $2 \times 2$ determinant (pp. 180, 265)
- The geometric interpretation of the scalar triple product (p. 266)
- How to draw a plane (p. 277)
- The geometric interpretation of linear dependence in $\mathbb{R}^3$ (p. 321)
- The geometry of linear transformations from $\mathbb{R}^2$ to $\mathbb{R}^2$ (pp. 495–503)
- Isometries of $\mathbb{R}^2$ (p. 521).

## Historical Emphasis

Mathematics becomes more interesting if one knows something about the historical development of the subject. To stimulate this interest I have included a number of small historical notes scattered throughout the text. In addition, I have written seven "focuses" that are longer and somewhat more detailed. They include:

- Carl Friedrich Gauss (p. 23)
- Sir William Rowan Hamilton (p. 54)
- Arthur Cayley and the Algebra of Matrices (p. 76)
- A Short History of Determinants (p. 209)
- Joseph Willard Gibbs and the Origins of Vector Analysis (p. 268)
- History of Mathematical Induction (p. A-6)
- George Dantzig and the History (and Future) of Linear Programming (p. 69 in the *Applications Supplement*)

## Features New to the Fifth Edition

This edition, like the ones that preceded it, contains hundreds of small changes that were suggested by users and reviewers of the last edition. The major additions and deletions in this fifth edition are:

- Self-Quiz Problems at the beginning of most problem sets
- Calculator boxes
- MATLAB tutorials and problems (as described below)
- A discussion of matrix multiplication as a linear combination of the columns of a matrix, and block matrix multiplication (in Section 1.6)
- A section on LU-factorizations of a matrix (Section 1.11)
- Two of the sections on numerical methods (iterative methods for solving a system of equations and for computing eigenvalues and eigenvectors) do not appear in this edition, because we felt that students solving problems numerically are much more likely to use software packages like MATLAB.

## MATLAB

A major addition to this text is the incorporation of more than 230 optional MATLAB problems, most with multiple parts, that follow most of the regular end-of-section problem sets. (MATLAB® is a registered trademark of The MathWorks, Inc.) MATLAB is a powerful, yet user-friendly software designed to handle wide-ranging problems involving matrix computations and linear algebra concepts. The problems, directly tied to examples and regular problems in the core text, will encourage the student to exploit the computational power of MATLAB and to explore linear algebra principles through conjecture and analysis.

The MATLAB problems appear first at the end of Section 1.3, where they are preceded by an introduction and short tutorial. The MATLAB problems in each section are designed to introduce the user to MATLAB commands as required by the problem set. There are numerous applications and project problems that demonstrate the real-world relevance of linear algebra; these could serve as group or short-paper assignments.

Many of the MATLAB problems are designed to encourage students to discover theorems in linear algebra. For example, a student who has generated several upper triangular matrices and computed their inverses will naturally conjecture that the inverse of an upper triangular matrix is upper triangular. The standard proof of this result is not at all trivial, but it will make more sense if the student "sees" that the result is plausible. Virtually every MATLAB problem set contains problems that lead to mathematical discoveries.

The problems have helpful hints that allow students to approach them without getting bogged down in the mechanics of MATLAB usage. Additionally, there are various "paper and pencil" parts that permit the student to exercise judgment and demonstrate the learning of concepts. The problems are carefully developed and located within the text to permit a gradual mastery of the software.

**Elementary Linear Algebra Toolbox (M-file disk)**

There are 10 "M-files," or short MATLAB programs, that are related to numerous MATLAB problems. The M-files are designated by M in the margin. A description of the use of the M-files is given in the appropriate problems. For example, see **lincomb.m** for visualizing linear combinations to understand their geometry (Section 3.1, MATLAB Problem 3 on p. 240), or **grafics.m** for computer graphics using matrices (Section 5.1, MATLAB Problem 1 on p. 474). The M-files are obtainable in disk form (either PC or Mac), *free of charge,* from The MathWorks, by mailing the card found in this text. (The MathWorks address can be found in Appendix 5 if the card is missing.)

Appendix 5 contains more information on MATLAB, including use of the M-file disk, availability of a MATLAB Primer, obtaining a record of work on results, graphics considerations, version 4.0 considerations, and special variable names.

As with the calculator boxes, we stress that the MATLAB material is *optional.* It can be assigned or not assigned as the instructor sees fit.

## A Note about Length

The fifth edition is about 130 pages longer than the fourth, despite the fact that it contains two fewer sections. The increased length is due primarily to the optional MATLAB material described above. We chose to place it in the text—rather than in a separate supplement—so that it could be integrated better and thereby be more effective.

Very little textual material was added (and some was deleted), so *Elementary Linear Algebra* is still designed to be covered in *one* semester. We expect that, when it is used, the MATLAB material will be covered in separate labs that supplement classroom work.

### Numbering

Numbering in this book is fairly standard. Within each section, examples, problems, theorems, and equations are numbered consecutively starting with 1. Reference to an example, problem, theorem, or equation outside the section is by chapter, section, and number. Thus, Example 4 in Section 2.5 is called Example 4 in that section but outside the section it is referred to as Example 2.5.4. In addition, page numbers are frequently provided to make referenced items easier to find.

## ORGANIZATION

The approach I have used in this text is gradual. Chapters 1 and 2 contain the basic computational material common to most elementary linear algebra texts. Chapter 1 discusses systems of linear equations, vectors, and matrices. It has been reorganized to cover all systems of equations material before introducing matrices. This presentation provides more motivation for the student and follows a more common course order. There is also a section (1.12) applying matrices to graph theory. Chapter 2 provides an introduction to determinants and includes a historical essay on the contributions to linear algebra of Leibniz and Cauchy (Section 2.3).

Even in this early material there are optional sections to challenge the student a bit more. For example, Section 2.3 provides a complete proof that $\det AB = \det A \det B$. The proof of this central result, using elementary matrices, is not often included in introductory texts.

Chapter 3 discusses vectors in the plane and in space. Many of the topics in this chapter are covered in a calculus sequence and may be familiar to students already. However, since much of linear algebra is concerned with a discussion of abstract vector spaces, students need a storehouse of concrete examples, which are most easily provided by the study of vectors in the plane and in space. The more difficult and abstract material in Chapters 4 and 5 is illustrated with examples from the concrete Chapter 3. Section 3.4 includes a historical essay on Gibbs and the origins of vector analysis.

Chapter 4 contains an introduction to general vector spaces and is, necessarily, more abstract than the earlier chapters. I have tried, however, to present the material as a natural extension of properties of vectors in the plane, which is really how the subject evolved. The fifth edition discusses linear combination and span (Section 4.4) before linear independence (Section 4.5) to motivate these topics more clearly. Chapter 4 also includes an interesting applications section (4.10) on least squares approximation.

At the end of Chapter 4 I have added an optional section (4.12) in which I prove that every vector space has a basis. In doing so I discuss partially ordered sets and Zorn's lemma. The material here is more difficult than anything else in the book and can easily be omitted. However, since linear algebra is often considered the first course in which proofs are as important as computations, I feel strongly that a proof of this fundamental result should be available to the more motivated student.

Chapter 5 continues the discussion begun in Chapter 4 with an introduction to linear transformations from one vector space to another. The chapter begins with two examples showing how such transformations can arise in a natural way. In Section 5.3 I have included a detailed description of the geometry of transformations from $\mathbb{R}^2$ to $\mathbb{R}^2$, including expansions, compressions, reflections, and shears. And Section 5.5 now includes a more detailed discussion of isometries of $\mathbb{R}^2$.

Chapter 6 describes the theory of eigenvalues and eigenvectors. These are introduced in Section 6.1, and a detailed biological application to population growth is given in Section 6.2. Sections 6.3, 6.4, and 6.5 all involve the diagonalization of a matrix, while Section 6.6 illustrates, for a few cases, how a matrix can be reduced to its Jordan canonical form. Section 6.7 discusses matrix differential equations and is the only section of the book that requires a knowledge of freshman calculus. This section provides an illustration of the usefulness of reducing a matrix to its Jordan canonical form (which is usually a diagonal matrix). In Section 6.8, I introduce two of my favorite results from matrix theory: the Cayley-Hamilton theorem and the Gershgorin circle theorem. The latter result, rarely discussed in an elementary linear algebra text, provides an easy way to estimate the eigenvalues of any matrix.

In Chapter 6, I had to make a difficult decision: whether or not to discuss complex eigenvalues and eigenvectors. I decided to include them because it seemed to be the only honest thing to do. Some of the "nicest" matrices have complex eigenvalues. To define an eigenvalue as a real number only may at first make things seem simpler, but it is certainly wrong. Moreover, for many applications involving eigenvalues (including some in Section 6.7), the most interesting models involve periodic phenomena, and these require complex eigenvalues. Complex numbers are not avoided in this book. For students who have not encountered them before, the few properties they need are discussed fully in Appendix 2.

The book has five appendices, the first on mathematical induction and the second on complex numbers. Some of the proofs in the book use mathematical induction, so Appendix 1 provides a brief introduction to this important technique for students who have not used it before.

Appendix 3 discusses the important notion of computational complexity, which, among other things, will help students understand why the writers of software packages choose certain algorithms over others. Appendix 4 presents one reasonably efficient method for solving systems of equations numerically. Finally, Appendix 5 includes a few technical details about the use of MATLAB in this book.

A word on chapter interdependence: The book is written sequentially. Each chapter depends on the ones that precede it, with one exception: Chapter 6 can be covered without most of the material in Chapter 5. Sections marked "Optional" can be omitted without any loss of continuity.

## SUPPLEMENTS

*Elementary Linear Algebra, fifth edition,* offers a powerful and diverse supporting package to suit all types of instruction and student backgrounds.

The **Applications Supplement** contains two applications chapters on linear programming, Markov chains and game theory with numerous examples and exercises.

A **Student Solutions Manual,** prepared by David L. Ragozin at the University of Washington, with the assistance of Andy Demetre and Fred Gylys-Colwell, contains complete detailed solutions to all odd-numbered problems and review exercises, including the graphing calculator and MATLAB problems.

An **Instructor's Solutions Manual,** also prepared by David L. Ragozin and colleagues, provides complete solutions for all the problems and review exercises in the text and is available free to adopters.

The **Elementary Linear Algebra Toolbox** is a disk of M-files related to some of the text's MATLAB Problems. The disk is available free by sending in the card found in the text or by contacting The MathWorks, Inc. (see Appendix 5).

Also available are the following manuals that can be used with any linear algebra text for lab activity:

**HP-48G/GX Calculator Enhancement for Science and Engineering Mathematics,** edited by Don LaTorre of Clemson University, addresses the use of high-level HP calculators in undergraduate mathematics. Topics include single-variable calculus; multivariable calculus, differential equations, linear algebra, advanced engineering mathematics; and probability and statistics.

**MATLAB Computer Lab Manual** by Karen Donnelly of Saint Joseph's College, Rennselaer, Indiana, contains computer lab exercises for using MATLAB. Each section lists objectives, prerequisites, and MATLAB features before the lab exercise is presented. The student is then encouraged to apply concepts interactively and create an edited diary session. Additional exercises follow the lab exercises that vary in length and difficulty, which the instructor can assign as desired.

## ACKNOWLEDGMENTS

I am grateful to many people who helped as this book was written. Some of the material here first appeared in *Mathematics for the Biological Sciences* (New York: Macmillan, 1974) written by James E. Turner and myself. I am grateful to Professor Turner for permission to use this material.

A great deal of this book was written while I was a research associate at University College London. I wish to thank the Mathematics Department of UCL for providing office facilities, mathematical suggestions, and, especially, friendship during my annual visits there.

The MATLAB material was written by Cecelia Laurie at the University of Alabama. I am very grateful to Professor Laurie for the truly outstanding way she used computer software to enhance the teaching process. This is a better book because of her efforts.

I'd also like to thank Cristina Palumbo of The MathWorks, Inc., for providing us with the most recent information on MATLAB.

The effectiveness of a mathematics textbook depends to a certain extent on the accuracy of the answers. There is nothing more frustrating to a student than to work to obtain the answers at the back of the book, only to discover that the given answer is incorrect. Four things were done to ensure that this doesn't happen. First, I solved or re-solved every odd-numbered problem and prepared an answer key. Sudhir Goel at Valdosta State University solved every new problem and checked my answers to the old ones. Cecelia Laurie prepared answers to MATLAB problems. Finally, David Ragozin at the University of Washington prepared the Student Solutions Manual, and in doing so, solved every odd-numbered problem in the book. Professors Goel and Ragozin then sent me the errors they found and these were corrected during typesetting. I am grateful to both of them for ensuring that the answers at the back of the book are as error free as humanly possible.

I'd like to thank those individuals who commented on the fourth edition and read the manuscript for this fifth edition. Without the comments and criticisms of those who actually teach the course, a linear algebra textbook author can easily lose touch with his audience. The reviewers for this fifth edition are:

Ken Armstrong, University of Manitoba
Thomas Cairns, University of Tulsa
Pat Collier, University of Wisconsin–Oshkosh
Roger Horn, University of Utah
Irving Katz, George Washington University
Gerald Leibowitz, University of Connecticut
Maurice Monahan, South Dakota State University
John Ratcliffe, Vanderbilt University
W. Vance Underhill, East Texas State University
Marvin Zeman, Southern Illinois University–Carbondale

I am grateful to the following experienced MATLAB users for reviewing the MATLAB problems:

Thomas Cairns, University of Tulsa
Karen Donnelly, Saint Joseph's College
Roger Horn, University of Utah
Irving Katz, George Washington University
Gary Klatt, University of Wisconsin–Whitewater

Special thanks go to the editorial and production staffs at Saunders College Publishing for the care and skill they brought to this product. My editor, Jay Ricci, made many helpful suggestions. In particular, it was he who insisted that MATLAB be incorporated into this book. Thanks, Jay. I very much appreciate the tremendous help given me from inception to final production by my outstanding developmental editor, Marc Sherman. Marc handled every detail with great care and tremendous competence. Finally, I am grateful to Kirsten Kauffman at York Production Services for her great skill and unflagging cheerfulness while helping to turn my manuscript into a bound book.

**Stanley I. Grossman**
Missoula, Montana
December 1993

# Contents

The *Applications Supplement* to accompany *Elementary Linear Algebra,* Fifth Edition, contains the following material:

# MATLAB Problems Contents

MATLAB Problem Sets and topics of special note are listed here.

## 1 Systems of Linear Equations and Matrices



## 2 Determinants

1

# Systems of Linear Equations and Matrices

## 1.1 INTRODUCTION

This is a book about linear algebra. If you look up the word "linear" in a dictionary, you will find something like the following: lin-e-ar (lin′ ē ər), adj. 1. of, consisting of, or using lines.† In mathematics, the word "linear" means a good deal more than that. Nevertheless, much of the theory of elementary linear algebra is in fact a generalization of properties of straight lines. As a review, here are some fundamental facts about straight lines:

**i.** The **slope** $m$ of a line passing through the points $(x_1, y_1)$ and $(x_2, y_2)$ is given by

$$m = \frac{y_2 - y_1}{x_2 - x_1} = \frac{\Delta y}{\Delta x} \qquad \text{if } x_1 \neq x_2$$

**ii.** If $x_2 - x_1 = 0$ and $y_2 \neq y_1$, then the line is vertical and the slope is said to be **undefined.**‡

**iii.** Any line (except one with undefined slope) can be described by writing its equation in the slope-intercept form $y = mx + b$, where $m$ is the slope of the line and $b$ is the $y$-intercept of the line (the value of $y$ at the point where the line crosses the $y$-axis).

**iv.** Two distinct lines are parallel if and only if they have the same slope.

**v.** If the equation of a line is written in the form $ax + by = c$ $(b \neq 0)$, then, as is easily computed, $m = -a/b$.

† Taken from the pocket edition of *The Random House Dictionary.*

‡ In some textbooks a vertical line is said to have "an infinite slope."

**vi.** If $m_1$ is the slope of line $L_1$, $m_2$ is the slope of line $L_2$, $m_1 \neq 0$, and $L_1$ and $L_2$ are perpendicular, then $m_2 = -1/m_1$.

**vii.** Lines parallel to the $x$-axis have a slope of zero.

**viii.** Lines parallel to the $y$-axis have an undefined slope.

In the next section we shall illustrate the relationship between solving systems of equations and finding points of intersection of pairs of straight lines.

## 1.2 TWO LINEAR EQUATIONS IN TWO UNKNOWNS

Consider the following system of two linear equations in the two unknowns $x$ and $y$:

$$\begin{aligned} a_{11}x + a_{12}y &= b_1 \\ a_{21}x + a_{22}y &= b_2 \end{aligned} \qquad (1)$$

where $a_{11}, a_{12}, a_{21}, a_{22}, b_1$, and $b_2$ are given numbers. Each of these equations is the equation of a straight line. A **solution** to system (1) is a pair of numbers, denoted by $(x, y)$, that satisfies (1). The questions that naturally arise are whether (1) has any solutions and if so, how many? We answer these questions after looking at some examples. In these examples we make use of two important facts from elementary algebra:

**Fact A** If $a = b$ and $c = d$, then $a + c = b + d$.

**Fact B** If $a = b$ and $c$ is any real number, then $ca = cb$.

Fact A states that if we add two equations together, we obtain a third, correct equation. Fact B states that if we multiply both sides of an equation by a constant, we obtain a second, valid equation. We shall assume that $c \neq 0$ for although the equation $0 = 0$ is correct, it is not very useful.

EXAMPLE 1 **A System with a Unique Solution** Consider the system

$$\begin{aligned} x - y &= 7 \\ x + y &= 5 \end{aligned} \qquad (2)$$

Adding the two equations together gives us, by Fact A, the following equation: $2x = 12$ (or $x = 6$). Then, from the second equation, $y = 5 - x = 5 - 6 = -1$. Thus the pair $(6, -1)$ satisfies system (2) and the way we found the solution shows that it is the only pair of numbers to do so. That is, system (2) has a **unique solution.**

EXAMPLE 2 **A System with an Infinite Number of Solutions** Consider the system

$$\begin{aligned} x - y &= 7 \\ 2x - 2y &= 14 \end{aligned} \qquad (3)$$

It is apparent that these two equations are equivalent. That is, any two numbers, $x$ and $y$, that satisfy the first equation also satisfy the second equation, and vice versa. To see this multiply the first by 2. This is permitted by Fact B. Then $x - y = 7$ or $y = x - 7$. Thus the pair $(x, x - 7)$ is a solution to system (3) for any real number $x$. That is, system (3) has an **infinite number of solutions.** For example, the following pairs are solutions: $(7, 0)$, $(0, -7)$, $(8, 1)$, $(1, -6)$, $(3, -4)$, and $(-2, -9)$.

EXAMPLE 3 **A System with No Solution** Consider the system

$$\begin{aligned} x - \phantom{2}y &= 7 \\ 2x - 2y &= 13 \end{aligned} \qquad \textbf{(4)}$$

Multiplying the first equation by 2 (which, again, is permitted by Fact B) gives us $2x - 2y = 14$. This contradicts the second equation. Thus system (4) has **no solution.**

A system with no solution is said to be **inconsistent.**

It is easy to explain, geometrically, what is going on in the preceding examples. First, we repeat that the equations in system (1) are both equations of straight lines. A solution to (1) is a point $(x, y)$ that lies on both lines. If the two lines are not parallel, then they intersect at a single point. If they are parallel, then they either never intersect (no points in common) or are the same line (infinite number of points in common). In Example 1 the lines have slopes of 1 and $-1$, respectively. Thus they are not parallel. They have the single point $(6, -1)$ in common. In Example 2 the lines are parallel (slope of 1) and coincident. In Example 3 the lines are parallel and distinct. These relationships are all illustrated in Figure 1.1.

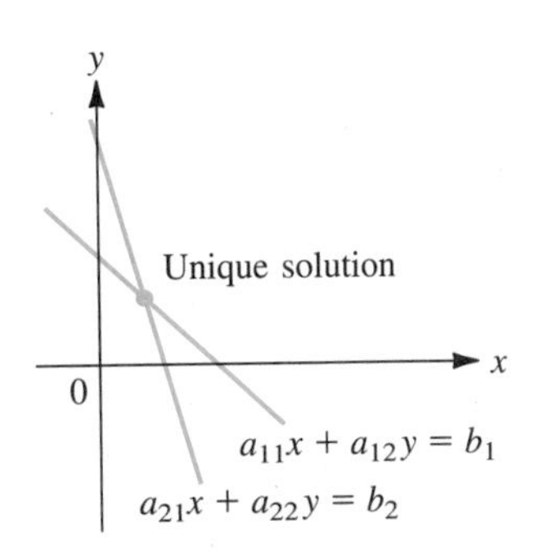


(a) Lines not parallel; one point of intersection

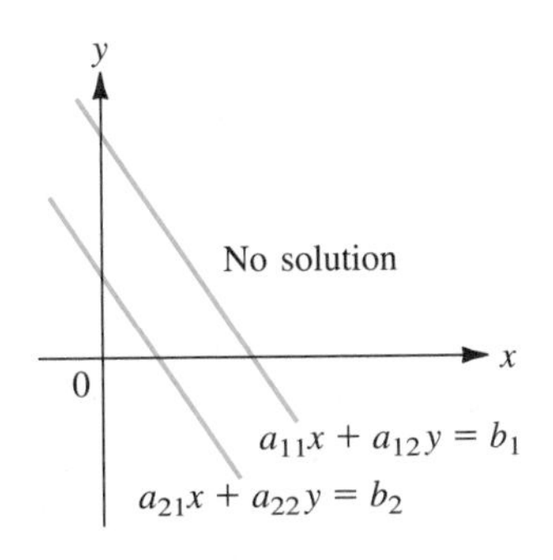


(b) Lines parallel; no points of intersection

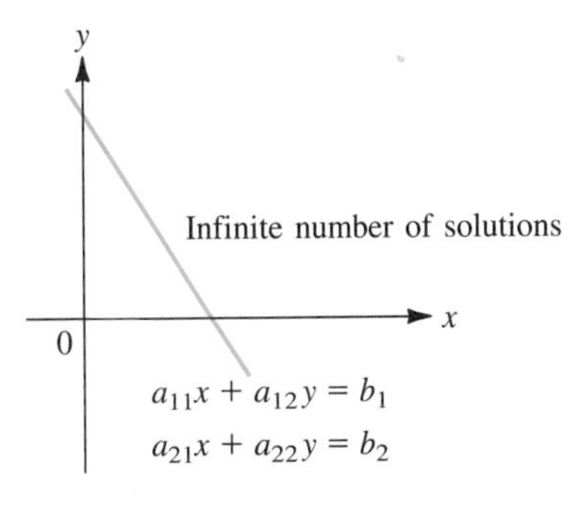


(c) Lines coincide; infinite number of points of intersection

**Figure 1.1** Two lines intersect at one point, no points, or (if they coincide) an infinite number of points

Let us now solve system (1) formally. We have

$$\begin{aligned} a_{11}x + a_{12}y &= b_1 \\ a_{21}x + a_{22}y &= b_2 \end{aligned} \qquad \textbf{(1)}$$

If $a_{12} = 0$, then $x = \dfrac{b_1}{a_{11}}$ and we can use the second equation to solve for $y$.

If $a_{22} = 0$, then $x = \dfrac{b_2}{a_{21}}$ and we can use the first equation to solve for $y$.

If $a_{12} = a_{22} = 0$, then system (1) contains only one unknown, $x$.

Thus we may assume that neither $a_{12}$ nor $a_{22}$ is zero.

Multiplying the first equation by $a_{22}$ and the second by $a_{12}$ yields

$$\begin{aligned} a_{11}a_{22}x + a_{12}a_{22}y &= a_{22}b_1 \\ a_{12}a_{21}x + a_{12}a_{22}y &= a_{12}b_2 \end{aligned} \qquad \textbf{(5)}$$

Equivalent systems

Before continuing we note that system (1) and system (5) are **equivalent.** By that we mean that any solution to system (1) is a solution to system (5) and vice versa. This follows immediately from Fact B assuming that the $c$ in Fact B is not zero. Next we subtract the second equation from the first to obtain

$$(a_{11}a_{22} - a_{12}a_{21})x = a_{22}b_1 - a_{12}b_2 \qquad \textbf{(6)}$$

At this point we must pause. If $a_{11}a_{22} - a_{12}a_{21} \neq 0$, then we can divide by it to obtain

$$x = \frac{a_{22}b_1 - a_{12}b_2}{a_{11}a_{22} - a_{12}a_{21}}$$

Then we can substitute this value of $x$ into system (1) to solve for $y$, and thus we have found the unique solution to the system.

We have shown the following:

> If $a_{11}a_{22} - a_{12}a_{21} \neq 0$, then
> the system (1) has a unique solution.

How does this statement relate to what we discussed earlier? In system (1) we see that the slope of the first line is $-a_{11}/a_{12}$ and the slope of the second is $-a_{21}/a_{22}$. In Problems 31, 32, and 33 you are asked to show that $a_{11}a_{22} - a_{12}a_{21} = 0$ if and only if the lines are parallel (have the same slope). So, if $a_{11}a_{22} - a_{12}a_{21} \neq 0$, the lines are not parallel and the system has a unique solution.

We now put the facts discussed above together in a theorem. This theorem will be generalized in later sections of this chapter and in subsequent chapters. We shall keep track of our progress by referring to the theorem as our "Summing Up Theorem." When all its parts have been proved, we shall see a remarkable relationship among several important concepts in linear algebra.

**THEOREM 1** **Summing Up Theorem—View 1** The system

$$a_{11}x + a_{12}y = b_1$$
$$a_{21}x + a_{22}y = b_2$$

of two equations in the two unknowns $x$ and $y$ has no solution, a unique solution, or an infinite number of solutions. It has:

**i.** A unique solution if and only if $a_{11}a_{22} - a_{12}a_{21} \neq 0$.

**ii.** No solution or an infinite number of solutions if and only if

$$a_{11}a_{22} - a_{12}a_{21} = 0.$$

In Section 1.3 we shall discuss systems of $m$ equations in $n$ unknowns and shall see that there is always either no solution, one solution, or an infinite number of solutions.

## PROBLEMS 1.2

### Self-Quiz

**I.** Which of the following is *not* true about the solution for a system of two linear equations in two unknowns?
- **a.** It is an ordered pair that satisfies both equations.
- **b.** Its graph consists of the point(s) of intersection of the graphs of the equations.
- **c.** Its graph is the $x$-intercept of the graphs of the equations.
- **d.** If the system is inconsistent, there is no solution.

**II.** Which of the following is true of an inconsistent system of two linear equations?
- **a.** There is no solution.
- **b.** The graph of the system is on the $y$-axis.
- **c.** The graph of the solution is one line.
- **d.** The graph of the solution is the point of intersection of two lines.

**III.** Which of the following is true of the system of equations below?

$$3x - 2y = 8$$
$$4x + y = 7$$

- **a.** The system is inconsistent.
- **b.** The solution is $(-1, 2)$.
- **c.** The solution is on the line $x = 2$.
- **d.** The equations are equivalent.

**IV.** Which of the following is a second equation for the system whose first equation is $x - 2y = -5$ if there are to be an infinite number of solutions for the system?
- **a.** $6y = 3x + 15$
- **b.** $6x - 3y = -15$
- **c.** $y = -\frac{1}{2}x + \frac{5}{2}$
- **d.** $\frac{3}{2}x = 3y + \frac{15}{2}$

**V.** The graph of which of the following systems is a pair of parallel lines?

**a.** $3x - 2y = 7$
$4y = 6x - 14$

**b.** $x - 2y = 7$
$3x = 4 + 6y$

**c.** $2x + 3y = 7$
$3x - 2y = 6$

**d.** $5x + y = 1$
$7y = 3x$

In Problems 1–12 find all solutions (if any) to the given systems. In each case calculate the value $a_{11}a_{22} - a_{12}a_{21}$.

**1.** $x - 3y = 4$
$-4x + 2y = 6$

**2.** $2x - y = -3$
$5x + 7y = 4$

**3.** $2x - 8y = 5$
$-3x + 12y = 8$

**4.** $2x - 8y = 6$
$-3x + 12y = -9$

**5.** $6x + y = 3$
$-4x - y = 8$

**6.** $3x + y = 0$
$2x - 3y = 0$

**7.** $4x - 6y = 0$
$-2x + 3y = 0$

**8.** $5x + 2y = 3$
$2x + 5y = 3$

**9.** $2x + 3y = 4$
$3x + 4y = 5$

**10.** $ax + by = c$
$ax - by = c$

**11.** $ax + by = c$
$bx + ay = c$

**12.** $ax - by = c$
$bx + ay = d$

**13.** Find conditions on $a$ and $b$ such that the system in Problem 10 has a unique solution.

**14.** Find conditions on $a$, $b$, and $c$ such that the system in Problem 11 has an infinite number of solutions.

**15.** Find conditions on $a$, $b$, $c$, and $d$ such that the system in Problem 12 has no solutions.

In Problems 16–21 find the point of intersection (if there is one) of the two lines.

**16.** $x - y = 7$; $2x + 3y = 1$

**17.** $y - 2x = 4$; $4x - 2y = 6$

**18.** $4x - 6y = 7$; $6x - 9y = 12$

**19.** $4x - 6y = 10$; $6x - 9y = 15$

**20.** $3x + y = 4$; $y - 5x = 2$

**21.** $3x + 4y = 5$; $6x - 7y = 8$

Let $L$ be a line and let $L_{\perp}$ denote the line perpendicular to $L$ that passes through a given point $P$. The **distance** from $L$ to $P$ is defined to be the distance† between $P$ and the point of intersection of $L$ and $L_{\perp}$. In Problems 22–27 find the distance between the given line and point.

**22.** $x - y = 6$; $(0, 0)$

**23.** $2x + 3y = -1$; $(0, 0)$

**24.** $3x + y = 7$; $(1, 2)$

**25.** $5x - 6y = 3$; $(2, \frac{16}{5})$

**26.** $2y - 5x = -2$; $(5, -3)$

**27.** $6y + 3x = 3$; $(8, -1)$

**28.** Find the distance between the line $2x - y = 6$ and the point of intersection of the lines $2x - 3y = 1$ and $3x + 6y = 12$.

---

## Answers to Self-Quiz

**I.** c **II.** a **III.** c **IV.** a **V.** b

---

†Recall that if $(x_1, y_1)$ and $(x_2, y_2)$ are two points in the $xy$-plane, then the distance $d$ between them is given by $d = \sqrt{(x_1 - x_2)^2 + (y_1 - y_2)^2}$.

***29.** Prove that the distance between the point $(x_1, y_1)$ and the line $ax + by = c$ is given by

$$d = \frac{|ax_1 + by_1 - c|}{\sqrt{a^2 + b^2}}$$

**30.** A zoo keeps birds (two-legged) and beasts (four-legged). If the zoo contains 60 heads and 200 feet, how many birds and how many beasts live there?

**31.** Suppose that $a_{11}a_{22} - a_{12}a_{21} = 0$. Show that the lines given in system (1) are parallel. Assume that $a_{11} \neq 0$ or $a_{12} \neq 0$ and $a_{21} \neq 0$ or $a_{22} \neq 0$.

**32.** If there is a unique solution to system (1), show that $a_{11}a_{22} - a_{12}a_{21} \neq 0$.

**33.** If $a_{11}a_{22} - a_{12}a_{21} \neq 0$, show that system (1) has a unique solution.

**34.** The Sunrise Porcelain Company manufactures ceramic cups and saucers. For each cup or saucer a worker measures a fixed amount of material and puts it into a forming machine, from which it is automatically glazed and dried. On the average, a worker needs 3 minutes to get the process started for a cup and 2 minutes for a saucer. The material for a cup costs 25¢ and the material for a saucer costs 20¢. If $44 is allocated daily for production of cups and saucers, how many of each can be manufactured in an 8-hour workday if a worker is working every minute and exactly $44 is spent on materials?

**35.** Answer the question of Problem 34 if the materials for a cup and saucer cost 15¢ and 10¢, respectively, and $24 is spent in an 8-hour day.

**36.** Answer the question of Problem 35 if $25 is spent in an 8-hour day.

**37.** An ice-cream shop sells only ice-cream sodas and milk shakes. It puts 1 ounce of syrup and 4 ounces of ice cream in an ice-cream soda, and 1 ounce of syrup and 3 ounces of ice cream in a milk shake. If the store used 4 gallons of ice cream and 5 quarts of syrup in a day, how many ice-cream sodas and milk shakes did it sell? [*Hint:* 1 quart = 32 ounces; 1 gallon = 128 ounces]

## 1.3 *m* EQUATIONS IN *n* UNKNOWNS: GAUSS-JORDAN AND GAUSSIAN ELIMINATION

In this section we describe a method for finding all solutions (if any) to a system of $m$ linear equations in $n$ unknowns. In doing so we shall see that, like the $2 \times 2$ case, such a system has no solutions, one solution, or an infinite number of solutions. Before launching into the general method, let us look at some simple examples. As variables, we use $x_1, x_2, x_3$, and so on instead of $x, y, z, \ldots$ because the subscripted notation is easier to generalize.

EXAMPLE 1 **Solving a System of Three Equations in Three Unknowns: Unique Solution**
Solve the system

$$\begin{aligned} 2x_1 + 4x_2 + 6x_3 &= 18 \\ 4x_1 + 5x_2 + 6x_3 &= 24 \\ 3x_1 + x_2 - 2x_3 &= 4 \end{aligned} \quad \textbf{(1)}$$

**Solution** Here we seek three numbers $x_1$, $x_2$, and $x_3$ such that the three equations in (1) are satisfied. Our method of solution will be to simplify the equations as we did in Section 1.2 so that solutions can be readily identified. We begin by dividing the first equation by 2. This gives us

$$\begin{aligned} x_1 + 2x_2 + 3x_3 &= 9 \\ 4x_1 + 5x_2 + 6x_3 &= 24 \\ 3x_1 + x_2 - 2x_3 &= 4 \end{aligned} \tag{2}$$

As we saw in Section 1.2, adding two equations together leads to a third, correct equation. This equation may replace either of the two equations used to obtain it in the system. We begin simplifying system (2) by multiplying both sides of the first equation in (2) by $-4$ and adding this new equation to the second equation. This gives us

$$\begin{aligned} -4x_1 - 8x_2 - 12x_3 &= -36 \\ 4x_1 + 5x_2 + 6x_3 &= 24 \\ \hline - 3x_2 - 6x_3 &= -12 \end{aligned}$$

The equation $-3x_2 - 6x_3 = -12$ is our new second equation and the system is now

$$\begin{aligned} x_1 + 2x_2 + 3x_3 &= 9 \\ - 3x_2 - 6x_3 &= -12 \\ 3x_1 + x_2 - 2x_3 &= 4 \end{aligned}$$

*Note.* We have *replaced* the equation $4x_1 + 5x_2 + 6x_3 = 24$ with the equation $-3x_2 - 6x_3 = -12$. Throughout this example and later ones we shall replace equations with simpler ones until a system with a readily identifiable solution is obtained.

We then multiply the first equation by $-3$ and add it to the third equation:

$$\begin{aligned} x_1 + 2x_2 + 3x_3 &= 9 \\ - 3x_2 - 6x_3 &= -12 \\ - 5x_2 - 11x_3 &= -23 \end{aligned} \tag{3}$$

Note that in system (3) the variable $x_1$ has been eliminated from the second and third equations. Next we divide the second equation by $-3$:

$$\begin{aligned} x_1 + 2x_2 + 3x_3 &= 9 \\ x_2 + 2x_3 &= 4 \\ - 5x_2 - 11x_3 &= -23 \end{aligned}$$

We multiply the second equation by $-2$ and add it to the first and then multiply the second equation by 5 and add it to the third:

$$\begin{aligned} x_1 \qquad - x_3 &= 1 \\ x_2 + 2x_3 &= 4 \\ - x_3 &= -3 \end{aligned}$$

We multiply the third equation by $-1$:

$$\begin{aligned} x_1 \qquad - x_3 &= 1 \\ x_2 + 2x_3 &= 4 \\ x_3 &= 3 \end{aligned}$$

Finally, we add the third equation to the first and then multiply the third equation by $-2$ and add it to the second to obtain the following system [which is equivalent to system (1)]:

$$\begin{aligned} x_1 \qquad\qquad &= 4 \\ x_2 \qquad &= -2 \\ x_3 &= 3 \end{aligned}$$

This is the unique solution to the system. We write it in the form $(4, -2, 3)$. The method we used here is called **Gauss-Jordan elimination.**†

Gauss-Jordan elimination

Before going on to another example, let us summarize what we have done in this example:

**i.** We divided to make the coefficient of $x_1$ in the first equation equal to 1.

**ii.** We "eliminated" the $x_1$ terms in the second and third equations. That is, we made the coefficients of these terms equal to zero by multiplying the first equation by appropriate numbers and then adding it to the second and third equations, respectively.

**iii.** We divided to make the coefficient of the $x_2$ term in the second equation equal to 1 and then proceeded to use the second equation to eliminate the $x_2$ terms in the first and third equations.

**iv.** We divided to make the coefficient of the $x_3$ term in the third equation equal to 1 and then proceeded to use the third equation to eliminate the $x_3$ terms in the first and second equations.

We emphasize that, at every step, we obtained systems that were equivalent. That is, each system had the same set of solutions as the one that preceded it. This follows from Facts A and B on page 2.

Matrix

Before solving other systems of equations, we introduce notation that makes it easier to write down each step in our procedure. A **matrix** is a rectangular array of numbers. We shall discuss matrices in great detail beginning in Section 1.5. For example, the coefficients of the variables $x_1$, $x_2$, $x_3$ in system (1) can be written as

---

†Named after the great German mathematician Karl Friedrich Gauss (1777–1855) and the German engineer Wilhelm Jordan (1844–1899). See the biographical sketch of Gauss on page 23. Jordan was an expert on geodesy-surveying with allowance for the earth's curvature. His work on solving systems of equations appeared in 1888 in his book *Handbuch der Vermessungskunde* (Handbook of Geodesy).

Coefficient matrix

the entries of a matrix $A$, called the **coefficient matrix** of the system:

$$A = \begin{pmatrix} 2 & 4 & 6 \\ 4 & 5 & 6 \\ 3 & 1 & -2 \end{pmatrix} \qquad \textbf{(4)}$$

$m \times n$ matrix

A matrix with $m$ rows and $n$ columns is called an $\boldsymbol{m \times n}$ **matrix.** The symbol $m \times n$ is read "$m$ by $n$." The study of matrices will take a large part of the remaining chapters of this text. We introduce them here for convenience of notation.

Augmented matrix

Using matrix notation, we can write system (1) as the **augmented matrix**

$$\left(\begin{array}{ccc|c} 2 & 4 & 6 & 18 \\ 4 & 5 & 6 & 24 \\ 3 & 1 & -2 & 4 \end{array}\right) \qquad \textbf{(5)}$$

We now introduce some terminology. We have seen that multiplying (or dividing) the sides of an equation by a nonzero number gives us a new, correct equation. Moreover, adding a multiple of one equation to another equation in a system gives us another correct equation. Finally, if we interchange two equations in a system of equations, we obtain an equivalent system. These three operations, when applied to the rows of the augmented matrix representation of a system of equations, are called **elementary row operations.**

To sum up, the three elementary row operations applied to the augmented matrix representation of a system of equations are:

**Elementary Row Operations**

**i.** Multiply (or divide) one row by a nonzero number.

**ii.** Add a multiple of one row to another row.

**iii.** Interchange two rows.

Row reduction

The process of applying elementary row operations to simplify an augmented matrix is called **row reduction.**

## Notation

1. $R_i \to cR_i$ stands for "replace the $i$th row by the $i$th row multiplied by $c$." [To multiply the $i$th row by $c$, multiply each number in the $i$th row by $c$.]
2. $R_j \to R_j + cR_i$ stands for "replace the $j$th row with the sum of the $j$th row and the $i$th row multiplied by $c$."
3. $R_i \rightleftarrows R_j$ stands for "interchange rows $i$ and $j$."
4. $A \to B$ indicates that the augmented matrices $A$ and $B$ are equivalent; that is, the systems they represent have the same solution.

In Example 1 we saw that by using the elementary row operations (*i*) and (*ii*) several times, we could obtain a system in which the solutions to the system were given explicitly. We now repeat the steps in Example 1, using the notation just introduced:

$$\left(\begin{array}{ccc|c} 2 & 4 & 6 & 18 \\ 4 & 5 & 6 & 24 \\ 3 & 1 & -2 & 4 \end{array}\right) \xrightarrow{R_1 \to \frac{1}{2}R_1} \left(\begin{array}{ccc|c} 1 & 2 & 3 & 9 \\ 4 & 5 & 6 & 24 \\ 3 & 1 & -2 & 4 \end{array}\right) \xrightarrow[R_3 \to R_3 - 3R_1]{R_2 \to R_2 - 4R_1} \left(\begin{array}{ccc|c} 1 & 2 & 3 & 9 \\ 0 & -3 & -6 & -12 \\ 0 & -5 & -11 & -23 \end{array}\right)$$

$$\xrightarrow{R_2 \to -\frac{1}{3}R_2} \left(\begin{array}{ccc|c} 1 & 2 & 3 & 9 \\ 0 & 1 & 2 & 4 \\ 0 & -5 & -11 & -23 \end{array}\right) \xrightarrow[R_3 \to R_3 + 5R_2]{R_1 \to R_1 - 2R_2} \left(\begin{array}{ccc|c} 1 & 0 & -1 & 1 \\ 0 & 1 & 2 & 4 \\ 0 & 0 & -1 & -3 \end{array}\right)$$

$$\xrightarrow{R_3 \to -R_3} \left(\begin{array}{ccc|c} 1 & 0 & -1 & 1 \\ 0 & 1 & 2 & 4 \\ 0 & 0 & 1 & 3 \end{array}\right) \xrightarrow[R_2 \to R_2 - 2R_3]{R_1 \to R_1 + R_3} \left(\begin{array}{ccc|c} 1 & 0 & 0 & 4 \\ 0 & 1 & 0 & -2 \\ 0 & 0 & 1 & 3 \end{array}\right)$$

Again, we can easily "see" the solution $x_1 = 4$, $x_2 = -2$, $x_3 = 3$.

**EXAMPLE 2** **Solving a System of Three Equations in Three Unknowns: Infinite Number of Solutions** Solve the system

$$\begin{aligned} 2x_1 + 4x_2 + 6x_3 &= 18 \\ 4x_1 + 5x_2 + 6x_3 &= 24 \\ 2x_1 + 7x_2 + 12x_3 &= 30 \end{aligned}$$

**Solution** To solve, we proceed as in Example 1, first writing the system as an augmented matrix:

$$\left(\begin{array}{ccc|c} 2 & 4 & 6 & 18 \\ 4 & 5 & 6 & 24 \\ 2 & 7 & 12 & 30 \end{array}\right)$$

We then obtain, successively,

$$\xrightarrow{R_1 \to \frac{1}{2}R_1} \left(\begin{array}{ccc|c} 1 & 2 & 3 & 9 \\ 4 & 5 & 6 & 24 \\ 2 & 7 & 12 & 30 \end{array}\right) \xrightarrow[R_3 \to R_3 - 2R_1]{R_2 \to R_2 - 4R_1} \left(\begin{array}{ccc|c} 1 & 2 & 3 & 9 \\ 0 & -3 & -6 & -12 \\ 0 & 3 & 6 & 12 \end{array}\right)$$

$$\xrightarrow{R_2 \to -\frac{1}{3}R_2} \left(\begin{array}{ccc|c} 1 & 2 & 3 & 9 \\ 0 & 1 & 2 & 4 \\ 0 & 3 & 6 & 12 \end{array}\right) \xrightarrow[R_3 \to R_3 - 3R_2]{R_1 \to R_1 - 2R_2} \left(\begin{array}{ccc|c} 1 & 0 & -1 & 1 \\ 0 & 1 & 2 & 4 \\ 0 & 0 & 0 & 0 \end{array}\right)$$

This is equivalent to the system of equations

$$\begin{aligned} x_1 \qquad + x_3 &= 1 \\ x_2 + 2x_3 &= 4 \end{aligned}$$

This is as far as we can go. There are now only two equations in the three unknowns $x_1, x_2, x_3$ and there are an infinite number of solutions. To see this let $x_3$ be chosen. Then $x_2 = 4 - 2x_3$ and $x_1 = 1 + x_3$. This will be a solution for any number $x_3$. We write these solutions in the form $(1 + x_3, 4 - 2x_3, x_3)$. For example, if $x_3 = 0$, we obtain the solution $(1, 4, 0)$. For $x_3 = 10$ we obtain the solution $(11, -16, 10)$.

**EXAMPLE 3** **An Inconsistent System** Solve the system

$$\begin{aligned} 2x_2 + 3x_3 &= 4 \\ 2x_1 - 6x_2 + 7x_3 &= 15 \\ x_1 - 2x_2 + 5x_3 &= 10 \end{aligned} \tag{6}$$

**Solution** The augmented matrix for this system is

$$\left(\begin{array}{ccc|c} 0 & 2 & 3 & 4 \\ 2 & -6 & 7 & 15 \\ 1 & -2 & 5 & 10 \end{array}\right)$$

We cannot make the 1,1 entry 1 as before because multiplying 0 by any real number results in 0. Instead, we may use elementary row operation (*iii*) to get a nonzero number in the 1,1 position. We can interchange row 1 with either of the other two rows; however, interchanging row 1 and row 3 puts a 1 in the 1,1 position. If we do this, we obtain the following:

$$\left(\begin{array}{ccc|c} 0 & 2 & 3 & 4 \\ 2 & -6 & 7 & 15 \\ 1 & -2 & 5 & 10 \end{array}\right) \xrightarrow{R_1 \rightleftarrows R_3} \left(\begin{array}{ccc|c} 1 & -2 & 5 & 10 \\ 2 & -6 & 7 & 15 \\ 0 & 2 & 3 & 4 \end{array}\right) \xrightarrow{R_2 \to R_2 - 2R_1} \left(\begin{array}{ccc|c} 1 & -2 & 5 & 10 \\ 0 & -2 & -3 & -5 \\ 0 & 2 & 3 & 4 \end{array}\right)$$

Let us pause here; the last two equations read

$$\begin{aligned} -2x_2 - 3x_3 &= -5 \\ 2x_2 + 3x_3 &= 4 \end{aligned}$$

which is impossible (if $-2x_2 - 3x_3 = -5$, then $2x_2 + 3x_3 = 5$, not 4). Thus there is no solution. We can proceed as in the last two examples to obtain a more standard form:

$$\xrightarrow{R_2 \to -\frac{1}{2}R_2} \left(\begin{array}{ccc|c} 1 & -2 & 5 & 10 \\ 0 & 1 & \frac{3}{2} & \frac{5}{2} \\ 0 & 2 & 3 & 4 \end{array}\right) \xrightarrow[R_3 \to R_3 - 2R_2]{R_1 \to R_1 + 2R_2} \left(\begin{array}{ccc|c} 1 & 0 & 8 & 15 \\ 0 & 1 & \frac{3}{2} & \frac{5}{2} \\ 0 & 0 & 0 & -1 \end{array}\right)$$

The last equation now reads $0x_1 + 0x_2 + 0x_3 = -1$, which is impossible since $0 \neq -1$. Thus system (6) has *no* solution. In this case the system is said to be *inconsistent.*

**DEFINITION 1** **Inconsistent and Consistent Systems** A system of linear equations is said to be **inconsistent** if it has no solutions. A system with at least one solution is said to be **consistent.**

Let us take another look at these three examples. In Example 1 we began with the coefficient matrix

$$A_1 = \begin{pmatrix} 2 & 4 & 6 \\ 4 & 5 & 6 \\ 3 & 1 & -2 \end{pmatrix}$$

In the process of row reduction $A_1$ was "reduced" to the matrix

$$R_1 = \begin{pmatrix} 1 & 0 & 0 \\ 0 & 1 & 0 \\ 0 & 0 & 1 \end{pmatrix}$$

In Example 2 we started with

$$A_2 = \begin{pmatrix} 2 & 4 & 6 \\ 4 & 5 & 6 \\ 2 & 7 & 12 \end{pmatrix}$$

and ended up with

$$R_2 = \begin{pmatrix} 1 & 0 & -1 \\ 0 & 1 & 2 \\ 0 & 0 & 0 \end{pmatrix}$$

In Example 3 we began with

$$A_3 = \begin{pmatrix} 0 & 2 & 3 \\ 2 & -6 & 7 \\ 1 & -2 & 5 \end{pmatrix}$$

and ended up with

$$R_3 = \begin{pmatrix} 1 & 0 & 8 \\ 0 & 1 & \frac{3}{2} \\ 0 & 0 & 0 \end{pmatrix}$$

The matrices $R_1$, $R_2$, and $R_3$ are called the *reduced row echelon forms* of the matrices $A_1$, $A_2$, and $A_3$, respectively. In general, we have the following definition:

**DEFINITION 2** **Reduced Row Echelon Form and Pivot** A matrix is in **reduced row echelon form** if the following four conditions hold:

**i.** All rows (if any) consisting entirely of zeros appear at the bottom of the matrix.

**ii.** The first nonzero number (starting from the left) in any row not consisting entirely of zeros is 1.

**iii.** If two successive rows do not consist entirely of zeros, then the first 1 in the lower row occurs farther to the right than the first 1 in the higher row.

**iv.** Any column containing the first 1 in a row has zeros everywhere else.

The first nonzero number in a row (if any) is called a **pivot** for that row.

*Note.* Condition (*iii*) can be restated as "the pivot in any row is to the right of the pivot in the row above it."

**EXAMPLE 4** **Five Matrices in Reduced Row Echelon Form** The following matrices are in reduced row echelon form:

$$\textbf{i. } \begin{pmatrix} 1 & 0 & 0 \\ 0 & 1 & 0 \\ 0 & 0 & 1 \end{pmatrix} \quad \textbf{ii. } \begin{pmatrix} 1 & 0 & 0 & 0 \\ 0 & 1 & 0 & 0 \\ 0 & 0 & 0 & 1 \end{pmatrix} \quad \textbf{iii. } \begin{pmatrix} 1 & 0 & 0 & 5 \\ 0 & 0 & 1 & 2 \end{pmatrix}$$

$$\textbf{iv. } \begin{pmatrix} 1 & 0 \\ 0 & 1 \end{pmatrix} \quad \textbf{v. } \begin{pmatrix} 1 & 0 & 2 & 5 \\ 0 & 1 & 3 & 6 \\ 0 & 0 & 0 & 0 \end{pmatrix}$$

The matrices in *i* and *ii* have three pivots; the other three matrices have two pivots.

**DEFINITION 3** **Row Echelon Form** A matrix is in **row echelon form** if conditions (*i*), (*ii*), and (*iii*) hold in Definition 2.

**EXAMPLE 5** **Five Matrices in Row Echelon Form** The following matrices are in row echelon form:

$$\textbf{i. } \begin{pmatrix} 1 & 2 & 3 \\ 0 & 1 & 5 \\ 0 & 0 & 1 \end{pmatrix} \quad \textbf{ii. } \begin{pmatrix} 1 & -1 & 6 & 4 \\ 0 & 1 & 2 & -8 \\ 0 & 0 & 0 & 1 \end{pmatrix}$$

$$\textbf{iii. } \begin{pmatrix} 1 & 0 & 2 & 5 \\ 0 & 0 & 1 & 2 \end{pmatrix} \quad \textbf{iv. } \begin{pmatrix} 1 & 2 \\ 0 & 1 \end{pmatrix} \quad \textbf{v. } \begin{pmatrix} 1 & 3 & 2 & 5 \\ 0 & 1 & 3 & 6 \\ 0 & 0 & 0 & 0 \end{pmatrix}$$

*Note.* The row echelon form of a matrix is not usually unique. That is, a matrix may be row equivalent to more than one matrix in row echelon form. For example

$$A = \begin{pmatrix} 1 & 3 & 2 & 5 \\ 0 & 1 & 3 & 6 \\ 0 & 0 & 0 & 0 \end{pmatrix} \xrightarrow{R_1 \to R_1 - R_2} \begin{pmatrix} 1 & 2 & -1 & -1 \\ 0 & 1 & 3 & 6 \\ 0 & 0 & 0 & 0 \end{pmatrix} = B$$

shows that the two matrices above, both of which are in row echelon form, are row equivalent. Thus, any matrix which has $A$ as a row echelon form also has $B$ as a row echelon form.

*Remark 1.* The difference between these two forms should be clear from the examples. In row echelon form, all the numbers below the first 1 in a row are zero. In reduced row echelon form, all the numbers above and below the first 1 in a row are zero. Thus, reduced row echelon form is more exclusive. That is, every matrix in reduced row echelon form is in row echelon form, but not conversely.

*Remark 2.* We can always reduce a matrix to reduced row echelon form or row echelon form by performing elementary row operations. We saw this reduction to reduced row echelon form in Examples 1, 2, and 3.

As we saw in Examples 1, 2, and 3, there is a strong connection between the reduced row echelon form of a matrix and the existence of a unique solution to the system. In Example 1 the reduced row echelon form of the coefficient matrix (that is, the first three columns of the augmented matrix) had a 1 in each row and there was a unique solution. In Examples 2 and 3 the reduced row echelon form of the coefficient matrix had a row of zeros and the system had either no solution or an infinite number of solutions. This turns out always to be true in any system with the same number of equations as unknowns. But before turning to the general case, let us discuss the usefulness of the row echelon form of a matrix. It is possible to solve the system in Example 1 by reducing the coefficient matrix to its row echelon form.

**EXAMPLE 6** **Solving a System by Gaussian Elimination** Solve the system of Example 1 by reducing the coefficient matrix to row echelon form.

**Solution** We begin as before:

$$\left(\begin{array}{ccc|c} 2 & 4 & 6 & 18 \\ 4 & 5 & 6 & 24 \\ 3 & 1 & -2 & 4 \end{array}\right) \xrightarrow{R_1 \to \frac{1}{2}R_1} \left(\begin{array}{ccc|c} 1 & 2 & 3 & 9 \\ 4 & 5 & 6 & 24 \\ 3 & 1 & -2 & 4 \end{array}\right)$$

$$\xrightarrow[R_3 \to R_3 - 3R_1]{R_2 \to R_2 - 4R_1} \left(\begin{array}{ccc|c} 1 & 2 & 3 & 9 \\ 0 & -3 & -6 & -12 \\ 0 & -5 & -11 & -23 \end{array}\right) \xrightarrow{R_2 \to -\frac{1}{3}R_2} \left(\begin{array}{ccc|c} 1 & 2 & 3 & 9 \\ 0 & 1 & 2 & 4 \\ 0 & -5 & -11 & -23 \end{array}\right)$$

So far, this process is identical to our earlier one. Now, however, we only make zero the number (−5) below the first 1 in the second row:

$$\xrightarrow{R_3 \to R_3 + 5R_2} \left(\begin{array}{ccc|c} 1 & 2 & 3 & 9 \\ 0 & 1 & 2 & 4 \\ 0 & 0 & -1 & -3 \end{array}\right) \xrightarrow{R_3 \to -R_3} \left(\begin{array}{ccc|c} 1 & 2 & 3 & 9 \\ 0 & 1 & 2 & 4 \\ 0 & 0 & 1 & 3 \end{array}\right)$$

Back substitution

The augmented matrix of the system (and the coefficient matrix) are now in row echelon form and we immediately see that $x_3 = 3$. We then use **back substitution** to solve for $x_2$ and then $x_1$. The second equation reads $x_2 + 2x_3 = 4$. Thus $x_2 + 2(3) = 4$ and $x_2 = -2$. Similarly, from the first equation we obtain $x_1 + 2(-2) + 3(3) = 9$ or $x_1 = 4$. Thus we again obtain the solution $(4, -2, 3)$. The method of solution just employed is called **Gaussian elimination.**

Gaussian elimination

We therefore have two methods for solving our sample systems of equations:

**i. Gauss-Jordan Elimination**
Row-reduce the coefficient matrix to reduced row echelon form using the procedure outlined on page 9.

**ii. Gaussian Elimination**
Row-reduce the coefficient matrix to row echelon form, solve for the last unknown, and then use back substitution to solve for the other unknowns.

Which method is more useful? It depends. In solving systems of equations on a computer, Gaussian elimination is the preferred method because it involves fewer elementary row operations. In fact, as we shall see in Appendix 3, to solve a system of $n$ equations in $n$ unknowns requires approximately $n^3/2$ multiplications and additions using Gauss-Jordan elimination but only $n^3/3$ multiplications and additions using Gaussian elimination. We shall discuss the numerical solution of systems of equations in Appendix 4. On the other hand, at times it is essential to obtain the reduced row echelon form of a matrix (one of these is discussed in Section 1.8). In these cases Gauss-Jordan elimination is the preferred method.

We now turn to the solution of a general system of $m$ equations in $n$ unknowns. Because of our need to do so in Section 1.8, we shall be solving most of the systems by Gauss-Jordan elimination. Keep in mind, however, that Gaussian elimination is sometimes the preferred approach.

The general $m \times n$ system of $m$ linear equations in $n$ unknowns is given by

$$\begin{array}{cccccc} a_{11}x_1 & + a_{12}x_2 & + a_{13}x_3 & + \cdots + & a_{1n}x_n & = b_1 \\ a_{21}x_1 & + a_{22}x_2 & + a_{23}x_3 & + \cdots + & a_{2n}x_n & = b_2 \\ a_{31}x_1 & + a_{32}x_2 & + a_{33}x_3 & + \cdots + & a_{3n}x_n & = b_3 \\ \vdots & \vdots & \vdots & \vdots & \vdots & \vdots \\ a_{m1}x_1 & + a_{m2}x_2 & + a_{m3}x_3 & + \cdots + & a_{mn}x_n & = b_m \end{array} \tag{7}$$

In system (7) all the $a$'s and $b$'s are given real numbers. The problem is to find all sets of $n$ numbers, denoted by $(x_1, x_2, x_3, \ldots, x_n)$, that satisfy every one of the $m$ equations in (7). The number $a_{ij}$ is the coefficient of the variable $x_j$ in the $i$th equation.

We may solve a system of $m$ equations in $n$ unknowns by either Gauss-Jordan or Gaussian elimination. We provide below one example where the numbers of equations and unknowns are not the same.

**EXAMPLE 7** **Solving a System of Two Equations in Four Unknowns** Solve the system

$$\begin{aligned} x_1 + 3x_2 - 5x_3 + x_4 &= 4 \\ 2x_1 + 5x_2 - 2x_3 + 4x_4 &= 6 \end{aligned}$$

**Solution** We write this system as an augmented matrix and row-reduce:

$$\left(\begin{array}{cccc|c} 1 & 3 & -5 & 1 & 4 \\ 2 & 5 & -2 & 4 & 6 \end{array}\right) \xrightarrow{R_2 \to R_2 - 2R_1} \left(\begin{array}{cccc|c} 1 & 3 & -5 & 1 & 4 \\ 0 & -1 & 8 & 2 & -2 \end{array}\right)$$

$$\xrightarrow{R_2 \to -R_2} \left(\begin{array}{cccc|c} 1 & 3 & -5 & 1 & 4 \\ 0 & 1 & -8 & -2 & 2 \end{array}\right) \xrightarrow{R_1 \to R_1 - 3R_2} \left(\begin{array}{cccc|c} 1 & 0 & 19 & 7 & -2 \\ 0 & 1 & -8 & -2 & 2 \end{array}\right)$$

This is as far as we can go. The coefficient matrix is in reduced row echelon form. There are evidently an infinite number of solutions. The variables $x_3$ and $x_4$ can be chosen arbitrarily. Then $x_2 = 2 + 8x_3 + 2x_4$ and $x_1 = -2 - 19x_3 - 7x_4$. All solutions are, therefore, represented by $(-2 - 19x_3 - 7x_4, 2 + 8x_3 + 2x_4, x_3, x_4)$. For example, if $x_3 = 1$ and $x_4 = 2$, we obtain the solution $(-35, 14, 1, 2)$.

As you will see if you do a lot of system solving, the computations can become very messy. It is a good rule of thumb to use a calculator or computer whenever the fractions become unpleasant. It should be noted, however, that if computations are carried out on a computer or calculator, "round-off" errors can be introduced. This problem is discussed in Appendix 3.

**EXAMPLE 8** **A Problem in Resource Management** A State Fish and Game Department supplies three types of food to a lake that supports three species of fish. Each fish of Species 1 consumes, each week, an average of 1 unit of Food 1, 1 unit of Food 2, and 2 units of Food 3. Each fish of Species 2 consumes, each week, an average of 3 units of Food 1, 4 units of Food 2, and 5 units of Food 3. For a fish of Species 3, the average weekly consumption is 2 units of Food 1, 1 unit of Food 2, and 5 units of Food 3. Each week 25,000 units of Food 1, 20,000 units of Food 2, and 55,000 units of Food 3 are supplied to the lake. If we assume that all food is eaten, how many fish of each species can coexist in the lake?

**Solution** We let $x_1$, $x_2$, and $x_3$ denote the numbers of fish of the three species being supported by the lake environment. Using the information in the problem, we see that $x_1$ fish of Species 1 consume $x_1$ units of Food 1, $x_2$ fish of Species 2 consume $3x_2$ units of Food 1, and $x_3$ fish of Species 3 consume $2x_3$ units of Food 1. Thus $x_1 + 3x_2 + 2x_3 = 25{,}000$ = total weekly supply of Food 1. Obtaining a similar equation for each of the other two foods, we are led to the following system:

$$\begin{aligned} x_1 + 3x_2 + 2x_3 &= 25{,}000 \\ x_1 + 4x_2 + x_3 &= 20{,}000 \\ 2x_1 + 5x_2 + 5x_3 &= 55{,}000 \end{aligned}$$

Upon solving, we obtain

$$\left(\begin{array}{ccc|c} 1 & 3 & 2 & 25{,}000 \\ 1 & 4 & 1 & 20{,}000 \\ 2 & 5 & 5 & 55{,}000 \end{array}\right)$$

$$\xrightarrow[R_3 \to R_3 - 2R_1]{R_2 \to R_2 - R_1} \left(\begin{array}{ccc|c} 1 & 3 & 2 & 25{,}000 \\ 0 & 1 & -1 & -5{,}000 \\ 0 & -1 & 1 & 5{,}000 \end{array}\right) \xrightarrow[R_3 \to R_3 + R_2]{R_1 \to R_1 - 3R_2} \left(\begin{array}{ccc|c} 1 & 0 & 5 & 40{,}000 \\ 0 & 1 & -1 & -5{,}000 \\ 0 & 0 & 0 & 0 \end{array}\right)$$

Thus, if $x_3$ is chosen arbitrarily, we have an infinite number of solutions given by $(40{,}000 - 5x_3,\ x_3 - 5{,}000,\ x_3)$. Of course, we must have $x_1 \geq 0$, $x_2 \geq 0$, and $x_3 \geq 0$. Since $x_2 = x_3 - 5{,}000 \geq 0$, we have $x_3 \geq 5{,}000$. This means that $0 \leq x_1 \leq 40{,}000 - 5(5{,}000) = 15{,}000$. Finally, since $40{,}000 - 5x_3 \geq 0$, we see that $x_3 \leq 8{,}000$. This means that the populations that can be supported by the lake with all food consumed are

$$\begin{aligned} x_1 &= 40{,}000 - 5x_3 \\ x_2 &= x_3 - 5{,}000 \\ 5{,}000 &\leq x_3 \leq 8{,}000 \end{aligned}$$

For example, if $x_3 = 6{,}000$, then $x_1 = 10{,}000$ and $x_2 = 1{,}000$.

*Note.* The system of equations does have an infinite number of solutions. However, the resource management problem has only a finite number of solutions because $x_1$, $x_2$, and $x_3$ must be positive integers and there are only 3001 integers in the interval [5000, 8000]. (You can't stock 5237.578 fish, for example.)

## Input-Output Analysis (Optional)

The next two examples show how systems of equations can arise in economic modeling.

EXAMPLE 9 **The Leontief Input-Output Model** A model that is often used in economics is the **Leontief input-output model.**† Suppose an economic system has $n$ industries. There are two kinds of demands on each industry. First, there is the *external* demand from outside the system. If the system is a country, for example, then the external demand could be from another country. Second, there is the demand placed on one industry by another industry in the same system. In the United States, for example, there is a demand on the output of the steel industry by the automobile industry.

Let $e_i$ represent the external demand placed on the $i$th industry. Let $a_{ij}$ represent the internal demand placed on the $i$th industry by the $j$th industry. More precisely, $a_{ij}$ represents the number of units of the output of industry $i$ needed to produce 1 unit of the output of industry $j$. Let $x_i$ represent the output of industry $i$. Now we assume that the output of each industry is equal to its demand (that is, there is no overproduction). The total demand is equal to the sum of the internal and external demands. To calculate the internal demand on industry 2, for example, we note that industry 1 needs $a_{21}$ units of the output of industry 2 to produce 1 unit of its output. If the output from industry 1 is $x_1$, then $a_{21}x_1$ is the total amount industry 1 needs from industry 2. Thus the total internal demand on industry 2 is $a_{21}x_1 + a_{22}x_2 + \cdots + a_{2n}x_n$.

We are led to the following system of equations obtained by equating the total demand with the output of each industry:

$$
\begin{array}{ccccccccccc}
a_{11}x_1 & + & a_{12}x_2 & + & \cdots & + & a_{1n}x_n & + & e_1 & = & x_1 \\
a_{21}x_1 & + & a_{22}x_2 & + & \cdots & + & a_{2n}x_n & + & e_2 & = & x_2 \\
\vdots & & \vdots & & & & \vdots & & \vdots & & \vdots \\
a_{n1}x_1 & + & a_{n2}x_2 & + & \cdots & + & a_{nn}x_n & + & e_n & = & x_n
\end{array}
\qquad (8)
$$

Or, rewriting (8) so it looks like system (7), we get

$$
\begin{array}{rcrcl}
(1 - a_{11})x_1 - a_{12}x_2 & - \cdots - & a_{1n}x_n & = & e_1 \\
-a_{21}x_1 + (1 - a_{22})x_2 & - \cdots - & a_{2n}x_n & = & e_2 \\
\vdots \qquad\quad \vdots & & \vdots & & \vdots \\
-a_{n1}x_1 - a_{n2}x_2 & - \cdots + & (1 - a_{nn})x_n & = & e_n
\end{array}
\qquad (9)
$$

System (9) of $n$ equations in $n$ unknowns is very important in economic analysis.

---

† Named after American economist Wassily W. Leontief. This model was used in his pioneering paper "Qualitative Input and Output Relations in the Economic System of the United States" in *Review of Economic Statistics* 18(1936):105–125. An updated version of this model appears in Leontief's book *Input-Output Analysis* (New York: Oxford University Press, 1966). Leontief won the Nobel Prize in economics in 1973 for his development of input-output analysis.

**EXAMPLE 10** **The Leontief Model Applied to an Economic System with Three Industries**
In an economic system with three industries, suppose that the external demands are, respectively, 10, 25, and 20. Suppose that $a_{11} = 0.2$, $a_{12} = 0.5$, $a_{13} = 0.15$, $a_{21} = 0.4$, $a_{22} = 0.1$, $a_{23} = 0.3$, $a_{31} = 0.25$, $a_{32} = 0.5$, and $a_{33} = 0.15$. Find the output in each industry such that supply exactly equals demand.

**Solution** Here $n = 3$, $1 - a_{11} = 0.8$, $1 - a_{22} = 0.9$, and $1 - a_{33} = 0.85$. Then system (9) is

$$\begin{aligned} 0.8x_1 - 0.5x_2 - 0.15x_3 &= 10 \\ -0.4x_1 + 0.9x_2 - \quad 0.3x_3 &= 25 \\ -0.25x_1 - 0.5x_2 + 0.85x_3 &= 20 \end{aligned}$$

Solving this system by Gauss-Jordan elimination on a calculator, while carrying five decimal places at every step, results in

$$\left(\begin{array}{ccc|c} 1 & 0 & 0 & 110.30442 \\ 0 & 1 & 0 & 118.74070 \\ 0 & 0 & 1 & 125.81787 \end{array}\right)$$

We conclude that the outputs needed for supply to equal demand (approximately) are $x_1 = 110$, $x_2 = 119$, and $x_3 = 126$.

## The Geometry of a System of Three Equations in Three Unknowns (Optional)

In Figure 1.1 on page 3 we saw that a system of two equations in two unknowns can be depicted as two straight lines. If the lines have a single point of intersection, then the system has a unique solution; if they coincide, then there are an infinite number of solutions; if they are parallel, there is no solution and the system is inconsistent.

Something similar happens when we have three equations in three unknowns.

As we shall see in Section 3.5, the graph of the equation $ax + by + cz = d$ in three-dimensional space is a plane.

Consider a system of three equations in three unknowns:

$$\begin{aligned} ax + by + cz &= d \\ ex + fy + gz &= h \\ jx + ky + \ell z &= m \end{aligned} \qquad \textbf{(10)}$$

where $a, b, c, d, e, f, g, h, j, k, \ell$, and $m$ are constants and at least one constant in each row is nonzero.

Each equation in (10) is the equation of a plane. *Every* solution $(x, y, z)$ to the system must be a point on *each one* of the three planes. There are six possibilities:

1. The three planes intersect at a single point. Then there is a unique solution to the system (see Figure 1.2).

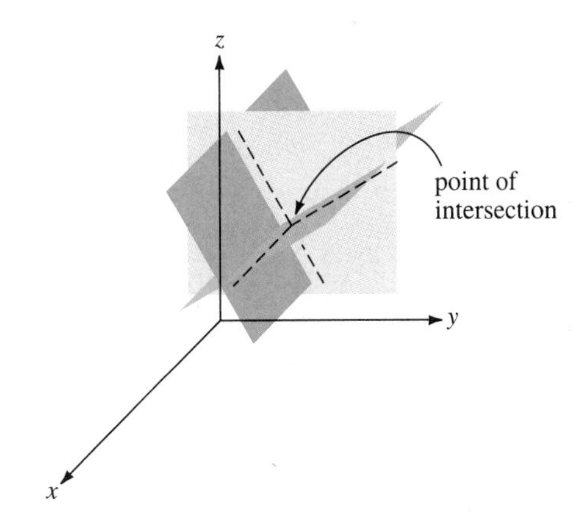


**Figure 1.2** Three planes intersect at a single point

2. The three planes intersect at the same line. Then every point on the line is a solution, and the system has an infinite number of solutions (see Figure 1.3).
3. All three planes coincide. Then every point on the plane is a solution, and there are an infinite number of solutions.
4. Two of the planes coincide and intersect the third plane at a line. Then every point on the line is a solution, and there are an infinite number of solutions (see Figure 1.4).

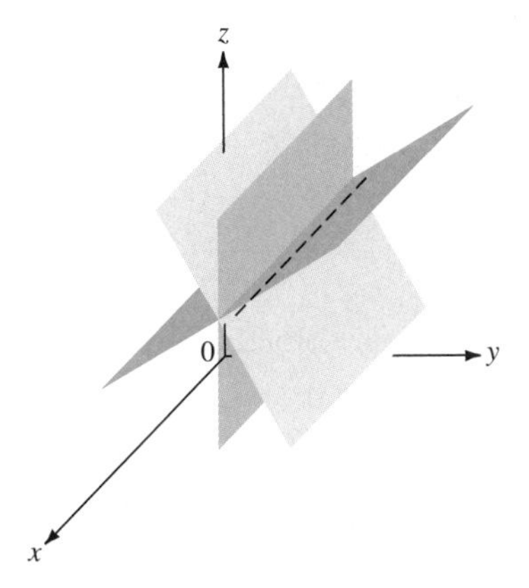


**Figure 1.3** Three planes intersect at the same line

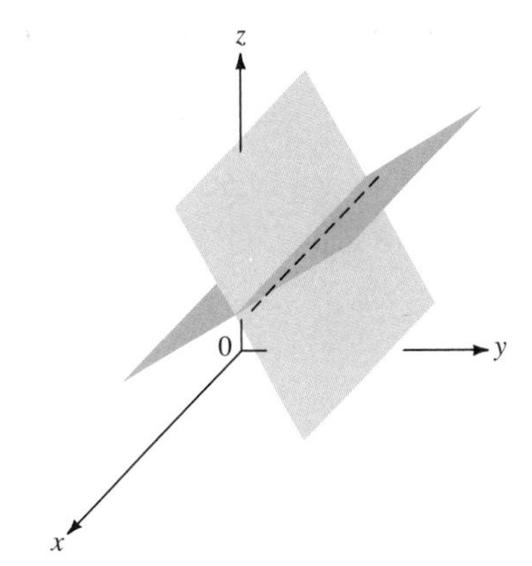


**Figure 1.4** Two planes intersect at a line

5. At least two of the planes are parallel and distinct. Then no points can be on both of them, and there is no solution. The system is inconsistent (see Figure 1.5).

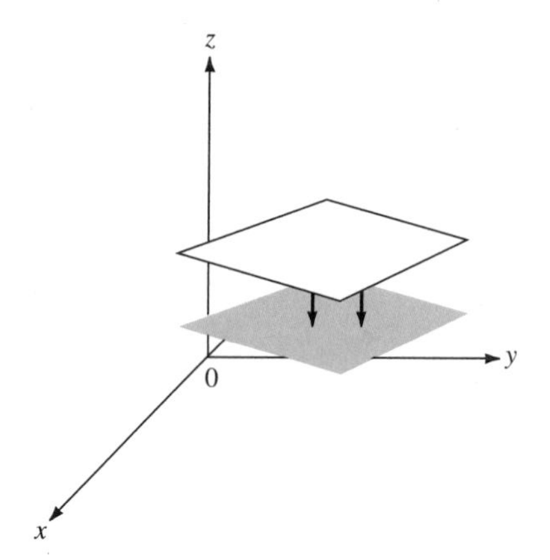


**Figure 1.5** Parallel planes have no points in common

6. Two of the planes meet at a line $L$. The third plane is parallel to $L$ (and does not contain it), so no point on the third plane is on both of the first two planes. There is no solution, and the system is inconsistent (see Figure 1.6).

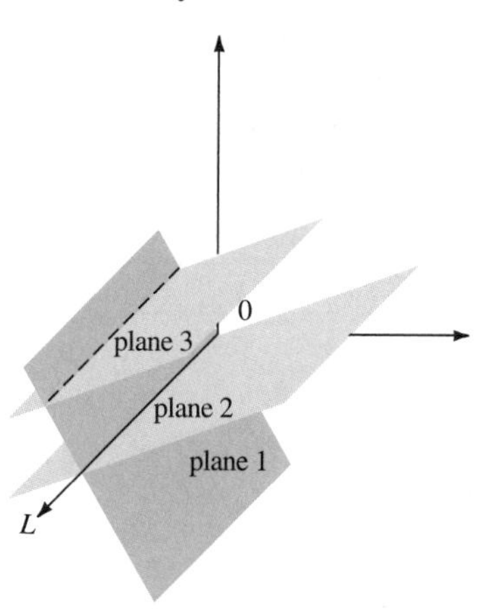


**Figure 1.6** Plane 3 is parallel to $L$, the line of intersection of planes 1 and 2

We see that in all cases the system has a unique solution, has an infinite number of solutions, or is inconsistent. Because it is so difficult to draw planes accurately, we shall say no more about them here. However, it is useful to see how ideas in the $xy$-plane can be extended to more complicated settings.

## Focus on . . .

# Carl Friedrich Gauss, 1777–1855

Carl Friedrich Gauss
*(Library of Congress)*

The greatest mathematician of the nineteenth century, Carl Friedrich Gauss is considered one of the three greatest mathematicians of all time—the others being Archimedes and Newton.

Gauss was born in Brunswick, Germany, in 1777. His father, a hard-working laborer who was exceptionally stubborn and did not believe in formal education, did what he could to keep Gauss from appropriate schooling. Fortunately for Carl (and for mathematics), his mother, while uneducated herself, encouraged her son in his studies and took considerable pride in his achievements until her death at the age of 97.

Gauss was a child prodigy. At the age of 3, he found an error in his father's bookkeeping. A famous story tells of Carl, age 10, as a student in the local Brunswick school. The teacher there was known to assign tasks to keep his pupils busy. One day he asked his students to add the numbers from 1 to 100. Almost at once, Carl placed his slate face down with the words "There it is." Afterwards, the teacher found that Gauss was the only one with the correct answer, 5050. Gauss had noticed that the numbers could be arranged in 50 pairs, each with the sum 101 ($1 + 100$, $2 + 99$, and so on), and $50 \times 101 = 5050$. Later in life, Gauss joked that he could add before he could speak.

When Gauss was 15, the Duke of Brunswick noticed him and became his patron. The duke helped him to enter Brunswick College in 1795 and, three years later, to enter the university at Göttingen. Undecided between careers in mathematics and philosophy, Gauss chose mathematics after two remarkable discoveries. First, he invented the method of least squares a decade before the result was published by Legendre. Second, a month before his nineteenth birthday, he solved a problem whose solution had been sought for more than two thousand years. Gauss showed how to construct, using compass and ruler, a regular polygon with the number of sides not a multiple of 2, 3, or 5.† On March 30, 1796, the day of this discovery, he began a diary, which contained as its first entry rules for construction of a 17-sided regular polygon. The diary, which contains 146 statements of results in only 19 pages, is one of the most important documents in the history of mathematics.

After a short period at Göttingen, Gauss went to the University of Helmstädt and, in 1798 at the age of 20, wrote his now famous doctoral dissertation. In it

† More generally, Gauss proved that a regular $n$-gon is constructible with compass and ruler if and only if $n$ has the form $n = 2^k p_2 \cdot p_3 \cdots p_m$ where $k \geq 0$ and the $p_i$ are distinct Fermat primes. The Fermat primes are primes that take the form $2^{2^n} + 1$. The first five Fermat primes are 3, 5, 17, 257, and 65,537.

he gave the first mathematically rigorous proof of the fundamental theorem of algebra—that every polynomial of degree $n$ has, counting multiplicities, exactly $n$ roots. Many mathematicians, including Euler, Newton, and Lagrange, had attempted to prove this result.

Gauss made a great number of discoveries in physics as well as in mathematics. For example, in 1801 he used a new procedure to calculate, from very little data, the orbit of the planetoid Ceres. In 1833, he invented the electromagnetic telegraph with his colleague Wilhelm Weber (1804–1891). While he did brilliant work in astronomy and electricity, however, it was Gauss's mathematical output that was astonishing. He made fundamental contributions to algebra and geometry. In 1811, he discovered a result that led to the development of complex variable theory by Cauchy. We encounter him here in the Gauss-Jordan method of elimination. Students of numerical analysis study Gaussian quadrature—a technique for numerical integration.

Gauss became a professor of mathematics at Göttingen in 1807 and remained at that post until his death in 1855. Even after his death, his mathematical spirit remained to haunt nineteenth-century mathematicians. Often it turned out that an important new result was discovered earlier by Gauss and could be found in his unpublished notes.

In his mathematical writings, Gauss was a perfectionist and is probably the last mathematician who knew everything in his subject. Claiming that a cathedral was not a cathedral until the last piece of scaffolding was removed, he endeavored to make each of his published works complete, concise, and polished. He used a seal that pictured a tree carrying only a few fruit together with the motto *pauca sed matura* (few, but ripe). But Gauss also believed that mathematics must reflect the real world. At his death, Gauss was honored by a commemorative medal on which was inscribed "George V. King of Hanover to the Prince of Mathematicians."

# PROBLEMS 1.3

## Self-Quiz

**I.** Which of the following systems has the coefficient matrix given at the right?

$$\begin{pmatrix} 3 & 2 & -1 \\ 0 & 1 & 5 \\ 2 & 0 & 1 \end{pmatrix}$$

**a.** $3x + 2y = -1$
$y = 5$
$2x = 1$

**b.** $3x + 2z = 10$
$2x + y = 0$
$-x + 5y + z = 5$

**c.** $3x = 2$
$2x + y = 0$
$-x + 5y = 1$

**d.** $3x + 2y - z = -3$
$y + 5z = 15$
$2x + z = 3$

**II.** Which of the following is an elementary row operation?

**a.** Replace a row with a nonzero multiple of that row.
**b.** Add a nonzero constant to each entry in a row.
**c.** Interchange two columns.
**d.** Replace a row with a sum of the row and a nonzero constant.

**III.** Which of the following is true about the given matrix?

$$\begin{pmatrix} 1 & 0 & 0 & 3 \\ 0 & 1 & 1 & 2 \\ 0 & 0 & 0 & 3 \\ 0 & 0 & 0 & 0 \end{pmatrix}$$

**a.** It is in row echelon form.
**b.** It is not in row echelon form because the fourth number in row 1 is not 1.
**c.** It is not in row echelon form because the first nonzero entry in row 3 is 3.
**d.** It is not in row echelon form because the last column contains a 0.

**IV.** Which of the following is true about the system given below?

$$\begin{aligned} x + y + z &= 3 \\ 2x + 2y + 2z &= 6 \\ 3x + 3y + 3z &= 10 \end{aligned}$$

**a.** It has the unique solution $x = 1$, $y = 1$, $z = 1$.
**b.** It is inconsistent.
**c.** It has an infinite number of solutions.

In Problems 1–20 use Gauss-Jordan or Gaussian elimination to find all solutions, if any, to the given systems.

**1.** $$\begin{aligned} x_1 - 2x_2 + 3x_3 &= 11 \\ 4x_1 + x_2 - x_3 &= 4 \\ 2x_1 - x_2 + 3x_3 &= 10 \end{aligned}$$

**2.** $$\begin{aligned} -2x_1 + x_2 + 6x_3 &= 18 \\ 5x_1 \qquad + 8x_3 &= -16 \\ 3x_1 + 2x_2 - 10x_3 &= -3 \end{aligned}$$

**3.** $$\begin{aligned} 3x_1 + 6x_2 - 6x_3 &= 9 \\ 2x_1 - 5x_2 + 4x_3 &= 6 \\ -x_1 + 16x_2 - 14x_3 &= -3 \end{aligned}$$

**4.** $$\begin{aligned} 3x_1 + 6x_2 - 6x_3 &= 9 \\ 2x_1 - 5x_2 + 4x_3 &= 6 \\ 5x_1 + 28x_2 - 26x_3 &= -8 \end{aligned}$$

**5.** $$\begin{aligned} x_1 + x_2 - x_3 &= 7 \\ 4x_1 - x_2 + 5x_3 &= 4 \\ 2x_1 + 2x_2 - 3x_3 &= 0 \end{aligned}$$

**6.** $$\begin{aligned} x_1 + x_2 - x_3 &= 7 \\ 4x_1 - x_2 + 5x_3 &= 4 \\ 6x_1 + x_2 + 3x_3 &= 18 \end{aligned}$$

**7.** $$\begin{aligned} x_1 + x_2 - x_3 &= 7 \\ 4x_1 - x_2 + 5x_3 &= 4 \\ 6x_1 + x_2 + 3x_3 &= 20 \end{aligned}$$

**8.** $$\begin{aligned} x_1 - 2x_2 + 3x_3 &= 0 \\ 4x_1 + x_2 - x_3 &= 0 \\ 2x_1 - x_2 + 3x_3 &= 0 \end{aligned}$$

**9.** $$\begin{aligned} x_1 + x_2 - x_3 &= 0 \\ 4x_1 - x_2 + 5x_3 &= 0 \\ 6x_1 + x_2 + 3x_3 &= 0 \end{aligned}$$

**10.** $$\begin{aligned} 2x_2 + 5x_3 &= 6 \\ x_1 \qquad - 2x_3 &= 4 \\ 2x_1 + 4x_2 \qquad &= -2 \end{aligned}$$

---

**Answers to Self-Quiz**

**I.** d **II.** a **III.** c **IV.** b

**11.** $\begin{aligned} x_1 + 2x_2 - x_3 &= 4 \\ 3x_1 + 4x_2 - 2x_3 &= 7 \end{aligned}$

**12.** $\begin{aligned} x_1 + 2x_2 - 4x_3 &= 4 \\ -2x_1 - 4x_2 + 8x_3 &= -8 \end{aligned}$

**13.** $\begin{aligned} x_1 + 2x_2 - 4x_3 &= 4 \\ -2x_1 - 4x_2 + 8x_3 &= -9 \end{aligned}$

**14.** $\begin{aligned} x_1 + 2x_2 - x_3 + x_4 &= 7 \\ 3x_1 + 6x_2 - 3x_3 + 3x_4 &= 21 \end{aligned}$

**15.** $\begin{aligned} 2x_1 + 6x_2 - 4x_3 + 2x_4 &= 4 \\ x_1 - x_3 + x_4 &= 5 \\ -3x_1 + 2x_2 - 2x_3 &= -2 \end{aligned}$

**16.** $\begin{aligned} x_1 - 2x_2 + x_3 + x_4 &= 2 \\ 3x_1 + 2x_3 - 2x_4 &= -8 \\ 4x_2 - x_3 - x_4 &= 1 \\ -x_1 + 6x_2 - 2x_3 &= 7 \end{aligned}$

**17.** $\begin{aligned} x_1 - 2x_2 + x_3 + x_4 &= 2 \\ 3x_1 + 2x_3 - 2x_4 &= -8 \\ 4x_2 - x_3 - x_4 &= 1 \\ 5x_1 + 3x_3 - x_4 &= -3 \end{aligned}$

**18.** $\begin{aligned} x_1 - 2x_2 + x_3 + x_4 &= 2 \\ 3x_1 + 2x_3 - 2x_4 &= -8 \\ 4x_2 - x_3 - x_4 &= 1 \\ 5x_1 + 3x_3 - x_4 &= 0 \end{aligned}$

**19.** $\begin{aligned} x_1 + x_2 &= 4 \\ 2x_1 - 3x_2 &= 7 \\ 3x_1 + 2x_2 &= 8 \end{aligned}$

**20.** $\begin{aligned} x_1 + x_2 &= 4 \\ 2x_1 - 3x_2 &= 7 \\ 3x_1 - 2x_2 &= 11 \end{aligned}$

In Problems 21–29 determine whether the given matrix is in row echelon form (but not in reduced row echelon form), reduced row echelon form, or neither.

**21.** $\begin{pmatrix} 1 & 1 & 0 \\ 0 & 1 & 0 \\ 0 & 0 & 1 \end{pmatrix}$ **22.** $\begin{pmatrix} 2 & 0 & 0 \\ 0 & 1 & 0 \\ 0 & 0 & -1 \end{pmatrix}$ **23.** $\begin{pmatrix} 1 & 0 & 1 & 0 \\ 0 & 1 & 1 & 0 \\ 0 & 0 & 0 & 0 \end{pmatrix}$

**24.** $\begin{pmatrix} 1 & 0 & 0 & 0 \\ 0 & 0 & 1 & 0 \\ 0 & 0 & 0 & 1 \end{pmatrix}$ **25.** $\begin{pmatrix} 0 & 1 & 0 & 0 \\ 1 & 0 & 0 & 0 \\ 0 & 0 & 0 & 0 \end{pmatrix}$ **26.** $\begin{pmatrix} 1 & 0 & 1 & 2 \\ 0 & 1 & 3 & 4 \end{pmatrix}$

**27.** $\begin{pmatrix} 1 & 0 \\ 0 & 1 \\ 0 & 0 \end{pmatrix}$ **28.** $\begin{pmatrix} 1 & 0 & 0 \\ 0 & 0 & 0 \\ 0 & 0 & 1 \end{pmatrix}$ **29.** $\begin{pmatrix} 1 & 0 & 0 & 4 \\ 0 & 1 & 0 & 5 \\ 0 & 1 & 1 & 6 \end{pmatrix}$

In Problems 30–35 use the elementary row operations to reduce the given matrices to row echelon form and reduced row echelon form.

**30.** $\begin{pmatrix} 1 & 1 \\ 2 & 3 \end{pmatrix}$ **31.** $\begin{pmatrix} -1 & 6 \\ 4 & 2 \end{pmatrix}$ **32.** $\begin{pmatrix} 1 & -1 & 1 \\ 2 & 4 & 3 \\ 5 & 6 & -2 \end{pmatrix}$

**33.** $\begin{pmatrix} 2 & -4 & 8 \\ 3 & 5 & 8 \\ -6 & 0 & 4 \end{pmatrix}$ **34.** $\begin{pmatrix} 2 & -4 & -2 \\ 3 & 1 & 6 \end{pmatrix}$ **35.** $\begin{pmatrix} 2 & -7 \\ 3 & 5 \\ 4 & -3 \end{pmatrix}$

**36.** In the Leontief input-output model of Example 9 suppose that there are three industries. Suppose further that $e_1 = 10$, $e_2 = 15$, $e_3 = 30$, $a_{11} = \frac{1}{3}$, $a_{12} = \frac{1}{2}$, $a_{13} = \frac{1}{6}$, $a_{21} = \frac{1}{4}$, $a_{22} = \frac{1}{4}$, $a_{23} = \frac{1}{8}$, $a_{31} = \frac{1}{12}$, $a_{32} = \frac{1}{3}$, and $a_{33} = \frac{1}{6}$. Find the output of each industry such that supply exactly equals demand.

**37.** In Example 8 assume that there are 15,000 units of the first food, 10,000 units of the second, and 35,000 units of the third supplied to the lake each week. Assuming that all three foods are consumed, what populations of the three species can coexist in the lake? Is there a unique solution?

**38.** A traveler who just returned from Europe spent \$30 a day for housing in England, \$20 a day in France, and \$20 a day in Spain. For food the traveler spent \$20 a day in England, \$30 a day in France, and \$20 a day in Spain. The traveler spent \$10 a day in each country for incidental expenses. The traveler's records of the trip indicate a total of \$340 spent for housing, \$320 for food, and \$140 for incidental expenses while traveling in these countries. Calculate the number of days the traveler spent in each of the countries or show that the records must be incorrect because the amounts spent are incompatible with each other.

**39.** An investor remarks to a stockbroker that all her stock holdings are in three companies, Delta Airlines, Hilton Hotels, and McDonald's, and that 2 days ago the value of her stocks went down \$350 but yesterday the value increased by \$600. The broker recalls that 2 days ago the price of Delta Airlines stock dropped by \$1 a share, Hilton Hotels dropped \$1.50, but the price of McDonald's stock rose by \$0.50. The broker also remembers that yesterday the price of Delta Airlines stock rose \$1.50, there was a further drop of \$0.50 a share in Hilton Hotels stock, and McDonald's stock rose \$1. Show that the broker does not have enough information to calculate the number of shares the investor owns of each company's stock, but that when the investor says that she owns 200 shares of McDonald's stock, the broker can calculate the number of shares of Delta Airlines and Hilton Hotels.

**40.** An intelligence agent knows that 60 aircraft, consisting of fighter planes and bombers, are stationed at a certain secret airfield. The agent wishes to determine how many of the 60 are fighter planes and how many are bombers. There is a type of rocket carried by both sorts of planes; the fighter carries six of these rockets, the bomber only two. The agent learns that 250 rockets are required to arm every plane at this airfield. Furthermore, the agent overhears a remark that there are twice as many fighter planes as bombers at the base (that is, the number of fighter planes minus twice the number of bombers equals zero). Calculate the number of fighter planes and bombers at the airfield or show that the agent's information must be incorrect because it is inconsistent.

**41.** Consider the system

$$\begin{aligned} 2x_1 - \phantom{5}x_2 + \phantom{2}3x_3 &= a \\ 3x_1 + \phantom{5}x_2 - \phantom{2}5x_3 &= b \\ -5x_1 - 5x_2 + 21x_3 &= c \end{aligned}$$

Show that the system is inconsistent if $c \neq 2a - 3b$.

**42.** Consider the system

$$\begin{aligned} 2x_1 + 3x_2 - \phantom{5}x_3 &= a \\ x_1 - \phantom{3}x_2 + 3x_3 &= b \\ 3x_1 + 7x_2 - 5x_3 &= c \end{aligned}$$

Find conditions on $a$, $b$, and $c$ such that the system is consistent.

***43.** Consider the general system of three linear equations in three unknowns:

$$\begin{aligned} a_{11}x_1 + a_{12}x_2 + a_{13}x_3 &= b_1 \\ a_{21}x_1 + a_{22}x_2 + a_{23}x_3 &= b_2 \\ a_{31}x_1 + a_{32}x_2 + a_{33}x_3 &= b_3 \end{aligned}$$

Find conditions on the coefficients $a_{ij}$ such that the system has a unique solution.

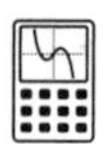

## CALCULATOR BOX

Both the TI-85 and the CASIO fx-7700 GB calculators can solve a system of $n$ equations in $n$ unknowns (up to a maximum of $n = 255$ on the TI-85 and $n = 9$ on the CASIO fx-7700 GB) when there is a unique solution. Consider the system

$$\begin{aligned} 3.8x_1 + 1.6x_2 + 0.9x_3 &= 3.72 \\ -0.7x_1 + 5.4x_2 + 1.6x_3 &= 3.16 \\ 1.5x_1 + 1.1x_2 - 3.2x_3 &= 43.78 \end{aligned}$$

### TI-85

**1.** Enter the augmented matrix. There are several ways to do this. The simplest way follows:

[[3.8, 1.6, .9, 3.72][−.7, 5.4, 1.6, 3.16][1.5, 1.1, −3.2, 43.78]]

STO▶ A ENTER

The last three keys store the augmented matrix in A.

**2.** Find the reduced row echelon form of A.

2nd MATRX F4 ⟨ops⟩ F5 ⟨rref⟩ ALPHA A ENTER

The result is

rref A

```
[[1  0  0    1.90081294721]
  0  1  0    4.19411081557]
  0  0  1  −11.3485183381]]
```

Thus $x_1 = 1.90081294721$, and so on.

*Note.* The TI-85 will give answers correct to 12 significant figures. However, in many applications it is inconvenient to have so many. You can control the number of decimal places of accuracy by pressing 2nd MODE and entering the number of decimal places of accuracy desired next to the symbol "Float." For example, by pressing ▼ to get to "Float," moving right to "3," and then pressing ENTER , only three decimal places will be displayed. Following the steps in the matrix A above then yields

rref A

```
[[1.000  0.000  0.000    1.901]
 [0.000  1.000  0.000    4.194]
 [0.000  0.000  1.000  −11.349]]
```

and thus, to three decimal places,

$$x_1 = 1.901, \qquad x_2 = 4.194, \text{ and } x_3 = -11.349$$

## CASIO fx-7700 GB

The CASIO does not compute the reduced row echelon form of a matrix so a more elaborate procedure must be followed.

**1.** Press MODE 0

**2.** Insert the number of equations and unknowns as follows:

F1 F6 F1 3 EXE 3 EXE

**3.** Enter the coefficient matrix by entering each number that is screened and then pressing EXE. A total of 9 numbers are entered, resulting in

$$\begin{array}{c|ccc} A & 1 & 2 & 3 \\ 1 & 3.8 & 1.6 & .9 \\ 2 & -0.7 & 5.4 & 1.6 \\ 3 & 1.5 & 1.1 & -3.2 \end{array}$$

**4.** Compute $A^{-1}$ and store it in $A$ [The symbol $A^{-1}$ is explained in Section 1.8] F4 F1 EXE. The result is

$$\begin{array}{c|ccc} A & 1 & 2 & 3 \\ 1 & 0.2377 & -0.076 & 0.0287 \\ 2 & -1\text{E-}03 & 0.1687 & 0.0837 \\ 3 & 0.1107 & 0.0222 & -0.27 \end{array}$$

**5.** Insert the number of equations for a second matrix B as follows:

MODE 0 F2 F6 F1 3 EXE 1 EXE

**6.** Enter the number on the right in the same way you entered numbers in Step 3. This results in

$$\begin{array}{c|c} B & 1 \\ 1 & 3.72 \\ 2 & 3.16 \\ 3 & 43.78 \end{array}$$

**7.** Compute $AB$ (really the inverse of the original $A$ times $B$) MODE 0 F5. The result is

$$\begin{array}{c|c} C & 1 \\ 1 & 1.9008 \\ 2 & 4.1941 \\ 3 & -11.34 \end{array}$$

and by pressing EXE twice you can see the fuller answer

$$\begin{bmatrix} 1.900812947 \\ 4.194110816 \\ -11.34851834 \end{bmatrix}$$

Thus $x = 1.900812947$, and so on.

*Note.* On the CASIO you will get an error message if the system either is inconsistent or has an infinite number of solutions.

In Problems 44–48 solve each system on a calculator.

**44.**
$$\begin{aligned} 2.6x_1 - 4.3x_2 + 9.6x_3 &= 21.62 \\ -8.5x_1 + 3.6x_2 + 9.1x_3 &= 14.23 \\ 12.3x_1 - 8.4x_2 - 0.6x_3 &= 12.61 \end{aligned}$$

**45.**
$$\begin{aligned} 2x_2 - x_3 - 4x_4 &= 2 \\ x_1 - x_2 + 5x_3 + 2x_4 &= -4 \\ 3x_1 + 3x_2 - 7x_3 - x_4 &= 4 \\ -x_1 - 2x_2 + 3x_3 &= -7 \end{aligned}$$

**46.**
$$\begin{aligned} 1.247x_1 - 2.583x_2 + 7.161x_3 + 8.275x_4 &= -1.205 \\ 3.472x_1 + 9.283x_2 + 11.275x_3 + 3.606x_4 &= 2.374 \\ -5.216x_1 - 12.816x_2 - 6.298x_3 + 1.877x_4 &= 21.206 \\ 6.812x_1 + 5.223x_2 - 9.725x_3 - 2.306x_4 &= -11.466 \end{aligned}$$

**47.**
$$\begin{aligned} 23.42x_1 - 16.89x_2 + 57.31x_3 + 82.6x_4 &= 2158.36 \\ -14.77x_1 + 38.29x_2 + 92.36x_3 - 4.36x_4 &= -1123.02 \\ -77.21x_1 + 71.26x_2 - 16.55x_3 + 43.09x_4 &= 3248.71 \\ 91.82x_1 + 81.43x_2 + 33.94x_3 + 57.22x_4 &= 235.25 \end{aligned}$$

**48.**
$$\begin{aligned} 6.1x_1 - 2.4x_2 + 23.3x_3 - 16.4x_4 - 8.9x_5 &= 121.7 \\ -14.2x_1 - 31.6x_2 - 5.8x_3 + 9.6x_4 + 23.1x_5 &= -87.7 \\ 10.5x_1 + 46.1x_2 - 19.6x_3 - 8.8x_4 - 41.2x_5 &= 10.8 \\ 37.3x_1 - 14.2x_2 + 62.0x_3 + 14.7x_4 - 9.6x_5 &= 61.3 \\ 0.8x_1 + 17.7x_2 - 47.5x_3 - 50.2x_4 + 29.8x_5 &= -27.8 \end{aligned}$$

## Further Exercises on the TI-85

The TI-85 has a command called ⟨ref⟩.

In Problems 49–53 compute the row echelon form (ref instead of rref) of each augmented matrix.

**49.** The matrix of Problem 45

**50.** The matrix of Problem 44

**51.** The matrix of Problem 47

**52.** The matrix of Problem 46

**53.** The matrix of Problem 48

In Problems 54–58 find all solutions, if any, to each system. Round off all answers to three decimal places. [*Hint:* First, obtain the reduced row echelon form of the augmented matrix.]

**54.**
$$\begin{aligned} 2.1x_1 + 4.2x_2 - 3.5x_3 &= 12.9 \\ -5.9x_1 + 2.7x_2 + 9.8x_3 &= -1.6 \end{aligned}$$

**55.**
$$\begin{aligned} -13.6x_1 + 71.8x_2 + 46.3x_3 &= -19.5 \\ 41.3x_1 - 75.0x_2 - 82.9x_3 &= 46.4 \\ 41.8x_1 + 65.4x_2 - 26.9x_3 &= 34.3 \end{aligned}$$

**56.**
$$\begin{aligned} -13.6x_1 + 71.8x_2 + 46.3x_3 &= 19.5 \\ 41.3x_1 - 75.0x_2 - 82.9x_3 &= 46.4 \\ 41.8x_1 + 65.4x_2 - 26.9x_3 &= 35.3 \end{aligned}$$

**57.**
$$\begin{aligned} 5x_1 - 2x_2 + 11x_3 - 16x_4 + 12x_5 &= 105 \\ -6x_1 + 8x_2 - 14x_3 - 9x_4 + 26x_5 &= -62 \\ 7x_1 - 18x_2 - 12x_3 + 21x_4 - 2x_5 &= 53 \end{aligned}$$

**58.**
$$\begin{aligned} 5x_1 - 2x_2 + 11x_3 - 16x_4 + 12x_5 &= 105 \\ -6x_1 + 8x_2 - 14x_3 - 9x_4 + 26x_5 &= -62 \\ 7x_1 - 18x_2 - 12x_3 + 21x_4 - 2x_5 &= 53 \\ -15x_1 + 42x_2 + 21x_3 - 17x_4 + 42x_5 &= -63 \end{aligned}$$

**59.**
$$\begin{aligned} 5x_1 - 2x_2 + 11x_3 - 16x_4 + 12x_5 &= 105 \\ -6x_1 + 8x_2 - 14x_3 - 9x_4 + 26x_5 &= -62 \\ 7x_1 - 18x_2 - 12x_3 + 21x_4 - 2x_5 &= 53 \\ -15x_1 + 42x_2 + 21x_3 - 17x_4 + 42x_5 &= 63 \end{aligned}$$

# MATLAB INTRODUCTION

## Examples of Basic MATLAB Commands

***MATLAB Is Case Sensitive.*** This means that **a** and **A** represent different variables.

***Entering Matrices.*** Entries in a row are separated by spaces, and columns are separated by ";":

| | |
|---|---|
| **A = [1 2 3;4 5 6;7 8 9]** | Produces the matrix $A = \begin{pmatrix} 1 & 2 & 3 \\ 4 & 5 & 6 \\ 7 & 8 & 9 \end{pmatrix}$ |
| **A = [1 2 3;**<br>**4 5 6;**<br>**7 8 9]** | Also produces the matrix $A$ above |
| **b = [3;6;1]** | Produces the matrix $b = \begin{pmatrix} 3 \\ 6 \\ 1 \end{pmatrix}$ |

***Notation for Forming Submatrices and Augmented Matrices.***

| | |
|---|---|
| **f = A(2,3)** | $f$ is the entry in second row, third column of $A$. |
| **d = A(3,:)** | $d$ is the third row of $A$. |
| **d = A(:,3)** | $d$ is the third column of $A$. |
| **C = A([2 4],:)** | $C$ is the matrix consisting of the second and fourth rows of $A$. |
| **C = [A b]** | Forms an augmented matrix $C = (A\vert b)$. |

***Performing Row Operations.***

| | |
|---|---|
| **A(2,:) = 3*A(2,:)** | $R_2 \to 3R_2$ |
| **A(2,:) = A(2,:)/4** | $R_2 \to \frac{1}{4}R_2$ |
| **A([2 3],:) = A([3 2],:)** | Interchanges rows 2 and 3 |
| **A(3,:) = A(3,:) + 3*A(2,:)** | $R_3 \to R_3 + 3R_2$ |

***Note.*** All of these commands change $A$. If you want to keep the original matrix $A$ and call the changed matrix $C$,

| | |
|---|---|
| **C = A** | |
| **C(2,:) = 3*C(2,:)** | |
| **C = rref(A)** | $C$ = reduced row echelon form of $A$ |

***Generating Random Matrices.***

| | |
|---|---|
| **A = rand(2,3)** | $2 \times 3$ matrix with entries between 0 and 1 |
| **A = 2*rand(2,3)−1** | $2 \times 3$ matrix with entries between $-1$ and 1 |
| **A = 4*(2*rand(2)−1)** | $2 \times 2$ matrix with entries between $-4$ and 4 |
| **A = round(10*rand(3))** | $3 \times 3$ matrix with integer entries between 0 and 10 |
| **A = 2*rand(3)−1 + i*(2*rand(3)−1)** | $3 \times 3$ matrix with complex entries $a + bi$, $a$ and $b$ between $-1$ and 1 |

## Other Useful Features and Commands

***Help.*** If you type in **help** followed by a MATLAB command, a description of how to use the command will appear.

***Examples.***

**help :** will give a description of how ":" can be used in MATLAB.

**help rref** will give a description of the rref command.

***Use of the Cursor Keys.*** For the MS-DOS version, using the up-arrow cursor key will display previous commands that you entered. You can use the cursor keys to locate a command, you can then modify the command, and by pressing the enter key, you can execute the modified command.

***Comment Statements.*** If you start an input line with the symbol %, MATLAB will interpret that line as a comment statement.

***Example.***

**% This is a comment.**

***Suppressing Display—Using ;.*** If you wish to perform a MATLAB command and do not want to see the results displayed, end the command with a ; (semicolon).

***For Long Entry Lines.*** To extend an input line, use "**. . .**".

**a = [1 2 3 4 5 6 7 8 . . .**
**9 10]** will produce $a = (1 \quad 2 \quad 3 \quad 4 \quad 5 \quad 6 \quad 7 \quad 8 \quad 9 \quad 10)$.

***For Display of Additional Digits.*** Normally MATLAB displays only 4 digits after the decimal point. Thus 4/3 is displayed as 1.3333. The command **format long** will cause all numbers to be displayed in full. So if **format long** is entered and then 4/3 is entered, the display will be 1.33333333333333. To return to the default display of 4 digits after the decimal point, enter the command **format short.**

## MATLAB Tutorial

1. Enter the matrices below in two different ways.

$$A = \begin{pmatrix} 2 & 2 & 3 & 4 & 5 \\ -6 & -1 & 2 & 0 & 7 \\ 1 & 2 & -1 & 3 & 4 \end{pmatrix} \qquad \mathbf{b} = \begin{pmatrix} -1 \\ 2 \\ 5 \end{pmatrix}$$

2. Form $C$ = augmented matrix $(A|b)$ for the $A$ and $\mathbf{b}$ above.
3. Form $D$, a random $3 \times 4$ matrix with entries between $-2$ and 2.
4. Form $B$, a random $4 \times 4$ matrix with integer entries between $-10$ and 10.
5. Form $K$, the matrix obtained from $B$ by interchanging rows 1 and 4 of $B$. Do not change $B$. (First, let $K = B$. Then change $K$.)
6. Perform the row operation $R_3 \to R_3 + (-1/2)R_1$ on the matrix $C$.
7. Enter the command **B([2 4],[1 3]).** Use a comment statement to describe the submatrix of $B$ that is produced.
8. Form $U$, the matrix consisting of just the third and fourth columns of $D$.

9. **(*MS-DOS*).** Use the up-arrow cursor to locate the command you used to perform the row operation in **6**. Modify the statement so that it performs the row operation $R_2 \to R_2 + 3R_1$ and then execute it.

10. Form $T$, a random $8 \times 7$ matrix with entries between 0 and 1. Type in the command **help :**. From the information given in the displayed description, determine how to use the ":" notation to form, as efficiently as possible, the matrix $S$ that consists of rows 3 through 8 of $T$.

11. Find the reduced row echelon form of $C$ using the **rref** command. Use it to write an equivalent system of equations.

## MATLAB 1.3

1. For each system in Problems 1, 2, 5, 8, and 16 in this section, enter the augmented matrix and use the **rref** command to find the reduced row echelon form. Show that each of these systems has a unique solution and that the solution is contained in the last column of the reduced row echelon form of the augmented matrix. Use the ":" notation to assign the variable **x** to the solution, that is, to the last column of the reduced row echelon form of the augmented matrix.

2. For each system in Problems 4, 7, 13, and 18 in this section, enter the augmented matrix and use the **rref** command to find the reduced row echelon form. Conclude that each of these systems has no solution.

3. The matrices below are augmented matrices from systems of equations that have infinitely many solutions.

   a. For each, enter the matrix and use the **rref** command to find the reduced row echelon form.

   **i.** $\left(\begin{array}{ccc|c} 3 & 5 & 1 & 0 \\ 4 & 2 & -8 & 0 \\ 8 & 3 & -18 & 0 \end{array}\right)$ **ii.** $\left(\begin{array}{cccc|c} 9 & 27 & 3 & 3 & 12 \\ 9 & 27 & 10 & 1 & 19 \\ 1 & 3 & 5 & 9 & 6 \end{array}\right)$

   **iii.** $\left(\begin{array}{ccccc|c} 1 & 0 & 1 & -2 & 7 & -4 \\ 1 & 4 & 21 & -2 & 2 & 5 \\ 3 & 0 & 3 & -6 & 7 & 2 \end{array}\right)$ **iv.** $\left(\begin{array}{ccccc|c} 6 & 4 & 7 & 5 & 15 & 9 \\ 8 & 5 & 9 & 10 & 10 & 8 \\ 4 & 5 & 7 & 7 & -1 & 7 \\ 8 & 3 & 7 & 6 & 22 & 8 \\ 3 & 2 & 7/2 & 9 & -12 & -2 \end{array}\right)$

   The rest of this problem involves pencil and paper work.

   b. For each reduced row echelon form, locate the pivots by circling them.

   c. For each reduced echelon form, write the equivalent system of equations.

   d. Solve each of these equivalent systems by choosing the arbitrary variables to be the variables that correspond to the columns that do not have pivots in the reduced row echelon form. (These variables are the natural variables to be chosen arbitrarily.)

4. The following systems represent the intersection of three planes in 3-space. Use the **rref** command as a tool to solve the systems. What conclusions can you make about the geometry of the planes?

   **i.** $$\begin{aligned} x_1 + 2x_2 + 3x_3 &= -1 \\ -3x_2 + x_3 &= 4 \\ 4x_1 + x_2 - 2x_3 &= 0 \end{aligned}$$

   **ii.** $$\begin{aligned} 2x_1 - x_2 + 4x_3 &= 5 \\ x_1 + 2x_2 - 3x_3 &= 6 \\ 4x_1 + 3x_2 - 2x_3 &= 9 \end{aligned}$$

**iii.** Same as (ii), except change the 9 to 17

**iv.**
$$\begin{aligned} 2x_1 - 4x_2 + 2x_3 &= 4 \\ 3x_1 - 6x_2 + 3x_3 &= 6 \\ -x_1 + 2x_2 - x_3 &= -2 \end{aligned}$$

5. Use MATLAB to reduce the augmented matrices below to reduced row echelon form step by step by performing the row operations. (See the examples of row operation commands in the MATLAB introduction on page 32.) Check your results using the **rref** command.

*Note.* If you originally call the matrix $A$, let D = A at the beginning and check **rref(D)**.

**i.** $\left(\begin{array}{ccc|c} 1 & 2 & -1 & 2 \\ 2 & 4 & 2 & 8 \\ 3 & 4 & -7 & 0 \end{array}\right)$ **ii.** $\left(\begin{array}{ccc|c} 1 & 2 & 3 & 2 \\ 3 & 4 & -1 & -3 \\ -2 & 1 & 0 & 4 \end{array}\right)$ **iii.** $\left(\begin{array}{ccccc|c} 1 & 2 & -2 & 0 & 1 & -2 \\ 2 & 4 & -1 & 0 & -4 & -19 \\ -3 & -6 & 12 & 2 & -12 & -8 \\ 1 & 2 & -2 & -4 & -5 & -34 \end{array}\right)$

See MATLAB Problem 1 in Section 1.5 for more work on performing row operations.

6. a. Let $A = \begin{pmatrix} 1 & 2 & -2 & 0 \\ 2 & 4 & -1 & 0 \\ -3 & -6 & 12 & 2 \\ 1 & 2 & -2 & -4 \end{pmatrix}$ $\mathbf{b} = \begin{pmatrix} 1 \\ -4 \\ -12 \\ 3 \end{pmatrix}$

Show that the system with augmented matrix $[A\ \mathbf{b}]$ has no solution.

b. Let **b = 2 * A(:,1) + A(:,2) + 3 * A(:,3) − 4 * A(:,4).** Recall that $A(:,1)$ is the first column of $A$. Hence we are taking a sum of multiples of the columns of $A$. Use **rref([A b])** to solve this system.

c. Use the up-arrow cursor key to bring back the **b** = 2 ∗ $A$(:,1) + etc. statement and edit in a new set of coefficients. Again, solve the system with augmented matrix $[A\ \mathbf{b}]$ for this new **b**. Repeat for two more choices of coefficients.

d. Would it be possible to put in coefficients so that there is no solution? Do you agree with the following conjecture: A system [$A$ **b**] has a solution if **b** is a sum of multiples of the columns of $A$? Why?

e. Test this conjecture for $A$ formed by:

**A = 2 * rand(5)−1**
**A(:,3) = 2 * A(:,1) − A(:,2)**

7. Suppose we wish to solve several systems of equations where the coefficient matrices (coefficients of the variables) are the same but there are different right-hand sides. Forming a larger augmented matrix will allow us to solve with multiple right-hand sides. Suppose $A$ is the coefficient matrix and **b** and **c** are two different right-hand sides; let **Aug = [A b c]** and find **rref(Aug).**

a. Solve the two systems below.

$$\begin{aligned} x_1 + x_2 + x_3 &= 4 \\ 2x_1 + 3x_2 + 4x_3 &= 9 \\ -2x_1 \qquad + 3x_3 &= -7 \end{aligned} \qquad \begin{aligned} x_1 + x_2 + x_3 &= 4 \\ 2x_1 + 3x_2 + 4x_3 &= 16 \\ -2x_1 \qquad + 3x_3 &= 11 \end{aligned}$$

b. Solve the three systems below.

$$\begin{aligned} 2x_1 + 3x_2 - 4x_3 &= 1 \\ x_1 + 2x_2 - 3x_3 &= 0 \\ -x_1 + 5x_2 - 11x_3 &= -7 \end{aligned} \qquad \begin{aligned} 2x_1 + 3x_2 - 4x_3 &= -1 \\ x_1 + 2x_2 - 3x_3 &= -1 \\ -x_1 + 5x_2 - 11x_3 &= -6 \end{aligned} \qquad \begin{aligned} 2x_1 + 3x_2 - 4x_3 &= 1 \\ x_1 + 2x_2 - 3x_3 &= 2 \\ -x_1 + 5x_2 - 11x_3 &= -7 \end{aligned}$$

c. Let $A$ be the coefficient matrix from part (a). Choose any three right-hand sides of your own. Solve.

d. We wish to make some observations about solutions to *square* systems, that is, systems with as many equations as variables. Answer the following questions by drawing on your observations from parts (a) through (c). (Pay particular attention to the form of the coefficient part of the rref.)

   **i.** Is it possible for a square system to have a unique solution with one right-hand side but infinitely many solutions with another right-hand side? Why or why not?

   **ii.** Is it possible for a square system to have a unique solution for one right-hand side but no solution for another right-hand side?

   **iii.** Is it possible for a square system to have infinitely many solutions for one right-hand side but no solution for another right-hand side? Why or why not?

8. **Heat Distribution** We have a rectangular plate whose edges are kept at certain temperatures. We are interested in finding the temperature at interior points. Consider the diagram below. We wish to find approximations for $T_1$ through $T_9$, the temperatures of the intermediate points. We assume that the temperature at an interior point is the average of the temperature of the four surrounding points—above, to the right, below, and to the left.

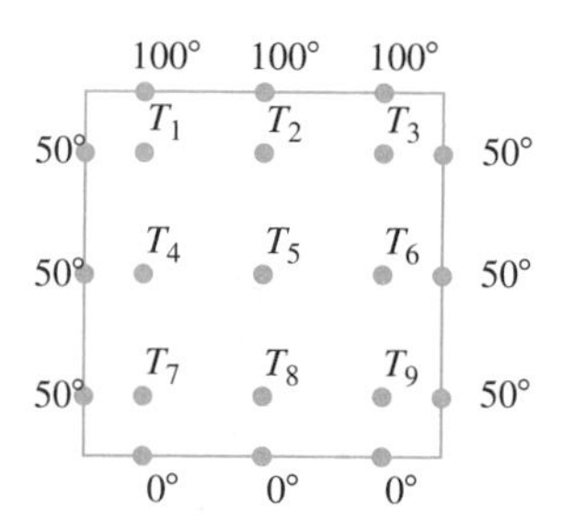


a. Using this assumption, set up a system of equations, first considering the point labeled $T_1$, then considering $T_2$, and so on. Rewrite so that all variables are on one side of the equation. For example, looking at $T_1$, we have

$$T_1 = (100 + T_2 + T_4 + 50)/4$$

which can be rewritten as $4T_1 - T_2 - T_4 = 150$.

Find the coefficient matrix and the augmented matrix. Describe the patterns you notice in the shape of the coefficient matrix. Such a matrix is called a **band matrix.** Do you see where that name comes from?

b. Solve the system using the **rref** command. Notice that you get a unique solution. Use the ":" notation to assign the variable name **x** to the solution.

c. Suppose $A$ is the coefficient matrix and **b** is the right-hand side for the above system. Enter the command **y = A\b.** (The slash here is called a **backslash.** It is NOT the divide slash.) Compare **y** and **x**.

9. **Leontief Input-Output Model**
   a. Refer to Example 10. Solve the system there using the **rref** command and the "\" command. Again, note that there is a unique solution.
   b. Suppose that there are three interdependent industries. The external demand for product 1 is 300,000; for product 2, 200,000; and for product 3, 200,000. Suppose the internal demands are given by

$$a_{11} = .2, \quad a_{12} = .1, \quad a_{13} = .3, \quad a_{21} = .15, \quad a_{22} = .25, \quad a_{23} = .25, \quad a_{31} = .1,$$
$$a_{32} = .05, \quad a_{33} = 0$$

   **i.** What does $a_{32} = .05$ tell you? What does $a_{33} = 0$ tell you?
   **ii.** Set up the appropriate augmented matrix for the system of equations to find $x_i$ = output of product $i$ for $i = 1, 2, 3$. FIRST REREAD EXAMPLE 10.
   **iii.** Solve the system using MATLAB. Interpret the solution; that is, how much of each product should be produced to have supply equal demand?
   **iv.** Suppose $x_i$ was measured in \$ (dollars of output) and you were interested in interpreting the solution to the penny. You would need more digits of the answer displayed than the default of four digits past the decimal point. Suppose you have assigned the variable name **x** to the solution. Enter the command **format long** (see page 33) and then the command **x**. This will display more digits. (When you are finished with this part, enter the command **format** to go back to the default.)

10. **Traffic Flow**
   a. Consider the diagram below of a grid of one-way streets with traffic entering and exiting intersections. Intersection $k$ is labeled as $[k]$. The arrows along each street indicate the direction of traffic flow. Let $x_i$ = the number of vehicles/hr traveling along street $i$. Assuming that the traffic that enters an intersection also leaves the intersection, set up a system of equations describing the traffic flow diagram. For example, at intersection [1]

$$x_1 + x_5 + 100 = \text{traffic in} = \text{traffic out} = x_3 + 300 \text{ yielding } x_1 - x_3 + x_5 = 200.$$

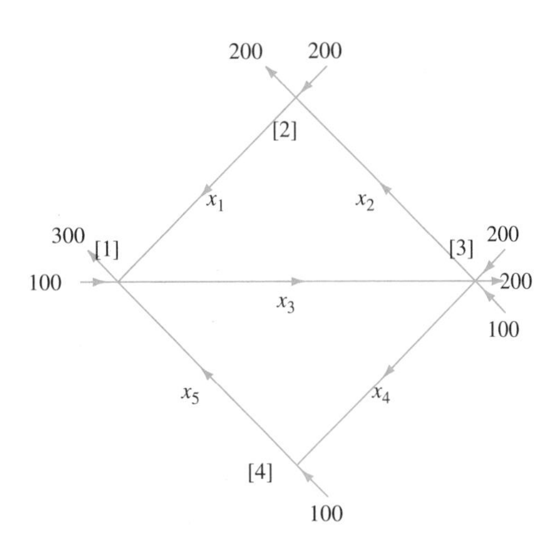

b. Solve the system using the **rref** command. There will be infinitely many solutions. Write the solutions in terms of the variables that are the natural variables to be chosen arbitrarily.

c. Suppose the street from [1] to [3] needs to be closed; that is, $x_3 = 0$. Can the street from [1] to [4] also be closed ($x_5 = 0$) without changing the one-way directions? If it cannot be closed, what is the smallest amount of traffic this street (from [1] to [4]) must be able to handle?

11. **Fitting Polynomials to Points** If we are given two points in the plane with distinct $x$-coordinates, there is a unique line $y = c_1x + c_2$ that passes through both points. If we are given three points in the plane with distinct $x$-coordinates, there is a unique parabola

$$y = c_1x^2 + c_2x + c_3$$

that passes through the three points. If we have $n + 1$ points in a plane with distinct $x$-coordinates, then there is a unique degree $n$ polynomial that passes through the $n + 1$ points:

$$y = c_1x^n + c_2x^{(n-1)} + \cdots + c_{n+1}$$

The coefficients $c_1, \ldots, c_{n+1}$ can be found by solving a system of linear equations.

*Example.*

$$P_1 = (2, 5) \qquad P_2 = (3, 10) \qquad P_3 = (4, -3)$$

We want to find $c_1$, $c_2$, and $c_3$ so that $y = c_1x^2 + c_2x + c_3$ passes through the points $P_1$, $P_2$, and $P_3$.

$$\begin{aligned} 5 &= c_1 2^2 + c_2 2 + c_3 \\ 10 &= c_1 3^2 + c_2 3 + c_3 \\ -3 &= c_1 4^2 + c_2 4 + c_3 \end{aligned}$$

Thus we have

$$A = \begin{pmatrix} 2^2 & 2 & 1 \\ 3^2 & 3 & 1 \\ 4^2 & 4 & 1 \end{pmatrix} \qquad \mathbf{b} = \begin{pmatrix} 5 \\ 10 \\ -3 \end{pmatrix}.$$

Solving the system for the coefficients, we get $\mathbf{c} = \begin{pmatrix} -9 \\ 50 \\ -59 \end{pmatrix}$, which tells us that the parabola that goes through each point is $y = -9x^2 + 50x - 59$. We say that the parabola *fits* the points.

a. For $P_1 = (1, -1)$ $P_2 = (3, 3)$ and $P_3 = (4, -2)$, set up the system of equations to find the coefficients of the parabola that fits the points. Set $A =$ the coefficient matrix and $\mathbf{b} =$ the right-hand side. Solve the system. In a comment statement, write the equation of the parabola that fits the points, that is, passes through the three points.

Enter **x = [1;3;4]** and enter **V = vander(x).** Compare $V$ to $A$.

b. For $P_1 = (0, 5)$ $P_2 = (1, -2)$ $P_3 = (3, 3)$ $P_4 = (4, -2)$, set up the system of equations, enter the augmented matrix and use MATLAB to solve the system.

In a comment statement, write down the equation of the cubic polynomial that fits the four points.

Let **x** be the column vector containing the $x$-coordinates of the points $P_1$ through $P_4$. Enter **x** and find **V = vander(x).** Compare $V$ to the coefficient matrix you found when setting up the system.

c. Using some graphics features of MATLAB, we can visualize the results by following the commands below. Follow these commands for the points in (a) and again for the points in (b).

Enter **x** as the column vector of $x$-coordinates of the points

Enter **y** as the column vector of $y$-coordinates of the points

Enter the following commands:

```
V = vander(x)
c = V\y
s = min(x):.01:max(x);
yy = polyval(c,s);
plot(x,y,'*',s,yy)
```

The first command gives the desired coefficient matrix.

The second solves the system for the coefficients of the polynomial.

The third creates a vector **s** containing many entries, each entry between the minimum and maximum $x$-coordinate, so that we can evaluate the polynomial at many points to create a good graph.

The fourth creates a vector **yy** that contains the $y$-coordinates obtained by evaluating the polynomial at all the entries of **s**.

The fifth produces a plot of the original points (with a "*" symbol) and a plot of the graph of the polynomial.

You should observe that the graph of the polynomial passes through the points (labeled with "*").

d. Generate **x = rand(7,1)** and **y = rand(7,1)** or generate a vector of $x$-coordinates and a vector of $y$-coordinates of your own choosing. Be sure that you change (or choose) the $x$-coordinates to ensure that they are distinct. Follow the commands in part (c) to visualize the polynomial fit.

## 1.4 HOMOGENEOUS SYSTEMS OF EQUATIONS

The general $m \times n$ system of linear equations [system (1.3.7), page 16] is called **homogeneous** if all the constants $b_1, b_2, \ldots, b_m$ are zero. That is, the general homogeneous system is given by

$$\begin{aligned} a_{11}x_1 + a_{12}x_2 + \cdots + a_{1n}x_n &= 0 \\ a_{21}x_1 + a_{22}x_2 + \cdots + a_{2n}x_n &= 0 \\ \vdots \qquad \vdots \qquad\quad \vdots \quad & \quad \vdots \\ a_{m1}x_1 + a_{m2}x_2 + \cdots + a_{mn}x_n &= 0 \end{aligned} \tag{1}$$

Homogeneous systems arise in a variety of ways. We shall see one of these in Section 4.4. In this section we solve some homogeneous systems—again by the method of Gauss-Jordan elimination.

For the general linear system there are three possibilities: no solution, one solution, or an infinite number of solutions. For the general homogeneous system the situation is simpler. Since $x_1 = x_2 = \cdots = x_n = 0$ is always a solution (called the **trivial solution** or **zero solution**), there are only two possibilities: Either the zero solution is the only solution or there are an infinite number of solutions in addition to the zero solution. Solutions other than the zero solution are called **nontrivial solutions.**

Trivial solution
Zero solution
Nontrivial solution

EXAMPLE 1 **A Homogeneous System with Only the Zero Solution** Solve the homogeneous system

$$\begin{aligned} 2x_1 + 4x_2 + 6x_3 &= 0 \\ 4x_1 + 5x_2 + 6x_3 &= 0 \\ 3x_1 + x_2 - 2x_3 &= 0 \end{aligned}$$

**Solution** This is the homogeneous version of the system in Example 1.3.1, page 7. Reducing successively, we obtain (after dividing the first equation by 2)

$$\left(\begin{array}{ccc|c} 1 & 2 & 3 & 0 \\ 4 & 5 & 6 & 0 \\ 3 & 1 & -2 & 0 \end{array}\right) \xrightarrow[R_3 \to R_3 - 3R_1]{R_2 \to R_2 - 4R_1} \left(\begin{array}{ccc|c} 1 & 2 & 3 & 0 \\ 0 & -3 & -6 & 0 \\ 0 & -5 & -11 & 0 \end{array}\right) \xrightarrow{R_2 \to -\frac{1}{3}R_2} \left(\begin{array}{ccc|c} 1 & 2 & 3 & 0 \\ 0 & 1 & 2 & 0 \\ 0 & -5 & -11 & 0 \end{array}\right)$$

$$\xrightarrow[R_3 \to R_3 + 5R_2]{R_1 \to R_1 - 2R_2} \left(\begin{array}{ccc|c} 1 & 0 & -1 & 0 \\ 0 & 1 & 2 & 0 \\ 0 & 0 & -1 & 0 \end{array}\right) \xrightarrow{R_3 \to -R_3} \left(\begin{array}{ccc|c} 1 & 0 & -1 & 0 \\ 0 & 1 & 2 & 0 \\ 0 & 0 & 1 & 0 \end{array}\right) \xrightarrow[R_2 \to R_2 - 2R_3]{R_1 \to R_1 + R_3} \left(\begin{array}{ccc|c} 1 & 0 & 0 & 0 \\ 0 & 1 & 0 & 0 \\ 0 & 0 & 1 & 0 \end{array}\right)$$

Thus the system has the unique solution (0, 0, 0). That is, the system has only the trivial solution.

EXAMPLE 2 **A Homogeneous System with an Infinite Number of Solutions** Solve the homogeneous system

$$\begin{aligned} x_1 + 2x_2 - x_3 &= 0 \\ 3x_1 - 3x_2 + 2x_3 &= 0 \\ -x_1 - 11x_2 + 6x_3 &= 0 \end{aligned}$$

**Solution** Using Gauss-Jordan elimination, we obtain, successively,

$$\left(\begin{array}{ccc|c} 1 & 2 & -1 & 0 \\ 3 & -3 & 2 & 0 \\ -1 & -11 & 6 & 0 \end{array}\right) \xrightarrow[R_3 \to R_3 + R_1]{R_2 \to R_2 - 3R_1} \left(\begin{array}{ccc|c} 1 & 2 & -1 & 0 \\ 0 & -9 & 5 & 0 \\ 0 & -9 & 5 & 0 \end{array}\right)$$

$$\xrightarrow{R_2 \to -\frac{1}{9}R_2} \left(\begin{array}{ccc|c} 1 & 2 & -1 & 0 \\ 0 & 1 & -\frac{5}{9} & 0 \\ 0 & -9 & 5 & 0 \end{array}\right) \xrightarrow[R_3 \to R_3 + 9R_2]{R_1 \to R_1 - 2R_2} \left(\begin{array}{ccc|c} 1 & 0 & \frac{1}{9} & 0 \\ 0 & 1 & -\frac{5}{9} & 0 \\ 0 & 0 & 0 & 0 \end{array}\right)$$

The augmented matrix is now in reduced row echelon form and, evidently, there are an infinite number of solutions given by $(-\frac{1}{9}x_3, \frac{5}{9}x_3, x_3)$. If $x_3 = 0$, for example, we obtain the trivial solution. If $x_3 = 1$, we obtain the solution $(-\frac{1}{9}, \frac{5}{9}, 1)$. If $x_3 = 9\pi$, we obtain the solution $(-\pi, 5\pi, 9\pi)$.

**EXAMPLE 3** **A Homogeneous System with More Unknowns Than Equations Has an Infinite Number of Solutions** Solve the system

$$\begin{aligned} x_1 + x_2 - x_3 &= 0 \\ 4x_1 - 2x_2 + 7x_3 &= 0 \end{aligned} \tag{2}$$

**Solution** Row-reducing, we obtain

$$\left(\begin{array}{rrr|r} 1 & 1 & -1 & 0 \\ 4 & -2 & 7 & 0 \end{array}\right) \xrightarrow{R_2 \to R_2 - 4R_1} \left(\begin{array}{rrr|r} 1 & 1 & -1 & 0 \\ 0 & -6 & 11 & 0 \end{array}\right)$$

$$\xrightarrow{R_2 \to -\frac{1}{6}R_2} \left(\begin{array}{rrr|r} 1 & 1 & -1 & 0 \\ 0 & 1 & -\frac{11}{6} & 0 \end{array}\right) \xrightarrow{R_1 \to R_1 - R_2} \left(\begin{array}{rrr|r} 1 & 0 & \frac{5}{6} & 0 \\ 0 & 1 & -\frac{11}{6} & 0 \end{array}\right)$$

Thus there are an infinite number of solutions given by $(-\frac{5}{6}x_3, \frac{11}{6}x_3, x_3)$. This may not be surprising because system (2) contains three unknowns and only two equations.

In general, if there are more unknowns than equations, the homogeneous system (1) will always have an infinite number of solutions. To see this, note that if there were only the trivial solution, then row reduction would lead us to the system

$$\begin{aligned} x_1 \qquad\qquad &= 0 \\ x_2 \qquad &= 0 \\ &\vdots \\ x_n &= 0 \end{aligned}$$

and, possibly, additional equations of the form $0 = 0$. But this system has at least as many equations as unknowns. Since row reduction does not change either the number of equations or the number of unknowns, we have a contradiction of our assumption that there were more unknowns than equations. Thus we have Theorem 1.

**THEOREM 1** The homogeneous system (1) has an infinite number of solutions if $n > m$.

# PROBLEMS 1.4

## Self-Quiz

**I.** Which of the following systems *must* have nontrivial solutions?

**a.** $a_{11}x_1 + a_{12}x_2 = 0$
$a_{21}x_1 + a_{22}x_2 = 0$

**b.** $a_{11}x_1 + a_{12}x_2 = 0$
$a_{21}x_1 + a_{22}x_2 = 0$
$a_{31}x_1 + a_{32}x_2 = 0$

**c.** $a_{11}x_1 + a_{12}x_2 + a_{13}x_3 = 0$
$a_{21}x_1 + a_{22}x_2 + a_{23}x_3 = 0$

**II.** For what value of $k$ will the following system have nontrivial solutions?

$$\begin{aligned} x + y + z &= 0 \\ 2x + 3y + 4z &= 0 \\ 3x + 4y + kz &= 0 \end{aligned}$$

**a.** 1 **b.** 2 **c.** 3 **d.** 4 **e.** 5 **f.** 0

In Problems 1–13 find all solutions to the homogeneous systems.

**1.** $$\begin{aligned} 2x_1 - x_2 &= 0 \\ 3x_1 + 4x_2 &= 0 \end{aligned}$$

**2.** $$\begin{aligned} x_1 - 5x_2 &= 0 \\ -x_1 + 5x_2 &= 0 \end{aligned}$$

**3.** $$\begin{aligned} x_1 + x_2 - x_3 &= 0 \\ 2x_1 - 4x_2 + 3x_3 &= 0 \\ 3x_1 + 7x_2 - x_3 &= 0 \end{aligned}$$

**4.** $$\begin{aligned} x_1 + x_2 - x_3 &= 0 \\ 2x_1 - 4x_2 + 3x_3 &= 0 \\ -x_1 - 7x_2 + 6x_3 &= 0 \end{aligned}$$

**5.** $$\begin{aligned} x_1 + x_2 - x_3 &= 0 \\ 2x_1 - 4x_2 + 3x_3 &= 0 \\ -5x_1 + 13x_2 - 10x_3 &= 0 \end{aligned}$$

**6.** $$\begin{aligned} 2x_1 + 3x_2 - x_3 &= 0 \\ 6x_1 - 5x_2 + 7x_3 &= 0 \end{aligned}$$

**7.** $$\begin{aligned} 4x_1 - x_2 &= 0 \\ 7x_1 + 3x_2 &= 0 \\ -8x_1 + 6x_2 &= 0 \end{aligned}$$

**8.** $$\begin{aligned} x_1 - x_2 + 7x_3 - x_4 &= 0 \\ 2x_1 + 3x_2 - 8x_3 + x_4 &= 0 \end{aligned}$$

**9.** $$\begin{aligned} x_1 - 2x_2 + x_3 + x_4 &= 0 \\ 3x_1 \qquad + 2x_3 - 2x_4 &= 0 \\ 4x_2 - x_3 - x_4 &= 0 \\ 5x_1 \qquad + 3x_3 - x_4 &= 0 \end{aligned}$$

**10.** $$\begin{aligned} -2x_1 \qquad\qquad + 7x_4 &= 0 \\ x_1 + 2x_2 - x_3 + 4x_4 &= 0 \\ 3x_1 \qquad - x_3 + 5x_4 &= 0 \\ 4x_1 + 2x_2 + 3x_3 \qquad &= 0 \end{aligned}$$

**11.** $$\begin{aligned} 2x_1 - x_2 &= 0 \\ 3x_1 + 5x_2 &= 0 \\ 7x_1 - 3x_2 &= 0 \\ -2x_1 + 3x_2 &= 0 \end{aligned}$$

**12.** $$\begin{aligned} x_1 - 3x_2 &= 0 \\ -2x_1 + 6x_2 &= 0 \\ 4x_1 - 12x_2 &= 0 \end{aligned}$$

**13.** $$\begin{aligned} x_1 + x_2 - x_3 &= 0 \\ 4x_1 - x_2 + 5x_3 &= 0 \\ -2x_1 + x_2 - 2x_3 &= 0 \\ 3x_1 + 2x_2 - 6x_3 &= 0 \end{aligned}$$

**Answers to Self-Quiz**

**I.** c **II.** e

**14.** Show that the homogeneous system

$$\begin{aligned} a_{11}x_1 + a_{12}x_2 &= 0 \\ a_{21}x_1 + a_{22}x_2 &= 0 \end{aligned}$$

has an infinite number of solutions if and only if $a_{11}a_{22} - a_{12}a_{21} = 0$.

**15.** Consider the system

$$\begin{aligned} 2x_1 - 3x_2 + 5x_3 &= 0 \\ -x_1 + 7x_2 - x_3 &= 0 \\ 4x_1 - 11x_2 + kx_3 &= 0 \end{aligned}$$

For what value of $k$ will the system have nontrivial solutions?

***16.** Consider the $3 \times 3$ homogeneous system

$$\begin{aligned} a_{11}x_1 + a_{12}x_2 + a_{13}x_3 &= 0 \\ a_{21}x_1 + a_{22}x_2 + a_{23}x_3 &= 0 \\ a_{31}x_1 + a_{32}x_2 + a_{33}x_3 &= 0 \end{aligned}$$

Find conditions on the coefficients $a_{ij}$ such that the zero solution is the only solution.

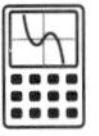

## CALCULATOR BOX

Homogeneous systems can be solved on the TI-85 but not on the CASIO fx-7700 GB. The reason is that the TI-85 computes the row or reduced row echelon form of a matrix while the CASIO does not. The CASIO can only solve a homogeneous system of $n$ equations in $n$ unknowns and then only when there is a unique solution (i.e., the zero solution). The following exercises are intended to be done on the TI-85.

### TI-85

In Problems 17–20 find all solutions to each system.

**17.**
$$\begin{aligned} 2.1x_1 + 4.2x_2 - 3.5x_3 &= 0 \\ -5.9x_1 + 2.7x_2 + 9.8x_3 &= 0 \end{aligned}$$

**18.**
$$\begin{aligned} -13.6x_1 + 71.8x_2 + 46.3x_3 &= 0 \\ 41.3x_1 - 75.0x_2 - 82.9x_3 &= 0 \\ 41.8x_1 + 65.4x_2 - 26.9x_3 &= 0 \end{aligned}$$

**19.**
$$\begin{aligned} 25x_1 - 16x_2 + 13x_3 + 33x_4 - 57x_5 &= 0 \\ -16x_1 + 3x_2 + x_3 \qquad\quad + 12x_5 &= 0 \\ -8x_2 \qquad\quad + 16x_4 - 26x_5 &= 0 \end{aligned}$$

**20.**
$$\begin{aligned} 5x_1 - 2x_2 + 11x_3 - 16x_4 + 12x_5 &= 0 \\ -6x_1 + 8x_2 - 14x_3 - 9x_4 + 26x_5 &= 0 \\ 7x_1 - 18x_2 - 12x_3 + 21x_4 - 2x_5 &= 0 \\ -x_1 + 11x_2 - 9x_3 + 13x_4 - 20x_5 &= 0 \end{aligned}$$

## MATLAB 1.4

1. a. Generate four random matrices with more columns (unknowns) than rows (equations).
   b. Use the **rref** command to find the reduced row echelon form of each.
   c. For each matrix, use the reduced row echelon form to write down the solutions to the associated homogeneous system. Verify Theorem 1, that is, there are always infinitely many solutions in this case.

   (To use MATLAB to generate random matrices, see the discussion preceding the MATLAB problems for Section 1.3.)

2. What can you conjecture about the solutions of a homogeneous system whose coefficient matrix has more rows (equations) than columns (unknowns)? Solve the homogeneous systems whose coefficient matrices are given below. Do these results confirm your conjecture?

$$\textbf{i.}\ \begin{pmatrix} 1 & 2 & 3 & 0 \\ -1 & 4 & 5 & -1 \\ 0 & 2 & -6 & 2 \\ 1 & 1 & 1 & 3 \\ 0 & 2 & 0 & 1 \end{pmatrix} \qquad \textbf{ii.}\ \begin{pmatrix} 1 & -1 & 3 \\ 2 & 1 & 3 \\ 0 & 2 & -2 \\ 4 & 4 & 4 \end{pmatrix}$$

3. **Balancing Chemical Reactions** In balancing a chemical reaction such as the one for photosynthesis

$$CO_2 + H_2O \rightarrow C_6H_{12}O_6 + O_2$$

we seek positive integers $x_1$, $x_2$, $x_3$, and $x_4$, which have no common divisor other than 1 so that in

$$x_1(CO_2) + x_2(H_2O) \rightarrow x_3(C_6H_{12}O_6) + x_4(O_2)$$

the number of atoms of each chemical element involved is the same on each side of the reaction. The number of atoms of a chemical element is indicated by the subscript; for example, in $CO_2$ there is one atom of C (carbon) and two atoms of O (oxygen). This leads to a homogeneous system of equations. Why would "balancing" lead to a homogeneous system?

$$\begin{aligned} &\text{C: } x_1 = 6x_3 && x_1 - 6x_3 = 0 \\ &\text{O: } 2x_1 + x_2 = 6x_3 + 2x_4 \quad \text{or} && 2x_1 + x_2 - 6x_3 - 2x_4 = 0 \\ &\text{H: } 2x_2 = 12x_3 && 2x_2 - 12x_3 = 0 \end{aligned}$$

This system has more unknowns than equations, so we expect infinitely many solutions. To solve the system, enter the augmented matrix, use the **rref** command, and write down the solution in terms of arbitrary variables. We will need to make a choice of the arbitrary variables so that $x_1$, $x_2$, $x_3$, and $x_4$ are all integers with no common divisor other than 1.

For the systems considered here, there will be one arbitrary variable corresponding to the last column of the rref of the coefficient matrix. To aid in finding the right choice of arbitrary variable to produce integers, use the ":" notation to assign the variable name **z** to the last column of the rref of the coefficient matrix. Enter the command **xx = rat(z,'s')**.

This will display the numbers in this column as fractions instead of as decimals. If you are using MATLAB 4.0, enter the command **format rat** and then dixplay **xx**. (Be sure to enter the command **format short** to return to the default display.)

a. Solve the photosynthesis system above and find $x_1$ through $x_4$, integers with no common divisor other than 1, that will balance the photosynthesis reaction.

b. Set up the homogeneous system of equations that will balance the reaction below:

$$Pb(N_3)_2 + Cr(MnO_4)_2 \rightarrow Cr_2O_3 + MnO_2 + Pb_3O_4 + NO$$

Solve the system and find $x_1$ through $x_6$, integers with no common divisor other than 1, that will balance the reaction.

# 1.5 VECTORS AND MATRICES

The study of vectors and matrices lies at the heart of linear algebra. The study of vectors began essentially with the work of the great Irish mathematician Sir William Rowan Hamilton (1805–1865).† His desire to find a way to represent certain objects in the plane and in space led to the discovery of what he called *quaternions*. This notion led to the development of what we now call *vectors*. Throughout Hamilton's life, and for the remainder of the nineteenth century, there was considerable debate over the usefulness of quaternions and vectors. At the end of the century the great British physicist Lord Kelvin wrote that quaternions, "although beautifully ingenious, have been an unmixed evil to those who have touched them in any way [and] vectors . . . have never been of the slightest use to any creature."

But Kelvin was wrong. Today nearly all branches of classical and modern physics are represented by means of the language of vectors. Vectors are also used with increasing frequency in the social and biological sciences.‡

On page 2 we described the solution to a system of two equations in two unknowns as a pair of numbers written $(x, y)$. In Example 1.3.1 on page 9 we wrote the solution of a system of three equations in three unknowns as the triple of numbers $(4, -2, 3)$. Both $(x, y)$ and $(4, -2, 3)$ are **vectors.**

**DEFINITION 1** ***n*-Component Row Vector** We define an ***n*-component row vector** to be an **ordered** set of $n$ numbers written as

$$(x_1, x_2, \ldots, x_n) \tag{1}$$

† See the biographical sketch of Hamilton on page 54.

‡ For interesting discussions of the development of modern vector analysis, consult the book by M. J. Crowe, *A History of Vector Analysis* (Notre Dame: University of Notre Dame Press, 1967) or Morris Kline's excellent book *Mathematical Thought from Ancient to Modern Times* (New York: Oxford University Press, 1972), Chapter 32.

**DEFINTION 2** ***n*-Component Column Vector** An ***n*-component column vector** is an **ordered** set of $n$ numbers written as

$$\begin{pmatrix} x_1 \\ x_2 \\ \vdots \\ x_n \end{pmatrix} \tag{2}$$

Components of a vector

In (1) or (2), $x_1$ is called the **first component** of the vector, $x_2$ is the **second component,** and so on. In general, $x_k$ is called the ***k*th component** of the vector.

For simplicity, we shall often refer to an $n$-component row vector as a **row vector** or an ***n*-vector.** Similarly, we shall use the term **column vector** (or $n$-vector) to denote an $n$-component column vector. Any vector whose entries are all zero is called a **zero vector.**

Zero vector

**EXAMPLE 1** **Four Vectors** The following are vectors:

**i.** (3, 6) is a row vector (or a 2-vector).

**ii.** $\begin{pmatrix} 2 \\ -1 \\ 5 \end{pmatrix}$ is a column vector (or a 3-vector).

**iii.** $(2, -1, 0, 4)$ is a row vector (or a 4-vector).

**iv.** $\begin{pmatrix} 0 \\ 0 \\ 0 \\ 0 \\ 0 \end{pmatrix}$ is a column vector and a zero vector.

**WARNING** The word "ordered" in the definition of a vector is essential. Two vectors with the same components written in different orders are *not* the same. Thus, for example, the row vectors (1, 2) and (2, 1) are not equal.

For the remainder of this text we shall denote vectors with boldface lowercase letters like **u**, **v**, **a**, **b**, **c**, and so on. A zero vector is denoted by **0**. Moreover, since it will usually be obvious whether a vector is a row or a column, we shall usually refer to row or column vectors simply as "vectors."

Vectors arise in a great number of ways. Suppose that the buyer for a manufacturing plant must order different quantities of steel, aluminum, oil, and paper. He can keep track of the quantities to be ordered with a single vector. The vector $\begin{pmatrix} 10 \\ 30 \\ 15 \\ 60 \end{pmatrix}$ indicates that he would order 10 units of steel, 30 units of aluminum, and so on.

*Remark.* We see here why the order in which the components of a vector are written is important. It is clear that the vectors $\begin{pmatrix} 30 \\ 15 \\ 60 \\ 10 \end{pmatrix}$ and $\begin{pmatrix} 10 \\ 30 \\ 15 \\ 60 \end{pmatrix}$ mean very different things to the buyer.

We now describe some properties of vectors. Since it would be repetitive to do so first for row vectors and then for column vectors, we shall give all definitions in terms of column vectors. Similar definitions hold for row vectors.

$\mathbb{R}$
$\mathbb{C}$

The components of all the vectors in this text are either real or complex numbers.† We denote the set of all real numbers by $\mathbb{R}$ and the set of all complex numbers by $\mathbb{C}$.

**The Space $\mathbb{R}^n$**

We use the symbol $\mathbb{R}^n$ to denote the set of all $n$-vectors $\begin{pmatrix} a_1 \\ a_2 \\ \vdots \\ a_n \end{pmatrix}$, where each $a_i$ is a real number.

$\mathbb{C}^n$

Similarly, we use the symbol $\mathbb{C}^n$ to denote the set of all $n$-vectors $\begin{pmatrix} c_1 \\ c_2 \\ \vdots \\ c_n \end{pmatrix}$, where each $c_i$ is a complex number. In Chapter 3 we shall discuss the sets $\mathbb{R}^2$ (vectors in the plane) and $\mathbb{R}^3$ (vectors in space). In Chapter 4 we shall examine arbitrary sets of vectors.

Vectors are really special types of matrices. Therefore, rather than discussing properties of vectors, we move on to the properties of matrices.

---

† A complex number is a number of the form $a + ib$, where $a$ and $b$ are real numbers and $i = \sqrt{-1}$. A description of complex numbers is given in Appendix 2. We shall not encounter complex vectors again until Chapter 4; they will be especially useful in Chapter 6. Therefore, unless otherwise stated, we assume, for the time being, that all vectors have real components.

**DEFINITION 3** **Matrix** An $m \times n$ **matrix** $A$ is a rectangular array of $mn$ numbers arranged in $m$ rows and $n$ columns:

$$A = \begin{pmatrix} a_{11} & a_{12} & \cdots & a_{1j} & \cdots & a_{1n} \\ a_{21} & a_{22} & \cdots & a_{2j} & \cdots & a_{2n} \\ \vdots & \vdots & & \vdots & & \vdots \\ a_{i1} & a_{i2} & \cdots & a_{ij} & \cdots & a_{in} \\ \vdots & \vdots & & \vdots & & \vdots \\ a_{m1} & a_{m2} & \cdots & a_{mj} & \cdots & a_{mn} \end{pmatrix} \tag{3}$$

The symbol $m \times n$ is read "$m$ by $n$." Unless stated otherwise, we shall always assume that the numbers in a matrix or vector are real. We call the row vector $(a_{i1}, a_{i2}, \ldots, a_{in})$ **row *i*** and the column vector $\begin{pmatrix} a_{1j} \\ a_{2j} \\ \vdots \\ a_{mj} \end{pmatrix}$ **column *j*.** The ***ij*th component** of $A$, denoted by $a_{ij}$, is the number appearing in the $i$th row and $j$th column of $A$. We shall sometimes write the matrix $A$ as $A = (a_{ij})$. Usually, matrices will be denoted by capital letters.

Rows and columns of a matrix

Component

Square matrix

Zero matrix

Size of a matrix

If $A$ is an $m \times n$ matrix with $m = n$, then $A$ is called a **square matrix.** An $m \times n$ matrix with all components equal to zero is called the $m \times n$ **zero matrix.**

An $m \times n$ matrix is said to have the **size** $m \times n$.

***Historical Note.*** The term "matrix" was first used in 1850 by the British mathematician James Joseph Sylvester (1814–1897) to distinguish matrices from determinants (which we shall discuss in Chapter 2). In fact, the term "matrix" was intended to mean "mother of determinants."

**EXAMPLE 2** **Five Matrices** Five matrices of different sizes are given below:

**i.** $\begin{pmatrix} 1 & 3 \\ 4 & 2 \end{pmatrix}$ is a $2 \times 2$ matrix (square).

**ii.** $\begin{pmatrix} -1 & 3 \\ 4 & 0 \\ 1 & -2 \end{pmatrix}$ is a $3 \times 2$ matrix.

**iii.** $\begin{pmatrix} -1 & 4 & 1 \\ 3 & 0 & 2 \end{pmatrix}$ is a $2 \times 3$ matrix.

**iv.** $\begin{pmatrix} 1 & 6 & -2 \\ 3 & 1 & 4 \\ 2 & -6 & 5 \end{pmatrix}$ is a $3 \times 3$ matrix (square).

**v.** $\begin{pmatrix} 0 & 0 & 0 & 0 \\ 0 & 0 & 0 & 0 \end{pmatrix}$ is the $2 \times 4$ zero matrix.

*Bracket Notation.* In some books matrices are given in square brackets rather than in parentheses. For example, the first two matrices in Example 2 can be written as

**i.** $A = \begin{bmatrix} 1 & 3 \\ 4 & 2 \end{bmatrix}$ **ii.** $A = \begin{bmatrix} -1 & 3 \\ 4 & 0 \\ 1 & -2 \end{bmatrix}$

In this text we shall use parentheses exclusively.

Throughout this book we refer to the $i$th row, the $j$th column, and the $ij$th component of a matrix for various numbers $i$ and $j$. We illustrate these ideas in the next example.

**EXAMPLE 3** **Finding Components of a Matrix** Find the 1,2, 3,1, and 2,2 components of

$$A = \begin{pmatrix} 1 & 6 & 4 \\ 2 & -3 & 5 \\ 7 & 4 & 0 \end{pmatrix}$$

**Solution** The 1,2 component is the number in the first row and the second column. We have shaded the first row and the second column; the 1,2 component is 6:

2nd column ↓

$$\text{1st row} \rightarrow \begin{pmatrix} 1 & 6 & 4 \\ 2 & -3 & 5 \\ 7 & 4 & 0 \end{pmatrix}$$

From the shaded matrices below, we see that the 3,1 component is 7 and the 2,2 component is $-3$:

1st column ↓ 2nd column ↓

$$\text{3rd row} \rightarrow \begin{pmatrix} 1 & 6 & 4 \\ 2 & -3 & 5 \\ 7 & 4 & 0 \end{pmatrix} \qquad \text{2nd row} \rightarrow \begin{pmatrix} 1 & 6 & 4 \\ 2 & -3 & 5 \\ 7 & 4 & 0 \end{pmatrix}$$

**DEFINITION 4** **Equality of Matrices** Two matrices $A = (a_{ij})$ and $B = (b_{ij})$ are **equal** if (1) they have the same size, and (2) corresponding components are equal.

**EXAMPLE 4** **Equal and Unequal Matrices** Are the following matrices equal?

**i.** $\begin{pmatrix} 4 & 1 & 5 \\ 2 & -3 & 0 \end{pmatrix}$ and $\begin{pmatrix} 1+3 & 1 & 2+3 \\ 1+1 & 1-4 & 6-6 \end{pmatrix}$

**ii.** $\begin{pmatrix} -2 & 0 \\ 1 & 3 \end{pmatrix}$ and $\begin{pmatrix} 0 & -2 \\ 1 & 3 \end{pmatrix}$

**iii.** $\begin{pmatrix} 1 & 0 \\ 0 & 1 \end{pmatrix}$ and $\begin{pmatrix} 1 & 0 & 0 \\ 0 & 1 & 0 \end{pmatrix}$

**Solution**

**i.** Yes; both matrices are $2 \times 3$, and $1 + 3 = 4$, $2 + 3 = 5$, $1 + 1 = 2$, $1 - 4 = -3$, and $6 - 6 = 0$.

**ii.** No; $-2 \neq 0$, so the matrices are unequal because, for example, the 1,1 components are unequal. This is true even though the two matrices contain the same numbers. *Corresponding* components must be equal. This means that the 1,1 component in $A$ must be equal to the 1,1 component in $B$, and so on.

**iii.** No; the first matrix is $2 \times 2$ and the second matrix is $2 \times 3$, so they do not have the same size.

### Vectors Are Matrices with One Row or One Column

Each vector is a special kind of matrix. Thus, for example, the $n$-component row vector $(a_1, a_2, \ldots, a_n)$ is a $1 \times n$ matrix, whereas the $n$-component column vector $\begin{pmatrix} a_1 \\ a_2 \\ \vdots \\ a_n \end{pmatrix}$ is an $n \times 1$ matrix.

Matrices, like vectors, arise in a great number of practical situations. For example, we saw on page 47 how the vector $\begin{pmatrix} 10 \\ 30 \\ 15 \\ 60 \end{pmatrix}$ could represent order quantities for four different products used by one manufacturer. Suppose that there were five different plants. Then the $4 \times 5$ matrix

$$Q = \begin{pmatrix} 10 & 20 & 15 & 16 & 25 \\ 30 & 10 & 20 & 25 & 22 \\ 15 & 22 & 18 & 20 & 13 \\ 60 & 40 & 50 & 35 & 45 \end{pmatrix}$$

could represent the orders for the four products in each of the five plants. We can see, for example, that plant 4 orders 25 units of the second product while plant 2 orders 40 units of the fourth product.

Matrices can be added and multiplied by real numbers.

**DEFINITION 5** **Addition of Matrices** Let $A = (a_{ij})$ and $B = (b_{ij})$ be two $m \times n$ matrices. Then the sum of $A$ and $B$ is the $m \times n$ matrix $A + B$ given by

$$A + B = (a_{ij} + b_{ij}) = \begin{pmatrix} a_{11} + b_{11} & a_{12} + b_{12} & \cdots & a_{1n} + b_{1n} \\ a_{21} + b_{21} & a_{22} + b_{22} & \cdots & a_{2n} + b_{2n} \\ \vdots & \vdots & & \vdots \\ a_{m1} + b_{m1} & a_{m2} + b_{m2} & \cdots & a_{mn} + b_{mn} \end{pmatrix} \quad (4)$$

That is, $A + B$ is the $m \times n$ matrix obtained by adding the corresponding components of $A$ and $B$.

**WARNING** The sum of two matrices is defined only when both matrices have the same size. Thus, for example, it is not possible to add together the matrices $\begin{pmatrix} 1 & 2 & 3 \\ 4 & 5 & 6 \end{pmatrix}$ and $\begin{pmatrix} -1 & 0 \\ 2 & -5 \\ 4 & 7 \end{pmatrix}$ or the matrices (vectors) $\begin{pmatrix} 1 \\ 2 \end{pmatrix}$ and $\begin{pmatrix} 1 \\ 2 \\ 3 \end{pmatrix}$. That is, they are incompatible under addition.

**EXAMPLE 5** **The Sum of Two Matrices**

$$\begin{pmatrix} 2 & 4 & -6 & 7 \\ 1 & 3 & 2 & 1 \\ -4 & 3 & -5 & 5 \end{pmatrix} + \begin{pmatrix} 0 & 1 & 6 & -2 \\ 2 & 3 & 4 & 3 \\ -2 & 1 & 4 & 4 \end{pmatrix} = \begin{pmatrix} 2 & 5 & 0 & 5 \\ 3 & 6 & 6 & 4 \\ -6 & 4 & -1 & 9 \end{pmatrix}$$

Scalars

When dealing with vectors, we shall refer to numbers as **scalars** (which may be real or complex depending on whether the vectors in question are real or complex).

***Historical Note.*** The term "scalar" originated with Hamilton. His definition of the quaternion included what he called a "real part" and an "imaginary part." In his paper "On Quaternions, or on a New System of Imaginaries in Algebra," in *Philosophical Magazine,* 3rd series, 25(1844):26–27, he wrote: "The algebraically *real* part may receive . . . all values contained on the one *scale* of progression of numbers from negative to positive infinity; we shall call it therefore the *scalar part,* or simply the *scalar* of the quaternion. . . ." In the same paper Hamilton went on to define the imaginary part of his quaternion as the *vector* part. Although this was not the first usage of the word "vector," it was the first time it was used in the context of the definitions in this section. It is fair to say that the paper from which the preceding quotation was taken marks the beginning of modern vector analysis.

**DEFINITION 6** **Multiplication of a Matrix by a Scalar** If $A = (a_{ij})$ is an $m \times n$ matrix and if $\alpha$ is a scalar, then the $m \times n$ matrix $\alpha A$ is given by

$$\alpha A = (\alpha a_{ij}) = \begin{pmatrix} \alpha a_{11} & \alpha a_{12} & \cdots & \alpha a_{1n} \\ \alpha a_{21} & \alpha a_{22} & \cdots & \alpha a_{2n} \\ \vdots & \vdots & & \vdots \\ \alpha a_{m1} & \alpha a_{m2} & \cdots & \alpha a_{mn} \end{pmatrix} \tag{5}$$

In other words, $\alpha A = (\alpha a_{ij})$ is the matrix obtained by multiplying each component of $A$ by $\alpha$. If $\alpha A = B = (b_{ij})$, then $b_{ij} = \alpha a_{ij}$ for $i = 1, 2, \ldots, m$ and $j = 1, 2, \ldots, n$.

**EXAMPLE 6** **Scalar Multiples of Matrices**

Let $A = \begin{pmatrix} 1 & -3 & 4 & 2 \\ 3 & 1 & 4 & 6 \\ -2 & 3 & 5 & 7 \end{pmatrix}$. Then $2A = \begin{pmatrix} 2 & -6 & 8 & 4 \\ 6 & 2 & 8 & 12 \\ -4 & 6 & 10 & 14 \end{pmatrix}$,

$-\frac{1}{3}A = \begin{pmatrix} -\frac{1}{3} & 1 & -\frac{4}{3} & -\frac{2}{3} \\ -1 & -\frac{1}{3} & -\frac{4}{3} & -2 \\ \frac{2}{3} & -1 & -\frac{5}{3} & -\frac{7}{3} \end{pmatrix}$, and $0A = \begin{pmatrix} 0 & 0 & 0 & 0 \\ 0 & 0 & 0 & 0 \\ 0 & 0 & 0 & 0 \end{pmatrix}$

**EXAMPLE 7** **The Sum of Scalar Multiples of Two Vectors**

Let $\mathbf{a} = \begin{pmatrix} 4 \\ 6 \\ 1 \\ 3 \end{pmatrix}$ and $\mathbf{b} = \begin{pmatrix} -2 \\ 4 \\ -3 \\ 0 \end{pmatrix}$. Calculate $2\mathbf{a} - 3\mathbf{b}$.

**Solution** $2\mathbf{a} - 3\mathbf{b} = 2\begin{pmatrix} 4 \\ 6 \\ 1 \\ 3 \end{pmatrix} + (-3)\begin{pmatrix} -2 \\ 4 \\ -3 \\ 0 \end{pmatrix} = \begin{pmatrix} 8 \\ 12 \\ 2 \\ 6 \end{pmatrix} + \begin{pmatrix} 6 \\ -12 \\ 9 \\ 0 \end{pmatrix} = \begin{pmatrix} 14 \\ 0 \\ 11 \\ 6 \end{pmatrix}$

The next theorem provides basic facts about matrix addition and scalar multiplication. We prove part (*iii*) and leave the remaining parts of the proof as an exercise (see Problems 41–43).

**THEOREM 1** Let $A$, $B$, and $C$ be $m \times n$ matrices and let $\alpha$ and $\beta$ be scalars. Then:

**i.** $A + 0 = A$

**ii.** $0A = 0$

**iii.** $A + B = B + A$ **(commutative law for matrix addition)**

**iv.** $(A + B) + C = A + (B + C)$ **(associative law for matrix addition)**

**v.** $\alpha(A + B) = \alpha A + \alpha B$ **(distributive law for scalar multiplication)**

**vi.** $1A = A$

**vii.** $(\alpha + \beta)A = \alpha A + \beta A$

*Note.* The zero in part $(i)$ of the theorem is the $m \times n$ zero matrix. In part $(ii)$ the zero on the left is a scalar while the zero on the right is the $m \times n$ zero matrix.

**Proof of (iii)**

$$\text{Let } A = \begin{pmatrix} a_{11} & a_{12} & \cdots & a_{1n} \\ a_{21} & a_{22} & \cdots & a_{2n} \\ \vdots & \vdots & & \vdots \\ a_{m1} & a_{m2} & \cdots & a_{mn} \end{pmatrix} \quad \text{and} \quad B = \begin{pmatrix} b_{11} & b_{12} & \cdots & b_{1n} \\ b_{21} & b_{22} & \cdots & b_{2n} \\ \vdots & \vdots & & \vdots \\ b_{m1} & b_{m2} & \cdots & b_{mn} \end{pmatrix}.$$

Then

$$A + B = \begin{pmatrix} a_{11} + b_{11} & a_{12} + b_{12} & \cdots & a_{1n} + b_{1n} \\ a_{21} + b_{21} & a_{22} + b_{22} & \cdots & a_{2n} + b_{2n} \\ \vdots & \vdots & & \vdots \\ a_{m1} + b_{m1} & a_{m2} + b_{m2} & \cdots & a_{mn} + b_{mn} \end{pmatrix}$$

$a + b = b + a$ for any real numbers $a$ and $b$

$$\overset{\downarrow}{=} \begin{pmatrix} b_{11} + a_{11} & b_{12} + a_{12} & \cdots & b_{1n} + a_{1n} \\ b_{21} + a_{21} & b_{22} + a_{22} & \cdots & b_{2n} + a_{2n} \\ \vdots & \vdots & & \vdots \\ b_{m1} + a_{m1} & b_{m2} + a_{m2} & \cdots & b_{mn} + a_{mn} \end{pmatrix} = B + A$$

■

**EXAMPLE 8** **Illustrating the Associative Law of Matrix Addition** To illustrate the associative law we note that

$$\left[\begin{pmatrix} 1 & 4 & -2 \\ 3 & -1 & 0 \end{pmatrix} + \begin{pmatrix} 2 & -2 & 3 \\ 1 & -1 & 5 \end{pmatrix}\right] + \begin{pmatrix} 3 & -1 & 2 \\ 0 & 1 & 4 \end{pmatrix}$$

$$= \begin{pmatrix} 3 & 2 & 1 \\ 4 & -2 & 5 \end{pmatrix} + \begin{pmatrix} 3 & -1 & 2 \\ 0 & 1 & 4 \end{pmatrix} = \begin{pmatrix} 6 & 1 & 3 \\ 4 & -1 & 9 \end{pmatrix}$$

Similarly,

$$\begin{pmatrix} 1 & 4 & -2 \\ 3 & -1 & 0 \end{pmatrix} + \left[\begin{pmatrix} 2 & -2 & 3 \\ 1 & -1 & 5 \end{pmatrix} + \begin{pmatrix} 3 & -1 & 2 \\ 0 & 1 & 4 \end{pmatrix}\right]$$

$$= \begin{pmatrix} 1 & 4 & -2 \\ 3 & -1 & 0 \end{pmatrix} + \begin{pmatrix} 5 & -3 & 5 \\ 1 & 0 & 9 \end{pmatrix} = \begin{pmatrix} 6 & 1 & 3 \\ 4 & -1 & 9 \end{pmatrix}$$

■

Example 8 illustrates the importance of the associative law of vector addition since if we wish to add together three or more matrices, we can only do so by adding them together two at a time. The associative law tells us that we can do this in two different ways and still come up with the same answer. If this were not the case, the sum of three or more matrices would be more difficult to define since we would have to specify whether we wanted $(A + B) + C$ or $A + (B + C)$ to be the definition of the sum $A + B + C$.

## Focus on . . .

## Sir William Rowan Hamilton, 1805–1865

Sir William Rowan Hamilton *(The Granger Collection)*

Born in Dublin in 1805, where he spent most of his life, William Rowan Hamilton was without question Ireland's greatest mathematician. Hamilton's father (an attorney) and mother died when he was a small boy. His uncle, a linguist, took over the boy's education. By his fifth birthday Hamilton could read English, Hebrew, Latin, and Greek. By his thirteenth birthday he had mastered not only the languages of continental Europe, but also Sanscrit, Chinese, Persian, Arabic, Malay, Hindi, Bengali, and several others as well. Hamilton liked to write poetry, both as a child and as an adult, and his friends included the great English poets Samuel Taylor Coleridge and William Wordsworth. Hamilton's poetry was considered so bad, however, that it is fortunate that he developed other interests—especially in mathematics.

Although he enjoyed mathematics as a young boy, Hamilton's interest was greatly enhanced by a chance meeting at the age of 15 with Zerah Colburn, the American lightning calculator. Shortly afterward, Hamilton began to read important mathematical books of the time. In 1823, at the age of 18, he discovered an error in Simon Laplace's *Mécanique céleste* and wrote an impressive paper on the subject. A year later he entered Trinity College in Dublin.

Hamilton's university career was astonishing. At the age of 21, while still an undergraduate, he had so impressed the faculty that he was appointed Royal Astronomer of Ireland and Professor of Astronomy at the University. Shortly thereafter he wrote what is now considered a classic work on optics. Using only mathematical theory, he predicted conical refraction in certain types of crystals. Later this theory was confirmed by physicists. Largely because of this work, Hamilton was knighted in 1835.

Hamilton's first great purely mathematical paper appeared in 1833. In this work he described an algebraic way to manipulate pairs of real numbers. This work gives rules that are used today to add, subtract, multiply, and divide complex numbers. At first, however, Hamilton was unable to devise a multiplication for triples or $n$-tuples of numbers for $n > 2$. For 10 years he pondered this problem, and it is said that he solved it in an inspiration while walking on the Brougham Bridge in Dublin in 1843. The key was to discard the familiar commutative property of multiplication. The new objects he created were called *quaternions,* which were the precursors of what we now call *vectors.* Today a tablet embedded in the stone of the bridge tells the story.

Here as he walked by
on the 16th of October 1843
Sir William Rowan Hamilton
in a flash of genius discovered
the fundamental formula for
quaternion multiplication
$i^2 = j^2 = k^2 = ijk = -1$
& cut it in a stone of this bridge

For the rest of his life, Hamilton spent most of his time developing the algebra of quaternions. He felt that they would have revolutionary significance in mathematical physics. His monumental work on this subject, *Treatise on Quaternions,* was published in 1853. Thereafter, he worked on an enlarged work, *Elements of Quaternions.* Although Hamilton died in 1865 before his *Elements* was completed, the work was published by his son in 1866.

Students of mathematics and physics know Hamilton in a variety of other contexts. In mathematical physics, for example, one encounters the Hamiltonian function, which often represents the total energy in a system, and the Hamilton-Jacobi differential equations of dynamics. In matrix theory, the Cayley-Hamilton theorem states that every matrix satisfies its own characteristic equation. We discuss this in Section 6.8.

Despite the great work he was doing, Hamilton's final years were a torment to him. His wife was a semi-invalid and he was plagued by alcoholism. It is therefore gratifying to point out that, during these last years, the newly formed American National Academy of Sciences elected Sir William Rowan Hamilton to be its first foreign associate.

# PROBLEMS 1.5

## Self-Quiz

**I.** Which of the following is true of the matrix

$$\begin{pmatrix} 1 & 2 & 3 \\ 7 & -1 & 0 \end{pmatrix}?$$

**a.** It is a square matrix.

**b.** If multiplied by the scalar $-1$, the product is $\begin{pmatrix} -1 & -2 & -3 \\ -7 & 1 & 0 \end{pmatrix}$.

**c.** It is a $3 \times 2$ matrix.

**d.** It is the sum of $\begin{pmatrix} 3 & 1 & 4 \\ 7 & 2 & 0 \end{pmatrix}$ and $\begin{pmatrix} -2 & 1 & 1 \\ 0 & 1 & 0 \end{pmatrix}$.

**II.** Which of the following is $2A - 4B$ if $A = (2 \quad 0 \quad 0)$ and $B = (3 \quad 1)$?

**a.** $(-8 \quad -4)$ **b.** $(5 \quad 0 \quad 1)$

**c.** $(16 \quad -4 \quad 0)$

**d.** This operation cannot be performed.

**III.** Which of the following is true when finding the difference of two matrices?

**a.** The matrices must have the same size.

**b.** The matrices must be square.

**c.** The matrices must be both row vectors or column vectors.

**d.** One matrix must be a row vector and the other must be a column vector.

**IV.** Which of the following would be the entires in the second column of matrix $B$, if

$$\begin{pmatrix} 3 & -4 & 0 \\ 2 & 8 & -1 \end{pmatrix} + B = \begin{pmatrix} 0 & 0 & 0 \\ 0 & 0 & 0 \end{pmatrix}?$$

**a.** $-2, -8, 1$ **b.** $4, -8$

**c.** $2, 8, -1$ **d.** $-4, 8$

**V.** Which of the following must be the second row of matrix $B$ if $3A - B = 2C$ for

$$A = \begin{pmatrix} 1 & -1 & 1 \\ 0 & 0 & 3 \\ 4 & 2 & 0 \end{pmatrix} \quad \text{and} \quad C = \begin{pmatrix} 1 & 0 & 0 \\ 0 & 1 & 0 \\ 0 & 0 & 1 \end{pmatrix}?$$

**a.** $-3, 2, 6$ **b.** $0, -2, 9$

**c.** $3, -2, 6$ **d.** $0, 2, -9$

### Answers to Self-Quiz

**I.** b **II.** d **III.** a **IV.** b **V.** b

In Problems 1–10 perform the indicated computation with $\mathbf{a} = \begin{pmatrix} -3 \\ 1 \\ 4 \end{pmatrix}$, $\mathbf{b} = \begin{pmatrix} 5 \\ -4 \\ 7 \end{pmatrix}$, and $\mathbf{c} = \begin{pmatrix} 2 \\ 0 \\ -2 \end{pmatrix}$.

**1.** $\mathbf{a} + \mathbf{b}$ **2.** $3\mathbf{b}$ **3.** $-2\mathbf{c}$

**4.** $\mathbf{b} + 3\mathbf{c}$ **5.** $2\mathbf{a} - 5\mathbf{b}$ **6.** $-3\mathbf{b} + 2\mathbf{c}$

**7.** $0\mathbf{c}$ **8.** $\mathbf{a} + \mathbf{b} + \mathbf{c}$ **9.** $3\mathbf{a} - 2\mathbf{b} + 4\mathbf{c}$

**10.** $3\mathbf{b} - 7\mathbf{c} + 2\mathbf{a}$

In Problems 11–20 perform the indicated computation with $\mathbf{a} = (3, -1, 4, 2)$, $\mathbf{b} = (6, 0, -1, 4)$, and $\mathbf{c} = (-2, 3, 1, 5)$.

**11.** $\mathbf{a} + \mathbf{c}$ **12.** $\mathbf{b} - \mathbf{a}$ **13.** $4\mathbf{c}$

**14.** $-2\mathbf{b}$ **15.** $2\mathbf{a} - \mathbf{c}$ **16.** $4\mathbf{b} - 7\mathbf{a}$

**17.** $\mathbf{a} + \mathbf{b} + \mathbf{c}$ **18.** $\mathbf{c} - \mathbf{b} + 2\mathbf{a}$ **19.** $3\mathbf{a} - 2\mathbf{b} + 4\mathbf{c}$

**20.** $\alpha\mathbf{a} + \beta\mathbf{b} + \gamma\mathbf{c}$

In Problems 21–32 perform the indicated computation with $A = \begin{pmatrix} 1 & 3 \\ 2 & 5 \\ -1 & 2 \end{pmatrix}$, $B = \begin{pmatrix} -2 & 0 \\ 1 & 4 \\ -7 & 5 \end{pmatrix}$, and $C = \begin{pmatrix} -1 & 1 \\ 4 & 6 \\ -7 & 3 \end{pmatrix}$.

**21.** $3A$ **22.** $A + B$ **23.** $A - C$

**24.** $2C - 5A$ **25.** $0B$ (0 is the scalar zero) **26.** $-7A + 3B$

**27.** $A + B + C$ **28.** $C - A - B$ **29.** $2A - 3B + 4C$

**30.** $7C - B + 2A$

**31.** Find a matrix $D$ such that $2A + B - D$ is the $3 \times 2$ zero matrix.

**32.** Find a matrix $E$ such that $A + 2B - 3C + E$ is the $3 \times 2$ zero matrix.

In Problems 33–40 perform the indicated computation with $A = \begin{pmatrix} 1 & -1 & 2 \\ 3 & 4 & 5 \\ 0 & 1 & -1 \end{pmatrix}$, $B = \begin{pmatrix} 0 & 2 & 1 \\ 3 & 0 & 5 \\ 7 & -6 & 0 \end{pmatrix}$, and $C = \begin{pmatrix} 0 & 0 & 2 \\ 3 & 1 & 0 \\ 0 & -2 & 4 \end{pmatrix}$.

**33.** $A - 2B$ **34.** $3A - C$ **35.** $A + B + C$

**36.** $2A - B + 2C$ **37.** $C - A - B$ **38.** $4C - 2B + 3A$

**39.** Find a matrix $D$ such that $A + B + C + D$ is the $3 \times 3$ zero matrix.

**40.** Find a matrix $E$ such that $3C - 2B + 8A - 4E$ is the $3 \times 3$ zero matrix.

**41.** Let $A = (a_{ij})$ be an $m \times n$ matrix and let $\overline{0}$ denote the $m \times n$ zero matrix. Use Definitions 5 and 6 to show that $0A = \overline{0}$ and $\overline{0} + A = A$. Similarly, show that $1A = A$.

**42.** If $A = (a_{ij})$, $B = (b_{ij})$, and $C = (c_{ij})$ are $m \times n$ matrices, compute $(A + B) + C$ and $A + (B + C)$ and show that they are equal.

**43.** If $\alpha$ and $\beta$ are scalars and $A$ and $B$ are $m \times n$ matrices, compute $\alpha(A + B)$ and $\alpha A + \alpha B$ and show that they are equal. Also, compute $(\alpha + \beta)A$ and $\alpha A + \beta A$ and show that they are equal.

**44.** Consider the "graph" joining the four points in Figure 1.7. Construct a $4 \times 4$ matrix having the property that $a_{ij} = 0$ if point $i$ is not connected (joined by a line) to point $j$ and $a_{ij} = 1$ if point $i$ is connected to point $j$.

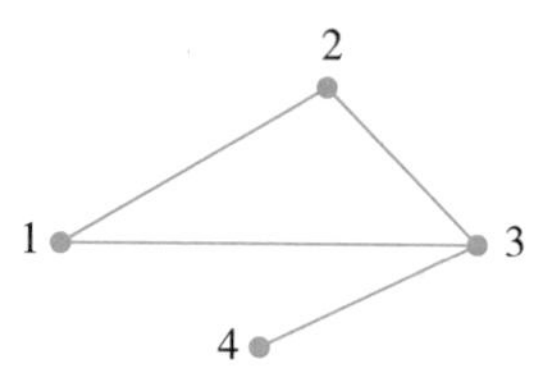


**Figure 1.7**

**45.** Do the same (this time constructing a $5 \times 5$ matrix) for the graph in Figure 1.8.

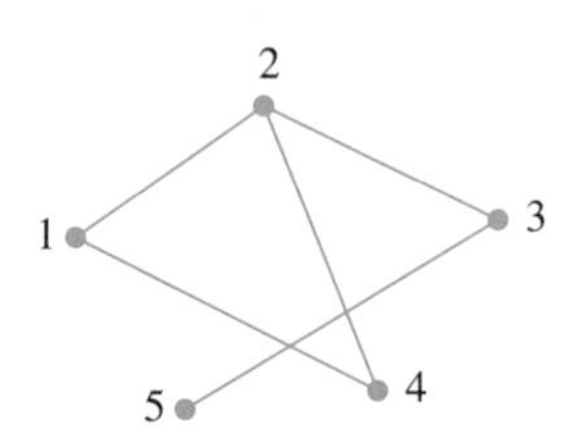


**Figure 1.8**

**46.** In the manufacture of a certain product, four raw materials are needed. The vector $\mathbf{d} = \begin{pmatrix} d_1 \\ d_2 \\ d_3 \\ d_4 \end{pmatrix}$ represents a given factory's demand for each of the four raw materials to produce 1 unit of its product. If $\mathbf{d}$ is the demand vector for factory 1 and $\mathbf{e}$ is the demand vector for factory 2, what is represented by the vectors $\mathbf{d} + \mathbf{e}$ and $2\mathbf{d}$?

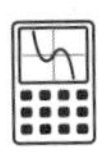

## CALCULATOR BOX

As we saw on page 28, you can enter matrices both on the TI-85 and on the CASIO fx-7700 GB. On the CASIO it is necessary first to specify the size (or dimension) of the matrix. As on page 29, you enter the size of an $m \times n$ matrix by the following sequence of key strokes:

MODE 0 F1 F6 F1 $m$ EXE $n$ EXE

Here $1 \le m \le 9$ and $1 \le n \le 9$.

On the TI-85 it is not necessary to specify the size in advance.

### CASIO fx-7700 GB

**For addition and scalar multiplication on the CASIO fx-7700 GB** To add two matrices of the same size, first enter one under A and the other under B. Then press the F3 key which, in the appropriate mode, is labeled "+."

To multiply $A$ by the scalar $k$, press $k$ $k$A. The $k$A key is accessed by pressing F1 when the calculator is in the appropriate mode.

### TI-85

**For addition and scalar multiplication on the TI-85** The easiest way to add two matrices of the same size is first to enter each matrix and to give each one a name (like $A$ and $B$). Then to obtain $A + B$, press

ALPHA A + ALPHA B ENTER

To obtain $k$A, press

$k$ ALPHA A ENTER

## MATLAB 1.5

1. This problem gives practice in working with matrix notation as well as in procedures used in future problems. In previous problems when performing the row operation $R_j \to R_j + cR_i$, you found the multiplier $c$ by observation. The multiplier $c$ can be computed accurately from the entries in the matrix.

   ***Example.***

$$A = \begin{pmatrix} a & b & c & d & e \\ 0 & 0 & f & g & h \\ 0 & 0 & i & j & k \end{pmatrix}$$

To create a 0 in the position that $i$ occupies, we need $R_3 \rightarrow R_3 + (-i/f)R_2$. Note that $f = A(2, 3)$ and $i = A(3, 3)$:

**c = −A(3,3)/A(2,3)**

In general, $c = -$entry to be zeroed/pivot used:

**A(3,:) = A(3,:) + c*A(2,:)**

a. For the matrix below, perform row operations $R_j \rightarrow R_j + cR_i$ to get the matrix in row echelon form (not reduced row echelon form), except that the pivot entry need not be 1. (Don't multiply or divide a row by a number to create 1's.) Find all multipliers using the matrix notation as above. For this matrix your multipliers will be nice numbers so that you can check yourself as you go along:

$$A = \begin{pmatrix} 1 & 2 & -2 & 0 & 1 \\ 2 & 4 & -1 & 0 & -4 \\ -3 & -6 & 12 & 2 & -12 \\ 1 & 2 & -2 & -4 & -5 \end{pmatrix}$$

b. Type in

**A = rand(4,5)**
**A(:,3) = 2*A(:,1) + 4*A(:,2)**

Follow the directions as in part (a). Be sure to compute the multipliers using matrix notation.

See MATLAB Problem 2 in Section 1.10 for a situation where we are interested in performing the type of reduction described above.

2. **MATLAB Feature** *Entering sparse matrices efficiently*

a. In Problem 45, you were asked to set up matrices for graphs where

$$a_{ij} = \begin{cases} 1 & \text{if point } i \text{ is connected to point } j \\ 0 & \text{otherwise} \end{cases}$$

For most such graphs, the matrix will consist mostly of 0's with some 1's. In MATLAB you can enter a matrix of all 0's and then modify it row by row.

Consider the graph below:

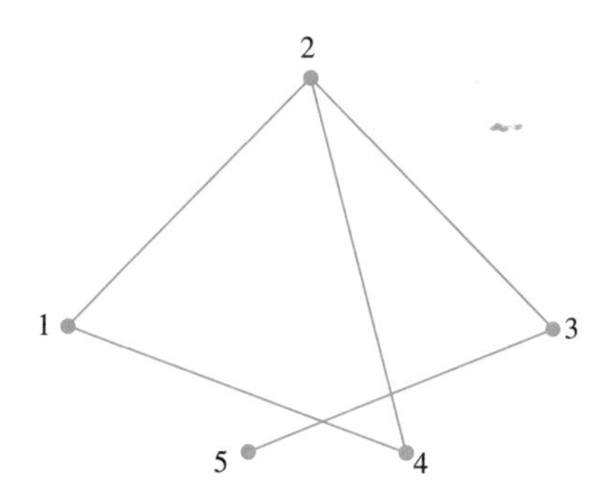

**a = zeros(5)**

**a(1,[2 4]) = [1 1]** (1 is connected to 2 and 4)

**a(2,[1 3 4]) = [1 1 1]** (2 is connected to 1, 3, and 4)

and so on

Finish entering the matrix above and check the result with your answer to Problem 45 in the text.

b. Consider the directed graph below.

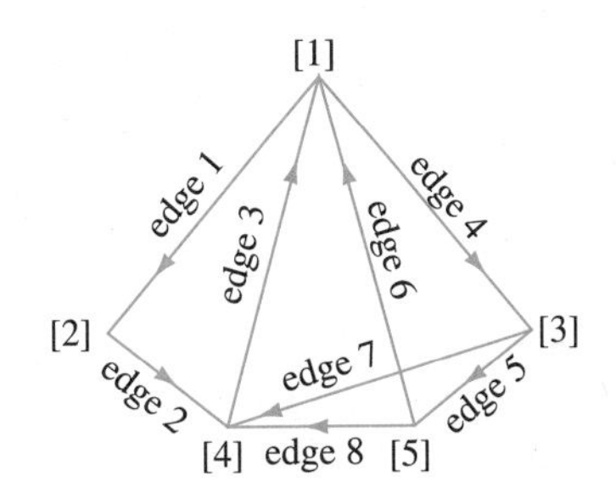


Define

$$a_{ij} = \begin{cases} 1 & \text{if edge } j \text{ goes into node } i \\ -1 & \text{if edge } j \text{ goes out of node } i \\ 0 & \text{otherwise} \end{cases}$$

What size will A be? Enter **A = zeros(n,m),** where $n$ is the number of rows and $m$ is the number of columns. We will modify $A$ column by column by looking at one edge at a time. For example,

**A([1 2],1) = [−1;1]** edge 1 goes out of [1] into [2]

**A([4 5],8) = [1;−1]** edge 8 goes out of [5] into [4]

Complete the process above to find $A$.

3. a. Enter any matrices $A$ and $B$ that are not of the same size. Find **A + B.** What does MATLAB tell you?

b. Enter any matrices $A$ and $B$ that are of the same size. Suppose $s$ is any scalar. From your knowledge of regular algebraic manipulations with numbers, what would you conjecture the relationship to be among **s * A**, **s * B**, and **s * (A + B)**? Using a comment statement, write your conjecture. Test your conjecture with three different choices of $s$. Test your conjecture with another choice of $A$ and another choice of $B$ and three choices of $s$. (To use MATLAB to generate random matrices, see the discussion preceding the MATLAB problems for Section 1.3.)

## 1.6 VECTOR AND MATRIX PRODUCTS

In this section we see how two matrices can be multiplied together. Quite obviously, we could define the product of two $m \times n$ matrices $A = (a_{ij})$ and $B = (b_{ij})$ to be the $m \times n$ matrix whose $ij$th component is $a_{ij}b_{ij}$. However, for just about all the important applications involving matrices, another kind of product is needed. Let us try to see why this is the case.

**EXAMPLE 1** **The Product of a Demand Vector and a Price Vector** Suppose that a manufacturer produces four items. The demand for the items is given by the **demand vector d** = (30 20 40 10) (a 1 × 4 matrix). The price per unit that the manufacturer receives for the items is given by the **price vector p** = $\begin{pmatrix} \$20 \\ \$15 \\ \$18 \\ \$40 \end{pmatrix}$ (a 4 × 1 matrix). If the demand is met, how much money will the manufacturer receive?

**Solution** Demand for the first item is 30, and the manufacturer receives \$20 for each of the first item sold. Thus (30)(20) = \$600 is received from the sales of the first item. By continuing this reasoning, we see that the total amount of money received is

$$\begin{aligned}(30)(20) + (20)(15) + (40)(18) + (10)(40) &= 600 + 300 + 720 + 400 \\ &= \$2020\end{aligned}$$

We write this result as

$$(30 \quad 20 \quad 40 \quad 10)\begin{pmatrix} 20 \\ 15 \\ 18 \\ 40 \end{pmatrix} = 2020$$

That is, we multiplied a 4-component row vector and a 4-component column vector to obtain a scalar (real number).

In the last example we multiplied a row vector by a column vector and obtained a scalar. In general, we have the following definition.

**DEFINITION 1** **Scalar Product** Let $\mathbf{a} = \begin{pmatrix} a_1 \\ a_2 \\ \vdots \\ a_n \end{pmatrix}$ and $\mathbf{b} = \begin{pmatrix} b_1 \\ b_2 \\ \vdots \\ b_n \end{pmatrix}$ be two $n$-vectors. Then the **scalar product** of **a** and **b**, denoted by **a · b**, is given by

$$\mathbf{a} \cdot \mathbf{b} = a_1 b_1 + a_2 b_2 + \cdots + a_n b_n \tag{1}$$

Because of the notation in (1), the scalar product of two vectors is often called the **dot product** or **inner product** of the vectors. Note that the scalar product of two $n$-vectors is a scalar (that is, a number).

**WARNING** When taking the scalar product of **a** and **b**, it is necessary that **a** and **b** have the same number of components.

We shall often be taking the scalar product of a row vector and column vector. In this case we have

**Scalar Product**

1 × $n$ row vector

$$(a_1, a_2, \ldots, a_n) \cdot \begin{pmatrix} b_1 \\ b_2 \\ \vdots \\ b_n \end{pmatrix} = a_1b_1 + a_2b_2 + \cdots + a_nb_n \quad (2)$$

$n$ × 1 column vector

This is a real number (a scalar)

**EXAMPLE 2** **The Scalar Product of Two Vectors** Let $\mathbf{a} = \begin{pmatrix} 1 \\ -2 \\ 3 \end{pmatrix}$ and $\mathbf{b} = \begin{pmatrix} 3 \\ -2 \\ 4 \end{pmatrix}$. Calculate $\mathbf{a} \cdot \mathbf{b}$.

**Solution** $\mathbf{a} \cdot \mathbf{b} = (1)(3) + (-2)(-2) + (3)(4) = 3 + 4 + 12 = 19$

**EXAMPLE 3** **The Scalar Product of Two Vectors** Let $\mathbf{a} = (2, -3, 4, -6)$ and $\mathbf{b} = \begin{pmatrix} 1 \\ 2 \\ 0 \\ 3 \end{pmatrix}$. Compute $\mathbf{a} \cdot \mathbf{b}$.

**Solution** Here $\mathbf{a} \cdot \mathbf{b} = (2)(1) + (-3)(2) + (4)(0) + (-6)(3) = 2 - 6 + 0 - 18 = -22$.

The next theorem follows directly from the definition of the scalar product. We prove part (*ii*) and leave the remaining parts as an exercise.

**THEOREM 1** Let **a**, **b**, and **c** be $n$-vectors and let $\alpha$ and $\beta$ be scalars. Then

**i.** $\mathbf{a} \cdot \mathbf{0} = 0$

**ii.** $\mathbf{a} \cdot \mathbf{b} = \mathbf{b} \cdot \mathbf{a}$ **(commutative law for scalar product)**

**iii.** $\mathbf{a} \cdot (\mathbf{b} + \mathbf{c}) = \mathbf{a} \cdot \mathbf{b} + \mathbf{a} \cdot \mathbf{c}$ **(distributive law for scalar product)**

**iv.** $(\alpha\mathbf{a}) \cdot \mathbf{b} = \alpha(\mathbf{a} \cdot \mathbf{b})$

**Proof of (ii)** Let $\mathbf{a} = \begin{pmatrix} a_1 \\ a_2 \\ \vdots \\ a_n \end{pmatrix}$ and $\mathbf{b} = \begin{pmatrix} b_1 \\ b_2 \\ \vdots \\ b_n \end{pmatrix}$.

Then

$ab = ba$ for any two numbers $a$ and $b$
↓

$$\mathbf{a} \cdot \mathbf{b} = a_1b_1 + a_2b_2 + \cdots + a_nb_n = b_1a_1 + b_2a_2 + \cdots + b_na_n = \mathbf{b} \cdot \mathbf{a}$$ ■

Note that there is *no* associative law for the scalar product. The expression $(\mathbf{a} \cdot \mathbf{b}) \cdot \mathbf{c} = \mathbf{a} \cdot (\mathbf{b} \cdot \mathbf{c})$ does not make sense because neither side of the equation is defined. For the left side, this follows from the fact that $\mathbf{a} \cdot \mathbf{b}$ is a scalar and the scalar product of the scalar $\mathbf{a} \cdot \mathbf{b}$ and the vector $\mathbf{c}$ is not defined.

We now define the product of two matrices.

**DEFINITION 2** **Product of Two Matrices** Let $A = (a_{ij})$ be an $m \times n$ matrix, and let $B = (b_{ij})$ be an $n \times p$ matrix. Then the **product** of $A$ and $B$ is an $m \times p$ matrix $C = (c_{ij})$, where

$$c_{ij} = (i\text{th row of } A) \cdot (j\text{th column of } B) \quad \textbf{(3)}$$

That is, the $ij$th element of $AB$ is the dot product of the $i$th row of $A$ and the $j$th column of $B$. If we write this out, we obtain

$$c_{ij} = a_{i1}b_{1j} + a_{i2}b_{2j} + \cdots + a_{in}b_{nj} \quad \textbf{(4)}$$

If the number of columns of $A$ is equal to the number of rows of $B$, then $A$ and $B$ are said to be **compatible under multiplication.**

**WARNING** Two matrices can be multiplied together only if the number of columns of the first matrix is equal to the number of rows of the second. Otherwise the vectors that are the $i$th row of $A$ and the $j$th column of $B$ will not have the same number of components, and the dot product in equation (3) will not be defined. That is, the matrices $A$ and $B$ will be **incompatible** under multiplication. To illustrate this, we write the matrices $A$ and $B$:

$$\begin{matrix} & & & & & & & & j\text{th column of } B \\ & & & & & & & & \downarrow \end{matrix}$$

$$i\text{th row of } A \rightarrow \begin{pmatrix} a_{11} & a_{12} & \cdots & a_{1n} \\ a_{21} & a_{22} & \cdots & a_{2n} \\ \vdots & \vdots & & \vdots \\ a_{i1} & a_{i2} & \cdots & a_{in} \\ \vdots & \vdots & & \vdots \\ a_{m1} & a_{m2} & \cdots & a_{mn} \end{pmatrix} \begin{pmatrix} b_{11} & b_{12} & \cdots & b_{1j} & \cdots & b_{1p} \\ b_{21} & b_{22} & \cdots & b_{2j} & \cdots & b_{2p} \\ \vdots & \vdots & & \vdots & & \vdots \\ b_{n1} & b_{n2} & \cdots & b_{nj} & \cdots & b_{np} \end{pmatrix}$$

The shaded row and column vectors must have the same number of components.

**EXAMPLE 4** **The Product of Two 2 × 2 Matrices** If $A = \begin{pmatrix} 1 & 3 \\ -2 & 4 \end{pmatrix}$ and $B = \begin{pmatrix} 3 & -2 \\ 5 & 6 \end{pmatrix}$, calculate $AB$ and $BA$.

**Solution** $A$ is a $2 \times 2$ matrix and $B$ is a $2 \times 2$ matrix, so $C = AB = (2 \times 2) \times (2 \times 2)$ is also a $2 \times 2$ matrix. If $C = (c_{ij})$, what is $c_{11}$? We know that

$$c_{11} = (\text{1st row of } A) \cdot (\text{1st column of } B)$$

Rewriting the matrices, we have

$$\begin{matrix} & & & \text{1st column of } B \\ & & & \downarrow \end{matrix}$$

$$\text{1st row of } A \rightarrow \begin{pmatrix} 1 & 3 \\ -2 & 4 \end{pmatrix} \begin{pmatrix} 3 & -2 \\ 5 & 6 \end{pmatrix}$$

Thus

$$c_{11} = (1 \quad 3)\begin{pmatrix} 3 \\ 5 \end{pmatrix} = 3 + 15 = 18$$

Similarly, to compute $c_{12}$ we have

$$\begin{matrix} & & & & \text{2nd column of } B \\ & & & & \downarrow \end{matrix}$$

$$\text{1st row of } A \rightarrow \begin{pmatrix} 1 & 3 \\ -2 & 4 \end{pmatrix} \begin{pmatrix} 3 & -2 \\ 5 & 6 \end{pmatrix}$$

and

$$c_{12} = (1 \quad 3) \begin{pmatrix} -2 \\ 6 \end{pmatrix} = -2 + 18 = 16$$

Continuing, we find that

$$c_{21} = (-2 \quad 4) \begin{pmatrix} 3 \\ 5 \end{pmatrix} = -6 + 20 = 14$$

and

$$c_{22} = (-2 \quad 4)\begin{pmatrix}-2\\6\end{pmatrix} = 4 + 24 = 28$$

Thus

$$C = AB = \begin{pmatrix}18 & 16\\14 & 28\end{pmatrix}$$

Similarly, leaving out the intermediate steps, we see that

$$C' = BA = \begin{pmatrix}3 & -2\\5 & 6\end{pmatrix}\begin{pmatrix}1 & 3\\-2 & 4\end{pmatrix} = \begin{pmatrix}3+4 & 9-8\\5-12 & 15+24\end{pmatrix} = \begin{pmatrix}7 & 1\\-7 & 39\end{pmatrix}$$

*Remark.* Example 4 illustrates an important fact: *Matrix products do not, in general, commute.* That is, $AB \neq BA$ in general. It sometimes happens that $AB = BA$, but this will be the exception, not the rule. If $AB = BA$, we say that $A$ and $B$ **commute.** In fact, as the next example illustrates, it may occur that $AB$ is defined while $BA$ is not. Thus we must be careful of *order* when multiplying two matrices together.

EXAMPLE 5 **The Product of a 2 × 3 and a 3 × 4 Matrix Is Defined While the Product of a 3 × 4 and a 2 × 3 Matrix Is Not Defined** Let

$$A = \begin{pmatrix}2 & 0 & -3\\4 & 1 & 5\end{pmatrix} \quad \text{and} \quad B = \begin{pmatrix}7 & -1 & 4 & 7\\2 & 5 & 0 & -4\\-3 & 1 & 2 & 3\end{pmatrix}. \text{ Calculate } AB.$$

**Solution** We first note that $A$ is a $2 \times 3$ matrix and $B$ is a $3 \times 4$ matrix. Hence the number of columns of $A$ equals the number of rows of $B$. The product $AB$ is therefore defined and is a $2 \times 4$ matrix. Let $AB = C = (c_{ij})$. Then

$$c_{11} = (2 \quad 0 \quad -3)\cdot\begin{pmatrix}7\\2\\-3\end{pmatrix} = 23 \qquad c_{12} = (2 \quad 0 \quad -3)\cdot\begin{pmatrix}-1\\5\\1\end{pmatrix} = -5$$

$$c_{13} = (2 \quad 0 \quad -3)\cdot\begin{pmatrix}4\\0\\2\end{pmatrix} = 2 \qquad c_{14} = (2 \quad 0 \quad -3)\cdot\begin{pmatrix}7\\-4\\3\end{pmatrix} = 5$$

$$c_{21} = (4 \quad 1 \quad 5)\cdot\begin{pmatrix}7\\2\\-3\end{pmatrix} = 15 \qquad c_{22} = (4 \quad 1 \quad 5)\cdot\begin{pmatrix}-1\\5\\1\end{pmatrix} = 6$$

$$c_{23} = (4 \quad 1 \quad 5)\cdot\begin{pmatrix}4\\0\\2\end{pmatrix} = 26 \qquad c_{24} = (4 \quad 1 \quad 5)\cdot\begin{pmatrix}7\\-4\\3\end{pmatrix} = 39$$

Hence $AB = \begin{pmatrix} 23 & -5 & 2 & 5 \\ 15 & 6 & 26 & 39 \end{pmatrix}$. This completes the problem. Note that the product $BA$ is *not* defined since the number of columns of $B$ (four) is not equal to the number of rows of $A$ (two).

EXAMPLE 6 **Direct and Indirect Contact with a Contagious Disease** In this example we show how matrix multiplication can be used to model the spread of a contagious disease. Suppose that four individuals have contracted such a disease. This group has contacts with six people in a second group. We can represent these contacts, called *direct contacts,* by a $4 \times 6$ matrix. An example of such a matrix is given below.

**Direct Contact Matrix:** First and second groups

$$A = \begin{pmatrix} 0 & 1 & 0 & 0 & 1 & 0 \\ 1 & 0 & 0 & 1 & 0 & 1 \\ 0 & 0 & 0 & 1 & 1 & 0 \\ 1 & 0 & 0 & 0 & 0 & 1 \end{pmatrix}$$

Here we set $a_{ij} = 1$ if the $i$th person in the first group has made contact with the $j$th person in the second group. For example, the 1 in the 2,4 position means that the second person in the first (infected) group has been in contact with the fourth person in the second group. Now suppose that a third group of five people has had a variety of direct contact with individuals of the second group. We can also represent this by a matrix.

**Direct Contact Matrix:** Second and third groups

$$B = \begin{pmatrix} 0 & 0 & 1 & 0 & 1 \\ 0 & 0 & 0 & 1 & 0 \\ 0 & 1 & 0 & 0 & 0 \\ 1 & 0 & 0 & 0 & 1 \\ 0 & 0 & 0 & 1 & 0 \\ 0 & 0 & 1 & 0 & 0 \end{pmatrix}$$

Note that $b_{64} = 0$, which means that the sixth person in the second group has had no contact with the fourth person in the third group.

The *indirect* or *second-order* contacts between the individuals in the first and third groups is represented by the $4 \times 5$ matrix $C = AB$. To see this, observe that a person in group 3 can be infected from someone in group 2, who in turn has been infected by someone in group 1. For example, since $a_{24} = 1$ and $b_{45} = 1$, we see that, indirectly, the fifth person in group 3 has contact (through the fourth person in group 2) with the second person in group 1. The total number of indirect contacts between the second person in group 1 and the fifth person in group 3 is given by

$$c_{25} = a_{21}b_{15} + a_{22}b_{25} + a_{23}b_{35} + a_{24}b_{45} + a_{25}b_{55} + a_{26}b_{65}$$
$$= 1 \cdot 1 + 0 \cdot 0 + 0 \cdot 0 + 1 \cdot 1 + 0 \cdot 0 + 1 \cdot 0 = 2$$

We now compute.

**Indirect Contact Matrix.** First and third groups

$$C = AB = \begin{pmatrix} 0 & 0 & 0 & 2 & 0 \\ 1 & 0 & 2 & 0 & 2 \\ 1 & 0 & 0 & 1 & 1 \\ 0 & 0 & 2 & 0 & 1 \end{pmatrix}$$

We observe that only the second person in group 3 has no indirect contacts with the disease. The fifth person in this group has $2 + 1 + 1 = 4$ indirect contacts.

We have seen that matrices do not, in general, commute. The next theorem shows that the associative law does hold.

**THEOREM 2** **Associative Law for Matrix Multiplication** Let $A = (a_{ij})$ be an $n \times m$ matrix, $B = (b_{ij})$ an $m \times p$ matrix, and $C = (c_{ij})$ a $p \times q$ matrix. Then the **associative law**

$$A(BC) = (AB)C \tag{5}$$

holds and $ABC$, defined by either side of (5), is an $n \times q$ matrix.

The proof of this theorem is not difficult, but it is somewhat tedious. It is best given using the summation notation. For that reason let us defer it until the end of the section.

From now on we shall write the product of three matrices simply as $ABC$. We can do this because $(AB)C = A(BC)$; thus we get the same answer no matter how the multiplication is carried out (provided that we do not commute any of the matrices).

The associative law can be extended to longer products. For example, suppose that $AB$, $BC$, and $CD$ are defined. Then

$$ABCD = A(B(CD)) = ((AB)C)D = A(BC)D = (AB)(CD) \tag{6}$$

There are two distributive laws for matrix multiplication.

**THEOREM 3** **Distributive Laws for Matrix Multiplication** If all the following sums and products are defined, then

$$A(B + C) = AB + AC \tag{7}$$

and

$$(A + B)C = AC + BC \tag{8}$$

The proofs are given at the end of the section.

## Matrix Multiplication as a Linear Combination of the Columns of $A$

Let $A$ be an $m \times n$ matrix and $\mathbf{x}$ an $n \times 1$ vector. Consider the product

$$A\mathbf{x} = \begin{pmatrix} a_{11} & a_{12} & \cdots & a_{1n} \\ a_{21} & a_{22} & \cdots & a_{2n} \\ \vdots & \vdots & & \vdots \\ a_{m1} & a_{m2} & \cdots & a_{mn} \end{pmatrix} \begin{pmatrix} x_1 \\ x_2 \\ \vdots \\ x_n \end{pmatrix} = \begin{pmatrix} a_{11}x_1 + a_{12}x_2 + \cdots + a_{1n}x_n \\ a_{21}x_1 + a_{22}x_2 + \cdots + a_{2n}x_n \\ \vdots \qquad\qquad \vdots \qquad\qquad \vdots \\ a_{m1}x_1 + a_{m2}x_2 + \cdots + a_{mn}x_n \end{pmatrix}$$

or

$$A\mathbf{x} = x_1 \begin{pmatrix} a_{11} \\ a_{21} \\ \vdots \\ a_{m1} \end{pmatrix} + x_2 \begin{pmatrix} a_{12} \\ a_{22} \\ \vdots \\ a_{m1} \end{pmatrix} + \cdots + x_n \begin{pmatrix} a_{1n} \\ a_{2n} \\ \vdots \\ a_{mn} \end{pmatrix} \tag{9}$$

Note that $\mathbf{c}_1 = \begin{pmatrix} a_{11} \\ a_{21} \\ \vdots \\ a_{m1} \end{pmatrix}$ is the first column of $A$, $\mathbf{c}_2 = \begin{pmatrix} a_{12} \\ a_{22} \\ \vdots \\ a_{m2} \end{pmatrix}$ is the second column of $A$, and so on. Thus we can write (9) as

$$A\mathbf{x} = x_1\mathbf{c}_1 + x_2\mathbf{c}_2 + \cdots + x_n\mathbf{c}_n \tag{10}$$

The right-hand side of expression (10) is called a **linear combination** of the vectors $\mathbf{c}_1, \mathbf{c}_2, \ldots, \mathbf{c}_n$. We will discuss linear combinations in detail in Section 4.4. Here we simply note the following very useful fact:

> The product of the $m \times n$ matrix $A$ and the column vector $\mathbf{x}$ is a linear combination of the columns of $A$.

Suppose now that $B$ is an $n \times p$ matrix. Let $C = AB$ and let $\mathbf{c}_1$ denote the first column of $C$. Then

$$\mathbf{c}_1 = \begin{pmatrix} c_{11} \\ c_{21} \\ \vdots \\ c_{m1} \end{pmatrix} = \begin{pmatrix} a_{11}b_{11} + a_{12}b_{21} + \cdots + a_{1n}b_{n1} \\ a_{21}b_{11} + a_{22}b_{21} + \cdots + a_{2n}b_{n1} \\ \vdots \qquad \vdots \qquad \vdots \\ a_{m1}b_{11} + a_{m2}b_{21} + \cdots + a_{mn}b_{n1} \end{pmatrix}$$

$$= b_{11}\begin{pmatrix} a_{11} \\ a_{21} \\ \vdots \\ a_{m1} \end{pmatrix} + b_{21}\begin{pmatrix} a_{12} \\ a_{22} \\ \vdots \\ a_{m2} \end{pmatrix} + \cdots + b_{n1}\begin{pmatrix} a_{1n} \\ a_{2n} \\ \vdots \\ a_{mn} \end{pmatrix}$$

equals a linear combination of the columns of $A$. Since this is true for every column of $C = AB$, we find that

> Each column in the product $AB$ is a linear combination of the columns of $A$.

EXAMPLE 7 **Writing the Columns of $AB$ as Linear Combinations of the Columns of $A$**

Let

$$A = \begin{pmatrix} 1 & -2 \\ 2 & 4 \\ 3 & 5 \end{pmatrix} \quad \text{and} \quad B = \begin{pmatrix} 1 & -1 \\ 2 & 7 \end{pmatrix}$$

Then $AB = \begin{pmatrix} -3 & -15 \\ 10 & 26 \\ 13 & 32 \end{pmatrix}$. Now

$$\begin{pmatrix} -3 \\ 10 \\ 13 \end{pmatrix} = 1\begin{pmatrix} 1 \\ 2 \\ 3 \end{pmatrix} + 2\begin{pmatrix} -2 \\ 4 \\ 5 \end{pmatrix} = \text{a linear combination of the columns of } A$$

and

$$\begin{pmatrix} -15 \\ 26 \\ 32 \end{pmatrix} = -1\begin{pmatrix} 1 \\ 2 \\ 3 \end{pmatrix} + 7\begin{pmatrix} -2 \\ 4 \\ 5 \end{pmatrix} = \text{a linear combination of the columns of } A$$

## Block Matrix Multiplication

These are situations when it is convenient to treat matrices as blocks of smaller matrices, called **submatrices,** and to multiply block by block rather than component by component. It turns out that multiplying in blocks is very similar to ordinary matrix multiplications.

EXAMPLE 8 **Multiplying Blocks** Consider the product

$$AB = \begin{pmatrix} 1 & -1 & 2 & 4 \\ 2 & 0 & 4 & 5 \\ 1 & 1 & 2 & -3 \\ -2 & 3 & 5 & 0 \end{pmatrix} \begin{pmatrix} 1 & 4 & 3 \\ 2 & -1 & 0 \\ -3 & 2 & 1 \\ 0 & 1 & 2 \end{pmatrix}$$

You should verify that this product is defined. We now partition these matrices by inserting dashed lines.

$$AB = \left(\begin{array}{cc:cc} 1 & -1 & 2 & 4 \\ 2 & 0 & 4 & 5 \\ \hdashline 1 & 1 & 2 & -3 \\ -2 & 3 & 5 & 0 \end{array}\right) \left(\begin{array}{cc:c} 1 & 4 & 3 \\ 2 & -1 & 0 \\ \hdashline -3 & 2 & 1 \\ 0 & 1 & 2 \end{array}\right) = \left(\begin{array}{c:c} C & D \\ \hdashline E & F \end{array}\right) \left(\begin{array}{c:c} G & H \\ \hdashline J & K \end{array}\right)$$

There are other ways to form the partition as well. Here $C = \begin{pmatrix} 1 & -1 \\ 2 & 0 \end{pmatrix}$, $K = \begin{pmatrix} 1 \\ 2 \end{pmatrix}$, and so on. Now, assuming that all the matrix sums and products are defined, we can multiply in the ordinary way to obtain

$$AB = \begin{pmatrix} C & D \\ E & F \end{pmatrix} \begin{pmatrix} G & H \\ J & K \end{pmatrix} = \left(\begin{array}{c:c} CG + DJ & CH + DK \\ \hdashline EG + FJ & EH + FK \end{array}\right)$$

Now

$$CG = \begin{pmatrix} 1 & -1 \\ 2 & 0 \end{pmatrix} \begin{pmatrix} 1 & 4 \\ 2 & -1 \end{pmatrix} = \begin{pmatrix} -1 & 5 \\ 2 & 8 \end{pmatrix}, \quad DJ = \begin{pmatrix} 2 & 4 \\ 4 & 5 \end{pmatrix} \begin{pmatrix} -3 & 2 \\ 0 & 1 \end{pmatrix} = \begin{pmatrix} -6 & 8 \\ -12 & 13 \end{pmatrix}$$

and

$$CG + DJ = \begin{pmatrix} -7 & 13 \\ -10 & 21 \end{pmatrix}.$$

Similarly,

$$EH = \begin{pmatrix} 1 & 1 \\ -2 & 3 \end{pmatrix} \begin{pmatrix} 3 \\ 0 \end{pmatrix} = \begin{pmatrix} 3 \\ -6 \end{pmatrix}, \quad FK = \begin{pmatrix} 2 & -3 \\ 5 & 0 \end{pmatrix} \begin{pmatrix} 1 \\ 2 \end{pmatrix} = \begin{pmatrix} -4 \\ 5 \end{pmatrix}$$

and

$$EH + FK = \begin{pmatrix} -1 \\ -1 \end{pmatrix}.$$

You should verify that $CH + DK = \begin{pmatrix} 13 \\ 20 \end{pmatrix}$ and $EG + FJ = \begin{pmatrix} -3 & 4 \\ -11 & -1 \end{pmatrix}$ so

$$AB = \left(\begin{array}{c:c} CG + DJ & CH + DK \\ \hdashline EG + FJ & EH + FK \end{array}\right) = \left(\begin{array}{cc:c} -7 & 13 & 13 \\ -10 & 21 & 20 \\ \hdashline -3 & 4 & -1 \\ -11 & -1 & -1 \end{array}\right) = \begin{pmatrix} -7 & 13 & 13 \\ -10 & 21 & 20 \\ -3 & 4 & -1 \\ -11 & -1 & -1 \end{pmatrix}$$

This is the same answer you would obtain if you multiplied $AB$ directly.

When two matrices are partitioned and, as in Example 8, all required products of submatrices are defined, then the two matrices are said to be partitioned **conformably.**

EXAMPLE 9 **Two Matrices Which Do Commute** Suppose that the matrices $A$ and $B$ are square and that $C = \begin{pmatrix} I & A \\ O & I \end{pmatrix}$ and $D = \begin{pmatrix} I & B \\ O & I \end{pmatrix}$ are partitioned conformably. Show that $C$ and $D$ commute. Here $O$ denotes a zero matrix and $I$ is a square matrix that has the property that $AI = IA = A$ whenever these products are defined (see page 99).

**Solution**

$$CD = \begin{pmatrix} I & A \\ O & I \end{pmatrix}\begin{pmatrix} I & B \\ O & I \end{pmatrix} = \begin{pmatrix} I^2 + A \cdot O & IB + AI \\ O \cdot I + I \cdot O & O \cdot B + I^2 \end{pmatrix} = \begin{pmatrix} I & B + A \\ O & I \end{pmatrix}$$

where $I^2 = I \cdot I$. Similarly

$$DC = \begin{pmatrix} I & B \\ O & I \end{pmatrix}\begin{pmatrix} I & A \\ O & I \end{pmatrix} = \begin{pmatrix} I^2 + B \cdot O & IA + BI \\ O \cdot I + I \cdot O & O \cdot A + I^2 \end{pmatrix} = \begin{pmatrix} I & A + B \\ O & I \end{pmatrix}$$

Since $B + A = A + B$, $CD = DC$ which means that the matrices commute.

In order to prove Theorems 2 and 3 and to discuss many other things in this book, we need to use the *summation notation.* If this is not familiar to you, then continue reading. Otherwise skip to the proofs of Theorems 2 and 3.

## The Σ Notation

A sum can be written,† with $N \geq M$,

Summation sign

$$a_M + a_{M+1} + a_{M+2} + \cdots + a_N = \sum_{k=M}^{N} a_k \tag{11}$$

Index of summation

which is read "the sum of the terms $a_k$ as $k$ goes from $M$ to $N$." In this context $\Sigma$ is called the **summation sign** and $k$ is called the **index of summation.**

† The Greek letter Σ (sigma) was first used to denote a sum by the Swiss mathematician Leonhard Euler (1707–1783).

**EXAMPLE 10** **Interpreting the Summation Notation** Write out the sum $\Sigma_{k=1}^{5} b_k$.

**Solution** Starting at $k = 1$ and ending at $k = 5$, we obtain

$$\sum_{k=1}^{5} b_k = b_1 + b_2 + b_3 + b_4 + b_5$$

**EXAMPLE 11** **Interpreting the Summation Notation** Write out the sum $\Sigma_{k=3}^{6} c_k$.

**Solution** Starting at $k = 3$ and ending at $k = 6$, we obtain

$$\sum_{k=3}^{6} c_k = c_3 + c_4 + c_5 + c_6$$

**EXAMPLE 12** **Interpreting the Summation Notation** Calculate $\Sigma_{-2}^{3} k^2$.

**Solution** Here $a_k = k^2$, and $k$ ranges from $-2$ to 3.

$$\sum_{-2}^{3} k^2 = (-2)^2 + (-1)^2 + (0)^2 + 1^2 + 2^2 + 3^2$$

$$= 4 + 1 + 0 + 1 + 4 + 9 = 19$$

*Note.* As in Example 12, the index of summation can take on negative integer values or zero.

**EXAMPLE 13** **Writing a Sum Using the Summation Notation** Write the sum $S_8 = 1 - 2 + 3 - 4 + 5 - 6 + 7 - 8$ by using the summation sign.

**Solution** Since $1 = (-1)^2$, $-2 = (-1)^3 \cdot 2$, $3 = (-1)^4 \cdot 3, \ldots$, we have

$$S_8 = \sum_{k=1}^{8} (-1)^{k+1} k$$

**EXAMPLE 14** **Writing the Scalar Product Using the Sigma Notation** Equation (1) for the scalar product can be written compactly using the summation notation:

$$\mathbf{a} \cdot \mathbf{b} = a_1 b_1 + a_2 b_2 + \cdots + a_n b_n = \sum_{i=1}^{n} a_i b_i$$

Formula (4) for the $ij$th component of the product $AB$ can be written

$$c_{ij} = a_{i1}b_{1j} + a_{i2}b_{2j} + \cdots + a_{in}b_{nj} = \sum_{k=1}^{n} a_{ik}b_{kj} \tag{12}$$

The sigma notation has a number of useful properties. For example,

$$\sum_{k=1}^{n} ca_k = ca_1 + ca_2 + ca_3 + \cdots + ca_n$$

$$= c(a_1 + a_2 + a_3 + \cdots + a_n) = c\sum_{k=1}^{n} a_k$$

This and other facts are summarized below.

**Facts About the Sigma Notation**

Let $\{a_n\}$ and $\{b_n\}$ be real sequences, and let $c$ be a real number. Then

$$\sum_{k=M}^{N} ca_k = c\sum_{k=M}^{N} a_k \tag{13}$$

$$\sum_{k=M}^{N} (a_k + b_k) = \sum_{k=M}^{N} a_k + \sum_{k=M}^{N} b_k \tag{14}$$

$$\sum_{k=M}^{N} (a_k - b_k) = \sum_{k=M}^{N} a_k - \sum_{k=M}^{N} b_k \tag{15}$$

$$\sum_{k=M}^{N} a_k = \sum_{k=M}^{m} a_k + \sum_{k=m+1}^{N} a_k \qquad \text{if } M < m < N \tag{16}$$

The proofs of these facts are left as exercises (see Problems 87–89).

We now use the summation notation to prove the associative and distributive laws.

**Proofs of Theorems 2 and 3**

**Associative Law** Since $A$ is $n \times m$ and $B$ is $m \times p$, $AB$ is $n \times p$. Thus $(AB)C = (n \times p) \times (p \times q)$ is an $n \times q$ matrix. Similarly, $BC$ is $m \times q$ and $A(BC)$ is $n \times q$ so that $(AB)C$ and $A(BC)$ are both of the same size. We must show that the $ij$th component of $(AB)C$ equals the $ij$th component of $A(BC)$. Define $D = (d_{ij}) = AB$. Then

$$d_{ij} \overset{\text{from (12)}}{=} \sum_{k=1}^{m} a_{ik}b_{kj}$$

The $ij$th component of $(AB)C = DC$ is

$$\sum_{l=1}^{p} d_{il}c_{lj} = \sum_{l=1}^{p}\left(\sum_{k=1}^{m} a_{ik}b_{kl}\right)c_{lj} = \sum_{k=1}^{m}\sum_{l=1}^{p} a_{ik}b_{kl}c_{lj}$$

Next we define $E = (e_{ij}) = BC$. Then

$$e_{kj} = \sum_{l=1}^{p} b_{kl}c_{lj}$$

and the $ij$th component of $A(BC) = AE$ is

$$\sum_{k=1}^{m} a_{ik}e_{kj} = \sum_{k=1}^{m}\sum_{l=1}^{p} a_{ik}b_{kl}c_{lj}$$

Thus the $ij$th component of $(AB)C$ is equal to the $ij$th component of $A(BC)$. This proves the associative law.

**Distributive Laws** We prove the first distributive law [equation (7)]. The proof of the second one [equation (8)] is virtually identical and is therefore omitted. Let $A$ be $n \times m$ and let $B$ and $C$ be $m \times p$. Then the $kj$th component of $B + C$ is $b_{kj} + c_{kj}$ and the $ij$th component of $A(B + C)$ is

$$\sum_{k=1}^{m} a_{ik}(b_{kj} + c_{kj}) \overset{\text{From (12)}}{=} \sum_{k=1}^{m} a_{ik}b_{kj} + \sum_{k=1}^{m} a_{ik}c_{kj} = ij\text{th component of } AB \text{ plus}$$

the $ij$th component of $AC$ and this proves equation (7).

## Focus on . . .

# Arthur Cayley and the Algebra of Matrices

Arthur Cayley
*(Library of Congress)*

The algebra of matrices, that is, the rules by which matrices can be added and multiplied, was developed by the English mathematician Arthur Cayley (1821–1895) in 1857. Cayley was born at Richmond, in Surrey (near London), and was educated at Trinity College, Cambridge, graduating in 1842. In that year he placed first in the very difficult test for the Smith's prize. For a period of several years he studied and practiced law, always being careful not to let his legal practice prevent him from working on mathematics. While a student of the bar he went to Dublin and attended Hamilton's lectures on quaternions. When the Sadlerian professorship was established at Cambridge in 1863, Cayley was offered the chair, which he accepted, thus giving up a lucrative future in the legal profession for the modest provision of an academic life. But then he could devote *all* of his time to mathematics.

Cayley ranks as the third most prolific writer of mathematics in the history of the subject, being surpassed only by Euler and Cauchy. He began publishing while still an undergraduate student at Cambridge, put out between 200 and 300 papers during his years of legal practice, and continued his prolific publication the rest of his long life. The massive *Collected Mathematical Papers* of Cayley contains 966 papers and fills 13 large quarto volumes averaging about 600 pages per volume. There is scarcely an area in pure mathematics that has not been touched and enriched by the genius of Cayley.

Besides developing matrix theory, Cayley made pioneering contributions to analytic geometry, the theory of determinants, higher-dimensional geometry, the theory of curves and surfaces, the study of binary forms, the theory of elliptic functions, and the development of invariant theory.

Cayley's mathematical style reflects his legal training, for his papers are severe, direct, methodical, and clear. He possessed a phenomenal memory and seemed never to forget anything he had once seen or read. He also possessed a singularly serene, even, and gentle temperament. He has been called "the mathematicians' mathematician."

Cayley developed an unusual avidity for novel reading. He read novels while traveling, while waiting for meetings to start, and at any odd moments that presented themselves. During his life he read thousands of novels, not only in English, but also in Greek, French, German, and Italian. He took great delight in painting, especially in water colors, and he exhibited a marked talent as a water colorist. He was also an ardent student of botany and nature in general.

Cayley was, in the true British tradition, an amateur mountain climber, and he made frequent trips to the Continent for long walks and mountain scaling. A story is told that he claimed the reason he undertook mountain climbing was that, though he found the ascent arduous and tiring, the grand feeling of exhilaration he attained when he conquered the peak was like that he experienced when he solved a difficult mathematics problem or completed an intricate mathematical theory, and it was easier for him to attain the desired feeling by climbing the mountain.

Matrices arose with Cayley in connection with linear transformations of the type

$$\begin{aligned} x' &= ax + by \\ y' &= cx + dy \end{aligned} \tag{17}$$

where $a, b, c, d$ are real numbers, and which may be thought of as functions that take the vector $(x, y)$ into the vector $(x', y')$. We shall discuss linear transformations in great detail in Chapter 5. Here we observe that the transformation (17) is completely determined by the four coefficients $a, b, c, d$, and so they can be symbolized by the square array

$$\begin{pmatrix} a & b \\ c & d \end{pmatrix}$$

which we have called a $2 \times 2$ matrix. Since two transformations like (17) are identical if and only if they possess the same coefficients, Cayley defined two matrices

$$\begin{pmatrix} a & b \\ c & d \end{pmatrix} \quad \text{and} \quad \begin{pmatrix} e & f \\ g & h \end{pmatrix}$$

to be equal if and only if $a = e$, $b = f$, $c = g$, and $d = h$.

Now suppose that the transformation (17) is followed by a second transformation

$$\begin{aligned} x'' &= ex' + fy' \\ y'' &= gx' + hy' \end{aligned} \tag{18}$$

Then

$$x'' = e(ax + by) + f(cx + dy) = (ea + fc)x + (eb + fd)y$$

and

$$y'' = g(ax + by) + h(cx + dy) = (ga + hc)x + (gb + hd)y$$

This led Cayley to the following definition for the product of two matrices:

$$\begin{pmatrix} e & f \\ g & h \end{pmatrix}\begin{pmatrix} a & b \\ c & d \end{pmatrix} = \begin{pmatrix} ea + fc & eb + fd \\ ga + hc & gb + hd \end{pmatrix}$$

which is, of course, a special case of the general definition of the matrix product we gave on page 64.

It is interesting to observe how, in mathematics, very simple observations can sometimes lead to important, and far reaching, definitions and theorems.

# PROBLEMS 1.6

## Self-Quiz

**I.** Which of the following is true of matrix multiplication of matrices $A$ and $B$?

**a.** It can be performed only if $A$ and $B$ are square matrices.

**b.** Each entry $c_{ij}$ is the product of $a_{ij}$ and $b_{ij}$.

**c.** $AB = BA$.

**d.** It can be performed only if the number of columns of $A$ is equal to the number of rows of $B$.

**II.** Which of the following would be the size of the product matrix $AB$ when a $2 \times 4$ matrix $A$ is multipled by a $4 \times 3$ matrix $B$?

**a.** $2 \times 3$ **b.** $3 \times 2$ **c.** $4 \times 4$

**d.** This product cannot be found.

**III.** Which of the following is true of matrices $A$ and $B$ if $AB$ is a column vector?

**a.** $B$ is a column vector.

**b.** $A$ is a row vector.

**c.** $A$ and $B$ are square matrices.

**d.** The number of rows in $A$ must equal the number of columns in $B$.

**IV.** Which of the following is true about a product $AB$ if $A$ is a $4 \times 5$ matrix?

**a.** $B$ must have four rows and the result will have five columns.

**b.** $B$ must have five columns and the result will be a square matrix.

**c.** $B$ must have four columns and the result will have five rows.

**d.** $B$ must have five rows and the result will have four rows.

In Problems 1–7 calculate the scalar product of the two vectors.

**1.** $\begin{pmatrix} 2 \\ 3 \\ -5 \end{pmatrix}; \begin{pmatrix} 3 \\ 0 \\ 4 \end{pmatrix}$

**2.** $(1, 2, -1, 0); (3, -7, 4, -2)$

**3.** $\begin{pmatrix} 5 \\ 7 \end{pmatrix}; \begin{pmatrix} 3 \\ -2 \end{pmatrix}$

**4.** $(8, 3, 1); (7, -4, 3)$

**5.** $(a, b); (c, d)$

**6.** $\begin{pmatrix} x \\ y \\ z \end{pmatrix}; \begin{pmatrix} y \\ z \\ x \end{pmatrix}$

**7.** $(-1, -3, 4, 5); (-1, -3, 4, 5)$

**8.** Let $\mathbf{a}$ be an $n$-vector. Show that $\mathbf{a} \cdot \mathbf{a} \geq 0$.

**9.** Find conditions on a vector $\mathbf{a}$ such that $\mathbf{a} \cdot \mathbf{a} = 0$.

---

### Answers to Self-Quiz

**I.** d **II.** a **III.** a **IV.** d

In Problems 10–14 perform the indicated computation with $\mathbf{a} = \begin{pmatrix} 1 \\ -2 \\ 4 \end{pmatrix}$, $\mathbf{b} = \begin{pmatrix} 0 \\ -3 \\ -7 \end{pmatrix}$, and $\mathbf{c} = \begin{pmatrix} 4 \\ -1 \\ 5 \end{pmatrix}$.

**10.** $(2\mathbf{a}) \cdot (3\mathbf{b})$ **11.** $\mathbf{a} \cdot (\mathbf{b} + \mathbf{c})$ **12.** $\mathbf{c} \cdot (\mathbf{a} - \mathbf{b})$

**13.** $(2\mathbf{b}) \cdot (3\mathbf{c} - 5\mathbf{a})$ **14.** $(\mathbf{a} - \mathbf{c}) \cdot (3\mathbf{b} - 4\mathbf{a})$

In Problems 15–29 perform the indicated computation.

**15.** $\begin{pmatrix} 2 & 3 \\ -1 & 2 \end{pmatrix}\begin{pmatrix} 4 & 1 \\ 0 & 6 \end{pmatrix}$ **16.** $\begin{pmatrix} 3 & -2 \\ 1 & 4 \end{pmatrix}\begin{pmatrix} -5 & 6 \\ 1 & 3 \end{pmatrix}$ **17.** $\begin{pmatrix} 1 & -1 \\ 1 & 1 \end{pmatrix}\begin{pmatrix} -1 & 0 \\ 2 & 3 \end{pmatrix}$

**18.** $\begin{pmatrix} -5 & 6 \\ 1 & 3 \end{pmatrix}\begin{pmatrix} 3 & -2 \\ 1 & 4 \end{pmatrix}$ **19.** $\begin{pmatrix} -4 & 5 & 1 \\ 0 & 4 & 2 \end{pmatrix}\begin{pmatrix} 3 & -1 & 1 \\ 5 & 6 & 4 \\ 0 & 1 & 2 \end{pmatrix}$ **20.** $\begin{pmatrix} 7 & 1 & 4 \\ 2 & -3 & 5 \end{pmatrix}\begin{pmatrix} 1 & 6 \\ 0 & 4 \\ -2 & 3 \end{pmatrix}$

**21.** $\begin{pmatrix} 1 & 6 \\ 0 & 4 \\ -2 & 3 \end{pmatrix}\begin{pmatrix} 7 & 1 & 4 \\ 2 & -3 & 5 \end{pmatrix}$ **22.** $\begin{pmatrix} 1 & 4 & -2 \\ 3 & 0 & 4 \end{pmatrix}\begin{pmatrix} 0 & 1 \\ 2 & 3 \end{pmatrix}$ **23.** $\begin{pmatrix} 1 & 4 & 6 \\ -2 & 3 & 5 \\ 1 & 0 & 4 \end{pmatrix}\begin{pmatrix} 2 & -3 & 5 \\ 1 & 0 & 6 \\ 2 & 3 & 1 \end{pmatrix}$

**24.** $\begin{pmatrix} 2 & -3 & 5 \\ 1 & 0 & 6 \\ 2 & 3 & 1 \end{pmatrix}\begin{pmatrix} 1 & 4 & 6 \\ -2 & 3 & 5 \\ 1 & 0 & 4 \end{pmatrix}$ **25.** $(1 \quad 4 \quad 0 \quad 2)\begin{pmatrix} 3 & -6 \\ 2 & 4 \\ 1 & 0 \\ -2 & 3 \end{pmatrix}$

**26.** $\begin{pmatrix} 3 & 2 & 1 & -2 \\ -6 & 4 & 0 & 3 \end{pmatrix}\begin{pmatrix} 1 \\ 4 \\ 0 \\ 2 \end{pmatrix}$ **27.** $\begin{pmatrix} 3 & -2 & 1 \\ 4 & 0 & 6 \\ 5 & 1 & 9 \end{pmatrix}\begin{pmatrix} 1 & 0 & 0 \\ 0 & 1 & 0 \\ 0 & 0 & 1 \end{pmatrix}$

**28.** $\begin{pmatrix} 1 & 0 & 0 \\ 0 & 1 & 0 \\ 0 & 0 & 1 \end{pmatrix}\begin{pmatrix} 3 & -2 & 1 \\ 4 & 0 & 6 \\ 5 & 1 & 9 \end{pmatrix}$ **29.** $\begin{pmatrix} a & b & c \\ d & e & f \\ g & h & j \end{pmatrix}\begin{pmatrix} 1 & 0 & 0 \\ 0 & 1 & 0 \\ 0 & 0 & 1 \end{pmatrix}$, where $a, b, c, d, e, f, g, h, j$ are real numbers.

**30.** Find a matrix $A = \begin{pmatrix} a & b \\ c & d \end{pmatrix}$ such that $A\begin{pmatrix} 2 & 3 \\ 1 & 2 \end{pmatrix} = \begin{pmatrix} 1 & 0 \\ 0 & 1 \end{pmatrix}$.

***31.** Let $a_{11}$, $a_{12}$, $a_{21}$, and $a_{22}$ be given real numbers such that $a_{11}a_{22} - a_{12}a_{21} \neq 0$. Find numbers $b_{11}$, $b_{12}$, $b_{21}$, and $b_{22}$ such that $\begin{pmatrix} a_{11} & a_{12} \\ a_{21} & a_{22} \end{pmatrix}\begin{pmatrix} b_{11} & b_{12} \\ b_{21} & b_{22} \end{pmatrix} = \begin{pmatrix} 1 & 0 \\ 0 & 1 \end{pmatrix}$.

**32.** Verify the associative law for multiplication for the matrices $A = \begin{pmatrix} 2 & -1 & 4 \\ 1 & 0 & 6 \end{pmatrix}$, $B = \begin{pmatrix} 1 & 0 & 1 \\ 2 & -1 & 2 \\ 3 & -2 & 0 \end{pmatrix}$, and $C = \begin{pmatrix} 1 & 6 \\ -2 & 4 \\ 0 & 5 \end{pmatrix}$.

**33.** As in Example 6, suppose that a group of people have contracted a contagious disease. These persons have contacts with a second group who in turn have contacts with a third group. Let $A = \begin{pmatrix} 1 & 0 & 1 & 0 \\ 0 & 1 & 1 & 0 \\ 1 & 0 & 0 & 1 \end{pmatrix}$ represent the contacts between the contagious group and the members of group 2, and let

$$B = \begin{pmatrix} 1 & 0 & 1 & 0 & 0 \\ 0 & 0 & 0 & 1 & 0 \\ 1 & 1 & 0 & 0 & 0 \\ 0 & 0 & 1 & 0 & 1 \end{pmatrix}$$

represent the contacts between groups 2 and 3. (a) How many people are in each group? (b) Find the matrix of indirect contacts between groups 1 and 3.

**34.** Answer the questions of Problem 33 for $A = \begin{pmatrix} 1 & 0 & 1 & 1 & 0 \\ 0 & 1 & 0 & 1 & 1 \end{pmatrix}$ and

$$B = \begin{pmatrix} 1 & 0 & 0 & 0 & 0 & 0 & 1 \\ 0 & 1 & 0 & 1 & 0 & 0 & 0 \\ 1 & 1 & 0 & 0 & 1 & 1 & 1 \\ 0 & 0 & 0 & 1 & 1 & 0 & 1 \\ 0 & 1 & 0 & 0 & 0 & 0 & 0 \end{pmatrix}$$

**ORTHOGONAL VECTORS**

Two vectors **a** and **b** are said to be **orthogonal** if $\mathbf{a} \cdot \mathbf{b} = 0$. In Problems 35–39 determine which pairs of vectors are orthogonal.†

**35.** $\begin{pmatrix} 2 \\ -3 \end{pmatrix}; \begin{pmatrix} 3 \\ 2 \end{pmatrix}$

**36.** $\begin{pmatrix} 2 \\ -3 \end{pmatrix}; \begin{pmatrix} -3 \\ 2 \end{pmatrix}$

**37.** $\begin{pmatrix} 1 \\ 4 \\ -7 \end{pmatrix}; \begin{pmatrix} 2 \\ 3 \\ 2 \end{pmatrix}$

**38.** (1, 0, 1, 0); (0, 1, 0, 1)

**39.** $\begin{pmatrix} a \\ 0 \\ b \\ 0 \\ c \end{pmatrix}; \begin{pmatrix} 0 \\ d \\ 0 \\ e \\ 0 \end{pmatrix}$

**40.** Determine a number $\alpha$ such that $(1, -2, 3, 5)$ is orthogonal to $(-4, \alpha, 6, -1)$.

**41.** Determine all numbers $\alpha$ and $\beta$ such that the vectors $\begin{pmatrix} 1 \\ -\alpha \\ 2 \\ 3 \end{pmatrix}$ and $\begin{pmatrix} 4 \\ 5 \\ -2\beta \\ 7 \end{pmatrix}$ are orthogonal.

**42.** Using the definition of the scalar product, prove Theorem 1.

† We shall be dealing extensively with orthogonal vectors in Chapters 3 and 4.

**43.** A manufacturer of custom-designed jewelry has orders for two rings, three pairs of earrings, five pins, and one necklace. The manufacturer estimates that it takes 1 hour of labor to make a ring, $1\frac{1}{2}$ hours to make a pair of earrings, $\frac{1}{2}$ hour for each pin, and 2 hours to make a necklace.

**a.** Express the manufacturer's orders as a row vector.

**b.** Express the hourly requirements for the various types of jewelry as a column vector.

**c.** Use the scalar product to calculate the total number of hours it will require to complete all the orders.

**44.** A tourist returned from a European trip with the following foreign currency: 1000 Austrian schillings, 20 British pounds, 100 French francs, 5000 Italian lire, and 50 German marks. In American money, a schilling was worth \$0.055, the pound \$1.80, the franc \$0.20, the lira \$0.001, and the mark \$0.40.

**a.** Express the quantity of each currency by means of a row vector.

**b.** Express the value of each currency in American money by means of a column vector.

**c.** Use the scalar product to compute how much the tourist's foreign currency was worth in American money.

**45.** A company pays its executives a salary and gives them shares of its stock as an annual bonus. Last year the president of the company received \$80,000 and 50 shares of stock, each of the three vice-presidents was paid \$45,000 and 20 shares of stock, and the treasurer was paid \$40,000 and 10 shares of stock.

**a.** Express the payments to the executives in money and stock by means of a $2 \times 3$ matrix.

**b.** Express the number of executives of each rank by means of a column vector.

**c.** Use matrix multiplication to calculate the total amount of money and the total number of shares of stock the company paid these executives last year.

**46.** Sales, unit gross profits, and unit taxes for sales of a large corporation are given in the following table:

| | Product | | | | | |
|---|---|---|---|---|---|---|
| | Sales of Item | | | | | |
| **Month** | **I** | **II** | **III** | **Item** | **Unit Profit (in hundreds of dollars)** | **Unit Taxes (in hundreds of dollars)** |
| **January** | **4** | **2** | **20** | **I** | **3.5** | **1.5** |
| **February** | **6** | **1** | **9** | **II** | **2.75** | **2** |
| **March** | **5** | **3** | **12** | **III** | **1.5** | **0.6** |
| **April** | **8** | **2.5** | **20** | | | |

Find a matrix that shows total profits and taxes in each of the 4 months.

**47.** Let $A$ be a square matrix. Then $A^2$ is defined simply as $AA$. Calculate $\begin{pmatrix} 2 & -1 \\ 4 & 6 \end{pmatrix}^2$.

**48.** Calculate $A^2$, where $A = \begin{pmatrix} 1 & -2 & 4 \\ 2 & 0 & 3 \\ 1 & 1 & 5 \end{pmatrix}$.

**49.** Calculate $A^3$, where $A = \begin{pmatrix} -1 & 2 \\ 3 & 4 \end{pmatrix}$.

**50.** Calculate $A^2$, $A^3$, $A^4$, and $A^5$, where

$$A = \begin{pmatrix} 0 & 1 & 0 & 0 \\ 0 & 0 & 1 & 0 \\ 0 & 0 & 0 & 1 \\ 0 & 0 & 0 & 0 \end{pmatrix}$$

**51.** Calculate $A^2$, $A^3$, $A^4$, and $A^5$, where

$$A = \begin{pmatrix} 0 & 1 & 0 & 0 & 0 \\ 0 & 0 & 1 & 0 & 0 \\ 0 & 0 & 0 & 1 & 0 \\ 0 & 0 & 0 & 0 & 1 \\ 0 & 0 & 0 & 0 & 0 \end{pmatrix}$$

**52.** An $n \times n$ matrix $A$ has the property that $AB$ is the zero matrix for any $n \times n$ matrix $B$. Prove that $A$ is the zero matrix.

**53.** A **probability matrix** is a square matrix having two properties: (*i*) every component is nonnegative ($\geq 0$) and (*ii*) the sum of the elements in each row is 1. The following are probability matrices:

$$P = \begin{pmatrix} \frac{1}{3} & \frac{1}{3} & \frac{1}{3} \\ \frac{1}{4} & \frac{1}{2} & \frac{1}{4} \\ 0 & 0 & 1 \end{pmatrix} \quad \text{and} \quad Q = \begin{pmatrix} \frac{1}{6} & \frac{1}{6} & \frac{2}{3} \\ 0 & 1 & 0 \\ \frac{1}{5} & \frac{1}{5} & \frac{3}{5} \end{pmatrix}$$

Show that $PQ$ is a probability matrix.

***54.** Let $P$ be a probability matrix. Show that $P^2$ is a probability matrix.

****55.** Let $P$ and $Q$ be probability matrices of the same size. Prove that $PQ$ is a probability matrix.

**56.** Prove formula (6) by using the associative law [equation (5)].

***57.** A round robin tennis tournament can be organized in the following way. Each of the $n$ players plays all the others, and the results are recorded in an $n \times n$ matrix $R$ as follows:

$$R_{ij} = \begin{cases} 1 & \text{if the } i\text{th player beats the } j\text{th player} \\ 0 & \text{if the } i\text{th player loses to the } j\text{th player} \\ 0 & \text{if } i = j \end{cases}$$

The $i$th player is then assigned the score

$$S_i = \sum_{j=1}^{n} R_{ij} + \frac{1}{2} \sum_{j=1}^{n} (R^2)_{ij}\dagger$$

† $(R^2)_{ij}$ is the $ij$th component of the matrix $R^2$.

**a.** In a tournament between four players

$$R = \begin{pmatrix} 0 & 1 & 0 & 0 \\ 0 & 0 & 1 & 1 \\ 1 & 0 & 0 & 0 \\ 1 & 0 & 1 & 0 \end{pmatrix}$$

Rank the players according to their scores.

**b.** Interpret the meaning of the score.

**58.** Let $O$ be the $m \times n$ zero matrix and let $A$ be an $n \times p$ matrix. Show that $OA = O_1$, where $O_1$ is the $m \times p$ zero matrix.

**59.** Verify the distributive law [equation (7)] for the matrices

$$A = \begin{pmatrix} 1 & 2 & 4 \\ 3 & -1 & 0 \end{pmatrix} \quad B = \begin{pmatrix} 2 & 7 \\ -1 & 4 \\ 6 & 0 \end{pmatrix} \quad C = \begin{pmatrix} -1 & 2 \\ 3 & 7 \\ 4 & 1 \end{pmatrix}$$

In Problems 60–64 multiply the matrices using the indicated blocks.

**60.** $$\left(\begin{array}{cc|cc} 2 & 3 & 1 & 5 \\ 0 & 1 & -4 & 2 \\ \hline 3 & 1 & 6 & 4 \end{array}\right)\left(\begin{array}{cc} 1 & 4 \\ -1 & 0 \\ \hline 2 & 3 \\ 1 & 5 \end{array}\right)$$

**61.** $$\left(\begin{array}{c} 1 \\ \hline 6 \\ \hline 2 \end{array}\right)\left(\begin{array}{c|c|c|c} 3 & 7 & 1 & 5 \end{array}\right)$$

**62.** $$\left(\begin{array}{cc|cc} 1 & 0 & -1 & 1 \\ 2 & 1 & -3 & 4 \\ \hline -2 & 1 & 4 & 6 \\ 0 & 2 & 3 & 5 \end{array}\right)\left(\begin{array}{cc|cc} 2 & 4 & 1 & 6 \\ 3 & 0 & -2 & 5 \\ \hline 2 & 1 & -1 & 0 \\ -2 & -4 & 1 & 3 \end{array}\right)$$

**63.** $$\left(\begin{array}{cc|cc} 1 & 0 & 0 & 0 \\ 0 & 1 & 0 & 0 \\ \hline 0 & 0 & a & b \\ 0 & 0 & c & d \end{array}\right)\left(\begin{array}{cc|cc} e & f & 0 & 0 \\ g & h & 0 & 0 \\ \hline 0 & 0 & 1 & 0 \\ 0 & 0 & 0 & 1 \end{array}\right)$$

**64.** $$\left(\begin{array}{cc|ccc} 1 & 0 & 2 & 3 & 1 \\ 0 & 1 & 5 & 2 & 6 \\ \hline 0 & 0 & -1 & 2 & 4 \\ 0 & 0 & 2 & 1 & 3 \end{array}\right)\left(\begin{array}{ccc} -1 & 1 & 4 \\ 0 & 4 & -3 \\ \hline 1 & 0 & 0 \\ 0 & 1 & 0 \\ 0 & 0 & 1 \end{array}\right)$$

**65.** Let $A = \begin{pmatrix} I & O \\ C & I \end{pmatrix}$ and $B = \begin{pmatrix} I & O \\ D & I \end{pmatrix}$. If $A$ and $B$ are partitioned conformably, show that $A$ and $B$ commute. Here $I$ is defined as in Example 9.

In Problems 66–73 evaluate the given sums.

**66.** $\sum_{k=1}^{4} 2k$ **67.** $\sum_{i=1}^{3} i^3$ **68.** $\sum_{k=0}^{6} 1$ **69.** $\sum_{k=1}^{8} 3^k$

**70.** $\sum_{i=2}^{5} \frac{1}{1+i}$ **71.** $\sum_{j=5}^{7} \frac{2j+3}{j-2}$ **72.** $\sum_{i=1}^{3}\sum_{j=1}^{4} ij$ **73.** $\sum_{k=1}^{3}\sum_{j=2}^{4} k^2 j^3$

In Problems 74–86 write each sum using the $\Sigma$ notation.

**74.** $1 + 2 + 4 + 8 + 16$

**75.** $1 - 3 + 9 - 27 + 81 - 243$

**76.** $\frac{2}{3} + \frac{3}{4} + \frac{4}{5} + \frac{5}{6} + \frac{6}{7} + \frac{7}{8} + \cdots + \frac{n}{n+1}$

**77.** $1 + 2^{1/2} + 3^{1/3} + 4^{1/4} + 5^{1/5} + \cdots + n^{1/n}$

**78.** $1 + x^3 + x^6 + x^9 + x^{12} + x^{15} + x^{18} + x^{21}$

**79.** $-1 + \frac{1}{a} - \frac{1}{a^2} + \frac{1}{a^3} - \frac{1}{a^4} + \frac{1}{a^5} - \frac{1}{a^6} + \frac{1}{a^7} - \frac{1}{a^8} + \frac{1}{a^9}$

**80.** $1 \cdot 3 + 3 \cdot 5 + 5 \cdot 7 + 7 \cdot 9 + 9 \cdot 11 + 11 \cdot 13 + 13 \cdot 15 + 15 \cdot 17$

**81.** $2^2 \cdot 4 + 3^2 \cdot 6 + 4^2 \cdot 8 + 5^2 \cdot 10 + 6^2 \cdot 12 + 7^2 \cdot 14$

**82.** $a_{11} + a_{12} + a_{13} + a_{21} + a_{22} + a_{23}$

**83.** $a_{11} + a_{12} + a_{21} + a_{22} + a_{31} + a_{32}$

**84.** $a_{21} + a_{22} + a_{23} + a_{24} + a_{31} + a_{32} + a_{33} + a_{34} + a_{41} + a_{42} + a_{43} + a_{44}$

**85.** $a_{31}b_{12} + a_{32}b_{22} + a_{33}b_{32} + a_{34}b_{42} + a_{35}b_{52}$

**86.** $a_{21}b_{11}c_{15} + a_{21}b_{12}c_{25} + a_{21}b_{13}c_{35} + a_{21}b_{14}c_{45}$
$+ a_{22}b_{21}c_{15} + a_{22}b_{22}c_{25} + a_{22}b_{23}c_{35} + a_{22}b_{24}c_{45}$
$+ a_{23}b_{31}c_{15} + a_{23}b_{32}c_{25} + a_{23}b_{33}c_{35} + a_{23}b_{34}c_{45}$

**87.** Prove formula (14) by writing out the terms in

$$\sum_{k=M}^{N} (a_k + b_k)$$

**88.** Prove formula (15).

[*Hint:* Use (13) to show that $\sum_{k=M}^{N} (-a_k) = -\sum_{k=M}^{N} a_n$. Then use (14).]

**89.** Prove formula (16).

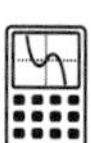

## CALCULATOR BOX

Matrix multiplication (of which the dot product is a special case) can be performed easily on both the TI-85 and the CASIO fx-7700 GB.

### TI-85

This can be done in two simple ways. If the matrices $A$ and $B$ are already entered, then the key sequence

ALPHA  A  ×  ALPHA  B  ENTER

will result in the product $AB$ being displayed.

If the product is not defined the calculator will display the message

ERROR 14 UNDEFINED

Alternatively, you may simply enter the matrices together. For example,

[[1,2][3,4]][[5,6][7,8]] ENTER results in [[19 22]]
[43 50]]

## CASIO fx-7700 GB

Enter the matrices $A$ and $B$. Then press MODE 0 F5 . If the product is undefined, the message Dim ERROR will be displayed.

In Problems 90–92 use a calculator to obtain each product.

**90.** $\begin{pmatrix} 1.23 & 4.69 & 5.21 \\ -1.08 & -3.96 & 8.57 \\ 6.28 & -5.31 & -4.27 \end{pmatrix}\begin{pmatrix} 9.61 & -2.30 \\ -8.06 & 0.69 \\ 2.67 & -5.23 \end{pmatrix}$

**91.** $\begin{pmatrix} 125 & 216 & 419 \\ 383 & 516 & 237 \\ 209 & 855 & 601 \\ 403 & 237 & 506 \end{pmatrix}\begin{pmatrix} 73 & 36 \\ 21 & 28 \\ 49 & 67 \end{pmatrix}$

**92.** $\begin{pmatrix} 23.2 & 56.3 & 19.6 & -31.4 \\ 18.9 & -9.6 & 17.4 & 51.2 \\ 30.8 & -17.9 & -14.4 & 28.6 \end{pmatrix}\begin{pmatrix} -0.071 & 0.068 \\ 0.051 & -0.023 \\ -0.011 & -0.082 \\ 0.053 & 0.065 \end{pmatrix}$

**93.** In Problem 55 you were asked to prove that the product of the two probability matrices is a probability matrix. Let

$$P = \begin{pmatrix} 0.23 & 0.16 & 0.57 & 0.04 \\ 0.15 & 0.09 & 0.34 & 0.42 \\ 0.66 & 0.22 & 0.11 & 0.01 \\ 0.07 & 0.51 & 0.20 & 0.22 \end{pmatrix} \quad \text{and} \quad Q = \begin{pmatrix} 0.112 & 0.304 & 0.081 & 0.503 \\ 0.263 & 0.015 & 0.629 & 0.093 \\ 0.402 & 0.168 & 0.039 & 0.391 \\ 0.355 & 0.409 & 0.006 & 0.230 \end{pmatrix}$$

**a.** Show that $P$ and $Q$ are probability matrices.
**b.** Compute $PQ$ and show that it is a probability matrix.

**94.** Let $A = \begin{pmatrix} 1 & 3 \\ 0 & 2 \end{pmatrix}$. Compute $A^2$, $A^5$, $A^{10}$, $A^{50}$, and $A^{100}$.

[*Hint:* On the TI-85, $A$^$n$ will compute $A^n$ for $0 \le n \le 255$; on the CASIO fx-7700 GB, you should enter $A$ into both memory $A$ and memory $B$. Then press MODE 0 F5 and $A^2$ will be displayed in memory C. Press F1 and $A^2$ will be stored in $A$. Then the product $AB$ will result in $A^2B = A^3$, and so on. Alternatively, store $A^2$ in $B$ as well. Then $AB$ results in $A^2A^2 = A^4$.]

**95.** Let $A = \begin{pmatrix} a & x & y \\ 0 & b & z \\ 0 & 0 & c \end{pmatrix}$. Based on your calculations in Problem 94, guess at the form of the diagonal components of $A^n$. Here $x$, $y$, and $z$ denote real numbers.

# MATLAB 1.6

*MATLAB Information.*

A matrix product $AB$ is found by **A*B.**

An integer power of a matrix, $A^n$, is found by **A^n,** where $n$ has previously been assigned some value.

We repeat some basic commands for generating random matrices: For a random $n \times m$ matrix with entries between $-c$ and $c$, **A = c*(2*rand(n,m)−1);** For a random $n \times m$ matrix with integer entries between $-c$ and $c$, **B = round(c*(2*rand(n,m)−1)).** To generate matrices with complex entries, generate $A$ and $B$ as above and let **C = A + i*B.** If a problem asks you to generate random matrices with certain entries, generate some complex as well as real matrices.

1. Enter any $3 \times 4$ matrix $A$ and $4 \times 2$ matrix $B$. Find **A*B** and **B*A.** Comment on the results.
2. Generate two random matrices, $A$ and $B$, with entries between $-10$ and 10. Find $AB$ and $BA$. Repeat for at least seven pairs of $A$ and $B$. How many of your pairs satisfied $AB = BA$? What can you conclude about the likelihood of $AB = BA$?
3. Enter matrices $A$, $B$, $\mathbf{x}$, and $\mathbf{z}$ below.

$$A = \begin{pmatrix} 2 & 9 & -23 & 0 \\ 0 & 4 & -12 & 4 \\ 7 & 5 & -1 & 1 \\ 7 & 8 & -10 & 4 \end{pmatrix} \quad \mathbf{b} = \begin{pmatrix} 34 \\ 24 \\ 15 \\ 33 \end{pmatrix} \quad \mathbf{z} = \begin{pmatrix} -2 \\ 3 \\ 1 \\ 0 \end{pmatrix} \quad \mathbf{x} = \begin{pmatrix} -5 \\ 10 \\ 2 \\ 2 \end{pmatrix}$$

   a. Show that $A\mathbf{x} = \mathbf{b}$ and $A\mathbf{z} = \mathbf{0}$.
   b. Based on your knowledge of regular algebraic manipulations and using the results of part (a), what would you conjecture that $A(\mathbf{x} + s\mathbf{z})$ would equal, where $s$ is any scalar? Test your conjecture by finding $A(\mathbf{x} + s\mathbf{z})$ for at least five choices of the scalar $s$.

4. a. Generate random matrices with integer entries, $A$ and $B$, such that the product $AB$ is defined. Modify $B$ such that two columns of $B$ are the same. (For example, **B(:,2) = B(:,3).**)
   b. Find $AB$ and look at its columns. What do you conjecture about the columns of $AB$ if $B$ has two columns that are the same?
   c. Test your conjecture by repeating the instructions above for three more choices of $A$ and $B$. (Don't make all your choices square matrices.)
   d. *(Paper and pencil)* Prove your conjecture, using the definition of matrix multiplication.

5. Generate a $5 \times 6$ random matrix $A$ with entries between $-10$ and 10 and generate a $6 \times 1$ random vector $\mathbf{x}$ with entries between $-10$ and 10. Find

   **A*x − (x(1)*A(:,1) + ··· + x(m)*A(:,m)).**

   Repeat for different random choices of $A$ and $\mathbf{x}$. How does this relate to expression (10) in this section in the text?

6. a. Let $A = \begin{pmatrix} a & b \\ c & d \end{pmatrix}$. Suppose $B = \begin{pmatrix} x_1 & x_2 \\ x_3 & x_4 \end{pmatrix}$.

   Set up the system of equations, with $x_1$ through $x_4$ as the unknowns, that arises from setting $AB = BA$. Verify that the system is a homogeneous system with coefficient matrix

$$R = \begin{pmatrix} 0 & -c & b & 0 \\ -b & a-d & 0 & b \\ c & 0 & d-a & -c \\ 0 & c & -b & 0 \end{pmatrix}$$

b. For $A = \begin{pmatrix} 1 & -1 \\ 5 & -4 \end{pmatrix}$, we wish to find a matrix $B$ such that $AB = BA$.

   **i.** Enter $R$ as above and solve the homogeneous system with coefficient matrix $R$ for $x_1$, $x_2$, $x_3$, and $x_4$. Explain why there are infinitely many solutions with one variable chosen arbitrarily.

   **ii.** Find **rat(rref(R),'s')** and use it to choose a value for the arbitrary variable so that $x_i$ is an integer. If you are using MATLAB 4.0, enter the commands **format rat** followed by **rref(R).**

   **iii.** Enter the resulting matrix $B = \begin{pmatrix} x_1 & x_2 \\ x_3 & x_4 \end{pmatrix}$ and verify that $AB = BA$.

   **iv.** Repeat *(iii)* for another choice of the arbitrary variable.

c. Repeat the process above for $A = \begin{pmatrix} 1 & 2 \\ 3 & 4 \end{pmatrix}$.

d. Repeat the process above for a $2 \times 2$ matrix $A$ of your choice.

7. Generate a pair, $A$ and $B$, of random $2 \times 2$ matrices with entries between $-10$ and 10. Find $C = (A + B)^2$ and $D = A^2 + 2AB + B^2$. Compare $C$ and $D$ (find $C - D$). Generate two more pairs of random $2 \times 2$ matrices and repeat the above. Enter the pair of matrices, $A$ and $B$, generated in MATLAB Problem 6(b) in this section and find $C - D$ as above. Enter the pair of matrices, $A$ and $B$, generated in MATLAB Problem 6(c) in this section and find $C - D$ as above. From this evidence, what can you conjecture about the statement $(A + B)^2 = A^2 + 2AB + B^2$? Prove your conjecture.

8. a. Enter **A = round(10*(2*rand(6,5)−1))**. Enter **E = [1 0 0 0 0 0]** and find **E*A.** Let **E = [0 0 1 0 0 0]** and find **E*A.** Describe how $EA$ is composed of parts of $A$ and how this depends on the position of the entry that equals 1 in the matrix $E$.

   b. Let **E = [2 0 0 0 0 0]** and find **E*A.** Let **E = [0 0 2 0 0 0]** and find **E*A.** Describe how $EA$ is composed of parts of $A$ and how this depends on the position of the entry that equals 2 in the matrix $E$.

   c. **i.** Let **E = [1 0 1 0 0 0]** and find **E*A.** Describe how $EA$ is composed of parts of $A$ and how the relationship depends on the position of the entries that are 1 in the matrix $E$.

   **ii.** Let **E = [2 0 1 0 0 0]** and find **E*A.** Describe how $EA$ is composed of parts of $A$ and how the relationship depends on the position of the nonzero entries in $E$.

   d. Suppose $A$ is $n \times m$ and $E$ is $1 \times n$, where the $k$th entry of $E$ equals some number $p$. From (a) and (b), formulate a conjecture about the relationship between $A$ and $EA$. Test your conjecture by generating a random $A$ (for some choice of $n$ and $m$), forming two different $E$ matrices (for some choice of $k$ and $p$), and finding $EA$ for each $E$. Repeat for another choice of $A$.

   e. Suppose $A$ is $n \times m$ and $E$ is $1 \times n$, where the $k$th entry of $E$ equals some number $p$ and the $j$th entry of $E$ equals some number $q$. From part (c), formulate a conjecture about the relationship between $A$ and $EA$. Test your conjecture by generating a random $A$, forming two different $E$ matrices of the form described and finding $EA$ for each $E$. Repeat for another choice of $A$.

f. Suppose $A$ is $n \times m$ and $F$ is $m \times 1$, where the $k$th entry of $F$ equals some number $p$ and the $j$th entry of $F$ equals some number $q$. Consider $AF$. Experiment as above to determine a conjecture about the relationship of $AF$ to $A$.

9. **Upper Triangular Matrices**

a. Let $A$ and $B$ be any two random $3 \times 3$ matrices. Let **UA = triu(A)** and **UB = triu(B).** The command **triu** forms upper triangular matrices. Find **UA*UB.** What property does the product have? Repeat for three more pairs of random $n \times n$ matrices, using different choices of $n$.

b. *(Pencil and paper)* From your observations, write a conjecture about the product of two upper triangular matrices. Prove your conjecture by using the definition of matrix multiplication.

c. What would you conjecture about the product of two lower triangular matrices? Test your conjecture for at least three pairs of lower triangular matrices. [*Hint:* Use **tril(A)** and **tril(B)** to generate lower triangular matrices from random matrices $A$ and $B$.]

10. **Nilpotent Matrices** A nonzero matrix $A$ is said to be **nilpotent** if there is some integer $k$ such that $A^k = 0$. The **index of nilpotency** is defined to be the smallest integer $k$ for which $A^k = 0$.

a. Generate a $5 \times 5$ random matrix $A$. Let **B = triu(A,1).** What does $B$ look like? Compute $B^2$, $B^3$, and so on; show that $B$ is nilpotent, and find its index of nilpotency.

b. Repeat the instructions from part (a) for **B = triu(A,2).**

c. Generate a $7 \times 7$ random matrix $A$. Repeat parts (a) and (b) using this $A$.

d. Using the experience gained from parts (a), (b), and (c) (and further exploration of the command **B = triu(A,j),** where $j$ is some integer), generate a $6 \times 6$ matrix $C$ that is nilpotent with the index of nilpotency of $C$ equal to 3.

11. **Block Matrices** If $A = \begin{pmatrix} a & b \\ c & d \end{pmatrix}$ and $B = \begin{pmatrix} e & f \\ g & h \end{pmatrix}$, then $AB = \begin{pmatrix} ae + bg & af + bh \\ ce + dg & cf + dh \end{pmatrix}$

Explain when this pattern holds true if $a$ through $h$ are matrices instead of numbers.

Generate eight $2 \times 2$ matrices $A$, $B$, $C$, $D$, $E$, $F$, $G$, and $H$. Find **AA = [A B; C D]** and **BB = [E F; G H].** Find **AA*BB** and compare with **K = [A*E+B*G A*F+B*H; C*E+D*G C*F+D*H]** (that is, find **AA*BB − K.**) Repeat for two more choices of $A$ through $H$.

12. **Outer Products** Generate a random $3 \times 4$ matrix $A$ and a random $4 \times 5$ matrix $B$. Compute

$$(\text{col } 1\ A)(\text{row } 1\ B) + (\text{col } 2\ A)(\text{row } 2\ B) + \cdots + (\text{col } 4\ A)(\text{row } 4\ B)$$

and label this expression $D$. Find $D - AB$. Describe the relationship between $D$ and $AB$. Repeat for a random matrix $A$ of size $5 \times 5$ and a random matrix $B$ of size $5 \times 6$. (Here the sum involved in calculating $D$ involves adding five products.)

13. **Contact Matrices** Consider four groups of people: Group 1 consists of $A1, A2$, and $A3$, group 2 consists of 5 people $B1$ through $B5$, group 3 consists of 8 people $C1$ through $C8$, and group 4 consists of 10 people $D1$ through $D10$.
    a. Given the following information, enter the three direct contact matrices. (See MATLAB Problem 2 from Section 1.5 for an efficient way to enter these matrices.)

    Contacts:

    ($A1$ with $B1$, $B2$) ($A2$ with $B2$, $B3$) ($A3$ with $B1$, $B4$, $B5$)
    ($B1$ with $C1$, $C3$, $C5$) ($B2$ with $C3$, $C4$, $C7$)
    ($B3$ with $C1$, $C5$, $C6$, $C8$) ($B4$ with $C8$) ($B5$ with $C5$, $C6$, $C7$)

    ($C1$ with $D1$, $D2$, $D3$) ($C2$ with $D3$, $D4$, $D6$) ($C3$ with $D8$, $D9$, $D10$)
    ($C4$ with $D4$, $D5$, $D7$) ($C5$ with $D1$, $D4$, $D6$, $D8$) ($C6$ with $D2$, $D4$)
    ($C7$ with $D1$, $D5$, $D9$) ($C8$ with $D1$, $D2$, $D4$, $D6$, $D7$, $D9$, $D10$)

    b. Find the indirect contact matrix for the contacts group 1 has with group 4. Which entries are zero? What is the significance of this? Interpret the $(1, 5)$ entry and the $(2, 4)$ entry of this indirect contact matrix.
    c. Which person in group 4 has the most indirect contacts with group 1? Which person has the fewest? Which person in group 1 is the "most dangerous" (for spreading a disease) for the people in group 4? Why?

    [*Hint:* There is a way to use matrix multiplication to compute the row and column sums. Use the vectors **d = ones(10,1)** and **e = ones(1,3).** Here the command **ones(n,m)** produces a matrix of size $n \times m$, where every entry equals 1.]

14. **Markov Chain** A market research firm has been studying the buying patterns for three competing products. The research firm has determined the percentage of the households that will change from buying one product to buying another product after 1 month. (Assume that each household buys one of these three products and that the percentages do not change from month to month.) This information is presented in matrix form:

    $$p_{ij} = \text{percentage who switch } \textit{from} \text{ product } j \textit{ to} \text{ product } i \text{ after 1 month}$$

    $$P = \begin{pmatrix} .8 & .2 & .05 \\ .05 & .75 & .05 \\ .15 & .05 & .9 \end{pmatrix} \qquad P \text{ is called a } \textbf{transition matrix.}$$

    For example, $P_{12} = .2$ means 20% of the households switch from buying product 2 to buying product 1 after one month and $P_{22} = .75$ means 75% of the households who were buying product 2 continue to buy product 2 after 1 month. Assume there is a total of 30,000 households.
    a. *(Pencil and paper)* Interpret the other entries in $P$.
    b. Let $\mathbf{x}$ be $3 \times 1$, where $x_k$ = the number of households that buy product $k$. What is the interpretation of $P\mathbf{x}$? of $P^2\mathbf{x} = P(P\mathbf{x})$?
    c. Suppose, initially,

    $$\mathbf{x} = \begin{pmatrix} 10000 \\ 10000 \\ 10000 \end{pmatrix}$$

Find $P^n\mathbf{x}$ for $n = 5, 10, 15, 20, 25, 30, 35, 40, 45$, and $50$. Describe the behavior of the vectors $P^n\mathbf{x}$ as $n$ gets larger. What is the interpretation of this?

d. Suppose, initially, that

$$\mathbf{x} = \begin{pmatrix} 0 \\ 30000 \\ 0 \end{pmatrix}$$

Repeat the instructions above. Compare results from (c) and (d).

e. Choose your own initial vector for $\mathbf{x}$, where the components of $\mathbf{x}$ add to 30000. Repeat the instructions and compare with previous results.

f. Compute $P^n$ and $30000P^n$ for the $n$'s given above. What do you notice about the columns of $P^n$? How do the columns of $30000P^n$ relate to the previous results of this problem?

g. Suppose a car rental agency has three offices. A car rented at one office can be returned to any of the three offices. Suppose

$$P = \begin{pmatrix} .8 & .1 & .1 \\ .05 & .75 & .1 \\ .15 & .15 & .8 \end{pmatrix}$$

is a transition matrix such that $P_{ij}$ = the percentage of cars rented at office $j$ and returned to office $i$ after one time interval. Suppose there is a total of 1000 cars. Based on your observations in previous parts of this problem, find the long-term distribution of cars, that is, the number of cars that will be at each office after a long period of time. How could a car rental agency use this information?

**PROJECT PROBLEM**

15. **Population Matrix** A population of fish is divided into five age groups, where group 1 represents the babies and group 5 is the oldest age group. The matrix $S$ below represents birth- and survival rates:

$$S = \begin{pmatrix} 0 & 0 & 2 & 2 & 0 \\ .4 & .2 & 0 & 0 & 0 \\ 0 & .5 & .2 & 0 & 0 \\ 0 & 0 & .5 & .2 & 0 \\ 0 & 0 & 0 & .4 & .1 \end{pmatrix}$$

$s_{1j}$ = number of fish born to each fish in group $j$ in one year
$s_{ij}$ = percentage of fish in group $j$ that survive to group $i$, where $i > 1$

For example, $s_{13} = 2$ says that each fish in group 3 has two babies in one year and $s_{21} = .4$ says that 40% of the fish in group 1 survive to group 2 one year later.

a. *(Paper and pencil)* Interpret the other entries of $S$.

b. *(Paper and pencil)* Let $\mathbf{x}$ be the $5 \times 1$ matrix such that $x_k$ = the number of fish in group $k$. Explain why $S^2\mathbf{x}$ represents the number of fish in each group 2 years later.

c. Let

$$\mathbf{x} = \begin{pmatrix} 5000 \\ 10000 \\ 20000 \\ 20000 \\ 5000 \end{pmatrix}$$

Find **floor(S^n*x)** for $n = 10, 20, 30, 40$, and 50. (The **floor** command rounds down to the nearest integer.) What is happening to the fish population over time? Is it growing or is it dying out? Explain.

d. Changes in birthrates or survival rates can affect the population growth. Change $s_{13}$ from 2 to 1 and repeat the commands as in part (c). Describe what appears to be happening to the population. Change $s_{13}$ back to 2 and change $s_{32}$ to .3 and repeat the commands as in part (c). Describe what appears to be happening to the population.

e. *(Paper and pencil)* Suppose we are interested in harvesting this fish population. Let **h** be the $5 \times 1$ vector, where $h_j$ = the number of fish harvested from group $j$ at the end of a year. Argue why **u = S*x−h** will give us the number of fish at the end of one year after the harvest and then why **w = S*u−h** will give us the number of fish at the end of 2 years after harvesting.

f. Change $s_{13}$ back to 2 and $s_{32}$ back to .5. Suppose we decide to harvest only mature fish, that is, fish from group 5. We will examine harvesting possibilities over a time span of 15 years. Let **h = [0;0;0;0;2000]**. To show that this is not a harvest that we can continue, enter the commands

```
u = S*x−h
u = S*u−h
```

Repeat the last command (using the up-arrow key) until you notice that there are negative fish after a harvest. For how many years can we harvest this amount?

g. Experiment with other harvests from group 5 to find the maximum amount of fish we can harvest in a given year in order to sustain this level of harvest for 15 years. (Enter **h = [0;0;0;0;n]** for some number $n$ and repeat the commands from part (f) as needed to represent 15 years of harvesting.) Write a description of your experimentation and results.

h. Experiment further to see if you can find a vector **h** representing harvests from groups 4 and 5 that would allow for more total fish to be harvested each year (and still sustain the harvest over 15 years). Write a description of your experimentation and your results.

## 1.7 MATRICES AND LINEAR SYSTEMS OF EQUATIONS

In Section 1.3, page 16, we discussed the following systems of $m$ linear equations in $n$ unknowns:

$$\begin{aligned} a_{11}x_1 + a_{12}x_2 + \cdots + a_{1n}x_n &= b_1 \\ a_{21}x_1 + a_{22}x_2 + \cdots + a_{2n}x_n &= b_2 \\ \vdots \qquad\quad \vdots \qquad\qquad\quad \vdots \quad &\quad \vdots \\ a_{m1}x_1 + a_{m2}x_2 + \cdots + a_{mn}x_n &= b_m \end{aligned} \tag{1}$$

Let

$$A = \begin{pmatrix} a_{11} & a_{12} & \cdots & a_{1n} \\ a_{21} & a_{22} & \cdots & a_{2n} \\ \vdots & \vdots & & \vdots \\ a_{m1} & a_{m2} & \cdots & a_{mn} \end{pmatrix}$$

be the coefficient matrix, $\mathbf{x}$ the vector $\begin{pmatrix} x_1 \\ x_2 \\ \vdots \\ x_n \end{pmatrix}$, and $\mathbf{b}$ the vector $\begin{pmatrix} b_1 \\ b_2 \\ \vdots \\ b_m \end{pmatrix}$. Since $A$ is an $m \times n$ matrix and $\mathbf{x}$ is an $n \times 1$ matrix, the matrix product $A\mathbf{x}$ is an $m \times 1$ matrix. It is not difficult to see that system (1) can be written as

**Matrix Representation of a Linear System of Equations**

$$A\mathbf{x} = \mathbf{b} \tag{2}$$

**EXAMPLE 1** **Writing a System in Matrix Representation** Consider the system

$$\begin{aligned} 2x_1 + 4x_2 + 6x_3 &= 18 \\ 4x_1 + 5x_2 + 6x_3 &= 24 \\ 3x_1 + \phantom{4}x_2 - 2x_3 &= 4 \end{aligned} \tag{3}$$

(See Example 1.3.1 on page 7.) This can be written as $A\mathbf{x} = \mathbf{b}$ with $A = \begin{pmatrix} 2 & 4 & 6 \\ 4 & 5 & 6 \\ 3 & 1 & -2 \end{pmatrix}$, $\mathbf{x} = \begin{pmatrix} x_1 \\ x_2 \\ x_3 \end{pmatrix}$, and $\mathbf{b} = \begin{pmatrix} 18 \\ 24 \\ 4 \end{pmatrix}$.

It is obviously easier to write out system (1) in the form $A\mathbf{x} = \mathbf{b}$. There are many other advantages, too. In Section 1.8 we shall see how a square system can be solved almost at once if we know a matrix called the *inverse* of $A$. Even without that, as we saw in Section 1.3, computations are much easier to write down by using an augmented matrix.

If $\mathbf{b} = \begin{pmatrix} 0 \\ 0 \\ \vdots \\ 0 \end{pmatrix}$ is the $m \times 1$ zero vector, then system (1) is homogeneous (see Section 1.4) and can be written

$$A\mathbf{x} = \mathbf{0} \quad \text{Matrix form of a homogeneous system of equations}$$

There is a fundamental relationship between homogeneous and nonhomogeneous systems. Let $A$ be an $m \times n$ matrix

$$\mathbf{x} = \begin{pmatrix} x_1 \\ x_2 \\ \vdots \\ x_n \end{pmatrix}, \quad \mathbf{b} = \begin{pmatrix} b_1 \\ b_2 \\ \vdots \\ b_m \end{pmatrix}, \quad \text{and} \quad \mathbf{0} = \begin{pmatrix} 0 \\ 0 \\ \vdots \\ 0 \end{pmatrix} \leftarrow m \text{ zeros}$$

The general linear, nonhomogeneous system can be written as

$$A\mathbf{x} = \mathbf{b} \tag{4}$$

Associated homogeneous system

With $A$ and $\mathbf{x}$ as in (4) and $\mathbf{b} \neq \mathbf{0}$, we define the **associated homogeneous system** by

$$A\mathbf{x} = \mathbf{0} \tag{5}$$

**THEOREM 1** Let $\mathbf{x}_1$ and $\mathbf{x}_2$ be solutions of the nonhomogeneous system (4). Then their difference, $\mathbf{x}_1 - \mathbf{x}_2$, is a solution of the related homogeneous system (5).

**Proof**

By the distributive law (7) on page 69
↓

$$A(\mathbf{x}_1 - \mathbf{x}_2) = A\mathbf{x}_1 - A\mathbf{x}_2 = \mathbf{b} - \mathbf{b} = \mathbf{0}$$ ■

**COROLLARY** Let $\mathbf{x}$ be a particular solution to the nonhomogeneous system (4) and let $\mathbf{y}$ be another solution to (4). Then there exists a solution $\mathbf{h}$ to the homogeneous system (5) such that

$$\mathbf{y} = \mathbf{x} + \mathbf{h} \tag{6}$$

**Proof** If $\mathbf{h}$ is defined by $\mathbf{h} = \mathbf{y} - \mathbf{x}$, then $\mathbf{h}$ solves (5) by Theorem 1 and $\mathbf{y} = \mathbf{x} + \mathbf{h}$. ■

Theorem 1 and its corollary are very useful. They tell us that

> In order to find all solutions to the nonhomogeneous system (4), it is sufficient to find *one* solution to (4) and all solutions to the associated homogeneous system (5).

*Remark.* A very similar result holds for solutions of homogeneous and nonhomogeneous linear differential equations (see Problems 23 and 24). One of the many nice things about mathematics is that seemingly very different topics are closely interrelated.

**EXAMPLE 2** **Writing an Infinite Number of Solutions as a Particular Solution to a Nonhomogeneous System Plus Solutions to the Homogeneous System** Find all solutions to the nonhomogeneous system

$$\begin{aligned} x + 2x_2 - \phantom{5}x_3 &= 2 \\ 2x_1 + 3x_2 + 5x_3 &= 5 \\ -x_1 - 3x_2 + 8x_3 &= -1 \end{aligned}$$

by using the result given above.

**Solution** First, we find one solution by row reduction:

$$\left(\begin{array}{rrr|r} 1 & 2 & -1 & 2 \\ 2 & 3 & 5 & 5 \\ -1 & -3 & 8 & -1 \end{array}\right) \xrightarrow[R_3 \to R_3 + R_1]{R_2 \to R_2 - 2R_1} \left(\begin{array}{rrr|r} 1 & 2 & -1 & 2 \\ 0 & -1 & 7 & 1 \\ 0 & -1 & 7 & 1 \end{array}\right)$$

$$\xrightarrow[R_3 \to R_3 - R_2]{R_1 \to R_1 + 2R_2} \left(\begin{array}{rrr|r} 1 & 0 & 13 & 4 \\ 0 & -1 & 7 & 1 \\ 0 & 0 & 0 & 0 \end{array}\right)$$

The equations corresponding to the first two rows of the last system are

$$x_1 = 4 - 13x_3, \quad \text{and} \quad x_2 = -1 + 7x_3$$

so the solutions are

$$\mathbf{x} = (x_1, x_2, x_3) = (4 - 13x_3, -1 + 7x_3, x_3) = \mathbf{x_p} + \mathbf{x_h}$$

where $\mathbf{x_p} = (4, -1, 0)$ is a particular solution and $\mathbf{x_h} = x_3(-13, 7, 1)$, $x_3$ a real number, is a solution to the associated homogeneous system. For example, $x_3 = 0$ yields the solution $(4, -1, 0)$, whereas $x_3 = 2$ gives the solution $(-22, 13, 2)$.

# PROBLEMS 1.7

## Self-Quiz

**I.** If the system $\begin{array}{rl} x \phantom{+2y} - z &= 2 \\ y + z &= 3 \\ x + 2y \phantom{- z} &= 4 \end{array}$ is written in the form $A\mathbf{x} = \mathbf{b}$, with $\mathbf{x} = \begin{pmatrix} x \\ y \\ z \end{pmatrix}$ and $\mathbf{b} = \begin{pmatrix} 2 \\ 3 \\ 4 \end{pmatrix}$, then $A =$ ______.

**a.** $\begin{pmatrix} 1 & 1 & -1 \\ 1 & 1 & 1 \\ 1 & 1 & 2 \end{pmatrix}$ **b.** $\begin{pmatrix} 1 & -1 & 0 \\ 0 & 1 & 1 \\ 1 & 2 & 0 \end{pmatrix}$ **c.** $\begin{pmatrix} 1 & 0 & -1 \\ 0 & 1 & 1 \\ 1 & 0 & 2 \end{pmatrix}$ **d.** $\begin{pmatrix} 1 & 0 & -1 \\ 0 & 1 & 1 \\ 1 & 2 & 0 \end{pmatrix}$

In Problems 1–6 write the given system in the form $A\mathbf{x} = \mathbf{b}$.

**1.** $\begin{array}{l} 2x_1 - x_2 = 3 \\ 4x_1 + 5x_2 = 7 \end{array}$

**2.** $\begin{array}{l} x_1 - x_2 + 3x_3 = 11 \\ 4x_1 + x_2 - x_3 = -4 \\ 2x_1 - x_2 + 3x_3 = 10 \end{array}$

**3.** $\begin{array}{l} 3x_1 + 6x_2 - 7x_3 = 0 \\ 2x_1 - x_2 + 3x_3 = 1 \end{array}$

**4.** $\begin{array}{l} 4x_1 - x_2 + x_3 - x_4 = -7 \\ 3x_1 + x_2 - 5x_3 + 6x_4 = 8 \\ 2x_1 - x_2 + x_3 \phantom{+ 6x_4} = 9 \end{array}$

**Answer to Self-Quiz**

**I.** d

**5.**
$$\begin{aligned} x_2 - x_3 &= 7 \\ x_1 \quad + x_3 &= 2 \\ 3x_1 + 2x_2 \quad &= -5 \end{aligned}$$

**6.**
$$\begin{aligned} 2x_1 + 3x_2 - x_3 &= 0 \\ -4x_1 + 2x_2 + x_3 &= 0 \\ 7x_1 + 3x_2 - 9x_3 &= 0 \end{aligned}$$

In Problems 7–15 write out the system of equations represented by the given augmented matrix.

**7.** $\left(\begin{array}{ccc|c} 1 & 1 & -1 & 7 \\ 4 & -1 & 5 & 4 \\ 6 & 1 & 3 & 20 \end{array}\right)$ **8.** $\left(\begin{array}{cc|c} 0 & 1 & 2 \\ 1 & 0 & 3 \end{array}\right)$ **9.** $\left(\begin{array}{ccc|c} 2 & 0 & 1 & 2 \\ -3 & 4 & 0 & 3 \\ 0 & 5 & 6 & 5 \end{array}\right)$

**10.** $\left(\begin{array}{ccc|c} 2 & 3 & 1 & 2 \\ 0 & 4 & 1 & 3 \\ 0 & 0 & 0 & 0 \end{array}\right)$ **11.** $\left(\begin{array}{cccc|c} 1 & 0 & 0 & 0 & 2 \\ 0 & 1 & 0 & 0 & 3 \\ 0 & 0 & 1 & 0 & -5 \\ 0 & 0 & 0 & 1 & 6 \end{array}\right)$ **12.** $\left(\begin{array}{ccc|c} 2 & 3 & 1 & 0 \\ 4 & -1 & 5 & 0 \\ 3 & 6 & -7 & 0 \end{array}\right)$

**13.** $\left(\begin{array}{ccc|c} 6 & 2 & 1 & 2 \\ -2 & 3 & 1 & 4 \\ 0 & 0 & 0 & 2 \end{array}\right)$ **14.** $\left(\begin{array}{ccc|c} 3 & 1 & 5 & 6 \\ 2 & 3 & 2 & 4 \end{array}\right)$ **15.** $\left(\begin{array}{cc|c} 7 & 2 & 1 \\ 3 & 1 & 2 \\ 6 & 9 & 3 \end{array}\right)$

**16.** Find a matrix $A$ and vectors $\mathbf{x}$ and $\mathbf{b}$ such that the system represented by the following augmented matrix can be written in the form $A\mathbf{x} = \mathbf{b}$ and solve the system.

$$\left(\begin{array}{ccc|c} 2 & 0 & 0 & 3 \\ 0 & 4 & 0 & 5 \\ 0 & 0 & -5 & 2 \end{array}\right)$$

In Problems 17–22 find all solutions to the given nonhomogeneous system by first finding one solution (if possible) and then finding all solutions to the associated homogeneous system.

**17.**
$$\begin{aligned} x_1 - 3x_2 &= 2 \\ -2x_1 + 6x_2 &= -4 \end{aligned}$$

**18.**
$$\begin{aligned} x_1 - x_2 + x_3 &= 6 \\ 3x_1 - 3x_2 + 3x_3 &= 18 \end{aligned}$$

**19.**
$$\begin{aligned} x_1 - x_2 - x_3 &= 2 \\ 2x_1 + x_2 + 2x_3 &= 4 \\ x_1 - 4x_2 - 5x_3 &= 2 \end{aligned}$$

**20.**
$$\begin{aligned} x_1 - x_2 - x_3 &= 2 \\ 2x_1 + x_2 + 2x_3 &= 4 \\ x_1 - 4x_2 - 5x_3 &= 2 \end{aligned}$$

**21.**
$$\begin{aligned} x_1 + x_2 - x_3 + 2x_4 &= 3 \\ 3x_1 + 2x_2 + x_3 - x_4 &= 5 \end{aligned}$$

**22.**
$$\begin{aligned} x_1 - x_2 + x_3 - x_4 &= -2 \\ -2x_1 + 3x_2 - x_3 + 2x_4 &= 5 \\ 4x_1 - 2x_2 + 2x_3 - 3x_4 &= 6 \end{aligned}$$

Calculus †**23.** Consider the linear, homogeneous second-order differential equation

$$y''(x) + a(x)y'(x) + b(x)y(x) = 0 \qquad \textbf{(7)}$$

where $a(x)$ and $b(x)$ are continuous and the unknown function $y$ is assumed to have a second derivative. Show that if $y_1$ and $y_2$ are solutions to (7), then $c_1y_1 + c_2y_2$ is a solution for any constants $y_1$ and $y_2$.

Calculus **24.** Suppose that $y_p$ and $y_q$ are solutions to the nonhomogeneous equation

$$y''(x) + a(x)y'(x) + b(x)y(x) = f(x) \qquad \textbf{(8)}$$

Show that $y_p - y_q$ is a solution to (7). Assume here that $f(x)$ is not the zero function.

† The symbol Calculus indicates that calculus is needed to solve the problem.

# MATLAB 1.7

*Note.* For generating random matrices, refer to the discussion preceding the MATLAB problems for Section 1.6.

1. a. Generate a random $3 \times 3$ matrix $A$ with entries between $-10$ and 10 and generate a random $3 \times 1$ vector **b** with entries between $-10$ and 10. Using MATLAB, solve the system with augmented matrix $[A\ \mathbf{b}]$ using **rref.** Use the ":" notation to place the solution in the variable **x**. Find $A$**x** and compare with **b** (find **A*x−b**). Find **y = x(1)*A(:,1) + x(2)*A(:,2) + x(3)*A(:,3)** and compare with **b** (find **y−b**). Repeat this for three more **b** vectors. What can you conclude about the relationship among $A$**x**, **y**, and **b**?

   b. Let

$$A = \begin{pmatrix} 4 & 9 & 17 & 5 \\ 2 & 1 & 5 & -1 \\ 5 & 9 & 19 & 4 \\ 9 & 5 & 23 & -4 \end{pmatrix} \qquad \mathbf{b} = \begin{pmatrix} 11 \\ 9 \\ 16 \\ 40 \end{pmatrix}$$

   i. Solve the system with augmented matrix $[A\ \mathbf{b}]$ using **rref.** If there are infinitely many solutions, make a choice for the arbitrary variables and find and enter the corresponding solution vector **x**.

   ii. Find $A$***x** and **y = x(1)*A(:,1) + x(2)*A(:,2) + x(3)*A(:,3) + x(4)*A(:,4),** and compare $A$**x**, **y**, and **b**.

   iii. Repeat with two other choices of the arbitrary variable.

   iv. What can you conclude about the relationship among $A$**x**, **y**, and **b**?

   c. What do parts (a) and (b) tell you about solutions to systems of equations? How does this relate to MATLAB Problem 6 in Section 1.3?

2. a. Assume that $A$ and **x** have real number entries. Using the definition of matrix multiplication, argue why $A\mathbf{x} = \mathbf{0}$ means that each row of $A$ is perpendicular to **x**. (Recall that two real vectors are perpendicular if their scalar product is zero.)

   b. Using the result from part (a), find all **x** so that **x** is perpendicular to the two vectors:

$$(1, 2, -3, 0, 4) \quad \text{and} \quad (4, -5, 2, 0, 1).$$

3. a. Recall MATLAB Problem 3 in Section 1.6. (Redo it.) How does this relate to the corollary of Theorem 1?

   b. Consider the $A$ and the **b** from MATLAB Problem 1(b) in this section.

   i. Verify that the system $[A\ \mathbf{b}]$ has infinitely many solutions.

   ii. Let **x = A\b.** Verify by using matrix multiplication that this produces a solution to the system with augmented matrix $[A\ \mathbf{b}]$. (Notice that MATLAB gives a warning. If there is not a unique solution, the MATLAB command "\" gives the least squares solution. In cases where a system has infinitely many solutions the least squares solution will be a solution.)

   iii. By considering **rref(A),** find four solutions to the homogeneous system $[A\ \mathbf{0}]$. Enter one at a time, calling it **z**, and verify by using matrix multiplication that $\mathbf{x} + \mathbf{z}$ is a solution to the system with augmented matrix $[A\ \mathbf{b}]$.

4. a. By looking at **rref(A)** for the $A$ below, argue why the system $[A\ \mathbf{b}]$ has a solution no matter what $4 \times 1$ vector $\mathbf{b}$ we choose.

$$A = \begin{pmatrix} 5 & 5 & 8 & 0 \\ 4 & 5 & 8 & 7 \\ 3 & 9 & 8 & 9 \\ 9 & 1 & 1 & 6 \end{pmatrix}$$

b. Conclude that every $\mathbf{b}$ is a linear combination of the columns of $A$. Generate three random $4 \times 1$ vectors $\mathbf{b}$ and, for each $\mathbf{b}$, find the coefficients needed to write $\mathbf{b}$ as a linear combination of the columns of $A$.

c. By looking at **rref(A)** for $A$ below, argue why there is some $4 \times 1$ vector $\mathbf{b}$ for which the system $[A\ \mathbf{b}]$ has no solution. Experiment to find a $\mathbf{b}$ vector for which there is no solution.

$$A = \begin{pmatrix} 5 & 5 & -5 & 0 \\ 4 & 5 & -6 & 7 \\ 3 & 9 & -15 & 9 \\ 9 & 1 & 7 & 6 \end{pmatrix}$$

d. How can you generate $\mathbf{b}$ vectors for which you will be guaranteed that there will be a solution? Decide on a procedure and describe it in a comment statement. Test out your procedure by forming three $\mathbf{b}$ vectors by your procedure and then solving the corresponding systems. (See MATLAB Problem 6 of Section 1.3.)

e. Prove that your procedure is valid using the theory developed in the text.

5. In this problem you will discover relationships between the reduced row echelon form of a matrix and information about linear combinations of the columns of $A$. The MATLAB part of the problem only involves computing some reduced row echelon forms. The theory is based on the facts that $A\mathbf{x} = \mathbf{0}$ means that $\mathbf{x}$ is a solution to the system $[A\ \mathbf{0}]$ and that

$$\mathbf{0} = x_1(\text{col } 1 \text{ of } A) + \cdots + x_n(\text{col } n \text{ of } A)$$

a. **i.** Let $A$ be the matrix in MATLAB Problem 4(c) in this section. Find **rref(A).** (The rest of this part involves paper and pencil work.)

**ii.** Find the solutions for the homogeneous system written in terms of the natural choices for the arbitrary variables.

**iii.** Set one arbitrary variable equal to 1 and the other arbitrary variables equal to 0 and find the other unknowns, producing a solution vector $\mathbf{x}$. For this $\mathbf{x}$, write out what the statement

$$\mathbf{0} = A\mathbf{x} = x_1(\text{col } 1 \text{ of } A) + \cdots + x_n(\text{col } n \text{ of } A)$$

tells you and solve for the column of $A$ corresponding to the arbitrary variable that you set equal to 1. Check your results.

**iv.** If there is more than one arbitrary value, set another arbitrary variable equal to 1 and the other arbitrary variables equal to 0 and repeat (*iii*). Continue in this fashion for each arbitrary variable.

**v.** Look back at **rref(A)** and see if you recognize any relationships between what you just discovered and the numbers in **rref(A).**

b. Let $A$ be the matrix in MATLAB Problem 1(b) in this section. Repeat the instructions above.

c. Let $A$ be a random $6 \times 6$ matrix. Modify $A$ so that

$$\mathbf{A(:,3) = 2*A(:,2) - 3*A(:,1)}$$
$$\mathbf{A(:,5) = -A(:,1) + 2*A(:,2) - 3*A(:,4)}$$
$$\mathbf{A(:,6) = A(:,2) + 4*A(:,4)}$$

Repeat the instructions above.

## 1.8 THE INVERSE OF A SQUARE MATRIX

In this section we define two kinds of matrices that are central to matrix theory. We begin with a simple example. Let $A = \begin{pmatrix} 2 & 5 \\ 1 & 3 \end{pmatrix}$ and $B = \begin{pmatrix} 3 & -5 \\ -1 & 2 \end{pmatrix}$. Then an easy computation shows that $AB = BA = I_2$, where $I_2 = \begin{pmatrix} 1 & 0 \\ 0 & 1 \end{pmatrix}$. The matrix $I_2$ is called the $2 \times 2$ *identity matrix*. The matrix $B$ is called the *inverse* of $A$ and is written $A^{-1}$.

**DEFINITION 1** **Identity Matrix** The $n \times n$ **identity matrix** $I_n$ is the $n \times n$ matrix with 1's down the **main diagonal**† and 0's everywhere else. That is,

$$I_n = (b_{ij}) \quad \text{where} \quad b_{ij} = \begin{cases} 1 & \text{if } i = j \\ 0 & \text{if } i \neq j \end{cases} \tag{1}$$

**EXAMPLE 1** **Two Identity Matrices**

$$I_3 = \begin{pmatrix} 1 & 0 & 0 \\ 0 & 1 & 0 \\ 0 & 0 & 1 \end{pmatrix} \quad \text{and} \quad I_5 = \begin{pmatrix} 1 & 0 & 0 & 0 & 0 \\ 0 & 1 & 0 & 0 & 0 \\ 0 & 0 & 1 & 0 & 0 \\ 0 & 0 & 0 & 1 & 0 \\ 0 & 0 & 0 & 0 & 1 \end{pmatrix}$$

† The main diagonal of $A = (a_{ij})$ consists of the components $a_{11}$, $a_{22}$, $a_{33}$, and so on. Unless otherwise stated, we shall refer to the main diagonal simply as the **diagonal.**

**THEOREM 1** Let $A$ be a square $n \times n$ matrix. Then

$$AI_n = I_nA = A$$

That is, $I_n$ commutes with every $n \times n$ matrix and leaves it unchanged after multiplication on the left or right.

*Note.* $I_n$ functions for $n \times n$ matrices the way the number 1 functions for real numbers (since $1 \cdot a = a \cdot 1 = a$ for every real number $a$).

**Proof** Let $c_{ij}$ be the $ij$th element of $AI_n$. Then

$$c_{ij} = a_{i1}b_{1j} + a_{i2}b_{2j} + \cdots + a_{ij}b_{jj} + \cdots + a_{in}b_{nj}$$

But from (1) this sum is equal to $a_{ij}$. Thus $AI_n = A$. In a similar fashion, we can show that $I_nA = A$, and this proves the theorem. ∎

*Notation.* From now on we shall write the identity matrix simply as $I$ since if $A$ is $n \times n$, the products $IA$ and $AI$ are defined only if $I$ is also $n \times n$.

**DEFINITION 2** **The Inverse of a Matrix** Let $A$ and $B$ be $n \times n$ matrices. Suppose that

$$AB = BA = I$$

Then $B$ is called the **inverse** of $A$ and is written as $A^{-1}$. We then have

$$AA^{-1} = A^{-1}A = I$$

If $A$ has an inverse, then $A$ is said to be **invertible.**

A square matrix that is not invertible is called **singular** and an invertible matrix is also called **nonsingular.**

*Remark 1.* From this definition it immediately follows that $(A^{-1})^{-1} = A$ if $A$ is invertible.

*Remark 2.* This definition does *not* state that every square matrix has an inverse. In fact, there are many square matrices that have no inverse. (See, for instance, Example 3 on page 103.)

In Definition 2 we defined *the* inverse of a matrix. This statement suggests that inverses are unique. This is indeed the case, as the following theorem shows.

**THEOREM 2** If a square matrix $A$ is invertible, then its inverse is unique.

**Proof** Suppose $B$ and $C$ are two inverses for $A$. We can show that $B = C$. By definition, we have $AB = BA = I$ and $AC = CA = I$. $B(AC) = (BA)C$ by the associative law of matrix multiplication. Then

$$B = BI = B(AC) = (BA)C = IC = C$$

Hence $B = C$ and the theorem is proved. ■

Another important fact about inverses is given below.

**THEOREM 3** Let $A$ and $B$ be invertible $n \times n$ matrices. Then $AB$ is invertible and

$$\boxed{(AB)^{-1} = B^{-1}A^{-1}}$$

**Proof** To prove this result, we refer to Definition 2. That is, $B^{-1}A^{-1} = (AB)^{-1}$ if and only if $B^{-1}A^{-1}(AB) = (AB)(B^{-1}A^{-1}) = I$. But this follows since

$$(B^{-1}A^{-1})(AB) \overset{\text{equation (6) on page 68}}{=} B^{-1}(A^{-1}A)B = B^{-1}IB = B^{-1}B = I$$

and

$$(AB)(B^{-1}A^{-1}) = A(BB^{-1})A^{-1} = AIA^{-1} = AA^{-1} = I.$$ ■

*Note.* It follows from Theorem 3 that $(ABC)^{-1} = C^{-1}B^{-1}A^{-1}$. See Problem 16.

Consider the system of $n$ equations in $n$ unknowns

$$A\mathbf{x} = \mathbf{b}$$

and suppose that $A$ is invertible. Then

$$A^{-1}A\mathbf{x} = A^{-1}\mathbf{b} \quad \text{we multiplied on the left by } A^{-1}$$
$$I\mathbf{x} = A^{-1}\mathbf{b} \quad A^{-1}A = I$$
$$\mathbf{x} = A^{-1}\mathbf{b} \quad I\mathbf{x} = \mathbf{x}$$

This is a solution to the system because

$$A\mathbf{x} = A(A^{-1}\mathbf{b}) = (AA^{-1})\mathbf{b} = I\mathbf{b} = \mathbf{b}$$

If $\mathbf{y}$ is the vector with $A\mathbf{y} = \mathbf{b}$, then the computation above shows that $\mathbf{y} = A^{-1}\mathbf{b}$.

That is, $\mathbf{y} = \mathbf{x}$. We have shown the following:

> If $A$ is invertible, the system $A\mathbf{x} = \mathbf{b}$ has the unique solution $\mathbf{x} = A^{-1}\mathbf{b}$. **(2)**

This is one of the reasons we study matrix inverses.

There are two basic questions that come to mind once we have defined the inverse of a matrix.

Question 1. Which matrices have inverses?

Question 2. If a matrix has an inverse, how can we compute it?

We answer both questions in this section. Let's begin by seeing what happens in the $2 \times 2$ case.

**EXAMPLE 2** **Finding the Inverse of a 2 × 2 Matrix** Let $A = \begin{pmatrix} 2 & -3 \\ -4 & 5 \end{pmatrix}$. Compute $A^{-1}$ if it exists.

**Solution** Suppose that $A^{-1}$ exists. We write $A^{-1} = \begin{pmatrix} x & y \\ z & w \end{pmatrix}$ and use the fact that $AA^{-1} = I$. Then

$$AA^{-1} = \begin{pmatrix} 2 & -3 \\ -4 & 5 \end{pmatrix}\begin{pmatrix} x & y \\ z & w \end{pmatrix} = \begin{pmatrix} 2x - 3z & 2y - 3w \\ -4x + 5z & -4y + 5w \end{pmatrix} = \begin{pmatrix} 1 & 0 \\ 0 & 1 \end{pmatrix}$$

The last two matrices can be equal only if each of their corresponding components are equal. This means that

$$\begin{aligned} 2x \qquad - 3z \qquad &= 1 && \textbf{(3)} \\ 2y \qquad - 3w &= 0 && \textbf{(4)} \\ -4x \qquad + 5z \qquad &= 0 && \textbf{(5)} \\ -4y \qquad + 5w &= 1 && \textbf{(6)} \end{aligned}$$

This is a system of four equations in four unknowns. Note that there are two equations involving $x$ and $z$ only [equations (3) and (5)] and two equations involving $y$ and $w$ only [equations (4) and (6)]. We write these two systems in augmented matrix form:

$$\left(\begin{array}{cc|c} 2 & -3 & 1 \\ -4 & 5 & 0 \end{array}\right) \qquad \textbf{(7)}$$

$$\left(\begin{array}{cc|c} 2 & -3 & 0 \\ -4 & 5 & 1 \end{array}\right) \qquad \textbf{(8)}$$

Now we know from Section 1.3 that if system (7) (in the variables $x$ and $z$) has a unique solution, then Gauss-Jordan elimination of (7) will result in

$$\left(\begin{array}{cc|c} 1 & 0 & x \\ 0 & 1 & z \end{array}\right)$$

where $(x, z)$ is the unique pair of numbers that satisfies $2x - 3z = 1$ and $-4x + 5z = 0$. Similarly, row reduction of (8) will result in

$$\left(\begin{array}{cc|c} 1 & 0 & y \\ 0 & 1 & w \end{array}\right)$$

where $(y, w)$ is the unique pair of numbers that satisfies $2y - 3w = 0$ and $-4y + 5w = 1$.

Since the coefficient matrices in (7) and (8) are the same, we can perform the row reductions on the two augmented matrices simultaneously by considering the new augmented matrix

$$\left(\begin{array}{cc|cc} 2 & -3 & 1 & 0 \\ -4 & 5 & 0 & 1 \end{array}\right) \tag{9}$$

If $A$ is invertible, then the system defined by (3), (4), (5), and (6) has a unique solution and, by what we said above, row reduction will result in

$$\left(\begin{array}{cc|cc} 1 & 0 & x & y \\ 0 & 1 & z & w \end{array}\right)$$

We now carry out the computation, noting that the matrix on the left in (9) is $A$ and the matrix on the right in (9) is $I$:

$$\left(\begin{array}{cc|cc} 2 & -3 & 1 & 0 \\ -4 & 5 & 0 & 1 \end{array}\right) \xrightarrow{R_1 \to \frac{1}{2}R_1} \left(\begin{array}{cc|cc} 1 & -\frac{3}{2} & \frac{1}{2} & 0 \\ -4 & 5 & 0 & 1 \end{array}\right)$$

$$\xrightarrow{R_2 \to R_2 + 4R_1} \left(\begin{array}{cc|cc} 1 & -\frac{3}{2} & \frac{1}{2} & 0 \\ 0 & -1 & 2 & 1 \end{array}\right)$$

$$\xrightarrow{R_2 \to -R_2} \left(\begin{array}{cc|cc} 1 & -\frac{3}{2} & \frac{1}{2} & 0 \\ 0 & 1 & -2 & -1 \end{array}\right)$$

$$\xrightarrow{R_1 \to R_1 + \frac{3}{2}R_2} \left(\begin{array}{cc|cc} 1 & 0 & -\frac{5}{2} & -\frac{3}{2} \\ 0 & 1 & -2 & -1 \end{array}\right)$$

Thus $x = -\frac{5}{2}$, $y = -\frac{3}{2}$, $z = -2$, $w = -1$, and $\begin{pmatrix} x & y \\ z & w \end{pmatrix} = \begin{pmatrix} -\frac{5}{2} & -\frac{3}{2} \\ -2 & -1 \end{pmatrix}$. We compute

$$\begin{pmatrix} 2 & -3 \\ -4 & 5 \end{pmatrix}\begin{pmatrix} -\frac{5}{2} & -\frac{3}{2} \\ -2 & -1 \end{pmatrix} = \begin{pmatrix} 1 & 0 \\ 0 & 1 \end{pmatrix}$$

and

$$\begin{pmatrix} -\frac{5}{2} & -\frac{3}{2} \\ -2 & -1 \end{pmatrix}\begin{pmatrix} 2 & -3 \\ -4 & 5 \end{pmatrix} = \begin{pmatrix} 1 & 0 \\ 0 & 1 \end{pmatrix}$$

Thus $A$ is invertible and $A^{-1} = \begin{pmatrix} -\frac{5}{2} & -\frac{3}{2} \\ -2 & -1 \end{pmatrix}$.

**EXAMPLE 3** **A 2 × 2 Matrix That Is Not Invertible** Let $A = \begin{pmatrix} 1 & 2 \\ -2 & -4 \end{pmatrix}$. Determine whether $A$ is invertible and, if it is, calculate its inverse.

**Solution** If $A^{-1} = \begin{pmatrix} x & y \\ z & w \end{pmatrix}$ exists, then

$$AA^{-1} = \begin{pmatrix} 1 & 2 \\ -2 & -4 \end{pmatrix}\begin{pmatrix} x & y \\ z & w \end{pmatrix} = \begin{pmatrix} x + 2z & y + 2w \\ -2x - 4z & -2y - 4w \end{pmatrix} = \begin{pmatrix} 1 & 0 \\ 0 & 1 \end{pmatrix}$$

This leads to the system

$$\begin{aligned} x \qquad + 2z \qquad &= 1 \\ y \qquad + 2w &= 0 \\ -2x \qquad - 4z \qquad &= 0 \\ -2y \qquad - 4w &= 1 \end{aligned} \qquad \textbf{(10)}$$

Using the same reasoning as in Example 1, we can write this system in the augmented matrix form $(A|I)$ and row-reduce:

$$\left(\begin{array}{cc|cc} 1 & 2 & 1 & 0 \\ -2 & -4 & 0 & 1 \end{array}\right) \xrightarrow{R_2 \to R_2 + 2R_1} \left(\begin{array}{cc|cc} 1 & 2 & 1 & 0 \\ 0 & 0 & 2 & 1 \end{array}\right)$$

This is as far as we can go. The last line reads $0 = 2$ or $0 = 1$, depending on which of the two systems of equations (in $x$ and $z$ or in $y$ and $w$) is being solved. Thus system (10) is inconsistent and $A$ is not invertible.

The last two examples illustrate a procedure that always works when you are trying to find the inverse of a matrix.

**Procedure for Computing the Inverse of a Square Matrix $A$**

*Step* 1. Write the augmented matrix $(A|I)$.

*Step* 2. Use row reduction to reduce the matrix $A$ to its reduced row echelon form.

*Step* 3. Decide if $A$ is invertible.

a. If the reduced row echelon form of $A$ is the identity matrix $I$, then $A^{-1}$ is the matrix to the right of the vertical bar.

b. If row reduction of $A$ leads to a row of zeros to the left of the vertical bar, then $A$ is not invertible.

*Remark.* We can rephrase (a) and (b) as follows:

> *An $n \times n$ matrix $A$ is invertible if and only if its reduced row echelon form is the identity matrix; that is, its reduced row echelon form has $n$ pivots.*

Let $A = \begin{pmatrix} a_{11} & a_{12} \\ a_{21} & a_{22} \end{pmatrix}$. Then we define

Determinant of a $2 \times 2$ matrix

$$\text{Determinant of } A = a_{11}a_{22} - a_{12}a_{21} \tag{11}$$

We abbreviate the determinant of $A$ by $\det A$.

**THEOREM 4** Let $A$ be a $2 \times 2$ matrix. Then

**i.** $A$ is invertible if and only if $\det A \neq 0$.

**ii.** If $\det A \neq 0$, then

$$A^{-1} = \frac{1}{\det A}\begin{pmatrix} a_{22} & -a_{12} \\ -a_{21} & a_{11} \end{pmatrix} \tag{12}$$

**Proof** First, suppose that $\det A \neq 0$ and let $B = (1/\det A)\begin{pmatrix} a_{22} & -a_{12} \\ -a_{21} & a_{11} \end{pmatrix}$. Then

$$BA = \frac{1}{\det A}\begin{pmatrix} a_{22} & -a_{12} \\ -a_{21} & a_{11} \end{pmatrix}\begin{pmatrix} a_{11} & a_{12} \\ a_{21} & a_{22} \end{pmatrix}$$

$$= \frac{1}{a_{11}a_{22} - a_{12}a_{21}}\begin{pmatrix} a_{22}a_{11} - a_{12}a_{21} & 0 \\ 0 & -a_{21}a_{12} + a_{11}a_{22} \end{pmatrix} = \begin{pmatrix} 1 & 0 \\ 0 & 1 \end{pmatrix} = I$$

Similarly, $AB = I$, which shows that $A$ is invertible and that $B = A^{-1}$. We still must show that if $A$ is invertible, then $\det A \neq 0$. To do so, we consider the system

$$\begin{aligned} a_{11}x_1 + a_{12}x_2 &= b_1 \\ a_{21}x_1 + a_{22}x_2 &= b_2 \end{aligned} \tag{13}$$

We do this because we know from our Summing Up Theorem (Theorem 1.2.1, page 5) that if this system has a unique solution, then $a_{11}a_{22} - a_{12}a_{21} \neq 0$. The system can be written in the form

$$A\mathbf{x} = \mathbf{b} \tag{14}$$

with $\mathbf{x} = \begin{pmatrix} x_1 \\ x_2 \end{pmatrix}$ and $\mathbf{b} = \begin{pmatrix} b_1 \\ b_2 \end{pmatrix}$. Then, since $A$ is invertible, we see from (2) that system (14) has a unique solution given by

$$\mathbf{x} = A^{-1}\mathbf{b}$$

But by Theorem 1.2.1 the fact that system (13) has a unique solution implies that $a_{11}a_{22} - a_{12}a_{21} = \det A \neq 0$. This completes the proof. ■

*Note.* Formula (12) can be obtained directly by applying our procedure for computing an inverse (see Problem 46).

**EXAMPLE 4** **Calculating the Inverse of a 2 × 2 Matrix** Let $A = \begin{pmatrix} 2 & -4 \\ 1 & 3 \end{pmatrix}$. Calculate $A^{-1}$ if it exists.

**Solution** We find that $\det A = (2)(3) - (-4)(1) = 10$; hence $A^{-1}$ exists. From equation (12) we get

$$A^{-1} = \frac{1}{10}\begin{pmatrix} 3 & 4 \\ -1 & 2 \end{pmatrix} = \begin{pmatrix} \frac{3}{10} & \frac{4}{10} \\ -\frac{1}{10} & \frac{2}{10} \end{pmatrix}$$

*Check*

$$A^{-1}A = \frac{1}{10}\begin{pmatrix} 3 & 4 \\ -1 & 2 \end{pmatrix}\begin{pmatrix} 2 & -4 \\ 1 & 3 \end{pmatrix} = \frac{1}{10}\begin{pmatrix} 10 & 0 \\ 0 & 10 \end{pmatrix} = \begin{pmatrix} 1 & 0 \\ 0 & 1 \end{pmatrix}$$

and

$$AA^{-1} = \begin{pmatrix} 2 & -4 \\ 1 & 3 \end{pmatrix}\begin{pmatrix} \frac{3}{10} & \frac{4}{10} \\ -\frac{1}{10} & \frac{2}{10} \end{pmatrix} = \begin{pmatrix} 1 & 0 \\ 0 & 1 \end{pmatrix}$$

■

**EXAMPLE 5** **A 2 × 2 Matrix That Is Not Invertible** Let $A = \begin{pmatrix} 1 & 2 \\ -2 & -4 \end{pmatrix}$. Calculate $A^{-1}$ if it exists.

**Solution** We find that $\det A = (1)(-4) - (2)(-2) = -4 + 4 = 0$ so that $A^{-1}$ does not exist, as we saw in Example 3. ■

The procedure described above for finding the inverse (if it exists) of a $2 \times 2$ matrix works for $n \times n$ matrices where $n > 2$. We illustrate this with a number of examples.

**EXAMPLE 6** **Calculating the Inverse of a 3 × 3 Matrix** Let $A = \begin{pmatrix} 2 & 4 & 6 \\ 4 & 5 & 6 \\ 3 & 1 & -2 \end{pmatrix}$ (see Example 1.3.1 on page 7). Calculate $A^{-1}$ if it exists.

**Solution** We first put $I$ next to $A$ in an augmented matrix form

$$\left(\begin{array}{ccc|ccc} 2 & 4 & 6 & 1 & 0 & 0 \\ 4 & 5 & 6 & 0 & 1 & 0 \\ 3 & 1 & -2 & 0 & 0 & 1 \end{array}\right)$$

and then carry out the row reduction.

$$\xrightarrow{R_1 \to \frac{1}{2}R_1} \left(\begin{array}{ccc|ccc} 1 & 2 & 3 & \frac{1}{2} & 0 & 0 \\ 4 & 5 & 6 & 0 & 1 & 0 \\ 3 & 1 & -2 & 0 & 0 & 1 \end{array}\right) \xrightarrow[R_3 \to R_3 - 3R_1]{R_2 \to R_2 - 4R_1} \left(\begin{array}{ccc|ccc} 1 & 2 & 3 & \frac{1}{2} & 0 & 0 \\ 0 & -3 & -6 & -2 & 1 & 0 \\ 0 & -5 & -11 & -\frac{3}{2} & 0 & 1 \end{array}\right)$$

$$\xrightarrow{R_2 \to -\frac{1}{3}R_2} \left(\begin{array}{ccc|ccc} 1 & 2 & 3 & \frac{1}{2} & 0 & 0 \\ 0 & 1 & 2 & \frac{2}{3} & -\frac{1}{3} & 0 \\ 0 & -5 & -11 & -\frac{3}{2} & 0 & 1 \end{array}\right) \xrightarrow[R_3 \to R_3 + 5R_2]{R_1 \to R_1 - 2R_2} \left(\begin{array}{ccc|ccc} 1 & 0 & -1 & -\frac{5}{6} & \frac{2}{3} & 0 \\ 0 & 1 & 2 & \frac{2}{3} & -\frac{1}{3} & 0 \\ 0 & 0 & -1 & \frac{11}{6} & -\frac{5}{3} & 1 \end{array}\right)$$

$$\xrightarrow{R_3 \to -R_3} \left(\begin{array}{ccc|ccc} 1 & 0 & -1 & -\frac{5}{6} & \frac{2}{3} & 0 \\ 0 & 1 & 2 & \frac{2}{3} & -\frac{1}{3} & 0 \\ 0 & 0 & 1 & -\frac{11}{6} & \frac{5}{3} & -1 \end{array}\right) \xrightarrow[R_2 \to R_2 - 2R_3]{R_1 \to R_1 + R_3} \left(\begin{array}{ccc|ccc} 1 & 0 & 0 & -\frac{8}{3} & \frac{7}{3} & -1 \\ 0 & 1 & 0 & \frac{13}{3} & -\frac{11}{3} & 2 \\ 0 & 0 & 1 & -\frac{11}{6} & \frac{5}{3} & -1 \end{array}\right)$$

Since $A$ has now been reduced to $I$, we have

$$A^{-1} = \begin{pmatrix} -\frac{8}{3} & \frac{7}{3} & -1 \\ \frac{13}{3} & -\frac{11}{3} & 2 \\ -\frac{11}{6} & \frac{5}{3} & -1 \end{pmatrix} = \frac{1}{6}\begin{pmatrix} -16 & 14 & -6 \\ 26 & -22 & 12 \\ -11 & 10 & -6 \end{pmatrix}$$

We factor out $\frac{1}{6}$ to make computations easier.

*Check*

$$A^{-1}A = \frac{1}{6}\begin{pmatrix} -16 & 14 & -6 \\ 26 & -22 & 12 \\ -11 & 10 & -6 \end{pmatrix}\begin{pmatrix} 2 & 4 & 6 \\ 4 & 5 & 6 \\ 3 & 1 & -2 \end{pmatrix} = \frac{1}{6}\begin{pmatrix} 6 & 0 & 0 \\ 0 & 6 & 0 \\ 0 & 0 & 6 \end{pmatrix} = I.$$

We can also verify that $AA^{-1} = I$.

**WARNING** It is easy to make numerical errors in computing $A^{-1}$. Therefore it is important to check the computations by verifying that $A^{-1}A = I$.

**EXAMPLE 7** **A 3 × 3 Matrix That Is Not Invertible** Let $A = \begin{pmatrix} 1 & -3 & 4 \\ 2 & -5 & 7 \\ 0 & -1 & 1 \end{pmatrix}$. Calculate $A^{-1}$ if it exists.

**Solution** Proceeding as before, we obtain, successively,

$$\left(\begin{array}{rrr|rrr} 1 & -3 & 4 & 1 & 0 & 0 \\ 2 & -5 & 7 & 0 & 1 & 0 \\ 0 & -1 & 1 & 0 & 0 & 1 \end{array}\right) \xrightarrow{R_2 \to R_2 - 2R_1} \left(\begin{array}{rrr|rrr} 1 & -3 & 4 & 1 & 0 & 0 \\ 0 & 1 & -1 & -2 & 1 & 0 \\ 0 & -1 & 1 & 0 & 0 & 1 \end{array}\right)$$

$$\xrightarrow[R_3 \to R_3 + R_2]{R_1 \to R_1 + 3R_2} \left(\begin{array}{rrr|rrr} 1 & 0 & 1 & -5 & 3 & 0 \\ 0 & 1 & -1 & -2 & 1 & 0 \\ 0 & 0 & 0 & -2 & 1 & 1 \end{array}\right)$$

This is as far as we can go. The matrix $A$ *cannot* be reduced to the identity matrix, and we can conclude that $A$ is *not* invertible.

There is another way to see the result of the last example. Let **b** be any 3-vector and consider the system $A\mathbf{x} = \mathbf{b}$. If we tried to solve this by Gaussian elimination, we would end up with an equation that reads $0 = c \neq 0$ as in Example 3, or $0 = 0$. That is, the system has either no solution or an infinite number of solutions. The one possibility ruled out is the case in which the system has a unique solution. But if $A^{-1}$ existed, then there would be a unique solution given by $\mathbf{x} = A^{-1}\mathbf{b}$. We are left to conclude that

> If row reduction of $A$ produces a row of zeros, then $A$ is *not* invertible.

**DEFINITION 3** **Row Equivalent Matrices** Suppose that by elementary row operations we can transform the matrix $A$ into the matrix $B$. Then $A$ and $B$ are said to be **row equivalent.**

The reasoning used above can be used to prove the following theorem (see Problem 47).

**THEOREM 5** Let $A$ be an $n \times n$ matrix.

**i.** $A$ is invertible if and only if $A$ is row equivalent to the identity matrix $I_n$; that is, the reduced row echelon form of $A$ is $I_n$.

**ii.** $A$ is invertible if and only if the system $A\mathbf{x} = \mathbf{b}$ has a unique solution for every $n$-vector **b**.

**iii.** If $A$ is invertible, then the unique solution of $A\mathbf{x} = \mathbf{b}$ is given by $\mathbf{x} = A^{-1}\mathbf{b}$.

**iv.** $A$ is invertible if and only if its row echelon form has $n$ pivots.

**EXAMPLE 8** **Using the Inverse of a Matrix to Solve a System of Equations** Solve the system

$$\begin{aligned} 2x_1 + 4x_2 + 3x_3 &= 6 \\ x_2 - x_3 &= -4 \\ 3x_1 + 5x_2 + 7x_3 &= 7 \end{aligned}$$

**Solution** This system can be written as $A\mathbf{x} = \mathbf{b}$, where $A = \begin{pmatrix} 2 & 4 & 3 \\ 0 & 1 & -1 \\ 3 & 5 & 7 \end{pmatrix}$ and $\mathbf{b} = \begin{pmatrix} 6 \\ -4 \\ 7 \end{pmatrix}$.

$$A^{-1} = \begin{pmatrix} 4 & -\frac{13}{3} & -\frac{7}{3} \\ -1 & \frac{5}{3} & \frac{2}{3} \\ -1 & \frac{2}{3} & \frac{2}{3} \end{pmatrix}$$

Thus the unique solution is given by

$$\mathbf{x} = \begin{pmatrix} x_1 \\ x_2 \\ x_3 \end{pmatrix} = A^{-1}\mathbf{b} = \begin{pmatrix} 4 & -\frac{13}{3} & -\frac{7}{3} \\ -1 & \frac{5}{3} & \frac{2}{3} \\ -1 & \frac{2}{3} & \frac{2}{3} \end{pmatrix} \begin{pmatrix} 6 \\ -4 \\ 7 \end{pmatrix} = \begin{pmatrix} 25 \\ -8 \\ -4 \end{pmatrix}$$

**EXAMPLE 9** **The Technology and Leontief Matrices: Modeling the 1958 American Economy** In the Leontief input-output model described in Example 1.3.9 on page 19, we obtained the system

$$\begin{aligned} a_{11}x_1 + a_{12}x_2 + \cdots + a_{1n}x_n + e_1 &= x_1 \\ a_{21}x_1 + a_{22}x_2 + \cdots + a_{2n}x_n + e_2 &= x_2 \\ \vdots \qquad \vdots \qquad\qquad \vdots \qquad \vdots \quad &\;\; \vdots \\ a_{n1}x_1 + a_{n2}x_2 + \cdots + a_{nn}x_n + e_n &= x_n \end{aligned} \tag{15}$$

which can be written as

$$A\mathbf{x} + \mathbf{e} = \mathbf{x} = I\mathbf{x}$$

or

$$(I - A)\mathbf{x} = \mathbf{e} \tag{16}$$

The matrix $A$ of internal demands is called the **technology matrix,** and the matrix $I - A$ is called the **Leontief matrix.** If the Leontief matrix is invertible, then systems (15) and (16) have unique solutions.

Leontief used his model to analyze the 1958 U.S. economy.† He divided the economy into 81 sectors and grouped them into six families of related sectors. For simplicity, we treat each family of sectors as a single sector so that we can treat the U.S. economy as an economy with six industries. These industries are listed in Table 1.1.

† *Scientific American* (April 1965): 26–27.

**Table 1.1**

| Sector | Examples |
|---|---|
| Final nonmetal (FN) | Furniture, processed food |
| Final metal (FM) | Household appliances, motor vehicles |
| Basic metal (BM) | Machine-shop products, mining |
| Basic nonmetal (BN) | Agriculture, printing |
| Energy (E) | Petroleum, coal |
| Services (S) | Amusements, real estate |

The input-output table, Table 1.2, gives internal demands in 1958 based on Leontief's figures. The units in the table are millions of dollars. Thus, for example, the number 0.173 in the 6,5 position means that in order to produce \$1 million worth of energy, it is necessary to provide \$0.173 million = \$173,000 worth of services. Similarly, the 0.037 in the 4,2 position means that in order to produce \$1 million worth of final metal, it is necessary to expend \$0.037 million = \$37,000 on basic nonmetal products.

**Table 1.2** Internal Demands in 1958 U.S. Economy

| | FN | FM | BM | BN | E | S |
|---|---|---|---|---|---|---|
| FN | 0.170 | 0.004 | 0 | 0.029 | 0 | 0.008 |
| FM | 0.003 | 0.295 | 0.018 | 0.002 | 0.004 | 0.016 |
| BM | 0.025 | 0.173 | 0.460 | 0.007 | 0.011 | 0.007 |
| BN | 0.348 | 0.037 | 0.021 | 0.403 | 0.011 | 0.048 |
| E | 0.007 | 0.001 | 0.039 | 0.025 | 0.358 | 0.025 |
| S | 0.120 | 0.074 | 0.104 | 0.123 | 0.173 | 0.234 |

Finally, Leontief estimated the external demands on the 1958 U.S. economy (in millions of dollars) as listed in Table 1.3.

**Table 1.3** External Demands on 1958 U.S. Economy (millions of dollars)

| | |
|---|---|
| FN | \$99,640 |
| FM | \$75,548 |
| BM | \$14,444 |
| BN | \$33,501 |
| E | \$23,527 |
| S | \$263,985 |

In order to run the U.S. economy in 1958 and to meet all external demands, how many units had to be produced in each of the six sectors?

**Solution** The technology matrix is given by

$$A = \begin{pmatrix} 0.170 & 0.004 & 0 & 0.029 & 0 & 0.008 \\ 0.003 & 0.295 & 0.018 & 0.002 & 0.004 & 0.016 \\ 0.025 & 0.173 & 0.460 & 0.007 & 0.011 & 0.007 \\ 0.348 & 0.037 & 0.021 & 0.403 & 0.011 & 0.048 \\ 0.007 & 0.001 & 0.039 & 0.025 & 0.358 & 0.025 \\ 0.120 & 0.074 & 0.104 & 0.123 & 0.173 & 0.234 \end{pmatrix} \quad \text{and} \quad \mathbf{e} = \begin{pmatrix} 99{,}640 \\ 75{,}548 \\ 14{,}444 \\ 33{,}501 \\ 23{,}527 \\ 263{,}985 \end{pmatrix}$$

To obtain the Leontief matrix, we subtract to obtain

$$I - A = \begin{pmatrix} 1 & 0 & 0 & 0 & 0 & 0 \\ 0 & 1 & 0 & 0 & 0 & 0 \\ 0 & 0 & 1 & 0 & 0 & 0 \\ 0 & 0 & 0 & 1 & 0 & 0 \\ 0 & 0 & 0 & 0 & 1 & 0 \\ 0 & 0 & 0 & 0 & 0 & 1 \end{pmatrix} - \begin{pmatrix} 0.170 & 0.004 & 0 & 0.029 & 0 & 0.008 \\ 0.003 & 0.295 & 0.018 & 0.002 & 0.004 & 0.016 \\ 0.025 & 0.173 & 0.460 & 0.007 & 0.011 & 0.007 \\ 0.348 & 0.037 & 0.021 & 0.403 & 0.011 & 0.048 \\ 0.007 & 0.001 & 0.039 & 0.025 & 0.358 & 0.025 \\ 0.120 & 0.074 & 0.104 & 0.123 & 0.173 & 0.234 \end{pmatrix}$$

The computation of the inverse of a $6 \times 6$ matrix is a tedious affair. The following results (rounded to three decimal places) were obtained using MATLAB:

$$(I - A)^{-1} \approx \begin{pmatrix} 1.234 & 0.014 & 0.007 & 0.064 & 0.006 & 0.017 \\ 0.017 & 1.436 & 0.056 & 0.014 & 0.019 & 0.032 \\ 0.078 & 0.467 & 1.878 & 0.036 & 0.044 & 0.031 \\ 0.752 & 0.133 & 0.101 & 1.741 & 0.065 & 0.123 \\ 0.061 & 0.045 & 0.130 & 0.083 & 1.578 & 0.059 \\ 0.340 & 0.236 & 0.307 & 0.315 & 0.376 & 1.349 \end{pmatrix}$$

Therefore the "ideal" output vector is given by

$$\mathbf{x} = (I - A)^{-1}\mathbf{e} \approx \begin{pmatrix} 131{,}033.21 \\ 120{,}458.90 \\ 80{,}680.56 \\ 178{,}732.04 \\ 66{,}929.26 \\ 431{,}562.04 \end{pmatrix}$$

This means that it would require approximately 131,033 units ($131,033 million worth) of final nonmetal products, 120,459 units of final metal products, 80,681 units of basic metal products, 178,732 units of basic nonmetal products, 66,929 units of energy, and 431,562 service units to run the U.S. economy and to meet the external demands in 1958.

In Section 1.2 we encountered the first form of our Summing Up Theorem (Theorem 1.2.1, page 5). We are now ready to improve upon it. The next theorem states that several statements involving inverse, uniqueness of solutions, row equivalence, and determinants are equivalent. At this point we can prove the equivalence of parts (*i*), (*ii*), (*iii*), (*iv*), and (*v*). We shall finish the proof after we have developed some basic theory about determinants (see Theorem 2.4.4 on page 216).

**THEOREM 6** **Summing Up Theorem—View 2** Let $A$ be an $n \times n$ matrix. Then the following six statements are equivalent. That is, each statement implies the other five (so that if one is true, all are true, and if one is false, all are false).

**i.** $A$ is invertible.

**ii.** The only solution to the homogeneous system $A\mathbf{x} = \mathbf{0}$ is the trivial solution ($\mathbf{x} = \mathbf{0}$).

**iii.** The system $A\mathbf{x} = \mathbf{b}$ has a unique solution for every $n$-vector $\mathbf{b}$.

**iv.** $A$ is row equivalent to the $n \times n$ identity matrix $I_n$; that is, the reduced row echelon form of $A$ is $I_n$.

**v.** The row echelon form of $A$ has $n$ pivots.

**vi.** $\det A \neq 0$. (So far $\det A$ is only defined if $A$ is a $2 \times 2$ matrix.)

**Proof** We have already seen that statements (*i*), (*iii*), (*iv*), and (*vi*) are equivalent [Theorem 5]. We shall show that (*ii*) and (*iv*) are equivalent. Suppose that (*ii*) holds. Then the reduced row echelon form of $A$ has $n$ pivots; otherwise at least one column of this form would have no pivot, and then the system $A\mathbf{x} = \mathbf{0}$ would have an infinite number of solutions because the variable corresponding to that column could be chosen arbitrarily (the coefficients in the column are zero). But if the reduced row echelon form of $A$ has $n$ pivots, then it is $I_n$.

Conversely, suppose that (*iv*) holds; that is, suppose that $A$ is row equivalent to $I_n$. Then by Theorem 5, part (*i*), $A$ is invertible and by Theorem 5, part (*iii*), the unique solution to $A\mathbf{x} = \mathbf{0}$ is $\mathbf{x} = A^{-1}\mathbf{0} = \mathbf{0}$. Thus (*ii*) and (*iv*) are equivalent. In Theorem 1.2.1 we showed that (*i*) and (*vi*) are equivalent in the $2 \times 2$ case. We shall prove the equivalence of (*i*) and (*vi*) in Section 2.4. ∎

*Remark.* If the row echelon form of $A$ has $n$ pivots, then it has the following appearance:

$$\begin{pmatrix} 1 & r_{12} & r_{13} & \cdots & r_{1n} \\ 0 & 1 & r_{23} & \cdots & r_{2n} \\ 0 & 0 & 1 & \cdots & r_{3n} \\ \vdots & \vdots & \vdots & & \vdots \\ 0 & 0 & 0 & \cdots & 1 \end{pmatrix} \qquad \textbf{(17)}$$

That is, $R$ is a matrix with 1's down the diagonal and 0's below it.

To verify that $B = A^{-1}$, we have to check that $AB = BA = I$. It turns out that only half this work has to be done.

**THEOREM 7** Let $A$ and $B$ be $n \times n$ matrices. Then $A$ is invertible and $B = A^{-1}$ if either (*i*) $BA = I$ or (*ii*) $AB = I$.

**Proof**

**i.** We assume that $BA = I$. Consider the homogeneous system $A\mathbf{x} = \mathbf{0}$. Multiplying both sides of this equation on the left by $B$, we obtain

$$BA\mathbf{x} = B\mathbf{0} \tag{18}$$

But $BA = I$ and $B\mathbf{0} = \mathbf{0}$, so (18) becomes $I\mathbf{x} = \mathbf{0}$ or $\mathbf{x} = \mathbf{0}$. This shows that $\mathbf{x} = \mathbf{0}$ is the only solution to $A\mathbf{x} = \mathbf{0}$ and by Theorem 6, parts (*i*) and (*ii*), this means that $A$ is invertible. We still have to show that $B = A^{-1}$. Let $A^{-1} = C$. Then $AC = I$. Thus

$$BAC = B(AC) = BI = B \quad \text{and} \quad BAC = (BA)C = IC = C$$

Hence $B = C$ and part (*i*) is proved.

**ii.** Let $AB = I$. Then from part (*i*) $A = B^{-1}$. From Definition 2 this means that $AB = BA = I$, which proves that $A$ is invertible and that $B = A^{-1}$. This completes the proof. ■

# PROBLEMS 1.8

## Self-Quiz

**I.** Which of the following is true?

**a.** Every square matrix has an inverse.

**b.** A square matrix has an inverse if its row reduction leads to a row of zeros.

**c.** A square matrix is invertible if it has an inverse.

**d.** A square matrix $B$ is the inverse of $A$ if $AI = B$.

**II.** Which of the following is true of a system of equations in matrix form?

**a.** It is of the form $A^{-1}\mathbf{x} = \mathbf{b}$.

**b.** If it has a unique solution, the solution will be $\mathbf{x} = A^{-1}\mathbf{b}$.

**c.** It will have a solution if $A$ is not invertible.

**d.** It will have a unique solution.

**III.** Which of the following is invertible?

**a.** $\begin{pmatrix} 1 & 3 \\ -3 & -9 \end{pmatrix}$ **b.** $\begin{pmatrix} 6 & -1 \\ 1 & -\frac{1}{6} \end{pmatrix}$

**c.** $\begin{pmatrix} 2 & -3 \\ 1 & -1 \end{pmatrix}$ **d.** $\begin{pmatrix} 1 & 0 \\ 2 & 0 \end{pmatrix}$

**IV.** Which of the following is true of an invertible matrix $A$?

**a.** The product of $A$ and $I$ is $A^{-1}$.

**b.** $A$ is a $2 \times 3$ matrix.

**c.** $A = A^{-1}$.

**d.** $A$ is a square matrix.

**V.** Which of the following is true of the system

$$4x - 5y = 3$$
$$6x - 7y = 4?$$

**a.** It has no solution because $\begin{pmatrix} 4 & -5 \\ 6 & -7 \end{pmatrix}$ is not invertible.

**b.** It has the solution $(-1, -\frac{1}{2})$.

**c.** If it had a solution, it would be found by solving $\begin{pmatrix} 4 & -5 \\ 6 & -7 \end{pmatrix}\begin{pmatrix} x \\ y \end{pmatrix} = \begin{pmatrix} 3 \\ 4 \end{pmatrix}$.

**d.** Its solution is $\begin{pmatrix} 4 & -5 \\ 6 & -7 \end{pmatrix}\begin{pmatrix} 3 \\ 4 \end{pmatrix}$.

In Problems 1–15 determine whether the given matrix is invertible. If it is, calculate the inverse.

**1.** $\begin{pmatrix} 2 & 1 \\ 3 & 2 \end{pmatrix}$

**2.** $\begin{pmatrix} -1 & 6 \\ 2 & -12 \end{pmatrix}$

**3.** $\begin{pmatrix} 0 & 1 \\ 1 & 0 \end{pmatrix}$

**4.** $\begin{pmatrix} 1 & 1 \\ 3 & 3 \end{pmatrix}$

**5.** $\begin{pmatrix} a & a \\ b & b \end{pmatrix}$

**6.** $\begin{pmatrix} 1 & 1 & 1 \\ 0 & 2 & 3 \\ 5 & 5 & 1 \end{pmatrix}$

**7.** $\begin{pmatrix} 3 & 2 & 1 \\ 0 & 2 & 2 \\ 0 & 0 & -1 \end{pmatrix}$

**8.** $\begin{pmatrix} 1 & 1 & 1 \\ 0 & 1 & 1 \\ 0 & 0 & 1 \end{pmatrix}$

**9.** $\begin{pmatrix} 1 & 6 & 2 \\ -2 & 3 & 5 \\ 7 & 12 & -4 \end{pmatrix}$

**10.** $\begin{pmatrix} 3 & 1 & 0 \\ 1 & -1 & 2 \\ 1 & 1 & 1 \end{pmatrix}$

**11.** $\begin{pmatrix} 2 & -1 & 4 \\ -1 & 0 & 5 \\ 19 & -7 & 3 \end{pmatrix}$

**12.** $\begin{pmatrix} 1 & 2 & 3 \\ 1 & 1 & 2 \\ 0 & 1 & 2 \end{pmatrix}$

**13.** $\begin{pmatrix} 1 & 1 & 1 & 1 \\ 1 & 2 & -1 & 2 \\ 1 & -1 & 2 & 1 \\ 1 & 3 & 3 & 2 \end{pmatrix}$

**14.** $\begin{pmatrix} 1 & 0 & 2 & 3 \\ -1 & 1 & 0 & 4 \\ 2 & 1 & -1 & 3 \\ -1 & 0 & 5 & 7 \end{pmatrix}$

**15.** $\begin{pmatrix} 1 & -3 & 0 & -2 \\ 3 & -12 & -2 & -6 \\ -2 & 10 & 2 & 5 \\ -1 & 6 & 1 & 3 \end{pmatrix}$

**16.** Show that if $A$, $B$, and $C$ are invertible matrices, then $ABC$ is invertible and $(ABC)^{-1} = C^{-1}B^{-1}A^{-1}$.

**17.** If $A_1, A_2, \ldots, A_m$ are invertible $n \times n$ matrices, show that $A_1A_2 \ldots A_m$ is invertible and calculate its inverse.

**18.** Show that the matrix $\begin{pmatrix} 3 & 4 \\ -2 & -3 \end{pmatrix}$ is its own inverse.

**19.** Show that the matrix $\begin{pmatrix} a_{11} & a_{12} \\ a_{21} & a_{22} \end{pmatrix}$ is its own inverse if $A = \pm I$ or if $a_{11} = -a_{22}$ and $a_{21}a_{12} = 1 - a_{11}^2$.

**20.** Find the output vector $\mathbf{x}$ in the Leontief input-output model if $n = 3$, $\mathbf{e} = \begin{pmatrix} 30 \\ 20 \\ 40 \end{pmatrix}$, and $A = \begin{pmatrix} \frac{1}{5} & \frac{1}{5} & 0 \\ \frac{2}{5} & \frac{2}{5} & \frac{3}{5} \\ \frac{1}{5} & \frac{1}{10} & \frac{2}{5} \end{pmatrix}$.

---

### Answers to Self-Quiz

**I.** c **II.** b **III.** c **IV.** d **V.** c

***21.** Suppose that $A$ is $n \times m$ and $B$ is $m \times n$ so that $AB$ is $n \times n$. Show that $AB$ is not invertible if $n > m$. [*Hint:* Show that there is a nonzero vector $\mathbf{x}$ such that $AB\mathbf{x} = \mathbf{0}$ and then apply Theorem 6.]

***22.** Use the methods of this section to find the inverses of the following matrices with complex entries:

**a.** $\begin{pmatrix} i & 2 \\ 1 & -i \end{pmatrix}$ **b.** $\begin{pmatrix} 1-i & 0 \\ 0 & 1+i \end{pmatrix}$ **c.** $\begin{pmatrix} 1 & i & 0 \\ -i & 0 & 1 \\ 0 & 1+i & 1-i \end{pmatrix}$

**23.** Show that for every real number $\theta$ the matrix $\begin{pmatrix} \sin\theta & \cos\theta & 0 \\ \cos\theta & -\sin\theta & 0 \\ 0 & 0 & 1 \end{pmatrix}$ is invertible and find its inverse.

**24.** Calculate the inverse of $A = \begin{pmatrix} 2 & 0 & 0 \\ 0 & 3 & 0 \\ 0 & 0 & 4 \end{pmatrix}$.

**25.** A square matrix $A = (a_{ij})$ is called **diagonal** if all its elements off the main diagonal are zero. That is, $a_{ij} = 0$ if $i \neq j$. (The matrix of Problem 24 is diagonal.) Show that a diagonal matrix is invertible if and only if each of its diagonal components is nonzero.

**26.** Let

$$A = \begin{pmatrix} a_{11} & 0 & \cdots & 0 \\ 0 & a_{22} & \cdots & 0 \\ \vdots & \vdots & \ddots & \vdots \\ 0 & 0 & \cdots & a_{nn} \end{pmatrix}$$

be a diagonal matrix such that each of its diagonal components is nonzero. Calculate $A^{-1}$.

**27.** Calculate the inverse of $A = \begin{pmatrix} 2 & 1 & -1 \\ 0 & 3 & 4 \\ 0 & 0 & 5 \end{pmatrix}$.

**28.** Show that the matrix $A = \begin{pmatrix} 1 & 0 & 0 \\ -2 & 0 & 0 \\ 4 & 6 & 1 \end{pmatrix}$ is not invertible.

***29.** A square matrix is called **upper (lower) triangular** if all its elements below (above) the main diagonal are zero. (The matrix of Problem 27 is upper triangular and the matrix of Problem 28 is lower triangular.) Show that an upper or lower triangular matrix is invertible if and only if each of its diagonal elements is nonzero.

***30.** Show that the inverse of an invertible upper triangular matrix is upper triangular. [*Hint:* First, prove the result for a $3 \times 3$ matrix.]

In Problems 31 and 32 a matrix is given. In each case show that the matrix is not invertible by finding a nonzero vector $\mathbf{x}$ such that $A\mathbf{x} = \mathbf{0}$.

**31.** $\begin{pmatrix} 2 & -1 \\ -4 & 2 \end{pmatrix}$ **32.** $\begin{pmatrix} 1 & -1 & 3 \\ 0 & 4 & -2 \\ 2 & -6 & 8 \end{pmatrix}$

**33.** A factory for the construction of quality furniture has two divisions: a machine shop where the parts of the furniture are fabricated and an assembly and finishing division where the parts are put together into the finished product. Suppose that there are 12 employees in the machine shop and 20 in the assembly and finishing division and that each employee works an 8-hour day. Suppose further that the factory produces only two products: chairs and tables. A chair requires $\frac{384}{17}$ hours of machine shop time and $\frac{480}{17}$ hours of assembly and finishing time. A table requires $\frac{240}{17}$ hours of machine shop time and $\frac{640}{17}$ hours of assembly and finishing time. Assuming that there is an unlimited demand for these products and that the manufacturer wishes to keep all employees busy, how many chairs and how many tables can this factory produce each day?

**34.** A witch's magic cupboard contains 10 ounces of ground four-leaf clovers and 14 ounces of powdered mandrake root. The cupboard will replenish itself automatically provided she uses up exactly all her supplies. A batch of love potion requires $3\frac{1}{13}$ ounces of ground four-leaf clovers and $2\frac{2}{13}$ ounces of powdered mandrake root. One recipe of a well-known (to witches) cure for the common cold requires $5\frac{5}{13}$ ounces of four-leaf clovers and $10\frac{10}{13}$ ounces of mandrake root. How much of the love potion and the cold remedy should the witch make in order to use up the supply in the cupboard exactly?

**35.** A farmer feeds his cattle a mixture of two types of feed. One standard unit of type A feed supplies a steer with 10% of its minimum daily requirement of protein and 15% of its requirement of carbohydrates. Type B feed contains 12% of the requirement of protein and 8% of the requirement of carbohydrates in a standard unit. If the farmer wishes to feed his cattle exactly 100% of their minimum daily requirement of protein and carbohydrates, how many units of each type of feed should he give a steer each day?

**36.** A much simplified version of an input-output table for the 1958 Israeli economy divides that economy into three sectors—agriculture, manufacturing, and energy—with the following result.†

| | **Agriculture** | **Manufacturing** | **Energy** |
|---|---|---|---|
| Agriculture | 0.293 | 0 | 0 |
| Manufacturing | 0.014 | 0.207 | 0.017 |
| Energy | 0.044 | 0.010 | 0.216 |

**a.** How many units of agricultural production are required to produce 1 unit of agricultural output?

**b.** How many units of agricultural production are required to produce 200,000 units of agricultural output?

**c.** How many units of agricultural product go into the production of 50,000 units of energy?

**d.** How many units of energy go into the production of 50,000 units of agricultural products?

† Wassily Leontief, *Input-Output Economics* (New York: Oxford University Press, 1966), 54–57.

**37.** Continuing Problem 36, exports (in thousands of Israeli pounds) in 1958 were:

| | |
|---|---|
| Agriculture | 13,213 |
| Manufacturing | 17,597 |
| Energy | 1,786 |

**a.** Compute the technology and Leontief matrices.
**b.** Determine the number of Israeli pounds worth of agricultural products, manufactured goods, and energy required to run this model of the Israeli economy and export the stated value of products.

In Problems 38–45 compute the row echelon form of the given matrix and use it to determine directly whether the given matrix is invertible.

**38.** The matrix of Problem 3.
**39.** The matrix of Problem 1.
**40.** The matrix of Problem 4.
**41.** The matrix of Problem 7.
**42.** The matrix of Problem 9.
**43.** The matrix of Problem 11.
**44.** The matrix of Problem 13.
**45.** The matrix of Problem 14.

**46.** Let $A = \begin{pmatrix} a_{11} & a_{12} \\ a_{21} & a_{22} \end{pmatrix}$ and assume that $a_{11}a_{22} - a_{12}a_{21} \neq 0$. Derive formula (12) by row-reducing the augmented matrix $\left(\begin{array}{cc|cc} a_{11} & a_{12} & 1 & 0 \\ a_{21} & a_{22} & 0 & 1 \end{array}\right)$.

**47.** Prove parts (*i*), (*ii*), and (*iv*) of Theorem 5.

**48.** Compute the inverse of $\begin{pmatrix} I & A \\ 0 & I \end{pmatrix}$ where $A$ is a square matrix [*Hint:* Review the multiplication of block matrices on page 70].†

**49.** Assuming that $A_{11}$ and $A_{22}$ are invertible, find the inverse of $\begin{pmatrix} A_{11} & 0 \\ A_{21} & A_{22} \end{pmatrix}$.

## CALCULATOR BOX

### TI-85

**To obtain an inverse on the TI-85,** give the matrix a name, say, $A$, and then press

ALPHA  A  2nd  $x^{-1}$  ENTER

If $A$ is not invertible, the message

ERROR 03 SINGULAR MAT

will be displayed.

† This problem and the next one were given by David Carlson in his article "Teaching Linear Algebra: Must the Fog Always Roll in?" in *The College Mathematics Journal,* 24(1), January 1993, 29–40.

## CASIO fx-7700 GB

**To obtain an inverse on the CASIO fx-7700 GB,** enter $A$. Then press F4 and $A^{-1}$ will be displayed (in memory $C$).

If $A$ is not invertible, the message Ma ERROR will be displayed.
In Problems 50–53 use a calculator to compute the inverse of the given matrix.

**50.** $\begin{pmatrix} 1.6 & 2.3 & 7.5 \\ -4.2 & 3.9 & 5.7 \\ -6.8 & -0.9 & 4.1 \end{pmatrix}$ **51.** $\begin{pmatrix} 20 & 37 & 11 \\ 26 & 49 & 10 \\ 57 & 98 & 36 \end{pmatrix}$

**52.** $\begin{pmatrix} -0.03 & 0.21 & 0.46 & -0.33 \\ -0.27 & 0.79 & 0.16 & 0.22 \\ 0.33 & 0.02 & 0 & -0.88 \\ 0.44 & -0.68 & 0.37 & 0.79 \end{pmatrix}$

**53.** $\begin{pmatrix} 23.46 & -59.62 & 38.36 & -44.21 \\ -59.32 & 77.01 & 91.38 & 50.02 \\ 36.38 & 67.92 & -81.31 & 15.06 \\ -61.31 & -70.80 & 43.59 & 71.22 \end{pmatrix}$

**54.** Show that the inverse of

$$\begin{pmatrix} 3 & 5 & -17 & 4 \\ 0 & 8 & 13 & 22 \\ 0 & 0 & 5 & -4 \\ 0 & 0 & 0 & -7 \end{pmatrix}$$

has zeros below the diagonal.

**55.** Do the same for the matrix

$$\begin{pmatrix} 23.1 & -42.1 & -63.7 & -19.4 & 23.8 \\ 0 & -14.5 & 36.2 & -15.9 & 61.3 \\ 0 & 0 & -37.2 & 64.8 & 23.5 \\ 0 & 0 & 0 & 91.2 & 13.8 \\ 0 & 0 & 0 & 0 & 46.9 \end{pmatrix}$$

**56.** The matrices in Problems 54 and 55 are called **upper triangular.** Using the results of those problems, make a conjecture about the inverse of an upper triangular matrix.

# MATLAB 1.8

*MATLAB Information.* The MATLAB command **eye(n)** forms the $n \times n$ identity matrix.

1. a. For $A = \begin{pmatrix} 1 & 2 & 3 \\ 2 & 5 & 4 \\ 1 & -1 & 10 \end{pmatrix}$, form **R = [A eye(3)].**

**i.** Find the reduced row echelon form of $R$. Use the ":" notation to assign the variable name $S$ to the matrix that consists of the last three columns of the reduced row echelon form of $R$.

**ii.** Find $SA$ and $AS$. Describe the relationship between $A$ and $S$.

**iii.** Compare $S$ with **inv(A)**.

b. Repeat the instructions above for **A = 2*rand(5)−1.** (Use **eye(5)** and let $S$ consist of the last five columns of the reduced row echelon form.)

2. Consider the matrices below.

**i.** $\frac{1}{13}\begin{pmatrix} 2 & 7 & 5 \\ 0 & 9 & 8 \\ 7 & 4 & 0 \end{pmatrix}$ **ii.** $\begin{pmatrix} 2 & -4 & 5 \\ 0 & 0 & 8 \\ 7 & -14 & 0 \end{pmatrix}$

**iii.** $\begin{pmatrix} 1 & 4 & -2 & 1 \\ 5 & 1 & 9 & 7 \\ 7 & 4 & 10 & 4 \\ 0 & 7 & -7 & 7 \end{pmatrix}$ **iv.** $\begin{pmatrix} 1 & 4 & 6 & 1 \\ 5 & 1 & 9 & 7 \\ 7 & 4 & 8 & 4 \\ 0 & 7 & 5 & 7 \end{pmatrix}$

**v.** $\frac{-1}{56}\begin{pmatrix} 1 & 2 & 3 & 4 & 5 \\ 0 & -1 & 2 & -1 & 2 \\ 1 & 0 & 0 & 2 & -1 \\ 1 & 1 & -1 & 1 & 1 \\ 0 & 0 & 0 & 0 & 4 \end{pmatrix}$ **vi.** $\begin{pmatrix} 1 & 2 & -1 & 7 & 5 \\ 0 & -1 & 2 & -3 & 2 \\ 1 & 0 & 3 & 1 & -1 \\ 1 & 1 & 1 & 4 & 1 \\ 0 & 0 & 0 & 0 & 4 \end{pmatrix}$

For each matrix $A$:

a. Use the **rref** command to test whether it is invertible and find **inv(A).**

b. If $A$ is not invertible, pay attention to the messages you get from MATLAB when you enter **inv(A).**

c. If $A$ is invertible, verify that **inv(A)** gives the inverse. Choose a random vector **b** of the right size, show that the system [$A$ **b**] has a unique solution using the **rref** command, assign the solution to the variable name **x**, and compare **x** with **y = inv(A)*b** (find **x−y**). Repeat this for one other **b** vector.

3. a. Let **A = round(10*(2*rand(5)−1)).** Let **B = A** but change one of the rows of $B$ to **B(3,:) = 3*B(1,:) + 5*B(2,:).** Show that $B$ is not invertible.

b. Let **B = A** and change a row of your choice to a linear combination of other rows of $B$. Show that $B$ is not invertible.

c. *(Paper and pencil)* By considering the process of reducing to the reduced row echelon form, prove that a matrix $B$ is not invertible if one row is a linear combination of other rows.

4. Let **A = round(10*(2*rand(7)−1)).**

Let **B = A** but **B(:,3) = 2*B(:,1)−B(:,2).**

Let **C = A** but **C(:,4) = C(:,1)+C(:,2)−C(:,3)** and **C(:,6) = 3*C(:,2).**

Let **D = A** but **D(:,2) = 3*D(:,1), D(:,4) = 2*D(:,1)−D(:,2)+4*D(:,3),**
**D(:,5) = D(:,2)−5*D(:,3).**

a. Find **rref** of $B$, $C$, and $D$. What do you conjecture about the invertibility of a matrix where some columns of the matrix are linear combinations of other columns?
b. Test your conjecture by generating another random matrix $E$ and modifying it by changing some columns to linear combinations of other columns.
c. For $B$, $C$, $D$, and $E$, look for patterns in the numbers of the **rref** that reflect the coefficients of the linear combinations. Describe these patterns.
d. How does this problem relate to MATLAB Problem 5 in Section 1.7?

5. **Special Types of Matrices**
   a. Generate five random upper triangular matrices with integer entries between $-10$ and 10. Use the **triu** command. For two of the generated matrices, change one diagonal entry to 0. (For example, if the matrix is called $A$, modify it by the command **A(2,2) = 0.**)
      **i.** Test each for invertibility. Describe a conjecture linking the diagonal terms of the upper triangular matrix and the property of being invertible or not. Test your conjecture with three more upper triangular matrices.
      **ii.** For each invertible matrix found in (i), find the inverse by using the **inv** command. What conjecture can you make about the form of the inverse of an upper triangular matrix? How are the diagonal entries of the inverse related to the diagonal entries of the original matrix? How does this observation relate to (i)?
      **iii.** *(Pencil and paper)* Assume $A$ is a $3 \times 3$ upper triangular matrix $\begin{pmatrix} a & b & c \\ 0 & d & e \\ 0 & 0 & f \end{pmatrix}$.
      Describe the steps involved in reducing the augmented matrix $[A\ I]$ ($I$ is the identity matrix) to reduced row echelon form and use this description to verify the conjectures about the inverses of upper triangular matrices that you made in (i) and (ii).
   b. Test the following matrices and others of the same general pattern for invertibility. Describe the results:

$$\begin{pmatrix} 1 & 2 & 3 \\ 4 & 5 & 6 \\ 7 & 8 & 9 \end{pmatrix} \quad \begin{pmatrix} 1 & 2 & 3 & 4 \\ 5 & 6 & 7 & 8 \\ 9 & 10 & 11 & 12 \\ 13 & 14 & 15 & 16 \end{pmatrix}$$

   c. In MATLAB Problem 11 in Section 1.3 we claimed that the system obtained by fitting a polynomial of degree $n$ to $n + 1$ points with distinct coordinates would yield a unique solution. What does this say about the coefficient matrix? Test out your conjecture: First enter a vector **x** with distinct coordinates and find **V = vander(x),** and then test $V$. Repeat for three more choices of the vector **x**.

6. Consider the matrices below.

$$A1 = \begin{pmatrix} 1 & 2 & 3 & 4 & 5 \\ 0 & -1 & 2 & -1 & 2 \\ 1 & 0 & 0 & 2 & -1 \\ 1 & 1 & -1 & 1 & 1 \\ 0 & 0 & 0 & 0 & 4 \end{pmatrix} \qquad A2 = \begin{pmatrix} 1 & 2 & -1 & 7 & 5 \\ 0 & -1 & 2 & -3 & 2 \\ 1 & 0 & 3 & 1 & -1 \\ 1 & 1 & 1 & 4 & 1 \\ 0 & 0 & 0 & 0 & 4 \end{pmatrix}$$

$$A3 = \begin{pmatrix} 3 & 9 & 5 & 5 & 1 \\ 4 & 9 & 5 & 3 & 2 \\ 2 & 1 & 3 & 1 & 3 \\ 5 & 9 & 10 & 9 & 4 \\ 0 & 0 & 0 & 0 & -5 \end{pmatrix} \qquad A4 = \begin{pmatrix} 1 & 2 & -3 & 4 & 5 \\ -2 & -5 & 8 & -8 & -9 \\ 1 & 2 & -2 & 7 & 9 \\ 1 & 1 & 0 & 6 & 12 \\ 2 & 4 & -6 & 8 & 11 \end{pmatrix}$$

$$A5 = \begin{pmatrix} 2 & -4 & 4 & 5 & -1 \\ 0 & 0 & 5 & 1 & -9 \\ 7 & -14 & 8 & 7 & -2 \\ 7 & -14 & 0 & 4 & 11 \\ 9 & -18 & 1 & 7 & 14 \end{pmatrix}$$

a. Using the **rref** command, test the invertibility of each matrix $A1$ through $A5$. Test the invertibility of $A1*A2$, $A1*A3$, $A1*A4$, $A1*A5$, $A2*A3$, $A2*A4$, $A2*A5$, $A3*A4$, $A3*A5$, and $A4*A5$. Make a conjecture concerning the relationship of the invertibility of two matrices and the invertibility of their product. Explain how the evidence supports your conjecture.

b. For each pair of matrices $A$ and $B$ above such that $AB$ is invertible, find

**inv(A*B) − inv(A)*inv(B)** and **inv(A*B) − inv(B)*inv(A)**

Conjecture a formula for $(AB)^{-1}$ in terms of $A^{-1}$ and $B^{-1}$. Explain.

7. **Perturbations: Matrices Close to a Noninvertible Matrix** Enter the matrix

$$A = \begin{pmatrix} 1 & 2 & 3 \\ 4 & 5 & 6 \\ 7 & 8 & 9 \end{pmatrix}$$

Verify that $A$ is not invertible. In what follows we change $A$ to an invertible matrix $C$ that is close to $A$ by modifying one of the entries of $A$:

$$C = \begin{pmatrix} 1 & 2 & 3 \\ 4 & 5 & 6 \\ 7 & 8 & 9+f \end{pmatrix}$$

where $f$ is a small number.

Before continuing, enter the command **format short e.** This command will cause numbers to be displayed in scientific notation. In MATLAB, for example, **1.e−5** represents $10^{-5}$.

a. Enter

**f = 1.e−5 ; C = A; C(3,3) = A(3,3) + f;**

Verify that $C$ is invertible and find **inv(C).**

b. Repeat for **f = 1.e−7** and **f = 1.e−10.**

c. Comment on the size of the entries of **inv(C)** (compared to the size of the entries of $C$) as $f$ gets smaller, that is, as $C$ gets closer to being noninvertible.

d. We will investigate the accuracy of solutions to systems where the coefficient matrix is close to being noninvertible. Note that if

$$C = \begin{pmatrix} 1 & 2 & 3 \\ 4 & 5 & 6 \\ 7 & 8 & 9+f \end{pmatrix} \quad \text{and} \quad \mathbf{b} = \begin{pmatrix} 6 \\ 15 \\ 24+f \end{pmatrix}$$

then $C\mathbf{x} = \mathbf{b}$, where $\mathbf{x} = \begin{pmatrix} 1 \\ 1 \\ 1 \end{pmatrix}$; that is, $\mathbf{x}$ is the exact solution. Enter **x = [1;1;1]**. For each $f$ used in (a) and (b) above, form $C$ and $\mathbf{b}$ and solve the system $C\mathbf{y} = \mathbf{b}$ using **inv(C)** (naming the solution **y**). Find **z = x−y.** How close is the computed solution **y** to the exact solution **x**? How does the accuracy change as $f$ gets smaller, that is, as $C$ gets closer to being noninvertible?

Before continuing with other problems, enter **format** to return to the default display.

8. This problem concerns the Leontief input-output model. Solve the problems using $(I - A)^{-1}$, where $A$ is the technology matrix describing the internal demands. Interpret the results. [MATLAB *Hint:* The $n \times n$ identity matrix $I$ can be generated by **eye(n).**]
   a. Problem 37 in this section.
   b. MATLAB Problem 9(b) in Section 1.3.

   Use **format long** if you want more digits of the answers displayed.

9. **Cryptography** One process for encoding a secret message is to use certain square matrices whose entries are integers with integer entries in the inverse. One takes a message, assigns a number to each letter (for example, A = 1, B = 2, and so on, and space equals 27), arranges the numbers in a matrix from left to right in each row where the number of entries in the row matches the size of the encoding matrix, multiplies this matrix by the encoding matrix *on the right,* transcribes the message to a string of numbers (reading left to right along each row), and sends the message.

   The person who is supposed to receive the message knows the encoding matrix. He or she arranges the encoded message in a matrix from left to right in each row where the number of entries in the row matches the size of the encoding matrix, multiplies *on the right* by the inverse of the encoding matrix, and then reads the decoded message (from left to right across each row).
   a. *(Paper and pencil)* If we arrange messages in a matrix reading left to right such that the number of entries in a row matches the size of the encoding matrix, why do we need to multiply on the right?

      Why would multiplying by the inverse decode the message (that is, undo the encoding)?
   b. You have received the following message that was encoded using the matrix $A$ below. Decode it. (Assume that A = 1, B = 2, and so on, and space equals 27.)

$$A = \begin{pmatrix} 1 & 2 & -3 & 4 & 5 \\ -2 & -5 & 8 & -8 & -9 \\ 1 & 2 & -2 & 7 & 9 \\ 1 & 1 & 0 & 6 & 12 \\ 2 & 4 & -6 & 8 & 11 \end{pmatrix}$$

***Message.*** 47, 49, −19, 257, 487, 10, −9, 63, 137, 236, 79, 142, −184, 372, 536, 59, 70, −40, 332, 588

*Note.* The first row of the matrix you need to construct is 47 49 −19 257 487. Now continue with the second row.

## 1.9 THE TRANSPOSE OF A MATRIX

Corresponding to every matrix is another matrix, which, as we shall see in Chapter 2, has properties very similar to those of the original matrix.

**DEFINITION 1** **Transpose** Let $A = (a_{ij})$ be an $m \times n$ matrix. Then the **transpose** of $A$, written $A^t$, is the $n \times m$ matrix obtained by interchanging the rows and columns of $A$. Succinctly, we may write $A^t = (a_{ji})$. In other words,

$$\text{if } A = \begin{pmatrix} a_{11} & a_{12} & \cdots & a_{1n} \\ a_{21} & a_{22} & \cdots & a_{2n} \\ \vdots & \vdots & & \vdots \\ a_{m1} & a_{m2} & \cdots & a_{mn} \end{pmatrix}, \quad \text{then} \quad A^t = \begin{pmatrix} a_{11} & a_{21} & \cdots & a_{m1} \\ a_{12} & a_{22} & \cdots & a_{m2} \\ \vdots & \vdots & & \vdots \\ a_{1n} & a_{2n} & \cdots & a_{mn} \end{pmatrix} \tag{1}$$

Simply put, the $i$th row of $A$ is the $i$th column of $A^t$ and the $j$th column of $A$ is the $j$th row of $A^t$.

**EXAMPLE 1** **Finding the Transposes of Three Matrices** Find the transposes of the matrices

$$A = \begin{pmatrix} 2 & 3 \\ 1 & 4 \end{pmatrix} \quad B = \begin{pmatrix} 2 & 3 & 1 \\ -1 & 4 & 6 \end{pmatrix} \quad C = \begin{pmatrix} 1 & 2 & -6 \\ 2 & -3 & 4 \\ 0 & 1 & 2 \\ 2 & -1 & 5 \end{pmatrix}$$

**Solution** Interchanging the rows and columns of each matrix, we obtain

$$A^t = \begin{pmatrix} 2 & 1 \\ 3 & 4 \end{pmatrix} \quad B^t = \begin{pmatrix} 2 & -1 \\ 3 & 4 \\ 1 & 6 \end{pmatrix} \quad C^t = \begin{pmatrix} 1 & 2 & 0 & 2 \\ 2 & -3 & 1 & -1 \\ -6 & 4 & 2 & 5 \end{pmatrix}$$

Note, for example, that 4 is the component in row 2 and column 3 of $C$ while 4 is the component in row 3 and column 2 of $C^t$. That is, the 2,3 element of $C$ is the 3,2 element of $C^t$.

**THEOREM 1** Suppose $A = (a_{ij})$ is an $n \times m$ matrix and $B = (b_{ij})$ is an $m \times p$ matrix. Then

**i.** $(A^t)^t = A$. **(2)**

**ii.** $(AB)^t = B^tA^t$ **(3)**

**iii.** If $A$ and $B$ are $n \times m$, then $(A + B)^t = A^t + B^t$. **(4)**

**iv.** If $A$ is invertible, then $A^t$ is invertible and $(A^t)^{-1} = (A^{-1})^t$. **(5)**

**Proof**

**i.** This follows directly from the definition of the transpose.

**ii.** First, we note that $AB$ is an $n \times p$ matrix, so $(AB)^t$ is $p \times n$. Also, $B^t$ is $p \times m$ and $A^t$ is $m \times n$, so $B^tA^t$ is $p \times n$. Thus both matrices in equation (3) have the same size. Now the $ij$th element of $AB$ is $\sum_{k=1}^{m} a_{ik}b_{kj}$, and this is the $ji$th element of $(AB)^t$. Let $C = B^t$ and $D = A^t$. Then the $ij$th element $c_{ij}$ of $C$ is $b_{ji}$ and the $ij$th element $d_{ij}$ of $D$ is $a_{ji}$. Thus the $ji$th element of $CD$ = the $ji$th element of $B^tA^t = \sum_{k=1}^{m} c_{jk}d_{ki} = \sum_{k=1}^{m} b_{kj}a_{ik} = \sum_{k=1}^{m} a_{ik}b_{kj}$ = the $ji$th element of $(AB)^t$. This completes the proof of part (*ii*).

**iii.** This part is left as an exercise (see Problem 11).

**iv.** Let $A^{-1} = B$. Then $AB = BA = I$ so that, from part (*ii*), $(AB)^t = B^tA^t = I^t = I$ and $(BA)^t = A^tB^t = I$. Thus $A^t$ is invertible and $B^t$ is the inverse of $A^t$; that is, $(A^t)^{-1} = B^t = (A^{-1})^t$. ■

The transpose plays an important role in matrix theory. We shall see in succeeding chapters that $A$ and $A^t$ have many properties in common. Since columns of $A^t$ are rows of $A$, we shall be able to use facts about the transpose to conclude that just about anything that is true about the rows of a matrix is true about its columns.

The following definition is central in matrix theory.

**DEFINITION 2** **Symmetric Matrix** The $n \times n$ (square) matrix $A$ is called **symmetric** if $A^t = A$. That is, the columns of $A$ are also the rows of $A$.

**EXAMPLE 2** **Four Symmetric Matrices** The following four matrices are symmetric:

$$I \qquad A = \begin{pmatrix} 1 & 2 \\ 2 & 3 \end{pmatrix} \qquad B = \begin{pmatrix} 1 & -4 & 2 \\ -4 & 7 & 5 \\ 2 & 5 & 0 \end{pmatrix} \qquad C = \begin{pmatrix} -1 & 2 & 4 & 6 \\ 2 & 7 & 3 & 5 \\ 4 & 3 & 8 & 0 \\ 6 & 5 & 0 & -4 \end{pmatrix}$$

■

We shall see the importance of real symmetric matrices in Chapters 5 and 6.

### Another Way to Write the Scalar Product

Let $\mathbf{a} = \begin{pmatrix} a_1 \\ a_2 \\ \vdots \\ a_n \end{pmatrix}$ and $\mathbf{b} = \begin{pmatrix} b_1 \\ b_2 \\ \vdots \\ b_n \end{pmatrix}$ be two $n$-component column vectors. Then from equation (1) on page 62,

$$\mathbf{a} \cdot \mathbf{b} = a_1b_1 + a_2b_2 + \cdots + a_nb_n$$

Now $\mathbf{a}$ is an $n \times 1$ matrix so $\mathbf{a}^t$ is a $1 \times n$ matrix and

$$\mathbf{a}^t = (a_1\ a_2 \cdots a_n)$$

Then $\mathbf{a}^t\mathbf{b}$ is a $1 \times 1$ matrix (or scalar), and by the definition of matrix multiplication

$$\mathbf{a}^t\mathbf{b} = (a_1\ a_2 \cdots a_n)\begin{pmatrix} b_1 \\ b_2 \\ \vdots \\ b_n \end{pmatrix} = a_1b_1 + a_2b_2 + \cdots + a_nb_n$$

Thus, if $\mathbf{a}$ and $\mathbf{b}$ are $n$-component column vectors, then

$$\boxed{\mathbf{a} \cdot \mathbf{b} = \mathbf{a}^t\mathbf{b}} \qquad (6)$$

Formula (6) will be useful to us later in the book.

## PROBLEMS 1.9

### Self-Quiz

**I.** If $A$ is a $3 \times 4$ matrix, then $A^t$ is a ______ matrix.
**a.** $4 \times 3$ **b.** $3 \times 4$ **c.** $3 \times 3$ **d.** $4 \times 4$

**II.** *True-False:* $A^t$ is defined only if $A$ is a square matrix.

**III.** *True-False:* If $A$ is an $n \times n$ matrix, then the main diagonal of $A^t$ is the same as the main diagonal of $A$.

**IV.** *True-False:* $[(A^t)^t]^t = A^t$

**V.** The transpose of $\begin{pmatrix} 1 & 2 & 3 \\ -1 & 0 & 0 \end{pmatrix}$ is ______.

**a.** $\begin{pmatrix} -1 & 1 \\ 2 & 0 \\ 3 & 0 \end{pmatrix}$ **b.** $\begin{pmatrix} 1 & -1 \\ 2 & 0 \\ 3 & 0 \end{pmatrix}$ **c.** $\begin{pmatrix} 1 & 0 \\ -1 & 3 \\ 2 & 0 \end{pmatrix}$ **d.** $\begin{pmatrix} -1 & -2 & -3 \\ 1 & 0 & 0 \end{pmatrix}$

### Answers to Self-Quiz

**I.** a **II.** False **III.** True **IV.** True **V.** b

In Problems 1–10 find the transpose of the given matrix.

**1.** $\begin{pmatrix} -1 & 4 \\ 6 & 5 \end{pmatrix}$ **2.** $\begin{pmatrix} 3 & 0 \\ 1 & 2 \end{pmatrix}$ **3.** $\begin{pmatrix} 2 & 3 \\ -1 & 2 \\ 1 & 4 \end{pmatrix}$ **4.** $\begin{pmatrix} 2 & -1 & 0 \\ 1 & 5 & 6 \end{pmatrix}$

**5.** $\begin{pmatrix} 1 & 2 & 3 \\ -1 & 0 & 4 \\ 1 & 5 & 5 \end{pmatrix}$ **6.** $\begin{pmatrix} 1 & 2 & 3 \\ 2 & 4 & -5 \\ 3 & -5 & 7 \end{pmatrix}$ **7.** $\begin{pmatrix} 1 & 0 & 1 & 0 \\ 0 & 1 & 0 & 1 \end{pmatrix}$ **8.** $\begin{pmatrix} 2 & -1 \\ 2 & 4 \\ 1 & 6 \\ 1 & 5 \end{pmatrix}$

**9.** $\begin{pmatrix} a & b & c \\ d & e & f \\ g & h & j \end{pmatrix}$ **10.** $\begin{pmatrix} 0 & 0 & 0 \\ 0 & 0 & 0 \end{pmatrix}$

**11.** Let $A$ and $B$ be $n \times m$ matrices. Show, using Definition 1, that $(A + B)^t = A^t + B^t$.

**12.** Find numbers $\alpha$ and $\beta$ such that $\begin{pmatrix} 2 & \alpha & 3 \\ 5 & -6 & 2 \\ \beta & 2 & 4 \end{pmatrix}$ is symmetric.

**13.** If $A$ and $B$ are symmetric $n \times n$ matrices, prove that $A + B$ is symmetric.

**14.** If $A$ and $B$ are symmetric $n \times n$ matrices, show that $(AB)^t = BA$.

**15.** Show that for any matrix $A$ the product matrix $AA^t$ is defined and is a symmetric matrix.

**16.** Show that every diagonal matrix (see Problem 1.8.25, page 114) is symmetric.

**17.** Show that the transpose of every upper triangular matrix (see Problem 1.8.29 on page 114) is lower triangular.

**18.** A square matrix is called **skew-symmetric** if $A^t = -A$ (that is, $a_{ij} = -a_{ji}$). Which of the following matrices are skew-symmetric?

**a.** $\begin{pmatrix} 1 & -6 \\ 6 & 0 \end{pmatrix}$ **b.** $\begin{pmatrix} 0 & -6 \\ 6 & 0 \end{pmatrix}$ **c.** $\begin{pmatrix} 2 & -2 & -2 \\ 2 & 2 & -2 \\ 2 & 2 & 2 \end{pmatrix}$ **d.** $\begin{pmatrix} 0 & 1 & -1 \\ -1 & 0 & 2 \\ 1 & -2 & 0 \end{pmatrix}$

**19.** Let $A$ and $B$ be $n \times n$ skew-symmetric matrices. Show that $A + B$ is skew-symmetric.

**20.** If $A$ is a real skew-symmetric matrix, show that every component on the main diagonal of $A$ is zero.

**21.** If $A$ and $B$ are skew-symmetric $n \times n$ matrices, show that $(AB)^t = BA$ so that $AB$ is symmetric if and only if $A$ and $B$ commute.

**22.** Let $A$ be an $n \times n$ matrix. Show that the matrix $\frac{1}{2}(A + A^t)$ is symmetric.

**23.** Let $A$ be an $n \times n$ matrix. Show that the matrix $\frac{1}{2}(A - A^t)$ is skew-symmetric.

***24.** Show that any square matrix can be written in a unique way as the sum of a symmetric matrix and a skew-symmetric matrix.

***25.** Let $A = \begin{pmatrix} a_{11} & a_{12} \\ a_{21} & a_{22} \end{pmatrix}$ be a matrix with real nonnegative entries having the properties that (*i*) $a_{11}^2 + a_{12}^2 = 1$ and $a_{21}^2 + a_{22}^2 = 1$ and (*ii*) $\begin{pmatrix} a_{11} \\ a_{12} \end{pmatrix} \cdot \begin{pmatrix} a_{21} \\ a_{22} \end{pmatrix} = 0$. Show that $A$ is invertible and that $A^{-1} = A^t$.

In Problems 26–29 compute $(A^t)^{-1}$ and $(A^{-1})^t$ and show that they are equal.

**26.** $A = \begin{pmatrix} 1 & 2 \\ 3 & 4 \end{pmatrix}$

**27.** $A = \begin{pmatrix} 2 & 1 \\ 3 & 2 \end{pmatrix}$

**28.** $A = \begin{pmatrix} 3 & 2 & 1 \\ 0 & 2 & 2 \\ 0 & 0 & -1 \end{pmatrix}$

**29.** $A = \begin{pmatrix} 1 & 1 & 1 \\ 0 & 2 & 3 \\ 5 & 5 & 1 \end{pmatrix}$

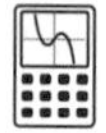

## CALCULATOR BOX

### TI-85

**To obtain $A^t$ on the TI-85,** press

2nd | MATRX | F3 ⟨MATH⟩ | ALPHA | A ⟨T⟩ | F2 | ENTER

### CASIO fx-7700 GB

**To obtain $A^t$ on the CASIO fx-7700 GB,** first display $A$. Then press F2 . $A^t$ will be stored in memory C.

# MATLAB 1.9

*MATLAB Information.* For most applications, to find the transpose of $A$, $A^t$, enter **A′**. Here the ′ is the single quote. If $A$ has complex entries, **A′** will produce the complex conjugate transpose; if you wish to find the transpose of $A$ (without complex conjugation), use **A.′**.

To generate random matrices, refer to the discussion preceding the MATLAB problems for Section 1.6.

1. Generate four pairs, $A$ and $B$, of random matrices such that $AB$ makes sense. Choose some square matrices and some nonsquare matrices. Find $(AB)^t - A^tB^t$ and $(AB)^t - B^tA^t$. Conjecture a formula for $(AB)^t$ in terms of the transposes of $A$ and $B$.
2. Refer back to MATLAB Problem 2 in Section 1.8. For each matrix there, check $A^t$ for invertibility and relate this to the invertibility of $A$. For those matrices for which it makes sense, compare **inv(A′)** with **inv(A)′**.
3. Generate four random square matrices of different sizes.
   a. For each matrix $A$, find **B = A′ + A.** Describe the patterns you notice in the form of these $B$ matrices.
   b. For each matrix $A$, let **C = A′ − A.** Describe the patterns you notice in the form of these $C$ matrices.
   c. Generate four random matrices of different sizes, some square and some nonsquare. For each matrix $F$ generated, find **G = F*F′.** Describe the patterns you notice in the form of these $G$ matrices.
   d. *(Paper and pencil)* Prove your observations from parts (a), (b), and (c) using properties of the transpose.

4. a. *(Paper and pencil)* If $A$ is a matrix with real entries, explain why solving the system $A^t\mathbf{x} = \mathbf{0}$ will produce all real vectors $\mathbf{x}$ such that $\mathbf{x}$ is perpendicular to all the *columns* of $A$.

   b. For each of the matrices $A$ below, find all real vectors $\mathbf{x}$ such that $\mathbf{x}$ is perpendicular to all the *columns* of $A$.

   **i.** $A = \begin{pmatrix} 2 & 0 & 1 \\ 0 & 2 & 1 \\ 1 & 1 & 1 \\ -1 & 1 & 1 \\ 1 & 1 & 1 \end{pmatrix}$ **ii.** $A = \begin{pmatrix} 2 & 4 & 5 \\ 0 & 5 & 7 \\ 7 & 8 & 0 \\ 7 & 0 & 4 \\ 9 & 1 & 1 \end{pmatrix}$

   **iii. A = rand(5,3)**

**PROJECT PROBLEM**

5. **Orthogonal Matrices** Let **A = 2*rand(4) − 1** and let **Q = orth(A).** $Q$ is an example of an *orthogonal* matrix. Orthogonal matrices have special properties which this problem will explore.

   a. Generate a pair of random $4 \times 1$ vectors $\mathbf{x}$ and $\mathbf{y}$. Compute the scalar product of $\mathbf{x}$ and $\mathbf{y}$; call it **s**. Compute the scalar product of $Q\mathbf{x}$ and $Q\mathbf{y}$; call it **r**. Find **s−r** and display by using **format short e.** Repeat for three more choices of $\mathbf{x}$ and $\mathbf{y}$.

   What conjecture can you make from comparing the scalar product of $\mathbf{x}$ with $\mathbf{y}$ to the scalar product of $Q\mathbf{x}$ with $Q\mathbf{y}$?

   b. Test your conjecture formulated in part (a). Generate three orthogonal matrices $Q$ of different sizes (using the **orth** command) and at least two choices of pairs of vectors $\mathbf{x}$ and $\mathbf{y}$ for each $Q$. Make at least one of the $Q$'s a complex matrix. For each $Q$ and pair $\mathbf{x}$ and $\mathbf{y}$, compare the scalar product of $\mathbf{x}$ with $\mathbf{y}$ to the scalar product of $Q\mathbf{x}$ with $Q\mathbf{y}$. Write a description of your process and results.

   c. For each generated $Q$, show that the length of each column of $Q$ equals 1 and that any two different columns of $Q$ are perpendicular to each other. (The length of a vector is given by the square root of the scalar product of a vector with itself: length = **sqrt(x′*x).** Two vectors are perpendicular if the scalar product of the two vectors is equal to zero.)

   d. For each $Q$, explore the relationship among **Q**, **Q′**, and **inv(Q).** Formulate a conjecture about this relationship. Describe your explorations and your thought process. Generate two more random orthogonal matrices of larger sizes and test your conjecture.

   e. *(Paper and pencil)* Use the property conjectured from part (d) (and other known properties) to prove the conjecture of part (b).

   Use the conjecture of part (b) to prove the observation in part (c). [*Hint:* Given a column of $Q$, choose an appropriate vector $\mathbf{x}$ such that $Q\mathbf{x}$ equals the given column.]

## 1.10 ELEMENTARY MATRICES AND MATRIX INVERSES

Let $A$ be an $m \times n$ matrix. Then, as we shall soon see, we can perform elementary row operations on $A$ by multiplying $A$ on the left by an appropriate matrix. The elementary row operations are:

**i.** Multiply the $i$th row by a nonzero number $c$. $R_i \to cR_i$

**ii.** Add a multiple of the $i$th row to the $j$th row. $R_j \to R_j + cR_i$

**iii.** Permute (interchange) the $i$th and $j$th rows. $R_i \rightleftarrows R_j$

**DEFINITION 1** **Elementary Matrix** An $n \times n$ (square) matrix $E$ is called an **elementary matrix** if it can be obtained from the $n \times n$ identity matrix $I_n$ by a *single* elementary row operation.

*Notation.* We denote an elementary matrix by $E$ or by $cR_i$, $R_j + cR_i$, or $P_{ij}$ depending on how the matrix is obtained from $I$. Here $P_{ij}$ is the matrix obtained by permuting the $i$th and $j$th rows of $I$.

**EXAMPLE 1** **Three Elementary Matrices** We obtain three elementary $3 \times 3$ matrices.

**i.** $$\begin{pmatrix} 1 & 0 & 0 \\ 0 & 1 & 0 \\ 0 & 0 & 1 \end{pmatrix} \xrightarrow{R_2 \to 5R_2} \begin{pmatrix} 1 & 0 & 0 \\ 0 & 5 & 0 \\ 0 & 0 & 1 \end{pmatrix} = 5R_2$$ Matrix obtained by multiplying the second row of $I$ by 5

**ii.** $$\begin{pmatrix} 1 & 0 & 0 \\ 0 & 1 & 0 \\ 0 & 0 & 1 \end{pmatrix} \xrightarrow{R_3 \to R_3 - 3R_1} \begin{pmatrix} 1 & 0 & 0 \\ 0 & 1 & 0 \\ -3 & 0 & 1 \end{pmatrix} = R_3 - 3R_1$$ Matrix obtained by multiplying the first row of $I$ by $-3$ and adding it to the third row

**iii.** $$\begin{pmatrix} 1 & 0 & 0 \\ 0 & 1 & 0 \\ 0 & 0 & 1 \end{pmatrix} \xrightarrow{R_2 \rightleftarrows R_3} \begin{pmatrix} 1 & 0 & 0 \\ 0 & 0 & 1 \\ 0 & 1 & 0 \end{pmatrix} = P_{23}$$ Matrix obtained by permuting the second and third rows of $I$

The proof of the following theorem is left as an exercise (see Problems 54–56).

**THEOREM 1** To perform an elementary row operation on a matrix $A$, multiply $A$ on the left by the appropriate elementary matrix.

**EXAMPLE 2** **Performing Elementary Row Operations by Multiplying by Elementary Matrices** Let $A = \begin{pmatrix} 1 & 3 & 2 & 1 \\ 4 & 2 & 3 & -5 \\ 3 & 1 & -2 & 4 \end{pmatrix}$. Perform the following elementary row operations on $A$ by multiplying $A$ on the left by an appropriate elementary matrix.

**i.** Multiply the second row by 5.

**ii.** Multiply the first row by $-3$ and add it to the third row.

**iii.** Permute the second and third rows.

**Solution** Since $A$ is a $3 \times 4$ matrix, each elementary matrix $E$ must be $3 \times 3$ since $E$ must be square and $E$ is multiplying $A$ on the left. We use the results of Example 1.

**i.** $$(5R_2)\,A = \begin{pmatrix} 1 & 0 & 0 \\ 0 & 5 & 0 \\ 0 & 0 & 1 \end{pmatrix}\begin{pmatrix} 1 & 3 & 2 & 1 \\ 4 & 2 & 3 & -5 \\ 3 & 1 & -2 & 4 \end{pmatrix} = \begin{pmatrix} 1 & 3 & 2 & 1 \\ 20 & 10 & 15 & -25 \\ 3 & 1 & -2 & 4 \end{pmatrix}$$

**ii.** $(R_3 - 3R_1)\,A = \begin{pmatrix} 1 & 0 & 0 \\ 0 & 1 & 0 \\ -3 & 0 & 1 \end{pmatrix}\begin{pmatrix} 1 & 3 & 2 & 1 \\ 4 & 2 & 3 & -5 \\ 3 & 1 & -2 & 4 \end{pmatrix} = \begin{pmatrix} 1 & 3 & 2 & 1 \\ 4 & 2 & 3 & -5 \\ 0 & -8 & -8 & 1 \end{pmatrix}$

**iii.** $(P_{23})\,A = \begin{pmatrix} 1 & 0 & 0 \\ 0 & 0 & 1 \\ 0 & 1 & 0 \end{pmatrix}\begin{pmatrix} 1 & 3 & 2 & 1 \\ 4 & 2 & 3 & -5 \\ 3 & 1 & -2 & 4 \end{pmatrix} = \begin{pmatrix} 1 & 3 & 2 & 1 \\ 3 & 1 & -2 & 4 \\ 4 & 2 & 3 & -5 \end{pmatrix}$

Consider the following three products, with $c \neq 0$:

$$\begin{pmatrix} 1 & 0 & 0 \\ 0 & c & 0 \\ 0 & 0 & 1 \end{pmatrix}\begin{pmatrix} 1 & 0 & 0 \\ 0 & 1/c & 0 \\ 0 & 0 & 1 \end{pmatrix} = \begin{pmatrix} 1 & 0 & 0 \\ 0 & 1 & 0 \\ 0 & 0 & 1 \end{pmatrix} \tag{1}$$

$$\begin{pmatrix} 1 & 0 & 0 \\ 0 & 1 & 0 \\ c & 0 & 1 \end{pmatrix}\begin{pmatrix} 1 & 0 & 0 \\ 0 & 1 & 0 \\ -c & 0 & 1 \end{pmatrix} = \begin{pmatrix} 1 & 0 & 0 \\ 0 & 1 & 0 \\ 0 & 0 & 1 \end{pmatrix} \tag{2}$$

$$\begin{pmatrix} 1 & 0 & 0 \\ 0 & 0 & 1 \\ 0 & 1 & 0 \end{pmatrix}\begin{pmatrix} 1 & 0 & 0 \\ 0 & 0 & 1 \\ 0 & 1 & 0 \end{pmatrix} = \begin{pmatrix} 1 & 0 & 0 \\ 0 & 1 & 0 \\ 0 & 0 & 1 \end{pmatrix} \tag{3}$$

Equations (1), (2), and (3) suggest that each elementary matrix is invertible and that its inverse is of the same type (Table 1.4). These facts follow from Theorem 1. Evidently, if the operations $R_j \to R_j + cR_i$ followed by $R_j \to R_j - cR_i$ are performed on the matrix $A$, the matrix $A$ is unchanged. Also, $R_i \to cR_i$ followed by $R_i \to \frac{1}{c}R_i$, and permuting the same two rows twice leave the matrix $A$ unchanged.

We have

$$(cR_i)^{-1} = \frac{1}{c}R_i \tag{4}$$

$$(R_j + cR_i)^{-1} = R_j - cR_i \tag{5}$$

$$(P_{ij})^{-1} = P_{ij} \tag{6}$$

Equation (6) indicates that

> Every elementary permutation matrix is its own inverse.

We summarize our results.

**Table 1.4**

| Elementary Matrix Type $E$ | Effect of Multiplying $A$ on the Left by $E$ | Symbolic Representation of Elementary Row Operation | When Multiplied on the Left, $E^{-1}$ Does the Following | Symbolic Representation of Inverse Operation |
|---|---|---|---|---|
| Multiplication | Multiplies $i$th row of $A$ by $c \neq 0$ | $cR_i$ | Multiplies $i$th row of $A$ by $\frac{1}{c}$ | $\frac{1}{c}R_i$ |
| Addition | Multiplies $i$th row of $A$ by $c$ and adds it to $j$th row | $R_j + cR_i$ | Multiplies $i$th row of $A$ by $-c$ and adds it to $j$th row | $R_j - cR_i$ |
| Permutation | Permutes the $i$th and $j$th rows of $A$ | $P_{ij}$ | Permutes the $i$th and $j$th rows of $A$ | $P_{ij}$ |

**THEOREM 2** Each elementary matrix is invertible. The inverse of an elementary matrix is a matrix of the same type.

*Note.* The inverse of an elementary matrix can be found by inspection. No computation is necessary.

**THEOREM 3** A square matrix is invertible if and only if it is the product of elementary matrices.

**Proof** Let $A = E_1E_2 \cdots E_m$ where each $E_i$ is an elementary matrix. By Theorem 2, each $E_i$ is invertible. Moreover, by Theorem 1.8.3 on page 100 $A$ is invertible† and

$$A^{-1} = E_m^{-1}E_{m-1}^{-1} \cdots E_2^{-1}E_1^{-1}$$

Conversely, suppose that $A$ is invertible. According to Theorem 1.8.6 (the Summing Up Theorem), $A$ is row equivalent to the identity matrix. This means that $A$ can be reduced to $I$ by a finite number, say, $m$, of elementary row operations. By Theorem 1 each such operation is accomplished by multiplying $A$ on the left by an elementary matrix. This means that there are elementary matrices $E_1, E_2, \ldots, E_m$

† Here we have used the generalization of Theorem 1.8.3 to more than two matrices. See, for example, Problem 1.8.16 on page 113.

such that

$$E_m E_{m-1} \cdots E_2 E_1 A = I,$$

Thus from Theorem 1.8.7 on page 112,

$$E_m E_{m-1} \cdots E_2 E_1 = A^{-1}$$

and since each $E_i$ is invertible by Theorem 2,

$$A = (A^{-1})^{-1} = (E_m E_{m-1} \cdots E_2 E_1)^{-1} = E_1^{-1} E_2^{-1} \cdots E_{m-1}^{-1} E_m^{-1} \tag{7}$$

Since the inverse of an elementary matrix is an elementary matrix, we have written $A$ as a product of elementary matrices and the proof is complete. ■

**EXAMPLE 3** **Writing an Invertible Matrix as the Product of Elementary Matrices** Show that the matrix $A = \begin{pmatrix} 2 & 4 & 6 \\ 4 & 5 & 6 \\ 3 & 1 & -2 \end{pmatrix}$ is invertible and write it as a product of elementary matrices.

**Solution** We have encountered this matrix before, first in Example 1.3.1 on page 7. To solve the problem, we reduce $A$ to $I$ and keep track of the elementary row operations. In Example 1.8.6 on page 106 we did reduce $A$ to $I$ by using the following operations:

① $\frac{1}{2}R_1$ ② $R_2 - 4R_1$ ③ $R_3 - 3R_1$ ④ $-\frac{1}{3}R_2$
⑤ $R_1 - 2R_2$ ⑥ $R_3 + 5R_2$ ⑦ $-R_3$ ⑧ $R_1 + R_3$
⑨ $R_2 - 2R_3$

$A^{-1}$ was obtained by starting with $I$ and applying these nine elementary row operations. Thus $A^{-1}$ is the product of nine elementary matrices:

$$A^{-1} = \underset{R_2 - 2R_3}{\begin{pmatrix} 1 & 0 & 0 \\ 0 & 1 & -2 \\ 0 & 0 & 1 \end{pmatrix}} \underset{R_1 + R_3}{\begin{pmatrix} 1 & 0 & 1 \\ 0 & 1 & 0 \\ 0 & 0 & 1 \end{pmatrix}} \underset{-R_3}{\begin{pmatrix} 1 & 0 & 0 \\ 0 & 1 & 0 \\ 0 & 0 & -1 \end{pmatrix}} \underset{R_3 + 5R_2}{\begin{pmatrix} 1 & 0 & 0 \\ 0 & 1 & 0 \\ 0 & 5 & 1 \end{pmatrix}} \underset{R_1 - 2R_2}{\begin{pmatrix} 1 & -2 & 0 \\ 0 & 1 & 0 \\ 0 & 0 & 1 \end{pmatrix}}$$

$$\times \underset{-\frac{1}{3}R_2}{\begin{pmatrix} 1 & 0 & 0 \\ 0 & -\frac{1}{3} & 0 \\ 0 & 0 & 1 \end{pmatrix}} \underset{R_3 - 3R_1}{\begin{pmatrix} 1 & 0 & 0 \\ 0 & 1 & 0 \\ -3 & 0 & 1 \end{pmatrix}} \underset{R_2 - 4R_1}{\begin{pmatrix} 1 & 0 & 0 \\ -4 & 1 & 0 \\ 0 & 0 & 1 \end{pmatrix}} \underset{\frac{1}{2}R_1}{\begin{pmatrix} \frac{1}{2} & 0 & 0 \\ 0 & 1 & 0 \\ 0 & 0 & 1 \end{pmatrix}}$$

Then $A = (A^{-1})^{-1} =$ the product of the inverses of the nine matrices in the opposite order:

$$\begin{pmatrix} 2 & 4 & 6 \\ 4 & 5 & 6 \\ 3 & 1 & -2 \end{pmatrix} = \underset{2R_1}{\begin{pmatrix} 2 & 0 & 0 \\ 0 & 1 & 0 \\ 0 & 0 & 1 \end{pmatrix}} \underset{R_2 + 4R_1}{\begin{pmatrix} 1 & 0 & 0 \\ 4 & 1 & 0 \\ 0 & 0 & 1 \end{pmatrix}} \underset{R_3 + 3R_1}{\begin{pmatrix} 1 & 0 & 0 \\ 0 & 1 & 0 \\ 3 & 0 & 1 \end{pmatrix}} \underset{-3R_2}{\begin{pmatrix} 1 & 0 & 0 \\ 0 & -3 & 0 \\ 0 & 0 & 1 \end{pmatrix}} \underset{R_1 + 2R_2}{\begin{pmatrix} 1 & 2 & 0 \\ 0 & 1 & 0 \\ 0 & 0 & 1 \end{pmatrix}}$$

$$\times \begin{pmatrix} 1 & 0 & 0 \\ 0 & 1 & 0 \\ 0 & -5 & 1 \end{pmatrix} \begin{pmatrix} 1 & 0 & 0 \\ 0 & 1 & 0 \\ 0 & 0 & -1 \end{pmatrix} \begin{pmatrix} 1 & 0 & -1 \\ 0 & 1 & 0 \\ 0 & 0 & 1 \end{pmatrix} \begin{pmatrix} 1 & 0 & 0 \\ 0 & 1 & 2 \\ 0 & 0 & 1 \end{pmatrix}$$
$$\quad R_3 - 5R_2 \qquad -R_3 \qquad R_1 - R_3 \qquad R_2 + 2R_3$$

We can use Theorem 3 to extend our Summing Up Theorem, last seen on page 111.

**THEOREM 4** **Summing Up Theorem—View 3** Let $A$ be an $n \times n$ matrix. Then the following seven statements are equivalent. That is, each one implies the other six (so that if one statement is true, all are true, and if one is false, all are false).

**i.** $A$ is invertible.

**ii.** The only solution to the homogeneous system $A\mathbf{x} = \mathbf{0}$ is the trivial solution ($\mathbf{x} = \mathbf{0}$).

**iii.** The system $A\mathbf{x} = \mathbf{b}$ has a unique solution for every $n$-vector $\mathbf{b}$.

**iv.** $A$ is row equivalent to the $n \times n$ identity matrix $I_n$; that is, the reduced row echelon form of $A$ is $I_n$.

**v.** $A$ can be written as a product of elementary matrices.

**vi.** The row echelon form of $A$ has $n$ pivots.

**vii.** $\det A \neq 0$ (so far, $\det A$ is defined only if $A$ is a $2 \times 2$ matrix).

There is one further result that will prove very useful in Section 2.3. First, we need a definition (given earlier in Problem 1.8.29 on page 114).

**DEFINITION 2** **Upper Triangular Matrix and Lower Triangular Matrix** A square matrix is called **upper (lower) triangular** if all its components below (above) the main diagonal are zero.

*Note.* $a_{ij}$ is below the main diagonal if $i > j$.

**EXAMPLE 4** **Two Upper Triangular and Two Lower Triangular Matrices** Matrices $U$ and $V$ are upper triangular while matrices $L$ and $M$ are lower triangular:

$$U = \begin{pmatrix} 2 & -3 & 5 \\ 0 & 1 & 6 \\ 0 & 0 & 2 \end{pmatrix} \qquad V = \begin{pmatrix} 1 & 5 \\ 0 & -2 \end{pmatrix}$$

$$L = \begin{pmatrix} 0 & 0 \\ 5 & 1 \end{pmatrix} \qquad M = \begin{pmatrix} 2 & 0 & 0 & 0 \\ -5 & 4 & 0 & 0 \\ 6 & 1 & 2 & 0 \\ 3 & 0 & 1 & 5 \end{pmatrix}$$

**THEOREM 5** Let $A$ be a square matrix. Then $A$ can be written as a product of elementary matrices and an upper triangular matrix $U$. In the product the elementary matrices are on the left and the upper triangular matrix is on the right.

**Proof** Gaussian elimination to solve the system $A\mathbf{x} = \mathbf{b}$ results in an upper triangular matrix. To see this, observe that Gaussian elimination will terminate when the matrix is in row echelon form—and the row echelon form of a square matrix is upper triangular. We denote the row echelon form of $A$ by $U$. Then $A$ is reduced to $U$ by a sequence of elementary row operations each of which can be obtained by multiplication by an elementary matrix. Thus

$$U = E_m E_{m-1} \cdots E_2 E_1 A$$

and

$$A = E_1^{-1} E_2^{-1} \cdots E_{m-1}^{-1} E_m^{-1} U$$

Since the inverse of an elementary matrix is an elementary matrix, we have written $A$ as the product of elementary matrices and $U$.

**EXAMPLE 5** **Writing a Matrix as the Product of Elementary Matrices and an Upper Triangular Matrix** Write the matrix

$$A = \begin{pmatrix} 3 & 6 & 9 \\ 2 & 5 & 1 \\ 1 & 1 & 8 \end{pmatrix}$$

as the product of elementary matrices and an upper triangular matrix.

**Solution** We row-reduce $A$ to obtain its row echelon form:

$$\begin{pmatrix} 3 & 6 & 9 \\ 2 & 5 & 1 \\ 1 & 1 & 8 \end{pmatrix} \xrightarrow{R_1 \to \frac{1}{3}R_1} \begin{pmatrix} 1 & 2 & 3 \\ 2 & 5 & 1 \\ 1 & 1 & 8 \end{pmatrix}$$

$$\xrightarrow[R_3 \to R_3 - R_1]{R_2 \to R_2 - 2R_1} \begin{pmatrix} 1 & 2 & 3 \\ 0 & 1 & -5 \\ 0 & -1 & 5 \end{pmatrix} \xrightarrow{R_3 \to R_3 + R_2} \begin{pmatrix} 1 & 2 & 3 \\ 0 & 1 & -5 \\ 0 & 0 & 0 \end{pmatrix} = U$$

Then working backward, we see that

$$U = \begin{pmatrix} 1 & 2 & 3 \\ 0 & 1 & -5 \\ 0 & 0 & 0 \end{pmatrix} = \underset{R_3 + R_2}{\begin{pmatrix} 1 & 0 & 0 \\ 0 & 1 & 0 \\ 0 & 1 & 1 \end{pmatrix}} \underset{R_3 - R_1}{\begin{pmatrix} 1 & 0 & 0 \\ 0 & 1 & 0 \\ -1 & 0 & 1 \end{pmatrix}}$$

$$\times \underset{R_2 - 2R_1}{\begin{pmatrix} 1 & 0 & 0 \\ -2 & 1 & 0 \\ 0 & 0 & 1 \end{pmatrix}} \underset{\frac{1}{3}R_1}{\begin{pmatrix} \frac{1}{3} & 0 & 0 \\ 0 & 1 & 0 \\ 0 & 0 & 1 \end{pmatrix}} \underset{A}{\begin{pmatrix} 3 & 6 & 9 \\ 2 & 5 & 1 \\ 1 & 1 & 8 \end{pmatrix}}$$

and, taking inverses of the four elementary matrices, we obtain

$$A = \begin{pmatrix} 3 & 6 & 9 \\ 2 & 5 & 1 \\ 1 & 1 & 8 \end{pmatrix} = \underset{3R_1}{\begin{pmatrix} 3 & 0 & 0 \\ 0 & 1 & 0 \\ 0 & 0 & 1 \end{pmatrix}} \underset{R_2 + 2R_1}{\begin{pmatrix} 1 & 0 & 0 \\ 2 & 1 & 0 \\ 0 & 0 & 1 \end{pmatrix}}$$

$$\times \underset{R_3 + R_1}{\begin{pmatrix} 1 & 0 & 0 \\ 0 & 1 & 0 \\ 1 & 0 & 1 \end{pmatrix}} \underset{R_3 - R_2}{\begin{pmatrix} 1 & 0 & 0 \\ 0 & 1 & 0 \\ 0 & -1 & 1 \end{pmatrix}} \underset{U}{\begin{pmatrix} 1 & 2 & 3 \\ 0 & 1 & -5 \\ 0 & 0 & 0 \end{pmatrix}}$$

## PROBLEMS 1.10

### Self-Quiz

#### *True-False*

**I.** The product of two elementary matrices is an elementary matrix.

**II.** The inverse of an elementary matrix is an elementary matrix.

**III.** Every matrix can be written as the product of elementary matrices.

**IV.** Every square matrix can be written as the product of elementary matrices.

**V.** Every invertible matrix can be written as the product of elementary matrices.

**VI.** Every square matrix can be written as the product of elementary matrices and an upper triangular matrix.

#### *Multiple Choice*

**VII.** The inverse of $\begin{pmatrix} 1 & 0 & 0 \\ 0 & 1 & 0 \\ 0 & 3 & 1 \end{pmatrix}$ is ________.

**a.** $\begin{pmatrix} 1 & 0 & 0 \\ 0 & 1 & 0 \\ 0 & -3 & 1 \end{pmatrix}$ **b.** $\begin{pmatrix} 1 & 0 & 0 \\ 0 & 1 & 0 \\ 0 & \frac{1}{3} & 1 \end{pmatrix}$ **c.** $\begin{pmatrix} 1 & -3 & 0 \\ 0 & 1 & 0 \\ 0 & 0 & 1 \end{pmatrix}$ **d.** $\begin{pmatrix} 1 & 0 & 0 \\ 0 & 1 & 0 \\ 0 & 3 & 1 \end{pmatrix}$

**VIII.** The inverse of $\begin{pmatrix} 1 & 0 & 0 \\ 0 & 1 & 0 \\ 0 & 0 & 4 \end{pmatrix}$ is ______.

**a.** $\begin{pmatrix} 1 & 0 & 0 \\ 0 & 1 & 0 \\ 0 & 0 & -4 \end{pmatrix}$ **b.** $\begin{pmatrix} 1 & 0 & 0 \\ 0 & 1 & 0 \\ 0 & 0 & \frac{1}{4} \end{pmatrix}$ **c.** $\begin{pmatrix} \frac{1}{4} & 0 & 0 \\ 0 & 1 & 0 \\ 0 & 0 & 1 \end{pmatrix}$ **d.** $\begin{pmatrix} 1 & 0 & 0 \\ 0 & 1 & 0 \\ 0 & 0 & 4 \end{pmatrix}$

**IX.** The inverse of $\begin{pmatrix} 0 & 1 & 0 \\ 1 & 0 & 0 \\ 0 & 0 & 1 \end{pmatrix}$ is ______.

**a.** $\begin{pmatrix} 0 & 1 & 0 \\ 1 & 0 & 0 \\ 0 & 0 & -1 \end{pmatrix}$ **b.** $\begin{pmatrix} 0 & -1 & 0 \\ -1 & 0 & 0 \\ 0 & 0 & -1 \end{pmatrix}$ **c.** $\begin{pmatrix} 1 & 0 & 0 \\ 0 & 0 & 1 \\ 0 & 1 & 0 \end{pmatrix}$ **d.** $\begin{pmatrix} 0 & 1 & 0 \\ 1 & 0 & 0 \\ 0 & 0 & 1 \end{pmatrix}$

In Problems 1–12 determine which matrices **are** elementary matrices.

**1.** $\begin{pmatrix} 0 & 1 \\ 1 & 0 \end{pmatrix}$ **2.** $\begin{pmatrix} 1 & 0 \\ 1 & 1 \end{pmatrix}$ **3.** $\begin{pmatrix} 0 & 1 \\ 1 & 1 \end{pmatrix}$

**4.** $\begin{pmatrix} 1 & 0 \\ 0 & 2 \end{pmatrix}$ **5.** $\begin{pmatrix} 3 & 0 \\ 0 & 3 \end{pmatrix}$ **6.** $\begin{pmatrix} 0 & 1 & 0 \\ 1 & 0 & 0 \\ 0 & 0 & 1 \end{pmatrix}$

**7.** $\begin{pmatrix} 0 & 1 & 0 \\ 0 & 0 & 1 \\ 1 & 0 & 0 \end{pmatrix}$ **8.** $\begin{pmatrix} 1 & 0 & 0 \\ 2 & 1 & 0 \\ 3 & 0 & 1 \end{pmatrix}$ **9.** $\begin{pmatrix} 1 & 0 & 0 \\ 2 & 1 & 0 \\ 0 & 0 & 1 \end{pmatrix}$

**10.** $\begin{pmatrix} 1 & 0 & 0 & 0 \\ 0 & 1 & 0 & 0 \\ 0 & 0 & 1 & 0 \\ 0 & 1 & 0 & 1 \end{pmatrix}$ **11.** $\begin{pmatrix} 1 & 0 & 0 & 0 \\ 1 & 1 & 0 & 0 \\ 0 & 0 & 1 & 0 \\ 0 & 0 & 1 & 1 \end{pmatrix}$ **12.** $\begin{pmatrix} 1 & -1 & 0 & 0 \\ 0 & 1 & 0 & 0 \\ 0 & 0 & 1 & 0 \\ 0 & 0 & 0 & 1 \end{pmatrix}$

In Problems 13–20 write the $3 \times 3$ elementary matrix that carries out the given row operation on a $3 \times 5$ matrix $A$ by left multiplication.

**13.** $R_2 \to 4R_2$ **14.** $R_2 \to R_2 + 2R_1$ **15.** $R_1 \to R_1 - 3R_2$

---

### Answers to Self-Quiz

**I.** False **II.** True **III.** False **IV.** False **V.** True **VI.** True **VII.** a
**VIII.** b **IX.** d

**16.** $R_1 \to R_1 + 4R_3$ **17.** $R_1 \rightleftarrows R_3$ **18.** $R_2 \rightleftarrows R_3$

**19.** $R_2 \to R_2 + R_3$ **20.** $R_3 \to -R_3$

In Problems 21–30 find the elementary matrix $E$ such that $EA = B$.

**21.** $A = \begin{pmatrix} 2 & 3 \\ -1 & 4 \end{pmatrix}$, $B = \begin{pmatrix} 2 & 3 \\ 2 & -8 \end{pmatrix}$

**22.** $A = \begin{pmatrix} 2 & 3 \\ -1 & 4 \end{pmatrix}$, $B = \begin{pmatrix} 2 & 3 \\ -5 & -2 \end{pmatrix}$

**23.** $A = \begin{pmatrix} 2 & 3 \\ -1 & 4 \end{pmatrix}$, $B = \begin{pmatrix} 0 & 11 \\ -1 & 4 \end{pmatrix}$

**24.** $A = \begin{pmatrix} 2 & 3 \\ -1 & 4 \end{pmatrix}$, $B = \begin{pmatrix} -1 & 4 \\ 2 & 3 \end{pmatrix}$

**25.** $A = \begin{pmatrix} 1 & 2 \\ 3 & 4 \\ 5 & 6 \end{pmatrix}$, $B = \begin{pmatrix} 5 & 6 \\ 3 & 4 \\ 1 & 2 \end{pmatrix}$

**26.** $A = \begin{pmatrix} 1 & 2 \\ 3 & 4 \\ 5 & 6 \end{pmatrix}$, $B = \begin{pmatrix} 1 & 2 \\ 0 & -2 \\ 5 & 6 \end{pmatrix}$

**27.** $A = \begin{pmatrix} 1 & 2 \\ 3 & 4 \\ 5 & 6 \end{pmatrix}$, $B = \begin{pmatrix} -1 & -2 \\ 3 & 4 \\ 5 & 6 \end{pmatrix}$

**28.** $A = \begin{pmatrix} 1 & 2 \\ 3 & 4 \\ 5 & 6 \end{pmatrix}$, $B = \begin{pmatrix} -5 & -6 \\ 3 & 4 \\ 5 & 6 \end{pmatrix}$

**29.** $A = \begin{pmatrix} 1 & 2 & 5 & 2 \\ 0 & -1 & 3 & 4 \\ 5 & 0 & -2 & 7 \end{pmatrix}$, $B = \begin{pmatrix} 1 & 2 & 5 & 2 \\ 0 & -1 & 3 & 4 \\ 0 & -10 & -27 & -3 \end{pmatrix}$

**30.** $A = \begin{pmatrix} 1 & 2 & 5 & 2 \\ 0 & -1 & 3 & 4 \\ 5 & 0 & -2 & 7 \end{pmatrix}$, $B = \begin{pmatrix} 1 & 0 & 11 & 10 \\ 0 & -1 & 3 & 4 \\ 5 & 0 & -2 & 7 \end{pmatrix}$

In Problems 31–40 find the inverse of the given elementary matrix.

**31.** $\begin{pmatrix} 0 & 1 \\ 1 & 0 \end{pmatrix}$ **32.** $\begin{pmatrix} 1 & 3 \\ 0 & 1 \end{pmatrix}$ **33.** $\begin{pmatrix} 1 & 0 \\ 0 & 4 \end{pmatrix}$

**34.** $\begin{pmatrix} 0 & 1 & 0 \\ 1 & 0 & 0 \\ 0 & 0 & 1 \end{pmatrix}$ **35.** $\begin{pmatrix} 1 & -2 & 0 \\ 0 & 1 & 0 \\ 0 & 0 & 1 \end{pmatrix}$ **36.** $\begin{pmatrix} 1 & 0 & 0 \\ 0 & 1 & 0 \\ -2 & 0 & 1 \end{pmatrix}$

**37.** $\begin{pmatrix} 1 & 0 & 0 \\ 0 & -\frac{1}{2} & 0 \\ 0 & 0 & 1 \end{pmatrix}$ **38.** $\begin{pmatrix} 1 & 0 & 1 & 0 \\ 0 & 1 & 0 & 0 \\ 0 & 0 & 1 & 0 \\ 0 & 0 & 0 & 1 \end{pmatrix}$ **39.** $\begin{pmatrix} 1 & 0 & 0 & 5 \\ 0 & 1 & 0 & 0 \\ 0 & 0 & 1 & 0 \\ 0 & 0 & 0 & 1 \end{pmatrix}$

**40.** $\begin{pmatrix} 1 & 0 & 0 & 0 \\ 0 & 1 & 0 & 0 \\ 0 & -3 & 1 & 0 \\ 0 & 0 & 0 & 1 \end{pmatrix}$

In Problems 41–48 show that each matrix is invertible and write it as a product of elementary matrices.

**41.** $\begin{pmatrix} 2 & 1 \\ 3 & 2 \end{pmatrix}$ **42.** $\begin{pmatrix} 1 & 2 \\ 3 & 4 \end{pmatrix}$ **43.** $\begin{pmatrix} 1 & 1 & 1 \\ 0 & 2 & 3 \\ 5 & 5 & 1 \end{pmatrix}$

**44.** $\begin{pmatrix} 3 & 2 & 1 \\ 0 & 2 & 2 \\ 0 & 0 & -1 \end{pmatrix}$

**45.** $\begin{pmatrix} 0 & -1 & 0 \\ 0 & 1 & -1 \\ 1 & 0 & 1 \end{pmatrix}$

**46.** $\begin{pmatrix} 2 & 0 & 4 \\ 0 & 1 & 1 \\ 3 & -1 & 1 \end{pmatrix}$

**47.** $\begin{pmatrix} 2 & 0 & 0 & 0 \\ 0 & 3 & 0 & 0 \\ 0 & 0 & -4 & 0 \\ 0 & 0 & 0 & 5 \end{pmatrix}$

**48.** $\begin{pmatrix} 2 & 1 & 0 & 0 \\ 0 & 2 & 1 & 0 \\ 0 & 0 & 2 & 1 \\ 0 & 0 & 0 & 2 \end{pmatrix}$

**49.** Let $A = \begin{pmatrix} a & b \\ 0 & c \end{pmatrix}$ where $ac \neq 0$. Write $A$ as a product of three elementary matrices and conclude that $A$ is invertible.

**50.** Let $A = \begin{pmatrix} a & b & c \\ 0 & d & e \\ 0 & 0 & f \end{pmatrix}$ where $adf \neq 0$. Write $A$ as a product of six elementary matrices and conclude that $A$ is invertible.

***51.** Let $A$ be an $n \times n$ upper triangular matrix. Prove that if each diagonal component of $A$ is nonzero, then $A$ is invertible [*Hint:* Look at Problems 49 and 50].

***52.** Show that if $A$ is an $n \times n$ upper triangular matrix with nonzero diagonal components, then $A^{-1}$ is upper triangular.

***53.** Use Theorem 1.9.1(*iv*) on page 122 and the result of Problem 52 to show that if $A$ is an $n \times n$ lower triangular matrix with nonzero diagonal components, then $A$ is invertible and $A^{-1}$ is lower triangular.

**54.** Show that if $P_{ij}$ is the $n \times n$ matrix obtained by permuting the $i$th and $j$th rows of $I_n$, then $P_{ij}A$ is the matrix obtained from $A$ by permuting its $i$th and $j$th rows.

**55.** Let $A_{ij}$ be the matrix with $c$ in the $ji$th position, 1's down the diagonal, and 0's everywhere else. Show that $A_{ij}A$ is the matrix obtained from $A$ by multiplying the $i$th row of $A$ by $c$ and adding it to the $j$th row.

**56.** Let $M_i$ be the matrix with $c$ in the $ii$ position, 1's in the other diagonal positions, and 0's everywhere else. Show that $M_iA$ is the matrix obtained from $A$ by multiplying the $i$th row of $A$ by $c$.

In Problems 57–62 write each square matrix as a product of elementary matrices and an upper triangular matrix.

**57.** $A = \begin{pmatrix} 1 & 2 \\ 2 & 4 \end{pmatrix}$

**58.** $A = \begin{pmatrix} 2 & -3 \\ -4 & 6 \end{pmatrix}$

**59.** $A = \begin{pmatrix} 0 & 0 \\ 1 & 0 \end{pmatrix}$

**60.** $A = \begin{pmatrix} 1 & -1 & 2 \\ 2 & 1 & 4 \\ 4 & -1 & 8 \end{pmatrix}$

**61.** $A = \begin{pmatrix} 1 & -3 & 3 \\ 0 & -3 & 1 \\ 1 & 0 & 2 \end{pmatrix}$

**62.** $A = \begin{pmatrix} 1 & 0 & 0 \\ 2 & 3 & 0 \\ -1 & 4 & 0 \end{pmatrix}$

## MATLAB 1.10

1. This problem explores the form of elementary matrices. Note that each elementary matrix can be obtained from the identity with one change. For example,

$$F = \begin{pmatrix} 1 & 0 & 0 \\ 0 & c & 0 \\ 0 & 0 & 1 \end{pmatrix} \quad \text{is the identity with } F(2,2) = c$$

In MATLAB, **F = eye(3); F(2,2) = c**

$$F = \begin{pmatrix} 1 & 0 & 0 \\ 0 & 1 & 0 \\ 0 & c & 1 \end{pmatrix} \quad \text{is the identity with } F(3,2) = c$$

In MATLAB, **F = eye(3); F(3,2) = c**

$$F = \begin{pmatrix} 1 & 0 & 0 \\ 0 & 0 & 1 \\ 0 & 1 & 0 \end{pmatrix} \quad \text{is the identity with row 2 and row 3 interchanged}$$

In MATLAB, **F = eye(3); F([2 3],:) = F([3 2],:)**

a. Enter **A = round(10*(2*rand(4)−1))**. In the manner described above, enter the matrices $F$ that represent the row operations below. Find **F*A** to test that $F$ performs the desired row operation.

**i.** $R_3 \to 4R_3$ **ii.** $R_1 \to R_1 - 3R_2$ **iii.** interchange $R_1$ and $R_4$

b. Find **inv(F)** for each $F$ from (a). For each $F$, explain why **inv(F)** is an elementary matrix and describe what row operation it represents. How is this row operation the "inverse" of the original row operation?

2. Given a matrix, we want to reduce to reduced row echelon form by multiplying by elementary matrices, keeping track of the product of the elementary matrices in the order in which they are used. For accuracy, you will want to compute your multipliers using matrix notation. (See MATLAB Problem 1 in Section 1.5 for computing multipliers and see MATLAB Problem 1 in this section for forming elementary matrices.)

a. Let $A = \begin{pmatrix} 7 & 2 & 3 \\ -1 & 0 & 4 \\ 2 & 1 & 1 \end{pmatrix}$

Enter the matrix above and store it in $A$. Enter **B = A.** This puts a copy of $A$ into $B$. We can then reduce $B$ so that $B$ will contain **rref(A)** and $A$ will still contain the original matrix.

**c = −B(2,1)/B(1,1)**

**F1 = eye(3); F1(2,1) = c**

**B = F1*B**

**F = F1**

**c = −B(3,1)/B(1,1)**

form **F2** with $c$ in the correct position

**B = F2*B**

**F = F2*F**

Continue in this fashion until **B** is in reduced row echelon form. If any pivot entry is zero, you will need to perform a row interchange by multiplying by an appropriate elementary matrix.

b. Find **F*A** and **A*F,** where $F$ is the product of the elementary matrices used and $A$ is the original matrix. What does this tell you about the relationship between $F$ and $A$? (Justify your answer.)

c. Find $D = F1^{-1}{*}F2^{-1}{*}\cdots{*}Fm^{-1}$, where $F1$ is the first elementary matrix used and $Fm$ is the last elementary matrix used. What is the relationship between $D$ and $A$? (Justify your answer.)

d. Repeat parts (a) through (c) for $A = \begin{pmatrix} 0 & 2 & 3 \\ 1 & 1 & 4 \\ 2 & 4 & 1 \end{pmatrix}$.

3. a. Let $A = \begin{pmatrix} 1 & 2 & 3 \\ 1 & 1 & 7 \\ 2 & 4 & 5 \end{pmatrix}$.

Perform row operations using multiplication by elementary matrices as described in MATLAB Problem 1 in this section, keeping track of the products of the elementary matrices but only performing row operations of the form $R_j \to R_j + cR_i$ until $A$ is reduced to upper triangular form. (Do not create 1's in the pivot positions.) Give each elementary matrix its own variable name and display each elementary matrix used and its inverse. Let the upper triangular form, which is the end result, be called $U$ and the product of all the elementary matrices used be called $F$.

b. Find $L = F1^{-1}{*}F2^{-1}{*}\cdots{*}Fm^{-1}$, where $F1$ is the first elementary matrix used and $Fm$ is the last elementary matrix used. What can you say about the form of $L$? What observations can you make about the entries of $L$, the entries of the elementary matrices, and the entries of the inverses of the elementary matrices? (Discuss the entries *and* their positions.)

Explain why we can say that $L$ holds all the information needed to reduce $A$ to upper triangular form.

c. Verify that $LU = A$. (Be sure that $A$ is the original matrix. Recall that $U$ is the end result of your reduction.) Prove that this should be true.

d. Repeat parts (a) through (c) for $A = \begin{pmatrix} 6 & 2 & 7 & 3 \\ 8 & 10 & 1 & 4 \\ 10 & 7 & 6 & 8 \\ 4 & 8 & 9 & 5 \end{pmatrix}$.

## 1.11 LU-FACTORIZATIONS OF A MATRIX

In this section we show how to write a square matrix as a product of triangular matrices. This factorization is useful to solve linear systems on a computer and can be used to prove important results about matrices.

In Section 1.3 we discussed **Gaussian elimination.** In that process we row-reduced a matrix to its row echelon form. Recall that the row echelon form of a square matrix is an upper triangular matrix with 1's or 0's down the main diagonal.

For example, the row echelon form of a 3 × 3 matrix looks like one of the following:

$$\begin{pmatrix}1&x&x\\0&1&x\\0&0&1\end{pmatrix} \text{ or } \begin{pmatrix}1&x&x\\0&1&x\\0&0&0\end{pmatrix} \text{ or } \begin{pmatrix}1&x&x\\0&0&1\\0&0&0\end{pmatrix} \text{ or } \begin{pmatrix}1&x&x\\0&0&0\\0&0&0\end{pmatrix} \text{ or }$$

$$\begin{pmatrix}0&1&x\\0&0&1\\0&0&0\end{pmatrix} \text{ or } \begin{pmatrix}0&0&1\\0&0&0\\0&0&0\end{pmatrix} \text{ or } \begin{pmatrix}0&0&0\\0&0&0\\0&0&0\end{pmatrix}$$

For our purposes in this section we want instead to row-reduce a matrix to an upper triangular matrix where the nonzero numbers on the diagonal are not necessarily 1's. We can do this simply by not insisting that each pivot be made equal to one.

**EXAMPLE 1** **Finding an LU-Factorization of a Matrix $A$** Row-reduce the matrix $A = \begin{pmatrix}2&3&2&4\\4&10&-4&0\\-3&-2&-5&-2\\-2&4&4&-7\end{pmatrix}$ into an upper triangular matrix and then write $A$ as a product of a lower triangular and an upper triangular matrix.

**Solution** We proceed as before; only this time we do not divide each diagonal component (pivot) by itself:

$$\begin{pmatrix}2&3&2&4\\4&10&-4&0\\-3&-2&-5&-2\\-2&4&4&-7\end{pmatrix} \xrightarrow[\substack{R_3 \to R_3 + \frac{3}{2}R_1\\ R_4 \to R_4 + R_1}]{R_2 \to R_2 - 2R_1} \begin{pmatrix}2&3&2&4\\0&4&-8&-8\\0&\frac{5}{2}&-2&4\\0&7&6&-3\end{pmatrix} \xrightarrow[R_4 \to R_4 - \frac{7}{4}R_2]{R_3 \to R_3 - \frac{5}{8}R_2} \begin{pmatrix}2&3&2&4\\0&4&-8&-8\\0&0&3&9\\0&0&20&11\end{pmatrix}$$

$$\xrightarrow{R_4 \to R_4 - \frac{20}{3}R_3} \begin{pmatrix}2&3&2&4\\0&4&-8&-8\\0&0&3&9\\0&0&0&-49\end{pmatrix} = U$$

Using elementary matrices as in Example 1.10.5 on page 133, we may write

$$U = \begin{pmatrix}1&0&0&0\\0&1&0&0\\0&0&1&0\\0&0&-\frac{20}{3}&1\end{pmatrix}\begin{pmatrix}1&0&0&0\\0&1&0&0\\0&0&1&0\\0&-\frac{7}{4}&0&1\end{pmatrix}\begin{pmatrix}1&0&0&0\\0&1&0&0\\0&-\frac{5}{8}&1&0\\0&0&0&1\end{pmatrix}$$

$$\times \begin{pmatrix}1&0&0&0\\0&1&0&0\\0&0&1&0\\1&0&0&1\end{pmatrix}\begin{pmatrix}1&0&0&0\\0&1&0&0\\\frac{3}{2}&0&1&0\\0&0&0&1\end{pmatrix}\begin{pmatrix}1&0&0&0\\-2&1&0&0\\0&0&1&0\\0&0&0&1\end{pmatrix}A$$

or

$$A = \begin{pmatrix} 1 & 0 & 0 & 0 \\ 2 & 1 & 0 & 0 \\ 0 & 0 & 1 & 0 \\ 0 & 0 & 0 & 1 \end{pmatrix} \begin{pmatrix} 1 & 0 & 0 & 0 \\ 0 & 1 & 0 & 0 \\ -\frac{3}{2} & 0 & 1 & 0 \\ 0 & 0 & 0 & 1 \end{pmatrix} \begin{pmatrix} 1 & 0 & 0 & 0 \\ 0 & 1 & 0 & 0 \\ 0 & 0 & 1 & 0 \\ -1 & 0 & 0 & 1 \end{pmatrix}$$

$$\times \begin{pmatrix} 1 & 0 & 0 & 0 \\ 0 & 1 & 0 & 0 \\ 0 & \frac{5}{8} & 1 & 0 \\ 0 & 0 & 0 & 1 \end{pmatrix} \begin{pmatrix} 1 & 0 & 0 & 0 \\ 0 & 1 & 0 & 0 \\ 0 & 0 & 1 & 0 \\ 0 & \frac{7}{4} & 0 & 1 \end{pmatrix} \begin{pmatrix} 1 & 0 & 0 & 0 \\ 0 & 1 & 0 & 0 \\ 0 & 0 & 1 & 0 \\ 0 & 0 & \frac{20}{3} & 1 \end{pmatrix} U$$

We have written $A$ as a product of six elementary matrices and an upper triangular matrix. Let $L$ denote the product of the elementary matrices. You should verify that

$$L = \begin{pmatrix} 1 & 0 & 0 & 0 \\ 2 & 1 & 0 & 0 \\ -\frac{3}{2} & \frac{5}{8} & 1 & 0 \\ -1 & \frac{7}{4} & \frac{20}{3} & 1 \end{pmatrix},$$

which is a lower triangular matrix with 1's down the diagonal. We can then write $A = LU$, where $L$ is lower triangular and $U$ is upper triangular. The diagonal components of $L$ are all equal to 1 and the diagonal components of $U$ are its pivots. This factorization is called the **LU-factorization of *A*.**

The procedure used in Example 1 can be carried out as long as no permutations are required in order to row-reduce $A$ into a triangular form. This is not always the case. For example, the first step in the row reduction of

$$\begin{pmatrix} 0 & 2 & 3 \\ 2 & -4 & 7 \\ 1 & -2 & 5 \end{pmatrix}$$

is to permute (interchange) the first and second rows or the first and third rows.

Let us assume for a moment that no such permutation is necessary. Then, as in Example 1, we may write $A = E_1 E_2 \cdots E_n U$, where $U$ is an upper triangular matrix and each elementary matrix is a lower triangular matrix with 1's on the diagonal. This follows from the fact that each $E$ is of the form $R_j + cR_i$. (There are no permutations and no multiplications of rows by constants.) Moreover, the numbers that are made zero in the row reduction are always *below* the diagonal so that in $R_j + cR_i$ it is always the case that $j > i$. Thus the $c$'s appear below the diagonal. The proof of the following theorem is not difficult (see Problems 24 and 25).

**THEOREM 1** The product of lower triangular matrices with 1's on the diagonal is a lower triangular matrix with 1's on the diagonal. Moreover, the product of two upper triangular matrices is upper triangular.

**THEOREM 2** **LU-Factorization Theorem** Let $A$ be a square $(n \times n)$ matrix and suppose that $A$ can be row-reduced to an upper triangular matrix $U$ without performing any permutations on the rows of $A$. Then there exists an invertible lower triangular matrix $L$ with 1's on the diagonal such that $A = LU$. If in addition $U$ has $n$ pivots (that is, $A$ is invertible), then this factorization is unique.

**Proof** $U$ and $L$ are obtained as in Example 1. We need only prove uniqueness in the case $A$ is invertible. Since $U$ has $n$ pivots, its row echelon form also has $n$ pivots (to see this, divide each row in $U$ by the pivot in that row). Then by the Summing Up Theorem on page 132, $U$ is invertible.

To show that $L$ is invertible, consider the equation $L\mathbf{x} = \mathbf{0}$:

$$\begin{pmatrix} 1 & 0 & \cdots & 0 \\ a & 1 & \cdots & 0 \\ \vdots & \vdots & & \vdots \\ a & a & \cdots & 1 \end{pmatrix} \begin{pmatrix} x_1 \\ x_2 \\ \vdots \\ x_n \end{pmatrix} = \begin{pmatrix} 0 \\ 0 \\ \vdots \\ 0 \end{pmatrix}$$

It follows that $x_1 = 0$, $ax_1 + x_2 = 0$, and so on which shows that $x_1 = x_2 = \cdots = x_n = 0$, and $L$ is invertible by the Summing Up Theorem. To show uniqueness, suppose that $A = L_1U_1 = L_2U_2$. Then

$$U_1U_2^{-1} = (L_1^{-1}L_1)(U_1U_2^{-1}) = L_1^{-1}(L_1U_1)U_2^{-1} = L_1^{-1}(L_2U_2)U_2^{-1} = (L_1^{-1}L_2)(U_2U_2^{-1}) = L_1^{-1}L_2$$

From the result of Problem 1.8.30 on page 114, $U_2^{-1}$ is upper triangular and $L_1^{-1}$ is lower triangular. Moreover, according to Theorem 1, $L_1^{-1}L_2$ is a lower triangular matrix with 1's on the diagonal while $U_1U_2^{-1}$ is upper triangular. The only way an upper and a lower triangular matrix can be equal is if they are both diagonal. Since $L_1^{-1}L_2$ has 1's down the diagonal, we see that

$$U_1U_2^{-1} = L_1^{-1}L_2 = I$$

from which it follows that $U_1 = U_2$ and $L_1 = L_2$. ■

## Using the *LU*-Factorization to Solve a System of Equations

Suppose we wish to solve the system $A\mathbf{x} = \mathbf{b}$, where $A$ is invertible. If $A$ satisfies the hypotheses of Theorem 2, we may write

$$LU\mathbf{x} = \mathbf{b}$$

Because $L$ is invertible, there exists a unique vector $\mathbf{y}$ such that $L\mathbf{y} = \mathbf{b}$. Because $U$ is also invertible, there exists a unique vector $\mathbf{x}$ such that $U\mathbf{x} = \mathbf{y}$. Then $A\mathbf{x} = L(U\mathbf{x}) = L\mathbf{y} = \mathbf{b}$ and our system is solved. Note that $L\mathbf{y} = \mathbf{b}$ can be solved directly by **forward substitution** and $U\mathbf{x} = \mathbf{b}$ can be solved directly by **back substitution.** We illustrate this by an example.

**EXAMPLE 2** **Using the *LU*-Factorization to Solve a System** Solve the system $A\mathbf{x} = \mathbf{b}$, where

$$A = \begin{pmatrix} 2 & 3 & 2 & 4 \\ 4 & 10 & -4 & 0 \\ -3 & -2 & -5 & -2 \\ -2 & 4 & 4 & -7 \end{pmatrix} \quad \text{and} \quad \mathbf{b} = \begin{pmatrix} 4 \\ -8 \\ -4 \\ -1 \end{pmatrix}$$

**Solution** From Example 1 we may write $A = LU$, where

$$L = \begin{pmatrix} 1 & 0 & 0 & 0 \\ 2 & 1 & 0 & 0 \\ -\frac{3}{2} & \frac{5}{8} & 1 & 0 \\ -1 & \frac{7}{4} & \frac{20}{3} & 1 \end{pmatrix} \quad \text{and} \quad U = \begin{pmatrix} 2 & 3 & 2 & 4 \\ 0 & 4 & -8 & -8 \\ 0 & 0 & 3 & 9 \\ 0 & 0 & 0 & -49 \end{pmatrix}$$

The system $L\mathbf{y} = \mathbf{b}$ yields the equations

$$\begin{aligned} y_1 \qquad\qquad\qquad &= 4 \\ 2y_1 + y_2 \qquad\qquad &= -8 \\ -\tfrac{3}{2}y_1 + \tfrac{5}{8}y_2 + y_3 \qquad &= -4 \\ -y_1 + \tfrac{7}{4}y_2 + \tfrac{20}{3}y_3 + y_4 &= -1 \end{aligned}$$

or

$$\begin{aligned} y_1 &= 4 \\ y_2 &= -8 - 2y_1 = -16 \\ y_3 &= -4 + \tfrac{3}{2}y_1 - \tfrac{5}{8}y_2 = 12 \\ y_4 &= -1 + y_1 - \tfrac{7}{4}y_2 - \tfrac{20}{3}y_3 = -49 \end{aligned}$$

We have just performed the forward substitution. Now from $U\mathbf{x} = \mathbf{y}$ we obtain

$$\begin{aligned} 2x_1 + 3x_2 + 2x_3 + 4x_4 &= 4 \\ 4x_2 - 8x_3 - 8x_4 &= -16 \\ 3x_3 + 9x_4 &= 12 \\ -49x_4 &= -49 \end{aligned}$$

or

$$\begin{aligned} x_4 &= 1 \\ 3x_3 &= 12 - 9x_4 = 3, \quad \text{so } x_3 = 1 \\ 4x_2 &= -16 + 8x_3 + 8x_4 = 0, \quad \text{so } x_2 = 0 \\ 2x_1 &= 4 - 3x_2 - 2x_3 - 4x_4 = -2, \quad \text{so } x_1 = -1 \end{aligned}$$

The solution is

$$\mathbf{x} = \begin{pmatrix} -1 \\ 0 \\ 1 \\ 1 \end{pmatrix}$$

## The Factorization $PA = LU$

Suppose that in order to row-reduce $A$ to a triangular matrix some permutations are required. An elementary permutation matrix is an elementary matrix associated with the row operation $R_i \rightleftarrows R_j$. Let us assume for the moment that we know before starting which permutations must be done. Each permutation is carried out by multiplying $A$ on the left by an elementary permutation matrix, denoted by $P_i$. Suppose that in the row reduction we perform $m$ permutations. Let

$$P = P_n P_{n-1} \cdots P_2 P_1$$

The product of *elementary* permutation matrices is called a **permutation matrix.** Alternatively, a permutation matrix is an $n \times n$ matrix whose rows are the rows of $I_n$, but not necessarily in the same order.

Now doing all $n$ permutations in advance is equivalent to multiplying $A$ on the left by $P$. That is,

> $PA$ is a matrix that can be row-reduced to an upper triangular matrix without performing any additional permutations.

**EXAMPLE 3** **A $PA = LU$-Factorization**

$$\text{Let } A = \begin{pmatrix} 0 & 2 & 3 \\ 2 & -4 & 7 \\ 1 & -2 & 5 \end{pmatrix}$$

To row-reduce $A$ to an upper triangular form, we first interchange rows 1 and 3 and then continue as follows:

$$\begin{pmatrix} 0 & 2 & 3 \\ 2 & -4 & 7 \\ 1 & -2 & 5 \end{pmatrix} \xrightarrow{R_1 \rightleftarrows R_3} \begin{pmatrix} 1 & -2 & 5 \\ 2 & -4 & 7 \\ 0 & 2 & 3 \end{pmatrix} \xrightarrow{R_2 \to R_2 - 2R_1} \begin{pmatrix} 1 & -2 & 5 \\ 0 & 0 & -3 \\ 0 & 2 & 3 \end{pmatrix} \xrightarrow{R_2 \rightleftarrows R_3} \begin{pmatrix} 1 & -2 & 5 \\ 0 & 2 & 3 \\ 0 & 0 & -3 \end{pmatrix}$$

In performing this row reduction we permuted twice. We first interchanged rows 1 and 3 and then interchanged rows 2 and 3:

$$P_1 = \begin{pmatrix} 0 & 0 & 1 \\ 0 & 1 & 0 \\ 1 & 0 & 0 \end{pmatrix} \quad \text{and} \quad P_2 = \begin{pmatrix} 1 & 0 & 0 \\ 0 & 0 & 1 \\ 0 & 1 & 0 \end{pmatrix}$$

Then

$$P = P_2 P_1 = \begin{pmatrix} 1 & 0 & 0 \\ 0 & 0 & 1 \\ 0 & 1 & 0 \end{pmatrix} \begin{pmatrix} 0 & 0 & 1 \\ 0 & 1 & 0 \\ 1 & 0 & 0 \end{pmatrix} = \begin{pmatrix} 0 & 0 & 1 \\ 1 & 0 & 0 \\ 0 & 1 & 0 \end{pmatrix}$$

and

$$PA = \begin{pmatrix} 0 & 0 & 1 \\ 1 & 0 & 0 \\ 0 & 1 & 0 \end{pmatrix} \begin{pmatrix} 0 & 2 & 3 \\ 2 & -4 & 7 \\ 1 & -2 & 5 \end{pmatrix} = \begin{pmatrix} 1 & -2 & 5 \\ 0 & 2 & 3 \\ 2 & -4 & 7 \end{pmatrix}.$$

This matrix can be row-reduced to an upper triangular form without permuting. We have

$$\begin{pmatrix} 1 & -2 & 5 \\ 0 & 2 & 3 \\ 2 & -4 & 7 \end{pmatrix} \xrightarrow{R_3 \to R_3 - 2R_1} \begin{pmatrix} 1 & -2 & 5 \\ 0 & 2 & 3 \\ 0 & 0 & -3 \end{pmatrix} = U$$

Thus, as in Example 1,

$$\begin{pmatrix} 1 & 0 & 0 \\ 0 & 1 & 0 \\ -2 & 0 & 1 \end{pmatrix} PA = U$$

or

$$PA = \begin{pmatrix} 1 & 0 & 0 \\ 0 & 1 & 0 \\ 2 & 0 & 1 \end{pmatrix} \begin{pmatrix} 1 & -2 & 5 \\ 0 & 2 & 3 \\ 0 & 0 & -3 \end{pmatrix} = LU$$

Generalizing the result of Example 3, we obtain the following theorem.

**THEOREM 3** Let $A$ be an invertible $n \times n$ matrix. Then there exists a permutation matrix $P$ such that

$$PA = LU$$

where $L$ is lower triangular with 1's on the diagonal and $U$ is upper triangular. For each choice of $P$ (and there may be more than one), the matrices $L$ and $U$ are unique.

*Note.* If we choose a different $P$, then we may obtain different matrices $L$ and $U$. For example, in Example 3 let

$$P^* = \begin{pmatrix} 0 & 1 & 0 \\ 1 & 0 & 0 \\ 0 & 0 & 1 \end{pmatrix}$$

(corresponding to permuting the first two rows in the first step)

You should verify that

$$P^*A = L_1U_1 = \begin{pmatrix} 1 & 0 & 0 \\ 0 & 1 & 0 \\ \frac{1}{2} & 0 & 1 \end{pmatrix}\begin{pmatrix} 2 & -4 & 7 \\ 0 & 2 & 3 \\ 0 & 0 & \frac{3}{2} \end{pmatrix}$$

## Solving a System by Using the $PA = LU$-Factorization

Consider the system $A\mathbf{x} = \mathbf{b}$ and suppose that $PA = LU$. Then

$$PA\mathbf{x} = P\mathbf{b}$$
$$LU\mathbf{x} = P\mathbf{b}$$

and we can solve this system the same way we solved the system in Example 2.

**EXAMPLE 4** **Solving a System Using the $PA = LU$-Factorization** Solve the system

$$\begin{aligned} 2x_2 + 3x_3 &= 7 \\ 2x_1 - 4x_2 + 7x_3 &= 9 \\ x_1 - 2x_2 + 5x_3 &= -6 \end{aligned}$$

**Solution** We can write this system as $A\mathbf{x} = \mathbf{b}$, where

$$A = \begin{pmatrix} 0 & 2 & 3 \\ 2 & -4 & 7 \\ 1 & -2 & 5 \end{pmatrix} \quad \text{and} \quad \mathbf{b} = \begin{pmatrix} 7 \\ 9 \\ -6 \end{pmatrix}$$

Then from Example 3

$$LU\mathbf{x} = PA\mathbf{x} = P\mathbf{b} = \begin{pmatrix} 0 & 0 & 1 \\ 1 & 0 & 0 \\ 0 & 1 & 0 \end{pmatrix}\begin{pmatrix} 7 \\ 9 \\ -6 \end{pmatrix} = \begin{pmatrix} -6 \\ 7 \\ 9 \end{pmatrix}$$

We seek a $\mathbf{y}$ such that $L\mathbf{y} = \begin{pmatrix} -6 \\ 7 \\ 9 \end{pmatrix}$. That is,

$$\begin{pmatrix} 1 & 0 & 0 \\ 0 & 1 & 0 \\ 2 & 0 & 1 \end{pmatrix}\begin{pmatrix} y_1 \\ y_2 \\ y_3 \end{pmatrix} = \begin{pmatrix} -6 \\ 7 \\ 9 \end{pmatrix}$$

Then $y_1 = -6$, $y_2 = 7$, and $2y_1 + y_3 = 9$, so $y_3 = 21$ and

$$\mathbf{y} = \begin{pmatrix} -6 \\ 7 \\ 21 \end{pmatrix}$$

Continuing, we seek an $\mathbf{x}$ such that $U\mathbf{x} = \begin{pmatrix} -6 \\ 7 \\ 21 \end{pmatrix}$; that is,

$$\begin{pmatrix} 1 & -2 & 5 \\ 0 & 2 & 3 \\ 0 & 0 & -3 \end{pmatrix}\begin{pmatrix} x_1 \\ x_2 \\ x_3 \end{pmatrix} = \begin{pmatrix} -6 \\ 7 \\ 21 \end{pmatrix}$$

So

$$\begin{aligned} x_1 - 2x_2 + 5x_3 &= -6 \\ 2x_2 + 3x_3 &= 7 \\ -3x_3 &= 21 \end{aligned}$$

Finally,

$$\begin{aligned} x_3 &= -7 \\ 2x_2 + 3(-7) &= 7, \text{ so } x_2 = 14 \\ x_1 - 2(14) + 5(-7) &= -6, \text{ so } x_1 = 57 \end{aligned}$$

The solution is

$$\mathbf{x} = \begin{pmatrix} 57 \\ 14 \\ -7 \end{pmatrix}$$

## An Easier Way to Find the *LU*-Factorization of a Matrix

Suppose that $A$ is a square matrix that can be row-reduced to an upper triangular matrix without performing permutations. Then there is an easy way to find the $LU$-factorization of $A$ without using row reduction. We illustrate this method in the following example.

**EXAMPLE 5** **An Easier Way to Obtain an *LU*-Factorization** Find the $LU$-factorization of

$$A = \begin{pmatrix} 2 & 3 & 2 & 4 \\ 4 & 10 & -4 & 0 \\ -3 & -2 & -5 & -2 \\ -2 & 4 & 4 & -7 \end{pmatrix}$$

**Solution** We solved this problem in Example 1. We now do it by using an easier method. If $A = LU$, we know that $A$ can be factored as follows:

$$A = \begin{pmatrix} 2 & 3 & 2 & 4 \\ 4 & 10 & -4 & 0 \\ -3 & -2 & -5 & -2 \\ -2 & 4 & 4 & -7 \end{pmatrix} = \begin{pmatrix} 1 & 0 & 0 & 0 \\ a & 1 & 0 & 0 \\ b & c & 1 & 0 \\ d & e & f & 1 \end{pmatrix}\begin{pmatrix} 2 & 3 & 2 & 4 \\ 0 & u & v & w \\ 0 & 0 & x & y \\ 0 & 0 & 0 & z \end{pmatrix} = LU$$

Note that the first row of $U$ is the same as the first row of $A$ because in reducing $A$ to a triangular form, we need not do anything to the first row of $A$.

We can obtain all the missing coefficients by simple matrix multiplication. The 2,1 component of $A$ is 4. Thus the dot product of the second row of $L$ and the first column of $U$ equals 4:

$$4 = 2a \quad \text{or} \quad a = 2$$

Thus

$$\begin{pmatrix} 2 & 3 & 2 & 4 \\ 4 & 10 & -4 & 0 \\ -3 & -2 & -5 & -2 \\ -2 & 4 & 4 & -7 \end{pmatrix} = \begin{pmatrix} 1 & 0 & 0 & 0 \\ 2 & 1 & 0 & 0 \\ \cancel{b}-\frac{3}{2} & \cancel{c}\,\frac{5}{8} & 1 & 0 \\ \cancel{d}-1 & \cancel{e}\,\frac{7}{4} & \cancel{f}\,\frac{20}{3} & 1 \end{pmatrix} \begin{pmatrix} 2 & 3 & 2 & 4 \\ 0 & \cancel{u}\,4 & \cancel{v}-8 & \cancel{w}-8 \\ 0 & 0 & \cancel{x}\,3 & \cancel{y}\,9 \\ 0 & 0 & 0 & \cancel{z}-49 \end{pmatrix}$$

Next we have:

2,2 component: $10 = 6 + u$, so $u = 4$

From now on we insert the newly found numbers in $L$ and $U$ above:

2,3 component: $-4 = 4 + v$, so $v = -8$

2,4 component: $0 = 8 + w$, so $w = -8$

3,1 component: $-3 = 2b$, so $b = -\frac{3}{2}$

3,2 component: $-2 = -\frac{9}{2} + 4c$, so $c = \frac{5}{8}$

3,3 component: $-5 = -3 - 5 + x$, so $x = 3$

3,4 component: $-2 = -6 - 5 + y$, so $y = 9$

4,1 component: $-2 = 2d$, so $d = -1$

4,2 component: $4 = -3 + 4e$, so $e = \frac{7}{4}$

4,3 component: $4 = -2 - 14 + 3f$, so $f = \frac{20}{3}$

4,4 component: $-7 = -4 - 14 + 60 + z$, so $z = -49$

The result is the factorization we obtained in Example 1 with considerably less work.

*Remark.* The technique illustrated in Example 5 is easily implemented on a computer.

WARNING

The technique in Example 5 works only if $A$ can be row-reduced to a triangular matrix without performing permutations. If permutations are necessary, you must first multiply $A$ on the left by an appropriate permutation matrix; then you can apply this process to get the $PA = LU$-factorization.

## *LU*-Factorization for Singular Matrices

If $A$ is a singular (not invertible) square matrix, then the row echelon form of $A$ will have at least one row of 0's, as will the triangular form of $A$. We may still be able to write $A = LU$ or $PA = LU$, but in this case $U$ will not be invertible and $L$ and $U$ may not be unique.

**EXAMPLE 6** **When $A$ Is Not Invertible, the $LU$-Factorization May Not Be Unique** Using the technique of Example 1 or Example 5, we obtain the factorization

$$A = \begin{pmatrix} 1 & 2 & 3 \\ -1 & -2 & -3 \\ 2 & 4 & 6 \end{pmatrix} = \begin{pmatrix} 1 & 0 & 0 \\ -1 & 1 & 0 \\ 2 & 0 & 1 \end{pmatrix} \begin{pmatrix} 1 & 2 & 3 \\ 0 & 0 & 0 \\ 0 & 0 & 0 \end{pmatrix} = LU$$

However, if we let $L_1 = \begin{pmatrix} 1 & 0 & 0 \\ -1 & 1 & 0 \\ 2 & x & 1 \end{pmatrix}$, then $A = L_1 U$ for any real number $x$. Thus in this case $A$ has an $LU$-factorization but it is not unique. You should verify that $A$ is not invertible.

On the other hand,

$$B = \begin{pmatrix} 1 & 2 & 3 \\ 2 & -1 & 4 \\ 3 & 1 & 7 \end{pmatrix} = \begin{pmatrix} 1 & 0 & 0 \\ 2 & 1 & 0 \\ 3 & 1 & 1 \end{pmatrix} \begin{pmatrix} 1 & 2 & 3 \\ 0 & -5 & -2 \\ 0 & 0 & 0 \end{pmatrix} = L'U'$$

and this factorization is unique—even though $B$ is not invertible. You should verify these facts.

This example shows that if a square matrix with an $LU$-factorization is not invertible, then its $LU$-factorization may or may not be unique.

## *LU*-Factorization for Nonsquare Matrices

It is sometimes possible to find $LU$-factorizations for matrices that are not square.

**THEOREM 4** **$LU$-Factorization for Nonsquare Matrices** Let $A$ be an $m \times n$ matrix. Suppose that $A$ can be reduced to its row echelon form without performing permutations. Then there exists an $m \times m$ lower triangular matrix $L$ with 1's on the diagonal and an $m \times n$ matrix $U$ with $U_{ij} = 0$ if $i > j$ such that $A = LU$.

*Note.* The condition $U_{ij} = 0$ if $i > j$ means that $U$ is upper triangular in the sense that all entries under the "diagonal" are 0. For example, a $3 \times 5$ $U$ that satisfies this condition has the form

$$U = \begin{pmatrix} d_1 & u_{12} & u_{13} & u_{14} & u_{15} \\ 0 & d_2 & u_{23} & u_{24} & u_{25} \\ 0 & 0 & d_3 & u_{34} & u_{35} \end{pmatrix} \tag{1}$$

while a $5 \times 3$ $U$ that satisfies this condition takes the form

$$U = \begin{pmatrix} d_1 & u_{12} & u_{13} \\ 0 & d_2 & u_{23} \\ 0 & 0 & d_3 \\ 0 & 0 & 0 \\ 0 & 0 & 0 \end{pmatrix} \tag{2}$$

We do not prove this theorem here, but rather, we illustrate it with two examples.

**EXAMPLE 7** **The *LU*-Factorization of a $4 \times 3$ Matrix** Find an *LU*-factorization for

$$A = \begin{pmatrix} 1 & 2 & 3 \\ -1 & -4 & 5 \\ 6 & -3 & 2 \\ 4 & 1 & -12 \end{pmatrix}$$

**Solution** We proceed as in Example 5 and set

$$\begin{pmatrix} 1 & 2 & 3 \\ -1 & -4 & 5 \\ 6 & -3 & 2 \\ 4 & 1 & -12 \end{pmatrix} = \begin{pmatrix} 1 & 0 & 0 & 0 \\ a & 1 & 0 & 0 \\ b & c & 1 & 0 \\ d & e & f & 1 \end{pmatrix} \begin{pmatrix} 1 & 2 & 3 \\ 0 & u & v \\ 0 & 0 & w \\ 0 & 0 & 0 \end{pmatrix} = LU$$

You should verify that this leads immediately to

$$L = \begin{pmatrix} 1 & 0 & 0 & 0 \\ -1 & 1 & 0 & 0 \\ 6 & \frac{15}{2} & 1 & 0 \\ 4 & \frac{7}{2} & \frac{13}{19} & 1 \end{pmatrix} \quad \text{and} \quad U = \begin{pmatrix} 1 & 2 & 3 \\ 0 & -2 & 8 \\ 0 & 0 & -76 \\ 0 & 0 & 0 \end{pmatrix}$$

**EXAMPLE 8** **The *LU*-Factorization of a $3 \times 4$ Matrix** Find an *LU*-factorization for

$$A = \begin{pmatrix} 3 & -1 & 4 & 2 \\ 1 & 2 & -3 & 5 \\ 2 & 4 & 1 & 5 \end{pmatrix}$$

**Solution** We write

$$\begin{pmatrix} 3 & -1 & 4 & 2 \\ 1 & 2 & -3 & 5 \\ 2 & 4 & 1 & 5 \end{pmatrix} = \begin{pmatrix} 1 & 0 & 0 \\ a & 1 & 0 \\ b & c & 1 \end{pmatrix}\begin{pmatrix} 3 & -1 & 4 & 2 \\ 0 & u & v & w \\ 0 & 0 & x & y \end{pmatrix}$$

Solving for the unknown variables as in Example 5 leads to

$$\begin{pmatrix} 3 & -1 & 4 & 2 \\ 1 & 2 & -3 & 5 \\ 2 & 4 & 1 & 5 \end{pmatrix} = \begin{pmatrix} 1 & 0 & 0 \\ \frac{1}{3} & 1 & 0 \\ \frac{2}{3} & 2 & 1 \end{pmatrix}\begin{pmatrix} 3 & -1 & 4 & 2 \\ 0 & \frac{7}{3} & -\frac{13}{3} & \frac{13}{3} \\ 0 & 0 & 7 & -5 \end{pmatrix}$$

*Note.* As in the case of a square, singular matrix, if a nonsquare matrix has an $LU$-factorization, it may or may not be unique.

## A Remark About Computers and the *LU*-Factorization

The TI-85, MATLAB, and other computer software systems will carry out the $PA$ = LU-factorization of a square matrix. However, the matrix $L$ so obtained is sometimes not a lower triangular matrix with 1's on the diagonal but may be a permutation of such a matrix. Alternatively, the system may give a lower triangular $L$ and a $U$ with 1's on the diagonal. The reason for this is that these systems use an $LU$-factorization to compute inverses and determinants and to solve systems of equations. Certain reorderings or permutations will minimize the accumulated round-off errors. We will say more about these errors and procedures in Appendixes 3 and 4.

In the meantime, you should be aware that the results you obtain on your calculator or computer will often be different from those you get by hand. In particular, if $A$ can be row-reduced to a triangular matrix without permutations, then when $PA = LU$, $P = I$. However, a different $P$ will often occur in your calculator. For example, if

$$A = \begin{pmatrix} 2 & 3 & 2 & 4 \\ 4 & 10 & -4 & 0 \\ -3 & -2 & -5 & -2 \\ -2 & 4 & 4 & -7 \end{pmatrix}$$

as in Examples 1 and 5, then MATLAB gives the factorization $A = LU$, where

$$L = \begin{pmatrix} \frac{1}{2} & -\frac{2}{9} & -\frac{40}{83} & 1 \\ 1 & 0 & 0 & 0 \\ -\frac{3}{4} & \frac{11}{18} & 1 & 0 \\ -\frac{1}{2} & 1 & 0 & 0 \end{pmatrix} \quad \text{and} \quad U = \begin{pmatrix} 4 & 10 & -4 & 0 \\ 0 & 9 & 2 & -7 \\ 0 & 0 & -\frac{83}{9} & \frac{41}{18} \\ 0 & 0 & 0 & \frac{294}{83} \end{pmatrix}$$

*Note.* A permutation of the rows of $L$ yields a lower triangular matrix with 1's on the diagonal.

# PROBLEMS 1.11

## Self-Quiz

### *True-False*

**I.** For every square matrix $A$ there exist invertible matrices $L$ and $U$ such that $A = LU$, where $L$ is lower triangular with 1's on the diagonal and $U$ is upper triangular.

**II.** For every invertible matrix $A$, there exist matrices $L$ and $U$ as in Problem I.

**III.** For every invertible matrix $A$ there exists a permutation matrix $P$ such that $PA = LU$, where $L$ and $U$ are as in Problem I.

**IV.** The product of permutation matrices is a permutation matrix.

In Problems 1–8 find a lower triangular matrix $L$ with 1's down the diagonal and an upper triangular matrix $U$ such that $A = LU$.

**1.** $\begin{pmatrix} 1 & 2 \\ 3 & 4 \end{pmatrix}$ **2.** $\begin{pmatrix} 1 & 2 \\ 0 & 3 \end{pmatrix}$ **3.** $\begin{pmatrix} -1 & 5 \\ 6 & 3 \end{pmatrix}$

**4.** $\begin{pmatrix} 1 & 4 & 6 \\ 2 & -1 & 3 \\ 3 & 2 & 5 \end{pmatrix}$ **5.** $\begin{pmatrix} 2 & 1 & 7 \\ 4 & 3 & 5 \\ 2 & 1 & 6 \end{pmatrix}$ **6.** $\begin{pmatrix} 3 & 9 & -2 \\ 6 & -3 & 8 \\ 4 & 6 & 5 \end{pmatrix}$

**7.** $\begin{pmatrix} 1 & 2 & -1 & 4 \\ 0 & -1 & 5 & 8 \\ 2 & 3 & 1 & 4 \\ 1 & -1 & 6 & 4 \end{pmatrix}$ **8.** $\begin{pmatrix} 2 & 3 & -1 & 6 \\ 4 & 7 & 2 & 1 \\ -2 & 5 & -2 & 0 \\ 0 & -4 & 5 & 2 \end{pmatrix}$

In Problems 9–16 solve the given systems using the $LU$-factorizations found in Problems 1–8. That is, solve $A\mathbf{x} = LU\mathbf{x} = \mathbf{b}$.

**9.** $A = \begin{pmatrix} 1 & 2 \\ 3 & 4 \end{pmatrix}$; $\mathbf{b} = \begin{pmatrix} -2 \\ 4 \end{pmatrix}$ **10.** $A = \begin{pmatrix} 1 & 2 \\ 0 & 3 \end{pmatrix}$; $\mathbf{b} = \begin{pmatrix} -1 \\ 4 \end{pmatrix}$

**11.** $A = \begin{pmatrix} -1 & 5 \\ 6 & 3 \end{pmatrix}$; $\mathbf{b} = \begin{pmatrix} 0 \\ 5 \end{pmatrix}$ **12.** $A = \begin{pmatrix} 1 & 4 & 6 \\ 2 & -1 & 3 \\ 3 & 2 & 5 \end{pmatrix}$; $\mathbf{b} = \begin{pmatrix} -1 \\ 7 \\ 2 \end{pmatrix}$

**13.** $A = \begin{pmatrix} 2 & 1 & 7 \\ 4 & 3 & 5 \\ 2 & 1 & 6 \end{pmatrix}$; $\mathbf{b} = \begin{pmatrix} 6 \\ 1 \\ 1 \end{pmatrix}$ **14.** $A = \begin{pmatrix} 3 & 9 & -2 \\ 6 & -3 & 8 \\ 4 & 6 & 5 \end{pmatrix}$; $\mathbf{b} = \begin{pmatrix} 3 \\ 10 \\ 4 \end{pmatrix}$

---

**Answers to Self-Quiz**

**I.** False **II.** False **III.** True **IV.** True

**15.** $A = \begin{pmatrix} 1 & 2 & -1 & 4 \\ 0 & -1 & 5 & 8 \\ 2 & 3 & 1 & 4 \\ 1 & -1 & 6 & 4 \end{pmatrix}$; $\mathbf{b} = \begin{pmatrix} 3 \\ -11 \\ 4 \\ -5 \end{pmatrix}$

**16.** $A = \begin{pmatrix} 2 & 3 & -1 & 6 \\ 4 & 7 & 2 & 1 \\ -2 & 5 & -2 & 0 \\ 0 & -4 & 5 & 2 \end{pmatrix}$; $\mathbf{b} = \begin{pmatrix} 1 \\ 0 \\ 0 \\ 4 \end{pmatrix}$

In Problems 17–23 (a) find a permutation matrix $P$ and lower and upper triangular matrices $L$ and $U$ such that $PA = LU$. (b) Use the result of part (a) to solve the system $A\mathbf{x} = \mathbf{b}$.

**17.** $A = \begin{pmatrix} 0 & 2 \\ 1 & 4 \end{pmatrix}$; $\mathbf{b} = \begin{pmatrix} 3 \\ -5 \end{pmatrix}$

**18.** $A = \begin{pmatrix} 0 & 2 & 4 \\ 1 & -1 & 2 \\ 0 & 3 & 2 \end{pmatrix}$; $\mathbf{b} = \begin{pmatrix} -1 \\ 2 \\ 4 \end{pmatrix}$

**19.** $A = \begin{pmatrix} 0 & 2 & 4 \\ 0 & 3 & 7 \\ 4 & 1 & 5 \end{pmatrix}$; $\mathbf{b} = \begin{pmatrix} -1 \\ 0 \\ 2 \end{pmatrix}$

**20.** $A = \begin{pmatrix} 0 & 5 & -1 \\ 2 & 3 & 5 \\ 4 & 6 & -7 \end{pmatrix}$; $\mathbf{b} = \begin{pmatrix} 10 \\ -3 \\ 5 \end{pmatrix}$

**21.** $A = \begin{pmatrix} 0 & 2 & 3 & 1 \\ 0 & 4 & -1 & 5 \\ 2 & 0 & 3 & 1 \\ 1 & -4 & 5 & 6 \end{pmatrix}$; $\mathbf{b} = \begin{pmatrix} 3 \\ -1 \\ 2 \\ 4 \end{pmatrix}$

**22.** $A = \begin{pmatrix} 0 & 0 & -2 & 3 \\ 5 & 0 & -6 & 4 \\ 2 & 0 & 1 & -2 \\ 0 & 4 & -2 & 5 \end{pmatrix}$; $\mathbf{b} = \begin{pmatrix} -2 \\ 4 \\ 5 \\ 7 \end{pmatrix}$

**23.** $A = \begin{pmatrix} 0 & -2 & 3 & 1 \\ 0 & 4 & -3 & 2 \\ 1 & 2 & -3 & 2 \\ -2 & -4 & 5 & -10 \end{pmatrix}$; $\mathbf{b} = \begin{pmatrix} 6 \\ 1 \\ 0 \\ 5 \end{pmatrix}$

**24.** Suppose that $L$ and $M$ are lower triangular with 1's on the diagonal. Show that $LM$ is lower triangular with 1's on the diagonal. [*Hint:* If $B = LM$, show that

$$b_{ii} = \sum_{k=1}^{n} l_{ik}m_{ki} = 1 \quad \text{and} \quad b_{ij} = \sum_{k=1}^{n} l_{ik}m_{kj} = 0 \qquad \text{if } j > i.]$$

**25.** Show that the product of two upper triangular matrices is upper triangular.

**26.** Show that $\begin{pmatrix} -1 & 2 & 1 \\ 2 & -4 & -2 \\ 4 & -8 & -4 \end{pmatrix}$ has more than one $LU$-factorization.

**27.** Do the same for the matrix $\begin{pmatrix} 3 & -3 & 2 & 5 \\ 2 & 1 & -6 & 0 \\ 5 & -2 & -4 & 5 \\ 1 & -4 & 8 & 5 \end{pmatrix}$

In Problems 28–32 find an $LU$-factorization for each singular matrix:

**28.** $\begin{pmatrix} 1 & 2 \\ 2 & 4 \end{pmatrix}$

**29.** $\begin{pmatrix} -1 & 2 & 3 \\ 2 & 1 & 7 \\ 1 & 3 & 10 \end{pmatrix}$

**30.** $\begin{pmatrix} -1 & 1 & 4 & 6 \\ 2 & -1 & 0 & 2 \\ 0 & 3 & 1 & 5 \\ 1 & 3 & 5 & 13 \end{pmatrix}$

**31.** $\begin{pmatrix} 2 & -1 & 1 & 7 \\ 3 & 2 & 1 & 6 \\ 1 & 3 & 0 & -1 \\ 4 & 5 & 1 & 5 \end{pmatrix}$

**32.** $\begin{pmatrix} 2 & -1 & 0 & 2 \\ 4 & -2 & 0 & 4 \\ -2 & 1 & 0 & -2 \\ 6 & -3 & 0 & 6 \end{pmatrix}$

In Problems 33–38 find an $LU$-factorization for each nonsquare matrix.

**33.** $\begin{pmatrix} 1 & 2 & 3 \\ -1 & 2 & 4 \end{pmatrix}$

**34.** $\begin{pmatrix} 2 & 1 \\ -1 & 4 \\ 6 & 0 \end{pmatrix}$

**35.** $\begin{pmatrix} 7 & 1 & 3 & 4 \\ -2 & 5 & 6 & 8 \end{pmatrix}$

**36.** $\begin{pmatrix} 4 & -1 & 2 & 1 \\ 2 & 1 & 6 & 5 \\ 3 & 2 & -1 & 7 \end{pmatrix}$

**37.** $\begin{pmatrix} 5 & 1 & 3 \\ -2 & 4 & 2 \\ 1 & 6 & 1 \\ -2 & 2 & 0 \\ 5 & -3 & 1 \end{pmatrix}$

**38.** $\begin{pmatrix} -1 & 2 & 1 \\ 1 & 6 & 5 \\ -2 & 3 & 7 \\ 1 & 0 & 2 \\ 4 & 1 & 5 \end{pmatrix}$

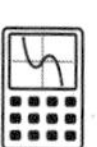

## CALCULATOR BOX

### TI-85

The $PA = LU$-factorization can be obtained on a TI-85 calculator. Once a matrix $A$ is entered, press the following key sequence:

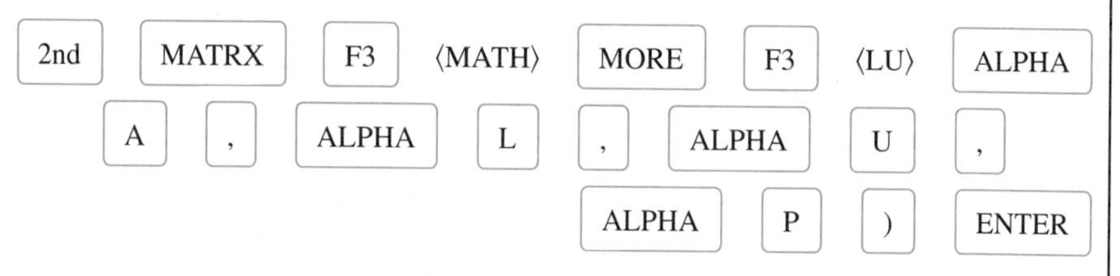


The word "Done" will be displayed. You can obtain the matrix $L$ by pressing ALPHA L ENTER and similarly for $U$ and $P$. However, the result obtained will not be the one obtained by hand. You will obtain a $U$ with 1's down the diagonal.

In Problems 39–44 find a $PA = LU$-factorization on a calculator.

**39.** $A = \begin{pmatrix} 0 & 2 & 5 \\ 3 & 1 & 7 \\ 2 & -1 & 9 \end{pmatrix}$

**40.** $A = \begin{pmatrix} 2 & -1 & 5 & 9 \\ 4 & 12 & 16 & -8 \\ 13 & 2 & 5 & 3 \\ 16 & 5 & -8 & 4 \end{pmatrix}$

**41.** $A = \begin{pmatrix} 0 & -7 & 4 & 1 \\ 5 & 3 & 9 & 2 \\ 2 & -1 & 0 & 4 \\ 16 & -5 & 11 & 8 \end{pmatrix}$

**42.** $A = \begin{pmatrix} 23 & 10 & 4 & -8 & 26 \\ 14 & 5 & 9 & -18 & 13 \\ 71 & -46 & 59 & 65 & -22 \\ 35 & 47 & -81 & 23 & -50 \\ 14 & 29 & 31 & 26 & 92 \end{pmatrix}$

**43.** $A = \begin{pmatrix} 0.21 & 0.32 & -0.34 & 0.37 \\ 0.91 & 0.23 & 0.16 & -0.20 \\ 0.46 & 0.08 & 0.33 & -0.59 \\ 0.83 & 0.71 & -0.68 & 0.77 \end{pmatrix}$

**44.** $A = \begin{pmatrix} 1 & 2 & 0 & 0 & 0 \\ 1 & 2 & 3 & 0 & 0 \\ 0 & 5 & 6 & 1 & 0 \\ 0 & 0 & 2 & 3 & 2 \\ 0 & 0 & 0 & 5 & 6 \end{pmatrix}$

## MATLAB 1.11

1. Following the steps outlined in MATLAB Problem 3 in Section 1.10, find the LU-decomposition for $A$ below; that is, find $L$ and $U$ and verify that $LU = A$. Here $U$ is not upper triangular but, rather, it is in row echelon form (except the pivots are not necessarily equal to 1):

$$A = \begin{pmatrix} 8 & 2 & -4 & 6 \\ 10 & 1 & -8 & 9 \\ 4 & 7 & 10 & 3 \end{pmatrix}$$

2. Solving systems (with unique solutions) by using the LU-decomposition is more efficient than other methods we have discussed.

   *MATLAB Information.* The command **x = A\b** solves the system [$A$ **b**] by finding the LU-decomposition and then doing the forward and backward substitutions. The command **flops** counts the number of (significant) floating point operations that are performed. The command **flops(0)** starts the count at 0.

   a. Choose **A = rand(5)** and **b = rand(5,1).** Enter

   **flops(0), rref([A b]), frref = flops**

   **flops(0), x = A\b, flu = flops**

   b. Repeat for three more choices of $A$ and **b** (use different sizes greater than 5).
   c. Comment on the comparison of the two flop counts, **frref** and **flu.**

3. MATLAB can find an LU-decomposition for you, but it may not be what you expect. There is almost always a permutation matrix $P$ involved.
   a. Let **A = 2*rand(3) − 1.** Enter **[L,U,P] = lu(A)** and check that $LU = PA$. Repeat for two more random square matrices of different sizes.

b. The reason there is almost always a $P$: To minimize round-off errors for better accuracy, rows are interchanged so that the largest entry (in absolute value) in a column (from rows not used yet) is in the pivot position.

Let **A = round(10*(2*rand(4)−1)).** For this $A$, find $L$, $U$, and $P$ using the **lu** command. Let **C = P*A.**

**i.** Reduce to upper triangular form by using row operations of the form $R_j \rightarrow R_j + c * R_i$. (Compute your multipliers by using matrix notation and perform the row operations by multiplying by elementary matrices.) (See MATLAB Problem 3 in Section 1.10.)

**ii.** Show that the reduction can proceed and that at each stage the pivot is the largest entry (in absolute value) of the entries in the column below the pivot position. Verify that the final result is the matrix $U$ produced from the **lu** command.

**iii.** Describe the relationship between the multipliers and their positions (in the elementary matrix that performs the row operation) and the entries of $L$ and their positions in $L$.

4. Enter a random $3 \times 3$ matrix $A$. Find $L$, $U$, and $P$ by using the **lu** command as in MATLAB Problem 3 in this section. Interpreting the information stored in $L$ as discovered in MATLAB Problem 3 in Section 1.10 (or as observed in MATLAB Problem 3 in this section), perform the indicated row operations to $PA$ and show that the end result is $U$. (Be sure to refer to a number from $L$ by using matrix notation and not the number displayed.)

## 1.12 GRAPH THEORY: AN APPLICATION OF MATRICES

In recent years much attention has been focused on a relatively new area of mathematical research called **graph theory.** Graphs, which we shall define shortly, are useful in studying the interrelationships among components in networks that arise in commerce, the social sciences, medicine, and many other areas. For example, graphs are useful in studying family ties in a tribal society, the spread of a communicable disease, or a network of commercial flights connecting a given number of major cities. Graph theory is a vast subject. In this section we merely give some definitions and show how close the relationship is between graph theory and matrix theory.

We now illustrate how a graph can arise in practice.

EXAMPLE 1 **Representing a Communication System by a Graph** Suppose we are studying a communication system linked by telephone wires. In this system there are five stations. In the following table we indicate available lines to and from the stations:

| Station | 1 | 2 | 3 | 4 | 5 |
|---|---|---|---|---|---|
| 1 | | ✓ | | | |
| 2 | ✓ | | | | ✓ |
| 3 | | | | ✓ | |
| 4 | | ✓ | ✓ | | |
| 5 | ✓ | | | ✓ | |

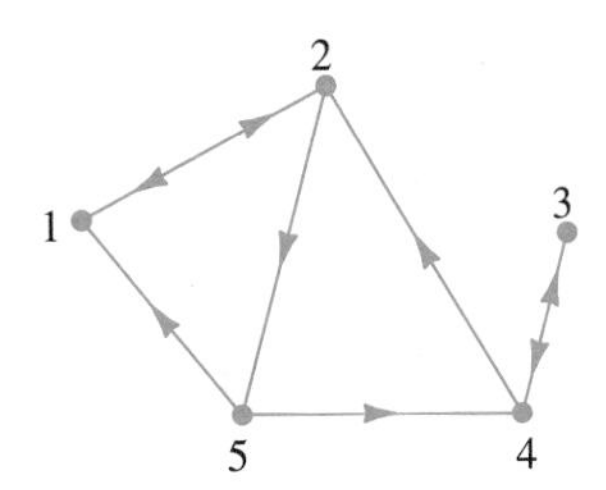


**Figure 1.9** The graph showing lines from one station to another

For example, the check in the 1,2 box indicates that there is a line from station 1 to station 2. The information in the table can be represented by a directed graph, as illustrated in Figure 1.9.

Directed graph
Vertices
Edges

In general, a **directed graph** is a collection of $n$ points called **vertices,** denoted by $V_1, V_2, \ldots, V_n$, together with a finite number of **edges** joining various pairs of vertices. Any directed graph can be represented by an $n \times n$ matrix where the number in the $ij$ position is the number of edges joining the $i$th vertex to the $j$th vertex.

**EXAMPLE 2** **The Matrix Representation of a Directed Graph** The matrix representation of the graph in Figure 1.9 is

$$A = \begin{pmatrix} 0 & 1 & 0 & 0 & 0 \\ 1 & 0 & 0 & 0 & 1 \\ 0 & 0 & 0 & 1 & 0 \\ 0 & 1 & 1 & 0 & 0 \\ 1 & 0 & 0 & 1 & 0 \end{pmatrix} \qquad \textbf{(1)}$$

**EXAMPLE 3** **The Matrix Representations of Two Directed Graphs** Find the matrix representations of the directed graphs in Figure 1.10.

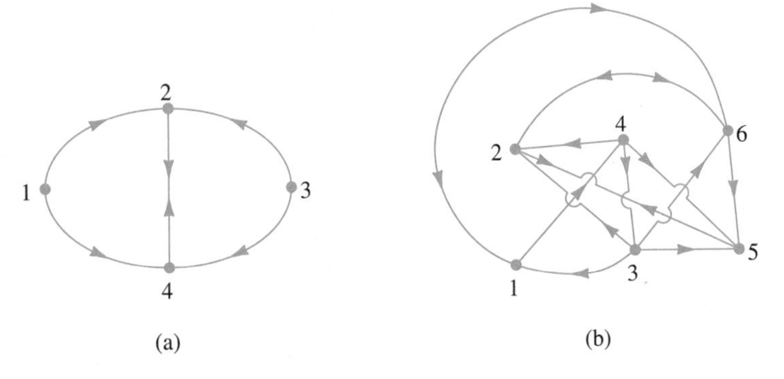


**Figure 1.10** Two directed graphs

**Solution**

$$\text{(a) } A = \begin{pmatrix} 0 & 1 & 0 & 1 \\ 0 & 0 & 0 & 1 \\ 0 & 1 & 0 & 1 \\ 0 & 1 & 0 & 0 \end{pmatrix} \qquad \text{(b) } A = \begin{pmatrix} 0 & 0 & 0 & 1 & 0 & 1 \\ 0 & 0 & 0 & 0 & 1 & 1 \\ 1 & 1 & 0 & 0 & 1 & 1 \\ 0 & 1 & 1 & 0 & 1 & 0 \\ 0 & 1 & 0 & 0 & 0 & 0 \\ 1 & 1 & 0 & 0 & 1 & 0 \end{pmatrix}$$

**EXAMPLE 4** **Obtaining a Graph from Its Matrix Representation** Sketch a graph represented by the matrix

$$A = \begin{pmatrix} 0 & 1 & 1 & 0 & 1 \\ 1 & 0 & 0 & 1 & 0 \\ 0 & 1 & 0 & 0 & 0 \\ 1 & 0 & 1 & 0 & 1 \\ 0 & 1 & 1 & 1 & 0 \end{pmatrix}$$

**Solution** Since $A$ is a $5 \times 5$ matrix, the graph has five vertices. See Figure 1.11.

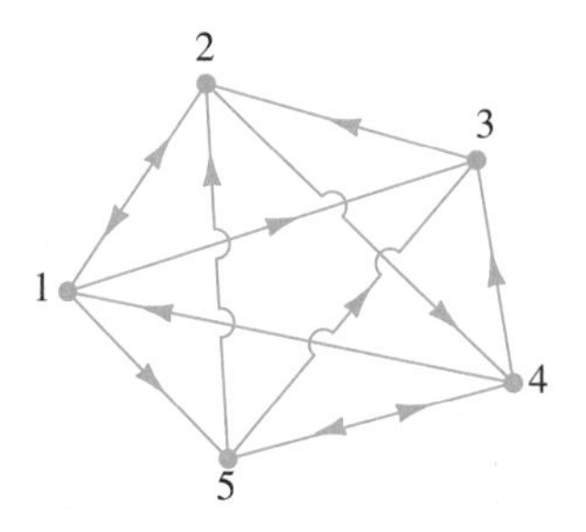


**Figure 1.11** The directed graph that is represented by $A$

*Remark.* In the examples considered we have had directed graphs that satisfy the following two conditions:

**i.** No vertex is connected to itself.

**ii.** At most, one edge leads from one vertex to another.

Incidence matrix

The matrix representing a directed graph satisfying these conditions is called an **incidence matrix.** In general, however, it is possible to have either a 1 on the main diagonal of a representing matrix (indicating an edge from a vertex to itself) or an integer bigger than 1 in the matrix (indicating more than one path from one vertex to another). To avoid more complicated (but treatable) situations, we have assumed, and shall continue to assume, that ($i$) and ($ii$) are satisfied.

**EXAMPLE 5** **A Directed Graph That Depicts Domination in a Group** Directed graphs are often used by sociologists to study group interactions. In many group situations certain individuals dominate others. This domination may be physical, intellectual, or emotional. To be more specific, we assume that in a certain setting involving six persons, a sociologist has been able to determine who dominates whom. (This may have been done by psychological testing, by filling out questionnaires, or simply by observation.) The directed graph in Figure 1.12 indicates the sociologist's findings.

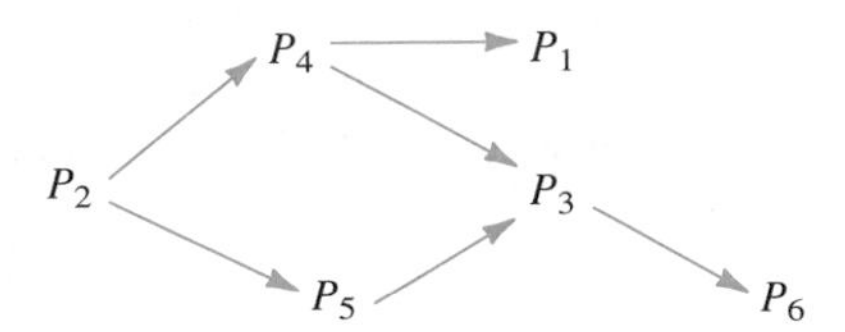


**Figure 1.12** The graph shows who dominates whom in the group

The matrix representation of this graph is

$$A = \begin{pmatrix} 0 & 0 & 0 & 0 & 0 & 0 \\ 0 & 0 & 0 & 1 & 1 & 0 \\ 0 & 0 & 0 & 0 & 0 & 1 \\ 1 & 0 & 1 & 0 & 0 & 0 \\ 0 & 0 & 1 & 0 & 0 & 0 \\ 0 & 0 & 0 & 0 & 0 & 0 \end{pmatrix}$$

There would be little point in introducing matrix representations of graphs if all we could do with them was to write them down. There are several facts that can be asked about graphs that may not be apparent. To illustrate this point, consider the graph in Figure 1.13.

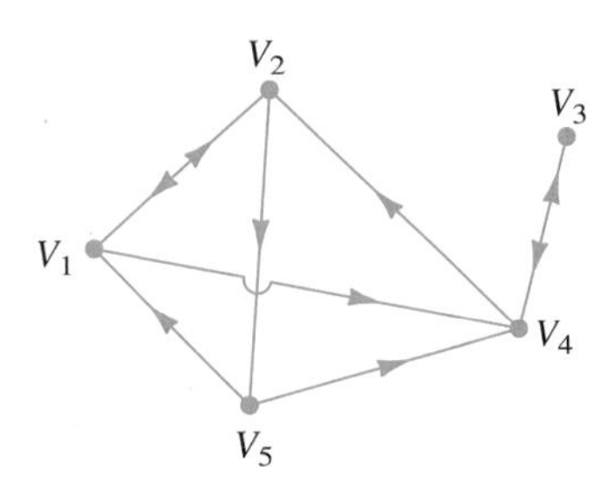


**Figure 1.13** There are paths from $V_1$ to $V_5$ even though there is no edge from $V_1$ to $V_5$. One such path is $V_1 \to V_2 \to V_5$

We can see that although there is no edge from $V_1$ to $V_5$, it is possible to send a message between these vertices. In fact, there are at least two ways to do this:

$$V_1 \rightarrow V_2 \rightarrow V_5 \tag{2}$$

and

$$V_1 \rightarrow V_4 \rightarrow V_2 \rightarrow V_5 \tag{3}$$

Path
Chain

A route from one vertex to another is called a **path** or **chain.** The path from $V_1$ to $V_5$ in (2) is called a **2-chain** because we are traversing two edges. Path (3) is called a **3-chain.** In general, a path traversing $n$ edges (and therefore passing through $n+1$ vertices) is called an ***n*-chain.** Now, returning to our graph, we see that we can also go from $V_1$ to $V_5$ along the 5-chain

$$V_1 \rightarrow V_4 \rightarrow V_3 \rightarrow V_4 \rightarrow V_2 \rightarrow V_5 \tag{4}$$

However, it would be somewhat foolish to do so since part of the path doesn't gain us anything. A path in which some vertex is encountered more than once is called **redundant.** The 5-chain (4) is redundant because vertex 4 is encountered twice.

It is of great interest to be able to determine the shortest path (if any) joining two vertices in a directed graph. There is a theorem that shows us how this can be done, but first we make an interesting observation. As we have seen, the matrix representation of the graph in Figure 1.9 is given by

$$A = \begin{pmatrix} 0 & 1 & 0 & 0 & 0 \\ 1 & 0 & 0 & 0 & 1 \\ 0 & 0 & 0 & 1 & 0 \\ 0 & 1 & 1 & 0 & 0 \\ 1 & 0 & 0 & 1 & 0 \end{pmatrix}$$

We compute

$$A^2 = \begin{pmatrix} 0 & 1 & 0 & 0 & 0 \\ 1 & 0 & 0 & 0 & 1 \\ 0 & 0 & 0 & 1 & 0 \\ 0 & 1 & 1 & 0 & 0 \\ 1 & 0 & 0 & 1 & 0 \end{pmatrix} \begin{pmatrix} 0 & 1 & 0 & 0 & 0 \\ 1 & 0 & 0 & 0 & 1 \\ 0 & 0 & 0 & 1 & 0 \\ 0 & 1 & 1 & 0 & 0 \\ 1 & 0 & 0 & 1 & 0 \end{pmatrix} = \begin{pmatrix} 1 & 0 & 0 & 0 & 1 \\ 1 & 1 & 0 & 1 & 0 \\ 0 & 1 & 1 & 0 & 0 \\ 1 & 0 & 0 & 1 & 1 \\ 0 & 2 & 1 & 0 & 0 \end{pmatrix}$$

Let us look more closely at the components of $A^2$. For example, the 1 in the 2,4 position is the dot product of the second row of $A$ and the fourth column of $A$:

$$(1 \quad 0 \quad 0 \quad 0 \quad 1) \begin{pmatrix} 0 \\ 0 \\ 1 \\ 0 \\ 1 \end{pmatrix} = 1$$

In the second row the last 1 represents the link

$$V_2 \rightarrow V_5$$

In the fourth column the last 1 represents the link

$$V_5 \to V_4$$

Multiplying these 1's together represents the 2-chain

$$V_2 \to V_5 \to V_4$$

Similarly, the 2 in the 5,2 position of $A^2$ is the dot product of the fifth row and second column of $A$:

$$(1 \quad 0 \quad 0 \quad 1 \quad 0)\begin{pmatrix} 1 \\ 0 \\ 0 \\ 1 \\ 0 \end{pmatrix} = 2$$

Reasoning as above, we see that this indicates the pair of 2-chains

$$V_5 \to V_1 \to V_2$$

and

$$V_5 \to V_4 \to V_2$$

If we generalize these facts, we can prove the following results:

**THEOREM 1** If $A$ is the incidence matrix of a directed graph, then the $ij$th component of $A^2$ gives the number of 2-chains from vertex $i$ to vertex $j$. ■

Using this theorem, we can show that the number of 3-chains joining vertex $i$ to vertex $j$ is the $ij$th component of $A^3$. In Example 2

$$A^3 = \begin{pmatrix} 1 & 1 & 0 & 1 & 0 \\ 1 & 2 & 1 & 0 & 1 \\ 1 & 0 & 0 & 1 & 1 \\ 1 & 2 & 1 & 1 & 0 \\ 2 & 0 & 0 & 1 & 2 \end{pmatrix}$$

For example, the two 3-chains from vertex 4 to vertex 2 are

$$V_4 \to V_3 \to V_4 \to V_2$$

and

$$V_4 \to V_2 \to V_1 \to V_2$$

Both of these are redundant. The two 3-chains from vertex 5 to vertex 1 are

$$V_5 \to V_4 \to V_2 \to V_1$$

and

$$V_5 \to V_1 \to V_2 \to V_1$$

The following theorem answers the question posed earlier about finding a smallest chain linking two vertices.

**THEOREM 2** Let $A$ be an incidence matrix of a directed graph. Let $a_{ij}^{(n)}$ denote the $ij$th component of $A^n$.

**i.** If $a_{ij}^{(n)} = k$, then there are exactly $k$ $n$-chains from vertex $i$ to vertex $j$.

**ii.** Moreover, if $a_{ij}^{(m)} = 0$ for all $m < n$ and $a_{ij}^{(n)} \neq 0$, then the shortest chain from vertex $i$ to vertex $j$ is an $n$-chain.

**EXAMPLE 6** **Computing Chains by Taking Powers of the Incidence Matrix** In Example 2 we have

$$A = \begin{pmatrix} 0 & 1 & 0 & 0 & 0 \\ 1 & 0 & 0 & 0 & 1 \\ 0 & 0 & 0 & 1 & 0 \\ 0 & 1 & 1 & 0 & 0 \\ 1 & 0 & 0 & 1 & 0 \end{pmatrix}, \quad A^2 = \begin{pmatrix} 1 & 0 & 0 & 0 & 1 \\ 1 & 1 & 0 & 1 & 0 \\ 0 & 1 & 1 & 0 & 0 \\ 1 & 0 & 0 & 1 & 1 \\ 0 & 2 & 1 & 0 & 0 \end{pmatrix}, \quad A^3 = \begin{pmatrix} 1 & 1 & 0 & 1 & 0 \\ 1 & 2 & 1 & 0 & 1 \\ 1 & 0 & 0 & 1 & 1 \\ 1 & 2 & 1 & 1 & 0 \\ 2 & 0 & 0 & 1 & 2 \end{pmatrix}$$

$$A^4 = \begin{pmatrix} 1 & 2 & 1 & 0 & 1 \\ 3 & 1 & 0 & 2 & 2 \\ 1 & 2 & 1 & 1 & 0 \\ 2 & 2 & 1 & 1 & 2 \\ 2 & 3 & 1 & 2 & 0 \end{pmatrix}, \quad \text{and} \quad A^5 = \begin{pmatrix} 3 & 1 & 0 & 2 & 2 \\ 3 & 5 & 2 & 2 & 1 \\ 2 & 2 & 1 & 1 & 2 \\ 4 & 3 & 1 & 3 & 2 \\ 3 & 4 & 2 & 1 & 3 \end{pmatrix}$$

Since $a_{13}^{(1)} = a_{13}^{(2)} = a_{13}^{(3)} = 0$ and $a_{13}^{(4)} = 1$, we see that the shortest path from vertex 1 to vertex 3 is a 4-chain. It is given by

$$V_1 \to V_2 \to V_5 \to V_4 \to V_3$$

*Note.* There are also five 5-chains (all of which are redundant) joining vertex 2 to itself.

**EXAMPLE 7** **Indirect Domination in a Group** In our example from sociology (Example 5) a chain (which is not an edge) represents indirect control of one person over another. That is, if Peter dominates Paul, who dominates Mary, then we can see that Peter exercises some control (albeit indirect) over Mary. To determine who has direct or indirect control over whom, we need only compute powers of the incidence matrix $A$. We have

$$A = \begin{pmatrix} 0 & 0 & 0 & 0 & 0 & 0 \\ 0 & 0 & 0 & 1 & 1 & 0 \\ 0 & 0 & 0 & 0 & 0 & 1 \\ 1 & 0 & 1 & 0 & 0 & 0 \\ 0 & 0 & 1 & 0 & 0 & 0 \\ 0 & 0 & 0 & 0 & 0 & 0 \end{pmatrix}, \quad A^2 = \begin{pmatrix} 0 & 0 & 0 & 0 & 0 & 0 \\ 1 & 0 & 2 & 0 & 0 & 0 \\ 0 & 0 & 0 & 0 & 0 & 0 \\ 0 & 0 & 0 & 0 & 0 & 1 \\ 0 & 0 & 0 & 0 & 0 & 1 \\ 0 & 0 & 0 & 0 & 0 & 0 \end{pmatrix}$$

and

$$A^3 = \begin{pmatrix} 0 & 0 & 0 & 0 & 0 & 0 \\ 0 & 0 & 0 & 0 & 0 & 2 \\ 0 & 0 & 0 & 0 & 0 & 0 \\ 0 & 0 & 0 & 0 & 0 & 0 \\ 0 & 0 & 0 & 0 & 0 & 0 \\ 0 & 0 & 0 & 0 & 0 & 0 \end{pmatrix}$$

As was apparent from the graph on page 159, these matrices show that person $P_2$ has direct or indirect control over everyone else. He or she has direct control over $P_4$ and $P_5$, second-order control over $P_1$ and $P_3$, and third-order control over $P_6$.

*Note.* In real situations things are much more complex. There may be hundreds of stations in a communications network or hundreds of individuals in a dominant-passive sociological study. In these cases matrices are essential for dealing with the huge amount of data that has to be analyzed.

# PROBLEMS 1.12

In Problems 1–4 find the matrix representation of the given directed graph.

**1.**

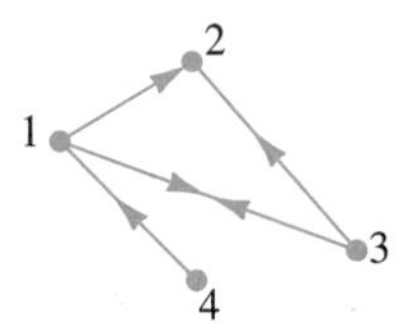


**2.**

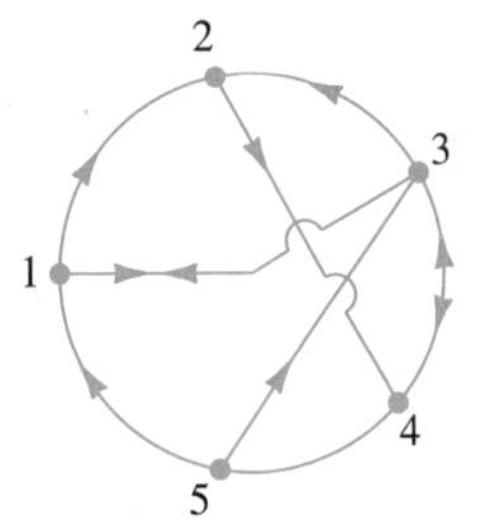


**3.**

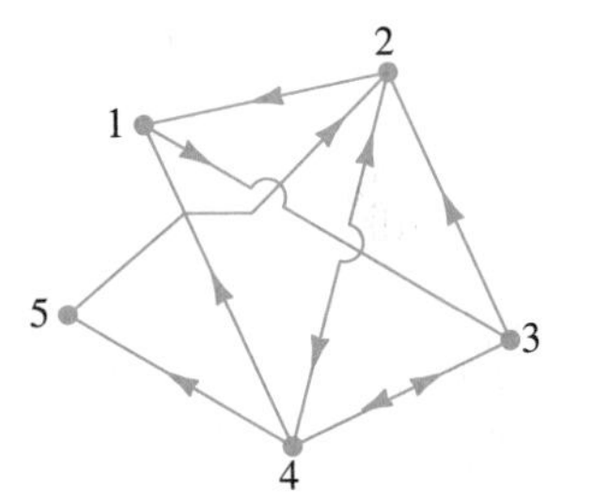


**4.**

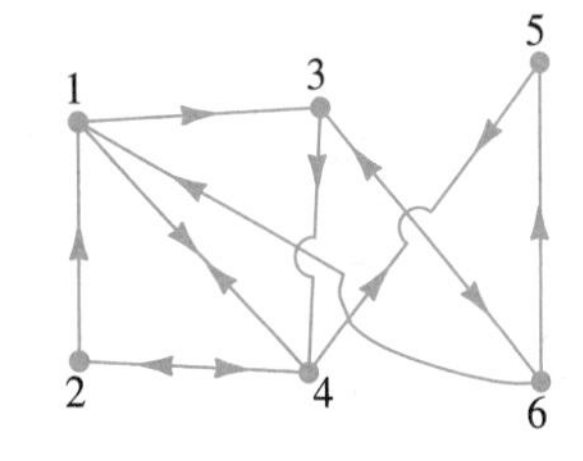

In Problems 5–7 draw graphs that are represented by the given matrices.

**5.** $\begin{pmatrix} 0 & 1 & 0 & 1 \\ 1 & 0 & 0 & 0 \\ 1 & 1 & 0 & 1 \\ 1 & 0 & 1 & 0 \end{pmatrix}$ **6.** $\begin{pmatrix} 0 & 1 & 0 & 1 & 0 \\ 0 & 0 & 1 & 1 & 1 \\ 1 & 1 & 0 & 0 & 0 \\ 1 & 0 & 0 & 0 & 0 \\ 0 & 1 & 1 & 1 & 0 \end{pmatrix}$ **7.** $\begin{pmatrix} 0 & 1 & 1 & 1 & 0 & 0 \\ 1 & 0 & 0 & 0 & 1 & 0 \\ 1 & 1 & 0 & 1 & 0 & 1 \\ 0 & 0 & 1 & 0 & 0 & 0 \\ 0 & 0 & 0 & 1 & 0 & 1 \\ 1 & 1 & 0 & 0 & 1 & 0 \end{pmatrix}$

**8.** Determine the number of 2-, 3-, and 4-chains linking the vertices in the graph of Problem 2.

**9.** Do the same for the graph of Problem 3.

**10.** Prove that the shortest path linking two vertices in a directed graph is not redundant.

**11.** If $A$ is the incidence matrix of a directed graph, show that $A + A^2$ represents the total number of 1-step or 2-step links between vertices.

**12.** Describe the direct and indirect dominance given by the following graph:

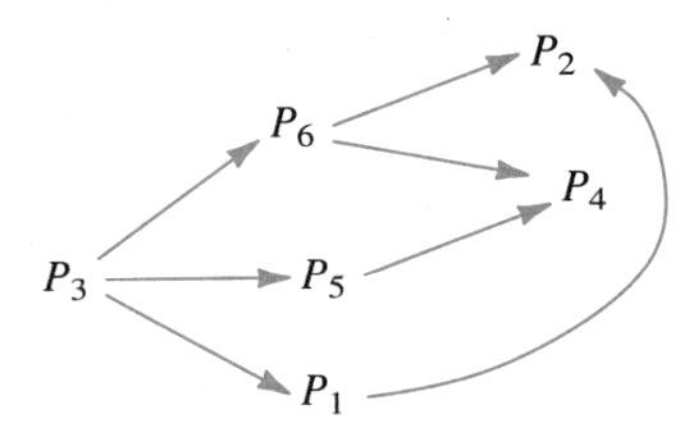


**Figure 1.18**

# SUMMARY

- An ***n*-component row vector** is an ordered set of $n$ numbers called **scalars,** written as $(x_1, x_2, \ldots, x_n)$. (p. 45)
- An ***n*-component column vector** is an ordered set of $n$ numbers written as (p. 46)

$$\begin{pmatrix} x_1 \\ x_2 \\ \vdots \\ x_n \end{pmatrix}$$

- A vector all of whose components are zero is called a **zero vector.** (p. 46)
- **Vector addition** and **multiplication by scalars** are defined by (p. 51)

$$\mathbf{a} + \mathbf{b} = \begin{pmatrix} a_1 + b_1 \\ a_2 + b_2 \\ \vdots \\ a_n + b_n \end{pmatrix} \quad \text{and} \quad \alpha\mathbf{a} = \begin{pmatrix} \alpha a_1 \\ \alpha a_2 \\ \vdots \\ \alpha a_n \end{pmatrix}$$

- An **$m \times n$ matrix** is a rectangular array of $mn$ numbers arranged in $m$ rows and $n$ columns:(pp. 9, 10, 48)

$$A = \begin{pmatrix} a_{11} & a_{12} \cdots a_{1n} \\ a_{21} & a_{22} \cdots a_{2n} \\ \vdots & \vdots \quad\;\; \vdots \\ a_{m1} & a_{m2} \cdots a_{mn} \end{pmatrix}$$

The matrix $A$ is also written as $A = (a_{ij})$.

- A matrix all of whose components are zero is called a **zero matrix.** (p. 48)
- If $A$ and $B$ are $m \times n$ matrices, then $A + B$ and $\alpha A$ ($\alpha$ a scalar) are $m \times n$ matrices: (pp. 51, 52)

The $ij$th component of $A + B$ is $a_{ij} + b_{ij}$.

The $ij$th component of $\alpha A$ is $\alpha a_{ij}$.

- The **scalar product** of two $n$-component vectors:

$$\mathbf{a} \cdot \mathbf{b} = (a_1, a_2, \cdots, a_n) \cdot \begin{pmatrix} b_1 \\ b_2 \\ \vdots \\ b_n \end{pmatrix} = a_1b_1 + a_2b_2 + \cdots + a_nb_n = \sum_{i=1}^{n} a_ib_i$$

(pp. 63, 64)

- ***Matrix products***

Let $A$ be an $m \times n$ matrix and let $B$ be an $n \times p$ matrix. Then $AB$ is an $m \times p$ matrix and (p. 64)

$$ij\text{th component of } AB = (i\text{th row of } A) \cdot (j\text{th column of } B)$$

$$= a_{i1}b_{1j} + a_{i2}b_{2j} + \cdots + a_{in}b_{nj} = \sum_{k=1}^{n} a_{ik}b_{kj}$$

- Matrix products do not in general commute; that is, it is usually the case that $AB \neq BA$. (p. 66)
- ***Associative law for matrix multiplication***

If $A$ is an $n \times m$ matrix, $B$ is $m \times p$ and $C$ is $p \times q$, then

$$A(BC) = (AB)C$$

and both $A(BC)$ and $(AB)C$ are $n \times q$ matrices. (p. 68)

- ***Distributive laws for matrix multiplication***

If all the products are defined, then (p. 69)

$$A(B + C) = AB + AC \quad \text{and} \quad (A + B)C = AC + BC$$

- The **coefficient matrix** of the linear system

$$\begin{aligned} a_{11}x_1 + a_{12}x_2 + \cdots + a_{1n}x_n &= b_1 \\ a_{21}x_1 + a_{22}x_2 + \cdots + a_{2n}x_n &= b_2 \\ \vdots \qquad\quad \vdots \qquad\qquad\quad \vdots \quad &\quad \vdots \\ a_{m1}x_1 + a_{m2}x_2 + \cdots + a_{mn}x_n &= b_m \end{aligned}$$

is the matrix (pp. 10, 16)

$$A = \begin{pmatrix} a_{11} & a_{12} & \cdots & a_{1n} \\ a_{21} & a_{22} & \cdots & a_{2n} \\ \vdots & \vdots & & \vdots \\ a_{m1} & a_{m2} & \cdots & a_{mn} \end{pmatrix}$$

- The linear system above can be written using the **augmented matrix** (pp. 10, 16)

$$\left(\begin{array}{cccc|c} a_{11} & a_{12} & \cdots & a_{1n} & b_1 \\ a_{21} & a_{22} & \cdots & a_{2n} & b_2 \\ \vdots & \vdots & & \vdots & \vdots \\ a_{m1} & a_{m2} & \cdots & a_{mn} & b_m \end{array}\right)$$

It can also be written as $A\mathbf{x} = \mathbf{b}$, where (p. 92)

$$\mathbf{x} = \begin{pmatrix} x_1 \\ x_2 \\ \vdots \\ x_n \end{pmatrix} \quad \text{and} \quad \mathbf{b} = \begin{pmatrix} b_1 \\ b_2 \\ \vdots \\ b_m \end{pmatrix}$$

- A matrix is in **reduced row echelon form** if the four conditions on page 14 hold. (p. 14)
- A matrix is in **row echelon form** if the first three conditions on page 14 hold. (p. 14)
- A **pivot** is the first nonzero component in the row of a matrix. (p. 14)
- The three **elementary row operations** are: (p. 10)

Multiply the $i$th row of a matrix by $c$: $R_i \to cR_i$, where $c \neq 0$.

Multiply the $i$th row by $c$ and add it to the $j$th row: $R_j \to R_j + cR_i$.

Permute the $i$th and $j$th rows: $R_i \rightleftarrows R_j$.

- The process of applying elementary row operations to a matrix is called **row reduction.** (p. 10)
- **Gauss-Jordan elimination** is the process of solving a system of equations by row-reducing its augmented matrix to its reduced row echelon form, using the process outlined on page 9. (pp. 9, 16)
- **Gaussian elimination** is the process of solving a system of equations by row-reducing its augmented matrix to row echelon form and using **back substitution.** (p. 16)
- A linear system that has one or more solutions is called **consistent.** (p. 10)
- A linear system that has no solution is called **inconsistent.** (pp. 3, 10)
- A linear system having solutions has either a unique solution or an infinite number of solutions. (p. 15)
- A **homogeneous** system of $m$ equations in $n$ unknowns is a linear system of the form (p. 39)

$$\begin{aligned} a_{11}x_1 + a_{12}x_2 + \cdots + a_{1n}x_n &= 0 \\ a_{21}x_1 + a_{22}x_2 + \cdots + a_{2n}x_n &= 0 \\ \vdots \qquad \vdots \qquad\quad \vdots \quad &\quad \vdots \\ a_{m1}x_1 + a_{m2}x_2 + \cdots + a_{mn}x_n &= 0 \end{aligned}$$

- A homogeneous linear system always has the **trivial solution** (or **zero solution**) (p. 40)

$$x_1 = x_2 = \cdots = x_n = 0$$

- Solutions other than the zero solution to a homogeneous linear system are called **nontrivial solutions.** (p. 40)

- The homogeneous linear system above has an infinite number of solutions if there are more unknowns than equations ($n > m$). (p. 41)
- The **$n \times n$ identity matrix, $I_n$,** is the $n \times n$ matrix with 1's down the **main diagonal** and 0's everywhere else. $I_n$ is usually denoted by $I$. (p. 98)
- If $A$ is a square matrix, then $AI = IA = A$. (p. 99)
- The $n \times n$ matrix $A$ is **invertible** if there is an $n \times n$ matrix $A^{-1}$ such that (p. 99)

$$AA^{-1} = A^{-1}A = I$$

In this case the $A^{-1}$ is called the **inverse** of $A$.

- If $A$ is invertible, its inverse is unique. (p. 100)
- If $A$ and $B$ are invertible $n \times n$ matrices, then $AB$ is invertible and (p. 100)

$$(AB)^{-1} = B^{-1}A^{-1}$$

- To determine whether an $n \times n$ matrix $A$ is invertible:

  **i.** Write the augmented matrix $(A|I)$. (p. 103)

  **ii.** Use row reduction to reduce $A$ to its reduced row echelon form.

  **iii. a.** If the reduced row echelon form of $A$ is $I$, then $A^{-1}$ will be the matrix to the right of the vertical bar.

  **b.** If the reduced row echelon form of $A$ contains a row of zeros, then $A$ is not invertible.

- The $2 \times 2$ matrix $A = \begin{pmatrix} a_{11} & a_{12} \\ a_{21} & a_{22} \end{pmatrix}$ is invertible if and only if (p. 104)

$$\text{determinant of } A = \det A = a_{11}a_{22} - a_{12}a_{21} \neq 0$$

In that case

$$A^{-1} = \frac{1}{\det A}\begin{pmatrix} a_{22} & -a_{12} \\ -a_{21} & a_{11} \end{pmatrix}$$

- Two matrices $A$ and $B$ are **row equivalent** if $A$ can be transformed into $B$ by row reduction. (p. 107)
- Let $A$ be an $n \times n$ matrix. If $AB = I$ or $BA = I$, then $A$ is invertible and $B = A^{-1}$. (p. 112)
- If $A = (a_{ij})$, then the **transpose of A,** written $A^t$, is given by $A^t = (a_{ji})$. That is, $A^t$ is obtained by interchanging the rows and columns of $A$. (p. 122)
- ***Facts about transpose***

  If all sums and products are defined and if $A$ is invertible, then (p. 122)

  $(A^t)^t = A$ $\quad (AB)^t = B^tA^t \quad (A+B)^t = A^t + B^t \quad$ If $A$ is invertible, then $(A^t)^{-1} = (A^{-1})^t$

- A square matrix $A$ is **symmetric** if $A^t = A$. (p. 123)
- An **elementary matrix** is a square matrix obtained by performing exactly one of the elementary row operations on the identity matrix. The three types of elementary matrices are: (p. 128)

| | |
|---|---|
| $cR_i$ | multiply the $i$th row of $I$ by $c$, $c \neq 0$ |
| $R_j + cR_i$ | multiply the $i$th row of $I$ by $c$ and add it to the $j$th row, $c \neq 0$ |
| $P_{ij}$ | permute the $i$th and $j$th rows |

- A square matrix is invertible if and only if it is the product of elementary matrices. (p. 130)

- Any square matrix can be written as the product of elementary matrices and one upper triangular matrix. (p. 133)

- ***LU*-factorization**

Suppose that the invertible matrix $A$ can be row-reduced to an upper triangular matrix without performing any permutations. Then there exist unique matrices $L$ and $U$ such that $L$ is lower triangular with 1's on the diagonal, $U$ is an invertible upper triangular matrix, and $A = LU$. (p. 142)

- ***Permutation matrix***

$E = P_{ij}$ is an **elementary permutation** matrix. A product of elementary permutation matrices is called a **permutation matrix.** (p. 144)

- ***PA = LU*-factorization**

Let $A$ be any $m \times n$ matrix. Then there exists a permutation matrix $P$ such that $PA = LU$, where $L$ and $U$ are as in the $LU$-factorization. $P$, $A$, and $U$ are not, in general, unique. (p. 145)

- ***Summing up theorem***

Let $A$ be an $n \times n$ matrix. Then the following are equivalent: (pp. 132, 145)

**i.** $A$ is invertible.

**ii.** The only solution to the homogeneous system $A\mathbf{x} = \mathbf{0}$ is the trivial solution ($\mathbf{x} = \mathbf{0}$).

**iii.** The system $A\mathbf{x} = \mathbf{b}$ has a unique solution for every $n$-vector $\mathbf{b}$.

**iv.** $A$ is row equivalent to the $n \times n$ identity matrix $I_n$.

**v.** $A$ can be written as a product of elementary matrices.

**vi.** $\det A \neq 0$ (so far, $\det A$ is defined only if $A$ is a $2 \times 2$ matrix).

**vii.** The row echelon form of $A$ has $n$ pivots.

**viii.** There exists a permutation matrix $P$, a lower triangular matrix $L$ with 1's on the diagonal, and an invertible upper triangular matrix $U$ such that $PA = LU$.

# REVIEW EXERCISES

In Exercises 1–14 find all solutions (if any) to the given systems.

**1.** $$\begin{aligned} 3x_1 + 6x_2 &= 9 \\ -2x_1 + 3x_2 &= 4 \end{aligned}$$

**2.** $$\begin{aligned} 3x_1 + 6x_2 &= 9 \\ 2x_1 + 4x_2 &= 6 \end{aligned}$$

**3.** $$\begin{aligned} 3x_1 - 6x_2 &= 9 \\ -2x_1 + 4x_2 &= 6 \end{aligned}$$

**4.** $$\begin{aligned} x_1 + x_2 + x_3 &= 2 \\ 2x_1 - x_2 + 2x_3 &= 4 \\ -3x_1 + 2x_2 + 3x_3 &= 8 \end{aligned}$$

**5.** $$\begin{aligned} x_1 + x_2 + x_3 &= 0 \\ 2x_1 - x_2 + 2x_3 &= 0 \\ -3x_1 + 2x_2 + 3x_3 &= 0 \end{aligned}$$

**6.** $$\begin{aligned} x_1 + x_2 + x_3 &= 2 \\ 2x_1 - x_2 + 2x_3 &= 4 \\ -x_1 + 4x_2 + x_3 &= 2 \end{aligned}$$

**7.** $$\begin{aligned} x_1 + x_2 + x_3 &= 2 \\ 2x_1 - x_2 + 2x_3 &= 4 \\ -x_1 + 4x_2 + x_3 &= 3 \end{aligned}$$

**8.** $$\begin{aligned} x_1 + x_2 + x_3 &= 0 \\ 2x_1 - x_2 + 2x_3 &= 0 \\ -x_1 + 4x_2 + x_3 &= 0 \end{aligned}$$

**9.** $$\begin{aligned} 2x_1 + x_2 - 3x_3 &= 0 \\ 4x_1 - x_2 + x_3 &= 0 \end{aligned}$$

**10.** $$\begin{aligned} x_1 + x_2 &= 0 \\ 2x_1 + x_2 &= 0 \\ 3x_1 + x_2 &= 0 \end{aligned}$$

**11.** $$\begin{aligned} x_1 + x_2 &= 1 \\ 2x_1 + x_2 &= 3 \\ 3x_1 + x_2 &= 4 \end{aligned}$$

**12.** $$\begin{aligned} x_1 + x_2 + x_3 + x_4 &= 4 \\ 2x_1 - 3x_2 - x_3 + 4x_4 &= 7 \\ -2x_1 + 4x_2 + x_3 - 2x_4 &= 1 \\ 5x_1 - x_2 + 2x_3 + x_4 &= -1 \end{aligned}$$

**13.** $$\begin{aligned} x_1 + x_2 + x_3 + x_4 &= 0 \\ 2x_1 - 3x_2 - x_3 + 4x_4 &= 0 \\ -2x_1 + 4x_2 + x_3 - 2x_4 &= 0 \\ 5x_1 - x_2 + 2x_3 + x_4 &= 0 \end{aligned}$$

**14.** $$\begin{aligned} x_1 + x_2 + x_3 + x_4 &= 0 \\ 2x_1 - 3x_2 - x_3 + 4x_4 &= 0 \\ -2x_1 + 4x_2 + x_3 - 2x_4 &= 0 \end{aligned}$$

In Exercises 15–22 perform the indicated computations.

**15.** $$3\begin{pmatrix} -2 & 1 \\ 0 & 4 \\ 2 & 3 \end{pmatrix}$$

**16.** $$\begin{pmatrix} 1 & 0 & 3 \\ 2 & -1 & 6 \end{pmatrix} + \begin{pmatrix} 2 & 0 & 4 \\ -2 & 5 & 8 \end{pmatrix}$$

**17.** $$5\begin{pmatrix} 2 & 1 & 3 \\ -1 & 2 & 4 \\ -6 & 1 & 5 \end{pmatrix} - 3\begin{pmatrix} -2 & 1 & 4 \\ 5 & 0 & 7 \\ 2 & -1 & 3 \end{pmatrix}$$

**18.** $$\begin{pmatrix} 2 & 3 \\ -1 & 4 \end{pmatrix}\begin{pmatrix} 5 & -1 \\ 2 & 7 \end{pmatrix}$$

**19.** $$\begin{pmatrix} 2 & 3 & 1 & 5 \\ 0 & 6 & 2 & 4 \end{pmatrix}\begin{pmatrix} 5 & 7 & 1 \\ 2 & 0 & 3 \\ 1 & 0 & 0 \\ 0 & 5 & 6 \end{pmatrix}$$

**20.** $$\begin{pmatrix} 2 & 3 & 5 \\ -1 & 6 & 4 \\ 1 & 0 & 6 \end{pmatrix}\begin{pmatrix} 0 & -1 & 2 \\ 3 & 1 & 2 \\ -7 & 3 & 5 \end{pmatrix}$$

**21.** $$\begin{pmatrix} 1 & 0 & 3 & -1 & 5 \\ 2 & 1 & 6 & 2 & 5 \end{pmatrix}\begin{pmatrix} 7 & 1 \\ 2 & 3 \\ -1 & 0 \\ 5 & 6 \\ 2 & 3 \end{pmatrix}$$

**22.** $$\begin{pmatrix} 1 & -1 & 2 \\ 3 & 5 & 6 \\ 2 & 4 & -1 \end{pmatrix}\begin{pmatrix} 2 \\ 1 \\ 3 \end{pmatrix}$$

In Exercises 23–26 determine whether the given matrix is in row echelon form (but not reduced row echelon form), reduced row echelon form, or neither.

**23.** $$\begin{pmatrix} 1 & 0 & 0 & 0 \\ 0 & 1 & 0 & 2 \\ 0 & 0 & 1 & 3 \end{pmatrix}$$

**24.** $$\begin{pmatrix} 1 & 8 & 1 & 0 \\ 0 & 1 & 5 & -7 \\ 0 & 0 & 1 & 4 \end{pmatrix}$$

**25.** $$\begin{pmatrix} 1 & 0 \\ 0 & 3 \\ 0 & 0 \end{pmatrix}$$

**26.** $$\begin{pmatrix} 1 & 0 & 2 & 0 \\ 0 & 1 & 3 & 0 \end{pmatrix}$$

In Exercises 27 and 28 reduce the matrix to row echelon form and reduced row echelon form.

**27.** $$\begin{pmatrix} 2 & 8 & -2 \\ 1 & 0 & -6 \end{pmatrix}$$

**28.** $$\begin{pmatrix} 1 & -1 & 2 & 4 \\ -1 & 2 & 0 & 3 \\ 2 & 3 & -1 & 1 \end{pmatrix}$$

In Exercises 29–33 calculate the row echelon form and the inverse of the given matrix (if the inverse exists).

**29.** $\begin{pmatrix} 2 & 3 \\ -1 & 4 \end{pmatrix}$ **30.** $\begin{pmatrix} -1 & 2 \\ 2 & -4 \end{pmatrix}$ **31.** $\begin{pmatrix} 1 & 2 & 0 \\ 2 & 1 & -1 \\ 3 & 1 & 1 \end{pmatrix}$

**32.** $\begin{pmatrix} -1 & 2 & 0 \\ 4 & 1 & -3 \\ 2 & 5 & -3 \end{pmatrix}$ **33.** $\begin{pmatrix} 2 & 0 & 4 \\ -1 & 3 & 1 \\ 0 & 1 & 2 \end{pmatrix}$

In Exercises 34–36 first write the system in the form $A\mathbf{x} = \mathbf{b}$, then calculate $A^{-1}$, and finally, use matrix multiplication to obtain the solution vector.

**34.** $\begin{aligned} x_1 - 3x_2 &= 4 \\ 2x_1 + 5x_2 &= 7 \end{aligned}$

**35.** $\begin{aligned} x_1 + 2x_2 \phantom{{}-x_3} &= 3 \\ 2x_1 + x_2 - x_3 &= -1 \\ 3x_1 + x_2 + x_3 &= 7 \end{aligned}$

**36.** $\begin{aligned} 2x_1 \phantom{{}+3x_2} + 4x_3 &= 7 \\ -x_1 + 3x_2 + x_3 &= -4 \\ x_2 + 2x_3 &= 5 \end{aligned}$

In Exercises 37–42 calculate the transpose of the given matrix and determine whether the matrix is symmetric or skew-symmetric.†

**37.** $\begin{pmatrix} 2 & 3 & 1 \\ -1 & 0 & 2 \end{pmatrix}$ **38.** $\begin{pmatrix} 4 & 6 \\ 6 & 4 \end{pmatrix}$ **39.** $\begin{pmatrix} 2 & 3 & 1 \\ 3 & -6 & -5 \\ 1 & -5 & 9 \end{pmatrix}$

**40.** $\begin{pmatrix} 0 & 5 & 6 \\ -5 & 0 & 4 \\ -6 & -4 & 0 \end{pmatrix}$ **41.** $\begin{pmatrix} 1 & -1 & 4 & 6 \\ -1 & 2 & 5 & 7 \\ 4 & 5 & 3 & -8 \\ 6 & 7 & -8 & 9 \end{pmatrix}$ **42.** $\begin{pmatrix} 0 & 1 & -1 & 1 \\ -1 & 0 & 1 & -2 \\ 1 & 1 & 0 & 1 \\ 1 & -2 & -1 & 0 \end{pmatrix}$

In Exercises 43–47 find a $3 \times 3$ elementary matrix that will carry out the given row operation.

**43.** $R_2 \to -2R_2$ **44.** $R_1 \to R_1 + 2R_2$ **45.** $R_3 \to R_3 - 5R_1$

**46.** $R_3 \rightleftarrows R_1$ **47.** $R_2 \to R_2 + \frac{1}{5}R_3$

In Exercises 48–50 find the inverse of the elementary matrix.

**48.** $\begin{pmatrix} 1 & 3 \\ 0 & 1 \end{pmatrix}$ **49.** $\begin{pmatrix} 0 & 1 & 0 \\ 1 & 0 & 0 \\ 0 & 0 & 1 \end{pmatrix}$ **50.** $\begin{pmatrix} 1 & 0 & 0 \\ 0 & 1 & 0 \\ 0 & 0 & -\frac{1}{3} \end{pmatrix}$

In Exercises 51 and 52 write the matrix as the product of elementary matrices.

**51.** $\begin{pmatrix} 2 & -1 \\ -1 & 1 \end{pmatrix}$ **52.** $\begin{pmatrix} 1 & 0 & 3 \\ 2 & 1 & -5 \\ 3 & 2 & 4 \end{pmatrix}$

---

† From Problem 1.9.18 on page 125 we have $A$ is skew-symmetric if $A^t = -A$.

In Exercises 53 and 54 write each matrix as the product of elementary matrices and one upper triangular matrix.

**53.** $\begin{pmatrix} 2 & -1 \\ -4 & 2 \end{pmatrix}$

**54.** $\begin{pmatrix} 1 & -2 & 3 \\ 2 & 0 & 4 \\ 1 & 2 & 1 \end{pmatrix}$

In Exercises 55 and 56 find the $LU$-factorization of $A$ and use it to solve $A\mathbf{x} = \mathbf{b}$.

**55.** $A = \begin{pmatrix} 1 & -2 & 5 \\ 2 & -5 & 7 \\ 4 & -3 & 8 \end{pmatrix}$; $\mathbf{b} = \begin{pmatrix} -1 \\ 2 \\ 5 \end{pmatrix}$

**56.** $A = \begin{pmatrix} 2 & 5 & -2 \\ 4 & 11 & 3 \\ 6 & -1 & 2 \end{pmatrix}$; $\mathbf{b} = \begin{pmatrix} 3 \\ 0 \\ 7 \end{pmatrix}$

In Exercises 57 and 58 find a permutation matrix $P$ and matrices $L$ and $U$ such that $PA = LU$ and use them to solve the system $A\mathbf{x} = \mathbf{b}$.

**57.** $A = \begin{pmatrix} 0 & -1 & 4 \\ 3 & 5 & 8 \\ 1 & 3 & -2 \end{pmatrix}$; $\mathbf{b} = \begin{pmatrix} 3 \\ -2 \\ -1 \end{pmatrix}$

**58.** $A = \begin{pmatrix} 0 & 3 & 2 \\ 1 & 2 & 4 \\ 2 & 6 & -5 \end{pmatrix}$; $\mathbf{b} = \begin{pmatrix} -2 \\ 8 \\ 10 \end{pmatrix}$

In Exercises 59 and 60 find the matrix that represents each graph.

**59.**

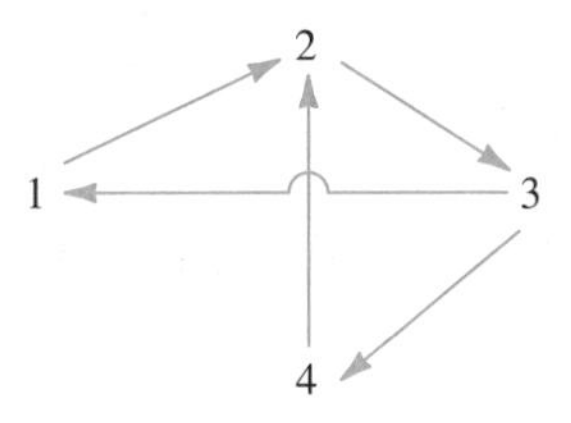


**60.**

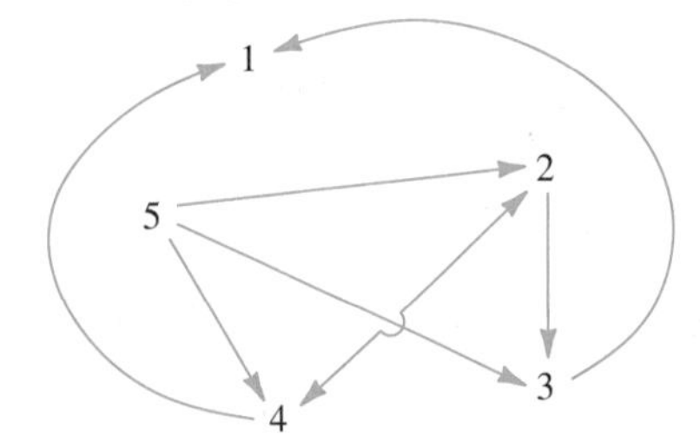


**61.** Draw the graph represented by the following matrix. $\begin{pmatrix} 0 & 0 & 1 & 1 & 0 \\ 0 & 0 & 0 & 1 & 1 \\ 1 & 0 & 0 & 0 & 0 \\ 0 & 1 & 1 & 0 & 1 \\ 1 & 0 & 1 & 0 & 0 \end{pmatrix}$

# 2

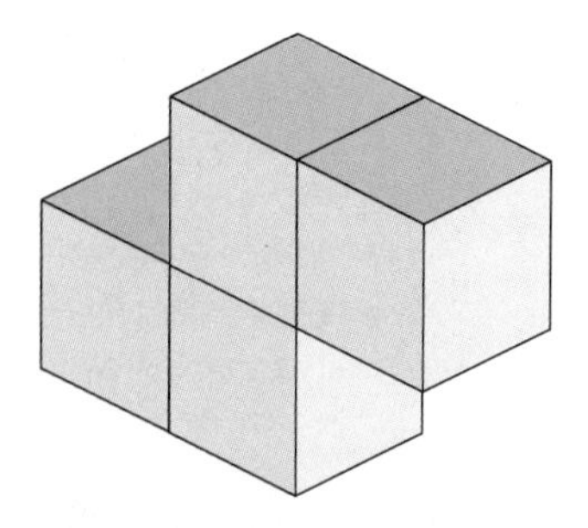

# Determinants

## 2.1 DEFINITIONS

Let $A = \begin{pmatrix} a_{11} & a_{12} \\ a_{21} & a_{22} \end{pmatrix}$ be a $2 \times 2$ matrix. In Section 1.8 on page 104 we defined the determinant of $A$ by

$$\det A = a_{11}a_{22} - a_{12}a_{21} \tag{1}$$

We shall often denote $\det A$ by

$$|A| \quad \text{or} \quad \begin{vmatrix} a_{11} & a_{12} \\ a_{21} & a_{22} \end{vmatrix} \tag{2}$$

*Remark.* You should not be confused by the use of absolute value bars here. $|A|$ denotes $\det A$ if $A$ is a square matrix. $|x|$ denotes the absolute value of $x$ if $x$ is a real or complex number.

We showed that $A$ is invertible if and only if $\det A \neq 0$. As we shall see, this important theorem is valid for $n \times n$ matrices.

In this chapter we develop some of the basic properties of determinants and see how they can be used to calculate matrix inverses and solve systems of $n$ linear equations in $n$ unknowns.

We define the determinant of an $n \times n$ matrix *inductively*. In other words, we use our knowledge of a $2 \times 2$ determinant to define a $3 \times 3$ determinant, use this to define a $4 \times 4$ determinant, and so on. We start by defining a $3 \times 3$ determinant.†

**DEFINITION 1** **3 × 3 Determinant** Let $A = \begin{pmatrix} a_{11} & a_{12} & a_{13} \\ a_{21} & a_{22} & a_{23} \\ a_{31} & a_{32} & a_{33} \end{pmatrix}$. Then

$$\det A = |A| = a_{11}\begin{vmatrix} a_{22} & a_{23} \\ a_{32} & a_{33} \end{vmatrix} - a_{12}\begin{vmatrix} a_{21} & a_{23} \\ a_{31} & a_{33} \end{vmatrix} + a_{13}\begin{vmatrix} a_{21} & a_{22} \\ a_{31} & a_{32} \end{vmatrix} \qquad \textbf{(3)}$$

Note the minus sign before the second term on the right side of (3).

**EXAMPLE 1** **Calculating a 3 × 3 Determinant** Let $A = \begin{pmatrix} 3 & 5 & 2 \\ 4 & 2 & 3 \\ -1 & 2 & 4 \end{pmatrix}$. Calculate $|A|$.

**Solution**

$$|A| = \begin{vmatrix} 3 & 5 & 2 \\ 4 & 2 & 3 \\ -1 & 2 & 4 \end{vmatrix} = 3\begin{vmatrix} 2 & 3 \\ 2 & 4 \end{vmatrix} - 5\begin{vmatrix} 4 & 3 \\ -1 & 4 \end{vmatrix} + 2\begin{vmatrix} 4 & 2 \\ -1 & 2 \end{vmatrix}$$
$$= 3 \cdot 2 - 5 \cdot 19 + 2 \cdot 10 = -69$$

**EXAMPLE 2** **Calculating a 3 × 3 Determinant** Calculate $\begin{vmatrix} 2 & -3 & 5 \\ 1 & 0 & 4 \\ 3 & -3 & 9 \end{vmatrix}$.

**Solution**

$$\begin{vmatrix} 2 & -3 & 5 \\ 1 & 0 & 4 \\ 3 & -3 & 9 \end{vmatrix} = 2\begin{vmatrix} 0 & 4 \\ -3 & 9 \end{vmatrix} - (-3)\begin{vmatrix} 1 & 4 \\ 3 & 9 \end{vmatrix} + 5\begin{vmatrix} 1 & 0 \\ 3 & -3 \end{vmatrix}$$
$$= 2 \cdot 12 + 3(-3) + 5(-3) = 0$$

† There are several ways to define a determinant and this is one of them. It is important to realize that "det" is a function that assigns a *number* to a *square* matrix.

There is another method for calculating $3 \times 3$ determinants. From equation (3) we have

$$\begin{vmatrix} a_{11} & a_{12} & a_{13} \\ a_{21} & a_{22} & a_{23} \\ a_{31} & a_{32} & a_{33} \end{vmatrix} = a_{11}(a_{22}a_{33} - a_{23}a_{32}) - a_{12}(a_{21}a_{33} - a_{23}a_{31}) + a_{13}(a_{21}a_{32} - a_{22}a_{31})$$

or

$$|A| = a_{11}a_{22}a_{33} + a_{12}a_{23}a_{31} + a_{13}a_{21}a_{32} - a_{13}a_{22}a_{31} - a_{12}a_{21}a_{33} - a_{11}a_{32}a_{23} \tag{4}$$

We write $A$ and adjoin its first two columns to it:

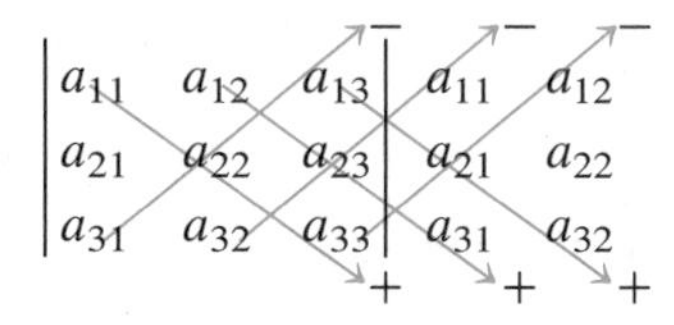

We then calculate the six products, put minus signs before the products with arrows pointing upward, and add. This gives the sum in equation (4).

**EXAMPLE 3** **Calculating a $3 \times 3$ Determinant by Using the New Method** Calculate $\begin{vmatrix} 3 & 5 & 2 \\ 4 & 2 & 3 \\ -1 & 2 & 4 \end{vmatrix}$ by using this new method.

**Solution** Writing $\begin{vmatrix} 3 & 5 & 2 \\ 4 & 2 & 3 \\ -1 & 2 & 4 \end{vmatrix} \begin{matrix} 3 & 5 \\ 4 & 2 \\ -1 & 2 \end{matrix}$ and multiplying as indicated, we obtain

$$\begin{aligned} |A| &= (3)(2)(4) + (5)(3)(-1) + (2)(4)(2) - (-1)(2)(2) - 2(3)(3) - (4)(4)(5) \\ &= 24 - 15 + 16 + 4 - 18 - 80 = -69 \end{aligned}$$

**WARNING** The method given above will *not* work for $n \times n$ determinants if $n > 3$. If you try something analogous for $4 \times 4$ or higher-order determinants, you will get the wrong answer.

Before defining $n \times n$ determinants, we first note that in equation (3) $\begin{pmatrix} a_{22} & a_{23} \\ a_{32} & a_{33} \end{pmatrix}$ is the matrix obtained by deleting the first row and first column of $A$; $\begin{pmatrix} a_{21} & a_{23} \\ a_{31} & a_{33} \end{pmatrix}$ is the matrix obtained by deleting the first row and second column of

$A$; and $\begin{pmatrix} a_{21} & a_{22} \\ a_{31} & a_{32} \end{pmatrix}$ is the matrix obtained by deleting the first row and third column of $A$. If we denote these three matrices by $M_{11}$, $M_{12}$, and $M_{13}$, respectively, and if $A_{11} = \det M_{11}$, $A_{12} = -\det M_{12}$, and $A_{13} = \det M_{13}$, then equation (3) can be written

$$\det A = |A| = a_{11}A_{11} + a_{12}A_{12} + a_{13}A_{13} \tag{5}$$

**DEFINITION 2** **Minor** Let $A$ be an $n \times n$ matrix and let $M_{ij}$ be the $(n-1) \times (n-1)$ matrix obtained from $A$ by deleting the $i$th row and $j$th column of $A$. $M_{ij}$ is called the ***ij*th minor** of $A$.

**EXAMPLE 4** **Finding Two Minors of a 3 × 3 Matrix** Let $A = \begin{pmatrix} 2 & -1 & 4 \\ 0 & 1 & 5 \\ 6 & 3 & -4 \end{pmatrix}$. Find $M_{13}$ and $M_{32}$.

**Solution** Deleting the first row and third column of $A$, we obtain $M_{13} = \begin{pmatrix} 0 & 1 \\ 6 & 3 \end{pmatrix}$. Similarly, by eliminating the third row and second column, we obtain $M_{32} = \begin{pmatrix} 2 & 4 \\ 0 & 5 \end{pmatrix}$.

**EXAMPLE 5** **Finding Two Minors of a 4 × 4 Matrix** Let $A = \begin{pmatrix} 1 & -3 & 5 & 6 \\ 2 & 4 & 0 & 3 \\ 1 & 5 & 9 & -2 \\ 4 & 0 & 2 & 7 \end{pmatrix}$. Find $M_{32}$ and $M_{24}$.

**Solution** Deleting the third row and second column of $A$, we find that $M_{32} = \begin{pmatrix} 1 & 5 & 6 \\ 2 & 0 & 3 \\ 4 & 2 & 7 \end{pmatrix}$; similarly, $M_{24} = \begin{pmatrix} 1 & -3 & 5 \\ 1 & 5 & 9 \\ 4 & 0 & 2 \end{pmatrix}$.

**DEFINITION 3** **Cofactor** Let $A$ be an $n \times n$ matrix. The ***ij*th cofactor** of $A$, denoted by $A_{ij}$, is given by

$$A_{ij} = (-1)^{i+j}|M_{ij}| \tag{6}$$

That is, the $ij$th cofactor of $A$ is obtained by taking the determinant of the $ij$th minor and multiplying it by $(-1)^{i+j}$. Note that

$$(-1)^{i+j} = \begin{cases} 1 & \text{if } i+j \text{ is even} \\ -1 & \text{if } i+j \text{ is odd} \end{cases}$$

*Remark.* Definition 3 makes sense because we are going to define an $n \times n$ determinant with the assumption that we already know what an $(n-1) \times (n-1)$ determinant is.

**EXAMPLE 6** **Finding Two Cofactors of a 4 × 4 Matrix** In Example 5 we have

$$A_{32} = (-1)^{3+2}|M_{32}| = -\begin{vmatrix} 1 & 5 & 6 \\ 2 & 0 & 3 \\ 4 & 2 & 7 \end{vmatrix} = -8$$

$$A_{24} = (-1)^{2+4}\begin{vmatrix} 1 & -3 & 5 \\ 1 & 5 & 9 \\ 4 & 0 & 2 \end{vmatrix} = -192$$

We now consider the general $n \times n$ matrix. Here

$$A = \begin{pmatrix} a_{11} & a_{12} & \cdots & a_{1n} \\ a_{21} & a_{22} & \cdots & a_{2n} \\ \vdots & \vdots & & \vdots \\ a_{n1} & a_{n2} & \cdots & a_{nn} \end{pmatrix} \tag{7}$$

**DEFINITION 4** ***n* × *n* Determinant** Let $A$ be an $n \times n$ matrix. Then the determinant of $A$, written $\det A$ or $|A|$, is given by

$$\det A = |A| = a_{11}A_{11} + a_{12}A_{12} + a_{13}A_{13} + \cdots + a_{1n}A_{1n} = \sum_{k=1}^{n} a_{1k}A_{1k} \tag{8}$$

The expression on the right side of (8) is called an **expansion by cofactors.**

*Remark.* In equation (8) we defined the determinant by expanding by cofactors in the first row of $A$. We shall see in the next section (Theorem 2.2.5) that we get the same answer if we expand by cofactors in any row or column.

**EXAMPLE 7** **Calculating the Determinant of a 4 × 4 Matrix** Calculate $\det A$, where

$$A = \begin{pmatrix} 1 & 3 & 5 & 2 \\ 0 & -1 & 3 & 4 \\ 2 & 1 & 9 & 6 \\ 3 & 2 & 4 & 8 \end{pmatrix}$$

**Solution**

$$\begin{vmatrix} 1 & 3 & 5 & 2 \\ 0 & -1 & 3 & 4 \\ 2 & 1 & 9 & 6 \\ 3 & 2 & 4 & 8 \end{vmatrix} = a_{11}A_{11} + a_{12}A_{12} + a_{13}A_{13} + a_{14}A_{14}$$

$$= 1\begin{vmatrix} -1 & 3 & 4 \\ 1 & 9 & 6 \\ 2 & 4 & 8 \end{vmatrix} - 3\begin{vmatrix} 0 & 3 & 4 \\ 2 & 9 & 6 \\ 3 & 4 & 8 \end{vmatrix} + 5\begin{vmatrix} 0 & -1 & 4 \\ 2 & 1 & 6 \\ 3 & 2 & 8 \end{vmatrix} - 2\begin{vmatrix} 0 & -1 & 3 \\ 2 & 1 & 9 \\ 3 & 2 & 4 \end{vmatrix}$$

$$= 1(-92) - 3(-70) + 5(2) - 2(-16) = 160$$

It is clear that calculating the determinant of an $n \times n$ matrix can be tedious. To calculate a $4 \times 4$ determinant, we must calculate four $3 \times 3$ determinants. To calculate a $5 \times 5$ determinant, we must calculate five $4 \times 4$ determinants—which is the same as calculating twenty $3 \times 3$ determinants. Fortunately, techniques exist for greatly simplifying these computations. Some of these methods are discussed in the next section. There are, however, some matrices whose determinants can easily be calculated. We begin by repeating a definition given on page 132.

**DEFINITION 5** **Triangular Matrix** A square matrix is called **upper triangular** if all its components below the diagonal are zero. It is **lower triangular** if all its components above the diagonal are zero. A matrix is called **diagonal** if all its elements not on the diagonal are zero; that is, $A = (a_{ij})$ is upper triangular if $a_{ij} = 0$ for $i > j$, lower triangular if $a_{ij} = 0$ for $i < j$, and diagonal if $a_{ij} = 0$ for $i \neq j$. Note that a diagonal matrix is both upper and lower triangular.

**EXAMPLE 8** **Six Triangular Matrices** The matrices $A = \begin{pmatrix} 2 & 1 & 7 \\ 0 & 2 & -5 \\ 0 & 0 & 1 \end{pmatrix}$ and

$B = \begin{pmatrix} -2 & 3 & 0 & 1 \\ 0 & 0 & 2 & 4 \\ 0 & 0 & 1 & 3 \\ 0 & 0 & 0 & -2 \end{pmatrix}$ are upper triangular; $C = \begin{pmatrix} 5 & 0 & 0 \\ 2 & 3 & 0 \\ -1 & 2 & 4 \end{pmatrix}$ and

$D = \begin{pmatrix} 0 & 0 \\ 1 & 0 \end{pmatrix}$ are lower triangular; $I$ (the identity matrix) and $E = \begin{pmatrix} 2 & 0 & 0 \\ 0 & -7 & 0 \\ 0 & 0 & -4 \end{pmatrix}$

are diagonal. Note that the matrix $E$ is both upper and lower triangular.

**EXAMPLE 9** **The Determinant of a Lower Triangular Matrix** The matrix

$$A = \begin{pmatrix} a_{11} & 0 & 0 & 0 \\ a_{21} & a_{22} & 0 & 0 \\ a_{31} & a_{32} & a_{33} & 0 \\ a_{41} & a_{42} & a_{43} & a_{44} \end{pmatrix}$$

is lower triangular. Compute $\det A$.

**Solution**

$$\begin{aligned} \det A &= a_{11}A_{11} + 0A_{12} + 0A_{13} + 0A_{14} = a_{11}A_{11} \\ &= a_{11} \begin{vmatrix} a_{22} & 0 & 0 \\ a_{32} & a_{33} & 0 \\ a_{42} & a_{43} & a_{44} \end{vmatrix} \\ &= a_{11}a_{22} \begin{vmatrix} a_{33} & 0 \\ a_{43} & a_{44} \end{vmatrix} \\ &= a_{11}a_{22}a_{33}a_{44} \end{aligned}$$

Example 9 can be generalized to prove the following theorem.

**THEOREM 1** Let $A = (a_{ij})$ be an upper or lower triangular $n \times n$ matrix. Then

$$\boxed{\det A = a_{11}a_{22}a_{33} \cdots a_{nn}} \quad \textbf{(9)}$$

That is: The determinant of a triangular matrix equals the product of its diagonal components.

**Proof** The lower triangular part of the theorem follows as in Example 9. We prove the upper triangular part by mathematical induction starting with $n = 2$. If $A$ is an upper triangular $2 \times 2$ matrix, then $A = \begin{pmatrix} a_{11} & a_{12} \\ 0 & a_{22} \end{pmatrix}$ and $\det A = a_{11}a_{22} - a_{12} \cdot 0 = a_{11}a_{22}$, so the theorem is true if $n = 2$. We assume it is true for $k = n - 1$ and prove it true for $k = n$. The determinant of an upper triangular $n \times n$ matrix is

$$\begin{vmatrix} a_{11} & a_{12} & a_{13} & \cdots & a_{1n} \\ 0 & a_{22} & a_{23} & \cdots & a_{2n} \\ 0 & 0 & a_{33} & \cdots & a_{3n} \\ \vdots & \vdots & \vdots & & \vdots \\ 0 & 0 & 0 & \cdots & a_{nn} \end{vmatrix} = a_{11} \begin{vmatrix} a_{22} & a_{23} & \cdots & a_{2n} \\ 0 & a_{23} & \cdots & a_{3n} \\ \vdots & \vdots & & \vdots \\ 0 & 0 & \cdots & a_{nn} \end{vmatrix} - a_{12} \begin{vmatrix} 0 & a_{23} & \cdots & a_{2n} \\ 0 & a_{23} & \cdots & a_{3n} \\ \vdots & \vdots & & \vdots \\ 0 & 0 & \cdots & a_{nn} \end{vmatrix}$$

$$+ a_{13} \begin{vmatrix} 0 & a_{22} & \cdots & a_{2n} \\ 0 & 0 & \cdots & a_{3n} \\ \vdots & \vdots & & \vdots \\ 0 & 0 & \cdots & a_{nn} \end{vmatrix} + \cdots + (-1)^{1+n} a_{1n} \begin{vmatrix} 0 & a_{22} & \cdots & a_{2,n-1} \\ 0 & 0 & \cdots & a_{3,n-1} \\ \vdots & \vdots & & \vdots \\ 0 & 0 & \cdots & 0 \end{vmatrix}$$

Each determinant above is the determinant of an upper triangular $(n - 1) \times (n - 1)$ matrix that, by the induction assumption, is equal to the product of its diagonal components. Each matrix except the first one has a column of zeros. So at least one of its diagonal components is zero. Thus each determinant except the first is zero. Finally,

$$\det A = a_{11} \begin{vmatrix} a_{22} & a_{23} & \cdots & a_{2n} \\ 0 & a_{33} & \cdots & a_{3n} \\ \vdots & \vdots & & \vdots \\ 0 & 0 & \cdots & a_{nn} \end{vmatrix} = a_{11}(a_{22}a_{33} \cdots a_{nn})$$

which shows that the theorem is true for $n \times n$ matrices.

EXAMPLE 10 **The Determinants of Six Triangular Matrices** The determinants of the six matrices in Example 8 are $|A| = 2 \cdot 2 \cdot 1 = 4$; $|B| = (-2)(0)(1)(-2) = 0$; $|C| = 5 \cdot 3 \cdot 4 = 60$; $|D| = 0$; $|I| = 1$; $|E| = (2)(-7)(-4) = 56$.

The following theorem will be very useful to us.

**THEOREM 2** Let $T$ be an upper triangular matrix. Then $T$ is invertible if and only if $\det T \neq 0$.

**Proof** Let

$$T = \begin{pmatrix} a_{11} & a_{12} & a_{13} & \cdots & a_{1n} \\ 0 & a_{22} & a_{23} & \cdots & a_{2n} \\ 0 & 0 & a_{33} & \cdots & a_{3n} \\ \vdots & \vdots & \vdots & & \vdots \\ 0 & 0 & 0 & \cdots & a_{nn} \end{pmatrix}$$

From Theorem 1,

$$\det T = a_{11}a_{22} \cdots a_{nn}$$

Thus $\det T \neq 0$ if and only if each of its diagonal components is nonzero.

If $\det T \neq 0$, then $T$ can be row-reduced to $I$ in the following way. For $i = 1, 2, \ldots, n$, divide the $i$th row of $T$ by $a_{ii} \neq 0$ to obtain

$$\begin{pmatrix} 1 & a'_{12} & \cdots & a'_{1n} \\ 0 & 1 & \cdots & a'_{2n} \\ \vdots & \vdots & & \vdots \\ 0 & 0 & \cdots & 1 \end{pmatrix}$$

This is the row echelon form of $T$, which has $n$ pivots, so by the Summing Up Theorem on page 132, $T$ is invertible.

Suppose that $\det T = 0$. Then at least one of the diagonal components of $T$ is zero. Let $a_{ii}$ be the first such component. Then $T$ can be written

$$T = \begin{pmatrix} a_{11} & a_{12} & \cdots & a_{1,i-1} & a_{1i} & a_{1,i+1} & \cdots & a_{1n} \\ 0 & a_{22} & \cdots & a_{2,i-1} & a_{2i} & a_{2,i+1} & \cdots & a_{2n} \\ \vdots & \vdots & & \vdots & \vdots & \vdots & & \vdots \\ 0 & 0 & \cdots & a_{i-1,i-1} & a_{i-1,i} & a_{i-1,i+1} & \cdots & a_{i-1,n} \\ 0 & 0 & \cdots & 0 & 0 & a_{i,i+1} & \cdots & a_{in} \\ 0 & 0 & \cdots & 0 & 0 & a_{i+1,i+1} & \cdots & a_{i+1,n} \\ \vdots & \vdots & & \vdots & \vdots & \vdots & \ddots & \vdots \\ 0 & 0 & \cdots & 0 & 0 & 0 & \cdots & a_{nn} \end{pmatrix}$$

When $T$ is reduced to its row echelon form, there will be no pivot in the $i$th column (explain why). Thus the row echelon form of $T$ has fewer than $n$ pivots, and from the Summing Up Theorem, we may conclude that $T$ is not invertible. ■

## Geometric Interpretation of the 2 × 2 Determinant

Let $A = \begin{vmatrix} a & b \\ c & d \end{vmatrix}$. In Figure 2.1 we plot the points $(a, c)$ and $(b, d)$ in the $xy$-plane

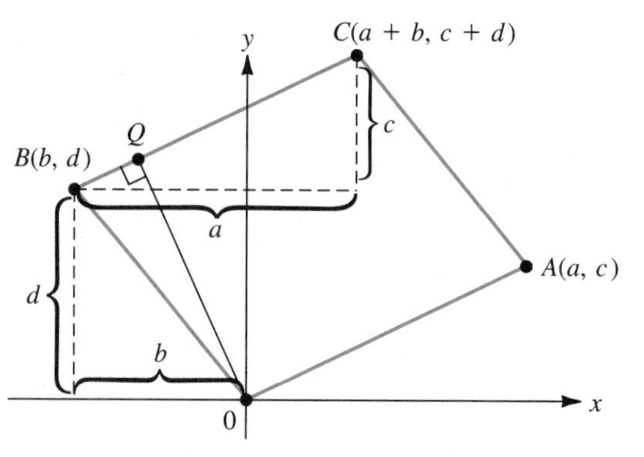


**Figure 2.1** $Q$ is on the line segment $BC$ and is also on a line perpendicular to $BC$ that passes through the origin. The area of the parallelogram is $\overline{0Q} \times \overline{0A}$.

and draw line segments from $(0, 0)$ to each of these points. We assume that the two lines are not collinear. This is the same as assuming that $(b, d)$ is not a multiple of $(a, c)$.

We define **the area generated by $A$** as the area of the parallelogram with three vertices at $(0, 0)$, $(a, c)$, and $(b, d)$.

**THEOREM 3** The area generated by $A = |\det A|$.†

**Proof** We assume that neither $a$ nor $c$ is zero. The proof for $a = 0$ or $c = 0$ is left as an exercise (see Problem 15).

The area of a parallelogram = Base × Height. The base of the parallelogram in Figure 2.1 has length $\overline{0A} = \sqrt{a^2 + c^2}$. The height of the parallelogram is $\overline{0Q}$, where $0Q$ is a line segment perpendicular to $BC$. From the figure we see that the coordinates of $C$, the fourth vertex of the parallelogram, are $x = a + b$ and $y = c + d$. Thus

$$\text{Slope of } BC = \frac{\Delta y}{\Delta x} = \frac{(c + d) - d}{(a + b) - b} = \frac{c}{a}$$

Then the equation of the line passing through $B$ and $C$ is

$$\frac{y - d}{x - b} = \frac{c}{a} \quad \text{or} \quad y = \frac{c}{a}x + d - \frac{bc}{a}$$

$$\text{Slope of } 0Q \underset{\text{Fact }(vi)\text{ on page 2}}{=} -\frac{1}{\text{slope of } BC} = -\frac{a}{c}$$

The equation of the line passing through $(0, 0)$ and $Q$ is

$$\frac{y - 0}{x - 0} = -\frac{a}{c} \quad \text{or} \quad y = -\frac{a}{c}x$$

$Q$ is the intersection of $BC$ and $0Q$, so it satisfies both equations. At the point of intersection we have

$$\frac{c}{a}x + d - \frac{bc}{a} = -\frac{a}{c}x$$

$$\left(\frac{c}{a} + \frac{a}{c}\right)x = \frac{bc}{a} - d$$

$$\frac{a^2 + c^2}{ac}x = \frac{bc - ad}{a}$$

$$x = \frac{ac(bc - ad)}{a(a^2 + c^2)} = \frac{c(bc - ad)}{a^2 + c^2} = -\frac{c(ad - bc)}{a^2 + c^2} = -\frac{c \det A}{a^2 + c^2}$$

† Here $|\det A|$ denotes the absolute value of the determinant of $A$.

and

$$y = -\frac{a}{c}x = -\frac{a}{c} \cdot -\frac{c \det A}{a^2 + c^2} = \frac{a \det A}{a^2 + c^2}$$

Thus $Q$ has coordinates $\left(\frac{-c \det A}{a^2 + c^2}, \frac{a \det A}{a^2 + c^2}\right)$

and

$$\overline{0Q} = \text{distance from } (0, 0) \text{ to } Q = \sqrt{\frac{c^2(\det A)^2}{(a^2 + c^2)^2} + \frac{a^2(\det A)^2}{(a^2 + c^2)^2}}$$

$$= \sqrt{\frac{(c^2 + a^2)(\det A)^2}{(c^2 + a^2)^2}} = \sqrt{\frac{(\det A)^2}{a^2 + c^2}} = \frac{|\det A|}{\sqrt{a^2 + c^2}}$$

Finally,

$$\text{Area of parallelogram} = \overline{0A} \times \overline{0Q} = \sqrt{a^2 + c^2} \times \frac{|\det A|}{\sqrt{a^2 + c^2}} = |\det A|$$

We shall be able to give a much simpler proof of this theorem when we discuss the cross product of two vectors in Section 3.4.

## PROBLEMS 2.1

### Self-Quiz

**I.** Which of the following is the cofactor of 3 in $\begin{pmatrix} 1 & 2 & 3 \\ 2 & -2 & 1 \\ 4 & 0 & 2 \end{pmatrix}$?

**a.** 8 **b.** −8
**c.** 3 **d.** 6
**e.** −10 **f.** 0

**II.** Which of the following is 0 for all $a$ and $b$?

**a.** $\begin{vmatrix} a & b \\ -b & a \end{vmatrix}$ **b.** $\begin{vmatrix} a & -b \\ -a & b \end{vmatrix}$ **c.** $\begin{vmatrix} a & a \\ b & -b \end{vmatrix}$

**d.** The determinants cannot be determined because values of $a$ and $b$ are not known.

**III.** If $A = \begin{pmatrix} 2 & -1 & 5 & 6 \\ 0 & 3 & 2 & 4 \\ 0 & 0 & -2 & 15 \\ 0 & 0 & 0 & 1 \end{pmatrix}$, then $\det A =$ ________.

**a.** 0 **b.** 12 **c.** −12 **d.** 6 **e.** −6

**IV.** Which of the following matrices are not invertible?

**a.** $\begin{pmatrix} 2 & 4 & 7 \\ 0 & 3 & 0 \\ 0 & 0 & 1 \end{pmatrix}$ **b.** $\begin{pmatrix} 2 & 4 & 7 \\ 0 & 0 & 3 \\ 0 & 0 & 1 \end{pmatrix}$

**c.** $\begin{pmatrix} 2 & -1 & 5 & 2 \\ 0 & 3 & 1 & 6 \\ 0 & 0 & 0 & 4 \\ 0 & 0 & 0 & 7 \end{pmatrix}$ **d.** $\begin{pmatrix} 2 & -1 & 5 & 2 \\ 0 & 3 & 1 & 6 \\ 0 & 0 & 4 & 0 \\ 0 & 0 & 0 & 7 \end{pmatrix}$

In Problems 1–10 calculate the determinant.

**1.** $\begin{vmatrix} 1 & 0 & 3 \\ 0 & 1 & 4 \\ 2 & 1 & 0 \end{vmatrix}$ **2.** $\begin{vmatrix} -1 & 1 & 0 \\ 2 & 1 & 4 \\ 1 & 5 & 6 \end{vmatrix}$ **3.** $\begin{vmatrix} 3 & -1 & 4 \\ 6 & 3 & 5 \\ 2 & -1 & 6 \end{vmatrix}$

**4.** $\begin{vmatrix} -1 & 0 & 6 \\ 0 & 2 & 4 \\ 1 & 2 & -3 \end{vmatrix}$ **5.** $\begin{vmatrix} -2 & 3 & 1 \\ 4 & 6 & 5 \\ 0 & 2 & 1 \end{vmatrix}$ **6.** $\begin{vmatrix} 5 & -2 & 1 \\ 6 & 0 & 3 \\ -2 & 1 & 4 \end{vmatrix}$

**7.** $\begin{vmatrix} 2 & 0 & 3 & 1 \\ 0 & 1 & 4 & 2 \\ 0 & 0 & 1 & 5 \\ 1 & 2 & 3 & 0 \end{vmatrix}$ **8.** $\begin{vmatrix} -3 & 0 & 0 & 0 \\ -4 & 7 & 0 & 0 \\ 5 & 8 & -1 & 0 \\ 2 & 3 & 0 & 6 \end{vmatrix}$ **9.** $\begin{vmatrix} -2 & 0 & 0 & 7 \\ 1 & 2 & -1 & 4 \\ 3 & 0 & -1 & 5 \\ 4 & 2 & 3 & 0 \end{vmatrix}$

**10.** $\begin{vmatrix} 2 & 3 & -1 & 4 & 5 \\ 0 & 1 & 7 & 8 & 2 \\ 0 & 0 & 4 & -1 & 5 \\ 0 & 0 & 0 & -2 & 8 \\ 0 & 0 & 0 & 0 & 6 \end{vmatrix}$

**11.** Show that if $A$ and $B$ are diagonal $n \times n$ matrices, then $\det AB = \det A \det B$.

***12.** Show that if $A$ and $B$ are lower triangular matrices, then $\det AB = \det A \det B$.

**13.** Show that, in general, it is not true that $\det(A + B) = \det A + \det B$.

**14.** Show that if $A$ is triangular, then $\det A \neq 0$ if and only if all the diagonal components of $A$ are nonzero.

**15.** Prove Theorem 3 when $A$ has coordinates $(0, c)$ or $(a, 0)$.

****16.** **More on the geometric interpretation of the determinant:** Let $\mathbf{u}_1$ and $\mathbf{u}_2$ be two 2-vectors and let $\mathbf{v}_1 = A\mathbf{u}_1$ and $\mathbf{v}_2 = A\mathbf{u}_2$. Show that (area generated by $\mathbf{v}_1$ and $\mathbf{v}_2$) = (area generated by $\mathbf{u}_1$ and $\mathbf{u}_2$) $|\det A|$

## Answers to Self-Quiz

**I.** a **II.** b **III.** c **IV.** b, c

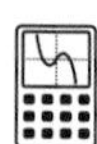

## CALCULATOR BOX

Both the TI-85 and the CASIO fx-7700 GB compute determinants quite easily.

### TI-85

First, enter a matrix and give it a name, say, $A$. Then press

[2nd] [MATRX] [F3] ⟨MATH⟩ [F1] ⟨det⟩ [ALPHA] [A] [ENTER]

### CASIO fx-7700 GB

Once $A$ is entered and displayed, press the [F3] key and $|A| = \det A$ will be displayed.

In Problems 17–20 find the determinant of each matrix on a calculator.

**17.** $\begin{pmatrix} 1 & -1 & 2 & 3 & 5 \\ 6 & 10 & -6 & 4 & 3 \\ 7 & -1 & 2 & -12 & 6 \\ 9 & 4 & 13 & 8 & 15 \\ 8 & 11 & -9 & -8 & 6 \end{pmatrix}$

**18.** $\begin{pmatrix} 1 & -1 & 4 & 6 \\ 2 & 9 & 16 & 4 \\ 37 & -6 & 0 & 23 \\ 14 & 4 & 6 & -11 \end{pmatrix}$

**19.** $\begin{pmatrix} -238 & -159 & 146 & 382 & -189 \\ -319 & 248 & -556 & 700 & 682 \\ 462 & 96 & -331 & 516 & -322 \\ 511 & 856 & 619 & 384 & 906 \\ 603 & -431 & -236 & 692 & -857 \end{pmatrix}$

**20.** $\begin{pmatrix} 0.62 & 0.37 & 0.42 & 0.56 & 0.33 \\ 0.29 & 0.46 & 0.33 & 0.48 & 0.97 \\ 0.81 & 0.37 & 0.91 & 0.33 & 0.77 \\ 0.35 & 0.62 & 0.73 & 0.98 & 0.18 \\ 0.29 & 0.08 & 0.46 & 0.71 & 0.29 \end{pmatrix}$

# MATLAB 2.1

*MATLAB Information* The command **det(A)** finds the determinant of $A$. As before, you can use MATLAB to generate random $n \times n$ matrices. For example,

**A = 2*rand(n)−1** (with entries between −1 and 1)

**A = 2*rand(n)−1 + i*(2*rand(n)−1)** (with real and imaginary entries between −1 and 1)

**A = round(10*(2*rand(n)−1))** (with integer entries between −10 and 10)

1. In this problem you will investigate the relationship between $\det(A)$ and the invertibility of $A$.

   a. For each of the matrices below, determine if $A$ is or is not invertible (using **rref**) and find $\det(A)$. How can you use **det(A)** to determine whether or not $A$ is invertible?

   **i.** $\begin{pmatrix} -6 & 4 & 0 \\ -9 & 9 & 7 \\ 4 & -2 & -9 \end{pmatrix}$ **ii.** $\begin{pmatrix} -9 & -2 & 2 & -8 \\ 1 & -9 & 9 & 3 \\ 3 & -2 & 7 & -2 \\ -10 & 4 & 1 & 4 \end{pmatrix}$ **iii.** $\begin{pmatrix} 23 & 19 & 11 \\ 5 & 1 & 5 \\ 9 & 9 & 3 \end{pmatrix}$

**iv.** $\begin{pmatrix} 8 & -3 & 5 & -9 & 5 \\ 5 & 3 & 8 & 3 & 0 \\ -5 & 5 & 0 & 8 & -5 \\ -9 & 10 & 1 & -5 & -5 \\ 5 & -3 & 2 & -1 & -3 \end{pmatrix}$ **v.** $\begin{pmatrix} 1 & 2 & -3 & 4 & 5 \\ -2 & -5 & 8 & -8 & -9 \\ 1 & 2 & -2 & 7 & 9 \\ 1 & 1 & 0 & 6 & 12 \\ 2 & 4 & -6 & 8 & 11 \end{pmatrix}$

**b.** Parts (i) and (ii) below test your conjecture from part (a) with several random matrices. (Choose at least four matrices in (i) of different sizes and at least four matrices in (ii). Include at least one matrix with complex entries for each part.)

**i.** Let $A$ be a random $n \times n$ matrix. Find $\det(A)$. Using previous knowledge, determine if $A$ is or is not invertible. How does this evidence support your conjecture?

**ii.** Let $B$ be a random $n \times n$ matrix, but for some $j$ of your choice, let **B(:,j)** equal a linear combination of some of the columns of $B$ (of your choice). For example, **B(:,3) = B(:,1) + 2*B(:,2).** Determine if $B$ is or is not invertible and find $\det(B)$. How does this evidence support your conjecture?

**2.** For six random $n \times n$ matrices $A$ with real entries (for different choices of $n$), compare **det(A)** with **det(A′)** where A′ denotes (in MATLAB) the transpose of $A$. Include at least two noninvertible matrices (see description in MATLAB Problem (1) (b) (ii) in this section). What does your comparison tell you? Repeat for matrices with complex entries.

**3.** Generate six pairs of $n \times n$ random matrices, $A$ and $B$. (Use different $n$'s.) For each pair, let $C = A + B$. Compare $\det(C)$ and $\det(A) + \det(B)$. What can you conclude about the statement

$$\det(A + B) = \det(A) + \det(B)$$

**4. a.** Using the pairs of matrices ($A$ and $B$) below, formulate a conjecture concerning **det(A*B)** in terms of the determinants of $A$ and $B$.

**i.** $A = \begin{pmatrix} 2 & 7 & 5 \\ 0 & 9 & 8 \\ 7 & 4 & 0 \end{pmatrix}$ $B = \begin{pmatrix} 1 & 4 & 2 \\ -1 & -2 & 1 \\ 1 & 6 & 6 \end{pmatrix}$

**ii.** $A = \begin{pmatrix} 2 & 7 & 5 \\ 0 & 9 & 8 \\ 7 & 4 & 0 \end{pmatrix}$ $B = \begin{pmatrix} 1 & 2 & 5 \\ 1 & -1 & 4 \\ 2 & 4 & 11 \end{pmatrix}$

**iii.** $A = \begin{pmatrix} 1 & 2 & 5 \\ 1 & -1 & 4 \\ 2 & 4 & 11 \end{pmatrix}$ $B = \begin{pmatrix} 1 & 4 & 2 \\ -1 & -2 & 1 \\ 1 & 6 & 6 \end{pmatrix}$

**iv.** $A = \begin{pmatrix} 10 & 6 & 4 & 1 \\ 1 & 1 & 0 & 0 \\ 2 & 7 & -5 & 9 \\ 3 & 6 & -3 & 4 \end{pmatrix}$ $B = \begin{pmatrix} 1 & 9 & 4 & 5 \\ 9 & 1 & 3 & 3 \\ 4 & 2 & 1 & 5 \\ 1 & 1 & 8 & 8 \end{pmatrix}$

**b.** Test your conjecture further by generating random $n \times n$ matrices. (Generate at least six pairs, with different choices for $n$. Include a pair in which one of the matrices is not invertible. Include matrices with complex entries.)

5. a. For the matrices below, formulate a conjecture relating **det(A)** and **det(inv(A))**.

i. $\begin{pmatrix} 2 & 2 \\ 1 & 2 \end{pmatrix}$ ii. $\begin{pmatrix} 2 & -1 \\ 1 & -2 \end{pmatrix}$ iii. $\begin{pmatrix} 2 & 1 & 2 \\ -2 & 0 & 3 \\ 2 & 1 & 4 \end{pmatrix}$ iv. $\begin{pmatrix} -1 & 1 & 2 \\ 1 & -2 & 1 \\ -2 & 2 & 9 \end{pmatrix}$

b. Test your conjecture with several (at least six) random $n \times n$ invertible matrices for different choices of $n$. Include matrices with complex entries.

c. *(Paper and pencil)* Prove your conjecture by using the definition of inverse (that is, consider $AA^{-1}$) and the property discovered in MATLAB Problem 4 in this section.

6. Let **A = 2*rand(6) − 1.**

a. Choose $i$, $j$, and $c$ and let $B$ be the matrix obtained by performing the row operation $R_j \rightarrow cR_i + R_j$ on $A$. Compare $\det(A)$ and $\det(B)$. Repeat for at least four more choices of $i$, $j$, and $c$. What can you conjecture about the relationship between the determinant of $A$ and the determinant of the matrix obtained from $A$ by performing the type of row operation above?

b. Follow the same instructions as for part (a) but for the row operation $R_i \rightarrow cR_i$.

c. Follow the same instructions as for part (a) but for the row operation that interchanges $R_i$ and $R_j$.

d. For each of the row operations you performed in (a), (b), and (c), find the elementary matrix $F$ such that $FA$ is the matrix obtained by performing the row operation on $A$. Find $\det F$. Explain the results obtained in parts (a), (b), and (c) by using your observation about $\det(F)$ and your conjecture from MATLAB Problem 4 in this section.

7. We know that if $A$ is an upper triangular matrix, then $\det(A)$ is the product of its diagonal entries. Consider the matrix $M$ below, where $A$, $B$, and $D$ are random $n \times n$ matrices and 0 stands for the $n \times n$ matrix consisting of all 0's:

$$M = \begin{pmatrix} A & B \\ 0 & D \end{pmatrix}$$

Can you conjecture a relationship between $\det(M)$ and the determinants of $A$, $B$, and $D$?

a. Enter random $n \times n$ matrices $A$, $B$, and $D$. Let **C = zeros(n).** Form the block matrix **M = [A B;C D].** Test your conjecture. (If you have not formulated a conjecture yet, find the determinants of $M$, $A$, $B$, and $D$ and look for patterns.) Repeat for several choices of $n$, $A$, $B$, and $D$.

b. Repeat the process above for

$$M = \begin{pmatrix} A & B & C \\ 0 & D & E \\ 0 & 0 & F \end{pmatrix}$$

where $A$, $B$, $C$, $D$, $E$, and $F$ are random $n \times n$ matrices and 0 stands for the $n \times n$ matrix consisting of all 0's (that is, **zeros(n)**).

M 8. *(Uses m-file ornt.m)* A geometric application for $2 \times 2$ determinants concerns orientation. If we travel around the edges of a parallelogram, we travel in either a clockwise or counterclockwise orientation. Mutliplication by a $2 \times 2$ matrix can affect orientation.

Given two vectors **u** and **v**, let's suppose we trace the parallelogram formed by starting at (0, 0), traveling to the end of **u**, then to the end of **u** + **v**, then to the end of **v**, and then back to (0, 0); and then we do the same for the parallelogram formed by $A$**u** and

$A\mathbf{v}$, where $A$ is a $2 \times 2$ matrix (where we first travel along $A\mathbf{u}$). When will the orientation (clockwise or counterclockwise) of the $A\mathbf{u}$ and $A\mathbf{v}$ parallelogram be reversed from the orientation of the $\mathbf{u}$ and $\mathbf{v}$ parallelogram?

On the accompanying disk there is an m-file called *ornt.m* that you can use to investigate this question. Enter **help ornt** for a description of what this m-file does.

For each problem below, enter $\mathbf{u}$, $\mathbf{v}$, and $A$. (Here $\mathbf{u}$ and $\mathbf{v}$ are $2 \times 1$ vectors and $A$ is a $2 \times 2$ matrix.) Find **det A.** Enter **ornt (u, v, A).** A graphics window will display both the $\mathbf{u}$ and $\mathbf{v}$ parallelogram and the $A\mathbf{u}$ and $A\mathbf{v}$ parallelogram with the orientations described on the screen. Is or is not the orientation changed? (To return to the command screen after viewing the graph, hit any key. If you wish to go from the command screen back to the graphics screen, enter the command **shg.** This stands for "show graph.") After doing the problems below, formulate a conjecture as to how you can use **det(A)** to determine whether or not the orientation will change. Test your conjecture on further examples (change $A$ and/or $\mathbf{u}$ and $\mathbf{v}$).

For each $A$ below, use **u = [1;0]** and **v = [0;1]** and then **u = [−2;1]** and **v = [1;3].**

a. $\begin{pmatrix} 1 & 1 \\ 1 & 2 \end{pmatrix}$ b. $\begin{pmatrix} 2 & 3 \\ 2 & 2 \end{pmatrix}$ c. $\begin{pmatrix} 1 & 0 \\ 3 & -1 \end{pmatrix}$ d. $\begin{pmatrix} 1 & 2 \\ 1 & 4 \end{pmatrix}$

e. Test your conjecture for more choices of $A$. Try different $\mathbf{u}$ and $\mathbf{v}$ also.

***Important Note.*** When you are finished with this problem, be sure to enter the command **clg** to clear the graphics screen before beginning other problems. If using MATLAB 4.0, enter **clf.**

## 2.2 PROPERTIES OF DETERMINANTS

There are some problems in mathematics that are theoretically simple but practically impossible. Think, for example, of the determinant of a $50 \times 50$ matrix. We can compute it by expanding by **cofactors.** This involves 50 $49 \times 49$ determinants that involve $50 \cdot 49$ $48 \times 48$ determinants that involve . . . that involve $50 \cdot 49 \cdot 48 \cdot 47 \ldots \cdot 3$ $2 \times 2$ determinants. Now $50 \cdot 49 \ldots \cdot 3 = 50!/2 \approx 1.5 \times 10^{64}$ $2 \times 2$ determinants. Suppose we had a computer that could compute 1 million $= 10^6$ $2 \times 2$ determinants each second. Then it would take about $1.5 \times 10^{58}$ seconds $\approx 4.8 \times 10^{50}$ *years* to finish the computation. (The universe is about 15 billion $= 1.5 \times 10^{10}$ years old according to the most recent version of the "big bang" theory). Evidently, while calculating a $50 \times 50$ determinant directly from the definition is theoretically straightforward, it is practically impossible.

Now a $50 \times 50$ matrix is not at all uncommon. Think of 50 stores each stocking 50 different products. In fact, $n \times n$ matrices with $n > 100$ arise frequently in applications. Fortunately, there are at least two ways to reduce significantly the amount of work needed to compute a determinant.

The first result we need is perhaps the most important theorem about determinants. This theorem states that the determinant of a product is equal to the product of the determinants.

**THEOREM 1** Let $A$ and $B$ be $n \times n$ matrices. Then

$$\boxed{\det AB = \det A \det B} \qquad \textbf{(8)}$$

That is: The determinant of the product is the product of the determinants.

*Remark.* Note that the product on the left is a product of matrices while the product on the right is a product of scalars.

**Proof** The proof, using elementary matrices, is given in Section 2.3. You are asked in Problem 38 to verify the result in the $2 \times 2$ case. ■

**EXAMPLE 1** **Illustration of Fact That det $AB$ = det $A$ det $B$** Verify Theorem 1 for

$$A = \begin{pmatrix} 1 & -1 & 2 \\ 3 & 1 & 4 \\ 0 & -2 & 5 \end{pmatrix} \quad \text{and} \quad B = \begin{pmatrix} 1 & -2 & 3 \\ 0 & -1 & 4 \\ 2 & 0 & -2 \end{pmatrix}$$

**Solution** $\det A = 16$ and $\det B = -8$. We calculate

$$AB = \begin{pmatrix} 1 & -1 & 2 \\ 3 & 1 & 4 \\ 0 & -2 & 5 \end{pmatrix}\begin{pmatrix} 1 & -2 & 3 \\ 0 & -1 & 4 \\ 2 & 0 & -2 \end{pmatrix} = \begin{pmatrix} 5 & -1 & -5 \\ 11 & -7 & 5 \\ 10 & 2 & -18 \end{pmatrix}$$

and $\det AB = -128 = (16)(-8) = \det A \det B$. ■

**WARNING** The determinant of the sum is *not* always equal to the sum of the determinants. That is for most pairs of matrices, $A$ and $B$,

$$\det(A + B) \neq \det A + \det B$$

For example, let $A = \begin{pmatrix} 1 & 2 \\ 3 & 4 \end{pmatrix}$ and $B = \begin{pmatrix} 3 & 0 \\ -2 & 2 \end{pmatrix}$. Then $A + B = \begin{pmatrix} 4 & 2 \\ 1 & 6 \end{pmatrix}$:

$$\det A = -2, \quad \det B = 6, \quad \text{and}$$
$$\det(A + B) = 22 \neq \det A + \det B = -2 + 6 = 4$$ □

Now let $A = LU$ be an $LU$-factorization of an $n \times n$ matrix (see page 142). Then from Theorem 1,

$$\det A = \det LU = \det L \det U$$

But $L$ is a lower triangular matrix with 1's on the diagonal, so

$$\det L = \text{product of its diagonal components} = 1$$

Similarly, since $U$ is upper triangular,

$$\det U = \text{product of its diagonal components}$$

We therefore have the following:

**THEOREM 2** If the square matrix $A$ has the $LU$-factorization $A = LU$ where $L$ has 1's on the diagonal, then

$$\det A = \det U = \text{product of the diagonal components of } U$$

**EXAMPLE 2** **Using the $LU$-Factorization to Compute the Determinant of a $4 \times 4$ Matrix**

Compute $\det A$, where $A = \begin{pmatrix} 2 & 3 & 2 & 4 \\ 4 & 10 & -4 & 0 \\ -3 & -2 & -5 & -2 \\ -2 & 4 & 4 & -7 \end{pmatrix}$

**Solution** From Example 1.11.1 on page 140, $A = LU$, where

$$U = \begin{pmatrix} 2 & 3 & 2 & 4 \\ 0 & 4 & -8 & -8 \\ 0 & 0 & 3 & 9 \\ 0 & 0 & 0 & -49 \end{pmatrix}$$

so $\det A = \det U = (2)(4)(3)(-49) = -1176$.

If $A$ cannot be reduced to a triangular form without performing any permutations, then, by Theorem 1.11.3 on page 145, there exists a permutation matrix $P$ such that

$$PA = LU$$

It is not difficult to prove that if $P$ is a permutation matrix, then $\det P = \pm 1$ (see Problem 42). Then

$$\begin{aligned} \det PA &= \det LU \\ \det P \det A &= \det L \det U = \det U \qquad \det L = 1 \\ \pm \det A &= \det U \\ \det A &= \pm \det U \end{aligned}$$

**THEOREM 3** If $PA = LU$, where $P$ is a permutation matrix and $L$ and $U$ are as before, then

$$\det A = \frac{\det U}{\det P} = \pm \det U$$

**EXAMPLE 3** **Using the $PA = LU$-Factorization to Compute the Determinant of a $3 \times 3$ Matrix**

Find $\det A$, where $A = \begin{pmatrix} 0 & 2 & 3 \\ 2 & -4 & 7 \\ 1 & -2 & 5 \end{pmatrix}$

**Solution** In Example 1.11.3 on page 141 we found that $PA = LU$, where

$$P = \begin{pmatrix} 0 & 0 & 1 \\ 1 & 0 & 0 \\ 0 & 1 & 0 \end{pmatrix} \quad \text{and} \quad U = \begin{pmatrix} 1 & -2 & 5 \\ 0 & 2 & 3 \\ 0 & 0 & -3 \end{pmatrix}$$

Now $\det P = 1$ and $\det U = (1)(2)(-3)$, so $\det A = \frac{-6}{1} = -6$.

We now state an important theorem about determinants.

**THEOREM 4** $\det A^t = \det A$

**Proof** Suppose that $A = LU$. Then $A^t = (LU)^t = U^t L^t$ by Theorem 1.9.1 (*ii*) on page 122. Next we compute

$$\det A = \det L \det U = \det U$$

$$\det A^t = \det U^t \det L^t = \det U^t = \det U = \det A \qquad \det L = 1$$

The last step followed from the facts that the transpose of a lower triangular matrix is upper triangular and vice versa, and transposing does not change the diagonal components of a matrix.

If $A$ cannot be written as $LU$, then there exists a permutation matrix $P$ such that $PA = LU$. By what we just proved,

$$\det PA = \det (PA)^t = \det(A^t P^t)$$

and, by Theorem 1,

$$\det P \det A = \det PA = \det(A^t P^t) = \det A^t \det P^t$$

It is not difficult to prove (see Problem 43) that if $P$ is a permutation matrix, then $\det P = \det P^t$. Since $\det P = \det P^t = \pm 1$, we conclude that $\det A = \det A^t$.

**EXAMPLE 4** **A Matrix and Its Transpose Have the Same Determinant**

Let $A = \begin{pmatrix} 1 & -1 & 2 \\ 3 & 1 & 4 \\ 0 & -2 & 5 \end{pmatrix}$. Then $A^t = \begin{pmatrix} 1 & 3 & 0 \\ -1 & 1 & -2 \\ 2 & 4 & 5 \end{pmatrix}$ and it is easy to verify that $|A| = |A^t| = 16$.

*Remark.* Since the rows of a matrix are the columns of its transpose, it follows that anything one can say about determinants that involves the rows is also true about the columns. The following properties about determinants comprise a second way to simplify the computation of determinants. The results are proved for rows. By what we have just said, the theorems are also true for columns.

We begin to describe these properties by stating a theorem from which many important results follow. The proof of this theorem is difficult and is deferred to the next section.

**THEOREM 5** **Basic Theorem** Let

$$A = \begin{pmatrix} a_{11} & a_{12} & \cdots & a_{1n} \\ a_{21} & a_{22} & \cdots & a_{2n} \\ \vdots & \vdots & & \vdots \\ a_{n1} & a_{n2} & \cdots & a_{nn} \end{pmatrix}$$

be an $n \times n$ matrix. Then

$$\det A = a_{i1}A_{i1} + a_{i2}A_{i2} + \cdots + a_{in}A_{in} = \sum_{k=1}^{n} a_{ik}A_{ik} \tag{1}$$

for $i = 1, 2, \ldots, n$. That is, **we can calculate det $A$ by expanding by cofactors in *any* row of $A$.** Furthermore,

$$\det A = a_{1j}A_{1j} + a_{2j}A_{2j} + \cdots + a_{nj}A_{nj} = \sum_{k=1}^{n} a_{kj}A_{kj} \tag{2}$$

Since the $j$th column of $A$ is $\begin{pmatrix} a_{1j} \\ a_{2j} \\ \vdots \\ a_{nj} \end{pmatrix}$, equation (2) indicates that **we can calculate det $A$ by expanding by cofactors in any column of $A$.** ■

**EXAMPLE 5** **Obtaining the Determinant by Expanding in the Second Row or the Third Column** For $A = \begin{pmatrix} 3 & 5 & 2 \\ 4 & 2 & 3 \\ -1 & 2 & 4 \end{pmatrix}$ we saw in Example 2.1.1 on page 173 that $\det A = -69$. Expanding in the second row, we obtain

$$\begin{aligned} \det A &= 4A_{21} + 2A_{22} + 3A_{23} \\ &= 4(-1)^{2+1}\begin{vmatrix} 5 & 2 \\ 2 & 4 \end{vmatrix} + 2(-1)^{2+2}\begin{vmatrix} 3 & 2 \\ -1 & 4 \end{vmatrix} + 3(-1)^{2+3}\begin{vmatrix} 3 & 5 \\ -1 & 2 \end{vmatrix} \\ &= -4(16) + 2(14) - 3(11) = -69 \end{aligned}$$

Similarly, if we expand in the third column, say, we obtain

$$\begin{aligned}\det A &= 2A_{13} + 3A_{23} + 4A_{33} \\ &= 2(-1)^{1+3}\begin{vmatrix} 4 & 2 \\ -1 & 2 \end{vmatrix} + 3(-1)^{2+3}\begin{vmatrix} 3 & 5 \\ -1 & 2 \end{vmatrix} + 4(-1)^{3+3}\begin{vmatrix} 3 & 5 \\ 4 & 2 \end{vmatrix} \\ &= 2(10) - 3(11) + 4(-14) = -69\end{aligned}$$

You should verify that we get the same answer if we expand in the third row or the first or second column.

We now list and prove some additional properties of determinants. In each case we assume that $A$ is an $n \times n$ matrix. We shall see that these properties can be used to reduce greatly the work involved in evaluating a determinant.

**Property 1** If any row or column of $A$ is a zero vector, then $\det A = 0$.

**Proof** Suppose the $i$th row of $A$ contains all zeros. That is, $a_{ij} = 0$ for $j = 1, 2, \ldots, n$. Then $\det A = a_{i1}A_{i1} + a_{i2}A_{i2} + \cdots + a_{in}A_{in} = 0 + 0 + \cdots + 0 = 0$. The same proof works if the $j$th column is the zero vector.

**EXAMPLE 6** **If $A$ Has a Row or Column of Zeros, Then $\det A = 0$** It is easy to verify that

$$\begin{vmatrix} 2 & 3 & 5 \\ 0 & 0 & 0 \\ 1 & -2 & 4 \end{vmatrix} = 0 \quad \text{and} \quad \begin{vmatrix} -1 & 3 & 0 & 1 \\ 4 & 2 & 0 & 5 \\ -1 & 6 & 0 & 4 \\ 2 & 1 & 0 & 1 \end{vmatrix} = 0$$

**Property 2** If the $i$th row or the $j$th column of $A$ is multiplied by a scalar $c$, then $\det A$ is multiplied by $c$. That is, if we call this new matrix $B$, then

$$|B| = \begin{vmatrix} a_{11} & a_{12} & \cdots & a_{1n} \\ a_{21} & a_{22} & \cdots & a_{2n} \\ \vdots & \vdots & & \vdots \\ ca_{i1} & ca_{i2} & \cdots & ca_{in} \\ \vdots & \vdots & & \vdots \\ a_{n1} & a_{n2} & \cdots & a_{nn} \end{vmatrix} = c\begin{vmatrix} a_{11} & a_{12} & \cdots & a_{1n} \\ a_{21} & a_{22} & \cdots & a_{2n} \\ \vdots & \vdots & & \vdots \\ a_{i1} & a_{i2} & \cdots & a_{in} \\ \vdots & \vdots & & \vdots \\ a_{n1} & a_{n2} & \cdots & a_{nn} \end{vmatrix} = c|A| \qquad (3)$$

**Proof** To prove (3) we expand in the $i$th row of $A$ to obtain

$$\begin{aligned}\det B &= ca_{i1}A_{i1} + ca_{i2}A_{i2} + \cdots + ca_{in}A_{in} \\ &= c(a_{i1}A_{i1} + a_{i2}A_{i2} + \cdots + a_{in}A_{in}) = c \det A\end{aligned}$$

A similar proof works for columns.

**EXAMPLE 7** **Illustration of Property 2** Let $A = \begin{pmatrix} 1 & -1 & 2 \\ 3 & 1 & 4 \\ 0 & -2 & 5 \end{pmatrix}$. Then $\det A = 16$. If we multiply the second row by 4, we have $B = \begin{pmatrix} 1 & -1 & 2 \\ 12 & 4 & 16 \\ 0 & -2 & 5 \end{pmatrix}$ and $\det B = 64 = 4 \det A$. If the third column is multiplied by $-3$, we obtain $C = \begin{pmatrix} 1 & -1 & -6 \\ 3 & 1 & -12 \\ 0 & -2 & -15 \end{pmatrix}$ and $\det C = -48 = -3 \det A$.

*Remark.* Using Property 2 we can prove (see Problem 28) the following interesting fact: For any scalar $\alpha$ and $n \times n$ matrix $A$, $\det \alpha A = \alpha^n \det A$.

**Property 3** Let

$$A = \begin{pmatrix} a_{11} & a_{12} & \cdots & a_{1j} & \cdots & a_{1n} \\ a_{21} & a_{22} & \cdots & a_{2j} & \cdots & a_{2n} \\ \vdots & \vdots & & \vdots & & \vdots \\ a_{n1} & a_{n2} & \cdots & a_{nj} & \cdots & a_{nn} \end{pmatrix}, \quad B = \begin{pmatrix} a_{11} & a_{12} & \cdots & \alpha_{1j} & \cdots & a_{1n} \\ a_{21} & a_{22} & \cdots & \alpha_{2j} & \cdots & a_{2n} \\ \vdots & \vdots & & \vdots & & \vdots \\ a_{n1} & a_{n2} & \cdots & \alpha_{nj} & \cdots & a_{nn} \end{pmatrix},$$

and

$$C = \begin{pmatrix} a_{11} & a_{12} & \cdots & a_{1j} + \alpha_{1j} & \cdots & a_{1n} \\ a_{21} & a_{22} & \cdots & a_{2j} + \alpha_{2j} & \cdots & a_{2n} \\ \vdots & \vdots & & \vdots & & \vdots \\ a_{n1} & a_{n2} & \cdots & a_{nj} + \alpha_{nj} & \cdots & a_{nn} \end{pmatrix}$$

Then

$$\boxed{\det C = \det A + \det B} \tag{4}$$

In other words, suppose that $A$, $B$, and $C$ are identical except for the $j$th column and that the $j$th column of $C$ is the sum of the $j$th columns of $A$ and $B$. Then $\det C = \det A + \det B$. The same statement is true for rows.

**Proof** We expand $\det C$ in the $j$th column to obtain

$$\begin{aligned} \det C &= (a_{1j} + \alpha_{1j})A_{1j} + (a_{2j} + \alpha_{2j})A_{2j} + \cdots + (a_{nj} + \alpha_{nj})A_{nj} \\ &= (a_{1j}A_{1j} + a_{2j}A_{2j} + \cdots + a_{nj}A_{nj}) \\ &\quad + (\alpha_{1j}A_{1j} + \alpha_{2j}A_{2j} + \cdots + \alpha_{nj}A_{nj}) = \det A + \det B \end{aligned}$$

**EXAMPLE 8** **Illustration of Property 3** Let $A = \begin{pmatrix} 1 & -1 & 2 \\ 3 & 1 & 4 \\ 0 & -2 & 5 \end{pmatrix}$, $B = \begin{pmatrix} 1 & -6 & 2 \\ 3 & 2 & 4 \\ 0 & 4 & 5 \end{pmatrix}$,

and $C = \begin{pmatrix} 1 & -1-6 & 2 \\ 3 & 1+2 & 4 \\ 0 & -2+4 & 5 \end{pmatrix} = \begin{pmatrix} 1 & -7 & 2 \\ 3 & 3 & 4 \\ 0 & 2 & 5 \end{pmatrix}$. Then $\det A = 16$, $\det B = 108$, and $\det C = 124 = \det A + \det B$.

**Property 4** Interchanging any two distinct rows (or columns) of $A$ has the effect of multiplying $\det A$ by $-1$.

**Proof** We prove the statement for rows and assume first that two adjacent rows are interchanged. That is, we assume that the $i$th and $(i+1)$st rows are interchanged. Let

$$A = \begin{pmatrix} a_{11} & a_{12} & \cdots & a_{1n} \\ a_{21} & a_{22} & \cdots & a_{2n} \\ \vdots & \vdots & & \vdots \\ a_{i1} & a_{i2} & \cdots & a_{in} \\ a_{i+1,1} & a_{i+1,2} & \cdots & a_{i+1,n} \\ \vdots & \vdots & & \vdots \\ a_{n1} & a_{n2} & & a_{nn} \end{pmatrix} \quad \text{and} \quad B = \begin{pmatrix} a_{11} & a_{12} & \cdots & a_{1n} \\ a_{21} & a_{22} & \cdots & a_{2n} \\ \vdots & \vdots & & \vdots \\ a_{i+1,1} & a_{i+1,2} & \cdots & a_{i+1,n} \\ a_{i1} & a_{i2} & \cdots & a_{in} \\ \vdots & \vdots & & \vdots \\ a_{n1} & a_{n2} & \cdots & a_{nn} \end{pmatrix}$$

Then, expanding $\det A$ in its $i$th row and $\det B$ in its $(i+1)$st row, we obtain

$$\begin{aligned} \det A &= a_{i1}A_{i1} + a_{i2}A_{i2} + \cdots + a_{in}A_{in} \\ \det B &= a_{i1}B_{i+1,1} + a_{i2}B_{i+1,2} + \cdots + a_{in}B_{i+1,n} \end{aligned} \tag{5}$$

Here $A_{ij} = (-1)^{i+j}|M_{ij}|$, where $M_{ij}$ is obtained by crossing off the $i$th row and $j$th column of $A$. Notice now that if we cross off the $(i+1)$st row and $j$th column of $B$, we obtain the same $M_{ij}$. Thus

$$B_{i+1,j} = (-1)^{i+1+j}|M_{ij}| = -(-1)^{i+j}|M_{ij}| = -A_{ij}$$

so that, from equations (5), $\det B = -\det A$.

Now suppose that $i < j$ and that the $i$th and $j$th rows are to be interchanged. We can do this by interchanging adjacent rows several times. It will take $j - i$ interchanges to move row $j$ into the $i$th row. Then row $i$ will be in the $(i+1)$st row and it will take an additional $j - i - 1$ interchanges to move row $i$ into the $j$th row. To illustrate, we interchange rows 2 and 6:†

---

† Note that all the numbers here refer to rows.

$$\begin{matrix} 1 & & 1 & & 1 & & 1 & & 1 & & 1 & & 1 & & 1 \\ 2 & & 2 & & 2 & & 2 & & 6 & & 6 & & 6 & & 6 \\ 3 & & 3 & & 3 & & 6 & & 2 & & 3 & & 3 & & 3 \\ 4 & \rightarrow & 4 & \rightarrow & 6 & \rightarrow & 3 & \rightarrow & 3 & \rightarrow & 2 & \rightarrow & 4 & \rightarrow & 4 \\ 5 & & 6 & & 4 & & 4 & & 4 & & 4 & & 2 & & 5 \\ 6 & & 5 & & 5 & & 5 & & 5 & & 5 & & 5 & & 2 \\ 7 & & 7 & & 7 & & 7 & & 7 & & 7 & & 7 & & 7 \end{matrix}$$

$6 - 2 = 4$ interchanges to move the 6 into the 2 position; $6 - 2 - 1 = 3$ interchanges to get the 2 into the 6 position

Finally, the total number of interchanges of adjacent rows is $(j - i) + (j - i - 1) = 2j - 2i - 1$, which is odd. Thus $\det A$ is multiplied by $-1$ an odd number of times, which is what we needed to show. ■

EXAMPLE 9 **Illustration of Property 4** Let $A = \begin{pmatrix} 1 & -1 & 2 \\ 3 & 1 & 4 \\ 0 & -2 & 5 \end{pmatrix}$. By interchanging the first and third rows, we obtain $B = \begin{pmatrix} 0 & -2 & 5 \\ 3 & 1 & 4 \\ 1 & -1 & 2 \end{pmatrix}$. By interchanging the first and second columns of $A$, we obtain $C = \begin{pmatrix} -1 & 1 & 2 \\ 1 & 3 & 4 \\ -2 & 0 & 5 \end{pmatrix}$. Then, by direct calculation, we find that $\det A = 16$ and $\det B = \det C = -16$. ■

**Property 5** If $A$ has two equal rows or columns, then $\det A = 0$.

**Proof** Suppose the $i$th and $j$th rows of $A$ are equal. By interchanging these rows, we get a matrix $B$ having the property that $\det B = -\det A$ (from Property 4). But since row $i$ = row $j$, interchanging them gives us the same matrix. Thus $A = B$ and $\det A = \det B = -\det A$. Thus $2 \det A = 0$, which can happen only if $\det A = 0$. ■

EXAMPLE 10 **Illustration of Property 5** By direct calculation, we can verify that for $A = \begin{pmatrix} 1 & -1 & 2 \\ 5 & 7 & 3 \\ 1 & -1 & 2 \end{pmatrix}$ [two equal rows] and $B = \begin{pmatrix} 5 & 2 & 2 \\ 3 & -1 & -1 \\ -2 & 4 & 4 \end{pmatrix}$ [two equal columns], $\det A = \det B = 0$. ■

**Property 6** If one row (column) of $A$ is a scalar multiple of another row (column), then $\det A = 0$.

**Proof** Let $(a_{j1}, a_{j2}, \ldots, a_{jn}) = c(a_{i1}, a_{i2}, \ldots, a_{in})$. Then from Property 2,

$$\det A = c\begin{vmatrix} a_{11} & a_{12} & \cdots & a_{1n} \\ a_{21} & a_{22} & \cdots & a_{2n} \\ \vdots & \vdots & & \vdots \\ a_{i1} & a_{i2} & \cdots & a_{in} \\ \vdots & \vdots & & \vdots \\ a_{i1} & a_{i2} & \cdots & a_{in} \\ \vdots & \vdots & & \vdots \\ a_{n1} & a_{n2} & \cdots & a_{nn} \end{vmatrix} \begin{matrix} \\ \\ \\ \\ \\ \leftarrow j\text{th row} \\ \\ \\ \end{matrix} = 0 \qquad \text{(from Property 5)}$$

**EXAMPLE 11** **Illustration of Property 6** $\begin{vmatrix} 2 & -3 & 5 \\ 1 & 7 & 2 \\ -4 & 6 & -10 \end{vmatrix} = 0$ since the third row is $-2$ times the first row.

**EXAMPLE 12** **Another Illustration of Property 6** $\begin{vmatrix} 2 & 4 & 1 & 12 \\ -1 & 1 & 0 & 3 \\ 0 & -1 & 9 & -3 \\ 7 & 3 & 6 & 9 \end{vmatrix} = 0$ since the fourth column is three times the second column.

**Property 7** If a scalar multiple of one row (column) of $A$ is added to another row (column) of $A$, then the determinant is unchanged.

**Proof** Let $B$ be the matrix obtained by adding $c$ times the $i$th row of $A$ to the $j$th row of $A$. Then

$$\det B = \begin{vmatrix} a_{11} & a_{12} & \cdots & a_{1n} \\ a_{21} & a_{22} & \cdots & a_{2n} \\ \vdots & \vdots & & \vdots \\ a_{i1} & a_{i2} & \cdots & a_{in} \\ \vdots & \vdots & & \vdots \\ a_{j1} + ca_{i1} & a_{j2} + ca_{i2} & \cdots & a_{jn} + ca_{in} \\ \vdots & \vdots & & \vdots \\ a_{n1} & a_{n2} & \cdots & a_{nn} \end{vmatrix}$$

$$\overset{\text{(from Property 3)}}{=}\begin{vmatrix} a_{11} & a_{12} & \cdots & a_{1n} \\ a_{21} & a_{22} & \cdots & a_{2n} \\ \vdots & \vdots & & \vdots \\ a_{i1} & a_{i2} & \cdots & a_{in} \\ \vdots & \vdots & & \vdots \\ a_{j1} & a_{j2} & \cdots & a_{jn} \\ \vdots & \vdots & & \vdots \\ a_{n1} & a_{n2} & \cdots & a_{nn} \end{vmatrix} + \begin{vmatrix} a_{11} & a_{12} & \cdots & a_{1n} \\ a_{21} & a_{22} & \cdots & a_{2n} \\ \vdots & \vdots & & \vdots \\ a_{i1} & a_{i2} & \cdots & a_{in} \\ \vdots & \vdots & & \vdots \\ ca_{i1} & ca_{i2} & \cdots & ca_{in} \\ \vdots & \vdots & & \vdots \\ a_{n1} & a_{n2} & \cdots & a_{nn} \end{vmatrix}$$

$$= \det A + 0 = \det A \qquad \text{(the zero comes from Property 6)}$$

**EXAMPLE 13** **Illustration of Property 7** Let $A = \begin{pmatrix} 1 & -1 & 2 \\ 3 & 1 & 4 \\ 0 & -2 & 5 \end{pmatrix}$. Then $\det A = 16$. If we multiply the third row by 4 and add it to the second row, we obtain a new matrix $B$ given by

$$B = \begin{pmatrix} 1 & -1 & 2 \\ 3+4(0) & 1+4(-2) & 4+5(4) \\ 0 & -2 & 5 \end{pmatrix} = \begin{pmatrix} 1 & -1 & 2 \\ 3 & -7 & 24 \\ 0 & -2 & 5 \end{pmatrix}$$

and $\det B = 16 = \det A$.

The properties discussed above make it much easier to evaluate high-order determinants. We simply "row-reduce" the determinant, using Property 7, until the determinant is in an easily evaluated form. The most common goal will be to use Property 7 repeatedly until either (1) the new determinant has a row (column) of zeros or one row (column) is a multiple of another row (column)—in which case the determinant is zero—or (2) the new matrix is triangular so that its determinant is the product of its diagonal elements.

**EXAMPLE 14** **Using the Properties of a Determinant to Calculate a 4 × 4 Determinant** Calculate

$$|A| = \begin{vmatrix} 1 & 3 & 5 & 2 \\ 0 & -1 & 3 & 4 \\ 2 & 1 & 9 & 6 \\ 3 & 2 & 4 & 8 \end{vmatrix}$$

**Solution** (See Example 2.1.7, page 177.)

There is already a zero in the first column, so it is simplest to reduce other elements in the first column to zero. We then continue to reduce, aiming for a triangular matrix.

Multiply the first row by $-2$ and add it to the third row and multiply the first row by $-3$ and add it to the fourth row.

$$|A| = \begin{vmatrix} 1 & 3 & 5 & 2 \\ 0 & -1 & 3 & 4 \\ 0 & -5 & -1 & 2 \\ 0 & -7 & -11 & 2 \end{vmatrix}$$

Multiply the second row by $-5$ and $-7$ and add it to the third and fourth rows, respectively.

$$= \begin{vmatrix} 1 & 3 & 5 & 2 \\ 0 & -1 & 3 & 4 \\ 0 & 0 & -16 & -18 \\ 0 & 0 & -32 & -26 \end{vmatrix}$$

Factor out $-16$ from the third row (using Property 2).

$$= -16\begin{vmatrix} 1 & 3 & 5 & 2 \\ 0 & -1 & 3 & 4 \\ 0 & 0 & 1 & \frac{9}{8} \\ 0 & 0 & -32 & -26 \end{vmatrix}$$

Multiply the third row by 32 and add it to the fourth row.

$$= -16\begin{vmatrix} 1 & 3 & 5 & 2 \\ 0 & -1 & 3 & 4 \\ 0 & 0 & 1 & \frac{9}{8} \\ 0 & 0 & 0 & 10 \end{vmatrix}$$

Now we have an upper triangular matrix and $|A| = -16(1)(-1)(1)(10) = (-16)(-10) = 160$.

**EXAMPLE 15** **Using the Properties to Calculate a 4 × 4 Determinant** Calculate

$$|A| = \begin{vmatrix} -2 & 1 & 0 & 4 \\ 3 & -1 & 5 & 2 \\ -2 & 7 & 3 & 1 \\ 3 & -7 & 2 & 5 \end{vmatrix}$$

**Solution** There are a number of ways to proceed here and it is not apparent which way will get us the answer most quickly. However, since there is already one zero in the first row, we begin our reduction in that row.

Multiply the second column by 2 and $-4$ and add it to the first and fourth columns, respectively.

$$|A| = \begin{vmatrix} 0 & 1 & 0 & 0 \\ 1 & -1 & 5 & 6 \\ 12 & 7 & 3 & -27 \\ -11 & -7 & 2 & 33 \end{vmatrix}$$

Interchange the first two columns.

$$= -\begin{vmatrix} 1 & 0 & 0 & 0 \\ -1 & 1 & 5 & 6 \\ 7 & 12 & 3 & -27 \\ -7 & -11 & 2 & 33 \end{vmatrix}$$

Multiply the second column by $-5$ and $-6$ and add it to the third and fourth columns, respectively.

$$= -\begin{vmatrix} 1 & 0 & 0 & 0 \\ -1 & 1 & 0 & 0 \\ 7 & 12 & -57 & -99 \\ -7 & -11 & 57 & 99 \end{vmatrix}$$

Since the fourth column is now a multiple of the third column (column 4 $= \frac{99}{57} \times$ column 3), we see that $|A| = 0$.

**EXAMPLE 16** **Using the Properties to Calculate a 5 × 5 Determinant** Calculate

$$|A| = \begin{vmatrix} 1 & -2 & 3 & -5 & 7 \\ 2 & 0 & -1 & -5 & 6 \\ 4 & 7 & 3 & -9 & 4 \\ 3 & 1 & -2 & -2 & 3 \\ -5 & -1 & 3 & 7 & -9 \end{vmatrix}$$

**Solution** Adding first row 2 and then row 4 to row 5, we obtain

$$|A| = \begin{vmatrix} 1 & -2 & 3 & -5 & 7 \\ 2 & 0 & -1 & -5 & 6 \\ 4 & 7 & 3 & -9 & 4 \\ 3 & 1 & -2 & -2 & 3 \\ 0 & 0 & 0 & 0 & 0 \end{vmatrix} = 0 \qquad \text{(from Property 1)}$$

This example illustrates the fact that a little looking before beginning the computations can simplify matters considerably.

There is one additional fact about determinants that we will find very useful.

**THEOREM 6** Let $A$ be an $n \times n$ matrix. Then

$$a_{i1}A_{j1} + a_{i2}A_{j2} + \cdots + a_{in}A_{jn} = 0 \qquad \text{if } i \neq j \tag{6}$$

*Note.* From Theorem 5 the sum in equation (6) equals $\det A$ if $i = j$.

**Proof** Let

$$B = \begin{pmatrix} a_{11} & a_{12} & \cdots & a_{1n} \\ a_{21} & a_{22} & \cdots & a_{2n} \\ \vdots & \vdots & & \vdots \\ a_{i1} & a_{i2} & \cdots & a_{in} \\ \vdots & \vdots & & \vdots \\ a_{i1} & a_{i2} & \cdots & a_{in} \\ \vdots & \vdots & & \vdots \\ a_{n1} & a_{n2} & \cdots & a_{nn} \end{pmatrix} \begin{matrix} \\ \\ \\ \\ \\ \leftarrow j\text{th row} \\ \\ \\ \end{matrix}$$

Then since two rows of $B$ are equal, $\det B = 0$. But $B = A$ except in the $j$th row. Thus if we calculate $\det B$ by expanding in the $j$th row of $B$, we obtain the sum in (6) and the theorem is proved. Note that when we expand in the $j$th row, the $j$th row is deleted in computing the cofactors of $B$. Thus $B_{jk} = A_{jk}$ for $k = 1, 2, \ldots, n$. ■

# PROBLEMS 2.2

## Self-Quiz

**I.** Which of the following is 0?

**a.** $\begin{vmatrix} 1 & 2 & 3 \\ 1 & 2 & 4 \\ 1 & 6 & 4 \end{vmatrix}$ **b.** $\begin{vmatrix} 1 & 2 & 7 \\ 2 & 3 & 8 \\ -1 & -2 & -7 \end{vmatrix}$

**c.** $\begin{vmatrix} 2 & 1 & 3 \\ -2 & 1 & 3 \\ 0 & 2 & 5 \end{vmatrix}$ **d.** $\begin{vmatrix} 1 & 0 & 0 \\ 0 & -1 & 0 \\ 0 & 0 & 4 \end{vmatrix}$

**II.** Which of the following is 0?

**a.** $\begin{vmatrix} 1 & 2 & 3 & 4 \\ -1 & 2 & -3 & 4 \\ 3 & -1 & 5 & 2 \\ 3 & 1 & 5 & 2 \end{vmatrix}$ **b.** $\begin{vmatrix} 1 & 3 & 0 & 1 \\ 0 & 2 & 1 & 4 \\ 3 & 1 & 0 & 2 \\ 0 & 0 & 0 & 5 \end{vmatrix}$

**c.** $\begin{vmatrix} 1 & 2 & 2 & 1 \\ -1 & 5 & -2 & 0 \\ 2 & 4 & 4 & 2 \\ 3 & 6 & 6 & 5 \end{vmatrix}$ **d.** $\begin{vmatrix} 2 & 1 & -1 & 1 \\ 2 & 1 & 1 & -1 \\ 3 & 0 & 0 & 2 \\ 0 & 3 & 2 & 0 \end{vmatrix}$

**III.** The determinant of $\begin{pmatrix} 1 & 2 & 3 \\ -1 & 2 & 4 \\ -1 & 2 & 5 \end{pmatrix}$ is ______.

**a.** 4 **b.** 10 **c.** −10 **d.** 8 **e.** 6

## Answers to Self-Quiz

**I.** b **II.** c **III.** a

In Problems 1–20 evaluate the determinant by using the methods of this section.

**1.** $\begin{vmatrix} 3 & -5 \\ 2 & 6 \end{vmatrix}$

**2.** $\begin{vmatrix} 4 & 1 \\ 0 & -3 \end{vmatrix}$

**3.** $\begin{vmatrix} -1 & 0 & 2 \\ 3 & 1 & 4 \\ 2 & 0 & -6 \end{vmatrix}$

**4.** $\begin{vmatrix} 2 & 1 & -1 \\ 3 & -2 & 0 \\ 5 & 1 & 6 \end{vmatrix}$

**5.** $\begin{vmatrix} -3 & 2 & 4 \\ 1 & -1 & 2 \\ -1 & 4 & 0 \end{vmatrix}$

**6.** $\begin{vmatrix} 0 & -2 & 3 \\ 1 & 2 & -3 \\ 4 & 0 & 5 \end{vmatrix}$

**7.** $\begin{vmatrix} -2 & 3 & 6 \\ 4 & 1 & 8 \\ -2 & 0 & 0 \end{vmatrix}$

**8.** $\begin{vmatrix} 2 & -1 & 3 \\ 4 & 0 & 6 \\ 5 & -2 & 3 \end{vmatrix}$

**9.** $\begin{vmatrix} 1 & -1 & 2 & 4 \\ 0 & -3 & 5 & 6 \\ 1 & 4 & 0 & 3 \\ 0 & 5 & -6 & 7 \end{vmatrix}$

**10.** $\begin{vmatrix} 2 & -3 & 1 & 4 \\ 0 & -2 & 0 & 0 \\ 3 & 7 & -1 & 2 \\ 4 & 1 & -3 & 8 \end{vmatrix}$

**11.** $\begin{vmatrix} 1 & 1 & -1 & 0 \\ -3 & 4 & 6 & 0 \\ 2 & 5 & -1 & 3 \\ 4 & 0 & 3 & 0 \end{vmatrix}$

**12.** $\begin{vmatrix} 3 & -1 & 2 & 1 \\ 4 & 3 & 1 & -2 \\ -1 & 0 & 2 & 3 \\ 6 & 2 & 5 & 2 \end{vmatrix}$

**13.** $\begin{vmatrix} 2 & 0 & 0 & 0 \\ 0 & 0 & 3 & 0 \\ 0 & -1 & 0 & 0 \\ 0 & 0 & 0 & 4 \end{vmatrix}$

**14.** $\begin{vmatrix} 0 & a & 0 & 0 \\ b & 0 & 0 & 0 \\ 0 & 0 & 0 & c \\ 0 & 0 & d & 0 \end{vmatrix}$

**15.** $\begin{vmatrix} 1 & 2 & 0 & 0 \\ 3 & -2 & 0 & 0 \\ 0 & 0 & 1 & -5 \\ 0 & 0 & 7 & 2 \end{vmatrix}$

**16.** $\begin{vmatrix} a & b & 0 & 0 \\ c & d & 0 & 0 \\ 0 & 0 & a & -b \\ 0 & 0 & c & d \end{vmatrix}$

**17.** $\begin{vmatrix} 2 & -1 & 0 & 4 & 1 \\ 3 & 1 & -1 & 2 & 0 \\ 3 & 2 & -2 & 5 & 1 \\ 0 & 0 & 4 & -1 & 6 \\ 3 & 2 & 1 & -1 & 1 \end{vmatrix}$

**18.** $\begin{vmatrix} 1 & -1 & 2 & 0 & 0 \\ 3 & 1 & 4 & 0 & 0 \\ 2 & -1 & 5 & 0 & 0 \\ 0 & 0 & 0 & 2 & 3 \\ 0 & 0 & 0 & -1 & 4 \end{vmatrix}$

**19.** $\begin{vmatrix} a & 0 & 0 & 0 & 0 \\ 0 & 0 & b & 0 & 0 \\ 0 & 0 & 0 & 0 & c \\ 0 & 0 & 0 & d & 0 \\ 0 & e & 0 & 0 & 0 \end{vmatrix}$

**20.** $\begin{vmatrix} 2 & 5 & -6 & 8 & 0 \\ 0 & 1 & -7 & 6 & 0 \\ 0 & 0 & 0 & 4 & 0 \\ 0 & 2 & 1 & 5 & 1 \\ 4 & -1 & 5 & 3 & 0 \end{vmatrix}$

In Problems 21–27 compute the determinant assuming that

$$\begin{vmatrix} a_{11} & a_{12} & a_{13} \\ a_{21} & a_{22} & a_{23} \\ a_{31} & a_{32} & a_{33} \end{vmatrix} = 8$$

**21.** $\begin{vmatrix} a_{31} & a_{32} & a_{33} \\ a_{21} & a_{22} & a_{23} \\ a_{11} & a_{12} & a_{13} \end{vmatrix}$

**22.** $\begin{vmatrix} a_{31} & a_{32} & a_{33} \\ a_{11} & a_{12} & a_{13} \\ a_{21} & a_{22} & a_{23} \end{vmatrix}$

**23.** $\begin{vmatrix} a_{11} & a_{12} & a_{13} \\ 2a_{21} & 2a_{22} & 2a_{23} \\ a_{31} & a_{32} & a_{33} \end{vmatrix}$

**24.** $\begin{vmatrix} -3a_{11} & -3a_{12} & -3a_{13} \\ 2a_{21} & 2a_{22} & 2a_{23} \\ 5a_{31} & 5a_{32} & 5a_{33} \end{vmatrix}$

**25.** $\begin{vmatrix} a_{11} & 2a_{13} & a_{12} \\ a_{21} & 2a_{23} & a_{22} \\ a_{31} & 2a_{33} & a_{32} \end{vmatrix}$

**26.** $\begin{vmatrix} a_{11} - a_{12} & a_{12} & a_{13} \\ a_{21} - a_{22} & a_{22} & a_{23} \\ a_{31} - a_{32} & a_{32} & a_{33} \end{vmatrix}$

**27.** $\begin{vmatrix} 2a_{11} - 3a_{21} & 2a_{12} - 3a_{22} & 2a_{13} - 3a_{23} \\ a_{31} & a_{32} & a_{33} \\ a_{21} & a_{22} & a_{23} \end{vmatrix}$

**28.** Using Property 2, show that if $\alpha$ is a scalar and $A$ is an $n \times n$ matrix, then $\det \alpha A = \alpha^n \det A$.

***29.** Show that

$$\begin{vmatrix} 1 + x_1 & x_2 & x_3 & \cdots & x_n \\ x_1 & 1 + x_2 & x_3 & \cdots & x_n \\ x_1 & x_2 & 1 + x_3 & \cdots & x_n \\ \vdots & \vdots & \vdots & & \vdots \\ x_1 & x_2 & x_3 & \cdots & 1 + x_n \end{vmatrix} = 1 + x_1 + x_2 + \cdots + x_n$$

***30.** A matrix is **skew-symmetric** if $A^t = -A$. If $A$ is an $n \times n$ skew-symmetric matrix, show that $\det A = (-1)^n \det A$.

**31.** Using the result of Problem 30, show that if $A$ is a skew-symmetric $n \times n$ matrix and $n$ is odd, then $\det A = 0$.

**32.** A matrix $A$ is called **orthogonal** if $A$ is invertible and $A^{-1} = A^t$. Show that if $A$ is orthogonal, then $\det A = \pm 1$.

****33.** Let $\Delta$ denote the triangle in the plane with vertices at $(x_1, y_1)$, $(x_2, y_2)$, and $(x_3, y_3)$. Show that the area of the triangle is given by

$$\text{Area of } \Delta = \pm\frac{1}{2}\begin{vmatrix} 1 & x_1 & y_1 \\ 1 & x_2 & y_2 \\ 1 & x_3 & y_3 \end{vmatrix}$$

Under what circumstances will this determinant equal zero?

****34.** Three lines, no two of which are parallel, determine a triangle in the plane. Suppose that the lines are given by

$$\begin{aligned} a_{11}x + a_{12}y + a_{13} &= 0 \\ a_{21}x + a_{22}y + a_{23} &= 0 \\ a_{31}x + a_{32}y + a_{33} &= 0 \end{aligned}$$

Show that the area determined by the lines is

$$\frac{\pm 1}{2A_{13}A_{23}A_{33}} \begin{vmatrix} A_{11} & A_{12} & A_{13} \\ A_{21} & A_{22} & A_{23} \\ A_{31} & A_{32} & A_{33} \end{vmatrix}$$

**35.** The $3 \times 3$ **Vandermonde† determinant** is given by

$$D_3 = \begin{vmatrix} 1 & 1 & 1 \\ a_1 & a_2 & a_3 \\ a_1^2 & a_2^2 & a_3^2 \end{vmatrix}$$

Show that $D_3 = (a_2 - a_1)(a_3 - a_1)(a_3 - a_2)$.

**36.** $D_4 = \begin{vmatrix} 1 & 1 & 1 & 1 \\ a_1 & a_2 & a_3 & a_4 \\ a_1^2 & a_2^2 & a_3^2 & a_4^2 \\ a_1^3 & a_2^3 & a_3^3 & a_4^3 \end{vmatrix}$ is the $4 \times 4$ Vandermonde determinant.

Show that $D_4 = (a_2 - a_1)(a_3 - a_1)(a_4 - a_1)(a_3 - a_2)(a_4 - a_2)(a_4 - a_3)$.

****37. a.** Define the $n \times n$ Vandermonde determinant $D_n$.

**b.** Show that $D_n = \prod_{\substack{i=1 \\ j>i}}^{n-1} (a_j - a_i)$, where $\prod$ stands for the word "product." Note that the product in Problem 36 can be written $\prod_{\substack{i=1 \\ j>i}}^{3} (a_j - a_i)$.

**38.** Let $A = \begin{pmatrix} a_{11} & a_{12} \\ a_{21} & a_{22} \end{pmatrix}$ and $B = \begin{pmatrix} b_{11} & b_{12} \\ b_{21} & b_{22} \end{pmatrix}$.

**a.** Write out the product $AB$.
**b.** Compute $\det A$, $\det B$, and $\det AB$.
**c.** Show that $\det AB = (\det A)(\det B)$.

**39.** The $n \times n$ matrix $A$ is called **nilpotent** if $A^k = 0$, the $n \times n$ zero matrix, for some integer $k \geq 1$. Show that the following matrices are nilpotent and find the smallest $k$ such that $A^k = 0$.

**a.** $\begin{pmatrix} 0 & 2 \\ 0 & 0 \end{pmatrix}$ **b.** $\begin{pmatrix} 0 & 1 & 3 \\ 0 & 0 & 4 \\ 0 & 0 & 0 \end{pmatrix}$

**40.** Show that if $A$ is nilpotent, then $\det A = 0$.

**41.** The matrix $A$ is called **idempotent** if $A^2 = A$. What are the possible values for $\det A$ if $A$ is idempotent?

**42.** Let $P$ be a permutation matrix. Show that $\det P = \pm 1$. [*Hint:* By the definition on page 144, $P = P_n P_{n-1} \cdots P_2 P_1$, where each $P_i$ is an elementary permutation matrix. Use Property 4 to show that $\det P_i = -1$ and then compute $\det P$ using Theorem 1].

**43.** Let $P$ be a permutation matrix. Show that $P^t$ is also a permutation matrix and that $\det P = \det P^t$ [*Hint:* If $P_i$ is an elementary permutation matrix, show that $P_i^t = P_i$.]

---

† A. T. Vandermonde (1735–1796) was a French mathematician.

## MATLAB 2.2

1. a. Let **A = round(10*(2*rand(n)−1))** for $n = 2$. Find **det(A).** Now find **det(2*A).** Repeat for $n = 3$ and $n = 4$.
   b. *(Paper and pencil)* Conjecture a formula for $\det(2A)$ in terms of $n$ and $\det(A)$. Conjecture a formula for $\det(kA)$ for general $k$.
   c. Use MATLAB to test your formula for $\det(3A)$.
   d. *(Paper and pencil)* Prove the formula by using properties you learned in this section.
2. For the matrices below, first find $\det(A)$. Then reduce $A$ to upper triangular form, $U$, by using row operations of the forms $R_j \rightarrow R_j + cR_i$ or interchange $R_i$ and $R_j$. Find $\det(U)$ and verify that $\det(A) = (-1)^k \det(U)$, where $k$ is the number of row interchanges performed in the reduction process.

   a. $A = \begin{pmatrix} 6 & 1 & 2 & 3 \\ -1 & 4 & 1 & 1 \\ 0 & 1 & -3 & 1 \\ 1 & 1 & 2 & 5 \end{pmatrix}$  b. $A = \begin{pmatrix} 0 & 1 & 2 \\ 3 & 4 & 5 \\ 1 & 2 & 3 \end{pmatrix}$

   c. For this matrix, before each row operation, interchange rows so that the entry in the pivot position is the largest in absolute value of the possible entries to use for that pivot:

$$A = \begin{pmatrix} 1 & 2 & 3 \\ 4 & 5 & 6 \\ -2 & 1 & 4 \end{pmatrix}$$

   d. Choose a random $n \times n$ matrix $A$ and reduce to upper triangular form by finding the LU-decomposition of $A$ by using the command **[L,U,P] = lu(A).** Use $P$ to determine the number of row interchanges performed and verify that $\det(A) = (-1)^k \det(U)$, where $k$ is the number of row interchanges. Describe the role of $\det(P)$. Repeat for two more choices of $A$.

## 2.3 PROOFS OF THREE IMPORTANT THEOREMS AND SOME HISTORY

Earlier in this book we cited three theorems that are central in the theory of matrices and determinants. The proofs of these theorems are more difficult than those proofs we have already given. Work through these proofs slowly; the reward will be a deeper understanding of some of the important ideas in linear algebra.

**THEOREM 1** **Basic Theorem** Let $A = (a_{ij})$ be an $n \times n$ matrix. Then

$$\det A = a_{11}A_{11} + a_{12}A_{12} + \cdots + a_{1n}A_{1n}$$

$$= a_{i1}A_{i1} + a_{i2}A_{i2} + \cdots + a_{in}A_{in} \quad (1)$$

$$= a_{1j}A_{1j} + a_{2j}A_{2j} + \cdots + a_{nj}A_{nj} \quad (2)$$

for $i = 1, 2, \ldots, n$ and $j = 1, 2, \ldots, n$.

*Note.* The first equality is Definition 2.1.4 of the determinant by cofactor expansion in the first row; the second equality says that the expansion by cofactors in any other row yields the determinant; the third equality says that expansion by cofactors in any column gives the determinant. By the remark on page 190, we need only prove the theorem for rows [equation (1)].

**Proof** We prove equality (1) by mathematical induction. For the $2 \times 2$ matrix $A = \begin{pmatrix} a_{11} & a_{12} \\ a_{21} & a_{22} \end{pmatrix}$, we first expand the first row by cofactors: $\det A = a_{11}A_{11} + a_{12}A_{12} = a_{11}(a_{22}) + a_{12}(-a_{21}) = a_{11}a_{22} - a_{12}a_{21}$. Similarly, expanding in the second row, we obtain $a_{21}A_{21} + a_{22}A_{22} = a_{21}(-a_{12}) + a_{22}(a_{11}) = a_{11}a_{22} - a_{12}a_{21}$. Thus we get the same result by expanding in any row of a $2 \times 2$ matrix, and this proves equality (1) in the $2 \times 2$ case.

We now assume that equality (1) holds for all $(n-1) \times (n-1)$ matrices. We must show that it holds for $n \times n$ matrices. Our procedure will be to expand by cofactors in the first and $i$th rows and show that the expansions are identical. If we expand in the first row, then a typical term in the cofactor expansion is

$$a_{1k}A_{1k} = (-1)^{1+k}a_{1k}|M_{1k}| \tag{3}$$

Note that this is the only place in the expansion of $|A|$ that the term $a_{1k}$ occurs since another typical term is $a_{1m}A_{1m} = (-1)^{1+m}|M_{1m}|$, $k \neq m$, and $M_{1m}$ is obtained by deleting the first row and $m$th column of $A$ (and $a_{1k}$ is in the first row of $A$). Since $M_{1k}$ is an $(n-1) \times (n-1)$ matrix, we can, by the induction hypothesis, calculate $|M_{1k}|$ by expanding in the $i$th row of $A$ (which is the $(i-1)$st row of $M_{1k}$). A typical term in this expansion is

$$a_{il} \text{ (cofactor of } a_{il} \text{ in } M_{1k}) \qquad (k \neq l) \tag{4}$$

For the reasons outlined above, this is the only term in the expansion of $|M_{1k}|$ in the $i$th row of $A$ that contains the term $a_{il}$. Substituting (4) into (3), we find that

$$(-1)^{1+k}a_{1k}a_{il} \text{ (cofactor of } a_{il} \text{ in } M_{1k}) \qquad (k \neq l) \tag{5}$$

is the only occurrence of the term $a_{1k}a_{il}$ in the cofactor expansion of $\det A$ in the first row.

Now if we expand by cofactors in the $i$th row of $A$ (where $i \neq 1$), a typical term is

$$(-1)^{i+l}a_{il}|M_{il}| \tag{6}$$

and a typical term in the expansion of $|M_{il}|$ in the first row of $M_{il}$ is

$$a_{1k} \text{ (cofactor of } a_{1k} \text{ in } M_{il}) \qquad (k \neq l) \tag{7}$$

Inserting (7) in (6), we find that the only occurrence of the term $a_{il}a_{1k}$ in the expansion of $\det A$ along its $i$th row is

$$(-1)^{i+l}a_{1k}a_{il}\text{(cofactor of } a_{1k} \text{ in } M_{il}) \qquad (k \neq l) \tag{8}$$

If we can show that the expressions in (5) and (8) are the same, then (1) will be proved, for the term in (5) is the only occurrence of $a_{1k}a_{il}$ in the first row expansion, the term in (8) is the only occurrence of $a_{1k}a_{il}$ in the $i$th row expansion, and $k$, $i$, and $l$ are arbitrary. This will show that the sums of the terms in the first and $i$th row expansions are the same.

Now let $M_{1i,kl}$ denote the $(n-2) \times (n-2)$ matrix obtained by deleting the first and $i$th rows and $k$th and $l$th columns of $A$. (This is called a **second-order minor** of $A$.) We first suppose that $k < l$. Then

$$M_{1k} = \begin{pmatrix} a_{21} & \cdots & a_{2,k-1} & a_{2,k+1} & \cdots & a_{2l} & \cdots & a_{2n} \\ \vdots & & \vdots & \vdots & & \vdots & & \vdots \\ a_{i1} & \cdots & a_{i,k-1} & a_{i,k+1} & \cdots & a_{il} & \cdots & a_{in} \\ \vdots & & \vdots & \vdots & & \vdots & & \vdots \\ a_{n1} & \cdots & a_{n,k-1} & a_{n,k+1} & \cdots & a_{nl} & \cdots & a_{nn} \end{pmatrix} \tag{9}$$

$$M_{il} = \begin{pmatrix} a_{11} & \cdots & a_{1k} & \cdots & a_{1,l-1} & a_{1,l+1} & \cdots & a_{1n} \\ \vdots & & \vdots & & \vdots & \vdots & & \vdots \\ a_{i-1,1} & \cdots & a_{i-1,k} & \cdots & a_{i-1,l-1} & a_{i-1,l+1} & \cdots & a_{i-1,n} \\ a_{i+1,1} & \cdots & a_{i+1,k} & \cdots & a_{i+1,l-1} & a_{i+1,l+1} & \cdots & a_{i+1,n} \\ \vdots & & \vdots & & \vdots & \vdots & & \vdots \\ a_{n1} & \cdots & a_{nk} & \cdots & a_{n,l-1} & a_{n,l+1} & \cdots & a_{nn} \end{pmatrix} \tag{10}$$

From (9) and (10) we see that

$$\text{Cofactor of } a_{il} \text{ in } M_{1k} = (-1)^{(i-1)+(l-1)}|M_{1i,kl}| \tag{11}$$

$$\text{Cofactor of } a_{1k} \text{ in } M_{il} = (-1)^{1+k}|M_{1i,kl}| \tag{12}$$

Thus (5) becomes

$$(-1)^{1+k}a_{1k}a_{il}(-1)^{(i-1)+(l-1)}|M_{1i,kl}| = (-1)^{i+k+l-1}a_{1k}a_{il}|M_{1i,kl}| \tag{13}$$

and (8) becomes

$$(-1)^{i+l}a_{1k}a_{il}(-1)^{1+k}|M_{1i,kl}| = (-1)^{i+k+l+1}a_{1k}a_{il}|M_{1i,kl}| \tag{14}$$

But $(-1)^{i+k+l-1} = (-1)^{i+k+l+1}$, so that the right-hand sides of equations (13) and (14) are equal. Hence expressions (5) and (8) are equal and (1) is proved in the case $k < l$. If $k > l$, then by similar reasoning, we find that

$$\text{Cofactor of } a_{il} \text{ in } M_{1k} = (-1)^{(i-1)+l}|M_{1i,kl}|$$
$$\text{Cofactor of } a_{1k} \text{ in } M_{il} = (-1)^{1+(k-1)}|M_{1i,kl}|$$

so that (5) becomes

$$(-1)^{1+k}a_{1k}a_{il}(-1)^{(i-1)+l}|M_{1i,kl}| = (-1)^{i+k+l}a_{1k}a_{il}|M_{1i,kl}|$$

and (8) becomes

$$(-1)^{i+l}a_{1k}a_{il}(-1)^{1+k-1}|M_{1i,kl}| = (-1)^{i+k+l}a_{1k}a_{il}|M_{1i,kl}|$$

This completes the proof of equation (1). ■

We now wish to prove that for any two $n \times n$ matrices $A$ and $B$, $\det AB = \det A \det B$. The proof is difficult and involves several steps. We shall make use of a number of facts about elementary matrices proved in Section 1.10.

We begin by computing the determinants of elementary matrices.

**LEMMA 1** Let $E$ be an elementary matrix:

**i.** If $E$ is the matrix representing the elementary row operation $R_i \rightleftarrows R_j$, then $\det E = -1$. **(15)**

**ii.** If $E$ is the matrix representing the elementary row operation $R_j \to R_j + cR_i$, then $\det E = 1$. **(16)**

**iii.** If $E$ is the matrix representing the elementary row operation $R_i \to cR_i$, then $\det E = c$. **(17)**

**Proof**

**i.** $\det I = 1$. $E$ is obtained from $I$ by interchanging the $i$th and $j$th rows of $I$. From Property 4 on page 194, $\det E = (-1) \det I = -1$.

**ii.** $E$ is obtained from $I$ by multiplying the $i$th row of $I$ by $c$ and adding it to the $j$th row. Thus by Property 7 on page 196 $\det E = \det I = 1$.

**iii.** $E$ is obtained from $I$ by multiplying the $i$th row of $I$ by $c$. Thus from Property 2 on page 192, $\det E = c \det I = c$. ■

**LEMMA 2** Let $B$ be an $n \times n$ matrix and let $E$ be an elementary matrix. Then

$$\det EB = \det E \det B \tag{18}$$

■

The proof of this lemma follows from Lemma 1 and the results relating elementary row operations to determinants discussed in Section 2.2. The steps in the proof are indicated in Problems 1 to 3.

The next theorem is a central result in matrix theory.

**THEOREM 2** Let $A$ be an $n \times n$ matrix. Then $A$ is invertible if and only if $\det A \neq 0$.

**Proof** From Theorem 1.10.5 on page 133, we know that there are elementary matrices $E_1, E_2, \ldots, E_m$ and an upper triangular matrix $T$ such that

$$A = E_1 E_2 \cdots E_m T \tag{19}$$

Using Lemma 2 $m$ times, we see that

$$\begin{aligned} \det A &= \det E_1 \det (E_2 E_3 \cdots E_m T) \\ &= \det E_1 \det E_2 \det(E_3 \cdots E_m T) \\ &\ \ \vdots \\ &= \det E_1 \det E_2 \cdots \det E_{m-1} \det(E_m T) \end{aligned}$$

or

$$\det A = \det E_1 \det E_2 \cdots \det E_{m-1} \det E_m \det T \qquad \textbf{(20)}$$

By Lemma 1, $\det E_i \neq 0$ for $i = 1, 2, \ldots, m$. We conclude that $\det A \neq 0$ if and only if $\det T \neq 0$.

Now suppose that $A$ is invertible. Then by using (19) and the fact that every elementary matrix is invertible, $E_m^{-1} \cdots E_1^{-1}A$ is the product of invertible matrices. Thus $T$ is invertible, and by Theorem 2.1.2 on page 179, $\det T \neq 0$. Thus $\det A \neq 0$.

If $\det A \neq 0$, then by (20), $\det T \neq 0$, so $T$ is invertible (by Theorem 2.1.2). Then the right side of (20) is the product of invertible matrices, and so $A$ is invertible. This completes the proof. ■

We can now, finally, prove the main result. Using the results already established, the proof is straightforward.

**THEOREM 3** Let $A$ and $B$ be $n \times n$ matrices. Then

$$\det AB = \det A \det B \qquad \textbf{(21)}$$

**Proof** *Case 1:* $\det A = \det B = 0$. Then by Theorem 2, $B$ is not invertible, so by Theorem 1.8.6, there is an $n$-vector $\mathbf{x} \neq \mathbf{0}$ such that $B\mathbf{x} = \mathbf{0}$. Then $(AB)\mathbf{x} = A(B\mathbf{x}) = A\mathbf{0} = \mathbf{0}$. Therefore, again by Theorem 1.8.6, $AB$ is not invertible. By Theorem 2,

$$0 = \det AB = 0 \cdot 0 = \det A \det B$$

*Case 2:* $\det A = 0$ and $\det B \neq 0$. $A$ is not invertible, so there is an $n$-vector $\mathbf{y} \neq \mathbf{0}$ such that $A\mathbf{y} = \mathbf{0}$. Since $\det B \neq 0$, $B$ is invertible and there is a unique vector $\mathbf{x} \neq \mathbf{0}$ such that $B\mathbf{x} = \mathbf{y}$. Then $AB\mathbf{x} = A(B\mathbf{x}) = A\mathbf{y} = \mathbf{0}$. Thus $AB$ is not invertible, so

$$\det AB = 0 = 0 \det B = \det A \det B$$

*Case 3:* $\det A \neq 0$. $A$ is invertible and can be written as a product of elementary matrices:

$$A = E_1 E_2 \cdots E_m$$

Then

$$AB = E_1 E_2 \cdots E_m B$$

Using the result of Lemma 2 repeatedly, we see that

$$\begin{aligned} \det AB &= \det(E_1 E_2 \cdots E_m B) \\ &= \det E_1 \det E_2 \cdots \det E_m \det B \\ &= \det(E_1 E_2 \cdots E_m) \det B \\ &= \det A \det B \end{aligned}$$

■

Focus on . . .

## A Short History of Determinants

Gottfried Wilhelm Leibniz
*(David Eugene Smith Collection, Rare Book and Manuscript Library, Columbia University)*

Determinants appeared in mathematical literature over a century before matrices. As pointed out in the note on page 48, the term *matrix* was coined by James Joseph Sylvester and was intended to mean "mother of determinants."

Some of the greatest mathematicians of the eighteenth and nineteenth centuries helped to develop properties of determinants. Most historians believe that the theory of determinants originated with the German mathematician Gottfried Wilhelm Leibniz (1646–1716), who, with Newton, was the co-inventor of calculus. Leibniz used determinants in 1693 in reference to systems of simultaneous linear equations. Some believe, however, that a Japanese mathematician, Seki Kōwa, did the same thing about 10 years earlier.

The most prolific contributor to the theory of determinants was the French mathematician Augustin-Louis Cauchy (1789–1857). Cauchy wrote an 84-page memoir in 1812 that contained the first proof of the theorem $\det AB = \det A \det B$. In 1840 Cauchy defined the characteristic equation of the matrix $A$ to be the polynomial equation $\det(A - \lambda I) = 0$. We shall discuss this equation in great detail in Chapter 6.

Augustin-Louis Cauchy
*(David Eugene Smith Collection, Rare Book and Manuscript Library, Columbia University)*

Cauchy made many other contributions to mathematics. In his 1829 calculus textbook *Leçons sur le calcul différential,* he gave the first reasonably clear definition of a limit.

Cauchy wrote extensively in both pure and applied mathematics. Only Euler wrote more. Cauchy contributed to many areas including real and complex function theory, probability theory, geometry, wave propagation theory, and infinite series.

Cauchy is credited with setting a new standard of rigor in mathematical publication. After Cauchy, it was much more difficult to publish a paper based on intuition; a strict adherence to formal proof was demanded.

The sheer volume of Cauchy's publication was overwhelming. When the French Academy of Sciences began publishing its journal *Comptes Rendus* in 1835, Cauchy sent his work there to be published. Soon the printing bill for Cauchy's work alone became so large that the Academy placed a limit of four pages on each published paper. This rule is still in force today.

Some other mathematicians are worthy of mention here. The expansion of a determinant by cofactors was first used by the French mathematician Pierre-Simon Laplace (1749–1827). Laplace is best known for the Laplace transform studied in applied mathematics courses.

A major contributor to determinant theory (second only to Cauchy) was the German mathematician Carl Gustav Jacobi (1804–1851). It was with him that the word "determinant" gained final acceptance. Jacobi first used the determinant applied to functions in the setting of the theory of functions of several variables. This determinant was later named the *Jacobian* by Sylvester. Students today study Jacobians in multivariable calculus classes.

Finally, no history of determinants would be complete without citing the text book *An Elementary Theory of Determinants,* written in 1867 by Charles Dodgson (1832–1898). In this book Dodgson gives conditions such that systems of equations have nontrivial solutions. These conditions are written in terms of the determinants of the minors of coefficient matrices. Charles Dodgson is better known by his pen name Lewis Carroll. Under this pen name he wrote his much better known book *Alice in Wonderland.*

## PROBLEMS 2.3

1. Let $E$ be the representation of $R_i \rightleftarrows R_j$ and let $B$ be an $n \times n$ matrix. Show that $\det EB = \det E \det B$. [*Hint:* Describe the matrix $EB$ and then use equation (15) and Property 4.]
2. Let $E$ be the representation of $R_j \to R_j + cR_i$ and let $B$ be an $n \times n$ matrix. Show that $\det EB = \det E \det B$. [*Hint:* Describe the matrix $EB$ and then use equation (16) and Property 7.]
3. Let $E$ be the representation of $R_i \to cR_i$ and let $B$ be an $n \times n$ matrix. Show that $\det EB = \det E \det B$. [*Hint:* Describe the matrix $EB$ and then use equation (7) and Property 2.]

## 2.4 DETERMINANTS AND INVERSES

In this section we see how matrix inverses can be calculated by using determinants. Moreover, we complete the task, begun in Chapter 1, of proving the important Summing Up Theorem (see Theorems 1.8.6 on page 111 and 1.10.4 on page 132), which shows the equivalence of various properties of matrices. We begin with a simple result.

**THEOREM 1** If $A$ is invertible, then $\det A \neq 0$ and

$$\boxed{\det A^{-1} = \frac{1}{\det A}} \tag{1}$$

**Proof** Suppose that $A$ is invertible. According to Theorem 2.3.2 on page 207, $\det A \neq 0$. From Theorem 2.2.1, page 188

$$1 = \det I = \det AA^{-1} = \det A \det A^{-1} \tag{2}$$

which implies that

$$\det A^{-1} = 1/\det A$$

Before using determinants to calculate inverses, we need to define the *adjoint* of a matrix $A = (a_{ij})$. Let $B = (A_{ij})$ be the matrix of cofactors of $A$. (Remember that a cofactor, defined on page 175, is a number.) Then

$$B = \begin{pmatrix} A_{11} & A_{12} & \cdots & A_{1n} \\ A_{21} & A_{22} & \cdots & A_{2n} \\ \vdots & \vdots & & \vdots \\ A_{n1} & A_{n2} & \cdots & A_{nn} \end{pmatrix} \tag{3}$$

**DEFINITION 1** **The Adjoint** Let $A$ be an $n \times n$ matrix and let $B$, given by (3), denote the matrix of its cofactors. Then the **adjoint** of $A$, written adj $A$, is the transpose of the $n \times n$ matrix $B$; that is,

$$\boxed{\operatorname{adj} A = B^t = \begin{pmatrix} A_{11} & A_{21} & \cdots & A_{n1} \\ A_{12} & A_{22} & \cdots & A_{n2} \\ \vdots & \vdots & & \vdots \\ A_{1n} & A_{2n} & \cdots & A_{nn} \end{pmatrix}} \tag{4}$$

*Remark.* In some books the term **adjugate** of $A$ is used instead of the term **adjoint** because adjoint has a second meaning in mathematics. In this book we will stick with the term adjoint.

**EXAMPLE 1** **Computing the Adjoint of a 3 × 3 Matrix** Let $A = \begin{pmatrix} 2 & 4 & 3 \\ 0 & 1 & -1 \\ 3 & 5 & 7 \end{pmatrix}$. Compute adj $A$.

**Solution** We have $A_{11} = \begin{vmatrix} 1 & -1 \\ 5 & 7 \end{vmatrix} = 12$, $A_{12} = -\begin{vmatrix} 0 & -1 \\ 3 & 7 \end{vmatrix} = -3$, $A_{13} = -3$, $A_{21} = -13$, $A_{22} = 5$, $A_{23} = 2$, $A_{31} = -7$, $A_{32} = 2$, and $A_{33} = 2$. Thus $B = \begin{pmatrix} 12 & -3 & -3 \\ -13 & 5 & 2 \\ -7 & 2 & 2 \end{pmatrix}$ and $\text{adj}\, A = B^t = \begin{pmatrix} 12 & -13 & -7 \\ -3 & 5 & 2 \\ -3 & 2 & 2 \end{pmatrix}$.

**EXAMPLE 2** **Computing the Adjoint of a 4 × 4 Matrix** Let

$$A = \begin{pmatrix} 1 & -3 & 0 & -2 \\ 3 & -12 & -2 & -6 \\ -2 & 10 & 2 & 5 \\ -1 & 6 & 1 & 3 \end{pmatrix}$$

Calculate adj $A$.

**Solution** This is more tedious since we have to compute sixteen 3 × 3 determinants. For example, we have $A_{12} = -\begin{vmatrix} 3 & -2 & -6 \\ -2 & 2 & 5 \\ -1 & 1 & 3 \end{vmatrix} = -1$, $A_{24} = \begin{vmatrix} 1 & -3 & 0 \\ -2 & 10 & 2 \\ -1 & 6 & 1 \end{vmatrix} = -2$, and $A_{43} = -\begin{vmatrix} 1 & -3 & -2 \\ 3 & -12 & -6 \\ -2 & 10 & 5 \end{vmatrix} = 3$. Completing these calculations, we find that

$$B = \begin{pmatrix} 0 & -1 & 0 & 2 \\ -1 & 1 & -1 & -2 \\ 0 & 2 & -3 & -3 \\ -2 & -2 & 3 & 2 \end{pmatrix}$$

and

$$\text{adj}\, A = B^t = \begin{pmatrix} 0 & -1 & 0 & -2 \\ -1 & 1 & 2 & -2 \\ 0 & -1 & -3 & 3 \\ 2 & -2 & -3 & 2 \end{pmatrix}$$

**EXAMPLE 3** **The Adjoint of a 2 × 2 Matrix** Let $A = \begin{pmatrix} a_{11} & a_{12} \\ a_{21} & a_{22} \end{pmatrix}$. Then

$$\text{adj } A = \begin{pmatrix} A_{11} & A_{21} \\ A_{12} & A_{22} \end{pmatrix} = \begin{pmatrix} a_{22} & -a_{12} \\ -a_{21} & a_{11} \end{pmatrix}.$$

**WARNING** In computing the adjoint of a matrix, do not forget to transpose the matrix of cofactors.

**THEOREM 2** Let $A$ be an $n \times n$ matrix. Then

$$(A)(\text{adj } A) = \begin{pmatrix} \det A & 0 & 0 & \cdots & 0 \\ 0 & \det A & 0 & \cdots & 0 \\ 0 & 0 & \det A & \cdots & 0 \\ \vdots & \vdots & \vdots & & \vdots \\ 0 & 0 & 0 & \cdots & \det A \end{pmatrix} = (\det A)I \qquad (5)$$

**Proof** Let $C = (c_{ij}) = (A)(\text{adj } A)$. Then

$$C = \begin{pmatrix} a_{11} & a_{12} & \cdots & a_{1n} \\ a_{21} & a_{22} & \cdots & a_{2n} \\ \vdots & \vdots & & \vdots \\ a_{n1} & a_{n2} & \cdots & a_{nn} \end{pmatrix} \begin{pmatrix} A_{11} & A_{21} & \cdots & A_{n1} \\ A_{12} & A_{22} & \cdots & A_{n2} \\ \vdots & \vdots & & \vdots \\ A_{1n} & A_{2n} & \cdots & A_{nn} \end{pmatrix} \qquad (6)$$

We have

$$\begin{aligned} c_{ij} &= (i\text{th row of } A) \cdot (j\text{th column of adj } A) \\ &= (a_{i1} \quad a_{i2} \cdots a_{in}) \cdot \begin{pmatrix} A_{j1} \\ A_{j2} \\ \vdots \\ A_{jn} \end{pmatrix} \end{aligned}$$

Thus

$$c_{ij} = a_{i1}A_{j1} + a_{i2}A_{j2} + \cdots + a_{in}A_{jn} \qquad (7)$$

Now if $i = j$, the sum in (7) equals $a_{i1}A_{i1} + a_{i2}A_{i2} + \cdots + a_{in}A_{in}$, which is the expansion of $\det A$ in the $i$th row of $A$. On the other hand, if $i \neq j$, then from Theorem 2.2.6 on page 199, the sum in (7) equals zero. Thus

$$c_{ij} = \begin{cases} \det A & \text{if } i = j \\ 0 & \text{if } i \neq j \end{cases}$$

This proves the theorem.

We can now state the main result.

**THEOREM 3** Let $A$ be an $n \times n$ matrix. Then $A$ is invertible if and only if $\det A \neq 0$. If $\det A \neq 0$, then

$$A^{-1} = \frac{1}{\det A} \operatorname{adj} A \qquad \textbf{(8)}$$

Note that Theorem 1.8.4 on page 104 for $2 \times 2$ matrices is a special case of this theorem.

**Proof** The first part of the theorem is Theorem 2.3.2. If $\det A \neq 0$, then we show that $(1/\det A)(\operatorname{adj} A)$ is the inverse of $A$ by multiplying it by $A$ and getting the identity matrix:

$$(A)\left(\frac{1}{\det A} \operatorname{adj} A\right) = \frac{1}{\det A}[A(\operatorname{adj} A)] \overset{\text{Theorem 2}}{=} \frac{1}{\det A}(\det A)I = I$$

But by Theorem 1.8.7 on page 112, if $AB = I$, then $B = A^{-1}$. Thus

$$(1/\det A) \operatorname{adj} A = A^{-1}$$

**EXAMPLE 4** **Using the Determinant and the Adjoint to Calculate an Inverse** Let $A = \begin{pmatrix} 2 & 4 & 3 \\ 0 & 1 & -1 \\ 3 & 5 & 7 \end{pmatrix}$. Determine whether $A$ is invertible and calculate $A^{-1}$ if it is.

**Solution** Since $\det A = 3 \neq 0$, we see that $A$ is invertible. From Example 1

$$\operatorname{adj} A = \begin{pmatrix} 12 & -13 & -7 \\ -3 & 5 & 2 \\ -3 & 2 & 2 \end{pmatrix}$$

Thus

$$A^{-1} = \frac{1}{3}\begin{pmatrix} 12 & -13 & -7 \\ -3 & 5 & 2 \\ -3 & 2 & 2 \end{pmatrix} = \begin{pmatrix} 4 & -\frac{13}{3} & -\frac{7}{3} \\ -1 & \frac{5}{3} & \frac{2}{3} \\ -1 & \frac{2}{3} & \frac{2}{3} \end{pmatrix}$$

*Check.*

$$A^{-1}A = \frac{1}{3}\begin{pmatrix} 12 & -13 & -7 \\ -3 & 5 & 2 \\ -3 & 2 & 2 \end{pmatrix}\begin{pmatrix} 2 & 4 & 3 \\ 0 & 1 & -1 \\ 3 & 5 & 7 \end{pmatrix} = \frac{1}{3}\begin{pmatrix} 3 & 0 & 0 \\ 0 & 3 & 0 \\ 0 & 0 & 3 \end{pmatrix} = I$$

**EXAMPLE 5** **Calculating the Inverse of a 4 × 4 Matrix Using the Determinant and the Adjoint** Let

$$A = \begin{pmatrix} 1 & -3 & 0 & -2 \\ 3 & -12 & -2 & -6 \\ -2 & 10 & 2 & 5 \\ -1 & 6 & 1 & 3 \end{pmatrix}$$

Determine whether $A$ is invertible, and if so, calculate $A^{-1}$.

**Solution** Using properties of determinants, we compute $\det A = -1 \neq 0$ and $A^{-1}$ exists.

By Example 2, we have

$$\text{adj } A = \begin{pmatrix} 0 & -1 & 0 & -2 \\ -1 & 1 & 2 & -2 \\ 0 & -1 & -3 & 3 \\ 2 & -2 & -3 & 2 \end{pmatrix}$$

Thus

$$A^{-1} = \frac{1}{-1}\begin{pmatrix} 0 & -1 & 0 & -2 \\ -1 & 1 & 2 & -2 \\ 0 & -1 & -3 & 3 \\ 2 & -2 & -3 & 2 \end{pmatrix} = \begin{pmatrix} 0 & 1 & 0 & 2 \\ 1 & -1 & -2 & 2 \\ 0 & 1 & 3 & -3 \\ -2 & 2 & 3 & -2 \end{pmatrix}$$

*Note 1.* As you have noticed, if $n > 3$ it is generally easier to compute $A^{-1}$ by row reduction than by using adj $A$ since, even for the $4 \times 4$ case, it is necessary to calculate 17 determinants (16 for the adjoint plus det $A$). Nevertheless, Theorem 3 is very important since, before you do any row reduction, the calculation of det $A$ (if it can be done easily) will tell you whether or not $A^{-1}$ exists.

*Note 2.* In many applications of matrix theory matrices are given symbolically (i.e., in terms of variables) rather than numerically. For example, we might have $A = \begin{pmatrix} x & y \\ z & w \end{pmatrix}$ rather than $\begin{pmatrix} 2 & -1 \\ 3 & 5 \end{pmatrix}$. In this case calculating $A^{-1}$ by using determinants will often be the most effective way to proceed. This is particularly true in certain engineering applications—like control theory.

We last saw our Summing Up Theorem (Theorems 1.2.1, 1.8.6, and 1.10.4) in Section 1.10. This is the theorem that ties together many of the concepts developed in the first chapters of this book.

**THEOREM 4** **Summing Up Theorem—View 4** Let $A$ be an $n \times n$ matrix. Then the following seven statements are equivalent. That is, each one implies the other six (so that if one is true, all of them are true):

**i.** $A$ is invertible.

**ii.** The only solution to the homogeneous system $A\mathbf{x} = \mathbf{0}$ is the trivial solution ($\mathbf{x} = \mathbf{0}$).

**iii.** The system $A\mathbf{x} = \mathbf{b}$ has a unique solution for every $n$-vector $\mathbf{b}$.

**iv.** $A$ is row equivalent to the $n \times n$ identity matrix $I_n$.

**v.** $A$ is the product of elementary matrices.

**vi.** The row echelon form of $A$ has $n$ pivots.

**vii.** $\det A \neq 0$.

**Proof** In Theorem 1.8.6 we proved the equivalence of parts (*i*), (*ii*), (*iii*), (*iv*), and (*vi*). In Theorem 1.10.3 we proved the equivalence of parts (*i*) and (*v*). Theorem 1 (or Theorem 2.3.2) proves the equivalence of (*i*) and (*vii*).

# PROBLEMS 2.4

## Self-Quiz

**I.** The determinant of $\begin{pmatrix} 1 & 2 & -1 & 4 \\ 2 & 3 & 2 & 4 \\ 5 & 1 & 0 & -3 \\ -4 & 3 & 1 & 6 \end{pmatrix}$ is $-149$. The 2,3 component of $A^{-1}$ is given by

**a.** $-\frac{1}{149}\begin{vmatrix} 1 & 2 & 4 \\ 5 & 1 & -3 \\ -4 & 3 & 6 \end{vmatrix}$, **b.** $\frac{1}{149}\begin{vmatrix} 1 & 2 & 4 \\ 5 & 1 & -3 \\ -4 & 3 & 6 \end{vmatrix}$,

**c.** $-\frac{1}{149}\begin{vmatrix} 1 & -1 & 4 \\ 2 & 2 & 4 \\ -4 & 1 & 6 \end{vmatrix}$, **d.** $\frac{1}{149}\begin{vmatrix} 1 & -1 & 4 \\ 2 & 2 & 4 \\ -4 & 1 & 6 \end{vmatrix}$

**II.** The determinant of $\begin{pmatrix} 3 & 7 & 2 \\ -1 & 5 & 8 \\ 6 & -4 & 4 \end{pmatrix}$ is 468. The 3,1 component of $A^{-1}$ is

**a.** $-\frac{26}{468}$ **b.** $\frac{26}{468}$ **c.** $\frac{46}{468}$ **d.** $-\frac{46}{468}$

**Answers to Self-Quiz**

**I.** d **II.** a

In Problems 1–12 use the methods of this section to determine whether the given matrix is invertible. If so, compute the inverse.

**1.** $\begin{pmatrix} 3 & 2 \\ 1 & 2 \end{pmatrix}$

**2.** $\begin{pmatrix} 3 & 6 \\ -4 & -8 \end{pmatrix}$

**3.** $\begin{pmatrix} 0 & 1 \\ 1 & 0 \end{pmatrix}$

**4.** $\begin{pmatrix} 1 & 1 & 1 \\ 0 & 2 & 3 \\ 5 & 5 & 1 \end{pmatrix}$

**5.** $\begin{pmatrix} 3 & 2 & 1 \\ 0 & 2 & 2 \\ 0 & 1 & -1 \end{pmatrix}$

**6.** $\begin{pmatrix} 1 & 1 & 1 \\ 0 & 1 & 1 \\ 0 & 0 & 1 \end{pmatrix}$

**7.** $\begin{pmatrix} 1 & 2 & 3 \\ 1 & 1 & 2 \\ 0 & 1 & 2 \end{pmatrix}$

**8.** $\begin{pmatrix} 3 & 1 & 0 \\ 1 & -1 & 2 \\ 1 & 1 & 1 \end{pmatrix}$

**9.** $\begin{pmatrix} 2 & -1 & 4 \\ -1 & 0 & 5 \\ 19 & -7 & 3 \end{pmatrix}$

**10.** $\begin{pmatrix} 1 & 6 & 2 \\ -2 & 3 & 5 \\ 7 & 12 & -4 \end{pmatrix}$

**11.** $\begin{pmatrix} 1 & 1 & 1 & 1 \\ 1 & 2 & -1 & 2 \\ 1 & -1 & 2 & 1 \\ 1 & 3 & 3 & 2 \end{pmatrix}$

**12.** $\begin{pmatrix} 1 & -3 & 0 & -2 \\ 3 & -12 & -2 & -6 \\ -2 & 10 & 2 & 5 \\ -1 & 6 & 1 & 3 \end{pmatrix}$

**13.** Use determinants to show that an $n \times n$ matrix $A$ is invertible if and only if $A^t$ is invertible.

**14.** For $A = \begin{pmatrix} 1 & 1 \\ 2 & 5 \end{pmatrix}$, verify that $\det A^{-1} = 1/\det A$.

**15.** For $A = \begin{pmatrix} 1 & -1 & 3 \\ 4 & 1 & 6 \\ 2 & 0 & -2 \end{pmatrix}$, verify that $\det A^{-1} = 1/\det A$.

**16.** For what values of $\alpha$ is the matrix $\begin{pmatrix} \alpha & -3 \\ 4 & 1-\alpha \end{pmatrix}$ not invertible?

**17.** For what values of $\alpha$ does the matrix $\begin{pmatrix} -\alpha & \alpha-1 & \alpha+1 \\ 1 & 2 & 3 \\ 2-\alpha & \alpha+3 & \alpha+7 \end{pmatrix}$ not have an inverse?

**18.** Suppose that the $n \times n$ matrix $A$ is not invertible. Show that $(A)(\text{adj } A)$ is the zero matrix.

**19.** Let $\theta$ be a real number. Show that $\begin{pmatrix} \cos\theta & \sin\theta \\ -\sin\theta & \cos\theta \end{pmatrix}$ is invertible and find its inverse.

## MATLAB 2.4

1. Generate a random $n \times m$ matrix by **A = 2*rand(n,m)−1** for some values of $n$ and $m$ such that $m > n$. Find the determinant of $A^tA$. What can you conclude about $A^tA$? Test your conjecture for three more choices of $A$. Is your conjecture valid if $m < n$?

2. Let **A = round(10*(2*rand(4)−1))**.

   a. Enter the command **flops(0)**. Compute adj$(A)$ by using MATLAB. (To get started, set **C = zeros(4); C(1,1) = det(A([2 3 4],[2 3 4])); C(1,2) = −det(A([2 3 4],[1 3 4]));** and so on. Don't forget the transpose.) Enter the command **s = flops.**

b. Enter the command **flops(0)**. Compute **D = det(A)*inv(A)**. Enter the command **ss = flops.**

c. How does adj$(A)$, computed in part (a), compare with $D$, computed in part (b)? Why would you expect that?

d. How do the flop counts compare? What did you learn from comparing the flop counts? (Recall that **flops** counts the number of floating point operations performed.)

3. We have shown that $A$ is not invertible if $\det(A) = 0$. A natural assumption is that if $A$ is close to being noninvertible, then $\det(A)$ will be close to 0.

Consider the matrix $C$ below. Verify that $C$ is not invertible. Enter **A = C; A(3,3) = C(3,3) + 1.e−10.** Verify that $A$ is invertible and note that $A$ is close to the noninvertible matrix $C$. Find **det(A).** What conclusion can you make about the "natural assumption" mentioned above?

$$C = 20 * \begin{pmatrix} 7 & 7 & -7 & 2 & 5 & 6 \\ 0 & 5 & -10 & 4 & 8 & 6 \\ 9 & 7 & -5 & 3 & 4 & 0 \\ 5 & 7 & -9 & 5 & 2 & 0 \\ 5 & 2 & 1 & 9 & 10 & 8 \\ 1 & 9 & -17 & 4 & 2 & 7 \end{pmatrix}$$

4. a. Enter a $5 \times 5$ upper triangular matrix $A$ with integer entries so that the determinant of $A$ is 1. Choose values of $c$ (integer), $i$, and $j$ and perform several row operations of the form $R_j \rightarrow R_j + cR_i$ so that the matrix is full, that is, as few zero entries as possible. Call the new matrix $A$ also.

b. Verify that $\det(A)$ is still equal to 1. Why would you expect this? Find inv$(A)$ and verify that it has integer entries. Why would you expect this?

**PROJECT PROBLEM**

c. Refer to MATLAB Problem 9 in Section 1.8 on encoding and decoding messages. This problem has you encode a message for your instructor by using the matrix $A$ created above.

**i.** Create a message for your instructor. Using numbers for letters as described in MATLAB Problem 9, Section 1.8, write the message in matrix form so that you can multiply by $A$ on the right to encode the message. (You may need to place extra spaces at the end of the message.)

**ii.** Use $A$ to encode the message.

**iii.** Hand in to your instructor the encoded message (as a string of numbers) and the matrix $A$.

## 2.5 CRAMER'S RULE (OPTIONAL)

In this section we examine an old method for solving systems with the same number of unknowns as equations. Consider the system of $n$ linear equations in $n$ unknowns.

$$\begin{aligned} a_{11}x_1 + a_{12}x_2 + \cdots + a_{1n}x_n &= b_1 \\ a_{21}x_1 + a_{22}x_2 + \cdots + a_{2n}x_n &= b_2 \\ \vdots \quad\quad \vdots \quad\quad\quad\quad \vdots \quad &\quad \vdots \\ a_{n1}x_1 + a_{n2}x_2 + \cdots + a_{nn}x_n &= b_n \end{aligned} \tag{1}$$

which can be written in the form

$$A\mathbf{x} = \mathbf{b} \tag{2}$$

If $\det A \neq 0$, then system (2) has a unique solution given by $\mathbf{x} = A^{-1}\mathbf{b}$. We can develop a method for finding that solution without row reduction and without computing $A^{-1}$.

Let $D = \det A$. We define $n$ new matrices:

$$A_1 = \begin{pmatrix} b_1 & a_{12} & \cdots & a_{1n} \\ b_2 & a_{22} & \cdots & a_{2n} \\ \vdots & \vdots & & \vdots \\ b_n & a_{n2} & \cdots & a_{nn} \end{pmatrix}, \quad A_2 = \begin{pmatrix} a_{11} & b_1 & \cdots & a_{1n} \\ a_{21} & b_2 & \cdots & a_{2n} \\ \vdots & \vdots & & \vdots \\ a_{n1} & b_n & \cdots & a_{nn} \end{pmatrix}, \ldots,$$

$$A_n = \begin{pmatrix} a_{11} & a_{12} & \cdots & b_1 \\ a_{21} & a_{22} & \cdots & b_2 \\ \vdots & \vdots & & \vdots \\ a_{n1} & a_{n2} & \cdots & b_n \end{pmatrix}$$

That is, $A_i$ is the matrix obtained by replacing the $i$th column of $A$ with $\mathbf{b}$. Finally, let $D_1 = \det A_1$, $D_2 = \det A_2, \ldots, D_n = \det A_n$.

**THEOREM 1** **Cramer's Rule** Let $A$ be an $n \times n$ matrix and suppose that $\det A \neq 0$. Then the unique solution to the system $A\mathbf{x} = \mathbf{b}$ is given by

$$\boxed{x_1 = \frac{D_1}{D}, x_2 = \frac{D_2}{D}, \ldots, x_i = \frac{D_i}{D}, \ldots, x_n = \frac{D_n}{D}} \tag{3}$$

**Proof** The solution to $A\mathbf{x} = \mathbf{b}$ is $\mathbf{x} = A^{-1}\mathbf{b}$. But

$$A^{-1}\mathbf{b} = \frac{1}{D}(\text{adj } A)\mathbf{b} = \frac{1}{D}\begin{pmatrix} A_{11} & A_{21} & \cdots & A_{n1} \\ A_{12} & A_{22} & \cdots & A_{n2} \\ \vdots & \vdots & & \vdots \\ A_{1n} & A_{2n} & \cdots & A_{nn} \end{pmatrix}\begin{pmatrix} b_1 \\ b_2 \\ \vdots \\ b_n \end{pmatrix} \tag{4}$$

Now $(\text{adj } A)\mathbf{b}$ is an $n$-vector, the $j$th component of which is

$$(A_{1j}\ A_{2j} \ldots A_{nj}) \cdot \begin{pmatrix} b_1 \\ b_2 \\ \vdots \\ b_n \end{pmatrix} = b_1A_{1j} + b_2A_{2j} + \cdots + b_nA_{nj} \tag{5}$$

Consider the matrix $A_j$:

$$A_j = \begin{pmatrix} a_{11} & a_{12} & \cdots & b_1 & \cdots & a_{1n} \\ a_{21} & a_{22} & \cdots & b_2 & \cdots & a_{2n} \\ \vdots & \vdots & & \vdots & & \vdots \\ a_{n1} & a_{n2} & \cdots & b_n & \cdots & a_{nn} \end{pmatrix} \tag{6}$$

$\uparrow$ $j$th column

If we expand the determinant of $A_j$ in its $j$th column, we obtain

$$\begin{aligned} D_j = b_1 \text{ (cofactor of } b_1) + b_2 \text{ (cofactor of } b_2) + \cdots \\ + b_n \text{ (cofactor of } b_n) \end{aligned} \tag{7}$$

But in order to find the cofactor of $b_i$, say, we delete the $i$th row and $j$th column of $A_j$ (since $b_i$ is in the $j$th column of $A_j$). But the $j$th column of $A_j$ is $\mathbf{b}$, and with this deleted, we simply have the $ij$ minor, $M_{ij}$, of $A$. Thus

$$\text{Cofactor of } b_i \text{ in } A_j = A_{ij}$$

so that (7) becomes

$$D_j = b_1 A_{1j} + b_2 A_{2j} + \cdots + b_n A_{nj} \tag{8}$$

But this is the same as the right side of (5). Thus the $i$th component of (adj $A$)$\mathbf{b}$ is $D_i$, and we have

$$\mathbf{x} = \begin{pmatrix} x_1 \\ x_2 \\ \vdots \\ x_n \end{pmatrix} = A^{-1}\mathbf{b} = \frac{1}{D}(\text{adj } A)\mathbf{b} = \frac{1}{D}\begin{pmatrix} D_1 \\ D_2 \\ \vdots \\ D_n \end{pmatrix} = \begin{pmatrix} D_1/D \\ D_2/D \\ \vdots \\ D_n/D \end{pmatrix}$$

and the proof is complete. ■

***Historical Note.*** Cramer's rule is named for the Swiss mathematician Gabriel Cramer (1704–1752). Cramer published the rule in 1750 in his *Introduction to the Analysis of Lines of Algebraic Curves.* Actually, there is much evidence to suggest that the rule was known as early as 1729 to Colin Maclaurin (1698–1746), who was probably the most outstanding British mathematician in the years following the death of Newton. Cramer's rule is one of the most famous results in the history of mathematics. For almost 200 years it was central in the teaching of algebra and the theory of equations. Because of the great number of computations involved, the rule is rarely used today. However, the result was very important in its time.

**EXAMPLE 1** **Solving a 3 × 3 System Using Cramer's Rule** Solve by using Cramer's rule the system

$$\begin{aligned} 2x_1 + 4x_2 + 6x_3 &= 18 \\ 4x_1 + 5x_2 + 6x_3 &= 24 \\ 3x_1 + x_2 - 2x_3 &= 4 \end{aligned} \tag{9}$$

**Solution** We have solved this before—using row reduction in Example 1.3.1 on page 7. We could also solve it by calculating $A^{-1}$ (Example 1.8.6, page 106) and then finding $A^{-1}\mathbf{b}$. We now solve it by using Cramer's rule. First, we have

$$D = \begin{vmatrix} 2 & 4 & 6 \\ 4 & 5 & 6 \\ 3 & 1 & -2 \end{vmatrix} = 6 \neq 0$$

so that system (9) has a unique solution. Then $D_1 = \begin{vmatrix} 18 & 4 & 6 \\ 24 & 5 & 6 \\ 4 & 1 & -2 \end{vmatrix} = 24$,

$D_2 = \begin{vmatrix} 2 & 18 & 6 \\ 4 & 24 & 6 \\ 3 & 4 & -2 \end{vmatrix} = -12$ and $D_3 = \begin{vmatrix} 2 & 4 & 18 \\ 4 & 5 & 24 \\ 3 & 1 & 4 \end{vmatrix} = 18$. Hence

$$x_1 = \frac{D_1}{D} = \frac{24}{6} = 4, \quad x_2 = \frac{D_2}{D} = -\frac{12}{6} = -2, \quad \text{and} \quad x_3 = \frac{D_3}{D} = \frac{18}{6} = 3.$$

**EXAMPLE 2** **Solving a 4 × 4 System Using Cramer's Rule** Show that the system

$$\begin{aligned} x_1 + 3x_2 + 5x_3 + 2x_4 &= 2 \\ -x_2 + 3x_3 + 4x_4 &= 0 \\ 2x_1 + x_2 + 9x_3 + 6x_4 &= -3 \\ 3x_1 + 2x_2 + 4x_3 + 8x_4 &= -1 \end{aligned} \tag{10}$$

has a unique solution and find it by using Cramer's rule.

**Solution** We saw in Example 2.2.14 on page 197 that

$$|A| = \begin{vmatrix} 1 & 3 & 5 & 2 \\ 0 & -1 & 3 & 4 \\ 2 & 1 & 9 & 6 \\ 3 & 2 & 4 & 8 \end{vmatrix} = 160 \neq 0$$

Thus the system has a unique solution. To find it we compute $D_1 = -464$; $D_2 = 280$; $D_3 = -56$; $D_4 = 112$. Thus $x_1 = D_1/D = -464/160$, $x_2 = D_2/D = 280/160$, $x_3 = D_3/D = -56/160$, and $x_4 = D_4/D = 112/160$. These solutions can be verified by direct substitution into system (10).

# PROBLEMS 2.5

## Self-Quiz

**I.** Consider the system

$$\begin{aligned} 2x - 3y + 4z &= 7 \\ 3x + 8y - z &= 2 \\ -5x - 12y + 6z &= 11 \end{aligned}$$

If $D = \begin{vmatrix} 2 & -3 & 4 \\ 3 & 8 & -1 \\ -5 & -12 & 6 \end{vmatrix}$, then $y =$ ________.

**a.** $\frac{1}{D}\begin{vmatrix} 7 & -3 & 4 \\ 2 & 8 & -1 \\ 11 & -12 & 6 \end{vmatrix}$

**b.** $\frac{1}{D}\begin{vmatrix} 2 & -3 & 7 \\ 3 & 8 & 2 \\ -5 & -12 & 11 \end{vmatrix}$

**c.** $\frac{1}{D}\begin{vmatrix} 2 & 7 & 4 \\ 3 & 2 & -1 \\ -5 & 11 & 6 \end{vmatrix}$

**d.** $\frac{1}{D}\begin{vmatrix} 2 & -7 & 4 \\ 3 & -2 & -1 \\ -5 & -1 & 6 \end{vmatrix}$

In Problems 1–9 solve the given system by using Cramer's rule.

**1.** $\begin{aligned} 2x_1 + 3x_2 &= -1 \\ -7x_1 + 4x_2 &= 47 \end{aligned}$

**2.** $\begin{aligned} 3x_1 - x_2 &= 0 \\ 4x_1 + 2x_2 &= 5 \end{aligned}$

**3.** $\begin{aligned} 2x_1 + x_2 + x_3 &= 6 \\ 3x_1 - 2x_2 - 3x_3 &= 5 \\ 8x_1 + 2x_2 + 5x_3 &= 11 \end{aligned}$

**4.** $\begin{aligned} x_1 + x_2 + x_3 &= 8 \\ 4x_2 - x_3 &= -2 \\ 3x_1 - x_2 + 2x_3 &= 0 \end{aligned}$

**5.** $\begin{aligned} 2x_1 + 2x_2 + x_3 &= 7 \\ x_1 + 2x_2 - x_3 &= 0 \\ -x_1 + x_2 + 3x_3 &= 1 \end{aligned}$

**6.** $\begin{aligned} 2x_1 + 5x_2 - x_3 &= -1 \\ 4x_1 + x_2 + 3x_3 &= 3 \\ -2x_1 + 2x_2 \phantom{+ 3x_3} &= 0 \end{aligned}$

**7.** $\begin{aligned} 2x_1 + x_2 - x_3 &= 4 \\ x_1 + \phantom{x_2 +} x_3 &= 2 \\ - x_2 + 5x_3 &= 1 \end{aligned}$

**8.** $\begin{aligned} x_1 + x_2 + x_3 + x_4 &= 6 \\ 2x_1 - x_3 - x_4 &= 4 \\ 3x_3 + 6x_4 &= 3 \\ x_1 - x_4 &= 5 \end{aligned}$

**9.** $\begin{aligned} x_1 - x_4 &= 7 \\ 2x_2 + x_3 &= 2 \\ 4x_1 - x_2 &= -3 \\ 3x_3 - 5x_4 &= 2 \end{aligned}$

---

### Answer to Self-Quiz

**I.** c

***10.** Consider the triangle in Figure 2.2.

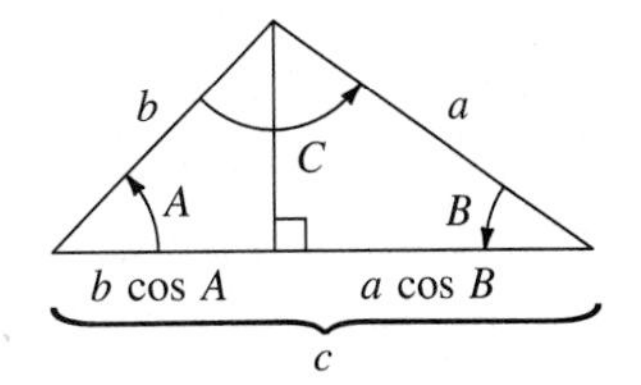


**Figure 2.2**

**a.** Show by using elementary trigonometry that

$$\begin{aligned} c\cos A \qquad\qquad\qquad & + a\cos C = b \\ b\cos A + a\cos B \qquad & \qquad\quad\;\; = c \\ c\cos B & + b\cos C = a \end{aligned}$$

**b.** If the system of part (*a*) is thought of as a system of three equations in the three unknowns $\cos A$, $\cos B$, and $\cos C$, show that the determinant of the system is nonzero.

**c.** Use Cramer's rule to solve for $\cos C$.

**d.** Use part (*c*) to prove the **law of cosines:** $c^2 = a^2 + b^2 - 2ab\cos C$.

## MATLAB 2.5

1. Generate a random $5 \times 5$ matrix $A$ and a random $5 \times 1$ matrix **b**.
   a. Enter the command **flops(0).** Solve the system $A\mathbf{x} = \mathbf{b}$ by using Cramer's rule. First find **d = det(A).** In the rest of the computations, use *d*; that is, do not recompute det(*A*). (To use Cramer's rule, you will need to form the matrix obtained from *A* by replacing the *i*th column of *A* by **b**: **C = A; C(:,i) = b.** Make efficient use of the up-arrow cursor key to repeat statements after some modification.) After finding each component of the solution, form a column vector containing the solution and call it **x**. Enter the command **s = flops.**
   b. Enter the command **flops(0).** Solve the system by using **z = A\b.** Enter the command **ss = flops.**
   c. Compare **x** and **z** by computing **x** − **z** and displaying it by using **format short e.** Compare the flop counts, **s** and **ss**. What did you learn from these comparisons?
   d. Repeat for a random $7 \times 7$ matrix. What further statements can you make about the flop counts?

## SUMMARY

- The **determinant** of a $2 \times 2$ matrix $A = \begin{pmatrix} a_{11} & a_{12} \\ a_{21} & a_{22} \end{pmatrix}$ is given by (p. 172)

$$\text{Determinant of } A = \det A = |A| = a_{11}a_{22} - a_{12}a_{21}$$

- **3 × 3 determinant**

$$\det\begin{pmatrix} a_{11} & a_{12} & a_{13} \\ a_{21} & a_{22} & a_{23} \\ a_{31} & a_{32} & a_{33} \end{pmatrix} = a_{11}\begin{vmatrix} a_{22} & a_{23} \\ a_{32} & a_{33} \end{vmatrix} - a_{12}\begin{vmatrix} a_{21} & a_{23} \\ a_{31} & a_{33} \end{vmatrix} + a_{13}\begin{vmatrix} a_{21} & a_{22} \\ a_{31} & a_{32} \end{vmatrix}$$ (p. 173)

- The ***ij*th minor** of the $n \times n$ matrix $A$, denoted by $M_{ij}$, is the $(n-1) \times (n-1)$ matrix obtained by crossing off the $i$th row and $j$th column of $A$. (p. 175)
- The ***ij*th cofactor** of $A$, denoted by $A_{ij}$, is given by

$$A_{ij} = (-i)^{i+j} \det M_{ij}$$ (p. 175)

- ***n × n determinant***

Let $A$ be an $n \times n$ matrix. Then (p. 176)

$$\det A = a_{11}A_{11} + a_{12}A_{12} + \cdots + A_{1n}A_{1n} = \sum_{k=1}^{n} a_{1k}A_{1k}$$

The sum above is called the **expansion of det *A* by cofactors in the first row.**

- If $A$ is an upper triangular, lower triangular, or diagonal $n \times n$ matrix with diagonal components $a_{11}, a_{22}, \ldots, a_{nn}$, then (p. 178)

$$\det A = a_{11}\, a_{22} \cdots a_{nn}$$

- If $A = LU$ is an $LU$-factorization of $A$, then $\det A = \det U$ (p. 189)
- If $PA = LU$ is an $LU$-factorization of $PA$, then $\det A = \det U/\det P = \pm\det U$ (p. 189)
- ***Basic Theorem***

If $A$ is an $n \times n$ matrix, then

$$\det A = a_{i1}A_{i1} + a_{i2}A_{i2} + \cdots + a_{in}A_{in} = \sum_{k=1}^{n} a_{ik}A_{ik}$$

and (pp. 191, 204)

$$\det A = a_{1j}A_{1j} + a_{2j}A_{2j} + \cdots + a_{nj}A_{nj} = \sum_{k=1}^{n} a_{kj}A_{kj}$$

for $i = 1, 2, \ldots, n$ and $j = 1, 2, \ldots, n$.

That is, the determinant of $A$ can be obtained by expanding in any row or column of $A$.

- If any row or column $A$ is the zero vector, then $\det A = 0$. (p. 192)
- If any row (column) of $A$ is multiplied by a scalar $c$, then $\det A$ is multiplied by $c$. (p. 192)
- If $A$ and $B$ are two $n \times n$ matrices that are equal except in the $j$th column ($i$th row) and $C$ is the matrix that is identical to $A$ and $B$ except that the $j$th column ($i$th row) of $C$ is the sum of the $j$th column of $A$ and the $j$th column of $B$ ($i$th row of $A$ and $i$th row of $B$), then $\det C = \det A + \det B$. (p. 193)
- Interchanging any two distinct rows or columns of $A$ has the effect of multiplying $\det A$ by $-1$. (p. 194)
- If any row (column) of $A$ is multiplied by a scalar and added to any other row (column) of $A$, then $\det A$ is unchanged. (p. 196)
- If one row (column) of $A$ is a multiple of another row (column) of $A$, then $\det A = 0$. (p. 196)
- $\det A = \det A^t$ (p. 190)

- The $n \times n$ matrix $A$ is invertible if and only if $\det A \neq 0$. (p. 207)
- $\det AB = \det A \det B$ (pp. 188, 208)
- If $A$ is invertible, then $\det A \neq 0$ and (p. 211)

$$\det A^{-1} = \frac{1}{\det A}$$

- Let $A$ be an $n \times n$ matrix. The **adjoint** or **adjugate** of $A$, denoted by adj $A$, is the $n \times n$ matrix whose $ij$th component is $A_{ji}$, the $ji$th cofactor of $A$. (p. 211)
- If $\det A \neq 0$, then $A$ is invertible and (p. 214)

$$A^{-1} = \frac{1}{\det A} \text{adj } A$$

- ***Summing Up Theorem***

  Let $A$ be an $n \times n$ matrix. Then the following seven statements are equivalent: (p. 216)

  **i.** $A$ is invertible.

  **ii.** The only solution to the homogeneous system $A\mathbf{x} = \mathbf{0}$ is the trivial solution ($\mathbf{x} = \mathbf{0}$).

  **iii.** The system $A\mathbf{x} = \mathbf{b}$ has a unique solution for every $n$-vector $\mathbf{b}$.

  **iv.** $A$ is row equivalent to the $n \times n$ identity matrix $I_n$.

  **v.** $A$ is the product of elementary matrices.

  **vi.** The row echelon form of $A$ has $n$ pivots.

  **vii.** $\det A \neq 0$.

- ***Cramer's Rule***

  Let $A$ be an $n \times n$ matrix with $\det A \neq 0$. Then the unique solution to the system $A\mathbf{x} = \mathbf{b}$ is given by (p. 219)

$$x_1 = \frac{D_1}{\det A},\ x_2 = \frac{D_2}{\det A},\ \ldots,\ x_n = \frac{D_n}{\det A}$$

  where $D_j$ is the determinant of the matrix obtained by replacing the $j$th column of $A$ by the column vector $\mathbf{b}$.

## REVIEW EXERCISES

In Exercises 1–8 calculate the determinant.

**1.** $\begin{vmatrix} -1 & 2 \\ 0 & 4 \end{vmatrix}$

**2.** $\begin{vmatrix} -3 & 5 \\ -7 & 4 \end{vmatrix}$

**3.** $\begin{vmatrix} 1 & -2 & 3 \\ 0 & 4 & 5 \\ 0 & 0 & 6 \end{vmatrix}$

**4.** $\begin{vmatrix} 5 & 0 & 0 \\ 6 & 2 & 0 \\ 10 & 100 & 6 \end{vmatrix}$

**5.** $\begin{vmatrix} 1 & -1 & 2 \\ 3 & 4 & 2 \\ -2 & 3 & 4 \end{vmatrix}$

**6.** $\begin{vmatrix} 3 & 1 & -2 \\ 4 & 0 & 5 \\ -6 & 1 & 3 \end{vmatrix}$

**7.** $\begin{vmatrix} 1 & -1 & 2 & 3 \\ 4 & 0 & 2 & 5 \\ -1 & 2 & 3 & 7 \\ 5 & 1 & 0 & 4 \end{vmatrix}$ **8.** $\begin{vmatrix} 3 & 15 & 17 & 19 \\ 0 & 2 & 21 & 60 \\ 0 & 0 & 1 & 50 \\ 0 & 0 & 0 & -1 \end{vmatrix}$

In Exercises 9–14 use determinants to calculate the inverse (if one exists).

**9.** $\begin{pmatrix} -3 & 4 \\ 2 & 1 \end{pmatrix}$ **10.** $\begin{pmatrix} 3 & -5 & 7 \\ 0 & 2 & 4 \\ 0 & 0 & -3 \end{pmatrix}$ **11.** $\begin{pmatrix} 1 & -1 & 2 \\ 3 & 1 & 4 \\ 5 & -1 & 8 \end{pmatrix}$

**12.** $\begin{pmatrix} 1 & 1 & 1 \\ 1 & 0 & 1 \\ 0 & 1 & 1 \end{pmatrix}$ **13.** $\begin{pmatrix} 2 & 1 & 0 & 0 \\ 0 & -1 & 3 & 0 \\ 1 & 0 & 0 & -2 \\ 3 & 0 & -1 & 0 \end{pmatrix}$ **14.** $\begin{pmatrix} 3 & -1 & 2 & 4 \\ 1 & 1 & 0 & 3 \\ -2 & 4 & 1 & 5 \\ 6 & -4 & 1 & 2 \end{pmatrix}$

In Exercises 15–18 solve the system by using Cramer's rule.

**15.** $\begin{aligned} 2x_1 - x_2 &= 3 \\ 3x_1 + 2x_2 &= 5 \end{aligned}$

**16.** $\begin{aligned} x_1 - x_2 + x_3 &= 7 \\ 2x_1 \qquad - 5x_3 &= 4 \\ 3x_2 - x_3 &= 2 \end{aligned}$

**17.** $\begin{aligned} 2x_1 + 3x_2 - x_3 &= 5 \\ -x_1 + 2x_2 + 3x_3 &= 0 \\ 4x_1 - x_2 + x_3 &= -1 \end{aligned}$

**18.** $\begin{aligned} x_1 \qquad - x_3 + x_4 &= 7 \\ 2x_2 + 2x_3 - 3x_4 &= -1 \\ 4x_1 - x_2 - x_3 \qquad &= 0 \\ -2x_1 + x_2 + 4x_3 \qquad &= 2 \end{aligned}$

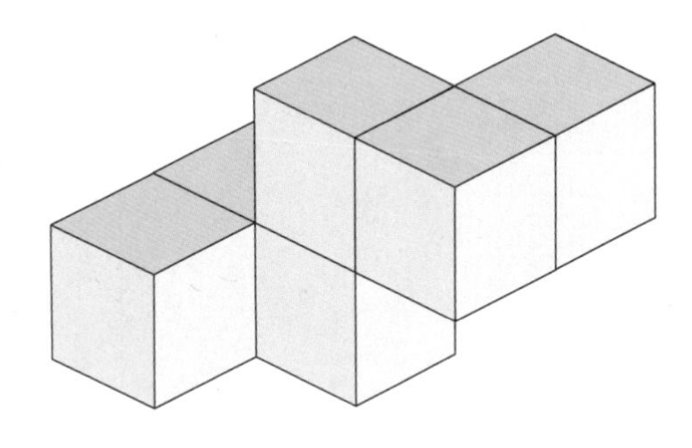

# 3

# Vectors in $\mathbb{R}^2$ and $\mathbb{R}^3$

In Section 1.5 we defined column and row vectors as ordered sets of $n$ real numbers or scalars. In the next chapter we shall define other kinds of sets of vectors, called *vector spaces.*

The study of arbitrary vector spaces is, initially, quite abstract. For that reason it is helpful to have a store of easily visualized vectors that we can use as examples.

In this chapter we discuss basic properties of vectors in the $xy$-plane and in real, three-dimensional space. Students who have studied multivariable calculus will have seen this material before. In that case it should be covered briefly, as a review. For others, coverage of this chapter will provide the examples that can make the material in Chapters 4 and 5 a great deal more comprehensible.

## 3.1 VECTORS IN THE PLANE

As defined in Section 1.5, $\mathbb{R}^2$ is the set of vectors $(x_1, x_2)$ with $x_1$ and $x_2$ real numbers. Since any point in the plane can be written in the form $(x, y)$, it is apparent that any point in the plane can be thought of as a vector in $\mathbb{R}^2$, and vice versa. Thus the terms "the plane" and "$\mathbb{R}^2$" are often used interchangeably. However, for a variety of physical applications (including the notions of force, velocity, acceleration, and momentum), it is important to think of a vector not as a point but as an entity having "length" and "direction." Now we shall see how this is done.

Directed line segment

Let $P$ and $Q$ be two points in the plane. Then the **directed line segment** from $P$ to $Q$, denoted by $\overrightarrow{PQ}$, is the straight-line segment that extends from $P$ to $Q$ (see Figure 3.1*a*). Note that the directed line segments $\overrightarrow{PQ}$ and $\overrightarrow{QP}$ are different since they point in opposite directions (Figure 3.1*b*).

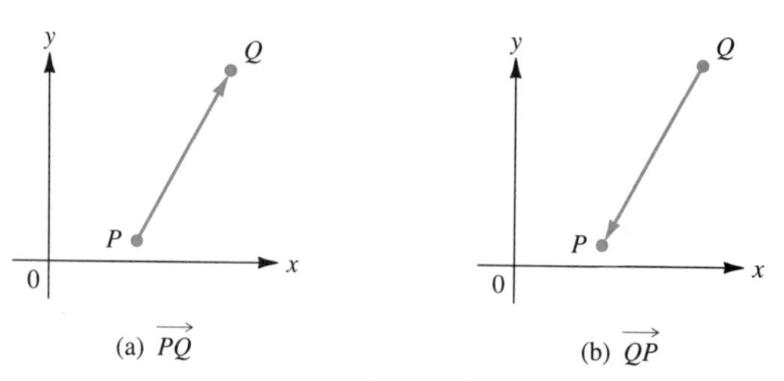


**Figure 3.1** The directed line segments $\overrightarrow{PQ}$ and $\overrightarrow{QP}$ point in opposite directions

Initial point
Terminal point

Equivalent directed line segments

The point $P$ in the directed line segment $\overrightarrow{PQ}$ is called the **initial point** of the segment and the point $Q$ is called the **terminal point.** The two major properties of a directed line segment are its magnitude (length) and its direction. If two directed line segments $\overrightarrow{PQ}$ and $\overrightarrow{RS}$ have the same magnitude and direction, we say that they are **equivalent** no matter where they are located with respect to the origin. The directed line segments in Figure 3.2 are all equivalent.

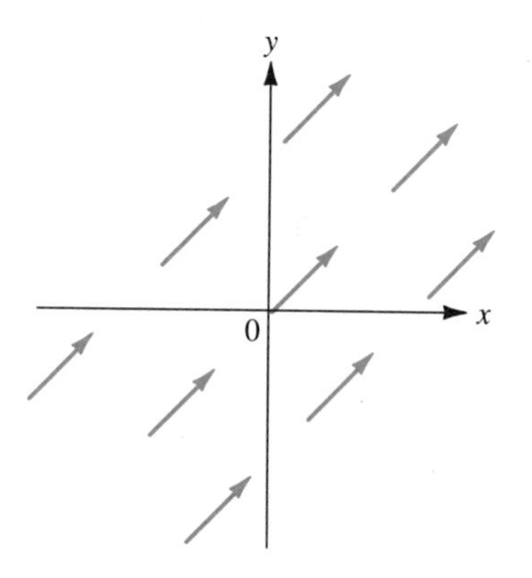


**Figure 3.2** A set of equivalent directed line segments

**DEFINITION 1** **Geometric Definition of a Vector** The set of all directed line segments equivalent to a given directed line segment is called a **vector.** Any directed line segment in that set is called a **representation** of the vector.

*Remark.* The directed line segments in Figure 3.2 are all representations of the same vector.

From Definition 1 we see that a given vector $\mathbf{v}$ can be represented in many different ways. Let $\overrightarrow{PQ}$ be a representation of $\mathbf{v}$. Then, without changing magnitude or direction, we can move $\overrightarrow{PQ}$ in a parallel way so that its initial point is shifted to

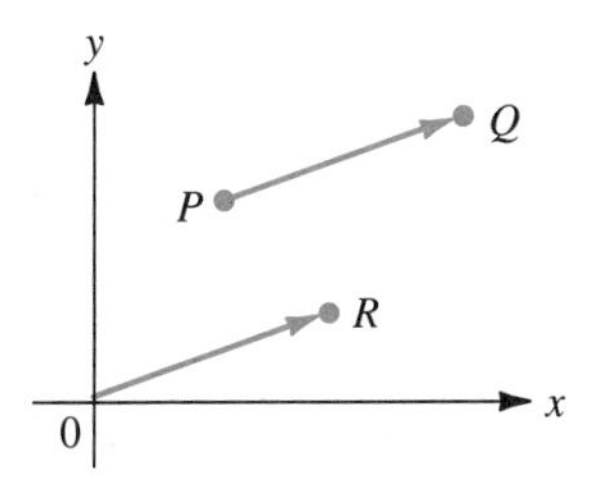


**Figure 3.3** We can move $\overrightarrow{PQ}$ to obtain an equivalent directed line segment with its initial point at the origin. Note that $\overrightarrow{OR}$ and $\overrightarrow{PQ}$ are parallel and have the same length.

the origin. We then obtain the directed line segment $\overrightarrow{0R}$, which is another representation of the vector **v** (see Figure 3.3). Now suppose that $R$ has the Cartesian coordinates $(a, b)$. Then we can describe the directed line segment $\overrightarrow{0R}$ by the coordinates $(a, b)$. That is, $\overrightarrow{0R}$ is the directed line segment with initial point $(0, 0)$ and terminal point $(a, b)$. Since one representation of a vector is as good as another, we can write the vector **v** as $(a, b)$.

**DEFINITION 2** **Algebraic Definition of a Vector** A **vector v** in the $xy$-plane is an ordered pair of real numbers $(a, b)$. The numbers $a$ and $b$ are called the **entries** or **components** of the vector **v**. The **zero vector** is the vector $(0, 0)$.

*Remark 1.* With this definition, a point in the $xy$-plane with coordinates $(a, b)$ can be thought of as a vector originating at the origin and terminating at $(a, b)$.

*Remark 2.* The zero vector has a magnitude of zero. Therefore, since the initial and terminal points coincide, we say that the zero vector has *no direction.*

*Remark 3.* We emphasize that Definitions 1 and 2 describe precisely the same objects. Each point of view (geometric and algebraic) has its advantages. Definition 2 is the definition of a 2-vector that we have been using all along.

Magnitude or length of a vector

Since a vector is really a set of equivalent line segments, we define the **magnitude** or **length** of a vector as the length of any one of its representations and its **direction** as the direction of any one of its representations. Using the representation $\overrightarrow{0R}$ and writing the vector $\mathbf{v} = (a, b)$, we find that

$$|\mathbf{v}| = \text{magnitude of } \mathbf{v} = \sqrt{a^2 + b^2} \tag{1}$$

This follows from the Pythagorean theorem (see Figure 3.4). We have used the notation $|\mathbf{v}|$ to denote the magnitude of **v**. Note that $|\mathbf{v}|$ is a *scalar.*

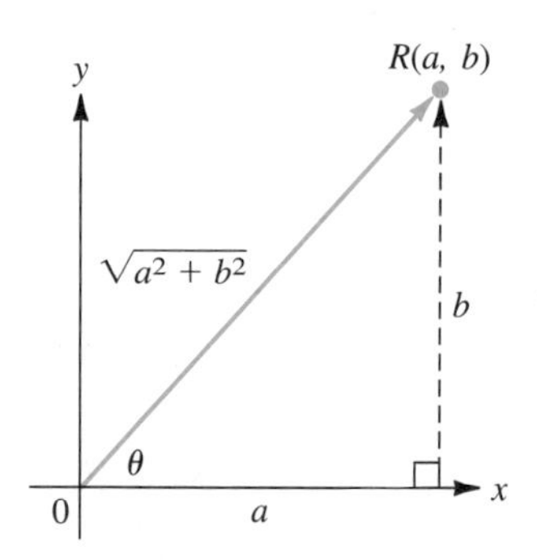


**Figure 3.4** The magnitude of a vector with $x$-coordinate $a$ and $y$-coordinate $b$ is $\sqrt{a^2 + b^2}$

EXAMPLE 1 **Calculating the Magnitudes of Six Vectors** Calculate the magnitudes of the vectors **(i)** $\mathbf{v} = (2, 2)$; **(ii)** $\mathbf{v} = (2, 2\sqrt{3})$; **(iii)** $\mathbf{v} = (-2\sqrt{3}, 2)$; **(iv)** $\mathbf{v} = (-3, -3)$; **(v)** $\mathbf{v} = (6, -6)$; **(vi)** $\mathbf{v} = (0, 3)$.

**Solution**

**i.** $|\mathbf{v}| = \sqrt{2^2 + 2^2} = \sqrt{8} = 2\sqrt{2}$

**ii.** $|\mathbf{v}| = \sqrt{2^2 + (2\sqrt{3})^2} = 4$

**iii.** $|\mathbf{v}| = \sqrt{(-2\sqrt{3})^2 + 2^2} = 4$

**iv.** $|\mathbf{v}| = \sqrt{(-3)^2 + (-3)^2} = \sqrt{18} = 3\sqrt{2}$

**v.** $|\mathbf{v}| = \sqrt{6^2 + (-6)^2} = \sqrt{72} = 6\sqrt{2}$

**vi.** $|\mathbf{v}| = \sqrt{0^2 + 3^2} = \sqrt{9} = 3$

Direction of a vector

We now define the **direction** of the vector $\mathbf{v} = (a, b)$ to be the angle $\theta$, measured in radians, that the vector makes with the positive $x$-axis. By convention, we choose $\theta$ such that $0 \leq \theta < 2\pi$. It follows from Figure 3.4 that if $a \neq 0$, then

$$\boxed{\tan \theta = \frac{b}{a}} \tag{2}$$

*Note:* $\tan \theta$ is periodic of period $\pi$, so if $a \neq 0$, there are always *two* numbers in $[0, 2\pi)$ such that $\tan \theta = b/a$. For example, $\tan \pi/4 = \tan 5\pi/4 = 1$. In order to determine $\theta$ uniquely, we need to determine the quadrant of $\mathbf{v}$, as we shall see in the next example.

EXAMPLE 2 **Calculating the Directions of Six Vectors** Calculate the directions of the vectors in Example 1.

**Solution** These six vectors are depicted in Figure 3.5.

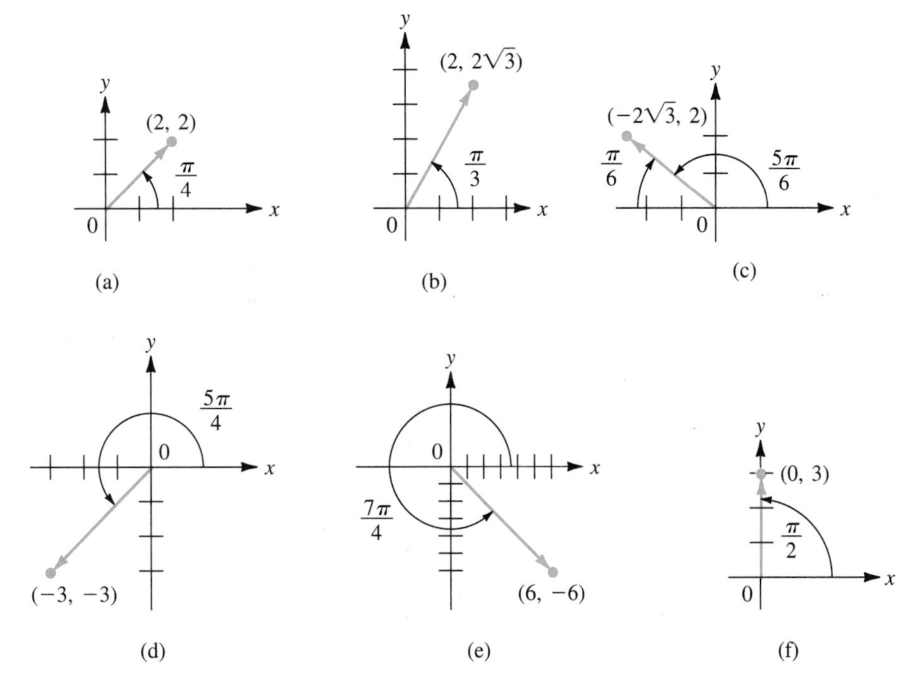


**Figure 3.5** The directions of six vectors

**i.** Here $\mathbf{v}$ is in the first quadrant and since $\tan\theta = 2/2 = 1$, $\theta = \pi/4$.

**ii.** Here $\theta = \tan^{-1} 2\sqrt{3}/2 = \tan^{-1}\sqrt{3} = \pi/3$ (since $\mathbf{v}$ is in the first quadrant).

**iii.** We see that $\mathbf{v}$ is in the second quadrant, and since $\tan^{-1} 2/2\sqrt{3} = \tan^{-1} 1/\sqrt{3} = \pi/6$, we see from Figure 3.5*c* that $\theta = \pi - (\pi/6) = 5\pi/6$.

**iv.** Here $\mathbf{v}$ is in the third quadrant, and since $\tan^{-1} 1 = \pi/4$, we find that $\theta = \pi + (\pi/4) = 5\pi/4$.

**v.** Since $\mathbf{v}$ is in the fourth quadrant and $\tan^{-1}(-1) = -\pi/4$, we get $\theta = 2\pi - (\pi/4) = 7\pi/4$.

**vi.** We cannot use equation (2) because $b/a$ is undefined. However, we see in Figure 3.5*f* that $\theta = \pi/2$.

In general, if $b > 0$

$$\text{Direction of } (0, b) = \frac{\pi}{2} \quad \text{and} \quad \text{direction of } (0, -b) = \frac{3\pi}{2} \qquad b > 0$$

In Section 1.5 we defined vector addition and scalar multiplication. What do these concepts mean geometrically? We start with scalar multiplication. If $\mathbf{v} =$

$(a, b)$, then $\alpha\mathbf{v} = (\alpha a, \alpha b)$. We find that

$$|\alpha\mathbf{v}| = \sqrt{\alpha^2a^2 + \alpha^2b^2} = |\alpha|\sqrt{a^2 + b^2} = |\alpha|\,|\mathbf{v}| \tag{3}$$

That is,

**Magnitude of $\alpha\mathbf{v}$**

Multiplying a vector by a nonzero scalar has the effect of multiplying the length of the vector by the absolute value of that scalar.

Moreover, if $\alpha > 0$, then $\alpha\mathbf{v}$ is in the same quadrant as $\mathbf{v}$, and therefore the direction of $\alpha\mathbf{v}$ is the *same* as the direction of $\mathbf{v}$ since $\tan^{-1}(\alpha b/\alpha a) = \tan^{-1}(b/a)$. If $\alpha < 0$, then $\alpha\mathbf{v}$ points in the direction opposite to that of $\mathbf{v}$. In other words,

**Direction of $\alpha\mathbf{v}$**

$$\begin{aligned} &\text{Direction of } \alpha\mathbf{v} = \text{direction of } \mathbf{v}, \text{ if } \alpha > 0 \\ &\text{Direction of } \alpha\mathbf{v} = (\text{direction of } \mathbf{v}) + \pi, \text{ if } \alpha < 0 \end{aligned} \tag{4}$$

**EXAMPLE 3** **Multiplying a Vector by a Scalar** Let $\mathbf{v} = (1, 1)$. Then $|\mathbf{v}| = \sqrt{1 + 1} = \sqrt{2}$ and $|2\mathbf{v}| = |(2, 2)| = \sqrt{2^2 + 2^2} = \sqrt{8} = 2\sqrt{2} = 2|\mathbf{v}|$. Further, $|-2\mathbf{v}| = \sqrt{(-2)^2 + (-2)^2} = 2\sqrt{2} = 2|\mathbf{v}|$. Moreover, the direction of $2\mathbf{v}$ is $\pi/4$, whereas the direction of $-2\mathbf{v}$ is $5\pi/4$ (see Figure 3.6).

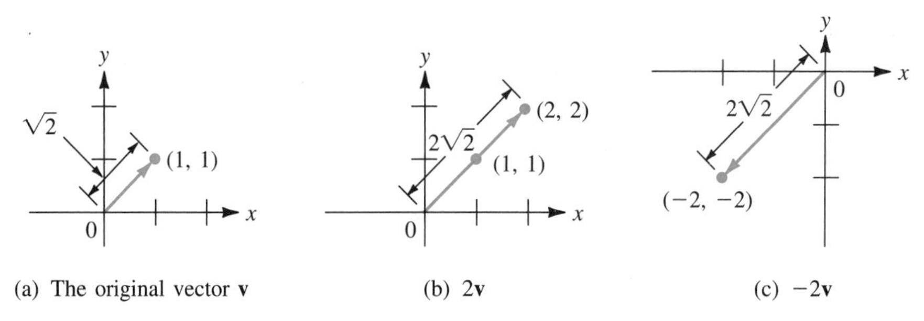


**Figure 3.6** The vector $2\mathbf{v}$ has the same direction as $\mathbf{v}$ and twice its magnitude. The vector $-2\mathbf{v}$ has the opposite direction of $\mathbf{v}$ and twice its magnitude

Now suppose we add the vectors $\mathbf{u} = (a_1, b_1)$ and $\mathbf{v} = (a_2, b_2)$ as in Figure 3.7. From the figure we see that the vector $\mathbf{u} + \mathbf{v} = (a_1 + a_2, b_1 + b_2)$ can be obtained by shifting the representation of the vector $\mathbf{v}$ so that its initial point coincides with the terminal point $(a_1, b_1)$ of the vector $\mathbf{u}$. We can therefore obtain the vector $\mathbf{u} + \mathbf{v}$

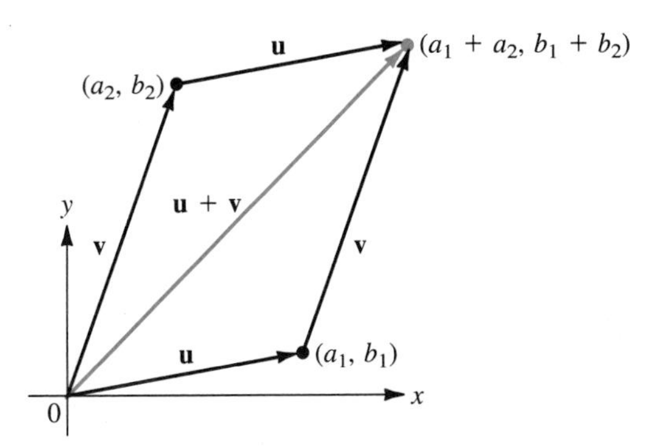


**Figure 3.7** The parallelogram rule for adding vectors

by drawing a parallelogram with one vertex at the origin and sides **u** and **v**. Then **u** + **v** is the vector that points from the origin along the diagonal of the parallelogram.

*Note.* Since a straight line is the shortest distance between two points, it immediately follows from Figure 3.7 that

**Triangle Inequality**

$$|\mathbf{u} + \mathbf{v}| \leq |\mathbf{u}| + |\mathbf{v}| \tag{5}$$

For reasons obvious from Figure 3.7, inequality (5) is called the **triangle inequality.**

We can also use Figure 3.7 to obtain a geometric representation of the vector **u** − **v**. Since **u** = **u** − **v** + **v**, the vector **u** − **v** is the vector that must be added to **v** to obtain **u**. This fact is illustrated in Figure 3.8*a*. A similar fact is illustrated in Figure 3.8*b*.

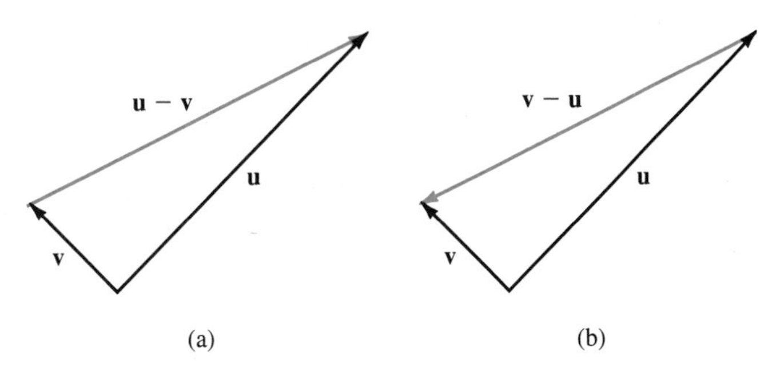


**Figure 3.8** The vectors **u** − **v** and **v** − **u** have the same magnitude but point in opposite directions

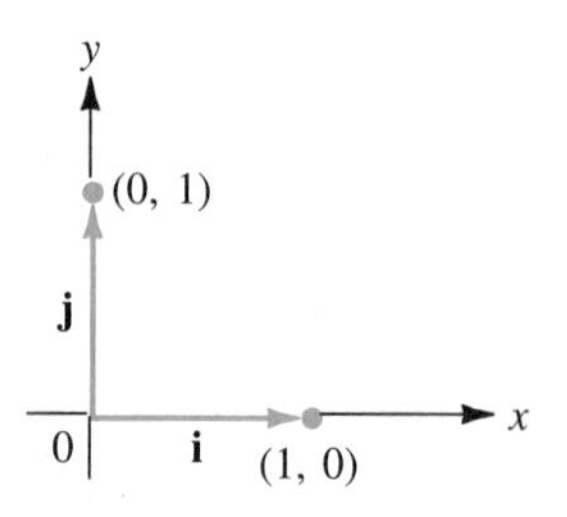


**Figure 3.9** The vectors **i** and **j**

There are two special vectors in $\mathbb{R}^2$ that allow us to represent other vectors in $\mathbb{R}^2$ in a convenient way. We denote the vector $(1, 0)$ by the symbol **i** and the vector $(0, 1)$ by the vector **j**. (See Figure 3.9.) If $\mathbf{v} = (a, b)$ is any vector in the plane, then since $(a, b) = a(1, 0) + b(0, 1)$, we may write

$$\boxed{\mathbf{v} = (a, b) = a\mathbf{i} + b\mathbf{j}} \tag{6}$$

With this representation we say that **v** is *resolved into its horizontal and vertical components.* The vectors **i** and **j** have two properties:

**i.** Neither one is a multiple of the other. (In the terminology of Chapter 4, they are *linearly independent.*)

**ii.** Any vector **v** can be written in terms of **i** and **j** as in equation (6).†

***Historical Note.*** The symbols **i** and **j** were first used by Hamilton. He defined his quaternion as a quantity of the form $a + b\mathbf{i} + c\mathbf{j} + d\mathbf{k}$, where $a$ is the "scalar part" and $b\mathbf{i} + c\mathbf{j} + d\mathbf{k}$ is the "vector part." In Section 3.3 we shall write vectors in space in the form $b\mathbf{i} + c\mathbf{j} + d\mathbf{k}$.

Under these two conditions **i** and **j** are said to form a **basis** in $\mathbb{R}^2$. We shall discuss bases in arbitrary vector spaces in Chapter 4.

We now define a kind of vector that is very useful in certain applications.

**DEFINITION 3** **Unit Vector** A **unit vector** is a vector that has length 1.

**EXAMPLE 4** **A Unit Vector** The vector $\mathbf{u} = (1/2)\mathbf{i} + (\sqrt{3}/2)\mathbf{j}$ is a unit vector since

$$|\mathbf{u}| = \sqrt{\left(\frac{1}{2}\right)^2 + \left(\frac{\sqrt{3}}{2}\right)^2} = \sqrt{\frac{1}{4} + \frac{3}{4}} = 1$$

† In equation (6) we say that **v** can be written as a *linear combination* of **i** and **j**. We shall discuss the notion of linear combination in Section 4.5.

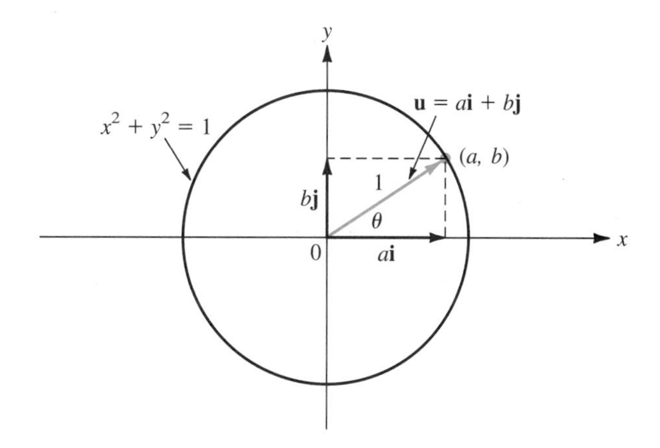


**Figure 3.10** The terminal point of a unit vector with initial point at the origin lies on the unit circle

Let $\mathbf{u} = a\mathbf{i} + b\mathbf{j}$ be a unit vector. Then $|\mathbf{u}| = \sqrt{a^2 + b^2} = 1$, so that $a^2 + b^2 = 1$ and $\mathbf{u}$ can be represented by a point on the unit circle (see Figure 3.10). If $\theta$ is the direction of $\mathbf{u}$, then we immediately see that $a = \cos\theta$ and $b = \sin\theta$. Thus any unit vector $\mathbf{u}$ can be written in the form

**Representing a Unit Vector**

$$\mathbf{u} = (\cos\theta)\mathbf{i} + (\sin\theta)\mathbf{j} \tag{7}$$

where $\theta$ is the direction of $\mathbf{u}$.

EXAMPLE 5 **Writing a Unit Vector as $(\cos\theta)\mathbf{i} + (\sin\theta)\mathbf{j}$** The unit vector $\mathbf{u} = (1/2)\mathbf{i} + (\sqrt{3}/2)\mathbf{j}$ of Example 4 can be written in the form of (7) with $\theta = \cos^{-1}(1/2) = \pi/3$.

We also have (see Problem 17)

> Let $\mathbf{v}$ be any nonzero vector. Then $\mathbf{u} = \mathbf{v}/|\mathbf{v}|$ is a unit vector having the same direction as $\mathbf{v}$.

EXAMPLE 6 **Finding a Unit Vector Having the Same Direction as a Given Nonzero Vector** Find a unit vector having the same direction as $\mathbf{v} = 2\mathbf{i} - 3\mathbf{j}$.

**Solution** Here $|\mathbf{v}| = \sqrt{4 + 9} = \sqrt{13}$, so $\mathbf{u} = \mathbf{v}/|\mathbf{v}| = (2/\sqrt{13})\mathbf{i} - (3/\sqrt{13})\mathbf{j}$ is the required unit vector.

We conclude this section with a summary of the properties of vectors.

**Table 3.1**

| Object | Intuitive Definition | Expression in terms of components if $\mathbf{u} = u_1\mathbf{i} + u_2\mathbf{j}$, $\mathbf{v} = v_1\mathbf{i} + v_2\mathbf{j}$, and $\mathbf{u} = (u_1, u_2)$, $\mathbf{v} = (v_1, v_2)$ |
|---|---|---|
| Vector $\mathbf{v}$ | An object having magnitude and direction | $v_1\mathbf{i} + v_2\mathbf{j}$ or $(v_1, v_2)$ |
| $\lvert\mathbf{v}\rvert$ | Magnitude (or length) of $\mathbf{v}$ | $\sqrt{v_1^2 + v_2^2}$ |
| $\alpha\mathbf{v}$ | ↗v ↗αv (In this sketch $\alpha = 2$) | $\alpha v_1\mathbf{i} + \alpha v_2\mathbf{j}$ or $(\alpha v_1, \alpha v_2)$ |
| $-\mathbf{v}$ | ↗v ↙−v | $-v_1\mathbf{i} - v_2\mathbf{j}$ or $(-v_1 - v_2)$ or $-(v_1, v_2)$ |
| $\mathbf{u} + \mathbf{v}$ | u + v, v, u | $(u_1 + v_1)\mathbf{i} + (u_2 + v_2)\mathbf{j}$ or $(u_1 + v_1, u_2 + v_2)$ |
| $\mathbf{u} - \mathbf{v}$ | u − v, v, u | $(u_1 - v_1)\mathbf{i} + (u_2 - v_2)\mathbf{j}$ or $(u_1 - v_1, u_2 - v_2)$ |

# PROBLEMS 3.1

## Self-Quiz

**I.** A *vector* is ________.
**a.** two points in the *xy*-plane
**b.** a line segment between two points
**c.** a directed line segment from one point to another
**d.** a collection of equivalent directed line segments

**II.** If $P = (3, -4)$ and $Q = (8, 6)$, the vector $\overrightarrow{PQ}$ has length ________.
**a.** $|3| + |-4|$ **b.** $(3)^2 + (-4)^2$
**c.** $(3 - 8)^2 + (-4 - 6)^2$ **d.** $\sqrt{(8 - 3)^2 + (6 - (-4))^2}$

**III.** The direction of the vector $(4, 8)$ is ________.
**a.** $\pi$ **b.** $\tan^{-1}(8 - 4)$ **c.** $(\frac{8}{4})\pi$ **d.** $\tan^{-1}(\frac{8}{4})$

**IV.** If $\mathbf{u} = (3, 4)$ and $\mathbf{v} = (5, 8)$, then $\mathbf{u} + \mathbf{v} =$ ________.
**a.** $(7, 13)$ **b.** $(8, 12)$ **c.** $(2, 4)$ **d.** $(15, 32)$

**V.** If $\mathbf{u} = (4, 3)$, then the unit vector with the same direction as $\mathbf{u}$ is ________.
**a.** $(0.4, 0.3)$ **b.** $(0.8, 0.6)$ **c.** $(\frac{4}{5}, \frac{3}{5})$ **d.** $(\frac{4}{7}, \frac{3}{7})$

**Answers to Self-Quiz**

**I.** d **II.** d **III.** d **IV.** b **V.** b = c

In Problems 1–12 find the magnitude and direction of the given vector.

**1.** $\mathbf{v} = (4, 4)$ **2.** $\mathbf{v} = (-4, 4)$ **3.** $\mathbf{v} = (4, -4)$

**4.** $\mathbf{v} = (-4, -4)$ **5.** $\mathbf{v} = (\sqrt{3}, 1)$ **6.** $\mathbf{v} = (1, \sqrt{3})$

**7.** $\mathbf{v} = (-1, \sqrt{3})$ **8.** $\mathbf{v} = (1, -\sqrt{3})$ **9.** $\mathbf{v} = (-1, -\sqrt{3})$

**10.** $\mathbf{v} = (1, 2)$ **11.** $\mathbf{v} = (-5, 8)$ **12.** $\mathbf{v} = (11, -14)$

**13.** Let $\mathbf{u} = (2, 3)$ and $\mathbf{v} = (-5, 4)$. Find: **(a)** $3\mathbf{u}$; **(b)** $\mathbf{u} + \mathbf{v}$; **(c)** $\mathbf{v} - \mathbf{u}$; **(d)** $2\mathbf{u} - 7\mathbf{v}$. Sketch these vectors.

**14.** Let $\mathbf{u} = 2\mathbf{i} - 3\mathbf{j}$ and $\mathbf{v} = -4\mathbf{i} + 6\mathbf{j}$. Find: **(a)** $\mathbf{u} + \mathbf{v}$; **(b)** $\mathbf{u} - \mathbf{v}$; **(c)** $3\mathbf{u}$; **(d)** $-7\mathbf{v}$; **(e)** $8\mathbf{u} - 3\mathbf{v}$; **(f)** $4\mathbf{v} - 6\mathbf{u}$. Sketch these vectors.

**15.** Show that the vectors $\mathbf{i}$ and $\mathbf{j}$ are unit vectors.

**16.** Show that the vector $(1/\sqrt{2})\mathbf{i} + (1/\sqrt{2})\mathbf{j}$ is a unit vector.

**17.** Show that if $\mathbf{v} = a\mathbf{i} + b\mathbf{j} \neq \mathbf{0}$, then $\mathbf{u} = (a/\sqrt{a^2 + b^2})\mathbf{i} + (b/\sqrt{a^2 + b^2})\mathbf{j}$ is a unit vector having the same direction as $\mathbf{v}$.

In Problems 18–21 find a unit vector having the same direction as the given vector.

**18.** $\mathbf{v} = 2\mathbf{i} + 3\mathbf{j}$ **19.** $\mathbf{v} = \mathbf{i} - \mathbf{j}$

**20.** $\mathbf{v} = -3\mathbf{i} + 4\mathbf{j}$ **21.** $\mathbf{v} = a\mathbf{i} + a\mathbf{j}$; $a \neq 0$.

**22.** If $\mathbf{v} = a\mathbf{i} + b\mathbf{j}$, show that $a/\sqrt{a^2 + b^2} = \cos\theta$ and $b/\sqrt{a^2 + b^2} = \sin\theta$, where $\theta$ is the direction of $\mathbf{v}$.

**23.** If $\mathbf{v} = 2\mathbf{i} - 3\mathbf{j}$, find $\sin\theta$ and $\cos\theta$.

**24.** If $\mathbf{v} = -3\mathbf{i} + 8\mathbf{j}$, find $\sin\theta$ and $\cos\theta$.

A vector $\mathbf{v}$ has a direction opposite to that of a vector $\mathbf{u}$ if direction $\mathbf{v}$ = direction $\mathbf{u} + \pi$. In Problems 25–28 find a unit vector $\mathbf{v}$ that has a direction opposite the direction of the given vector $\mathbf{u}$.

**25.** $\mathbf{u} = \mathbf{i} + \mathbf{j}$ **26.** $\mathbf{u} = 2\mathbf{i} - 3\mathbf{j}$

**27.** $\mathbf{u} = -3\mathbf{i} + 4\mathbf{j}$ **28.** $\mathbf{u} = -2\mathbf{i} + 3\mathbf{j}$

**29.** Let $\mathbf{u} = 2\mathbf{i} - 3\mathbf{j}$ and $\mathbf{v} = -\mathbf{i} + 2\mathbf{j}$. Find a unit vector having the same direction as: **(a)** $\mathbf{u} + \mathbf{v}$; **(b)** $2\mathbf{u} - 3\mathbf{v}$; **(c)** $3\mathbf{u} + 8\mathbf{v}$.

**30.** Let $P = (c, d)$ and $Q = (c + a, d + b)$. Show that the magnitude of $\overrightarrow{PQ}$ is $\sqrt{a^2 + b^2}$.

**31.** Show that the direction of $\overrightarrow{PQ}$ in Problem 30 is the same as the direction of the vector $(a, b)$. [*Hint:* If $R = (a, b)$, show that the line passing through the points $P$ and $Q$ is parallel to the line passing through the points 0 and $R$.]

In Problems 32–35 find a vector $\mathbf{v}$ having the given magnitude and direction.

**32.** $|\mathbf{v}| = 3$; $\theta = \pi/6$ **33.** $|\mathbf{v}| = 8$; $\theta = \pi/3$

**34.** $|\mathbf{v}| = 1$; $\theta = \pi/4$ **35.** $|\mathbf{v}| = 6$; $\theta = 2\pi/3$.

***36.** Show algebraically (that is, strictly from the definitions of vector addition and magnitude) that for any two vectors $\mathbf{u}$ and $\mathbf{v}$, $|\mathbf{u} + \mathbf{v}| \leq |\mathbf{u}| + |\mathbf{v}|$.

**37.** Show that if neither $\mathbf{u}$ nor $\mathbf{v}$ is the zero vector, then $|\mathbf{u} + \mathbf{v}| = |\mathbf{u}| + |\mathbf{v}|$ if and only if $\mathbf{u}$ is a positive scalar multiple of $\mathbf{v}$.

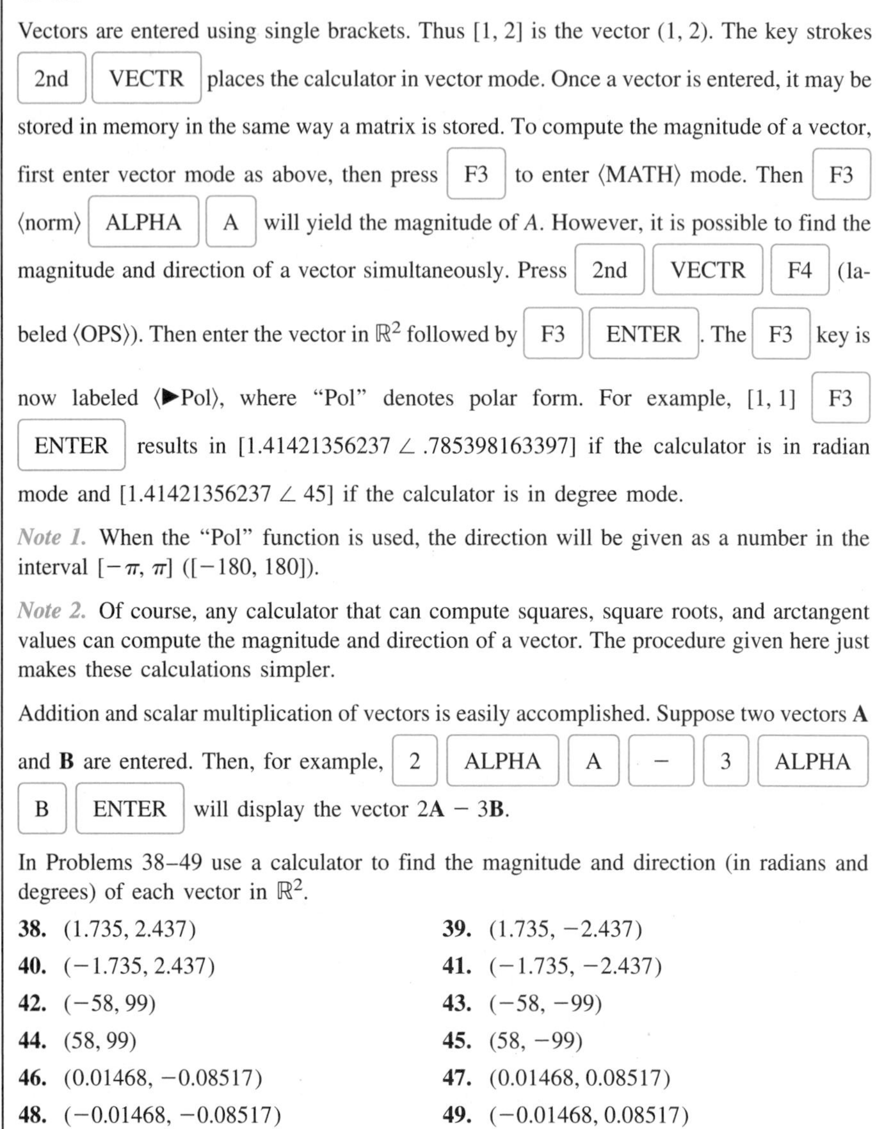

## CALCULATOR BOX

Many vector operations can be carried out on the TI-85.

### TI-85

Vectors are entered using single brackets. Thus [1, 2] is the vector (1, 2). The key strokes [2nd] [VECTR] places the calculator in vector mode. Once a vector is entered, it may be stored in memory in the same way a matrix is stored. To compute the magnitude of a vector, first enter vector mode as above, then press [F3] to enter ⟨MATH⟩ mode. Then [F3] ⟨norm⟩ [ALPHA] [A] will yield the magnitude of $A$. However, it is possible to find the magnitude and direction of a vector simultaneously. Press [2nd] [VECTR] [F4] (labeled ⟨OPS⟩). Then enter the vector in $\mathbb{R}^2$ followed by [F3] [ENTER]. The [F3] key is now labeled ⟨▶Pol⟩, where "Pol" denotes polar form. For example, [1, 1] [F3] [ENTER] results in [1.41421356237 ∠ .785398163397] if the calculator is in radian mode and [1.41421356237 ∠ 45] if the calculator is in degree mode.

*Note 1.* When the "Pol" function is used, the direction will be given as a number in the interval $[-\pi, \pi]$ ([−180, 180]).

*Note 2.* Of course, any calculator that can compute squares, square roots, and arctangent values can compute the magnitude and direction of a vector. The procedure given here just makes these calculations simpler.

Addition and scalar multiplication of vectors is easily accomplished. Suppose two vectors **A** and **B** are entered. Then, for example, [2] [ALPHA] [A] [−] [3] [ALPHA] [B] [ENTER] will display the vector 2**A** − 3**B**.

In Problems 38–49 use a calculator to find the magnitude and direction (in radians and degrees) of each vector in $\mathbb{R}^2$.

**38.** (1.735, 2.437)

**39.** (1.735, −2.437)

**40.** (−1.735, 2.437)

**41.** (−1.735, −2.437)

**42.** (−58, 99)

**43.** (−58, −99)

**44.** (58, 99)

**45.** (58, −99)

**46.** (0.01468, −0.08517)

**47.** (0.01468, 0.08517)

**48.** (−0.01468, −0.08517)

**49.** (−0.01468, 0.08517)

# MATLAB 3.1

*MATLAB Information.* Enter a vector as a $2 \times 1$ or $3 \times 1$ matrix. Addition and scalar multiplication is the same as for matrices:

*Scalar product* of **u** and **v**: **u′*v**

*Magnitude (length)* of **v**: **sqrt(v′*v)** or **norm(v)**

*Direction of v:* See Example 2 and use the fact that $\tan^{-1}(c)$ is found by **atan(*c*).**

*Graphics:* Several of the problems use graphics displays. Specific directions are given in each problem. Some general information that is useful for DOS systems: To return to the command screen after viewing a graph, you can hit any key. To view the graphics screen if you are on the command screen, you enter the command **shg.**

1. a. Use MATLAB to verify the results obtained by paper and pencil for the magnitude and direction of the vectors in the odd problems of Problems 1–12 in this section.

   ***Note.*** $\sqrt{3}$ is found by **sqrt(3).**

   b. Use MATLAB to find the magnitude and direction of the vectors in the even problems of Problems 38–49 in this section.

2. Linear combinations of vectors will be important in further work. This problem describes one way to visualize linear combinations of vectors in the plane. (See also MATLAB Problem 3 below.)

   a. We wish to plot several linear combinations of two given vectors on the same set of axes. Each vector will be represented by a line from (0, 0) to the head of the vector. Let **u** and **v** be given $2 \times 1$ matrices (vectors). We wish to plot several vectors **z**, where $\mathbf{z} = a\mathbf{u} + b\mathbf{v}$ for $-1 \leq a,b \leq 1$ to aid in understanding the geometry of linear combination. Read the note on *Graphics* preceding the MATLAB problems above.

   Enter **u** and **v** of your choice, with **u** and **v** not parallel. Enter the following:

```
w = u + v; ww = u − v; aa = [u′ v′ w′ ww′]; M = max(abs(aa));
axis('square'); axis([−M M −M M])
plot([0 v(1)],[0 v(2)],[0 u(1)],[0 u(2)])
hold on
grid
```

   So far you will see **u** and **v** plotted.

```
a = 1; b = 1;
z = a*u + b*v;
plot([0 z(1)],[0 z(2)],'c5')
```

   If you are using MATLAB 4.0, enter the **axis** statements *after* each plot statement.

Repeat the last three command lines above five more times, but modify the choices of $a$ and $b$ with $0 \le a,b \le 1$. (Recall the use of the up-arrow cursor.) Observe the geometry of each linear combination as it is plotted.

What would the graphics screen look like if many more choices of $a$ and $b$ as above were plotted?

Repeat the last three command lines above six times with the following changes: Change **'c5'** to **'c6'** and make at least six choices of $a$ and $b$ for $0 \le a \le 1$ and $-1 \le b \le 0$. Let the first choice be $a = 1$ and $b = -1$. Observe the geometry and answer the question as above.

Repeat the last three command lines above six times with the following changes: Change **'c5'** to **'c7'** and make at least six choices of $a$ and $b$ for $-1 \le a \le 0$ and $0 \le b \le 1$. Let the first choice be $a = -1$ and $b = 1$. Observe the geometry and answer the question as above.

Repeat the last three command lines above six times with the following changes: Change **'c5'** to **'c8'** and make at least six choices of $a$ and $b$ for $-1 \le a, b \le 1$. Let the first choice be $a = -1$ and $b = -1$. Observe the geometry and answer the question as above.

What would the graphics screen look like if more and more linear combinations were plotted?

When you are finished with this problem enter the command **hold off.**

**b.** Following the directions above, explore what happens if you start with **u** and **v** parallel.

When you are finished with this problem, enter the command **hold off.**

M

**3.** *(Uses the m-file lincomb.m)* Given two nonparallel vectors in the plane, we can write any other vector in the plane as a linear combination of these two vectors. The m-file *lincomb.m,* contained on the accompanying disk, is an aid for visualization. Enter the command **help lincomb** for a description of this m-file.

Let **u** and **v** be $2 \times 1$ vectors that are not parallel. Let **w = 5*(2*rand(2,1) – 1).** Enter **lincomb (u,v,w).** You will first see **u**, **v**, and **w** plotted. Hit any key and the geometry of **w** written as a linear combination of **u** and **v** will be displayed. Repeat for several choices of **w**, **u**, and **v**.

## 3.2 THE SCALAR PRODUCT AND PROJECTIONS IN $\mathbb{R}^2$

In Section 1.6 we defined the scalar product of two vectors. If $\mathbf{u} = (a_1, b_1)$ and $\mathbf{v} = (a_2, b_2)$, then

$$\boxed{\mathbf{u} \cdot \mathbf{v} = a_1a_2 + b_1b_2} \tag{1}$$

We now see how the scalar product can be interpreted geometrically.

**DEFINITION 1** **Angle Between Vectors** Let $\mathbf{u}$ and $\mathbf{v}$ be two nonzero vectors. Then the **angle $\varphi$ between $\mathbf{u}$ and $\mathbf{v}$** is defined to be the smallest nonnegative angle† between the representations of $\mathbf{u}$ and $\mathbf{v}$ that have the origin as their initial points. If $\mathbf{u} = \alpha\mathbf{v}$ for some scalar $\alpha$, then we define $\varphi = 0$ if $\alpha > 0$ and $\varphi = \pi$ if $\alpha < 0$.

This definition is illustrated in Figure 3.11. Note that $\varphi$ can always be chosen to be a nonnegative angle in the interval $[0, \pi]$.

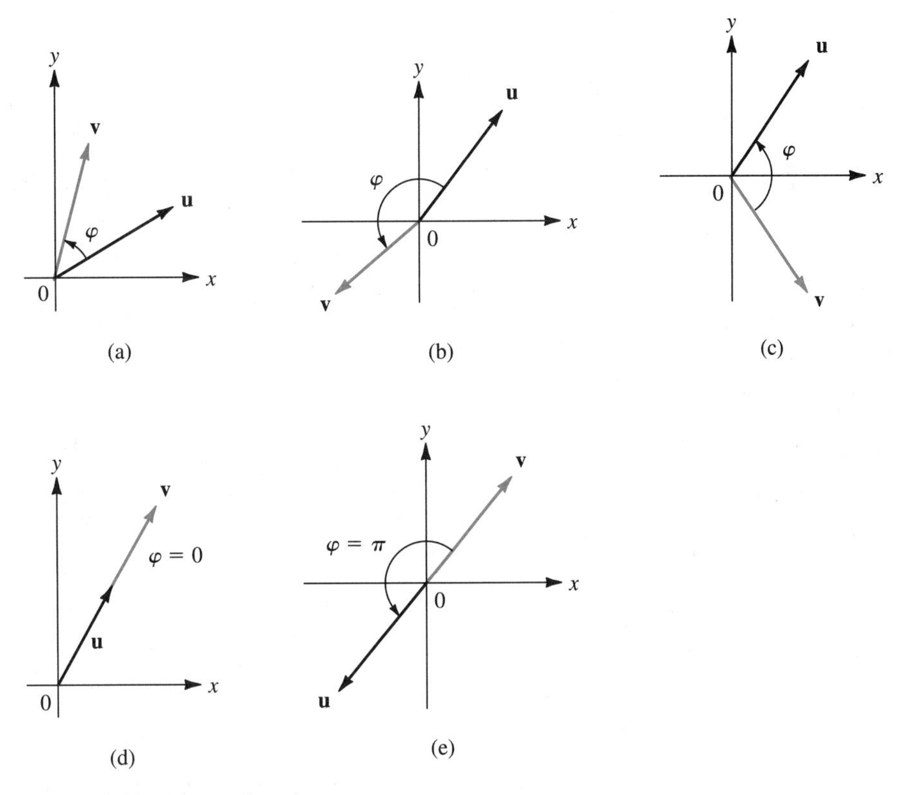


**Figure 3.11** The angle $\varphi$ between two vectors

**THEOREM 1** Let $\mathbf{v}$ be a vector. Then

$$|\mathbf{v}|^2 = \mathbf{v} \cdot \mathbf{v} \tag{2}$$

† This angle will be in the interval $[0, \pi]$.

**Proof** Let $\mathbf{v} = (a, b)$. Then

$$|\mathbf{v}|^2 = a^2 + b^2$$

and

$$\mathbf{v} \cdot \mathbf{v} = (a, b) \cdot (a, b) = a \cdot a + b \cdot b = a^2 + b^2 = |\mathbf{v}|^2$$

**THEOREM 2** Let $\mathbf{u}$ and $\mathbf{v}$ be two nonzero vectors. If $\varphi$ is the angle between them, then

$$\cos \varphi = \frac{\mathbf{u} \cdot \mathbf{v}}{|\mathbf{u}|\,|\mathbf{v}|} \tag{3}$$

**Proof** The law of cosines (see Problem 2.5.10, page 223) states that in the triangle of Figure 3.12

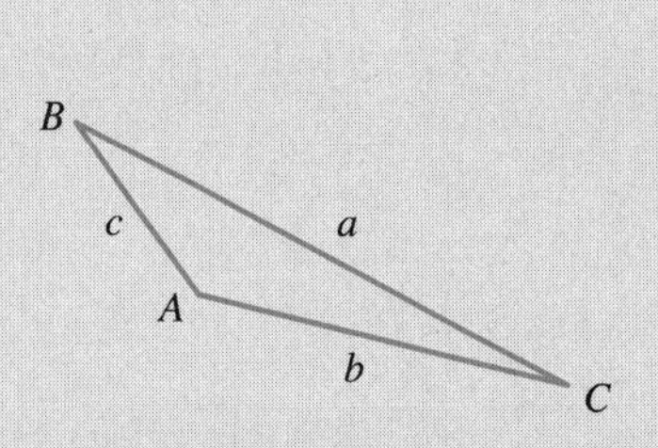


**Figure 3.12** A triangle with sides $a$, $b$, and $c$

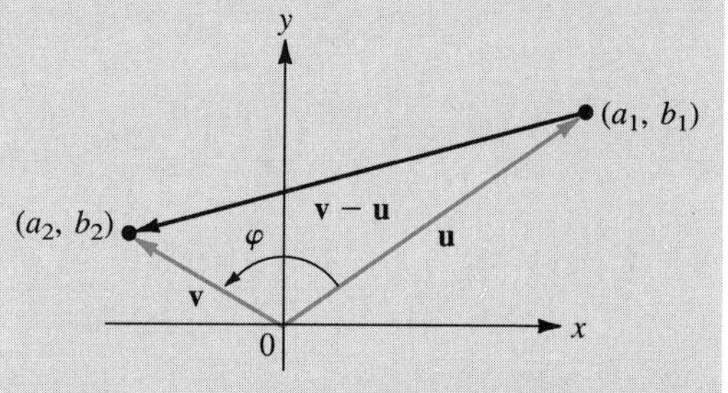

**Figure 3.13** A triangle with sides $|\mathbf{u}|$, $|\mathbf{v}|$, and $|\mathbf{v} - \mathbf{u}|$

$$c^2 = a^2 + b^2 - 2ab \cos C$$

We now place the representations of $\mathbf{u}$ and $\mathbf{v}$ with initial points at the origin so that $\mathbf{u} = (a_1, b_1)$ and $\mathbf{v} = (a_2, b_2)$ (see Figure 3.13). Then from the law of cosines, $|\mathbf{v} - \mathbf{u}|^2 = |\mathbf{v}|^2 + |\mathbf{u}|^2 - 2|\mathbf{u}|\,|\mathbf{v}| \cos \varphi$. But

$$\begin{aligned} |\mathbf{v} - \mathbf{u}|^2 &\overset{\text{from (2)}}{=} (\mathbf{v} - \mathbf{u}) \cdot (\mathbf{v} - \mathbf{u}) \overset{\text{Theorem 1 } (iii) \text{ on page 63}}{=} \mathbf{v} \cdot \mathbf{v} - 2\mathbf{u} \cdot \mathbf{v} + \mathbf{u} \cdot \mathbf{u} \\ &= |\mathbf{v}|^2 - 2\mathbf{u} \cdot \mathbf{v} + |\mathbf{u}|^2 \end{aligned}$$

Thus, after subtracting $|\mathbf{v}|^2 + |\mathbf{u}|^2$ from both sides, we obtain $-2\mathbf{u} \cdot \mathbf{v} = -2|\mathbf{u}|\,|\mathbf{v}| \cos \varphi$, from which the theorem follows.

*Remark.* Using Theorem 1, we could define the scalar product $\mathbf{u} \cdot \mathbf{v}$ by

$$\boxed{\mathbf{u} \cdot \mathbf{v} = |\mathbf{u}|\,|\mathbf{v}| \cos \varphi}$$

**EXAMPLE 1** **Computing the Angle Between Two Vectors** Find the angle between the vectors $\mathbf{u} = 2\mathbf{i} + 3\mathbf{j}$ and $\mathbf{v} = -7\mathbf{i} + \mathbf{j}$.

**Solution** $\mathbf{u} \cdot \mathbf{v} = -14 + 3 = -11$, $|\mathbf{u}| = \sqrt{2^2 + 3^2} = \sqrt{13}$, and $|\mathbf{v}| = \sqrt{(-7)^2 + 1^2} = \sqrt{50}$. Hence

$$\cos \varphi = \frac{\mathbf{u} \cdot \mathbf{v}}{|\mathbf{u}|\,|\mathbf{v}|} = \frac{-11}{\sqrt{13}\sqrt{50}} = \frac{-11}{\sqrt{650}} \approx -0.431455497\dagger$$

so

$$\varphi = \cos^{-1}(-0.431455497) \approx 2.0169\ddagger \ (\approx 115.6°)$$

*Note.* Since $0 \le \varphi \le \pi$, $\cos^{-1}(\cos \varphi) = \varphi$

**DEFINITION 2** **Parallel Vectors** Two nonzero vectors $\mathbf{u}$ and $\mathbf{v}$ are **parallel** if the angle between them is zero or $\pi$. Note that parallel vectors have the same or opposite directions.

**EXAMPLE 2** **Two Parallel Vectors** Show that the vectors $\mathbf{u} = (2, -3)$ and $\mathbf{v} = (-4, 6)$ are parallel.

**Solution**

$$\cos \varphi = \frac{\mathbf{u} \cdot \mathbf{v}}{|\mathbf{u}|\,|\mathbf{v}|} = \frac{-8 - 18}{\sqrt{13}\sqrt{52}} = \frac{-26}{\sqrt{13}(2\sqrt{13})} = \frac{-26}{2(13)} = -1$$

Hence $\varphi = \pi$ (so that $\mathbf{u}$ and $\mathbf{v}$ have opposite directions).

**THEOREM 3** If $\mathbf{u} \neq \mathbf{0}$, then $\mathbf{v} = \alpha\mathbf{u}$ for some nonzero constant $\alpha$ if and only if $\mathbf{u}$ and $\mathbf{v}$ are parallel.

**Proof** The proof is left as an exercise (see Problem 35).

**DEFINITION 3** **Orthogonal Vectors** The nonzero vectors $\mathbf{u}$ and $\mathbf{v}$ are called **orthogonal** (or **perpendicular**) if the angle between them is $\pi/2$.

---

† These numbers, like others in the text, were obtained with a hand calculator.

‡ When doing this computation yourself, make certain that your calculator is set to radian mode.

**EXAMPLE 3** **Two Orthogonal Vectors** Show that the vectors $\mathbf{u} = 3\mathbf{i} - 4\mathbf{j}$ and $\mathbf{v} = 4\mathbf{i} + 3\mathbf{j}$ are orthogonal.

**Solution** $\mathbf{u} \cdot \mathbf{v} = 3 \cdot 4 - 4 \cdot 3 = 0$. This implies that $\cos \varphi = (\mathbf{u} \cdot \mathbf{v})/(|\mathbf{u}|\,|\mathbf{v}|) = 0$. Since $\varphi$ is in the interval $[0, \pi]$, $\varphi = \pi/2$. ■

**THEOREM 4** The nonzero vectors $\mathbf{u}$ and $\mathbf{v}$ are orthogonal if and only if $\mathbf{u} \cdot \mathbf{v} = 0$.

**Proof** This proof is also left as an exercise (see Problem 36). ■

A number of interesting problems involves the notion of the projection of one vector along another. Before defining this, we prove the following theorem.

**THEOREM 5** Let $\mathbf{v}$ be a nonzero vector. Then for any other vector $\mathbf{u}$, the vector

$$\mathbf{w} = \mathbf{u} - \frac{(\mathbf{u} \cdot \mathbf{v})}{|\mathbf{v}|^2}\mathbf{v}$$

is orthogonal to $\mathbf{v}$.

**Proof**

$$\mathbf{w} \cdot \mathbf{v} = \left[\mathbf{u} - \frac{(\mathbf{u} \cdot \mathbf{v})\mathbf{v}}{|\mathbf{v}|^2}\right] \cdot \mathbf{v} = \mathbf{u} \cdot \mathbf{v} - \frac{(\mathbf{u} \cdot \mathbf{v})(\mathbf{v} \cdot \mathbf{v})}{|\mathbf{v}|^2}$$

$$= \mathbf{u} \cdot \mathbf{v} - \frac{(\mathbf{u} \cdot \mathbf{v})|\mathbf{v}|^2}{|\mathbf{v}|^2} = \mathbf{u} \cdot \mathbf{v} - \mathbf{u} \cdot \mathbf{v} = 0$$ ■

The vectors $\mathbf{u}$, $\mathbf{v}$, and $\mathbf{w}$ are illustrated in Figure 3.14.

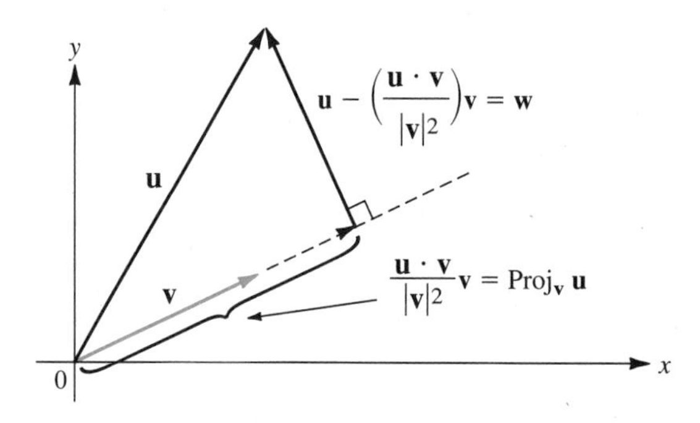


**Figure 3.14** The vector $\mathbf{w} = \mathbf{u} - \dfrac{\mathbf{u} \cdot \mathbf{v}}{|\mathbf{v}|^2}\mathbf{v}$ is orthogonal to $\mathbf{v}$

**DEFINITION 4** **Projection** Let **u** and **v** be nonzero vectors. Then the **projection** of **u** on **v** is a vector, denoted by $\text{proj}_{\mathbf{v}}\,\mathbf{u}$, which is defined by

$$\text{proj}_{\mathbf{v}}\,\mathbf{u} = \frac{\mathbf{u} \cdot \mathbf{v}}{|\mathbf{v}|^2}\mathbf{v} \tag{4}$$

The **component** of **u** in the direction **v** is $\dfrac{\mathbf{u} \cdot \mathbf{v}}{|\mathbf{v}|}$ **(5)**

It is a scalar.

Note that $\mathbf{v}/|\mathbf{v}|$ is a unit vector in the direction of **v**.

*Remark 1.* From Figures 3.14 and 3.15 and the fact that $\cos\varphi = (\mathbf{u} \cdot \mathbf{v})/(|\mathbf{u}|\,|\mathbf{v}|)$, we find that

> **v** and $\text{proj}_{\mathbf{v}}\,\mathbf{u}$ have (*i*) the same direction if $\mathbf{u} \cdot \mathbf{v} > 0$ and (*ii*) opposite directions if $\mathbf{u} \cdot \mathbf{v} < 0$.

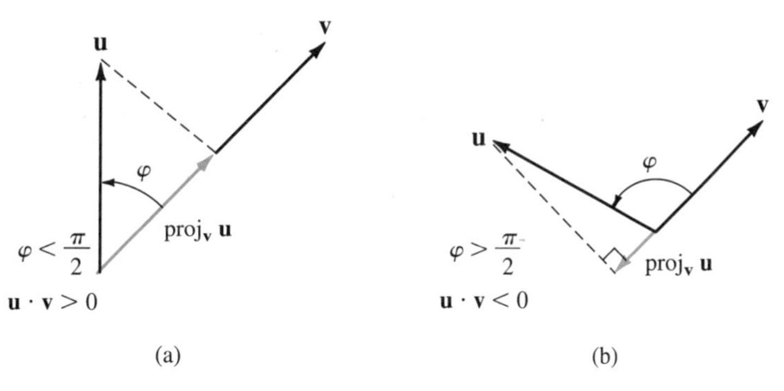


**Figure 3.15** (a) **v** and $\text{proj}_{\mathbf{v}}\,\mathbf{u}$ have the same direction if $\mathbf{u} \cdot \mathbf{v} > 0$. (b) **v** and $\text{proj}_{\mathbf{v}}\,\mathbf{u}$ have opposite directions if $\mathbf{u} \cdot \mathbf{v} < 0$

*Remark 2.* $\text{Proj}_{\mathbf{v}}\,\mathbf{u}$ can be thought of as the "**v**-component" of the vector **u**.

*Remark 3.* If **u** and **v** are orthogonal, then $\mathbf{u} \cdot \mathbf{v} = 0$ so that $\text{proj}_{\mathbf{v}}\,\mathbf{u} = \mathbf{0}$.

*Remark 4.* An alternative definition of projection is: If **u** and **v** are nonzero vectors, then $\text{proj}_{\mathbf{v}}\,\mathbf{u}$ is the unique vector having the following properties:

> **i.** $\text{Proj}_{\mathbf{v}}\, \mathbf{u}$ is parallel to $\mathbf{v}$.
> **ii.** $\mathbf{u} - \text{proj}_{\mathbf{v}}\, \mathbf{u}$ is orthogonal to $\mathbf{v}$.

EXAMPLE 4 **Calculating a Projection** Let $\mathbf{u} = 2\mathbf{i} + 3\mathbf{j}$ and $\mathbf{v} = \mathbf{i} + \mathbf{j}$. Calculate $\text{proj}_{\mathbf{v}}\, \mathbf{u}$.

**Solution** $\text{Proj}_{\mathbf{v}}\, \mathbf{u} = (\mathbf{u} \cdot \mathbf{v})\mathbf{v}/|\mathbf{v}|^2 = [5/(\sqrt{2})^2]\mathbf{v} = (5/2)\mathbf{i} + (5/2)\mathbf{j}$ (see Figure 3.16).

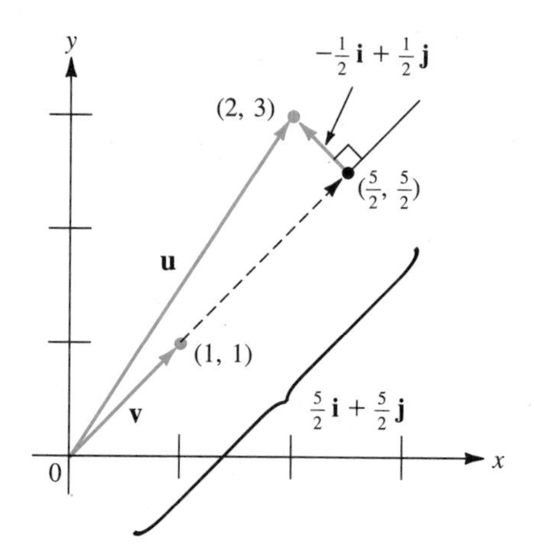


**Figure 3.16** The projection of (2, 3) on (1, 1) is $(\frac{5}{2}, \frac{5}{2})$

EXAMPLE 5 **Calculating a Projection** Let $\mathbf{u} = 2\mathbf{i} - 3\mathbf{j}$ and $\mathbf{v} = \mathbf{i} + \mathbf{j}$. Find $\text{proj}_{\mathbf{v}}\, \mathbf{u}$.

**Solution** Here $(\mathbf{u} \cdot \mathbf{v})/|\mathbf{v}|^2 = -\frac{1}{2}$; hence $\text{proj}_{\mathbf{v}}\, \mathbf{u} = -\frac{1}{2}\mathbf{i} - \frac{1}{2}\mathbf{j}$ (see Figure 3.17).

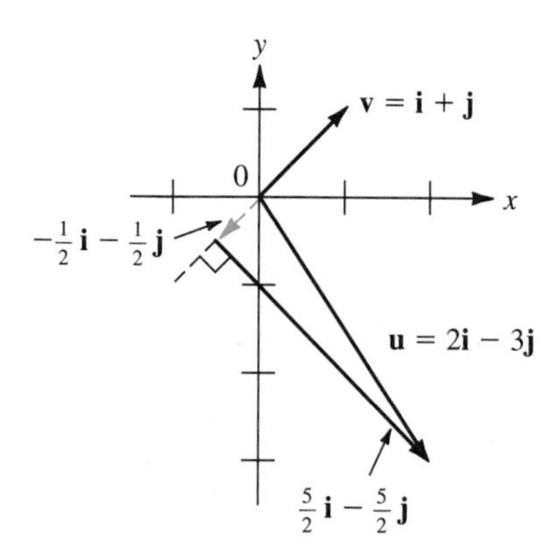


**Figure 3.17** The projection of $2\mathbf{i} - 3\mathbf{j}$ on $\mathbf{i} + \mathbf{j}$ is $-\frac{1}{2}\mathbf{i} - \frac{1}{2}\mathbf{j}$

# PROBLEMS 3.2

## Self-Quiz

**I.** $\mathbf{i}\cdot\mathbf{j} =$ ______.

**a.** 1 **b.** $\sqrt{(0-1)^2+(1-0)^2}$

**c.** 0 **d.** $\mathbf{i}+\mathbf{j}$

**II.** $(3, 4)\cdot(3, 2) =$ ______.

**a.** $(3+3)(4+2) = 36$

**b.** $(3)(3)+(4)(2) = 17$

**c.** $(3-3)(2-4) = 0$

**d.** $(3)(3)-(4)(2) = 1$

**III.** The cosine of the angle between $\mathbf{i}+\mathbf{j}$ and $\mathbf{i}-\mathbf{j}$ is ______.

**a.** $0\mathbf{i}+0\mathbf{j}$ **b.** 0

**c.** $\sqrt{2}$ **d.** $1/\sqrt{2}+0$

**IV.** The vectors $2\mathbf{i}-12\mathbf{j}$ and $3\mathbf{i}+(\frac{1}{2})\mathbf{j}$ are ______.

**a.** neither parallel nor orthogonal

**b.** parallel

**c.** orthogonal

**d.** identical

**V.** $\text{Proj}_{\mathbf{w}}\,\mathbf{u} =$ ______.

**a.** $\dfrac{\mathbf{u}\cdot\mathbf{w}}{|\mathbf{w}|}$ **b.** $\dfrac{\mathbf{w}}{|\mathbf{w}|}$

**c.** $\dfrac{\mathbf{u}\cdot\mathbf{w}}{|\mathbf{w}|}\dfrac{\mathbf{w}}{|\mathbf{w}|}$ **d.** $\dfrac{\mathbf{u}\cdot\mathbf{w}}{|\mathbf{u}|}\dfrac{\mathbf{u}}{|\mathbf{u}|}$

In Problems 1–8 calculate both the scalar product of the two vectors and the cosine of the angle between them.

**1.** $\mathbf{u}=\mathbf{i}+\mathbf{j}$; $\mathbf{v}=\mathbf{i}-\mathbf{j}$

**2.** $\mathbf{u}=3\mathbf{i}$; $\mathbf{v}=-7\mathbf{j}$

**3.** $\mathbf{u}=-5\mathbf{i}$; $\mathbf{v}=18\mathbf{j}$

**4.** $\mathbf{u}=\alpha\mathbf{i}$; $\mathbf{v}=\beta\mathbf{j}$; $\alpha$, $\beta$ real

**5.** $\mathbf{u}=2\mathbf{i}+5\mathbf{j}$; $\mathbf{v}=5\mathbf{i}+2\mathbf{j}$

**6.** $\mathbf{u}=2\mathbf{i}+5\mathbf{j}$; $\mathbf{v}=5\mathbf{i}-2\mathbf{j}$

**7.** $\mathbf{u}=-3\mathbf{i}+4\mathbf{j}$; $\mathbf{v}=-2\mathbf{i}-7\mathbf{j}$

**8.** $\mathbf{u}=4\mathbf{i}+5\mathbf{j}$; $\mathbf{v}=5\mathbf{i}-4\mathbf{j}$

**9.** Show that for any real numbers $\alpha$ and $\beta$, the vectors $\mathbf{u}=\alpha\mathbf{i}+\beta\mathbf{j}$ and $\mathbf{v}=\beta\mathbf{i}-\alpha\mathbf{j}$ are orthogonal.

**10.** Let $\mathbf{u}$, $\mathbf{v}$, and $\mathbf{w}$ denote three arbitrary vectors. Explain why the product $\mathbf{u}\cdot\mathbf{v}\cdot\mathbf{w}$ is *not defined.*

In Problems 11–16 determine whether the given vectors are orthogonal, parallel, or neither. Then sketch each pair.

**11.** $\mathbf{u}=3\mathbf{i}+5\mathbf{j}$; $\mathbf{v}=-6\mathbf{i}-10\mathbf{j}$

**12.** $\mathbf{u}=2\mathbf{i}+3\mathbf{j}$; $\mathbf{v}=6\mathbf{i}-4\mathbf{j}$

**13.** $\mathbf{u}=2\mathbf{i}+3\mathbf{j}$; $\mathbf{v}=6\mathbf{i}+4\mathbf{j}$

**14.** $\mathbf{u}=2\mathbf{i}+3\mathbf{j}$; $\mathbf{v}=-6\mathbf{i}+4\mathbf{j}$

---

**Answers to Self-Quiz**

**I.** c **II.** b **III.** b **IV.** c **V.** c

**15.** $\mathbf{u} = 7\mathbf{i}$; $\mathbf{v} = -23\mathbf{j}$

**16.** $\mathbf{u} = 2\mathbf{i} - 6\mathbf{j}$; $\mathbf{v} = -\mathbf{i} + 3\mathbf{j}$

**17.** Let $\mathbf{u} = 3\mathbf{i} + 4\mathbf{j}$ and $\mathbf{v} = \mathbf{i} + \alpha\mathbf{j}$. Determine $\alpha$ such that:

**a.** $\mathbf{u}$ and $\mathbf{v}$ are orthogonal.

**b.** $\mathbf{u}$ and $\mathbf{v}$ are parallel.

**c.** The angle between $\mathbf{u}$ and $\mathbf{v}$ is $\pi/4$.

**d.** The angle between $\mathbf{u}$ and $\mathbf{v}$ is $\pi/3$.

**18.** Let $\mathbf{u} = -2\mathbf{i} + 5\mathbf{j}$ and $\mathbf{v} = \alpha\mathbf{i} - 2\mathbf{j}$. Determine $\alpha$ such that

**a.** $\mathbf{u}$ and $\mathbf{v}$ are orthogonal.

**b.** $\mathbf{u}$ and $\mathbf{v}$ are parallel.

**c.** The angle between $\mathbf{u}$ and $\mathbf{v}$ is $2\pi/3$.

**d.** The angle between $\mathbf{u}$ and $\mathbf{v}$ is $\pi/3$.

**19.** In Problem 17 show that there is no value of $\alpha$ for which $\mathbf{u}$ and $\mathbf{v}$ have opposite directions.

**20.** In Problem 18 show that there is no value of $\alpha$ for which $\mathbf{u}$ and $\mathbf{v}$ have the same direction.

In Problems 21–30 calculate $\text{proj}_{\mathbf{v}}\,\mathbf{u}$.

**21.** $\mathbf{u} = 3\mathbf{i}$; $\mathbf{v} = \mathbf{i} + \mathbf{j}$

**22.** $\mathbf{u} = -5\mathbf{j}$; $\mathbf{v} = \mathbf{i} + \mathbf{j}$

**23.** $\mathbf{u} = 2\mathbf{i} + \mathbf{j}$; $\mathbf{v} = \mathbf{i} - 2\mathbf{j}$

**24.** $\mathbf{u} = 2\mathbf{i} + 3\mathbf{j}$; $\mathbf{v} = 4\mathbf{i} + \mathbf{j}$

**25.** $\mathbf{u} = \mathbf{i} + \mathbf{j}$; $\mathbf{v} = 2\mathbf{i} - 3\mathbf{j}$

**26.** $\mathbf{u} = \mathbf{i} + \mathbf{j}$; $\mathbf{v} = 2\mathbf{i} + 3\mathbf{j}$

**27.** $\mathbf{u} = \alpha\mathbf{i} + \beta\mathbf{j}$; $\mathbf{v} = \mathbf{i} + \mathbf{j}$; $\alpha$, $\beta$ real and positive

**28.** $\mathbf{u} = \mathbf{i} + \mathbf{j}$; $\mathbf{v} = \alpha\mathbf{i} + \beta\mathbf{j}$, $\alpha$, $\beta$ real and positive

**29.** $\mathbf{u} = \alpha\mathbf{i} - \beta\mathbf{j}$; $\mathbf{v} = \mathbf{i} + \mathbf{j}$; $\alpha$, $\beta$ real and positive with $\alpha > \beta$

**30.** $\mathbf{u} = \alpha\mathbf{i} - \beta\mathbf{j}$; $\mathbf{v} = \mathbf{i} + \mathbf{j}$; $\alpha$, $\beta$ real and positive with $\alpha < \beta$.

**31.** Let $\mathbf{u} = a_1\mathbf{i} + b_1\mathbf{j}$ and $\mathbf{v} = a_2\mathbf{i} + b_2\mathbf{j}$. Give a condition on $a_1$, $b_1$, $a_2$, and $b_2$ that will ensure that $\mathbf{v}$ and $\text{proj}_{\mathbf{v}}\,\mathbf{u}$ have the same direction.

**32.** In Problem 31 give a condition that will ensure that $\mathbf{v}$ and $\text{proj}_{\mathbf{v}}\,\mathbf{u}$ have opposite directions.

**33.** Let $P = (2, 3)$, $Q = (5, 7)$, $R = (2, -3)$, and $S = (1, 2)$. Calculate $\text{proj}_{\overrightarrow{PQ}}\overrightarrow{RS}$ and $\text{proj}_{\overrightarrow{RS}}\overrightarrow{PQ}$.

**34.** Let $P = (-1, 3)$, $Q = (2, 4)$, $R = (-6, -2)$, and $S = (3, 0)$. Calculate $\text{proj}_{\overrightarrow{PQ}}\overrightarrow{RS}$ and $\text{proj}_{\overrightarrow{RS}}\overrightarrow{PQ}$.

**35.** Prove that the nonzero vectors $\mathbf{u}$ and $\mathbf{v}$ are parallel if and only if $\mathbf{v} = \alpha\mathbf{u}$ for some constant $\alpha$. [*Hint:* Show that $\cos\varphi = \pm 1$ if and only if $\mathbf{v} = \alpha\mathbf{u}$.]

**36.** Prove that $\mathbf{u}$ and $\mathbf{v}$ are orthogonal if and only if $\mathbf{u} \cdot \mathbf{v} = 0$.

**37.** Show that the vector $\mathbf{v} = a\mathbf{i} + b\mathbf{j}$ is orthogonal to the line $ax + by + c = 0$.

**38.** Show that the vector $\mathbf{u} = b\mathbf{i} - a\mathbf{j}$ is parallel to the line $ax + by + c = 0$.

**39.** A triangle has vertices $(1, 3)$, $(4, -2)$, and $(-3, 6)$. Find the cosine of each of its angles.

**40.** A triangle has vertices $(a_1, b_1)$, $(a_2, b_2)$, and $(a_3, b_3)$. Find a formula for the cosines of each of its angles.

***41.** The **Cauchy-Schwarz inequality** states that for any real numbers $a_1$, $a_2$, $b_1$, and $b_2$,

$$\left|\sum_{k=1}^{2} a_k b_k\right| \le \left(\sum_{k=1}^{2} a_k^2\right)^{1/2}\left(\sum_{k=1}^{2} b_k^2\right)^{1/2}$$

Use the scalar product to prove this formula. Under what circumstances can the inequality be replaced by an equality?

*__42.__ Prove that the shortest distance between a point and a line is measured along a line through the point and perpendicular to the line.

**43.** Find the distance between $P = (2, 3)$ and the line through the points $Q = (-1, 7)$ and $R = (3, 5)$.

**44.** Find the distance between $(3, 7)$ and the line along the vector $\mathbf{v} = 2\mathbf{i} - 3\mathbf{j}$ passing through the origin.

**45.** Let $A$ be a $2 \times 2$ matrix such that each column is a unit vector and the two columns are orthogonal. Show that $A$ is invertible and that $A^{-1} = A^t$. ($A$ is called an **orthogonal** matrix.)

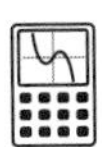

## CALCULATOR BOX

The dot product can be found easily on both the TI-85 and the CASIO fx-7700 GB.

### TI-85

Press [2nd] [VECTR] [F3] ⟨MATH⟩ [F4] ⟨dot⟩. "dot(" will be displayed. For example, dot (followed by [3, −1], [4, 5]) [ENTER] results in 7. Similarly, pressing [2nd] [VECTR] [F3] ⟨MATH⟩ [F2] ⟨unit V⟩ results in unit V. If you then enter a vector and press ENTER, a unit vector with the same direction as the entered vector will be displayed.

A projection, too, can be easily computed. If U and V are entered, then (dot (U, V)/dot (V, V)) ∗ V will yield $\text{proj}_V$ U.

### CASIO fx-7700 GB

The dot product can be computed by matrix multiplication. For example, if U is a $1 \times 2$ matrix (row vector) and V is a $2 \times 1$ matrix (column vector), then UV equals U · V. The Casio does not deal directly with the dot product.

In Problems 46–50 use a calculator to find a unit vector having the same direction as the given vector.

**46.** $(0.231, 0.816)$ **47.** $(-91, 48)$ **48.** $(1295, -7238)$

**49.** $(-5.2361, -18.6163)$ **50.** $(-20192, 58116)$

In Problems 51–54 use a calculator to find the projection of $\mathbf{u}$ on $\mathbf{v}$ and sketch $\mathbf{u}$, $\mathbf{v}$, and $\text{Proj}_{\mathbf{v}}\,\mathbf{u}$.

**51.** $\mathbf{u} = (3.28, -5.19)$, $\mathbf{v} = (-6.17, -11.526)$

**52.** $\mathbf{u} = (0.01629, -0.03556)$, $\mathbf{v} = (0.08171, 0.00119)$

**53.** $\mathbf{u} = (-5723, 4296)$, $\mathbf{v} = (17171, -9816)$

**54.** $\mathbf{u} = (37155, 42136)$, $\mathbf{v} = (25516, 72385)$

## MATLAB 3.2

1. For the pairs of vectors in Problems 21–26, verify the projection vectors computed with paper and pencil by computing using MATLAB. (Refer to the MATLAB information preceding MATLAB problems for Section 3.1.)

2. *(Uses the m-file prjtn.m)* This problem concerns the visualization of projections. It uses the m-file *prjtn.m* on the accompanying disk. Enter the command **help prjtn** for a description of this m-file.

   For the pairs of vectors **u** and **v** below:

   a. Input **u** and **v** as $2 \times 1$ matrices and compute **p** = the projection of **u** onto **v**.

   b. Enter the command **prjtn(u,v).** (This file displays **u** and **v** on the graphics screen. Hit any key and a perpendicular is dropped from the head of **u** onto the line determined by **v**. Hit any key and the projection vector is indicated.)

   c. While observing the graphics screen, verify that the **p** vector plotted is the vector computed in (a). Locate the vector (parallel to) **u** − **p**. What is the geometrical relationship between **u** − **p** and **v**?

   **i.** **u** = [2;1] **v** = [3;0]  **ii.** **u** = [2;3] **v** = [−3;0]

   **iii.** **u** = [2;1] **v** = [−1;2]  **iv.** **u** = [2;3] **v** = [−1;−2]

   **v.** Choose your own **u** and **v** (at least three choices).

## 3.3 VECTORS IN SPACE

We have seen how any point in a plane can be represented as an ordered pair of real numbers. Analogously, any point in space can be represented by an **ordered triple** of real numbers

$$(a, b, c) \tag{1}$$

$\mathbb{R}^3$
Origin
*x*-axis
*y*-axis
*z*-axis

Vectors of the form (1) constitute $\mathbb{R}^3$. To represent a point in space, we begin by choosing a point in $\mathbb{R}^3$. We call this point the **origin,** denoted by 0. Then we draw three mutually perpendicular axes, which we label the ***x*-axis,** the ***y*-axis,** and the ***z*-axis.** These axes can be selected in a variety of ways, but the most common selection has the $x$- and $y$-axes drawn horizontally with the $z$-axis vertical. On each axis, we choose a positive direction and measure distance along each axis as the number of units in this positive direction measured from the origin.

The two basic systems of drawing these axes are depicted in Figure 3.18. If the axes are placed as in Figure 3.18*a*, then the system is called a **right-handed system;** if they are placed as in Figure 3.18*b*, the system is a **left-handed system.** In the figures the arrows indicate the positive directions on the axes. The reason for this choice of terms is as follows: In a right-handed system, if you place your right hand so that your index finger points in the positive direction of the $x$-axis while your middle finger points in the positive direction of the $y$-axis, then your thumb will point in the positive direction of the $z$-axis. This concept is illustrated in Figure 3.19. For a left-handed system, the same rule works for your left hand. For the remainder of this text, we shall follow common practice and depict the coordinate axes using a right-handed system.

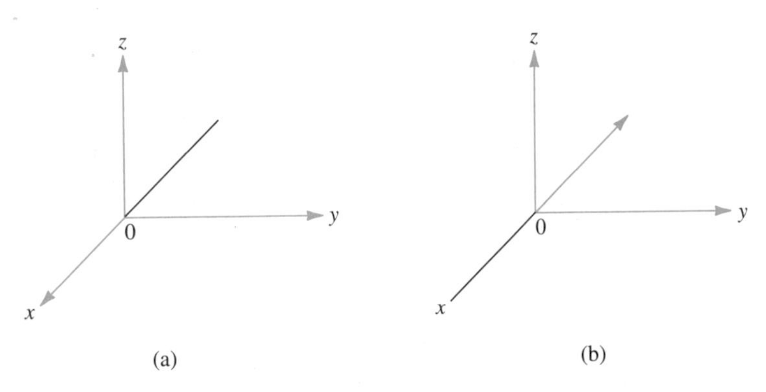


**Figure 3.18** (a) A right-handed system. (b) A left-handed system

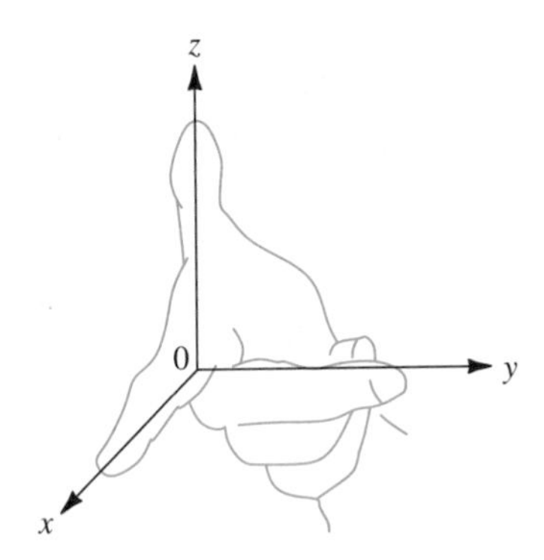


**Figure 3.19** A right hand indicates directions in a right-handed system

Coordinate planes

The three axes in our system determine three **coordinate planes,** which we call the $xy$-plane, the $xz$-plane, and the $yz$-plane. The $xy$-plane contains the $x$- and $y$-axes and is simply the plane with which we have been dealing to this point in most of this book. The $xz$- and $yz$-planes can be thought of in a similar way.

Having built our structure of coordinate axes and planes, we can describe any point $P$ in $\mathbb{R}^3$ in a unique way:

$$\boxed{P = (x, y, z)} \quad \textbf{(2)}$$

where the first coordinate $x$ is the directed distance from the $yz$-plane to $P$ (measured in the positive direction of the $x$-axis and along a line parallel to the $x$-axis), the second coordinate $y$ is the directed distance from the $xz$-plane to $P$ (measured in the positive direction of the $y$-axis and along a line parallel to the $y$-axis), and the third coordinate $z$ is the directed distance from the $xy$-plane to $P$ (measured in the positive direction of the $z$-axis and along a line parallel to the $z$-axis).

In this system the three coordinate planes divide $\mathbb{R}^3$ into eight **octants,** just as in $\mathbb{R}^2$ the two coordinate axes divide the plane into four quadrants. The first octant is always chosen to be the one in which the three coordinates are positive.

Cartesian coordinate system in $\mathbb{R}^3$

The coordinate system we have just established is often referred to as the **rectangular coordinate system** or the **Cartesian coordinate system.** Once we are comfortable with the notion of depicting a point in this system, we can extend many of our ideas from the plane.

**THEOREM 1** Let $P = (x_1, y_1, z_1)$ and $Q = (x_2, y_2, z_2)$ be two points in space. Then the distance $\overline{PQ}$ between $P$ and $Q$ is given by

$$\overline{PQ} = \sqrt{(x_1 - x_2)^2 + (y_1 - y_2)^2 + (z_1 - z_2)^2} \tag{3}$$

You are asked to prove this result in Problem 39.

**EXAMPLE 1** **Calculating the Distance Between Two Points in $\mathbb{R}^3$** Calculate the distance between the points $(3, -1, 6)$ and $(-2, 3, 5)$.

**Solution**

$$\overline{PQ} = \sqrt{[3 - (-2)]^2 + (-1 - 3)^2 + (6 - 5)^2} = \sqrt{42}$$

In Sections 3.1 and 3.2 we developed geometric properties of vectors in the plane. Given the similarity between the coordinate systems in $\mathbb{R}^2$ and $\mathbb{R}^3$, it should come as no surprise that vectors in $\mathbb{R}^2$ and $\mathbb{R}^3$ have very similar structures. We now develop the notion of a vector in space. The development will closely follow the development in the last two sections, and therefore some of the details will be omitted.

Directed line segment

Vector in $\mathbb{R}^3$

Let $P$ and $Q$ be two distinct points in $\mathbb{R}^3$. Then the **directed line segment** $\overrightarrow{PQ}$ is the straight-line segment that extends from $P$ to $Q$. Two directed line segments are **equivalent** if they have the same magnitude and direction. A **vector** in $\mathbb{R}^3$ is the set of all directed line segments equivalent to a given directed line segment, and any directed line segment $\overrightarrow{PQ}$ in that set is called a **representation** of the vector.

So far our definitions are identical. For convenience, we choose $P$ to be the origin so that the vector $\mathbf{v} = \overrightarrow{0Q}$ can be described by the coordinates $(x, y, z)$ of the point $Q$. Then the **magnitude** of $\mathbf{v} = |\mathbf{v}| = \sqrt{x^2 + y^2 + z^2}$ (from Theorem 1).

**EXAMPLE 2** **Calculating the Magnitude of a Vector in $\mathbb{R}^3$** Let $\mathbf{v} = (1, 3, -2)$. Find $|\mathbf{v}|$.

**Solution**

$$|\mathbf{v}| = \sqrt{1^2 + 3^2 + (-2)^2} = \sqrt{14}.$$

Let $\mathbf{u} = (x_1, y_1, z_1)$ and $\mathbf{v} = (x_2, y_2, z_2)$ be two vectors and let $\alpha$ be a real number (scalar). Then we define

**Vector Addition and Scalar Multiplication in $\mathbb{R}^3$**

$$\mathbf{u} + \mathbf{v} = (x_1 + x_2, y_1 + y_2, z_1 + z_2)$$

and

$$\alpha\mathbf{u} = (\alpha x_1, \alpha y_1, \alpha z_1)$$

This is the same definition of vector addition and scalar multiplication we had before; it is illustrated in Figure 3.20.

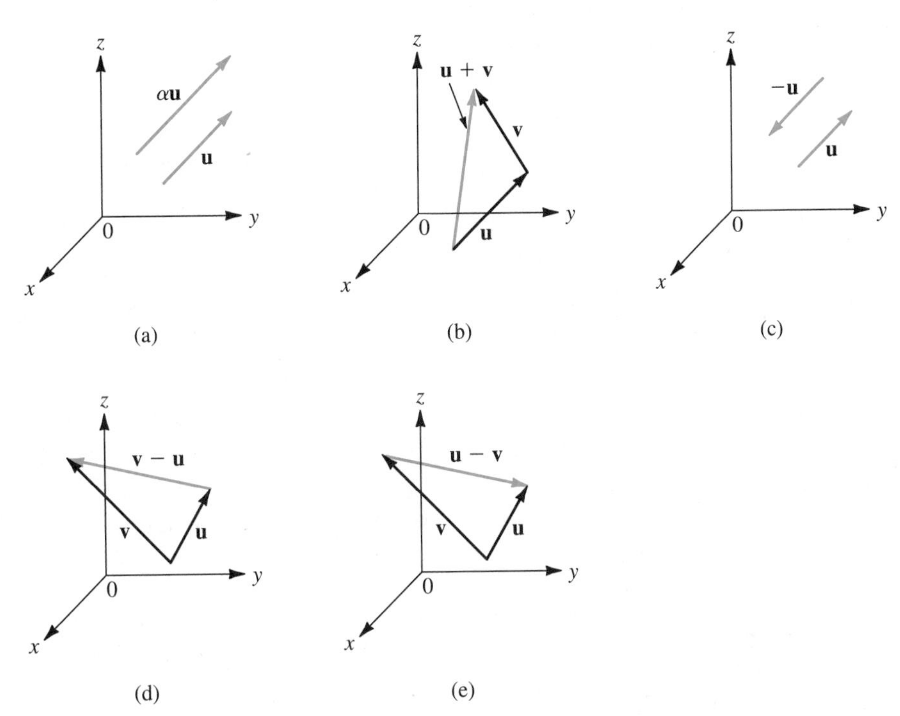


**Figure 3.20** Illustration of vector addition and scalar multiplication in $\mathbb{R}^3$

Unit vector

A **unit vector u** is a vector with magnitude 1. If $\mathbf{v}$ is any nonzero vector, then $\mathbf{u} = \mathbf{v}/|\mathbf{v}|$ is a unit vector having the same direction as $\mathbf{v}$.

**EXAMPLE 3** **Finding a Unit Vector in $\mathbb{R}^3$** Find a unit vector having the same direction as $\mathbf{v} = (2, 4, -3)$.

**Solution** Since $\mathbf{v} = \sqrt{2^2 + 4^2 + (-3)^2} = \sqrt{29}$, we have

$$\mathbf{u} = (2/\sqrt{29}, 4/\sqrt{29}, -3/\sqrt{29})$$

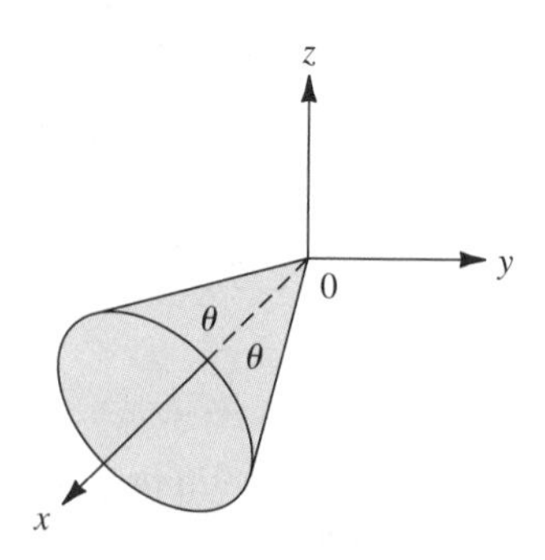


**Figure 3.21** All the vectors that lie on this cone make the angle $\theta$ with the positive $x$-axis

We can now formally define the direction of a vector in $\mathbb{R}^3$. We cannot define it as the angle $\theta$ that the vector makes with the positive $x$-axis, since, for example, if $0 < \theta < \pi/2$, then there are an *infinite number* of vectors making the angle $\theta$ with the positive $x$-axis, and these vectors together form a cone (see Figure 3.21).

**DEFINITION 1** **Direction in $\mathbb{R}^3$** The **direction** of a nonzero vector $\mathbf{v}$ in $\mathbb{R}^3$ is defined as the unit vector $\mathbf{u} = \mathbf{v}/|\mathbf{v}|$.

*Remark.* We could have defined the direction of a vector $\mathbf{v}$ in $\mathbb{R}^2$ in this way. For if $\mathbf{u} = \mathbf{v}/|\mathbf{v}|$, then $\mathbf{u} = (\cos\theta, \sin\theta)$, where $\theta$ is the direction of $\mathbf{v}$.

It would still be satisfying to define the direction of a vector in terms of some angles. Let $\mathbf{v}$ be the vector $\overrightarrow{0P}$ depicted in Figure 3.22. We define $\alpha$ to be the angle between $\mathbf{v}$ and the positive $x$-axis, $\beta$ the angle between $\mathbf{v}$ and the positive $y$-axis, and $\gamma$ the angle between $\mathbf{v}$ and the positive $z$-axis. The angles $\alpha$, $\beta$, and $\gamma$ are called the **direction angles** of the vector $\mathbf{v}$. Then, from Figure 3.22,

Direction angles

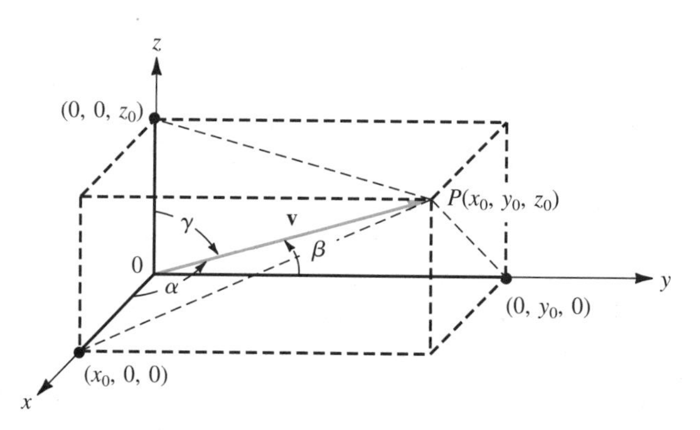


**Figure 3.22** The vector $\mathbf{v}$ makes the angle $\alpha$ with the positive $x$-axis, $\beta$ with the positive $y$-axis, and $\gamma$ with the positive $z$-axis

$$\cos\alpha = \frac{x_0}{|\mathbf{v}|} \qquad \cos\beta = \frac{y_0}{|\mathbf{v}|} \qquad \cos\gamma = \frac{z_0}{|\mathbf{v}|} \tag{4}$$

If $\mathbf{v}$ is a unit vector, then $|\mathbf{v}| = 1$ and

$$\cos\alpha = x_0 \qquad \cos\beta = y_0 \qquad \cos\gamma = z_0 \tag{5}$$

Direction cosines

By definition, each of these three angles lies in the interval $[0, \pi]$. The cosines of these angles are called the **direction cosines** of the vector $\mathbf{v}$. Note, from equations (4), that

$$\cos^2\alpha + \cos^2\beta + \cos^2\gamma = \frac{x_0^2 + y_0^2 + z_0^2}{|\mathbf{v}|^2} = \frac{x_0^2 + y_0^2 + z_0^2}{x_0^2 + y_0^2 + z_0^2} = 1 \tag{6}$$

If $\alpha$, $\beta$, and $\gamma$ are any three numbers between zero and $\pi$ such that condition (6) is satisfied, then they uniquely determine a unit vector given by $\mathbf{u} = (\cos\alpha, \cos\beta, \cos\gamma)$.

Direction numbers

*Remark.* If $\mathbf{v} = (a, b, c)$ and $|\mathbf{v}| \neq 1$, then the numbers $a$, $b$, and $c$ are called **direction numbers** of the vector $\mathbf{v}$.

**EXAMPLE 4** **Finding the Direction Cosines of a Vector in $\mathbb{R}^3$** Find the direction cosines of the vector $\mathbf{v} = (4, -1, 6)$.

**Solution** The direction of $\mathbf{v}$ is $\mathbf{v}/|\mathbf{v}| = \mathbf{v}/\sqrt{53} = (4/\sqrt{53}, -1/\sqrt{53}, 6/\sqrt{53})$. Then $\cos\alpha = 4/\sqrt{53} \approx 0.5494$, $\cos\beta = -1/\sqrt{53} \approx -0.1374$, and $\cos\gamma = 6/\sqrt{53} \approx 0.8242$. From these we use a table of cosines or a hand calculator to obtain $\alpha \approx 56.7° \approx 0.989$ rad, $\beta \approx 97.9° \approx 1.71$ rad, and $\gamma = 34.5° \approx 0.602$ rad. The vector, along with the angles $\alpha$, $\beta$, and $\gamma$, is sketched in Figure 3.23.

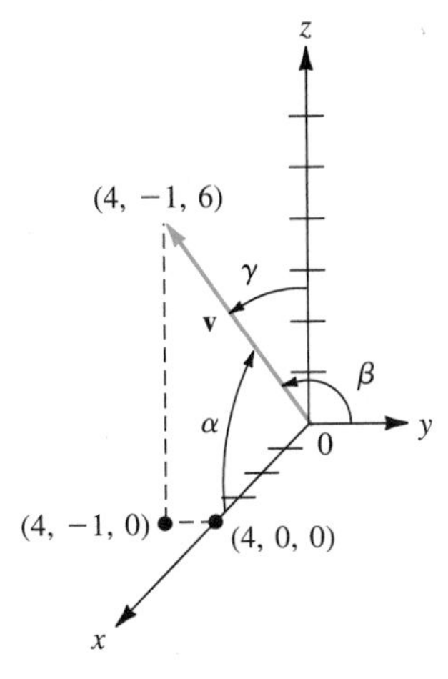


**Figure 3.23** The direction cosines of $(4, -1, 6)$ are $\cos\alpha$, $\cos\beta$, and $\cos\gamma$

**EXAMPLE 5** **Finding a Vector in $\mathbb{R}^3$ Given Its Magnitude and Direction Cosines** Find a vector **v** of magnitude 7 whose direction cosines are $1/\sqrt{6}$, $1/\sqrt{3}$, and $1/\sqrt{2}$.

**Solution** Let $\mathbf{u} = (1/\sqrt{6}, 1/\sqrt{3}, 1/\sqrt{2})$. Then **u** is a unit vector since $|\mathbf{u}| = 1$. Thus the direction of **v** is given by **u** and $\mathbf{v} = |\mathbf{v}|\,\mathbf{u} = 7\mathbf{u} = (7/\sqrt{6}, 7/\sqrt{3}, 7/\sqrt{2})$.

*Note.* We can solve this problem because $(1/\sqrt{6})^2 + (1/\sqrt{3})^2 + (1/\sqrt{2})^2 = 1$.

It is interesting to note that if **v** in $\mathbb{R}^2$ is a unit vector and we write $\mathbf{v} = (\cos\theta)\mathbf{i} + (\sin\theta)\mathbf{j}$, where $\theta$ is the direction of **v**, then $\cos\theta$ and $\sin\theta$ are the direction cosines of **v**. Here $\alpha = \theta$ and we define $\beta$ to be the angle that **v** makes with the $y$-axis (see Figure 3.24). Then $\beta = (\pi/2) - \alpha$ so that $\cos\beta = \cos(\pi/2 - \alpha) = \sin\alpha$ and **v** can be written in the "direction cosine" form

$$\mathbf{v} = \cos\alpha\mathbf{i} + \cos\beta\mathbf{j}$$

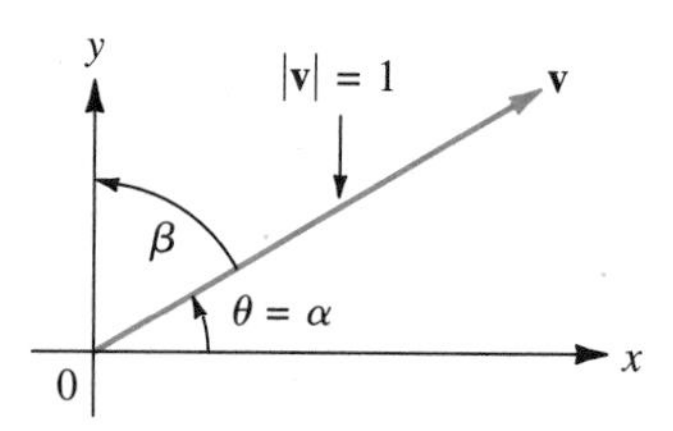


**Figure 3.24** If $\beta = \dfrac{\pi}{2} - \theta = \dfrac{\pi}{2} - \alpha$ and **v** is a unit vector, then $\mathbf{v} = \cos\theta\,\mathbf{i} + \sin\theta\,\mathbf{j} = \cos\alpha\mathbf{i} + \cos\beta\mathbf{j}$

In Section 3.1 we saw how any vector in the plane can be written in terms of the basis vectors **i** and **j**. To extend this idea to $\mathbb{R}^3$, we define

$$\mathbf{i} = (1, 0, 0) \qquad \mathbf{j} = (0, 1, 0) \qquad \mathbf{k} = (0, 0, 1) \tag{7}$$

Here **i**, **j**, and **k** are unit vectors. The vector **i** lies along the $x$-axis, **j** along the $y$-axis, and **k** along the $z$-axis. These are sketched in Figure 3.25. If $\mathbf{v} = (x, y, z)$ is any

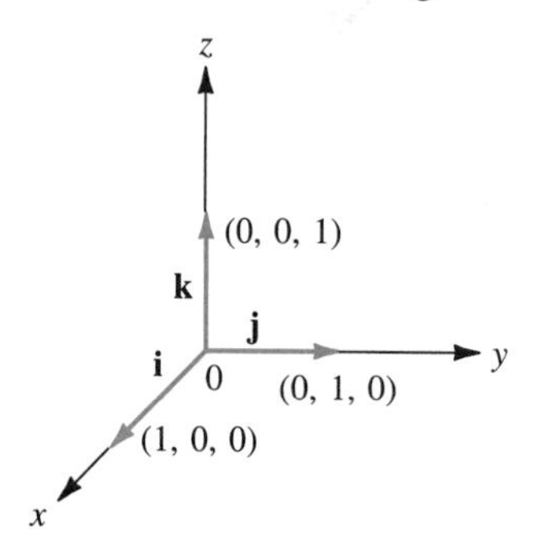


**Figure 3.25** The basis vectors **i**, **j**, and **k** in $\mathbb{R}^3$

vector in $\mathbb{R}^3$, then

$$\mathbf{v} = (x, y, z) = (x, 0, 0) + (0, y, 0) + (0, 0, z) = x\mathbf{i} + y\mathbf{j} + z\mathbf{k}$$

That is: *Any vector* $\mathbf{v}$ *in* $\mathbb{R}^3$ *can be written in a unique way in terms of the vectors* $\mathbf{i}$, $\mathbf{j}$, *and* $\mathbf{k}$.

The definition of the scalar product in $\mathbb{R}^3$ is, of course, the definition we have already seen in Section 1.6. Note that $\mathbf{i} \cdot \mathbf{i} = 1, \mathbf{j} \cdot \mathbf{j} = 1, \mathbf{k} \cdot \mathbf{k} = 1, \mathbf{i} \cdot \mathbf{j} = 0, \mathbf{j} \cdot \mathbf{k} = 0$, and $\mathbf{i} \cdot \mathbf{k} = 0$.

**THEOREM 2** If $\varphi$ denotes the smallest positive angle between two nonzero vectors $\mathbf{u}$ and $\mathbf{v}$, we have

$$\cos \varphi = \frac{\mathbf{u} \cdot \mathbf{v}}{|\mathbf{u}|\,|\mathbf{v}|} = \frac{\mathbf{u}}{|\mathbf{u}|} \cdot \frac{\mathbf{v}}{|\mathbf{v}|} \tag{8}$$

**Proof** The proof is almost identical to the proof of Theorem 3.2.2 on page 242 and is left as an exercise (see Problem 40). ■

**EXAMPLE 6** **Computing the Cosine of the Angle Between Two Vectors in $\mathbb{R}^3$** Calculate the cosine of the angle between $\mathbf{u} = 3\mathbf{i} - \mathbf{j} + 2\mathbf{k}$ and $\mathbf{v} = 4\mathbf{i} + 3\mathbf{j} - \mathbf{k}$.

**Solution** $\mathbf{u} \cdot \mathbf{v} = 7$, $|\mathbf{u}| = \sqrt{14}$, and $|\mathbf{v}| = \sqrt{26}$ so that $\cos \varphi = 7/\sqrt{(14)(26)} = 7/\sqrt{364} \approx 0.3669$ and $\varphi \approx 68.5° \approx 1.2$ rad. ■

**DEFINITION 2** **Parallel and Orthogonal Vectors** Two nonzero vectors $\mathbf{u}$ and $\mathbf{v}$ are:

**i.** **Parallel** if the angle between them is zero or $\pi$

**ii.** **Orthogonal** (or **perpendicular**) if the angle between them is $\pi/2$

**THEOREM 3**

**i.** If $\mathbf{u} \neq \mathbf{0}$, then $\mathbf{u}$ and $\mathbf{v}$ are parallel if and only if $\mathbf{v} = \alpha\mathbf{u}$ for some scalar $\alpha \neq 0$.

**ii.** If $\mathbf{u}$ and $\mathbf{v}$ are nonzero, then $\mathbf{u}$ and $\mathbf{v}$ are orthogonal if and only if $\mathbf{u} \cdot \mathbf{v} = 0$.

**Proof** Again, the proof is easy and is left as an exercise (see Problem 41). ■

We now turn to the definition of the projection of one vector on another. First, we state the theorem that is the analog of Theorem 3.2.5 (and that has an identical proof).

**THEOREM 4** Let $\mathbf{v}$ be a nonzero vector. Then for any other vector $\mathbf{u}$,

$$\mathbf{w} = \mathbf{u} - \frac{\mathbf{u} \cdot \mathbf{v}}{|\mathbf{v}|^2}\mathbf{v}$$

is orthogonal to $\mathbf{v}$.

**DEFINITION 3** **Projection** Let $\mathbf{u}$ and $\mathbf{v}$ be nonzero vectors. Then the **projection** of $\mathbf{u}$ on $\mathbf{v}$, denoted by $\text{proj}_{\mathbf{v}}\,\mathbf{u}$, is defined by

$$\text{proj}_{\mathbf{v}}\,\mathbf{u} = \frac{\mathbf{u} \cdot \mathbf{v}}{|\mathbf{v}|^2}\mathbf{v} \tag{9}$$

The **component** of $\mathbf{u}$ in the direction $\mathbf{v}$ is given by $\dfrac{\mathbf{u} \cdot \mathbf{v}}{|\mathbf{v}|}$. **(10)**

**EXAMPLE 7** **Calculating a Projection in $\mathbb{R}^3$** Let $\mathbf{u} = 2\mathbf{i} + 3\mathbf{j} + \mathbf{k}$ and $\mathbf{v} = \mathbf{i} + 2\mathbf{j} - 6\mathbf{k}$. Find $\text{proj}_{\mathbf{v}}\,\mathbf{u}$.

**Solution** Here $(\mathbf{u} \cdot \mathbf{v})/|\mathbf{v}|^2 = 2/41$ and $\text{proj}_{\mathbf{v}}\,\mathbf{u} = \frac{2}{41}\mathbf{i} + \frac{4}{41}\mathbf{j} - \frac{12}{41}\mathbf{k}$. The component of $\mathbf{u}$ in the direction $\mathbf{v}$ is $(\mathbf{u} \cdot \mathbf{v})/|\mathbf{v}| = 2/\sqrt{41}$.

Note that, as in the planar case, $\text{proj}_{\mathbf{v}}\,\mathbf{u}$ is a vector that has the same direction as $\mathbf{v}$ if $\mathbf{u} \cdot \mathbf{v} > 0$ and the direction opposite to that of $\mathbf{v}$ if $\mathbf{u} \cdot \mathbf{v} < 0$.

# PROBLEMS 3.3

## Self-Quiz

**I.** *True-False.* The common practice followed in this text is to display the *xyz*-axes for $\mathbb{R}^3$ as a right-handed system.

**II.** The distance between the points (1, 2, 3) and (3, 5, −1) is ________.
**a.** $\sqrt{(1+2+3)^2 + (3+5-1)^2}$
**b.** $\sqrt{2^2 + 3^2 + 2^2}$
**c.** $\sqrt{2^2 + 3^2 + 4^2}$
**d.** $\sqrt{4^2 + 7^2 + 2^2}$

**III.** The point (0.3, 0.5, 0.2) is ________ the unit sphere.
**a.** tangent to **b.** on
**c.** inside **d.** outside

**IV.** $(x-3)^2+(y+5)^2+z^2=81$ is the equation of the sphere with ______.
**a.** center 81 and radius $(-3, 5, 0)$
**b.** radius 81 and center $(-3, 5, 0)$
**c.** radius $-9$ and center $(3, -5, 0)$
**d.** radius 9 and center $(3, -5, 0)$

**V.** $\mathbf{j}-(4\mathbf{k}-3\mathbf{i})=$ ______.
**a.** $(1, -4, -3)$
**b.** $(1, -4, 3)$
**c.** $(-3, 1, -4)$
**d.** $(3, 1, -4)$

**VI.** $(\mathbf{i}+3\mathbf{k}-\mathbf{j})\cdot(\mathbf{k}-4\mathbf{j}+2\mathbf{i})=$ ______.
**a.** $2+4+3=9$
**b.** $(1+3-1)(1-4+2)=-3$
**c.** $1-12-2=-13$
**d.** $2-4-3=-5$

**VII.** The unit vector in the same direction as $2\mathbf{i}-2\mathbf{j}+\mathbf{k}$ is ______.
**a.** $\mathbf{i}-\mathbf{j}+\mathbf{k}$
**b.** $\frac{1}{5}(2\mathbf{i}-2\mathbf{j}+\mathbf{k})$
**c.** $\frac{1}{3}(2\mathbf{i}-2\mathbf{j}+\mathbf{k})$
**d.** $\frac{1}{3}(2\mathbf{i}+2\mathbf{j}+\mathbf{k})$

**VIII.** The component of **u** in the direction **w** is
**a.** $\dfrac{\mathbf{u}\cdot\mathbf{w}}{|\mathbf{w}|}$
**b.** $\dfrac{\mathbf{w}}{|\mathbf{w}|}$
**c.** $\dfrac{\mathbf{u}\cdot\mathbf{w}}{|\mathbf{w}|}\dfrac{\mathbf{w}}{|\mathbf{w}|}$
**d.** $\dfrac{\mathbf{u}\cdot\mathbf{w}}{|\mathbf{w}|}\dfrac{\mathbf{u}}{|\mathbf{u}|}$

In Problems 1–3 find the distance between the two points.

**1.** $(3, -4, 3)$; $(3, 2, 5)$
**2.** $(3, -4, 7)$; $(3, -4, 9)$
**3.** $(-2, 1, 3)$; $(4, 1, 3)$

In Problems 4–17 find the magnitude and the direction cosines of the given vector.

**4.** $\mathbf{v}=3\mathbf{j}$
**5.** $\mathbf{v}=-3\mathbf{i}$
**6.** $\mathbf{v}=4\mathbf{i}-\mathbf{j}$
**7.** $\mathbf{v}=\mathbf{i}+2\mathbf{k}$
**8.** $\mathbf{v}=\mathbf{i}-\mathbf{j}+\mathbf{k}$
**9.** $\mathbf{v}=\mathbf{i}+\mathbf{j}-\mathbf{k}$
**10.** $\mathbf{v}=-\mathbf{i}+\mathbf{j}+\mathbf{k}$
**11.** $\mathbf{v}=\mathbf{i}-\mathbf{j}-\mathbf{k}$
**12.** $\mathbf{v}=-\mathbf{i}+\mathbf{j}-\mathbf{k}$
**13.** $\mathbf{v}=-\mathbf{i}-\mathbf{j}+\mathbf{k}$
**14.** $\mathbf{v}=-\mathbf{i}-\mathbf{j}-\mathbf{k}$
**15.** $\mathbf{v}=2\mathbf{i}+5\mathbf{j}-7\mathbf{k}$
**16.** $\mathbf{v}=-3\mathbf{i}-3\mathbf{j}+8\mathbf{k}$
**17.** $\mathbf{v}=-2\mathbf{i}-3\mathbf{j}-4\mathbf{k}$

**18.** The three direction angles of a certain unit vector are the same and are between zero and $\pi/2$. What is the vector?

**19.** Find a vector of magnitude 12 that has the same direction as the vector in Problem 18.

**20.** Show that there is no unit vector whose direction angles are $\pi/6$, $\pi/3$, and $\pi/4$.

**21.** Let $P=(2, 1, 4)$ and $Q=(3, -2, 8)$. Find a unit vector in the direction $\overrightarrow{PQ}$.

**22.** Let $P=(-3, 1, 7)$ and $Q=(8, 1, 7)$. Find a unit vector whose direction is opposite that of $\overrightarrow{PQ}$.

---

### Answers to Self-Quiz

**I.** True **II.** c **III.** c **IV.** d **V.** d **VI.** a **VII.** c **VIII.** a

**23.** In Problem 22 find all points $R$ such that $\overrightarrow{PR} \perp \overrightarrow{PQ}$.

***24.** Show that the set of points that satisfy the condition of Problem 23 and the condition $|\overrightarrow{PR}| = 1$ forms a circle.

**25. Triangle Inequality** If $\mathbf{u}$ and $\mathbf{v}$ are in $\mathbb{R}^3$, show that $|\mathbf{u} + \mathbf{v}| \leq |\mathbf{u}| + |\mathbf{v}|$.

**26.** Under what circumstances can the inequality in Problem 25 be replaced by an equals sign?

In Problems 27–38 let $\mathbf{u} = 2\mathbf{i} - 3\mathbf{j} + 4\mathbf{k}$, $\mathbf{v} = -2\mathbf{i} - 3\mathbf{j} + 5\mathbf{k}$, $\mathbf{w} = \mathbf{i} - 7\mathbf{j} + 3\mathbf{k}$, and $\mathbf{t} = 3\mathbf{i} + 4\mathbf{j} + 5\mathbf{k}$.

**27.** Calculate $\mathbf{u} + \mathbf{v}$.

**28.** Calculate $2\mathbf{u} - 3\mathbf{v}$.

**29.** Calculate $\mathbf{t} + 3\mathbf{w} - \mathbf{v}$.

**30.** Calculate $2\mathbf{u} - 7\mathbf{w} + 5\mathbf{v}$.

**31.** Calculate $2\mathbf{v} + 7\mathbf{t} - \mathbf{w}$.

**32.** Calculate $\mathbf{u} \cdot \mathbf{v}$.

**33.** Calculate $|\mathbf{w}|$.

**34.** Calculate $\mathbf{u} \cdot \mathbf{w} - \mathbf{w} \cdot \mathbf{t}$.

**35.** Calculate the angle between $\mathbf{u}$ and $\mathbf{w}$.

**36.** Calculate the angle between $\mathbf{t}$ and $\mathbf{w}$.

**37.** Calculate $\text{proj}_{\mathbf{u}}\, \mathbf{v}$.

**38.** Calculate $\text{proj}_{\mathbf{t}}\, \mathbf{w}$.

**39.** Prove Theorem 1. [*Hint:* Use the Pythagorean theorem twice in Figure 3.26.]

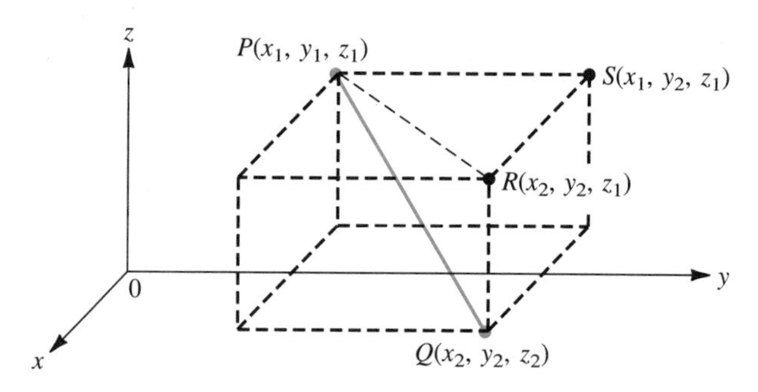


**Figure 3.26**

**40.** Prove Theorem 2.

**41.** Prove Theorem 3.

**42.** Prove Theorem 4.

## CALCULATOR BOX

The calculator instructions given in Sections 3.1 and 3.2 for vectors in $\mathbb{R}^2$ carry over into $\mathbb{R}^3$, with one exception.

### TI-85

The only difference is that the operation ⟨◀Pol⟩ applies only to $\mathbb{R}^2$. The direction of a vector A in $\mathbb{R}^3$ is obtained by using the operation ⟨unit V⟩ ALPHA A ENTER. All other operations are the same.

Solve the following problems on a calculator.

In Problems 43–46 find the magnitude and direction of each vector.

**43.** $(0.2316, 0.4179, -0.5213)$

**44.** $(-2356, -8194, 3299)$

**45.** $(17.3, 78.4, 28.6)$

**46.** $(0.0136, -0.0217, -0.0448)$

In Problems 47–50 compute $\text{proj}_{\mathbf{v}}\,\mathbf{u}$.

**47.** $\mathbf{u} = (-15, 27, 83)$; $\mathbf{v} = (-84, -77, 51)$

**48.** $\mathbf{u} = (-0.346, -0.517, -0.824)$; $\mathbf{v} = (-0.517, 0.811, 0.723)$

**49.** $\mathbf{u} = (5241, -3199, 2386)$; $\mathbf{v} = (1742, 8233, 9416)$

**50.** $\mathbf{u} = (0.24, 0.036, 0.055)$; $\mathbf{v} = (0.088, -0.064, 0.037)$

## 3.4 THE CROSS PRODUCT OF TWO VECTORS

To this point the only product of vectors that we have considered has been the scalar or dot product. We now define a new product, called the *cross product* (or *vector product*), which is defined *only* in $\mathbb{R}^3$.

***Historical Note.*** The cross product was defined by Hamilton in one of a series of papers published in *Philosophical Magazine* between the years 1844 and 1850.

**DEFINITION 1** **Cross Product** Let $\mathbf{u} = a_1\mathbf{i} + b_1\mathbf{j} + c_1\mathbf{k}$ and $\mathbf{v} = a_2\mathbf{i} + b_2\mathbf{j} + c_2\mathbf{k}$. Then the **cross product (vector product)** of $\mathbf{u}$ and $\mathbf{v}$, denoted by $\mathbf{u} \times \mathbf{v}$, is a new vector defined by

$$\mathbf{u} \times \mathbf{v} = (b_1c_2 - c_1b_2)\mathbf{i} + (c_1a_2 - a_1c_2)\mathbf{j} + (a_1b_2 - b_1a_2)\mathbf{k} \tag{1}$$

*Note that the result of the cross product is a vector, whereas the result of the scalar product is a scalar.*

Here the cross product seems to have been defined somewhat arbitrarily. There are obviously many ways to define a vector product. Why was this definition chosen? We answer that question in this section by demonstrating some of the properties of the cross product and illustrating some of its uses.

**EXAMPLE 1** **Calculating the Cross Product of Two Vectors** Let $\mathbf{u} = \mathbf{i} - \mathbf{j} + 2\mathbf{k}$ and $\mathbf{v} = 2\mathbf{i} + 3\mathbf{j} - 4\mathbf{k}$. Calculate $\mathbf{w} = \mathbf{u} \times \mathbf{v}$.

**Solution** Using formula (1), we obtain

$$\begin{aligned}\mathbf{w} &= [(-1)(-4) - (2)(3)]\mathbf{i} + [(2)(2) - (1)(-4)]\mathbf{j} + [(1)(3) - (-1)(2)]\mathbf{k} \\ &= -2\mathbf{i} + 8\mathbf{j} + 5\mathbf{k}\end{aligned}$$

*Note.* In this example $\mathbf{u} \cdot \mathbf{w} = (\mathbf{i} - \mathbf{j} + 2\mathbf{k}) \cdot (-2\mathbf{i} + 8\mathbf{j} + 5\mathbf{k}) = -2 - 8 + 10 = 0$. Similarly, $\mathbf{v} \cdot \mathbf{w} = 0$. That is, $\mathbf{u} \times \mathbf{v}$ is orthogonal to both $\mathbf{u}$ and $\mathbf{v}$. As we shall shortly see, the cross product of $\mathbf{u}$ and $\mathbf{v}$ is always orthogonal to $\mathbf{u}$ and $\mathbf{v}$. ■

Before continuing our discussion of the uses of the cross product, we observe that there is an easy way to calculate $\mathbf{u} \times \mathbf{v}$ by using determinants.

**THEOREM 1**

$$\mathbf{u} \times \mathbf{v} = \begin{vmatrix} \mathbf{i} & \mathbf{j} & \mathbf{k} \\ a_1 & b_1 & c_1 \\ a_2 & b_2 & c_2 \end{vmatrix}^{\dagger}$$

**Proof**

$$\begin{aligned}\begin{vmatrix} \mathbf{i} & \mathbf{j} & \mathbf{k} \\ a_1 & b_1 & c_1 \\ a_2 & b_2 & c_2 \end{vmatrix} &= \mathbf{i}\begin{vmatrix} b_1 & c_1 \\ b_2 & c_2 \end{vmatrix} - \mathbf{j}\begin{vmatrix} a_1 & c_1 \\ a_2 & c_2 \end{vmatrix} + \mathbf{k}\begin{vmatrix} a_1 & b_1 \\ a_2 & b_2 \end{vmatrix} \\ &= (b_1c_2 - c_1b_2)\mathbf{i} + (c_1a_2 - a_1c_2)\mathbf{j} + (a_1b_2 - b_1a_2)\mathbf{k}\end{aligned}$$

which is equal to $\mathbf{u} \times \mathbf{v}$ according to Definition 1. ■

**EXAMPLE 2** **Using Theorem 1 to Calculate a Cross Product** Calculate $\mathbf{u} \times \mathbf{v}$, where $\mathbf{u} = 2\mathbf{i} + 4\mathbf{j} - 5\mathbf{k}$ and $\mathbf{v} = -3\mathbf{i} - 2\mathbf{j} + \mathbf{k}$.

**Solution**

$$\begin{aligned}\mathbf{u} \times \mathbf{v} = \begin{vmatrix} \mathbf{i} & \mathbf{j} & \mathbf{k} \\ 2 & 4 & -5 \\ -3 & -2 & 1 \end{vmatrix} &= (4 - 10)\mathbf{i} - (2 - 15)\mathbf{j} + (-4 + 12)\mathbf{k} \\ &= -6\mathbf{i} + 13\mathbf{j} + 8\mathbf{k}\end{aligned}$$ ■

The following theorem summarizes some properties of the cross product. Its proof is left as an exercise (see Problems 32–35).

† This is not really a determinant because **i**, **j**, and **k** are not numbers. However, using determinant notation, Theorem 1 helps us remember how to calculate a cross product.

**THEOREM 2** Let **u**, **v**, and **w** be vectors in $\mathbb{R}^3$ and let $\alpha$ be a scalar. Then:

**i.** $\mathbf{u} \times \mathbf{0} = \mathbf{0} \times \mathbf{u} = \mathbf{0}$.

**ii.** $\mathbf{u} \times \mathbf{v} = -(\mathbf{v} \times \mathbf{u})$ **(anticommutative property for the vector product).**

**iii.** $(\alpha\mathbf{u}) \times \mathbf{v} = \alpha(\mathbf{u} \times \mathbf{v})$.

**iv.** $\mathbf{u} \times (\mathbf{v} + \mathbf{w}) = (\mathbf{u} \times \mathbf{v}) + (\mathbf{u} \times \mathbf{w})$ **(distributive property for the vector product).**

**v.** $(\mathbf{u} \times \mathbf{v}) \cdot \mathbf{w} = \mathbf{u} \cdot (\mathbf{v} \times \mathbf{w})$. (This is called the **scalar triple product** of **u**, **v**, and **w**.)

**vi.** $\mathbf{u} \cdot (\mathbf{u} \times \mathbf{v}) = \mathbf{v} \cdot (\mathbf{u} \times \mathbf{v}) = 0$. (That is, $\mathbf{u} \times \mathbf{v}$ is orthogonal to both **u** and **v**.)

**vii.** If neither **u** nor **v** is the zero vector, then **u** and **v** are parallel if and only if $\mathbf{u} \times \mathbf{v} = \mathbf{0}$.

Part (*vi*) is the most commonly used part of this theorem. We restate it below:

> The cross product $\mathbf{u} \times \mathbf{v}$ is orthogonal to both **u** and **v**.

We now know that $\mathbf{u} \times \mathbf{v}$ is a vector orthogonal to **u** and **v**. But there are always *two* unit vectors orthogonal to **u** and **v** (see Figure 3.27). The vectors **n** and $-\mathbf{n}$ (**n** stands for **normal**) are both orthogonal to **u** and **v**. Which one is in the direction of $\mathbf{u} \times \mathbf{v}$? The answer is given by the **right-hand rule.** If the right hand is placed so that the index finger points in the direction of **u** while the middle finger points in the direction of **v**, then the thumb points in the direction of $\mathbf{u} \times \mathbf{v}$ (see Figure 3.28).

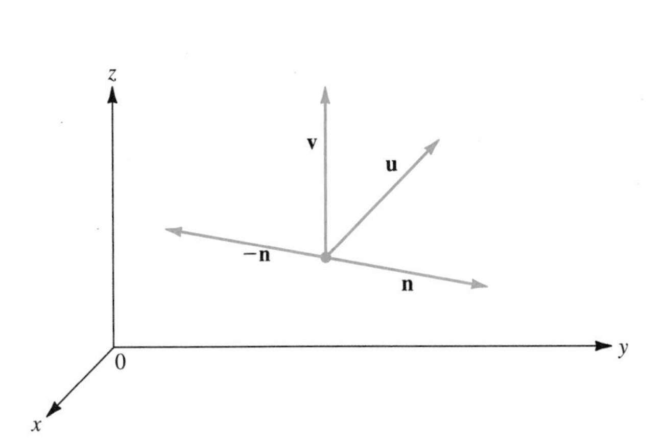


**Figure 3.27** There are exactly two vectors **n** and $-\mathbf{n}$ that are orthogonal to two nonparallel vectors **u** and **v** in $\mathbb{R}^3$

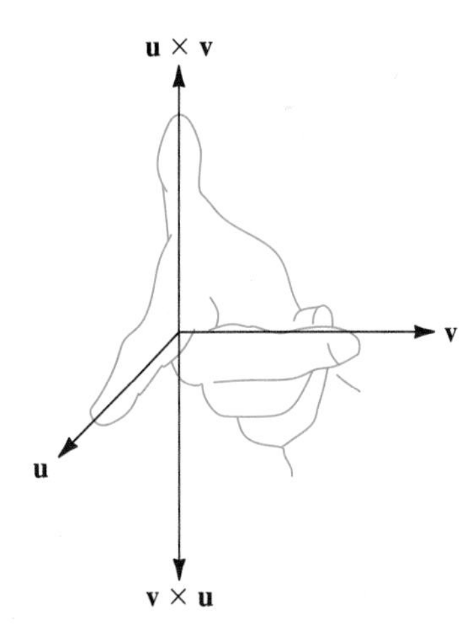


**Figure 3.28** The direction of $\mathbf{u} \times \mathbf{v}$ can be determined using the right-hand rule

Having discussed the direction of the vector $\mathbf{u} \times \mathbf{v}$, we now turn to a discussion of its magnitude.

**THEOREM 3** If $\varphi$ is the angle between $\mathbf{u}$ and $\mathbf{v}$, then

$$|\mathbf{u} \times \mathbf{v}| = |\mathbf{u}|\,|\mathbf{v}| \sin \varphi \tag{2}$$

**Proof** It is not difficult to show (by comparing coordinates) that $|\mathbf{u} \times \mathbf{v}|^2 = |\mathbf{u}|^2|\mathbf{v}|^2 - (\mathbf{u} \cdot \mathbf{v})^2$ (see Problem 31). Then since $(\mathbf{u} \cdot \mathbf{v})^2 = |\mathbf{u}|^2|\mathbf{v}|^2 \cos^2 \varphi$ (from Theorem 3.3.2, page 257),

$$\begin{aligned} |\mathbf{u} \times \mathbf{v}|^2 &= |\mathbf{u}|^2|\mathbf{v}|^2 - |\mathbf{u}|^2|\mathbf{v}|^2 \cos^2 \varphi = |\mathbf{u}|^2|\mathbf{v}|^2 (1 - \cos^2 \theta) \\ &= |\mathbf{u}|^2|\mathbf{v}|^2 \sin^2 \varphi \end{aligned}$$

and the theorem follows after taking the square root of both sides. Note that $\sin \varphi \geq 0$ because $0 \leq \varphi \leq \pi$. ■

There is an interesting geometric interpretation of Theorem 3. The vectors $\mathbf{u}$ and $\mathbf{v}$ are sketched in Figure 3.29 and can be thought of as two adjacent sides of a parallelogram. Then from elementary geometry, we see that

$$\text{Area of the parallelogram having } \mathbf{u} \text{ and } \mathbf{v} \text{ as adjacent sides} = |\mathbf{u}|\,|\mathbf{v}| \sin \varphi = |\mathbf{u} \times \mathbf{v}| \tag{3}$$

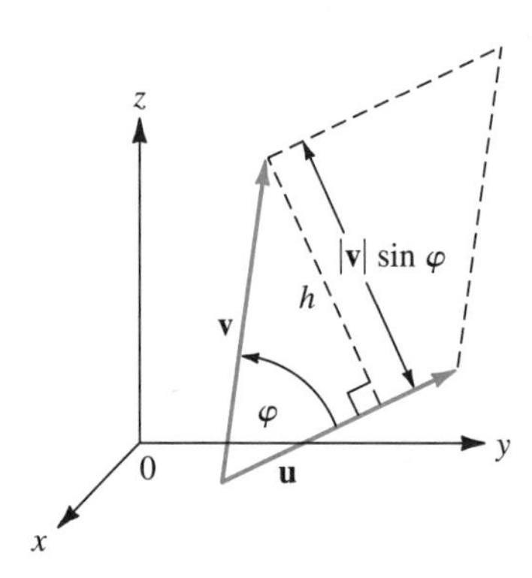


**Figure 3.29** $\varphi$ is the angle between $\mathbf{u}$ and $\mathbf{v}$. $\dfrac{h}{|\mathbf{v}|} = \sin \varphi$ so that $h = |\mathbf{v}| \sin \varphi$

**EXAMPLE 3** **Finding the Area of a Parallelogram in $\mathbb{R}^3$** Find the area of the parallelogram with consecutive vertices at $P = (1, 3, -2)$, $Q = (2, 1, 4)$, and $R = (-3, 1, 6)$.

**Solution** The parallelogram is sketched in Figure 3.30. We have

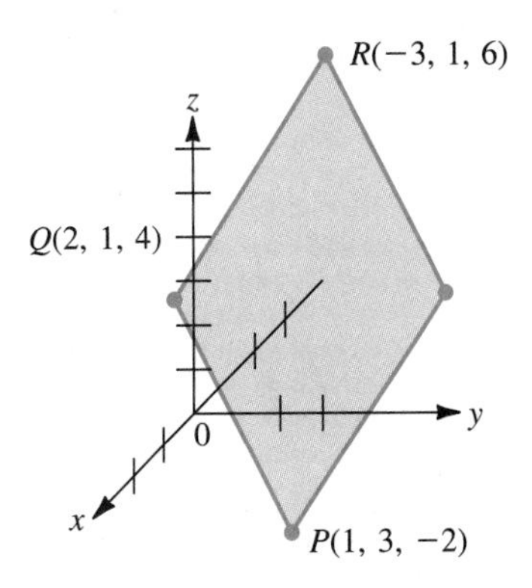


**Figure 3.30** A parallelogram in $\mathbb{R}^3$

$$\text{Area} = |\overrightarrow{PQ} \times \overrightarrow{QR}| = |(\mathbf{i} - 2\mathbf{j} + 6\mathbf{k}) \times (-5\mathbf{i} + 2\mathbf{k})|$$

$$= \begin{vmatrix} \mathbf{i} & \mathbf{j} & \mathbf{k} \\ 1 & -2 & 6 \\ -5 & 0 & 2 \end{vmatrix} = |-4\mathbf{i} - 32\mathbf{j} - 10\mathbf{k}| = \sqrt{1140} \text{ square units}$$

## Geometric Interpretation of 2 × 2 Determinants (Revisited)

In Section 2.1 (page 180) we discussed the geometric meaning of a $2 \times 2$ determinant. We look at the same problem now. Using the cross product, we obtain the result of Section 2.1 much more easily. Let $A$ be a $2 \times 2$ matrix and let $\mathbf{u}$ and $\mathbf{v}$ be two 2-vectors. Let $\mathbf{u} = \begin{pmatrix} u_1 \\ u_2 \end{pmatrix}$ and $\mathbf{v} = \begin{pmatrix} v_1 \\ v_2 \end{pmatrix}$. These vectors are given in Figure 3.31.

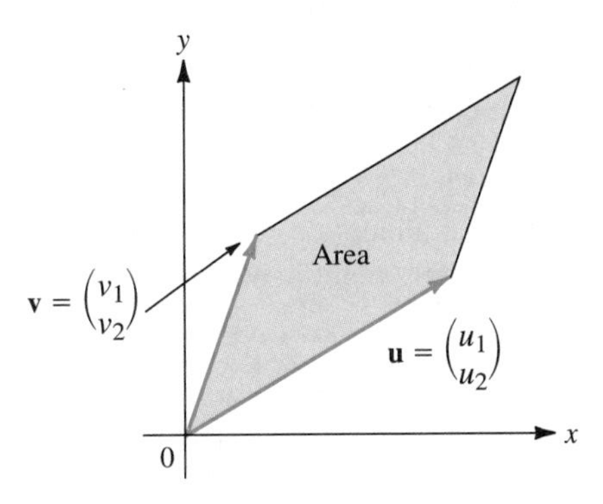


**Figure 3.31** The area of the shaded region is the area generated by **u** and **v**

The **area generated** by **u** and **v** is defined to be the area of the parallelogram given in the figure. We can think of **u** and **v** as vectors in $\mathbb{R}^3$ lying in the $xy$-plane. Then

$$\mathbf{u} = \begin{pmatrix} u_1 \\ u_2 \\ 0 \end{pmatrix}, \quad \mathbf{v} = \begin{pmatrix} v_1 \\ v_2 \\ 0 \end{pmatrix}, \quad \text{and}$$

$$\text{Area generated by } \mathbf{u} \text{ and } \mathbf{v} = |\mathbf{u} \times \mathbf{v}| = \begin{vmatrix} \mathbf{i} & \mathbf{j} & \mathbf{k} \\ u_1 & u_2 & 0 \\ v_1 & v_2 & 0 \end{vmatrix}$$

$$= |(u_1v_2 - u_2v_1)\mathbf{k}| = |u_1v_2 - u_2v_1|\dagger$$

Now let $A = \begin{pmatrix} a_{11} & a_{12} \\ a_{21} & a_{22} \end{pmatrix}$, $\mathbf{u}' = A\mathbf{u}$, and $\mathbf{v}' = A\mathbf{v}$. Then $\mathbf{u}' = \begin{pmatrix} a_{11}u_1 + a_{12}u_2 \\ a_{21}u_1 + a_{22}u_2 \end{pmatrix}$ and $\mathbf{v}' = \begin{pmatrix} a_{11}v_1 + a_{12}v_2 \\ a_{21}v_1 + a_{22}v_2 \end{pmatrix}$. What is the area generated by $\mathbf{u}'$ and $\mathbf{v}'$? Following the preceding steps, we calculate

Area generated by $\mathbf{u}'$ and $\mathbf{v}'$ =

$$|\mathbf{u}' \times \mathbf{v}'| = \left\| \begin{matrix} \mathbf{i} & \mathbf{j} & \mathbf{k} \\ a_{11}u_1 + a_{12}u_2 & a_{21}u_1 + a_{22}u_2 & 0 \\ a_{11}v_1 + a_{12}v_2 & a_{21}v_1 + a_{22}v_2 & 0 \end{matrix} \right\|$$

$$= |(a_{11}u_1 + a_{12}u_2)(a_{21}v_1 + a_{22}v_2) - (a_{21}u_1 + a_{22}u_2)(a_{11}v_1 + a_{12}v_2)|$$

It is simple algebra to verify that the last expression is equal to

$$|(a_{11}a_{22} - a_{12}a_{21})(u_1v_2 - u_2v_1)| = \pm\det A \text{ (area generated by } \mathbf{u} \text{ and } \mathbf{v})$$

Thus (in this context): *The determinant has the effect of multiplying area.* In Problem 39 you are asked to show that in a certain sense a $3 \times 3$ determinant has the effect of multiplying volume.

## Geometric Interpretation of the Scalar Triple Product

Let **u**, **v**, and **w** be the three vectors that are not in the same plane. Then they form the sides of a **parallelepiped** in space (see Figure 3.32). Let us compute its volume. The base of the parallelepiped is a parallelogram. Its area, from (3), is equal to $|\mathbf{u} \times \mathbf{v}|$.

The vector $\mathbf{u} \times \mathbf{v}$ is orthogonal to both **u** and **v**, and is therefore orthogonal to the parallelogram determined by **u** and **v**. The height of the parallelepiped, $h$, is measured along a vector orthogonal to the parallelogram.

---

† Note that this is the absolute value of $\det \begin{pmatrix} u_1 & v_1 \\ u_2 & v_2 \end{pmatrix}$.

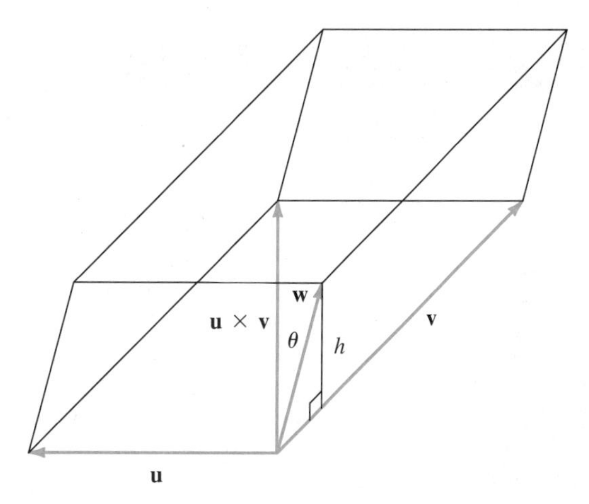


**Figure 3.32** Three vectors **u**, **v**, and **w** that are not in the same plane determine a parallelepiped in $\mathbb{R}^3$

From our discussion of projections on page 245, we see that $h$ is the absolute value of the component of **w** in the (orthogonal) direction **u** × **v**. Thus from equation (10) on page 258

$$h = \text{component of } \mathbf{w} \text{ in the direction } \mathbf{u} \times \mathbf{v} = \left| \frac{\mathbf{w} \cdot (\mathbf{u} \times \mathbf{v})}{|\mathbf{u} \times \mathbf{v}|} \right|$$

Thus

$$\begin{aligned} \text{Volume of parallelepiped} &= \text{area of base} \times \text{height} \\ &= |\mathbf{u} \times \mathbf{v}| \left[ \frac{|\mathbf{w} \cdot (\mathbf{u} \times \mathbf{v})|}{|\mathbf{u} \times \mathbf{v}|} \right] = |\mathbf{w} \cdot (\mathbf{u} \times \mathbf{v})| \end{aligned}$$

That is,

> The volume of the parallelepiped determined by the three vectors **u**, **v**, and **w** is equal to $|(\mathbf{u} \times \mathbf{v}) \cdot \mathbf{w}|$ = absolute value of the scalar triple product of **u**, **v**, and **w**. **(4)**

Focus on . . .

## Josiah Willard Gibbs and the Origins of Vector Analysis

Josiah Willard Gibbs *(The Granger Collection, New York)*

As we have already noted, the study of vectors originated with Hamilton's invention of quaternions. Quaternions were developed by Hamilton and others as mathematical tools for the exploration of physical space. But the results were disappointing because quaternions proved to be too complicated for quick mastery and easy application. Fortunately, there was a solution. Quaternions contained a scalar part and a vector part and difficulties arose when these parts were treated simultaneously. Scientists soon learned that many problems could be dealt with by considering the vector part separately, and the study of vector analysis began.

This work was due principally to the American physicist Josiah Willard Gibbs (1839–1903). As a native of New Haven, Connecticut, Gibbs studied mathematics and physics at Yale University, receiving a doctorate in physics in 1863. He then studied mathematics and physics further in Paris, Berlin, and Heidelberg. In 1871 he was appointed professor of mathematical physics at Yale. He was a highly original physicist who published widely in mathematical physics. Gibbs' book *Vector Analysis* appeared in 1881 and again in 1884. In 1902 he published his *Elementary Principles of Statistical Mechanics.* Students of applied mathematics encounter the curious **Gibbs' phenomenon** of Fourier series.

Gibbs' pioneering book *Vector Analysis* was actually a small pamphlet, printed for private distribution—primarily for the use of his students. Nevertheless, it created great excitement among those who were looking for an alternative to quaternions, and the book soon became widely known. The material was finally turned into a standard book by E. B. Wilson. The book, *Vector Analysis* by Gibbs and Wilson was based on Gibbs' lectures. It was published in 1901.

Gibbs' work is encountered by every student of elementary physics. In introductory physics a vector in space is regarded as a directed line segment, or arrow. Gibbs gave definitions of equality, addition, and multiplication of vectors; these are essentially the definitions given in this chapter. In particular, the vector part of a quaternion was written as $a\mathbf{i} + b\mathbf{j} + c\mathbf{k}$, and this is one way we now depict vectors in $\mathbb{R}^3$.

Gibbs defined the scalar product initially only for the vectors **i**, **j**, **k**:

$$\mathbf{i}\cdot\mathbf{i} = \mathbf{j}\cdot\mathbf{j} = \mathbf{k}\cdot\mathbf{k} = 1$$
$$\mathbf{i}\cdot\mathbf{j} = \mathbf{j}\cdot\mathbf{i} = \mathbf{i}\cdot\mathbf{k} = \mathbf{k}\cdot\mathbf{i} = \mathbf{j}\cdot\mathbf{k} = \mathbf{k}\cdot\mathbf{j} = 0$$

The more general definition followed soon thereafter. Gibbs applied the scalar product in a problem involving force. (Remember, he was first a physicist.) If **F** is a force vector of magnitude $|F|$ acting in the direction of the segment $\overrightarrow{0Q}$ (see Figure 3.33), then the effectiveness of this force in pushing an object along the

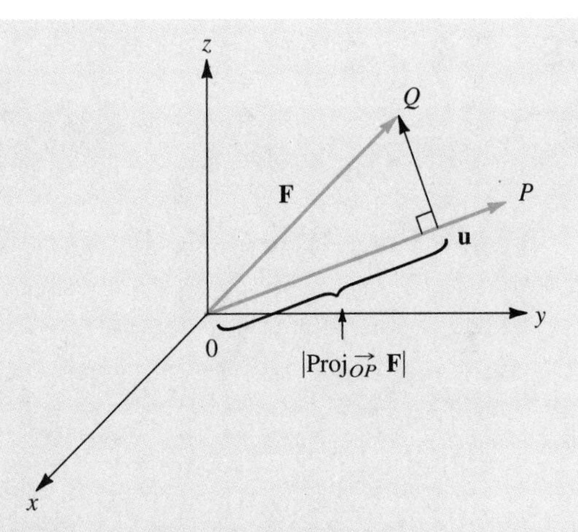


**Figure 3.33** The effectiveness of **F** in the direction of $\overrightarrow{0P}$ is the component of **F** in the direction $\overrightarrow{0P}(= \mathbf{u})$ if $\mathbf{u} = 1$

segment $\overrightarrow{0P}$ (i.e., along the vector **u**) is given by $\mathbf{F} \cdot \mathbf{u}$. If $|\mathbf{u}| = 1$, then $\mathbf{F} \cdot \mathbf{u}$ is the component of **F** in the direction **u**. The cross product, too, has physical significance. Suppose that a force vector **F** acts at a point $P$ in space in the direction $\overrightarrow{PQ}$. If **u** denotes the vector represented by $\overrightarrow{0P}$, then the moment of force exerted by **F** around the origin is the vector $\mathbf{u} \times \mathbf{F}$ (see Figure 3.34).

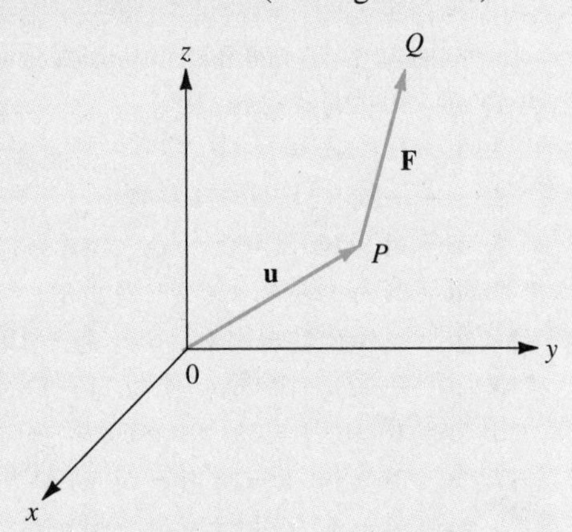


**Figure 3.34** The vector $\mathbf{u} \times \mathbf{F}$ is the moment of force around the origin

Both the scalar and cross products of vectors appear prominently in physical applications involving multivariable calculus. These include the famous Maxwell equations of electromagnetism.

In studying mathematics at the end of the twentieth century, we must not lose sight of the fact that much of modern mathematics was developed to solve real-world problems. Vectors were developed by Gibbs and others to make it easier to analyze physical phenomena. In that role they have been hugely successful.

# PROBLEMS 3.4

## Self-Quiz

**I.** $\mathbf{i} \times \mathbf{k} - \mathbf{k} \times \mathbf{i} =$ ______.

a. $\mathbf{0}$ b. $\mathbf{j}$ c. $2\mathbf{j}$ d. $-2\mathbf{j}$

**II.** $\mathbf{i} \cdot (\mathbf{j} \times \mathbf{k}) =$ ______.

a. 0 b. $\mathbf{0}$ c. 1 d. $\mathbf{i} - \mathbf{j} + \mathbf{k}$

**III.** $\mathbf{i} \times \mathbf{j} \times \mathbf{k}$ ______.

a. $= 0$ b. $= \mathbf{0}$ c. $= 1$ d. is undefined

**IV.** $(\mathbf{i} + \mathbf{j}) \times (\mathbf{j} + \mathbf{k}) =$ ______.

a. 0 b. $\mathbf{0}$ c. 1 d. $\mathbf{i} - \mathbf{j} + \mathbf{k}$

**V.** The sine of the angle between vectors **u** and **w** is ______.

a. $\dfrac{|\mathbf{u} \times \mathbf{w}|}{|\mathbf{u}||\mathbf{w}|}$ b. $\dfrac{|\mathbf{u} \times \mathbf{w}|}{|\mathbf{u} \cdot \mathbf{w}|}$

c. $\dfrac{|\mathbf{u} \cdot \mathbf{w}|}{|\mathbf{u} \times \mathbf{w}|}$ d. $|\mathbf{u} \times \mathbf{w}| - |\mathbf{u} \cdot \mathbf{w}|$

**VI.** $\mathbf{u} \times \mathbf{u} =$ ______.

a. $|\mathbf{u}|^2$ b. 1 c. $\mathbf{0}$ d. 0

In Problems 1–20 find the cross product $\mathbf{u} \times \mathbf{v}$.

**1.** $\mathbf{u} = \mathbf{i} - 2\mathbf{j}$; $\mathbf{v} = 3\mathbf{k}$

**2.** $\mathbf{u} = 3\mathbf{i} - 7\mathbf{j}$; $\mathbf{v} = \mathbf{i} + \mathbf{k}$

**3.** $\mathbf{u} = \mathbf{i} - \mathbf{j}$; $\mathbf{v} = \mathbf{j} + \mathbf{k}$

**4.** $\mathbf{u} = -7\mathbf{k}$; $\mathbf{v} = \mathbf{j} + 2\mathbf{k}$

**5.** $\mathbf{u} = -2\mathbf{i} + 3\mathbf{j}$; $\mathbf{v} = 7\mathbf{i} + 4\mathbf{k}$

**6.** $\mathbf{u} = a\mathbf{i} + b\mathbf{j}$; $\mathbf{v} = c\mathbf{i} + d\mathbf{j}$

**7.** $\mathbf{u} = a\mathbf{i} + b\mathbf{k}$; $\mathbf{v} = c\mathbf{i} + d\mathbf{k}$

**8.** $\mathbf{u} = a\mathbf{j} + b\mathbf{k}$; $\mathbf{v} = c\mathbf{i} + d\mathbf{k}$

**9.** $\mathbf{u} = 2\mathbf{i} - 3\mathbf{j} + \mathbf{k}$; $\mathbf{v} = \mathbf{i} + 2\mathbf{j} + \mathbf{k}$

**10.** $\mathbf{u} = 3\mathbf{i} - 4\mathbf{j} + 2\mathbf{k}$; $\mathbf{v} = 6\mathbf{i} - 3\mathbf{j} + 5\mathbf{k}$

**11.** $\mathbf{u} = -3\mathbf{i} - 2\mathbf{j} + \mathbf{k}$; $\mathbf{v} = 6\mathbf{i} + 4\mathbf{j} - 2\mathbf{k}$

**12.** $\mathbf{u} = \mathbf{i} + 7\mathbf{j} - 3\mathbf{k}$; $\mathbf{v} = -\mathbf{i} - 7\mathbf{j} + 3\mathbf{k}$

**13.** $\mathbf{u} = \mathbf{i} - 7\mathbf{j} - 3\mathbf{k}$; $\mathbf{v} = -\mathbf{i} + 7\mathbf{j} - 3\mathbf{k}$

**14.** $\mathbf{u} = 2\mathbf{i} - 3\mathbf{j} + 5\mathbf{k}$; $\mathbf{v} = 3\mathbf{i} - \mathbf{j} - \mathbf{k}$

**15.** $\mathbf{u} = 10\mathbf{i} + 7\mathbf{j} - 3\mathbf{k}$; $\mathbf{v} = -3\mathbf{i} + 4\mathbf{j} - 3\mathbf{k}$

**16.** $\mathbf{u} = 2\mathbf{i} + 4\mathbf{j} - 6\mathbf{k}$; $\mathbf{v} = -\mathbf{i} - \mathbf{j} + 3\mathbf{k}$

**17.** $\mathbf{u} = 2\mathbf{i} - \mathbf{j} + \mathbf{k}$; $\mathbf{v} = 4\mathbf{i} + 2\mathbf{j} + 2\mathbf{k}$

**18.** $\mathbf{u} = 3\mathbf{i} - \mathbf{j} + 8\mathbf{k}$; $\mathbf{v} = \mathbf{i} + \mathbf{j} - 4\mathbf{k}$

**19.** $\mathbf{u} = a\mathbf{i} + a\mathbf{j} + a\mathbf{k}$; $\mathbf{v} = b\mathbf{i} + b\mathbf{j} + b\mathbf{k}$

**20.** $\mathbf{u} = a\mathbf{i} + b\mathbf{j} + c\mathbf{k}$; $\mathbf{v} = a\mathbf{i} + b\mathbf{j} - c\mathbf{k}$

**21.** Find two unit vectors orthogonal to both $\mathbf{u} = 2\mathbf{i} - 3\mathbf{j}$ and $\mathbf{v} = 4\mathbf{j} + 3\mathbf{k}$.

**22.** Find two unit vectors orthogonal to both $\mathbf{u} = \mathbf{i} + \mathbf{j} + \mathbf{k}$ and $\mathbf{v} = \mathbf{i} - \mathbf{j} - \mathbf{k}$.

**23.** Use the cross product to find the sine of the angle $\varphi$ between the vectors $\mathbf{u} = 2\mathbf{i} + \mathbf{j} - \mathbf{k}$ and $\mathbf{v} = -3\mathbf{i} - 2\mathbf{j} + 4\mathbf{k}$.

## Answers to Self-Quiz

**I.** d **II.** c **III.** b = zero vector [*Note:* $\mathbf{i} \times \mathbf{j} \times \mathbf{k}$ *is* defined because $(\mathbf{i} \times \mathbf{j}) \times \mathbf{k} = \mathbf{0} = \mathbf{i} \times (\mathbf{j} \times \mathbf{k})$.] **IV.** d **V.** a **VI.** c = zero vector

**24.** Use the scalar product to calculate the cosine of the angle $\varphi$ between the vectors of Problem 23. Then show that for the values you have calculated, $\sin^2 \varphi + \cos^2 \varphi = 1$.

In Problems 25–30 find the area of the parallelogram with the given adjacent vertices.

**25.** $(1, -2, 3)$; $(2, 0, 1)$; $(0, 4, 0)$

**26.** $(-2, 1, 1)$; $(2, 2, 3)$; $(-1, -2, 4)$

**27.** $(-2, 1, 0)$; $(1, 4, 2)$; $(-3, 1, 5)$

**28.** $(7, -2, -3)$; $(-4, 1, 6)$; $(5, -2, 3)$

**29.** $(a, 0, 0)$; $(0, b, 0)$; $(0, 0, c)$

**30.** $(a, b, 0)$; $(a, 0, b)$; $(0, a, b)$

**31.** Show that $|\mathbf{u} \times \mathbf{v}|^2 = |\mathbf{u}|^2|\mathbf{v}|^2 - (\mathbf{u} \cdot \mathbf{v})^2$. [*Hint:* Write out in terms of components.]

**32.** Use Properties 1, 4, 2, and 3 (in that order) in Section 2.2 to prove parts (*i*), (*ii*), (*iii*), and (*iv*) of Theorem 2.

**33.** Prove Theorem 2($v$) by writing out the components of each side of the equality.

**34.** Prove Theorem 2($vi$). [*Hint:* Use parts ($ii$) and ($v$) and the fact that the scalar product is commutative to show that $\mathbf{u} \cdot (\mathbf{u} \times \mathbf{v}) = -\mathbf{u} \cdot (\mathbf{u} \times \mathbf{v})$.]

**35.** Prove Theorem 2($vii$). [*Hint:* Use Theorem 3.3.3 on page 257, Property 6 on page 196, and equation (2).]

**36.** Show that if $\mathbf{u} = (a_1, b_1, c_1)$, $\mathbf{v} = (a_2, b_2, c_2)$, and $\mathbf{w} = (a_3, b_3, c_3)$, then

$$\mathbf{u} \cdot (\mathbf{v} \times \mathbf{w}) = \begin{vmatrix} a_1 & b_1 & c_1 \\ a_2 & b_2 & c_2 \\ a_3 & b_3 & c_3 \end{vmatrix}$$

**37.** Calculate the volume of the parallelepiped determined by the vectors $\mathbf{i} - \mathbf{j}$, $3\mathbf{i} + 2\mathbf{k}$, $-7\mathbf{j} + 3\mathbf{k}$.

**38.** Calculate the volume of the parallelepiped determined by the vectors $\overrightarrow{PQ}$, $\overrightarrow{PR}$, and $\overrightarrow{PS}$, where $P = (2, 1, -1)$, $Q = (-3, 1, 4)$, $R = (-1, 0, 2)$, and $S = (-3, -1, 5)$.

****39.** The **volume generated** by three vectors $\mathbf{u}$, $\mathbf{v}$, and $\mathbf{w}$ in $\mathbb{R}^3$ is defined to be the volume of the parallelepiped whose sides are $\mathbf{u}$, $\mathbf{v}$, and $\mathbf{w}$ (as in Figure 3.32). Let $A$ be a $3 \times 3$ matrix and let $\mathbf{u}_1 = A\mathbf{u}$, $\mathbf{v}_1 = A\mathbf{v}$, and $\mathbf{w}_1 = A\mathbf{w}$. Show that

$$\text{Volume generated by } \mathbf{u}_1, \mathbf{v}_1, \mathbf{w}_1 = (\pm\det A)(\text{volume generated by } \mathbf{u}, \mathbf{v}, \mathbf{w})$$

This shows that just as the determinant of a $2 \times 2$ matrix multiplies area, the determinant of a $3 \times 3$ matrix multiplies volume.

**40.** Let $A = \begin{pmatrix} 2 & 3 & 1 \\ 4 & -1 & 5 \\ 1 & 0 & 6 \end{pmatrix}$, $\mathbf{u} = \begin{pmatrix} 2 \\ -1 \\ 0 \end{pmatrix}$, $\mathbf{v} = \begin{pmatrix} 1 \\ 0 \\ 4 \end{pmatrix}$, and $\mathbf{w} = \begin{pmatrix} -1 \\ 3 \\ 2 \end{pmatrix}$.

**a.** Calculate the volume generated by $\mathbf{u}$, $\mathbf{v}$, and $\mathbf{w}$.

**b.** Calculate the volume generated by $A\mathbf{u}$, $A\mathbf{v}$, and $A\mathbf{w}$.

**c.** Calculate $\det A$.

**d.** Show that [volume in part ($b$)] $= (\pm\det A) \times$ [volume in part ($a$)].

**41.** The **triple cross product** of three vectors in $\mathbb{R}^3$ is defined to be the vector $\mathbf{u} \times (\mathbf{v} \times \mathbf{w})$. Show that

$$\mathbf{u} \times (\mathbf{v} \times \mathbf{w}) = (\mathbf{u} \cdot \mathbf{w})\mathbf{v} - (\mathbf{u} \cdot \mathbf{v})\mathbf{w}$$

**CALCULATOR BOX**

The cross product of two vectors can be found directly on some calculators.

**TI-85**

Enter two vectors and give them names, say, A and B. Press [2nd] [VECTR] [F3]

⟨MATH⟩ [F1] ⟨cross⟩ and "cross (" will be displayed. Then press

[ALPHA] [A] [,] [ALPHA] [B] [)] [ENTER]

and A × B will be displayed.

In Problems 42–45 compute $\mathbf{u} \times \mathbf{v}$ on a calculator.

**42.** $\mathbf{u} = (-15, 27, 83)$; $\mathbf{v} = (-84, -77, 51)$

**43.** $\mathbf{u} = (-0.346, -0.517, -0.824)$; $\mathbf{v} = (-0.517, 0.811, 0.723)$

**44.** $\mathbf{u} = (5241, -3199, 2386)$; $\mathbf{v} = (1742, 8233, 9416)$

**45.** $\mathbf{u} = (0.024, 0.036, 0.055)$; $\mathbf{v} = (0.088, -0.064, 0.037)$

## MATLAB 3.4

1. Using MATLAB, compute the cross products of the vectors for Problems 1, 2, 3, 4, and 10 in this section. Verify your answers by computing the scalar products of the results with the individual vectors. (What should these scalar products be?) The cross product $\mathbf{u} \times \mathbf{v}$ is defined to be the $3 \times 1$ vector given by **[u(2)*v(3)−u(3)*v(2);−u(1)*v(3)+u(3)*v(1);u(1)*v(2)−v(1)*u(2)].**

2. a. Choose **u**, **v**, and **w** to be random $3 \times 1$ vectors (use **2*rand(3,1)−1**). Compute $\mathbf{u} \cdot (\mathbf{v} \times \mathbf{w})$, the scalar product of **u** with $\mathbf{v} \times \mathbf{w}$. Let $B = [\mathbf{u}\ \mathbf{v}\ \mathbf{w}]$. Find $\det(B)$. Compare $\det(B)$ with the scalar product. Do this for several choices of **u**, **v**, and **w**. Formulate a conjecture and then prove it (paper and pencil).

   b. Choose **u**, **v**, and **w** to be random $3 \times 1$ vectors and choose $A$ to be a random $3 \times 3$ matrix. Let **A = round(10*(2*rand(3)−1))**. Compute $|\mathbf{u} \cdot (\mathbf{v} \times \mathbf{w})|$, $|A\mathbf{u} \cdot (A\mathbf{v} \times A\mathbf{w})|$, and $|\det(A)|$. (In MATLAB, **abs(a)** gives $|\mathbf{a}|$.) Do this for several choices of $A$ until you can formulate a conjecture relating the three quantities computed. Test your conjecture for further random choices of $A$.

   Based on your conjecture, what geometrical significance does $|\det(A)|$ have?

   c. *(Paper and pencil)* Using (a), show that $A\mathbf{u} \cdot (A\mathbf{v} \times A\mathbf{w}) = \det([A\mathbf{u}\ A\mathbf{v}\ A\mathbf{w}])$, where $A$ is a $3 \times 3$ matrix. Argue why $[A\mathbf{u}\ A\mathbf{v}\ A\mathbf{w}] = AB$, where $B = [\mathbf{u}\ \mathbf{v}\ \mathbf{w}]$. Now prove the conjecture you made in part (b).

## 3.5 LINES AND PLANES IN SPACE

In the plane $\mathbb{R}^2$ we can find the equation of a line if we know either two points on the line or one point and the slope of the line. In $\mathbb{R}^3$ our intuition tells us that the basic ideas are the same. Since two points determine a line, we should be able to calculate the equation of a line in space if we know two points on it. Alternatively, if we know one point and the direction of a line, we should also be able to find its equation.

We begin with two points $P = (x_1, y_1, z_1)$ and $Q = (x_2, y_2, z_2)$ on a line $L$. One vector parallel to $L$ is the vector with representation $\overrightarrow{PQ}$. Thus

$$\mathbf{v} = (x_2 - x_1)\mathbf{i} + (y_2 - y_1)\mathbf{j} + (z_2 - z_1)\mathbf{k} \tag{1}$$

is a vector parallel to $L$. Now let $R = (x, y, z)$ be another point on the line. Then $\overrightarrow{PR}$ is parallel to $\overrightarrow{PQ}$, which is parallel to $\mathbf{v}$ so that, by Theorem 3.3.3 on page 257,

$$\overrightarrow{PR} = t\mathbf{v} \tag{2}$$

for some real number $t$. Now look at Figure 3.35. From this figure we have (in each of the three possible cases)

$$\overrightarrow{0R} = \overrightarrow{0P} + \overrightarrow{PR} \tag{3}$$

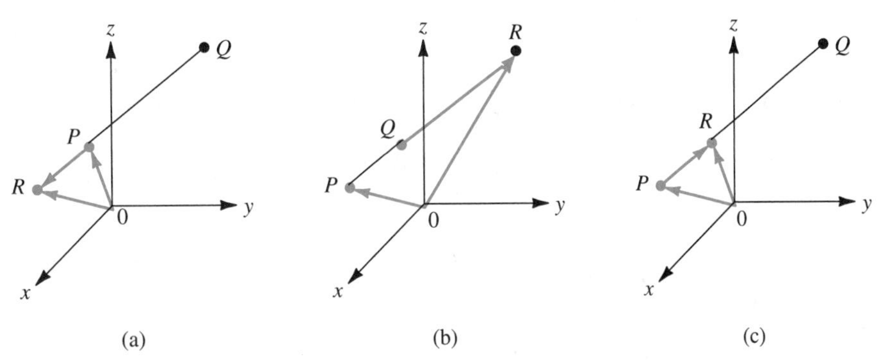


**Figure 3.35** In all three cases $\overrightarrow{0R} = \overrightarrow{0P} + \overrightarrow{PR}$

And, by combining (2) and (3), we get

$$\boxed{\overrightarrow{0R} = \overrightarrow{0P} + t\mathbf{v}} \tag{4}$$

Vector equation of a line

Equation (4) is called the **vector equation** of the line $L$. For if $R$ is on $L$, then (4) is satisfied for some real number $t$. Conversely, if (4) is satisfied, then, reversing our steps, we see that $\overrightarrow{PR}$ is parallel to $\mathbf{v}$, which means that $R$ is on $L$.

If we write out the components of equation (4), we obtain

$$x\mathbf{i} + y\mathbf{j} + z\mathbf{k} = x_1\mathbf{i} + y_1\mathbf{j} + z_1\mathbf{k} + t(x_2 - x_1)\mathbf{i} + t(y_2 - y_1)\mathbf{j} + t(z_2 - z_1)\mathbf{k}$$

or

$$\begin{aligned} x &= x_1 + t(x_2 - x_1) \\ y &= y_1 + t(y_2 - y_1) \\ z &= z_1 + t(z_2 - z_1) \end{aligned} \tag{5}$$

Parametric equations of a line

Equations (5) are called **parametric equations** of a line.

Finally, solving for $t$ in (5), and defining $x_2 - x_1 = a$, $y_2 - y_1 = b$, and $z_2 - z_1 = c$, we find that if $abc \neq 0$,

$$\frac{x - x_1}{a} = \frac{y - y_1}{b} = \frac{z - z_1}{c} \tag{6}$$

Symmetric equations of a line

Equations (6) are called **symmetric equations** of the line. Here $a$, $b$, and $c$ are direction numbers of the vector $\mathbf{v}$. Of course, equations (6) are valid only if $a$, $b$, and $c$ are nonzero.

**EXAMPLE 1** **Determining Equations of a Line** Find a vector equation, parametric equations, and symmetric equations of the line $L$ passing through the points $P = (2, -1, 6)$ and $Q = (3, 1, -2)$.

**Solution** First, we calculate $\mathbf{v} = (3 - 2)\mathbf{i} + [1 - (-1)]\mathbf{j} + (-2 - 6)\mathbf{k} = \mathbf{i} + 2\mathbf{j} - 8\mathbf{k}$. Then, from (4), if $R = (x, y, z)$ is on the line, we obtain $\overrightarrow{0R} = x\mathbf{i} + y\mathbf{j} + z\mathbf{k} = \overrightarrow{0P} + t\mathbf{v} = 2\mathbf{i} - \mathbf{j} + 6\mathbf{k} + t(\mathbf{i} + 2\mathbf{j} - 8\mathbf{k})$ or

$$x = 2 + t \qquad y = -1 + 2t \qquad z = 6 - 8t \qquad \text{parametric equations}$$

Finally, since $a = 1$, $b = 2$, and $c = -8$, we find the symmetric equations

$$\frac{x - 2}{1} = \frac{y + 1}{2} = \frac{z - 6}{-8} \qquad \text{symmetric equations} \tag{7}$$

To check these equations, we verify that $(2, -1, 6)$ and $(3, 1, -2)$ are indeed on the line. We have [after plugging these points into (7)]

$$\frac{2 - 2}{1} = \frac{-1 + 1}{2} = \frac{6 - 6}{-8} = 0$$

$$\frac{3 - 2}{1} = \frac{1 + 1}{2} = \frac{-2 - 6}{-8} = 1$$

Other points on the line can be found. If $t = 3$, for example, we obtain

$$3 = \frac{x - 2}{1} = \frac{y + 1}{2} = \frac{z - 6}{-8}$$

which yield the point $(5, 5, -18)$.

**EXAMPLE 2** **Finding Symmetric Equations of a Line** Find symmetric equations of the line passing through the point $(1, -2, 4)$ and parallel to the vector $\mathbf{v} = \mathbf{i} + \mathbf{j} - \mathbf{k}$.

**Solution** We use formula (6) with $P = (x_1, y_1, z_1) = (1, -2, 4)$ and $\mathbf{v}$ as above so that $a = 1$, $b = 1$, and $c = -1$. This gives us

$$\frac{x-1}{1} = \frac{y+2}{1} = \frac{z-4}{-1}$$

What happens if one of the direction numbers $a$, $b$, or $c$ is zero?

**EXAMPLE 3** **Finding Symmetric Equations of a Line When One Direction Number Is Zero** Find symmetric equations of the line containing the points $P = (3, 4, -1)$ and $Q = (-2, 4, 6)$.

**Solution** Here $\mathbf{v} = -5\mathbf{i} + 7\mathbf{k}$ and $a = -5$, $b = 0$, $c = 7$. Then a parametric representation of the line is $x = 3 - 5t$, $y = 4$, and $z = -1 + 7t$. Solving for $t$, we find that

$$\frac{x-3}{-5} = \frac{z+1}{7} \quad \text{and} \quad y = 4$$

The equation $y = 4$ is the equation of a plane parallel to the $xz$-plane, so we have obtained an equation of a line in that plane.

**EXAMPLE 4** **Finding Symmetric Equations of a Line When Two Direction Numbers Are Zero** Find symmetric equations of the line passing through the points $P = (2, 3, -2)$ and $Q = (2, -1, -2)$.

**Solution** Here $\mathbf{v} = -4\mathbf{j}$ so that $a = 0$, $b = -4$, and $c = 0$. A parametric representation of the line is, by equation (5), given by $x = 2$, $y = 3 - 4t$, $z = -2$. Now $x = 2$ is the equation of a plane parallel to the $yz$-plane, whereas $z = -2$ is the equation of a plane parallel to the $xy$-plane. Their intersection is the line $x = 2$, $z = -2$, which is parallel to the $y$-axis and passes through the point $(2, 0, -2)$. In fact, the equation $y = 3 - 4t$ says, essentially, that $y$ can take on any value (while $x$ and $z$ remain fixed).

**WARNING** **The parametric or symmetric equations of a line are *not* unique. To see this, simply start with two other points on the line.**

**EXAMPLE 5** **Illustration That Symmetric Equations of a Line Are Not Unique** In Example 1 the line whose equations we found contains the point $(5, 5, -18)$. Choose $P = (5, 5, -18)$ and $Q = (3, 1, -2)$. We find that $\mathbf{v} = -2\mathbf{i} - 4\mathbf{j} + 16\mathbf{k}$ so that $x = 5 - 2t$, $y = 5 - 4t$, and $z = -18 + 16t$. (Note that if $t = \frac{3}{2}$, we obtain $(x, y, z) =$

(2, −1, 6).) The symmetric equations are now

$$\frac{x-5}{-2} = \frac{y-5}{-4} = \frac{z+18}{16}$$

Note that $(-2, -4, -16) = -2(1, 2, -8)$.

The equation of a line in space is obtained by specifying a point on the line and a vector *parallel* to this line. We can derive the equation of a plane in space by specifying a point in the plane and a vector orthogonal to every vector in the plane. This orthogonal vector is called a **normal vector** to the plane and is denoted by **n** (see Figure 3.36).

Normal vector

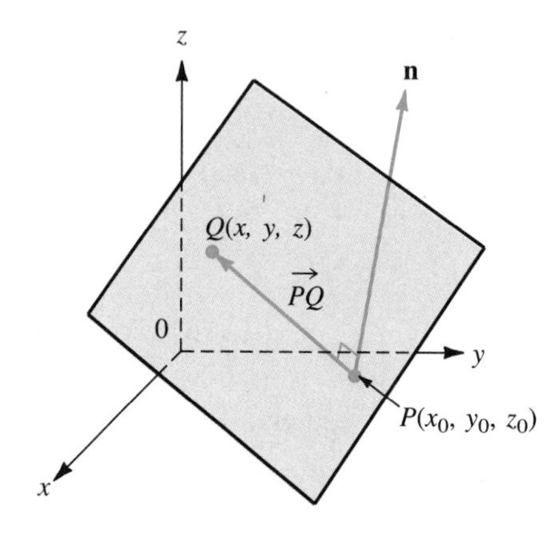


**Figure 3.36** The vector **n** is orthogonal to every vector in the plane

**DEFINITION 1**

**Plane** Let $P$ be a point in space and let **n** be a given nonzero vector. Then the set of all points $Q$ for which $\overrightarrow{PQ} \cdot \mathbf{n} = 0$ constitutes a **plane** in $\mathbb{R}^3$.

*Notation.* We usually denote a plane by the symbol $\pi$.

Let $P = (x_0, y_0, z_0)$ be a fixed point on a plane with normal vector $\mathbf{n} = a\mathbf{i} + b\mathbf{j} + c\mathbf{k}$. If $Q = (x, y, z)$ is any other point on the plane, then $\overrightarrow{PQ} = (x - x_0)\mathbf{i} + (y - y_0)\mathbf{j} + (z - z_0)\mathbf{k}$. Since $\overrightarrow{PQ} \perp \mathbf{n}$, we have $\overrightarrow{PQ} \cdot \mathbf{n} = 0$. But this implies that

$$a(x - x_0) + b(y - y_0) + c(z - z_0) = 0 \quad \textbf{(8)}$$

A more common way to write the equation of a plane is easily derived from (8):

**Standard Equation of a Plane**

$$ax + by + cz = d \quad \textbf{(9)}$$

where $d = ax_0 + by_0 + cz_0 = \overrightarrow{0P} \cdot \mathbf{n}$

**EXAMPLE 6** **Finding an Equation of the Plane Passing Through a Given Point with Given Normal Vector** Find the plane $\pi$ passing through the point (2, 5, 1) having the normal vector $\mathbf{n} = \mathbf{i} - 2\mathbf{j} + 3\mathbf{k}$.

**Solution** From (8) we immediately obtain $(x - 2) - 2(y - 5) + 3(z - 1) = 0$ or

$$x - 2y + 3z = -5 \tag{10}$$

The three coordinate planes are represented as follows:

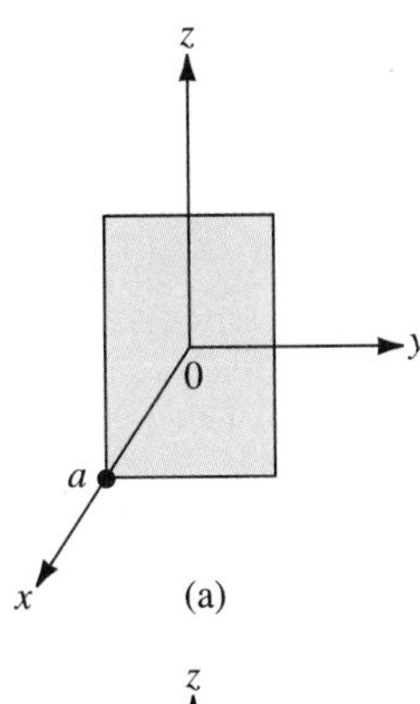


**i.** The *xy-plane* passes through the origin (0, 0, 0), and any vector lying along the $z$-axis is normal to it. The simplest such vector is $\mathbf{k}$. Thus, from (8), we obtain $0(x - 0) + 0(y - 0) + 1(z - 0) = 0$, which yields

$$z = 0 \tag{11}$$

as the equation of the $xy$-plane. (This result should not be very surprising.)

**ii.** The *xz-plane* has the equation

$$y = 0 \tag{12}$$

**iii.** The *yz-plane* has the equation

$$x = 0 \tag{13}$$

## How to Draw a Plane

It is not difficult to draw a plane.

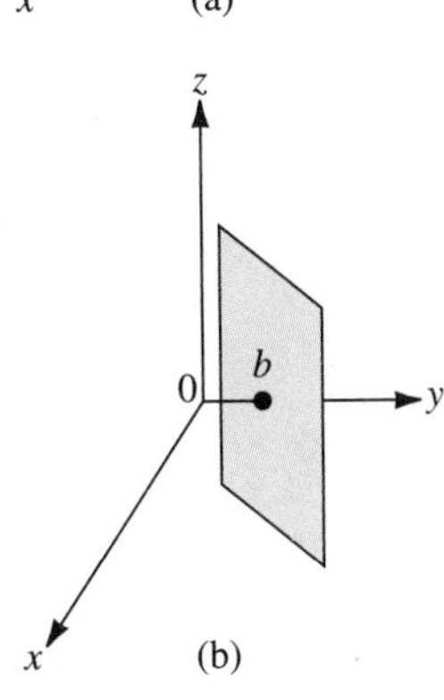


**Case 1.** *The plane is parallel to a coordinate plane.* If the plane is parallel to one of the coordinate planes, then the equation of the plane is one of the following:

$x = a$ (parallel to the $yz$-plane)
$y = b$ (parallel to the $xz$-plane)
$z = c$ (parallel to the $xy$-plane)

We draw each plane as a rectangle with sides parallel to the two other coordinate axes. We sketch three of these planes in Figure 3.37.

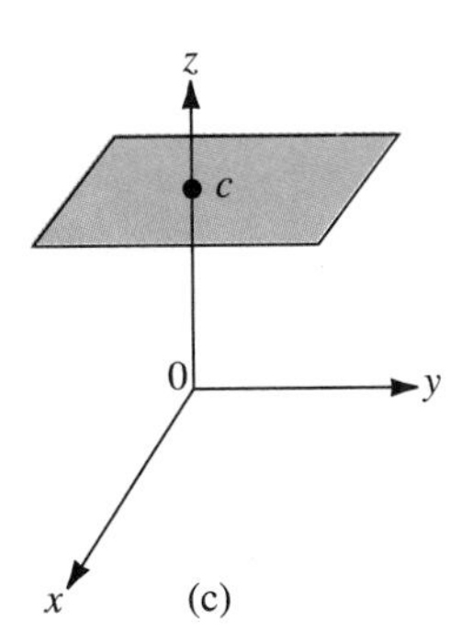


**Figure 3.37** Three planes parallel to a coordinate plane

**Case 2.** *The plane intersects each coordinate axis.* Suppose an equation of the plane is

$$ax + by + cz = d \qquad \text{with } abc \neq 0.$$

The $x$-intercept is the point $\left(\frac{d}{a}, 0, 0\right)$, the $y$-intercept is the point $\left(0, \frac{d}{b}, 0\right)$, and the $z$-intercept is the point $\left(0, 0, \frac{d}{c}\right)$.

**Step 1.** Plot the three intercepts.

**Step 2.** Join the three intercepts to form a triangle.

**Step 3.** By drawing two parallel lines, draw a parallelogram with one diagonal being the third side of the triangle.

**Step 4.** Expand the parallelogram by drawing four parallel lines.

We illustrate this process by drawing the plane $x + 2y + 3z = 6$ in Figure 3.38. The intercepts are (6, 0, 0), (0, 3, 0), and (0, 0, 2).

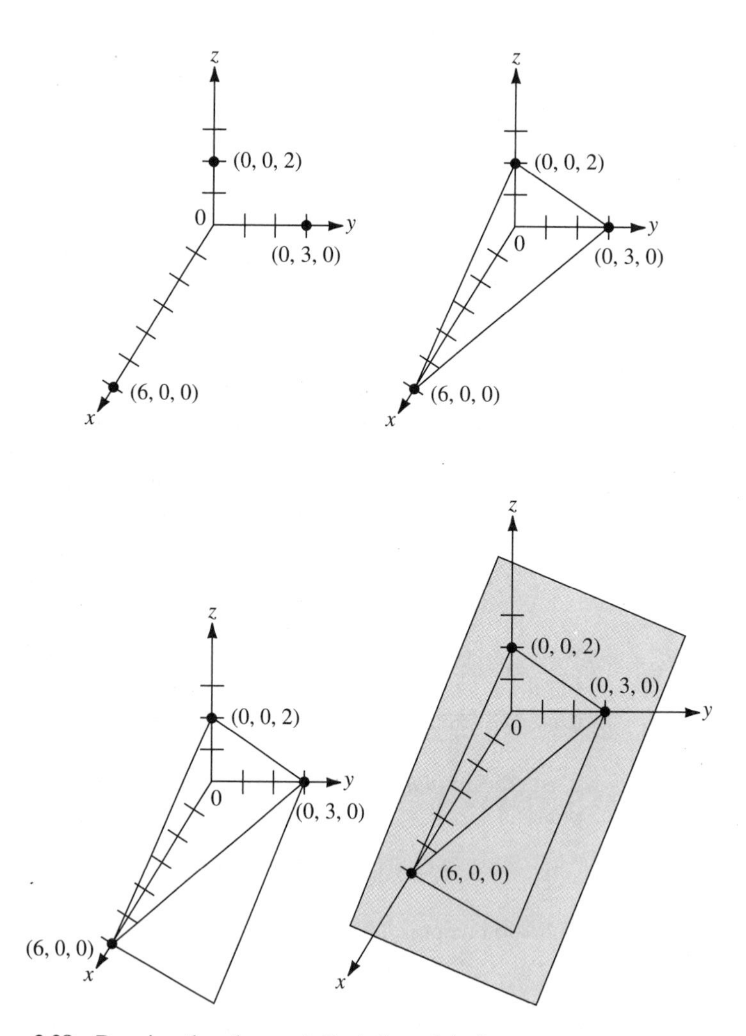


**Figure 3.38** Drawing the plane $x + 2y + 3z = 6$ in four steps

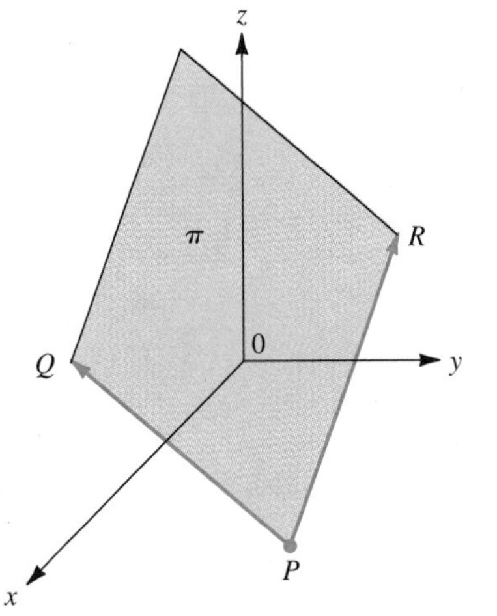

**Figure 3.39** The points $P$, $Q$, and $R$ determine a plane as long as they are not collinear

Three non-collinear points determine a plane since they determine two nonparallel vectors that intersect at a point (see Figure 3.39).

**EXAMPLE 7** **Finding the Equation of a Plane Passing Through Three Given Points** Find the equation of the plane passing through the points $P = (1, 2, 1)$, $Q = (-2, 3, -1)$, and $R = (1, 0, 4)$.

**Solution** The vectors $\overrightarrow{PQ} = -3\mathbf{i} + \mathbf{j} - 2\mathbf{k}$ and $\overrightarrow{QR} = 3\mathbf{i} - 3\mathbf{j} + 5\mathbf{k}$ lie on the plane and are therefore orthogonal to the normal vector so that

$$\mathbf{n} = \overrightarrow{PQ} \times \overrightarrow{QR} = \begin{vmatrix} \mathbf{i} & \mathbf{j} & \mathbf{k} \\ -3 & 1 & -2 \\ 3 & -3 & 5 \end{vmatrix} = -\mathbf{i} + 9\mathbf{j} + 6\mathbf{k}$$

and we obtain, using the point $P$ in equation (8),

$$\pi: \quad -(x-1) + 9(y-2) + 6(z-1) = 0$$

or

$$-x + 9y + 6z = 23$$

Note that if we choose another point, say, $Q$, we get the equation $-(x+2) + 9(y-3) + 6(z+1) = 0$, which reduces to $-x + 9y + 6z = 23$. This plane is sketched in Figure 3.40.

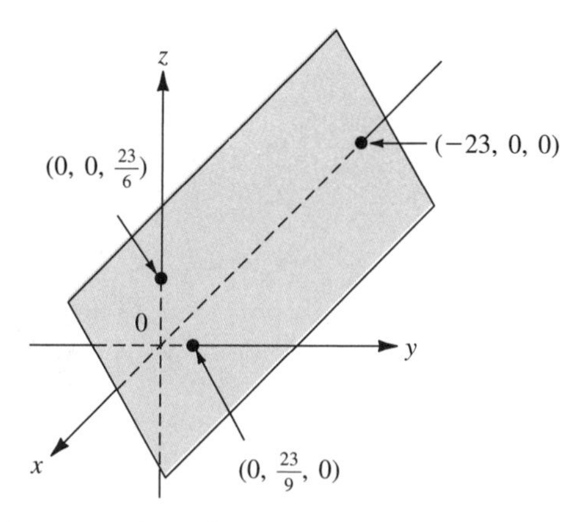


**Figure 3.40** The plane $-x + 9y + 6z = 23$

**DEFINITION 2** **Parallel Planes** Two planes are **parallel**† if their normal vectors are parallel; that is, if the cross product of their normal vectors is zero.

Two parallel planes are drawn in Figure 3.41.

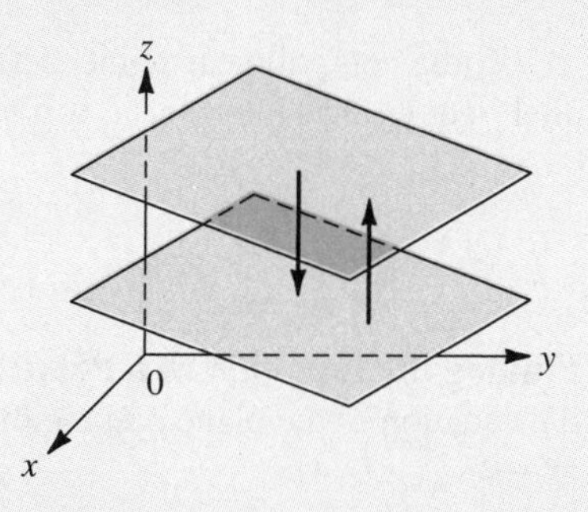


**Figure 3.41** Two parallel planes

**EXAMPLE 8** **Two Parallel Planes** The planes $\pi_1$: $2x + 3y - z = 3$ and $\pi_2$: $-4x - 6y + 2z = 8$ are parallel since $\mathbf{n}_1 = 2\mathbf{i} + 3\mathbf{j} - \mathbf{k}$, $\mathbf{n}_2 = -4\mathbf{i} - 6\mathbf{j} + 2\mathbf{k} = -2\mathbf{n}_1$ (and $\mathbf{n}_1 \times \mathbf{n}_2 = \mathbf{0}$).

If two planes are not parallel, then they intersect in a straight line.

**EXAMPLE 9** **Finding Points of Intersection of Planes** Find all points of intersection of the planes $2x - y - z = 3$ and $x + 2y + 3z = 7$.

†Note that two parallel planes could be coincident. For example, the planes $x + y + z = 1$ and $2x + 2y + 2z = 2$ are coincident (the same).

**Solution** The coordinates of any point $(x, y, z)$ on the line of intersection of these two planes must satisfy the equations $x + 2y + 3z = 7$ and $2x - y - z = 3$. Solving this system of two equations in three unknowns by row reduction, we obtain, successively,

$$\left(\begin{array}{ccc|c} 1 & 2 & 3 & 7 \\ 2 & -1 & -1 & 3 \end{array}\right) \xrightarrow{R_2 \to R_2 - 2R_1} \left(\begin{array}{ccc|c} 1 & 2 & 3 & 7 \\ 0 & -5 & -7 & -11 \end{array}\right)$$

$$\xrightarrow{R_2 \to -\frac{1}{5}R_2} \left(\begin{array}{ccc|c} 1 & 2 & 3 & 7 \\ 0 & 1 & \frac{7}{5} & \frac{11}{5} \end{array}\right) \xrightarrow{R_1 \to R_1 - 2R_2} \left(\begin{array}{ccc|c} 1 & 0 & \frac{1}{5} & \frac{13}{5} \\ 0 & 1 & \frac{7}{5} & \frac{11}{5} \end{array}\right)$$

Thus $y = \frac{11}{5} - (\frac{7}{5})z$ and $x = \frac{13}{5} - (\frac{1}{5})z$. Finally, setting $z = t$, we obtain a parametric representation of the line of intersection: $x = \frac{13}{5} - \frac{1}{5}t$, $y = \frac{11}{5} - \frac{7}{5}t$, and $z = t$.

We can derive an interesting fact from Theorem 2 *(vi)* on page 263.

If **w** is in the plane of **u** and **v**, then **w** is perpendicular to **u** × **v**, which means that **w** · (**u** × **v**) = 0. Conversely, if (**u** × **v**) · **w** = 0, then **w** is perpendicular to (**u** × **v**) so that **w** is in the plane determined by **u** and **v**. We conclude that

> Three vectors **u**, **v**, and **w** are coplanar if and only if their scalar triple product is zero.

# PROBLEMS 3.5

## Self-Quiz

**I.** The line through the points (1, 2, 4) and (5, 10, 15) satisfies the equation ________.

**a.** $(x, y, z) = (1, 2, 4) + t(4, 8, 11)$

**b.** $\dfrac{x-1}{4} = \dfrac{y-2}{8} = \dfrac{z-4}{11}$

**c.** $(x, y, z) = (5, 10, 15) + s(4, 8, 11)$

**d.** $\dfrac{x-5}{4} = \dfrac{y-10}{8} = \dfrac{z-15}{11}$

**II.** The line through the point $(7, 3, -4)$ and parallel to the vector **i** + 5**j** + 2**k** satisfies the equation ________.

**a.** $\dfrac{x-7}{1} = \dfrac{y-3}{5} = \dfrac{z+4}{2}$

**b.** $(x, y, z) = (1, 5, 2) + t(7, 3, -4)$

**c.** $\dfrac{x-7}{8} = \dfrac{y-3}{8} = \dfrac{z+4}{-2}$

**d.** $(x, y, z) = (7, 3, -4) + s(8, 8, -2)$

**III.** The vector equation $(x, y, z) - (3, 5, -7) = t(-1, 4, 8)$ describes ________.

**a.** the line through $(-1, 4, 8)$ and parallel to 3**i** + 5**j** − 7**k**

**b.** the line through $(-3, -5, 7)$ and parallel to −**i** + 4**j** + 8**k**

**c.** the line through $(3, 5, -7)$ and perpendicular to −**i** + 4**j** + 8**k**

**d.** the line through $(3, 5, -7)$ and parallel to −**i** + 4**j** + 8**k**

**IV.** The plane passing through (5, −4, 3) that is orthogonal to **j** satisfies ______.

**a.** $y = -4$

**b.** $(x - 5) + (z - 3) = 0$

**c.** $x + y + z = 4$

**d.** $5x - 4y + 3z = -4$

**V.** The plane passing through (5, −4, 3) that is orthogonal to **i** + **j** + **k** satisfies ______.

**a.** $y = -4$

**b.** $(x - 5)/1 = (y + 4)/1 = (z - 3)/1$

**c.** $x + y + z = 4$

**d.** $5x - 4y + 3z = -4$

**VI.** The vector ______ is orthogonal to the plane satisfying $2(x - 3) - 3(y + 2) + 5(z - 5) = 0$.

**a.** $-3\mathbf{i} + 2\mathbf{j} - 5\mathbf{k}$

**b.** $2\mathbf{i} - 3\mathbf{j} + 5\mathbf{k}$

**c.** $(2 - 3)\mathbf{i} + (-3 + 2)\mathbf{j} + (5 - 5)\mathbf{k}$

**d.** $(2)(-3)\mathbf{i} + (-3)(2)\mathbf{j} + (5)(-5)\mathbf{k}$

**VII.** The plane satisfying $6x + 18y - 12z = 17$ is ______ to the plane $-5x - 15y + 10z = 29$.

**a.** identical

**b.** parallel

**c.** orthogonal

**d.** neither parallel nor orthogonal

In Problems 1–14 find a vector equation, parametric equations, and symmetric equations of the indicated line.

**1.** Containing (2, 1, 3) and (1, 2, −1)

**2.** Containing (1, −1, 1) and (−1, 1, −1)

**3.** Containing (−4, 1, 3) and (−4, 0, 1)

**4.** Containing (2, 3, −4) and (2, 0, −4)

**5.** Containing (1, 2, 3) and (3, 2, 1)

**6.** Containing (7, 1, 3) and (−1, −2, 3)

**7.** Containing (2, 2, 1) and parallel to $2\mathbf{i} - \mathbf{j} - \mathbf{k}$

**8.** Containing (−1, −6, 2) and parallel to $4\mathbf{i} + \mathbf{j} - 3\mathbf{k}$

**9.** Containing (−1, −2, 5) and parallel to $-3\mathbf{j} + 7\mathbf{k}$

**10.** Containing (−2, 3, −2) and parallel to $4\mathbf{k}$

**11.** Containing $(a, b, c)$ and parallel to $d\mathbf{i} + e\mathbf{j}$

**12.** Containing $(a, b, c)$ and parallel to $d\mathbf{k}$

---

**Answers to Self-Quiz**

**I.** a, b, c, d **II.** a **III.** d **IV.** a **V.** c **VI.** b **VII.** b

**13.** Containing $(4, 1, -6)$ and parallel to $(x - 2)/3 = (y + 1)/6 = (z - 5)/2$

**14.** Containing $(3, 1, -2)$ and parallel to $(x + 1)/3 = (y + 3)/2 = (z - 2)/(-4)$

**15.** Let $L_1$ be given by

$$\frac{x - x_1}{a_1} = \frac{y - y_1}{b_1} = \frac{z - z_1}{c_1}$$

and $L_2$ be given by

$$\frac{x - x_1}{a_2} = \frac{y - y_1}{b_2} = \frac{z - z_1}{c_2}$$

Show that $L_1$ is orthogonal to $L_2$ if and only if $a_1a_2 + b_1b_2 + c_1c_2 = 0$.

**16.** Show that the lines

$$L_1\colon\ \frac{x - 3}{2} = \frac{y + 1}{4} = \frac{z - 2}{-1} \quad \text{and} \quad L_2\colon\ \frac{x - 3}{5} = \frac{y + 1}{-2} = \frac{z - 3}{2}$$

are orthogonal.

**17.** Show that the lines

$$L_1\colon\ \frac{x - 1}{1} = \frac{y + 3}{2} = \frac{z + 3}{3} \quad \text{and} \quad L_2\colon\ \frac{x - 3}{3} = \frac{y - 1}{6} = \frac{z - 8}{9}$$

are parallel.

Lines in $\mathbb{R}^3$ that do not have the same direction need not have a point in common.

**18.** Show that the lines $L_1\colon\ x = 1 + t, y = -3 + 2t, z = -2 - t$ and $L_2\colon\ x = 17 + 3s, y = 4 + s, z = -8 - s$ have the point $(2, -1, -3)$ in common.

**19.** Show that the lines $L_1\colon\ x = 2 - t, y = 1 + t, z = -2t$ and $L_2\colon\ x = 1 + s, y = -2s, z = 3 + 2s$ do *not* have a point in common.

**20.** Let $L$ be given in its vector form $\overrightarrow{0R} = \overrightarrow{0P} + t\mathbf{v}$. Find a number $t$ such that $\overrightarrow{0R}$ is perpendicular to $\mathbf{v}$.

**21.** Use the result of Problem 20 to find the distance between the line $L$ (containing $P$ and parallel to $\mathbf{v}$) and the origin when

**a.** $P = (2, 1, -4)$; $\mathbf{v} = \mathbf{i} + \mathbf{j} + \mathbf{k}$

**b.** $P = (1, 2, -3)$; $\mathbf{v} = 3\mathbf{i} - \mathbf{j} - \mathbf{k}$

**c.** $P = (-1, 4, 2)$; $\mathbf{v} = -\mathbf{i} + \mathbf{j} + 2\mathbf{k}$

In Problems 22–25 find a line $L$ orthogonal to the two given lines and passing through the given point.

**22.** $\dfrac{x + 2}{-3} = \dfrac{y - 1}{4} = \dfrac{z}{-5}$; $\dfrac{x - 3}{7} = \dfrac{y + 2}{-2} = \dfrac{z - 8}{3}$; $(1, -3, 2)$

**23.** $\dfrac{x - 2}{-4} = \dfrac{y + 3}{-7} = \dfrac{z + 1}{3}$; $\dfrac{x + 2}{3} = \dfrac{y - 5}{-4} = \dfrac{z + 3}{-2}$; $(-4, 7, 3)$

**24.** $x = 3 - 2t, y = 4 + 3t, z = -7 + 5t$; $x = -2 + 4s, y = 3 - 2s, z = 3 + s$; $(-2, 3, 4)$

**25.** $x = 4 + 10t, y = -4 - 8t, z = 3 + 7t$; $x = -2t, y = 1 + 4t, z = -7 - 3t$; $(4, 6, 0)$

***26.** Calculate the distance between the lines

$$L_1\colon \frac{x-2}{3} = \frac{y-5}{2} = \frac{z-1}{-1} \quad \text{and} \quad L_2\colon \frac{x-4}{-4} = \frac{y-5}{4} = \frac{z+2}{1}$$

[*Hint:* The distance is measured along a vector **v** that is perpendicular to both $L_1$ and $L_2$. Let $P$ be a point on $L_1$ and $Q$ a point on $L_2$. Then the length of the projection of $\overrightarrow{PQ}$ on **v** is the distance between the lines, measured along a vector that is perpendicular to them both.]

***27.** Find the distance between the lines

$$L_1\colon \frac{x+2}{3} = \frac{y-7}{-4} = \frac{z-2}{4} \quad \text{and} \quad L_2\colon \frac{x-1}{-3} = \frac{y+2}{4} = \frac{z+1}{1}$$

In Problems 28–41 find the equation of the plane.

**28.** $P = (0, 0, 0)$; $\mathbf{n} = \mathbf{i}$

**29.** $P = (0, 0, 0)$; $\mathbf{n} = \mathbf{j}$

**30.** $P = (0, 0, 0)$; $\mathbf{n} = \mathbf{k}$

**31.** $P = (1, 2, 3)$; $\mathbf{n} = \mathbf{i} + \mathbf{j}$

**32.** $P = (1, 2, 3)$; $\mathbf{n} = \mathbf{i} + \mathbf{k}$

**33.** $P = (1, 2, 3)$; $\mathbf{n} = \mathbf{j} + \mathbf{k}$

**34.** $P = (2, -1, 6)$; $\mathbf{n} = 3\mathbf{i} - \mathbf{j} + 2\mathbf{k}$

**35.** $P = (-4, -7, 5)$; $\mathbf{n} = -3\mathbf{i} - 4\mathbf{j} + \mathbf{k}$

**36.** $P = (-3, 11, 2)$; $\mathbf{n} = 4\mathbf{i} + \mathbf{j} - 7\mathbf{k}$

**37.** $P = (3, -2, 5)$; $\mathbf{n} = 2\mathbf{i} - 7\mathbf{j} - 8\mathbf{k}$

**38.** Containing $(1, 2, -4)$, $(2, 3, 7)$, and $(4, -1, 3)$

**39.** Containing $(-7, 1, 0)$, $(2, -1, 3)$, and $(4, 1, 6)$

**40.** Containing $(1, 0, 0)$, $(0, 1, 0)$, and $(0, 0, 1)$

**41.** Containing $(2, 3, -2)$, $(4, -1, -1)$, and $(3, 1, 2)$

Two planes are **orthogonal** if their normal vectors are orthogonal. In Problems 42–46 determine whether the given planes are parallel, orthogonal, coincident (that is, the same), or none of these.

**42.** $\pi_1\colon\ x + y + z = 2$; $\pi_2\colon\ 2x + 2y + 2z = 4$

**43.** $\pi_1\colon\ x - y + z = 3$; $\pi_2\colon\ -3x + 3y - 3z = -9$

**44.** $\pi_1\colon\ 2x - y + z = 3$; $\pi_2\colon\ x + y - z = 7$

**45.** $\pi_1\colon\ 2x - y + z = 3$; $\pi_2\colon\ x + y + z = 3$

**46.** $\pi_1\colon\ 3x - 2y + 7z = 4$; $\pi_2\colon\ -2x + 4y + 2z = 16$

In Problems 47–49 find the equation of the set of all points of intersection of the two planes.

**47.** $\pi_1\colon\ x - y + z = 2$; $\pi_2\colon\ 2x - 3y + 4z = 7$

**48.** $\pi_1\colon\ 3x - y + 4z = 3$; $\pi_2\colon\ -4x - 2y + 7z = 8$

**49.** $\pi_1\colon\ -2x - y + 17z = 4$; $\pi_2\colon\ 2x - y - z = -7$

***50.** Let $\pi$ be a plane, $P$ a point on the plane, **n** a vector normal to the plane, and $Q$ a point not on the plane (see Figure 3.42). Show that the perpendicular distance $D$ from $Q$ to the plane is given by

$$D = |\text{proj}_{\mathbf{n}}\, \overrightarrow{PQ}| = \frac{|\overrightarrow{PQ} \cdot \mathbf{n}|}{|\mathbf{n}|}$$

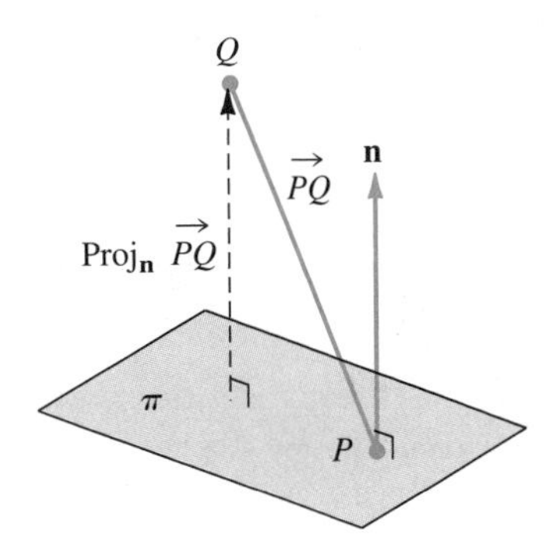


**Figure 3.42**

In Problems 51–53 find the distance from the given point to the given plane.

**51.** $(4, 0, 1)$; $2x - y + 8z = 3$

**52.** $(-7, -2, -1)$; $-2x + 8z = -5$

**53.** $(-3, 0, 2)$; $-3x + y + 5z = 0$

**54.** Prove that the distance between the plane $ax + by + cz = d$ and the point $(x_0, y_0, z_0)$ is given by

$$D = \frac{|ax_0 + by_0 + cz_0 - d|}{\sqrt{a^2 + b^2 + c^2}}$$

The **angle between two planes** is defined to be the acute† angle between their normal vectors. In Problems 55–57 find the angle between the two planes.

**55.** The two planes of Problem 47

**56.** The two planes of Problem 48

**57.** The two planes of Problem 49

***58.** Let **u** and **v** be two nonparallel, nonzero vectors in a plane $\pi$. Show that if **w** is any other vector in $\pi$, then there exist scalars $\alpha$ and $\beta$ such that $\mathbf{w} = \alpha\mathbf{u} + \beta\mathbf{v}$. This is called the **parametric representation** of the plane $\pi$. [*Hint:* Draw a parallelogram in which $\alpha\mathbf{u}$ and $\beta\mathbf{v}$ form adjacent sides and the diagonal vector is **w**.]

***59.** Three vectors **u**, **v**, and **w** are called **coplanar** if they all lie in the same plane $\pi$. Show that if **u**, **v**, and **w** all pass through the origin, then they are coplanar if and only if the scalar triple product equals zero: $\mathbf{u} \cdot (\mathbf{v} \times \mathbf{w}) = 0$.

In Problems 60–64 determine whether the three given position vectors (that is, one end point at the origin) are coplanar. If they are coplanar, find the equation of the plane containing them.

**60.** $\mathbf{u} = 2\mathbf{i} - 3\mathbf{j} + 4\mathbf{k}$; $\mathbf{v} = 7\mathbf{i} - 2\mathbf{j} + 3\mathbf{k}$; $\mathbf{w} = 9\mathbf{i} - 5\mathbf{j} + 7\mathbf{k}$

**61.** $\mathbf{u} = -3\mathbf{i} + \mathbf{j} + 8\mathbf{k}$; $\mathbf{v} = -2\mathbf{i} - 3\mathbf{j} + 5\mathbf{k}$; $\mathbf{w} = 2\mathbf{i} + 14\mathbf{j} - 4\mathbf{k}$

**62.** $\mathbf{u} = 2\mathbf{i} + \mathbf{j} - 2\mathbf{k}$; $\mathbf{v} = 2\mathbf{i} - \mathbf{j} - 2\mathbf{k}$; $\mathbf{w} = 2\mathbf{i} - \mathbf{j} + 2\mathbf{k}$

**63.** $\mathbf{u} = 3\mathbf{i} - 2\mathbf{j} + \mathbf{k}$; $\mathbf{v} = \mathbf{i} + \mathbf{j} - 5\mathbf{k}$; $\mathbf{w} = -\mathbf{i} + 5\mathbf{j} - 16\mathbf{k}$

**64.** $\mathbf{u} = 2\mathbf{i} - \mathbf{j} - \mathbf{k}$; $\mathbf{v} = 4\mathbf{i} + 3\mathbf{j} + 2\mathbf{k}$; $\mathbf{w} = 6\mathbf{i} + 7\mathbf{j} + 5\mathbf{k}$

† Recall that an acute angle $\alpha$ is an angle between 0° and 90°; that is, $\alpha \in (0, \pi/2)$.

## SUMMARY

- The **directed line segment** extending from $P$ to $Q$ in $\mathbb{R}^2$ or $\mathbb{R}^3$, denoted by $\overrightarrow{PQ}$, is the straight-line segment that extends from $P$ to $Q$. (pp. 227, 252)
- Two directed line segments in $\mathbb{R}^2$ or $\mathbb{R}^3$ are **equivalent** if they have the same magnitude (length) and direction. (pp. 228, 252)
- ***Geometric definition of a vector***

  A **vector** in $\mathbb{R}^2(\mathbb{R}^3)$ is the set of all directed line segments in $\mathbb{R}^2(\mathbb{R}^3)$ equivalent to a given directed line segment. One representation of a vector has its initial point at the origin and is denoted by $\overrightarrow{0R}$. (pp. 228, 252)
- ***Algebraic definition of a vector***

  A **vector v** in the $xy$-plane ($\mathbb{R}^2$) is an ordered pair of real numbers $(a, b)$. The numbers $a$ and $b$ are called the **components** of the vector **v**. The **zero vector** is the vector $(0, 0)$. In $\mathbb{R}^3$, a vector **v** is an **ordered triple** of real numbers $(a, b, c)$. The **zero vector** in $\mathbb{R}^3$ is the vector $(0, 0, 0)$. (pp. 229, 252)
- The geometric and algebraic definitions of a vector in $\mathbb{R}^2[\mathbb{R}^3]$ are related in the following way: If $\mathbf{v} = (a, b)[(a, b, c)]$, then one representation of $\mathbf{v}$ is $\overrightarrow{0R}$, where $R = (a, b)[R = (a, b, c)]$. (p. 229)
- If $\mathbf{v} = (a, b)$, then the **magnitude of v**, denoted by $|\mathbf{v}|$, is given by $|\mathbf{v}| = \sqrt{a^2 + b^2}$. If $\mathbf{v} = (a, b, c)$, then $|\mathbf{v}| = \sqrt{a^2 + b^2 + c^2}$. (pp. 229, 252)
- If **v** is a vector in $\mathbb{R}^2$, then the **direction of v** is the angle in $[0, 2\pi)$ that any representation of **v** makes with the positive $x$-axis. (p. 230)
- ***Triangle inequality***

  In $\mathbb{R}^2$ or $\mathbb{R}^3$

  $$|\mathbf{u} + \mathbf{v}| \le |\mathbf{u}| + |\mathbf{v}| \qquad \text{(p. 233)}$$
- In $\mathbb{R}^2$ let $\mathbf{i} = (1, 0)$ and $\mathbf{j} = (0, 1)$. Then $\mathbf{v} = (a, b)$ can be written $\mathbf{v} = a\mathbf{i} + b\mathbf{j}$. (p. 234)
- In $\mathbb{R}^3$ let $\mathbf{i} = (1, 0, 0)$, $\mathbf{j} = (0, 1, 0)$, and $\mathbf{k} = (0, 0, 1)$. Then $\mathbf{v} = (a, b, c)$ can be written (p. 257)

  $$\mathbf{v} = a\mathbf{i} + b\mathbf{j} + c\mathbf{k}$$
- A **unit vector u** in $\mathbb{R}^2$ or $\mathbb{R}^3$ is a vector that satisfies $|\mathbf{u}| = 1$. In $\mathbb{R}^2$ a unit vector can be written (pp. 234, 235)

  $$\mathbf{u} = (\cos\theta)\mathbf{i} + (\sin\theta)\mathbf{j}$$

  where $\theta$ is the direction of **u**.
- Let $\mathbf{u} = (a_1, b_1)$ and $\mathbf{v} = (a_2, b_2)$. Then the **scalar product** or **dot product** of **u** and **v**, written $\mathbf{u} \cdot \mathbf{v}$, is given by (p. 241)

  $$\mathbf{u} \cdot \mathbf{v} = a_1 a_2 + b_1 b_2$$

  If $\mathbf{u} = (a_1, b_1, c_1)$ and $\mathbf{v} = (a_2, b_2, c_2)$, then

  $$\mathbf{u} \cdot \mathbf{v} = a_1 a_2 + b_1 b_2 + c_1 c_2$$
- The **angle** $\varphi$ between two vectors **u** and **v** in $\mathbb{R}^2$ or $\mathbb{R}^3$ is the unique number in $[0, \pi]$ that satisfies (pp. 241, 242)

  $$\cos\varphi = \frac{\mathbf{u} \cdot \mathbf{v}}{|\mathbf{u}|\,|\mathbf{v}|}$$

- Two vectors in $\mathbb{R}^2$ or $\mathbb{R}^3$ are **parallel** if the angle between them is 0 or $\pi$. They are parallel if and only if one is a scalar multiple of the other. (pp. 243, 257)
- Two vectors in $\mathbb{R}^2$ or $\mathbb{R}^3$ are **orthogonal** if the angle between them is $\pi/2$. They are orthogonal if and only if their scalar product is zero. (pp. 243, 244, 257)
- Let **u** and **v** be two nonzero vectors in $\mathbb{R}^2$ or $\mathbb{R}^3$. Then the **projection** of **u** on **v** is a vector, denoted by $\text{proj}_{\mathbf{v}}\,\mathbf{u}$, which is defined by (pp. 245, 258)

$$\text{proj}_{\mathbf{v}}\,\mathbf{u} = \frac{\mathbf{u}\cdot\mathbf{v}}{|\mathbf{v}|^2}\mathbf{v}$$

The vector $\dfrac{\mathbf{u}\cdot\mathbf{v}}{|\mathbf{v}|}$ is called the **component** of **u** in the direction **v**.

- $\text{Proj}_{\mathbf{v}}\,\mathbf{u}$ is parallel to **v** and $\mathbf{u} - \text{proj}_{\mathbf{v}}\,\mathbf{u}$ is orthogonal to **v**. (pp. 246, 258)
- The **direction** of a vector **v** in $\mathbb{R}^3$ is the unit vector (p. 254)

$$\mathbf{u} = \frac{\mathbf{v}}{|\mathbf{v}|}$$

- If $\mathbf{v} = (a, b, c)$, then $\cos\alpha = \dfrac{a}{|\mathbf{v}|}$, $\cos\beta \dfrac{b}{|\mathbf{v}|}$ and $\cos\gamma = \dfrac{c}{|\mathbf{v}|}$ are called the **direction cosines** of **v**. (p. 255)
- Let $\mathbf{u} = a_1\mathbf{i} + b_1\mathbf{j} + c_1\mathbf{k}$ and $\mathbf{v} = a_2\mathbf{i} + b_2\mathbf{j} + c_2\mathbf{k}$. Then the **cross product** or **vector product** of **u** and **v**, denoted by $\mathbf{u}\times\mathbf{v}$, is given by (pp. 261, 262)

$$\mathbf{u}\times\mathbf{v} = \begin{vmatrix} \mathbf{i} & \mathbf{j} & \mathbf{k} \\ a_1 & b_1 & c_1 \\ a_2 & b_2 & c_2 \end{vmatrix}$$

- ***Properties of the cross product*** (p. 263)

  **i.** $\mathbf{u}\times\mathbf{0} = \mathbf{0}\times\mathbf{u} = \mathbf{0}$.

  **ii.** $\mathbf{u}\times\mathbf{v} = -\mathbf{v}\times\mathbf{u}$.

  **iii.** $(\alpha\mathbf{u})\times\mathbf{v} = \alpha(\mathbf{u}\times\mathbf{v})$.

  **iv.** $\mathbf{u}\times(\mathbf{v}+\mathbf{w}) = (\mathbf{u}\times\mathbf{v}) + (\mathbf{u}\times\mathbf{w})$.

  **v.** $(\mathbf{u}\times\mathbf{v})\cdot\mathbf{w} = \mathbf{u}\cdot(\mathbf{v}\times\mathbf{w})$ (the **scalar triple product**).

  **vi.** $\mathbf{u}\times\mathbf{v}$ is orthogonal to both **u** and **v**.

  **vii.** If neither **u** nor **v** is the zero vector, then **u** and **v** are parallel, if and only if $\mathbf{u}\times\mathbf{v} = \mathbf{0}$.

- If $\varphi$ is the angle between **u** and **v**, then (p. 264)

  $|\mathbf{u}\times\mathbf{v}| = |\mathbf{u}|\,|\mathbf{v}|\sin\varphi$ = area of the parallelogram with sides **u** and **v**

- Let $P = (x_1, y_1, z_1)$ and $Q = (x_2, y_2, z_2)$ be two points on a line $L$ in $\mathbb{R}^3$. Let $\mathbf{v} = (x_2 - x_1)\mathbf{i} + (y_2 - y_1)\mathbf{j} + (z_2 - z_1)\mathbf{k}$ and let $a = x_2 - x_1$, $b = y_2 - y_1$, and $c = z_2 - z_1$.

  **Vector equation of a line:** $\overrightarrow{0R} = \overrightarrow{0P} + t\mathbf{v}$ (p. 273)

  **Parametric equations of a line:** (p. 274)

$$x = x_1 + at$$
$$y = y_1 + bt$$
$$z = z_1 + ct$$

  **Symmetric equations of a line:** $\dfrac{x - x_1}{a} = \dfrac{y - y_1}{b} = \dfrac{z - z_1}{c}$, if $a$, $b$, and $c$ are nonzero. (p. 274)

- Let $P$ be a point in $\mathbb{R}^3$ and let $\mathbf{n}$ be a given nonzero vector. Then the set of all points $Q$ for which $\overrightarrow{PQ} \cdot \mathbf{n} = 0$ constitutes a plane in $\mathbb{R}^3$. The vector $\mathbf{n}$ is called the **normal vector** of the plane. (p. 276)
- If $\mathbf{n} = a\mathbf{i} + b\mathbf{j} + c\mathbf{k}$ and $P = (x_0, y_0, z_0)$, then the equation of the plane can be written (p. 276)

$$ax + by + cz = d$$

where

$$d = ax_0 + by_0 + cz_0 = \overrightarrow{0P} \cdot \mathbf{n}$$

- The ***xy*-plane** has the equation $z = 0$; the ***xz*-plane** has the equation $y = 0$; the ***yz*-plane** has the equation $x = 0$. (p. 277)
- Two planes are **parallel** if their normal vectors are parallel. If two planes are not parallel, then they intersect in a straight line. (p. 280)

# REVIEW EXERCISES

In Exercises 1–6 find the magnitude and direction of the given vector.

**1.** $\mathbf{v} = (3, 3)$ **2.** $\mathbf{v} = -3\mathbf{i} + 3\mathbf{j}$ **3.** $\mathbf{v} = (2, -2\sqrt{3})$

**4.** $\mathbf{v} = (\sqrt{3}, 1)$ **5.** $\mathbf{v} = -12\mathbf{i} - 12\mathbf{j}$ **6.** $\mathbf{v} = \mathbf{i} + 4\mathbf{j}$

In Exercises 7–10 write the vector $\mathbf{v}$ that is represented by $\overrightarrow{PQ}$ in the form $a\mathbf{i} + b\mathbf{j}$. Sketch $\overrightarrow{PQ}$ and $\mathbf{v}$.

**7.** $P = (2, 3)$; $Q = (4, 5)$ **8.** $P = (1, -2)$; $Q = (7, 12)$

**9.** $P = (-1, -6)$; $Q = (3, -4)$ **10.** $P = (-1, 3)$; $Q = (3, -1)$

**11.** Let $\mathbf{u} = (2, 1)$ and $\mathbf{v} = (-3, 4)$. Find **(a)** $5\mathbf{u}$; **(b)** $\mathbf{u} - \mathbf{v}$; **(c)** $-8\mathbf{u} + 5\mathbf{v}$.

**12.** Let $\mathbf{u} = -4\mathbf{i} + \mathbf{j}$ and $\mathbf{v} = -3\mathbf{i} - 4\mathbf{j}$. Find **(a)** $-3\mathbf{v}$; **(b)** $\mathbf{u} + \mathbf{v}$; **(c)** $3\mathbf{u} - 6\mathbf{v}$.

In Exercises 13–19 find a unit vector having the same direction as the given vector.

**13.** $\mathbf{v} = \mathbf{i} + \mathbf{j}$ **14.** $\mathbf{v} = -\mathbf{i} + \mathbf{j}$ **15.** $\mathbf{v} = 2\mathbf{i} + 5\mathbf{j}$

**16.** $\mathbf{v} = -7\mathbf{i} + 3\mathbf{j}$ **17.** $\mathbf{v} = 3\mathbf{i} + 4\mathbf{j}$ **18.** $\mathbf{v} = -2\mathbf{i} - 2\mathbf{j}$

**19.** $\mathbf{v} = a\mathbf{i} - a\mathbf{j}$

**20.** If $\mathbf{v} = 4\mathbf{i} - 7\mathbf{j}$, find $\sin\theta$ and $\cos\theta$, where $\theta$ is the direction of $\mathbf{v}$.

**21.** Find a unit vector with direction opposite to that of $\mathbf{v} = 5\mathbf{i} + 2\mathbf{j}$.

**22.** Find two unit vectors orthogonal to $\mathbf{v} = \mathbf{i} - \mathbf{j}$.

**23.** Find a unit vector with direction opposite to that of $\mathbf{v} = 10\mathbf{i} - 7\mathbf{j}$.

In Exercises 24–27 find a vector $\mathbf{v}$ having the given magnitude and direction.

**24.** $|\mathbf{v}| = 2$; $\theta = \pi/3$ **25.** $|\mathbf{v}| = 1$; $\theta = \pi/2$

**26.** $|\mathbf{v}| = 4$; $\theta = \pi$ **27.** $|\mathbf{v}| = 7$; $\theta = 5\pi/6$

In Exercises 28–31 calculate the scalar product of the two vectors and the cosine of the angle between them.

**28.** $\mathbf{u} = \mathbf{i} - \mathbf{j}$; $\mathbf{v} = \mathbf{i} + 2\mathbf{j}$ **29.** $\mathbf{u} = -4\mathbf{i}$; $\mathbf{v} = 11\mathbf{j}$

**30.** $\mathbf{u} = 4\mathbf{i} - 7\mathbf{j}$; $\mathbf{v} = 5\mathbf{i} + 6\mathbf{j}$ **31.** $\mathbf{u} = -\mathbf{i} - 2\mathbf{j}$; $\mathbf{v} = 4\mathbf{i} + 5\mathbf{j}$

In Exercises 32–37 determine whether the given vectors are orthogonal, parallel, or neither. Then sketch each pair.

**32.** $\mathbf{u} = 2\mathbf{i} - 6\mathbf{j}$; $\mathbf{v} = -\mathbf{i} + 3\mathbf{j}$

**33.** $\mathbf{u} = 4\mathbf{i} - 5\mathbf{j}$; $\mathbf{v} = 5\mathbf{i} - 4\mathbf{j}$

**34.** $\mathbf{u} = 4\mathbf{i} - 5\mathbf{j}$; $\mathbf{v} = -5\mathbf{i} + 4\mathbf{j}$

**35.** $\mathbf{u} = -7\mathbf{i} - 7\mathbf{j}$; $\mathbf{v} = \mathbf{i} + \mathbf{j}$

**36.** $\mathbf{u} = -7\mathbf{i} - 7\mathbf{j}$; $\mathbf{v} = -\mathbf{i} + \mathbf{j}$

**37.** $\mathbf{u} = -7\mathbf{i} - 7\mathbf{j}$; $\mathbf{v} = -\mathbf{i} - \mathbf{j}$

**38.** Let $\mathbf{u} = 2\mathbf{i} + 3\mathbf{j}$ and $\mathbf{v} = 4\mathbf{i} + \alpha\mathbf{j}$. Determine $\alpha$ such that

**a.** $\mathbf{u}$ and $\mathbf{v}$ are orthogonal.

**b.** $\mathbf{u}$ and $\mathbf{v}$ are parallel.

**c.** The angle between $\mathbf{u}$ and $\mathbf{v}$ is $\pi/4$.

**d.** The angle between $\mathbf{u}$ and $\mathbf{v}$ is $\pi/6$.

In Exercises 39–44 calculate $\text{proj}_{\mathbf{v}}\,\mathbf{u}$.

**39.** $\mathbf{u} = 14\mathbf{i}$; $\mathbf{v} = \mathbf{i} + \mathbf{j}$

**40.** $\mathbf{u} = 14\mathbf{i}$, $\mathbf{v} = \mathbf{i} - \mathbf{j}$

**41.** $\mathbf{u} = 3\mathbf{i} - 2\mathbf{j}$; $\mathbf{v} = 3\mathbf{i} + 2\mathbf{j}$

**42.** $\mathbf{u} = 3\mathbf{i} + 2\mathbf{j}$; $\mathbf{v} = \mathbf{i} - 3\mathbf{j}$

**43.** $\mathbf{u} = 2\mathbf{i} - 5\mathbf{j}$; $\mathbf{v} = -3\mathbf{i} - 7\mathbf{j}$

**44.** $\mathbf{u} = 4\mathbf{i} - 5\mathbf{j}$; $\mathbf{v} = -3\mathbf{i} - \mathbf{j}$

**45.** Let $P = (3, -2)$, $Q = (4, 7)$, $R = (-1, 3)$, and $S = (2, -1)$. Calculate $\text{proj}_{\overrightarrow{PQ}}\,\overrightarrow{RS}$ and $\text{proj}_{\overrightarrow{RS}}\,\overrightarrow{PQ}$.

In Exercises 46–48 find the distance between the two given points.

**46.** $(4, -1, 7)$; $(-5, 1, 3)$

**47.** $(-2, 4, -8)$; $(0, 0, 6)$

**48.** $(2, -7, 0)$; $(0, 5, -8)$

In Exercises 49–51 find the magnitude and the direction cosines of the given vector.

**49.** $\mathbf{v} = 3\mathbf{j} + 11\mathbf{k}$

**50.** $\mathbf{v} = \mathbf{i} - 2\mathbf{j} - 3\mathbf{k}$

**51.** $\mathbf{v} = -4\mathbf{i} + \mathbf{j} + 6\mathbf{k}$

**52.** Find a unit vector in the direction of $\overrightarrow{PQ}$, where $P = (3, -1, 2)$ and $Q = (-4, 1, 7)$.

**53.** Find a unit vector whose direction is opposite that of $\overrightarrow{PQ}$, where $P = (1, -3, 0)$ and $Q = (-7, 1, -4)$.

In Exercises 54–61 let $\mathbf{u} = \mathbf{i} - 2\mathbf{j} + 3\mathbf{k}$, $\mathbf{v} = -3\mathbf{i} + 2\mathbf{j} + 5\mathbf{k}$, and $\mathbf{w} = 2\mathbf{i} - 4\mathbf{j} + \mathbf{k}$. Calculate

**54.** $\mathbf{u} - \mathbf{v}$

**55.** $3\mathbf{v} + 5\mathbf{w}$

**56.** $\text{proj}_{\mathbf{v}}\,\mathbf{w}$

**57.** $\text{proj}_{\mathbf{w}}\,\mathbf{u}$

**58.** $2\mathbf{u} - 4\mathbf{v} + 7\mathbf{w}$

**59.** $\mathbf{u} \cdot \mathbf{w} - \mathbf{w} \cdot \mathbf{v}$

**60.** The angle between $\mathbf{u}$ and $\mathbf{v}$

**61.** The angle between $\mathbf{v}$ and $\mathbf{w}$

In Exercises 62–65 find the cross product $\mathbf{u} \times \mathbf{v}$.

**62.** $\mathbf{u} = 3\mathbf{i} - \mathbf{j}$; $\mathbf{v} = 2\mathbf{i} + 4\mathbf{k}$

**63.** $\mathbf{u} = 7\mathbf{j}$; $\mathbf{v} = \mathbf{i} - \mathbf{k}$

**64.** $\mathbf{u} = 4\mathbf{i} - \mathbf{j} + 7\mathbf{k}$; $\mathbf{v} = -7\mathbf{i} + \mathbf{j} - 2\mathbf{k}$

**65.** $\mathbf{u} = -2\mathbf{i} + 3\mathbf{j} - 4\mathbf{k}$; $\mathbf{v} = -3\mathbf{i} + \mathbf{j} - 10\mathbf{k}$

**66.** Find two unit vectors orthogonal to both $\mathbf{u} = \mathbf{i} - \mathbf{j} + 3\mathbf{k}$ and $\mathbf{v} = -2\mathbf{i} - 3\mathbf{j} + 4\mathbf{k}$.

**67.** Calculate the area of the parallelogram with the adjacent vertices $(1, 4, -2)$, $(-3, 1, 6)$, and $(1, -2, 3)$.

In Exercises 68–71 find a vector equation, parametric equations, and symmetric equations of the given line.

**68.** Containing $(3, -1, 4)$ and $(-1, 6, 2)$

**69.** Containing $(-4, 1, 0)$ and $(3, 0, 7)$

**70.** Containing $(3, 1, 2)$ and parallel to $3\mathbf{i} - \mathbf{j} - \mathbf{k}$

**71.** Containing $(1, -2, -3)$ and parallel to $(x + 1)/5 = (y - 2)/(-3) = (z - 4)/2$

**72.** Show that the lines $L_1$: $x = 3 - 2t$, $y = 4 + t$, $z = -2 + 7t$ and $L_2$: $x = -3 + s$, $y = 2 - 4s$, $z = 1 + 6s$ have no points of intersection.

**73.** Find the distance from the origin to the line passing through the point $(3, 1, 5)$ and having the direction $\mathbf{v} = 2\mathbf{i} - \mathbf{j} + \mathbf{k}$.

**74.** Find the equation of the line passing through $(-1, 2, 4)$ and orthogonal to $L_1$: $(x - 1)/4 = (y + 6)/3 = z/(-2)$ and $L_2$: $(x + 3)/5 = (y - 1)/1 = (z + 3)/4$.

In Exercises 75–77 find the equation of the plane containing the given point and orthogonal to the given normal vector.

**75.** $P = (1, 3, -2)$; $\mathbf{n} = \mathbf{i} + \mathbf{k}$

**76.** $P = (1, -4, 6)$; $\mathbf{n} = 2\mathbf{j} - 3\mathbf{k}$

**77.** $P = (-4, 1, 6)$; $\mathbf{n} = 2\mathbf{i} - 3\mathbf{j} + 5\mathbf{k}$

**78.** Find the equation of the plane containing the points $(-2, 4, 1)$, $(3, -7, 5)$, and $(-1, -2, -1)$.

**79.** Find all points of intersection of the planes $\pi_1$: $-x + y + z = 3$ and $\pi_2$: $-4x + 2y - 7z = 5$.

**80.** Find all points of intersection of the planes $\pi_1$: $-4x + 6y + 8z = 12$ and $\pi_2$: $2x - 3y - 4z = 5$.

**81.** Find all points of intersection of the planes $\pi_1$: $3x - y + 4z = 8$ and $\pi_2$: $-3x - y - 11z = 0$.

**82.** Find the distance from $(1, -2, 3)$ to the plane $2x - y - z = 6$.

**83.** Find the angle between the planes of Exercise 79.

**84.** Show that the position vectors $\mathbf{u} = \mathbf{i} - 2\mathbf{j} + \mathbf{k}$, $\mathbf{v} = 3\mathbf{i} + 2\mathbf{j} - 3\mathbf{k}$, and $\mathbf{w} = 9\mathbf{i} - 2\mathbf{j} - 3\mathbf{k}$ are coplanar and find the equation of the plane containing them.

4

# Vector Spaces

## 4.1 INTRODUCTION

As we saw in the last chapter, the sets $\mathbb{R}^2$ (vectors in the plane) and $\mathbb{R}^3$ (vectors in space) have a number of nice properties. We can add two vectors in $\mathbb{R}^2$ and obtain another vector in $\mathbb{R}^2$. Under addition, vectors in $\mathbb{R}^2$ commute and obey the associative law. If $\mathbf{x} \in \mathbb{R}^2$, then $\mathbf{x} + \mathbf{0} = \mathbf{x}$ and $\mathbf{x} + (-\mathbf{x}) = \mathbf{0}$. We can multiply vectors in $\mathbb{R}^2$ by scalars and obtain a number of distributive laws. The same properties also hold in $\mathbb{R}^3$.

The sets $\mathbb{R}^2$ and $\mathbb{R}^3$ together with the operations of vector addition and scalar multiplication are called *vector spaces*. Intuitively, we can say that a vector space is a set of objects together with two operations that obey the rules described in the previous paragraph.

In this chapter we make a seemingly great leap from the concrete world of solving equations and dealing with easily visualized vectors to the abstract world of arbitrary vector spaces. There is a great advantage in doing so. Once we have established a fact about vector spaces in general, we can apply that fact to *every* vector space. Otherwise, we would have to prove that fact again and again, once for each new vector space we encounter (and there is an endless supply of them). But as you will see, many of the abstract theorems we shall prove are really no more difficult than the ones already encountered.

## 4.2 DEFINITION AND BASIC PROPERTIES

**DEFINITION 1** **Real Vector Space** A **real vector space** $V$ is a set of objects, called **vectors,** together with two operations called **addition** and **scalar multiplication** that satisfy the ten axioms listed below.

*Notation.* If $\mathbf{x}$ and $\mathbf{y}$ are in $V$ and if $\alpha$ is a real number, then we write $\mathbf{x} + \mathbf{y}$ for the sum of $\mathbf{x}$ and $\mathbf{y}$ and $\alpha\mathbf{x}$ for the scalar product of $\alpha$ and $\mathbf{x}$.

Before we list the properties satisfied by vectors in a vector space, two things should be mentioned. First, while it might be helpful to think of $\mathbb{R}^2$ or $\mathbb{R}^3$ when dealing with a vector space, it often occurs that a vector space may appear to be very different from these comfortable spaces. (We shall see this shortly.) Second, Definition 1 gives a definition of a *real* vector space. The word "real" means that the scalars we use are real numbers. It would be just as easy to define a *complex* vector space by using complex numbers instead of real ones. This book deals primarily with real vector spaces, but generalizations to other sets of scalars present little difficulty.

**AXIOMS OF A VECTOR SPACE**

**i.** If $\mathbf{x} \in V$ and $\mathbf{y} \in V$, then $\mathbf{x} + \mathbf{y} \in V$ **(closure under addition).**

**ii.** For all $\mathbf{x}$, $\mathbf{y}$, and $\mathbf{z}$ in $V$, $(\mathbf{x} + \mathbf{y}) + \mathbf{z} = \mathbf{x} + (\mathbf{y} + \mathbf{z})$ **(associative law of vector addition).**

**iii.** There is a vector $\mathbf{0} \in V$ such that for all $\mathbf{x} \in V$, $\mathbf{x} + \mathbf{0} = \mathbf{0} + \mathbf{x} = \mathbf{x}$ (**0** is called the **zero vector** or **additive identity**).

**iv.** If $\mathbf{x} \in V$, there is a vector $-\mathbf{x}$ in $V$ such that $\mathbf{x} + (-\mathbf{x}) = \mathbf{0}$ ($-\mathbf{x}$ is called the **additive inverse** of $\mathbf{x}$).

**v.** If $\mathbf{x}$ and $\mathbf{y}$ are in $V$, then $\mathbf{x} + \mathbf{y} = \mathbf{y} + \mathbf{x}$ **(commutative law of vector addition).**

**vi.** If $\mathbf{x} \in V$ and $\alpha$ is a scalar, then $\alpha\mathbf{x} \in V$ **(closure under scalar multiplication).**

**vii.** If $\mathbf{x}$ and $\mathbf{y}$ are in $V$ and $\alpha$ is a scalar, then $\alpha(\mathbf{x} + \mathbf{y}) = \alpha\mathbf{x} + \alpha\mathbf{y}$ **(first distributive law).**

**viii.** If $\mathbf{x} \in V$ and $\alpha$ and $\beta$ are scalars, then $(\alpha + \beta)\mathbf{x} = \alpha\mathbf{x} + \beta\mathbf{x}$ **(second distributive law).**

**ix.** If $\mathbf{x} \in V$ and $\alpha$ and $\beta$ are scalars, then $\alpha(\beta\mathbf{x}) = (\alpha\beta)\mathbf{x}$ **(associative law of scalar multiplication).**

**x.** For every vector $\mathbf{x} \in V$, $1\mathbf{x} = \mathbf{x}$ ■

*Note.* It is not difficult to show that the additive identity and additive inverses in a vector space are unique (see Problems 21 and 22).

EXAMPLE 1 **The Space $\mathbb{R}^n$** Let $V = \mathbb{R}^n = \left\{ \begin{pmatrix} x_1 \\ x_2 \\ \vdots \\ x_n \end{pmatrix} : x_i \in \mathbb{R} \text{ for } i = 1, 2, \ldots, n \right\}$.

Each vector in $\mathbb{R}^n$ is an $n \times 1$ matrix. By the definition of matrix addition given on page 51, $\mathbf{x} + \mathbf{y}$ is an $n \times 1$ matrix if both $\mathbf{x}$ and $\mathbf{y}$ are $n \times 1$ matrices. Setting $\mathbf{0} = \begin{pmatrix} 0 \\ 0 \\ \vdots \\ 0 \end{pmatrix}$ and $-\mathbf{x} = \begin{pmatrix} -x_1 \\ -x_2 \\ \vdots \\ -x_n \end{pmatrix}$, we see that axioms (*ii*) to (*x*) follow from the definition of vector (matrix) addition and Theorem 1.5.1 on page 53.

*Note.* The vectors in $\mathbb{R}^n$ can be written as row vectors as well as column vectors.

EXAMPLE 2 **A Trivial Vector Space** Let $V = \{0\}$. That is, $V$ consists of the single number 0. Since $0 + 0 = 1 \cdot 0 = 0 + (0 + 0) = (0 + 0) + 0 = 0$, we see that $V$ is a vector space. It is often referred to as a **trivial** vector space.

EXAMPLE 3 **A Set That Is Not a Vector Space** Let $V = \{1\}$. That is, $V$ consists of the single number 1. This is *not* a vector space since it violates axiom (*i*)—the closure axiom. To see this, we simply note that $1 + 1 = 2 \notin V$. It also violates other axioms as well. However, all we need to show is that it violates at least one of the ten axioms in order to prove that $V$ is not a vector space.

*Note.* Checking all ten axioms can be tedious. From now on we shall check only those axioms that are not immediately obvious.

EXAMPLE 4 **The Set of Points in $\mathbb{R}^2$ That Lie on a Line Passing Through the Origin Constitutes a Vector Space** Let

$$V = \{(x, y): \quad y = mx, \text{ where } m \text{ is a fixed real number and } x \text{ is an arbitrary real number}\}.$$

That is, $V$ consists of all points lying on the line $y = mx$ passing through the origin with slope $m$. To show that $V$ is a vector space, we verify that each of the axioms holds. Note that we have written vectors in $\mathbb{R}^2$ as rows rather than columns. This makes no essential difference.

**i.** Suppose that $\mathbf{x} = (x_1, y_1)$ and $\mathbf{y} = (x_2, y_2)$ are in $V$. Then $y_1 = mx_1$, $y_2 = mx_2$, and

$$\mathbf{x} + \mathbf{y} = (x_1, y_1) + (x_2, y_2) = (x_1, mx_1) + (x_2, mx_2) = (x_1 + x_2, mx_1 + mx_2)$$
$$= (x_1 + x_2, m(x_1 + x_2)) \in V$$

Thus axiom $(i)$ is satisfied.

**ii.** Suppose $(x, y) \in V$. Then $y = mx$ and $-(x, y) = -(x, mx) = (-x, m(-x))$, so $-(x, y)$ belongs to $V$ as well and $(x, mx) + (-x, m(-x)) = (x - x, m(x - x)) = (0, 0)$.

Every vector in $V$ is a vector in $\mathbb{R}^2$ and $\mathbb{R}^2$ is a vector space, as shown in Example 1. Since $(0, 0) = \mathbf{0}$ is in $V$, (explain why) all the other properties follow from Example 1. Thus $V$ is a vector space.

EXAMPLE 5 **The Set of Points in $\mathbb{R}^2$ Lying on a Line Not Passing Through the Origin Does Not Constitute a Vector Space** Let $V = \{(x, y)\colon\ y = 2x + 1, x \in \mathbb{R}\}$. That is, $V$ is the set of points lying on the line $y = 2x + 1$. $V$ is *not* a vector space because closure under addition is violated, as in Example 3. To see this, let us suppose that $(x_1, y_1)$ and $(x_2, y_2)$ are in $V$. Then

$$(x_1, y_1) + (x_2, y_2) = (x_1 + x_2, y_1 + y_2)$$

If this last vector were in $V$, we would have

$$y_1 + y_2 = 2(x_1 + x_2) + 1 = 2x_1 + 2x_2 + 1$$

But $y_1 = 2x_1 + 1$ and $y_2 = 2x_2 + 1$ so that

$$y_1 + y_2 = (2x_1 + 1) + (2x_2 + 1) = 2x_1 + 2x_2 + 2$$

Hence we conclude that

$$(x_1 + x_2, y_1 + y_2) \notin V \qquad \text{if } (x_1, y_1) \in V \quad \text{and} \quad (x_2, y_2) \in V$$

For example, $(0, 1)$ and $(3, 7)$ are in $V$, but $(0, 1) + (3, 7) = (3, 8)$ is not in $V$ because $8 \neq 2 \cdot 3 + 1$. An easier way to see that $V$ is not a vector space is to observe that $\mathbf{0} = (0, 0)$ is not in $V$ because $0 \neq 2 \cdot 0 + 1$. It is not difficult to show that the set of points in $\mathbb{R}^2$ lying on any line that does not pass through $(0, 0)$ does not constitute a vector space.

EXAMPLE 6 **The Set of Points in $\mathbb{R}^3$ Lying on a Plane Passing Through the Origin Constitutes a Vector Space** Let $V = \{(x, y, z);\ ax + by + cz = 0\}$. That is, $V$ is the set of points in $\mathbb{R}^3$ lying on the plane with normal vector $(a, b, c)$ and passing through the origin. As in Example 4, we write vectors as rows rather than columns.

Suppose $(x_1, y_1, z_1)$ and $(x_2, y_2, z_2)$ are in $V$. Then $(x_1, y_1, z_1) + (x_2, y_2, z_2) = (x_1 + x_2, y_1 + y_2, z_1 + z_2) \in V$ because

$$a(x_1 + x_2) + b(y_1 + y_2) + c(z_1 + z_2)$$
$$= (ax_1 + by_1 + cz_1) + (ax_2 + by_2 + cz_2) = 0 + 0 = 0$$

hence axiom $(i)$ is satisfied. The other axioms are easily verified. Thus the set of points lying on a plane in $\mathbb{R}^3$ that passes through the origin constitutes a vector space.

**EXAMPLE 7** **The Vector Space $P_n$** Let $V = P_n$, the set of polynomials with real coefficients of degree less than or equal to $n$.† If $p \in P_n$, then

$$p(x) = a_n x^n + a_{n-1}x^{n-1} + \cdots + a_1 x + a_0$$

where each $a_i$ is real. The sum $p(x) + q(x)$ is defined in the usual way: If $q(x) = b_n x^n + b_{n-1}x^{n-1} + \cdots + b_1 x + b_0$, then

$$p(x) + q(x) = (a_n + b_n)x^n + (a_{n-1} + b_{n-1})x^{n-1} + \cdots + (a_1 + b_1)x + (a_0 + b_0)$$

Clearly, the sum of two polynomials of degree less than or equal to $n$ is another polynomial with degree less than or equal to $n$, so axiom ($i$) is satisfied. Properties ($ii$) and ($v$) to ($x$) are obvious. If we define the zero polynomial $\mathbf{0} = 0x^n + 0x^{n-1} + \cdots + 0x + 0$, then clearly $\mathbf{0} \in P_n$ and axiom ($iii$) is satisfied. Finally, letting $-p(x) = -a_n x^n - a_{n-1}x^{n-1} - \cdots - a_1 x - a_0$, we see that axiom ($iv$) holds, so $P_n$ is a real vector space.

**EXAMPLE 8** ‡ Calculus **The Vector Spaces $C[0, 1]$ and $C[a, b]$** Let $V = C[0, 1]$ = the set of real-valued continuous functions defined on the interval $[0, 1]$. We define

$$(f + g)x = f(x) + g(x) \quad \text{and} \quad (\alpha f)(x) = \alpha[f(x)]$$

Since the sum of continuous functions is continuous, axiom ($i$) is satisfied and the other axioms are easily verified with $\mathbf{0}$ = the zero function and $(-f)(x) = -f(x)$. Similarly, $C[a, b]$, the set of real-valued functions defined and continuous on $[a, b]$, constitutes a vector space.

**EXAMPLE 9** **The Vector Space $M_{34}$** Let $V = M_{34}$ denote the set of $3 \times 4$ matrices with real components. Then with the usual sum and scalar multiplication of matrices, it is again easy to verify that $M_{34}$ is a vector space with $\mathbf{0}$ as the $3 \times 4$ zero matrix. If $A = (a_{ij})$ is in $M_{34}$, then $-A = (-a_{ij})$ is also in $M_{34}$.

**EXAMPLE 10** **The Vector Space $M_{mn}$** In an identical manner we see that $M_{mn}$, the set of $m \times n$ matrices with real components, forms a vector space for any positive integers $m$ and $n$.

**EXAMPLE 11** **A Set of Invertible Matrices Might Not Form a Vector Space** Let $S_3$ denote the set of invertible $3 \times 3$ matrices. Define the "sum" $A \oplus B$ by $A \oplus B = AB$.§ If $A$ and $B$ are invertible, then $AB$ is invertible (by Theorem 1.8.3, page 100) so that axiom ($i$) is satisfied. Axiom ($ii$) is simply the associative law for matrix multiplication (Theorem 1.6.2, page 68); axioms ($iii$) and ($iv$) are satisfied with $\mathbf{0} = I_3$ and $-A = A^{-1}$. However, $AB \neq BA$ in general (see page 66), so axion ($v$) fails and $S_3$ is not a vector space.

---

†Constant functions (including the function $f(x) = 0$) are said to be polynomials of **degree zero.**

‡ Calculus This symbol is used throughout the book to indicate that the problem or example uses calculus.

§We use a plus with a circle around it to avoid confusion with the ordinary plus sign that denotes matrix addition.

**EXAMPLE 12** **A Set of Points in a Half Plane Might Not Form a Vector Space** Let $V = \{(x, y)\colon\ y \geq 0\}$. $V$ consists of the points in $\mathbb{R}^2$ in the upper half plane (the first two quadrants). If $y_1 \geq 0$ and $y_2 \geq 0$, then $y_1 + y_2 \geq 0$; hence if $(x_1, y_1) \in V$ and $(x_2, y_2) \in V$, then $(x_1 + x_2, y_1 + y_2) \in V$. $V$ is not a vector space, however, since the vector $(1, 1)$, for example, does not have an inverse in $V$ because $(-1, -1) \notin V$. Moreover, axiom (*vi*) fails since if $(x, y) \in V$, then $\alpha(x, y) \notin V$ if $\alpha < 0$.

**EXAMPLE 13** **The Space $\mathbb{C}^n$** Let $V = \mathbb{C}^n = \{(c_1, c_2, \ldots, c_n)\colon c_i$ is a complex number for $i = 1, 2, \ldots, n\}$ and the set of scalars is the set of complex numbers. It is not difficult to verify that $\mathbb{C}^n$, too, is a vector space.

As these examples suggest, there are many different kinds of vector spaces and many kinds of sets that are *not* vector spaces. Before leaving this section, let us prove some elementary results about vector spaces.

**THEOREM 1** Let $V$ be a vector space. Then

**i.** $\alpha\mathbf{0} = \mathbf{0}$ for every scalar $\alpha$.

**ii.** $0 \cdot \mathbf{x} = \mathbf{0}$ for every $\mathbf{x} \in V$.

**iii.** If $\alpha\mathbf{x} = \mathbf{0}$, then $\alpha = 0$ or $\mathbf{x} = \mathbf{0}$ (or both).

**iv.** $(-1)\mathbf{x} = -\mathbf{x}$ for every $\mathbf{x} \in V$.

**Proof**

**i.** By axiom (*iii*), $\mathbf{0} + \mathbf{0} = \mathbf{0}$; and from axiom (*vii*),

$$\alpha\mathbf{0} = \alpha(\mathbf{0} + \mathbf{0}) = \alpha\mathbf{0} + \alpha\mathbf{0} \tag{1}$$

Adding $-\alpha\mathbf{0}$ to both sides of (1) and using the associative law (axiom *ii*), we obtain

$$\begin{aligned} \alpha\mathbf{0} + (-\alpha\mathbf{0}) &= [\alpha\mathbf{0} + \alpha\mathbf{0}] + (-\alpha\mathbf{0}) \\ \mathbf{0} &= \alpha\mathbf{0} + [\alpha\mathbf{0} + (-\alpha\mathbf{0})] \\ \mathbf{0} &= \alpha\mathbf{0} + \mathbf{0} \\ \mathbf{0} &= \alpha\mathbf{0} \end{aligned}$$

**ii.** Essentially the same proof as used in part (*i*) works. We start with $0 + 0 = 0$ and use axiom (*viii*) to see that $0\mathbf{x} = (0 + 0)\mathbf{x} = 0\mathbf{x} + 0\mathbf{x}$ or $0\mathbf{x} + (-0\mathbf{x}) = 0\mathbf{x} + [0\mathbf{x} + (-0\mathbf{x})]$ or $\mathbf{0} = 0\mathbf{x} + \mathbf{0} = 0\mathbf{x}$.

**iii.** Let $\alpha\mathbf{x} = \mathbf{0}$. If $\alpha \neq 0$, we multiply both sides of the equation by $1/\alpha$ to obtain $(1/\alpha)(\alpha\mathbf{x}) = (1/\alpha)\mathbf{0} = \mathbf{0}$ [by part (*i*)]. But $(1/\alpha)(\alpha\mathbf{x}) = 1\mathbf{x} = \mathbf{x}$ (by axiom *ix*), so $\mathbf{x} = \mathbf{0}$.

**iv.** We start with the fact that $1 + (-1) = 0$. Then, using part (*ii*), we obtain

$$\mathbf{0} = 0\mathbf{x} = [1 + (-1)]\mathbf{x} = 1\mathbf{x} + (-1)\mathbf{x} = \mathbf{x} + (-1)\mathbf{x} \tag{2}$$

We add $-\mathbf{x}$ to both sides of (2) to obtain

$$\begin{aligned}-\mathbf{x} = \mathbf{0} + (-\mathbf{x}) &= \mathbf{x} + (-1)\mathbf{x} + (-\mathbf{x}) = \mathbf{x} + (-\mathbf{x}) + (-1)\mathbf{x} \\ &= \mathbf{0} + (-1)\mathbf{x} = (-1)\mathbf{x}\end{aligned}$$

Thus $-\mathbf{x} = (-1)\mathbf{x}$. Note that we were able to reverse the order of addition in the preceding equation by using the commutative law (axiom $v$). ■

*Remark.* Part $(iii)$ of Theorem 1 is not so obvious as it seems. There are familiar situations in which $xy = 0$ does not imply that either $x$ or $y$ is zero. As an example, we look at the multiplication of $2 \times 2$ matrices. If $A = \begin{pmatrix} 0 & 1 \\ 0 & 0 \end{pmatrix}$ and $B = \begin{pmatrix} 0 & -2 \\ 0 & 0 \end{pmatrix}$, then neither $A$ nor $B$ is zero, although, as is easily verified, the product $AB = 0$, the zero matrix.

# PROBLEMS 4.2

## Self-Quiz

### *True-False*

**I.** The set of vectors $\begin{pmatrix} x \\ y \end{pmatrix}$ in $\mathbb{R}^2$ with $y = -3x$ is a real vector space.

**II.** The set of vectors $\begin{pmatrix} x \\ y \end{pmatrix}$ in $\mathbb{R}^2$ with $y = -3x + 1$ is a real vector space.

**III.** The set of invertible $5 \times 5$ matrices forms a vector space (with "+" defined as ordinary matrix addition).

**IV.** The set of constant multiples of the $2 \times 2$ identity matrix is a real vector space (with "+" defined as above).

**V.** The set of $n \times n$ identity matrices for $n = 2, 3, 4, \ldots$ is a real vector space (with "+" defined as above).

**VI.** The set of vectors $\begin{pmatrix} x \\ y \\ z \end{pmatrix}$ in $\mathbb{R}^3$ with $2x - y - 12z = 0$ is a real vector space.

**VII.** The set of vectors $\begin{pmatrix} x \\ y \\ z \end{pmatrix}$ in $\mathbb{R}^3$ with $2x - y - 12z = 1$ is a real vector space.

**VIII.** The set of polynomials of degree 3 is a real vector space (with "+" defined as ordinary polynomial addition).

## Answers to Self-Quiz

**I.** True **II.** False **III.** False **IV.** True **V.** False **VI.** True
**VII.** False **VIII.** False

In Problems 1–20 determine whether the given set is a vector space. If it is not, list the axioms that do not hold.

**1.** The set of diagonal $n \times n$ matrices under the usual matrix addition and the usual scalar multiplication.

**2.** The set of diagonal $n \times n$ matrices under multiplication (that is, $A \oplus B = AB$).

**3.** $\{(x, y)\colon\ y \leq 0;\ x, y \text{ real}\}$ with the usual addition and scalar multiplication of vectors.

**4.** The vectors in the plane lying in the first quadrant.

**5.** The set of vectors in $\mathbb{R}^3$ of the form $(x, x, x)$.

**6.** The set of polynomials of degree 4 under the operations of Example 7.

**7.** The set of $n \times n$ symmetric matrices (see Section 1.9) under the usual addition and scalar multiplication.

**8.** The set of $2 \times 2$ matrices having the form $\begin{pmatrix} 0 & a \\ b & 0 \end{pmatrix}$ under the usual addition and scalar multiplication.

**9.** The set of matrices of the form $\begin{pmatrix} 1 & \alpha \\ \beta & 1 \end{pmatrix}$ with the matrix operations of addition and scalar multiplication.

**10.** The set consisting of the single vector $(0, 0)$ under the usual operations in $\mathbb{R}^2$.

**11.** The set of polynomials of degree $\leq n$ with zero constant term.

**12.** The set of polynomials of degree $\leq n$ with positive constant term $a_0$.

**13.** The set of real valued continuous functions defined on $[0, 1]$ with $f(0) = 0$ and $f(1) = 0$ under the operations of Example 8.

**14.** The set of points in $\mathbb{R}^3$ lying on a line passing through the origin.

**15.** The set of points in $\mathbb{R}^3$ lying on the line $x = t + 1,\ y = 2t,\ z = t - 1$.

**16.** $\mathbb{R}^2$ with addition defined by $(x_1, y_1) + (x_2, y_2) = (x_1 + x_2 + 1, y_1 + y_2 + 1)$ and ordinary scalar multiplication.

**17.** The set of Problem 16 with scalar multiplication defined by $\alpha(x, y) = (\alpha + \alpha x - 1, \alpha + \alpha y - 1)$.

**18.** The set consisting of one object with addition defined by *object* + *object* = *object* and scalar multiplication defined by $\alpha$(*object*) = *object*.

† Calculus **19.** The set of differentiable functions defined on $[0, 1]$ with the operations of Example 8.

***20.** The set of real numbers of the form $a + b\sqrt{2}$, where $a$ and $b$ are rational numbers, under the usual addition of real numbers and with scalar multiplication defined only for rational scalars.

**21.** Show that in a vector space the additive identity element is unique.

**22.** Show that in a vector space each vector has a unique additive inverse.

**23.** If $\mathbf{x}$ and $\mathbf{y}$ are vectors in a vector space $V$, show that there is a unique vector $\mathbf{z} \in V$ such that $\mathbf{x} + \mathbf{z} = \mathbf{y}$.

**24.** Show that the set of positive real numbers forms a vector space under the operations $x + y = xy$ and $\alpha x = x^{\alpha}$.

---

† Calculus This symbol is used throughout the book to indicate that the problem or example uses calculus.

* Calculus **25.** Consider the homogeneous second-order differential equation

$$y''(x) + a(x)y'(x) + b(x)y(x) = 0$$

where $a(x)$ and $b(x)$ are continuous functions. Show that the set of solutions to the equation is a vector space under the usual rules for adding functions and multiplying them by real numbers.

## MATLAB 4.2

1. *(Uses m-file vctrsp.m)* The m-file *vctrsp.m* is a demonstration that illustrates the geometry of some of the vector space properties of vectors in $\mathbb{R}^2$. Enter **help vctrsp** for a description.

   Enter the vectors **x**, **y**, and **z** below and then enter the command **vctrsp(x,y,z).** The demonstration will illustrate the geometry of the commutative and associative properties of vector addition and the distributive property of scalar multiplication over vector addition. When asked to "input a scalar value for *a*," enter just the value for *a* (NO spaces) and hit the enter key.
   a. $\mathbf{x} = [3;0]$, $\mathbf{y} = [2;2]$, $\mathbf{z} = [-2;4]$. Use $a = 2$, $a = \frac{1}{2}$, and $a = -2$.
   b. $\mathbf{x} = [-5;5]$, $\mathbf{y} = [0;-4]$, $\mathbf{z} = [4;4]$. Use $a = 2$, $a = \frac{1}{3}$, and $a = -\frac{3}{2}$.
   c. Your own choices for **x**, **y**, **z**, and/or $a$

2. a. Choose some values for $n$ and $m$ and generate three random matrices of size $n \times m$, called $X$, $Y$, and $Z$. Generate two random scalars $a$ and $b$. (For example, **a = 2*rand(1)−1.**) Verify each vector space property for these example matrices and scalars. To show $A = B$, show that $A - B = 0$; for property (*iii*), decide how to generate the additive identity for $n \times m$ matrices.) Repeat for three more choices of $X$, $Y$, $Z$, $a$, and $b$ (for the same $n$ and $m$).
   b. *(Paper and pencil)* Prove the vector space properties for $M_{nm}$, the $n \times m$ matrices.
   c. *(Paper and pencil)* What is the difference between parts (a) and (b)?

## 4.3 SUBSPACES

From Example 4.2.1, page 293, we know that $\mathbb{R}^2 = \{(x, y): x \in \mathbb{R} \text{ and } y \in \mathbb{R}\}$ is a vector space. In Example 4.2.4, page 293, we saw that $V = \{(x, y): y = mx\}$ is also a vector space. Moreover, it is clear that $V \subset \mathbb{R}^2$. That is, $\mathbb{R}^2$ has a subset that is also a vector space. In fact, all vector spaces have subsets that are also vector spaces. We examine these important subsets in this section.

**DEFINITION 1** **Subspace** Let $H$ be a nonempty subset of a vector space $V$ and suppose that $H$ is itself a vector space under the operations of addition and scalar multiplication defined on $V$. Then $H$ is said to be a **subspace** of $V$.

We can say that the subspace $H$ **inherits** the operations from the "parent" vector space $V$.

We encounter many examples of subspaces in this chapter. But first we prove a result that makes it relatively easy to determine whether a subset of $V$ is indeed a subspace of $V$.

**THEOREM 1** A nonempty subset $H$ of the vector space $V$ is a subspace of $V$ if the two closure rules hold:

**Rules for Checking Whether a Nonempty Subset Is a Subspace**

**i.** If $\mathbf{x} \in H$ and $\mathbf{y} \in H$, then $\mathbf{x} + \mathbf{y} \in H$.

**ii.** If $\mathbf{x} \in H$, then $\alpha\mathbf{x} \in H$ for every scalar $\alpha$.

**Proof** Evidently, if $H$ is a vector space, then the two closure rules must hold. Conversely, to show that $H$ is a vector space, we must show that axioms (*i*) to (*x*) on page 292 hold under the operations of vector addition and scalar multiplication defined in $V$. The two closure operations [axioms (*i*) and (*vi*)] hold by hypothesis. Since vectors in $H$ are also in $V$, the associative, commutative, distributive, and multiplicative identity laws [axioms (*ii*), (*v*), (*vii*), (*viii*), (*ix*), and (*x*)] hold. Let $\mathbf{x} \in H$. Then $0\mathbf{x} \in H$ by hypothesis (*ii*). But by Theorem 4.2.1, page 296, (part *ii*), $0\mathbf{x} = \mathbf{0}$. Thus $\mathbf{0} \in H$ and axiom (*iii*) holds. Finally, by part (*ii*), $(-1)\mathbf{x} \in H$ for every $\mathbf{x} \in H$. By Theorem 4.2.1 (part *iv*), $-\mathbf{x} = (-1)\mathbf{x} \in H$ so that axiom (*iv*) also holds and the proof is complete. ■

This theorem shows that to test whether $H$ is a subspace of $V$, it is sufficient to verify that

$\mathbf{x} + \mathbf{y}$ and $\alpha\mathbf{x}$ are in $H$ when $\mathbf{x}$ and $\mathbf{y}$ are in $H$ and $\alpha$ is a scalar.

The preceding proof contains a fact that is important enough to mention explicitly:

Every subspace of a vector space $V$ contains $\mathbf{0}$. **(1)**

This fact will often make it easy to see that a particular subset of $V$ is *not* a subspace of $V$. That is, if a subset does not contain $\mathbf{0}$, then it is not a subspace. Note that the zero vector in $H$, a subspace of $V$, is the same as the zero vector in $V$.

We now give some examples of subspaces.

**EXAMPLE 1** **The Trivial Subspace** For any vector space $V$, the subset $\{\mathbf{0}\}$ consisting of the zero vector alone is a subspace since $\mathbf{0} + \mathbf{0} = \mathbf{0}$ and $\alpha\mathbf{0} = \mathbf{0}$ for every real number $\alpha$ [part ($i$) of Theorem 4.2.1]. It is called the **trivial subspace.**

**EXAMPLE 2** **A Vector Space Is a Subspace of Itself** For every vector space $V$, $V$ is a subspace of itself.

The first two examples show that every vector space $V$ contains two subspaces $\{\mathbf{0}\}$ and $V$ (which coincide if $V = \{\mathbf{0}\}$). It is more interesting to find other subspaces.

Proper subspaces

Subspaces other than $\{\mathbf{0}\}$ and $\mathbf{V}$ are called **proper subspaces.**

**EXAMPLE 3** **A Proper Subspace of $\mathbb{R}^2$** Let $H = \{(x, y)\colon\ y = mx\}$ (see Example 4.2.4, page 293). Then as we have already mentioned, $H$ is a subspace of $\mathbb{R}^2$. As we shall see in Section 4.6 (Problem 15 on page 345), if $H$ is any proper subspace of $\mathbb{R}^2$, then $H$ consists of the set of points lying on a straight line through the origin; that is, a set of points lying on a straight line passing through the origin is the only kind of proper subspace of $\mathbb{R}^2$.

**EXAMPLE 4** **A Proper Subspace of $\mathbb{R}^3$** Let $H = \{(x, y, z)\colon\ x = at, y = bt, \text{ and } z = ct;\ a, b, c, t \text{ real}\}$. Then $H$ consists of the vectors in $\mathbb{R}^3$ lying on a straight line passing through the origin. To see that $H$ is a subspace of $\mathbb{R}^3$, let $\mathbf{x} = (at_1, bt_1, ct_1) \in H$ and $\mathbf{y} = (at_2, bt_2, ct_2) \in H$. Then

$$\mathbf{x} + \mathbf{y} = (a(t_1 + t_2),\ b(t_1 + t_2),\ c(t_1 + t_2)) \in H$$

and

$$\alpha\mathbf{x} = (a(\alpha t_1),\ b(\alpha t_2),\ c(\alpha t_3)) \in H.$$

Thus $H$ is a subspace of $\mathbb{R}^3$.

**EXAMPLE 5** **Another Proper Subspace of $\mathbb{R}^3$** Let $\pi = \{(x, y, z)\colon\ ax + by + cz = 0;\ a, b, c \text{ real}\}$. Then as we saw in Example 4.2.6, page 294, $\pi$ is a vector space; thus $\pi$ is a subspace of $\mathbb{R}^3$.

We shall prove in Section 4.6 that sets of vectors lying on lines and planes through the origin are the only proper subspaces of $\mathbb{R}^3$.

Before studying more examples, we note that *not every vector space has proper subspaces.*

**EXAMPLE 6** **$\mathbb{R}$ Has No Proper Subspace** Let $H$ be a subspace of $\mathbb{R}$.† If $H \neq \{\mathbf{0}\}$, then $H$ contains a nonzero real number $\alpha$. Then, by axiom $(vi)$, $1 = (1/\alpha)\alpha \in H$ and $\beta 1 = \beta \in H$ for every real number $\beta$. Thus if $H$ is not the trivial subspace, then $H = \mathbb{R}$. That is, $\mathbb{R}$ has *no* proper subspace.

**EXAMPLE 7** **Some Proper Subspaces of $P_n$** If $P_n$ denotes the vector space of polynomials of degree $\leq n$ (Example 4.2.7, page 295), and if $0 \leq m < n$, then $P_m$ is a proper subspace of $P_n$, as is easily verified.

**EXAMPLE 8** **A Proper Subspace of $M_{mn}$** Let $M_{mn}$ (Example 4.2.10, page 295) denote the vector space of $m \times n$ matrices with real components and let $H = \{A \in M_{mn}: a_{11} = 0\}$. By the definition of matrix addition and scalar multiplication, it is clear that the two closure axioms hold so that $H$ is a subspace.

**EXAMPLE 9** **A Subset That Is Not a Subspace of $M_{nn}$** Let $V = M_{nn}$ (the $n \times n$ matrices) and let $H = \{A \in M_{nn}: A \text{ is invertible}\}$. Then $H$ is not a subspace since the $n \times n$ zero matrix is not in $H$.

**EXAMPLE 10** Calculus

**A Proper Subspace of $C[0, 1]$** $P_n[0, 1]$‡ $\subset C[0, 1]$ (see Example 4.2.8, page 295) because every polynomial is continuous and $P_n$ is a vector space for every integer $n$ so that each $P_n[0, 1]$ is a subspace of $C[0, 1]$.

**EXAMPLE 11** Calculus

**$C^1[0, 1]$ Is a Proper Subspace of $C[0, 1]$** Let $C^1[0, 1]$ denote the set of functions with continuous first derivatives defined on $[0, 1]$. Since every differentiable function is continuous, we have $C^1[0, 1] \subset C[0, 1]$. Since the sum of two differentiable functions is differentiable and a constant multiple of a differentiable function is differentiable, we see that $C^1[0, 1]$ is a subspace of $C[0, 1]$. It is a proper subspace because not every continuous function is differentiable.

**EXAMPLE 12** Calculus

**Another Proper Subspace of $C[0, 1]$** If $f \in C[0, 1]$, then $\int_0^1 f(x)\,dx$ exists. Let $H = \{f \in C[0, 1]: \int_0^1 f(x)\,dx = 0\}$. If $f \in H$ and $g \in H$, then $\int_0^1 [f(x) + g(x)]\,dx = \int_0^1 f(x)\,dx + \int_0^1 g(x)\,dx = 0 + 0 = 0$ and $\int_0^1 \alpha f(x)\,dx = \alpha \int_0^1 f(x)\,dx = 0$. Thus $f + g$ and $\alpha f$ are in $H$ for every real number $\alpha$. This shows that $H$ is a proper subspace of $C[0, 1]$.

As the last three examples illustrate, a vector space can have a great number and variety of proper subspaces. Before leaving this section, we prove an interesting fact about subspaces.

---

† Note that $\mathbb{R}$ is a real vector space; that is, $\mathbb{R}$ is a vector space where the scalars are taken to be the reals. This is Example 4.2.1, page 293, with $n = 1$.

‡ $P_n[0, 1]$ denotes the set of polynomials defined on the interval $[0, 1]$ of degree $\leq n$.

**THEOREM 2** Let $H_1$ and $H_2$ be subspaces of a vector space $V$. Then $H_1 \cap H_2$ is a subspace of $V$.

**Proof** Note that $H_1 \cap H_2$ is nonempty because it contains $\mathbf{0}$. Let $\mathbf{x}_1 \in H_1 \cap H_2$ and $\mathbf{x}_2 \in H_1 \cap H_2$. Then since $H_1$ and $H_2$ are subspaces, $\mathbf{x}_1 + \mathbf{x}_2 \in H_1$ and $\mathbf{x}_1 + \mathbf{x}_2 \in H_2$. This means that $\mathbf{x}_1 + \mathbf{x}_2 \in H_1 \cap H_2$. Similarly, $\alpha\mathbf{x}_1 \in H_1 \cap H_2$. Thus the two closure axioms are satisfied and $H_1 \cap H_2$ is a subspace. ∎

**EXAMPLE 13** **The Intersection of Two Subspaces of $\mathbb{R}^3$ Is a Subspace** In $\mathbb{R}^3$ let $H_1 = \{(x, y, z)\colon 2x - y - z = 0\}$ and $H_2 = \{(x, y, z)\colon x + 2y + 3z = 0\}$. Then $H_1$ and $H_2$ consist of vectors lying on planes through the origin and are, by Example 5, subspaces of $\mathbb{R}^3$. $H_1 \cap H_2$ is the intersection of the two planes that we compute as in Example 9 in Section 3.5:

$$\begin{aligned} x + 2y + 3z &= 0 \\ 2x - \phantom{2}y - \phantom{3}z &= 0 \end{aligned}$$

or, row-reducing, we have

$$\left(\begin{array}{ccc|c} 1 & 2 & 3 & 0 \\ 2 & -1 & -1 & 0 \end{array}\right) \longrightarrow \left(\begin{array}{ccc|c} 1 & 2 & 3 & 0 \\ 0 & -5 & -7 & 0 \end{array}\right) \longrightarrow$$

$$\left(\begin{array}{ccc|c} 1 & 2 & 3 & 0 \\ 0 & 1 & \frac{7}{5} & 0 \end{array}\right) \longrightarrow \left(\begin{array}{ccc|c} 1 & 0 & \frac{1}{5} & 0 \\ 0 & 1 & \frac{7}{5} & 0 \end{array}\right)$$

Thus all solutions to the homogeneous system are given by $(-\frac{1}{5}z, -\frac{7}{5}z, z)$. Setting $z = t$, we obtain the parametric equations of a line $L$ in $\mathbb{R}^3$: $x = -\frac{1}{5}t$, $y = -\frac{7}{5}t$, $z = t$. As we saw in Example 4, the set of vectors on $L$ constitutes a subspace of $\mathbb{R}^3$. ∎

***Remark.*** It is not necessarily true that if $H_1$ and $H_2$ are subspaces of $V$, then $H_1 \cup H_2$ is a subspace of $V$ (it may or may not be). For example, $H_1 = \{(x, y)\colon y = 2x\}$ and $\{(x, y)\colon y = 3x\}$ are subspaces of $\mathbb{R}^2$, but $H_1 \cup H_2$ is not a subspace. To see this, observe that $(1, 2) \in H_1$ and $(1, 3) \in H_2$ so that both $(1, 2)$ and $(1, 3)$ are in $H_1 \cup H_2$. But $(1, 2) + (1, 3) = (2, 5) \notin H_1 \cup H_2$ because $(2, 5) \notin H_1$ and $(2, 5) \notin H_2$. Thus $H_1 \cup H_2$ is not closed under addition, and is therefore not a subspace.

## PROBLEMS 4.3

### Self-Quiz

#### *True-False*

**I.** The set of vectors of the form $\begin{pmatrix} x \\ y \\ 1 \end{pmatrix}$ is a subspace of $\mathbb{R}^3$.

**II.** The set of vectors of the form $\begin{pmatrix} x \\ 0 \\ z \end{pmatrix}$ is a subspace of $\mathbb{R}^3$.

**III.** The set of diagonal $3 \times 3$ matrices is a subspace of $M_{33}$.

**IV.** The set of upper triangular $3 \times 3$ matrices is a subspace of $M_{33}$.

**V.** The set of triangular $3 \times 3$ matrices is a subspace of $M_{33}$.

**VI.** Let $H$ be a subspace of $M_{22}$. Then $\begin{pmatrix} 0 & 0 \\ 0 & 0 \end{pmatrix}$ must be in $H$.

**VII.** Let $H = \left\{ \begin{pmatrix} x \\ y \\ z \end{pmatrix} : 2x + 3y - z = 0 \right\}$ and $K = \left\{ \begin{pmatrix} x \\ y \\ z \end{pmatrix} : x - 2y + 5z = 0 \right\}$, then $H \cup K$ is a subspace of $\mathbb{R}^3$.

**VIII.** If $H$ and $K$ are as in Problem VII, then $H \cap K$ is a subspace of $\mathbb{R}^3$.

**IX.** The set of polynomials of degree 2 is a subspace of $P_3$.

In Problems 1–20 determine whether the given subset $H$ of the vector space $V$ is a subspace of $V$.

**1.** $V = \mathbb{R}^2$; $H = \{(x, y): y \geq 0\}$

**2.** $V = \mathbb{R}^2$; $H = \{(x, y): x = y\}$

**3.** $V = \mathbb{R}^3$; $H =$ the $xy$-plane

**4.** $V = \mathbb{R}^2$; $H = \{(x, y): x^2 + y^2 \leq 1\}$

**5.** $V = M_{nn}$; $H = \{D \in M_{nn}: D \text{ is diagonal}\}$

**6.** $V = M_{nn}$; $H = \{T \in M_{nn}: T \text{ is upper triangular}\}$

**7.** $V = M_{nn}$; $H = \{S \in M_{nn}: S \text{ is symmetric}\}$

**8.** $V = M_{mn}$; $H = \{A \in M_{mn}: a_{ij} = 0\}$

**9.** $V = M_{22}$; $H = \left\{ A \in M_{22}: A = \begin{pmatrix} a & b \\ -b & c \end{pmatrix} \right\}$

**10.** $V = M_{22}$; $H = \left\{ A \in M_{22}: A = \begin{pmatrix} a & 1 + a \\ 0 & 0 \end{pmatrix} \right\}$

**11.** $V = M_{22}$; $H = \left\{ A \in M_{22}: A = \begin{pmatrix} 0 & a \\ b & 0 \end{pmatrix} \right\}$

**12.** $V = P_4$; $H = \{p \in P_4: \deg p = 4\}$

**13.** $V = P_4$; $H = \{p \in P_4: p(0) = 0\}$

**14.** $V = P_n$; $H = \{p \in P_n: p(0) = 0\}$

**15.** $V = P_n$; $H = \{p \in P_n: p(0) = 1\}$

**16.** $V = C[0, 1]$; $H = \{f \in C[0, 1]: f(0) = f(1) = 0\}$

**17.** $V = C[0, 1]$; $H = \{f \in C[0, 1]: f(0) = 2\}$

Calculus **18.** $V = C^1[0, 1]$; $H = \{f \in C^1[0, 1]: f'(0) = 0\}$

Calculus **19.** $V = C[a, b]$, where $a$ and $b$ are real numbers and $a < b$; $H = \{f \in C[a, b]: \int_a^b f(x)\, dx = 0\}$

Calculus **20.** $V = C[a, b]$; $H = \{f \in C[a, b]: \int_a^b f(x)\, dx = 1\}$

---

### Answers to Self-Quiz

**I.** False **II.** True **III.** True **IV.** True **V.** False **VI.** True
**VII.** False **VIII.** True **IX.** False

21. Let $V = M_{22}$; let $H_1 = \{A \in M_{22}\colon a_{11} = 0\}$ and $H_2 = \left\{A \in M_{22}\colon A = \begin{pmatrix} -b & a \\ a & b \end{pmatrix}\right\}$.
    **a.** Show that $H_1$ and $H_2$ are subspaces.
    **b.** Describe the subset $H = H_1 \cap H_2$ and show that it is a subspace.

Calculus 22. If $V = C[0, 1]$, let $H_1$ denote the subspace of Example 10 and $H_2$ denote the subspace of Example 11. Describe the set $H_1 \cap H_2$ and show that it is a subspace.

23. Let $A$ be an $n \times m$ matrix and let $H = \{\mathbf{x} \in \mathbb{R}^m\colon A\mathbf{x} = \mathbf{0}\}$. Show that $H$ is a subspace of $\mathbb{R}^m$. $H$ is called the **null space** of the matrix $A$.

24. In Problem 23 let $H = \{\mathbf{x} \in \mathbb{R}^m\colon A\mathbf{x} \neq \mathbf{0}\}$. Show that $H$ is not a subspace of $\mathbb{R}^m$.

25. Let $H = \{(x, y, z, w)\colon ax + by + cz + dw = 0\}$, where $a$, $b$, $c$, and $d$ are real numbers not all zero. Show that $H$ is a proper subspace of $\mathbb{R}^4$. $H$ is called a **hyperplane** in $\mathbb{R}^4$ that passes through the origin.

26. Let $H = \{(x_1, x_2, \ldots, x_n)\colon a_1x_1 + a_2x_2 + \cdots + a_nx_n = 0\}$, where $a_1, a_2, \ldots, a_n$ are real numbers that are not all zero. Show that $H$ is a proper subspace of $\mathbb{R}^n$. $H$, as in Problem 25, is called a **hyperplane** in $\mathbb{R}^n$.

27. Let $H_1$ and $H_2$ be subspaces of a vector space $V$. Let $H_1 + H_2 = \{\mathbf{v}\colon \mathbf{v} = \mathbf{v}_1 + \mathbf{v}_2$ with $\mathbf{v}_1 \in H_1$ and $\mathbf{v}_2 \in H_2\}$. Show that $H_1 + H_2$ is a subspace of $V$.

28. Let $\mathbf{v}_1$ and $\mathbf{v}_2$ be two vectors in $\mathbb{R}^2$. Show that $H = \{\mathbf{v}\colon \mathbf{v} = a\mathbf{v}_1 + b\mathbf{v}_2;\ a, b \text{ real}\}$ is a subspace of $\mathbb{R}^2$.

*29. In Problem 28 show that if $\mathbf{v}_1$ and $\mathbf{v}_2$ are not collinear, then $H = \mathbb{R}^2$.

*30. Let $\mathbf{v}_1, \mathbf{v}_2, \ldots, \mathbf{v}_n$ be arbitrary vectors in a vector space $V$. Let $H = \{\mathbf{v} \in V\colon \mathbf{v} = a_1\mathbf{v}_1 + a_2\mathbf{v}_2 + \cdots + a_n\mathbf{v}_n$, where $a_1, a_2, \ldots, a_n$ are scalars$\}$. Show that $H$ is a subspace of $V$. $H$ is called the subspace **spanned** by the vectors $\mathbf{v}_1, \mathbf{v}_2, \ldots, \mathbf{v}_n$.

## MATLAB 4.3

1. a. Generate a random $4 \times 4$ matrix $A$ and let **S = triu(A) + triu(A)'**. Verify that $S$ is symmetric.
   b. Using part (a), generate two random $4 \times 4$ real symmetric matrices, $S$ and $T$, and a random scalar, $a$. Verify that $aS$ and $S + T$ are also symmetric. Repeat for four more choices of $S$, $T$, and $a$.
   c. Why can we say that we have gathered evidence that the subset of $4 \times 4$ symmetric matrices is a subspace of $M_{44}$?
   d. *(Paper and pencil)* Prove that the subset of $n \times n$ symmetric matrices is a subspace of $M_{nn}$.

## 4.4 LINEAR COMBINATION AND SPAN

We have seen that every vector $\mathbf{v} = (a, b, c)$ in $\mathbb{R}^3$ can be written in the form

$$\mathbf{v} = a\mathbf{i} + b\mathbf{j} + c\mathbf{k}$$

In this case we say that $\mathbf{v}$ is a *linear combination* of the three vectors $\mathbf{i}$, $\mathbf{j}$, and $\mathbf{k}$. More generally, we have the following definition.

**DEFINITION 1** **Linear Combination** Let $\mathbf{v}_1, \mathbf{v}_2, \ldots, \mathbf{v}_n$ be vectors in a vector space $V$. Then any vector of the form

$$a_1\mathbf{v}_1 + a_2\mathbf{v}_2 + \cdots + a_n\mathbf{v}_n \tag{1}$$

where $a_1, a_2, \ldots, a_n$ are scalars is called a **linear combination** of $\mathbf{v}_1, \mathbf{v}_2, \ldots, \mathbf{v}_n$.

**EXAMPLE 1** **A Linear Combination in $\mathbb{R}^3$** In $\mathbb{R}^3$, $\begin{pmatrix} -7 \\ 7 \\ 7 \end{pmatrix}$ is a linear combination of $\begin{pmatrix} -1 \\ 2 \\ 4 \end{pmatrix}$ and $\begin{pmatrix} 5 \\ -3 \\ 1 \end{pmatrix}$ since $\begin{pmatrix} -7 \\ 7 \\ 7 \end{pmatrix} = 2\begin{pmatrix} -1 \\ 2 \\ 4 \end{pmatrix} - \begin{pmatrix} 5 \\ -3 \\ 1 \end{pmatrix}$.

**EXAMPLE 2** **A Linear Combination in $M_{23}$** In $M_{23}$, $\begin{pmatrix} -3 & 2 & 8 \\ -1 & 9 & 3 \end{pmatrix} = 3\begin{pmatrix} -1 & 0 & 4 \\ 1 & 1 & 5 \end{pmatrix} + 2\begin{pmatrix} 0 & 1 & -2 \\ -2 & 3 & -6 \end{pmatrix}$, which shows that $\begin{pmatrix} -3 & 2 & 8 \\ -1 & 9 & 3 \end{pmatrix}$ is a linear combination of $\begin{pmatrix} -1 & 0 & 4 \\ 1 & 1 & 5 \end{pmatrix}$ and $\begin{pmatrix} 0 & 1 & -2 \\ -2 & 3 & -6 \end{pmatrix}$.

**EXAMPLE 3** **Linear Combinations in $P_n$** In $P_n$ every polynomial can be written as a linear combination of the "monomials" $1, x, x^2, \ldots, x^n$.

**DEFINITION 2** **Span** The vectors $\mathbf{v}_1, \mathbf{v}_2, \ldots, \mathbf{v}_n$ in a vector space $V$ are said to **span** $V$ if every vector in $V$ can be written as a linear combination of them. That is, for every $\mathbf{v} \in V$ there are scalars $a_1, a_2, \ldots, a_n$ such that

$$\mathbf{v} = a_1\mathbf{v}_1 + a_2\mathbf{v}_2 + \cdots + a_n\mathbf{v}_n \tag{2}$$

**EXAMPLE 4** **Sets of Vectors That Span $\mathbb{R}^2$ and $\mathbb{R}^3$** We saw in Section 3.1 that the vectors $\mathbf{i} = \begin{pmatrix} 1 \\ 0 \end{pmatrix}$ and $\mathbf{j} = \begin{pmatrix} 0 \\ 1 \end{pmatrix}$ span $\mathbb{R}^2$. In Section 3.3 we saw that $\mathbf{i} = \begin{pmatrix} 1 \\ 0 \\ 0 \end{pmatrix}$, $\mathbf{j} = \begin{pmatrix} 0 \\ 1 \\ 0 \end{pmatrix}$, and $\mathbf{k} = \begin{pmatrix} 0 \\ 0 \\ 1 \end{pmatrix}$ span $\mathbb{R}^3$.

We now look briefly at spanning sets of some other vector spaces.

**EXAMPLE 5** **$n + 1$ Vectors That Span $P_n$** From Example 3 it follows that the monomials $1, x, x^2, \ldots, x^n$ span $P_n$.

**EXAMPLE 6** **Four Vectors That Span $M_{22}$** Since $\begin{pmatrix} a & b \\ c & d \end{pmatrix} = a\begin{pmatrix} 1 & 0 \\ 0 & 0 \end{pmatrix} + b\begin{pmatrix} 0 & 1 \\ 0 & 0 \end{pmatrix} + c\begin{pmatrix} 0 & 0 \\ 1 & 0 \end{pmatrix} + d\begin{pmatrix} 0 & 0 \\ 0 & 1 \end{pmatrix}$, we see that $\begin{pmatrix} 1 & 0 \\ 0 & 0 \end{pmatrix}$, $\begin{pmatrix} 0 & 1 \\ 0 & 0 \end{pmatrix}$, $\begin{pmatrix} 0 & 0 \\ 1 & 0 \end{pmatrix}$, and $\begin{pmatrix} 0 & 0 \\ 0 & 1 \end{pmatrix}$ span $M_{22}$.

**EXAMPLE 7** **No Finite Set of Polynomials Spans $P$** Let $P$ denote the vector space of polynomials. Then no *finite* set of polynomials spans $P$. To see this, suppose that $p_1, p_2, \ldots, p_m$ are polynomials. Let $p_k$ be the polynomial of largest degree in this set and let $N = \deg p_k$. Then the polynomial $p(x) = x^{N+1}$ cannot be written as a linear combination of $p_1, p_2, \ldots, p_m$. For example if $N = 3$, then $x^4 \neq c_0 + c_1x + c_2x^2 + c_3x^3$ for any scalars $c_0$, $c_1$, $c_2$, and $c_3$.

We now turn to another way of finding subspaces of a vector space $V$.

**DEFINITION 3** **Span of a Set of Vectors** Let $\mathbf{v}_1, \mathbf{v}_2, \ldots, \mathbf{v}_k$ be $k$ vectors in a vector space $V$. The **span** of $\{\mathbf{v}_1, \mathbf{v}_2, \ldots, \mathbf{v}_k\}$ is the set of linear combinations of $\mathbf{v}_1, \mathbf{v}_2, \ldots, \mathbf{v}_k$. That is,

$$\text{span}\,\{\mathbf{v}_1, \mathbf{v}_2, \ldots, \mathbf{v}_k\} = \{\mathbf{v}\colon \mathbf{v} = a_1\mathbf{v}_1 + a_2\mathbf{v}_2 + \cdots + a_k\mathbf{v}_k\} \tag{3}$$

where $a_1, a_2, \ldots, a_k$ are arbitrary scalars.

**THEOREM 1** If $\mathbf{v}_1, \mathbf{v}_2, \ldots, \mathbf{v}_k$ are vectors in a vector space $V$, then span $\{\mathbf{v}_1, \mathbf{v}_2, \ldots, \mathbf{v}_k\}$ is a subspace of $V$.

**Proof** The proof is easy and is left as an exercise (see Problem 16).

**EXAMPLE 8** **The Span of Two Vectors in $\mathbb{R}^3$** Let $\mathbf{v}_1 = (2, -1, 4)$ and $\mathbf{v}_2 = (4, 1, 6)$. Then $H = \text{span}\,\{\mathbf{v}_1, \mathbf{v}_2\} = \{\mathbf{v}\colon \mathbf{v} = a_1(2, -1, 4) + a_2(4, 1, 6)\}$. What does $H$ look like? If $\mathbf{v} = (x, y, z) \in H$, then we have $x = 2a_1 + 4a_2$, $y = -a_1 + a_2$, and $z = 4a_1 + 6a_2$. If we think of $(x, y, z)$ as being fixed, then we can view these equations as a system of three equations in the two unknowns $a_1$, $a_2$. We solve this system in the usual way:

$$\left(\begin{array}{cc|c} -1 & 1 & y \\ 2 & 4 & x \\ 4 & 6 & z \end{array}\right) \xrightarrow{R_1 \to -R_1} \left(\begin{array}{cc|c} 1 & -1 & -y \\ 2 & 4 & x \\ 4 & 6 & z \end{array}\right) \xrightarrow[R_3 \to R_3 - 4R_1]{R_2 \to R_2 - 2R_1} \left(\begin{array}{cc|c} 1 & -1 & -y \\ 0 & 6 & x + 2y \\ 0 & 10 & z + 4y \end{array}\right)$$

$$\xrightarrow{R_2 \to \frac{1}{6}R_2} \left(\begin{array}{cc|c} 1 & -1 & -y \\ 0 & 1 & (x + 2y)/6 \\ 0 & 10 & z + 4y \end{array}\right) \xrightarrow[R_3 \to R_3 - 10R_2]{R_1 \to R_1 + R_2} \left(\begin{array}{cc|c} 1 & 0 & x/6 - 2y/3 \\ 0 & 1 & x/6 + y/3 \\ 0 & 0 & -5x/3 + 2y/3 + z \end{array}\right)$$

From Chapter 1 we see that the system has a solution only if $-5x/3 + 2y/3 + z = 0$; or multiplying through by $-3$, if

$$5x - 2y - 3z = 0 \tag{4}$$

Equation (4) is the equation of a plane in $\mathbb{R}^3$ passing through the origin.

The last example can be generalized to prove the following interesting fact:

> *The span of two nonzero vectors in $\mathbb{R}^3$ that are not parallel is a plane passing through the origin.*

For a suggested proof see Problems 19 and 20.

We can give a geometric interpretation of this result. Look at the vectors in Figure 4.1. We know (from Section 3.1) the geometric interpretation of the vectors

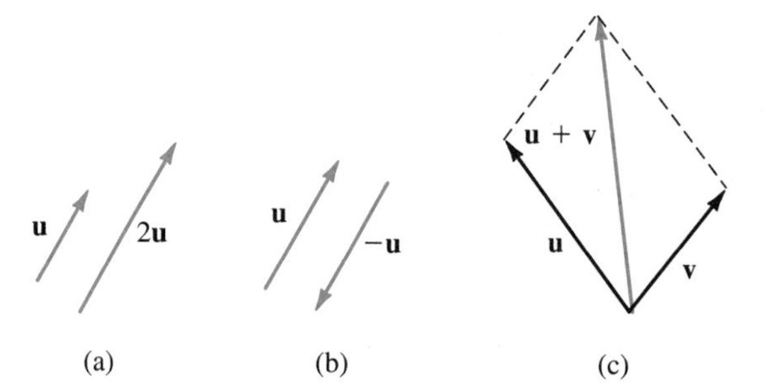


**Figure 4.1** $\mathbf{u} + \mathbf{v}$ is obtained from the parallelogram rule

$2\mathbf{u}$, $-\mathbf{u}$, and $\mathbf{u} + \mathbf{v}$, for example. Using these, we see that any other vector in the plane of $\mathbf{u}$ and $\mathbf{v}$ can be obtained as a linear combination of $\mathbf{u}$ and $\mathbf{v}$. Figure 4.2 shows how in four different situations a third vector $\mathbf{w}$ in the plane of $\mathbf{u}$ and $\mathbf{v}$ can be written as $\alpha\mathbf{u} + \beta\mathbf{v}$ for appropriate choices of the numbers $\alpha$ and $\beta$.

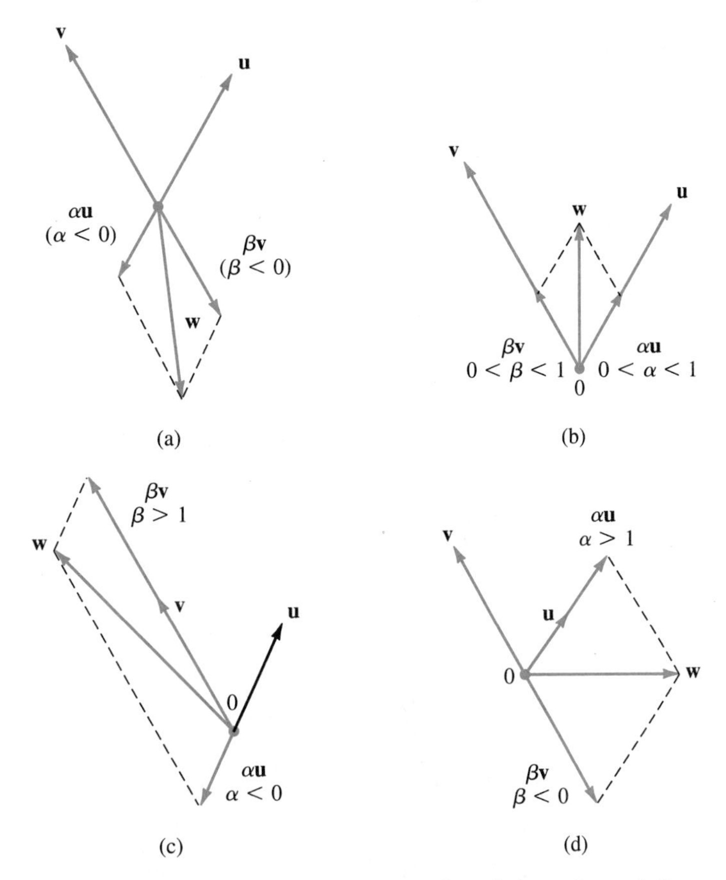


**Figure 4.2** In each case $\mathbf{w} = \alpha\mathbf{u} + \beta\mathbf{v}$ for appropriate choices of $\alpha$ and $\beta$

*Remark.* In Definitions 2 and 3 we used the word "span" in two different ways: as a verb and as a noun. We emphasize that

verb
↓

A set of vectors $\mathbf{v}_1, \mathbf{v}_2, \ldots, \mathbf{v}_n$ *span* $V$ if every vector in $V$ can be written as a linear combination of $\mathbf{v}_1, \mathbf{v}_2, \ldots, \mathbf{v}_n$;

but

noun
↓

The *span* of the $n$ vectors $\mathbf{v}_1, \mathbf{v}_2, \ldots, \mathbf{v}_k$ is the set of linear combinations of these vectors.

These two concepts are different—even though they use the same word.

We close this section by citing a useful result. Its proof is not difficult and is left as an exercise (see Problem 21).

**THEOREM 2** Let $\mathbf{v}_1, \mathbf{v}_2, \ldots, \mathbf{v}_n, \mathbf{v}_{n+1}$ be $n + 1$ vectors that are in a vector space $V$. If $\mathbf{v}_1, \mathbf{v}_2, \ldots, \mathbf{v}_n$ span $V$, then $\mathbf{v}_1, \mathbf{v}_2, \ldots, \mathbf{v}_n, \mathbf{v}_{n+1}$ also span $V$. That is, the addition of one (or more) vectors to a spanning set yields another spanning set. ■

# PROBLEMS 4.4

## Self-Quiz

**I.** Which of the following pairs of vectors could *not* possibly span $\mathbb{R}^2$?

**a.** $\begin{pmatrix}1\\1\end{pmatrix}, \begin{pmatrix}-3\\-3\end{pmatrix}$ **b.** $\begin{pmatrix}1\\1\end{pmatrix}, \begin{pmatrix}2\\2\end{pmatrix}$ **c.** $\begin{pmatrix}1\\1\end{pmatrix}, \begin{pmatrix}-1\\1\end{pmatrix}$ **d.** $\begin{pmatrix}1\\3\end{pmatrix}, \begin{pmatrix}0\\0\end{pmatrix}$ **e.** $\begin{pmatrix}1\\3\end{pmatrix}, \begin{pmatrix}3\\1\end{pmatrix}$

**II.** Which of the following sets of polynomials span $P_2$?

**a.** $1, x^2$ **b.** $3, 2x, -x^2$ **c.** $1 + x, 2 + 2x, x^2$ **d.** $1, 1 + x, 1 + x^2$

## *True-False*

**III.** $\begin{pmatrix}3\\5\end{pmatrix}$ is in the span of $\left\{\begin{pmatrix}1\\1\end{pmatrix}, \begin{pmatrix}2\\4\end{pmatrix}\right\}$.

**IV.** $\begin{pmatrix}1\\2\\3\end{pmatrix}$ is in the span of $\left\{\begin{pmatrix}2\\0\\4\end{pmatrix}, \begin{pmatrix}-1\\0\\3\end{pmatrix}\right\}$.

**V.** $\{1, x, x^2, x^3, \ldots, x^{10,000}\}$ spans $P$.

**VI.** $\left\{\begin{pmatrix}1 & 0\\0 & 0\end{pmatrix}, \begin{pmatrix}0 & 1\\0 & 0\end{pmatrix}, \begin{pmatrix}0 & 0\\1 & 0\end{pmatrix}, \begin{pmatrix}0 & 0\\0 & 1\end{pmatrix}\right\}$ spans $M_{22}$.

**VII.** span $\left\{\begin{pmatrix}1\\2\\-1\\3\end{pmatrix}, \begin{pmatrix}7\\1\\0\\4\end{pmatrix}, \begin{pmatrix}-8\\0\\8\\2\end{pmatrix}\right\}$ is a subspace of $\mathbb{R}^3$.

**VIII.** span $\left\{\begin{pmatrix}1\\2\\-1\\3\end{pmatrix}, \begin{pmatrix}7\\1\\0\\4\end{pmatrix}, \begin{pmatrix}-8\\0\\8\\2\end{pmatrix}\right\}$ is a subspace of $\mathbb{R}^4$.

**IX.** If $\left\{\begin{pmatrix}1\\2\end{pmatrix}, \begin{pmatrix}2\\3\end{pmatrix}\right\}$ spans $\mathbb{R}^2$, then $\left\{\begin{pmatrix}1\\2\end{pmatrix}, \begin{pmatrix}2\\3\end{pmatrix}, \begin{pmatrix}-2\\-3\end{pmatrix}\right\}$ also spans $\mathbb{R}^2$.

## Answers to Self-Quiz

**I.** a, b, d **II.** b, d **III.** True **IV.** False **V.** False **VI.** True
**VII.** False **VIII.** True **IX.** True

In Problems 1–13 determine whether the given set of vectors spans the given vector space.

**1.** In $\mathbb{R}^2$: $\begin{pmatrix}1\\2\end{pmatrix}, \begin{pmatrix}3\\4\end{pmatrix}$

**2.** In $\mathbb{R}^2$: $\begin{pmatrix}1\\1\end{pmatrix}, \begin{pmatrix}2\\1\end{pmatrix}, \begin{pmatrix}2\\2\end{pmatrix}$

**3.** In $\mathbb{R}^2$: $\begin{pmatrix}1\\1\end{pmatrix}, \begin{pmatrix}2\\2\end{pmatrix}, \begin{pmatrix}5\\5\end{pmatrix}$

**4.** In $\mathbb{R}^3$: $\begin{pmatrix}1\\2\\3\end{pmatrix}, \begin{pmatrix}-1\\2\\3\end{pmatrix}, \begin{pmatrix}5\\2\\3\end{pmatrix}$

**5.** In $\mathbb{R}^3$: $\begin{pmatrix}1\\1\\1\end{pmatrix}, \begin{pmatrix}0\\1\\1\end{pmatrix}, \begin{pmatrix}0\\0\\1\end{pmatrix}$

**6.** In $\mathbb{R}^3$: $\begin{pmatrix}2\\0\\1\end{pmatrix}, \begin{pmatrix}3\\1\\2\end{pmatrix}, \begin{pmatrix}1\\1\\1\end{pmatrix}, \begin{pmatrix}7\\3\\5\end{pmatrix}$

**7.** In $\mathbb{R}^3$: $(1, -1, 2), (1, 1, 2), (0, 0, 1)$

**8.** In $\mathbb{R}^3$: $(1, -1, 2), (-1, 1, 2), (0, 0, 1)$

**9.** In $P_2$: $1 - x,\ 3 - x^2$

**10.** In $P_2$: $1 - x,\ 3 - x^2,\ x$

**11.** In $M_{22}$: $\begin{pmatrix}2 & 1\\0 & 0\end{pmatrix}, \begin{pmatrix}0 & 0\\2 & 1\end{pmatrix}, \begin{pmatrix}3 & -1\\0 & 0\end{pmatrix}, \begin{pmatrix}0 & 0\\3 & 1\end{pmatrix}$

**12.** In $M_{22}$: $\begin{pmatrix}1 & 0\\1 & 0\end{pmatrix}, \begin{pmatrix}1 & 2\\0 & 0\end{pmatrix}, \begin{pmatrix}4 & -1\\3 & 0\end{pmatrix}, \begin{pmatrix}-2 & 5\\6 & 0\end{pmatrix}$

**13.** In $M_{23}$: $\begin{pmatrix}1 & 0 & 0\\0 & 0 & 0\end{pmatrix}, \begin{pmatrix}0 & 1 & 0\\0 & 0 & 0\end{pmatrix}, \begin{pmatrix}0 & 0 & 1\\0 & 0 & 0\end{pmatrix},$
$\begin{pmatrix}0 & 0 & 0\\1 & 0 & 0\end{pmatrix}, \begin{pmatrix}0 & 0 & 0\\0 & 1 & 0\end{pmatrix}, \begin{pmatrix}0 & 0 & 0\\0 & 0 & 1\end{pmatrix}$

**14.** Show that two polynomials cannot span $P_2$.

***15.** If $p_1, p_2, \ldots, p_m$ span $P_n$, show that $m \geq n + 1$.

**16.** Show that if $\mathbf{u}$ and $\mathbf{v}$ are in span $\{\mathbf{v}_1, \mathbf{v}_2, \ldots, \mathbf{v}_k\}$, then $\mathbf{u} + \mathbf{v}$ and $\alpha\mathbf{u}$ are in span $\{\mathbf{v}_1, \mathbf{v}_2, \ldots, \mathbf{v}_k\}$ [*Hint:* Using the definition of span write $\mathbf{u} + \mathbf{v}$ and $\alpha\mathbf{u}$ as linear combinations of $\mathbf{v}_1, \mathbf{v}_2, \ldots, \mathbf{v}_k$.]

**17.** Show that the infinite set $\{1, x, x^2, x^3, \ldots\}$ spans $P$, the vector space of polynomials.

**18.** Let $H$ be a subspace of $V$ containing $\mathbf{v}_1, \mathbf{v}_2, \ldots, \mathbf{v}_n$. Show that span $\{\mathbf{v}_1, \mathbf{v}_2, \ldots, \mathbf{v}_n\} \subseteq H$. That is, span $\{\mathbf{v}_1, \mathbf{v}_2, \ldots, \mathbf{v}_n\}$ is the *smallest* subspace of $V$ containing $\mathbf{v}_1, \mathbf{v}_2, \ldots, \mathbf{v}_n$.

**19.** Let $\mathbf{v}_1 = (x_1, y_1, z_1)$ and $\mathbf{v}_2 = (x_2, y_2, z_2)$ be in $\mathbb{R}^3$. Show that if $\mathbf{v}_2 = c\mathbf{v}_1$, then span $\{\mathbf{v}_1, \mathbf{v}_2\}$ is a line passing through the origin.

****20.** In Problem 19 assume that $\mathbf{v}_1$ and $\mathbf{v}_2$ are not parallel. Show that $H =$ span $\{\mathbf{v}_1, \mathbf{v}_2\}$ is a plane passing through the origin. What is the equation of that plane? [*Hint:* If $(x, y, z) \in H$, write $\mathbf{v} = a_1\mathbf{v}_1 + a_2\mathbf{v}_2$ and find a condition relating $x$, $y$, and $z$ such that the resulting $3 \times 2$ system has a solution.]

**21.** Prove Theorem 2. [*Hint:* If $\mathbf{v} \in V$, write $\mathbf{v}$ as a linear combination of $\mathbf{v}_1, \mathbf{v}_2, \ldots, \mathbf{v}_n, \mathbf{v}_{n+1}$ with the coefficient of $\mathbf{v}_{n+1}$ equal to zero.]

**22.** Show that $M_{22}$ can be spanned by invertible matrices.

***23.** Let $\{\mathbf{u}_1, \mathbf{u}_2, \ldots, \mathbf{u}_n\}$ and $\{\mathbf{v}_1, \mathbf{v}_2, \ldots, \mathbf{v}_n\}$ be $2n$ vectors in a vector space $V$. Suppose that

$$\begin{aligned}
\mathbf{v}_1 &= a_{11}\mathbf{u}_1 + a_{12}\mathbf{u}_2 + \cdots + a_{1n}\mathbf{u}_n \\
\mathbf{v}_2 &= a_{21}\mathbf{u}_1 + a_{22}\mathbf{u}_2 + \cdots + a_{2n}\mathbf{u}_n \\
\vdots \quad & \quad \vdots \qquad\quad \vdots \qquad\qquad\quad \vdots \\
\mathbf{v}_n &= a_{n1}\mathbf{u}_1 + a_{n2}\mathbf{u}_2 + \cdots + a_{nn}\mathbf{u}_n
\end{aligned}$$

Show that if

$$\begin{vmatrix} a_{11} & a_{12} & \cdots & a_{1n} \\ a_{21} & a_{22} & \cdots & a_{2n} \\ \vdots & \vdots & & \vdots \\ a_{n1} & a_{n2} & \cdots & a_{nn} \end{vmatrix} \neq 0$$

then span $\{\mathbf{u}_1, \mathbf{u}_2, \ldots, \mathbf{u}_n\}$ = span $\{\mathbf{v}_1, \mathbf{v}_2, \ldots, \mathbf{v}_n\}$.

## MATLAB 4.4

1. **Visualization of Linear Combinations**

   a. Rework MATLAB Problems 2 and 3 from Section 3.1.

   M

   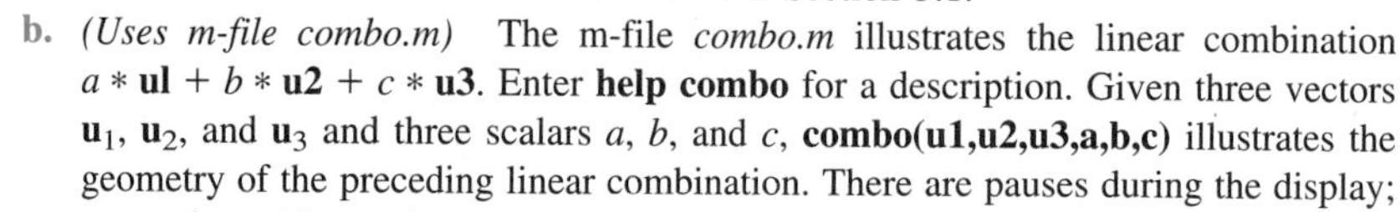

   b. *(Uses m-file combo.m)* The m-file *combo.m* illustrates the linear combination $a * \mathbf{ul} + b * \mathbf{u2} + c * \mathbf{u3}$. Enter **help combo** for a description. Given three vectors $\mathbf{u}_1$, $\mathbf{u}_2$, and $\mathbf{u}_3$ and three scalars $a$, $b$, and $c$, **combo(u1,u2,u3,a,b,c)** illustrates the geometry of the preceding linear combination. There are pauses during the display; to continue, hit any key.

   **i.** **ul** = [1;2], **u2** = [−2;3], **u3** = [5;4], $a = -2$, $b = 2$, $c = -1$

   **ii.** **ul** = [1;1], **u2** = [−1;1], **u3** = [3;0], $a = 2$, $b = -1$, $c = .5$

   **iii.** Your own choices

2. a. *(Paper and pencil)* To say that $\mathbf{w}$ is in span $\{\mathbf{u}, \mathbf{v}\}$ means that there exist scalars $c_1$ and $c_2$ such that $\mathbf{w} = c_1\mathbf{u} + c_2\mathbf{v}$. For the sets of vectors below, write down $\mathbf{w} = c_1\mathbf{u} + c_2\mathbf{v}$, interpret it as a system of equations for the unknowns $c_1$ and $c_2$, verify that the augmented matrix for the system is $[\mathbf{u}\ \mathbf{v}|\mathbf{w}]$, and solve the system.

   **i.** $\mathbf{u} = \begin{pmatrix} 1 \\ 2 \end{pmatrix}$ $\quad \mathbf{v} = \begin{pmatrix} -1 \\ 3 \end{pmatrix}$ $\quad \mathbf{w} = \begin{pmatrix} 3 \\ 1 \end{pmatrix}$

   **ii.** $\mathbf{u} = \begin{pmatrix} 2 \\ 4 \end{pmatrix}$ $\quad \mathbf{v} = \begin{pmatrix} -1 \\ 2 \end{pmatrix}$ $\quad \mathbf{w} = \begin{pmatrix} -1 \\ 6 \end{pmatrix}$

   **iii.** $\mathbf{u} = \begin{pmatrix} 1 \\ -1 \end{pmatrix}$ $\quad \mathbf{v} = \begin{pmatrix} 2 \\ 1 \end{pmatrix}$ $\quad \mathbf{w} = \begin{pmatrix} \frac{8}{3} \\ -\frac{5}{3} \end{pmatrix}$

   M

   b. *(Uses the m-file lincomb.m)* Verify results (and observe the geometry) by first entering the vectors **u**, **v**, and **w** and then entering **lincomb(u,v,w)** for each of the sets of vectors in part (a).

3. a. *(Paper and pencil)* To say that $\mathbf{w}$ is in span $\{\mathbf{v}_1, \mathbf{v}_2, \mathbf{v}_3\}$ means that there exist scalars $c_1$, $c_2$, and $c_3$ such that $\mathbf{w} = c_1\mathbf{v}_1 + c_2\mathbf{v}_2 + c_3\mathbf{v}_3$. For the sets of vectors below, write down $\mathbf{w} = c_1\mathbf{v}_1 + c_2\mathbf{v}_2 + c_3\mathbf{v}_3$, interpret it as a system of equations for the unknowns $c_1$, $c_2$, and $c_3$, verify that the augmented matrix for the system is $[\mathbf{v}_1\ \mathbf{v}_2\ \mathbf{v}_3|\mathbf{w}]$, and solve the system. Note that there will be infinitely many solutions.

   **i.** $\mathbf{v}_1 = \begin{pmatrix} 1 \\ 1 \end{pmatrix}$ $\quad \mathbf{v}_2 = \begin{pmatrix} -1 \\ 1 \end{pmatrix}$ $\quad \mathbf{v}_3 = \begin{pmatrix} 3 \\ 0 \end{pmatrix}$ $\quad \mathbf{w} = \begin{pmatrix} 1 \\ -4 \end{pmatrix}$

**ii.** $\mathbf{v}_1 = \begin{pmatrix} 1 \\ 2 \end{pmatrix}$ $\mathbf{v}_2 = \begin{pmatrix} -2 \\ 3 \end{pmatrix}$ $\mathbf{v}_3 = \begin{pmatrix} 5 \\ 4 \end{pmatrix}$ $\mathbf{w} = \begin{pmatrix} -4 \\ -1 \end{pmatrix}$

**b.** *(Paper and pencil)* This part and part (c) explore the "meaning" of having infinitely many solutions. For each of the sets of vectors in part (a):

**i.** Let $c_3 = 0$ and solve for $c_2$ and $c_1$. Write $\mathbf{w}$ as a linear combination of $\mathbf{v}_1$ and $\mathbf{v}_2$.

**ii.** Let $c_2 = 0$ and solve for $c_1$ and $c_3$. Write $\mathbf{w}$ as a linear combination of $\mathbf{v}_1$ and $\mathbf{v}_3$.

**iii.** Let $c_1 = 0$ and solve for $c_2$ and $c_3$. Write $\mathbf{w}$ as a linear combination of $\mathbf{v}_2$ and $\mathbf{v}_3$.

M

**c.** *(Uses the m-file combine2.m)* Enter **help combine2** for a description. For each of the sets of vectors in part (a), enter the vectors $\mathbf{v_1}$, $\mathbf{v_2}$, $\mathbf{v_3}$, and $\mathbf{w}$ and then enter **combine2(v1,v2,v3,w).** The geometry of the observations in part (b) is demonstrated.

*Note.* It is important to notice that no two of the vectors $\mathbf{v}_1$, $\mathbf{v}_2$, or $\mathbf{v}_3$ are parallel.

**4. a.** *(Paper and pencil)* For the set of vectors $\{\mathbf{v}_1, \mathbf{v}_2, \mathbf{v}_3\}$ and the vector $\mathbf{w}$ in (i) in part (c), write down the equations expressing $\mathbf{w} = c_1\mathbf{v}_1 + c_2\mathbf{v}_2 + c_3\mathbf{v}_3$ as a system of equations with $c_1$, $c_2$, and $c_3$ as the unknowns. Write the augmented matrix for this system of equations and verify that it is $[\mathbf{v}_1\ \mathbf{v}_2\ \mathbf{v}_3|\mathbf{w}]$. Explain why $\mathbf{w}$ is a linear combination of $\mathbf{v}_1$, $\mathbf{v}_2$, and $\mathbf{v}_3$ if and only if the system has a solution.

**b.** For each set of vectors $\{\mathbf{v}_1, \ldots, \mathbf{v}_k\}$ and $\mathbf{w}$ in part (c), find the augmented matrix $[\mathbf{v}_1\ \mathbf{v}_2 \cdots \mathbf{v}_k|\mathbf{w}]$ and solve the corresponding system using the **rref** command. Form $\mathbf{c} = \begin{pmatrix} c_1 \\ \vdots \\ c_k \end{pmatrix}$, a solution to the system if a solution exists.

**c.** For each case worked in part (b), write a conclusion saying whether or not $\mathbf{w}$ is a linear combination of $\{\mathbf{v}_1, \ldots, \mathbf{v}_k\}$ and why. If it is a linear combination, verify that $\mathbf{w} = c_1\mathbf{v}_1 + \cdots + c_k\mathbf{v}_k$, where $c_1, \ldots, c_k$ are the components of the solution vector $\mathbf{c}$ in part (b).

**i.** $\left\{\begin{pmatrix} 4 \\ 2 \\ 9 \end{pmatrix}, \begin{pmatrix} 7 \\ 1 \\ -8 \end{pmatrix}, \begin{pmatrix} 3 \\ -2 \\ 4 \end{pmatrix}\right\}$ $\mathbf{w} = \begin{pmatrix} 3 \\ -3 \\ 25 \end{pmatrix}$

**ii.** $\left\{\begin{pmatrix} 4 \\ 2 \\ 9 \end{pmatrix}, \begin{pmatrix} 7 \\ 0 \\ 13 \end{pmatrix}, \begin{pmatrix} 3 \\ -2 \\ 4 \end{pmatrix}\right\}$ $\mathbf{w} = \begin{pmatrix} 3 \\ -3 \\ 25 \end{pmatrix}$

**iii.** $\left\{\begin{pmatrix} 8 \\ 5 \\ -5 \\ -9 \end{pmatrix}, \begin{pmatrix} 5 \\ -3 \\ 3 \\ 5 \end{pmatrix}, \begin{pmatrix} 10 \\ -3 \\ -5 \\ 10 \end{pmatrix}\right\}$ $\mathbf{w} = \begin{pmatrix} 10.5 \\ 2 \\ -14 \\ 3.5 \end{pmatrix}$

**iv.** same set as (iii); $\mathbf{w} = \begin{pmatrix} 1 \\ 1 \\ 1 \\ 1 \end{pmatrix}$

**v.** $\left\{\begin{pmatrix} 4 \\ 5 \\ 3 \\ -9 \end{pmatrix}, \begin{pmatrix} 3 \\ 8 \\ -5 \\ -1 \end{pmatrix}, \begin{pmatrix} 5 \\ 2 \\ 11 \\ -17 \end{pmatrix}, \begin{pmatrix} -3 \\ -7 \\ 0 \\ 8 \end{pmatrix}\right\}$ $\mathbf{w} = \begin{pmatrix} -19 \\ -9 \\ -46 \\ 74 \end{pmatrix}$

**vi.** same set as (i); $\mathbf{w} = \begin{pmatrix} 1 \\ 1 \\ 1 \end{pmatrix}$

**vii.** $\left\{ \begin{pmatrix} 1 \\ 2 \end{pmatrix}, \begin{pmatrix} -1 \\ 0 \end{pmatrix}, \begin{pmatrix} 1 \\ -1 \end{pmatrix} \right\}$ $\mathbf{w} = \begin{pmatrix} 3 \\ 2 \end{pmatrix}$

5. **a.** For $\{\mathbf{v}_1, \ldots, \mathbf{v}_k\}$ below, let $A = [\mathbf{v}_1\ \mathbf{v}_2 \cdots \mathbf{v}_k]$ and find **rref(A).** Argue why there would be a solution to the system $[A|\mathbf{w}]$ for any $\mathbf{w}$ in the indicated $\mathbb{R}^n$. Explain why you can conclude that the set spans all of that $\mathbb{R}^n$.

**i.** $\mathbb{R}^3$ $\left\{ \begin{pmatrix} 4 \\ 2 \\ 9 \end{pmatrix}, \begin{pmatrix} 7 \\ 1 \\ -8 \end{pmatrix}, \begin{pmatrix} 3 \\ -2 \\ 4 \end{pmatrix} \right\}$

**ii.** $\mathbb{R}^3$ $\left\{ \begin{pmatrix} 9 \\ -9 \\ 5 \end{pmatrix}, \begin{pmatrix} 5 \\ 7 \\ -7 \end{pmatrix}, \begin{pmatrix} -10 \\ 4 \\ 7 \end{pmatrix}, \begin{pmatrix} 3 \\ 5 \\ 5 \end{pmatrix} \right\}$

**b.** For $\{\mathbf{v}_1, \ldots, \mathbf{v}_k\}$ below, let $A = [\mathbf{v}_1\ \mathbf{v}_2 \cdots \mathbf{v}_k]$ and find **rref(A).** Argue why there would be some $\mathbf{w}$ in the indicated $\mathbb{R}^n$ for which there is no solution to the system $[A|\mathbf{w}]$. Experiment using MATLAB to find such a $\mathbf{w}$. Explain why you can conclude that the set does not span all of that $\mathbb{R}^n$.

**i.** $\mathbb{R}^4$ $\left\{ \begin{pmatrix} 10 \\ 0 \\ -5 \\ -8 \end{pmatrix}, \begin{pmatrix} 9 \\ -9 \\ 0 \\ -2 \end{pmatrix}, \begin{pmatrix} -4 \\ 8 \\ 1 \\ -1 \end{pmatrix} \right\}$

**ii.** $\mathbb{R}^4$ $\left\{ \begin{pmatrix} 4 \\ 5 \\ 3 \\ -9 \end{pmatrix}, \begin{pmatrix} 3 \\ 8 \\ -5 \\ -1 \end{pmatrix}, \begin{pmatrix} 5 \\ 2 \\ 11 \\ -17 \end{pmatrix}, \begin{pmatrix} 3 \\ -7 \\ 0 \\ 8 \end{pmatrix} \right\}$

**iii.** $\mathbb{R}^3$ $\left\{ \begin{pmatrix} 9 \\ -9 \\ 5 \end{pmatrix}, \begin{pmatrix} 5 \\ 7 \\ 7 \end{pmatrix}, \begin{pmatrix} 14 \\ -2 \\ 12 \end{pmatrix}, \begin{pmatrix} -4 \\ 16 \\ 2 \end{pmatrix} \right\}$

6. Consider the matrices in MATLAB Problem 2 in Section 1.8. For each matrix, test for invertibility. For each matrix, decide if the columns of $A$ would span all of $\mathbb{R}^n$ or not (the size of the matrix is $n \times n$). Write a conjecture concerning the relationship between the invertibility of an $n \times n$ matrix and whether the columns of the matrix span all of $\mathbb{R}^n$.

7. Recall from previous problems that $\mathbf{w} = c_1\mathbf{v}_1 + \cdots + c_k\mathbf{v}_k$; that is, $\mathbf{w}$ is in span $\{\mathbf{v}_1, \ldots, \mathbf{v}_k\}$ whenever $\mathbf{c} = \begin{pmatrix} c_1 \\ \vdots \\ c_k \end{pmatrix}$ is a solution to the system of equations whose augmented matrix is $[\mathbf{v}_1 \cdots \mathbf{v}_k|\mathbf{w}]$.

**a.** For the following set of vectors, show that any $\mathbf{w}$ in $\mathbb{R}^4$ will be in the span of the set of vectors but that there will be infinitely many ways to write $\mathbf{w}$ as a linear combination of the set of vectors; that is, there will be infinitely many choices for coefficients $c_1, \ldots, c_k$.

$$\left\{\begin{pmatrix}3\\-7\\4\\-2\end{pmatrix},\begin{pmatrix}-2\\0\\-7\\2\end{pmatrix},\begin{pmatrix}7\\2\\9\\1\end{pmatrix},\begin{pmatrix}14\\-5\\27\\-5\end{pmatrix},\begin{pmatrix}1\\-5\\0\\-1\end{pmatrix}\right\}$$

b. For each **w** given below:
   i. Solve the system to find the coefficients to write **w** as a linear combination of the set of vectors and write the solutions in terms of the natural arbitrary variables (that is, the variables corresponding to the columns in the **rref** without pivots).
   ii. Set the arbitrary variables equal to zero and write **w** as a linear combination of the vectors in the set.
   iii. Verify that **w** equals the linear combination you found:

$$\mathbf{w}=\begin{pmatrix}23\\-15\\33\\-5\end{pmatrix}\qquad \mathbf{w}=\begin{pmatrix}-13\\18\\-45\\18\end{pmatrix}$$

c. From your results in part (b), which vectors of the original set were not needed in writing the **w**'s as linear combinations of the set of vectors? Why? How could you recognize them by looking at the reduced row echelon form of the matrix whose columns are the set of vectors?

d. Consider the subset of the original vectors obtained by deleting the unneeded vectors. Show that each unneeded vector is in the span of this subset of vectors. Argue why any vector **w** in $\mathbb{R}^4$ will be in the span of this subset of vectors and that the coefficients for the linear combination will be unique.

e. Repeat parts (a) through (d) for the following set of vectors and the given **w**'s in $\mathbb{R}^3$.

$$\left\{\begin{pmatrix}10\\8\\-5\end{pmatrix},\begin{pmatrix}0\\2\\7\end{pmatrix},\begin{pmatrix}-10\\-4\\19\end{pmatrix},\begin{pmatrix}-6\\-7\\1\end{pmatrix},\begin{pmatrix}32\\32\\-5\end{pmatrix}\right\}\qquad \mathbf{w}=\begin{pmatrix}26\\31\\17\end{pmatrix}\qquad \mathbf{w}=\begin{pmatrix}2\\20\\52\end{pmatrix}$$

8. **Application** A concrete company stocks three basic concrete mix blends, given below. Quantities are measured in grams and each "unit" of mix weighs 60 grams. Custom mixes can be formulated by blending combinations of the three basic mixes; hence possible custom mixes belong to the span of the three vectors representing the three basic mixes.

| | A | B | C |
|---|---|---|---|
| Cement | 20 | 18 | 12 |
| Water | 10 | 10 | 10 |
| Sand | 20 | 25 | 15 |
| Gravel | 10 | 5 | 15 |
| Fly ash | 0 | 2 | 8 |

a. Can you make a custom mix consisting of 1000 grams of cement, 200 grams of water, 1000 grams of sand, 500 grams of gravel, and 300 grams of fly ash? Why or why not? If so, how many units each of mix—mix $A$, mix $B$, and mix $C$—are needed to formulate the custom blend?

b. Suppose you wish to make 5000 grams of concrete that has a water to cement ratio of 2 to 3, with 1250 grams of cement. If 1500 grams of sand and 1000 grams of gravel are to be included in the specification, find the amount of fly ash to make 5000 grams of concrete. Can this be formulated as a custom blend? If so, how many units of each of the three mixes are needed to formulate the custom blend?

*Note.* This problem is taken from Deborah P. Levinson, "Teaching Elementary Linear Algebra with MATLAB to Engineering Students," in *Proceedings of the Fifth International Conference on Technology in Collegiate Mathematics,* 1992.

9. We can represent polynomials as vectors by focusing only on the coefficients. Let $p(x) = 5x^3 + 4x^2 + 3x + 1$. We can represent $p(x)$ as the vector $\mathbf{v} = \begin{pmatrix} 1 \\ 3 \\ 4 \\ 5 \end{pmatrix}$. In this representation, the first component is the constant term, the second component is the coefficient of the $x$ term, the third component is the coefficient of the $x^2$ term, and the fourth component is the coefficient of the $x^3$ term.

a. *(Paper and pencil)* Explain why $\mathbf{u} = \begin{pmatrix} -5 \\ 3 \\ 0 \\ 1 \end{pmatrix}$ represents the polynomial $q(x) = x^3 + 3x - 5$.

b. Find the polynomial $r(x) = 2p(x) - 3q(x)$. Find the vector $\mathbf{w} = 2\mathbf{v} - 3\mathbf{u}$ and explain how $\mathbf{w}$ represents $r(x)$.

For parts (c) through (e), first represent each polynomial by a vector as described above. Then answer the questions concerning span as you would if they were asked about a set of vectors.

c. In $P_2$, is $p(x) = 2x - 1$ in the span of $\{-5x^2 - 2, -6x^2 - 9x + 8, -x^2 - 7x + 9\}$? If so, write $p(x)$ as a linear combination of the polynomials in the set. Does the set of polynomials span all of $P_2$? Why or why not?

d. In $P_3$, is $p(x) = x^3 + 3x^2 + 29x - 17$ in the span of $\{-2x^3 - 7x^2 + 8x - 8, 7x^3 + 9x^2 + 3x + 5, -7x^3 + 6x^2 - x - 3\}$? If so, write $p(x)$ as a linear combination of the polynomials in the set. Does the set of polynomials span all of $P_3$? Why or why not?

e. Does the following set of polynomials span $P_3$? Why or why not?

$$\{x^3 - x + 2, x^3 + x^2 + 3x + 1, 2x^3 + x^2 + 2x + 1, -x^2 + 1\}$$

10. Suppose $A = \begin{pmatrix} a_1 & c_1 & e_1 \\ b_1 & d_1 & f_1 \end{pmatrix}$ and $B = \begin{pmatrix} a_2 & c_2 & e_2 \\ b_2 & d_2 & f_2 \end{pmatrix}$.

Let $\mathbf{v} = \begin{pmatrix} a_1 \\ b_1 \\ c_1 \\ d_1 \\ e_1 \\ f_1 \end{pmatrix}$ and $\mathbf{w} = \begin{pmatrix} a_2 \\ b_2 \\ c_2 \\ d_2 \\ e_2 \\ f_2 \end{pmatrix}$. Note that $\mathbf{v}$ represents $A$ in the sense that it is constructed from $A$ by starting with the (1, 1) entry of $A$, listing the entries of the first column in order, continuing the listing with the entries of the second column of $A$ in order, and finishing the listing with entries of the third column of $A$. Note also that $\mathbf{w}$ represents $B$ in the same way.

a. *(Paper and pencil)* Write out the matrix $C = A - 2B$. Write down the vector that represents $C$ in the manner described above and verify that this vector equals $\mathbf{v} - 2\mathbf{w}$.

For parts (b) through (d), first represent each matrix by a vector as described above. Then answer the questions concerning span as you would if they were asked about a set of vectors.

b. Is $\begin{pmatrix} 1 & 3 \\ 29 & -17 \end{pmatrix}$ in the span of the set of matrices below? If so, write it as a linear combination:

$$\left\{ \begin{pmatrix} -2 & -7 \\ 8 & -8 \end{pmatrix}, \begin{pmatrix} 7 & 9 \\ 3 & 5 \end{pmatrix}, \begin{pmatrix} -7 & 6 \\ -1 & -3 \end{pmatrix} \right\}$$

Does this set span all of $M_{22}$? Why or why not?

c. Is $\begin{pmatrix} 4 & 7 & -10 \\ -2 & -6 & 1 \end{pmatrix}$ in the span of the matrices below? If so, write it as a linear combination.

$$\left\{ \begin{pmatrix} 6 & 5 & -1 \\ 9 & 3 & -1 \end{pmatrix}, \begin{pmatrix} 6 & 4 & 4 \\ 10 & 9 & 7 \end{pmatrix}, \begin{pmatrix} -4 & 1 & 0 \\ -8 & -2 & 2 \end{pmatrix}, \begin{pmatrix} 8 & -1 & 5 \\ 7 & 4 & 6 \end{pmatrix}, \begin{pmatrix} 4 & 5 & -10 \\ 8 & 0 & -1 \end{pmatrix}, \begin{pmatrix} -9 & 4 & 0 \\ 3 & 4 & -6 \end{pmatrix} \right\}$$

Does this set span all of $M_{23}$? Why or why not?

d. Does the following set of matrices span all of $M_{22}$? Why or why not?

$$\left\{ \begin{pmatrix} 1 & 0 \\ -1 & 2 \end{pmatrix}, \begin{pmatrix} 1 & 1 \\ 3 & 1 \end{pmatrix}, \begin{pmatrix} 2 & 1 \\ 2 & 1 \end{pmatrix}, \begin{pmatrix} 0 & -1 \\ 0 & 1 \end{pmatrix} \right\}$$

## 4.5 LINEAR INDEPENDENCE

In the study of linear algebra, one of the central ideas is that of the linear dependence or independence of vectors. In this section we define what we mean by linear independence and show how it is related to the theory of homogeneous systems of equations and determinants.

Is there a special relationship between the vectors $\mathbf{v}_1 = \begin{pmatrix} 1 \\ 2 \end{pmatrix}$ and $\mathbf{v}_2 = \begin{pmatrix} 2 \\ 4 \end{pmatrix}$? Of course, we see that $\mathbf{v}_2 = 2\mathbf{v}_1$ or, writing this equation in another way,

$$2\mathbf{v}_1 - \mathbf{v}_2 = \mathbf{0} \tag{1}$$

That is, the zero vector can be written as a nontrivial linear combination of $\mathbf{v}_1$ and $\mathbf{v}_2$ (i.e., where the coefficients in the linear combination are not both zero). What is special about the vectors $\mathbf{v}_1 = \begin{pmatrix} 1 \\ 2 \\ 3 \end{pmatrix}$, $\mathbf{v}_2 = \begin{pmatrix} -4 \\ 1 \\ 5 \end{pmatrix}$, and $\mathbf{v}_3 = \begin{pmatrix} -5 \\ 8 \\ 19 \end{pmatrix}$? This question is more difficult to answer at first glance. It is easy to verify, however, that $\mathbf{v}_3 = 3\mathbf{v}_1 + 2\mathbf{v}_2$, or rewriting, we obtain

$$3\mathbf{v}_1 + 2\mathbf{v}_2 - \mathbf{v}_3 = \mathbf{0} \tag{2}$$

Now we have written the zero vector as a linear combination of $\mathbf{v}_1$, $\mathbf{v}_2$, and $\mathbf{v}_3$. It appears that the two vectors in equation (1) and the three vectors in (2) are more closely related than an arbitrary pair of 2-vectors or an arbitrary triple of 3-vectors. In each case we say that the vectors are *linearly dependent.* In general, we have the following important definition.

**DEFINITION 1** **Linear Dependence and Independence** Let $\mathbf{v}_1, \mathbf{v}_2, \ldots, \mathbf{v}_n$ be $n$ vectors in a vector space $V$. Then the vectors are said to be **linearly dependent** if there exist $n$ scalars $c_1, c_2, \ldots, c_n$ *not all zero* such that

$$c_1\mathbf{v}_1 + c_2\mathbf{v}_2 + \cdots + c_n\mathbf{v}_n = \mathbf{0} \tag{3}$$

If the vectors are not linearly dependent, they are said to be **linearly independent.**

Putting this another way, $\mathbf{v}_1, \mathbf{v}_2, \ldots, \mathbf{v}_n$ are linearly independent if the equation $c_1\mathbf{v}_1 + c_2\mathbf{v}_2 + \cdots + c_n\mathbf{v}_n = \mathbf{0}$ holds only for $c_1 = c_2 = \cdots = c_n = 0$. They are linearly dependent if the zero vector in $V$ can be written as a linear combination of $\mathbf{v}_1, \mathbf{v}_2, \ldots, \mathbf{v}_n$ with not all the coefficients equal to zero.

*Note.* We say that the *vectors* $\mathbf{v}_1, \mathbf{v}_2, \ldots, \mathbf{v}_n$ *are* linearly independent (or dependent), or that *the set* of vectors $\{\mathbf{v}_1, \mathbf{v}_2, \ldots, \mathbf{v}_n\}$ *is* linearly independent (or dependent). That is, we use these two phrases interchangeably.

How do we determine whether a set of vectors is linearly dependent or independent? The case for 2-vectors is easy.

**THEOREM 1** Two vectors in a vector space $V$ are linearly dependent if and only if one is a scalar multiple of the other.

**Proof** First, suppose that $\mathbf{v}_2 = c\mathbf{v}_1$ for some scalar $c \neq 0$. Then $c\mathbf{v}_1 - \mathbf{v}_2 = \mathbf{0}$ and $\mathbf{v}_1$ and $\mathbf{v}_2$ are linearly dependent. On the other hand, suppose that $\mathbf{v}_1$ and $\mathbf{v}_2$ are linearly dependent. Then there are constants $c_1$ and $c_2$, not both zero, such that $c_1\mathbf{v}_1 + c_2\mathbf{v}_2 = \mathbf{0}$. If $c_1 \neq 0$, then, dividing by $c_1$, we obtain $\mathbf{v}_1 + (c_2/c_1)\mathbf{v}_2 = \mathbf{0}$ or

$$\mathbf{v}_1 = \left(-\frac{c_2}{c_1}\right)\mathbf{v}_2$$

That is, $\mathbf{v}_1$ is a scalar multiple of $\mathbf{v}_2$. If $c_1 = 0$, then $c_2 \neq 0$, and hence $\mathbf{v}_2 = \mathbf{0} = 0\mathbf{v}_1$. ■

**EXAMPLE 1** **Two Linearly Dependent Vectors in $\mathbb{R}^4$** The vectors $\mathbf{v}_1 = \begin{pmatrix} 2 \\ -1 \\ 0 \\ 3 \end{pmatrix}$ and $\mathbf{v}_2 = \begin{pmatrix} -6 \\ 3 \\ 0 \\ -9 \end{pmatrix}$ are linearly dependent since $\mathbf{v}_2 = -3\mathbf{v}_1$. ■

**EXAMPLE 2** **Two Linearly Independent Vectors in $\mathbb{R}^3$** The vectors $\begin{pmatrix} 1 \\ 2 \\ 4 \end{pmatrix}$ and $\begin{pmatrix} 2 \\ 5 \\ -3 \end{pmatrix}$ are linearly independent; if they were not, we would have $\begin{pmatrix} 2 \\ 5 \\ -3 \end{pmatrix} = c\begin{pmatrix} 1 \\ 2 \\ 4 \end{pmatrix} = \begin{pmatrix} c \\ 2c \\ 4c \end{pmatrix}$. Then $2 = c$, $5 = 2c$, and $-3 = 4c$, which is clearly impossible for any number $c$. ■

**EXAMPLE 3** **Determining Whether Three Vectors in $\mathbb{R}^3$ Are Linearly Dependent or Independent** Determine whether the vectors $\begin{pmatrix} 1 \\ -2 \\ 3 \end{pmatrix}$, $\begin{pmatrix} 2 \\ -2 \\ 0 \end{pmatrix}$, and $\begin{pmatrix} 0 \\ 1 \\ 7 \end{pmatrix}$ are linearly dependent or independent.

**Solution** Suppose that $c_1\begin{pmatrix} 1 \\ -2 \\ 3 \end{pmatrix} + c_2\begin{pmatrix} 2 \\ -2 \\ 0 \end{pmatrix} + c_3\begin{pmatrix} 0 \\ 1 \\ 7 \end{pmatrix} = \mathbf{0} = \begin{pmatrix} 0 \\ 0 \\ 0 \end{pmatrix}$. Then multiplying through and adding, we have $\begin{pmatrix} c_1 + 2c_2 \\ -2c_1 - 2c_2 + c_3 \\ 3c_1 + 7c_3 \end{pmatrix} = \begin{pmatrix} 0 \\ 0 \\ 0 \end{pmatrix}$. This yields a homogeneous system of three equations in the three unknowns $c_1$, $c_2$, and $c_3$:

$$\begin{aligned} c_1 + 2c_2 \qquad &= 0 \\ -2c_1 - 2c_2 + c_3 &= 0 \\ 3c_1 \qquad + 7c_3 &= 0 \end{aligned} \tag{4}$$

Thus the vectors will be linearly dependent if and only if system (4) has nontrivial solutions. We write system (4) using an augmented matrix and then row-reduce. The reduced row echelon form of $\left(\begin{array}{ccc|c} 1 & 2 & 0 & 0 \\ -2 & -2 & 1 & 0 \\ 3 & 0 & 7 & 0 \end{array}\right)$ is $\left(\begin{array}{ccc|c} 1 & 0 & 0 & 0 \\ 0 & 1 & 0 & 0 \\ 0 & 0 & 1 & 0 \end{array}\right)$. The last system of equations reads $c_1 = 0$, $c_2 = 0$, $c_3 = 0$. Hence (4) has no nontrivial solutions and the given vectors are linearly independent. ■

**EXAMPLE 4** **Determining Whether Three Vectors in $\mathbb{R}^3$ Are Linearly Dependent or Independent** Determine whether the vectors $\begin{pmatrix} 1 \\ -3 \\ 0 \end{pmatrix}$, $\begin{pmatrix} 3 \\ 0 \\ 4 \end{pmatrix}$, and $\begin{pmatrix} 11 \\ -6 \\ 12 \end{pmatrix}$ are linearly dependent or independent.

**Solution** The equation $c_1\begin{pmatrix} 1 \\ -3 \\ 0 \end{pmatrix} + c_2\begin{pmatrix} 3 \\ 0 \\ 4 \end{pmatrix} + c_3\begin{pmatrix} 11 \\ -6 \\ 12 \end{pmatrix} = \begin{pmatrix} 0 \\ 0 \\ 0 \end{pmatrix}$ leads to the homogeneous system

$$\begin{aligned} c_1 + 3c_2 + 11c_3 &= 0 \\ -3c_1 \qquad - 6c_3 &= 0 \\ 4c_2 + 12c_3 &= 0 \end{aligned} \tag{5}$$

Writing system (5) in augmented-matrix form and row-reducing, we obtain

$$\left(\begin{array}{ccc|c} 1 & 3 & 11 & 0 \\ -3 & 0 & -6 & 0 \\ 0 & 4 & 12 & 0 \end{array}\right) \longrightarrow \left(\begin{array}{ccc|c} 1 & 3 & 11 & 0 \\ 0 & 9 & 27 & 0 \\ 0 & 4 & 12 & 0 \end{array}\right)$$

$$\longrightarrow \left(\begin{array}{ccc|c} 1 & 3 & 11 & 0 \\ 0 & 1 & 3 & 0 \\ 0 & 4 & 12 & 0 \end{array}\right) \longrightarrow \left(\begin{array}{ccc|c} 1 & 0 & 2 & 0 \\ 0 & 1 & 3 & 0 \\ 0 & 0 & 0 & 0 \end{array}\right)$$

We can stop here since the theory of Section 1.4 shows us that system (5) has an infinite number of solutions. For example, the last augmented matrix reads

$$\begin{aligned} c_1 \qquad + 2c_3 &= 0 \\ c_2 + 3c_3 &= 0 \end{aligned}$$

If we choose $c_3 = 1$, we have $c_2 = -3$ and $c_1 = -2$ so that, as is easily verified, $-2\begin{pmatrix} 1 \\ -3 \\ 0 \end{pmatrix} - 3\begin{pmatrix} 3 \\ 0 \\ 4 \end{pmatrix} + \begin{pmatrix} 11 \\ -6 \\ 12 \end{pmatrix} = \begin{pmatrix} 0 \\ 0 \\ 0 \end{pmatrix}$ and the vectors are linearly dependent.

## Geometric Interpretation of Linear Dependence in $\mathbb{R}^3$

In Example 3 we found three vectors in $\mathbb{R}^3$ that were linearly independent. In Example 4 we found three vectors that were dependent. What does this mean geometrically?

Suppose that **u**, **v**, and **w** are three linearly dependent vectors in $\mathbb{R}^3$. We may treat the vectors as if they have an end point at the origin. Then there are constants $c_1$, $c_2$, and $c_3$, not all zero, such that

$$c_1\mathbf{u} + c_2\mathbf{v} + c_3\mathbf{w} = \mathbf{0} \tag{6}$$

Suppose that $c_3 \neq 0$ (a similar result holds if $c_1 \neq 0$ or $c_2 \neq 0$). Then we may divide both sides of (6) by $c_3$ and rearrange terms to obtain

$$\mathbf{w} = -\frac{c_1}{c_3}\mathbf{u} - \frac{c_2}{c_3}\mathbf{v} = A\mathbf{u} + B\mathbf{v}$$

where $A = -c_1/c_3$ and $B = -c_2/c_3$. We now show that **u**, **v**, and **w** are coplanar. We compute

$$\begin{aligned} \mathbf{w} \cdot (\mathbf{u} \times \mathbf{v}) &= (A\mathbf{u} + B\mathbf{v}) \cdot (\mathbf{u} \times \mathbf{v}) = A[\mathbf{u} \cdot (\mathbf{u} \times \mathbf{v})] + B[\mathbf{v} \cdot (\mathbf{u} \times \mathbf{v})] \\ &= A \cdot 0 + B \cdot 0 = 0 \end{aligned}$$

because **u** and **v** are both orthogonal to $\mathbf{u} \times \mathbf{v}$ (see page 263). Let $\mathbf{n} = \mathbf{u} \times \mathbf{v}$. If $\mathbf{n} = \mathbf{0}$, then by Theorem 3.4.2 (*vii*) **u** and **v** are parallel (and collinear). Thus **u**, **v**, and **w** lie in any plane that contains both **u** and **v** and are therefore coplanar. If $\mathbf{n} \neq \mathbf{0}$, then **u** and **v** lie in the plane consisting of those vectors passing through the origin that are orthogonal to **n**. But **w** is in the same plane because $\mathbf{w} \cdot \mathbf{n} = \mathbf{w} \cdot (\mathbf{u} \times \mathbf{v}) = 0$. This shows that **u**, **v**, and **w** are coplanar.

In Problem 59 you are asked to show that if **u**, **v**, and **w** are coplanar, then they are linearly dependent. We conclude that

> Three vectors in $\mathbb{R}^3$ are linearly dependent if and only if they are coplanar.

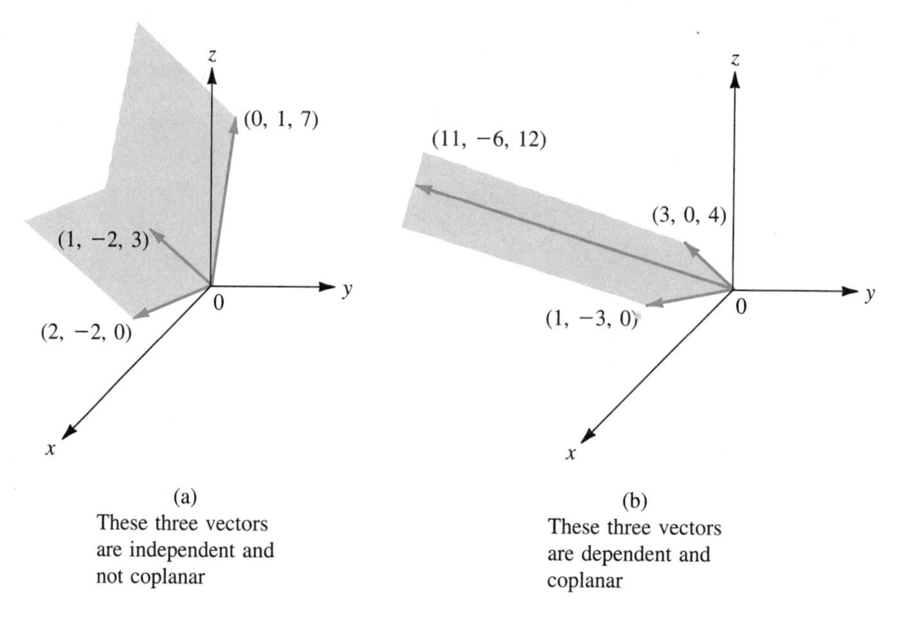


**Figure 4.3** Two sets of three vectors

Figure 4.3 illustrates this fact using the vectors in Examples 3 and 4.

The theory of homogeneous systems can tell us something about the linear dependence or independence of vectors.

**THEOREM 2** A set of $n$ vectors in $\mathbb{R}^m$ is always linearly dependent if $n > m$.

**Proof** Let $\mathbf{v}_1, \mathbf{v}_2, \ldots, \mathbf{v}_n$ be $n$ vectors in $\mathbb{R}^m$ and let us try to find constants $c_1, c_2, \ldots, c_n$ not all zero such that

$$c_1\mathbf{v}_1 + c_2\mathbf{v}_2 + \cdots + c_n\mathbf{v}_n = \mathbf{0} \tag{7}$$

Let $\mathbf{v}_1 = \begin{pmatrix} a_{11} \\ a_{21} \\ \vdots \\ a_{m1} \end{pmatrix}$, $\mathbf{v}_2 = \begin{pmatrix} a_{12} \\ a_{22} \\ \vdots \\ a_{m2} \end{pmatrix}, \ldots, \mathbf{v}_n = \begin{pmatrix} a_{1n} \\ a_{2n} \\ \vdots \\ a_{mn} \end{pmatrix}$. Then equation (7) becomes

$$\begin{aligned} a_{11}c_1 + a_{12}c_2 + \cdots + a_{1n}c_n &= 0 \\ a_{21}c_1 + a_{22}c_2 + \cdots + a_{2n}c_n &= 0 \\ \vdots \qquad \vdots \qquad\quad \vdots \quad &\;\; \vdots \\ a_{m1}c_1 + a_{m2}c_2 + \cdots + a_{mn}c_n &= 0 \end{aligned} \tag{8}$$

But system (8) is system (1.4.1) on page 41, and according to Theorem 1.4.1, this system has an infinite number of solutions if $n > m$. Thus there are scalars $c_1, c_2, \ldots, c_n$ not all zero that satisfy (8) and the vectors $\mathbf{v}_1, \mathbf{v}_2, \ldots, \mathbf{v}_n$ are therefore linearly dependent. ■

**EXAMPLE 5** **Four Vectors in $\mathbb{R}^3$ Are Linearly Dependent** The vectors $\begin{pmatrix} 2 \\ -3 \\ 4 \end{pmatrix}$, $\begin{pmatrix} 4 \\ 7 \\ -6 \end{pmatrix}$, $\begin{pmatrix} 18 \\ -11 \\ 4 \end{pmatrix}$, and $\begin{pmatrix} 2 \\ -7 \\ 3 \end{pmatrix}$ are linearly dependent since they comprise a set of four 3-vectors.

There is a very important (and obvious) corollary to Theorem 2.

**COROLLARY** A set of linearly independent vectors in $\mathbb{R}^n$ contains at most $n$ vectors.

*Note.* We can rephrase the corollary as follows: If we have $n$ linearly independent $n$-vectors, then we cannot include any more vectors without making the set linearly dependent.

From system (8) we can make another important observation whose proof is left as an exercise (see Problem 27).

**THEOREM 3** Let

$$A = \begin{pmatrix} a_{11} & a_{12} & \cdots & a_{1n} \\ a_{21} & a_{22} & \cdots & a_{2n} \\ \vdots & \vdots & & \vdots \\ a_{m1} & a_{m2} & \cdots & a_{mn} \end{pmatrix}$$

Then the columns of $A$, considered as vectors, are linearly dependent if and only if system (8), which can be written $A\mathbf{c} = \mathbf{0}$, has nontrivial solutions. Here $\mathbf{c} = \begin{pmatrix} c_1 \\ c_2 \\ \vdots \\ c_n \end{pmatrix}$.

**EXAMPLE 6** **Writing Solutions to a Homogeneous System as Linear Combinations of Linearly Independent Solution Vectors** Consider the homogeneous system

$$\begin{aligned} x_1 + 2x_2 - x_3 + 2x_4 &= 0 \\ 3x_1 + 7x_2 + x_3 + 4x_4 &= 0 \end{aligned} \qquad \textbf{(9)}$$

We solve this by row reduction:

$$\left(\begin{array}{cccc|c} 1 & 2 & -1 & 2 & 0 \\ 3 & 7 & 1 & 4 & 0 \end{array}\right) \longrightarrow \left(\begin{array}{cccc|c} 1 & 2 & -1 & 2 & 0 \\ 0 & 1 & 4 & -2 & 0 \end{array}\right)$$

$$\longrightarrow \left(\begin{array}{cccc|c} 1 & 0 & -9 & 6 & 0 \\ 0 & 1 & 4 & -2 & 0 \end{array}\right)$$

The last system is

$$\begin{aligned} x_1 \qquad - 9x_3 + 6x_4 &= 0 \\ x_2 + 4x_3 - 2x_4 &= 0 \end{aligned}$$

We see that this system has an infinite number of solutions, which we write as a linear combination of column vectors:

$$\begin{pmatrix} x_1 \\ x_2 \\ x_3 \\ x_4 \end{pmatrix} = \begin{pmatrix} 9x_3 - 6x_4 \\ -4x_3 + 2x_4 \\ x_3 \\ x_4 \end{pmatrix} = x_3 \begin{pmatrix} 9 \\ -4 \\ 1 \\ 0 \end{pmatrix} + x_4 \begin{pmatrix} -6 \\ 2 \\ 0 \\ 1 \end{pmatrix} \tag{10}$$

Note that $\begin{pmatrix} 9 \\ -4 \\ 1 \\ 0 \end{pmatrix}$ and $\begin{pmatrix} -6 \\ 2 \\ 0 \\ 1 \end{pmatrix}$ are linearly independent solutions to (9) because neither one is a multiple of the other. (You should verify that they are solutions.) Since $x_3$ and $x_4$ are arbitrary real numbers, we see from (10) that the set of all solutions to system (9) is the subspace of $\mathbb{R}^4$ spanned by these two linearly independent solution vectors.

The next two theorems follow directly from Theorem 3.

**THEOREM 4** Let $\mathbf{v}_1, \mathbf{v}_2, \ldots, \mathbf{v}_n$ be $n$ vectors in $\mathbb{R}^n$ and let $A$ be the $n \times n$ matrix whose columns are $\mathbf{v}_1, \mathbf{v}_2, \ldots, \mathbf{v}_n$. Then $\mathbf{v}_1, \mathbf{v}_2, \ldots, \mathbf{v}_n$ are linearly independent if and only if the only solution to the homogeneous system $A\mathbf{x} = \mathbf{0}$ is the trivial solution $\mathbf{x} = \mathbf{0}$.

**Proof** This is Theorem 3 in the case $m = n$.

**THEOREM 5** Let $A$ be an $n \times n$ matrix. Then $\det A \neq 0$ if and only if the columns of $A$ are linearly independent.

**Proof** From Theorem 4 and the Summing Up Theorem (see page 216). Columns of $A$ are linearly independent $\Leftrightarrow$ $\mathbf{0}$ is the only solution to $A\mathbf{x} = \mathbf{0}$ $\Leftrightarrow$ $\det A \neq 0$. Here, $\Leftrightarrow$ stands for the words "if and only if."

Theorem 5 enables us to extend our Summing Up Theorem.

**THEOREM 6** **Summing Up Theorem—View 5** Let $A$ be an $n \times n$ matrix. Then the following eight statements are equivalent; that is, each one implies the other seven (so that if one is true, all are true).

**i.** $A$ is invertible.

**ii.** The only solution to the homogeneous system $A\mathbf{x} = \mathbf{0}$ is the trivial solution ($\mathbf{x} = \mathbf{0}$).

**iii.** The system $A\mathbf{x} = \mathbf{b}$ has a unique solution for every $n$-vector $\mathbf{b}$.

**iv.** $A$ is row equivalent to the $n \times n$ identity matrix $I_n$.

**v.** $A$ can be written as the product of elementary matrices.

**vi.** The row echelon form of $A$ has $n$ pivots.

**vii.** $\det A \neq 0$.

**viii.** The columns (and rows) of $A$ are linearly independent.

**Proof** The only part not proved is that the rows of $A$ are linearly independent $\Leftrightarrow$ $\det A \neq 0$. The columns are independent $\Leftrightarrow$ $\det A \neq 0$ $\Leftrightarrow$ $\det A^t = \det A \neq 0$ (see Theorem 2.2.4 on page 190) $\Leftrightarrow$ the columns of $A^t$ are linearly independent. But the columns of $A^t$ are the rows of $A$. This completes the proof. ∎

The following theorem ties together the ideas of linear independence and spanning sets in $\mathbb{R}^n$.

**THEOREM 7** Any set of $n$ linearly independent vectors in $\mathbb{R}^n$ spans $\mathbb{R}^n$.

**Proof** Let $\mathbf{v}_1 = \begin{pmatrix} a_{11} \\ a_{21} \\ \vdots \\ a_{n1} \end{pmatrix}$, $\mathbf{v}_2 = \begin{pmatrix} a_{12} \\ a_{22} \\ \vdots \\ a_{n2} \end{pmatrix}$, . . . , $\mathbf{v}_n = \begin{pmatrix} a_{1n} \\ a_{2n} \\ \vdots \\ a_{nn} \end{pmatrix}$ be linearly independent and let $\mathbf{v} = \begin{pmatrix} x_1 \\ x_2 \\ \vdots \\ x_n \end{pmatrix}$ be a vector in $\mathbb{R}^n$. We must show that there exist scalars $c_1, c_2, \ldots, c_n$ such that

$$\mathbf{v} = c_1\mathbf{v}_1 + c_2\mathbf{v}_2 + \cdots + c_n\mathbf{v}_n$$

That is,

$$\begin{pmatrix} x_1 \\ x_2 \\ \vdots \\ x_n \end{pmatrix} = c_1\begin{pmatrix} a_{11} \\ a_{21} \\ \vdots \\ a_{n1} \end{pmatrix} + c_2\begin{pmatrix} a_{12} \\ a_{22} \\ \vdots \\ a_{n2} \end{pmatrix} + \cdots + c_n\begin{pmatrix} a_{1n} \\ a_{2n} \\ \vdots \\ a_{nn} \end{pmatrix} \tag{11}$$

In (11) we multiply through, add, and equate components to obtain a system of $n$ equations in the $n$ unknowns $c_1, c_2, \ldots, c_n$:

$$\begin{aligned} a_{11}c_1 + a_{12}c_2 + \cdots + a_{1n}c_n &= x_1 \\ a_{21}c_1 + a_{22}c_2 + \cdots + a_{2n}c_n &= x_2 \\ \vdots \qquad \vdots \qquad\qquad \vdots \quad &\quad \vdots \\ a_{n1}c_1 + a_{n2}c_2 + \cdots + a_{nn}c_n &= x_n \end{aligned} \tag{12}$$

We can write (12) as $A\mathbf{c} = \mathbf{v}$, where

$$A = \begin{pmatrix} a_{11} & a_{12} & \cdots & a_{1n} \\ a_{21} & a_{22} & \cdots & a_{2n} \\ \vdots & \vdots & & \vdots \\ a_{n1} & a_{n2} & \cdots & a_{nn} \end{pmatrix} \quad \text{and} \quad \mathbf{c} = \begin{pmatrix} c_1 \\ c_2 \\ \vdots \\ c_n \end{pmatrix}$$

But $\det A \neq 0$ since the columns of $A$ are linearly independent. So system (12) has a unique solution $\mathbf{c}$ by Theorem 6 and the theorem is proved. ■

***Remark.*** This proof not only shows that $\mathbf{v}$ can be written as a linear combination of the independent vectors $\mathbf{v}_1, \mathbf{v}_2, \ldots, \mathbf{v}_n$, but also that this can be done in *only one way* (since the solution vector $\mathbf{c}$ is unique).

**EXAMPLE 7** **Three Vectors in $\mathbb{R}^3$ Span $\mathbb{R}^3$ If Their Determinant Is Nonzero** The vectors $(2, -1, 4)$, $(1, 0, 2)$, and $(3, -1, 5)$ span $\mathbb{R}^3$ because $\begin{vmatrix} 2 & 1 & 3 \\ -1 & 0 & -1 \\ 4 & 2 & 5 \end{vmatrix} = -1 \neq 0$ so that they are independent. ■

Every example we have done so far has been in the space $\mathbb{R}^n$. This is not so much of a restriction as it seems. In Section 5.4 (Theorem 6) we shall show that many different looking vector spaces have essentially the same properties. For example, we shall see that the space $P_n$ is essentially the same as the space $\mathbb{R}^{n+1}$. We shall say that two such vector spaces are *isomorphic.*

This very powerful result will have to wait until Chapter 5. In the meantime, we shall do some examples in spaces other than $\mathbb{R}^n$.

**EXAMPLE 8** **Three Linearly Independent Matrices in $M_{23}$** In $M_{23}$ let $A_1 = \begin{pmatrix} 1 & 0 & 2 \\ 3 & 1 & -1 \end{pmatrix}$, $A_2 = \begin{pmatrix} -1 & 1 & 4 \\ 2 & 3 & 0 \end{pmatrix}$, and $A_3 = \begin{pmatrix} -1 & 0 & 1 \\ 1 & 2 & 1 \end{pmatrix}$. Determine whether $A_1$, $A_2$, and $A_3$ are linearly dependent or independent.

**Solution** Suppose that $c_1A_1 + c_2A_2 + c_3A_3 = 0$. Then

$$\begin{aligned}\begin{pmatrix}0&0&0\\0&0&0\end{pmatrix} &= c_1\begin{pmatrix}1&0&2\\3&1&-1\end{pmatrix} + c_2\begin{pmatrix}-1&1&4\\2&3&0\end{pmatrix} + c_3\begin{pmatrix}-1&0&1\\1&2&1\end{pmatrix}\\ &= \begin{pmatrix}c_1-c_2-c_3 & c_2 & 2c_1+4c_2+c_3\\ 3c_1+2c_2+c_3 & c_1+3c_2+2c_3 & -c_1+c_3\end{pmatrix}\end{aligned}$$

This gives us a homogeneous system of six equations in the three unknowns $c_1$, $c_2$, and $c_3$, and it is quite easy to verify that the only solution is $c_1 = c_2 = c_3 = 0$. Thus the three matrices are linearly independent.

**EXAMPLE 9** **Four Linearly Independent Polynomials in $P_3$** In $P_3$ determine whether the polynomials 1, $x$, $x^2$, and $x^3$ are linearly dependent or independent.

**Solution** Suppose that $c_1 + c_2x + c_3x^2 + c_4x^3 = 0$. This must hold for every real number $x$. In particular, if $x = 0$, we obtain $c_1 = 0$. Then setting $x = 1, -1, 2$, we obtain, successively,

$$\begin{aligned} c_2 + c_3 + c_4 &= 0\\ -c_2 + c_3 - c_4 &= 0\\ 2c_2 + 4c_3 + 8c_4 &= 0\end{aligned}$$

The determinant of this homogeneous system is

$$\begin{vmatrix}1&1&1\\-1&1&-1\\2&4&8\end{vmatrix} = 12 \neq 0$$

so that the system has the unique solution $c_2 = c_3 = c_4 = 0$ and the four polynomials are linearly independent. We can see this in another way. We know that any polynomial of degree 3 has at most three real roots. But if $c_1 + c_2x + c_3x^2 + c_4x^3 = 0$ for some nonzero constants $c_1$, $c_2$, $c_3$, and $c_4$ and for every real number $x$, then we have constructed a cubic polynomial for which every real number is a root. This is impossible.

**EXAMPLE 10** **Three Linearly Dependent Polynomials in $P_2$** In $P_2$ determine whether the polynomials $x - 2x^2$, $x^2 - 4x$, and $-7x + 8x^2$ are linearly dependent or independent.

**Solution** Let $c_1(x - 2x^2) + c_2(x^2 - 4x) + c_3(-7x + 8x^2) = 0$. Then rearranging terms, we obtain

$$\begin{aligned}(c_1 - 4c_2 - 7c_3)x &= 0\\ (-2c_1 + c_2 + 8c_3)x^2 &= 0\end{aligned}$$

These equations hold for every $x$ if and only if

$$c_1 - 4c_2 - 7c_3 = 0$$

and

$$-2c_1 + c_2 + 8c_3 = 0$$

But by Theorem 1.4.1, page 41 this system of two equations in three unknowns has an infinite number of solutions. This shows that the polynomials are linearly dependent.

If we solve this homogeneous system, we obtain, successively,

$$\left(\begin{array}{ccc|c} 1 & -4 & -7 & 0 \\ -2 & 1 & 8 & 0 \end{array}\right) \longrightarrow \left(\begin{array}{ccc|c} 1 & -4 & -7 & 0 \\ 0 & -7 & -6 & 0 \end{array}\right)$$

$$\longrightarrow \left(\begin{array}{ccc|c} 1 & -4 & -7 & 0 \\ 0 & 1 & \frac{6}{7} & 0 \end{array}\right) \longrightarrow \left(\begin{array}{ccc|c} 1 & 0 & -\frac{25}{7} & 0 \\ 0 & 1 & \frac{6}{7} & 0 \end{array}\right)$$

Thus $c_3$ can be chosen arbitrarily, $c_1 = \frac{25}{7}c_3$ and $c_2 = -\frac{6}{7}c_3$. If $c_3 = 7$, for example, then $c_1 = 25$, $c_2 = -6$, and we have

$$25(x - 2x^2) - 6(x^2 - 4x) + 7(-7x + 8x^2) = 0.$$

## PROBLEMS 4.5

### Self-Quiz

**I.** Which of the following pairs of vectors are linearly independent?

**a.** $\begin{pmatrix} 1 \\ 1 \end{pmatrix}, \begin{pmatrix} 1 \\ -1 \end{pmatrix}$ **b.** $\begin{pmatrix} 2 \\ 3 \end{pmatrix}, \begin{pmatrix} 3 \\ 2 \end{pmatrix}$ **c.** $\begin{pmatrix} 11 \\ 0 \end{pmatrix}, \begin{pmatrix} 0 \\ 4 \end{pmatrix}$

**d.** $\begin{pmatrix} -3 \\ -11 \end{pmatrix}, \begin{pmatrix} -6 \\ 11 \end{pmatrix}$ **e.** $\begin{pmatrix} -2 \\ 4 \end{pmatrix}, \begin{pmatrix} 2 \\ 4 \end{pmatrix}$

**II.** Which of the following pairs of vectors span $\mathbb{R}^2$?

**a.** $\begin{pmatrix} 1 \\ 1 \end{pmatrix}, \begin{pmatrix} 1 \\ -1 \end{pmatrix}$ **b.** $\begin{pmatrix} 2 \\ 3 \end{pmatrix}, \begin{pmatrix} 3 \\ 2 \end{pmatrix}$ **c.** $\begin{pmatrix} 11 \\ 0 \end{pmatrix}, \begin{pmatrix} 0 \\ 4 \end{pmatrix}$

**d.** $\begin{pmatrix} -3 \\ -11 \end{pmatrix}, \begin{pmatrix} -6 \\ 11 \end{pmatrix}$ **e.** $\begin{pmatrix} -2 \\ 4 \end{pmatrix}, \begin{pmatrix} 2 \\ 4 \end{pmatrix}$

**III.** Which of the following sets of vectors *must* be linearly dependent?

**a.** $\begin{pmatrix} a \\ b \\ c \end{pmatrix}, \begin{pmatrix} d \\ e \\ f \end{pmatrix}$ **b.** $\begin{pmatrix} a \\ b \end{pmatrix}, \begin{pmatrix} c \\ d \end{pmatrix}, \begin{pmatrix} e \\ f \end{pmatrix}$ **c.** $\begin{pmatrix} a \\ b \\ c \end{pmatrix}, \begin{pmatrix} d \\ e \\ f \end{pmatrix}, \begin{pmatrix} g \\ h \\ i \end{pmatrix}$

**d.** $\begin{pmatrix} a \\ b \\ c \end{pmatrix}, \begin{pmatrix} d \\ e \\ f \end{pmatrix}, \begin{pmatrix} g \\ h \\ i \end{pmatrix}, \begin{pmatrix} j \\ k \\ l \end{pmatrix}$

Here $a, b, c, d, e, f, g, h, i, j, k$, and $l$ are real numbers.

### *True-False*

**IV.** If $\mathbf{v}_1, \mathbf{v}_2, \ldots, \mathbf{v}_n$ are linearly independent, then $\mathbf{v}_1, \mathbf{v}_2, \ldots, \mathbf{v}_n, \mathbf{v}_{n+1}$ are also linearly independent.

**V.** If $\mathbf{v}_1, \mathbf{v}_2, \ldots, \mathbf{v}_n$ are linearly dependent, then $\mathbf{v}_1, \mathbf{v}_2, \ldots, \mathbf{v}_n, \mathbf{v}_{n+1}$ are linearly dependent.

**VI.** If $A$ is a $3 \times 3$ matrix and $\det A = 0$, then the rows of $A$ are linearly dependent vectors in $\mathbb{R}^3$.

**VII.** The polynomials 3, $2x$, $-x^3$, and $3x^4$ are linearly independent in $P_4$.

**VIII.** The matrices $\begin{pmatrix} 1 & 0 \\ 0 & 0 \end{pmatrix}$, $\begin{pmatrix} 0 & 1 \\ 0 & 0 \end{pmatrix}$, $\begin{pmatrix} 0 & 1 \\ 1 & 0 \end{pmatrix}$, and $\begin{pmatrix} 2 & 3 \\ -5 & 0 \end{pmatrix}$ are linearly independent in $M_{22}$.

In Problems 1–22 determine whether the given set of vectors is linearly dependent or independent.

**1.** $\begin{pmatrix} 1 \\ 2 \end{pmatrix}; \begin{pmatrix} -1 \\ -3 \end{pmatrix}$

**2.** $\begin{pmatrix} 2 \\ -1 \\ 4 \end{pmatrix}; \begin{pmatrix} 4 \\ -2 \\ 7 \end{pmatrix}$

**3.** $\begin{pmatrix} 2 \\ -1 \\ 4 \end{pmatrix}; \begin{pmatrix} 4 \\ -2 \\ 8 \end{pmatrix}$

**4.** $\begin{pmatrix} -2 \\ 3 \end{pmatrix}; \begin{pmatrix} 4 \\ 7 \end{pmatrix}$

**5.** $\begin{pmatrix} -3 \\ 2 \end{pmatrix}; \begin{pmatrix} 1 \\ 10 \end{pmatrix}; \begin{pmatrix} 4 \\ -5 \end{pmatrix}$

**6.** $\begin{pmatrix} 1 \\ 0 \\ 1 \end{pmatrix}; \begin{pmatrix} 0 \\ 1 \\ 1 \end{pmatrix}; \begin{pmatrix} 1 \\ 1 \\ 0 \end{pmatrix}$

**7.** $\begin{pmatrix} 1 \\ 0 \\ 1 \end{pmatrix}; \begin{pmatrix} 0 \\ 1 \\ 0 \end{pmatrix}; \begin{pmatrix} 0 \\ 0 \\ 1 \end{pmatrix}$

**8.** $\begin{pmatrix} -3 \\ 4 \\ 2 \end{pmatrix}; \begin{pmatrix} 7 \\ -1 \\ 3 \end{pmatrix}; \begin{pmatrix} 1 \\ 2 \\ 8 \end{pmatrix}$

**9.** $\begin{pmatrix} -3 \\ 4 \\ 2 \end{pmatrix}; \begin{pmatrix} 7 \\ -1 \\ 3 \end{pmatrix}; \begin{pmatrix} 1 \\ 1 \\ 8 \end{pmatrix}$

**10.** $\begin{pmatrix} 1 \\ -2 \\ 1 \\ 1 \end{pmatrix}; \begin{pmatrix} 3 \\ 0 \\ 2 \\ -2 \end{pmatrix}; \begin{pmatrix} 0 \\ 4 \\ -1 \\ -1 \end{pmatrix}; \begin{pmatrix} 5 \\ 0 \\ 3 \\ -1 \end{pmatrix}$

**11.** $\begin{pmatrix} 1 \\ -2 \\ 1 \\ 1 \end{pmatrix}; \begin{pmatrix} 3 \\ 0 \\ 2 \\ -2 \end{pmatrix}; \begin{pmatrix} 0 \\ 4 \\ -1 \\ 1 \end{pmatrix}; \begin{pmatrix} 5 \\ 0 \\ 3 \\ -1 \end{pmatrix}$

**12.** $\begin{pmatrix} 1 \\ -1 \\ 2 \end{pmatrix}; \begin{pmatrix} 4 \\ 0 \\ 0 \end{pmatrix}; \begin{pmatrix} -2 \\ 3 \\ 5 \end{pmatrix}; \begin{pmatrix} 7 \\ 1 \\ 2 \end{pmatrix}$

**13.** In $P_2$: $1 - x$, $x$

**14.** In $P_2$: $-x$, $x^2 - 2x$, $3x + 5x^2$

**15.** In $P_2$: $1 - x$, $1 + x$, $x^2$

**16.** In $P_3$: $x$, $x^2 - x$, $x^3 - x$

**17.** In $P_3$: $2x$, $x^3 - 3$, $1 + x - 4x^3$, $x^3 + 18x - 9$

---

## Answers to Self-Quiz

**I.** All of them **II.** All of them **III.** b, d **IV.** False **V.** True **VI.** True **VII.** True **VIII.** False

**18.** In $M_{22}$: $\begin{pmatrix} 2 & -1 \\ 4 & 0 \end{pmatrix}, \begin{pmatrix} 0 & -3 \\ 1 & 5 \end{pmatrix}, \begin{pmatrix} 4 & 1 \\ 7 & -5 \end{pmatrix}$

**19.** In $M_{22}$: $\begin{pmatrix} 1 & -1 \\ 0 & 6 \end{pmatrix}, \begin{pmatrix} -1 & 0 \\ 3 & 1 \end{pmatrix}, \begin{pmatrix} 1 & 1 \\ -1 & 2 \end{pmatrix}, \begin{pmatrix} 0 & 1 \\ 1 & 0 \end{pmatrix}$

**20.** In $M_{22}$: $\begin{pmatrix} -1 & 0 \\ 1 & 2 \end{pmatrix}, \begin{pmatrix} 2 & 3 \\ 7 & -4 \end{pmatrix}, \begin{pmatrix} 8 & -5 \\ 7 & 6 \end{pmatrix}, \begin{pmatrix} 4 & -1 \\ 2 & 3 \end{pmatrix}, \begin{pmatrix} 2 & 3 \\ -1 & 4 \end{pmatrix}$

***21.** In $C[0, 1]$: $\sin x$, $\cos x$

***22.** In $C[0, 1]$: $x$, $\sqrt{x}$, $\sqrt[3]{x}$

**23.** Determine a condition on the numbers $a$, $b$, $c$, and $d$ such that the vectors $\begin{pmatrix} a \\ b \end{pmatrix}$ and $\begin{pmatrix} c \\ d \end{pmatrix}$ are linearly dependent.

***24.** Find a condition on the numbers $a_{ij}$ such that the vectors $\begin{pmatrix} a_{11} \\ a_{21} \\ a_{31} \end{pmatrix}, \begin{pmatrix} a_{12} \\ a_{22} \\ a_{32} \end{pmatrix}$, and $\begin{pmatrix} a_{13} \\ a_{23} \\ a_{33} \end{pmatrix}$ are linearly dependent.

**25.** For what value(s) of $\alpha$ will the vectors $\begin{pmatrix} 1 \\ 2 \\ 3 \end{pmatrix}, \begin{pmatrix} 2 \\ -1 \\ 4 \end{pmatrix}, \begin{pmatrix} 3 \\ \alpha \\ 4 \end{pmatrix}$ be linearly dependent?

**26.** For what value(s) of $\alpha$ are the vectors $\begin{pmatrix} 2 \\ -3 \\ 1 \end{pmatrix}, \begin{pmatrix} -4 \\ 6 \\ -2 \end{pmatrix}, \begin{pmatrix} \alpha \\ 1 \\ 2 \end{pmatrix}$ linearly dependent? [*Hint:* Look carefully.]

**27.** Prove Theorem 3. [*Hint:* Look closely at system (8).]

**28.** Prove that if the vectors $\mathbf{v}_1, \mathbf{v}_2, \ldots, \mathbf{v}_n$ are linearly dependent vectors in $\mathbb{R}^m$ and if $\mathbf{v}_{n+1}$ is any other vector in $\mathbb{R}^m$, then the set $\mathbf{v}_1, \mathbf{v}_2, \ldots, \mathbf{v}_n, \mathbf{v}_{n+1}$ is linearly dependent.

**29.** Show that if $\mathbf{v}_1, \mathbf{v}_2, \ldots, \mathbf{v}_n$ $(n \geq 2)$ are linearly independent, then so too are $\mathbf{v}_1, \mathbf{v}_2, \ldots, \mathbf{v}_k$, where $k < n$.

**30.** Show that if the nonzero vectors $\mathbf{v}_1$ and $\mathbf{v}_2$ in $\mathbb{R}^n$ are orthogonal (see page 80), then the set $\{\mathbf{v}_1, \mathbf{v}_2\}$ is linearly independent.

***31.** Suppose that $\mathbf{v}_1$ is orthogonal to $\mathbf{v}_2$ and $\mathbf{v}_3$ and that $\mathbf{v}_2$ is orthogonal to $\mathbf{v}_3$. If $\mathbf{v}_1$, $\mathbf{v}_2$, and $\mathbf{v}_3$ are nonzero, show that the set $\{\mathbf{v}_1, \mathbf{v}_2, \mathbf{v}_3\}$ is linearly independent.

**32.** Let $A$ be a square $(n \times n)$ matrix whose columns are the vectors $\mathbf{v}_1, \mathbf{v}_2, \ldots, \mathbf{v}_n$. Show that $\mathbf{v}_1, \mathbf{v}_2, \ldots, \mathbf{v}_n$ are linearly independent if and only if the row echelon form of $A$ does not contain a row of zeros.

In Problems 33–37 write the solutions to the given homogeneous systems in terms of one or more linearly independent vectors.

**33.** $x_1 + x_2 + x_3 = 0$

**34.**
$$\begin{aligned} x_1 - x_2 + 7x_3 - x_4 &= 0 \\ 2x_1 + 3x_2 - 8x_3 + x_4 &= 0 \end{aligned}$$

**35.**
$$\begin{aligned} x_1 + 2x_2 - x_3 &= 0 \\ 2x_1 + 5x_2 + 4x_3 &= 0 \end{aligned}$$

**36.**
$$\begin{aligned} x_1 + x_2 + x_3 - x_4 - x_5 &= 0 \\ -2x_1 + 3x_2 + x_3 + 4x_4 - 6x_5 &= 0 \end{aligned}$$

**37.** $x_1 + 2x_2 - 3x_3 + 5x_4 = 0$

**38.** Let $\mathbf{u} = (1, 2, 3)$.

- **a.** Let $H = \{\mathbf{v} \in \mathbb{R}^3: \mathbf{u} \cdot \mathbf{v} = 0\}$. Show that $H$ is a subspace of $\mathbb{R}^3$.
- **b.** Find two linearly independent vectors in $H$. Call them $\mathbf{x}$ and $\mathbf{y}$.
- **c.** Compute $\mathbf{w} = \mathbf{x} \times \mathbf{y}$.
- **d.** Show that $\mathbf{u}$ and $\mathbf{w}$ are linearly dependent.
- **e.** Give a geometric interpretation of parts (*a*) and (*c*) and explain why (*d*) must be true.

*Remark.* If $V = \{\mathbf{v} \in \mathbb{R}^3: \mathbf{v} = \alpha\mathbf{u}$ for some real number $\alpha\}$, then $V$ is a subspace of $\mathbb{R}^3$ and $H$ is called the **orthogonal complement** of $V$.

**39.** Choose a vector $\mathbf{u} \neq \mathbf{0}$ in $\mathbb{R}^3$. Repeat the steps of Problem 38, starting with the vector you have chosen.

**40.** Show that any four polynomials in $P_2$ are linearly dependent.

**41.** Show that two polynomials cannot span $P_2$.

***42.** Show that any $n + 2$ polynomials in $P_n$ are linearly dependent.

**43.** Show that any subset of a set of linearly independent vectors is linearly independent. [*Note:* This generalizes Problem 29.]

**44.** Show that any seven matrices in $M_{32}$ are linearly dependent.

***45.** Prove that any $mn + 1$ matrices in $M_{mn}$ are linearly dependent.

**46.** Let $S_1$ and $S_2$ be two finite, linearly independent sets in a vector space $V$. Show that $S_1 \cap S_2$ is a linearly independent set.

***47.** Show that in $P_n$ the polynomials $1, x, x^2, \ldots, x^n$ are linearly independent. [*Hint:* This is certainly true if $n = 1$. Assume that $1, x, x^2, \ldots, x^{n-1}$ are linearly independent and show how this implies that $1, x, x^2, \ldots, x^n$ are also independent. This will complete the proof by mathematical induction.]

**48.** Let $\{\mathbf{v}_1, \mathbf{v}_2, \ldots, \mathbf{v}_n\}$ be a linearly independent set. Show that the vectors $\mathbf{v}_1, \mathbf{v}_1 + \mathbf{v}_2, \mathbf{v}_1 + \mathbf{v}_2 + \mathbf{v}_3, \ldots, \mathbf{v}_1 + \mathbf{v}_2 + \cdots + \mathbf{v}_n$ are linearly independent.

**49.** Let $S = \{\mathbf{v}_1, \mathbf{v}_2, \ldots, \mathbf{v}_n\}$ be a linearly dependent set of nonzero vectors in a vector space $V$. Show that at least one of the vectors in $S$ can be written as a linear combination of the vectors that precede it. That is, show that there is an integer $k \leq n$ and scalars $a_1, a_2, \ldots, a_{k-1}$ such that $\mathbf{v}_k = a_1\mathbf{v}_1 + a_2\mathbf{v}_2 + \cdots + a_{k-1}\mathbf{v}_{k-1}$.

**50.** Let $\{\mathbf{v}_1, \mathbf{v}_2, \ldots, \mathbf{v}_n\}$ be a set of vectors having the property that the set $\{\mathbf{v}_i, \mathbf{v}_j\}$ is linearly dependent when $i \neq j$. Show that each vector in the set is a multiple of a single vector in the set.

Calculus **51.** Let $f$ and $g$ be in $C^1[0,1]$. Then the **Wronskian†** of $f$ and $g$ is defined by

$$W(f, g)(x) = \begin{vmatrix} f(x) & g(x) \\ f'(x) & g'(x) \end{vmatrix}$$

Show that if $f$ and $g$ are linearly dependent, then $W(f, g)(x) = 0$ for every $x \in [0, 1]$.

---

†Named after the Polish mathematician Józef Maria Hoene-Wroński (1778–1853). Hoene-Wroński spent most of his adult life in France. He worked on the theory of determinants and was also known for his critical writings in the philosophy of mathematics.

Calculus **52.** Determine a suitable definition for the Wronskian of the functions $f_1, f_2, \ldots, f_n \in C^{(n-1)}[0, 1]$.†

**53.** Suppose that **u**, **v**, and **w** are linearly independent. Prove or disprove: $\mathbf{u} + \mathbf{v}$, $\mathbf{u} + \mathbf{w}$, and $\mathbf{v} + \mathbf{w}$ are linearly independent.

**54.** For what real values of $c$ are the vectors $(1 - c, 1 + c)$ and $(1 + c, 1 - c)$ linearly independent?

**55.** Show that the vectors $(1, a, a^2)$, $(1, b, b^2)$, and $(1, c, c^2)$ are linearly independent if $a \neq b$, $a \neq c$, and $b \neq c$.

**56.** Let $\{\mathbf{v}_1, \mathbf{v}_2, \ldots, \mathbf{v}_n\}$ be a linearly independent set and suppose that $\mathbf{v} \notin \operatorname{span}\{\mathbf{v}_1, \mathbf{v}_2, \ldots, \mathbf{v}_n\}$. Show that $\{\mathbf{v}_1, \mathbf{v}_2, \ldots, \mathbf{v}_n, \mathbf{v}\}$ is a linearly independent set.

**57.** Find a set of three linearly independent vectors in $\mathbb{R}^3$ that contains the vectors $\begin{pmatrix} 2 \\ 1 \\ 2 \end{pmatrix}$ and $\begin{pmatrix} -1 \\ 3 \\ 4 \end{pmatrix}$. $\left[ \textit{Hint: } \text{Find a vector } \mathbf{v} \notin \operatorname{span}\left\{ \begin{pmatrix} 2 \\ 1 \\ 2 \end{pmatrix}, \begin{pmatrix} -1 \\ 3 \\ 4 \end{pmatrix} \right\}. \right]$

**58.** Find a set of three linearly independent vectors in $P_2$ that contains the polynomials $1 - x^2$ and $1 + x^2$.

**59.** Suppose that $\mathbf{u} = \begin{pmatrix} u_1 \\ u_2 \\ u_3 \end{pmatrix}$, $\mathbf{v} = \begin{pmatrix} v_1 \\ v_2 \\ v_3 \end{pmatrix}$, and $\mathbf{w} = \begin{pmatrix} w_1 \\ w_2 \\ w_3 \end{pmatrix}$ are coplanar.

**a.** Show that there exist constants $a$, $b$, and $c$ not all zero such that

$$\begin{aligned} au_1 + bu_2 + cu_3 &= 0 \\ av_1 + bv_2 + cv_3 &= 0 \\ aw_1 + bw_2 + cw_3 &= 0 \end{aligned}$$

**b.** Explain why

$$\det \begin{pmatrix} u_1 & u_2 & u_3 \\ v_1 & v_2 & v_3 \\ w_1 & w_2 & w_3 \end{pmatrix} = 0$$

**c.** Use Theorem 3 to show that **u**, **v**, and **w** are linearly dependent.

† $C^{(n-1)}[0, 1]$ is the set of functions whose $(n - 1)$st derivatives are defined and continuous on $[0, 1]$.

## MATLAB 4.5

1. Use **rref** to check the independence or dependence of the sets of vectors in Problems 1 through 12 in this section. Explain your conclusions.

2. a. For text Problems 7 and 9, argue why the vectors given are not coplanar.
   b. Argue why the sets of vectors given below are coplanar.

   i. $\left\{\begin{pmatrix}1\\2\\1\end{pmatrix}, \begin{pmatrix}2\\1\\3\end{pmatrix}, \begin{pmatrix}3\\3\\4\end{pmatrix}\right\}$ ii. $\left\{\begin{pmatrix}1\\2\\1\end{pmatrix}, \begin{pmatrix}-1\\0\\1\end{pmatrix}, \begin{pmatrix}2\\6\\4\end{pmatrix}\right\}$

3. Choose $m$ and $n$ with $m > n$ and let **A = 2*rand(n,m)−1.** Determine the independence or dependence of the columns of $A$. Repeat for four more choices of $n$ and $m$. Write a conjecture about the linear independence of the columns of a matrix that has more columns than rows. Prove your conjecture.

4. Consider the matrices in MATLAB Problem 2 in Section 1.8. Test each $A$ for invertibility, linear independence of the columns of $A$, and linear independence of the rows of $A$ (consider $A'$). Write a conjecture relating invertibility of $A^t$ to linear independence of the columns of $A$ and to linear independence of the rows of $A$. Prove your conjecture in terms of properties of the reduced row echelon form.

5. a. *(Paper and pencil)* If $A$ is $n \times m$ and $\mathbf{z}$ is $m \times 1$, explain why $\mathbf{w} = A\mathbf{z}$ is in the span of the columns of $A$.
   b. For each set of vectors $\{\mathbf{v}_1, \ldots, \mathbf{v}_k\}$ below, generate a random vector $\mathbf{w}$ that is in the span of the set of vectors [use part (a)]. Test the linear independence or dependence of the set of vectors $\{\mathbf{v}_1, \ldots, \mathbf{v}_k, \mathbf{w}\}$. Repeat for three more choices of $\mathbf{w}$.

   i. $\left\{\begin{pmatrix}8\\7\\8\end{pmatrix}, \begin{pmatrix}1\\-7\\-1\end{pmatrix}, \begin{pmatrix}10\\-6\\-1\end{pmatrix}\right\}$ ii. $\left\{\begin{pmatrix}1\\0\\1\\1\end{pmatrix}, \begin{pmatrix}-1\\2\\3\\1\end{pmatrix}, \begin{pmatrix}2\\-1\\0\\4\end{pmatrix}\right\}$

   iii. $\left\{\begin{pmatrix}4\\3\\2\\0\\2\end{pmatrix}, \begin{pmatrix}10\\2\\8\\1\\4\end{pmatrix}, \begin{pmatrix}6\\2\\8\\2\\10\end{pmatrix}, \begin{pmatrix}3\\2\\1\\2\\6\end{pmatrix}\right\}$

   c. Write a conclusion to the following: If $\mathbf{w}$ is in the span of $\{\mathbf{v}_1, \ldots, \mathbf{v}_k\}$, then . . .

6. a. Recall the sets of vectors in MATLAB Problems 3 and 7 in Section 4.4. For $\mathbf{w}$ in the span of these sets of vectors, there were infinitely many ways to write $\mathbf{w}$ as a linear combination of the vectors. Verify that each of these sets of vectors is linearly dependent.
   b. *(Paper and pencil)* Prove the following statement: For vectors in $\mathbb{R}^n$ such that $\mathbf{w} = c_1\mathbf{v}_1 + \cdots + c_k\mathbf{v}_k$ has a solution, then there are infinitely many solutions for $c_1, \ldots, c_k$ if and only if $\{\mathbf{v}_1, \ldots, \mathbf{v}_k\}$ is linearly dependent. [*Hint:* Think of the reduced row echelon form.]

7. a. Choose $n$ and $m$ with $m \le n$ and let **A = 2*rand(n,m)−1.** Verify that the columns of $A$ are linearly independent. Change $A$ so that some column(s) of $A$ is (are) linear combinations of other columns of $A$. (For example, **B = A; B(:,3) = 3*B(:,1)−2*B(:,2).**) Verify that the columns of $B$ are dependent. Repeat for another

choice of linear combination(s). Which columns of **rref(B)** do not have pivots? How is this related to your linear combination?

b. Repeat part (a) for four more choices of $n$, $m$, and $A$.

c. Write a conclusion to the following: if a column of $A$ is a linear combination of other columns of $A$, then . . .

d. Redo MATLAB Problem 5 in Section 1.7. Verify for each matrix $A$ in that problem that the columns of $A$ are dependent.

e. Write a conclusion to the following: If the columns of $A$ are linearly dependent, then . . .

f. *(Paper and pencil)* Prove your conclusions.

8. a. From MATLAB Problem 7 in this section and MATLAB Problem 5 in Section 1.7, we can conclude that if the columns of $A$ are dependent, then the columns of $A$ corresponding to columns without pivots in **rref(A)** can be written as linear combinations of the columns of $A$ corresponding to columns with pivots in **rref(A).** Following the process outlined in MATLAB Problem 5 in Section 1.7, determine which columns of the matrices below are linear combinations of other columns; write these columns as linear combinations, and verify, using MATLAB, that the linear combinations are correct.

i. $\begin{pmatrix} 1 & 0 & 2 \\ 2 & 3 & 1 \\ -1 & 1 & -3 \end{pmatrix}$ ii. $\begin{pmatrix} 10 & 0 & -10 & -6 & 32 \\ 8 & 2 & -4 & -7 & 32 \\ -5 & 7 & 19 & 1 & -5 \end{pmatrix}$

iii. $\begin{pmatrix} 7 & 6 & 11 & 3 & 5 \\ 8 & 1 & -5 & -20 & 9 \\ 7 & 6 & 11 & 3 & 8 \\ 8 & 2 & -2 & -16 & 6 \\ 7 & 3 & 2 & -9 & 7 \end{pmatrix}$ iv. $\begin{pmatrix} 1 & 3 & 1 & 1 & 3 \\ -2 & 4 & 0 & 1 & -1 \\ 0 & -2 & -3 & 1 & 9 \\ 1 & 1 & 2 & 1 & 5 \end{pmatrix}$

b. *(Paper and pencil)* Do Problem 49 in Section 4.5.

9. a. Show that the following sets of vectors are independent but that there is a vector in their respective $\mathbb{R}^n$ that is not in the span of the set.

i. $\mathbb{R}^2$ $\begin{pmatrix} -1 \\ 2 \end{pmatrix}$

ii. $\mathbb{R}^4$ See part (b) (ii) of MATLAB Problem 5 in this section.

iii. $\mathbb{R}^5$ See part (b) (iii) of MATLAB Problem 5 in this section.

b. Show that the following sets of vectors span all of their respective $\mathbb{R}^n$ but are not linearly independent.

i. $\mathbb{R}^2$ $\left\{\begin{pmatrix} -1 \\ 2 \end{pmatrix}, \begin{pmatrix} 3 \\ -1 \end{pmatrix}, \begin{pmatrix} -1 \\ 0 \end{pmatrix}\right\}$ ii. $\mathbb{R}^3$ $\left\{\begin{pmatrix} 1 \\ 0 \\ 1 \end{pmatrix}, \begin{pmatrix} -1 \\ 2 \\ 3 \end{pmatrix}, \begin{pmatrix} 2 \\ -1 \\ 0 \end{pmatrix}, \begin{pmatrix} 1 \\ 1 \\ 4 \end{pmatrix}\right\}$,

iii. $\mathbb{R}^4$ $\left\{\begin{pmatrix} 4 \\ -1 \\ 3 \\ 1 \end{pmatrix}, \begin{pmatrix} 3 \\ 2 \\ 2 \\ 2 \end{pmatrix}, \begin{pmatrix} 0 \\ 1 \\ 2 \\ 2 \end{pmatrix}, \begin{pmatrix} 7 \\ 2 \\ 7 \\ 5 \end{pmatrix}, \begin{pmatrix} 1 \\ 1 \\ -1 \\ 0 \end{pmatrix}, \begin{pmatrix} 1 \\ 1 \\ 1 \\ 1 \end{pmatrix}\right\}$

c. Are either of the situations in parts (a) or (b) possible if we consider a set of $n$ vectors in $\mathbb{R}^n$? Why or why not? Give examples using MATLAB.

d. *(Paper and pencil)* Write a conclusion relating linear independence to spanning all of $\mathbb{R}^n$ for a set of $m$ vectors in $\mathbb{R}^n$. Consider $m < n$, $m = n$, and $m > n$. Prove your statements by considering properties of the reduced row echelon form of the matrix whose columns are the set of vectors.

**10.** a. Verify that each set of vectors below is linearly independent.

**i.** $\left\{ \begin{pmatrix} 1 \\ 0 \\ 1 \\ 1 \end{pmatrix}, \begin{pmatrix} -1 \\ 2 \\ 3 \\ 1 \end{pmatrix}, \begin{pmatrix} 2 \\ -1 \\ 0 \\ 4 \end{pmatrix} \right\}$ **ii.** $\left\{ \begin{pmatrix} 4 \\ 3 \\ 2 \\ 0 \end{pmatrix}, \begin{pmatrix} 10 \\ 2 \\ 8 \\ 1 \end{pmatrix}, \begin{pmatrix} 6 \\ 2 \\ 8 \\ 2 \end{pmatrix}, \begin{pmatrix} 3 \\ 2 \\ 1 \\ 2 \end{pmatrix} \right\}$

**iii.** $\left\{ \begin{pmatrix} -1 \\ 0 \\ 2 \\ 3 \end{pmatrix}, \begin{pmatrix} -1 \\ 0 \\ 1 \\ 5 \end{pmatrix} \right\}$

**iv.** Generate four random vectors in $\mathbb{R}^4$ using the **rand** command. Check for independence. (Keep generating sets until you get an independent set.)

b. Form an invertible $4 \times 4$ matrix $A$. For each set of linearly independent vectors $\{\mathbf{v}_1, \ldots, \mathbf{v}_k\}$ from part (a), check the independence or dependence of $\{A\mathbf{v}_1, A\mathbf{v}_2, \ldots, A\mathbf{v}_k\}$ to determine which sets $\{A\mathbf{v}_1, A\mathbf{v}_2, \ldots, A\mathbf{v}_k\}$ are independent.

c. Form a $4 \times 4$ matrix $A$ that is not invertible. (For example, given an invertible $A$, change one of the columns of $A$ to a linear combination of other columns of $A$.) For each set of linearly independent vectors $\{\mathbf{v}_1, \ldots, \mathbf{v}_k\}$ from part (a), check the independence or dependence of $\{A\mathbf{v}_1, A\mathbf{v}_2, \ldots, A\mathbf{v}_k\}$ to determine which sets $\{A\mathbf{v}_1, A\mathbf{v}_2, \ldots, A\mathbf{v}_k\}$ are independent.

d. Write a conjecture describing when multiplication by a square matrix preserves independence of a set of vectors.

**11.** Use MATLAB to check the independence or dependence of the sets of polynomials in Problems 13 through 17 in this section. If the set is dependent, write the dependent polynomials as linear combinations of other polynomials in the set and verify the linear combinations. (See MATLAB Problem 9 in Section 4.4 and MATLAB Problem 8 in Section 4.5.)

**12.** Use MATLAB to check the independence or dependence of the sets of matrices in Problems 18 through 20 in Section 4.5. If the set is dependent, write the dependent matrices as linear combinations of other matrices in the set and verify the linear combinations. (See MATLAB Problem 10 in Section 4.4 and MATLAB Problem 8 in Section 4.5.)

**13.** a. Generate a set of five random matrices in $M_{22}$ and show that the set is linearly dependent. Repeat for two more sets.

b. Generate a set of seven random matrices in $M_{23}$ and show that the set is linearly dependent. Repeat for two more sets.

c. For $M_{42}$, how many matrices are needed in a set to guarantee that the set is dependent? Test your conjecture by generating sets of random matrices. Demonstrate that sets with fewer matrices are not necessarily dependent.

d. *(Paper and pencil)* Work Problems 44 and 45 in this section.

**14. Cycles in Digraphs and Linear Independence** For a directed graph *(digraph),* the node-edge incidence matrix is defined as

$$a_{ij} = \begin{cases} 1 & \text{if edge } j \text{ goes into node } i \\ -1 & \text{if edge } j \text{ goes out of node } i \\ 0 & \text{otherwise.} \end{cases}$$

Therefore, each column corresponds to an edge of the digraph.

**a.** For the digraph below, set up the node-edge incidence matrix $A$. (To enter $A$ efficiently, see MATLAB Problem 2 in Section 1.5.)

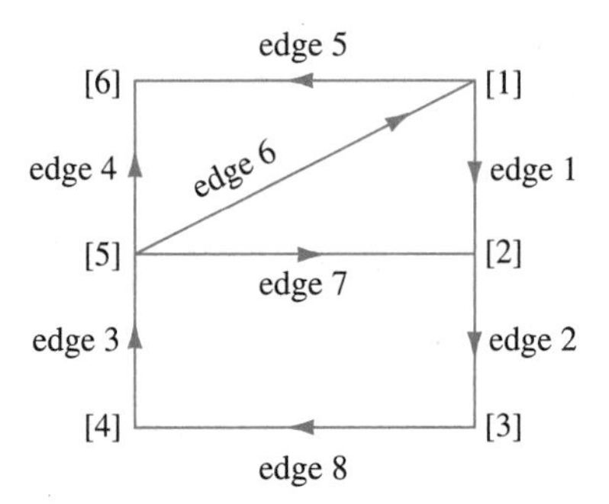


**b.** Find a closed loop *(undirected cycle)* in the digraph and note which edges it involves. Check the independence or dependence of the columns of $A$ that correspond to these edges. (For example, following edge 1, then opposite to edge 7, then along edge 4, and then opposite to edge 5 forms a cycle. Form the matrix **[A(:,1) A(:,7) A(:,4) A(:,5)]** and check for independence.)

Find as many other closed loops as you can recognize and test the independence/dependence of the corresponding columns of $A$.

**c.** Consider a subset of edges that contains no closed loops. Test the independence/dependence of the corresponding columns of $A$.

**d.** Repeat parts (a) through (c) for the digraph below.

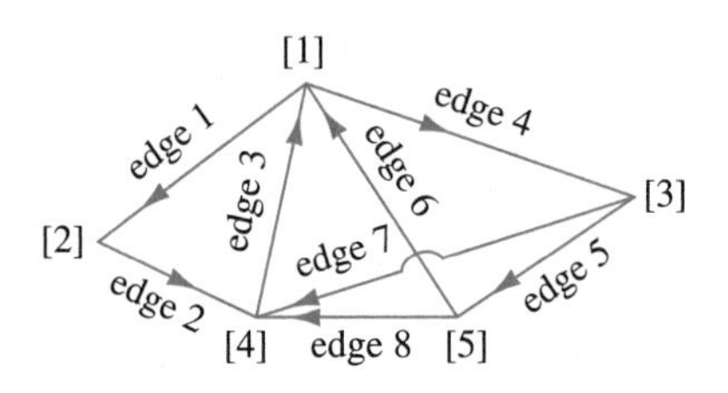


**e.** Write a conclusion about the relationship between undirected cycles in digraphs and the linear independence/dependence of the columns of the node-edge incidence matrix of the digraph.

*Note.* This problem was inspired by a lecture given by Gilbert Strang at the University of New Hampshire, June 1991.

## 4.6 BASIS AND DIMENSION

We have seen that in $\mathbb{R}^2$ it is convenient to write vectors as a linear combination of the vectors $\mathbf{i} = \begin{pmatrix} 1 \\ 0 \end{pmatrix}$ and $\mathbf{j} = \begin{pmatrix} 0 \\ 1 \end{pmatrix}$. In $\mathbb{R}^3$ we wrote vectors in terms of $\begin{pmatrix} 1 \\ 0 \\ 0 \end{pmatrix}$, $\begin{pmatrix} 0 \\ 1 \\ 0 \end{pmatrix}$, and $\begin{pmatrix} 0 \\ 0 \\ 1 \end{pmatrix}$. We now generalize this idea.

**DEFINITION 1** **Basis** A finite set of vectors $\{\mathbf{v}_1, \mathbf{v}_2, \ldots, \mathbf{v}_n\}$ is a **basis** for a vector space $V$ if

**i.** $\{\mathbf{v}_1, \mathbf{v}_2, \ldots, \mathbf{v}_n\}$ is linearly independent.

**ii.** $\{\mathbf{v}_1, \mathbf{v}_2, \ldots, \mathbf{v}_n\}$ spans $V$.

We have already seen quite a few examples of bases. In Theorem 4.5.7, for instance, we saw that any set of $n$ linearly independent vectors in $\mathbb{R}^n$ spans $\mathbb{R}^n$. Thus

*Every set of $n$ linearly independent vectors in $\mathbb{R}^n$ is a basis in $\mathbb{R}^n$.*

In $\mathbb{R}^n$ we define

$$\mathbf{e}_1 = \begin{pmatrix} 1 \\ 0 \\ 0 \\ \vdots \\ 0 \end{pmatrix}, \mathbf{e}_2 = \begin{pmatrix} 0 \\ 1 \\ 0 \\ \vdots \\ 0 \end{pmatrix}, \mathbf{e}_3 = \begin{pmatrix} 0 \\ 0 \\ 1 \\ \vdots \\ 0 \end{pmatrix}, \ldots, \mathbf{e}_n = \begin{pmatrix} 0 \\ 0 \\ 0 \\ \vdots \\ 1 \end{pmatrix}$$

Standard basis

Then since the $\mathbf{e}_i$'s are the columns of the identity matrix (which has determinant 1), $\{\mathbf{e}_1, \mathbf{e}_2, \ldots, \mathbf{e}_n\}$ is a linearly independent set and therefore constitutes a basis in $\mathbb{R}^n$. This special basis is called the **standard basis** in $\mathbb{R}^n$. We now find bases for some other spaces.

**EXAMPLE 1** **Standard Basis for $P_n$** By Example 4.5.9, page 327, the polynomials $1, x, x^2, x^3$ are linearly independent in $P_3$. By Example 4.4.3, page 306, these polynomials span $P_3$. Thus $\{1, x, x^2, x^3\}$ is a basis for $P_3$. In general, the monomials $\{1, x, x^2, x^3, \ldots, x^n\}$ constitute a basis for $P_n$. This is called the **standard basis** for $P_n$.

**EXAMPLE 2** **Standard Basis for $M_{22}$** We saw in Example 4.4.6, page 307, that $\begin{pmatrix} 1 & 0 \\ 0 & 0 \end{pmatrix}$, $\begin{pmatrix} 0 & 1 \\ 0 & 0 \end{pmatrix}$, $\begin{pmatrix} 0 & 0 \\ 1 & 0 \end{pmatrix}$, and $\begin{pmatrix} 0 & 0 \\ 0 & 1 \end{pmatrix}$ span $M_{22}$. If $\begin{pmatrix} c_1 & c_2 \\ c_3 & c_4 \end{pmatrix} = c_1\begin{pmatrix} 1 & 0 \\ 0 & 0 \end{pmatrix} + c_2\begin{pmatrix} 0 & 1 \\ 0 & 0 \end{pmatrix} + c_3\begin{pmatrix} 0 & 0 \\ 1 & 0 \end{pmatrix} + c_4\begin{pmatrix} 0 & 0 \\ 0 & 1 \end{pmatrix} = \begin{pmatrix} 0 & 0 \\ 0 & 0 \end{pmatrix}$, then, obviously, $c_1 = c_2 = c_3 = c_4 = 0$. Thus these four matrices are linearly independent and form a basis for $M_{22}$. This is called the **standard basis** for $M_{22}$.

**EXAMPLE 3** **A Basis for a Subspace of $\mathbb{R}^3$** Find a basis for the set of vectors lying on the plane

$$\pi = \left\{ \begin{pmatrix} x \\ y \\ z \end{pmatrix} : 2x - y + 3z = 0 \right\}$$

**Solution** We saw in Example 4.2.6 that $\pi$ is a vector space. To find a basis, we first note that if $x$ and $z$ are chosen arbitrarily and if $\begin{pmatrix} x \\ y \\ z \end{pmatrix} \in \pi$, then $y = 2x + 3z$. Thus vectors in $\pi$ have the form

$$\begin{pmatrix} x \\ 2x + 3z \\ z \end{pmatrix} = \begin{pmatrix} x \\ 2x \\ 0 \end{pmatrix} + \begin{pmatrix} 0 \\ 3z \\ z \end{pmatrix} = x\begin{pmatrix} 1 \\ 2 \\ 0 \end{pmatrix} + z\begin{pmatrix} 0 \\ 3 \\ 1 \end{pmatrix}$$

which shows that $\begin{pmatrix} 1 \\ 2 \\ 0 \end{pmatrix}$ and $\begin{pmatrix} 0 \\ 3 \\ 1 \end{pmatrix}$ span $\pi$. Since these two vectors are obviously linearly independent (because one is not a multiple of the other), they form a basis for $\pi$.

If $\mathbf{v}_1, \mathbf{v}_2, \ldots, \mathbf{v}_n$ is a basis for $V$, then any other vector $\mathbf{v} \in V$ can be written $\mathbf{v} = c_1\mathbf{v}_1 + c_2\mathbf{v}_2 + \cdots + c_n\mathbf{v}_n$. Can it be written in another way as a linear combination of the $\mathbf{v}_i$'s? The answer is *no*. (See the remark following the proof of Theorem 4.5.7, page 326, in the case $V = \mathbb{R}^n$.)

**THEOREM 1** If $\{\mathbf{v}_1, \mathbf{v}_2, \ldots, \mathbf{v}_n\}$ is a basis for $V$ and if $\mathbf{v} \in V$, then there exists a *unique* set of scalars $c_1, c_2, \ldots, c_n$ such that $\mathbf{v} = c_1\mathbf{v}_1 + c_2\mathbf{v}_2 + \cdots + c_n\mathbf{v}_n$.

**Proof** At least one such set of scalars exists because $\{\mathbf{v}_1, \mathbf{v}_2, \ldots, \mathbf{v}_n\}$ spans $V$. Suppose then that $\mathbf{v}$ can be written in two ways as a linear combination of the basis vectors.

That is, suppose that

$$\mathbf{v} = c_1\mathbf{v}_1 + c_2\mathbf{v}_2 + \cdots + c_n\mathbf{v}_n = d_1\mathbf{v}_1 + d_2\mathbf{v}_2 + \cdots + d_n\mathbf{v}_n$$

Then subtracting, we obtain the equation

$$(c_1 - d_1)\mathbf{v}_1 + (c_2 - d_2)\mathbf{v}_2 + \cdots + (c_n - d_n)\mathbf{v}_n = \mathbf{0}$$

But since the $\mathbf{v}_i$'s are linearly independent, this equation can hold only if $c_1 - d_1 = c_2 - d_2 = \cdots = c_n - d_n = 0$. Thus $c_1 = d_1, c_2 = d_2, \ldots, c_n = d_n$ and the theorem is proved.

We have seen that a vector space may have many bases. A question naturally arises: Do all bases contain the same number of vectors? In $\mathbb{R}^3$ the answer is certainly yes. To see this, we note that any three linearly independent vectors in $\mathbb{R}^3$ form a basis. But fewer than three vectors cannot form a basis since, as we saw in Section 4.4, the span of two linearly independent vectors in $\mathbb{R}^3$ is a plane in $\mathbb{R}^3$—and a plane is not all of $\mathbb{R}^3$. Similarly, a set of four or more vectors in $\mathbb{R}^3$ cannot be linearly independent; for if the first three vectors in the set are linearly independent, then they form a basis, and therefore all other vectors in the set can be written as a linear combination of the first three. Thus all bases in $\mathbb{R}^3$ contain three vectors. The next theorem tells us that the answer to the question posed above is *yes* for all vector spaces.

**THEOREM 2** If $\{\mathbf{u}_1, \mathbf{u}_2, \ldots, \mathbf{u}_m\}$ and $\{\mathbf{v}_1, \mathbf{v}_2, \ldots, \mathbf{v}_n\}$ are bases for the vector space $V$, then $m = n$; that is, any two bases in a vector space $V$ have the same number of vectors.

**Proof†** Let $S_1 = \{\mathbf{u}_1, \ldots, \mathbf{u}_m\}$ and $S_2 = \{\mathbf{v}_1, \ldots, \mathbf{v}_n\}$ be two bases for $V$. We must show that $m = n$. We prove this by showing that if $m > n$, then $S_1$ is a linearly dependent set, which contradicts the hypothesis that $S_1$ is a basis. This will show that $m \le n$. The same proof will then show that $n \le m$, and this will prove the theorem. Hence all we must show is that if $m > n$, then $S_1$ is dependent. Since $S_2$ constitutes a basis, we can write each $\mathbf{u}_i$ as a linear combination of the $\mathbf{v}_i$'s. We have

$$\begin{aligned}
\mathbf{u}_1 &= a_{11}\mathbf{v}_1 + a_{12}\mathbf{v}_2 + \cdots + a_{1n}\mathbf{v}_n \\
\mathbf{u}_2 &= a_{21}\mathbf{v}_1 + a_{22}\mathbf{v}_2 + \cdots + a_{2n}\mathbf{v}_n \\
&\vdots \\
\mathbf{u}_m &= a_{m1}\mathbf{v}_1 + a_{m2}\mathbf{v}_2 + \cdots + a_{mn}\mathbf{v}_n
\end{aligned} \tag{1}$$

To show that $S_1$ is dependent, we must find scalars $c_1, c_2, \ldots, c_m$, not all zero, such that

$$c_1\mathbf{u}_1 + c_2\mathbf{u}_2 + \cdots + c_m\mathbf{u}_m = \mathbf{0} \tag{2}$$

† This proof is given for vector spaces with bases containing a finite number of vectors. We also treat the scalars as though they were real numbers. However, the proof works in the complex case as well.

Inserting (1) into (2), we obtain

$$c_1(a_{11}\mathbf{v}_1 + a_{12}\mathbf{v}_2 + \cdots + a_{1n}\mathbf{v}_n) + c_2(a_{21}\mathbf{v}_1 + a_{22}\mathbf{v}_2 + \cdots + a_{2n}\mathbf{v}_n) + \cdots + c_m(a_{m1}\mathbf{v}_1 + a_{m2}\mathbf{v}_2 + \cdots + a_{mn}\mathbf{v}_n) = \mathbf{0} \quad (3)$$

Equation (3) can be rewritten as

$$(a_{11}c_1 + a_{21}c_2 + \cdots + a_{m1}c_m)\mathbf{v}_1 + (a_{12}c_1 + a_{22}c_2 + \cdots + a_{m2}c_m)\mathbf{v}_2 + \cdots + (a_{1n}c_1 + a_{2n}c_2 + \cdots + a_{mn}c_m)\mathbf{v}_n = \mathbf{0} \quad (4)$$

But since $\mathbf{v}_1, \mathbf{v}_2, \ldots, \mathbf{v}_n$ are linearly independent, we must have

$$\begin{aligned} a_{11}c_1 + a_{21}c_2 + \cdots + a_{m1}c_m &= 0 \\ a_{12}c_1 + a_{22}c_2 + \cdots + a_{m2}c_m &= 0 \\ \vdots \qquad \vdots \qquad\quad \vdots \quad &\ \ \vdots \\ a_{1n}c_1 + a_{2n}c_2 + \cdots + a_{mn}c_m &= 0 \end{aligned} \quad (5)$$

System (5) is a homogeneous system of $n$ equations in the $m$ unknowns $c_1, c_2, \ldots, c_m$, and since $m > n$, Theorem 1.4.1, page 41, tells us that the system has an infinite number of solutions. Thus there are scalars $c_1, c_2, \ldots, c_m$, not all zero, such that (2) is satisfied, and therefore $S_1$ is a linearly dependent set. This contradiction proves that $m \le n$; by exchanging the roles of $S_1$ and $S_2$, we can show that $n \le m$ and the proof is complete. ■

Because of this theorem, we can define one of the central concepts in linear algebra.

**DEFINITION 2** **Dimension** If the vector space $V$ has a finite basis, then the **dimension** of $V$ is the number of vectors in every basis and $V$ is called a **finite dimensional vector space.** Otherwise $V$ is called an **infinite dimensional vector space.** If $V = \{\mathbf{0}\}$, then $V$ is said to be **zero dimensional.**

***Notation.*** We write the dimension of $V$ as dim $V$.

***Remark.*** We have not proved that every vector space has a basis. This very difficult proof appears in Section 4.12. But we do not need this fact for Definition 2 to make sense; for *if* $V$ has a finite basis, then $V$ is finite dimensional. Otherwise $V$ is infinite dimensional. Thus, in order to show that $V$ is infinite dimensional, it is only necessary to show that $V$ does not have a finite basis. We can do this by showing that $V$ contains an infinite number of linearly independent vectors (see Example 7 below).

**EXAMPLE 4** **The Dimension of $\mathbb{R}^n$** Since $n$ linearly independent vectors in $\mathbb{R}^n$ constitute a basis, we see that

$$\dim \mathbb{R}^n = n$$

■

**EXAMPLE 5** **The Dimension of $P_n$** By Example 1 and Problem 4.5.47, page 331, the polynomials $\{1, x, x^2, \ldots, x^n\}$ constitute a basis in $P_n$. Thus dim $P_n = n + 1$.

**EXAMPLE 6** **The Dimension of $M_{mn}$** In $M_{mn}$ let $A_{ij}$ be the $m \times n$ matrix with a 1 in the $ij$th position and a zero everywhere else. It is easy to show that the $A_{ij}$ for $i = 1, 2, \ldots, m$ and $j = 1, 2, \ldots, n$ form a basis for $M_{mn}$. Thus dim $M_{mn} = mn$.

**EXAMPLE 7** ***P* Is Infinite Dimensional** In Example 4.4.7, page 307, we saw that no finite set of polynomials spans $P$. Thus $P$ has no finite basis, and is therefore an infinite dimensional vector space.

There are a number of theorems that tell us something about the dimension of a vector space.

**THEOREM 3** Suppose that dim $V = n$. If $\mathbf{u}_1, \mathbf{u}_2, \ldots, \mathbf{u}_m$ is a set of $m$ linearly independent vectors in $V$, then $m \leq n$.

**Proof** Let $\mathbf{v}_1, \mathbf{v}_2, \ldots, \mathbf{v}_n$ be a basis for $V$. If $m > n$, then as in the proof of Theorem 2, we can find constants $c_1, c_2, \ldots, c_m$ not all zero such that equation (2) is satisfied. This would contradict the linear independence of the $\mathbf{u}_i$'s. Thus $m \leq n$.

**THEOREM 4** Let $H$ be a subspace of the finite dimensional vector space $V$. Then $H$ is finite dimensional and

$$\boxed{\dim H \leq \dim V} \tag{6}$$

**Proof** Let dim $V = n$. Any set of linearly independent vectors in $H$ is also a linearly independent set in $V$. By Theorem 3, any linearly independent set in $H$ can contain at most $n$ vectors. If $H = \{\mathbf{0}\}$, then dim $H = 0$. If $H \neq \{\mathbf{0}\}$, let $\mathbf{v}_1$ be a vector in $H$ and let $H_1 = \text{span}\ \{\mathbf{v}_1\}$. If $H_1 = H$, dim $H = 1$ and we are done. If not, pick a $\mathbf{v}_2 \in H$ such that $\mathbf{v}_2 \notin H_1$ and let $H_2 = \text{span}\ \{\mathbf{v}_1, \mathbf{v}_2\}$, and so on. We continue until we have found linearly independent vectors $\mathbf{v}_1, \mathbf{v}_2, \ldots, \mathbf{v}_k$ such that $H = \text{span}\ \{\mathbf{v}_1, \mathbf{v}_2, \ldots, \mathbf{v}_k\}$. The process has to terminate because we can find at most $n$ linearly independent vectors in $H$. Thus dim $H = k \leq n$.

Theorem 4 has some interesting consequences. We give two of them here.

**EXAMPLE 8** Calculus

**$C[0, 1]$ and $C^1[0, 1]$ Are Infinite Dimensional** Let $P[0, 1]$ denote the set of polynomials defined on the interval $[0, 1]$. Then $P[0, 1] \subset C[0, 1]$. If $C[0, 1]$ were finite dimensional, then $P[0, 1]$ would be finite dimensional also. But by Example 7 this is not the case. Hence $C[0, 1]$ is infinite dimensional. Similarly, since $P[0, 1] \subset C^1[0, 1]$ (since every polynomial is differentiable), we also see that $C^1[0, 1]$ is infinite dimensional.

In general,

> *Any vector space containing an infinite dimensional subspace is infinite dimensional.*

**EXAMPLE 9**

**The Subspaces of $\mathbb{R}^3$** We can use Theorem 4 to find *all* subspaces of $\mathbb{R}^3$. Let $H$ be a subspace of $\mathbb{R}^3$. Then there are four possibilities: $H = \{\mathbf{0}\}$, $\dim H = 1$, $\dim H = 2$, and $\dim H = 3$. If $\dim H = 3$, then $H$ contains a basis of three linearly independent vectors $\mathbf{v}_1, \mathbf{v}_2, \mathbf{v}_3$ in $\mathbb{R}^3$. But then $\mathbf{v}_1, \mathbf{v}_2, \mathbf{v}_3$ also form a basis for $\mathbb{R}^3$. Thus $H = \text{span}\,\{\mathbf{v}_1, \mathbf{v}_2, \mathbf{v}_3\} = \mathbb{R}^3$. Hence the only way to get a *proper* subspace of $\mathbb{R}^3$ is to have $\dim H = 1$ or $\dim H = 2$. If $\dim H = 1$, then $H$ has a basis consisting of the one vector $\mathbf{v} = (a, b, c)$. Let $\mathbf{x}$ be in $H$. Then $\mathbf{x} = t(a, b, c)$ for some real number $t$ [since $(a, b, c)$ spans $H$]. If $\mathbf{x} = (x, y, z)$, this means that $x = at$, $y = bt$, $z = ct$. But this is the equation of a line in $\mathbb{R}^3$ passing through the origin with direction vector $(a, b, c)$.

Now suppose $\dim H = 2$ and let $\mathbf{v}_1 = (a_1, b_1, c_1)$ and $\mathbf{v}_2 = (a_2, b_2, c_2)$ be a basis for $H$. If $\mathbf{x} = (x, y, z) \in H$, then there exist real numbers $s$ and $t$ such that $\mathbf{x} = s\mathbf{v}_1 + t\mathbf{v}_2$ or $(x, y, z) = s(a_1, b_1, c_1) + t(a_2, b_2, c_2)$. Then

$$\begin{aligned} x &= sa_1 + ta_2 \\ y &= sb_1 + tb_2 \\ z &= sc_1 + tc_2 \end{aligned} \tag{7}$$

Let $\mathbf{v}_3 = (\alpha, \beta, \gamma) = \mathbf{v}_1 \times \mathbf{v}_2$. Then from Theorem 3.4.2, on page 263, part (*vi*), we have $\mathbf{v}_3 \cdot \mathbf{v}_1 = 0$ and $\mathbf{v}_3 \cdot \mathbf{v}_2 = 0$. Now we calculate

$$\begin{aligned} \alpha x + \beta y + \gamma z &= \alpha(sa_1 + ta_2) + \beta(sb_1 + tb_2) + \gamma(sc_1 + tc_2) \\ &= (\alpha a_1 + \beta b_1 + \gamma c_1)s + (\alpha a_2 + \beta b_2 + \gamma c_2)t \\ &= (\mathbf{v}_3 \cdot \mathbf{v}_1)s + (\mathbf{v}_3 \cdot \mathbf{v}_2)t = 0 \end{aligned}$$

Thus, if $(x, y, z) \in H$, then $\alpha x + \beta y + \gamma z = 0$, which shows that $H$ is a plane passing through the origin with normal vector $\mathbf{v}_3 = \mathbf{v}_1 \times \mathbf{v}_2$. Therefore we have proved that

> *The only proper subspaces of $\mathbb{R}^3$ are sets of vectors lying on a single line or a single plane passing through the origin.*

**EXAMPLE 10** **Solution Space and Null Space** Let $A$ be an $m \times n$ matrix and let $S = \{\mathbf{x} \in \mathbb{R}^n: A\mathbf{x} = \mathbf{0}\}$. Let $\mathbf{x}_1 \in S$ and $\mathbf{x}_2 \in S$; then $A(\mathbf{x}_1 + \mathbf{x}_2) = A\mathbf{x}_1 + A\mathbf{x}_2 = \mathbf{0} + \mathbf{0} = \mathbf{0}$ and $A(\alpha\mathbf{x}_1) = \alpha(A\mathbf{x}_1) = \alpha\mathbf{0} = \mathbf{0}$, so that $S$ is a subspace of $\mathbb{R}^n$ and $\dim S \le n$. $S$ is called the **solution space** of the homogeneous system $A\mathbf{x} = \mathbf{0}$. It is also called the **null space** of the matrix $A$.

**EXAMPLE 11** **Finding a Basis for the Solution Space of a Homogeneous System** Find a basis for (and the dimension of) the solution space $S$ of the homogeneous system

$$\begin{aligned} x + 2y - z &= 0 \\ 2x - y + 3z &= 0 \end{aligned}$$

**Solution** Here $A = \begin{pmatrix} 1 & 2 & -1 \\ 2 & -1 & 3 \end{pmatrix}$. Since $A$ is a $2 \times 3$ matrix, $S$ is a subspace of $\mathbb{R}^3$. Row-reducing, we find, successively,

$$\left(\begin{array}{ccc|c} 1 & 2 & -1 & 0 \\ 2 & -1 & 3 & 0 \end{array}\right) \longrightarrow \left(\begin{array}{ccc|c} 1 & 2 & -1 & 0 \\ 0 & -5 & 5 & 0 \end{array}\right)$$

$$\longrightarrow \left(\begin{array}{ccc|c} 1 & 2 & -1 & 0 \\ 0 & 1 & -1 & 0 \end{array}\right) \longrightarrow \left(\begin{array}{ccc|c} 1 & 0 & 1 & 0 \\ 0 & 1 & -1 & 0 \end{array}\right)$$

Then $y = z$ and $x = -z$ so that all solutions are of the form $\begin{pmatrix} -z \\ z \\ z \end{pmatrix}$. Thus $\begin{pmatrix} -1 \\ 1 \\ 1 \end{pmatrix}$ is a basis for $S$ and $\dim S = 1$. Note that $S$ is the set of vectors lying on the straight line $x = -t$, $y = t$, $z = t$.

**EXAMPLE 12** **Finding a Basis for the Solution Space of a Homogeneous System** Find a basis for the solution space $S$ of the system

$$\begin{aligned} 2x - y + 3z &= 0 \\ 4x - 2y + 6z &= 0 \\ -6x + 3y - 9z &= 0 \end{aligned}$$

**Solution** Row-reducing, we obtain

$$\left(\begin{array}{ccc|c} 2 & -1 & 3 & 0 \\ 4 & -2 & 6 & 0 \\ -6 & 3 & -9 & 0 \end{array}\right) \longrightarrow \left(\begin{array}{ccc|c} 2 & -1 & 3 & 0 \\ 0 & 0 & 0 & 0 \\ 0 & 0 & 0 & 0 \end{array}\right)$$

giving the single equation $2x - y + 3z = 0$. $S$ is a plane, and by Example 3 one basis is given by $\begin{pmatrix} 1 \\ 2 \\ 0 \end{pmatrix}$ and $\begin{pmatrix} 0 \\ 3 \\ 1 \end{pmatrix}$. So $\dim S = 2$.

Before leaving this section, we prove a result that is very useful in finding a basis for an arbitrary vector space. We have seen that $n$ linearly independent vectors in $\mathbb{R}^n$ constitute a basis for $\mathbb{R}^n$. This fact holds in *any* finite dimensional vector space.

**THEOREM 5** Any $n$ linearly independent vectors in a vector space $V$ of dimension $n$ constitute a basis for $V$.

**Proof** Let $\mathbf{v}_1, \mathbf{v}_2, \ldots, \mathbf{v}_n$ be the $n$ vectors. If they span $V$, then they constitute a basis. If they do not, then there is a vector $\mathbf{u} \in V$ such that $\mathbf{u} \notin \text{span}\,\{\mathbf{v}_1, \mathbf{v}_2, \ldots, \mathbf{v}_n\}$. This means that the $n + 1$ vectors $\mathbf{v}_1, \mathbf{v}_2, \ldots, \mathbf{v}_n, \mathbf{u}$ are linearly independent. To see this, note that if

$$c_1\mathbf{v}_1 + c_2\mathbf{v}_2 + \cdots + c_n\mathbf{v}_n + c_{n+1}\mathbf{u} = 0 \tag{8}$$

then $c_{n+1} = 0$, for if not we could write $\mathbf{u}$ as a linear combination of $\mathbf{v}_1, \mathbf{v}_2, \ldots, \mathbf{v}_n$ by dividing equation (8) by $c_{n+1}$ and putting all terms except $\mathbf{u}$ on the right-hand side. But if $c_{n+1} = 0$, then (8) reads

$$c_1\mathbf{v}_1 + c_2\mathbf{v}_2 + \cdots + c_n\mathbf{v}_n = 0$$

which means that $c_1 = c_2 = \cdots = c_n = 0$ since the $\mathbf{v}_i$'s are linearly independent. Now let $W = \text{span}\,\{\mathbf{v}_1, \mathbf{v}_2, \ldots, \mathbf{v}_n, \mathbf{u}\}$. Then as all the vectors in brackets are in $V$, $W$ is a subspace of $V$. Since $\mathbf{v}_1, \mathbf{v}_2, \ldots, \mathbf{v}_n, \mathbf{u}$ are linearly independent, they form a basis for $W$. Thus $\dim W = n + 1$. But from Theorem 4, $\dim W \leq n$. This contradiction shows that there is *no* vector $\mathbf{u} \in V$ such that $\mathbf{u} \notin \text{span}\,\{\mathbf{v}_1, \mathbf{v}_2, \ldots, \mathbf{v}_n\}$. Thus $\mathbf{v}_1, \mathbf{v}_2, \ldots, \mathbf{v}_n$ span $V$, and therefore constitute a basis for $V$. ■

# PROBLEMS 4.6

## Self-Quiz

### *True-False*

**I.** Any three vectors in $\mathbb{R}^3$ form a basis for $\mathbb{R}^3$.

**II.** Any three linearly independent vectors in $\mathbb{R}^3$ form a basis for $\mathbb{R}^3$.

**III.** A basis in a vector space is unique.

**IV.** Let $H$ be a proper subspace of $\mathbb{R}^4$. It is possible to find four linearly independent vectors in $H$.

**V.** Let $H = \left\{\begin{pmatrix} x \\ y \\ z \end{pmatrix} : 2x + 11y - 17z = 0\right\}$. Then $\dim H = 2$.

**VI.** Let $\{\mathbf{v}_1, \mathbf{v}_2, \ldots, \mathbf{v}_n\}$ be a basis for the vector space $V$. Then it is *not* possible to find a vector $\mathbf{v} \in V$ such that $\mathbf{v} \notin \text{span}\,\{\mathbf{v}_1, \mathbf{v}_2, \ldots, \mathbf{v}_n\}$.

**VII.** $\left\{\begin{pmatrix} 2 & 0 \\ 0 & 0 \end{pmatrix}, \begin{pmatrix} 0 & 3 \\ 0 & 0 \end{pmatrix}, \begin{pmatrix} 0 & 0 \\ -7 & 0 \end{pmatrix}, \begin{pmatrix} 0 & 0 \\ 0 & 12 \end{pmatrix}\right\}$ is a basis for $M_{22}$.

### Answers to Self-Quiz

**I.** False **II.** True **III.** False **IV.** False **V.** True **VI.** True **VII.** True

In Problems 1–10 determine whether the given set of vectors is a basis for the given vector space.

**1.** In $P_2$: $1 - x^2, x$

**2.** In $P_2$: $-3x, 1 + x^2, x^2 - 5$

**3.** In $P_2$: $x^2 - 1, x^2 - 2, x^2 - 3$

**4.** In $P_3$: $1, 1 + x, 1 + x^2, 1 + x^3$

**5.** In $P_3$: $3, x^3 - 4x + 6, x^2$

**6.** In $M_{22}$: $\begin{pmatrix} 3 & 1 \\ 0 & 0 \end{pmatrix}, \begin{pmatrix} 3 & 2 \\ 0 & 0 \end{pmatrix}, \begin{pmatrix} -5 & 1 \\ 0 & 6 \end{pmatrix}, \begin{pmatrix} 0 & 1 \\ 0 & -7 \end{pmatrix}$

**7.** In $M_{22}$: $\begin{pmatrix} a & 0 \\ 0 & 0 \end{pmatrix}, \begin{pmatrix} 0 & b \\ 0 & 0 \end{pmatrix}, \begin{pmatrix} 0 & 0 \\ c & 0 \end{pmatrix}, \begin{pmatrix} 0 & 0 \\ 0 & d \end{pmatrix}$, where $abcd \neq 0$

**8.** In $M_{22}$: $\begin{pmatrix} -1 & 0 \\ 3 & 1 \end{pmatrix}, \begin{pmatrix} 2 & 1 \\ 1 & 4 \end{pmatrix}, \begin{pmatrix} -6 & 1 \\ 5 & 8 \end{pmatrix}, \begin{pmatrix} 7 & -2 \\ 1 & 0 \end{pmatrix}, \begin{pmatrix} 0 & 1 \\ 0 & 0 \end{pmatrix}$

**9.** $H = \{(x, y) \in \mathbb{R}^2\colon x + y = 0\}$; $(1, -1)$

**10.** $H = \{(x, y) \in \mathbb{R}^2\colon x + y = 0\}$; $(1, -1), (-3, 3)$

**11.** Find a basis in $\mathbb{R}^3$ for the set of vectors in the plane $2x - y - z = 0$.

**12.** Find a basis in $\mathbb{R}^3$ for the set of vectors in the plane $3x - 2y + 6z = 0$.

**13.** Find a basis in $\mathbb{R}^3$ for the set of vectors on the line $x/2 = y/3 = z/4$.

**14.** Find a basis in $\mathbb{R}^3$ for the set of vectors on the line $x = 3t$, $y = -2t$, $z = t$.

**15.** Show that the only proper subspaces of $\mathbb{R}^2$ are straight lines passing through the origin.

**16.** In $\mathbb{R}^4$ let $H = \{(x, y, z, w)\colon ax + by + cz + dw = 0\}$, where $abcd \neq 0$.

**a.** Show that $H$ is a subspace of $\mathbb{R}^4$.

**b.** Find a basis for $H$.

**c.** What is dim $H$?

***17.** In $\mathbb{R}^n$ a **hyperplane** containing **0** is a subspace of dimension $n - 1$. If $H$ is a hyperplane in $\mathbb{R}^n$ that contains **0**, show that

$$H = \{(x_1, x_2, \ldots, x_n)\colon a_1x_1 + a_2x_2 + \cdots + a_nx_n = 0\}$$

where $a_1, a_2, \ldots, a_n$ are fixed real numbers, not all of which are zero.

**18.** In $\mathbb{R}^5$ find a basis for the hyperplane

$$H = \{(x_1, x_2, x_3, x_4, x_5)\colon 2x_1 - 3x_2 + x_3 + 4x_4 - x_5 = 0\}$$

In Problems 19–23 find a basis for the solution space of the given homogeneous system.

**19.** $\begin{aligned} x - y &= 0 \\ -2x + 2y &= 0 \end{aligned}$

**20.** $\begin{aligned} x - 2y &= 0 \\ 3x + y &= 0 \end{aligned}$

**21.** $\begin{aligned} x - y - z &= 0 \\ 2x - y + z &= 0 \end{aligned}$

**22.** $\begin{aligned} x - 3y + z &= 0 \\ -2x + 2y - 3z &= 0 \\ 4x - 8y + 5z &= 0 \end{aligned}$

**23.** $\begin{aligned} 2x - 6y + 4z &= 0 \\ -x + 3y - 2z &= 0 \\ -3x + 9y - 6z &= 0 \end{aligned}$

**24.** Find a basis for $D_3$, the vector space of diagonal $3 \times 3$ matrices. What is the dimension of $D_3$?

**25.** What is the dimension of $D_n$, the space of diagonal $n \times n$ matrices?

**26.** Let $S_{nn}$ denote the vector space of symmetric $n \times n$ matrices. Show that $S_{nn}$ is a subspace of $M_{nn}$ and that dim $S_{nn} = [n(n + 1)]/2$.

**27.** Suppose that $\mathbf{v}_1, \mathbf{v}_2, \ldots, \mathbf{v}_m$ are linearly independent vectors in a vector space $V$ of dimension $n$ and $m < n$. Show that $\{\mathbf{v}_1, \mathbf{v}_2, \ldots, \mathbf{v}_m\}$ can be enlarged to a basis for $V$. That is, there exist vectors $\mathbf{v}_{m+1}, \mathbf{v}_{m+2}, \ldots, \mathbf{v}_n$ such that $\{\mathbf{v}_1, \mathbf{v}_2, \ldots, \mathbf{v}_n\}$ is a basis. [*Hint:* Look at the proof of Theorem 5.]

**28.** Let $\{\mathbf{v}_1, \mathbf{v}_2, \ldots, \mathbf{v}_n\}$ be a basis for $V$. Let $\mathbf{u}_1 = \mathbf{v}_1$, $\mathbf{u}_2 = \mathbf{v}_1 + \mathbf{v}_2$, $\mathbf{u}_3 = \mathbf{v}_1 + \mathbf{v}_2 + \mathbf{v}_3$, $\ldots, \mathbf{u}_n = \mathbf{v}_1 + \mathbf{v}_2 + \cdots + \mathbf{v}_n$. Show that $\{\mathbf{u}_1, \mathbf{u}_2, \ldots, \mathbf{u}_n\}$ is also a basis for $V$.

**29.** Show that if $\{\mathbf{v}_1, \mathbf{v}_2, \ldots, \mathbf{v}_n\}$ spans $V$, then $\dim V \le n$. [*Hint:* Use the result of Problem 4.5.49.]

**30.** Let $H$ and $K$ be subspaces of $V$ such that $H \subseteq K$ and $\dim H = \dim K < \infty$. Show that $H = K$.

**31.** Let $H$ and $K$ be subspaces of $V$ and define $H + K = \{\mathbf{h} + \mathbf{k}\colon \mathbf{h} \in H \text{ and } \mathbf{k} \in K\}$.

**a.** Show that $H + K$ is a subspace of $V$.

**b.** If $H \cap K = \{\mathbf{0}\}$, show that $\dim (H + K) = \dim H + \dim K$.

***32.** If $H$ is a subspace of the finite dimensional vector space $V$, show that there exists a unique subspace $K$ of $V$ such that **(a)** $H \cap K = \{\mathbf{0}\}$ and **(b)** $H + K = V$.

**33.** Show that two vectors $\mathbf{v}_1$ and $\mathbf{v}_2$ in $\mathbb{R}^2$ with end points at the origin are collinear if and only if dim span $\{\mathbf{v}_1, \mathbf{v}_2\} = 1$.

**34.** Show that three vectors $\mathbf{v}_1$, $\mathbf{v}_2$, and $\mathbf{v}_3$ in $\mathbb{R}^3$ with end points at the origin are coplanar if and only if dim span $\{\mathbf{v}_1, \mathbf{v}_2, \mathbf{v}_3\} \le 2$.

**35.** Show that any $n$ vectors that span an $n$-dimensional space $V$ form a basis for $V$. [*Hint:* Show that if the $n$ vectors are not linearly independent, then $\dim V < n$.]

***36.** Show that every subspace of a finite dimensional vector space has a basis.

**37.** Find two bases for $\mathbb{R}^4$ that contain (1, 0, 1, 0) and (0, 1, 0, 1) and have no other vectors in common.

**38.** For what values of the real number $a$ do the vectors $(a, 1, 0)$, $(1, 0, a)$, and $(1 + a, 1, a)$ constitute a basis for $\mathbb{R}^3$?

# MATLAB 4.6

The problems in this section concentrate on working with bases for *all* of $\mathbb{R}^n$ (or all of $P_n$ or all of $M_{nm}$). The problems in Section 4.7 will concentrate on bases of subspaces.

1. **a.** Verify that the sets given in part (b) form a basis for the indicated vector space. Explain how each of the properties of the definition of a basis is satisfied.

   **b.** Generate a random vector in the given vector space. Show that it is a linear combination of the basis vectors with unique coefficients for the linear combination. Repeat for two more random vectors.

   **i.** $\mathbb{R}^3$ $\left\{\begin{pmatrix} 8.25 \\ 7 \\ 8 \end{pmatrix}, \begin{pmatrix} 1.01 \\ -7 \\ -1 \end{pmatrix}, \begin{pmatrix} 10 \\ -6.5 \\ -1 \end{pmatrix}\right\}$

   **ii.** $\mathbb{R}^5$ $\left\{\begin{pmatrix} 1 \\ -1 \\ 0 \\ 2 \\ 1 \end{pmatrix}, \begin{pmatrix} 1 \\ 0 \\ 3 \\ -1 \\ 1 \end{pmatrix}, \begin{pmatrix} 2 \\ 4 \\ -1 \\ 3 \\ 1 \end{pmatrix}, \begin{pmatrix} -1 \\ -1 \\ 2 \\ -1 \\ 1 \end{pmatrix}, \begin{pmatrix} 1 \\ 1 \\ 1 \\ 1 \\ 1 \end{pmatrix}\right\}$

**iii.** $M_{22}$ $\left\{\begin{pmatrix}1 & -1\\ 1.2 & 2.1\end{pmatrix}, \begin{pmatrix}2 & 1\\ -1 & 1\end{pmatrix}, \begin{pmatrix}1 & 3\\ -2 & 0\end{pmatrix}, \begin{pmatrix}-1.5 & 4\\ 4.3 & 5\end{pmatrix}\right\}$

(See MATLAB Problem 10 in Section 4.4.)

**iv.** $P_4$ $\{x^4 - x^3 + 2x + 1,\ \ x^4 + 3x^2 - x + 4,\ \ 2x^4 + 4x^3 - x^2 + 3x + 5,\ \ x^4 + x^3 - 2x^2 + x,\ \ x^4 + x^3 + x^2 + x + 1\}$

(See MATLAB Problem 9 in Section 4.4.)

2. For the sets of vectors in MATLAB Problem 9(b) in Section 4.5, show that the sets span their respective $\mathbb{R}^n$ but do not form a basis. For each set, generate a random vector $\mathbf{w}$ in the respective $\mathbb{R}^n$ and verify that $\mathbf{w}$ is a linear combination of the set of vectors but that the coefficients of the linear combination are not unique. Repeat for two more vectors $\mathbf{w}$.

3. For each of the bases in MATLAB Problem 1 in this section:
   a. Delete a vector from the set and show that the new set is not a basis, describing which property of basis is not satisfied. Repeat (delete a different vector).
   b. Generate a random vector $\mathbf{w}$ in the vector space. Add $\mathbf{w}$ to the set of vectors. Show that the new set is not a basis, describing which property of basis is not satisfied. Repeat with another $\mathbf{w}$.
   c. *(Paper and pencil)* Write a proof, based on reduced row echelon form, that a basis for $\mathbb{R}^n$ must contain exactly $n$ vectors and a proof that a basis for $P_n$ must contain exactly $n + 1$ vectors.

4. a. The dimension of $M_{32}$ is 6. Generate five random matrices in $M_{32}$ and show that they do not form a basis for $M_{32}$, describing which property of basis is not satisfied. Generate seven random matrices in $M_{32}$ and show that they do not form a basis for $M_{32}$, describing which property of basis is not satisfied.
   b. *(Paper and pencil)* Write a proof, based on reduced row echelon form, that the dimension of $M_{nm}$ is $nm$, the product of $n$ and $m$.

5. Consider the matrices in MATLAB Problem 2 in Section 1.8 and each matrix whose columns are the vectors in each set of vectors given in MATLAB Problem 1(b) (i) and (ii) in this section.
   a. Determine for each matrix $A$ (say, its size is $n \times n$) whether it is invertible and whether the columns of $A$ form a basis for $\mathbb{R}^n$.
   b. Write a conjecture relating the property of invertibility to the property of having the columns form a basis.
   c. *(Paper and pencil)* Prove your conjecture.

6. a. *(Paper and pencil)* Suppose $\{\mathbf{v}_1, \ldots, \mathbf{v}_5\}$ is a basis for $\mathbb{R}^5$. Suppose $\mathbf{w}_1 = A\mathbf{v}_1$, $\mathbf{w}_2 = A\mathbf{v}_2, \ldots, \mathbf{w}_5 = A\mathbf{v}_5$, for some $n \times 5$ matrix $A$. Answer the questions below to complete the description of how to find $A\mathbf{w}$ for any $\mathbf{w}$ knowing only what $A$ does to a basis.
      **i.** Given any $\mathbf{w}$ in $\mathbb{R}^5$, argue why $\mathbf{w} = c_1\mathbf{v}_1 + \cdots + c_5\mathbf{v}_5$, where $c_1, \ldots, c_5$ are unique.
      **ii.** Show that $A\mathbf{w} = c_1\mathbf{w}_1 + \cdots + c_5\mathbf{w}_5$.
      **iii.** Argue why $A\mathbf{w} = [\mathbf{w}_1\ \mathbf{w}_2\ \mathbf{w}_3\ \mathbf{w}_4\ \mathbf{w}_5]\begin{pmatrix}c_1\\ c_2\\ c_3\\ c_4\\ c_5\end{pmatrix}$

**b.** Let $\{\mathbf{v}_1, \ldots, \mathbf{v}_5\}$ be the basis for $\mathbb{R}^5$ given in MATLAB Problem 1(b)(ii) in this section. Suppose

$$A\mathbf{v}_1 = \begin{pmatrix} 5 \\ 5 \\ 3 \end{pmatrix} \quad A\mathbf{v}_2 = \begin{pmatrix} 7 \\ 5 \\ 7 \end{pmatrix} \quad A\mathbf{v}_3 = \begin{pmatrix} 36 \\ 25 \\ 13 \end{pmatrix} \quad A\mathbf{v}_4 = \begin{pmatrix} -10 \\ -2 \\ -1 \end{pmatrix} \quad A\mathbf{v}_5 = \begin{pmatrix} 5 \\ 9 \\ 5 \end{pmatrix}$$

Find $A\mathbf{w}$, where

**i.** $\mathbf{w} = \begin{pmatrix} 0 \\ -10 \\ 9 \\ -6 \\ -4 \end{pmatrix}$

**ii.** **w = 2*rand(5,1)−1**

**c.** Repeat (b) for

$$A\mathbf{v}_1 = \begin{pmatrix} 1 \\ 0 \\ 0 \\ 0 \\ 0 \end{pmatrix} \quad A\mathbf{v}_2 = \begin{pmatrix} 0 \\ 1 \\ 0 \\ 0 \\ 0 \end{pmatrix} \quad A\mathbf{v}_3 = \begin{pmatrix} 0 \\ 0 \\ 1 \\ 0 \\ 0 \end{pmatrix} \quad A\mathbf{v}_4 = \begin{pmatrix} 0 \\ 0 \\ 0 \\ 1 \\ 0 \end{pmatrix} \quad A\mathbf{v}_5 = \begin{pmatrix} 0 \\ 0 \\ 0 \\ 0 \\ 1 \end{pmatrix}$$

## 4.7 THE RANK, NULLITY, ROW SPACE, AND COLUMN SPACE OF A MATRIX

In Section 4.5 we introduced the notion of linear independence. We showed that if $A$ is an invertible $n \times n$ matrix, then the columns and rows of $A$ form sets of linearly independent vectors. However, if $A$ is not invertible (so that $\det A = 0$), or if $A$ is not a square matrix, then these results tell us nothing about the number of linearly independent rows or columns of $A$. In this section we fill in this gap. We also show how a basis for the span of a set of vectors can be obtained by row reduction.

Let $A$ be an $m \times n$ matrix and let

**The Null Space of a Matrix**

$$N_A = \{\mathbf{x} \in \mathbb{R}^n\colon\ A\mathbf{x} = \mathbf{0}\} \tag{1}$$

Then as we saw in Example 4.6.10 on page 343, $N_A$ is a subspace of $\mathbb{R}^n$.

**DEFINITION 1** **Null Space and Nullity of a Matrix** $N_A$ is called the **null space** of $A$ and $\nu(A) = \dim N_A$ is called the **nullity** of $A$. If $N_A$ contains only the zero vector, then $\nu(A) = 0$.

*Note.* The null space of a matrix is also called its **kernel.**

**EXAMPLE 1** **The Null Space and Nullity of a 2 × 3 Matrix** Let $A = \begin{pmatrix} 1 & 2 & -1 \\ 2 & -1 & 3 \end{pmatrix}$. Then as we saw in Example 4.6.11, on page 343, $N_A$ is spanned by $\begin{pmatrix} -1 \\ 1 \\ 1 \end{pmatrix}$, so $\nu(A) = 1$. ■

**EXAMPLE 2** **The Null Space and Nullity of a 3 × 3 Matrix** Let $A = \begin{pmatrix} 2 & -1 & 3 \\ 4 & -2 & 6 \\ -6 & 3 & -9 \end{pmatrix}$.

Then by Example 4.6.12, on page 343, $\left\{\begin{pmatrix} 1 \\ 2 \\ 0 \end{pmatrix}, \begin{pmatrix} 0 \\ 3 \\ 1 \end{pmatrix}\right\}$ is a basis for $N_A$, so $\nu(A) = 2$. ■

**THEOREM 1** Let $A$ be an $n \times n$ matrix. Then $A$ is invertible if and only if $\nu(A) = 0$.

**Proof** By our Summing Up Theorem [Theorem 4.5.6, page 325, parts (*i*) and (*ii*)], $A$ is invertible if and only if the homogeneous system $A\mathbf{x} = \mathbf{0}$ has only the trivial solution $\mathbf{x} = \mathbf{0}$. But from equation (1) this means that $A$ is invertible if and only if $N_A = \{\mathbf{0}\}$. Thus $A$ is invertible if and only if $\nu(A) = \dim N_A = 0$. ■

**DEFINITION 2** **Range of a Matrix** Let $A$ be an $m \times n$ matrix. Then the **range** of $A$, denoted by Range $A$, is given by

$$\text{Range } A = \{\mathbf{y} \in \mathbb{R}^m \colon A\mathbf{x} = \mathbf{y} \text{ for some } \mathbf{x} \in \mathbb{R}^n\} \tag{2}$$

**THEOREM 2** Let $A$ be an $m \times n$ matrix. Then Range $A$ is a subspace of $\mathbb{R}^m$.

**Proof** Suppose that $\mathbf{y}_1$ and $\mathbf{y}_2$ are in Range $A$. Then there are vectors $\mathbf{x}_1$ and $\mathbf{x}_2$ in $\mathbb{R}^n$ such that $\mathbf{y}_1 = A\mathbf{x}_1$ and $\mathbf{y}_2 = A\mathbf{x}_2$. Therefore

$$A(\alpha\mathbf{x}_1) = \alpha A\mathbf{x}_1 = \alpha\mathbf{y}_1 \quad \text{and} \quad A(\mathbf{x}_1 + \mathbf{x}_2) = A\mathbf{x}_1 + A\mathbf{x}_2 = \mathbf{y}_1 + \mathbf{y}_2$$

so $\alpha\mathbf{y}_1$ and $\mathbf{y}_1 + \mathbf{y}_2$ are in Range $A$. Thus, from Theorem 4.3.1, Range $A$ is a subspace of $\mathbb{R}^m$. ■

**DEFINITION 3** **Rank of a Matrix** Let $A$ be an $m \times n$ matrix. Then the **rank** of $A$, denoted by $\rho(A)$, is given by

$$\boxed{\rho(A) = \dim \text{Range } A}$$

We shall give two definitions and a theorem that make the calculation of rank relatively easy.

**DEFINITION 4** **Row and Column Space of a Matrix** If $A$ is an $m \times n$ matrix, let $\{\mathbf{r}_1, \mathbf{r}_2, \ldots, \mathbf{r}_m\}$ denote the rows of $A$ and let $\{\mathbf{c}_1, \mathbf{c}_2, \ldots, \mathbf{c}_n\}$ denote the columns of $A$. Then we define

$$\boxed{R_A = \textbf{row space of } A = \text{span } \{\mathbf{r}_1, \mathbf{r}_2, \ldots, \mathbf{r}_m\}} \quad \textbf{(3)}$$

and

$$\boxed{C_A = \textbf{column space of } A = \text{span } \{\mathbf{c}_1, \mathbf{c}_2, \ldots, \mathbf{c}_n\}} \quad \textbf{(4)}$$

*Note.* $R_A$ is a subspace of $\mathbb{R}^n$ and $C_A$ is a subspace of $\mathbb{R}^m$.

We have introduced a lot of notation in just three pages.

Before giving an example, we show that two of these four spaces are the same.

**THEOREM 3** For any matrix $A$, $C_A = \text{Range } A$. That is, the range of a matrix is equal to its column space.

**Proof** To show that $C_A = \text{Range } A$, we show that $\text{Range } A \subseteq C_A$ and $C_A \subseteq \text{Range } A$

**i.** We show that $\textbf{Range } \boldsymbol{A} \subseteq \boldsymbol{C_A}$. Suppose that $\mathbf{y} \in \text{Range } A$. Then there is a vector $\mathbf{x}$ such that $\mathbf{y} = A\mathbf{x}$. But as we observed in Section 1.6, page 69, $A\mathbf{x}$ can be written as a linear combination of the columns of $A$. Thus $\mathbf{y} \in C_A$, so $\text{Range } A \subseteq C_A$.

**ii.** We show that $\boldsymbol{C_A} \subseteq \textbf{Range } \boldsymbol{A}$. Suppose that $\mathbf{y} \in C_A$. Then $\mathbf{y}$ can be written as a linear combination of the columns of $A$ as in equation (1.6.9) on page 69. Let $\mathbf{x}$ be the column vector of coefficients in this linear combination. Then as in equation (1.6.9), $\mathbf{y} = A\mathbf{x}$, so $\mathbf{y} \in \text{Range } A$, which proves that $C_A \subseteq \text{Range } A$. ■

**EXAMPLE 3** **Finding $N_A$, $\nu(A)$, Range $A$, $\rho(A)$, $R_A$, and $C_A$ for a $2 \times 3$ Matrix** Let $A = \begin{pmatrix} 1 & 2 & -1 \\ 2 & -1 & 3 \end{pmatrix}$. $A$ is a $2 \times 3$ matrix.

**i.** *The null space of* $A = N_A = \{\mathbf{x} \in \mathbb{R}^3\colon\ A\mathbf{x} = \mathbf{0}\}$. As we saw in Example 1,

$$N_A = \text{span}\left\{\begin{pmatrix} -1 \\ 1 \\ 1 \end{pmatrix}\right\}.$$

**ii.** *The nullity of* $A = \nu(A) = \dim N_A = 1$.

**iii.** We know that *Range* $A = C_A$. The first two columns of $A$ are linearly independent vectors in $\mathbb{R}^2$, and thereby form a basis for $\mathbb{R}^2$. Thus Range $A = C_A = \mathbb{R}^2$

**iv.** $\rho(A) = \dim \text{Range } A = \dim \mathbb{R}^2 = 2$

**v.** *The row space of* $A = R_A = \text{span}\ \{(1, 2, -1), (2, -1, 3)\}$. Since these two vectors are linearly independent, we see that $R_A$ is a two-dimensional subspace of $\mathbb{R}^3$. From Example 4.6.9, on page 342, we observe that $R_A$ is a plane passing through the origin.

In Example 3(*iv*) we observe that $\rho(A) = \dim R_A = 2$. This is no coincidence.

**THEOREM 4** If $A$ is an $m \times n$ matrix, then

$$\dim R_A = \dim C_A = \dim \text{Range } A = \rho(A)$$

**Proof** As usual, we denote the $ij$th component of $A$ by $a_{ij}$. We must show that $\dim R_A = \dim C_A$. We denote the rows of $A$ by $\mathbf{r}_1, \mathbf{r}_2, \ldots, \mathbf{r}_m$, and let $k = \dim R_A$. Let $S = \{\mathbf{s}_1, \mathbf{s}_2, \ldots, \mathbf{s}_k\}$ be a basis for $R_A$. Then every row of $A$ can be written as a linear combination of the vectors in $S$, and we have, for some constants $\alpha_{ij}$,

$$\begin{aligned}
\mathbf{r}_1 &= \alpha_{11}\mathbf{s}_1 + \alpha_{12}\mathbf{s}_2 + \cdots + \alpha_{1k}\mathbf{s}_k \\
\mathbf{r}_2 &= \alpha_{21}\mathbf{s}_1 + \alpha_{22}\mathbf{s}_2 + \cdots + \alpha_{2k}\mathbf{s}_k \\
&\ \ \vdots \\
\mathbf{r}_m &= \alpha_{m1}\mathbf{s}_1 + \alpha_{m2}\mathbf{s}_2 + \cdots + \alpha_{mk}\mathbf{s}_k
\end{aligned} \tag{5}$$

Now the $j$th component of $\mathbf{r}_i$ is $a_{ij}$. Thus if we equate the $j$th components of both sides of (5) and set $\mathbf{s}_i = (s_{i1}, s_{i2}, \ldots, s_{in})$, we obtain

$$\begin{aligned}
a_{1j} &= \alpha_{11}s_{1j} + \alpha_{12}s_{2j} + \cdots + \alpha_{1k}s_{kj} \\
a_{2j} &= \alpha_{21}s_{1j} + \alpha_{22}s_{2j} + \cdots + \alpha_{2k}s_{kj} \\
&\ \ \vdots \\
a_{mj} &= \alpha_{m1}s_{1j} + \alpha_{m2}s_{2j} + \cdots + \alpha_{mk}s_{kj}
\end{aligned}$$

or

$$\begin{pmatrix} a_{1j} \\ a_{2j} \\ \vdots \\ a_{mj} \end{pmatrix} = s_{1j}\begin{pmatrix} \alpha_{11} \\ \alpha_{21} \\ \vdots \\ \alpha_{m1} \end{pmatrix} + s_{2j}\begin{pmatrix} \alpha_{12} \\ \alpha_{22} \\ \vdots \\ \alpha_{m2} \end{pmatrix} + \cdots + s_{kj}\begin{pmatrix} \alpha_{1k} \\ \alpha_{2k} \\ \vdots \\ \alpha_{mk} \end{pmatrix} \tag{6}$$

Let $\boldsymbol{\alpha}_i$ denote the vector $\begin{pmatrix} \alpha_{1i} \\ \alpha_{2i} \\ \vdots \\ \alpha_{mi} \end{pmatrix}$. Then since the left-hand side of (6) is the $j$th column of $A$, we see that we can write every column of $A$ as a linear combination of $\boldsymbol{\alpha}_1, \boldsymbol{\alpha}_2, \ldots, \boldsymbol{\alpha}_k$, which means that the vectors $\boldsymbol{\alpha}_1, \boldsymbol{\alpha}_2, \ldots, \boldsymbol{\alpha}_k$ span $C_A$ and

$$\dim C_A \le k = \dim R_A \tag{7}$$

But equation (7) holds for any matrix $A$. In particular, it holds for $A^t$. But $C_{A^t} = R_A$ and $R_{A^t} = C_A$. Since from (7) $\dim C_{A^t} \le \dim R_{A^t}$, we have

$$\dim R_A \le \dim C_A \tag{8}$$

Combining (7) and (8) completes the proof. ■

**EXAMPLE 4** **Finding Range $A$ and $\rho(A)$ for a 3 × 3 Matrix** Find a basis for Range $A$ and determine the rank of $A = \begin{pmatrix} 2 & -1 & 3 \\ 4 & -2 & 6 \\ -6 & 3 & -9 \end{pmatrix}$.

**Solution** Since $\mathbf{r}_2 = 2\mathbf{r}_1$ and $\mathbf{r}_3 = -3\mathbf{r}_1$, we see that $\rho(A) = \dim R_A = 1$. Thus any column in $C_A$ is a basis for $C_A = \text{Range } A$. For example, $\begin{pmatrix} 2 \\ 4 \\ -6 \end{pmatrix}$ is a basis for Range $A$. ■

The following theorem will simplify our computations of range, rank, and nullity.

**THEOREM 5** If $A$ is row equivalent to $B$, then $R_A = R_B$, $\rho(A) = \rho(B)$, and $\nu(A) = \nu(B)$.

**Proof** Recall from Definition 1.8.3, page 107, that $A$ is row equivalent to $B$ if $A$ can be "reduced" to $B$ by elementary row operations. Suppose that $C$ is the matrix obtained by performing an elementary row operation on $A$. We first show that $R_A = R_C$. Since $B$ is obtained by performing several elementary row operations on $A$, our first result, applied several times, will imply that $R_A = R_B$.

*Case 1:* Interchange two rows of $A$. Then $R_A = R_C$ because the rows of $A$ and $C$ are the same (just written in a different order).

*Case 2:* Multiply the $i$th row of $A$ by $c \neq 0$. If the rows of $A$ are $\{\mathbf{r}_1, \mathbf{r}_2, \ldots, \mathbf{r}_i, \ldots, \mathbf{r}_m\}$, then the rows of $C$ are $\{\mathbf{r}_1, \mathbf{r}_2, \ldots, c\mathbf{r}_i, \ldots, \mathbf{r}_m\}$. Obviously, $c\mathbf{r}_i = c(\mathbf{r}_i)$ and $\mathbf{r}_i = (1/c)\mathbf{r}_i$. Thus each row of $C$ is a multiple of one row of $A$ and vice versa. This means that each row of $C$ is in the span of the rows of $A$ and vice versa. We have

$$R_A \subseteq R_C \quad \text{and} \quad R_C \subseteq R_A, \qquad \text{so } R_C = R_A$$

*Case 3:* Multiply the $i$th row of $A$ by $c \neq 0$ and add it to the $j$th row. Now the rows of $C$ are $\{\mathbf{r}_1, \mathbf{r}_2, \ldots, \mathbf{r}_i, \ldots, \mathbf{r}_j + c\mathbf{r}_i, \ldots, \mathbf{r}_m\}$. Here

$$\mathbf{r}_j = \underbrace{(\mathbf{r}_j + c\mathbf{r}_i)}_{j\text{th row of } C} - \underset{\substack{\uparrow \\ i\text{th row of } C}}{c\mathbf{r}_i}$$

so each row of $A$ can be written as a linear combination of the rows of $C$ and vice versa. Then as before,

$$R_A \subseteq R_C \quad \text{and} \quad R_C \subseteq R_A, \qquad \text{so } R_C = R_A$$

We have shown that $R_A = R_B$. Hence $\rho(R_A) = \rho(R_B)$. Finally, the set of solutions to $A\mathbf{x} = \mathbf{0}$ does not change under elementary row operations. Thus $N_A = N_B$, so $\nu(A) = \nu(B)$. ■

Theorem 5 is very important. It tells us, for example, that the rank and row space of a matrix are the same as the rank and row space of the row echelon form of the matrix. It is not difficult to prove the following (see Problem 43).

**THEOREM 6** The rank of a matrix is equal to the number of pivots in its row echelon form. ■

EXAMPLE 5 **Finding $\rho(A)$ and $R_A$ for a $3 \times 3$ Matrix** Determine the rank and row space of $A = \begin{pmatrix} 1 & -1 & 3 \\ 2 & 0 & 4 \\ -1 & -3 & 1 \end{pmatrix}$. The row echelon form of $A$ is $\begin{pmatrix} 1 & -1 & 3 \\ 0 & 1 & -1 \\ 0 & 0 & 0 \end{pmatrix} = B$. Since $B$ has two pivots, $\rho(A) = \dim R_A = 2$. A basis for $R_A$ consists of the first two rows of $B$:

$$R_A = \text{span}\,\{(1, -1, 3), (0, 1, -1)\}$$

■

Theorem 5 is useful when we want to find a basis for the span of a set of vectors.

**EXAMPLE 6** **Finding a Basis for the Span of Four Vectors in $\mathbb{R}^3$** Find a basis for the space spanned by

$$\mathbf{v}_1 = \begin{pmatrix} 1 \\ 2 \\ -3 \end{pmatrix}, \quad \mathbf{v}_2 = \begin{pmatrix} -2 \\ 0 \\ 4 \end{pmatrix}, \quad \mathbf{v}_3 = \begin{pmatrix} 0 \\ 4 \\ -2 \end{pmatrix}, \quad \mathbf{v}_4 = \begin{pmatrix} -2 \\ -4 \\ 6 \end{pmatrix}$$

**Solution** We write the vectors as rows of a matrix $A$ and then reduce the matrix to row echelon form. The resulting matrix will have the same row space as $A$. The row echelon form of $\begin{pmatrix} 1 & 2 & -3 \\ -2 & 0 & 4 \\ 0 & 4 & -2 \\ -2 & -4 & 6 \end{pmatrix}$ is $\begin{pmatrix} 1 & 2 & -3 \\ 0 & 1 & -\frac{1}{2} \\ 0 & 0 & 0 \\ 0 & 0 & 0 \end{pmatrix}$, which has two pivots.

Thus a basis for span $\{\mathbf{v}_1, \mathbf{v}_2, \mathbf{v}_3, \mathbf{v}_4\}$ is $\left\{ \begin{pmatrix} 1 \\ 2 \\ -3 \end{pmatrix}, \begin{pmatrix} 0 \\ 1 \\ -\frac{1}{2} \end{pmatrix} \right\}$. For example,

$$\begin{pmatrix} -2 \\ 0 \\ 4 \end{pmatrix} = -2 \begin{pmatrix} 1 \\ 2 \\ -3 \end{pmatrix} + 4 \begin{pmatrix} 0 \\ 1 \\ -\frac{1}{2} \end{pmatrix}$$

There is a relatively easy way to find the null space of a matrix.

**EXAMPLE 7** **Finding the Null Space of a 4 × 4 Matrix** Find the null space of

$$A = \begin{pmatrix} 1 & 2 & -4 & 3 \\ 2 & 5 & 6 & -8 \\ 0 & -1 & -14 & 14 \\ 3 & 6 & -12 & 9 \end{pmatrix}$$

**Solution** The reduced row echelon form of $A$ is

$$U = \begin{pmatrix} 1 & 0 & -32 & 31 \\ 0 & 1 & 14 & -14 \\ 0 & 0 & 0 & 0 \\ 0 & 0 & 0 & 0 \end{pmatrix}$$

By the same reasoning as in the proof of Theorem 5, the solutions to $A\mathbf{x} = \mathbf{0}$ are the same as the solutions to $U\mathbf{x} = \mathbf{0}$. If $\mathbf{x} = \begin{pmatrix} x_1 \\ x_2 \\ x_3 \\ x_4 \end{pmatrix}$, then $U\mathbf{x} = \mathbf{0}$ results in

$$\begin{aligned} x_1 \quad - 32x_3 + 31x_4 &= 0 \\ x_2 + 14x_3 - 14x_4 &= 0 \end{aligned}$$

or

$$\begin{aligned} x_1 &= 32x_3 - 31x_4 \\ x_2 &= -14x_3 + 14x_4 \end{aligned}$$

So if $\mathbf{x} \in N_A$, then

$$\mathbf{x} = \begin{pmatrix} 32x_3 - 31x_4 \\ -14x_3 + 14x_4 \\ x_3 \\ x_4 \end{pmatrix} = x_3 \begin{pmatrix} 32 \\ -14 \\ 1 \\ 0 \end{pmatrix} + x_4 \begin{pmatrix} -31 \\ 14 \\ 0 \\ 1 \end{pmatrix}$$

basis for $N_A$

That is, $N_A = \text{span}\left\{ \begin{pmatrix} 32 \\ -14 \\ 1 \\ 0 \end{pmatrix}, \begin{pmatrix} -31 \\ 14 \\ 0 \\ 1 \end{pmatrix} \right\}$.

The procedure used in Example 7 can always be used to find the null space of a matrix.

We make an interesting geometric observation:

> Every vector in the row space of a real matrix is orthogonal to every vector in its null space.

We write this in shorthand notation as $R_A \perp N_A$. To see why this is so, consider the equation $A\mathbf{x} = \mathbf{0}$. If $A$ is an $m \times n$ matrix, then we have

$$\begin{pmatrix} a_{11} & a_{12} & \cdots & a_{1n} \\ a_{21} & a_{22} & \cdots & a_{2n} \\ \vdots & \vdots & & \vdots \\ a_{m1} & a_{m2} & \cdots & a_{mn} \end{pmatrix} \begin{pmatrix} x_1 \\ x_2 \\ \vdots \\ x_n \end{pmatrix} = \begin{pmatrix} 0 \\ 0 \\ \vdots \\ 0 \end{pmatrix}$$

If $\mathbf{r}_i$ denotes the $i$th row of $A$, we see from the equation above that $\mathbf{r}_i \cdot \mathbf{x} = 0$ for $i = 1, 2, \ldots, m$. Thus if $\mathbf{x} \in N_A$, then $\mathbf{r}_i \perp \mathbf{x}$ for $i = 1, 2, \ldots, m$. But if $\mathbf{y} \in R_A$, then $\mathbf{y} = c_1\mathbf{r}_2 + \cdots + c_m\mathbf{r}_m$ for some constants $c_1, c_2, \ldots, c_m$. Then

$$\mathbf{y} \cdot \mathbf{x} = (c_1\mathbf{r}_1 + c_2\mathbf{r}_2 + \cdots + c_m\mathbf{r}_m) \cdot \mathbf{x} = c_1\mathbf{r}_1 \cdot \mathbf{x} + c_2\mathbf{r}_2 \cdot \mathbf{x} + \cdots + c_m\mathbf{r}_m \cdot \mathbf{x} = 0$$

which proves our claim.

In Example 7, $R_A = \text{span}\,\{(1, 0, -32, 31), (0, 1, 14, -14)\}$ and $N_A = \text{span}\left\{\begin{pmatrix} 32 \\ -14 \\ 1 \\ 0 \end{pmatrix}, \begin{pmatrix} -31 \\ 14 \\ 0 \\ 1 \end{pmatrix}\right\}$. You should verify that basis vectors for $R_A$ are orthogonal to basis vectors for $N_A$.

The next theorem gives the relationship between rank and nullity.

**THEOREM 7** Let $A$ be an $m \times n$ matrix. Then

$$\boxed{\rho(A) + \nu(A) = n}$$

That is, the rank of $A$ plus the nullity of $A$ equals the number of columns of $A$.

**Proof** We assume that $k = \rho(A)$ and that the first $k$ columns of $A$ are linearly independent. Let $\mathbf{c}_i$ $(i > k)$ denote any other column of $A$. Since $\mathbf{c}_1, \mathbf{c}_2, \ldots, \mathbf{c}_k$ form a basis for $C_A$, we have, for some scalars $a_1, a_2, \ldots, a_k$,

$$\mathbf{c}_i = a_1\mathbf{c}_1 + a_2\mathbf{c}_2 + \cdots + a_k\mathbf{c}_k$$

Thus by adding $-a_1\mathbf{c}_1, -a_2\mathbf{c}_2, \ldots, -a_k\mathbf{c}_k$ successively to the $i$th column of $A$, we obtain a new $m \times n$ matrix $B$ with $\rho(B) = \rho(A)$ and $\nu(B) = \nu(A)$ with the $i$th column of $B = \mathbf{0}$.† We do this to all other columns of $A$ (except the first $k$) to obtain the matrix

$$D = \begin{pmatrix} a_{11} & a_{12} & \cdots & a_{1k} & 0 & 0 & \cdots & 0 \\ a_{21} & a_{22} & \cdots & a_{2k} & 0 & 0 & \cdots & 0 \\ \vdots & \vdots & & \vdots & \vdots & \vdots & & \vdots \\ a_{m1} & a_{m2} & \cdots & a_{mk} & 0 & 0 & \cdots & 0 \end{pmatrix}$$

where $\rho(D) = \rho(A)$ and $\nu(D) = \nu(A)$. By possibly rearranging the rows of $D$, we can assume that the first $k$ rows of $D$ are independent. Then we do the same thing to the rows (i.e., add multiples of the first $k$ rows to the last $m - k$ rows) to obtain a new matrix:

† This follows by considering $A^t$ (the columns of $A$ are the rows of $A^t$).

$$F = \begin{pmatrix} a_{11} & a_{12} & \cdots & a_{1k} & 0 & \cdots & 0 \\ a_{21} & a_{22} & \cdots & a_{2k} & 0 & \cdots & 0 \\ \vdots & \vdots & & \vdots & \vdots & & \vdots \\ a_{k1} & a_{k2} & \cdots & a_{kk} & 0 & \cdots & 0 \\ 0 & 0 & \cdots & 0 & 0 & \cdots & 0 \\ \vdots & \vdots & & \vdots & \vdots & & \vdots \\ 0 & 0 & \cdots & 0 & 0 & \cdots & 0 \end{pmatrix}$$

where $\rho(F) = \rho(A)$ and $\nu(F) = \nu(A)$. It is now obvious that if $i > k$, then $F\mathbf{e}_i = \mathbf{0}$,† so $E_k = \{\mathbf{e}_{k+1}, \mathbf{e}_{k+2}, \ldots, \mathbf{e}_n\}$ is a linearly independent set of $n - k$ vectors in $N_F$. We now show that $E_k$ spans $N_F$. Let the vector $\mathbf{x} \in N_F$ have the form

$$\mathbf{x} = \begin{pmatrix} x_1 \\ x_2 \\ \vdots \\ x_k \\ \vdots \\ x_n \end{pmatrix}$$

Then

$$\mathbf{0} = F\mathbf{x} = \begin{pmatrix} a_{11}x_1 + a_{12}x_2 + \cdots + a_{1k}x_k \\ a_{21}x_1 + a_{22}x_2 + \cdots + a_{2k}x_k \\ \vdots \qquad\qquad \vdots \qquad\qquad \vdots \\ a_{k1}x_1 + a_{k2}x_2 + \cdots + a_{kk}x_k \\ 0 \\ \vdots \\ 0 \end{pmatrix} = \begin{pmatrix} 0 \\ 0 \\ \vdots \\ 0 \end{pmatrix}$$

The determinant of the matrix of the $k \times k$ homogeneous system given above is nonzero, since the rows of this matrix are linearly independent. Thus the only solution to the system is $x_1 = x_2 = \cdots = x_k = 0$. Thus $\mathbf{x}$ has the form

$$(0, 0, \ldots, 0, x_{k+1}, x_{k+2}, \ldots, x_n) = x_{k+1}\mathbf{e}_{k+1} + x_{k+2}\mathbf{e}_{k+2} + \cdots + x_n\mathbf{e}_n$$

This means that $E_k$ spans $N_F$ so that $\nu(F) = n - k = n - \rho(F)$. This completes the proof. ■

*Note.* We know that $\rho(A)$ equals the number of pivots in the row echelon form of $A$ equals the number of columns of the row echelon form of $A$ that contain pivots. Then, from Theorem 7, $\nu(A)$ = number of columns of the row echelon form of $A$ that do not contain pivots.

---

† Recall that $\mathbf{e}_i$ is the vector with a 1 in the $i$th position and a zero everywhere else.

**EXAMPLE 8** **Illustration that $\rho(A) + \nu(A) = n$** For $A = \begin{pmatrix} 1 & 2 & -1 \\ 2 & -1 & 3 \end{pmatrix}$ we calculated (in Examples 1 and 3) that $\rho(A) = 2$ and $\nu(A) = 1$; this illustrates that $\rho(A) + \nu(A) = n\ (=3)$.

**EXAMPLE 9** **Illustration That $\rho(A) + \nu(A) = n$** For $A = \begin{pmatrix} 1 & -1 & 3 \\ 2 & 0 & 4 \\ -1 & -3 & 1 \end{pmatrix}$, calculate $\nu(A)$.

**Solution** In Example 5 we found that $\rho(A) = 2$. Thus $\nu(A) = 3 - 2 = 1$. You can show this directly by solving the system $A\mathbf{x} = \mathbf{0}$ to find that $N_A = \text{span}\left\{\begin{pmatrix} -2 \\ 1 \\ 1 \end{pmatrix}\right\}$.

**THEOREM 8** Let $A$ be an $n \times n$ matrix. Then $A$ is invertible if and only if $\rho(A) = n$.

**Proof** By Theorem 1, $A$ is invertible if and only if $\nu(A) = 0$. But by Theorem 7, $\rho(A) = n - \nu(A)$. Thus $A$ is invertible if and only if $\rho(A) = n - 0 = n$.

We next show how the notion of rank can be used to determine whether a linear system of equations has solutions or is inconsistent. Again, we consider the system of $m$ equations in $n$ unknowns:

$$\begin{aligned} a_{11}x_1 + a_{12}x_2 + \cdots + a_{1n}x_n &= b_1 \\ a_{21}x_1 + a_{22}x_2 + \cdots + a_{2n}x_n &= b_2 \\ \vdots \qquad \vdots \qquad\qquad \vdots \qquad &\quad \vdots \\ a_{m1}x_1 + a_{m2}x_2 + \cdots + a_{mn}x_n &= b_m \end{aligned} \tag{9}$$

which we write as $A\mathbf{x} = \mathbf{b}$. We use the symbol $(A, \mathbf{b})$ to denote the $m \times (n + 1)$ augmented matrix obtained (as in Section 1.3) by adjoining the vector $\mathbf{b}$ to $A$.

**THEOREM 9** The system $A\mathbf{x} = \mathbf{b}$ has at least one solution if and only if $\mathbf{b} \in C_A$. This will occur if and only if $A$ and the augmented matrix $(A, \mathbf{b})$ have the same rank.

**Proof** If $\mathbf{c}_1, \mathbf{c}_2, \ldots, \mathbf{c}_n$ are the columns of $A$, then we can write system (9) as

$$x_1\mathbf{c}_1 + x_2\mathbf{c}_2 + \cdots + x_n\mathbf{c}_n = \mathbf{b} \tag{10}$$

System (10) will have a solution if and only if $\mathbf{b}$ can be written as a linear combination of the columns of $A$. That is, to have a solution we must have $\mathbf{b} \in C_A$. If $\mathbf{b} \in C_A$, then $(A, \mathbf{b})$ has the same number of linearly independent columns as $A$ so that $A$ and $(A, \mathbf{b})$ have the same rank. If $\mathbf{b} \notin C_A$, then $\rho(A, \mathbf{b}) = \rho(A) + 1$ and the system has no solutions. This completes the proof.

**EXAMPLE 10** **Using Theorem 9 to Determine Whether a System Has Solutions** Determine whether the system

$$\begin{aligned} 2x_1 + 4x_2 + \ \ 6x_3 &= 18 \\ 4x_1 + 5x_2 + \ \ 6x_3 &= 24 \\ 2x_1 + 7x_2 + 12x_3 &= 40 \end{aligned}$$

has solutions.

**Solution** Let $A = \begin{pmatrix} 2 & 4 & 6 \\ 4 & 5 & 6 \\ 2 & 7 & 12 \end{pmatrix}$. The row echelon form of $A$ is $\begin{pmatrix} 1 & 2 & 3 \\ 0 & 1 & 2 \\ 0 & 0 & 0 \end{pmatrix}$ and $\rho(A) = 2$. The row echelon form of the augmented matrix $(A, \mathbf{b}) = \left(\begin{array}{ccc|c} 2 & 4 & 6 & 18 \\ 4 & 5 & 6 & 24 \\ 2 & 7 & 12 & 40 \end{array}\right)$ is $\left(\begin{array}{ccc|c} 1 & 2 & 3 & 9 \\ 0 & 1 & 2 & 4 \\ 0 & 0 & 0 & 1 \end{array}\right)$, which has three pivots, so $\rho(A, \mathbf{b}) = 3$ and the system has no solution.

**EXAMPLE 11** **Using Theorem 9 to Determine Whether a System Has Solutions** Determine whether the system

$$\begin{aligned} x_1 - x_2 + 2x_3 &= 4 \\ 2x_1 + x_2 - 3x_3 &= -2 \\ 4x_1 - x_2 + \ \ x_3 &= 6 \end{aligned}$$

has solutions.

**Solution** Let $A = \begin{pmatrix} 1 & -1 & 2 \\ 2 & 1 & -3 \\ 4 & -1 & 1 \end{pmatrix}$. Then $\det A = 0$, so $\rho(A) < 3$. Since the first column is not a multiple of the second, we see that the first two columns are linearly independent; hence $\rho(A) = 2$. To compute $\rho(A, \mathbf{b})$, we row-reduce:

$$\left(\begin{array}{ccc|c} 1 & -1 & 2 & 4 \\ 2 & 1 & -3 & -2 \\ 4 & -1 & 1 & 6 \end{array}\right) \longrightarrow \left(\begin{array}{ccc|c} 1 & -1 & 2 & 4 \\ 0 & 3 & -7 & -10 \\ 0 & 3 & -7 & -10 \end{array}\right)$$

We see that $\rho(A, \mathbf{b}) = 2$ and there are an infinite number of solutions to the system. (If there were a unique solution, we would have $\det A \neq 0$.)

The results of this section allow us to improve on our Summing Up Theorem—last seen in Section 4.5 page 325.

**THEOREM 10** **Summing Up Theorem—View 6** Let $A$ be an $n \times n$ matrix. Then the following ten statements are equivalent: that is, each one implies the other nine, (so if one is true, all are true.)

**i.** $A$ is invertible.

**ii.** The only solution to the homogeneous system $A\mathbf{x} = \mathbf{0}$ is the trivial solution ($\mathbf{x} = \mathbf{0}$).

**iii.** The system $A\mathbf{x} = \mathbf{b}$ has a unique solution for every $n$-vector $\mathbf{b}$.

**iv.** $A$ is row equivalent to the $n \times n$ identity matrix $I_n$.

**v.** $A$ can be written as the product of elementary matrices.

**vi.** The row echelon form of $A$ has $n$ pivots.

**vii.** The rows (and columns) of $A$ are linearly independent.

**viii.** $\det A \neq 0$.

**ix.** $\nu(A) = 0$.

**x.** $\rho(A) = n$.

Moreover, if one of the above fails to hold, then for every vector $\mathbf{b} \in \mathbb{R}^n$, the system $A\mathbf{x} = \mathbf{b}$ has either no solution or an infinite number of solutions. It has an infinite number of solutions if and only if $\rho(A) = \rho((A, \mathbf{b}))$.

# PROBLEMS 4.7

## Self-Quiz

**I.** The rank of the matrix $\begin{pmatrix} 1 & 2 & 3 & 4 \\ 0 & 2 & -1 & 5 \\ 0 & 0 & 3 & 7 \end{pmatrix}$ is ________.

**a.** 1 **b.** 2 **c.** 3 **d.** 4

**II.** The nullity of the matrix in Problem I is ________.

**a.** 1 **b.** 2 **c.** 3 **d.** 4

**III.** If a $5 \times 7$ matrix has nullity 2, then its rank is ________.

**a.** 5 **b.** 3 **c.** 2 **d.** 7

**e.** It cannot be determined without further information.

**IV.** The rank of the matrix $\begin{pmatrix} 1 & 2 \\ -2 & -4 \\ 3 & 6 \end{pmatrix}$ is ________.

**a.** 1 **b.** 2 **c.** 3

**V.** The nullity of the matrix in Problem 4 is ________.

**a.** 0 **b.** 1 **c.** 2 **d.** 3

**VI.** If $A$ is a $4 \times 4$ matrix and $\det A = 0$, then the maximum possible value for $\rho(A)$ is ________.

**a.** 1 **b.** 2 **c.** 3 **d.** 4

**VII.** In Problem IV dim $C_A =$ ________.
**a.** 1 **b.** 2 **c.** 3

**VIII.** In Problem I dim $R_A =$ ________.
**a.** 1 **b.** 2 **c.** 3 **d.** 4

### *True-False*

**IX.** In any $m \times n$ matrix, $C_A = R_A$

**X.** In any $m \times n$ matrix, $C_A =$ Range $A$

In Problems 1–15 find the rank and nullity of the given matrix.

**1.** $\begin{pmatrix} 1 & 2 \\ 3 & 4 \end{pmatrix}$ **2.** $\begin{pmatrix} 1 & -1 & 2 \\ 3 & 1 & 0 \end{pmatrix}$ **3.** $\begin{pmatrix} -1 & 3 & 2 \\ 2 & -6 & -4 \end{pmatrix}$

**4.** $\begin{pmatrix} 1 & -1 & 2 \\ 3 & 1 & 4 \\ -1 & 0 & 4 \end{pmatrix}$ **5.** $\begin{pmatrix} 1 & -1 & 2 \\ 3 & 1 & 4 \\ 5 & -1 & 8 \end{pmatrix}$ **6.** $\begin{pmatrix} -1 & 2 & 1 \\ 2 & -4 & -2 \\ -3 & 6 & 3 \end{pmatrix}$

**7.** $\begin{pmatrix} 1 & -1 & 2 & 3 \\ 0 & 1 & 4 & 3 \\ 1 & 0 & 6 & 6 \end{pmatrix}$ **8.** $\begin{pmatrix} 1 & -1 & 2 & 3 \\ 0 & 1 & 4 & 3 \\ 1 & 0 & 6 & 5 \end{pmatrix}$ **9.** $\begin{pmatrix} 2 & 3 \\ -1 & 1 \\ 4 & 7 \end{pmatrix}$

**10.** $\begin{pmatrix} 1 & -1 & 2 & 3 \\ 0 & 1 & 0 & 1 \\ 1 & 0 & 1 & 0 \\ 0 & 0 & 0 & 1 \end{pmatrix}$ **11.** $\begin{pmatrix} 1 & -1 & 2 & 1 \\ -1 & 0 & 1 & 2 \\ 1 & -2 & 5 & 4 \\ 2 & -1 & 1 & -1 \end{pmatrix}$ **12.** $\begin{pmatrix} 1 & -1 & 2 & 3 \\ -2 & 2 & -4 & -6 \\ 2 & -2 & 4 & 6 \\ 3 & -3 & 6 & 9 \end{pmatrix}$

**13.** $\begin{pmatrix} -1 & -1 & 0 & 0 \\ 0 & 0 & 2 & 3 \\ 4 & 0 & -2 & 1 \\ 3 & -1 & 0 & 4 \end{pmatrix}$ **14.** $\begin{pmatrix} 3 & 0 & 0 \\ 0 & 0 & 0 \\ 0 & 0 & 6 \end{pmatrix}$ **15.** $\begin{pmatrix} 1 & 2 & 3 \\ 0 & 0 & 4 \\ 0 & 0 & 6 \end{pmatrix}$

In Problems 16–22 find a basis for the range and null space of the given matrix.

**16.** The matrix of Problem 2
**17.** The matrix of Problem 5
**18.** The matrix of Problem 6
**19.** The matrix of Problem 8
**20.** The matrix of Problem 11
**21.** The matrix of Problem 12
**22.** The matrix of Problem 13

In Problems 23–26 find a basis for the span of the given set of vectors.

**23.** $\begin{pmatrix} 1 \\ 4 \\ -2 \end{pmatrix}, \begin{pmatrix} 2 \\ 1 \\ 2 \end{pmatrix}, \begin{pmatrix} -1 \\ 3 \\ -4 \end{pmatrix}$

---

**Answers to Self-Quiz**

**I.** c **II.** a **III.** a **IV.** a **V.** b **VI.** c **VII.** a **VIII.** c
**IX.** False **X.** True

**24.** $(1, -2, 3), (2, -1, 4), (3, -3, 3), (2, 1, 0)$

**25.** $(1, -1, 1, -1), (2, 0, 0, 1), (4, -2, 2, 1), (7, -3, 3, -1)$

**26.** $\begin{pmatrix} 1 \\ 0 \\ 0 \\ 1 \end{pmatrix}, \begin{pmatrix} 0 \\ 1 \\ 1 \\ 0 \end{pmatrix}, \begin{pmatrix} 1 \\ -2 \\ -2 \\ 1 \end{pmatrix}, \begin{pmatrix} 0 \\ 2 \\ 2 \\ 1 \end{pmatrix}$

In Problems 27–30 use Theorem 9 to determine whether the given system has any solutions.

**27.**
$$\begin{aligned} x_1 + x_2 - x_3 &= 7 \\ 4x_1 - x_2 + 5x_3 &= 4 \\ 6x_1 + x_2 + 3x_3 &= 20 \end{aligned}$$

**28.**
$$\begin{aligned} x_1 + x_2 - x_3 &= 7 \\ 4x_1 - x_2 + 5x_3 &= 4 \\ 6x_1 + x_2 + 3x_3 &= 18 \end{aligned}$$

**29.**
$$\begin{aligned} x_1 - 2x_2 + x_3 + x_4 &= 2 \\ 3x_1 \qquad + 2x_3 - 2x_4 &= -8 \\ 4x_2 - x_3 - x_4 &= 1 \\ 5x_1 \qquad + 3x_3 - x_4 &= -3 \end{aligned}$$

**30.**
$$\begin{aligned} x_1 - 2x_2 + x_3 + x_4 &= 2 \\ 3x_1 \qquad + 2x_3 - 2x_4 &= -8 \\ 4x_2 - x_3 - x_4 &= 1 \\ 5x_1 \qquad + 3x_3 - x_4 &= 0 \end{aligned}$$

**31.** Show that the rank of a diagonal matrix is equal to the number of nonzero components on the diagonal.

**32.** Let $A$ be an upper triangular $n \times n$ matrix with zeros on the diagonal. Show that $\rho(A) < n$.

**33.** Show that for any matrix $A$, $\rho(A) = \rho(A^t)$.

**34.** Show that if $A$ is an $m \times n$ matrix and $m < n$, then **(a)** $\rho(A) \leq m$ and **(b)** $\nu(A) \geq n - m$.

**35.** Let $A$ be an $m \times n$ matrix and let $B$ and $C$ be invertible $m \times m$ and $n \times n$ matrices, respectively. Prove that $\rho(A) = \rho(BA) = \rho(AC)$. That is, multiplying a matrix by an invertible matrix does not change its rank.

**36.** Let $A$ and $B$ be $m \times n$ and $n \times p$ matrices, respectively. Show that $\rho(AB) \leq \min(\rho(A), \rho(B))$.

**37.** Let $A$ be a $5 \times 7$ matrix with rank 5. Show that the linear system $A\mathbf{x} = \mathbf{b}$ has at least one solution for every 5-vector $\mathbf{b}$.

***38.** Let $A$ and $B$ be $m \times n$ matrices. Show that if $\rho(A) = \rho(B)$, then there exist invertible matrices $C$ and $D$ such that $B = CAD$.

**39.** If $B = CAD$, where $C$ and $D$ are invertible, prove that $\rho(A) = \rho(B)$.

**40.** Suppose that any $k$ rows of $A$ are linearly independent while any $k + 1$ rows of $A$ are linearly dependent. Show that $\rho(A) = k$.

**41.** If $A$ is an $n \times n$ matrix, show that $\rho(A) < n$ if and only if there is a vector $\mathbf{x} \in \mathbb{R}^n$ such that $\mathbf{x} \neq \mathbf{0}$ and $A\mathbf{x} = \mathbf{0}$.

**42.** Let $A$ be an $m \times n$ matrix. Suppose that for every $\mathbf{y} \in \mathbb{R}^m$ there is an $\mathbf{x} \in \mathbb{R}^n$ such that $A\mathbf{x} = \mathbf{y}$. Show that $\rho(A) = m$.

**43.** Prove that the rank of a matrix equals the number of pivots in its row echelon form. [*Hint:* Show that if the row echelon form has $k$ pivots, then the row echelon form has exactly $k$ linearly independent rows.]

## CALCULATOR BOX

### TI-85

There is one clear way to determine the rank, range, and row space of a matrix on the TI-85; find the row echelon form or reduced row echelon form of the matrix. For example, suppose the matrix

$$A = \begin{pmatrix} 1 & 3 & 4 & 1 \\ 4 & 2 & 6 & -6 \\ 3 & 5 & 8 & -1 \end{pmatrix}$$

is entered. Then as on page 28, press the following keys:

[2nd] [MATRX] [F4] ⟨ops⟩ [F5]

⟨rref⟩ [ALPHA] [A] [ENTER]

The result is

rref A

[[1 0 1 −2]
[0 1 1 1]
[0 0 0 0]]

Clearly, $\rho(A) = 2$, $R_A = \text{span}\,\{(1, 0, 1, -2), (0, 1, 1, 1)\}$; since $\rho(A) = 2$, $A$ has two linearly independent columns so

$$C_A = \text{Range } A = \text{span}\left\{\begin{pmatrix}1\\4\\3\end{pmatrix}, \begin{pmatrix}3\\2\\5\end{pmatrix}\right\}, \quad \text{and} \quad \nu(A) = 4 - 2 = 2$$

In Problems 44–47 use a calculator to find the rank, range, row space, and nullity of the given matrix.

**44.** $\begin{pmatrix} 0.37 & 0.48 & -0.70 & -1.16 \\ 0.46 & -0.39 & 2.09 & 0.83 \\ 0.52 & 0.87 & -1.57 & 1.04 \\ 0.67 & 0.35 & 0.29 & -0.33 \end{pmatrix}$

**45.** $\begin{pmatrix} 187 & -46 & 512 & 653 & 512 \\ -35 & 51 & -223 & -207 & -325 \\ 257 & -148 & 958 & 1067 & 1162 \end{pmatrix}$

**46.** $\begin{pmatrix} 37 & 81 & -29 & 58 & 33 & -19 & 102 \\ -48 & 91 & 306 & 38 & 205 & 0 & -58 \\ 53 & 215 & -47 & -11 & -38 & 423 & 99 \\ -85 & 10 & 335 & -20 & 172 & 19 & -160 \\ -80 & 316 & 594 & 7 & 339 & 442 & -119 \\ -71 & 46 & -416 & -83 & 201 & -88 & 144 \end{pmatrix}$

**47.** $\begin{pmatrix} .0284 & -.0311 & -.0207 & .0431 & .0615 \\ -.0511 & -.1216 & -.1811 & .0904 & .0310 \\ -.0965 & -.4270 & -.5847 & .3574 & .2160 \\ .0795 & .0905 & .1604 & -.4730 & .0305 \\ -.0110 & -.3365 & -.4243 & .3101 & .5210 \end{pmatrix}$

## MATLAB 4.7

1. For each of the matrices below:
   a. Find a basis for the null space by following Example 7. This will involve solving an appropriate homogeneous system of equations.
   b. Verify that the set of vectors obtained for each problem is an independent set.
   c. *(Paper and pencil)* If the set of vectors is to be a basis for the null space, it must also be shown that every vector in the null space can be written as a linear combination of the basis vectors. Show that every vector in the null space, that is, every solution to the homogeneous system solved in part (a), can be written as a linear combination of the vectors found in part (a).
   d. For each problem, find the dimension of the null space. Explain. How does the dimension relate to the number of arbitrary variables that arose in the solution of the homogeneous system solved in part (a)?

   **i.–vi.** Problems 7, 8, and 10 through 13 in Section 4.7.

   **vii.** $\begin{pmatrix} -6 & -2 & -18 & -2 & -10 \\ -9 & 0 & -18 & 4 & -5 \\ 4 & 7 & 29 & 2 & 13 \end{pmatrix}$

2. a. **i.** For Problem 13 in this section, find the basis for the null space following Example 7.
   **ii.** Let **R = rref(A).** Verify that the basis consists of the one vector **B = [−R(1,4);−R(2,4);−R(3,4);1].**
   **iii.** Verify that **A*B** = 0. Why would you expect this?

   b. **i.** For the matrix $A = \begin{pmatrix} -6 & -2 & -18 & -2 & -10 \\ -9 & 0 & -18 & 4 & -5 \\ 4 & 7 & 29 & 2 & 13 \end{pmatrix}$ find the basis for the null space.
   **ii.** Let **R = rref(A)** and let

   **B = [[−R(1,3);−R(2,3);1;0;0] [−R(1,5);−R(2,5);0;−R(3,5);1]]**

   Verify that the columns of $B$ are the basis vectors that you found in part (b) (i).
   **iii.** Verify that **A*B** = 0 and explain why that should be so.

   c. For the following matrices $A$, find **R = rref(A)** and find the basis for the null space by forming the matrix $B$ as illustrated in the examples in parts (a) and (b). Verify that **A*B** = 0. (To help you recognize the procedure for finding $B$: For example, in part (b), columns 3 and 5 of $R$ did not have pivots indicating that $x_3$ and $x_5$ were arbitrary variables. Columns 3 and 5 of $R$ are *not* vectors in the null space, but a basis for the null space can be found using the numbers in columns 3 and 5 appropriately. Note that the third and fifth positions in the basis vectors are either 1 or 0.)

   **i.** $A = \begin{pmatrix} -9 & 3 & 8 & -5 & -1 \\ 5 & 0 & -5 & -5 & -3 \\ -7 & 0 & 8 & 8 & 9 \end{pmatrix}$

   **ii.** **A = rand(4,6);A(:,4) = 1/3*A(:,2) − 2/7*A(:,3)**

3. a. MATLAB has a command **null(A)** that will produce a basis for the null space of $A$. (It produces an orthonormal basis. See Section 4.9 for a definition of orthonormal.)
   **i.** For each of the matrices $A$ in MATLAB Problem 2 in this section, find **N = null(A).** Find $B$, the matrix whose columns form a basis for the null space, by using the procedure of Example 7.

ii. How many vectors are in each basis? What property does this confirm?

iii. By considering **rref([B N])** and **rref([N B]),** verify that every vector in the basis for the null space determined by the **null** command is a linear combination of the basis vectors found in the columns of B, and that every column vector in $B$ is a linear combination of the basis vectors found using the **null** command. Explain your reasoning and process. Explain why such a statement should be true.

b. The algorithm used by MATLAB's **null** command is numerically more stable than the process involving **rref;** that is, **null** is better at minimizing the buildup of round-off error. For the matrix $A$ below, find **N = null(A)** and find **B** as in part (a). Find **A*B** and **A*N** and discuss how this provides some evidence for the statement made at the beginning of part (b).

$$A = \begin{pmatrix} 1 & -2 & 5 & 1 & 9 \\ -3 & 6 & 6 & 3.56 & 3 \\ 4.2 & -8.4 & -10 & 4 & -1 \end{pmatrix}$$

4. **Geometric Application of Null Space**

a. *(Paper and pencil)* Argue why a basis for the null space of an $m \times n$ matrix $A$ will be a basis for the subspace of all vectors in $\mathbb{R}^n$ perpendicular (orthogonal) to the *rows* of $A$.

b. Find a basis for the plane formed by all vectors perpendicular to $\begin{pmatrix} -1 \\ 2 \\ 3 \end{pmatrix}$

c. Find a basis for the line perpendicular to the plane spanned by $\left\{ \begin{pmatrix} 2 \\ -3 \\ 1 \end{pmatrix}, \begin{pmatrix} -1 \\ 0 \\ \frac{1}{2} \end{pmatrix} \right\}$.

Compare your answer to the cross product of the two vectors.

d. Find a basis for the subspace of all vectors perpendicular to

$$\left\{ \begin{pmatrix} 1 \\ 2 \\ -3 \\ 1 \\ 2 \end{pmatrix}, \begin{pmatrix} 0 \\ 1 \\ 5 \\ -1 \\ 1 \end{pmatrix}, \begin{pmatrix} -2 \\ 3 \\ 1 \\ 4 \\ 0 \end{pmatrix} \right\}.$$

5. **Application of Null Space to Systems of Equations** Let

$$A = \begin{pmatrix} 0 & 8 & -6 & -5 & 4 & -4 \\ 9 & 2 & 4 & -10 & 9 & 8 \\ 5 & 7 & -7 & -2 & -5 & 3 \\ 1 & -7 & -8 & -9 & -6 & -7 \end{pmatrix} \quad \mathbf{b} = \begin{pmatrix} 46 \\ 29 \\ 0 \\ -15 \end{pmatrix} \quad \mathbf{x} = \begin{pmatrix} 1 \\ 2 \\ -1 \\ 0 \\ 4 \\ -2 \end{pmatrix}$$

a. Show that $\mathbf{x}$ is a solution to the system $[A\ \mathbf{b}]$. (Use matrix multiplication.)

b. Find a basis for the null space of $A$, forming a matrix whose columns are the vectors in the basis.

c. Generate a random vector **w** that is a linear combination of the basis vectors found in part (b). (Use matrix multiplication.) Show that $\mathbf{z} = \mathbf{x} + \mathbf{w}$ is a solution to the system $[A\ \mathbf{b}]$. Repeat for another choice of **w**.

6. For the following sets of vectors:
   a. Let $A$ be the matrix whose *rows* are the vectors. Find **rref(A).** Use the ":" command to find the matrix $C$ that consists only of the nonzero rows of **rref(A).** Let **B = C′.** Explain why the columns of $B$ are a basis for the span of the vectors. (See Example 6.)
   b. Verify that the basis found is linearly independent.
   c. Verify that each vector in the original set is a unique linear combination of the vectors in the basis. Describe any patterns you discover in the coefficients of the linear combinations.

i. $\left\{\begin{pmatrix}1\\-2\\3\end{pmatrix}, \begin{pmatrix}-2\\4\\-6\end{pmatrix}, \begin{pmatrix}1\\0\\1\end{pmatrix}\right\}$ ii. $\left\{\begin{pmatrix}1\\-1\\0\\3\\-1\\4\end{pmatrix}, \begin{pmatrix}2\\0\\1\\7\\2\\\frac{1}{2}\end{pmatrix}, \begin{pmatrix}3\\5\\1\\4\\1\\5\end{pmatrix}\right\}$

iii. $\left\{\begin{pmatrix}1\\2\\-1\\3\\1\end{pmatrix}, \begin{pmatrix}-1\\0\\1\\2\\0\end{pmatrix}, \begin{pmatrix}5\\4\\-5\\0\\2\end{pmatrix}, \begin{pmatrix}1\\2\\3\\-2\\0\end{pmatrix}, \begin{pmatrix}6\\8\\-2\\3\\3\end{pmatrix}\right\}$

7. a. *(Paper and pencil)* Suppose you want to find the basis for the range (column space) of a real matrix $A$. Explain how you could use **rref(A′)** to do this.
   b. For the matrices below, find a basis for the range, forming a matrix whose columns are the basis vectors. Verify that each column of the original matrix is a unique linear combination of the vectors in the basis.
   **i.–iv.** Matrices of Problems 7 and 11 through 13 in this section.
   **v. A = round(10*(2*rand(5)−1));A(:,2) = .5*A(:,1);**
   **A(:,4) = A(:,1)−1/3*A(:,3)**

8. a. For each of the matrices in MATLAB Problem 7 in this section, find **rref(A)** and **rref(A′).**
   b. Find a basis of the column space of $A$ and hence the dimension of the column space of $A$.
   c. Find a basis of the row space of $A$ and hence the dimension of the row space of $A$.
   d. Write a conjecture relating the dimension of the column space of $A$ to the dimension of the row space of $A$.
   e. What do **rref(A)** and **rref(A′)** have in common and how does this relate to part (d)?

9. This problem explains another way to find a basis for a span of vectors so that the basis consists of a subset of the original set of vectors.
   a. Recall (or do) MATLAB Problems 3 and 7 in Section 4.4. If $A$ is the matrix whose columns are the vectors in a given set, conclude that the columns of $A$ corresponding to columns without pivots in the reduced row echelon form are not needed in forming the span of the original set of vectors.

b. For the sets of vectors in MATLAB Problem 6 in this section, let $A$ be the matrix whose *columns* are the vectors in the given set.
   **i.** Using **rref(A)** to decide which vectors of the original set can be deleted (not needed), form a matrix $B$ that is a submatrix of the original $A$ [NOT **rref**($A$)] consisting of the minimal number of vectors from the original set needed to form the span.
   **ii.** Verify that the subset chosen (the columns of the submatrix) is linearly independent.
   **iii.** Verify that the number of vectors is the same as the number of vectors in the basis determined in MATLAB Problem 6 in this section.
   **iv.** Verify that each vector in the basis found in MATLAB Problem 6 is a unique linear combination of the basis found in this problem and that each vector of the basis found in this problem is a unique linear combination of the basis found in MATLAB Problem 6. [*Hint.* If $C$ is the matrix whose columns are the basis vectors found in MATLAB Problem 6, look at **rref([B C])** and **rref([C B])**.]

c. Follow the directions of part (b) for the column space of the matrices in MATLAB Problem 7 in this section.

**10.** Suppose $\{\mathbf{v}_1, \ldots, \mathbf{v}_k\}$ is a set of linearly independent vectors in $\mathbb{R}^n$. Suppose we wish to add some vectors to the set to create a basis for all of $\mathbb{R}^n$ that contains the original set. For each of the sets of vectors below:

a. Let $A$ be the matrix such that the $i$th column of $A$ equals $\mathbf{v}_i$. Form the matrix $B = [A\ I]$, where $I$ is the $n \times n$ identity matrix. Verify that the columns of $B$ span all of $\mathbb{R}^n$.

b. Follow the procedure outlined in MATLAB Problem 9 in this section to find a basis for the column space of $B$. Verify that the basis so obtained is a basis for $\mathbb{R}^n$ and contains the original set of vectors.
   **i.** Generate three random vectors, $\{\mathbf{v}_1, \mathbf{v}_2, \mathbf{v}_3\}$, in $\mathbb{R}^5$ using MATLAB. (First, check that they are linearly independent.)
   **ii.** In $\mathbb{R}^4$, $\mathbf{v}_1 = \begin{pmatrix} 1 \\ 2 \\ 3 \\ 1 \end{pmatrix} \quad \mathbf{v}_2 = \begin{pmatrix} 2 \\ 8 \\ 9 \\ 3 \end{pmatrix} \quad \mathbf{v}_3 = \begin{pmatrix} -1 \\ 1 \\ -3 \\ -1 \end{pmatrix}.$

c. *(Paper and pencil)* Explain why this procedure will always yield a basis for $\mathbb{R}^n$ containing the original set of linearly independent vectors.

**11.** The MATLAB command **orth(A)** will produce a basis for the range (column space) of the matrix $A$. (It produces an orthonormal basis.) For each of the matrices in MATLAB Problem 7 in this section, use **orth(A)** to find a basis for the column space of $A$. Verify that this basis contains the same number of vectors as the basis found in MATLAB Problem 7, and show that each vector in the basis found using **orth** is a linear combination of the basis found in MATLAB Problem 7. Also show that each vector in the basis found in MATLAB Problem 7 is a linear combination of the basis found using **orth.**

**12.** Find a basis for the span of the following sets:

a. In $P_3$: $\{-x^3 + 4x + 3, -x^3 - 1, x^2 - 2x, 3x^2 + x + 4\}$ [See MATLAB Problem 4.4.9]

b. In $M_{22}$: $\left\{\begin{pmatrix} -6 & -9 \\ 4 & 4 \end{pmatrix}, \begin{pmatrix} -2 & 0 \\ 7 & -9 \end{pmatrix}, \begin{pmatrix} -18 & -18 \\ 29 & -19 \end{pmatrix}, \begin{pmatrix} -2 & 4 \\ 2 & 9 \end{pmatrix}\right\}$
   [See MATLAB Problem 4.4.10]

13. a. Choose a value for $n \geq 4$ and generate a random $n \times n$ matrix $A$ using MATLAB. Find **rref(A)** and find **rank(A).** (The command **rank(A)** finds the rank of $A$.) Verify that $A$ is invertible.
    b. Let **B = A** and change one column of $B$ to be a linear combination of previous columns of $B$. Find **rref(B)** and **rank(B).** Verify that $B$ is not invertible.
    c. Let $B$ be as in part (b) after the change and change another column of $B$ to be a linear combination of previous columns of $B$. Find **rref(B)** and **rank(B).** Verify that $B$ is not invertible.
    d. Repeat for four more choices of $A$. (Use some different values of $n$.)
    e. Based on the evidence gathered so far, conjecture a relationship between **rank(A)** and the number of pivots in **rref(A).**
    f. Conjecture a relationship among **rank(A),** the size of $A$, and the invertibility of $A$.
    d. Create a $5 \times 5$ matrix with rank 2 and a $6 \times 6$ matrix with rank 4.

14. a. Generate three random real $n \times m$ matrices of different sizes, with $m$ not equal to $n$. Find **rank(A)** and **rank(A′).**
    b. Choose a value for $n$ and generate three real $n \times n$ matrices, each with a different rank. (See MATLAB Problem 13 in this section.) Find **rank(A)** and **rank(A′).** Repeat for another value of $n$.
    c. Describe the relationship between **rank(A)** and **rank(A′).**
    d. Describe the relationship between this problem and MATLAB Problem 8 in this section.

15. Consider the systems of equations in MATLAB Problems 1 through 3 in Section 1.3. For two of the systems from each problem, find the rank of the coefficient matrix and the rank of the augmented matrix. Formulate a conjecture relating these ranks to whether or not the system has a solution. Test your conjecture on some of the other systems in MATLAB Problems 1 through 3 in Section 1.3. Prove the conjecture.

16. **Rank Explorations for Special Matrices**
    a. Magic Square Matrices The command **magic(n)** will generate a magic square of size $n \times n$. (A magic square has the property that every column sum and row sum equals the same number.) Generate three magic square matrices for each of $n = 3, \ldots, 9$ and find the rank of each. How does the size of the matrix affect the rank? Write a description of the patterns you discovered.

***Note.*** This problem was inspired by a talk given by Cleve Moler at the University of New Hampshire in 1991.

   b. Explore the rank of $\begin{pmatrix} 1 & 2 & 3 \\ 4 & 5 & 6 \\ 7 & 8 & 9 \end{pmatrix}$, $\begin{pmatrix} 1 & 2 & 3 & 4 \\ 5 & 6 & 7 & 8 \\ 9 & 10 & 11 & 12 \\ 13 & 14 & 15 & 16 \end{pmatrix}$, and the next two matrices in this pattern. Write a description of the behavior of the rank of such matrices. Prove your conclusion. [*Hint.* Look at row $j + 1$ − row $j$.]
   c. Generate a random $n \times 1$ vector **u** and a random $n \times 1$ vector **v**. Form **A = u*v′**, an $n \times n$ matrix. Find the rank of $A$. Repeat for three more choices of **u** and **v**. Write a description of the rank of matrices formed in this manner.

17. **Rank and Products of Matrices**
    a. Choose a value for $n$ and let $A$ be an $n \times n$ invertible matrix. [*Hint.* Look for invertible matrices encountered in previous problems or generate a random matrix using the **rand** command. Check the invertibility.] Generate four $n \times m$ matrices,

some square and some not square, of differing ranks. (See MATLAB Problem 13 in this section to help create matrices of certain ranks.) Keep a record of the rank of each of these matrices. For each $B$ (one of these matrices), let **C = A*B.** Find **rank (C).** Relate rank($C$) to rank($B$). Complete the following statement: If $A$ is invertible and $B$ has rank $k$, then $AB$ has rank ______. Describe how this problem relates to MATLAB Problem 10 in Section 4.5.

b. Generate a $6 \times 6$ matrix $A$ with rank 4. Generate four matrices of size $6 \times m$ with different ranks, some bigger than 4 and some less than 4. For each $B$ (one of these four matrices), find **rank(A*B)** and relate it to the ranks of $A$ and $B$.

c. Repeat part (b) with $A$ as a $5 \times 7$ matrix with rank 3 and the $B$ matrices $7 \times m$.

d. Formulate a conjecture relating rank($AB$) to rank($A$) and rank($B$).

e. Let

$$A = \begin{pmatrix} 1 & -1 & 0 \\ 2 & 0 & 2 \\ 3 & 1 & 4 \end{pmatrix} \qquad B = \begin{pmatrix} 1 & -3 & 2 \\ 1 & -3 & 2 \\ -1 & 3 & -2 \end{pmatrix}$$

Find rank($A$), rank($B$), and rank($AB$). Modify your conjecture from part (d). [*Hint.* Think about inequalities.]

**PROJECT PROBLEM**

18. **Cycles in Digraphs** Directed graphs, such as those below, are used to describe various physical situations. One such situation concerns electric circuits where current flows along the edges. In applying Kirchhoff's laws to determine the current along each edge, one must examine closed loops in the diagram for voltage drops. Not every possible closed loop needs to be examined, however, since some loops can be formed from other loops. Thus one needs to examine a "basis" of closed loops, that is, a minimal number of loops that will generate all other loops.

Diagrams such as those below are called directed graphs, **digraphs** for short. A closed loop in a directed graph is called an **undirected cycle.**

a. Any digraph has an associated matrix called the **node-edge incidence matrix.** It is defined as

$$a_{ij} = \begin{cases} 1 & \text{if edge } j \text{ enters node } i \\ -1 & \text{if edge } j \text{ exits node } i \\ 0 & \text{otherwise} \end{cases}$$

It is easy to set up (or enter using MATLAB) a node-edge incidence matrix by looking at one edge at a time. (See MATLAB Problem 2 in Section 1.5.)

Enter the node-edge incidence matrix $A$ for the digraph below. Note that each edge corresponds to a column of $A$ and that $A$ will be an $n \times m$ matrix, where $n$ is the number of nodes and $m$ is the number of edges.

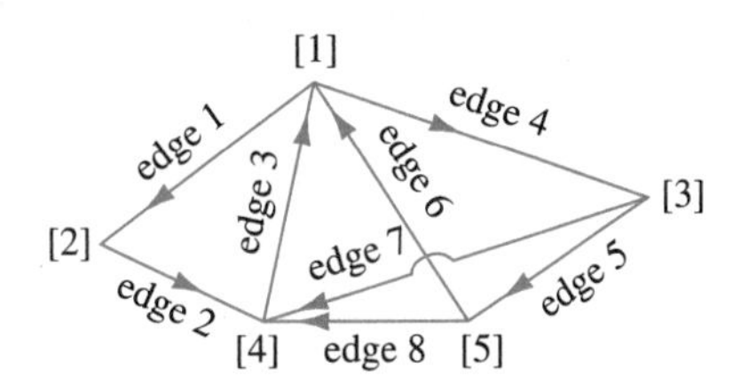


b. A cycle (closed loop) can be represented by an $m \times 1$ vector where each entry in the vector corresponds to a coefficient in front of an edge. For example, one cycle in the digraph above is found by: start at node [3], follow edge 5, then edge 8, then go opposite to edge 7. This can be expressed as edge 5 + edge 8 − edge 7, which can be represented by the $m \times 1$ vector: $(0\,0\,0\,0\,1\,0\,-1\,1)^t$.

**i.** Verify that this vector is in the null space of $A$, the node-edge incidence matrix.

**ii.** Form the vector corresponding to the loop that goes from node [1] to node [2] to node [4] to node [3] and back to node [1]. Verify that it is in the null space of $A$.

c. Verify that $\mathbf{x} = (1\,1\,2\,0\,0\,-1\,0\,1)^t$ is in the null space of $A$. Show that this vector corresponds to the loop traced out by starting at node [1], then following edge 1 + edge 2 + edge 3 − edge 6 + edge 8 + edge 3.

d. Find a basis for the null space of $A$.

e. For each vector in the basis, identify the loop corresponding to that vector by writing the edges in the order they are followed. Draw it, labeling edges and nodes.

f. Form a linear combination of these basis vectors (of the null space of $A$), using coefficients of 1 or −1. Identify the loop that this linear combination describes by writing the edges in the order in which they are traced, such as was done in part (c). (Also, draw the loop.)

Repeat for another linear combination.

g. Identify a loop in the digraph that is not one of the loops in the basis for the null space or one of the loops described in part (f). Write down the corresponding vector in the null space of $A$. Find the coefficients needed to write the vector as a linear combination of the basis vectors for the null space. Draw (or otherwise describe) your loop and the basis loops that are involved in the linear combination and show how your loop is made up of these basis loops.

Repeat for another loop.

h. For the digraph below, enter the node-edge incidence matrix and repeat parts (d) through (g) for this digraph. The label $e_i$ refers to edge $i$.

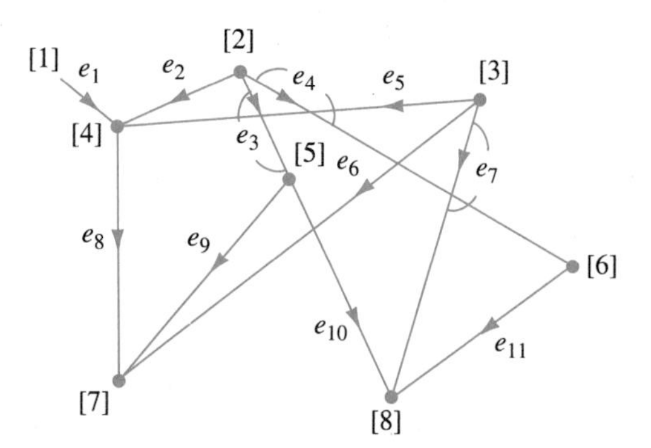


***Note.*** This problem was inspired by a talk given by Gilbert Strang at the University of New Hampshire in June 1991.

**PROJECT PROBLEM**

19. **Subspace Sum and Intersection** Let $V$ and $W$ be subspaces of $\mathbb{R}^n$. The intersection is defined as

$$U = V \cap W = \{\mathbf{z} \text{ in } \mathbb{R}^n | \mathbf{z} \text{ is in } V \text{ and } \mathbf{z} \text{ is in } W\}.$$

The subspace sum is defined as

$$S = V + W = \{\mathbf{z} | \mathbf{z} = \mathbf{v} + \mathbf{w} \text{ for some } \mathbf{v} \text{ in } V \text{ and some } \mathbf{w} \text{ in } W\}.$$

**Suppose** $\{\mathbf{v}_1, \ldots, \mathbf{v}_k\}$ is a basis for $V$ and $\{\mathbf{w}_1, \ldots, \mathbf{w}_m\}$ is a basis for $W$.

a. *(Paper and pencil)* Verify that $U$ and $S$ are subspaces.

b. *(Paper and pencil)* Verify that $\{\mathbf{v}_1, \ldots, \mathbf{v}_k, \mathbf{w}_1, \ldots, \mathbf{w}_m\}$ spans $S$, the subspace sum.

c. For each of the $V$ and $W$ pairs of bases given below, find a basis for $S = V + W$ and find the dimension of $S$. Do some checking of your answer by generating a random vector in $S$ (generate random vectors in $V$ and $W$ and add) and by showing that the vector is a linear combination of the basis vectors that you found.

**i.** Basis for $V = \left\{ \begin{pmatrix} 1 \\ 2 \\ 3 \\ 4 \\ 1 \\ 0 \end{pmatrix}, \begin{pmatrix} -1 \\ 0 \\ 1 \\ 2 \\ 1 \\ 1 \end{pmatrix} \right\}$ For $W = \left\{ \begin{pmatrix} 0 \\ 1 \\ 2 \\ 3 \\ -1 \\ -1 \end{pmatrix}, \begin{pmatrix} 5 \\ 4 \\ 2 \\ 3 \\ 1 \\ 2 \end{pmatrix}, \begin{pmatrix} 0 \\ 0 \\ 1 \\ -2 \\ 1 \\ 1 \end{pmatrix} \right\}$

**ii.** Basis for $V = \left\{ \begin{pmatrix} 1 \\ 2 \\ 3 \\ 4 \\ 1 \\ 0 \end{pmatrix}, \begin{pmatrix} -1 \\ 0 \\ 1 \\ 2 \\ 1 \\ 1 \end{pmatrix}, \begin{pmatrix} 0 \\ 1 \\ 2 \\ 3 \\ -1 \\ -1 \end{pmatrix} \right\}$

For $W = \left\{ \begin{pmatrix} -1 \\ 2 \\ 1 \\ 3 \\ -1 \\ 2 \end{pmatrix}, \begin{pmatrix} 4 \\ 3 \\ 5 \\ 4 \\ 2 \\ -8 \end{pmatrix}, \begin{pmatrix} 10 \\ 13 \\ 18 \\ 20 \\ -1 \\ -19 \end{pmatrix} \right\}$

**iii.** Basis for $V = \left\{ \begin{pmatrix} 1 \\ 2 \\ 3 \\ 4 \\ 1 \\ 0 \end{pmatrix}, \begin{pmatrix} -1 \\ 0 \\ 1 \\ 2 \\ 1 \\ 1 \end{pmatrix}, \begin{pmatrix} 0 \\ 1 \\ 2 \\ 3 \\ 1 \\ -1 \end{pmatrix}, \begin{pmatrix} -1 \\ 2 \\ 1 \\ 3 \\ -1 \\ 2 \end{pmatrix} \right\}$

Basis for $W = \left\{ \begin{pmatrix} 4 \\ 3 \\ 5 \\ 4 \\ 2 \\ 8 \end{pmatrix}, \begin{pmatrix} 0 \\ 0 \\ 1 \\ -2 \\ 1 \\ 1 \end{pmatrix}, \begin{pmatrix} 0 \\ -1 \\ 1 \\ 4 \\ 2 \\ -3 \end{pmatrix}, \begin{pmatrix} -2 \\ -8 \\ 0 \\ 8 \\ 8 \\ 9 \end{pmatrix} \right\}$

d. *(Paper and pencil)* Let $\overline{V}$ be the matrix $[\mathbf{v}_1 \ldots \mathbf{v}_k]$ and let $\overline{W}$ be the matrix $[\mathbf{w}_1 \ldots \mathbf{w}_m]$. Let $A$ be the matrix $[\overline{V}\,\overline{W}]$. Suppose $\mathbf{p}$ is a $(k + m) \times 1$ vector *in the null space of* $A$. Let $\mathbf{p} = \begin{pmatrix} \mathbf{a} \\ \mathbf{b} \end{pmatrix}$, where $\mathbf{a}$ is $k \times 1$ and $\mathbf{b}$ is $m \times 1$.

Show that $\overline{V}\mathbf{a} = -\overline{W}\mathbf{b}$. Letting $\mathbf{z} = \overline{V}\mathbf{a}$, explain why you can thus conclude that $\mathbf{z}$ is in $U$, the intersection of $V$ and $W$.

e. *(Paper and pencil)* Conversely, suppose $\mathbf{z}$ is in $U$, the intersection of $V$ and $W$. Explain why $\mathbf{z} = \overline{V}\mathbf{x}$ for some $\mathbf{x}$ and $\mathbf{z} = \overline{W}\mathbf{y}$ for some $\mathbf{y}$. Argue why the vector $\begin{pmatrix} \mathbf{x} \\ -\mathbf{y} \end{pmatrix}$ is in the null space of $A$.

f. *(Paper and pencil)* Explain why you can conclude that $U$, the intersection, is equal to

$$\left\{\overline{V}\mathbf{a} \mid \begin{pmatrix} \mathbf{a} \\ \mathbf{b} \end{pmatrix} \text{ is in the null space of } A\right\}$$

Conclude that if $\{\mathbf{s}_1, \ldots, \mathbf{s}_q\}$ is a basis for the null space of $A$ and each $\mathbf{s}_i = \begin{pmatrix} \mathbf{a}_i \\ \mathbf{b}_i \end{pmatrix}$ where $\mathbf{a}_i$ is $k \times 1$ and $\mathbf{b}_i$ is $m \times 1$, then $\{\overline{V}\mathbf{a}_1, \ldots, \overline{V}\mathbf{a}_q\}$ spans $U$.

g. Using information from part (f), find a basis for $U = V \cap W$ for the pairs of bases for $V$ and $W$ given in part (c). For each pair, find the dimension of $U$.

Do some checking of your answer: Verify that the set of vectors that you found is linearly independent and show that a random linear combination of vectors in the set is in $V$ *and* in $W$.

h. From your work above, conjecture a relationship among the dimensions of $V$, $W$, $U$, and $S$.

## 4.8 CHANGE OF BASIS

In $\mathbb{R}^2$ we wrote vectors in terms of the standard basis $\mathbf{i} = \begin{pmatrix} 1 \\ 0 \end{pmatrix}$, $\mathbf{j} = \begin{pmatrix} 0 \\ 1 \end{pmatrix}$. In $\mathbb{R}^n$ we defined the standard basis $\{\mathbf{e}_1, \mathbf{e}_2, \ldots, \mathbf{e}_n\}$. In $P_n$ we defined the standard basis to be $\{1, x, x^2, \ldots, x^n\}$. These bases are most commonly used because it is relatively easy to work with them. But it sometimes happens that some other basis is more convenient. There are infinitely many bases to choose from since in an $n$-dimensional vector space *any* $n$ linearly independent vectors form a basis. In this section we shall see how to change from one basis to another by computing a certain matrix.

We start with a simple example. Let $\mathbf{u}_1 = \begin{pmatrix} 1 \\ 0 \end{pmatrix}$ and $\mathbf{u}_2 = \begin{pmatrix} 0 \\ 1 \end{pmatrix}$. Then $B_1 = \{\mathbf{u}_1, \mathbf{u}_2\}$ is the standard basis in $\mathbb{R}^2$. Let $\mathbf{v}_1 = \begin{pmatrix} 1 \\ 3 \end{pmatrix}$ and $\mathbf{v}_2 = \begin{pmatrix} -1 \\ 2 \end{pmatrix}$. Since $\mathbf{v}_1$ and $\mathbf{v}_2$ are linearly independent (because $\mathbf{v}_1$ is not a multiple of $\mathbf{v}_2$), $B_2 = \{\mathbf{v}_1, \mathbf{v}_2\}$ is a second basis in $\mathbb{R}^2$. Let $\mathbf{x} = \begin{pmatrix} x_1 \\ x_2 \end{pmatrix}$ be a vector in $\mathbb{R}^2$. This notation means that

$$\mathbf{x} = \begin{pmatrix} x_1 \\ x_2 \end{pmatrix} = x_1\begin{pmatrix} 1 \\ 0 \end{pmatrix} + x_2\begin{pmatrix} 0 \\ 1 \end{pmatrix} = x_1\mathbf{u}_1 + x_2\mathbf{u}_2$$

That is, $\mathbf{x}$ is written in terms of the vectors in the basis $B_1$. To emphasize this fact, we write

$$(\mathbf{x})_{B_1} = \begin{pmatrix} x_1 \\ x_2 \end{pmatrix}$$

Since $B_2$ is another basis in $\mathbb{R}^2$, there are scalars $c_1$ and $c_2$ such that

$$\mathbf{x} = c_1\mathbf{v}_1 + c_2\mathbf{v}_2 \tag{1}$$

Once these scalars are found, we write

$$(\mathbf{x})_{B_2} = \begin{pmatrix} c_1 \\ c_2 \end{pmatrix}$$

to indicate that $\mathbf{x}$ is now expressed in terms of the vectors in $B_2$. To find the numbers $c_1$ and $c_2$, we write the old basis vectors ($\mathbf{u}_1$ and $\mathbf{u}_2$) in terms of the new basis vectors ($\mathbf{v}_1$ and $\mathbf{v}_2$). It is easy to verify that

$$\mathbf{u}_1 = \begin{pmatrix} 1 \\ 0 \end{pmatrix} = \tfrac{2}{5}\begin{pmatrix} 1 \\ 3 \end{pmatrix} - \tfrac{3}{5}\begin{pmatrix} -1 \\ 2 \end{pmatrix} = \tfrac{2}{5}\mathbf{v}_1 - \tfrac{3}{5}\mathbf{v}_2 \tag{2}$$

and

$$\mathbf{u}_2 = \begin{pmatrix} 0 \\ 1 \end{pmatrix} = \tfrac{1}{5}\begin{pmatrix} 1 \\ 3 \end{pmatrix} + \tfrac{1}{5}\begin{pmatrix} -1 \\ 2 \end{pmatrix} = \tfrac{1}{5}\mathbf{v}_1 + \tfrac{1}{5}\mathbf{v}_2 \tag{3}$$

That is,

$$(\mathbf{u}_1)_{B_2} = \begin{pmatrix} \frac{2}{5} \\ -\frac{3}{5} \end{pmatrix} \quad \text{and} \quad (\mathbf{u}_2)_{B_2} = \begin{pmatrix} \frac{1}{5} \\ \frac{1}{5} \end{pmatrix}$$

Then

$$\begin{aligned} \mathbf{x} = x_1\mathbf{u}_1 + x_2\mathbf{u}_2 &\overset{\text{from (2) and (3)}}{=} x_1(\tfrac{2}{5}\mathbf{v}_1 - \tfrac{3}{5}\mathbf{v}_2) + x_2(\tfrac{1}{5}\mathbf{v}_1 + \tfrac{1}{5}\mathbf{v}_2) \\ &= (\tfrac{2}{5}x_1 + \tfrac{1}{5}x_2)\mathbf{v}_1 + (-\tfrac{3}{5}x_1 + \tfrac{1}{5}x_2)\mathbf{v}_2 \end{aligned}$$

Thus, from (1),

$$\begin{aligned} c_1 &= \tfrac{2}{5}x_1 + \tfrac{1}{5}x_2 \\ c_2 &= -\tfrac{3}{5}x_1 + \tfrac{1}{5}x_2 \end{aligned}$$

or

$$(\mathbf{x})_{B_2} = \begin{pmatrix} c_1 \\ c_2 \end{pmatrix} = \begin{pmatrix} \frac{2}{5}x_1 + \frac{1}{5}x_2 \\ -\frac{3}{5}x_1 + \frac{1}{5}x_2 \end{pmatrix} = \begin{pmatrix} \frac{2}{5} & \frac{1}{5} \\ -\frac{3}{5} & \frac{1}{5} \end{pmatrix}\begin{pmatrix} x_1 \\ x_2 \end{pmatrix}$$

For example, if $(\mathbf{x})_{B_1} = \begin{pmatrix} 3 \\ -4 \end{pmatrix}$, then

$$(\mathbf{x})_{B_2} = \begin{pmatrix} \frac{2}{5} & \frac{1}{5} \\ -\frac{3}{5} & \frac{1}{5} \end{pmatrix}\begin{pmatrix} 3 \\ -4 \end{pmatrix} = \begin{pmatrix} \frac{2}{5} \\ -\frac{13}{5} \end{pmatrix}$$

***Check.***

$$\begin{aligned} \tfrac{2}{5}\mathbf{v}_1 - \tfrac{13}{5}\mathbf{v}_2 &= \tfrac{2}{5}\begin{pmatrix} 1 \\ 3 \end{pmatrix} - \tfrac{13}{5}\begin{pmatrix} -1 \\ 2 \end{pmatrix} = \begin{pmatrix} \frac{2}{5} + \frac{13}{5} \\ \frac{6}{5} - \frac{26}{5} \end{pmatrix} = \begin{pmatrix} 3 \\ -4 \end{pmatrix} = 3\begin{pmatrix} 1 \\ 0 \end{pmatrix} - 4\begin{pmatrix} 0 \\ 1 \end{pmatrix} \\ &= 3\mathbf{u}_1 - 4\mathbf{u}_2 \end{aligned}$$

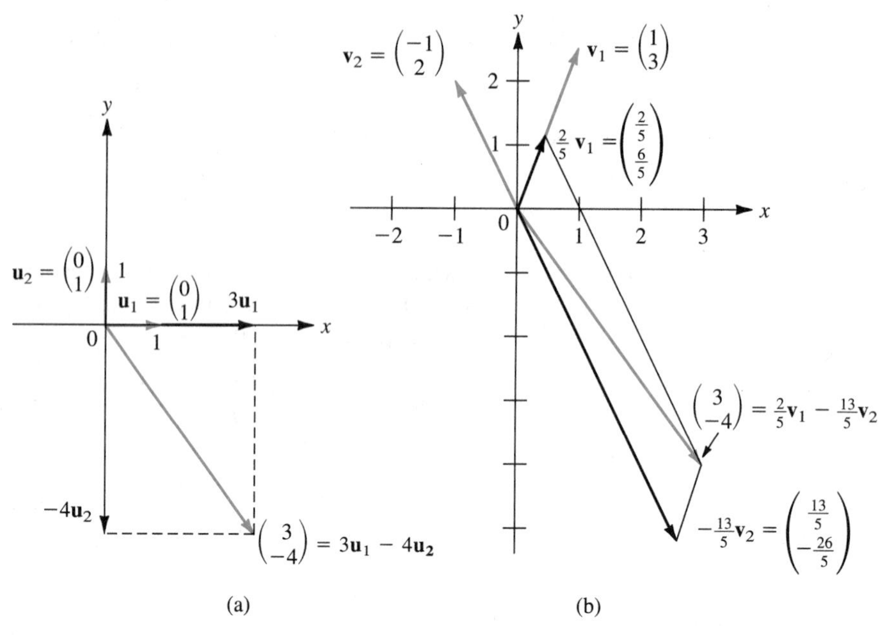


**Figure 4.4** (a) Writing $\begin{pmatrix} 3 \\ -4 \end{pmatrix}$ in terms of standard basis $\left\{\begin{pmatrix} 1 \\ 0 \end{pmatrix}, \begin{pmatrix} 0 \\ 1 \end{pmatrix}\right\}$.
(b) Writing $\begin{pmatrix} 3 \\ -4 \end{pmatrix}$ in terms of basis $\left\{\begin{pmatrix} 1 \\ 3 \end{pmatrix}, \begin{pmatrix} -1 \\ 2 \end{pmatrix}\right\}$

The matrix $A = \begin{pmatrix} \frac{2}{5} & \frac{1}{5} \\ -\frac{3}{5} & \frac{1}{5} \end{pmatrix}$ is called the **transition matrix** from $B_1$ to $B_2$, and we have shown that

$$(\mathbf{x})_{B_2} = A(\mathbf{x})_{B_1} \tag{4}$$

We illustrate the two bases $\left\{\begin{pmatrix} 1 \\ 0 \end{pmatrix}, \begin{pmatrix} 0 \\ 1 \end{pmatrix}\right\}$ and $\left\{\begin{pmatrix} 1 \\ 3 \end{pmatrix}, \begin{pmatrix} -1 \\ 2 \end{pmatrix}\right\}$ in Figure 4.4.

This example can be easily generalized, but first we need to extend our notation. Let $B_1 = \{\mathbf{u}_1, \mathbf{u}_2, \ldots, \mathbf{u}_n\}$ and $B_2 = \{\mathbf{v}_1, \mathbf{v}_2, \ldots, \mathbf{v}_n\}$ be two bases for an $n$-dimensional real vector space $V$. Let $\mathbf{x} \in V$. Then $\mathbf{x}$ can be written in terms of both bases:

$$\mathbf{x} = b_1\mathbf{u}_1 + b_2\mathbf{u}_2 + \cdots + b_n\mathbf{u}_n \tag{5}$$

and

$$\mathbf{x} = c_1\mathbf{v}_1 + c_2\mathbf{v}_2 + \cdots + c_n\mathbf{v}_n \tag{6}$$

where the $b_i$'s and $c_i$'s are real numbers. We then write $(\mathbf{x})_{B_1} = \begin{pmatrix} b_1 \\ b_2 \\ \vdots \\ b_n \end{pmatrix}$ to denote the representation of $\mathbf{x}$ in terms of the basis $B_1$. This is unambiguous because the coefficients $b_i$ in (5) are unique by Theorem 4.6.1, page 338. Likewise $(\mathbf{x})_{B_2} = \begin{pmatrix} c_1 \\ c_2 \\ \vdots \\ c_n \end{pmatrix}$ has a similar meaning. Suppose that $\mathbf{w}_1 = a_1\mathbf{u}_1 + a_2\mathbf{u}_2 + \cdots + a_n\mathbf{u}_n$ and $\mathbf{w}_2 = b_1\mathbf{u}_1 + b_2\mathbf{u}_2 + \cdots + b_n\mathbf{u}_n$. Then $\mathbf{w}_1 + \mathbf{w}_2 = (a_1 + b_1)\mathbf{u}_1 + (a_2 + b_2)\mathbf{u}_2 + \cdots + (a_n + b_n)\mathbf{u}_n$, so that

$$(\mathbf{w}_1 + \mathbf{w}_2)_{B_1} = (\mathbf{w}_1)_{B_1} + (\mathbf{w}_2)_{B_1}$$

That is, in the new notation we can add vectors just as we add vectors in $\mathbb{R}^n$. The coefficients of the "sum" vector are the sums of the coefficients of the two individual vectors. Moreover, it is easy to show that

$$\alpha(\mathbf{w})_{B_1} = (\alpha\mathbf{w})_{B_1}$$

Now since $B_2$ is a basis, each $\mathbf{u}_j$ in $B_1$ can be written as a linear combination of the $\mathbf{v}_i$'s. Thus there exists a unique set of scalars $a_{1j}, a_{2j}, \ldots, a_{nj}$ such that for $j = 1, 2, \ldots, n$

$$\mathbf{u}_j = a_{1j}\mathbf{v}_1 + a_{2j}\mathbf{v}_2 + \cdots + a_{nj}\mathbf{v}_n \tag{7}$$

or

$$(\mathbf{u}_j)_{B_2} = \begin{pmatrix} a_{1j} \\ a_{2j} \\ \vdots \\ a_{nj} \end{pmatrix} \tag{8}$$

**DEFINITION 1** **Transition Matrix** The $n \times n$ matrix $A$ whose columns are given by (8) is called the **transition matrix** from basis $B_1$ to basis $B_2$. That is,

$$A = \begin{pmatrix} a_{11} & a_{12} & a_{13} & \cdots & a_{1n} \\ a_{21} & a_{22} & a_{23} & \cdots & a_{2n} \\ \vdots & \vdots & \vdots & & \vdots \\ a_{n1} & a_{n2} & a_{n3} & \cdots & a_{nn} \end{pmatrix} \tag{9}$$

$$\begin{matrix} \uparrow & \uparrow & \uparrow & & \uparrow \\ (\mathbf{u}_1)_{B_2} & (\mathbf{u}_2)_{B_2} & (\mathbf{u}_3)_{B_2} & \cdots & (\mathbf{u}_n)_{B_2} \end{matrix}$$

*Note.* If you change the order in which you write the vectors in a basis, then you must also change the order of the columns in the transition matrix.

**THEOREM 1** Let $B_1$ and $B_2$ be bases for a vector space $V$. Let $A$ be the transition matrix from $B_1$ to $B_2$. Then for every $\mathbf{x} \in V$

$$\boxed{(\mathbf{x})_{B_2} = A(\mathbf{x})_{B_1}} \tag{10}$$

**Proof** We use the representation of $\mathbf{x}$ given in (5) and (6):

$$\begin{aligned}
\mathbf{x} &\overset{\text{from (5)}}{=} b_1\mathbf{u}_1 + b_2\mathbf{u}_2 + \cdots + b_n\mathbf{u}_n \\
&\overset{\text{from (7)}}{=} b_1(a_{11}\mathbf{v}_1 + a_{21}\mathbf{v}_2 + \cdots + a_{n1}\mathbf{v}_n) + b_2(a_{12}\mathbf{v}_1 + a_{22}\mathbf{v}_2 + \cdots + a_{n2}\mathbf{v}_n) \\
&\quad + \cdots + b_n(a_{1n}\mathbf{v}_1 + a_{2n}\mathbf{v}_2 + \cdots + a_{nn}\mathbf{v}_n) \\
&= (a_{11}b_1 + a_{12}b_2 + \cdots + a_{1n}b_n)\mathbf{v}_1 + (a_{21}b_1 + a_{22}b_2 + \cdots + a_{2n}b_n)\mathbf{v}_2 + \cdots \\
&\quad + (a_{n1}b_1 + a_{n2}b_2 + \cdots + a_{nn}b_n)\mathbf{v}_n \\
&\overset{\text{from (6)}}{=} c_1\mathbf{v}_1 + c_2\mathbf{v}_2 + \cdots + c_n\mathbf{v}_n
\end{aligned} \tag{11}$$

Thus

$$(\mathbf{x})_{B_2} = \begin{pmatrix} c_1 \\ c_2 \\ \vdots \\ c_n \end{pmatrix} \overset{\text{from (11)}}{=} \begin{pmatrix} a_{11}b_1 + a_{12}b_2 + \cdots + a_{1n}b_n \\ a_{21}b_1 + a_{22}b_2 + \cdots + a_{2n}b_n \\ \vdots \\ a_{n1}b_1 + a_{n2}b_2 + \cdots + a_{nn}b_n \end{pmatrix}$$

$$= \begin{pmatrix} a_{11} & a_{12} & \cdots & a_{1n} \\ a_{21} & a_{22} & \cdots & a_{2n} \\ \vdots & \vdots & & \vdots \\ a_{n1} & a_{n2} & \cdots & a_{nn} \end{pmatrix} \begin{pmatrix} b_1 \\ b_2 \\ \vdots \\ b_n \end{pmatrix} = A(\mathbf{x})_{B_1} \tag{12}$$

■

Before doing any further examples, we prove a theorem that is very useful for computations.

**THEOREM 2** If $A$ is the transition matrix from $B_1$ to $B_2$, then $A^{-1}$ is the transition matrix from $B_2$ to $B_1$.

**Proof** Let $C$ be the transition matrix from $B_2$ to $B_1$. Then from (10) we have

$$(\mathbf{x})_{B_1} = C(\mathbf{x})_{B_2} \tag{13}$$

But $(\mathbf{x})_{B_2} = A(\mathbf{x})_{B_1}$, and substituting this into (13) yields

$$(\mathbf{x})_{B_1} = CA(\mathbf{x})_{B_1} \tag{14}$$

We leave it as an exercise (see Problem 39) to show that (14) can hold for every $\mathbf{x}$ in $V$ only if $CA = I$. Thus from Theorem 1.8.7 on page 112, $C = A^{-1}$, and the theorem is proven. ■

*Remark.* This theorem makes it especially easy to find the transition matrix from the standard basis $B_1 = \{\mathbf{e}_1, \mathbf{e}_2, \ldots, \mathbf{e}_n\}$ in $\mathbb{R}^n$ to any other basis in $\mathbb{R}^n$. Let $B_2 = \{\mathbf{v}_1, \mathbf{v}_2, \ldots, \mathbf{v}_n\}$ be any other basis. Let $C$ be the matrix whose columns are the vectors $\mathbf{v}_1, \mathbf{v}_2, \ldots, \mathbf{v}_n$. Then $C$ is the transition matrix from $B_2$ to $B_1$ since each vector $\mathbf{v}_i$ is already written in terms of the standard basis. For example,

$$\begin{pmatrix} 1 \\ 3 \\ -2 \\ 4 \end{pmatrix}_{B_1} = \begin{pmatrix} 1 \\ 3 \\ -2 \\ 4 \end{pmatrix} = 1\begin{pmatrix} 1 \\ 0 \\ 0 \\ 0 \end{pmatrix} + 3\begin{pmatrix} 0 \\ 1 \\ 0 \\ 0 \end{pmatrix} - 2\begin{pmatrix} 0 \\ 0 \\ 1 \\ 0 \end{pmatrix} + 4\begin{pmatrix} 0 \\ 0 \\ 0 \\ 1 \end{pmatrix}$$

Thus the transition matrix from $B_1$ to $B_2$ is $C^{-1}$.

**Procedure for Finding the Transition Matrix from the Standard Basis to the Basis $B_2 = \{\mathbf{v}_1, \mathbf{v}_2, \ldots, \mathbf{v}_n\}$**

**i.** Write the matrix $C$ whose columns are $\mathbf{v}_1, \mathbf{v}_2, \ldots, \mathbf{v}_n$.

**ii.** Compute $C^{-1}$. This is the required transition matrix.

*Note.* As on page 376, the transition matrix is unique relative to the order in which we write the basis vectors in $B_2$.

**EXAMPLE 1** **Writing Vectors in $\mathbb{R}^3$ in Terms of a New Basis** In $\mathbb{R}^3$ let $B_1 = \{\mathbf{i}, \mathbf{j}, \mathbf{k}\}$ and let $B_2 = \left\{\begin{pmatrix} 1 \\ 0 \\ 2 \end{pmatrix}, \begin{pmatrix} 3 \\ -1 \\ 0 \end{pmatrix}, \begin{pmatrix} 0 \\ 1 \\ -2 \end{pmatrix}\right\}$. If $\mathbf{x} = \begin{pmatrix} x \\ y \\ z \end{pmatrix} \in \mathbb{R}^3$, write $\mathbf{x}$ in terms of the vectors in $B_2$.

**Solution** We first verify that $B_2$ is a basis. This is evident since $\begin{vmatrix} 1 & 3 & 0 \\ 0 & -1 & 1 \\ 2 & 0 & -2 \end{vmatrix} = 8 \neq 0$.

Since $\mathbf{u}_1 = \begin{pmatrix} 1 \\ 0 \\ 0 \end{pmatrix}$, $\mathbf{u}_2 = \begin{pmatrix} 0 \\ 1 \\ 0 \end{pmatrix}$, and $\mathbf{u}_3 = \begin{pmatrix} 0 \\ 0 \\ 1 \end{pmatrix}$, we immediately see that the transition matrix, $C$, from $B_2$ to $B_1$ is given by

$$C = \begin{pmatrix} 1 & 3 & 0 \\ 0 & -1 & 1 \\ 2 & 0 & -2 \end{pmatrix}$$

Thus from Theorem 2 the transition matrix $A$ from $B_1$ to $B_2$ is

$$A = C^{-1} = \tfrac{1}{8}\begin{pmatrix} 2 & 6 & 3 \\ 2 & -2 & -1 \\ 2 & 6 & -1 \end{pmatrix}$$

For example, if $(\mathbf{x})_{B_1} = \begin{pmatrix} 1 \\ -2 \\ 4 \end{pmatrix}$, then

$$(\mathbf{x})_{B_2} = \tfrac{1}{8}\begin{pmatrix} 2 & 6 & 3 \\ 2 & -2 & -1 \\ 2 & 6 & -1 \end{pmatrix}\begin{pmatrix} 1 \\ -2 \\ 4 \end{pmatrix} = \tfrac{1}{8}\begin{pmatrix} 2 \\ 2 \\ -14 \end{pmatrix} = \begin{pmatrix} \frac{1}{4} \\ \frac{1}{4} \\ -\frac{7}{4} \end{pmatrix}$$

As a check, note that

$$\tfrac{1}{4}\begin{pmatrix} 1 \\ 0 \\ 2 \end{pmatrix} + \tfrac{1}{4}\begin{pmatrix} 3 \\ -1 \\ 0 \end{pmatrix} - \tfrac{7}{4}\begin{pmatrix} 0 \\ 1 \\ -2 \end{pmatrix} = \begin{pmatrix} 1 \\ -2 \\ 4 \end{pmatrix} = 1\begin{pmatrix} 1 \\ 0 \\ 0 \end{pmatrix} - 2\begin{pmatrix} 0 \\ 1 \\ 0 \end{pmatrix} + 4\begin{pmatrix} 0 \\ 0 \\ 1 \end{pmatrix}$$

**EXAMPLE 2** **Writing Polynomials in $P_2$ in Terms of a New Basis** In $P_2$ the standard basis is $B_1 = \{1, x, x^2\}$. Another basis is $B_2 = \{4x - 1, 2x^2 - x, 3x^2 + 3\}$. If $p = a_0 + a_1x + a_2x^2$, write $p$ in terms of the polynomials in $B_2$.

**Solution** We first verify that $B_2$ is a basis. If $c_1(4x - 1) + c_2(2x^2 - x) + c_3(3x^2 + 3) = 0$ for all $x$, then rearranging terms we obtain

$$(-c_1 + 3c_3)1 + (4c_1 - c_2)x + (2c_2 + 3c_3)x^2 = 0$$

But since $\{1, x, x^2\}$ is a linearly independent set, we must have

$$\begin{aligned} -c_1 \qquad\quad + 3c_3 &= 0 \\ 4c_1 - \quad c_2 \qquad\quad &= 0 \\ 2c_2 + 3c_3 &= 0 \end{aligned}$$

The determinant of this homogeneous system is $\begin{vmatrix} -1 & 0 & 3 \\ 4 & -1 & 0 \\ 0 & 2 & 3 \end{vmatrix} = 27 \neq 0$, which means that $c_1 = c_2 = c_3 = 0$ is the only solution. Now $(4x - 1)_{B_1} = \begin{pmatrix} -1 \\ 4 \\ 0 \end{pmatrix}$, $(2x^2 - x)_{B_1} = \begin{pmatrix} 0 \\ -1 \\ 2 \end{pmatrix}$, and $(3 + 3x^2)_{B_1} = \begin{pmatrix} 3 \\ 0 \\ 3 \end{pmatrix}$. Hence

$$C = \begin{pmatrix} -1 & 0 & 3 \\ 4 & -1 & 0 \\ 0 & 2 & 3 \end{pmatrix}$$

is the transition matrix from $B_2$ to $B_1$ so that

$$A = C^{-1} = \tfrac{1}{27}\begin{pmatrix} -3 & 6 & 3 \\ -12 & -3 & 12 \\ 8 & 2 & 1 \end{pmatrix}$$

is the transition matrix from $B_1$ to $B_2$. Since $(a_0 + a_1x + a_2x^2)_{B_1} = \begin{pmatrix} a_0 \\ a_1 \\ a_2 \end{pmatrix}$, we have

$$(a_0 + a_1x + a_2x^2)_{B_2} = \tfrac{1}{27}\begin{pmatrix} -3 & 6 & 3 \\ -12 & -3 & 12 \\ 8 & 2 & 1 \end{pmatrix}\begin{pmatrix} a_0 \\ a_1 \\ a_2 \end{pmatrix}$$

$$= \begin{pmatrix} \frac{1}{27}[-3a_0 + 6a_1 + 3a_2] \\ \frac{1}{27}[-12a_0 - 3a_1 + 12a_2] \\ \frac{1}{27}[8a_0 + 2a_1 + a_2] \end{pmatrix}$$

For example, if $p(x) = 5x^2 - 3x + 4$, then

$$(5x^2 - 3x + 4)_{B_2} = \tfrac{1}{27}\begin{pmatrix} -3 & 6 & 3 \\ -12 & -3 & 12 \\ 8 & 2 & 1 \end{pmatrix}\begin{pmatrix} 4 \\ -3 \\ 5 \end{pmatrix} = \begin{pmatrix} -\frac{15}{27} \\ \frac{21}{27} \\ \frac{31}{27} \end{pmatrix}$$

or

check this ↓

$$5x^2 - 3x + 4 = -\tfrac{15}{27}(4x - 1) + \tfrac{21}{27}(2x^2 - x) + \tfrac{31}{27}(3x^2 + 3)$$

**EXAMPLE 3** **Converting from One Basis to Another in $\mathbb{R}^2$** Let $B_1 = \left\{\begin{pmatrix}3\\1\end{pmatrix}, \begin{pmatrix}2\\-1\end{pmatrix}\right\}$ and $B_2 = \left\{\begin{pmatrix}2\\4\end{pmatrix}, \begin{pmatrix}-5\\3\end{pmatrix}\right\}$ be two bases in $\mathbb{R}^2$. If $(\mathbf{x})_{B_1} = \begin{pmatrix}b_1\\b_2\end{pmatrix}$, write $\mathbf{x}$ in terms of the vectors in $B_2$.

**Solution** This problem is a bit more difficult because neither basis is the standard basis. We must write the vectors in $B_1$ as linear combinations of the vectors in $B_2$. That is, we must find constants $a_{11}$, $a_{21}$, $a_{12}$, $a_{22}$ such that

$$\begin{pmatrix}3\\1\end{pmatrix} = a_{11}\begin{pmatrix}2\\4\end{pmatrix} + a_{21}\begin{pmatrix}-5\\3\end{pmatrix} \quad \text{and} \quad \begin{pmatrix}2\\-1\end{pmatrix} = a_{12}\begin{pmatrix}2\\4\end{pmatrix} + a_{22}\begin{pmatrix}-5\\3\end{pmatrix}$$

This leads to the following systems:

$$\begin{aligned} 2a_{11} - 5a_{21} &= 3 \\ 4a_{11} + 3a_{21} &= 1 \end{aligned} \quad \text{and} \quad \begin{aligned} 2a_{12} - 5a_{22} &= 2 \\ 4a_{12} + 3a_{22} &= -1 \end{aligned}$$

The solutions are $a_{11} = \frac{7}{13}$, $a_{21} = -\frac{5}{13}$, $a_{12} = \frac{1}{26}$, and $a_{22} = -\frac{5}{13}$. Thus

$$A = \frac{1}{26}\begin{pmatrix}14 & 1\\-10 & -10\end{pmatrix}$$

and

$$(\mathbf{x})_{B_2} = \frac{1}{26}\begin{pmatrix}14 & 1\\-10 & -10\end{pmatrix}\begin{pmatrix}b_1\\b_2\end{pmatrix} = \begin{pmatrix}\frac{1}{26}(14b_1 + b_2)\\-\frac{10}{26}(b_1 + b_2)\end{pmatrix}$$

For example, let $\mathbf{x} = \begin{pmatrix}7\\4\end{pmatrix}$ (in standard basis). Then

$$\begin{pmatrix}7\\4\end{pmatrix}_{B_1} = b_1\begin{pmatrix}3\\1\end{pmatrix} + b_2\begin{pmatrix}2\\-1\end{pmatrix} = 3\begin{pmatrix}3\\1\end{pmatrix} - \begin{pmatrix}2\\-1\end{pmatrix}$$

so that

$$\begin{pmatrix}7\\4\end{pmatrix}_{B_1} = \begin{pmatrix}3\\-1\end{pmatrix}$$

and

$$\begin{pmatrix}7\\4\end{pmatrix}_{B_2} = \frac{1}{26}\begin{pmatrix}14 & 1\\-10 & -10\end{pmatrix}\begin{pmatrix}3\\-1\end{pmatrix} = \begin{pmatrix}\frac{41}{26}\\-\frac{20}{26}\end{pmatrix}$$

That is,

$$\begin{pmatrix}7\\4\end{pmatrix} \overset{\text{check!}}{=} \frac{41}{26}\begin{pmatrix}2\\4\end{pmatrix} - \frac{20}{26}\begin{pmatrix}-5\\3\end{pmatrix}$$

Using the notation of this section, we can derive a convenient way to determine whether a given set of vectors in any finite dimensional real vector space is linearly dependent or independent.

**THEOREM 3** Let $B_1 = \{\mathbf{v}_1, \mathbf{v}_2, \ldots, \mathbf{v}_n\}$ be a basis for the $n$-dimensional vector space $V$. Suppose that

$$(\mathbf{x}_1)_{B_1} = \begin{pmatrix} a_{11} \\ a_{21} \\ \vdots \\ a_{n1} \end{pmatrix}, (\mathbf{x}_2)_{B_1} = \begin{pmatrix} a_{12} \\ a_{22} \\ \vdots \\ a_{n2} \end{pmatrix}, \ldots, (\mathbf{x}_n)_{B_1} = \begin{pmatrix} a_{1n} \\ a_{2n} \\ \vdots \\ a_{nn} \end{pmatrix}$$

Let

$$A = \begin{pmatrix} a_{11} & a_{12} & \cdots & a_{1n} \\ a_{21} & a_{22} & \cdots & a_{2n} \\ \vdots & \vdots & & \vdots \\ a_{n1} & a_{n2} & \cdots & a_{nn} \end{pmatrix}$$

Then $\mathbf{x}_1, \mathbf{x}_2, \ldots, \mathbf{x}_n$ are linearly independent if and only if $\det A \neq 0$.

**Proof** Let $\mathbf{a}_1, \mathbf{a}_2, \ldots, \mathbf{a}_n$ denote the columns of $A$. Suppose that

$$c_1\mathbf{x}_1 + c_2\mathbf{x}_2 + \cdots + c_n\mathbf{x}_n = \mathbf{0} \tag{15}$$

Then using the addition defined on page 375, we may write (15) as

$$(c_1\mathbf{a}_1 + c_2\mathbf{a}_2 + \cdots + c_n\mathbf{a}_n)_{B_1} = (\mathbf{0})_{B_1} \tag{16}$$

Equation (16) gives two representations of the zero vector in $V$ in terms of the basis vectors in $B_1$. Since the representation of a vector in terms of basis vectors is unique (by Theorem 4.6.1, page 338) we conclude that

$$c_1\mathbf{a}_1 + c_2\mathbf{a}_2 + \cdots + c_n\mathbf{a}_n = \mathbf{0} \tag{17}$$

where the zero on the right-hand side is the zero vector in $\mathbb{R}^n$. But this proves the theorem since equation (17) involves the columns of $A$, which are linearly independent if and only if $\det A \neq 0$. ∎

EXAMPLE 4 **Determining Whether Three Polynomials in $P_2$ Are Linearly Dependent or Independent** In $P_2$ determine whether the polynomials $3 - x$, $2 + x^2$, and $4 + 5x - 2x^2$ are linearly dependent or independent.

**Solution** Using the basis $B_1 = \{1, x, x^2\}$, we have $(3 - x)_{B_1} = \begin{pmatrix} 3 \\ -1 \\ 0 \end{pmatrix}$, $(2 + x^2)_{B_1} = \begin{pmatrix} 2 \\ 0 \\ 1 \end{pmatrix}$, and $(4 + 5x - 2x^2)_{B_1} = \begin{pmatrix} 4 \\ 5 \\ -2 \end{pmatrix}$. Then $\det A = \begin{vmatrix} 3 & 2 & 4 \\ -1 & 0 & 5 \\ 0 & 1 & -2 \end{vmatrix} = -23 \neq 0$, so the polynomials are independent.

**EXAMPLE 5** **Determining Whether Four 2 × 2 Matrices Are Linearly Dependent or Independent** In $M_{22}$ determine whether the matrices $\begin{pmatrix} 1 & 2 \\ 3 & 6 \end{pmatrix}, \begin{pmatrix} -1 & 3 \\ -1 & 1 \end{pmatrix}, \begin{pmatrix} 2 & -1 \\ 0 & 1 \end{pmatrix}$, and $\begin{pmatrix} 1 & 4 \\ 4 & 9 \end{pmatrix}$ are linearly dependent or independent.

**Solution** Using the standard basis $B_1 = \left\{ \begin{pmatrix} 1 & 0 \\ 0 & 0 \end{pmatrix}, \begin{pmatrix} 0 & 1 \\ 0 & 0 \end{pmatrix}, \begin{pmatrix} 0 & 0 \\ 1 & 0 \end{pmatrix}, \begin{pmatrix} 0 & 0 \\ 0 & 1 \end{pmatrix} \right\}$, we obtain

$$\det A = \begin{vmatrix} 1 & -1 & 2 & 1 \\ 2 & 3 & -1 & 4 \\ 3 & -1 & 0 & 4 \\ 6 & 1 & 1 & 9 \end{vmatrix} = 0$$

so the matrices are dependent. Note that $\det A = 0$ because the fourth row of $A$ is the sum of the first three rows of $A$. Note also that

$$-29\begin{pmatrix} 1 & 2 \\ 3 & 6 \end{pmatrix} - 7\begin{pmatrix} -1 & 3 \\ -1 & 1 \end{pmatrix} + \begin{pmatrix} 2 & -1 \\ 0 & 1 \end{pmatrix} + 20\begin{pmatrix} 1 & 4 \\ 4 & 9 \end{pmatrix} = \begin{pmatrix} 0 & 0 \\ 0 & 0 \end{pmatrix}$$

which illustrates that the four matrices are linearly dependent.

## PROBLEMS 4.8

### Self-Quiz

**I.** The transition matrix in $\mathbb{R}^2$ from the basis $\left\{ \begin{pmatrix} 1 \\ 0 \end{pmatrix}, \begin{pmatrix} 0 \\ 1 \end{pmatrix} \right\}$ to the basis $\left\{ \begin{pmatrix} 2 \\ 3 \end{pmatrix}, \begin{pmatrix} -3 \\ -4 \end{pmatrix} \right\}$ is ________.

**a.** $\begin{pmatrix} 2 & -3 \\ 3 & -4 \end{pmatrix}$ **b.** $\begin{pmatrix} 2 & 3 \\ -3 & -4 \end{pmatrix}$ **c.** $\begin{pmatrix} -4 & 3 \\ -3 & 2 \end{pmatrix}$ **d.** $\begin{pmatrix} -4 & -3 \\ 3 & 2 \end{pmatrix}$

**II.** The transition matrix in $\mathbb{R}^2$ from the basis $\left\{ \begin{pmatrix} 2 \\ 3 \end{pmatrix}, \begin{pmatrix} -3 \\ -4 \end{pmatrix} \right\}$ to the basis $\left\{ \begin{pmatrix} 1 \\ 0 \end{pmatrix}, \begin{pmatrix} 0 \\ 1 \end{pmatrix} \right\}$ is ________.

**a.** $\begin{pmatrix} 2 & -3 \\ 3 & -4 \end{pmatrix}$ **b.** $\begin{pmatrix} 2 & 3 \\ -3 & -4 \end{pmatrix}$ **c.** $\begin{pmatrix} -4 & 3 \\ -3 & 2 \end{pmatrix}$ **d.** $\begin{pmatrix} -4 & -3 \\ 3 & 2 \end{pmatrix}$

**III.** The transition matrix in $P_1$ from the basis $\{1, x\}$ to the basis $\{2 + 3x, -4 + 5x\}$ is ______.

**a.** $\begin{pmatrix} 2 & 3 \\ -4 & 5 \end{pmatrix}$ **b.** $\begin{pmatrix} 2 & -4 \\ 3 & 5 \end{pmatrix}$ **c.** $\frac{1}{22}\begin{pmatrix} 5 & -3 \\ 4 & 2 \end{pmatrix}$ **d.** $\frac{1}{22}\begin{pmatrix} 5 & 4 \\ -3 & 2 \end{pmatrix}$

In Problems 1–5 write $\begin{pmatrix} x \\ y \end{pmatrix} \in \mathbb{R}^2$ in terms of the given basis.

**1.** $\begin{pmatrix} 1 \\ 1 \end{pmatrix}, \begin{pmatrix} 1 \\ -1 \end{pmatrix}$ **2.** $\begin{pmatrix} 2 \\ -3 \end{pmatrix}, \begin{pmatrix} 3 \\ -2 \end{pmatrix}$ **3.** $\begin{pmatrix} 5 \\ 7 \end{pmatrix}, \begin{pmatrix} 3 \\ -4 \end{pmatrix}$ **4.** $\begin{pmatrix} -1 \\ -2 \end{pmatrix}, \begin{pmatrix} -1 \\ 2 \end{pmatrix}$

**5.** $\begin{pmatrix} a \\ c \end{pmatrix}, \begin{pmatrix} b \\ d \end{pmatrix}$, where $ad - bc \neq 0$

In Problems 6–10 write $\begin{pmatrix} x \\ y \\ z \end{pmatrix} \in \mathbb{R}^3$ in terms of the given basis.

**6.** $\begin{pmatrix} 1 \\ 0 \\ 0 \end{pmatrix}, \begin{pmatrix} 0 \\ 0 \\ 1 \end{pmatrix}, \begin{pmatrix} 1 \\ 1 \\ 1 \end{pmatrix}$ **7.** $\begin{pmatrix} 1 \\ 0 \\ 0 \end{pmatrix}, \begin{pmatrix} 1 \\ 1 \\ 0 \end{pmatrix}, \begin{pmatrix} 1 \\ 1 \\ 1 \end{pmatrix}$ **8.** $\begin{pmatrix} 1 \\ 0 \\ -1 \end{pmatrix}, \begin{pmatrix} -1 \\ 1 \\ 0 \end{pmatrix}, \begin{pmatrix} 0 \\ 1 \\ 1 \end{pmatrix}$

**9.** $\begin{pmatrix} 2 \\ 1 \\ 3 \end{pmatrix}, \begin{pmatrix} -1 \\ 4 \\ 5 \end{pmatrix}, \begin{pmatrix} 3 \\ -2 \\ -4 \end{pmatrix}$ **10.** $\begin{pmatrix} a \\ 0 \\ 0 \end{pmatrix}, \begin{pmatrix} b \\ d \\ 0 \end{pmatrix}, \begin{pmatrix} c \\ e \\ f \end{pmatrix}$, where $adf \neq 0$

In Problems 11–13 write the polynomial $a_0 + a_1x + a_2x^2$ in $P_2$ in terms of the given basis.

**11.** $1, x - 1, x^2 - 1$ **12.** $6, 2 + 3x, 3 + 4x + 5x^2$ **13.** $x + 1, x - 1, x^2 - 1$

**14.** In $M_{22}$ write the matrix $\begin{pmatrix} 2 & -1 \\ 4 & 6 \end{pmatrix}$ in terms of the basis $\left\{\begin{pmatrix} 1 & 1 \\ -1 & 0 \end{pmatrix}, \begin{pmatrix} 2 & 0 \\ 3 & 1 \end{pmatrix}, \begin{pmatrix} 0 & 1 \\ -1 & 0 \end{pmatrix}, \begin{pmatrix} 0 & -2 \\ 0 & 4 \end{pmatrix}\right\}$.

**15.** In $P_3$ write the polynomial $2x^3 - 3x^2 + 5x - 6$ in terms of the basis polynomials $1$, $1 + x$, $x + x^2$, $x^2 + x^3$.

**16.** In $P_3$ write the polynomial $4x^2 - x + 5$ in terms of the basis polynomials $1$, $1 - x$, $(1 - x)^2$, $(1 - x)^3$.

**17.** In $\mathbb{R}^2$ suppose that $(\mathbf{x})_{B_1} = \begin{pmatrix} 2 \\ -1 \end{pmatrix}$, where $B_1 = \left\{\begin{pmatrix} 1 \\ 1 \end{pmatrix}, \begin{pmatrix} 2 \\ 3 \end{pmatrix}\right\}$. Write $\mathbf{x}$ in terms of the basis $B_2 = \left\{\begin{pmatrix} 0 \\ 3 \end{pmatrix}, \begin{pmatrix} 5 \\ -1 \end{pmatrix}\right\}$.

---

### Answers to Self-Quiz

**I.** c **II.** a **III.** d

**18.** In $\mathbb{R}^2$, $(\mathbf{x})_{B_1} = \begin{pmatrix} 4 \\ -1 \end{pmatrix}$, where $B_1 = \left\{\begin{pmatrix} 2 \\ -5 \end{pmatrix}, \begin{pmatrix} 7 \\ 3 \end{pmatrix}\right\}$. Write $\mathbf{x}$ in terms of $B_2 = \left\{\begin{pmatrix} -2 \\ 1 \end{pmatrix}, \begin{pmatrix} -3 \\ 2 \end{pmatrix}\right\}$.

**19.** In $\mathbb{R}^3$, $(\mathbf{x})_{B_1} = \begin{pmatrix} 2 \\ -1 \\ 4 \end{pmatrix}$, where $B_1 = \left\{\begin{pmatrix} 1 \\ -1 \\ 0 \end{pmatrix}, \begin{pmatrix} 0 \\ 1 \\ -1 \end{pmatrix}, \begin{pmatrix} 1 \\ 0 \\ 1 \end{pmatrix}\right\}$. Write $\mathbf{x}$ in terms of $B_2 = \left\{\begin{pmatrix} 3 \\ 0 \\ 0 \end{pmatrix}, \begin{pmatrix} 1 \\ 2 \\ -1 \end{pmatrix}, \begin{pmatrix} 0 \\ 1 \\ 5 \end{pmatrix}\right\}$.

**20.** In $P_2$, $(\mathbf{x})_{B_1} = \begin{pmatrix} 2 \\ 1 \\ 3 \end{pmatrix}$, where $B_1 = \{1 - x, 3x, x^2 - x - 1\}$. Write $\mathbf{x}$ in terms of $B_2 = \{3 - 2x, 1 + x, x + x^2\}$.

In Problems 21–28 use Theorem 2 to determine whether the given set of vectors is linearly dependent or independent.

**21.** In $P_2$: $2 + 3x + 5x^2$, $1 - 2x + x^2$, $-1 + 6x^2$

**22.** In $P_2$: $-3 + x^2$, $2 - x + 4x^2$, $4 + 2x$

**23.** In $P_2$: $x + 4x^2$, $-2 + 2x$, $2 + x + 12x^2$

**24.** In $P_2$: $-2 + 4x - 2x^2$, $3 + x$, $6 + 8x$

**25.** In $P_3$: $1 + x^2$, $-1 - 3x + 4x^2 + 5x^3$, $2 + 5x - 6x^3$, $4 + 6x + 3x^2 + 7x^3$

**26.** In $M_{22}$: $\begin{pmatrix} 2 & 0 \\ 3 & 4 \end{pmatrix}, \begin{pmatrix} -3 & -2 \\ 7 & 1 \end{pmatrix}, \begin{pmatrix} 1 & 0 \\ -1 & -3 \end{pmatrix}, \begin{pmatrix} 11 & 2 \\ -5 & -5 \end{pmatrix}$

**27.** In $M_{22}$: $\begin{pmatrix} 1 & -3 \\ 2 & 4 \end{pmatrix}, \begin{pmatrix} 1 & 4 \\ 5 & 0 \end{pmatrix}, \begin{pmatrix} -1 & 6 \\ -1 & 3 \end{pmatrix}, \begin{pmatrix} 0 & 0 \\ 3 & 0 \end{pmatrix}$

**28.** In $M_{22}$: $\begin{pmatrix} a & 0 \\ 0 & 0 \end{pmatrix}, \begin{pmatrix} b & c \\ 0 & 0 \end{pmatrix}, \begin{pmatrix} d & e \\ f & 0 \end{pmatrix}, \begin{pmatrix} g & h \\ j & k \end{pmatrix}$, where $acfk \neq 0$

**29.** In $P_n$ let $p_1, p_2, \ldots, p_{n+1}$ be $n + 1$ polynomials such that $p_i(0) = 0$ for $i = 1, 2, \ldots, n + 1$. Show that the polynomials are linearly dependent.

*Calculus **30.** In Problem 29, instead of $p_i(0) = 0$ suppose that $p_i^{(j)} = 0$ for $i = 1, 2, \ldots, n + 1$, and for some $j$ with $1 \leq j \leq n$, and $p_i^{(j)}$ denotes the $j$th derivative of $p_i$. Show that the polynomials are linearly dependent in $P_n$.

**31.** In $M_{mn}$ let $A_1, A_2, \ldots, A_{mn}$ be $mn$ matrices each of whose components in the 1, 1 position is zero. Show that the matrices are linearly dependent.

***32.** Suppose the $x$- and $y$-axes in the plane are rotated counterclockwise through an angle of $\theta$ (measure in degrees or radians). This gives us new axes which we denote $(x', y')$. What are the $x$- and $y$-coordinates of the now rotated basis vectors $\mathbf{i}$ and $\mathbf{j}$?

**33.** Show that the "change of coordinates" matrix in Problem 32 is given by $A^{-1} = \begin{pmatrix} \cos\theta & \sin\theta \\ -\sin\theta & \cos\theta \end{pmatrix}$.

**34.** If in Problems 32 and 33, $\theta = \pi/6 = 30°$, write the vector $\begin{pmatrix} -4 \\ 3 \end{pmatrix}$ in terms of the new coordinate axes $x'$ and $y'$.

**35.** If $\theta = \pi/4 = 45°$, write $\begin{pmatrix} 2 \\ -7 \end{pmatrix}$ in terms of the new coordinate axes.

**36.** If $\theta = 2\pi/3 = 120°$, write $\begin{pmatrix} 4 \\ 5 \end{pmatrix}$ in terms of the new coordinate axes.

**37.** Let $C = (c_{ij})$ be an $n \times n$ invertible matrix and let $B_1 = \{\mathbf{v}_1, \mathbf{v}_2, \ldots, \mathbf{v}_n\}$ be a basis for a vector space $V$. Let

$$\mathbf{c}_1 = \begin{pmatrix} c_{11} \\ c_{21} \\ \vdots \\ c_{n1} \end{pmatrix}_{B_1}, \mathbf{c}_2 = \begin{pmatrix} c_{12} \\ c_{22} \\ \vdots \\ c_{n2} \end{pmatrix}_{B_1}, \ldots, \mathbf{c}_n = \begin{pmatrix} c_{1n} \\ c_{2n} \\ \vdots \\ c_{nn} \end{pmatrix}_{B_1}$$

Show that $B_2 = \{\mathbf{c}_1, \mathbf{c}_2, \ldots, \mathbf{c}_n\}$ is a basis for $V$.

**38.** Let $B_1$ and $B_2$ be bases for the $n$-dimensional vector space $V$ and let $C$ be the transition matrix from $B_1$ to $B_2$. Show that $C^{-1}$ is the transition matrix from $B_2$ to $B_1$.

**39.** Show that $(\mathbf{x})_{B_1} = CA(\mathbf{x})_{B_1}$ for every $\mathbf{x}$ in a vector space $V$ if and only if $CA = I$ [*Hint:* Let $\mathbf{x}_i$ be the $i$th vector in $B_1$. Then $(\mathbf{x}_i)_{B_1}$ has a 1 in the $i$th position and a 0 everywhere else. What can you say about $CA(\mathbf{x}_i)_{B_1}$?]

## MATLAB 4.8

1. Let $B = \{\mathbf{v}_1, \mathbf{v}_2\}$, where $\mathbf{v}_1 = \begin{pmatrix} 1 \\ 1 \end{pmatrix}$ and $\mathbf{v}_2 = \begin{pmatrix} -1 \\ 1 \end{pmatrix}$. Note that $B$ is a basis for $\mathbb{R}^2$. For $\mathbf{w}$ in $\mathbb{R}^2$, $(\mathbf{w})_B = \begin{pmatrix} a \\ b \end{pmatrix}$ means that $\mathbf{w} = a\mathbf{v}_1 + b\mathbf{v}_2$.

M

a. For the **w**'s below, write the system of equations to find $(\mathbf{w})_B$, that is, to find $a$ and $b$ and solve by hand. Verify by calling **lincomb(v1,v2,w).** (Use the m-file *lincomb.m* on the accompanying disk.)

**i.** $\mathbf{w} = \begin{pmatrix} 1 \\ 2 \end{pmatrix}$  **ii.** $\mathbf{w} = \begin{pmatrix} -3 \\ 4 \end{pmatrix}$

b. *(Paper and pencil)* In general, explain why $\begin{pmatrix} a \\ b \end{pmatrix}$ is a solution to the system whose augmented matrix is $[\mathbf{v}_1\ \mathbf{v}_2 \mid \mathbf{w}]$.

2. Let $B = \left\{ \begin{pmatrix} 1 \\ 2 \\ 1 \\ 0 \end{pmatrix}, \begin{pmatrix} 2 \\ 5 \\ 3 \\ -2 \end{pmatrix}, \begin{pmatrix} 3 \\ 5 \\ 3 \\ 2 \end{pmatrix}, \begin{pmatrix} 4 \\ 8 \\ 9 \\ 1 \end{pmatrix} \right\}$ and $\mathbf{w} = \begin{pmatrix} 1 \\ 2 \\ -3 \\ 1 \end{pmatrix}$. We will refer to the $i$th vector in $B$ as $\mathbf{v}_i$.

a. Verify that $B$ is a basis for $\mathbb{R}^4$.

b. *(Paper and pencil)* Write the system of equations to find $(\mathbf{w})_B = \begin{pmatrix} x_1 \\ x_2 \\ x_3 \\ x_4 \end{pmatrix}$, the coordinates of $\mathbf{w}$ with respect to $B$. Show that $[\mathbf{v}_1\ \mathbf{v}_2\ \mathbf{v}_3\ \mathbf{v}_4|\mathbf{w}]$ is the augmented matrix for the system.

c. Solve the system for $(\mathbf{w})_B$. Verify that $\mathbf{w} = A(\mathbf{w})_B$, where $A = [\mathbf{v}_1\ \mathbf{v}_2\ \mathbf{v}_3\ \mathbf{v}_4]$.

d. For the bases $B = \{\mathbf{v}_1, \mathbf{v}_2, \mathbf{v}_3, \mathbf{v}_4\}$ below and the given $\mathbf{w}$'s, find $(\mathbf{w})_B$ and verify that $\mathbf{w} = A(\mathbf{w})_B$, where $A = [\mathbf{v}_1\ \mathbf{v}_2\ \mathbf{v}_3\ \mathbf{v}_4]$.

**i.** $B = \left\{ \begin{pmatrix} 1 \\ 1 \\ 1 \\ .5 \end{pmatrix}, \begin{pmatrix} 2 \\ 3 \\ 2 \\ 1 \end{pmatrix}, \begin{pmatrix} 3 \\ 2 \\ 4 \\ 1.5 \end{pmatrix}, \begin{pmatrix} 4 \\ 4 \\ 10 \\ 2.5 \end{pmatrix} \right\}$

$\mathbf{w} =$ **round(10*(2*rand(4,1)−1))**

**ii.** For $B$, generate four $4 \times 1$ random vectors. (Check that they form a basis.) For $\mathbf{w}$, generate a random $4 \times 1$ vector.

3. Let $B = \{\mathbf{v}_1, \mathbf{v}_2, \mathbf{v}_3, \mathbf{v}_4\}$ as in MATLAB Problem 2(a) in this section. Let

$$\mathbf{w}_1 = \begin{pmatrix} 1 \\ 0 \\ 0 \\ 0 \end{pmatrix} \quad \mathbf{w}_2 = \begin{pmatrix} 0 \\ 1 \\ 0 \\ 0 \end{pmatrix} \quad \mathbf{w}_3 = \begin{pmatrix} 0 \\ 0 \\ 1 \\ 0 \end{pmatrix} \quad \mathbf{w}_4 = \begin{pmatrix} 0 \\ 0 \\ 0 \\ 1 \end{pmatrix}$$

a. *(Paper and pencil)* Argue why if you find the **rref** of the matrix $[\mathbf{v}_1\ \mathbf{v}_2\ \mathbf{v}_3\ \mathbf{v}_4\ \mathbf{w}_1\ \mathbf{w}_2\ \mathbf{w}_3\ \mathbf{w}_4] = [\mathbf{v}_1\ \mathbf{v}_2\ \mathbf{v}_3\ \mathbf{v}_4$ **eye(4)**], then the 5th column of the **rref** is $(\mathbf{w}_1)_B$, the 6th column is $(\mathbf{w}_2)_B$, and so on.

b. Find $(\mathbf{w}_1)_B$, $(\mathbf{w}_2)_B$, $(\mathbf{w}_3)_B$, and $(\mathbf{w}_4)_B$. Form $C$, the matrix whose $i$th column equals $(\mathbf{w}_i)_B$. Verify that $C$ is equal to the inverse of $A = [\mathbf{v}_1\ \mathbf{v}_2\ \mathbf{v}_3\ \mathbf{v}_4]$. Use observations from part (a) to explain why.

c. Let $\mathbf{w} = \begin{pmatrix} 1 \\ -2 \\ 3 \\ 4 \end{pmatrix}$. Note that $\mathbf{w} = 1\mathbf{w}_1 + -2\mathbf{w}_2 + 3\mathbf{w}_3 + 4\mathbf{w}_4$

**i.** Solve $[A|\mathbf{w}] = [\mathbf{v}_1\ \mathbf{v}_2\ \mathbf{v}_3\ \mathbf{v}_4 \mid \mathbf{w}]$ to find $(\mathbf{w})_B$.

**ii.** Verify that $C\mathbf{w} = A^{-1}\mathbf{w} = (\mathbf{w})_B$. [Here $C$ is the matrix from part (b).]

**iii.** *(Paper and pencil)* $C$ is called the transition matrix from what to what? Using subpart (ii), and recalling what the columns of $C$ are, explain why

$$(\mathbf{w})_B = 1(\mathbf{w}_1)_B - 2(\mathbf{w}_2)_B + 3(\mathbf{w}_3)_B + 4(\mathbf{w}_4)_B$$

d. Repeat part (c) for the $B$ and $\mathbf{w}$ in MATLAB Problem 2 (d) (i) in this section.

4. a. Read MATLAB Problem 9 in Section 4.4. Explain why what was done there was finding coordinates of a polynomial in terms of the standard basis for polynomials.

b. Work Problems 14 through 16 in this section.

5. Let $B = \{\mathbf{v}_1, \mathbf{v}_2, \mathbf{v}_3\} = \left\{ \begin{pmatrix} 1 \\ 1 \\ 1 \end{pmatrix}, \begin{pmatrix} 2 \\ 3 \\ 3 \end{pmatrix}, \begin{pmatrix} -3 \\ 2 \\ 3 \end{pmatrix} \right\}$

   Let $C = \{\mathbf{w}_1, \mathbf{w}_2, \mathbf{w}_3\} = \left\{ \begin{pmatrix} 1 \\ 2 \\ 1 \end{pmatrix}, \begin{pmatrix} -1 \\ -1 \\ 0 \end{pmatrix}, \begin{pmatrix} 2 \\ 9 \\ 8 \end{pmatrix} \right\}$

   a. Verify that $B$ and $C$ are bases for $\mathbb{R}^3$. Let $W = [\mathbf{w}_1\ \mathbf{w}_2\ \mathbf{w}_3]$ and let $V = [\mathbf{v}_1\ \mathbf{v}_2\ \mathbf{v}_3]$.
   b. *(Paper and pencil)* Write the three systems of equations needed to express each vector in $B$ as a linear combination of vectors in $C$. Explain why the solutions to these systems can be found by solving the system(s) with extended augmented matrix $[\mathbf{w}_1\ \mathbf{w}_2\ \mathbf{w}_3|\mathbf{v}_1\ \mathbf{v}_2\ \mathbf{v}_3]$.
   c. Solve the system(s) to find $(\mathbf{v}_1)_C$, $(\mathbf{v}_2)_C$, and $(\mathbf{v}_3)_C$ and form the matrix $D = [(\mathbf{v}_1)_C\ (\mathbf{v}_2)_C\ (\mathbf{v}_3)_C]$.
   d. Let $\mathbf{x} = \begin{pmatrix} 1 \\ -2 \\ -3 \end{pmatrix}$. Find $(\mathbf{x})_B$ and $(\mathbf{x})_C$. Verify that $(\mathbf{x})_C = D(\mathbf{x})_B$.

      Repeat for a random $3 \times 1$ vector $\mathbf{x}$.
   e. With $W$ and $V$ as in part (a), find $W^{-1}V$ and compare it with $D$.
   f. Repeat parts (a) through (e) with

   $$B = \left\{ \begin{pmatrix} 1 \\ 2 \\ 1 \\ 0 \end{pmatrix}, \begin{pmatrix} 2 \\ 5 \\ 3 \\ -2 \end{pmatrix}, \begin{pmatrix} 3 \\ 5 \\ 3 \\ 2 \end{pmatrix}, \begin{pmatrix} 4 \\ 8 \\ 9 \\ 1 \end{pmatrix} \right\}, \qquad C = \left\{ \begin{pmatrix} 1 \\ 1 \\ 1 \\ .5 \end{pmatrix}, \begin{pmatrix} 2 \\ 3 \\ 2 \\ 1 \end{pmatrix}, \begin{pmatrix} 3 \\ 2 \\ 4 \\ 1.5 \end{pmatrix}, \begin{pmatrix} 4 \\ 4 \\ 10 \\ 2.5 \end{pmatrix} \right\},$$

   where $\mathbf{x}$ is a random $4 \times 1$ vector.
   g. *(Paper and pencil)* Explain why $W^{-1}V = D$ in two ways:
      **i.** Based on the process of solving $[W|V]$ to find $D$.
      **ii.** By interpreting $W^{-1}$ and $V$ as transition matrices involving the standard basis.

6. Using the facts learned in MATLAB Problem 5 in this section:
   a. Work Problems 18 through 20 in the text.
   b. Generate a random basis $B$ for $\mathbb{R}^5$ and a random basis $C$ for $\mathbb{R}^5$. Find the transition matrix, $T$, from $B$ to $C$. Verify your answer by generating a random vector $\mathbf{x}$ in $\mathbb{R}^5$, finding $(\mathbf{x})_B$ and $(\mathbf{x})_C$ and showing that $T(\mathbf{x})_B = (\mathbf{x})_C$.

7. Let $B$ and $C$ be as in MATLAB Problem 5(a) in this section. Let $D$ be the basis:

   $$\left\{ \begin{pmatrix} 2 \\ 8 \\ 5 \end{pmatrix}, \begin{pmatrix} 4 \\ 7 \\ 3 \end{pmatrix}, \begin{pmatrix} .5 \\ 1 \\ .5 \end{pmatrix} \right\}$$

   a. Find $T$, the transition matrix from $B$ to $C$. Find $S$, the transition matrix from $C$ to $D$. Find $K$, the transition matrix from $B$ to $D$.
   b. Conjecture a way to find $K$ from $T$ and $S$. Test your conjecture. Explain your reasoning.
   c. Repeat parts (a) and (b) for three random bases ($B$, $C$, and $D$) for $\mathbb{R}^4$.

8. Let $B = \{\mathbf{v}_1, \mathbf{v}_2, \mathbf{v}_3\} = \left\{\begin{pmatrix}1\\1\\1\end{pmatrix}, \begin{pmatrix}2\\3\\3\end{pmatrix}, \begin{pmatrix}-3\\2\\3\end{pmatrix}\right\}$. Let $A = \begin{pmatrix}5 & -6 & 4\\3 & -19 & 19\\3 & -24 & 24\end{pmatrix}$.
   a. Verify that $A\mathbf{v}_1 = 3\mathbf{v}_1$, $A\mathbf{v}_2 = 2\mathbf{v}_2$, and $A\mathbf{v}_3 = 5\mathbf{v}_3$.
   b. Suppose $\mathbf{x} = -1\mathbf{v}_1 + 2\mathbf{v}_2 + 4\mathbf{v}_3$. Note that $(\mathbf{x})_B = \begin{pmatrix}-1\\2\\4\end{pmatrix}$. Find $\mathbf{z} = A\mathbf{x}$, then find $(\mathbf{z})_B$, and verify that $(\mathbf{z})_B = D(\mathbf{x})_B$, where $D = \begin{pmatrix}3 & 0 & 0\\0 & 2 & 0\\0 & 0 & 5\end{pmatrix}$.
   c. Let $\mathbf{x} = a\mathbf{v}_1 + b\mathbf{v}_2 + c\mathbf{v}_3$. Repeat part (b) for three choices of $a$, $b$, and $c$.
   d. Let $V = [\mathbf{v}_1\ \mathbf{v}_2\ \mathbf{v}_3]$. Show that $A = VDV^{-1}$.
   e. Repeat parts (a) through (d) for

$$B = \left\{\begin{pmatrix}1\\2\\1\end{pmatrix}, \begin{pmatrix}-1\\-1\\0\end{pmatrix}, \begin{pmatrix}2\\9\\8\end{pmatrix}\right\} \qquad A = \begin{pmatrix}37 & -33 & 28\\48.5 & -44.5 & 38.5\\12 & -12 & 11\end{pmatrix}$$

   Verify $A\mathbf{v}_1 = -\mathbf{v}_1$, $A\mathbf{v}_2 = 4\mathbf{v}_2$, and $A\mathbf{v}_3 = .5\mathbf{v}_3$ and use

$$D = \begin{pmatrix}-1 & 0 & 0\\0 & 4 & 0\\0 & 0 & .5\end{pmatrix}$$

   f. *(Paper and pencil)* Suppose $B = \{\mathbf{v}_1, \mathbf{v}_2, \mathbf{v}_3\}$ is a basis and $A\mathbf{v}_1 = r\mathbf{v}_1$, $A\mathbf{v}_2 = s\mathbf{v}_2$, and $A\mathbf{v}_3 = t\mathbf{v}_3$. Suppose $\mathbf{x} = a\mathbf{v}_1 + b\mathbf{v}_2 + c\mathbf{v}_3$. Prove that $(\mathbf{z})_B = D(\mathbf{x})_B$, where $\mathbf{z} = A\mathbf{x}$ and $D = \begin{pmatrix}r & 0 & 0\\0 & s & 0\\0 & 0 & t\end{pmatrix}$.

   Using this fact and thinking in terms of transition matrices, explain why $A = VDV^{-1}$, where $V = [\mathbf{v}_1\ \mathbf{v}_2\ \mathbf{v}_3]$.

9. **Change of Basis by Rotation in $\mathbb{R}^2$** Let $\mathbf{e}_1$ and $\mathbf{e}_2$ denote the standard basis for $\mathbb{R}^2$, where $\mathbf{e}_1$ is a unit vector along the $x$-axis and $\mathbf{e}_2$ is a unit vector along the $y$-axis. If we rotate the axes counterclockwise by an angle $\theta$ around the origin, then $\mathbf{e}_1$ rotates to a vector $\mathbf{v}_1$ and $\mathbf{e}_2$ rotates to a vector $\mathbf{v}_2$ such that $\{\mathbf{v}_1, \mathbf{v}_2\}$ is a basis for $\mathbb{R}^2$.
   a. *(Paper and pencil)* Show that

$$\mathbf{v}_1 = \begin{pmatrix}\cos(\theta)\\ \sin(\theta)\end{pmatrix} \quad \text{and} \quad \mathbf{v}_2 = \begin{pmatrix}-\sin(\theta)\\ \cos(\theta)\end{pmatrix}$$

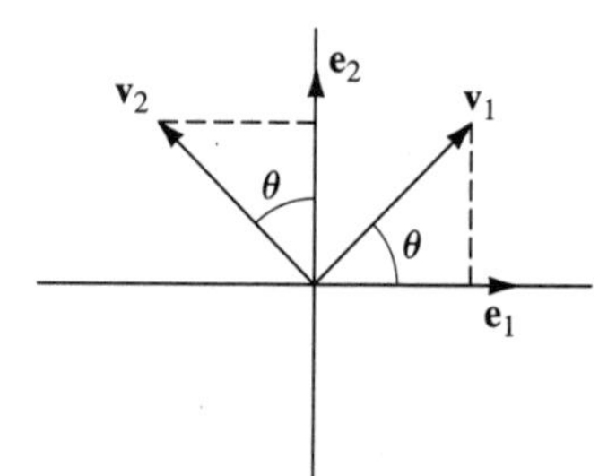

b. Let $V = [\mathbf{v}_1\ \mathbf{v}_2]$. Then $\mathbf{v}_1 = V\mathbf{e}_1$ and $\mathbf{v}_2 = V\mathbf{e}_2$. We will explore the geometry of $\mathbf{w} = a\mathbf{v}_1 + b\mathbf{v}_2$, that is, the geometry of linear combinations in terms of the new basis. We are interested in the relationship of the linear combinations to the rotation.

Suppose $\mathbf{x} = a\mathbf{e}_1 + b\mathbf{e}_2$. Then $\mathbf{w} = a\mathbf{v}_1 + b\mathbf{v}_2 = V\mathbf{x}$ represents the vector $\mathbf{x}$ rotated counterclockwise by an angle $\theta$ around the origin.

The MATLAB coding below helps you visualize this geometry. It plots vectors as line segments starting at the origin. The vector $\mathbf{x}$ is plotted in red and the vector $\mathbf{w}$ is plotted in blue. Observe how $\mathbf{w}$ (the blue vector) is the rotation counterclockwise by $\theta$ of $\mathbf{v}$ (the red vector). If you are using MATLAB 4.0, enter the **plot** command first and then the two **axis** commands. View the graph after the **axis** commands.

*Caution.* A hard copy of the graph produced by a screen-dump will not show equal lengths or show right angles as right angles.

```
a = 1;b = 2;                          (looking at v1 + 2v2)
x = [a;b];M = sqrt(x'*x);
th = pi/2;                            (assuming the angle θ = π/2)
v1 = [cos(th);sin(th)]
v2 = [−sin(th);cos(th)]
V = [v1 v2]
w = V*x
axis('square')
axis([ −M M −M M])
plot([0 x(1)], [0 x(2)],'r', [0 w(1)], [0 w(2)], 'b')
grid
```

Repeat the preceding instructions, modifying the values for $a$ and $b$.

Repeat the preceding instructions for $\theta = -\pi/2,\ \pi/4,\ -\pi/4,\ 2\pi/3$, and an angle of your choice. For each angle, make two choices of $a$ and $b$. When you are finished with this part enter the command **axis** to clear the axis scale. If using MATLAB 4.0, use **axis('auto')**.

c. Let's say that a basis has **orientation given by $\boldsymbol{\theta}$** if it is a basis obtained by rotating the standard basis counterclockwise about the origin by an angle of $\theta$.

Suppose $\{\mathbf{v}_1, \mathbf{v}_2\}$ is a basis with orientation given by $\theta$. Suppose $\mathbf{v}_1$ and $\mathbf{v}_2$ represent directions of sensors for some tracking device. The tracking device records the location of an object as coordinates with respect to the basis $\{\mathbf{v}_1, \mathbf{v}_2\}$. If two devices have different orientations, how can one device use information gathered by another? This involves translating coordinates in terms of one basis into coordinates in terms of another basis.

**i.** Suppose $B = \{\mathbf{v}_1, \mathbf{v}_2\}$ is a basis with orientation given by $\pi/4$ and $C = \{\mathbf{w}_1, \mathbf{w}_2\}$ is a basis with orientation given by $2\pi/3$. Find the transition matrix $T$ from the basis $B$ to the basis $C$. Find the transition matrix $S$ from the basis $C$ to the basis $B$. (*Note.* Lines 3, 4, and 5 in the MATLAB code in part (b) give an example of finding a basis with orientation $\pi/2$.)

**ii.** Suppose the device with orientation given by $\pi/4$ locates an object in its coordinates as [.5; 3]. Find the coordinates of the object with respect to the device oriented by $2\pi/3$. Explain your process. Check your result by finding the standard coordinates of the object using the [.5; 3] coordinates for the first basis $B$ and finding the standard coordinates of the object using the coordinates found for the second basis $C$.

**iii.** Suppose the device with orientation given by $2\pi/3$ locates an object in its coordinates as [2; −1.4]. Find the coordinates of the object with respect to the device oriented by $\pi/4$. Explain your process. Check your answer as in subpart (ii).

**iv.** The m-file *rotcoor.m* on the accompanying disk helps visualize the process above. The format is **rotcoor(E,F,c),** where each of $E$ and $F$ is a $2 \times 2$ matrix whose columns form a basis for $\mathbb{R}^2$ and $\mathbf{c}$ is a $2 \times 1$ matrix representing the coordinates of a vector with respect to the basis given by $E$. Two images will be displayed on the screen: One shows the vector as a linear combination of the basis vectors given by $E$ and the other shows the vector as a linear combination of the basis vectors given by $F$. After each use of **rotcoor,** be sure to enter the command **clg** or **clf** if using MATLAB 4.0.

Use this m-file to visualize the results from subparts (ii) and (iii). Check your answers from subparts (ii) and (iii) using the information in the displays. For example, in (ii), $E$ will be the basis for the orientation of $\pi/4$, $F$ the basis for the orientation of $2\pi/3$, and $\mathbf{c} = [.5; 3]$.

**10. Change of Basis by Rotations in $\mathbb{R}^3$; Pitch, Yaw, Roll**

a. *(Paper and pencil)* In $\mathbb{R}^3$, one can rotate counterclockwise around the $x$-axis, the $y$-axis, or the $z$-axis (the $x$-, $y$-, and $z$-axes form a right-handed coordinate system). Let $\mathbf{e}_1$, $\mathbf{e}_2$, and $\mathbf{e}_3$ be the unit vectors in the standard basis in the direction of positive $x$, positive $y$, and positive $z$, respectively.

**i.** A rotation counterclockwise by an angle $\theta$ around the $z$-axis will produce a basis $\{\mathbf{v}, \mathbf{w}, \mathbf{e}_3\}$, where $\mathbf{v}$ is the vector obtained by rotating $\mathbf{e}_1$ and $\mathbf{w}$ is the vector by rotating $\mathbf{e}_2$. Using the diagrams below as guides, show that

$$\mathbf{v} = \begin{pmatrix} \cos(\theta) \\ \sin(\theta) \\ 0 \end{pmatrix} \quad \text{and} \quad \mathbf{w} = \begin{pmatrix} -\sin(\theta) \\ \cos(\theta) \\ 0 \end{pmatrix}$$

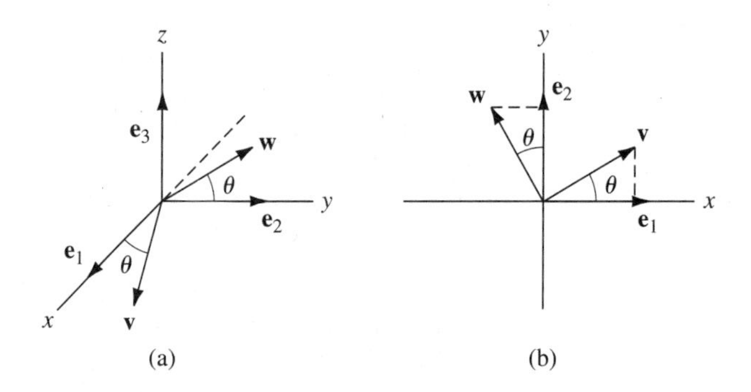


Let $Y = [\mathbf{v}\ \mathbf{w}\ \mathbf{e}_3]$. Interpret $Y$ as a transition matrix.

**ii.** A rotation counterclockwise by an angle $\alpha$ around the $x$-axis will produce a basis $\{\mathbf{e}_1, \mathbf{v}, \mathbf{w}\}$, where $\mathbf{v}$ is the vector obtained by rotating $\mathbf{e}_2$ and $\mathbf{w}$ is the vector obtained by rotating $\mathbf{e}_3$. Using the diagrams below as a guide, show that

$$\mathbf{v} = \begin{pmatrix} 0 \\ \cos(\alpha) \\ \sin(\alpha) \end{pmatrix} \quad \text{and} \quad \mathbf{w} = \begin{pmatrix} 0 \\ -\sin(\alpha) \\ \cos(\alpha) \end{pmatrix}$$

(a) (b)

Let $R = [\mathbf{e}_1\ \mathbf{v}\ \mathbf{w}]$. Interpret $R$ as a transition matrix.

**iii.** A rotation counterclockwise by an angle $\varphi$ around the $y$-axis will produce a basis $\{\mathbf{v}, \mathbf{e}_2, \mathbf{w}\}$, where $\mathbf{v}$ is the vector obtained by rotating $\mathbf{e}_1$ and $\mathbf{w}$ is the vector obtained by rotating $\mathbf{e}_3$. Using the diagrams below as a guide, show that

$$\mathbf{v} = \begin{pmatrix} \cos(\varphi) \\ 0 \\ -\sin(\varphi) \end{pmatrix} \quad \text{and} \quad \mathbf{w} = \begin{pmatrix} \sin(\varphi) \\ 0 \\ \cos(\varphi) \end{pmatrix}$$

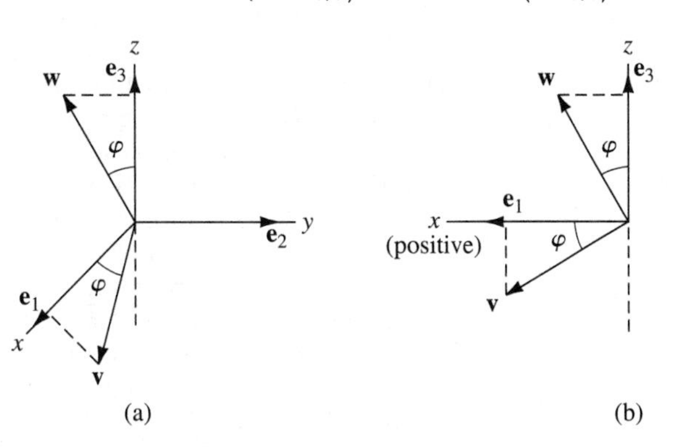


(a) (b)

Let $P = [\mathbf{v}\ \mathbf{e}_2\ \mathbf{w}]$. Interpret $P$ as a transition matrix.

**b.** *(Paper and pencil)* Assume $Y$ is a matrix obtained as in part (a) (i) for an angle $\theta$, $R$ is a matrix obtained as in part (a) (ii) for an angle $\alpha$, and $P$ is a matrix obtained as in part (a) (iii) for an angle $\varphi$.

The matrices $Y$, $R$, and $P$ for any given angles have geometric interpretations similar to the geometric interpretation of a rotation matrix in $\mathbb{R}^2$. Let $M$ be any one of these rotation matrices. Let $\mathbf{u} = a\mathbf{e}_1 + b\mathbf{e}_2 + c\mathbf{e}_3$. Then $\mathbf{r} = M\mathbf{u}$ will give the standard coordinates of the vector obtained by rotating the vector $\mathbf{u}$.

Using this geometric interpretation, explain why the matrix $YR$ would represent rotation by $\alpha$ counterclockwise around the $x$-axis followed by rotation by $\theta$ counterclockwise around the $z$-axis.

What matrix would represent rotation by $\theta$ counterclockwise around the $z$-axis followed by rotation by $\alpha$ counterclockwise around the $x$-axis? Do you expect this matrix to give the same result as the matrix in the previous paragraph? Why or why not?

c. The rotations we have been talking about are useful for describing the **attitude** of a spacecraft (or airplane). The attitude is the rotational orientation of the spacecraft about its center. Here we assume that a spacecraft has a set of axes through its center of mass such that the $x$- and $y$-axes are at right angles to each other (such as one axis running from the back to the front of the craft and the other from side to side) and a $z$-axis perpendicular to the $x$- and $y$-axes to form a right-handed system.

Corrections in attitude can be made by performing rotations of the form described in part (a). Without some form of attitude control, a satellite begins to spin wildly. A rotation around the $z$-axis is called a **yaw** maneuver, a rotation around the $x$-axis is called a **roll** maneuver, and a rotation around the $y$-axis is called a **pitch** maneuver.

Suppose the craft's set of axes are initially aligned with a fixed reference system (axes representing a standard basis). The attitude of a craft can be given by a matrix whose columns are unit vectors in the directions of the axes associated with the craft.

**i.** Find the matrix that represents the craft's attitude after performing a pitch maneuver of angle $\pi/4$, followed by a roll maneuver of angle $-\pi/3$, and then followed by a yaw maneuver of angle $\pi/2$.

**ii.** Perform the same maneuvers in a different order and compare the attitudes. (Describe the order of the maneuvers.)

**iii.** Repeat for another choice of angles for each type of maneuver; that is, find the attitudes from performing the maneuvers in two different orders (describing the orders) and compare the attitudes.

d. Suppose two satellites with different attitudes wish to transfer information to each other. Each satellite records information in terms of its coordinate system; that is, it records information as coordinates with respect to the basis of unit vectors defining its axes system. Besides the adjustment for different locations (which is a simple translation), a transfer of information requires using a transition matrix from the coordinates of one of the satellites to the coordinates of the other.

**i.** Consider two spacecrafts where one is oriented as in part (c) (i) and the other is oriented as in part (c) (ii). Suppose the first craft records the location of an object as **p = [.2;.3;−1]**. Translate this information into the coordinate system of the second craft. Check the result by finding the standard coordinates of the object using the first craft's reading and then finding the standard coordinates of the object using the second craft's adjusted reading.

**ii.** Repeat for two spacecrafts whose orientations were generated in part (c) (iii).

e. **Optional** Suppose your craft has an attitude matrix given by **A = orth(rand(3)).** Experiment with pitch, yaw, and roll maneuvers to realign the craft with the fixed reference system (standard basis).

**PROJECT PROBLEM**

11. Combine MATLAB Problems 9 and 10 in this section.

## 4.9 ORTHONORMAL BASES AND PROJECTIONS IN $\mathbb{R}^n$

In $\mathbb{R}^n$ we saw that $n$ linearly independent vectors constitute a basis. The most commonly used basis is the standard basis $E = \{\mathbf{e}_1, \mathbf{e}_2, \ldots, \mathbf{e}_n\}$. These vectors have two properties:

**i.** $\mathbf{e}_i \cdot \mathbf{e}_j = 0 \quad \text{if } i \neq j$

**ii.** $\mathbf{e}_i \cdot \mathbf{e}_i = 1$

**DEFINITION 1**

**Orthonormal Set in $\mathbb{R}^n$** A set of vectors $S = \{\mathbf{u}_1, \mathbf{u}_2, \ldots, \mathbf{u}_k\}$ in $\mathbb{R}^n$ is said to be an **orthonormal set** if

$$\mathbf{u}_i \cdot \mathbf{u}_j = 0 \quad \text{if } i \neq j \tag{1}$$

$$\mathbf{u}_i \cdot \mathbf{u}_i = 1 \tag{2}$$

If only equation (1) is satisfied, the set is called **orthogonal.**

Since we shall be working with the scalar product extensively in this section, let us recall some basic facts (see Theorem 1.6.1, page 63). Without mentioning them again explicitly, we shall use these facts often in the rest of this section.

If $\mathbf{u}$, $\mathbf{v}$, and $\mathbf{w}$ are in $\mathbb{R}^n$ and $\alpha$ is a real number, then

$$\mathbf{u} \cdot \mathbf{v} = \mathbf{v} \cdot \mathbf{u} \tag{3}$$

$$(\mathbf{u} + \mathbf{v}) \cdot \mathbf{w} = \mathbf{u} \cdot \mathbf{w} + \mathbf{v} \cdot \mathbf{w} \tag{4}$$

$$\mathbf{u} \cdot (\mathbf{v} + \mathbf{w}) = \mathbf{u} \cdot \mathbf{v} + \mathbf{u} \cdot \mathbf{w} \tag{5}$$

$$(\alpha\mathbf{u}) \cdot \mathbf{v} = \alpha(\mathbf{u} \cdot \mathbf{v}) \tag{6}$$

$$\mathbf{u} \cdot (\alpha\mathbf{v}) = \alpha(\mathbf{u} \cdot \mathbf{v}) \tag{7}$$

We now give another useful definition.

**DEFINITION 2**

**Length or Norm of a Vector** If $\mathbf{v} \in \mathbb{R}^n$, then the **length** or **norm** of $\mathbf{v}$, written $|\mathbf{v}|$, is given by

$$|\mathbf{v}| = \sqrt{\mathbf{v} \cdot \mathbf{v}} \tag{8}$$

*Note.* If $\mathbf{v} = (x_1, x_2, \ldots, x_n)$, then $\mathbf{v} \cdot \mathbf{v} = x_1^2 + x_2^2 + \cdots + x_n^2$. This means that

$$\mathbf{v} \cdot \mathbf{v} \geq 0 \quad \text{and} \quad \mathbf{v} \cdot \mathbf{v} = 0 \quad \text{if and only if } \mathbf{v} = \mathbf{0} \tag{9}$$

Thus we can take the square root in (8), and we have

$$|\mathbf{v}| = \sqrt{\mathbf{v} \cdot \mathbf{v}} \geq 0 \qquad \text{for every } \mathbf{v} \in \mathbb{R}^n \tag{10}$$

$$|\mathbf{v}| = 0 \qquad \text{if and only if } \mathbf{v} = \mathbf{0} \tag{11}$$

**EXAMPLE 1** **The Norm of a Vector in $\mathbb{R}^2$** Let $\mathbf{v} = (x, y) \in \mathbb{R}^2$. Then $|\mathbf{v}| = \sqrt{x^2 + y^2}$ conforms to our usual definition of length of a vector in the plane (see Equation 3.1.1, page 394).

**EXAMPLE 2** **The Norm of a Vector in $\mathbb{R}^3$** If $\mathbf{v} = (x, y, z) \in \mathbb{R}^3$, then $|\mathbf{v}| = \sqrt{x^2 + y^2 + z^2}$ as in Section 3.3.

**EXAMPLE 3** **The Norm of a Vector in $\mathbb{R}^5$** If $\mathbf{v} = (2, -1, 3, 4, -6) \in \mathbb{R}^5$, then $|\mathbf{v}| = \sqrt{4 + 1 + 9 + 16 + 36} = \sqrt{66}$.

We can now restate Definition 1:

> A set of vectors is orthonormal if any pair of them is orthogonal and each has length 1.

Orthonormal sets of vectors are reasonably easy to work with. We shall see an example of this characteristic in Chapter 5. Now we prove that any finite orthogonal set of nonzero vectors is linearly independent.

**THEOREM 1** If $S = \{\mathbf{v}_1, \mathbf{v}_2, \ldots, \mathbf{v}_k\}$ is an orthogonal set of nonzero vectors, then $S$ is linearly independent.

**Proof** Suppose that $c_1\mathbf{v}_1 + c_2\mathbf{v}_2 + \cdots + c_n\mathbf{v}_k = \mathbf{0}$. Then for any $i = 1, 2, \ldots, k$

$$\begin{aligned} 0 = \mathbf{0} \cdot \mathbf{v}_i &= (c_1\mathbf{v}_1 + c_2\mathbf{v}_2 + \cdots + c_i\mathbf{v}_i + \cdots + c_k\mathbf{v}_k) \cdot \mathbf{v}_i \\ &= c_1(\mathbf{v}_1 \cdot \mathbf{v}_i) + c_2(\mathbf{v}_2 \cdot \mathbf{v}_i) + \cdots + c_i(\mathbf{v}_i \cdot \mathbf{v}_i) + \cdots + c_k(\mathbf{v}_k \cdot \mathbf{v}_i) \\ &= c_1 0 + c_2 0 + \cdots + c_i\,|\mathbf{v}_i|^2 + \cdots + c_k 0 = c_i\,|\mathbf{v}_i|^2 \end{aligned}$$

Since $\mathbf{v}_i \neq 0$ by hypothesis, $|\mathbf{v}_i|^2 > 0$ and we have $c_i = 0$. This is true for $i = 1, 2, \ldots, k$ and the proof is complete.

We now see how *any* basis in $\mathbb{R}^n$ can be "turned into" an orthonormal basis. The method described below is called the **Gram-Schmidt orthonormalization process.**†

**THEOREM 2** **Gram-Schmidt Orthonormalization Process** Let $H$ be an $m$-dimensional subspace of $\mathbb{R}^n$. Then $H$ has an orthonormal basis.‡

**Proof** Let $S = \{\mathbf{v}_1, \mathbf{v}_2, \ldots, \mathbf{v}_m\}$ be a basis for $H$. We shall prove the theorem by constructing an orthonormal basis from the vectors in $S$. Before giving the steps in this construction, we note the simple fact that a linearly independent set of vectors does *not* contain the zero vector (see Problem 21).

**Step 1. Choosing the First Unit Vector** Let

$$\mathbf{u}_1 = \frac{\mathbf{v}_1}{|\mathbf{v}_1|} \tag{12}$$

Then

$$\mathbf{u}_1 \cdot \mathbf{u}_1 = \left(\frac{\mathbf{v}_1}{|\mathbf{v}_1|}\right) \cdot \left(\frac{\mathbf{v}_1}{|\mathbf{v}_1|}\right) = \left(\frac{1}{|\mathbf{v}_1|^2}\right)(\mathbf{v}_1 \cdot \mathbf{v}_1) = 1$$

so that $|\mathbf{u}_1| = 1$.

**Step 2. Choosing a Second Vector Orthogonal to $\mathbf{u}_1$** In Section 3.2 (Theorem 5 on page 244) we saw that, in $\mathbb{R}^2$, the vector $\mathbf{w} = \mathbf{u} - \dfrac{\mathbf{u} \cdot \mathbf{v}}{|\mathbf{v}|^2}\mathbf{v}$ is orthogonal to $\mathbf{v}$. Here $\dfrac{\mathbf{u} \cdot \mathbf{v}}{|\mathbf{v}|^2}\mathbf{v}$ is the projection of $\mathbf{u}$ on $\mathbf{v}$. This is illustrated in Figure 4.5.

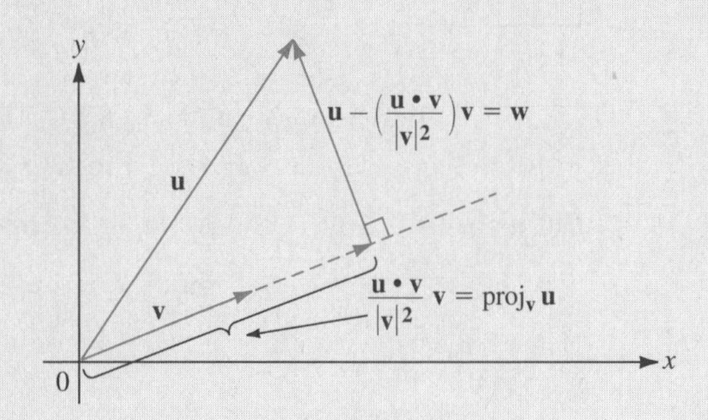


**Figure 4.5** The vector $\mathbf{w} = \mathbf{u} - \dfrac{\mathbf{u} \cdot \mathbf{v}}{|\mathbf{v}|^2}\mathbf{v}$ is orthogonal to $\mathbf{v}$

---

† Jörgen Pederson Gram (1850–1916) was a Danish actuary who was very interested in the science of measurement. Erhardt Schmidt (1876–1959) was a German mathematician.

‡ Note that $H$ may be $\mathbb{R}^n$ in this theorem. That is, $\mathbb{R}^n$ itself has an orthonormal basis.

It turns out that the vector $\mathbf{w}$ given above is orthogonal to $\mathbf{v}$ when $\mathbf{w}$ and $\mathbf{v}$ are in $\mathbb{R}^n$ for any $n \geq 2$. Note here that as $\mathbf{u}_1$ is a unit vector, $\dfrac{\mathbf{v} \cdot \mathbf{u}}{|\mathbf{u}_1|^2}\mathbf{u}_1 = (\mathbf{v} \cdot \mathbf{u}_1)\mathbf{u}_1$ for any vector $\mathbf{v}$.

Let

$$\mathbf{v}_2' = \mathbf{v}_2 - (\mathbf{v}_2 \cdot \mathbf{u}_1)\mathbf{u}_1 \tag{13}$$

Then

$$\mathbf{v}_2' \cdot \mathbf{u}_1 = \mathbf{v}_2 \cdot \mathbf{u}_1 - (\mathbf{v}_2 \cdot \mathbf{u}_1)(\mathbf{u}_1 \cdot \mathbf{u}_1) = \mathbf{v}_2 \cdot \mathbf{u}_1 - (\mathbf{v}_2 \cdot \mathbf{u}_1)1 = 0$$

so that $\mathbf{v}_2'$ is orthogonal to $\mathbf{u}_1$. Moreover, by Theorem 1, $\mathbf{u}_1$ and $\mathbf{v}_2'$ are linearly independent. $\mathbf{v}_2' \neq \mathbf{0}$ because otherwise $\mathbf{v}_2 = (\mathbf{v}_2 \cdot \mathbf{u}_1)\mathbf{u}_1 = \dfrac{(\mathbf{v}_2 \cdot \mathbf{u}_1)}{|\mathbf{v}_1|}\mathbf{v}_1$, contradicting the independence of $\mathbf{v}_1$ and $\mathbf{v}_2$.

**Step 3. Choosing a Second Unit Vector** Let

$$\mathbf{u}_2 = \frac{\mathbf{v}_2'}{|\mathbf{v}_2'|} \tag{14}$$

Then clearly $\{\mathbf{u}_1, \mathbf{u}_2\}$ is an orthonormal set.

Suppose now that the vectors $\mathbf{u}_1, \mathbf{u}_2, \ldots, \mathbf{u}_k$ $(k < m)$ have been constructed and form an orthonormal set. We show how to construct $\mathbf{u}_{k+1}$.

**Step 4. Continuing the Process** Let

$$\mathbf{v}_{k+1}' = \mathbf{v}_{k+1} - (\mathbf{v}_{k+1} \cdot \mathbf{u}_1)\mathbf{u}_1 - (\mathbf{v}_{k+1} \cdot \mathbf{u}_2)\mathbf{u}_2 - \cdots - (\mathbf{v}_{k+1} \cdot \mathbf{u}_k)\mathbf{u}_k \tag{15}$$

Then for $i = 1, 2, \ldots, k$

$$\begin{aligned}\mathbf{v}_{k+1}' \cdot \mathbf{u}_i = \mathbf{v}_{k+1} \cdot \mathbf{u}_i &- (\mathbf{v}_{k+1} \cdot \mathbf{u}_1)(\mathbf{u}_1 \cdot \mathbf{u}_i) - (\mathbf{v}_{k+1} \cdot \mathbf{u}_2)(\mathbf{u}_2 \cdot \mathbf{u}_i)\\ &- \cdots - (\mathbf{v}_{k+1} \cdot \mathbf{u}_i)(\mathbf{u}_i \cdot \mathbf{u}_i) - \cdots - (\mathbf{v}_{k+1} \cdot \mathbf{u}_k)(\mathbf{u}_k \cdot \mathbf{u}_i)\end{aligned}$$

But $\mathbf{u}_j \cdot \mathbf{u}_i = 0$ if $j \neq i$ and $\mathbf{u}_i \cdot \mathbf{u}_i = 1$. Thus

$$\mathbf{v}_{k+1}' \cdot \mathbf{u}_i = \mathbf{v}_{k+1} \cdot \mathbf{u}_i - \mathbf{v}_{k+1} \cdot \mathbf{u}_i = 0$$

Hence $\{\mathbf{u}_1, \mathbf{u}_2, \ldots, \mathbf{u}_k, \mathbf{v}_{k+1}'\}$ is an orthogonal, linearly independent set and $\mathbf{v}_{k+1}' \neq \mathbf{0}$.

**Step 5.** Let $\mathbf{u}_{k+1} = \mathbf{v}_{k+1}'/|\mathbf{v}_{k+1}'|$. Then clearly $\{\mathbf{u}_1, \mathbf{u}_2, \ldots, \mathbf{u}_k, \mathbf{u}_{k+1}\}$ is an orthonormal set, and we continue in this manner until $k + 1 = m$ and the proof is complete.

*Note.* Since each $\mathbf{u}_i$ is a linear combination of $\mathbf{v}_i$'s, span $\{\mathbf{u}_1, \mathbf{u}_2, \ldots, \mathbf{u}_k\}$ is a subspace of span $\{\mathbf{v}_1, \mathbf{v}_2, \ldots, \mathbf{v}_n\}$ and since each space has dimension $k$, the spaces are equal. ■

**EXAMPLE 4** **Constructing an Orthonormal Basis in $\mathbb{R}^3$** Construct an orthonormal basis in $\mathbb{R}^3$ starting with the basis $\{\mathbf{v}_1, \mathbf{v}_2, \mathbf{v}_3\} = \left\{\begin{pmatrix}1\\1\\0\end{pmatrix}, \begin{pmatrix}0\\1\\1\end{pmatrix}, \begin{pmatrix}1\\0\\1\end{pmatrix}\right\}$.

**Solution** We have $|\mathbf{v}_1| = \sqrt{2}$, so $\mathbf{u}_1 = \begin{pmatrix}1/\sqrt{2}\\1/\sqrt{2}\\0\end{pmatrix}$. Then

$$\mathbf{v}_2' = \mathbf{v}_2 - (\mathbf{v}_2 \cdot \mathbf{u}_1)\mathbf{u}_1 = \begin{pmatrix}0\\1\\1\end{pmatrix} - \frac{1}{\sqrt{2}}\begin{pmatrix}1/\sqrt{2}\\1/\sqrt{2}\\0\end{pmatrix} = \begin{pmatrix}0\\1\\1\end{pmatrix} - \begin{pmatrix}\frac{1}{2}\\\frac{1}{2}\\0\end{pmatrix} = \begin{pmatrix}-\frac{1}{2}\\\frac{1}{2}\\1\end{pmatrix}$$

Since $|\mathbf{v}_2'| = \sqrt{3/2}$, $\mathbf{u}_2 = \sqrt{2/3}\begin{pmatrix}-\frac{1}{2}\\\frac{1}{2}\\1\end{pmatrix} = \begin{pmatrix}-1/\sqrt{6}\\1/\sqrt{6}\\2/\sqrt{6}\end{pmatrix}$. Continuing, we have

$$\begin{aligned}\mathbf{v}_3' &= \mathbf{v}_3 - (\mathbf{v}_3 \cdot \mathbf{u}_1)\mathbf{u}_1 - (\mathbf{v}_3 \cdot \mathbf{u}_2)\mathbf{u}_2 \\ &= \begin{pmatrix}1\\0\\1\end{pmatrix} - \frac{1}{\sqrt{2}}\begin{pmatrix}1/\sqrt{2}\\1/\sqrt{2}\\0\end{pmatrix} - \frac{1}{\sqrt{6}}\begin{pmatrix}-1/\sqrt{6}\\1/\sqrt{6}\\2/\sqrt{6}\end{pmatrix} = \begin{pmatrix}1\\0\\1\end{pmatrix} - \begin{pmatrix}\frac{1}{2}\\\frac{1}{2}\\0\end{pmatrix} - \begin{pmatrix}-\frac{1}{6}\\\frac{1}{6}\\\frac{2}{6}\end{pmatrix} = \begin{pmatrix}\frac{2}{3}\\-\frac{2}{3}\\\frac{2}{3}\end{pmatrix}\end{aligned}$$

Finally, $|\mathbf{v}_3'| = \sqrt{12/9} = 2/\sqrt{3}$ so that $\mathbf{u}_3 = \frac{\sqrt{3}}{2}\begin{pmatrix}2/3\\-2/3\\2/3\end{pmatrix} = \begin{pmatrix}1/\sqrt{3}\\-1/\sqrt{3}\\1/\sqrt{3}\end{pmatrix}$. Thus an orthonormal basis in $\mathbb{R}^3$ is $\left\{\begin{pmatrix}1/\sqrt{2}\\1/\sqrt{2}\\0\end{pmatrix}, \begin{pmatrix}-1/\sqrt{6}\\1/\sqrt{6}\\2/\sqrt{6}\end{pmatrix}, \begin{pmatrix}1/\sqrt{3}\\-1/\sqrt{3}\\1/\sqrt{3}\end{pmatrix}\right\}$. This result should be checked.

**EXAMPLE 5** **Finding an Orthonormal Basis for a Subspace of $\mathbb{R}^3$** Find an orthonormal basis for the set of vectors in $\mathbb{R}^3$ lying on the plane $\pi = \left\{\begin{pmatrix}x\\y\\z\end{pmatrix}: 2x - y + 3z = 0\right\}$.

**Solution** As we saw in Example 4.6.3, page 338, a basis for this two-dimensional subspace is $\mathbf{v}_1 = \begin{pmatrix}1\\2\\0\end{pmatrix}$ and $\mathbf{v}_2 = \begin{pmatrix}0\\3\\1\end{pmatrix}$. Then $|\mathbf{v}_1| = \sqrt{5}$ and $\mathbf{u}_1 = \mathbf{v}_1/|\mathbf{v}_1| = \begin{pmatrix}1/\sqrt{5}\\2/\sqrt{5}\\0\end{pmatrix}$. Continuing, we define

$$\mathbf{v}_2' = \mathbf{v}_2 - (\mathbf{v}_2 \cdot \mathbf{u}_1)\mathbf{u}_1$$
$$= \begin{pmatrix} 0 \\ 3 \\ 1 \end{pmatrix} - \frac{6}{\sqrt{5}} \begin{pmatrix} 1/\sqrt{5} \\ 2/\sqrt{5} \\ 0 \end{pmatrix} = \begin{pmatrix} 0 \\ 3 \\ 1 \end{pmatrix} - \begin{pmatrix} \frac{6}{5} \\ \frac{12}{5} \\ 0 \end{pmatrix} = \begin{pmatrix} -\frac{6}{5} \\ \frac{3}{5} \\ 1 \end{pmatrix}$$

Finally, $|\mathbf{v}_2'| = \sqrt{70/25} = \sqrt{70}/5$ so that $\mathbf{u}_2 = \mathbf{v}_2'/|\mathbf{v}_2'| = \frac{5}{\sqrt{70}} \begin{pmatrix} -\frac{6}{5} \\ \frac{3}{5} \\ 1 \end{pmatrix} = \begin{pmatrix} -6/\sqrt{70} \\ 3/\sqrt{70} \\ 5/\sqrt{70} \end{pmatrix}$. Thus an orthonormal basis is $\left\{ \begin{pmatrix} 1/\sqrt{5} \\ 2/\sqrt{5} \\ 0 \end{pmatrix}, \begin{pmatrix} -6/\sqrt{70} \\ 3/\sqrt{70} \\ 5/\sqrt{70} \end{pmatrix} \right\}$. To check this answer, we note that (1) the vectors are orthogonal, (2) each has length 1, and (3) each satisfies $2x - y + 3z = 0$.

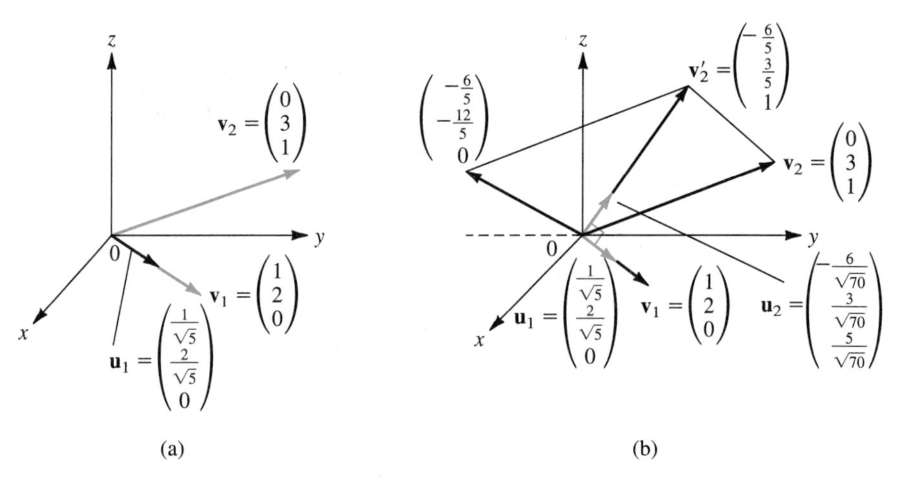


**Figure 4.6** The vectors $\mathbf{u}_1$ and $\mathbf{u}_2$ form an orthonormal basis for the plane spanned by the vectors $\mathbf{v}_1$ and $\mathbf{v}_2$

In Figure 4.6*a* we draw the vectors $\mathbf{v}_1$, $\mathbf{v}_2$, and $\mathbf{u}_1$. In Figure 4.6*b* we draw the vector $-\begin{pmatrix} \frac{6}{5} \\ \frac{12}{5} \\ 0 \end{pmatrix} = \begin{pmatrix} -\frac{6}{5} \\ -\frac{12}{5} \\ 0 \end{pmatrix}$ and add it to $\mathbf{v}_2$ using the parallelogram rule to obtain $\mathbf{v}_2' = \begin{pmatrix} -\frac{6}{5} \\ \frac{3}{5} \\ 1 \end{pmatrix}$. Finally, $\mathbf{u}_2$ lies 1 unit along the vector $\mathbf{v}_2'$.

We now define a new kind of matrix that will be very useful in later chapters.

**DEFINITION 3** **Orthogonal Matrix** An $n \times n$ matrix $Q$ is called **orthogonal** if $Q$ is invertible and

$$Q^{-1} = Q^t \tag{16}$$

Note that if $Q^{-1} = Q^t$, then $Q^tQ = I$.

Orthogonal matrices are not difficult to construct, according to the next theorem.

**THEOREM 3** The $n \times n$ matrix $Q$ is orthogonal if and only if the columns of $Q$ form an orthonormal basis for $\mathbb{R}^n$.

**Proof** Let

$$Q = \begin{pmatrix} a_{11} & a_{12} & \cdots & a_{1n} \\ a_{21} & a_{22} & \cdots & a_{2n} \\ \vdots & \vdots & & \vdots \\ a_{n1} & a_{n2} & \cdots & a_{nn} \end{pmatrix}$$

Then

$$Q^t = \begin{pmatrix} a_{11} & a_{21} & \cdots & a_{n1} \\ a_{12} & a_{22} & \cdots & a_{n2} \\ \vdots & \vdots & & \vdots \\ a_{1n} & a_{2n} & \cdots & a_{nn} \end{pmatrix}$$

Let $B = (b_{ij}) = Q^tQ$. Then

$$b_{ij} = a_{1i}a_{1j} + a_{2i}a_{2j} + \cdots + a_{ni}a_{nj} = \mathbf{c}_i \cdot \mathbf{c}_j \tag{17}$$

where $\mathbf{c}_i$ denotes the $i$th column of $Q$. If the columns of $Q$ are orthonormal, then

$$b_{ij} = \begin{cases} 0 & \text{if } i \neq j \\ 1 & \text{if } i = j \end{cases} \tag{18}$$

That is, $B = I$. Conversely, if $Q^t = Q^{-1}$, then $B = I$ so that (18) holds and (17) shows that the columns of $Q$ are orthonormal. This completes the proof. ■

**EXAMPLE 6** **An Orthogonal Matrix** From Example 4, the vectors $\begin{pmatrix} 1/\sqrt{2} \\ 1/\sqrt{2} \\ 0 \end{pmatrix}$, $\begin{pmatrix} 1/\sqrt{3} \\ 1/\sqrt{3} \\ 1/\sqrt{3} \end{pmatrix}$, $\begin{pmatrix} 1/\sqrt{3} \\ -1/\sqrt{3} \\ 1/\sqrt{3} \end{pmatrix}$ form an orthonormal basis in $\mathbb{R}^3$. Thus the matrix $Q =$

$\begin{pmatrix} 1/\sqrt{2} & -1/\sqrt{6} & 1/\sqrt{3} \\ 1/\sqrt{2} & 1/\sqrt{6} & -1/\sqrt{3} \\ 0 & 2/\sqrt{6} & 1/\sqrt{3} \end{pmatrix}$ is an orthogonal matrix. To check this, we note that

$$Q^tQ = \begin{pmatrix} 1/\sqrt{2} & 1/\sqrt{2} & 0 \\ -1/\sqrt{6} & 1/\sqrt{6} & 2/\sqrt{6} \\ 1/\sqrt{3} & -1/\sqrt{3} & 1/\sqrt{3} \end{pmatrix} \begin{pmatrix} 1/\sqrt{2} & -1/\sqrt{6} & 1/\sqrt{3} \\ 1/\sqrt{2} & 1/\sqrt{6} & -1/\sqrt{3} \\ 0 & 2/\sqrt{6} & 1/\sqrt{3} \end{pmatrix} = \begin{pmatrix} 1 & 0 & 0 \\ 0 & 1 & 0 \\ 0 & 0 & 1 \end{pmatrix}$$

In the proof of Theorem 2 we defined $\mathbf{v}_2' = \mathbf{v}_2 - (\mathbf{v}_2 \cdot \mathbf{u}_1)\mathbf{u}_1$. But as we have seen, $(\mathbf{v}_2 \cdot \mathbf{u}_1)\mathbf{u}_1 = \text{proj}_{\mathbf{u}_1} \mathbf{v}_2$ (since $|\mathbf{u}_1|^2 = 1$). We now extend this notion from projection onto a vector to projection onto a subspace.

**DEFINITION 4** **Orthogonal Projection** Let $H$ be a subspace of $\mathbb{R}^n$ with orthonormal basis $\{\mathbf{u}_1, \mathbf{u}_2, \ldots, \mathbf{u}_k\}$. If $\mathbf{v} \in \mathbb{R}^n$, then the **orthogonal projection** of $\mathbf{v}$ onto $H$, denoted by $\text{proj}_H \mathbf{v}$, is given by

$$\text{proj}_H \mathbf{v} = (\mathbf{v} \cdot \mathbf{u}_1)\mathbf{u}_1 + (\mathbf{v} \cdot \mathbf{u}_2)\mathbf{u}_2 + \cdots + (\mathbf{v} \cdot \mathbf{u}_k)\mathbf{u}_k \tag{19}$$

Note that $\text{proj}_H \mathbf{v} \in H$.

**EXAMPLE 7** **The Orthogonal Projection of a Vector onto a Plane** Find $\text{proj}_\pi \mathbf{v}$, where $\pi$ is the plane $\left\{\begin{pmatrix} x \\ y \\ z \end{pmatrix}: 2x - y + 3z = 0\right\}$ and $\mathbf{v}$ is the vector $\begin{pmatrix} 3 \\ -2 \\ 4 \end{pmatrix}$.

**Solution** From Example 5, an orthonormal basis for $\pi$ is $\mathbf{u}_1 = \begin{pmatrix} 1/\sqrt{5} \\ 2/\sqrt{5} \\ 0 \end{pmatrix}$ and $\mathbf{u}_2 = \begin{pmatrix} -6/\sqrt{70} \\ 3/\sqrt{70} \\ 5/\sqrt{70} \end{pmatrix}$. Then

$$\text{proj}_\pi \mathbf{v} = \left[\begin{pmatrix} 3 \\ -2 \\ 4 \end{pmatrix} \cdot \begin{pmatrix} 1/\sqrt{5} \\ 2/\sqrt{5} \\ 0 \end{pmatrix}\right] \begin{pmatrix} 1/\sqrt{5} \\ 2/\sqrt{5} \\ 0 \end{pmatrix} + \left[\begin{pmatrix} 3 \\ -2 \\ 4 \end{pmatrix} \cdot \begin{pmatrix} -6/\sqrt{70} \\ 3/\sqrt{70} \\ 5/\sqrt{70} \end{pmatrix}\right] \begin{pmatrix} -6/\sqrt{70} \\ 3/\sqrt{70} \\ 5/\sqrt{70} \end{pmatrix}$$

$$= -\frac{1}{\sqrt{5}} \begin{pmatrix} 1/\sqrt{5} \\ 2/\sqrt{5} \\ 0 \end{pmatrix} - \frac{4}{\sqrt{70}} \begin{pmatrix} -6/\sqrt{70} \\ 3/\sqrt{70} \\ 5/\sqrt{70} \end{pmatrix} = \begin{pmatrix} -\frac{1}{5} \\ -\frac{2}{5} \\ 0 \end{pmatrix} + \begin{pmatrix} \frac{24}{70} \\ -\frac{12}{70} \\ -\frac{20}{70} \end{pmatrix} = \begin{pmatrix} \frac{1}{7} \\ -\frac{4}{7} \\ -\frac{2}{7} \end{pmatrix}$$

The notion of projection gives us a convenient way to write a vector in $\mathbb{R}^n$ in terms of an orthonormal basis.

**THEOREM 4** Let $B = \{\mathbf{u}_1, \mathbf{u}_2, \ldots, \mathbf{u}_n\}$ be an orthonormal basis for $\mathbb{R}^n$ and let $\mathbf{v} \in \mathbb{R}^n$. Then

$$\boxed{\mathbf{v} = (\mathbf{v} \cdot \mathbf{u}_1)\mathbf{u}_1 + (\mathbf{v} \cdot \mathbf{u}_2)\mathbf{u}_2 + \cdots + (\mathbf{v} \cdot \mathbf{u}_n)\mathbf{u}_n} \tag{20}$$

That is, $\mathbf{v} = \text{proj}_{\mathbb{R}^n} \mathbf{v}$.

**Proof** Since $B$ is a basis, we can write $\mathbf{v}$ in a unique way as $\mathbf{v} = c_1\mathbf{u}_1 + c_2\mathbf{u}_2 + \cdots + c_n\mathbf{u}_n$. Then

$$\mathbf{v} \cdot \mathbf{u}_i = c_1(\mathbf{u}_1 \cdot \mathbf{u}_i) + c_2(\mathbf{u}_2 \cdot \mathbf{u}_i) + \cdots + c_i(\mathbf{u}_i \cdot \mathbf{u}_i) + \cdots + c_n(\mathbf{u}_n \cdot \mathbf{u}_i) = c_i$$

since the $\mathbf{u}_i$'s are orthonormal. Since this is true for $i = 1, 2, \ldots, n$, the proof is complete. ■

**EXAMPLE 8** **Writing a Vector in Terms of an Orthonormal Basis** Write the vector $\begin{pmatrix} 2 \\ -1 \\ 3 \end{pmatrix}$ in $\mathbb{R}^3$ in terms of the orthonormal basis $\left\{\begin{pmatrix} 1/\sqrt{2} \\ 1/\sqrt{2} \\ 0 \end{pmatrix}, \begin{pmatrix} -1/\sqrt{6} \\ 1/\sqrt{6} \\ 2/\sqrt{6} \end{pmatrix}, \begin{pmatrix} 1/\sqrt{3} \\ -1/\sqrt{3} \\ 1/\sqrt{3} \end{pmatrix}\right\}$.

**Solution**

$$\begin{pmatrix} 2 \\ -1 \\ 3 \end{pmatrix} = \left[\begin{pmatrix} 2 \\ -1 \\ 3 \end{pmatrix} \cdot \begin{pmatrix} 1/\sqrt{2} \\ 1/\sqrt{2} \\ 0 \end{pmatrix}\right]\begin{pmatrix} 1/\sqrt{2} \\ 1/\sqrt{2} \\ 0 \end{pmatrix} + \left[\begin{pmatrix} 2 \\ -1 \\ 3 \end{pmatrix} \cdot \begin{pmatrix} -1/\sqrt{6} \\ 1/\sqrt{6} \\ 2/\sqrt{6} \end{pmatrix}\right]\begin{pmatrix} -1/\sqrt{6} \\ 1/\sqrt{6} \\ 2/\sqrt{6} \end{pmatrix}$$

$$+ \left[\begin{pmatrix} 2 \\ -1 \\ 3 \end{pmatrix} \cdot \begin{pmatrix} 1/\sqrt{3} \\ -1/\sqrt{3} \\ 1/\sqrt{3} \end{pmatrix}\right]\begin{pmatrix} 1/\sqrt{3} \\ -1/\sqrt{3} \\ 1/\sqrt{3} \end{pmatrix}$$

$$= \frac{1}{\sqrt{2}}\begin{pmatrix} 1/\sqrt{2} \\ 1/\sqrt{2} \\ 0 \end{pmatrix} + \frac{3}{\sqrt{6}}\begin{pmatrix} -1/\sqrt{6} \\ 1/\sqrt{6} \\ 2/\sqrt{6} \end{pmatrix} + \frac{6}{\sqrt{3}}\begin{pmatrix} 1/\sqrt{3} \\ -1/\sqrt{3} \\ 1/\sqrt{3} \end{pmatrix}$$

■

Before continuing, we need to know that an orthogonal projection is well defined. By this we mean that the definition of $\text{proj}_H \mathbf{v}$ is independent of the orthonormal basis chosen in $H$. The following theorem takes care of this problem.

**THEOREM 5** Let $H$ be a subspace of $\mathbb{R}^n$. Suppose that $H$ has two orthonormal bases, $\{\mathbf{u}_1, \mathbf{u}_2, \ldots, \mathbf{u}_k\}$ and $\{\mathbf{w}_1, \mathbf{w}_2, \ldots, \mathbf{w}_k\}$. Let $\mathbf{v}$ be a vector in $\mathbb{R}^n$. Then

$$(\mathbf{v}\cdot\mathbf{u}_1)\mathbf{u}_1 + (\mathbf{v}\cdot\mathbf{u}_2)\mathbf{u}_2 + \cdots + (\mathbf{v}\cdot\mathbf{u}_k)\mathbf{u}_k$$
$$= (\mathbf{v}\cdot\mathbf{w}_1)\mathbf{w}_1 + (\mathbf{v}\cdot\mathbf{w}_2)\mathbf{w}_2 + \cdots + (\mathbf{v}\cdot\mathbf{w}_k)\mathbf{w}_k \qquad \textbf{(21)}$$

**Proof** Choose vectors $\mathbf{u}_{k+1}, \mathbf{u}_{k+2}, \ldots, \mathbf{u}_n$ such that $B_1 = \{\mathbf{u}_1, \mathbf{u}_2, \ldots, \mathbf{u}_k, \mathbf{u}_{k+1}, \ldots, \mathbf{u}_n\}$ is an orthonormal basis for $\mathbb{R}^n$ (this can be done as in the proof of Theorem 2).† Then $B_2 = \{\mathbf{w}_1, \mathbf{w}_2, \ldots, \mathbf{w}_k, \mathbf{u}_{k+1}, \mathbf{u}_{k+2}, \ldots, \mathbf{u}_n\}$ is also an orthonormal basis for $\mathbb{R}^n$. To see this, note first that none of the vectors $\mathbf{u}_{k+1}, \mathbf{u}_{k+2}, \ldots, \mathbf{u}_n$ can be written as a linear combination of $\mathbf{w}_1, \mathbf{w}_2, \ldots, \mathbf{w}_k$ because none of these vectors is in $H$ and $\{\mathbf{w}_1, \mathbf{w}_2, \ldots, \mathbf{w}_k\}$ is a basis for $H$. Thus $B_2$ is a basis for $\mathbb{R}^n$ because it contains $n$ linearly independent vectors. The orthonormality of the vectors in $B_2$ follows from the way these vectors were chosen ($\mathbf{u}_{k+j}$ is orthogonal to every vector in $H$ for $j = 1, 2, \ldots, n-k$). Let $\mathbf{v}$ be a vector in $\mathbb{R}^n$. Then from Theorem 4 [equation (20)]

$$\mathbf{v} = (\mathbf{v}\cdot\mathbf{u}_1)\mathbf{u}_1 + (\mathbf{v}\cdot\mathbf{u}_2)\mathbf{u}_2 + \cdots + (\mathbf{v}\cdot\mathbf{u}_k)\mathbf{u}_k + (\mathbf{v}\cdot\mathbf{u}_{k+1})\mathbf{u}_{k+1} + \cdots + (\mathbf{v}\cdot\mathbf{u}_n)\mathbf{u}_n$$
$$= (\mathbf{v}\cdot\mathbf{w}_1)\mathbf{w}_1 + (\mathbf{v}\cdot\mathbf{w}_2)\mathbf{w}_2 + \cdots + (\mathbf{v}\cdot\mathbf{w}_k)\mathbf{w}_k + (\mathbf{v}\cdot\mathbf{u}_{k+1})\mathbf{u}_{k+1} + \cdots + (\mathbf{v}\cdot\mathbf{u}_n)\mathbf{u}_n \qquad \textbf{(22)}$$

Equation (21) now follows from equation (22). ■

**DEFINITION 5** **Orthogonal Complement** Let $H$ be a subspace of $\mathbb{R}^n$. Then the **orthogonal complement** of $H$, denoted by $H^\perp$, is given by

$$H^\perp = \{\mathbf{x} \in \mathbb{R}^n\colon\ \mathbf{x}\cdot\mathbf{h} = 0 \quad \text{for every } \mathbf{h} \in H\}$$

**THEOREM 6** If $H$ is a subspace of $\mathbb{R}^n$, then

**i.** $H^\perp$ is a subspace of $\mathbb{R}^n$.

**ii.** $H \cap H^\perp = \{\mathbf{0}\}$.

**iii.** $\dim H^\perp = n - \dim H$.

**Proof** **i.** If $\mathbf{x}$ and $\mathbf{y}$ are in $H^\perp$ and if $\mathbf{h} \in H$, then $(\mathbf{x}+\mathbf{y})\cdot\mathbf{h} = \mathbf{x}\cdot\mathbf{h} + \mathbf{y}\cdot\mathbf{h} = 0 + 0 = 0$ and $(\alpha\mathbf{x}\cdot\mathbf{h}) = \alpha(\mathbf{x}\cdot\mathbf{h}) = 0$, so $H^\perp$ is a subspace.

† First, we must find vectors $\mathbf{v}_{k+1}, \mathbf{v}_{k+2}, \ldots, \mathbf{v}_n$ such that $\{\mathbf{u}_1, \ldots, \mathbf{u}_k, \mathbf{v}_{k+1}, \ldots, \mathbf{v}_n\}$ is a basis for $\mathbb{R}^2$. This can be done as in the proof of Theorem 4.6.4 on page 341; see also Problem 4.6.27.

**ii.** If $\mathbf{x} \in H \cap H^\perp$, then $\mathbf{x} \cdot \mathbf{x} = 0$, so $\mathbf{x} = \mathbf{0}$, which shows that $H \cap H^\perp = \{\mathbf{0}\}$.

**iii.** Let $\{\mathbf{u}_1, \mathbf{u}_2, \ldots, \mathbf{u}_k\}$ be an orthonormal basis for $H$. By the result of Problem 4.6.27, page 346, this can be expanded into a basis $B$ for $\mathbb{R}^n$: $B = \{\mathbf{u}_1, \mathbf{u}_2, \ldots, \mathbf{u}_k, \mathbf{v}_{k+1}, \ldots, \mathbf{v}_n\}$. Using the Gram-Schmidt process, we can turn $B$ into an orthonormal basis for $\mathbb{R}^n$. As in the proof of Theorem 2, the already orthonormal $\mathbf{u}_1, \mathbf{u}_2, \ldots, \mathbf{u}_k$ will remain unchanged in the process, and we obtain the orthonormal basis $B_1 = \{\mathbf{u}_1, \mathbf{u}_2, \ldots, \mathbf{u}_k, \mathbf{u}_{k+1}, \ldots, \mathbf{u}_n\}$. To complete the proof, we need only show that $\{\mathbf{u}_{k+1}, \ldots, \mathbf{u}_n\}$ is a basis for $H^\perp$. Since the $\mathbf{u}_i$'s are independent, we must show that they span $H^\perp$. Let $\mathbf{x} \in H^\perp$; then by Theorem 4

$$\mathbf{x} = (\mathbf{x} \cdot \mathbf{u}_1)\mathbf{u}_1 + (\mathbf{x} \cdot \mathbf{u}_2)\mathbf{u}_2 + \cdots + (\mathbf{x} \cdot \mathbf{u}_k)\mathbf{u}_k + (\mathbf{x} \cdot \mathbf{u}_{k+1})\mathbf{u}_{k+1} + \cdots + (\mathbf{x} \cdot \mathbf{u}_n)\mathbf{u}_n$$

But $(\mathbf{x} \cdot \mathbf{u}_i) = 0$ for $i = 1, 2, \ldots, k$, since $\mathbf{x} \in H^\perp$ and $\mathbf{u}_i \in H$. Thus $\mathbf{x} = (\mathbf{x} \cdot \mathbf{u}_{k+1})\mathbf{u}_{k+1} + \cdots + (\mathbf{x} \cdot \mathbf{u}_n)\mathbf{u}_n$. This shows that $\{\mathbf{u}_{k+1}, \ldots, \mathbf{u}_n\}$ is a basis for $H^\perp$, which means that $\dim H^\perp = n - k$. ■

The spaces $H$ and $H^\perp$ allow us to "decompose" any vector in $\mathbb{R}^n$.

**THEOREM 7** **Projection Theorem** Let $H$ be a subspace of $\mathbb{R}^n$ and let $\mathbf{v} \in \mathbb{R}^n$. Then there exists a unique pair of vectors $\mathbf{h}$ and $\mathbf{p}$ such that $\mathbf{h} \in H$, $\mathbf{p} \in H^\perp$, and $\mathbf{v} = \mathbf{h} + \mathbf{p}$. In particular, $\mathbf{h} = \text{proj}_H \mathbf{v}$ and $\mathbf{p} = \text{proj}_{H^\perp} \mathbf{v}$ so that

$$\boxed{\mathbf{v} = \mathbf{h} + \mathbf{p} = \text{proj}_H \mathbf{v} + \text{proj}_{H^\perp} \mathbf{v}} \tag{23}$$

**Proof** Let $\mathbf{h} = \text{proj}_H \mathbf{v}$ and let $\mathbf{p} = \mathbf{v} - \mathbf{h}$. By Definition 4 we have $\mathbf{h} \in H$. We now show that $\mathbf{p} \in H^\perp$. Let $\{\mathbf{u}_1, \mathbf{u}_2, \ldots, \mathbf{u}_k\}$ be an orthonormal basis for $H$. Then

$$\mathbf{h} = (\mathbf{v} \cdot \mathbf{u}_1)\mathbf{u}_1 + (\mathbf{v} \cdot \mathbf{u}_2)\mathbf{u}_2 + \cdots + (\mathbf{v} \cdot \mathbf{u}_k)\mathbf{u}_k$$

Let $\mathbf{x}$ be a vector in $H$. There exist constants $\alpha_1, \alpha_2, \ldots, \alpha_k$ such that

$$\mathbf{x} = \alpha_1\mathbf{u}_1 + \alpha_2\mathbf{u}_2 + \cdots + \alpha_k\mathbf{u}_k$$

Then

$$\mathbf{p} \cdot \mathbf{x} = (\mathbf{v} - \mathbf{h}) \cdot \mathbf{x} = [\mathbf{v} - (\mathbf{v} \cdot \mathbf{u}_1)\mathbf{u}_1 - (\mathbf{v} \cdot \mathbf{u}_2)\mathbf{u}_2 - \cdots - (\mathbf{v} \cdot \mathbf{u}_k)\mathbf{u}_k] \cdot [\alpha_1\mathbf{u}_1 + \alpha_2\mathbf{u}_2 + \cdots + \alpha_k\mathbf{u}_k] \tag{24}$$

Since $\mathbf{u}_i \cdot \mathbf{u}_j = \begin{cases} 0, & i \neq j \\ 1, & i = j \end{cases}$, it is easy to verify that the scalar product in (24) is given by

$$\mathbf{p} \cdot \mathbf{x} = \sum_{i=1}^{k} \alpha_i(\mathbf{v} \cdot \mathbf{u}_i) - \sum_{i=1}^{k} \alpha_i(\mathbf{v} \cdot \mathbf{u}_i) = 0$$

Thus $\mathbf{p} \cdot \mathbf{x} = 0$ for every $\mathbf{x} \in H$, which means that $\mathbf{p} \in H^\perp$. To show that $\mathbf{p} = \text{proj}_{H^\perp} \mathbf{v}$, we extend $\{\mathbf{u}_1, \mathbf{u}_2, \ldots, \mathbf{u}_k\}$ to an orthonormal basis for $\mathbb{R}^n$: $\{\mathbf{u}_1, \mathbf{u}_2, \ldots, \mathbf{u}_k, \mathbf{u}_{k+1}, \ldots, \mathbf{u}_n\}$. Then $\{\mathbf{u}_{k+1}, \ldots, \mathbf{u}_n\}$ is a basis for $H^\perp$ and, by Theorem 4,

$$\begin{aligned} \mathbf{v} &= (\mathbf{v} \cdot \mathbf{u}_1)\mathbf{u}_1 + (\mathbf{v} \cdot \mathbf{u}_2)\mathbf{u}_2 + \cdots + (\mathbf{v} \cdot \mathbf{u}_k)\mathbf{u}_k + (\mathbf{v} \cdot \mathbf{u}_{k+1})\mathbf{u}_{k+1} \\ &\quad + \cdots + (\mathbf{v} \cdot \mathbf{u}_n)\mathbf{u}_n \\ &= \text{proj}_H \mathbf{v} + \text{proj}_{H^\perp} \mathbf{v} \qquad \text{(by Definition 4)} \end{aligned}$$

This proves equation (23). To prove uniqueness, suppose that $\mathbf{v} = \mathbf{h}_1 - \mathbf{p}_1 = \mathbf{h}_2 - \mathbf{p}_2$, where $\mathbf{h}_1, \mathbf{h}_2 \in H$ and $\mathbf{p}_1, \mathbf{p}_2 \in H^\perp$. Then $\mathbf{h}_1 - \mathbf{h}_2 = \mathbf{p}_1 - \mathbf{p}_2$. But $\mathbf{h}_1 - \mathbf{h}_2 \in H$ and $\mathbf{p}_1 - \mathbf{p}_2 \in H^\perp$, so $\mathbf{h}_1 - \mathbf{h}_2 \in H \cap H^\perp = \{\mathbf{0}\}$. Thus $\mathbf{h}_1 - \mathbf{h}_2 = \mathbf{0}$ and $\mathbf{p}_1 - \mathbf{p}_2 = \mathbf{0}$, which completes the proof. ■

**EXAMPLE 9** **Decomposing a Vector in $\mathbb{R}^3$** In $\mathbb{R}^3$ let $\pi = \left\{\begin{pmatrix} x \\ y \\ z \end{pmatrix} : 2x - y + 3z = 0\right\}$. Write the vector $\begin{pmatrix} 3 \\ -2 \\ 4 \end{pmatrix}$ as $\mathbf{h} + \mathbf{p}$, where $\mathbf{h} \in \pi$ and $\mathbf{p} \in \pi^\perp$.

**Solution** An orthonormal basis for $\pi$ is $B_1 = \left\{\begin{pmatrix} 1/\sqrt{5} \\ 2/\sqrt{5} \\ 0 \end{pmatrix}, \begin{pmatrix} -6/\sqrt{70} \\ 3/\sqrt{70} \\ 5/\sqrt{70} \end{pmatrix}\right\}$, and from Example 7, $\mathbf{h} = \text{proj}_\pi \mathbf{v} = \begin{pmatrix} \frac{1}{7} \\ -\frac{4}{7} \\ -\frac{2}{7} \end{pmatrix} \in \pi$. Then

$$\mathbf{p} = \mathbf{v} - \mathbf{h} = \begin{pmatrix} 3 \\ -2 \\ 4 \end{pmatrix} - \begin{pmatrix} \frac{1}{7} \\ -\frac{4}{7} \\ -\frac{2}{7} \end{pmatrix} = \begin{pmatrix} \frac{20}{7} \\ -\frac{10}{7} \\ \frac{30}{7} \end{pmatrix} \in \pi^\perp$$

Note that $\mathbf{p} \cdot \mathbf{h} = 0$. ■

The following theorem is very useful in statistics and other applied areas. We shall provide one application of this theorem in the next section and apply an extended version of this result in Section 4.11.

**THEOREM 8** **Norm Approximation Theorem** Let $H$ be a subspace of $\mathbb{R}^n$ and let $\mathbf{v}$ be a vector in $\mathbb{R}^n$. Then $\text{proj}_H \mathbf{v}$ is the best approximation to $\mathbf{v}$ in $H$ in the following sense: If $\mathbf{h}$ is any other vector in $H$, then

$$|\mathbf{v} - \text{proj}_H \mathbf{v}| < |\mathbf{v} - \mathbf{h}| \qquad \textbf{(25)}$$

**Proof** From Theorem 7, $\mathbf{v} - \text{proj}_H \mathbf{v} \in H^\perp$. We write

$$\mathbf{v} - \mathbf{h} = (\mathbf{v} - \text{proj}_H \mathbf{v}) + (\text{proj}_H \mathbf{v} - \mathbf{h})$$

The first term on the right is in $H^\perp$, while the second is in $H$, so

$$(\mathbf{v} - \text{proj}_H \mathbf{v}) \cdot (\text{proj}_H \mathbf{v} - \mathbf{h}) = 0 \qquad \textbf{(26)}$$

Now

$$\begin{aligned}
|\mathbf{v} - \mathbf{h}|^2 &= (\mathbf{v} - \mathbf{h}) \cdot (\mathbf{v} - \mathbf{h}) \\
&= [(\mathbf{v} - \text{proj}_H \mathbf{v}) + (\text{proj}_H \mathbf{v} - \mathbf{h})] \cdot [(\mathbf{v} - \text{proj}_H \mathbf{v}) + (\text{proj}_H \mathbf{v} - \mathbf{h})] \\
&= |\mathbf{v} - \text{proj}_H \mathbf{v}|^2 + 2(\mathbf{v} - \text{proj}_H \mathbf{v}) \cdot (\text{proj}_H \mathbf{v} - \mathbf{h}) + |\text{proj}_H \mathbf{v} - \mathbf{h}|^2 \\
&= |\mathbf{v} - \text{proj}_H \mathbf{v}|^2 + |\text{proj}_H \mathbf{v} - \mathbf{h}|^2
\end{aligned}$$

But $|\text{proj}_H \mathbf{v} - \mathbf{h}|^2 > 0$ because $\mathbf{h} \neq \text{proj}_H \mathbf{v}$. Hence

$$|\mathbf{v} - \mathbf{h}|^2 > |\mathbf{v} - \text{proj}_H \mathbf{v}|^2$$

or

$$|\mathbf{v} - \mathbf{h}| > |\mathbf{v} - \text{proj}_H \mathbf{v}|$$ ■

**Orthogonal Bases in $\mathbb{R}^3$ with Integer Coefficients and Integer Norms**

It is sometimes useful to construct an orthogonal basis of vectors where the coordinates and norm of each vector is an integer. For example,

$$\left\{ \begin{pmatrix} 2 \\ 2 \\ -1 \end{pmatrix}, \begin{pmatrix} 2 \\ -1 \\ 2 \end{pmatrix}, \begin{pmatrix} -1 \\ 2 \\ 2 \end{pmatrix} \right\}$$

constitutes an orthogonal basis in $\mathbb{R}^3$ where each vector has norm 3. As another example,

$$\left\{ \begin{pmatrix} 12 \\ 4 \\ -3 \end{pmatrix}, \begin{pmatrix} 0 \\ 3 \\ 4 \end{pmatrix}, \begin{pmatrix} -25 \\ 48 \\ -36 \end{pmatrix} \right\}$$

is an orthogonal basis in $\mathbb{R}^3$ whose vectors have norms 13, 5, and 65, respectively. Finding bases like this in $\mathbb{R}^3$ turns out to be not so difficult as you might imagine. A discussion of this topic appears in the interesting paper "Orthogonal Bases of $\mathbb{R}^3$ with Integer Coordinates and Integer Lengths" by Anthony Osborne and Hans Liebeck in *The American Mathematical Monthly,* Volume 96, Number 1, January 1989, pp. 49–53.

We close this section with an important theorem.

**THEOREM 9** **Cauchy-Schwarz Inequality in $\mathbb{R}^n$** Let $\mathbf{u}$ and $\mathbf{v}$ be vectors in $\mathbb{R}^n$. Then

**i.** $|\mathbf{u}\cdot\mathbf{v}| \le |\mathbf{u}|\,|\mathbf{v}|$. **(27)**

**ii.** $|\mathbf{u}\cdot\mathbf{v}| = |\mathbf{u}|\,|\mathbf{v}|$ if and only if $\mathbf{u}=\mathbf{0}$ or $\mathbf{v}=\lambda\mathbf{u}$ for some real number $\lambda$.

**Proof** **i.** If $\mathbf{u}=\mathbf{0}$ or $\mathbf{v}=\mathbf{0}$ (or both), then (27) holds (both sides are equal to 0). We assume that $\mathbf{u}\ne\mathbf{0}$ and $\mathbf{v}\ne\mathbf{0}$. Then

$$0 \le \left|\frac{\mathbf{u}}{|\mathbf{u}|}-\frac{\mathbf{v}}{|\mathbf{v}|}\right|^2 = \left(\frac{\mathbf{u}}{|\mathbf{u}|}-\frac{\mathbf{v}}{|\mathbf{v}|}\right)\cdot\left(\frac{\mathbf{u}}{|\mathbf{u}|}-\frac{\mathbf{v}}{|\mathbf{v}|}\right) = \frac{\mathbf{u}\cdot\mathbf{u}}{|\mathbf{u}|^2}-\frac{2\mathbf{u}\cdot\mathbf{v}}{|\mathbf{u}|\,|\mathbf{v}|}+\frac{\mathbf{v}\cdot\mathbf{v}}{|\mathbf{v}|^2}$$
$$= \frac{|\mathbf{u}|^2}{|\mathbf{u}|^2}-\frac{2\mathbf{u}\cdot\mathbf{v}}{|\mathbf{u}|\,|\mathbf{v}|}+\frac{|\mathbf{v}|^2}{|\mathbf{v}|^2} = 2-\frac{2\mathbf{u}\cdot\mathbf{v}}{|\mathbf{u}|\,|\mathbf{v}|}$$

Thus $\dfrac{2\mathbf{u}\cdot\mathbf{v}}{|\mathbf{u}|\,|\mathbf{v}|}\le 2$, so $\dfrac{\mathbf{u}\cdot\mathbf{v}}{|\mathbf{u}|\,|\mathbf{v}|}\le 1$ and $\mathbf{u}\cdot\mathbf{v}\le|\mathbf{u}|\,|\mathbf{v}|$. Similarly, starting with $0 \le \left|\dfrac{\mathbf{u}}{|\mathbf{u}|}+\dfrac{\mathbf{v}}{|\mathbf{v}|}\right|^2$, we end up with $\dfrac{\mathbf{u}\cdot\mathbf{v}}{|\mathbf{u}|\,|\mathbf{v}|}\ge -1$ or $\mathbf{u}\cdot\mathbf{v}\ge -|\mathbf{u}|\,|\mathbf{v}|$. Putting these together, we obtain

$$-|\mathbf{u}|\,|\mathbf{v}| \le \mathbf{u}\cdot\mathbf{v} \le |\mathbf{u}|\,|\mathbf{v}| \quad\text{or}\quad |\mathbf{u}\cdot\mathbf{v}| \le |\mathbf{u}|\,|\mathbf{v}|$$

**ii.** If $\mathbf{u}=\lambda\mathbf{v}$, then $|\mathbf{u}\cdot\mathbf{v}| = |\lambda\mathbf{v}\cdot\mathbf{v}| = |\lambda|\,|\mathbf{v}|^2$ and $|\mathbf{u}|\,|\mathbf{v}| = |\lambda\mathbf{v}|\,|\mathbf{v}| = |\lambda|\,|\mathbf{v}|\,|\mathbf{v}| = |\lambda|\,|\mathbf{v}|^2 = |\mathbf{u}\cdot\mathbf{v}|$. Conversely, suppose that $|\mathbf{u}\cdot\mathbf{v}| = |\mathbf{u}|\,|\mathbf{v}|$ with $\mathbf{u}\ne 0$ and $\mathbf{v}\ne\mathbf{0}$. Then $\left|\dfrac{\mathbf{u}\cdot\mathbf{v}}{|\mathbf{u}|\,|\mathbf{v}|}\right| = 1$, so $\dfrac{\mathbf{u}\cdot\mathbf{v}}{|\mathbf{u}|\,|\mathbf{v}|} = \pm 1$

*Case 1:* $\dfrac{\mathbf{u}\cdot\mathbf{v}}{|\mathbf{u}|\,|\mathbf{v}|} = 1$. Then

$$\left|\frac{\mathbf{u}}{|\mathbf{u}|}-\frac{\mathbf{v}}{|\mathbf{v}|}\right|^2 = \left(\frac{\mathbf{u}}{|\mathbf{u}|}-\frac{\mathbf{v}}{|\mathbf{v}|}\right)\cdot\left(\frac{\mathbf{u}}{|\mathbf{u}|}-\frac{\mathbf{v}}{|\mathbf{v}|}\right) \overset{\text{as in (i)}}{=} 2-\frac{2\mathbf{u}\cdot\mathbf{v}}{|\mathbf{u}|\,|\mathbf{v}|} = 2-2 = 0.$$

Thus

$$\frac{\mathbf{u}}{|\mathbf{u}|} = \frac{\mathbf{v}}{|\mathbf{v}|} \quad\text{or}\quad \mathbf{u} = \frac{|\mathbf{u}|}{|\mathbf{v}|}\mathbf{v} = \lambda\mathbf{v}$$

*Case 2:* $\dfrac{\mathbf{u}\cdot\mathbf{v}}{|\mathbf{u}|\,|\mathbf{v}|} = -1$. Then

$$\left|\frac{\mathbf{u}}{|\mathbf{u}|} + \frac{\mathbf{v}}{|\mathbf{v}|}\right|^2 = 2 + \frac{2\mathbf{u}\cdot\mathbf{v}}{|\mathbf{u}|\,|\mathbf{v}|} = 2 - 2 = 0$$

so

$$\frac{\mathbf{u}}{|\mathbf{u}|} = -\frac{\mathbf{v}}{|\mathbf{v}|} \quad\text{and}\quad \mathbf{u} = -\frac{|\mathbf{u}|}{|\mathbf{v}|}\mathbf{v} = \lambda\mathbf{v}$$

## PROBLEMS 4.9

### Self-Quiz

#### *True-False*

**I.** The set $\{(1, 1), (1, -1)\}$ is an orthonormal set in $\mathbb{R}^2$.

**II.** The set $\left\{\left(\frac{1}{\sqrt{2}}, \frac{1}{\sqrt{2}}\right), \left(\frac{1}{\sqrt{2}}, \frac{-1}{\sqrt{2}}\right)\right\}$ is an orthonormal set in $\mathbb{R}^2$.

**III.** Every basis in $\mathbb{R}^n$ can be turned into an orthonormal basis by using the Gram-Schmidt orthonormalization process.

**IV.** The matrix $\begin{pmatrix} 1 & 1 \\ 1 & -1 \end{pmatrix}$ is orthogonal.

**V.** The matrix $\begin{pmatrix} 1/\sqrt{2} & 1/\sqrt{2} \\ 1/\sqrt{2} & -1/\sqrt{2} \end{pmatrix}$ is orthogonal.

#### *Multiple Choice*

**VI.** For which of the following matrices is $Q^{-1}$ equal to $Q^t$?

**a.** $\begin{pmatrix} 1 & 6 \\ 3 & -2 \end{pmatrix}$ **b.** $\begin{pmatrix} 1/\sqrt{10} & 6/\sqrt{40} \\ 3/\sqrt{10} & 2/\sqrt{40} \end{pmatrix}$

**c.** $\begin{pmatrix} 1/\sqrt{10} & 6/\sqrt{40} \\ 3/\sqrt{10} & -2/\sqrt{40} \end{pmatrix}$ **d.** $\begin{pmatrix} 1 & 6 \\ 3 & 2 \end{pmatrix}$

In Problems 1–13 construct an orthonormal basis for the given vector space or subspace.

**1.** In $\mathbb{R}^2$, starting with the basis vectors $\begin{pmatrix} 1 \\ 1 \end{pmatrix}, \begin{pmatrix} -1 \\ 1 \end{pmatrix}$

**2.** $H = \{(x, y) \in \mathbb{R}^2\colon x + y = 0\}$

**3.** $H = \{(x, y) \in \mathbb{R}^2\colon ax + by = 0\}$

**4.** In $\mathbb{R}^2$, starting with $\begin{pmatrix} a \\ b \end{pmatrix}, \begin{pmatrix} c \\ d \end{pmatrix}$, where $ad - bc \neq 0$.

### Answers to Self-Quiz

**I.** False **II.** True **III.** True **IV.** False **V.** True **VI.** c

**5.** $\pi = \{(x, y, z)\colon 2x - y - z = 0\}$

**6.** $\pi = \{(x, y, z)\colon 3x - 2y + 6z = 0\}$

**7.** $L = \{(x, y, z)\colon x/2 = y/3 = z/4\}$

**8.** $L = \{(x, y, z)\colon x = 3t,\ y = -2t,\ z = t;\ t \text{ real}\}$

**9.** $H = \{(x, y, z, w) \in \mathbb{R}^4\colon 2x - y + 3z - w = 0\}$

**10.** $\pi = \{(x, y, z)\colon ax + by + cz = 0\}$, where $abc \neq 0$

**11.** $L = \{(x, y, z)\colon x/a = y/b = z/c\}$, where $abc \neq 0$.

**12.** $H = \{x_1, x_2, x_3, x_4, x_5) \in \mathbb{R}^5\colon 2x_1 - 3x_2 + x_3 + 4x_4 - x_5 = 0\}$

**13.** $H$ is the solution space of

$$\begin{aligned} x - 3y + \phantom{3}z &= 0 \\ -2x + 2y - 3z &= 0 \\ 4x - 8y + 5z &= 0 \end{aligned}$$

***14.** Find an orthonormal basis in $\mathbb{R}^4$ that includes the vectors

$$\mathbf{u}_1 = \begin{pmatrix} 1/\sqrt{2} \\ 0 \\ 1/\sqrt{2} \\ 0 \end{pmatrix} \quad \text{and} \quad \mathbf{u}_2 = \begin{pmatrix} -\frac{1}{2} \\ \frac{1}{2} \\ \frac{1}{2} \\ -\frac{1}{2} \end{pmatrix}$$

[*Hint:* First find two vectors $\mathbf{v}_3$ and $\mathbf{v}_4$ to complete the basis.]

**15.** Show that $Q = \begin{pmatrix} \frac{2}{3} & \frac{1}{3} & \frac{2}{3} \\ \frac{1}{3} & \frac{2}{3} & -\frac{2}{3} \\ -\frac{2}{3} & \frac{2}{3} & \frac{1}{3} \end{pmatrix}$ is an orthogonal matrix.

**16.** Show that if $P$ and $Q$ are orthogonal $n \times n$ matrices, then $PQ$ is orthogonal.

**17.** Verify the result of Problem 16 with

$$P = \begin{pmatrix} 1/\sqrt{2} & -1/\sqrt{2} \\ 1/\sqrt{2} & 1/\sqrt{2} \end{pmatrix} \quad \text{and} \quad Q = \begin{pmatrix} 1/3 & -\sqrt{8}/3 \\ \sqrt{8}/3 & 1/3 \end{pmatrix}$$

**18.** Show that if $Q$ is a symmetric orthogonal matrix, then $Q^2 = I$.

**19.** Show that if $Q$ is orthogonal, then $\det Q = \pm 1$.

**20.** Show that for any real number $t$, the matrix $A = \begin{pmatrix} \sin t & \cos t \\ \cos t & -\sin t \end{pmatrix}$ is orthogonal.

**21.** Let $\{\mathbf{v}_1, \mathbf{v}_2, \ldots, \mathbf{v}_k\}$ be a linearly independent set of vectors in $\mathbb{R}^n$. Prove that $\mathbf{v}_i \neq \mathbf{0}$ for $i = 1, 2, \ldots, k$. [*Hint:* If $\mathbf{v}_i = \mathbf{0}$, then it is easy to find constants $c_1, c_2, \ldots, c_k$ with $c_i \neq 0$ such that $c_1\mathbf{v}_1 + c_2\mathbf{v}_2 + \cdots + c_k\mathbf{v}_k = \mathbf{0}$.]

In Problems 22–28 a subspace $H$ and a vector $\mathbf{v}$ are given. **(a)** Compute $\text{proj}_H \mathbf{v}$; **(b)** find an orthonormal basis for $H^\perp$; **(c)** write $\mathbf{v}$ as $\mathbf{h} + \mathbf{p}$, where $\mathbf{h} \in H$ and $\mathbf{p} \in H^\perp$.

**22.** $H = \left\{\begin{pmatrix} x \\ y \end{pmatrix} \in \mathbb{R}^2\colon x + y = 0\right\}$; $\mathbf{v} = \begin{pmatrix} -1 \\ 2 \end{pmatrix}$

**23.** $H = \left\{\begin{pmatrix} x \\ y \end{pmatrix} \in \mathbb{R}^2\colon ax + by = 0\right\}$; $\mathbf{v} = \begin{pmatrix} a \\ b \end{pmatrix}$

**24.** $H = \left\{\begin{pmatrix} x \\ y \\ z \end{pmatrix} \in \mathbb{R}^3\colon ax + by + cz = 0\right\}$; $\mathbf{v} = \begin{pmatrix} a \\ b \\ c \end{pmatrix}$, $\mathbf{v} \neq \mathbf{0}$

**25.** $H = \left\{ \begin{pmatrix} x \\ y \\ z \end{pmatrix} \in \mathbb{R}^3 \colon 3x - 2y + 6z = 0 \right\}$; $\mathbf{v} = \begin{pmatrix} -3 \\ 1 \\ 4 \end{pmatrix}$

**26.** $H = \left\{ \begin{pmatrix} x \\ y \\ z \end{pmatrix} \in \mathbb{R}^3 \colon x/2 = y/3 = z/4 \right\}$; $\mathbf{v} = \begin{pmatrix} 1 \\ 1 \\ 1 \end{pmatrix}$

**27.** $H = \left\{ \begin{pmatrix} x \\ y \\ z \\ w \end{pmatrix} \in \mathbb{R}^4 \colon 2x - y + 3z - w = 0 \right\}$; $\mathbf{v} = \begin{pmatrix} 1 \\ -1 \\ 2 \\ 3 \end{pmatrix}$

**28.** $H = \left\{ \begin{pmatrix} x \\ y \\ z \\ w \end{pmatrix} \in \mathbb{R}^4 \colon x = y \text{ and } w = 3y \right\}$; $\mathbf{v} = \begin{pmatrix} -1 \\ 2 \\ 3 \\ 1 \end{pmatrix}$

**29.** Let $\mathbf{u}_1$ and $\mathbf{u}_2$ be two orthonormal vectors in $\mathbb{R}^n$. Show that $|\mathbf{u}_1 - \mathbf{u}_2| = \sqrt{2}$.

**30.** If $\mathbf{u}_1, \mathbf{u}_2, \ldots, \mathbf{u}_n$ are orthonormal, show that

$$|\mathbf{u}_1 + \mathbf{u}_2 + \cdots + \mathbf{u}_n|^2 = |\mathbf{u}_1|^2 + |\mathbf{u}_2|^2 + \cdots + |\mathbf{u}_n|^2 = n$$

**31.** Find a condition on the numbers $a$ and $b$ such that $\left\{ \begin{pmatrix} a \\ b \end{pmatrix}, \begin{pmatrix} b \\ -a \end{pmatrix} \right\}$ and $\left\{ \begin{pmatrix} a \\ b \end{pmatrix}, \begin{pmatrix} -b \\ a \end{pmatrix} \right\}$ form orthonormal bases in $\mathbb{R}^2$.

**32.** Show that *any* orthonormal basis in $\mathbb{R}^2$ has one of the forms of the bases in Problem 31.

**33.** Using the Cauchy-Schwarz inequality, prove that if $|\mathbf{u} + \mathbf{v}| = |\mathbf{u}| + |\mathbf{v}|$, then $\mathbf{u}$ and $\mathbf{v}$ are linearly dependent.

**34.** Using the Cauchy-Schwarz inequality, prove the **triangle inequality:**

$$|\mathbf{u} + \mathbf{v}| \le |\mathbf{u}| + |\mathbf{v}|$$

[*Hint:* Expand $|\mathbf{u} + \mathbf{v}|^2$.]

***35.** Suppose that $\mathbf{x}_1, \mathbf{x}_2, \ldots, \mathbf{x}_k$ are vectors in $\mathbb{R}^n$ (not all zero) and

$$|\mathbf{x}_1 + \mathbf{x}_2 + \cdots + \mathbf{x}_k| = |\mathbf{x}_1| + |\mathbf{x}_2| + \cdots + |\mathbf{x}_k|$$

Show that $\dim \operatorname{span} \{\mathbf{x}_1, \mathbf{x}_2, \ldots, \mathbf{x}_k\} = 1$. [*Hint:* Use the results of Problems 33 and 34.]

**36.** Let $\{\mathbf{u}_1, \mathbf{u}_2, \ldots, \mathbf{u}_n\}$ be an orthonormal basis in $\mathbb{R}^n$ and let $\mathbf{v}$ be a vector in $\mathbb{R}^n$. Prove that $|\mathbf{v}|^2 = |\mathbf{v} \cdot \mathbf{u}_1|^2 + |\mathbf{v} \cdot \mathbf{u}_2|^2 + \cdots + |\mathbf{v} \cdot \mathbf{u}_n|^2$. This equality is called **Parseval's equality** in $\mathbb{R}^n$.

**37.** Show that for any subspace $H$ of $\mathbb{R}^n$, $(H^\perp)^\perp = H$.

**38.** Let $H_1$ and $H_2$ be two subspaces of $\mathbb{R}^n$ and suppose that $H_1^\perp = H_2^\perp$. Show that $H_1 = H_2$.

**39.** If $H_1$ and $H_2$ are subspaces of $\mathbb{R}^n$, show that if $H_1 \subset H_2$, then $H_2^\perp \subset H_1^\perp$.

**40.** Prove the **generalized Pythagorean theorem:** Let $\mathbf{u}$ and $\mathbf{v}$ be vectors in $\mathbb{R}^n$ with $\mathbf{u} \perp \mathbf{v}$. Then

$$|\mathbf{u} + \mathbf{v}|^2 = |\mathbf{u}|^2 + |\mathbf{v}|^2$$

## CALCULATOR BOX

### TI-85

On page 238 we indicated how to find the length or norm of a vector in $\mathbb{R}^2$ on the TI-85. On page 249 we showed how to find the dot product of two vectors in $\mathbb{R}^2$. The same procedures work equally well in $\mathbb{R}^n$. For example, the key sequence

[2nd] [VECTR] [F3] ⟨MATH⟩ [F3]

⟨norm⟩ [1, 3, 5, −8] [ENTER]

results in 9.94987437107 (which is $\sqrt{99}$ to 12 significant figures). Mimicking the procedure on page 249 will yield $\mathbf{a} \cdot \mathbf{b}$, where both $\mathbf{a}$ and $\mathbf{b}$ are in $\mathbb{R}^n$ for any $n > 2$.

# MATLAB 4.9

***MATLAB Reminder*** $\mathbf{u} \cdot \mathbf{v}$ is computed by **u'*v** or **v'*u.** $|\mathbf{v}|$ is computed by **sqrt(v'*v)** or **norm(v).** $\text{proj}_{\mathbf{v}}\mathbf{u}$ is computed by **((u'*v)/(v'*v))*v** (the vector projection of **u** onto **v**).

1. Find orthonormal bases for the span of each set of vectors below by the Gram-Schmidt process. Check your answers by verifying that the set of vectors obtained is orthonormal and that each vector in the original set is a linear combination of the set of vectors you obtained.

   a. $\left\{\begin{pmatrix}-1\\2\\-1\end{pmatrix}, \begin{pmatrix}3\\4\\0\end{pmatrix}\right\}$ b. $\left\{\begin{pmatrix}0\\-2\\-3\\-3\\1\end{pmatrix}, \begin{pmatrix}3\\-5\\0\\0\\5\end{pmatrix}, \begin{pmatrix}2\\1\\4\\1\\3\end{pmatrix}\right\}$

   c. $\left\{\begin{pmatrix}-1\\2\\0\\1\end{pmatrix}, \begin{pmatrix}1\\-1\\2\\2\end{pmatrix}, \begin{pmatrix}1\\-2\\3\\1\end{pmatrix}, \begin{pmatrix}-1\\2\\-1\\4\end{pmatrix}\right\}$

   d. Generate four random vectors in $\mathbb{R}^6$.

2. Find an orthonormal basis for

$$H = \left\{\begin{pmatrix}x\\y\\z\\w\end{pmatrix} \;\middle|\; x - y + 3z + w = 0\right\}$$

***Hint.*** First, find a basis for $H$ by finding a basis for solutions to $A\mathbf{x} = \mathbf{0}$, where $A = (1\ -1\ 3\ 1)$, and then apply the Gram-Schmidt process.

3. a. *(Paper and pencil)* Suppose $\mathbf{v} = \begin{pmatrix} a \\ b \end{pmatrix}$ and $\mathbf{z} = \begin{pmatrix} -b \\ a \end{pmatrix}$. Suppose $\mathbf{v}_1 = \mathbf{v}/|\mathbf{v}|$ and $\mathbf{v}_2 = \mathbf{z}/|\mathbf{z}|$. Show that $\{\mathbf{v}_1, \mathbf{v}_2\}$ forms an orthonormal basis for $\mathbb{R}^2$ as long as $a$ and $b$ are not both zero.

   b. For $\mathbf{v} = \begin{pmatrix} 1 \\ 2 \end{pmatrix}$, form $\mathbf{v}_1$ and $\mathbf{v}_2$ as in part (a). Let $\mathbf{w} = \begin{pmatrix} -3 \\ 4 \end{pmatrix}$. Compute $\mathbf{p}_1$, the vector projection of $\mathbf{w}$ onto $\mathbf{v}_1$ and $\mathbf{p}_2$, the vector projection of $\mathbf{w}$ onto $\mathbf{v}_2$. Recall the geometry of a projection by using the m-file *prjtn.m* on the accompanying disk. Use the commands **prjtn(w,v1)** and **prjtn(w,v2).** (On the graphics display, $\mathbf{w}$ will be labeled $U$ and $\mathbf{v}_1$ or $\mathbf{v}_2$ will be labeled $V$.)

M

   c. Verify that $\mathbf{w} = \mathbf{p}_1 + \mathbf{p}_2 = (\mathbf{w} \cdot \mathbf{v}_1)\mathbf{v}_1 + (\mathbf{w} \cdot \mathbf{v}_2)\mathbf{v}_2$. Enter the command **lincomb(v1,v2,w).** (The m-file *lincomb.m* is on the accompanying disk.)

   Describe how the geometry of projection and the geometry of linear combination is reflected in the graph displayed.

***Caution.*** A hard copy produced by a screen dump will NOT preserve equal lengths or right angles.

To check that the numbers displayed on the graphics screen are **w · v1** and **w · v2**, enter the commands **rat(w'*v1,'s')** and **rat(w'*v2,'s').** In MATLAB 4.0, use **format rat.**

   d. Repeat parts (b) and (c) for $\mathbf{v} = \begin{pmatrix} 1 \\ 2 \end{pmatrix}$ and $\mathbf{w} = \begin{pmatrix} 4 \\ 2 \end{pmatrix}$.

   e. Repeat parts (b) and (c) for $\mathbf{v}$ and $\mathbf{w}$ of your choice.

   f. *(Paper and pencil)* Explain how this problem also illustrates Theorem 7 in this section, where $H$ is span $\{\mathbf{v}\}$.

4. a. Let $\mathbf{v}$ be a vector of length 1 in the direction of $\begin{pmatrix} 2 \\ 1 \end{pmatrix}$ (divide the vector by its length).

   Let $\mathbf{w} = \begin{pmatrix} 3 \\ 5 \end{pmatrix}$, find $\mathbf{p}$, the vector projection of $\mathbf{w}$ onto $\mathbf{v}$, and compute $|\mathbf{w} - \mathbf{p}|$.

   b. Choose any scalar value for $c$; let $\mathbf{z} = c\mathbf{v}$, and verify that $|\mathbf{w} - \mathbf{z}| \geq |\mathbf{w} - \mathbf{p}|$. Repeat for three more values of $c$. Explain how this relates to Theorem 8, where $H$ is span $\{\mathbf{v}\}$.

   c. Repeat parts (a) and (b) with $\mathbf{w} = \begin{pmatrix} -3 \\ 2 \end{pmatrix}$.

   d. Repeat parts (a) and (b) for a $\mathbf{v}$ and a $\mathbf{w}$ of your choice.

   e. *(Paper and pencil)* In the schematic diagrams below, label $\mathbf{p}$, the vector projection of $\mathbf{w}$ onto $\mathbf{v}$, and label $\mathbf{w} - \mathbf{p}$ and $\mathbf{w} - \mathbf{z}$. Explain how these diagrams illustrate the geometry of Theorem 8, where $H$ is the subspace span $\{\mathbf{v}\}$.

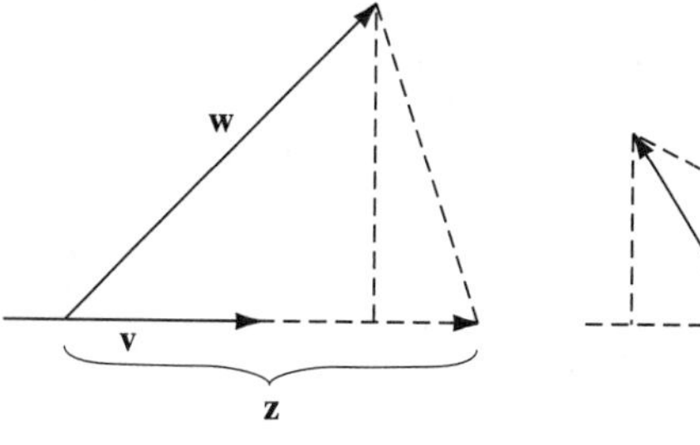


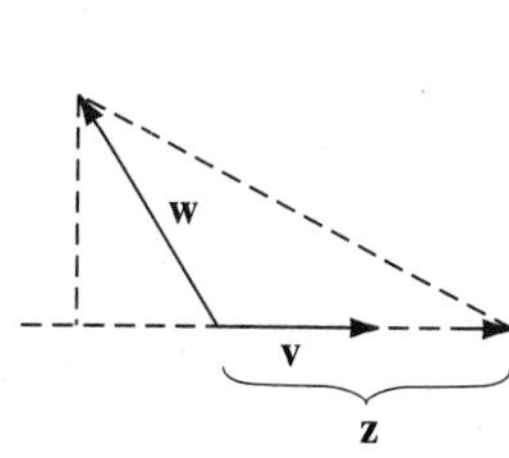

5. **Projection onto a Plane in $\mathbb{R}^3$**

a. Let $\mathbf{v}_1 = \begin{pmatrix} -1 \\ 2 \\ 3 \end{pmatrix}$ and $\mathbf{v}_2 = \begin{pmatrix} 0 \\ 1 \\ 2 \end{pmatrix}$.

Find an orthonormal basis $\{\mathbf{z}_1, \mathbf{z}_2\}$ for the plane given by span $\{\mathbf{v}_1, \mathbf{v}_2\}$ using the Gram-Schmidt process.

b. *(Paper and pencil)* Verify that $\mathbf{z} = \begin{pmatrix} -1 \\ -2 \\ 1 \end{pmatrix}$ is perpendicular to both $\mathbf{v}_1$ and $\mathbf{v}_2$ and hence perpendicular to $H = \text{span}\ \{\mathbf{v}_1, \mathbf{v}_2\}$. Let $\mathbf{n} = \mathbf{z}/|\mathbf{z}|$. Explain why $\mathbf{n}$ is an orthonormal basis for $H^\perp$.

c. Definition 4 says that the projection of a vector $\mathbf{w}$ onto $H$ is given by $\text{proj}_H\mathbf{w} = (\mathbf{w} \cdot \mathbf{z}_1)\mathbf{z}_1 + (\mathbf{w} \cdot \mathbf{z}_2)\mathbf{z}_2$. Theorem 7 says that $\mathbf{w} = \text{proj}_H\mathbf{w} + \text{proj}_{H^\perp}\mathbf{w}$, which can be rewritten to say that $\text{proj}_H\mathbf{w} = \mathbf{w} - \text{proj}_{H^\perp}\mathbf{w}$.

For four $3 \times 1$ vectors $\mathbf{w}$ of your choice, compute $\text{proj}_H\mathbf{w}$ both ways and compare. (*Note.* Since $H^\perp$ is one dimensional, $\text{proj}_{H^\perp}\mathbf{w}$ equals the vector projection of $\mathbf{w}$ onto $\mathbf{n}$.)

d. *(Paper and pencil)* The following diagram illustrates the geometry of $\text{proj}_H\mathbf{w} = \mathbf{w} - \text{proj}_{H^\perp}\mathbf{w}$. On the diagram, label $\mathbf{h} = \text{proj}_{H^\perp}\mathbf{w}$, sketch $\mathbf{w} - \mathbf{h}$, and verify that it is parallel to $\mathbf{p}$, the projection of $\mathbf{w}$ onto the plane.

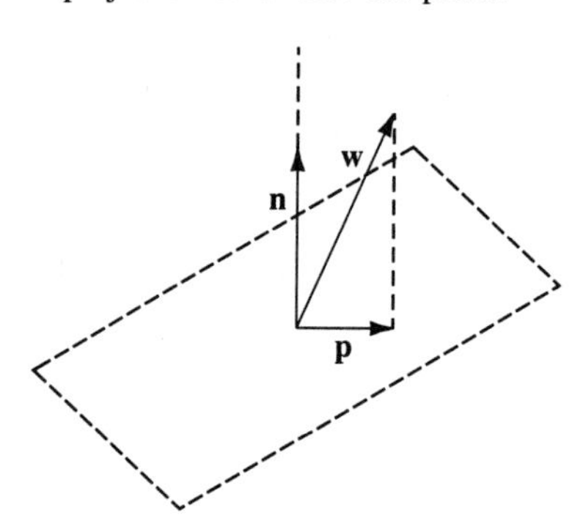


6. For vectors $\mathbf{v}_1, \ldots, \mathbf{v}_k$, if you form the matrix $A = [\mathbf{v}_1 \cdots \mathbf{v}_k]$, then the MATLAB command **B = orth(A)** will produce a matrix $B$ whose columns form an orthonormal basis for the subspace $H$ = range of $A$ = span $\{\mathbf{v}_1, \ldots, \mathbf{v}_k\}$.

a. Let $\{\mathbf{v}_1, \mathbf{v}_2, \mathbf{v}_3\}$ be the set of vectors in MATLAB Problem 1(b) in this section. Find $A$ and $B$ as described above. Verify that the columns of $B$ are orthonormal.

b. Let $\mathbf{x}$ be a random $3 \times 1$ vector and find $A\mathbf{x}$. Explain why $A\mathbf{x}$ is in $H$.

Theorem 4 says that if $\mathbf{w}$ is in $H$, then $\mathbf{w} = (\mathbf{w} \cdot \mathbf{u}_1)\mathbf{u}_1 + \cdots + (\mathbf{w} \cdot \mathbf{u}_k)\mathbf{u}_k$, where $\{\mathbf{u}_1, \ldots, \mathbf{u}_k\}$ is an orthonormal basis for $H$. Verify this for $\mathbf{w} = A\mathbf{x}$ using the fact that $\mathbf{u}_i$ is column $i$ of $B$.

c. Repeat the directions of parts (a) and (b) for $\{\mathbf{v}_1, \mathbf{v}_2, \mathbf{v}_3, \mathbf{v}_4\}$, where each $\mathbf{v}_i$ is a random $6 \times 1$ vector and $\mathbf{x}$ is a random $4 \times 1$ vector.

7. Generate four random vectors in $\mathbb{R}^6$, $\mathbf{v}_1, \mathbf{v}_2, \mathbf{v}_3$, and $\mathbf{v}_4$. Let $H$ denote span $\{\mathbf{v}_1, \mathbf{v}_2, \mathbf{v}_3, \mathbf{v}_4\}$. Let $A = [\mathbf{v}_1\ \mathbf{v}_2\ \mathbf{v}_3\ \mathbf{v}_4]$ and **B = orth(A).** Let $\mathbf{u}_i$ denote the $i$th column of $B$.

a. Let $\mathbf{w}$ be a random $6 \times 1$ vector. Find the projection of $\mathbf{w}$ onto $H$, $\mathbf{p} = \text{proj}_H\mathbf{w}$, using Definition 4.

Compute $\mathbf{z} = \begin{pmatrix} \mathbf{w} \cdot \mathbf{u}_1 \\ \mathbf{w} \cdot \mathbf{u}_2 \\ \mathbf{w} \cdot \mathbf{u}_3 \\ \mathbf{w} \cdot \mathbf{u}_4 \end{pmatrix}$. Verify that $\mathbf{z} = B^t\mathbf{w}$ and $\mathbf{p} = BB^t\mathbf{w}$. Repeat for another choice of $\mathbf{w}$.

b. Let $\mathbf{x}$ be a random $4 \times 1$ vector and form $\mathbf{h} = A\mathbf{x}$. Hence $\mathbf{h}$ is in $H$. Compare $|\mathbf{w} - \mathbf{p}|$ and $|\mathbf{w} - \mathbf{h}|$. Repeat for three more choices of $\mathbf{x}$. Write an interpretation of your observations.

c. Let $\mathbf{z} = 2\mathbf{v}_1 - 3\mathbf{v}_3 + \mathbf{v}_4$. Then $H = \text{span}\ \{\mathbf{v}_1, \mathbf{v}_2, \mathbf{v}_3, \mathbf{z}\}$. (Here $H$ is the subspace described in previous parts of this problem.) Why? Let $C = [\mathbf{v}_1\ \mathbf{v}_2\ \mathbf{v}_3\ \mathbf{z}]$ and **D = orth(C).** Then the columns of $D$ will be another orthonormal basis for $H$.

Let $\mathbf{w}$ be a random $6 \times 1$ vector. Compute the projection of $\mathbf{w}$ onto $H$ using $B$ and the projection of $\mathbf{w}$ onto $H$ using $D$. Compare the results. Repeat for two more choices of $\mathbf{w}$. Write an interpretation of your observations.

d. *(Paper and pencil)* If $\{\mathbf{u}_1, \ldots, \mathbf{u}_k\}$ is an orthonormal basis for a subspace $H$ and $B$ is the matrix $[\mathbf{u}_1 \cdots \mathbf{u}_k]$, prove that the projection of $\mathbf{w}$ onto $H$ equals $BB^t\mathbf{w}$.

8. a. *(Paper and pencil)* If $A$ is a real matrix, explain why the null space of $A^t$ is perpendicular to the range of $A$; that is, if $H = \text{range}\ A$, then null space $(A^t) = H^\perp$.

b. Let $A$ be a random real $7 \times 4$ matrix. Let **B = orth(A)** and let **C = null(A′).** (Then the columns of $B$ form an orthonormal basis for $H = \text{range}\ A$ and the columns of $C$ form an orthonormal basis for $H^\perp$.) Verify that the columns of $C$ are orthonormal.

c. Let $\mathbf{w}$ be a random $7 \times 1$ vector. Find $\mathbf{h}$, the projection of $\mathbf{w}$ onto $H$ and $\mathbf{p}$, the projection of $\mathbf{w}$ onto $H^\perp$. (See MATLAB Problem 7 in this section.) Verify that $\mathbf{w} = \mathbf{p} + \mathbf{h}$. Repeat for three more choices of $\mathbf{w}$.

d. Verify that $BB^t + CC^t = I$, where $I$ is the identity matrix.

e. *(Paper and pencil)* Prove the relationship in part (d).

9. a. *(Paper and pencil)* Suppose $\{\mathbf{u}_1, \ldots, \mathbf{u}_n\}$ is an orthonormal basis for $\mathbb{R}^n$ and $B$ is the matrix $[\mathbf{u}_1 \cdots \mathbf{u}_n]$. Let $\mathbf{v}$ be a vector in $\mathbb{R}^n$. Using Theorem 4, explain why the coordinates of $\mathbf{v}$ with respect to the basis $\{\mathbf{u}_1, \ldots, \mathbf{u}_n\}$ can be found by $B^t\mathbf{v}$.

b. *(Paper and pencil)* Recall that if $\theta$ is the angle between $\mathbf{u}$ and $\mathbf{w}$, then $\cos(\theta) = \mathbf{u} \cdot \mathbf{w}/|\mathbf{u}||\mathbf{w}|$. Assume $|\mathbf{w}| = 1$. Using Theorem 4, prove that the coordinates of $\mathbf{w}$ with respect to an orthonormal basis can be interpreted as the cosines of the angles that $\mathbf{w}$ makes with each of the basis vectors; that is, the coordinate of $\mathbf{w}$ that is the coefficient of the $i$th basis vector equals the cosine of the angle between $\mathbf{w}$ and the $i$th basis vector.

c. Verify this interpretation by finding the angles between the given vector $\mathbf{w}$ and the given orthonormal basis $\{\mathbf{v}_1, \mathbf{v}_2\}$ for $\mathbb{R}^2$. First, sketch by hand to decide what you expect the angles to be. (Use the MATLAB command **acos.** Enter **help acos** for a description. To change an angle from radian measure to degree measure, multiply by $180/\pi$.)

**i.** $\mathbf{w}$ = vector of length 1 in the direction of $\begin{pmatrix} 1 \\ 1 \end{pmatrix}$

$$\mathbf{v}_1 = \begin{pmatrix} 1 \\ 0 \end{pmatrix} \quad \mathbf{v}_2 = \begin{pmatrix} 0 \\ 1 \end{pmatrix}$$

**ii.** $\mathbf{w} = \begin{pmatrix} -1 \\ 0 \end{pmatrix}$

$$\mathbf{v}_1 = \text{vector of length 1 in the direction of} \begin{pmatrix} 1 \\ 1 \end{pmatrix}$$

$$\mathbf{v}_2 = \text{vector of length 1 in the direction of} \begin{pmatrix} -1 \\ 1 \end{pmatrix}$$

d. Verify that $\left\{ \begin{pmatrix} \frac{2}{3} \\ \frac{2}{3} \\ -\frac{1}{3} \end{pmatrix}, \begin{pmatrix} \frac{2}{3} \\ -\frac{1}{3} \\ \frac{2}{3} \end{pmatrix}, \begin{pmatrix} -\frac{1}{3} \\ \frac{2}{3} \\ \frac{2}{3} \end{pmatrix} \right\}$ is an orthonormal basis for $\mathbb{R}^3$. Let $\mathbf{s} = \begin{pmatrix} 1 \\ 1 \\ 1 \end{pmatrix}$.
Find the angles between $\mathbf{s}$ and each of the basis vectors. First form $\mathbf{w} = \mathbf{s}/|\mathbf{s}|$. The angles between $\mathbf{w}$ and each of the basis vectors will be the same as the angles between $\mathbf{s}$ and each of the basis vectors. Repeat for another choice of $\mathbf{s}$.

10. Verify that the following are orthogonal matrices.

a. $\left(\frac{1}{\sqrt{2}}\right)\begin{pmatrix} 1 & 1 \\ 1 & -1 \end{pmatrix} = B$

b. $\left(\frac{1}{14}\right)\begin{pmatrix} -4 & -6 & 12 \\ 6 & -12 & -4 \\ 12 & 4 & 6 \end{pmatrix} = B_1$

c. $\left(\frac{1}{39}\right)\begin{pmatrix} -13 & 14 & -34 \\ -26 & -29 & -2 \\ -26 & 22 & 19 \end{pmatrix} = B_2$

d. **orth(rand3))** $= B_3$

e. $[\mathbf{u}_1\ \mathbf{u}_2\ \mathbf{u}_3] = B_4$, where $\{\mathbf{u}_1, \mathbf{u}_2, \mathbf{u}_3\}$ is the basis obtained by applying the Gram-Schmidt process to $\left\{ \begin{pmatrix} -1 \\ 2 \\ 3 \end{pmatrix}, \begin{pmatrix} 0 \\ 1 \\ 1 \end{pmatrix}, \begin{pmatrix} -1 \\ 2 \\ 4 \end{pmatrix} \right\}$.

11. a. Verify that each of the following are orthogonal matrices: $B_1B_2$, $B_1B_3$, $B_2B_4$, and $B_3B_4$, where $B_1$, $B_2$, $B_3$, and $B_4$ are as in MATLAB Problem 10 above.

b. *(Paper and pencil)* Work Problem 16.

12. a. Find the inverse of each matrix in MATLAB Problem 10 in this section and verify that the inverses are orthogonal.

b. *(Paper and pencil)* Prove that the inverse of an orthogonal matrix is an orthogonal matrix.

13. a. Find the determinant of each of the matrices in MATLAB Problem 10. Formulate a conjecture about the determinant of an orthogonal matrix.

b. *(Paper and pencil)* Prove your conjecture.

c. Review (or do) MATLAB Problem 2 in Section 3.4. Suppose $\mathbf{u}$, $\mathbf{v}$, and $\mathbf{w}$ are vectors in $\mathbb{R}^3$ that form a parallelepiped. If $Q$ is a $3 \times 3$ orthogonal matrix, explain why $Q\mathbf{u}$, $Q\mathbf{v}$, and $Q\mathbf{w}$ form a parallelepiped with the same volume as the parallelepiped formed by $\mathbf{u}$, $\mathbf{v}$, and $\mathbf{w}$.

14. **Orthogonal Matrices: Length and Angle** Recall that if $\theta$ is the angle between $\mathbf{u}$ and $\mathbf{w}$, then $\cos(\theta) = \mathbf{u} \cdot \mathbf{w}/|\mathbf{u}||\mathbf{w}|$.

a. Let $Q$ be the orthogonal matrix $B_1$ in MATLAB Problem 10. Choose random vectors $\mathbf{v}$ and $\mathbf{w}$. Compute and compare the length of $\mathbf{v}$ and the length of $Q\mathbf{v}$. Compute and compare the cosine of the angle between **v** and **w** and the cosine of the angle between $Q$**v** and $Q$**w**. Repeat for a total of three choices of **v** and **w**.

b. Repeat part (a) for another orthogonal matrix from MATLAB Problem 10. Repeat part (a) for **Q = orth(2*rand(5)−1).** (First, check that this $Q$ is orthogonal.) Write an interpretation of your observations from parts (a) and (b).

c. Let **Q = orth(2*rand(6)−1).** Verify that $Q$ is an orthogonal matrix and hence that the columns of $Q$ form an orthonormal basis for $\mathbb{R}^6$.

Let **x** and **z** be random $6 \times 1$ vectors. Find **xx**, the coordinates of **x** with respect to the basis given by the columns of $Q$. Find **zz**, the coordinates of **z** with respect to the basis given by the columns of $Q$.

Compare $|\mathbf{x} - \mathbf{z}|$ with $|\mathbf{xx} - \mathbf{zz}|$. Repeat for another pair of vectors **x** and **z** and write a description of your observations.

d. Part (c) has some important ramifications. In any computations or measurements, errors are introduced. An important issue in designing numerical algorithms is the concern about compounding errors. We can interpret $|\mathbf{x} - \mathbf{z}|$ as an error; for example, **x** may represent the true values and **z** may represent an approximation. Explain how the observations from part (c) tell you that the process of changing to coordinates of an orthonormal basis does not compound (increase) any error that is already present. Why would changing the coordinates back to standard coordinates also not increase error?

e. *(Paper and pencil)* If $Q$ is an orthogonal matrix and **v** and **w** are vectors, prove that $Q\mathbf{v} \cdot Q\mathbf{w} = \mathbf{v} \cdot \mathbf{w}$. Use this to prove that $|Q\mathbf{v}| = |\mathbf{v}|$ and that the cosine of the angle between $Q$**v** and $Q$**w** equals the cosine of the angle between **v** and **w**.

f. *(Paper and pencil)* Prove your observation from part (c). (First, explain why finding the coordinates of a vector **x** with respect to the columns of $Q$ is the same as multiplying **x** by an orthogonal matrix.)

15. **Rotation Matrices** You will need to have done MATLAB Problems 9 and 10 in Section 4.8. You can do parts (a) and (b) for $\mathbb{R}^2$ if you have only done MATLAB Problem 9 in Section 4.8.

a. Consider the rotation matrix $V$ in MATLAB Problem 9(b) in Section 4.8 and the rotation matrices $P$, $Y$, and $R$ in MATLAB Problem 10(a) in Section 4.8. Choose a value for a rotation angle, for example, $\pi/4$, and verify that each of the matrices $V$, $P$, $Y$, and $R$ (using your choice of angle) is an orthogonal matrix. Repeat for two more choices of angle.

b. *(Paper and pencil)* Since an $n \times n$ rotation matrix is orthogonal, the columns of the matrix form an orthonormal basis for $\mathbb{R}^n$. Why? Why would you expect this from the geometry?

c. *(Paper and pencil)* Recall from MATLAB Problem 10 in Section 4.8 that spacecraft attitude is found by doing pitch, yaw, and roll maneuvers in some order. This yields an attitude matrix that is formed by the product of some $P$, $Y$, and $R$ rotation matrices. Explain why the attitude matrix will be an orthogonal matrix.

d. Assume that a satellite was initially oriented by pitch, yaw, and roll maneuvers so that its attitude matrix is orthogonal. The control center (oriented along standard coordinates) periodically checks the attitude of the satellite by having it send readings (in the satellite's coordinates) of objects at locations known to the control center.

A particular satellite sends the following readings (which are adjusted to take into account the different locations of the control center and the satellite):

$$\mathbf{v}_1 = \begin{pmatrix} .7017 \\ -.7017 \\ 0 \end{pmatrix} \quad \text{for an object at} \begin{pmatrix} 1 \\ 0 \\ 0 \end{pmatrix} \text{(standard coordinates)}$$

$$\mathbf{v}_2 = \begin{pmatrix} .2130 \\ .2130 \\ .9093 \end{pmatrix} \quad \text{for an object at} \begin{pmatrix} 0 \\ 1 \\ 0 \end{pmatrix} \text{(standard coordinates)}$$

$$\mathbf{v}_3 = \begin{pmatrix} .1025 \\ -.4125 \\ .0726 \end{pmatrix} \quad \text{for an object at} \begin{pmatrix} 0 \\ 0 \\ 1 \end{pmatrix} \text{(standard coordinates)}$$

Explain why the control center knows that something is wrong with the satellite. [*Hint:* First explain why the matrix $[\mathbf{v}_1\ \mathbf{v}_2\ \mathbf{v}_3]$ should equal $A^{-1}I$, where $I = \begin{pmatrix} 1 & 0 & 0 \\ 0 & 1 & 0 \\ 0 & 0 & 1 \end{pmatrix}$ and $A$ is the attitude matrix of the satellite. Recall that the readings are the coordinates of $\begin{pmatrix} 1 \\ 0 \\ 0 \end{pmatrix}$, $\begin{pmatrix} 0 \\ 1 \\ 0 \end{pmatrix}$, and $\begin{pmatrix} 0 \\ 0 \\ 1 \end{pmatrix}$ with respect to the coordinate system of the satellite given by $A$, the attitude matrix. What kind of matrices should $A$ and $A^{-1}$ be?]

**e.** Suppose a spacecraft is oriented by a pitch maneuver with an angle of $\pi/4$, followed by a roll maneuver with an angle of $-\pi/3$, followed by a yaw maneuver with an angle of $\pi/6$. Find the attitude matrix.

Find the angles between each of the coordinate axes of the spacecraft and the standard $x$-axis, that is, the angles between the columns of the attitude matrix and the vector $\begin{pmatrix} 1 \\ 0 \\ 0 \end{pmatrix}$. Find the angles between each of the coordinate axes of the spacecraft and the standard $y$-axis and the angles between each of the coordinate axes of the spacecraft and the standard $z$-axis. (See MATLAB Problem 9 in this section.) Explain your procedure.

**16. a.** Let $\mathbf{x}$ be a random $3 \times 1$ vector. Let $\mathbf{v} = \mathbf{x}/|\mathbf{x}|$. Find the matrix $H = I - 2\mathbf{v}\mathbf{v}^t$, where $I$ is the $3 \times 3$ identity. Verify that $H$ is orthogonal. Repeat for two more choices of $\mathbf{x}$. (Recall that the **eye** command will create an identity matrix.)

**b.** Repeat part (a) for $\mathbf{x}$, a random $n \times 1$ vector for two different choices of $n$. (Here $I$ will be the $n \times n$ identity.)

**c.** *(Paper and pencil)* If $\mathbf{v}$ is a vector of length 1 in $\mathbb{R}^n$, prove that $H = I - 2\mathbf{v}\mathbf{v}^t$ is an orthogonal matrix.

**d.** **Geometry** Matrices constructed above are called **elementary reflectors.** Let $\mathbf{v}$ be a vector of length 1 in $\mathbb{R}^2$ and construct $H$ as above. Let $\mathbf{x}$ be any vector in $\mathbb{R}^2$. Then $H\mathbf{x}$ is the reflection of $\mathbf{x}$ across the line perpendicular to $\mathbf{v}$.

The MATLAB code below illustrates this geometry. The vector $\mathbf{z}$ that is computed is $\mathbf{x} - \text{proj}_{\mathbf{v}}\mathbf{x}$ and hence will be a vector perpendicular to $\mathbf{v}$. Thus $\mathbf{z}$ represents the line perpendicular to $\mathbf{v}$. This line is represented by a dotted white line. The line determined by $\mathbf{v}$ is represented by a dashed white line. The original vector $\mathbf{x}$ is plotted in red and the reflected vector $\mathbf{h}$ is plotted in blue. The lines of the code preceding the plot statement are needed to set the axis perspective properly so that equal lengths appear equal and right angles appear as right angles. When you are finished with this part, enter the command **axis** to clear the axis scale. If using MATLAB 4.0 use **axis('auto')**.

***Caution.*** A hard copy produced by a screen dump will NOT preserve equal lengths or right angles.

***Note.*** If you are using MATLAB 4.0, enter the **plot** command before the two **axis** statements. Observe the final graphics screen after entering the two **axis** statements.

Enter 2 × 1 vectors **vv** and **x**:

```
v = vv/norm(vv);
z = x − x'*v*v;
H = eye(2)−2*v*v';
h = H*x;
aa = [x', z', h', −z', v', −v'];
m = min(aa); M = max(aa);
axis([m M m M])
axis('square')
plot([0 z(1)], [0 z(2)], 'w:', [0 −z(1)], [0 −z(2)], 'w:', . . .
   [0 v(1)], [0 v(2)], 'w--', [0 −v(1)], [0 −v(2)], 'w--', . . .
   [0 x(1)], [0 x(2)], 'r', [0 h(1)], [0 h(2)], 'b')
```

Suggested choices are:

```
vv = [0;1]    x = [3;3]
vv = [1;1]    x = [−1;2]
vv = [1;1]    x = [4;2]
```

e. By observing the geometry, conjecture a relationship between $H$ and $H^{-1}$. Test your conjecture for four $H$ matrices generated as in parts (a) or (b).

**PROJECT PROBLEM**

17. Work MATLAB Problems 9 and 10 in Section 4.8 and MATLAB Problem 15 in this section.

## 4.10 LEAST SQUARES APPROXIMATION

In many problems in the biological, physical, and social sciences it is useful to describe the relationship among the variables of the problem by means of a mathematical expression. Thus, for example, we may describe the relationship among cost, revenue, and profit by means of the simple formula

$$P = R - C$$

In a different vein, we may represent the relationship among the acceleration due to gravity, the time an object has been falling, and the height of the object by the physical law

$$s = s_0 - v_0 t - \tfrac{1}{2} g t^2$$

where $s_0$ is the initial height of the object and $v_0$ is its initial velocity.

Unfortunately, formulas like the ones above do not come easily. It is usually the task of the scientist or economist to sort through large amounts of data in order to find relationships among the variables in the problem. A common way to do this is to fit a curve among the various data points. This curve may be a straight line or a quadratic or a cubic, and so on. The object is to find the curve of the given type that "best" fits the given data. In this section we show how to do this when there are two variables in the problem. In every case we assume that there are $n$ data points $(x_1, y_1), (x_2, y_2), \ldots, (x_n, y_n)$.

In Figure 4.7 we can indicate three of the curves that can be used to fit data.

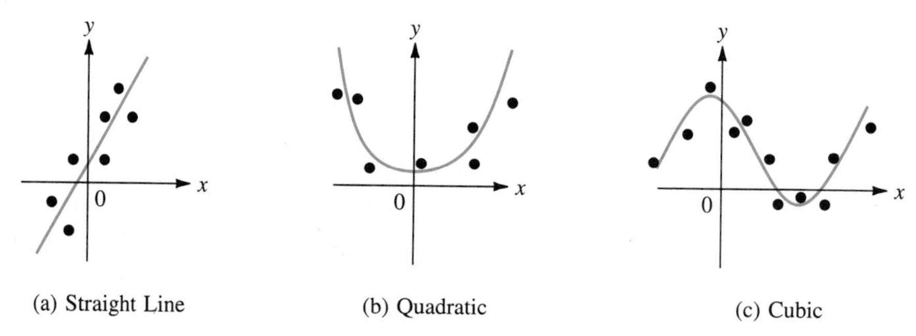


**Figure 4.7** Three curves in the $xy$-plane

## Straight-Line Approximation

Before continuing, we must be clear as to what we mean by the "best fit." Suppose we seek a straight line of the form $y = b + mx$ that best represents the $n$ data points $(x_1, y_1), (x_2, y_2), \ldots, (x_n, y_n)$.

Figure 4.8 illustrates what is going on (using three data points). From the figure we see that if we assume that the $x$- and $y$-variables are related by the formula $y = b + mx$, then, for example, for $x = x_1$ the corresponding $y$-value is $b + mx_1$. This is different from the "true" $y$-value $y = y_1$.

In $\mathbb{R}^2$ the distance between the points $(a_1, b_1)$ and $(a_2, b_2)$ is given by $d = \sqrt{(a_1 - a_2)^2 + (b_1 - b_2)^2}$. Therefore, in determining how to choose the line $y = b + mx$ that best approximates the given data, it is reasonable to use the criterion of choosing the line that minimizes the sum of the squares of the differences between the $y$-values of the points and the corresponding $y$-values on the line. Note that since the distance between $(x_1, y_1)$ and $(x_1, b + mx_1)$ is $y_1 - (b + mx_1)$, our problem (for $n$ data points) can be stated as follows:

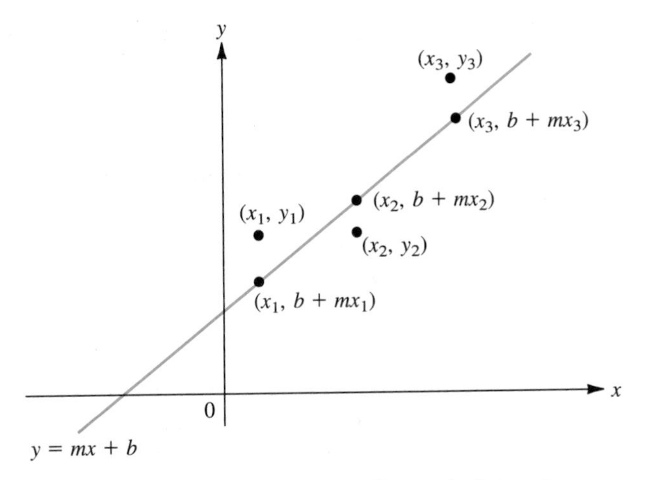


**Figure 4.8** Points on the straight line have coordinates $(x, b + mx)$

> **The Least Squares Problem for a Line**
>
> Find numbers $m$ and $b$ such that the sum
>
> $$[y_1 - (b + mx_1)]^2 + [y_2 - (b + mx_2)]^2 + \cdots + [y_n - (b + mx_n)]^2 \tag{1}$$
>
> is a minimum. For this choice of $m$ and $b$, the line $y = mx + b$ is called the **least squares straight-line approximation to the data points** $(x_1, y_1), (x_2, y_2), \ldots, (x_n, y_n)$.

Having defined the problem, we now seek a method for finding the least squares approximation. This is most easily done by writing everything in matrix form. If the points $(x_1, y_1), (x_2, y_2), \ldots, (x_n, y_n)$ all lie on the line $y = b + mx$ (that is, if they are collinear), then we have

$$\begin{aligned} y_1 &= b + mx_1 \\ y_2 &= b + mx_2 \\ &\vdots \\ y_n &= b + mx_n \end{aligned}$$

or

$$\mathbf{y} = A\mathbf{u} \tag{2}$$

where

$$\mathbf{y} = \begin{pmatrix} y_1 \\ y_2 \\ \vdots \\ y_n \end{pmatrix}, \quad A = \begin{pmatrix} 1 & x_1 \\ 1 & x_2 \\ \vdots & \vdots \\ 1 & x_n \end{pmatrix}, \quad \text{and} \quad \mathbf{u} = \begin{pmatrix} b \\ m \end{pmatrix} \tag{3}$$

If the points are not collinear, then $\mathbf{y} - A\mathbf{u} \neq \mathbf{0}$ and the problem becomes

> **Vector Form of the Least Squares Problem**
>
> Find a vector $\mathbf{u}$ such that the Euclidean norm
>
> $$|\mathbf{y} - A\mathbf{u}| \tag{4}$$
>
> is a minimum.

Note that in $\mathbb{R}^2$, $|(x, y)| = \sqrt{x^2 + y^2}$; in $\mathbb{R}^3$, $|(x, y, z)| = \sqrt{x^2 + y^2 + z^2}$, etc. Thus minimizing (4) is equivalent to minimizing the sum of the squares in (1).

Finding the minimizing vector $\mathbf{u}$ is not so difficult as it seems. Since $A$ is an $n \times 2$ matrix and $\mathbf{u}$ is a $2 \times 1$ matrix, the vector $A\mathbf{u}$ is a vector in $\mathbb{R}^n$ that belongs to the range of $A$. The range of $A$ is a subspace of $\mathbb{R}^n$ of dimension at most two (since at most two of the columns of $A$ are linearly independent). Thus by the norm approximation theorem in $\mathbb{R}^n$ (Theorem 8 on page 405), (4) is a minimum when

$$A\mathbf{u} = \text{proj}_H \, \mathbf{y}$$

where $H$ is the range of $A$. We illustrate this graphically in the case $n = 3$.

In $\mathbb{R}^3$ the range of $A$ will be a plane or a line passing through the origin (since these are the only subspaces of $\mathbb{R}^3$ of dimension one or two). Look at Figure 4.9. We denote the minimizing vector by $\bar{\mathbf{u}}$. It follows from the figure (and the Pythagorean theorem) that $|\mathbf{y} - A\mathbf{u}|$ is minimized when $\mathbf{y} - A\mathbf{u}$ is orthogonal to the range of $A$.

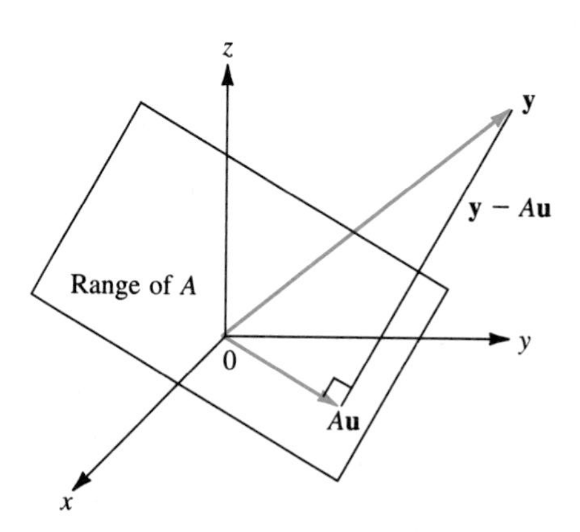


**Figure 4.9** $\mathbf{y} - A\mathbf{u}$ is orthogonal to $A\mathbf{u}$

That is, if $\bar{\mathbf{u}}$ is the minimizing vector, then for every vector $\mathbf{u} \in \mathbb{R}^2$

$$A\mathbf{u} \perp (\mathbf{y} - A\bar{\mathbf{u}}) \tag{5}$$

Using the definition of the scalar product in $\mathbb{R}^n$, we find that (5) becomes

$$A\mathbf{u} \cdot (\mathbf{y} - A\overline{\mathbf{u}}) = 0$$

$$(A\mathbf{u})^t(\mathbf{y} - A\overline{\mathbf{u}}) = 0 \qquad \text{formula (6) on page 124}$$

$$(\mathbf{u}^tA^t)(\mathbf{y} - A\overline{\mathbf{u}}) = 0 \qquad \text{Theorem 1 }(ii)\text{ on page 122}$$

or

$$\mathbf{u}^t(A^t\mathbf{y} - A^tA\overline{\mathbf{u}}) = 0 \tag{6}$$

Equation (6) can hold for every $\mathbf{u} \in \mathbb{R}^2$ only if

$$A^t\mathbf{y} - A^tA\overline{\mathbf{u}} = 0 \tag{7}$$

Solving (7) for $\overline{\mathbf{u}}$, we obtain

**Solution to the Least Square Problem for a Straight-Line Fit**

If $A$ and $\mathbf{y}$ are as in (3), then the line $y = mx + b$ gives the best straight-line fit (in the least squares sense) to the data points $(x_1, y_1), (x_2, y_2), \ldots, (x_n, y_n)$ when $\begin{pmatrix} b \\ m \end{pmatrix} = \overline{\mathbf{u}}$ and

$$\overline{\mathbf{u}} = (A^tA)^{-1}A^t\mathbf{y} \tag{8}$$

Here we have assumed that $A^tA$ is invertible. This is always the case when the $n$ data points are not collinear. The proof of this fact is left to the end of the section.

**EXAMPLE 1** **Finding the Best Straight-Line Fit to Four Points** Find the best straight-line fit to the data points (1, 4), (−2, 5), (3, −1), and (4, 1).

**Solution** Here

$$A = \begin{pmatrix} 1 & 1 \\ 1 & -2 \\ 1 & 3 \\ 1 & 4 \end{pmatrix}, \qquad A^t = \begin{pmatrix} 1 & 1 & 1 & 1 \\ 1 & -2 & 3 & 4 \end{pmatrix} \quad \text{and} \quad \mathbf{y} = \begin{pmatrix} 4 \\ 5 \\ -1 \\ 1 \end{pmatrix}$$

Then

$$A^tA = \begin{pmatrix} 4 & 6 \\ 6 & 30 \end{pmatrix}, \qquad (A^tA)^{-1} = \tfrac{1}{84}\begin{pmatrix} 30 & -6 \\ -6 & 4 \end{pmatrix} \quad \text{and}$$

$$\overline{\mathbf{u}} = (A^tA)^{-1}A^t\mathbf{y} = \tfrac{1}{84}\begin{pmatrix} 30 & -6 \\ -6 & 4 \end{pmatrix}\begin{pmatrix} 1 & 1 & 1 & 1 \\ 1 & -2 & 3 & 4 \end{pmatrix}\begin{pmatrix} 4 \\ 5 \\ -1 \\ 1 \end{pmatrix}$$

$$= \frac{1}{84}\begin{pmatrix} 30 & -6 \\ -6 & 4 \end{pmatrix}\begin{pmatrix} 9 \\ -5 \end{pmatrix} = \frac{1}{84}\begin{pmatrix} 300 \\ -74 \end{pmatrix} \approx \begin{pmatrix} 3.57 \\ -0.88 \end{pmatrix}$$

Therefore the best straight-line fit is given by

$$y = 3.57 - 0.88x$$

This line and the four data points are sketched in Figure 4.10.

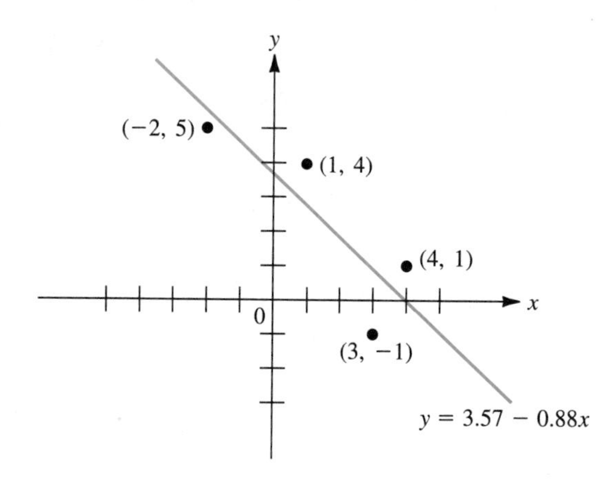


**Figure 4.10** The best straight-line fit to the four points is $y = 3.57 - 0.88x$

## Quadratic Approximation

Here we wish to fit a quadratic to our $n$ data points. Recall that a quadratic in $x$ is any expression of the form

$$y = a + bx + cx^2 \tag{9}$$

Equation (9) is the equation of a parabola in the plane. If the $n$ data points were on the parabola, we would have

$$\begin{aligned} y_1 &= a + bx_1 + cx_1^2 \\ y_2 &= a + bx_2 + cx_2^2 \\ &\vdots \\ y_n &= a + bx_n + cx_n^2 \end{aligned} \tag{10}$$

For

$$\mathbf{y} = \begin{pmatrix} y_1 \\ y_2 \\ \vdots \\ y_n \end{pmatrix}, \quad A = \begin{pmatrix} 1 & x_1 & x_1^2 \\ 1 & x_2 & x_2^2 \\ \vdots & \vdots & \vdots \\ 1 & x_n & x_n^2 \end{pmatrix} \quad \text{and} \quad \mathbf{u} = \begin{pmatrix} a \\ b \\ c \end{pmatrix} \tag{11}$$

(10) can be rewritten as

$$\mathbf{y} = A\mathbf{u}$$

as before. If the data points do not all lie on the same parabola, then $\mathbf{y} - A\mathbf{u} \neq \mathbf{0}$ for any vector $\mathbf{u}$, and our problem is, again,

> Find a vector $\mathbf{u}$ in $\mathbb{R}^3$ such that $|\mathbf{y} - A\mathbf{u}|$ is a minimum.

Using reasoning similar to that used earlier, we can show that if at least three of the $x_i$'s are distinct, then $A^tA$ is invertible and the minimizing vector $\bar{\mathbf{u}}$ is given by

$$\boxed{\bar{\mathbf{u}} = (A^tA)^{-1}A^t\mathbf{y}} \qquad \textbf{(12)}$$

**EXAMPLE 2** **Finding the Best Quadratic Fit to Four Points** Find the best quadratic fit to the data points of Example 1.

**Solution** Here

$$A = \begin{pmatrix} 1 & 1 & 1 \\ 1 & -2 & 4 \\ 1 & 3 & 9 \\ 1 & 4 & 16 \end{pmatrix}, \qquad A^t = \begin{pmatrix} 1 & 1 & 1 & 1 \\ 1 & -2 & 3 & 4 \\ 1 & 4 & 9 & 16 \end{pmatrix} \quad \text{and} \quad \mathbf{y} = \begin{pmatrix} 4 \\ 5 \\ -1 \\ 1 \end{pmatrix}$$

Then

$$A^tA = \begin{pmatrix} 4 & 6 & 30 \\ 6 & 30 & 84 \\ 30 & 84 & 354 \end{pmatrix}, \qquad (A^tA)^{-1} = \tfrac{1}{4752}\begin{pmatrix} 3564 & 396 & -396 \\ 396 & 516 & -156 \\ -396 & -156 & 84 \end{pmatrix}$$

and

$$\bar{\mathbf{u}} = (A^tA)^{-1}A^t\mathbf{y} = \tfrac{1}{4752}\begin{pmatrix} 3565 & 396 & -396 \\ 396 & 516 & -156 \\ -396 & -156 & 84 \end{pmatrix}\begin{pmatrix} 1 & 1 & 1 & 1 \\ 1 & -2 & 3 & 4 \\ 1 & 4 & 9 & 16 \end{pmatrix}\begin{pmatrix} 4 \\ 5 \\ -1 \\ 1 \end{pmatrix}$$

$$= \tfrac{1}{4752}\begin{pmatrix} 3564 & 396 & -396 \\ 396 & 516 & -156 \\ -396 & -156 & 84 \end{pmatrix}\begin{pmatrix} 9 \\ -5 \\ 31 \end{pmatrix} = \tfrac{1}{4752}\begin{pmatrix} 17820 \\ -3852 \\ -180 \end{pmatrix} \approx \begin{pmatrix} 3.75 \\ -0.81 \\ -0.04 \end{pmatrix}$$

Thus the best quadratic fit to the data is given by the parabola

$$\mathbf{y} = 3.75 - 0.81x - 0.04x^2$$

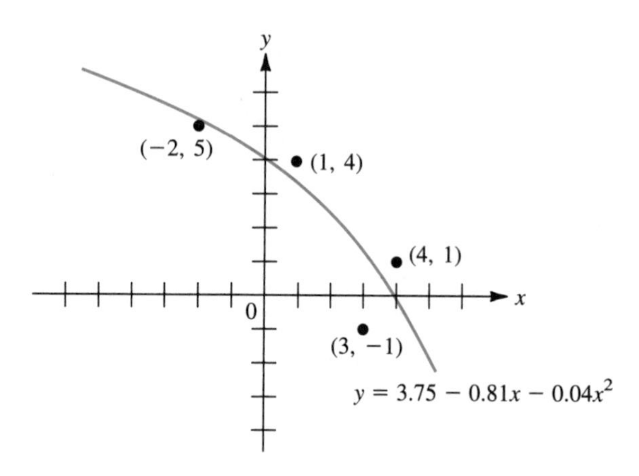


**Figure 4.11** The quadratic $y = 3.75 - 0.81x - 0.04x^2$ is the best quadratic fit to the four points

The parabola and data points are sketched in Figure 4.11.

*Note.* If $n$ is large, then computing $(A^tA)^{-1}$ may lead to large numerical errors. In this case it is much more efficient to find $\overline{\mathbf{u}}$ by solving the system $(A^tA)\overline{\mathbf{u}} = A^t\mathbf{y}$ by Gaussian elimination. In fact, solving $A^tA\overline{\mathbf{u}} = A^ty$ by Gaussian elimination is almost always more efficient than computing $(A^tA)^{-1}$ when $n > 3$.

EXAMPLE 3 **Finding the Best Quadratic Fit to Five Data Points Can Provide an Estimate for $g$** The method of curve-fitting can be used to measure physical constants. Suppose, for example, that an object is dropped from a height of 200 meters. The following measurements are taken:

| Elapsed Time | Height (in meters) |
|---|---|
| 0 | 200 |
| 1 | 195 |
| 2 | 180 |
| 4 | 120 |
| 6 | 25 |

If an object at an initial height of 200 meters is dropped from rest, then its height after $t$ seconds is given by

$$s = 200 - \tfrac{1}{2}gt^2$$

To estimate $g$, we can fit a quadratic to the five data points given above. The coefficients of the $t^2$ term will, if our measurements are accurate, be a reasonable approximation to the number $-\frac{1}{2}g$. Using the earlier notation, we have

$$A = \begin{pmatrix} 1 & 0 & 0 \\ 1 & 1 & 1 \\ 1 & 2 & 4 \\ 1 & 4 & 16 \\ 1 & 6 & 36 \end{pmatrix}, \qquad A^t = \begin{pmatrix} 1 & 1 & 1 & 1 & 1 \\ 0 & 1 & 2 & 4 & 6 \\ 0 & 1 & 4 & 16 & 36 \end{pmatrix} \quad \text{and} \quad \mathbf{y} = \begin{pmatrix} 200 \\ 195 \\ 180 \\ 120 \\ 25 \end{pmatrix}$$

Then

$$A^tA = \begin{pmatrix} 5 & 13 & 57 \\ 13 & 57 & 289 \\ 57 & 289 & 1569 \end{pmatrix}, \qquad (A^tA)^{-1} = \tfrac{1}{7504}\begin{pmatrix} 5912 & -3924 & 508 \\ -3924 & 4596 & -704 \\ 508 & -704 & 116 \end{pmatrix}$$

and

$$\overline{\mathbf{u}} = \tfrac{1}{7504}\begin{pmatrix} 5912 & -3924 & 508 \\ -3924 & 4596 & -704 \\ 508 & -704 & 116 \end{pmatrix}\begin{pmatrix} 1 & 1 & 1 & 1 & 1 \\ 0 & 1 & 2 & 4 & 6 \\ 0 & 1 & 4 & 16 & 36 \end{pmatrix}\begin{pmatrix} 200 \\ 195 \\ 180 \\ 120 \\ 25 \end{pmatrix}$$

$$= \tfrac{1}{7504}\begin{pmatrix} 5912 & -3924 & 508 \\ -3924 & 4596 & -704 \\ 508 & -704 & 116 \end{pmatrix}\begin{pmatrix} 720 \\ 1185 \\ 3735 \end{pmatrix} = \tfrac{1}{7504}\begin{pmatrix} 1504080 \\ -8460 \\ -35220 \end{pmatrix} \approx \begin{pmatrix} 200.44 \\ -1.13 \\ -4.69 \end{pmatrix}$$

Thus the data points are fitted by the quadratic

$$s(t) = 200.44 - 1.13t - 4.69t^2$$

and we have $\frac{1}{2}g \approx 4.69$ or

$$g \approx 2(4.69) = 9.38 \text{ m/sec}^2$$

This is reasonably close to the correct value 9.81 m/sec$^2$. To obtain a more accurate approximation for $g$, we would need to obtain more accurate observations. Note that the term $-1.13t$ represents an initial (downward) velocity of 1.13 m/sec.

We note here that higher-order polynomial approximations are carried out in a virtually identical manner. For details, see Problems 7 and 9.

We conclude this section by proving the result which guarantees that equation (8) will always be valid except when the data points lie on the same vertical line.

**THEOREM 1** Let $(x_1, y_1), (x_2, y_2), \ldots, (x_n, y_n)$ be $n$ points in $\mathbb{R}^2$, and suppose that not all the $x_i$'s are equal. Then if $A$ is given as in (3), the matrix $A^tA$ is an invertible $2 \times 2$ matrix.

*Note.* If $x_1 = x_2 = x_3 = \cdots = x_n$, then all the data points lie on the vertical line $x = x_1$, and the best linear approximation is, of course, this line.

**Proof** We have

$$A = \begin{pmatrix} 1 & x_1 \\ 1 & x_2 \\ \vdots & \vdots \\ 1 & x_n \end{pmatrix}$$

Since not all the $x_i$'s are equal, the columns of $A$ are linearly independent. Now

$$A^tA = \begin{pmatrix} 1 & 1 & \cdots & 1 \\ x_1 & x_2 & \cdots & x_n \end{pmatrix} \begin{pmatrix} 1 & x_1 \\ 1 & x_2 \\ \vdots & \vdots \\ 1 & x_n \end{pmatrix} = \begin{pmatrix} n & \sum_{i=1}^{n} x_i \\ \sum_{i=1}^{n} x_i & \sum_{i=1}^{n} x_i^2 \end{pmatrix}$$

If $A^tA$ is not invertible, then $\det A^tA = 0$. This means that

$$n \sum_{i=1}^{n} x_i^2 = \left( \sum_{i=1}^{n} x_i \right)^2 \tag{13}$$

Let $\mathbf{u} = \begin{pmatrix} 1 \\ 1 \\ \vdots \\ 1 \end{pmatrix}$ and $\mathbf{x} = \begin{pmatrix} x_1 \\ x_2 \\ \vdots \\ x_n \end{pmatrix}$. Then

$$|\mathbf{u}|^2 = \mathbf{u} \cdot \mathbf{u} = n, \qquad |\mathbf{x}|^2 = \sum_{i=1}^{n} x_i^2, \quad \text{and} \quad \mathbf{u} \cdot \mathbf{x} = \sum_{i=1}^{n} x_i$$

so that equation (13) can be restated as

$$|\mathbf{u}|^2 |\mathbf{x}|^2 = |\mathbf{u} \cdot \mathbf{x}|^2$$

or taking square roots, we obtain

$$|\mathbf{u} \cdot \mathbf{x}| = |\mathbf{u}|\,|\mathbf{x}|$$

Now the Cauchy-Schwartz inequality (page 406) states that $|\mathbf{u} \cdot \mathbf{x}| \le |\mathbf{u}|\,|\mathbf{x}|$ with equality if and only if $\mathbf{x}$ is a constant multiple of $\mathbf{u}$. But $\mathbf{u}$ and $\mathbf{x}$ are the columns of $A$ that are linearly independent, by hypothesis. This contradiction proves the theorem. ■

# PROBLEMS 4.10

## Self-Quiz

**I.** The least squares line for the data points (2, 1), (−1, 2), and (3, −5) will minimize

**a.** $[2 - (b + m)]^2 + [-1 - (b + 2m)]^2 + [3 - (b - 5m)]^2$
**b.** $[1 - (b + 2m)]^2 + [2 - (b - m)]^2 + [-5 - (b + 3m)]^2$
**c.** $[1 - (b + 2m)]^2 + |2 - (b - m)| + |-5 - (b + 3m)|$
**d.** $[1 - (b + 2)]^2 + [2 - (b - 1)]^2 + [-5 - (b + 3)]^2$

In Problems 1–3 find the best straight-line fit to the given data points.

**1.** (1, 3), (−2, 4), (7, 0)

**2.** (−3, 7), (4, 9)

**3.** (1, −3), (4, 6), (−2, 5), (3, −1)

In Problems 4–6 find the best quadratic fit to the given data points.

**4.** (2, −5), (3, 0), (1, 1), (4, −2)

**5.** (−7, 3), (2, 8), (1, 5)

**6.** (1, −1), (3, −6), (5, 2), (−3, 1), (7, 4)

**7.** The general cubic is given by

$$a + bx + cx^2 + dx^3$$

Show that the best cubic approximation to $n$ data points is given by

$$\bar{\mathbf{u}} = \begin{pmatrix} a \\ b \\ c \\ d \end{pmatrix} = (A^tA)^{-1}A^t\mathbf{y}$$

where $\mathbf{y}$ is as before, and

$$A = \begin{pmatrix} 1 & x_1 & x_1^2 & x_1^3 \\ 1 & x_2 & x_2^2 & x_2^3 \\ \vdots & \vdots & \vdots & \vdots \\ 1 & x_n & x_n^2 & x_n^3 \end{pmatrix}$$

**8.** Find the best cubic approximation to the data points (3, −2), (0, 3), (−1, 4), (2, −2), and (1, 2).

**9.** The general $k$th-degree polynomial is given by

$$a_0 + a_1x + a_2x^2 + \cdots + a_kx^k$$

---

**Answer to Self-Quiz**

**I.** b

Show that the best $k$th-degree fit to $n$ data points is given by

$$\bar{\mathbf{u}} = \begin{pmatrix} a_0 \\ a_1 \\ \vdots \\ a_k \end{pmatrix} = (A^tA)^{-1}A^t\mathbf{y}$$

where

$$A = \begin{pmatrix} 1 & x_1 & x_1^2 & \cdots & x_1^k \\ 1 & x_2 & x_2^2 & \cdots & x_2^k \\ \vdots & \vdots & \vdots & \cdots & \vdots \\ 1 & x_n & x_n^2 & \cdots & x_n^k \end{pmatrix}$$

**10.** The points $(1, 5.52)$, $(-1, 15.52)$, $(3, 11.28)$, and $(-2, 26.43)$ all lie on a parabola.

**a.** Find the parabola.

**b.** Show that $|\mathbf{y} - A\bar{\mathbf{u}}| = 0$.

**11.** A manufacturer buys large quantities of a certain machine replacement part. He finds that his cost depends on the number of cases bought at the same time and that the cost per unit decreases as the number of cases bought increases. He assumes that cost is a quadratic function of volume, and from past invoices he obtains the following table:

| Number of Cases Bought | Total Cost (dollars) |
|---|---|
| 10 | 150 |
| 30 | 260 |
| 50 | 325 |
| 100 | 500 |
| 175 | 670 |

Find his total cost function.

**12.** A person throws a ball straight into the air. Its height is given by $s(t) = s_0 + v_0t + \frac{1}{2}gt^2$. The following measurements are taken:

| Elapsed Time (seconds) | Height (feet) |
|---|---|
| 1 | 57 |
| 1.5 | 67 |
| 2.5 | 68 |
| 4 | 9.5 |

Using these data, estimate

**a.** The height at which the ball was released

**b.** Its initial velocity

**c.** $g$ (in ft/sec$^2$)

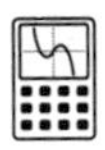

## CALCULATOR BOX

Finding the least squares approximating line is an important problem in statistics. In the statistical context, the procedure for doing so is called **linear regression.** Finding the best quadratic fit is called **quadratic regression,** and so on. Linear regression is a very commonly used statistical tool and virtually all graphing calculators will compute the values $m$ and $b$ once the data points are entered.

### TI-85

All statistical calculations are performed by pressing the STAT key. We will recompute the regression line for the data points in Example 1: (1, 4), (−2, 5), (3, −1), and (4, 1). Press STAT F2 ⟨EDIT⟩ ENTER ENTER. This has the effect of naming the x-coordinates and $y$-coordinates **x Stat** and **y Stat,** respectively. Then enter $x_1 = 1, y_1 = 4, x_2 = -2, y_2 = 5, x_3 = 3, y_3 = -1, x_4 = 4, y_4 = 1$ EXIT.† The data are now entered. Now press F1 ⟨CALC⟩ ENTER ENTER:

**x = x Stat** **y = y Stat**

will be displayed. Now press F2 ⟨LINR⟩ and the following appears:

LinR
**a** = 3.57142857143
**b** = −.880952380952
corr = −.846391670113
**n** = 4

The regression line on the TI-85 is given as $y = a + bx$, so the line is

$$y = 3.57142857143 - 0.880952380952x$$

which is what we obtained (with fewer decimal places) in Example 1. The "corr" stands for "correlation coefficient," which is best explained in a statistics course. The $n = 4$ stands for four data points. To get the best quadratic fit with the same data, start as before until

**x = x** Stat **y = y** Stat

is displayed. Then press MORE F1 ⟨P2REG⟩ and the following is displayed:

P2Reg
**n** = 4
PRegC = {−.037878787879 − .810606060606 3.75}

which denotes the polynomial

$$y = -.037878787879x^2 - .810606060606x + 3.75$$

which is what we obtained in Example 2.

---

† If some data already appear, press the F5 key ⟨CLR$xy$⟩ in the EDIT mode to clear these data.

We can also get third- and fourth-degree polynomial regression curves by pressing [F2] ⟨P3REG⟩ or [F3] ⟨P4REG⟩ instead of [F1] in the last step.

### CASIO fx-7700 GB

The linear regression line can be computed, but not the quadratic regression curve. To begin, press [MODE] [SHIFT] [2] and something like the following appears:

RUN / LIN − REG
S-data: NON-
S-graph: NON-
G-type: REC/CON
angle: Deg
display: Nrm 1

Then press [MODE] [4] to enter the linear regression mode. Enter the data as follows:

1,4 [F1] −2,5 [F1] 3,−1 [F1] 4,1 [F1] [F6] ⟨REG⟩

The linear regression line is given as $y = A + Bx$. Press [F1] ⟨A⟩ [ENTER] and 3.571428571 is displayed; then [F2] ⟨B⟩ [ENTER] yields −0.880952381.

In Problems 13–16 find, to eight decimal places, the regression line for the given data.

**13.** (57, 84); (43, 91), (71, 36); (83, 24); (108, 15); (141, 8)

**14.** (0.32, 14.16); (−0.29, 51.3); (0.58, −13.4); (0.71, −29.8); (0.44, 19.6); (0.88, −46.5)

**15.** (461, 982); (511, 603); (846, 429); (599, 1722); (806, 2415); (1508, 3295); (2409, 5002)

**16.** (−0.0162, −0.0315); (−0.0515, −0.0813); (0.0216, −0.0339); (0.0628, −0.0616); (0.0855, −0.0919); (0.1163, −0.2105); (0.1316, −0.3002); (−0.4416, −0.8519)

In Problems 17–20 find the quadratic regression curve for the given data.

**17.** The data of Problem 13.

**18.** The data of Problem 14.

**19.** The data of Problem 15.

**20.** The data of Problem 16.

## MATLAB 4.10

1. Consider the data set (1, 2), (2, .5), (−1, 4), (3.5, −1), (2.2, .4), and (4, −2). Let **x** be a $6 \times 1$ vector containing the $x$-coordinates and let **y** be a $6 \times 1$ vector containing the $y$-coordinates.
   a. Enter **A = [ones(6, 1) x]** and explain why $A$ is the matrix used in finding the least squares line fit to these data.

b. Find the least squares solution $\mathbf{u} = (A^tA)^{-1}A^t\mathbf{y}$. Find $\mathbf{v} = \mathbf{A\backslash y}$ and compare with $\mathbf{u}$. (The backslash "\" command in MATLAB will find the least squares solution to an overdetermined full rank system.)

c. Find $|\mathbf{y} - A\mathbf{u}|$. Choose **w = u + [.1;−.5]**, find $|\mathbf{y} - A\mathbf{w}|$, and compare with $|\mathbf{y} - A\mathbf{u}|$. Repeat for two more choices of **w**. Explain what part of the theory of least squares approximation this illustrates.

d. The theory of least squares approximation claims that $A\mathbf{u} = \text{proj}_H\mathbf{y}$, where $H$ is the range of $A$ and $\mathbf{u}$ is the least squares solution. Find $\text{proj}_H\mathbf{y}$ by using **B = orth(A)** as in MATLAB Problem 7(a) in Section 4.9. Verify that $A\mathbf{u} = \text{proj}_H\mathbf{y}$.

e. Visualizing the data and the least squares line fit can be useful. The following MATLAB code finds the coefficients for the least squares line fit, generates several $x$-coordinate values (the vector **s**), evaluates the equation of the line at these values, plots the original data set with white *, and plots the least squares line.

***Note.*** Of course, plotting a line does not require evaluating the equation at several values, so finding the vector **s** is not really necessary. However, to plot higher-degree polynomial (or exponential) fits, one needs to evaluate the function at several $x$-coordinate values. The generation of **s** is included here to provide a model MATLAB code that will need only slight modification for other types of fit.

```
u = A\y
s = min(x):(max(x)−min(x))/100:max(x);
fit = u(1)+u(2)*s;
plot(x,y,'w*',s,fit)
```

Does the least squares line seem to be a reasonable fit to these data?

f. Use the equation of the least squares line to approximate a $y$-value for $x = 2.9$.

2. Consider the data in Problem 11 in this section. Let **x** be a $5 \times 1$ vector containing the values for the number of cases bought. Let **y** be the $5 \times 1$ vector containing the values for the corresponding total cost.

a. The problem asks for a least squares quadratic fit. Enter **A = [ones(5,1) x x.^2]** and explain why this is the matrix used for a least squares quadratic fit.

***Note.*** The period (.) before the "^" symbol is important. It tells MATLAB to square each component of the vector **x**.

b. Follow the same directions as in parts (b) through (e) in MATLAB Problem 1 above; except, for part (b), choose **w** to be a $3 \times 1$ vector, for example **w = u + [.1;−.2;−.05]**; for part (e), use **fit = u(1) + u(2)*s + u(3)*s.^2;**.

c. Using the equation of the least squares quadratic, estimate the total cost for 75 cases and estimate the total cost for 200 cases.

3. Work Problem 12 in this section.

4. It is important to look at plots of the data and the least squares solution. A least squares solution can be affected greatly by one or two data points. Some data points could be very different from the rest of the data. These are called **outliers.** Outliers could indicate errors in the data or unusual behavior that could be investigated further.

a. Let **x** and **y** be vectors representing the data points in MATLAB Problem 1 in this section. We will add the point $(1.5, -3.8)$ to the data set. Let $r = 1.5$ and $t = -3.8$. Form **xx = [x;r]** and **yy = [y;t]**.

**i.** Enter the command **plot(xx,yy,'w*')**, locate the additional data point, and explain why it could be considered an outlier.

**ii.** We will plot the least squares line fit to the original data and the least squares line fit to the extended data on the same graph so that they can be compared.

Find **u**, the least squares line solution for the data in **x** and **y**. Find **uu**, the least squares line solution for the data in **xx** and **yy**. Form **s** as in MATLAB Problem 1(e) above using **xx** instead of **x**. Find **fit** as in MATLAB Problem 1(e) above using **u** and find **fit1** using **uu**. Enter the command:

**plot(x, y, 'w*', r, t, 'wo', s, fit, 'r', s, fit1, 'b')**

This command will plot the original data points with white * and the outlier as a white o. The least squares line fit to the original data will be red and the least squares line fit to the extended data will be blue.

**iii.** Write a description of the effect that the outlier has on the least squares line fit. Which line do you think better represents the data?

b. Repeat part (a) for $r = 4.9$ and $t = 4.5$.

5. a. For the data in Calculator Problem 16:

Find $A$, the matrix for a least squares line fit, and then find **u**, the least squares solution.

Find $B$, the matrix for a least squares quadratic fit, and then find **v**, the least squares solution.

Find $|\mathbf{y} - A\mathbf{u}|$ and $|\mathbf{y} - B\mathbf{v}|$.

Plot the data and both least squares curves on the same graph: Generate **s** and **fit** as in MATLAB Problem 1(e) above and generate **fitq = v(1) + v(2)*s + v(3)*s.^2;**. Then enter **plot(x, y, 'w*', s, fit, 'r', s, fitq, 'b')**.

Discuss whether a line or a quadratic is a better fit. Justify your conclusion using the work above.

b. Repeat part (a) for Calculator Problem 14.

6. The following data on fuel efficiency in mi/gal (miles per gallon, mpg) for U.S. passenger cars comes from the *World Almanac*.

| Year | Average mpg for U.S. Passenger Cars |
|---|---|
| 1980 | 15.5 |
| 1981 | 15.9 |
| 1982 | 16.7 |
| 1983 | 17.1 |
| 1984 | 17.8 |
| 1985 | 18.2 |
| 1986 | 18.3 |
| 1987 | 19.2 |
| 1988 | 20.0 |

a. Find a least squares line fit and plot. (Let $x = 0$ represent 1980, $x = 8$ represent 1988, and so on.) Discuss whether the line appears to be a reasonable fit to the data.

b. Assuming the trend continues, use the equation of the line to predict the year that the average mpg will be 25.

7. A manufacturing designer employs your professional services for advice on an experiment that she has run. She is interested in knowing what effect temperature has on the strength of her product. Because of the costs involved, the designer was limited in the amount of data that could be obtained:

| Temperature | Strength Level |
|---|---|
| 600 | 40 |
| 600 | 44 |
| 700 | 48 |
| 700 | 46 |
| 700 | 50 |
| 900 | 48 |
| 950 | 46 |
| 950 | 45 |

Find a least squares line fit *and* a least squares quadratic fit. Plot both of them. Using this analysis, discuss whether you think there is evidence that temperature has an effect on the strength and, if so, discuss what temperature you would recommend to produce the strongest product. (Larger values of strength level indicate a stronger product.)

M

8. On the accompanying disk there is an m-file called *mile.m* that contains data from the *World Almanac* for record times in the mile run and the year that the records were made (from 1880 to 1985).

Enter the command **mile**. This will load the data variables on the file. Nothing will be displayed. The data values for the year are stored in the variable **xm** and the record times are stored in the variable **ym**. To display the data, enter **[xm ym]**.

The values in **xm** are between 80 and 185, where 80 represents the year 1880 and 185 represents the year 1985. The record times in **ym** are in seconds. There are 37 data points.

a. Find a least squares line fit and plot. Is the line a reasonable fit?
b. From the slope of the line, determine the average number of seconds per year that the record time has decreased.
c. If the trend were to continue, predict when the 3-minute mile barrier will be broken; that is, when the record time will be 3 minutes or less. Do you think the trend will continue?

9. **Population Growth** Population growth is often said to be exponential. Least squares line fits can still be valuable if used in conjunction with *reexpression* of the data values. If $x$ and $p$ have an exponential relationship, this means that $p = Ae^{kx}$ for some constants $A$ and $k$. Using properties of logarithms, we find that $\ln(p) = \ln(A) + kx$. Note that $x$ and $\ln(p)$ have a linear relationship.

Thus if we expect an exponential relationship, we reexpress the data $(x, p)$ in terms of the data $(x, \ln(p))$ and find a least squares solution for the reexpressed data. This yields $\ln(p) = mx + b$, and hence $p = e^{mx+b}$ is the exponential fit.

a. Below is population data for the United States for each decade from 1800 to 1900.

| Year | Population (in millions) |
|---|---|
| 1800 | 5.3 |
| 1810 | 7.2 |
| 1820 | 9.6 |
| 1830 | 12.9 |
| 1840 | 17.1 |
| 1850 | 23.2 |
| 1860 | 31.4 |
| 1870 | 38.6 |
| 1880 | 50.2 |
| 1890 | 62.9 |
| 1900 | 76.2 |

Enter **x = [0:10]′**. (Thus the $x$-values are such that $x = 0$ represents 1800 and $x = 10$ represents 1900.) Let **p** be the vector of the corresponding population values. Enter **y = log(p)**;

i. Find the least squares line fit to the data in **x** and **y**. Find **s** and **fit** as in MATLAB Problem 1(e) above. Enter

```
fite = exp(fit);
plot(x, p, '*w', s, fite)
```

Here **exp(fit)** will find the exponential $e^{\text{fit}}$. Does it appear that population growth is exponential?

ii. Assuming that the population continued to grow at the same rate, use the least squares solution to predict the population in 1950. (Find the $y$-value using the least squares line solution and then find the population $p$ using $p = e^y$.)

b. Below is population data for the United States from 1900 to 1980.

| Year | Population (in millions) |
|---|---|
| 1910 | 92.2 |
| 1920 | 106.0 |
| 1930 | 123.2 |
| 1940 | 132.2 |
| 1950 | 151.3 |
| 1960 | 179.3 |
| 1970 | 203.3 |
| 1980 | 226.5 |

i. From these data and your projection of the population in 1950 from part (a), explain why it appears that the rate of growth has slowed in this second century.

ii. Find a least squares exponential fit to these data and plot, following the steps in part (a). For **y**, be sure to use the logarithms of the population values. Is the population growth still exponential?

**iii.** Explain how the coefficients in the least squares solutions from part (a) and part (b) (ii) show that the growth rate has slowed.

**iv.** Assuming the population growth continues as in recent years, predict the population in the year 2000 using the exponential fit from part (b) (ii).

10. **Mineral Geology** Geologists study the composition of rocks and minerals in formations to gather information about the formations. Studying metamorphic rocks and determining such things as the temperature and pressure at which they were formed will provide useful information about the conditions present at the time they were formed. One common mineral is garnet. The Fe-Mg distribution coefficient of a garnet is known to be highly dependent on the temperature at which the garnet was formed. (Here the Fe-Mg distribution coefficient is related to the proportions of iron (Fe) and magnesium (Mg) in the garnet.) However, the amount of calcium (Ca) in a garnet also affects the Fe-Mg distribution coefficient. Corrections to estimates of temperature can be made if the relationship between the amount of calcium present and the Fe-Mg coefficient of the garnet can be determined. The following data were gathered from garnet samples in the Esplanade Range in British Columbia.

| Mole Fraction of Ca | Fe-Mg Distribution Coefficient |
|---|---|
| .1164 | .12128 |
| .0121 | .17185 |
| .0562 | .13365 |
| .0931 | .1485 |
| .0664 | .12637 |
| .1728 | .10406 |
| .1793 | .10703 |
| .1443 | .1189 |
| .1824 | .09952 |

Find the least squares line solution and plot. Use the mole fraction of Ca for the $x$-coordinates and the Fe-Mg distribution coefficient for the $y$-coordinates. Do the data appear to be linearly related? Write the equation of the least squares line.

**PROJECT PROBLEM**

11. **Petroleum Geology** Rock formations are formed in layers. Folds in the rock can be caused by compression deformations. In simple folds, called **detachment anticlines,** when lower layers are compressed, fractures occur and the rock pushes up above its original formation level (called the **regional datum level**). The schematic diagram below represents a cross section.

Oil and gas can be trapped in the part of the fold where fractures occur. There is a level below which no compression has occurred, hence no fracturing, and therefore no oil or gas. This level is called the **detachment level.** It is of interest to estimate the depth of the detachment level since an oil company could then reasonably conclude that it would not pay to drill any deeper than the detachment level to find oil.

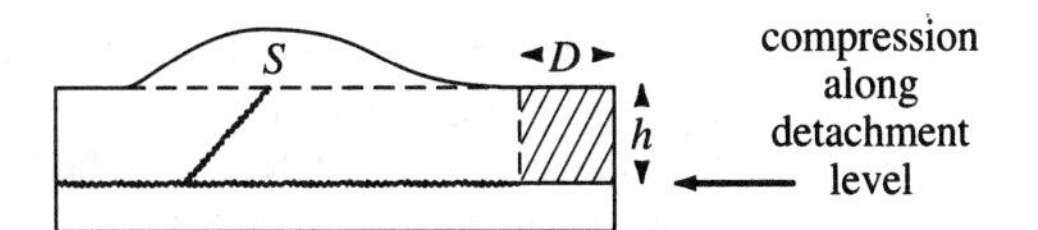

If we assume that a fold has uniform cross sections, then conservation of the volume of the rock would imply that the area of the rock above the regional datum level (labeled $S$ in the diagram) must equal the area of the rock compressed (represented by the shaded area in the diagram). Thus $S = Dh$, where $h$ is the depth to detachment level and $D$ is called the **displacement.** Note that $S$ is linearly related to $h$.

Using seismic images of cross sections, geologists can approximate the excess area ($S$) above regional datum levels at various locations in the fold. A recent method proposed to estimate both the depth to detachment and the displacement involves least squares. The process involves measuring the excess areas ($y$-coordinates) and measuring the depth to some arbitrary fixed reference level ($x$-coordinates). The relationship between the excess area and depth to the reference level will be linear and, in fact, will just be a translation of the line that relates excess area to the depth of detachment. Hence the slope of the line will be an approximation to $D$, the displacement. The depth to detachment will correspond to the $x$-coordinate of the point on the line for which the excess area is 0 (zero) since there is no compression just below this level, and hence no rock is pushed up.

a. The data below were obtained from measurements at various regional datum levels and locations in the Tip Top field, a producing oil field located at the front of the central Wyoming thrust belt.

| **Distance to Reference Level (km)** | **Excess Area ($km^2$)** |
|---|---|
| 3.13 | 2.19 |
| 2.68 | 1.88 |
| 2.50 | 1.73 |
| 2.08 | 1.56 |
| 1.69 | 1.53 |
| 1.37 | 1.39 |
| 1.02 | 1.12 |
| .76 | .96 |
| .53 | .69 |

i. Find the least squares line fit and plot. Does a linear relationship seem reasonable; that is, does it seem reasonable that this fold could be a deformation anticline?

ii. Find the approximation to the displacement and to the depth to detachment. Write a report summarizing the advice you would give to the oil company based on this analysis.

b. There are other kinds of folds, a common one being a **fault-bend fold.** Here there are two levels of interest, an upper and a lower detachment level. Between the lower and upper levels, the excess rock is pushed up. Above the upper level, some excess rock is pushed up and some is pushed (displaced) horizontally. This different structure has different implications for the potential of trapping oil. A careful look at the data and a modified least squares process can indicate the presence of this kind of fold.

For this type of fold, the relationship between depth to detachment and excess area consists of two lines, with the top line having a smaller slope. This would be reflected in the data of the excess area versus the depth to reference level by noticing

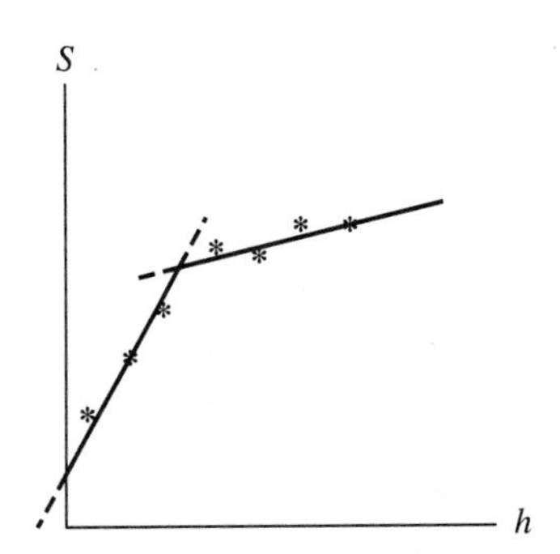


that the data points fall into two natural subsets. A least squares line would be fit to each subset of data. This would be called a **piecewise line fit.** These lines would be translations of the relationship between the excess area and the depth to detachment.

Where the lower line intersects the $h$-axis would be the lower detachment level. The $h$-coordinate of the intersection point of the lines would be the elevation of the upper detachment level above the reference level. The difference in the slopes of the two lines represents the horizontal displacment of rock along the upper detachment level.

For the Tip-Top field data above, we wish to explore if it could be reasonable to interpret the fold as a fault-bend fold.

**i.** First, find the least squares line fit to the whole set of data and find $|\mathbf{y} - A\mathbf{u}|^2$, where $A$ is the matrix used in the least squares fit and $\mathbf{u}$ is the least squares solution. Recall that $|\mathbf{y} - A\mathbf{u}|^2$ measures the sum of the squares of the distances from each of the $y$-values of the data points to the corresponding $y$-value of the least squares line.

**ii.** Next, plot the data points and determine what might be a natural grouping of the points into two line segments. Determine which data values belong in each group. Fit a least squares line to each group and determine $|\mathbf{y} - A\mathbf{u}|^2$ for each group. Add these lengths together to obtain a number that represents the sum of the squares of the distances from each of the $y$-values of the data points to the $y$-value of the piecewise line fit. Compare this with the number obtained in subpart (i) above. Is this piecewise fit better?

**iii.** Continue to experiment with different groupings of the data. Is there one for which the piecewise fit is better?

**iv.** For the best piecewise fit, determine the information that this provides about the detachment levels and the horizontal displacement. [See the paragraph preceding subpart (i).]

c. Write a report to the oil company summarizing your conclusions and recommendations.

***Note.*** The method described above comes from an article entitled "Excess Area and Depth to Detachment" by Jean-Luc Epard and Richard Groshong, Jr., *American Association of Petroleum Geologists Bulletin,* August 1993. (The article also discusses how a quadratic fit to data of excess area versus the depth to reference level would indicate a shear compression.)

**PROJECT PROBLEM** M

12. On the accompanying disk there is a file of data from an introductory astronomy course containing three midterm exam scores and a final exam score. (These are real data.) The professor wants to know if giving a comprehensive final exam significantly affects students' grades for the course.

If you enter the command **astest,** a $110 \times 4$ matrix called $D$ will be loaded. The first three columns of $D$ contain the midterm exam scores and the last column of $D$ contains the final exam scores for 110 students. Each midterm score is out of 30 points as is the final exam score.

a. Let **x = (D(:,1) + D(:,2) + D(:,3))/3;.** Let **y = D(:,4);.** We thus have that **x** contains the average of the midterm exam scores and **y** contains the final exam scores. Find the least squares line fit and plot (use **s = [0:30]**). Does it appear that a line is a reasonable fit to the data?

b. The professor wants to explore further if using the $y$-value of the least squares line for the final exam score (instead of the actual final exam score) has much effect on the students' overall course grade.

   **i.** One measure of the goodness of fit of the line would be to see how many students would receive the same letter grade for the final exam if the $y$-value of the least squares line were used instead of the true final exam score. This can be explored graphically. On the same plot as the data and the least squares line, plot the line given by the points whose $x$-coordinates are in the vector **s** and whose $y$-coordinates are 3 units above the least squares line (use **fit + 3**) and the line whose $y$-coordinates are $-3$ units below the least squares line (use **fit − 3**). (Here **fit** is the vector of $y$-values found by evaluating the least squares line at the values in **s**.)

   Explain why any student represented by a data point within these lines would receive the same letter grade for the final exam if the $y$-value of the least squares line were used. (Remember that the final exam score is out of 30 points. Assume that letter grades are assigned on the basis of 100–90 is an A, 80–89 is a B, and so on.)

   How many students are outside the lines? What is the percentage of students whose letter grade on the final exam would be different if the $y$-value of the least squares line is used?

   **ii.** A possibly better measure of the goodness of fit of the line for our purposes would be to compare the overall grade average using the $y$-value of the least squares line for the final exam score with the overall grade average using the true final exam score. Assume the midterm exams count 60% and the final exam counts 40%. Compute the vector **true,** containing the overall grade averages for each student using the true final exam score. Compute the vector **approx,** containing the overall grade averages for each student using the $y$-value of the least squares line for the final exam score. (First, compute a vector containing the $y$-values of the least squares line for each of the students; that is, for each of the $x$-values in **x**.) Find **true − approx** and determine how many of the components have an absolute value greater than or equal to 3. This will be the number of students whose overall grade would be different if the $y$-value of the least squares line is substituted for the final exam score. Find the percentage of students whose grade would be different.

   **iii.** Write a description to the professor about your findings. Address the concern of the professor about whether a final exam significantly affects students' grades.

c. Experiment to see if other kinds of averages of the midterm grades (for example, having the last exam count more than the first two) might produce a better least squares fit.

## 4.11 INNER PRODUCT SPACES AND PROJECTIONS

This section makes use of a knowledge of elementary properties of complex numbers (summarized in Appendix 2) and some familiarity with material in the first year of calculus.

In Section 1.6 we saw how we could multiply two vectors in $\mathbb{R}^n$ to get a scalar. This scalar product is also called an *inner product.* Other vector spaces have inner products defined on them as well. Before giving a general definition, we note that in $\mathbb{R}^n$ the inner product of two vectors is a real scalar. In other spaces (see Example 2 below) the inner product gives us a complex scalar. To include all cases, therefore, we assume in the following definition that the inner product of two vectors is a complex number.

**DEFINITION 1** **Inner Product Space** A complex vector space $V$ is called an **inner product space** if for every ordered pair of vectors $\mathbf{u}$ and $\mathbf{v}$ in $V$, there is a unique complex number $(\mathbf{u}, \mathbf{v})$, called the **inner product** of $\mathbf{u}$ and $\mathbf{v}$, such that if $\mathbf{u}$, $\mathbf{v}$, and $\mathbf{w}$ are in $V$ and $\alpha \in \mathbb{C}$, then

**i.** $(\mathbf{v}, \mathbf{v}) \geq 0$

**ii.** $(\mathbf{v}, \mathbf{v}) = 0$ if and only if $\mathbf{v} = \mathbf{0}$

**iii.** $(\mathbf{u}, \mathbf{v} + \mathbf{w}) = (\mathbf{u}, \mathbf{v}) + (\mathbf{u}, \mathbf{w})$

**iv.** $(\mathbf{u} + \mathbf{v}, \mathbf{w}) = (\mathbf{u}, \mathbf{w}) + (\mathbf{v}, \mathbf{w})$

**v.** $(\mathbf{u}, \mathbf{v}) = \overline{(\mathbf{v}, \mathbf{u})}$

**vi.** $(\alpha\mathbf{u}, \mathbf{v}) = \alpha(\mathbf{u}, \mathbf{v})$

**vii.** $(\mathbf{u}, \alpha\mathbf{v}) = \overline{\alpha}(\mathbf{u}, \mathbf{v})$

The bar in conditions ($v$) and ($vii$) denotes the complex conjugate.

*Note.* If $(\mathbf{u}, \mathbf{v})$ is real, then $\overline{(\mathbf{u}, \mathbf{v})} = (\mathbf{u}, \mathbf{v})$ and we can remove the bar in ($v$).

EXAMPLE 1 **An Inner Product in $\mathbb{R}^n$** $\mathbb{R}^n$ is an inner product space with $(\mathbf{u}, \mathbf{v}) = \mathbf{u} \cdot \mathbf{v}$. Conditions ($iii$–$vii$) are contained in Theorem 1.6.1 on page 63. Conditions ($i$) and ($ii$) are included in the result (4.9.9) on page 393.

EXAMPLE 2 **An Inner Product in $\mathbb{C}^n$** We defined the space $\mathbb{C}^n$ in Example 4.2.13, page 296. Let $\mathbf{x} = (x_1, x_2, \ldots, x_n)$ and $\mathbf{y} = (y_1, y_2, \ldots, y_n)$ be in $\mathbb{C}^n$. (Remember—this means that the $x_i$'s and $y_i$'s are complex numbers.) Then we define

$$(\mathbf{x}, \mathbf{y}) = x_1\overline{y}_1 + x_2\overline{y}_2 + \cdots + x_n\overline{y}_n \tag{1}$$

To show that equation (1) defines an inner product, we need some facts about complex numbers. If these are unfamiliar, refer to Appendix 2. For (*i*),

$$(\mathbf{x}, \mathbf{x}) = x_1\bar{x}_1 + x_2\bar{x}_2 + \cdots + x_n\bar{x}_n = |x_1|^2 + |x_2|^2 + \cdots + |x_n|^2$$

Thus (*i*) and (*ii*) are satisfied since $|x_i|$ is a real number. Conditions (*iii*) and (*iv*) follow from the fact that $z_1(z_2 + z_3) = z_1z_2 + z_1z_3$ for any complex numbers $z_1$, $z_2$, and $z_3$. Condition (*v*) follows from the fact that $\overline{z_1z_2} = \overline{z_1}\,\overline{z_2}$ and $\bar{\bar{z}}_1 = z_1$ so that $\overline{x_1\bar{y}_1} = \bar{x}_1y_1$. Condition (*vi*) is obvious. For (*vii*), $(\mathbf{u}, \alpha\mathbf{v}) = \overline{(\alpha\mathbf{v}, \mathbf{u})} = (\overline{\alpha\mathbf{v}}, \overline{\mathbf{u}}) = \bar{\alpha}(\overline{\mathbf{v}}, \overline{\mathbf{u}}) = \bar{\alpha}(\mathbf{u}, \mathbf{v})$. Here we used (*vi*) and (*v*).

**EXAMPLE 3** **The Inner Product of Two Vectors in $\mathbb{C}^3$** In $\mathbb{C}^3$ let $\mathbf{x} = (1 + i, -3, 4 - 3i)$ and $\mathbf{y} = (2 - i, -i, 2 + i)$. Then

$$\begin{aligned}(\mathbf{x}, \mathbf{y}) &= (1 + i)(\overline{2 - i}) + (-3)(\overline{-i}) + (4 - 3i)(\overline{2 + i}) \\ &= (1 + i)(2 + i) + (-3)(i) + (4 - 3i)(2 - i) \\ &= (1 + 3i) - 3i + (5 - 10i) = 6 - 10i\end{aligned}$$

**EXAMPLE 4** Calculus **An Inner Product in $C[a, b]$** Suppose that $a < b$; let $V = C[a, b]$, the space of real-valued functions that are continuous on the interval $[a, b]$, and define

$$(f, g) = \int_a^b f(t)g(t)\,dt \tag{2}$$

We shall see that this is also an inner product.†

(*i*) $(f, f) = \int_a^b f^2(t)\,dt \geq 0$. It is a basic theorem of calculus that if $f \in C[a, b]$, $f \geq 0$ on $[a, b]$, and $\int_a^b f(t)\,dt = 0$, then $f = 0$ on $[a, b]$. This proves (*i*) and (*ii*). (*iii–vii*) follow from basic facts about definite integrals.

*Note.* In $C[a, b]$ the scalars are assumed to be real numbers and the functions are real-valued so that we do not have to worry about complex conjugates. However, if the functions are complex-valued, then we can still define an inner product. See Problem 27 for details.

**EXAMPLE 5** Calculus **The Inner Product of Two Functions in $C[0, 1]$** Let $f(t) = t^2 \in C[0, 1]$ and $g(t) = (4 - t) \in C[0, 1]$. Then

$$(f, g) = \int_0^1 t^2(4 - t)\,dt = \int_0^1 (4t^2 - t^3)\,dt = \left(\frac{4t^3}{3} - \frac{t^4}{4}\right)\Big|_0^1 = \frac{13}{12}$$

† This is not the only way to define an inner product on $C[a, b]$, but it is the most common one.

**DEFINITION 2** Let $V$ be an inner product space and suppose that $\mathbf{u}$ and $\mathbf{v}$ are in $V$. Then

**i.** $\mathbf{u}$ and $\mathbf{v}$ are **orthogonal** if $(\mathbf{u}, \mathbf{v}) = 0$.

**ii.** The **norm** of $\mathbf{u}$, denoted by $\|\mathbf{u}\|$, is given by

$$\|\mathbf{u}\| = \sqrt{(\mathbf{u}, \mathbf{u})} \tag{3}$$

*Note 1.* We use double bars instead of single bars here to avoid confusion with absolute value. For example, in Example 7 $\|\sin t\|$ denotes the norm of $\sin t$ as a "vector" in $C[0, 2\pi]$, while $|\sin t|$ denotes the absolute value of the function $\sin t$.

*Note 2.* Equation (3) makes sense since $(\mathbf{u}, \mathbf{u}) \geq 0$.

**EXAMPLE 6** **Two Orthogonal Vectors in $\mathbb{C}^2$** In $\mathbb{C}^2$ the vectors $(3, -i)$ and $(2, 6i)$ are orthogonal because

$$((3, -i), (2, 6i)) = 3 \cdot \overline{2} + (-i)(\overline{6i}) = 6 + (-i)(-6i) = 6 - 6 = 0$$

Also $\|(3, -i)\| = \sqrt{3 \cdot 3 + (-i)(\overline{i})} = \sqrt{10}$. ■

**EXAMPLE 7** **Two Orthogonal Functions in $C[0, 2\pi]$** In $C[0, 2\pi]$ the functions $\sin t$ and $\cos t$ are orthogonal since

$$(\sin t, \cos t) = \int_0^{2\pi} \sin t \cos t \, dt = \frac{1}{2}\int_0^{2\pi} \sin 2t \, dt = -\frac{\cos 2t}{4}\Big|_0^{2\pi} = 0$$

Also,

$$\begin{aligned} \|\sin t\| &= (\sin t, \sin t)^{1/2} \\ &= \left[\int_0^{2\pi} \sin^2 t \, dt\right]^{1/2} \\ &= \left[\frac{1}{2}\int_0^{2\pi} (1 - \cos 2t) \, dt\right]^{1/2} \\ &= \left[\frac{1}{2}\left(t - \frac{\sin 2t}{2}\right)\Big|_0^{2\pi}\right]^{1/2} \\ &= \sqrt{\pi} \end{aligned}$$ ■

If you look at the proofs of Theorems 4.9.1 and 4.9.2 on pages 394 and 395, you will see that no use was made of the fact that $V = \mathbb{R}^n$. The same theorems are true in any inner product space $V$. We list them for convenience after giving a definition.

**DEFINITION 3** **Orthonormal Set** The set of vectors $\{\mathbf{v}_1, \mathbf{v}_2, \ldots, \mathbf{v}_n\}$ is an **orthonormal set** in $V$ if

$$(\mathbf{v}_i, \mathbf{v}_j) = 0 \qquad \text{for } i \neq j \tag{4}$$

and

$$\|\mathbf{v}_i\| = \sqrt{(\mathbf{v}_i, \mathbf{v}_i)} = 1 \tag{5}$$

If only (4) holds, the set is said to be **orthogonal.**

**THEOREM 1** Any finite orthogonal set of nonzero vectors in an inner product space is linearly independent.

**THEOREM 2** Any finite, linearly independent set in an inner product space can be made into an orthonormal set by the Gram-Schmidt process. In particular, any finite dimensional inner product space has an orthonormal basis.

**EXAMPLE 8** Calculus

**An Orthonormal Basis in $P_2[0, 1]$** Construct an orthonormal basis for $P_2[0, 1]$.

**Solution** We start with the standard basis $\{1, x, x^2\}$. Since $P_2[0, 1]$ is a subspace of $C[0, 1]$, we may use the inner product of Example 4. Since $\int_0^1 1^2\,dx = 1$, we let $\mathbf{u}_1 = 1$. Then $\mathbf{v}_1' = \mathbf{v}_2 - (\mathbf{v}_2, \mathbf{u}_1)\mathbf{u}_1$. Here $(\mathbf{v}_2, \mathbf{u}_1) = \int_0^1 (x \cdot 1)\,dx = \frac{1}{2}$. Thus $\mathbf{v}_2' = x - \frac{1}{2} \cdot 1 = x - \frac{1}{2}$. Next we compute

$$\|x - \tfrac{1}{2}\| = \left[\int_0^1 (x - \tfrac{1}{2})^2\,dx\right]^{1/2} = \left[\int_0^1 (x^2 - x + \tfrac{1}{4})\,dx\right]^{1/2} = \frac{1}{\sqrt{12}} = \frac{1}{2\sqrt{3}}$$

Hence $\mathbf{u}_2 = 2\sqrt{3}(x - \frac{1}{2}) = \sqrt{3}(2x - 1)$. Then

$$\mathbf{v}_3' = \mathbf{v}_3 - (\mathbf{v}_3, \mathbf{u}_1)\mathbf{u}_1 - (\mathbf{v}_3, \mathbf{u}_2)\mathbf{u}_2$$

We have $(\mathbf{v}_3, \mathbf{u}_1) = \int_0^1 x^2\,dx = \frac{1}{3}$ and

$$(\mathbf{v}_3, \mathbf{u}_2) = \sqrt{3}\int_0^1 x^2(2x - 1)\,dx = \sqrt{3}\int_0^1 (2x^3 - x^2)\,dx = \frac{\sqrt{3}}{6}$$

Thus

$$\mathbf{v}_3' = x^2 - \tfrac{1}{3} - \frac{\sqrt{3}}{6}[\sqrt{3}(2x - 1)] = x^2 - x + \tfrac{1}{6}$$

and

$$\|\mathbf{v}_3'\| = \left[\int_0^1 (x^2 - x + \tfrac{1}{6})^2\,dx\right]^{1/2}$$
$$= \left[\int_0^1 \left(x^4 - 2x^3 + \tfrac{4}{3}x^2 - \frac{x}{3} + \tfrac{1}{36}\right) dx\right]^{1/2}$$
$$= \left[\left(\frac{x^5}{5} - \frac{x^4}{2} + \frac{4x^3}{9} - \frac{x^2}{6} + \frac{x}{36}\right)\Big|_0^1\right]^{1/2}$$
$$= \frac{1}{\sqrt{180}} = \frac{1}{6\sqrt{5}}$$

Thus $\mathbf{u}_3 = 6\sqrt{5}(x^2 - x + \frac{1}{6}) = \sqrt{5}(6x^2 - 6x + 1)$. Finally, an orthonormal basis is $\{1, \sqrt{3}(2x - 1), \sqrt{5}(6x^2 - 6x + 1)\}$.

EXAMPLE 9

Calculus

**An Infinite Orthonormal Set in $C[0, 2\pi]$** In $C[0, 2\pi]$ the infinite set

$$S = \left\{\frac{1}{\sqrt{2\pi}}, \frac{1}{\sqrt{\pi}}\sin x, \frac{1}{\sqrt{\pi}}\cos x, \frac{1}{\sqrt{\pi}}\sin 2x, \frac{1}{\sqrt{\pi}}\cos 2x, \ldots, \frac{1}{\sqrt{\pi}}\sin nx, \frac{1}{\sqrt{\pi}}\cos nx, \ldots\right\}$$

is an orthonormal set. This follows since if $m \neq n$, then

$$\int_0^{2\pi} \sin mx \cos nx\,dx = \int_0^{2\pi} \sin mx \sin nx\,dx = \int_0^{2\pi} \cos mx \cos nx\,dx = 0$$

To prove one of these, we note that

$$\int_0^{2\pi} \sin mx \cos nx\,dx = \frac{1}{2}\int_0^{2\pi} [\sin(m+n)x + \sin(m-n)x]\,dx$$
$$= -\frac{1}{2}\left[\frac{\cos(m+n)x}{m+n} + \frac{\cos(m-n)x}{m-n}\right]\Big|_0^{2\pi}$$
$$= 0$$

since $\cos x$ is periodic of period $2\pi$. We have seen that $|\sin x| = \sqrt{\pi}$. Thus $\|(1/\sqrt{\pi})\sin x\| = 1$. The other facts follow in a similar fashion. This example provides a situation in which we have an *infinite* orthonormal set. In fact, although this is far beyond us in this elementary text, it is true that some functions in $C[0, 2\pi]$ can be written as linear combinations of functions in $S$. Suppose $f \in C[0, 2\pi]$. Then if we write $f$ as an infinite linear combination of the vectors in $S$, we obtain what is called the **Fourier series representation** of $f$.

**DEFINITION 4** **Orthogonal Projection** Let $H$ be a subspace of an inner product space $V$ with an orthonormal basis $\{\mathbf{u}_1, \mathbf{u}_2, \ldots, \mathbf{u}_k\}$. If $\mathbf{v} \in V$, then the **orthogonal projection** of $\mathbf{v}$ onto $H$, denoted by $\text{proj}_H \mathbf{v}$, is given by

$$\text{proj}_H \mathbf{v} = (\mathbf{v}, \mathbf{u}_1)\mathbf{u}_1 + (\mathbf{v}, \mathbf{u}_2)\mathbf{u}_2 + \cdots + (\mathbf{v}, \mathbf{u}_k)\mathbf{u}_k \tag{6}$$

The following theorems have proofs that are identical to their $\mathbb{R}^n$ counterparts proved in Section 4.9.

**THEOREM 3** Let $H$ be a subspace of the finite dimensional inner product space $V$. Suppose that $H$ has two orthonormal bases $\{\mathbf{u}_1, \mathbf{u}_2, \ldots, \mathbf{u}_k\}$ and $\{\mathbf{w}_1, \mathbf{w}_2, \ldots, \mathbf{w}_k\}$. Let $\mathbf{v} \in V$. Then

$$(\mathbf{v}, \mathbf{u}_1)\mathbf{u}_1 + (\mathbf{v}, \mathbf{u}_2)\mathbf{u}_2 + \cdots + (\mathbf{v}, \mathbf{u}_k)\mathbf{u}_k = (\mathbf{v}, \mathbf{w}_1)\mathbf{w}_1 + (\mathbf{v}, \mathbf{w}_2)\mathbf{w}_2 + \cdots + (\mathbf{v}, \mathbf{w}_k)\mathbf{w}_k$$

**DEFINITION 5** **Orthogonal Complement** Let $H$ be a subspace of the inner product space $V$. Then the **orthogonal complement** of $H$, denoted by $H^\perp$, is given by

$$H^\perp = \{\mathbf{x} \in V\colon\ (\mathbf{x}, \mathbf{h}) = 0 \quad \text{for every } \mathbf{h} \in H\} \tag{7}$$

**THEOREM 4** If $H$ is a subspace of the inner product space $V$, then

**i.** $H^\perp$ is a subspace of $V$.

**ii.** $H \cap H^\perp = \{\mathbf{0}\}$.

**iii.** $\dim H^\perp = n - \dim H$ if $\dim V = n < \infty$.

**THEOREM 5** **Projection Theorem** Let $H$ be a finite dimensional subspace of the inner product space $V$ and suppose that $\mathbf{v} \in V$. Then there exists a unique pair of vectors $\mathbf{h}$ and $\mathbf{p}$ such that $\mathbf{h} \in H$, $\mathbf{p} \in H^\perp$, and

$$\mathbf{v} = \mathbf{h} + \mathbf{p} \tag{8}$$

where $\mathbf{h} = \text{proj}_H \mathbf{v}$.

If $V$ is finite dimensional, then $\mathbf{p} = \text{proj}_{H^\perp} \mathbf{v}$.

*Remark.* If you look at the proof of Theorem 4.9.7, you will notice that (8) holds even if $V$ is infinite dimensional. The only difference is that if $V$ is infinite dimensional, then $H^\perp$ is infinite dimensional (since $H$ is finite dimensional), and so $\text{proj}_{H^\perp}\,\mathbf{v}$ is not defined.

**THEOREM 6** **Norm Approximation Theorem** Let $H$ be a finite dimensional subspace of the inner product space $V$ and let $\mathbf{v}$ be a vector in $V$. Then, in $H$, $\text{proj}_H\,\mathbf{v}$ is the best approximation to $\mathbf{v}$ in the following sense: If $\mathbf{h}$ is any other vector in $H$, then

$$\|\mathbf{v} - \text{proj}_H\,\mathbf{v}\| < \|\mathbf{v} - \mathbf{h}\| \tag{9}$$

**EXAMPLE 10** Calculus

**Computing a Projection onto $P_2[0, 1]$** Since $P_2[0, 1]$ is a finite dimensional subspace of $C[0, 1]$, we can talk about $\text{proj}_{P_2[0,1]} f$ if $f \in C[0, 1]$. If $f(x) = e^x$, for example, we compute $\text{proj}_{P_2[0,1]}\, e^x$. Since $\{\mathbf{u}_1, \mathbf{u}_2, \mathbf{u}_3\} = \{1, \sqrt{3}(2x - 1), \sqrt{5}(6x^2 - 6x + 1)\}$ is an orthonormal basis in $P_2[0, 1]$ by Example 8, we have

$$\begin{aligned}\text{proj}_{P_2[0,1]}\, e^x &= (e^x, 1)1 + (e^x, \sqrt{3}(2x-1))\sqrt{3}(2x-1)\\ &\quad + (e^x, \sqrt{5}(6x^2 - 6x + 1))\sqrt{5}(6x^2 - 6x + 1)\end{aligned}$$

We shall spare you the computations. Using the fact that $\int_0^1 e^x\,dx = e - 1$, $\int_0^1 xe^x\,dx = 1$, and $\int_0^1 x^2e^x\,dx = e - 2$, we obtain $(e^x, 1) = e - 1$, $(e^x, \sqrt{3}(2x - 1)) = \sqrt{3}(3 - e)$, and $(e^x, \sqrt{5}(6x^2 - 6x + 1)) = \sqrt{5}(7e - 19)$. Finally,

$$\begin{aligned}\text{proj}_{P_2[0,1]}\, e^x &= (e - 1) + \sqrt{3}(3 - e)\sqrt{3}(2x - 1)\\ &\quad + \sqrt{5}(7e - 19)(\sqrt{5})(6x^2 - 6x + 1)\\ &= (e - 1) + (9 - 3e)(2x - 1)\\ &\quad + 5(7e - 19)(6x^2 - 6x + 1)\\ &\approx 1.01 + 0.85x + 0.84x^2\end{aligned}$$

We conclude this section with an application of the norm approximation theorem.

**Mean Square Approximation of a Continuous Function**

Let $f \in C[a, b]$. We wish to approximate $f$ by an $n$th-degree polynomial. What is the polynomial that does this with the smallest error?

In order to answer this question, we must define what we mean by *error*. There are many different ways to define error. Three are given below:

$$\textbf{Maximum error} = \max|f(x) - g(x)| \quad \text{for } x \in [a, b] \tag{10}$$

$$\textbf{Area error} = \int_a^b |f(x) - g(x)|\,dx \tag{11}$$

$$\textbf{Mean square error} = \int_a^b |f(x) - g(x)|^2\,dx \tag{12}$$

**EXAMPLE 11** Calculus

**Computing Errors** Let $f(x) = x^2$ and $g(x) = x^3$ on $[0, 1]$. On $[0, 1]$, $x^2 \geq x^3$, so $|x^2 - x^3| = x^2 - x^3$. Then

**i.** Maximum error $= \max (x^2 - x^3)$. To compute this, we compute $d/dx\,(x^2 - x^3) = 2x - 3x^2 = x(2 - 3x) = 0$ when $x = 0$ and $x = 2/3$. The maximum error occurs when $x = 2/3$ and is given by $[(\frac{2}{3})^2 - (\frac{2}{3})^3] = \frac{4}{9} - \frac{8}{27} = \frac{4}{27} \approx 0.148$.

**ii.** Area error $= \int_0^1 (x^2 - x^3)\,dx = (x^3/3 - x^4/4)|_0^1 = \frac{1}{3} - \frac{1}{4} = \frac{1}{12} \approx 0.083$. This is sketched in Figure 4.12.

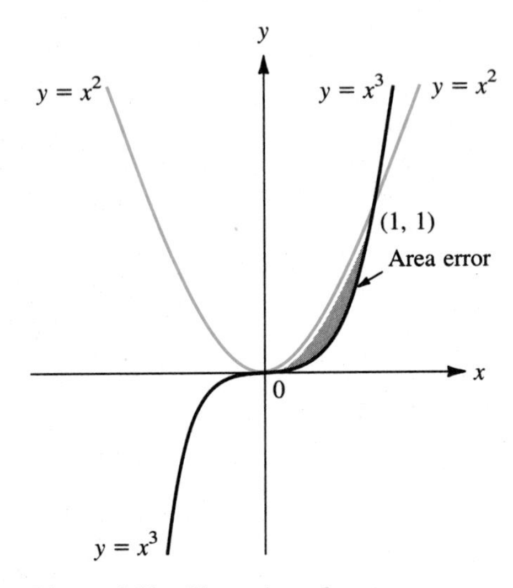


**Figure 4.12** Illustration of area error

**iii.** Mean square error $= \int_0^1 (x^2 - x^3)^2\,dx = \int_0^1 (x^4 - 2x^5 + x^6)\,dx = (x^5/5 - x^6/3 + x^7/7)|_0^1 = \frac{1}{5} - \frac{1}{3} + \frac{1}{7} = \frac{1}{105} \approx 0.00952$

Each of the three measurements of error is useful. The mean square error is used in statistics and other applications. We can use the norm approximation theorem to find the unique $n$th-degree polynomial that approximates a given continuous function with the smallest mean square error.

From Example 4, $C[a, b]$ is an inner product space with

$$(f, g) = \int_a^b f(t)g(t)\,dt \tag{13}$$

For every positive integer $n$, $P_n[a, b]$—the space of $n$th-degree polynomials defined on $[a, b]$—is a finite dimensional subspace of $C[a, b]$. We compute, for $f \in C[a, b]$ and $p_n \in P_n$,

$$\|f - p_n\|^2 = (f - p_n, f - p_n) = \int_a^b [(f(t) - p_n(t))(\overline{f(t) - p_n(t)})]\, dt$$
$$= \int_a^b |f(t) - p_n(t)|^2\, dt = \text{mean square error}$$

Thus by Theorem 6

> The $n$th-degree polynomial that approximates a continuous function with the smallest mean square error is given by
> $$p_n = \mathbf{proj}_{P_n} f$$ **(14)**

EXAMPLE 12 **The Best Mean Square Quadratic Approximation to $e^x$** From Example 10, the second-degree polynomial that best approximates $e^x$ on [0, 1] in the mean square sense is given by

$$p_2(x) \approx 1.01 + 0.85x + 0.84x^2$$

# PROBLEMS 4.11

## Self-Quiz

**I.** In $C[0, 1]$, $(x, x^3) =$ ______.

**a.** $\frac{1}{2}$ **b.** $\frac{1}{3}$ **c.** $\frac{1}{4}$ **d.** $\frac{1}{5}$ **e.** $\frac{1}{6}$

**II.** In $C[0, 1]$, $\|x^2\|^2 =$ ______.

**a.** $\frac{1}{2}$ **b.** $\frac{1}{3}$ **c.** $\frac{1}{4}$ **d.** $\frac{1}{5}$ **e.** $\frac{1}{6}$

**III.** In $\mathbb{C}^2$, $((1 + i, 2 - 3i), (2 - i, -1 + 2i)) =$ ______.

**a.** $-7 + 2i$ **b.** $7 + 8i$ **c.** $4 - 3i$ **d.** $4 + 3i$ **e.** $-2 + 5i$

**IV.** In $\mathbb{C}^2$, $\|(1 + i, 2 - 3i)\| =$ ______.

**a.** $-5 - 10i$ **b.** 15 **c.** $\sqrt{15}$ **d.** 7 **e.** $\sqrt{7}$

### *True-False*

**V.** If $H$ is a finite dimensional subspace of the inner product space $V$ and if $\mathbf{v} \in V$, then there are vectors $\mathbf{h} \in H$ and $\mathbf{p} \in H^\perp$ such that $\mathbf{v} = \mathbf{h} + \mathbf{p}$.

**VI.** In Problem V, $\mathbf{h} = \mathbf{proj}_H\, \mathbf{v}$ and $\mathbf{p} = \mathbf{proj}_{H^\perp}\, \mathbf{v}$.

---

Answers to Self-Quiz

**I.** d **II.** d **III.** a **IV.** c **V.** True **VI.** False (True only if $V$ is finite dimensional.)

**1.** Let $D_n$ denote the set of $n \times n$ diagonal matrices with real components under the usual matrix operations. If $A$ and $B$ are in $D_n$, define

$$(A, B) = a_{11}b_{11} + a_{22}b_{22} + \cdots + a_{nn}b_{nn}$$

Prove that $D_n$ is an inner product space.

**2.** If $A \in D_n$, show that $\|A\| = 1$ if and only if $a_{11}^2 + a_{22}^2 + \cdots + a_{nn}^2 = 1$.

**3.** Find an orthonormal basis for $D_n$.

**4.** Find an orthonormal basis for $D_2$ starting with $A = \begin{pmatrix} 2 & 0 \\ 0 & 1 \end{pmatrix}$ and $B = \begin{pmatrix} -3 & 0 \\ 0 & 4 \end{pmatrix}$.

**5.** In $\mathbb{C}^2$ find an orthonormal basis starting with the basis $(1, i)$, $(2 - i, 3 + 2i)$.

Calculus **6.** Find an orthonormal basis for $P_3[0, 1]$.

Calculus **7.** Find an orthonormal basis for $P_2[-1, 1]$. The polynomials you obtain are called **normalized Legendre polynomials.**

*Calculus **8.** Find an orthonormal basis for $P_2[a, b]$, $a < b$.

**9.** If $A = (a_{ij})$ is a real $n \times n$ matrix, the **trace** of $A$, written tr $A$, is the sum of the diagonal components of $A$: $\text{tr } A = a_{11} + a_{22} + \cdots + a_{nn}$. In $M_{nn}$ define $(A, B) = \text{tr } (AB^t)$. Prove that with the preceding inner product, $M_{nn}$ is an inner product space.

**10.** If $A \in M_{nn}$, show that $\|A\|^2 = \text{tr } (AA^t)$ is the sum of the squares of the elements of $A$. [*Note:* Here $\|A\| = (A, A)^{1/2}$; use the notation of Problem 9.]

**11.** Find an orthonormal basis for $M_{22}$.

**12.** We can think of the complex plane as a vector space over the reals with basis vectors 1, $i$. If $z = a + ib$ and $w = c + id$, define $(z, w) = ac + bd$. Show that this is an inner product and that $\|z\|$ is the usual length of a complex number.

**13.** Let $a$, $b$, and $c$ be three distinct real numbers. Let $p$ and $q$ be in $P_2$ and define $(p, q) = p(a)q(a) + p(b)q(b) + p(c)q(c)$.

**a.** Prove that $(p, q)$ is an inner product in $P_2$.

**b.** Is $(p, q) = p(a)q(a) + p(b)q(b)$ an inner product?

**14.** In $\mathbb{R}^2$, if $\mathbf{x} = \begin{pmatrix} x_1 \\ x_2 \end{pmatrix}$ and $\mathbf{y} = \begin{pmatrix} y_1 \\ y_2 \end{pmatrix}$, let $(\mathbf{x}, \mathbf{y})_* = x_1y_1 + 3x_2y_2$. Show that $(x, y)$ is an inner product on $\mathbb{R}^2$.

**15.** With the inner product of Problem 14, calculate $\left\| \begin{pmatrix} 2 \\ -3 \end{pmatrix} \right\|_*$.

**16.** In $\mathbb{R}$ let $(\mathbf{x}, \mathbf{y}) = x_1y_1 - x_2y_2$. Is this an inner product? If not, why not?

***17.** Let $V$ be an inner product space. Prove that $|(\mathbf{u}, \mathbf{v})| \le |\mathbf{u}|\,|\mathbf{v}|$. This is called the **Cauchy-Schwarz inequality.** [*Hint:* See Theorem 9 in Section 4.9.]

***18.** Using the result of Problem 17, prove that $\|\mathbf{u} + \mathbf{v}\| \le \|\mathbf{u}\| + \|\mathbf{v}\|$. This is called the **triangle inequality.**

Calculus **19.** In $P_3[0, 1]$ let $H$ be the subspace spanned by $\{1, x^2\}$. Find $H^\perp$.

*Calculus **20.** In $C[-1, 1]$ let $H$ be the subspace of even functions. Show that $H^\perp$ consists of odd functions. [*Hint:* $f$ is odd if $f(-x) = -f(x)$ and is even if $f(-x) = f(x)$.]

*Calculus **21.** $H = P_2[0, 1]$ is a subspace of $P_3[0, 1]$. Write the polynomial $1 + 2x + 3x^2 - x^3$ as $h(x) + p(x)$, where $h(x) \in H$ and $p(x) \in H^\perp$.

***22.** Find a second-degree polynomial that best approximates $\sin \frac{\pi}{2} x$ on the interval $[0, 1]$ in the mean square sense.

***23.** Solve Problem 22 for the function $\cos \frac{\pi}{2} x$.

**24.** Let $A$ be an $m \times n$ matrix with complex entries. Then the **conjugate transpose** of $A$, denoted by $A^*$, is defined by $(A^*)_{ij} = \overline{a_{ji}}$. Compute $A^*$ if

$$A = \begin{pmatrix} 1 - 2i & 3 + 4i \\ 2i & -6 \end{pmatrix}$$

**25.** Let $A$ be an invertible $n \times n$ matrix with complex entries. $A$ is called **unitary** if $A^{-1} = A^*$. Show that the following matrix is unitary:

$$A = \begin{pmatrix} \frac{1}{\sqrt{2}} & -\frac{1}{2} + \frac{i}{2} \\ \frac{1}{\sqrt{2}} & \frac{1}{2} - \frac{i}{2} \end{pmatrix}$$

***26.** Show that an $n \times n$ matrix with complex entries is unitary if and only if the columns of $A$ constitute an orthonormal basis for $\mathbb{C}^n$.

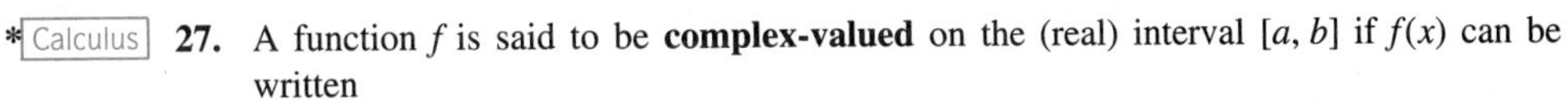

*Calculus **27.** A function $f$ is said to be **complex-valued** on the (real) interval $[a, b]$ if $f(x)$ can be written

$$f(x) = f_1(x) + f_2(x)i, \qquad x \in [a, b]$$

where $f_1$ and $f_2$ are real-valued functions. The complex-valued function $f$ is **continuous** if $f_1$ and $f_2$ are continuous. Let $CV[a, b]$ denote the set of complex-valued functions that are continuous on $[a, b]$. For $f$ and $g$ in $CV[a, b]$, define

$$(f, g) = \int_a^b f(x)\overline{g(x)}\, dx \tag{15}$$

Show that (15) defines an inner product in $CV[a, b]$.

Calculus **28.** Show that $f(x) = \sin x + i \cos x$ and $g(x) = \sin x - i \cos x$ are **orthogonal** in $CV[0, \pi]$.

Calculus **29.** Compute $\|\sin x + i \cos x\|$ in $CV[0, \pi]$.

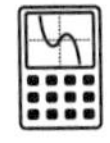

## CALCULATOR BOX

Many of the computations in this section can be carried out on most graphing calculators. In particular, most graphing calculators can perform complex arithmetic and approximate definite integrals.

### TI-85

To compute a definite integral, press [2nd] [CALC] to get into calculus mode. Then [F5] ⟨fnint⟩ will display fnInt(.To compute $\int_a^b f(x)\, dx$, enter $f(x), x, a, b)$ [ENTER] and

and the value of the integral will be displayed. For example, fnInt ($X$^4 − 2$X$^5 + $X$^6, $X$, 0, 1) ENTER results in .009523809524. That is, $\int_0^1 (x^4 - 2x^5 + x^6)\,dx \approx$ 0.009523809524 (see Example 11).

**CASIO fx-7700 GB**

Press SHIFT $\int dx$ to display ∫ (. Then to compute the integral above, enter $X$ (the X, θ, T key) $\mathbf{x}^{\mathbf{y}}$4 − 2$X$$\mathbf{x}^{\mathbf{y}}$5 + $X$$\mathbf{x}^{\mathbf{y}}$6, 0, 1, 8) to obtain 0.00952381. The 8 indicates that the interval [0, 1] has been divided into $2^8 = 256$ subintervals. You can enter an integer from 1 to 9; the larger the integer, the more accuracy you get. However, when $n = 8$ or 9, the computation may take a relatively long time.

## MATLAB 4.11

In MATLAB, if a matrix $A$ has complex entries, then **A′** will produce the complex conjugate transpose. Hence if **u** and **v** are vectors in $\mathbb{C}^n$, they can be represented by $n \times 1$ matrices with complex entries and $(\mathbf{u}, \mathbf{v})$ is computed by **v′*u** and $|\mathbf{u}|$ is computed by **norm(u)** or **sqrt(u′*u)**.

The variable **i** in MATLAB is built-in to represent the imaginary number $\sqrt{-1}$. MATLAB will recognize **i** as such as long as you have not used the variable **i** for any other purpose.

For a given $n$, to generate a random vector in $\mathbb{C}^n$, enter

**v = 2*rand(n,1)−1 + i*(2*rand(n,1)−1)**

1. Generate four random vectors in $\mathbb{C}^4$. Find the orthonormal basis for the span of these vectors using the Gram-Schmidt process. Check your answers by verifying that the set of orthonormal vectors obtained from the Gram-Schmidt process is orthonormal and that each vector in the original set is a linear combination of the set of vectors you obtained.

2. a. Let $\{\mathbf{u}_1, \mathbf{u}_2, \mathbf{u}_3, \mathbf{u}_4\}$ be the set of orthonormal vectors obtained in MATLAB Problem 1 above. Let $A$ be the matrix $[\mathbf{u}_1\ \mathbf{u}_2\ \mathbf{u}_3\ \mathbf{u}_4]$. Let **w** be a random vector in $\mathbb{C}^4$. Verify that

$$\mathbf{w} = (\mathbf{w}, \mathbf{u}_1)\mathbf{u}_1 + \cdots + (\mathbf{w}, \mathbf{u}_4)\mathbf{u}_4$$

Repeat for another choice of **w**.

b. *(Paper and pencil)* What general property of an orthonormal basis for $\mathbb{C}^n$ is expressed by part (a)? Write a description of how to find the coordinates of a vector in $\mathbb{C}^n$ with respect to an orthonormal basis.

3. Generate four random vectors in $\mathbb{C}^6$, $\mathbf{v}_1$, $\mathbf{v}_2$, $\mathbf{v}_3$, and $\mathbf{v}_4$. Let $H$ denote span $\{\mathbf{v}_1, \mathbf{v}_2, \mathbf{v}_3, \mathbf{v}_4\}$. Let $A = [\mathbf{v}_1\ \mathbf{v}_2\ \mathbf{v}_3\ \mathbf{v}_4]$ and **B = orth(A).** Let $\mathbf{u}_i$ denote the $i$th column of $B$.

a. Let **w** be a random vector in $\mathbb{C}^6$. Find the projection of **w** onto $H$, $\mathbf{p} = \text{proj}_H\mathbf{w}$.

Compute $\mathbf{z} = \begin{pmatrix} (\mathbf{w}, \mathbf{u}_1) \\ (\mathbf{w}, \mathbf{u}_2) \\ (\mathbf{w}, \mathbf{u}_3) \\ (\mathbf{w}, \mathbf{u}_4) \end{pmatrix}$. Verify that **z** = **B′*w** and **p = B*B*′*w.** Repeat for another choice of **w**.

b. Let **x** be a random vector in $\mathbb{C}^4$ and form $\mathbf{h} = A\mathbf{x}$. Hence **h** is in $H$. Compare $|\mathbf{w} - \mathbf{p}|$ and $|\mathbf{w} - \mathbf{h}|$. Repeat for three more choices of **x**. Write an interpretation of your observations.

c. Let $\mathbf{z} = 2\mathbf{v}_1 - 3\mathbf{v}_3 + \mathbf{v}_4$. Then $H = \text{span}\,\{\mathbf{v}_1, \mathbf{v}_2, \mathbf{v}_3, \mathbf{z}\}$. (Here $H$ is the subspace described in previous parts of this problem.) Why? Let $C = [\mathbf{v}_1\ \mathbf{v}_2\ \mathbf{v}_3\ \mathbf{z}]$ and **D = orth(C).** Then the columns of $D$ will be another orthonormal basis for $H$.

Let **w** be a random vector in $\mathbb{C}^6$. Compute the projection of **w** onto $H$ using $B$ and the projection of **w** onto $H$ using $D$. Compare the results. Repeat for two more choices of **w**. Write an interpretation of your observations.

4. a. *(Paper and pencil)* Explain why the null space of **A′** is orthogonal to the range of $A$; that is, if $H = \text{range } A$, then null space of $\mathbf{A}' = H^\perp$.

b. Let $A$ be a random $7 \times 4$ matrix with complex entries. (Let **A = 2*rand(7,4)−1 + i*(2*rand(7,4)−1).**) Let **B = orth(A)** and let **C = null(A′).** (Then the columns of $B$ form an orthonormal basis for $H = \text{range } A$ and the columns of $C$ form an orthonormal basis for $H^\perp$.) Verify that the columns of $C$ are orthonormal.

c. Let **w** be a random vector in $\mathbb{C}^7$. Find **h**, the projection of **w** onto $H$, and **p**, the projection of **w** onto $H^\perp$. Verify that $\mathbf{w} = \mathbf{p} + \mathbf{h}$. Repeat for three more choices of **w**.

5. If $Q$ is an $n \times n$ matrix with complex entries, then $Q$ is a **unitary** matrix if **Q′*Q = eye(n).** You can generate a random unitary matrix $Q$ by generating a random complex matrix $A$ and then letting **Q = orth(A).**

a. Generate two random unitary $4 \times 4$ matrices as described above. Verify that they satisfy the property of being unitary and verify that the columns form an orthonormal basis for $\mathbb{C}^4$.

b. Verify that the inverse of each matrix is unitary.

c. Verify that the product of the matrices is unitary.

d. Generate a random vector **v** in $\mathbb{C}^4$. Verify that each unitary matrix preserves length; that is, $|Q\mathbf{v}| = |\mathbf{v}|$.

e. Repeat parts (a) through (d) for two random $6 \times 6$ unitary matrices.

## 4.12 THE FOUNDATIONS OF VECTOR SPACE THEORY: THE EXISTENCE OF A BASIS (Optional)

In this section we prove one of the most important results in linear algebra: **Every vector space has a basis.** The proof is more difficult than any other proof in this book; it involves concepts that are part of the foundation of mathematics. It will take some hard work to go through the details of this proof. However, after you have done so, you should have a deeper appreciation of a fundamental mathematical idea.

We begin with some definitions.

**DEFINITION 1** **Partial Ordering** Let $S$ be a set. A **partial ordering** on $S$ is a relation, denoted by $\leq$, which is defined for some of the ordered pairs of elements of $S$ and satisfies three conditions:

**i.** $x \leq x$ for all $x \in S$ **reflexive law**

**ii.** If $x \leq y$ and $y \leq x$, then $x = y$ **antisymmetric law**

**iii.** If $x \leq y$ and $y \leq z$, then $x \leq z$ **transitive law**

It may be the case that there are elements $x$ and $y$ in $S$ such that neither $x \leq y$ nor $y \leq x$. However, if for every $x, y \in S$, either $x \leq y$ or $y \leq x$, then the ordering is said to be a **total ordering.** If $x \leq y$ or $y \leq x$, then $x$ and $y$ are said to be **comparable.**

*Notation.* $x < y$ means $x \leq y$ and $x \neq y$.

EXAMPLE 1 **A Partial Ordering on $\mathbb{R}$** The real numbers are partially ordered by $\leq$ where $\leq$ stands for "less than or equal to." The ordering here is a total ordering.

EXAMPLE 2 **A Partial Order on a Set of Subsets** Let $S$ be a set and let $P(S)$, called the **power set** of $S$, denote the set of all subsets of $S$.

We say that $A \leq B$ if $A \subseteq B$. The inclusion relation is a partial ordering on $P(S)$. This is easy to prove. We have

**i.** $A \subseteq A$ for every set $A$.

**ii.** $A \subseteq B$ and $B \subseteq A$ if and only if $A = B$.

**iii.** Suppose $A \subseteq B$ and $B \subseteq C$. If $x \in A$, then $x \in B$, so $x \in C$. This means that $A \subseteq C$.

Except in unusual circumstances (for example, if $S$ contains only one element), the ordering will not be a total ordering. This is illustrated in Figure 4.13.

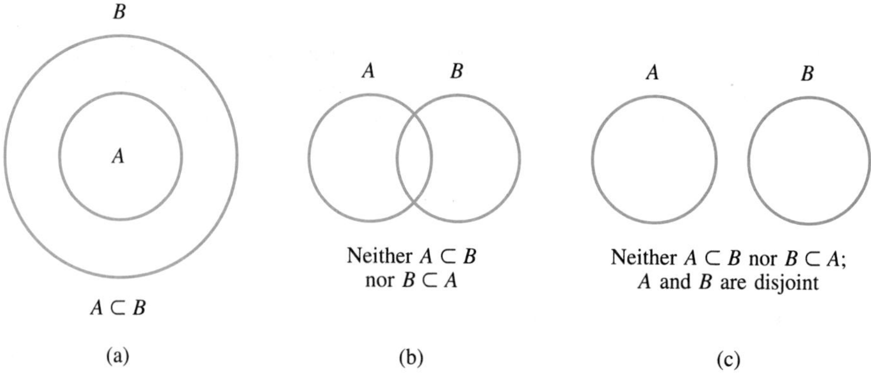


**Figure 4.13** Three possibilities for set inclusion

**DEFINITION 2** **Chain, Upper Bound, and Maximal Element** Let $S$ be a set partially ordered by $\leq$.

**i.** A subset $T$ of $S$ is called a **chain** if it is totally ordered; that is, if $x$ and $y$ are distinct elements of $T$, then $x \leq y$ or $y \leq x$.

**ii.** Let $C$ be a subset of $S$. An element $u \in S$ is an **upper bound** for $C$ if $c \leq u$ for every element $c \in C$.

**iii.** The element $m \in S$ is a **maximal element** for $S$ if there is no $s \in S$ with $m < s$.

*Remark 1.* In (*ii*) the upper bound for $C$ must be comparable with every element in $C$ but it need not be in $C$ (although it must be in $S$). For example, the number 1 is an upper bound for the set (0, 1) but is not in (0, 1). Any number greater than 1 is an upper bound. However, there is no number in (0, 1) that is an upper bound for (0, 1).

*Remark 2.* If $m$ is a maximal element for $S$, it is not necessarily the case that $s \leq m$ for every $s \in S$. In fact, $m$ may be comparable to very few elements of $S$. The only condition for maximality is that there is no element of $S$ "larger than" $m$.

**EXAMPLE 3** **A Chain of Subsets of $\mathbb{R}^2$** Let $S = \mathbb{R}^2$. Then $P(S)$ consists of subsets of the $xy$-plane. Let $D_r = \{(x, y)\colon x^2 + y^2 < r^2\}$; that is, $D_r$ is an open disk of radius $r$—the interior of the circle of radius $r$ centered at the origin. Let

$$T = \{D_r\colon r > 0\}$$

Clearly, $T$ is a chain. For if $D_{r_1}$ and $D_{r_2}$ are in $T$, then

$$D_{r_1} \subseteq D_{r_2} \text{ if } r_1 \leq r_2 \quad \text{and} \quad D_{r_2} \subseteq D_{r_1} \text{ if } r_2 \leq r_1$$

Before going further, we need some new notation. Let $V$ be a vector space. We have seen that a linear combination of vectors in $V$ is a finite sum $\Sigma_{i=1}^{n} \alpha_i \mathbf{v}_i = \alpha_1\mathbf{v}_1 + \alpha_2\mathbf{v}_2 + \cdots + \alpha_n\mathbf{v}_n$. If you have studied power series, you have seen infinite sums of the form $\Sigma_{n=0}^{\infty} a_n x^n$. For example,

$$e^x = \sum_{n=0}^{\infty} \frac{x^n}{n!} = 1 + x + \frac{x^2}{2!} + \frac{x^3}{3!} + \cdots$$

Here we need a different kind of sum. Let $C$ be a set of vectors in $V$.† For each $\mathbf{v} \in C$, let $\alpha_{\mathbf{v}}$ denote a scalar (the set of scalars is given in the definition of $V$). Then

† $C$ is not necessarily a subspace of $V$.

when we write

$$\mathbf{x} = \sum_{\mathbf{v} \in C} \alpha_{\mathbf{v}} \mathbf{v} \tag{1}$$

it will be understood that only a finite number of the scalars $\alpha_{\mathbf{v}}$ are nonzero and that all the terms with $\alpha_{\mathbf{v}} = 0$ are to be left out of the summation. We can describe the sum in (1) as follows:

> For each $\mathbf{v} \in C$, assign a scalar $\alpha_{\mathbf{v}}$ and form the product $\alpha_{\mathbf{v}} \mathbf{v}$. Then $\mathbf{x}$ is the sum of the finite subset of the vectors $\alpha_{\mathbf{v}} \mathbf{v}$ for which $\alpha_{\mathbf{v}} \neq 0$.

**DEFINITION 3** **Linear Combination, Spanning Set, Linear Independence, and Basis**

**i.** Let $C$ be a subset of a vector space $V$. Then any vector that can be written in form (1) is called a **linear combination** of vectors in $C$. The set of linear combinations of vectors in $C$ is denoted by $\mathrm{L}(C)$.

**ii.** The set $C$ is said to **span** the vector space $V$ if $V \subseteq \mathrm{L}(C)$.

**iii.** A subset $C$ of a vector space $V$ is said to be **linearly independent** if

$$\sum_{\mathbf{v} \in C} \alpha_{\mathbf{v}} \mathbf{v} = \mathbf{0}$$

holds only when $\alpha_{\mathbf{v}} = 0$ for every $\mathbf{v} \in C$.

**iv.** The subset $B$ of a vector space $V$ is a **basis** for $V$ if it spans $V$ and is linearly independent.

*Remark.* If $C$ contains only a finite number of vectors, then these definitions are precisely the ones we have seen earlier in this chapter.

**THEOREM 1** Let $B$ be a linearly independent subset of a vector space $V$. Then $B$ is a basis if and only if it is maximal; that is, if $B \subsetneqq D$, then $D$ is linearly dependent.

**Proof** Suppose that $B$ is a basis and that $B \subsetneqq D$. Choose $\mathbf{x}$ such that $\mathbf{x} \in D$ but $\mathbf{x} \notin B$. Since $B$ is a basis, $\mathbf{x}$ can be written as a linear combination of vectors in $B$:

$$\mathbf{x} = \sum_{\mathbf{v} \in B} \alpha_{\mathbf{v}} \mathbf{v}$$

If $\alpha_{\mathbf{v}} = 0$ for every $\mathbf{v}$, then $\mathbf{x} = \mathbf{0}$ and $D$ is dependent. Otherwise $\alpha_{\mathbf{v}} \neq 0$ for some $\mathbf{v}$, and so the sum

$$\mathbf{x} - \sum_{\mathbf{v} \in B} \alpha_{\mathbf{v}} \mathbf{v} = \mathbf{0}$$

shows that $D$ is dependent. Thus $B$ is maximal.

Conversely, suppose that $B$ is maximal. Let $\mathbf{x}$ be a vector in $V$ that is not in $B$. Let $D = B \cup \{\mathbf{x}\}$. Then $D$ is dependent (since $B$ is maximal) and there is an equation

$$\sum_{\mathbf{v} \in B} \alpha_{\mathbf{v}} \mathbf{v} + \beta \mathbf{x} = \mathbf{0}$$

in which not every coefficient is zero. But $\beta \neq 0$ since otherwise we would obtain a contradiction of the linear independence of $B$. Thus we can write

$$\mathbf{x} = -\beta^{-1} \sum_{\mathbf{v} \in B} \alpha_{\mathbf{v}} \mathbf{v}\dagger$$

Thus $B$ is a spanning set and is therefore a basis for $V$. ■

Where is this all leading? Perhaps you can see the general direction. We have defined ordering on sets and maximal elements. We have shown that a linearly independent set is a basis if it is maximal. We now are lacking only a result that can help us prove the existence of a maximal element. That result is one of the basic assumptions of mathematics.

Many of you studied Euclidean geometry in high school. There, you had perhaps your first contact with mathematical proof. To prove things, Euclid made certain assumptions, which he called *axioms*. For example, he assumed that the shortest distance between two points is a straight line. Starting with these axioms, he, and students of geometry, were able to prove a number of theorems.

In all branches of mathematics it is necessary to have some axioms. If we assume nothing, we can prove nothing. To complete our proof, we need the following axiom:

**AXIOM** **Zorn's Lemma**‡ If $S$ is a nonempty, partially ordered set such that every nonempty chain has an upper bound, then $S$ has a maximal element. ■

† If the scalars are real or complex numbers, then $\beta^{-1} = 1/\beta$.

‡ Max A. Zorn (1906–1993) spent a number of years at Indiana University where he was a Professor Emeritus until his death on March 9, 1993. He published his famous result in 1935 ("A Remark on Method in Transfinite Algebra," *Bulletin of the American Mathematical Society* 41 (1935): 667–670).

*Remark.* The **axiom of choice** says, roughly, that given a number (finite or infinite) of nonempty sets, there is a function that chooses one element from each set. This axiom is equivalent to Zorn's lemma; that is, if you assume the axiom of choice, then you can prove Zorn's lemma and vice versa. For a proof of this equivalence and other interesting results, see the excellent book *Naive Set Theory* by Paul R. Halmos (New York: Van Nostrand, 1960), especially page 63.

We can now, finally, state and prove our main result.

**THEOREM 2** Every vector space $V$ has a basis.

**Proof** We show that $V$ has a maximal linearly independent subset. We do this in steps.

**i.** Let $S$ be the collection of all linearly independent subsets partially ordered by inclusion.

**ii.** A chain in $S$ is a subset $T$ of $S$ such that if $A$ and $B$ are in $T$, either $A \subseteq B$ or $B \subseteq A$.

**iii.** Let $T$ be a chain. Define

$$M(T) = \bigcup_{A \in T} A$$

Clearly, $M(T)$ is a subset of $V$ and $A \subseteq M(T)$ for every $A \in T$. We want to show that $M(T)$ is an upper bound for $T$. Since $A \subseteq M(T)$ for every $A \in T$, we need only show that $M(T) \in S$; that is, we must show that $M(T)$ is linearly independent.

**iv.** Suppose $\sum_{\mathbf{v} \in M(T)} \alpha_{\mathbf{v}} \mathbf{v} = \mathbf{0}$, where only a finite number of the $\alpha_{\mathbf{v}}$'s are nonzero. Denote these scalars by $\alpha_1, \alpha_2, \ldots, \alpha_n$ and the corresponding vectors by $\mathbf{v}_1, \mathbf{v}_2, \ldots, \mathbf{v}_n$. For each $i$, $i = 1, 2, \ldots, n$ there is a set $A_i \in T$ such that $\mathbf{v}_i \in A_i$ (because each $\mathbf{v}_i$ is in $M(T)$ and $M(T)$ is the union of sets in $T$). But $T$ is totally ordered, so one of the $A_i$'s contains all the others (see Problem 3); call it $A_k$. (We can only draw this conclusion because $\{A_1, A_2, \ldots, A_n\}$ is finite.) Thus $A_i \subseteq A_k$ for $i = 1, 2, \ldots, n$ and $\mathbf{v}_1, \mathbf{v}_2, \ldots, \mathbf{v}_n \in A_k$. Since $A_k$ is linearly independent and $\sum_{i=1}^{n} \alpha_i \mathbf{v}_i = \mathbf{0}$, it follows that $\alpha_1 = \alpha_2 = \cdots = \alpha_n = 0$. Thus $M(T)$ is linearly independent.

**v.** $S$ is nonempty because $\emptyset \in S$ ($\emptyset$ denotes the empty set). We have shown that every chain $T$ in $S$ has an upper bound $M(T)$, which is in $S$. By Zorn's lemma, $S$ has a maximal element. But $S$ consists of all linearly independent subsets of $V$. The maximal element $B \in S$ is therefore a maximal linearly independent subset of $V$. Thus by Theorem 1, $B$ is a basis for $V$. ■

## PROBLEMS 4.12

1. Show that every linearly independent set in a vector space $V$ can be expanded to a basis.
2. Show that every spanning set in a vector space $V$ has a subset that is a basis.
3. Let $A_1, A_2, \ldots, A_n$ be $n$ sets in a chain $T$. Show that one of the sets contains all the others. [*Hint:* Because $T$ is a chain, either $A_1 \subseteq A_2$ or $A_2 \subseteq A_1$. Thus the result is true if $n = 2$. Complete the proof by mathematical induction.]

## SUMMARY

- A **real vector space** $V$ is a set of objects, called **vectors,** together with two operations called **addition** (denoted by $\mathbf{x} + \mathbf{y}$) and **scalar multiplication** (denoted by $\alpha\mathbf{x}$) that satisfy the following ten axioms: (p. 292)

  **i.** If $\mathbf{x} \in V$ and $\mathbf{y} \in V$, then $\mathbf{x} + \mathbf{y} \in V$ (closure under addition).

  **ii.** For all $\mathbf{x}$, $\mathbf{y}$, and $\mathbf{z}$ in $V$, $(\mathbf{x} + \mathbf{y}) + \mathbf{z} = \mathbf{x} + (\mathbf{y} + \mathbf{z})$ (associative law of vector addition).

  **iii.** There is a vector $\mathbf{0} \in V$ such that for all $\mathbf{x} \in V$, $\mathbf{x} + \mathbf{0} = \mathbf{0} + \mathbf{x} = \mathbf{x}$ ($\mathbf{0}$ is called the additive identity).

  **iv.** If $\mathbf{x} \in V$, there is a vector $-\mathbf{x}$ in $V$ such that $\mathbf{x} + (-\mathbf{x}) = \mathbf{0}$ ($-\mathbf{x}$ is called the additive inverse of $\mathbf{x}$).

  **v.** If $\mathbf{x}$ and $\mathbf{y}$ are in $V$, then $\mathbf{x} + \mathbf{y} = \mathbf{y} + \mathbf{x}$ (commutative law of vector addition).

  **vi.** If $\mathbf{x} \in V$ and $\alpha$ is a scalar, then $\alpha\mathbf{x} \in V$ (closure under scalar multiplication).

  **vii.** If $\mathbf{x}$ and $\mathbf{y}$ are in $V$ and $\alpha$ is a scalar, then $\alpha(\mathbf{x} + \mathbf{y}) = \alpha\mathbf{x} + \alpha\mathbf{y}$ (first distributive law).

  **viii.** If $\mathbf{x} \in V$ and $\alpha$ and $\beta$ are scalars, then $(\alpha + \beta)\mathbf{x} = \alpha\mathbf{x} + \beta\mathbf{x}$ (second distributive law).

  **ix.** If $\mathbf{x} \in V$ and $\alpha$ and $\beta$ are scalars, then $\alpha(\beta\mathbf{x}) = \alpha\beta\mathbf{x}$ (associative law of scalar multiplication).

  **x.** For every vector $\mathbf{x} \in V$, $1\mathbf{x} = \mathbf{x}$ (the scalar 1 is called a multiplicative identity).

- **The space** $\mathbb{R}^n = \{(x_1, x_2, \ldots, x_n)\colon x_i \in \mathbb{R} \text{ for } i = 1, 2, \ldots, n\}$. (p. 293)
- **The space** $P_n$ = {polynomials of degree less than or equal to $n$}. (p. 295)
- **The space** $C[a, b]$ = {real-valued functions that are continuous on the interval $[a, b]$}. (p. 295)
- **The space** $M_{mn}$ = {$m \times n$ matrices with real coefficients}. (p. 295)
- **The space** $\mathbb{C}^n = \{(c_1, c_2, \ldots, c_n)\colon c_i \in \mathbb{C} \text{ for } i = 1, 2, \ldots, n\}$. $\mathbb{C}$ denotes the set of complex numbers. (p. 296)
- A **subspace** $H$ of a vector space $V$ is a subset of $V$ that is itself a vector space. (p. 299)
- A nonempty subset $H$ of a vector space $V$ is a subspace of $V$ if the following two rules hold: (p. 300)

  **i.** If $\mathbf{x} \in H$ and $\mathbf{y} \in H$, then $\mathbf{x} + \mathbf{y} \in H$.

  **ii.** If $\mathbf{x} \in H$, then $\alpha\mathbf{x} \in H$ for every scalar $\alpha$.

- A **proper subspace** of a vector space $V$ is a subspace of $V$ other than $\{\mathbf{0}\}$ or $V$. (p. 301)
- A **linear combination** of the vectors $\mathbf{v}_1, \mathbf{v}_2, \ldots, \mathbf{v}_n$ in a vector space $V$ is a sum of the form (p. 306)

$$\alpha_1\mathbf{v}_1 + \alpha_2\mathbf{v}_2 + \cdots + \alpha_n\mathbf{v}_n$$

  where $\alpha_1, \alpha_2, \ldots, \alpha_n$ are scalars.

- The vectors $\mathbf{v}_1, \mathbf{v}_2, \ldots, \mathbf{v}_n$ in a vector space $V$ are said to **span** $V$ if every vector in $V$ can be written as a linear combination of $\mathbf{v}_1, \mathbf{v}_2, \ldots, \mathbf{v}_n$. (p. 306)
- The **span of a set of vectors** $\mathbf{v}_1, \mathbf{v}_2, \ldots, \mathbf{v}_k$ in a vector space $V$ is the set of linear combinations of $\mathbf{v}_1, \mathbf{v}_2, \ldots, \mathbf{v}_k$. (p. 307)
- span $\{\mathbf{v}_1, \mathbf{v}_2, \ldots, \mathbf{v}_k\}$ is a subspace of $V$. (p. 307)
- ***Linear dependence and independence***

  The vectors $\mathbf{v}_1, \mathbf{v}_2, \ldots, \mathbf{v}_n$ in a vector space $V$ are said to be **linearly dependent** if there exist scalars $c_1, c_2, \ldots, c_n$ not all zero such that (p. 318)

$$c_1\mathbf{v}_1 + c_2\mathbf{v}_2 + \cdots + c_n\mathbf{v}_n = \mathbf{0}$$

  If the vectors are not linearly dependent, they are said to be **linearly independent.**
- Two vectors in a vector space $V$ are linearly dependent if and only if one is a scalar multiple of the other. (p. 319)
- Any set of $n$ linearly independent vectors in $\mathbb{R}^n$ spans $\mathbb{R}^n$. (p. 325)
- A set of $n$ vectors in $\mathbb{R}^m$ is linearly dependent if $n > m$. (p. 322)
- ***Basis***

  A set of vectors $\mathbf{v}_1, \mathbf{v}_2, \ldots, \mathbf{v}_n$ is a **basis** for a vector space $V$ if (p. 337)

  **i.** $\{\mathbf{v}_1, \mathbf{v}_2, \ldots, \mathbf{v}_n\}$ is linearly independent

  **ii.** $\{\mathbf{v}_1, \mathbf{v}_2, \ldots, \mathbf{v}_n\}$ spans $V$
- Every set of $n$ linearly independent vectors in $\mathbb{R}^n$ is a basis in $\mathbb{R}^n$ (p. 337)
- The **standard basis** in $\mathbb{R}^n$ consists of the $n$ vectors (p. 337)

$$\mathbf{e}_1 = \begin{pmatrix} 1 \\ 0 \\ 0 \\ \vdots \\ 0 \end{pmatrix}, \mathbf{e}_2 = \begin{pmatrix} 0 \\ 1 \\ 0 \\ \vdots \\ 0 \end{pmatrix}, \mathbf{e}_3 = \begin{pmatrix} 0 \\ 0 \\ 1 \\ \vdots \\ 0 \end{pmatrix}, \ldots, \mathbf{e}_n = \begin{pmatrix} 0 \\ 0 \\ 0 \\ \vdots \\ 1 \end{pmatrix}$$

- ***Dimension***

  If the vector space $V$ has a finite basis, then the **dimension** of $V$ is the number of vectors in every basis and $V$ is called a **finite dimensional vector space.** Otherwise $V$ is called an **infinite dimensional vector space.** If $V = \{\mathbf{0}\}$, then $V$ is said to be **zero dimensional.** (p. 340)

  We write the dimension of $V$ as dim $V$.
- If $H$ is a subspace of the finite dimensional space $V$, then $\dim H \le \dim V$. (p. 341)
- The only proper subspaces of $\mathbb{R}^3$ are sets of vectors lying on a single line or a single plane passing through the origin. (p. 342)
- The **null space** of an $n \times n$ matrix $A$ is the subspace of $\mathbb{R}^n$ given by (p. 348)

$$N_A = \{\mathbf{x} \in \mathbb{R}^n\colon A\mathbf{x} = \mathbf{0}\}$$

- The **nullity** of an $n \times n$ matrix $A$ is the dimension of $N_A$ and is denoted by $\nu(A)$. (p. 348)
- Let $A$ be an $m \times n$ matrix. The **range of A,** denoted by Range $A$, is the subspace of $\mathbb{R}^m$ given by (p. 349)

$$\text{Range } A = \{\mathbf{y} \in \mathbb{R}^m\colon A\mathbf{x} = \mathbf{y} \text{ for some } \mathbf{x} \in \mathbb{R}^n\}$$

- The **rank of A,** denoted by $\rho(A)$, is the dimension of Range $A$. (p. 350)

- The **row space of $A$,** denoted by $R_A$, is the span of the rows of $A$ and is a subspace of $\mathbb{R}^n$. (p. 350)
- The **column space of $A$,** denoted by $C_A$, is the span of the columns of $A$ and is a subspace of $\mathbb{R}^m$. (p. 350)
- If $A$ is an $m \times n$ matrix, then

$$C_A = \text{Range } A \quad \text{and} \quad \dim R_A = \dim C_A = \dim \text{Range } A = \rho(A) \qquad \text{(pp. 350, 351)}$$

Moreover,

$$\rho(A) + \nu(A) = n \qquad \text{(p. 356)}$$

- The system $A\mathbf{x} = \mathbf{b}$ has at least one solution if and only if $\rho(A) = \rho(A, \mathbf{b})$, where $(A, \mathbf{b})$ is the augmented matrix obtained by adjoining the column vector $\mathbf{b}$ to $A$. (p. 358)

- ***Summing up theorem***

Let $A$ be an $n \times n$ matrix. Then the following are equivalent: (p. 360)

**i.** $A$ is invertible.
**ii.** The only solution to the homogeneous system $A\mathbf{x} = \mathbf{0}$ is the trivial solution ($\mathbf{x} = \mathbf{0}$).
**iii.** The system $A\mathbf{x} = \mathbf{b}$ has a unique solution for every $n$-vector $\mathbf{b}$.
**iv.** $A$ is row equivalent to the $n \times n$ identity matrix $I_n$.
**v.** $A$ can be written as the product of elementary matrices.
**vi.** The row echelon form of $A$ has $n$ pivots.
**vii.** The columns (and rows) of $A$ are linearly independent.
**viii.** $\det A \neq 0$.
**ix.** $\nu(A) = 0$
**x.** $\rho(A) = n$.

- Let $B_1 = \{\mathbf{u}_1, \mathbf{u}_2, \ldots, \mathbf{u}_n\}$ and $B_2 = \{\mathbf{v}_1, \mathbf{v}_2, \ldots, \mathbf{u}_n\}$ be two bases for the vector space $V$. If $\mathbf{x} \in V$ and (pp. 372–373)

$$\mathbf{x} = b_1\mathbf{u}_1 + b_2\mathbf{u}_2 + \cdots + b_n\mathbf{u}_n = c_1\mathbf{v}_1 + c_2\mathbf{v}_2 + c_n\mathbf{v}_n$$

then we write $(\mathbf{x})_{B_1} = \begin{pmatrix} b_1 \\ b_2 \\ \vdots \\ b_n \end{pmatrix}$ and $(\mathbf{x})_{B_2} = \begin{pmatrix} c_1 \\ c_2 \\ \vdots \\ c_n \end{pmatrix}$.

Suppose that $(\mathbf{u}_j)_{B_2} = \begin{pmatrix} a_{1j} \\ a_{2j} \\ \vdots \\ a_{nj} \end{pmatrix}$. Then the **transition matrix** from $B_1$ to $B_2$ is the $n \times n$ matrix (pp. 375, 376)

$$A = \begin{pmatrix} a_{11} & a_{12} & \cdots & a_{1n} \\ a_{21} & a_{22} & \cdots & a_{2n} \\ \vdots & \vdots & & \vdots \\ a_{n1} & a_{n2} & \cdots & a_{nn} \end{pmatrix}$$

Moreover, $(\mathbf{x})_{B_2} = A(\mathbf{x})_{B_1}$.

- If $A$ is the transition matrix from $B_1$ to $B_2$, then $A^{-1}$ is the transition matrix from $B_2$ to $B_1$. (p. 377)
- If $(\mathbf{x}_j)_{B_1} = \begin{pmatrix} a_{1j} \\ a_{2j} \\ \vdots \\ a_{nj} \end{pmatrix}$ for $j = 1, 2, \ldots, n$, then $\mathbf{x}_1, \mathbf{x}_2, \ldots, \mathbf{x}_n$ are linearly independent if and only if det $A \neq 0$, where (p. 381)

$$A = \begin{pmatrix} a_{11} & a_{12} & \cdots & a_{1n} \\ a_{21} & a_{22} & \cdots & a_{2n} \\ \vdots & \vdots & & \vdots \\ a_{n1} & a_{n2} & \cdots & a_{nn} \end{pmatrix}$$

- The vectors $\mathbf{u}_1, \mathbf{u}_2, \ldots, \mathbf{u}_k$ in $\mathbb{R}^n$ form an **orthogonal set** if $\mathbf{u}_i \cdot \mathbf{u}_j = 0$ for $i \neq j$. If, in addition, $\mathbf{u}_i \cdot \mathbf{u}_i = 1$ for $i = 1, 2, \ldots, k$, the set is said to be **orthonormal.** (p. 393)
- $|\mathbf{v}| = |\mathbf{v} \cdot \mathbf{v}|^{1/2}$ is called the **length** or **norm** of $\mathbf{v}$. (p. 393)
- Every subspace of $\mathbb{R}^n$ has an orthonormal basis. **The Gram-Schmidt orthonormalization process** can be used to construct such a basis. (p. 395)
- An **orthogonal matrix** is an invertible $n \times n$ matrix $Q$ such that $Q^{-1} = Q^t$. (p. 399)
- An $n \times n$ matrix is orthogonal if and only if its columns form an orthonormal basis for $\mathbb{R}^n$. (p. 399)
- Let $H$ be a subspace of $\mathbb{R}^n$ with orthonormal basis $\{\mathbf{u}_1, \mathbf{u}_2, \ldots, \mathbf{u}_k\}$. If $\mathbf{v} \in \mathbb{R}^n$, then the **orthogonal projection** of $\mathbf{v}$ onto $H$, denoted by $\text{proj}_H \mathbf{v}$, is given by (p. 400)

$$\text{proj}_H \mathbf{v} = (\mathbf{v} \cdot \mathbf{u}_1)\mathbf{u}_1 + (\mathbf{v} \cdot \mathbf{u}_2)\mathbf{u}_2 + \cdots + (\mathbf{v} \cdot \mathbf{u}_k)\mathbf{u}_k$$

- Let $H$ be a subspace of $\mathbb{R}^n$. Then the **orthogonal complement** of $H$, denoted by $H^\perp$, is given by (p. 402)

$$H^\perp = \{\mathbf{x} \in \mathbb{R}^n\text{: } \mathbf{x} \cdot \mathbf{h} = 0 \quad \text{for every } \mathbf{h} \in H\}$$

- ***Projection theorem***

  Let $H$ be a subspace of $\mathbb{R}^n$ and let $\mathbf{v} \in \mathbb{R}^n$. Then there exists a unique pair of vectors $\mathbf{h}$ and $\mathbf{p}$ such that $\mathbf{h} \in H$, $\mathbf{p} \in H^\perp$, and (p. 403)

$$\mathbf{v} = \mathbf{h} + \mathbf{p} = \text{proj}_H \mathbf{v} + \text{proj}_{H^\perp} \mathbf{v}$$

- ***Norm approximation theorem***

  Let $H$ be a subspace of $\mathbb{R}^n$ and let $\mathbf{v}$ be a vector in $\mathbb{R}^n$. Then, in $H$, $\text{proj}_H \mathbf{v}$ is the best approximation to $\mathbf{v}$ in the following sense: If $\mathbf{h}$ is any other vector in $H$, then (p. 405)

$$|\mathbf{v} - \text{proj}_H \mathbf{v}| < |\mathbf{v} - \mathbf{h}|$$

- Let $(x_1, y_1), (x_2, y_2), \ldots, (x_n, y_n)$ be a set of data points. If we wish to represent these data by the straight line $y = mx + b$, then the **least squares problem** is to find the $m$ and $b$ that minimizes the sum of squares (p. 419)

$$[y_1 - (b + mx_1)]^2 + [y_2 - (b + mx_2)]^2 + \cdots + [y_n - (b + mx_n)]^2$$

  The solution to this problem is to set (p. 421)

$$\begin{pmatrix} b \\ m \end{pmatrix} = \mathbf{u} = (A^tA)^{-1}A^t\mathbf{y}$$

where

$$\mathbf{y} = \begin{pmatrix} y_1 \\ y_2 \\ \vdots \\ y_n \end{pmatrix} \quad \text{and} \quad A = \begin{pmatrix} 1 & x_1 \\ 1 & x_2 \\ \vdots & \vdots \\ 1 & x_n \end{pmatrix}$$

Similar results apply when we attempt to represent the data using a polynomial of degree > 1.

- ***Inner product space***

The complex vector space $V$ is called an **inner product space** if for every pair of vectors $\mathbf{u}$ and $\mathbf{v}$ in $V$ there is a unique complex number $(\mathbf{u}, \mathbf{v})$, called the **inner product** of $\mathbf{u}$ and $\mathbf{v}$, such that if $\mathbf{u}$, $\mathbf{v}$, and $\mathbf{w}$ are in $V$ and $\alpha \in \mathbb{C}$, then (p. 439)

**i.** $(\mathbf{v}, \mathbf{v}) \geq 0$

**ii.** $(\mathbf{v}, \mathbf{v}) = 0$ if and only if $\mathbf{v} = \mathbf{0}$

**iii.** $(\mathbf{u}, \mathbf{v} + \mathbf{w}) = (\mathbf{u}, \mathbf{v}) + (\mathbf{u}, \mathbf{w})$

**iv.** $(\mathbf{u} + \mathbf{v}, \mathbf{w}) = (\mathbf{u}, \mathbf{w}) + (\mathbf{v}, \mathbf{w})$

**v.** $(\mathbf{u}, \mathbf{v}) = \overline{(\mathbf{v}, \mathbf{u})}$

**vi.** $(\alpha\mathbf{u}, \mathbf{v}) = \alpha(\mathbf{u}, \mathbf{v})$

**vii.** $(\mathbf{u}, \alpha\mathbf{v}) = \overline{\alpha}(\mathbf{u}, \mathbf{v})$

- ***Inner product in $\mathbb{C}^n$*** (p. 439)

$$(\mathbf{x}, \mathbf{y}) = x_1\overline{y}_1 + x_2\overline{y}_2 + \cdots + x_n\overline{y}_n$$

- Let $V$ be an inner product space and suppose that $\mathbf{u}$ and $\mathbf{v}$ are in $V$. Then (p. 441)

$\mathbf{u}$ and $\mathbf{v}$ are **orthogonal** if $(\mathbf{u}, \mathbf{v}) = 0$

- The **norm** of $\mathbf{u}$, denoted by $\|\mathbf{u}\|$, is given by

$$\|\mathbf{u}\| = \sqrt{(\mathbf{u}, \mathbf{u})}$$

- ***Orthonormal set***

The set of vectors $\{\mathbf{v}_1, \mathbf{v}_2, \ldots, \mathbf{v}_n\}$ is an **orthonormal set** in $V$ if (p. 442)

$$(\mathbf{v}_i, \mathbf{v}_j) = 0 \qquad \text{for } i \neq j$$

and

$$\|\mathbf{v}_i\| = \sqrt{(\mathbf{v}_i, \mathbf{v}_i)} = 1$$

If only the first condition holds, then the set is said to be **orthogonal.**

- ***Orthogonal projection***

Let $H$ be a subspace of an inner product space $V$ with an orthonormal basis $\{\mathbf{u}_1, \mathbf{u}_2, \ldots, \mathbf{u}_k\}$. If $\mathbf{v} \in V$, then the **orthogonal projection** of $\mathbf{v}$ onto $H$, denoted by $\text{proj}_H\, \mathbf{v}$, is given by (p. 444)

$$\text{proj}_H\, \mathbf{v} = (\mathbf{v}, \mathbf{u}_1)\mathbf{u}_1 + (\mathbf{v}, \mathbf{u}_2)\mathbf{u}_2 + \cdots + (\mathbf{v}, \mathbf{u}_k)\mathbf{u}_k$$

- ***Orthogonal complement***

  Let $H$ be a subspace of the inner product space $V$. Then the **orthogonal complement** of $H$, denoted by $H^\perp$, is given by (p. 444)

  $$H^\perp = \{\mathbf{x} \in V: (\mathbf{x}, \mathbf{h}) = 0 \quad \text{for every } \mathbf{h} \in H\}$$

- If $H$ is a subspace of the inner product space $V$, then

  **i.** $H^\perp$ is a subspace of $\mathbf{v}$. (p. 444)

  **ii.** $H \cap H^\perp = \{\mathbf{0}\}$.

  **iii.** $\dim H^\perp = n - \dim H$ if $\dim V = n < \infty$.

- ***Projection theorem***

  Let $H$ be a finite dimensional subspace of the inner product space $V$ and suppose that $\mathbf{v} \in V$. Then there exists a unique pair of vectors $\mathbf{h}$ and $\mathbf{p}$ such that $\mathbf{h} \in H$, $\mathbf{p} \in H^\perp$, and (p. 444)

  $$\mathbf{v} = \mathbf{h} + \mathbf{p}$$

  where $\mathbf{h} = \text{proj}_H \mathbf{v}$.

  If $V$ is finite dimensional, then $\mathbf{p} = \text{proj}_{H^\perp} \mathbf{v}$.

- ***Norm approximation theorem***

  Let $H$ be a finite dimensional subspace of the inner product space $V$ and let $\mathbf{v}$ be a vector in $V$. Then, in $H$, $\text{proj}_H \mathbf{v}$ is the best approximation to $\mathbf{v}$ in the following sense: If $\mathbf{h}$ is any other vector in $H$, then (p. 445)

  $$|\mathbf{v} - \text{proj}_H \mathbf{v}| < |\mathbf{v} - \mathbf{h}|$$

# REVIEW EXERCISES

In Exercises 1–10 determine whether the given set is a vector space. If so, determine its dimension. If it is finite dimensional, find a basis for it.

**1.** The vectors $(x, y, z)$ in $\mathbb{R}^3$ satisfying $x + 2y - z = 0$

**2.** The vectors $(x, y, z)$ in $\mathbb{R}^3$ satisfying $x + 2y - z \leq 0$

**3.** The vectors $(x, y, z, w)$ in $\mathbb{R}^4$ satisfying $x + y + z + w = 0$

**4.** The vectors in $\mathbb{R}^3$ satisfying $x - 2 = y + 3 = z - 4$

**5.** The set of upper triangular $n \times n$ matrices under the operations of matrix addition and scalar multiplication

**6.** The set of polynomials of degree $\leq 5$

**7.** The set of polynomials of degree 5

**8.** The set of $3 \times 2$ matrices $A = (a_{ij})$, with $a_{12} = 0$, under the operations of matrix addition and scalar multiplication

**9.** The set in Exercise 8 except that $a_{12} = 1$

**10.** The set $S = \{f \in C[0, 2]: f(2) = 0\}$

In Exercises 11–19 determine whether the given set of vectors is linearly dependent or independent.

**11.** $\begin{pmatrix} 2 \\ 3 \end{pmatrix}; \begin{pmatrix} 4 \\ -6 \end{pmatrix}$ **12.** $\begin{pmatrix} 2 \\ 3 \end{pmatrix}; \begin{pmatrix} 4 \\ 6 \end{pmatrix}$ **13.** $\begin{pmatrix} 1 \\ -1 \\ 2 \end{pmatrix}; \begin{pmatrix} 3 \\ 0 \\ 1 \end{pmatrix}; \begin{pmatrix} 0 \\ 0 \\ 0 \end{pmatrix}$

**14.** $\begin{pmatrix} 1 \\ -4 \\ 2 \end{pmatrix}; \begin{pmatrix} 0 \\ 2 \\ -1 \end{pmatrix}; \begin{pmatrix} 2 \\ -10 \\ 5 \end{pmatrix}$ **15.** $\begin{pmatrix} 1 \\ 0 \\ 0 \\ 0 \end{pmatrix}; \begin{pmatrix} 0 \\ 1 \\ 0 \\ 0 \end{pmatrix}; \begin{pmatrix} 0 \\ 0 \\ 1 \\ 0 \end{pmatrix}; \begin{pmatrix} 0 \\ 0 \\ 0 \\ 1 \end{pmatrix}$

**16.** In $P_3$: $1,\ 2 - x^2,\ 3 - x,\ 7x^2 - 8x$ **17.** In $P_3$: $1,\ 2 + x^3,\ 3 - x,\ 7x^2 - 8x$

**18.** In $M_{22}$: $\begin{pmatrix} 1 & -1 \\ 0 & 0 \end{pmatrix}, \begin{pmatrix} 1 & 1 \\ 0 & 0 \end{pmatrix}, \begin{pmatrix} 0 & 0 \\ 1 & 1 \end{pmatrix}, \begin{pmatrix} 0 & 0 \\ 1 & -1 \end{pmatrix}$

**19.** In $M_{22}$: $\begin{pmatrix} 1 & 1 \\ 0 & 0 \end{pmatrix}, \begin{pmatrix} 1 & -1 \\ 0 & 0 \end{pmatrix}, \begin{pmatrix} 0 & 0 \\ 1 & 1 \end{pmatrix}, \begin{pmatrix} 0 & 0 \\ 1 & -1 \end{pmatrix}$

**20.** Using determinants, determine whether each set of vectors is linearly dependent or independent.

**a.** $\begin{pmatrix} 1 \\ 5 \\ 2 \end{pmatrix}; \begin{pmatrix} 3 \\ 0 \\ 4 \end{pmatrix}; \begin{pmatrix} -5 \\ 5 \\ 6 \end{pmatrix}$ **b.** $(2, 1, 4);\ (3, -2, 6);\ (-1, -4, -2)$

In Exercises 21–26 find a basis for the given vector space and determine its dimension.

**21.** The vectors in $\mathbb{R}^3$ lying on the plane $2x + 3y - 4z = 0$

**22.** $H = \{(x, y)\colon 2x - 3y = 0\}$ **23.** $\{\mathbf{v} \in \mathbb{R}^4\colon 3x - y - z + w = 0\}$

**24.** $\{p \in P_3\colon p(0) = 0\}$ **25.** The set of diagonal $4 \times 4$ matrices

**26.** $M_{32}$

In Exercises 27–32 find the null space, range, nullity, and rank of the given matrix.

**27.** $A = \begin{pmatrix} 1 & -2 \\ -2 & 4 \end{pmatrix}$ **28.** $A = \begin{pmatrix} 1 & -1 & 3 \\ 2 & 0 & 4 \\ 0 & -2 & 2 \end{pmatrix}$ **29.** $A = \begin{pmatrix} 1 & -1 & 2 \\ 0 & 1 & 4 \\ 1 & -1 & 0 \end{pmatrix}$

**30.** $A = \begin{pmatrix} 2 & 4 & -2 \\ -1 & -2 & 1 \end{pmatrix}$ **31.** $A = \begin{pmatrix} 2 & 3 \\ -1 & 2 \\ 4 & 6 \end{pmatrix}$ **32.** $A = \begin{pmatrix} 1 & -1 & 2 & 3 \\ 0 & 1 & -1 & 0 \\ 1 & -2 & 3 & 3 \\ 2 & -3 & 5 & 6 \end{pmatrix}$

In Exercises 33–36 write the given vector in terms of the given basis vectors.

**33.** In $\mathbb{R}^2$: $\mathbf{x} = \begin{pmatrix} 2 \\ -1 \end{pmatrix}; \begin{pmatrix} 1 \\ 2 \end{pmatrix}, \begin{pmatrix} -1 \\ 2 \end{pmatrix}$ **34.** In $\mathbb{R}^3$: $\mathbf{x} = \begin{pmatrix} -3 \\ 4 \\ 2 \end{pmatrix}; \begin{pmatrix} 1 \\ 0 \\ 1 \end{pmatrix}, \begin{pmatrix} 1 \\ 1 \\ 0 \end{pmatrix}, \begin{pmatrix} 0 \\ 2 \\ 3 \end{pmatrix}$

**35.** In $P_2$: $\mathbf{x} = 4 + x^2;\ 1 + x^2,\ 1 + x,\ 1$

**36.** In $M_{22}$: $\mathbf{x} = \begin{pmatrix} 3 & 1 \\ 0 & 1 \end{pmatrix}$; $\begin{pmatrix} 1 & 1 \\ 0 & 0 \end{pmatrix}, \begin{pmatrix} 1 & -1 \\ 0 & 0 \end{pmatrix}, \begin{pmatrix} 0 & 0 \\ 1 & 1 \end{pmatrix}, \begin{pmatrix} 0 & 0 \\ 1 & -1 \end{pmatrix}$

In Exercises 37–40 find an orthonormal basis for the given vector space.

**37.** $\mathbb{R}^2$ starting with the basis $\begin{pmatrix} 2 \\ 3 \end{pmatrix}, \begin{pmatrix} -1 \\ 4 \end{pmatrix}$

**38.** $\{(x, y, z) \in \mathbb{R}^3\colon x - y - z = 0\}$

**39.** $\{(x, y, z) \in \mathbb{R}^3\colon x = y = z\}$

**40.** $\{(x, y, z, w) \in \mathbb{R}^4\colon x = z \text{ and } y = w\}$

In Exercises 41–43: **(a)** compute $\text{proj}_H \mathbf{v}$; **(b)** find an orthonormal basis for $H^\perp$; **(c)** write $\mathbf{v}$ as $\mathbf{h} + \mathbf{p}$, where $\mathbf{h} \in H$ and $\mathbf{p} \in H^\perp$.

**41.** $H$ is the subspace of Problem 38; $\mathbf{v} = \begin{pmatrix} -1 \\ 2 \\ 4 \end{pmatrix}$.

**42.** $H$ is the subspace of Problem 39; $\mathbf{v} = \begin{pmatrix} 1 \\ 0 \\ -1 \end{pmatrix}$.

**43.** $H$ is the subspace of Problem 40; $\mathbf{v} = \begin{pmatrix} 1 \\ 0 \\ 0 \\ 1 \end{pmatrix}$.

Calculus **44.** Find an orthonormal basis for $P_2[0, 2]$.

Calculus **45.** Use the result of Exercise 44 to find the polynomial that gives the best mean square approximation to $e^x$ on the interval $[0, 2]$.

**46.** Find the best straight-line fit to the points $(2, 5)$, $(-1, -3)$, $(1, 0)$

**47.** Find the best quadratic fit to the points in Exercise 46.

# 5

# Linear Transformations

## 5.1 DEFINITION AND EXAMPLES

In this chapter we discuss a special class of functions, called *linear transformations* that occur with great frequency in linear algebra and other branches of mathematics. They are also important in a wide variety of applications. Before defining a linear transformation, let us study two simple examples to see what can happen.

EXAMPLE 1 **Reflection About the $x$-Axis** In $\mathbb{R}^2$ define a function $T$ by the formula $T\begin{pmatrix} x \\ y \end{pmatrix} = \begin{pmatrix} x \\ -y \end{pmatrix}$. Geometrically, $T$ takes a vector in $\mathbb{R}^2$ and reflects it about the $x$-axis. This is illustrated in Figure 5.1. Once we have given our basic definition, we shall see that $T$ is a linear transformation from $\mathbb{R}^2$ into $\mathbb{R}^2$.

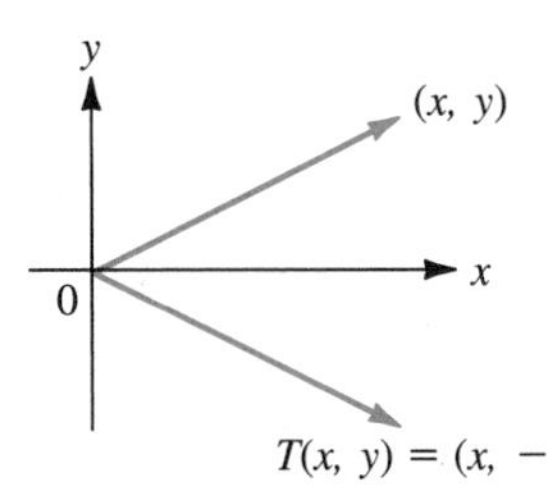


**Figure 5.1** The vector $(x, -y)$ is the reflection about the $x$-axis of the vector $(x, y)$

**EXAMPLE 2** **Transforming a Production Vector into a Raw Material Vector** A manufacturer makes four different products, each of which requires three raw materials. We denote the four products by $P_1$, $P_2$, $P_3$, and $P_4$ and denote the raw materials by $R_1$, $R_2$, and $R_3$. The accompanying table gives the number of units of each raw material required to manufacture 1 unit of each product.

**Needed to Produce 1 Unit of**

| | | $P_1$ | $P_2$ | $P_3$ | $P_4$ |
|---|---|---|---|---|---|
| **Number of Units of Raw Material** | $R_1$ | 2 | 1 | 3 | 4 |
| | $R_2$ | 4 | 2 | 2 | 1 |
| | $R_3$ | 3 | 3 | 1 | 2 |

A natural question arises: If certain numbers of the four products are produced, how many units of each raw material are needed? We let $p_1$, $p_2$, $p_3$, and $p_4$ denote the number of items of the four products manufactured and let $r_1$, $r_2$, and $r_3$ denote the number of units of the three raw materials needed. Then we define

$$\mathbf{p} = \begin{pmatrix} p_1 \\ p_2 \\ p_3 \\ p_4 \end{pmatrix} \qquad \mathbf{r} = \begin{pmatrix} r_1 \\ r_2 \\ r_3 \end{pmatrix} \qquad A = \begin{pmatrix} 2 & 1 & 3 & 4 \\ 4 & 2 & 2 & 1 \\ 3 & 3 & 1 & 2 \end{pmatrix}$$

For example, suppose that $\mathbf{p} = \begin{pmatrix} 10 \\ 30 \\ 20 \\ 50 \end{pmatrix}$. How many units of $R_1$ are needed to produce these numbers of units of the four products? From the table we find that

$$\begin{aligned} r_1 &= p_1 \cdot 2 + p_2 \cdot 1 + p_3 \cdot 3 + p_4 \cdot 4 \\ &= 10 \cdot 2 + 30 \cdot 1 + 20 \cdot 3 + 50 \cdot 4 = 310 \text{ units} \end{aligned}$$

Similarly,

$$r_2 = 10 \cdot 4 + 30 \cdot 2 + 20 \cdot 2 + 50 \cdot 1 = 190 \text{ units}$$

and

$$r_3 = 10 \cdot 3 + 30 \cdot 3 + 20 \cdot 1 + 50 \cdot 2 = 240 \text{ units}$$

In general, we see that

$$\begin{pmatrix} 2 & 1 & 3 & 4 \\ 4 & 2 & 2 & 1 \\ 3 & 3 & 1 & 2 \end{pmatrix} \begin{pmatrix} p_1 \\ p_2 \\ p_3 \\ p_4 \end{pmatrix} = \begin{pmatrix} r_1 \\ r_2 \\ r_3 \end{pmatrix}$$

or $$A\mathbf{p} = \mathbf{r}$$

We can look at this in another way. If $\mathbf{p}$ is called the **production vector** and $\mathbf{r}$ the **raw material vector,** we define the function $T$ by $\mathbf{r} = T\mathbf{p} = A\mathbf{p}$. That is, $T$ is the function that "transforms" the production vector into the raw material vector. It is defined by ordinary matrix multiplication. As we shall see, this function is also a linear transformation.

Before defining a linear transformation, let us say a bit about functions. In Section 1.7 we wrote a system of equations as

$$A\mathbf{x} = \mathbf{b}$$

where $A$ is an $m \times n$ matrix, $\mathbf{x} \in \mathbb{R}^n$ and $\mathbf{b} \in \mathbb{R}^m$. We were asked to find $\mathbf{x}$ when $A$ and $\mathbf{b}$ were known. However, we can look at this equation in another way: Suppose $A$ is given. Then the equation $A\mathbf{x} = \mathbf{b}$ "says": Give me an $\mathbf{x}$ in $\mathbb{R}^n$ and I'll give you a $\mathbf{b}$ in $\mathbb{R}^m$; that is, $A$ represents a *function* with domain $\mathbb{R}^n$ and range in $\mathbb{R}^m$.

The function defined above has the properties that $A(\alpha\mathbf{x}) = \alpha A\mathbf{x}$ if $\alpha$ is a scalar and $A(\mathbf{x} + \mathbf{y}) = A\mathbf{x} + A\mathbf{y}$. This property characterizes linear transformations.

**DEFINITION 1** **Linear Transformation** Let $V$ and $W$ be real vector spaces. A **linear transformation** $T$ from $V$ into $W$ is a function that assigns to each vector $\mathbf{v} \in V$ a unique vector $T\mathbf{v} \in W$ and that satisfies, for each $\mathbf{u}$ and $\mathbf{v}$ in $V$ and each scalar $\alpha$,

$$T(\mathbf{u} + \mathbf{v}) = T\mathbf{u} + T\mathbf{v} \quad \textbf{(1)}$$

and

$$T(\alpha\mathbf{v}) = \alpha T\mathbf{v} \quad \textbf{(2)}$$

## Three Notes on Notation

1. We write $T\colon V \to W$ to indicate that $T$ takes the real vector space $V$ into the real vector space $W$. That is, $T$ is a function with $V$ as its domain and a subset of $W$ as its range.
2. We write $T\mathbf{v}$ and $T(\mathbf{v})$ interchangeably. They denote the same thing; each is read "$T$ of $\mathbf{v}$." This is analogous to the functional notation $f(x)$, which is read "$f$ of $x$."
3. Many of the definitions and theorems in this chapter hold for complex vector spaces as well. (These are vector spaces where the scalars are complex numbers.) However, except briefly in Section 5.5, we deal only with real vector spaces and will, therefore, drop the word "real" in our discussions of vector spaces and linear transformations.

*Terminology.* Linear transformations are often called **linear operators.**

**EXAMPLE 3** **A Linear Transformation from $\mathbb{R}^2$ to $\mathbb{R}^3$** Let $T: \mathbb{R}^2 \to \mathbb{R}^3$ be defined by $T\begin{pmatrix} x \\ y \end{pmatrix} = \begin{pmatrix} x+y \\ x-y \\ 3y \end{pmatrix}$. For example, $T\begin{pmatrix} 2 \\ -3 \end{pmatrix} = \begin{pmatrix} -1 \\ 5 \\ -9 \end{pmatrix}$. Then

$$T\left[\begin{pmatrix} x_1 \\ y_1 \end{pmatrix} + \begin{pmatrix} x_2 \\ y_2 \end{pmatrix}\right] = T\begin{pmatrix} x_1 + x_2 \\ y_1 + y_2 \end{pmatrix} = \begin{pmatrix} x_1 + x_2 + y_1 + y_2 \\ x_1 + x_2 - y_1 - y_2 \\ 3y_1 + 3y_2 \end{pmatrix}$$

$$= \begin{pmatrix} x_1 + y_1 \\ x_1 - y_1 \\ 3y_1 \end{pmatrix} + \begin{pmatrix} x_2 + y_2 \\ x_2 - y_2 \\ 3y_2 \end{pmatrix}$$

But

$$\begin{pmatrix} x_1 + y_1 \\ x_1 - y_1 \\ 3y_1 \end{pmatrix} = T\begin{pmatrix} x_1 \\ y_1 \end{pmatrix} \quad \text{and} \quad \begin{pmatrix} x_2 + y_2 \\ x_2 - y_2 \\ 3y_2 \end{pmatrix} = T\begin{pmatrix} x_2 \\ y_2 \end{pmatrix}$$

Thus

$$T\left[\begin{pmatrix} x_1 \\ y_1 \end{pmatrix} + \begin{pmatrix} x_2 \\ y_2 \end{pmatrix}\right] = T\begin{pmatrix} x_1 \\ y_1 \end{pmatrix} + T\begin{pmatrix} x_2 \\ y_2 \end{pmatrix}$$

Similarly,

$$T\left[\alpha\begin{pmatrix} x \\ y \end{pmatrix}\right] = T\begin{pmatrix} \alpha x \\ \alpha y \end{pmatrix} = \begin{pmatrix} \alpha x + \alpha y \\ \alpha x - \alpha y \\ 3\alpha y \end{pmatrix} = \alpha\begin{pmatrix} x + y \\ x - y \\ 3y \end{pmatrix} = \alpha T\begin{pmatrix} x \\ y \end{pmatrix}$$

Thus $T$ is a linear transformation.

**EXAMPLE 4** **The Zero Transformation** Let $V$ and $W$ be vector spaces and define $T: V \to W$ by $T\mathbf{v} = \mathbf{0}$ for every $\mathbf{v}$ in $V$. Then $T(\mathbf{v}_1 + \mathbf{v}_2) = \mathbf{0} = \mathbf{0} + \mathbf{0} = T\mathbf{v}_1 + T\mathbf{v}_2$ and $T(\alpha\mathbf{v}) = \mathbf{0} = \alpha\mathbf{0} = \alpha T\mathbf{v}$. Here $T$ is called the **zero transformation.**

**EXAMPLE 5** **The Identity Transformation** Let $V$ be a vector space and define $I: V \to V$ by $I\mathbf{v} = \mathbf{v}$ for every $\mathbf{v}$ in $V$. Here $I$ is obviously a linear transformation. It is called the **identity transformation** or **identity operator.**

**EXAMPLE 6** **A Reflection Transformation** Let $T: \mathbb{R}^2 \to \mathbb{R}^2$ be defined by $T\begin{pmatrix} x \\ y \end{pmatrix} = \begin{pmatrix} -x \\ y \end{pmatrix}$. It is easy to verify that $T$ is linear. Geometrically, $T$ takes a vector in $\mathbb{R}^2$ and reflects it about the $y$-axis (see Figure 5.2).

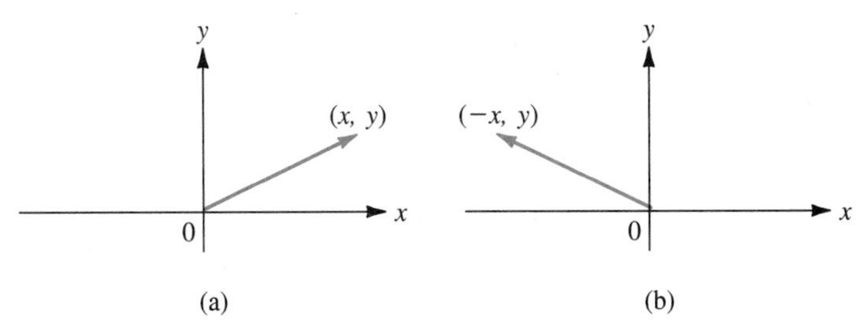


**Figure 5.2** The vector $(-x, y)$ is the reflection about the $y$-axis of the vector $(x, y)$

**EXAMPLE 7** **A Transformation from $\mathbb{R}^n \to \mathbb{R}^m$ Given by Multiplication by an $m \times n$ Matrix** Let $A$ be an $m \times n$ matrix and define $T$: $\mathbb{R}^n \to \mathbb{R}^m$ by $T\mathbf{x} = A\mathbf{x}$. Since $A(\mathbf{x} + \mathbf{y}) = A\mathbf{x} + A\mathbf{y}$ and $A(\alpha\mathbf{x}) = \alpha A\mathbf{x}$ if $\mathbf{x}$ and $\mathbf{y}$ are in $\mathbb{R}^n$, we see that $T$ is a linear transformation. Thus: *Every $m \times n$ matrix $A$ can be used to define a linear transformation from $\mathbb{R}^n$ into $\mathbb{R}^m$.* In Section 5.3 we shall see that a certain converse is true: *Every linear transformation between finite dimensional vector spaces can be represented by a matrix.*

**EXAMPLE 8** **A Rotation Transformation** Suppose the vector $\mathbf{v} = \begin{pmatrix} x \\ y \end{pmatrix}$ in the $xy$-plane is rotated through an angle of $\theta$ (measured in degrees or radians) in the counterclockwise direction. Call the new rotated vector $\mathbf{v}' = \begin{pmatrix} x' \\ y' \end{pmatrix}$. Then as in Figure 5.3, if $r$ denotes the length of $\mathbf{v}$ (which is unchanged by rotation),

$$x = r \cos \alpha \qquad y = r \sin \alpha$$
$$x' = r \cos (\theta + \alpha) \qquad y' = r \sin (\theta + \alpha)†$$

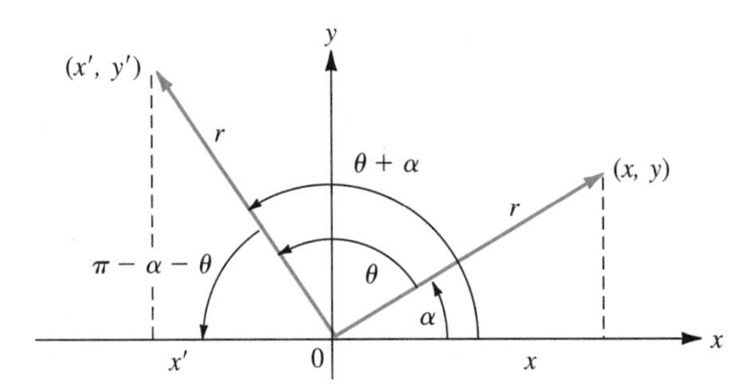


**Figure 5.3** $(x', y')$ is obtained by rotating $(x, y)$ through the angle $\theta$

† These follow from the standard definitions of $\cos \theta$ and $\sin \theta$ as the $x$- and $y$-coordinates of a point on the unit circle. If $(x, y)$ is a point on the circle centered at the origin of radius $r$, then $x = r \cos \varphi$ and $y = r \sin \varphi$, where $\varphi$ is the angle the vector $(x, y)$ makes with the positive $x$-axis.

But $r\cos(\theta + \alpha) = r\cos\theta\cos\alpha - r\sin\theta\sin\alpha$, so that

$$x' = x\cos\theta - y\sin\theta \tag{3}$$

Similarly, $r\sin(\theta + \alpha) = r\sin\theta\cos\alpha + r\cos\theta\sin\alpha$ or

$$y' = x\sin\theta + y\cos\theta \tag{4}$$

Let

$$A_\theta = \begin{pmatrix} \cos\theta & -\sin\theta \\ \sin\theta & \cos\theta \end{pmatrix} \tag{5}$$

Then from (3) and (4), we see that $A_\theta\begin{pmatrix} x \\ y \end{pmatrix} = \begin{pmatrix} x' \\ y' \end{pmatrix}$. The linear transformation $T$: $\mathbb{R}^2 \to \mathbb{R}^2$ defined by $T\mathbf{v} = A_\theta\mathbf{v}$, where $A_\theta$ is given by (5), is called a **rotation transformation.**

EXAMPLE 9 **An Orthogonal Projection Transformation** Let $H$ be a subspace of $\mathbb{R}^n$. We define the **orthogonal projection transformation** $P$: $V \to H$ by

$$P\mathbf{v} = \text{proj}_H\,\mathbf{v} \tag{6}$$

Let $\{\mathbf{u}_1, \mathbf{u}_2, \ldots, \mathbf{u}_k\}$ be an orthonormal basis for $H$. Then from Definition 4.9.4 on page 400 we have

$$P\mathbf{v} = (\mathbf{v}\cdot\mathbf{u}_1)\mathbf{u}_1 + (\mathbf{v}\cdot\mathbf{u}_2)\mathbf{u}_2 + \cdots + (\mathbf{v}\cdot\mathbf{u}_k)\mathbf{u}_k \tag{7}$$

Since $(\mathbf{v}_1 + \mathbf{v}_2)\cdot\mathbf{u} = \mathbf{v}_1\cdot\mathbf{u} + \mathbf{v}_2\cdot\mathbf{u}$ and $(\alpha\mathbf{v})\cdot\mathbf{u} = \alpha(\mathbf{v}\cdot\mathbf{u})$, we see that $P$ is a linear transformation.

EXAMPLE 10 **Two Projection Operators** Let $T$: $\mathbb{R}^3 \to \mathbb{R}^3$ be defined by $T\begin{pmatrix} x \\ y \\ z \end{pmatrix} = \begin{pmatrix} x \\ y \\ 0 \end{pmatrix}$. Then $T$ is the projection operator taking a vector in three-dimensional space and projecting it into the $xy$-plane. Similarly, $T\begin{pmatrix} x \\ y \\ z \end{pmatrix} = \begin{pmatrix} x \\ 0 \\ z \end{pmatrix}$ projects a vector in space into the $xz$-plane. These two transformations are depicted in Figure 5.4.

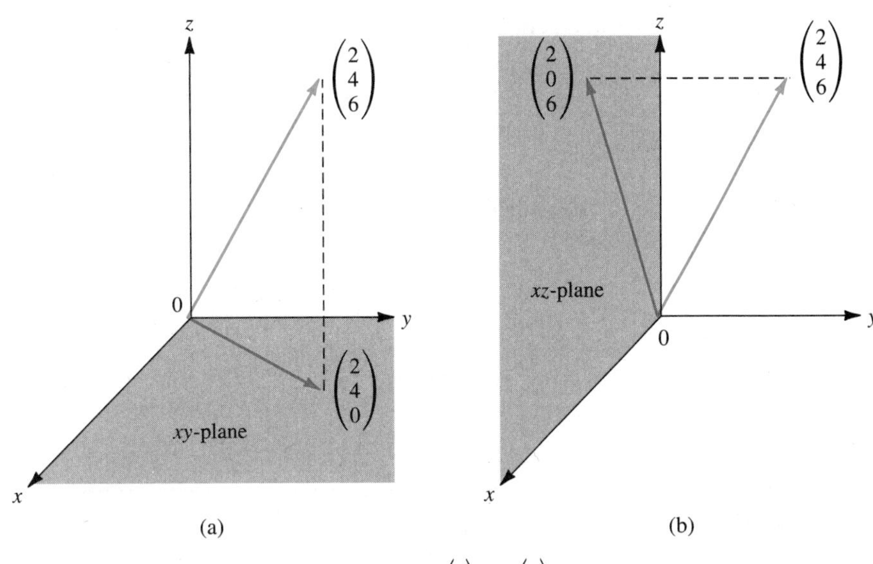


**Figure 5.4** (a) Projection onto $xy$-plane: $T\begin{pmatrix}x\\y\\z\end{pmatrix} = \begin{pmatrix}x\\y\\0\end{pmatrix}$.

(b) Projection onto $xz$-plane: $T\begin{pmatrix}x\\y\\z\end{pmatrix} = \begin{pmatrix}x\\0\\z\end{pmatrix}$

**EXAMPLE 11** **A Transpose Operator** Define $T$: $M_{mn} \to M_{nm}$ by $T(A) = A^t$. Since $(A + B)^t = A^t + B^t$ and $(\alpha A)^t = \alpha A^t$, we see that $T$, called the **transpose operator,** is a linear transformation.

**EXAMPLE 12** Calculus

**An Integral Operator** Let $J$: $C[0, 1] \to \mathbb{R}$ be defined by $Jf = \int_0^1 f(x)\,dx$. Since $\int_0^1 [f(x) + g(x)]\,dx = \int_0^1 f(x)\,dx + \int_0^1 g(x)\,dx$ and $\int_0^1 \alpha f(x)\,dx = \alpha \int_0^1 f(x)\,dx$ if $f$ and $g$ are continuous, we see that $J$ is linear. For example, $J(x^3) = \frac{1}{4}$. $J$ is called an **integral operator.**

**EXAMPLE 13** Calculus

**A Differential Operator** Let $D$: $C^1[0, 1] \to C[0, 1]$ be defined by $Df = f'$. Since $(f + g)' = f' + g'$ and $(\alpha f)' = \alpha f'$ if $f$ and $g$ are differentiable, we see that $D$ is linear. $D$ is called a **differential operator.**

**WARNING**

Not every transformation that looks linear actually is linear. For example, define $T$: $\mathbb{R} \to \mathbb{R}$ by $Tx = 2x + 3$. Then the graph of $\{(x, Tx): x \in \mathbb{R}\}$ is a straight line in the $xy$-plane. But $T$ is not linear since $T(x + y) = 2(x + y) + 3 = 2x + 2y + 3$ and $Tx + Ty = (2x + 3) + (2y + 3) = 2x + 2y + 6$. The only linear transformations from $\mathbb{R}$ to $\mathbb{R}$ are functions of the form $f(x) = mx$ for some real number $m$. Thus among all functions whose graphs are lines, the only ones that are linear are those that pass through the origin. In algebra and calculus a **linear function** with domain $\mathbb{R}$ is defined as a function having the form $f(x) = mx + b$. Thus we can say that a linear function is a linear transformation from $\mathbb{R}$ to $\mathbb{R}$ if and only if $b$ (the $y$-intercept) is zero.

**EXAMPLE 14** **A Transformation That Is Not Linear** Let $T$: $C[0, 1] \to \mathbb{R}$ be defined by $Tf = f(0) + 1$. Then $T$ is not linear. To see this, we compute

$$T(f + g) = (f + g)(0) + 1 = f(0) + g(0) + 1$$
$$Tf + Tg = [f(0) + 1] + [g(0) + 1] = f(0) + g(0) + 2$$

This provides another example of a transformation that might look linear but in fact is not.

# PROBLEMS 5.1

## Self-Quiz

### *True-False*

**I.** If $T$ is a linear transformation, then $T(3\mathbf{x}) = 3T\mathbf{x}$.

**II.** If $T$ is a linear transformation, then $T(\mathbf{x} + \mathbf{y}) = T\mathbf{x} + T\mathbf{y}$.

**III.** If $T$ is a linear transformation, then $T(\mathbf{xy}) = T\mathbf{x}T\mathbf{y}$

**IV.** If $A$ is a $4 \times 5$ matrix, then $T\mathbf{x} = A\mathbf{x}$ is a linear transformation from $\mathbb{R}^4$ to $\mathbb{R}^5$.

**V.** If $A$ is a $4 \times 5$ matrix, then $T\mathbf{x} = A\mathbf{x}$ is a linear transformation from $\mathbb{R}^5$ to $\mathbb{R}^4$.

In Problems 1–29 determine whether the given transformation from $V$ to $W$ is linear.

**1.** $T$: $\mathbb{R}^2 \to \mathbb{R}^2$; $T\begin{pmatrix} x \\ y \end{pmatrix} = \begin{pmatrix} x \\ 0 \end{pmatrix}$

**2.** $T$: $\mathbb{R}^2 \to \mathbb{R}^2$; $T\begin{pmatrix} x \\ y \end{pmatrix} = \begin{pmatrix} 1 \\ y \end{pmatrix}$

**3.** $T$: $\mathbb{R}^3 \to \mathbb{R}^2$; $T\begin{pmatrix} x \\ y \\ z \end{pmatrix} = \begin{pmatrix} x \\ y \end{pmatrix}$

**4.** $T$: $\mathbb{R}^3 \to \mathbb{R}^2$; $T\begin{pmatrix} x \\ y \\ z \end{pmatrix} = \begin{pmatrix} 0 \\ y \end{pmatrix}$

**5.** $T$: $\mathbb{R}^3 \to \mathbb{R}^2$; $T\begin{pmatrix} x \\ y \\ z \end{pmatrix} = \begin{pmatrix} 1 \\ z \end{pmatrix}$

**6.** $T$: $\mathbb{R}^2 \to \mathbb{R}^2$; $T\begin{pmatrix} x \\ y \end{pmatrix} = \begin{pmatrix} x^2 \\ y^2 \end{pmatrix}$

**7.** $T$: $\mathbb{R}^2 \to \mathbb{R}^2$; $T\begin{pmatrix} x \\ y \end{pmatrix} = \begin{pmatrix} y \\ x \end{pmatrix}$

**8.** $T$: $\mathbb{R}^2 \to \mathbb{R}^2$; $T\begin{pmatrix} x \\ y \end{pmatrix} = \begin{pmatrix} x + y \\ x - y \end{pmatrix}$

**9.** $T$: $\mathbb{R}^2 \to \mathbb{R}$; $T\begin{pmatrix} x \\ y \end{pmatrix} = xy$

---

**Answers to Self-Quiz**

**I.** True **II.** True **III.** False **IV.** False **V.** True

**10.** $T$: $\mathbb{R}^n \to \mathbb{R}$; $T\begin{pmatrix} x_1 \\ x_2 \\ \vdots \\ x_n \end{pmatrix} = x_1 + x_2 + \cdots + x_n$

**11.** $T$: $\mathbb{R} \to \mathbb{R}^n$; $T(x) = \begin{pmatrix} x \\ x \\ \vdots \\ x \end{pmatrix}$

**12.** $T$: $\mathbb{R}^4 \to \mathbb{R}^2$; $T\begin{pmatrix} x \\ y \\ z \\ w \end{pmatrix} = \begin{pmatrix} x + z \\ y + w \end{pmatrix}$

**13.** $T$: $\mathbb{R}^4 \to \mathbb{R}^2$; $T = \begin{pmatrix} x \\ y \\ z \\ w \end{pmatrix} = \begin{pmatrix} xz \\ yw \end{pmatrix}$

**14.** $T$: $M_{nn} \to M_{nn}$; $T(A) = AB$, where $B$ is a fixed $n \times n$ matrix

**15.** $T$: $M_{nn} \to M_{nn}$; $T(A) = A^tA$

**16.** $T$: $M_{mn} \to M_{mp}$; $T(A) = AB$, where $B$ is a fixed $n \times p$ matrix

**17.** $T$: $D_n \to D_n$; $T(D) = D^2$ ($D_n$ is the set of $n \times n$ diagonal matrices)

**18.** $T$: $D_n \to D_n$; $T(D) = I + D$

**19.** $T$: $P_2 \to P_1$; $T(a_0 + a_1x + a_2x^2) = a_0 + a_1x$

**20.** $T$: $P_2 \to P_1$; $T(a_0 + a_1x + a_2x^2) = a_1 + a_2x$

**21.** $T$: $\mathbb{R} \to P_n$; $T(a) = a + ax + ax^2 + \cdots + ax^n$

**22.** $T$: $P_2 \to P_4$; $T(p(x)) = [p(x)]^2$

**23.** $T$: $C[0, 1] \to C[0, 1]$; $Tf(x) = f^2(x)$

**24.** $T$: $C[0, 1] \to C[0, 1]$; $Tf(x) = f(x) + 1$

Calculus **25.** $T$: $C[0, 1] \to \mathbb{R}$; $Tf = \int_0^1 f(x)g(x)\,dx$, where $g$ is a fixed function in $C[0, 1]$

Calculus **26.** $T$: $C^1[0, 1] \to C[0, 1]$; $Tf = (fg)'$, where $g$ is a fixed function in $C^1[0, 1]$

**27.** $T$: $C[0, 1] \to C[1, 2]$; $Tf(x) = f(x - 1)$

**28.** $T$: $C[0, 1] \to \mathbb{R}$; $Tf = f(\frac{1}{2})$

**29.** $T$: $M_{nn} \to \mathbb{R}$; $T(A) = \det A$

**30.** Let $T$: $\mathbb{R}^2 \to \mathbb{R}^2$ be given by $T(x, y) = (-x, -y)$. Describe $T$ geometrically.

**31.** Let $T$ be a linear transformation from $\mathbb{R}^2 \to \mathbb{R}^3$ such that $T\begin{pmatrix} 1 \\ 0 \end{pmatrix} = \begin{pmatrix} 1 \\ 2 \\ 3 \end{pmatrix}$ and $T\begin{pmatrix} 0 \\ 1 \end{pmatrix} = \begin{pmatrix} -4 \\ 0 \\ 5 \end{pmatrix}$. Find: **(a)** $T\begin{pmatrix} 2 \\ 4 \end{pmatrix}$ and **(b)** $T\begin{pmatrix} -3 \\ 7 \end{pmatrix}$.

**32.** In Example 8:

**a.** Find the rotation matrix $A_\theta$ when $\theta = \pi/6$.

**b.** What happens to the vector $\begin{pmatrix} -3 \\ 4 \end{pmatrix}$ if it is rotated through an angle of $\pi/6$ in the counterclockwise direction?

**33.** Let $A_\theta = \begin{pmatrix} \cos\theta & -\sin\theta & 0 \\ \sin\theta & \cos\theta & 0 \\ 0 & 0 & 1 \end{pmatrix}$. Describe geometrically the linear transformation $T$: $\mathbb{R}^3 \to \mathbb{R}^3$ given by $T\mathbf{x} = A_\theta\mathbf{x}$.

**34.** Answer the questions in Problem 33 for $A_\theta = \begin{pmatrix} \cos\theta & 0 & -\sin\theta \\ 0 & 1 & 0 \\ \sin\theta & 0 & \cos\theta \end{pmatrix}$.

**35.** Suppose that in a real vector space $V$, $T$ satisfies $T(\mathbf{x}+\mathbf{y}) = T\mathbf{x} + T\mathbf{y}$ and $T(\alpha\mathbf{x}) = \alpha T\mathbf{x}$ for $\alpha \geq 0$. Show that $T$ is linear.

**36.** Find a linear transformation $T$: $M_{33} \to M_{22}$.

**37.** If $T$ is a linear transformation from $V$ to $W$, show that $T(\mathbf{x}-\mathbf{y}) = T\mathbf{x} - T\mathbf{y}$.

**38.** If $T$ is a linear transformation from $V$ to $W$, show that $T\mathbf{0} = \mathbf{0}$. Are the two zero vectors here the same?

**39.** Let $V$ be an inner product space and let $\mathbf{u}_0 \in V$ be fixed. Let $T$: $V \to \mathbb{R}$ (or $\mathbb{C}$) be defined by $T\mathbf{v} = (\mathbf{v}, \mathbf{u}_0)$. Show that $T$ is linear.

***40.** Show that if $V$ is a complex inner product space and $T$: $V \to \mathbb{C}$ is defined by $T\mathbf{v} = (\mathbf{u}_0, \mathbf{v})$ for a fixed vector $\mathbf{u}_0 \in V$, then $T$ is not linear.

**41.** Let $V$ be an inner product space with the finite dimensional subspace $H$. Let $\{\mathbf{u}_1, \mathbf{u}_2, \ldots, \mathbf{u}_k\}$ be a basis for $H$. Show that $T$: $V \to H$ defined by $T\mathbf{v} = (\mathbf{v}, \mathbf{u}_1)\mathbf{u}_1 + (\mathbf{v}, \mathbf{u}_2)\mathbf{u}_2 + \cdots + (\mathbf{v}, \mathbf{u}_k)\mathbf{u}_k$ is a linear transformation.

**42.** Let $V$ and $W$ be vector spaces. Let $L(V, W)$ denote the set of linear transformations from $V$ to $W$. If $T_1$ and $T_2$ are in $L(V, W)$, define $\alpha T_1$ and $T_1 + T_2$ by $(\alpha T_1)\mathbf{v} = \alpha(T_1\mathbf{v})$ and $(T_1 + T_2)\mathbf{v} = T_1\mathbf{v} + T_2\mathbf{v}$. Prove that $L(V, W)$ is a vector space.

# MATLAB 5.1

## MATLAB Information: Obtaining Graphics Printout

If you are using MATLAB on a system that has access to extended memory or if you are using MATLAB 4.0, enter **help print** for appropriate directions. If your system does not have access to extended memory, enter **help meta** for information about saving a graphics screen to a file. Then read the MATLAB manual for information about using the program **gpp.** If you are using the student edition of MATLAB, the only way to obtain a graphics printout is by performing a screen dump (Shift-PrtScr keys). (Before beginning MATLAB, you must be sure you have loaded the graphics package for your operating system.)

*Warning.* Output from a screen dump does not necessarily preserve the aspect ratio of the screen; thus right angles may not appear as right angles and equal lengths may not appear equal.

1. **Computer Graphics: Creating a figure** A figure to be graphed is described using a matrix containing key points in the figure and a matrix containing information about which points are to be connected with line segments.

**The points matrix.** The points matrix is a $2 \times n$ matrix, where $n$ is the number of points; the first row contains the $x$-coordinates of the points and the second row contains the $y$-coordinates of the points.

**The lines matrix.** The lines matrix is a $2 \times m$ matrix, where $m$ is the number of lines. Each entry is the number of a column of the points matrix. The information indicates that the two points referred to in a column of the lines matrix should be connected by a line segment.

For example, to describe the first rectangle below:

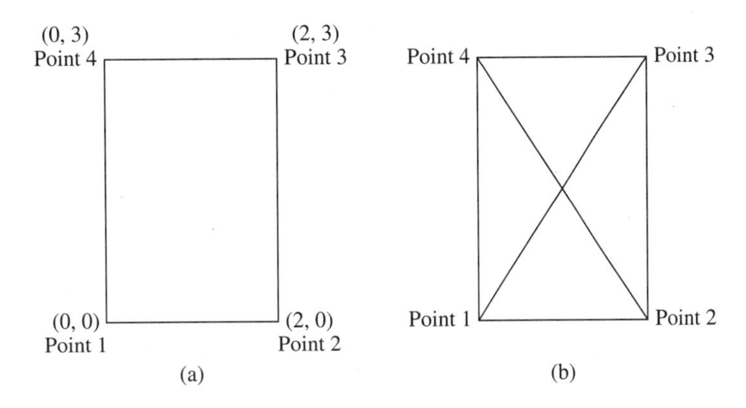


(a) (b)

$$pts = \begin{pmatrix} 0 & 2 & 2 & 0 \\ 0 & 0 & 3 & 3 \end{pmatrix}$$

$$lns = \begin{pmatrix} 1 & 2 & 3 & 4 \\ 2 & 3 & 4 & 1 \end{pmatrix}$$

The matrix *lns* says that point 1, (0, 0), (column 1 of *pts*) is connected to point 2, (2, 0), (column 2 of *pts*); point 2 is connected to point 3, (2, 3), (column 3 of *pts*); point 3 is connected to point 4, (0, 3), (column 4 of *pts*); and point 4 is connected to point 1.

If the figure is the second rectangle above with the diagonals drawn from corner to corner, the *pts* matrix would be the same, and

$$lns = \begin{pmatrix} 1 & 2 & 3 & 4 & 1 & 2 \\ 2 & 3 & 4 & 1 & 3 & 4 \end{pmatrix}.$$

M

To plot the figure after entering the matrices *pts* and *lns*, you will use the m-file *grafics.m* on the accompanying disk. The syntax for using *grafics* is **grafics(pts, lns, clr, sym, M):**

*pts* = the points matrix

*lns* = the lines matrix

*clr* = color options; for example **'r'** represents red; enter **help plot** for a description of further color options.

*sym* = **'*'** or **'o'** or **'+'** or **'x'** or **'.'**

The points in the points matrix will be plotted individually using the chosen symbol.

*M* is some positive number, usually an integer. This sets the scale on the axes of the graphics screen to be $-M \leq x \leq M$ and $-M \leq y \leq M$.

For example, **grafics(pts, lns, 'b', '+', 10)** will plot the rectangle given by the first set of matrices, *pts* and *lns*, above in blue, with the points (the corners of the rectangle) plotted with a "+" symbol and the axes scale of $-10 \leq x \leq 10$ and $-10 \leq y \leq 10$.

a. Enter the following matrices:

$$pts = \begin{pmatrix} 0 & 3 & 3 & 8 & 8 & 11 & 11 & 15 & 15 & 11 & 8 & 8 & 0 & 10 \\ 0 & 0 & 3 & 3 & 0 & 0 & 7 & 7 & 10 & 10 & 12 & 7 & 7 & 9 \end{pmatrix}$$

$$lns = \begin{pmatrix} 1 & 2 & 3 & 4 & 5 & 6 & 7 & 8 & 9 & 10 & 11 & 12 & 13 \\ 2 & 3 & 4 & 5 & 6 & 7 & 8 & 9 & 10 & 11 & 12 & 13 & 1 \end{pmatrix}$$

Now enter the command **grafics(pts, lns, 'r', '*', 20)**
Describe in words the figure that is produced and describe other features of the graphics screen.

b. Design your own figure. Form a points and lines matrix for it and plot it using the m-file *grafics.m.*

2. Suppose $T:\mathbb{R}^2 \to \mathbb{R}^2$ is a linear transformation (such as rotation about the origin) and we wish to plot the image of a figure after applying the transformation to it.

a. *(Paper and pencil)* Consider points $P_1$ and $P_2$ in the plane. Let $\mathbf{x}$ be the vector starting at the origin and ending at $P_1$ and let $\mathbf{y}$ be the vector starting at the origin and ending at $P_2$. Explain why the vector $\mathbf{z} = \mathbf{x} - \mathbf{y}$ is parallel to the line segment from $P_1$ to $P_2$.

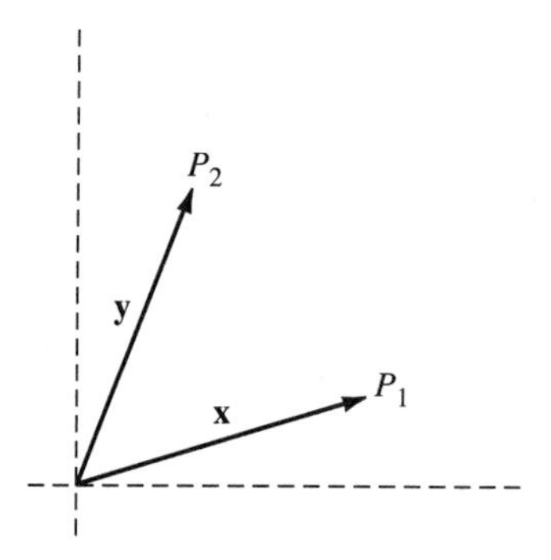


Let $T:\mathbb{R}^2 \to \mathbb{R}^2$ be a linear transformation. Then the endpoint of $T\mathbf{x}$ will be the point in the transformed image that came from $P_1$ and the endpoint of $T\mathbf{y}$ will be the point in the transformed image that came from $P_2$. Thus $T\mathbf{x} - T\mathbf{y}$ will be parallel to the line segment between the transformed image of $P_1$ and the transformed image of $P_2$. Explain why we can conclude from the linearity of $T$ that the line segment between $P_1$ and $P_2$, represented by $\mathbf{x} - \mathbf{y}$, is transformed into the line segment between the transformed image of $P_1$ and the transformed image of $P_2$, represented by $T\mathbf{x} - T\mathbf{y}$.

Part (a) implies that, to graph the image of a figure after applying a linear transformation $T$, we need only apply the transformation to the points matrix; the lines matrix of the transformed image will be the same. Any linear transformation $T:\mathbb{R}^2 \to \mathbb{R}^2$ can be represented by multiplication by a $2 \times 2$ matrix $A$. Hence the points matrix of the transformed image will be $A * pts$, where *pts* is the points matrix of the original figure.

b. We wish to plot, on the same set of axes, the figure given by the points and lines matrices given in MATLAB Problem 1(a) in this section and its transformed image after applying a rotation transformation. Recall that the matrix for the linear transformation that rotates *counterclockwise* around the origin by an angle $\theta$ is given by

$$A = \begin{pmatrix} \cos(\theta) & -\sin(\theta) \\ \sin(\theta) & \cos(\theta) \end{pmatrix}.$$

The following commands will plot the original figure (in red) and its rotation by an angle of $\pi/2$ *clockwise* around the origin (in blue):

```
th = -pi/2; A = [cos(th) -sin(th);sin(th) cos(th)]
grafics(pts, lns, 'r', '*', 20)
hold on
grafics(A*pts, lns, 'b', '*', 20)
hold off
```

Note the use of the command **hold on** to enable both figures to be plotted on the same set of axes. Using the **hold off** command is a good habit; strictly speaking, it is not needed here since the m-file *grafics.m* contains a final **hold off** command.

*Interpretation.* On the plot, identify the four points of the original figure that lie along the bottom (along the $x$-axis). Identify the points into which they are transformed. Identify any line segments among these points in the original figure and identify the corresponding line segments in the transformed figure. Verify that these line segments of the transformed figure are indeed rotations by $\pi/2$ clockwise of the line segments of the original figure. Do the same for the two points of the original figure that lie along the $y$-axis.

c. On the same set of axes, plot the original figure (the one used in previous parts of this problem) and the transformed image after rotation around the origin by $2\pi/3$ *counterclockwise.* Interpret as indicated in part (b).

d. On the same set of axes, plot your figure from MATLAB Problem 1(b) in this section and the transformed image after rotation around the origin by an angle of your choice.

3. Consider the figure whose points and lines matrices are given in MATLAB Problem 1(a) in this section.

M

a. Using the m-file *grafics* on the accompanying disk, on the same set of axes, plot the original figure and the figure after applying the transformation given by multiplication by the matrix $A$, where

$$A = \begin{pmatrix} 2 & 0 \\ 0 & 2 \end{pmatrix}$$

Make an appropriate choice of the parameter $M$ in the call to *grafics* so that both figures fit nicely on the graphics screen. (You may need to experiment with the choice of the parameter $M$. After determining an appropriate $M$, enter **hold off,** and repeat the sequence of commands needed to plot both images on the same set of axes.)

Describe the geometry of the transformation.

b. Repeat part (a) for each transformation given below:

$$A = \begin{pmatrix} 2 & 0 \\ 0 & 1 \end{pmatrix}, \quad A = \begin{pmatrix} 1 & 0 \\ 0 & 2 \end{pmatrix}$$

c. *(Paper and pencil)* Describe the geometry of $T:\mathbb{R}^2 \to \mathbb{R}^2$ given by $T(\mathbf{x}) = A\mathbf{x}$ where

$$A = \begin{pmatrix} r & 0 \\ 0 & s \end{pmatrix}$$

for $r > 0$ and $s > 0$.

## 5.2 PROPERTIES OF LINEAR TRANSFORMATIONS: RANGE AND KERNEL

In this section we develop some of the basic properties of linear transformations.

**THEOREM 1** Let $T: V \to W$ be a linear transformation. Then for all vectors $\mathbf{u}, \mathbf{v}, \mathbf{v}_1, \mathbf{v}_2, \ldots, \mathbf{v}_n$ in $V$ and all scalars $\alpha_1, \alpha_2, \ldots, \alpha_n$:

**i.** $T(\mathbf{0}) = \mathbf{0}$

**ii.** $T(\mathbf{u} - \mathbf{v}) = T\mathbf{u} - T\mathbf{v}$

**iii.** $T(\alpha_1\mathbf{v}_1 + \alpha_2\mathbf{v}_2 + \cdots + \alpha_n\mathbf{v}_n) = \alpha_1 T\mathbf{v}_1 + \alpha_2 T\mathbf{v}_2 + \cdots + \alpha_n T\mathbf{v}_n$

*Note.* In part ($i$) the $\mathbf{0}$ on the left is the zero vector in $V$, whereas the $\mathbf{0}$ on the right is the zero vector in $W$.

**Proof** **i.** $T(\mathbf{0}) = T(\mathbf{0} + \mathbf{0}) = T(\mathbf{0}) + T(\mathbf{0})$. Thus

$$\mathbf{0} = T(\mathbf{0}) - T(\mathbf{0}) = T(\mathbf{0}) + T(\mathbf{0}) - T(\mathbf{0}) = T(\mathbf{0})$$

**ii.** $T(\mathbf{u} - \mathbf{v}) = T[\mathbf{u} + (-1)\mathbf{v}] = T\mathbf{u} + T[(-1)\mathbf{v}] = T\mathbf{u} + (-1)T\mathbf{v} = T\mathbf{u} - T\mathbf{v}$.

**iii.** We prove this part by induction (see Appendix 1). For $n = 2$ we get $T(\alpha_1\mathbf{v}_1 + \alpha_2\mathbf{v}_2) = T(\alpha_1\mathbf{v}_1) + T(\alpha_2\mathbf{v}_2) = \alpha_1 T\mathbf{v}_1 + \alpha_2 T\mathbf{v}_2$. Thus the equation holds for $n = 2$. We assume that it holds for $n = k$ and prove it for $n = k + 1$: $T(\alpha_1\mathbf{v}_1 + \alpha_2\mathbf{v}_2 + \cdots + \alpha_k\mathbf{v}_k + \alpha_{k+1}\mathbf{v}_{k+1}) = T(\alpha_1\mathbf{v}_1 + \alpha_2\mathbf{v}_2 + \cdots + \alpha_k\mathbf{v}_k) + T(\alpha_{k+1}\mathbf{v}_{k+1})$, and using the equation in part ($iii$) for $n = k$, this is equal to $(\alpha_1 T\mathbf{v}_1 + \alpha_2 T\mathbf{v}_2 + \cdots + \alpha_k T\mathbf{v}_k) + \alpha_{k+1}T\mathbf{v}_{k+1}$, which is what we wanted to show. This completes the proof. ∎

*Remark.* Note that parts ($i$) and ($ii$) of Theorem 1 are special cases of part ($iii$).

An important fact about linear transformations is that they are completely determined by what they do to basis vectors.

**THEOREM 2** Let $V$ be a finite dimensional vector space with basis $B = \{\mathbf{v}_1, \mathbf{v}_2, \ldots, \mathbf{v}_n\}$. Let $\mathbf{w}_1, \mathbf{w}_2, \ldots, \mathbf{w}_n$ be $n$ vectors in $W$. Suppose that $T_1$ and $T_2$ are two linear transformations from $V$ to $W$ such that $T_1\mathbf{v}_i = T_2\mathbf{v}_i = \mathbf{w}_i$ for $i = 1, 2, \ldots, n$. Then for any vector $\mathbf{v} \in V$, $T_1\mathbf{v} = T_2\mathbf{v}$. That is, $T_1 = T_2$.

**Proof** Since $B$ is a basis for $V$, there exists a unique set of scalars $\alpha_1, \alpha_2, \ldots, \alpha_n$ such that $\mathbf{v} = \alpha_1\mathbf{v}_1 + \alpha_2\mathbf{v}_2 + \cdots + \alpha_n\mathbf{v}_n$. Then, from part ($iii$) of Theorem 1,

$$\begin{aligned} T_1\mathbf{v} = T_1(\alpha_1\mathbf{v}_1 + \alpha_2\mathbf{v}_2 + \cdots + \alpha_n\mathbf{v}_n) &= \alpha_1 T_1\mathbf{v}_1 + \alpha_2 T_1\mathbf{v}_2 + \cdots + \alpha_n T_1\mathbf{v}_n \\ &= \alpha_1\mathbf{w}_1 + \alpha_2\mathbf{w}_2 + \cdots + \alpha_n\mathbf{w}_n \end{aligned}$$

Similarly,

$$T_2\mathbf{v} = T_2(\alpha_1\mathbf{v}_1 + \alpha_2\mathbf{v}_2 + \cdots + \alpha_n\mathbf{v}_n) = \alpha_1T_2\mathbf{v}_1 + \alpha_2T_2\mathbf{v}_2 + \cdots + \alpha_nT_2\mathbf{v}_n$$
$$= \alpha_1\mathbf{w}_1 + \alpha_2\mathbf{w}_2 + \cdots + \alpha_n\mathbf{w}_n$$

Thus $T_1\mathbf{v} = T_2\mathbf{v}$.

Theorem 2 tells us that if $T$: $V \rightarrow W$ and $V$ is finite dimensional, then we need to know only what $T$ does to basis vectors in $V$. That is, if you know the image of each basis vector, you can determine the image of any vector in $V$. This determines $T$ completely. To see this, let $\mathbf{v}_1, \mathbf{v}_2, \ldots, \mathbf{v}_n$ be a basis in $V$ and let $\mathbf{v}$ be another vector in $V$. Then as in the proof of Theorem 2,

$$T\mathbf{v} = \alpha_1T\mathbf{v}_1 + \alpha_2T\mathbf{v}_2 + \cdots + \alpha_nT\mathbf{v}_n$$

Thus we can compute $T\mathbf{v}$ for any vector $\mathbf{v} \in V$ if we know $T\mathbf{v}_1, T\mathbf{v}_2, \ldots, T\mathbf{v}_n$.

**EXAMPLE 1** **If You Know What a Linear Transformation Does to Basis Vectors, Then You Know What It Does to Any Other Vector** Let $T$ be a linear transformation from $\mathbb{R}^3$ into $\mathbb{R}^2$ and suppose that $T\begin{pmatrix}1\\0\\0\end{pmatrix} = \begin{pmatrix}2\\3\end{pmatrix}$, $T\begin{pmatrix}0\\1\\0\end{pmatrix} = \begin{pmatrix}-1\\4\end{pmatrix}$, and $T\begin{pmatrix}0\\0\\1\end{pmatrix} = \begin{pmatrix}5\\-3\end{pmatrix}$. Compute $T\begin{pmatrix}3\\-4\\5\end{pmatrix}$.

**Solution** We have $\begin{pmatrix}3\\-4\\5\end{pmatrix} = 3\begin{pmatrix}1\\0\\0\end{pmatrix} - 4\begin{pmatrix}0\\1\\0\end{pmatrix} + 5\begin{pmatrix}0\\0\\1\end{pmatrix}$.

Thus

$$T\begin{pmatrix}3\\-4\\5\end{pmatrix} = 3T\begin{pmatrix}1\\0\\0\end{pmatrix} - 4T\begin{pmatrix}0\\1\\0\end{pmatrix} + 5T\begin{pmatrix}0\\0\\1\end{pmatrix}$$
$$= 3\begin{pmatrix}2\\3\end{pmatrix} - 4\begin{pmatrix}-1\\4\end{pmatrix} + 5\begin{pmatrix}5\\-3\end{pmatrix} = \begin{pmatrix}6\\9\end{pmatrix} + \begin{pmatrix}4\\-16\end{pmatrix} + \begin{pmatrix}25\\-15\end{pmatrix} = \begin{pmatrix}35\\-22\end{pmatrix}$$

Another question arises: If $\mathbf{w}_1, \mathbf{w}_2, \ldots, \mathbf{w}_n$ are $n$ vectors in $W$, does there exist a linear transformation $T$ such that $T\mathbf{v}_i = \mathbf{w}_i$ for $i = 1, 2, \ldots, n$? The answer is yes, as the next theorem shows.

**THEOREM 3** Let $V$ be a finite dimensional vector space with basis $B = \{\mathbf{v}_1, \mathbf{v}_2, \ldots, \mathbf{v}_n\}$. Let $W$ be a vector space containing the $n$ vectors $\mathbf{w}_1, \mathbf{w}_2, \ldots, \mathbf{w}_n$. Then there exists a unique linear transformation $T: V \to W$ such that $T\mathbf{v}_i = \mathbf{w}_i$ for $i = 1, 2, \ldots, n$.

**Proof** Define a function $T$ as follows:

**i.** $T\mathbf{v}_i = \mathbf{w}_i$

**ii.** If $\mathbf{v} = \alpha_1\mathbf{v}_1 + \alpha_2\mathbf{v}_2 + \cdots + \alpha_n\mathbf{v}_n$, then

$$T\mathbf{v} = \alpha_1\mathbf{w}_1 + \alpha_2\mathbf{w}_2 + \cdots + \alpha_n\mathbf{w}_n \tag{1}$$

Because $B$ is a basis for $V$, $T$ is defined for every $\mathbf{v} \in V$; and since $W$ is a vector space, $T\mathbf{v} \in W$. Thus it only remains to show that $T$ is linear. But this follows directly from equation (1). For if $\mathbf{u} = \alpha_1\mathbf{v}_1 + \alpha_2\mathbf{v}_2 + \cdots + \alpha_n\mathbf{v}_n$ and $\mathbf{v} = \beta_1\mathbf{v}_1 + \beta_2\mathbf{v}_2 + \cdots + \beta_n\mathbf{v}_n$, then

$$\begin{aligned} T(\mathbf{u} + \mathbf{v}) &= T[(\alpha_1 + \beta_1)\mathbf{v}_1 + (\alpha_2 + \beta_2)\mathbf{v}_2 + \cdots + (\alpha_n + \beta_n)\mathbf{v}_n] \\ &= (\alpha_1 + \beta_1)\mathbf{w}_1 + (\alpha_2 + \beta_2)\mathbf{w}_2 + \cdots + (\alpha_n + \beta_n)\mathbf{w}_n \\ &= (\alpha_1\mathbf{w}_1 + \alpha_2\mathbf{w}_2 + \cdots + \alpha_n\mathbf{w}_n) + (\beta_1\mathbf{w}_1 + \beta_2\mathbf{w}_2 + \cdots + \beta_n\mathbf{w}_n) \\ &= T\mathbf{u} + T\mathbf{v} \end{aligned}$$

Similarly, $T(\alpha\mathbf{v}) = \alpha T\mathbf{v}$, so $T$ is linear. The uniqueness of $T$ follows from Theorem 2 and the theorem is proved. ■

*Remark.* In Theorems 2 and 3 the vectors $\mathbf{w}_1, \mathbf{w}_2, \ldots, \mathbf{w}_n$ need not be independent and, in fact, need not even be distinct. Moreover, we emphasize that the theorems are true if $V$ is any finite dimensional vector space, not just $\mathbb{R}^n$. Note also that $W$ does not have to be finite dimensional.

**EXAMPLE 2** **Defining a Linear Transformation from $\mathbb{R}^2$ into a Subspace of $\mathbb{R}^3$** Find a linear transformation from $\mathbb{R}^2$ into the plane

$$W = \left\{ \begin{pmatrix} x \\ y \\ z \end{pmatrix} : 2x - y + 3z = 0 \right\}$$

**Solution** From Example 4.6.3, page 338, we know that $W$ is a two-dimensional subspace of $\mathbb{R}^3$ with basis vectors $\mathbf{w}_1 = \begin{pmatrix} 1 \\ 2 \\ 0 \end{pmatrix}$ and $\mathbf{w}_2 = \begin{pmatrix} 0 \\ 3 \\ 1 \end{pmatrix}$. Using the standard basis in $\mathbb{R}^2$, $\mathbf{v}_1 = \begin{pmatrix} 1 \\ 0 \end{pmatrix}$ and $\mathbf{v}_2 = \begin{pmatrix} 0 \\ 1 \end{pmatrix}$, we define the linear transformation $T$ by $T\begin{pmatrix} 1 \\ 0 \end{pmatrix} = \begin{pmatrix} 1 \\ 2 \\ 0 \end{pmatrix}$

and $T\begin{pmatrix}0\\1\end{pmatrix} = \begin{pmatrix}0\\3\\1\end{pmatrix}$. Then as the discussion following Theorem 2 shows, $T$ is completely determined. For example,

$$T\begin{pmatrix}5\\-7\end{pmatrix} = T\left[5\begin{pmatrix}1\\0\end{pmatrix} - 7\begin{pmatrix}0\\1\end{pmatrix}\right] = 5T\begin{pmatrix}1\\0\end{pmatrix} - 7T\begin{pmatrix}0\\1\end{pmatrix} = 5\begin{pmatrix}1\\2\\0\end{pmatrix} - 7\begin{pmatrix}0\\3\\1\end{pmatrix} = \begin{pmatrix}5\\-11\\-7\end{pmatrix}$$

More generally,

$$T\begin{pmatrix}x\\y\end{pmatrix} = \begin{pmatrix}x\\2x+3y\\y\end{pmatrix}$$

We now turn to two important definitions in the theory of linear transformations.

**DEFINITION 1** **Kernel and Range of a Linear Transformation** Let $V$ and $W$ be vector spaces and let $T$: $V \rightarrow W$ be a linear transformation. Then

**i.** The **kernel** of $T$, denoted by ker $T$, is given by

$$\ker T = \{\mathbf{v} \in V\colon\ T\mathbf{v} = \mathbf{0}\} \tag{2}$$

**ii.** The **range** of $T$, denoted by range $T$, is given by

$$\text{Range } T = \{\mathbf{w} \in W\colon\ \mathbf{w} = T\mathbf{v} \text{ for some } \mathbf{v} \in V\} \tag{3}$$

*Remark 1.* Note that ker $T$ is nonempty because, by Theorem 1, $T(\mathbf{0}) = \mathbf{0}$ so that $\mathbf{0} \in \ker T$ for any linear transformation $T$. We shall be interested in finding other vectors in $V$ that get "mapped to zero." Again, note that when we write $T(\mathbf{0}) = \mathbf{0}$, the $\mathbf{0}$ on the left is in $V$ and the $\mathbf{0}$ on the right is in $W$.

Image

*Remark 2.* Range $T$ is simply the set of "images" of vectors in $V$ under the transformation $T$. In fact, if $\mathbf{w} = T\mathbf{v}$, we say that $\mathbf{w}$ is the **image** of $\mathbf{v}$ under $T$.

Before giving examples of kernels and ranges, we prove a theorem that will be very useful.

**THEOREM 4** If $T$: $V \to W$ is a linear transformation, then

i. $\ker T$ is a subspace of $V$.

ii. range $T$ is a subspace of $W$.

**Proof** i. Let $\mathbf{u}$ and $\mathbf{v}$ be in $\ker T$; then $T(\mathbf{u} + \mathbf{v}) = T\mathbf{u} + T\mathbf{v} = \mathbf{0} + \mathbf{0} = \mathbf{0}$ and $T(\alpha\mathbf{u}) = \alpha T\mathbf{u} = \alpha\mathbf{0} = \mathbf{0}$ so that $\mathbf{u} + \mathbf{v}$ and $\alpha\mathbf{u}$ are in $\ker T$.

ii. Let $\mathbf{w}$ and $\mathbf{x}$ be in range $T$. Then $\mathbf{w} = T\mathbf{u}$ and $\mathbf{x} = T\mathbf{v}$ for two vectors $\mathbf{u}$ and $\mathbf{v}$ in $V$. This means that $T(\mathbf{u} + \mathbf{v}) = T\mathbf{u} + T\mathbf{v} = \mathbf{w} + \mathbf{x}$ and $T(\alpha\mathbf{u}) = \alpha T\mathbf{u} = \alpha\mathbf{w}$. Thus $\mathbf{w} + \mathbf{x}$ and $\alpha\mathbf{w}$ are in range $T$. ■

**EXAMPLE 3** **Kernel and Range of the Zero Transformation** Let $T\mathbf{v} = \mathbf{0}$ for every $\mathbf{v} \in V$. ($T$ is the zero transformation.) Then $\ker T = V$ and range $T = \{\mathbf{0}\}$. ■

**EXAMPLE 4** **Kernel and Range of the Identity Transformation** Let $T\mathbf{v} = \mathbf{v}$ for every $\mathbf{v} \in V$. ($T$ is the identity transformation.) Then $\ker T = \{\mathbf{0}\}$ and range $T = V$. ■

The zero and identity transformations provide two extremes. In the first, everything is in the kernel. In the second, only the zero vector is in the kernel. The cases in between are more interesting.

**EXAMPLE 5** **Kernel and Range of a Projection Operator** Let $T$: $\mathbb{R}^3 \to \mathbb{R}^3$ be defined by $T\begin{pmatrix} x \\ y \\ z \end{pmatrix} = \begin{pmatrix} x \\ y \\ 0 \end{pmatrix}$. That is (see Example 5.1.10, page 470), $T$ is the projection operator from $\mathbb{R}^3$ into the $xy$-plane. If $T\begin{pmatrix} x \\ y \\ z \end{pmatrix} = \begin{pmatrix} x \\ y \\ 0 \end{pmatrix} = \mathbf{0} = \begin{pmatrix} 0 \\ 0 \\ 0 \end{pmatrix}$, then $x = y = 0$. Thus $\ker T = \{(x, y, z)\colon\ x = y = 0\}$ = the $z$-axis, and range $T = \{(x, y, z)\colon\ z = 0\}$ = the $xy$-plane. Note that $\dim \ker T = 1$ and dim range $T = 2$. ■

**DEFINITION 2** **Nullity and Rank of a Linear Transformation** If $T$ is a linear transformation from $V$ to $W$, then we define

$$\textbf{Nullity of } T = \nu(T) = \dim \ker T \tag{4}$$

$$\textbf{Rank of } T = \rho(T) = \dim \text{range } T \tag{5}$$

***Remark.*** In Section 4.7 we defined the rank, range, null space, and nullity of a matrix. According to Example 5.1.7, every $m \times n$ matrix $A$ gives rise to a linear transformation $T: \mathbb{R}^n \to \mathbb{R}^m$ defined by $T\mathbf{x} = A\mathbf{x}$. Evidently, $\ker T = N_A$, range $T$ = range $A = C_A$, $\nu(T) = \nu(A)$ and $\rho(T) = \rho(A)$. Thus we see that the definitions of kernel, range, nullity, and rank of a linear transformation are extensions of the null space, range, nullity, and rank of a matrix.

**EXAMPLE 6** **Kernel and Nullity of a Projection Operator** Let $H$ be a subspace of $\mathbb{R}^n$ and define $T\mathbf{v} = \text{proj}_H \mathbf{v}$. Clearly, range $T = H$. From Theorem 4.9.7, page 403, we can write any $\mathbf{v} \in V$ as $\mathbf{v} = \mathbf{h} + \mathbf{p} = \text{proj}_H \mathbf{v} + \text{proj}_{H^\perp}\mathbf{v}$. If $T\mathbf{v} = \mathbf{0}$, then $\mathbf{h} = \mathbf{0}$, which means that $\mathbf{v} = \mathbf{p} \in H^\perp$. Thus $\ker T = H^\perp$, $\rho(T) = \dim H$, and $\nu(T) = \dim H^\perp = n - \rho(T)$.

**EXAMPLE 7** **The Kernel and Range of a Transpose Operator** Let $V = M_{mn}$ and define $T: M_{mn} \to M_{nm}$ by $T(A) = A^t$ (see Example 5.1.11, page 471). If $TA = A^t = 0$, then $A^t$ is the $n \times m$ zero matrix so that $A$ is the $m \times n$ zero matrix. Thus $\ker T = \{\mathbf{0}\}$, and clearly range $T = M_{nm}$. This means that $\nu(T) = 0$ and $\rho(T) = nm$.

**EXAMPLE 8** **The Kernel and Range of a Transformation from $P_3$ to $P_2$** Let $T: P_3 \to P_2$ be defined by $T(p) = T(a_0 + a_1x + a_2x^2 + a_3x^3) = a_0 + a_1x + a_2x^2$. Then if $T(p) = 0$, $a_0 + a_1x + a_2x^2 = 0$ for every $x$, which implies that $a_0 = a_1 = a_2 = 0$. Thus $\ker T = \{p \in P_3\colon p(x) = a_3x^3\}$ and range $T = P_2$, $\nu(T) = 1$, and $\rho(T) = 3$.

**EXAMPLE 9** Calculus **The Kernel and Range of an Integral Operator** Let $V = C[0, 1]$ and define $J: C[0, 1] \to \mathbb{R}$ by $Jf = \int_0^1 f(x)\,dx$ (see Example 5.1.12, page 471). Then $\ker J = \{f \in C[0, 1]\colon \int_0^1 f(x)\,dx = 0\}$. Let $\alpha$ be a real number. Then the constant function $f(x) = \alpha$ for $x \in [0, 1]$ is in $C[0, 1]$ and $\int_0^1 \alpha\,dx = \alpha$. Since this is true for every real number $\alpha$, we have range $J = \mathbb{R}$.

In the next section we shall see how every linear transformation from one finite dimensional vector space to another can be represented by a matrix. This will enable us to compute the kernel and range of any linear transformation between finite dimensional vector spaces by finding the null space and range of a corresponding matrix.

# PROBLEMS 5.2

## Self-Quiz

### True-False

**I.** Let $T$: $V \to W$ be a linear transformation. It is sometimes possible to find three different vectors $\mathbf{v}_1 \in V$, $\mathbf{v}_2 \in V$, and $\mathbf{w} \in W$ such that $T\mathbf{v}_1 = T\mathbf{v}_2 = \mathbf{w}$.

**II.** If $T\mathbf{v}_1 = T\mathbf{v}_2 = \mathbf{w}$ as in Problem I, then $\mathbf{v}_1 - \mathbf{v}_2 \in \ker T$.

**III.** If $T$ is a linear transformation from $V$ into $W$, then the range of $T$ is $W$.

**IV.** Let $\mathbf{v}_1, \mathbf{v}_2, \ldots, \mathbf{v}_n$ be a basis for $\mathbb{R}^n$ and let $\mathbf{w}_1, \mathbf{w}_2, \ldots, \mathbf{w}_n$ be a basis for $P_{n-1}$. Then there may be two linear transformation $S$ and $T$ such that $T\mathbf{v}_i = \mathbf{w}_i$ for $i = 1, 2, \ldots, n$.

**V.** If $T$: $\mathbb{R}^2 \to \mathbb{R}^2$ is a linear transformation and $T\begin{pmatrix}0\\0\end{pmatrix} = \begin{pmatrix}0\\0\end{pmatrix}$, then $T$ is the zero transformation.

**VI.** There exists a linear transformation $T$ from $\mathbb{R}^5 \to \mathbb{R}^5$ with $\rho(T) = \nu(T)$.

**VII.** Suppose that $T$: $M_{22} \to M_{22}$ with $\rho(T) = 4$. If $TA = \begin{pmatrix}0 & 0\\0 & 0\end{pmatrix}$, then $A = \begin{pmatrix}0 & 0\\0 & 0\end{pmatrix}$.

In Problems 1–10 find the kernel, range, rank, and nullity of the given linear transformation.

**1.** $T$: $\mathbb{R}^2 \to \mathbb{R}^2$; $T\begin{pmatrix}x\\y\end{pmatrix} = \begin{pmatrix}x\\0\end{pmatrix}$

**2.** $T$: $\mathbb{R}^3 \to \mathbb{R}^2$; $T\begin{pmatrix}x\\y\\z\end{pmatrix} = \begin{pmatrix}z\\y\end{pmatrix}$

**3.** $T$: $\mathbb{R}^2 \to \mathbb{R}$; $T\begin{pmatrix}x\\y\end{pmatrix} = x + y$

**4.** $T$: $\mathbb{R}^4 \to \mathbb{R}^2$; $T\begin{pmatrix}x\\y\\z\\w\end{pmatrix} = \begin{pmatrix}x+z\\y+w\end{pmatrix}$

**5.** $T$: $M_{22} \to M_{22}$; $T(A) = AB$, where $B = \begin{pmatrix}1 & 2\\0 & 1\end{pmatrix}$

**6.** $T$: $\mathbb{R} \to P_3$; $T(a) = a + ax + ax^2 + ax^3$

* **7.** $T$: $M_{nn} \to M_{nn}$; $T(A) = A^t + A$

Calculus **8.** $T$: $C^1[0, 1] \to C[0, 1]$; $Tf = f'$

**9.** $T$: $C[0, 1] \to \mathbb{R}$; $Tf = f(\frac{1}{2})$

**10.** $T$: $\mathbb{R}^2 \to \mathbb{R}^2$; $T$ is a rotation through an angle of $\pi/3$

**11.** Let $T$: $V \to W$ be a linear transformation, let $\{\mathbf{v}_1, \mathbf{v}_2, \ldots, \mathbf{v}_n\}$ be a basis for $V$, and suppose that $T\mathbf{v}_i = \mathbf{0}$ for $i = 1, 2, \ldots, n$. Show that $T$ is the zero transformation.

---

**Answers to Self-Quiz**

**I.** True **II.** True **III.** False **IV.** False **V.** False **VI.** False
**VII.** True

**12.** In Problem 11 suppose that $W = V$ and $T\mathbf{v}_i = \mathbf{v}_i$ for $i = 1, 2, \ldots, n$. Show that $T$ is the identity operator.

**13.** Let $T$: $V \to \mathbb{R}^3$. Prove that range $T$ is either **(a)** $\{\mathbf{0}\}$, **(b)** a line through the origin, **(c)** a plane through the origin, or **(d)** $\mathbb{R}^3$.

**14.** Let $T$: $\mathbb{R}^3 \to V$. Show that ker $T$ is one of four spaces listed in Problem 13.

**15.** Find all linear transformations from $\mathbb{R}^2$ into $\mathbb{R}^2$ such that the line $y = 0$ is carried into the line $x = 0$.

**16.** Find all linear transformations from $\mathbb{R}^2$ into $\mathbb{R}^2$ that carry the line $y = ax$ into the line $y = bx$.

**17.** Find a linear transformation $T$ from $\mathbb{R}^3 \to \mathbb{R}^3$ such that

$$\ker T = \{(x, y, z): 2x - y + z = 0\}.$$

**18.** Find a linear transformation $T$ from $\mathbb{R}^3 \to \mathbb{R}^3$ such that

$$\text{range } T = \{(x, y, z): 2x - y + z = 0\}.$$

**19.** Let $T$: $M_{nn} \to M_{nn}$ be defined by $TA = A - A^t$. Show that $\ker T = \{$symmetric $n \times n$ matrices$\}$ and range of $T = \{$skew-symmetric $n \times n$ matrices$\}$.

*Calculus **20.** Let $T$: $C^1[0, 1] \to C[0, 1]$ be defined by $Tf(x) = xf'(x)$. Find the kernel and range of $T$.

***21.** In Problem 5.1.42 you were asked to show that the set of linear transformations from a vector space $V$ to a vector space $W$, denoted by $L(V, W)$, is a vector space. Suppose that $\dim V = n < \infty$ and $\dim W = m < \infty$. Find $\dim L(V, W)$.

**22.** Let $H$ be a subspace of $V$ where $\dim H = k$ and $\dim V = n$. Let $U$ be the subset of $L(V, V)$ having the property that if $T \in L(V, V)$, then $T\mathbf{h} = \mathbf{0}$ for every $\mathbf{h} \in H$.

**a.** Prove that $U$ is a subspace of $L(V, V)$.

**b.** Find dim $U$.

***23.** Let $S$ and $T$ be in $L(V, V)$ such that $ST$ is the zero transformation. Prove or disprove: $TS$ is the zero transformation.

## 5.3 THE MATRIX REPRESENTATION OF A LINEAR TRANSFORMATION

If $A$ is an $m \times n$ matrix and $T$: $\mathbb{R}^n \to \mathbb{R}^m$ is defined by $T\mathbf{x} = A\mathbf{x}$, then, as we saw in Example 5.1.7 on page 469, $T$ is a linear transformation. We shall now see that for *every* linear transformation from $\mathbb{R}^n$ into $\mathbb{R}^m$ there exists an $m \times n$ matrix $A$ such that $T\mathbf{x} = A\mathbf{x}$ for every $\mathbf{x} \in \mathbb{R}^n$. This fact is extremely useful. As we saw in the remark on page 483, if $T\mathbf{x} = A\mathbf{x}$, then $\ker T = N_A$ and range $T = R_A$. Moreover, $\nu(T) = \dim \ker T = \nu(A)$ and $\rho(T) = \dim \text{range } T = \rho(A)$. Thus we can determine the kernel, range, nullity, and rank of a linear transformation from $\mathbb{R}^n \to \mathbb{R}^m$ by determining the null space and range of a corresponding matrix. Moreover, once we know that $T\mathbf{x} = A\mathbf{x}$, we can evaluate $T\mathbf{x}$ for any $\mathbf{x}$ in $\mathbb{R}^n$ by simple matrix multiplication.

But this is not all. As we shall see, any linear transformation between finite dimensional vector spaces can be represented by a matrix.

**THEOREM 1** Let $T: \mathbb{R}^n \rightarrow \mathbb{R}^m$ be a linear transformation. Then there exists a unique $m \times n$ matrix $A_T$ such that

$$T\mathbf{x} = A_T\mathbf{x} \qquad \text{for every } \mathbf{x} \in \mathbb{R}^n \tag{1}$$

**Proof** Let $\mathbf{w}_1 = T\mathbf{e}_1, \mathbf{w}_2 = T\mathbf{e}_2, \ldots, \mathbf{w}_n = T\mathbf{e}_n$. Let $A_T$ be the matrix whose columns are $\mathbf{w}_1, \mathbf{w}_2, \ldots, \mathbf{w}_n$ and let $A_T$ also denote the transformation from $\mathbb{R}^n \rightarrow \mathbb{R}^m$, which multiplies a vector in $\mathbb{R}^n$ on the left by $A_T$. If

$$\mathbf{w}_i = \begin{pmatrix} a_{1i} \\ a_{2i} \\ \vdots \\ a_{mi} \end{pmatrix} \qquad \text{for } i = 1, 2, \ldots, n$$

then

$$A_T\mathbf{e}_i = \begin{pmatrix} a_{11} & a_{12} & \cdots & a_{1i} & \cdots & a_{1n} \\ a_{21} & a_{22} & \cdots & a_{2i} & \cdots & a_{2n} \\ \vdots & \vdots & & \vdots & & \vdots \\ a_{m1} & a_{m2} & \cdots & a_{mi} & \cdots & a_{mn} \end{pmatrix} \begin{pmatrix} 0 \\ 0 \\ \vdots \\ 1 \\ 0 \\ \vdots \\ 0 \end{pmatrix} = \begin{pmatrix} a_{1i} \\ a_{2i} \\ \vdots \\ a_{mi} \end{pmatrix} = \mathbf{w}_i$$

*i*th position

Thus $A_T\mathbf{e}_i = \mathbf{w}_i$ for $i = 1, 2, \ldots, n$. By Theorem 5.2.2 on page 478, $T$ and the transformation $A_T$ are the same because they agree on basis vectors.

We can now show that $A_T$ is unique. Suppose that $T\mathbf{x} = A_T\mathbf{x}$ and $T\mathbf{x} = B_T\mathbf{x}$ for every $\mathbf{x} \in \mathbb{R}^n$. Then $A_T\mathbf{x} = B_T\mathbf{x}$, or setting $C_T = A_T - B_T$, we have $C_T\mathbf{x} = \mathbf{0}$ for every $\mathbf{x} \in \mathbb{R}^n$. In particular, $C_T\mathbf{e}_i = \mathbf{0}$ for $i = 1, 2, \ldots, n$. But as we see from the proof of the first part of the theorem, $C_T\mathbf{e}_i$ is the $i$th column of $C_T$. Thus each of the $n$ columns of $C_T$ is the $m$-zero vector and $C_T = 0$, the $m \times n$ zero matrix. This shows that $A_T = B_T$ and the theorem is proved. ∎

*Remark 1.* In this theorem we assumed that every vector in $\mathbb{R}^n$ and $\mathbb{R}^m$ is written in terms of the standard basis vectors in those spaces. If we choose other bases for $\mathbb{R}^n$ and $\mathbb{R}^m$, we shall, of course, get a different matrix $A_T$. See, for instance, Example 4.8.1 on page 377 or Example 8 below.

*Remark 2.* The proof of the theorem shows us that $A_T$ is easily obtained as the matrix whose columns are the vectors $T\mathbf{e}_i$.

**DEFINITION 1** **Transformation Matrix** The matrix $A_T$ in Theorem 1 is called the **transformation matrix** corresponding to $T$ or the **matrix representation** of $T$.

*Note.* The transformation matrix $A_T$ is defined by using the standard bases in both $\mathbb{R}^n$ and $\mathbb{R}^m$. If you use other bases, you will get a different transformation matrix. See Theorem 3 on page 490.

In Section 5.2 we defined the range, rank, kernel, and nullity of a linear transformation. In Section 4.7 we defined the range, rank, null space, and nullity of a matrix. The proof of the following theorem follows from Theorem 1 and is left as an exercise (see Problem 36).

**THEOREM 2** Let $A_T$ be the transformation matrix corresponding to the linear transformation $T$. Then

**i.** range $T$ = Range $A = C_{A_T}$

**ii.** $\rho(T) = \rho(A_T)$

**iii.** $\ker T = N_{A_T}$

**iv.** $\nu(T) = \nu(A_T)$

**EXAMPLE 1** **The Matrix Representation of a Projection Transformation** Find the transformation matrix $A_T$ corresponding to the projection of a vector in $\mathbb{R}^3$ onto the $xy$-plane.

**Solution** Here $T\begin{pmatrix} x \\ y \\ z \end{pmatrix} = \begin{pmatrix} x \\ y \\ 0 \end{pmatrix}$. In particular, $T\begin{pmatrix} 1 \\ 0 \\ 0 \end{pmatrix} = \begin{pmatrix} 1 \\ 0 \\ 0 \end{pmatrix}$, $T\begin{pmatrix} 0 \\ 1 \\ 0 \end{pmatrix} = \begin{pmatrix} 0 \\ 1 \\ 0 \end{pmatrix}$, and $T\begin{pmatrix} 0 \\ 0 \\ 1 \end{pmatrix} = \begin{pmatrix} 0 \\ 0 \\ 0 \end{pmatrix}$. Thus $A_T = \begin{pmatrix} 1 & 0 & 0 \\ 0 & 1 & 0 \\ 0 & 0 & 0 \end{pmatrix}$. Note that $A_T\begin{pmatrix} x \\ y \\ z \end{pmatrix} = \begin{pmatrix} 1 & 0 & 0 \\ 0 & 1 & 0 \\ 0 & 0 & 0 \end{pmatrix}\begin{pmatrix} x \\ y \\ z \end{pmatrix} = \begin{pmatrix} x \\ y \\ 0 \end{pmatrix}$.

**EXAMPLE 2** **The Matrix Representation of a Transformation from $\mathbb{R}^3$ to $\mathbb{R}^4$** Let $T$: $\mathbb{R}^3 \to \mathbb{R}^4$ be defined by

$$T\begin{pmatrix} x \\ y \\ z \end{pmatrix} = \begin{pmatrix} x - y \\ y + z \\ 2x - y - z \\ -x + y + 2z \end{pmatrix}$$

Find $A_T$, $\ker T$, range $T$, $\nu(T)$, and $\rho(T)$.

**Solution** $T\begin{pmatrix}1\\0\\0\end{pmatrix}=\begin{pmatrix}1\\0\\2\\-1\end{pmatrix}$, $T\begin{pmatrix}0\\1\\0\end{pmatrix}=\begin{pmatrix}-1\\1\\-1\\1\end{pmatrix}$, and $T\begin{pmatrix}0\\0\\1\end{pmatrix}=\begin{pmatrix}0\\1\\-1\\2\end{pmatrix}$. Thus

$$A_T=\begin{pmatrix}1&-1&0\\0&1&1\\2&-1&-1\\-1&1&2\end{pmatrix}$$

Note (as a check) that

$$\begin{pmatrix}1&-1&0\\0&1&1\\2&-1&-1\\-1&1&2\end{pmatrix}\begin{pmatrix}x\\y\\z\end{pmatrix}=\begin{pmatrix}x-y\\x+z\\2x-y-z\\-x+y+2z\end{pmatrix}$$

Next we compute the kernel and range of $A$. The row echelon form of $\begin{pmatrix}1&-1&0\\0&1&1\\2&-1&-1\\-1&1&2\end{pmatrix}$ is $\begin{pmatrix}1&-1&0\\0&1&1\\0&0&1\\0&0&0\end{pmatrix}$. The row echelon form has three pivots so

$$\rho(A)=3 \quad\text{and}\quad \underset{\text{since } \rho(A)+\nu(A)=3}{\nu(A)=3-3=0}$$

This means that $\ker T=\{\mathbf{0}\}$, range $T=\text{span}\left\{\begin{pmatrix}1\\0\\2\\-1\end{pmatrix},\begin{pmatrix}-1\\1\\-1\\1\end{pmatrix},\begin{pmatrix}0\\1\\-1\\2\end{pmatrix}\right\}$, $\nu(T)=0$, and $\rho(T)=3$.

## EXAMPLE 3 The Matrix Representation of a Transformation from $\mathbb{R}^3$ to $\mathbb{R}^3$

Let $T$: $\mathbb{R}^3\to\mathbb{R}^3$ be defined by $T\begin{pmatrix}x\\y\\z\end{pmatrix}=\begin{pmatrix}2x-y+3z\\4x-2y+6z\\-6x+3y-9z\end{pmatrix}$. Find $A_T$, $\ker T$, range $T$, $\nu(T)$, and $\rho(T)$.

**Solution** Since $T\begin{pmatrix}1\\0\\0\end{pmatrix}=\begin{pmatrix}2\\4\\-6\end{pmatrix}$, $T\begin{pmatrix}0\\1\\0\end{pmatrix}=\begin{pmatrix}-1\\-2\\3\end{pmatrix}$, and $T\begin{pmatrix}0\\0\\1\end{pmatrix}=\begin{pmatrix}3\\6\\-9\end{pmatrix}$, we have

$$A_T = \begin{pmatrix} 2 & -1 & 3 \\ 4 & -2 & 6 \\ -6 & 3 & -9 \end{pmatrix}$$

From Example 4.7.4 on page 352 we see that $\rho(A) \overset{\text{Theorem 2}(ii)}{=} \rho(T) = 1$ and range $T = \text{span}\left\{\begin{pmatrix} 2 \\ 4 \\ -6 \end{pmatrix}\right\}$. Then $\nu(T) = 2$.

To find $N_A \overset{\text{Theorem 2}(iii)}{=} \ker T$, we row-reduce to solve the system $A\mathbf{x} = \mathbf{0}$:

$$\left(\begin{array}{ccc|c} 2 & -1 & 3 & 0 \\ 4 & -2 & 6 & 0 \\ -6 & 3 & -9 & 0 \end{array}\right) \longrightarrow \left(\begin{array}{ccc|c} 2 & -1 & 3 & 0 \\ 0 & 0 & 0 & 0 \\ 0 & 0 & 0 & 0 \end{array}\right)$$

This means that $\begin{pmatrix} x \\ y \\ z \end{pmatrix} \in N_A$ if $2x - y + 3z = 0$ or $y = 2x + 3z$. First setting $x = 1$, $z = 0$ and then $x = 0$, $z = 1$, we obtain a basis for $N_A$:

$$\ker T = N_A = \text{span}\left\{\begin{pmatrix} 1 \\ 2 \\ 0 \end{pmatrix}, \begin{pmatrix} 0 \\ 3 \\ 1 \end{pmatrix}\right\}$$

**EXAMPLE 4** **The Matrix Representation of a Zero Transformation** It is easy to verify that if $T$ is the zero transformation from $\mathbb{R}^n \to \mathbb{R}^m$, then $A_T$ is the $m \times n$ zero matrix. Similarly, if $T$ is the identity transformation from $\mathbb{R}^n \to \mathbb{R}^n$, then $A_T = I_n$.

**EXAMPLE 5** **The Matrix Representation of a Rotation Transformation** We saw in Example 5.1.8 on page 469 that if $T$ is the function that rotates every vector in $\mathbb{R}^2$ through an angle of $\theta$, then $A_T = \begin{pmatrix} \cos\theta & -\sin\theta \\ \sin\theta & \cos\theta \end{pmatrix}$.

We now generalize the notion of matrix representation to arbitrary finite dimensional vector spaces.

**THEOREM 3** Let $V$ be an $n$-dimensional vector space, $W$ be an $m$-dimensional vector space, and $T: V \to W$ be a linear transformation. Let $B_1 = \{\mathbf{v}_1, \mathbf{v}_2, \ldots, \mathbf{v}_n\}$ be a basis for $V$ and let $B_2 = \{\mathbf{w}_1, \mathbf{w}_2, \ldots, \mathbf{w}_m\}$ be a basis for $W$. Then there is a unique $m \times n$ matrix $A_T$ such that

$$\boxed{(T\mathbf{x})_{B_2} = A_T(\mathbf{x})_{B_1}} \tag{2}$$

***Remark 1.*** The notation in (2) is the notation of Section 4.8 (see page 375). If $\mathbf{x} \in V = c_1\mathbf{v}_1 + c_2\mathbf{v}_2 + \cdots + c_n\mathbf{v}_n$, then $(\mathbf{x})_{B_1} = \begin{pmatrix} c_1 \\ c_2 \\ \vdots \\ c_n \end{pmatrix}$. Let $\mathbf{c} = \begin{pmatrix} c_1 \\ c_2 \\ \vdots \\ c_n \end{pmatrix}$. Then $A_T\mathbf{c}$ is an $m$-vector that we denote by $\mathbf{d} = \begin{pmatrix} d_1 \\ d_2 \\ \vdots \\ d_m \end{pmatrix}$. Equation (2) says that $(T\mathbf{x})_{B_2} = \begin{pmatrix} d_1 \\ d_2 \\ \vdots \\ d_m \end{pmatrix}$.

That is,

$$T\mathbf{x} = d_1\mathbf{w}_1 + d_2\mathbf{w}_2 + \cdots + d_m\mathbf{w}_m$$

***Remark 2.*** As in Theorem 1, the uniqueness of $A_T$ is relative to the bases $B_1$ and $B_2$. If we change the bases, we change $A_T$ (see Examples 8 and 9, and Theorem 5). If the standard bases are used, then this $A_T$ is the $A_T$ in Definition 1.

**Proof** Let $T\mathbf{v}_1 = \mathbf{y}_1, T\mathbf{v}_2 = \mathbf{y}_2, \ldots, T\mathbf{v}_n = \mathbf{y}_n$. Since $\mathbf{y}_i \in W$, we have for $i = 1, 2, \ldots, n$

$$\mathbf{y}_i = a_{1i}\mathbf{w}_1 + a_{2i}\mathbf{w}_2 + \cdots + a_{mi}\mathbf{w}_m$$

for some (unique) set of scalars $a_{1i}, a_{2i}, \ldots, a_{mi}$, and we write

$$(\mathbf{y}_1)_{B_2} = \begin{pmatrix} a_{11} \\ a_{21} \\ \vdots \\ a_{m1} \end{pmatrix}, (\mathbf{y}_2)_{B_2} = \begin{pmatrix} a_{12} \\ a_{22} \\ \vdots \\ a_{m2} \end{pmatrix}, \ldots, (\mathbf{y}_n)_{B_2} = \begin{pmatrix} a_{1n} \\ a_{2n} \\ \vdots \\ a_{mn} \end{pmatrix}$$

This means, for example, that $\mathbf{y}_1 = a_{11}\mathbf{w}_1 + a_{21}\mathbf{w}_2 + \cdots + a_{m1}\mathbf{w}_m$. We now define

$$A_T = \begin{pmatrix} a_{11} & a_{12} & \cdots & a_{1n} \\ a_{21} & a_{22} & \cdots & a_{2n} \\ \vdots & \vdots & & \vdots \\ a_{m1} & a_{m2} & \cdots & a_{mn} \end{pmatrix}$$

Since

$$(\mathbf{v}_1)_{B_1} = \begin{pmatrix} 1 \\ 0 \\ \vdots \\ 0 \end{pmatrix}, \quad (\mathbf{v}_2)_{B_1} = \begin{pmatrix} 0 \\ 1 \\ 0 \\ \vdots \\ 0 \end{pmatrix}, \ldots, (\mathbf{v}_n)_{B_1} = \begin{pmatrix} 0 \\ 0 \\ \vdots \\ 1 \end{pmatrix}$$

we have, as in the proof of Theorem 1,

$$A_T(\mathbf{v}_i)_{B_1} = \begin{pmatrix} a_{11} & a_{12} & \cdots & a_{1n} \\ a_{21} & a_{22} & \cdots & a_{2n} \\ \vdots & \vdots & & \vdots \\ a_{i1} & a_{i2} & \cdots & a_{in} \\ \vdots & \vdots & & \vdots \\ a_{m1} & a_{m2} & \cdots & a_{mn} \end{pmatrix} \begin{pmatrix} 0 \\ 0 \\ \vdots \\ 1 \\ 0 \\ \vdots \\ 0 \end{pmatrix} = \begin{pmatrix} a_{1i} \\ a_{2i} \\ \vdots \\ a_{mi} \end{pmatrix} = (\mathbf{y}_i)_{B_2}$$

($i$th position: the entry 1)

If $\mathbf{x}$ is in $V$, then

$$\mathbf{x} = c_1\mathbf{v}_1 + c_2\mathbf{v}_2 + \cdots + c_n\mathbf{v}_n$$

$$(\mathbf{x})_{B_1} = \begin{pmatrix} c_1 \\ c_2 \\ \vdots \\ c_n \end{pmatrix}$$

and

$$A_T(\mathbf{x})_{B_1} = \begin{pmatrix} a_{11} & a_{12} & \cdots & a_{1n} \\ a_{21} & a_{22} & \cdots & a_{2n} \\ \vdots & \vdots & & \vdots \\ a_{m1} & a_{m2} & \cdots & a_{mn} \end{pmatrix} \begin{pmatrix} c_1 \\ c_2 \\ \vdots \\ c_n \end{pmatrix} = \begin{pmatrix} a_{11}c_1 + a_{12}c_2 + \cdots + a_{1n}c_n \\ a_{21}c_1 + a_{22}c_2 + \cdots + a_{2n}c_n \\ \vdots \qquad \vdots \qquad \vdots \\ a_{m1}c_1 + a_{m2}c_2 + \cdots + a_{mn}c_n \end{pmatrix}$$

$$= c_1\begin{pmatrix} a_{11} \\ a_{21} \\ \vdots \\ a_{m1} \end{pmatrix} + c_2\begin{pmatrix} a_{12} \\ a_{22} \\ \vdots \\ a_{m2} \end{pmatrix} + \cdots + c_n\begin{pmatrix} a_{1n} \\ a_{2n} \\ \vdots \\ a_{mn} \end{pmatrix}$$

$$= c_1(\mathbf{y}_1)_{B_2} + c_2(\mathbf{y}_2)_{B_2} + \cdots + c_n(\mathbf{y}_n)_{B_2}$$

Similarly, $T\mathbf{x} = T(c_1\mathbf{v}_1 + c_2\mathbf{v}_2 + \cdots + c_n\mathbf{v}_n) = c_1T\mathbf{v}_1 + c_2T\mathbf{v}_2 + \cdots + c_nT\mathbf{v}_n = c_1\mathbf{y}_1 + c_2\mathbf{y}_2 + \cdots + c_n\mathbf{y}_n$, so $T(\mathbf{x})_{B_2} = (c_1\mathbf{y}_1 + c_2\mathbf{y}_2 + \cdots + c_n\mathbf{y}_n)_{B_2} = c_1(\mathbf{y}_1)_{B_2} + c_2(\mathbf{y}_2)_{B_2} + \cdots + c_n(\mathbf{y}_n)_{B_2} = A_T(\mathbf{x})_{B_1}$. Thus $(T\mathbf{x})_{B_2} = A_T(\mathbf{x})_{B_1}$. The proof of uniqueness is exactly as in the proof of uniqueness in Theorem 1. ■

The following useful result follows immediately from Theorem 4.7.7 on on page 356 and generalizes Theorem 2. Its proof is left as an exercise (see Problem 37).

**THEOREM 4** Let $V$ and $W$ be finite dimensional vector spaces with dim $V = n$. Let $T\colon V \to W$ be a linear transformation and let $A_T$ be a matrix representation of $T$ with respect to the bases $B_1$ in $V$ and $B_2$ in $W$. Then

**i.** $\rho(T) = \rho(A_T)$

**ii.** $\nu(T) = \nu(A_T)$

**iii.** $\nu(T) + \rho(T) = n$

*Note.* $(i)$ and $(ii)$ imply that $\rho(A_T)$ and $\nu(A_T)$ are independent of the bases $B_1$ and $B_2$.

**EXAMPLE 6** **The Matrix Representation of a Transformation from $P_2$ to $P_3$**
Let $T\colon P_2 \to P_3$ be defined by $(Tp)(x) = xp(x)$. Find $A_T$ and use it to determine the kernel and range of $T$.

**Solution** Using the standard basis $B_1 = \{1, x, x^2\}$ in $P_2$ and $B_2 = \{1, x, x^2, x^3\}$ in $P_3$, we have

$(T(1))_{B_2} = (x)_{B_2} = \begin{pmatrix} 0 \\ 1 \\ 0 \\ 0 \end{pmatrix}$, $(T(x))_{B_2} = (x^2)_{B_2} = \begin{pmatrix} 0 \\ 0 \\ 1 \\ 0 \end{pmatrix}$, and $(T(x^2))_{B_2} = (x^3)_{B_2} = \begin{pmatrix} 0 \\ 0 \\ 0 \\ 1 \end{pmatrix}$. Thus $A_T = \begin{pmatrix} 0 & 0 & 0 \\ 1 & 0 & 0 \\ 0 & 1 & 0 \\ 0 & 0 & 1 \end{pmatrix}$. Clearly, $\rho(A) = 3$ and a basis for $R_A$ is $\left\{\begin{pmatrix} 0 \\ 1 \\ 0 \\ 0 \end{pmatrix}, \begin{pmatrix} 0 \\ 0 \\ 1 \\ 0 \end{pmatrix}, \begin{pmatrix} 0 \\ 0 \\ 0 \\ 1 \end{pmatrix}\right\}$. Therefore range $T = \text{span}\,\{x, x^2, x^3\}$. Since $\nu(A) = 3 - \rho(A) = 0$, we see that $\ker T = \{\mathbf{0}\}$.

**EXAMPLE 7** **The Matrix Representation of a Transformation from $P_3$ to $P_2$**
Let $T\colon P_3 \to P_2$ be defined by $T(a_0 + a_1x + a_2x^2 + a_3x^3) = a_1 + a_2x^2$. Compute $A_T$ and use it to find the kernel and range of $T$.

**Solution** Using the standard bases $B_1 = (1, x, x^2, x^3\}$ in $P_3$ and $B_2 = \{1, x, x^2\}$ in $P_2$, we immediately see that $(T(1))_{B_2} = \begin{pmatrix} 0 \\ 0 \\ 0 \end{pmatrix}$, $(T(x))_{B_2} = \begin{pmatrix} 1 \\ 0 \\ 0 \end{pmatrix}$, $(T(x^2))_{B_2} = \begin{pmatrix} 0 \\ 0 \\ 1 \end{pmatrix}$, and

$(T(x^3))_{B_2} = \begin{pmatrix} 0 \\ 0 \\ 0 \end{pmatrix}$. Thus $A_T = \begin{pmatrix} 0 & 1 & 0 & 0 \\ 0 & 0 & 0 & 0 \\ 0 & 0 & 1 & 0 \end{pmatrix}$. Clearly, $\rho(A) = 2$ and a basis for $R_A$ is $\left\{ \begin{pmatrix} 1 \\ 0 \\ 0 \end{pmatrix}, \begin{pmatrix} 0 \\ 0 \\ 1 \end{pmatrix} \right\}$ so that range $T = \text{span } \{1, x^2\}$. Then $\nu(A) = 4 - 2 = 2$; and if $A_T \begin{pmatrix} a_0 \\ a_1 \\ a_2 \\ a_3 \end{pmatrix} = \begin{pmatrix} 0 \\ 0 \\ 0 \end{pmatrix}$, then $a_1 = 0$ and $a_2 = 0$. Hence $a_0$ and $a_3$ are arbitrary and a basis for $N_A$ is $\left\{ \begin{pmatrix} 1 \\ 0 \\ 0 \\ 0 \end{pmatrix}, \begin{pmatrix} 0 \\ 0 \\ 0 \\ 1 \end{pmatrix} \right\}$ so that a basis for ker $T$ is $\{1, x^3\}$. ■

In all the examples of this section we have obtained the matrix $A_T$ by using the standard basis in each vector space. However, Theorem 3 holds for any bases in $V$ and $W$. The next example illustrates this.

**EXAMPLE 8** **Finding a Matrix Representation Relative to Two Nonstandard Bases in $\mathbb{R}^2$**

Let $T$: $\mathbb{R}^2 \rightarrow \mathbb{R}^2$ be defined by $T\begin{pmatrix} x \\ y \end{pmatrix} = \begin{pmatrix} x + y \\ x - y \end{pmatrix}$. Using the bases $B_1 = B_2 = \left\{ \begin{pmatrix} 1 \\ -1 \end{pmatrix}, \begin{pmatrix} -3 \\ 2 \end{pmatrix} \right\}$, compute $A_T$.

**Solution** We have $T\begin{pmatrix} 1 \\ -1 \end{pmatrix} = \begin{pmatrix} 0 \\ 2 \end{pmatrix}$ and $T\begin{pmatrix} -3 \\ 2 \end{pmatrix} = \begin{pmatrix} -1 \\ -5 \end{pmatrix}$. Since $\begin{pmatrix} 0 \\ 2 \end{pmatrix} = -6\begin{pmatrix} 1 \\ -1 \end{pmatrix} - 2\begin{pmatrix} -3 \\ 2 \end{pmatrix}$, we find that $\begin{pmatrix} 0 \\ 2 \end{pmatrix}_{B_2} = \begin{pmatrix} -6 \\ -2 \end{pmatrix}$. Similarly, $\begin{pmatrix} -1 \\ -5 \end{pmatrix} = 17\begin{pmatrix} 1 \\ -1 \end{pmatrix} + 6\begin{pmatrix} -3 \\ 2 \end{pmatrix}$ so that $\begin{pmatrix} -1 \\ -5 \end{pmatrix}_{B_2} = \begin{pmatrix} 17 \\ 6 \end{pmatrix}$. Thus $A_T = \begin{pmatrix} -6 & 17 \\ -2 & 6 \end{pmatrix}$. To compute $T\begin{pmatrix} -4 \\ 7 \end{pmatrix}$, for example, we first write $\begin{pmatrix} -4 \\ 7 \end{pmatrix} = -13\begin{pmatrix} 1 \\ -1 \end{pmatrix} - 3\begin{pmatrix} -3 \\ 2 \end{pmatrix}$, so $\begin{pmatrix} -4 \\ 7 \end{pmatrix}_{B_1} = \begin{pmatrix} -13 \\ -3 \end{pmatrix}$. Then $\left(T\begin{pmatrix} -4 \\ 7 \end{pmatrix}\right)_{B_2} = A_T\begin{pmatrix} -4 \\ 7 \end{pmatrix}_{B_1} = A_T\begin{pmatrix} -13 \\ -3 \end{pmatrix} = \begin{pmatrix} -6 & 17 \\ -2 & 6 \end{pmatrix}\begin{pmatrix} -13 \\ -3 \end{pmatrix} = \begin{pmatrix} 27 \\ 8 \end{pmatrix}$. Hence $T\begin{pmatrix} -4 \\ 7 \end{pmatrix} = 27\begin{pmatrix} 1 \\ -1 \end{pmatrix} + 8\begin{pmatrix} -3 \\ 2 \end{pmatrix} = \begin{pmatrix} 3 \\ -11 \end{pmatrix}$. Note that $T\begin{pmatrix} -4 \\ 7 \end{pmatrix} = \begin{pmatrix} -4 + 7 \\ -4 - 7 \end{pmatrix} = \begin{pmatrix} 3 \\ -11 \end{pmatrix}$, which verifies our calculations. ■

To avoid confusion, we shall, unless explicitly stated otherwise, always compute the matrix $A_T$ with respect to the standard basis.† If $T$: $V \to V$ is a linear transformation and some other basis $B$ is used, then we refer to $A_T$ as *the transformation matrix of $T$ with respect to the basis $B$*. Thus in the last example $A_T = \begin{pmatrix} -6 & 17 \\ -2 & 6 \end{pmatrix}$ is the transformation matrix of $T$ with respect to the basis $\left\{\begin{pmatrix} 1 \\ -1 \end{pmatrix}, \begin{pmatrix} -3 \\ 2 \end{pmatrix}\right\}$.

Before leaving this section, we must answer an obvious question. Why bother to use a basis other than the standard basis since the computations are, as in Example 8, a good deal more complicated? The answer is that it is often possible to find a basis $B^*$ in $\mathbb{R}^n$ so that the transformation matrix with respect to $B^*$ is a diagonal matrix. Diagonal matrices are very easy to work with, and as we shall see in Chapter 6, there are numerous advantages to writing a matrix in a diagonal form.

**EXAMPLE 9** **The Matrix Representation of a Linear Transformation Relative to Two Nonstandard Bases in $\mathbb{R}^2$ May Be Diagonal** Let $T$: $\mathbb{R}^2 \to \mathbb{R}^2$ be defined by $T\begin{pmatrix} x \\ y \end{pmatrix} = \begin{pmatrix} 12x + 10y \\ -15x - 13y \end{pmatrix}$. Find $A_T$ with respect to the basis $B_1 = B_2 = \left\{\begin{pmatrix} 1 \\ -1 \end{pmatrix}, \begin{pmatrix} 2 \\ -3 \end{pmatrix}\right\}$.

**Solution** $T\begin{pmatrix} 1 \\ -1 \end{pmatrix} = \begin{pmatrix} 2 \\ -2 \end{pmatrix}$ and $T\begin{pmatrix} 2 \\ -3 \end{pmatrix} = \begin{pmatrix} -6 \\ 9 \end{pmatrix}$. Then $\begin{pmatrix} 2 \\ -2 \end{pmatrix} = 2\begin{pmatrix} 1 \\ -1 \end{pmatrix} + 0\begin{pmatrix} 2 \\ -3 \end{pmatrix}$, so $\begin{pmatrix} 2 \\ -2 \end{pmatrix}_{B_2} = \begin{pmatrix} 2 \\ 0 \end{pmatrix}$. Similarly, $\begin{pmatrix} -6 \\ 9 \end{pmatrix} = 0\begin{pmatrix} 1 \\ -1 \end{pmatrix} - 3\begin{pmatrix} 2 \\ -3 \end{pmatrix}$, so $\begin{pmatrix} -6 \\ 9 \end{pmatrix}_{B_2} = \begin{pmatrix} 0 \\ -3 \end{pmatrix}$. Thus $A_T = \begin{pmatrix} 2 & 0 \\ 0 & -3 \end{pmatrix}$.

There is another way to solve this problem. The vectors $\begin{pmatrix} 1 \\ -1 \end{pmatrix}$ and $\begin{pmatrix} 2 \\ -3 \end{pmatrix}$ are written in terms of the standard basis $S = \left\{\begin{pmatrix} 1 \\ 0 \end{pmatrix}, \begin{pmatrix} 0 \\ 1 \end{pmatrix}\right\}$. That is, $\begin{pmatrix} 1 \\ -1 \end{pmatrix} = 1\begin{pmatrix} 1 \\ 0 \end{pmatrix} + (-1)\begin{pmatrix} 0 \\ 1 \end{pmatrix}$ and $\begin{pmatrix} 2 \\ -3 \end{pmatrix} = 2\begin{pmatrix} 1 \\ 0 \end{pmatrix} + (-3)\begin{pmatrix} 0 \\ 1 \end{pmatrix}$. Thus the matrix $A = \begin{pmatrix} 1 & 2 \\ -1 & -3 \end{pmatrix}$ is the matrix whose first and second columns represent the expansions of the vectors in $B_1$ in terms of the standard basis. Then from the procedure outlined on page 377, the matrix $A^{-1} = \begin{pmatrix} 3 & 2 \\ -1 & -1 \end{pmatrix}$ is the transition matrix from $S$ to $B_1$. Similarly, the matrix $A$ is the transition matrix from $B_1$ to $S$ (see Problem 4.8.38, page 385). Now

† That is, in any space where we have defined a standard basis.

suppose that $\mathbf{x}$ is written in terms of $B_1$. Then $A\mathbf{x}$ is the same vector now written in terms of $S$. Let $C = \begin{pmatrix} 12 & 10 \\ -15 & -13 \end{pmatrix}$. Then $CA\mathbf{x} = T(A\mathbf{x})$ is the image of $A\mathbf{x}$ written in terms of $S$. Finally, since we want $T(A\mathbf{x})$ in terms of $B_1$ (that was the problem), we multiply on the left by the transition matrix $A^{-1}$ to obtain $(T\mathbf{x})_{B_1} = (A^{-1}CA)(\mathbf{x})_{B_1}$. That is,

$$A_T = A^{-1}CA = \begin{pmatrix} 3 & 2 \\ -1 & -1 \end{pmatrix}\begin{pmatrix} 12 & 10 \\ -15 & -13 \end{pmatrix}\begin{pmatrix} 1 & 2 \\ -1 & -3 \end{pmatrix} = \begin{pmatrix} 3 & 2 \\ -1 & -1 \end{pmatrix}\begin{pmatrix} 2 & -6 \\ -2 & 9 \end{pmatrix} = \begin{pmatrix} 2 & 0 \\ 0 & -3 \end{pmatrix}$$

as before. We summarize this result below.

**THEOREM 5** Let $T$: $\mathbb{R}^n \to \mathbb{R}^m$ be a linear transformation. Suppose that $C$ is the transformation matrix of $T$ with respect to the standard bases $S_n$ and $S_m$ in $\mathbb{R}^n$ and $\mathbb{R}^m$, respectively. Let $A_1$ be the transition matrix from $S_n$ to the basis $B_1$ in $\mathbb{R}^n$ and let $A_2$ be the transition matrix from $S_m$ to the basis $B_2$ in $\mathbb{R}^m$. If $A_T$ denotes the transformation matrix of $T$ with respect to the bases $B_1$ and $B_2$, then

$$\boxed{A_T = A_2^{-1}CA_1} \tag{3}$$

In Example 9 we saw that by looking at the linear transformation $T$ with respect to a new basis, the transformation matrix $A_T$ turned out to be a diagonal matrix. We shall return to this "diagonalizing" procedure in Section 6.3. Given a linear transformation from $\mathbb{R}^n$ to $\mathbb{R}^n$, we shall see that it is often possible to find a basis $B$ such that the transformation matrix of $T$ with respect to $B$ will be diagonal.

## The Geometry of Linear Transformations from $\mathbb{R}^2$ to $\mathbb{R}^2$

Let $T$: $\mathbb{R}^2 \to \mathbb{R}^2$ be a linear transformation with matrix representation $A_T$. We now show that if $A_T$ is invertible, then $T$ can be written as a succession of one or more of four special transformations, called **expansions, compressions, reflections,** and **shears.**

**Expansions Along the $x$- or $y$-Axis.** An **expansion along the $x$-axis** is a linear transformation that multiplies the $x$-coordinate of a vector in $\mathbb{R}^2$ by a constant $c > 1$. That is,

$$T\begin{pmatrix} x \\ y \end{pmatrix} = \begin{pmatrix} cx \\ y \end{pmatrix}$$

Then $T\begin{pmatrix}1\\0\end{pmatrix}=\begin{pmatrix}c\\0\end{pmatrix}$ and $T\begin{pmatrix}0\\1\end{pmatrix}=\begin{pmatrix}0\\1\end{pmatrix}$, so if $A_T=\begin{pmatrix}c&0\\0&1\end{pmatrix}$, we have

$$T\begin{pmatrix}x\\y\end{pmatrix}=A\begin{pmatrix}x\\y\end{pmatrix}=\begin{pmatrix}c&0\\0&1\end{pmatrix}\begin{pmatrix}x\\y\end{pmatrix}=\begin{pmatrix}cx\\y\end{pmatrix}$$

Similarly, **an expansion along the $y$-axis** is a linear transformation that multiplies the $y$-coordinate of every vector in $\mathbb{R}^2$ by a constant $c > 1$. As above,

If $T\begin{pmatrix}x\\y\end{pmatrix}=\begin{pmatrix}x\\cy\end{pmatrix}$, then the matrix representation of $T$ is $A_T=\begin{pmatrix}1&0\\0&c\end{pmatrix}$ so that $\begin{pmatrix}1&0\\0&c\end{pmatrix}\begin{pmatrix}x\\y\end{pmatrix}=\begin{pmatrix}x\\cy\end{pmatrix}$.

In Figure 5.5 we depict an expansion along each axis.

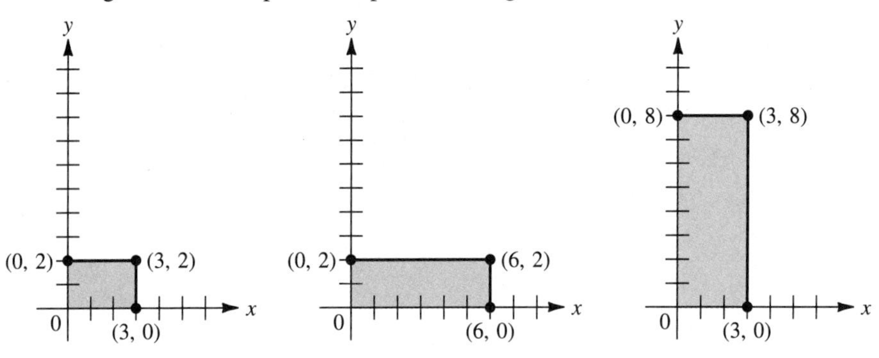


**Figure 5.5** Two expansions

**Compression Along the $x$-Axis or $y$-Axis.** A **compression** along the $x$- or $y$-axis is a linear transformation that multiplies the $x$- or $y$-coordinate of a vector in $\mathbb{R}^2$ by a positive constant $c < 1$. The matrix representations of a compression are the same as for an expansion except that for a compression $0 < c < 1$, whereas for an expansion $c > 1$. Two compressions are illustrated in Figure 5.6.

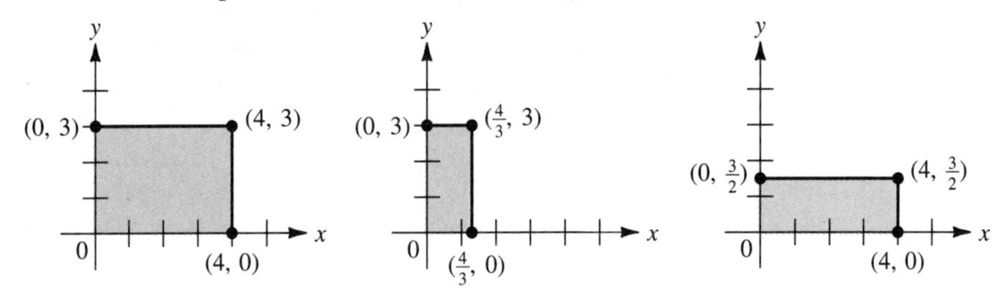


**Figure 5.6** Two compressions

**Reflections.** There are three kinds of reflections that will be of interest to us. In Example 5.1.1 on page 465 we saw that the transformation

$$T\begin{pmatrix} x \\ y \end{pmatrix} = \begin{pmatrix} x \\ -y \end{pmatrix}$$

reflects a vector in $\mathbb{R}^2$ about the $x$-axis (see Figure 5.1). In Example 5.1.6 on page 468 we saw that the transformation

$$T\begin{pmatrix} x \\ y \end{pmatrix} = \begin{pmatrix} -x \\ y \end{pmatrix}$$

reflects a vector in $\mathbb{R}^2$ about the $y$-axis (see Figure 5.2). Now

$$\begin{pmatrix} 1 & 0 \\ 0 & -1 \end{pmatrix}\begin{pmatrix} x \\ y \end{pmatrix} = \begin{pmatrix} x \\ -y \end{pmatrix} \quad \text{and} \quad \begin{pmatrix} -1 & 0 \\ 0 & 1 \end{pmatrix}\begin{pmatrix} x \\ y \end{pmatrix} = \begin{pmatrix} -x \\ y \end{pmatrix}$$

so $\begin{pmatrix} 1 & 0 \\ 0 & -1 \end{pmatrix}$ is the matrix representation of the reflection about the $x$-axis and $\begin{pmatrix} -1 & 0 \\ 0 & 1 \end{pmatrix}$ is the matrix representation of the reflection about the $y$-axis. Finally, the mapping $T\begin{pmatrix} x \\ y \end{pmatrix} = \begin{pmatrix} y \\ x \end{pmatrix}$, which interchanges $x$ and $y$, has the effect of reflecting a vector in $\mathbb{R}^2$ **about the line $y = x$** (see Figure 5.7).

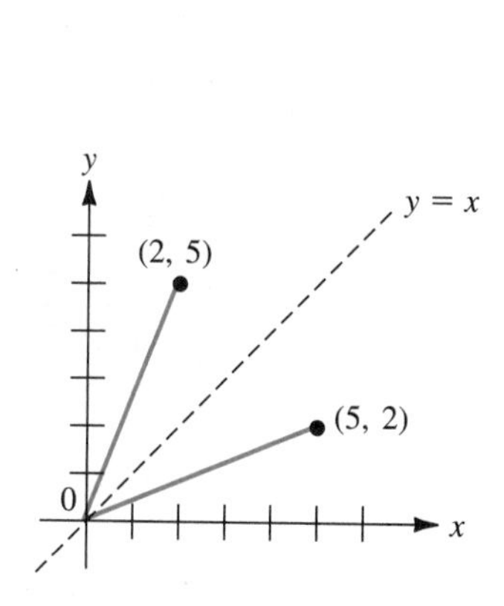


(a) (2, 5) is obtained by reflecting (5, 2) about the line $y = x$

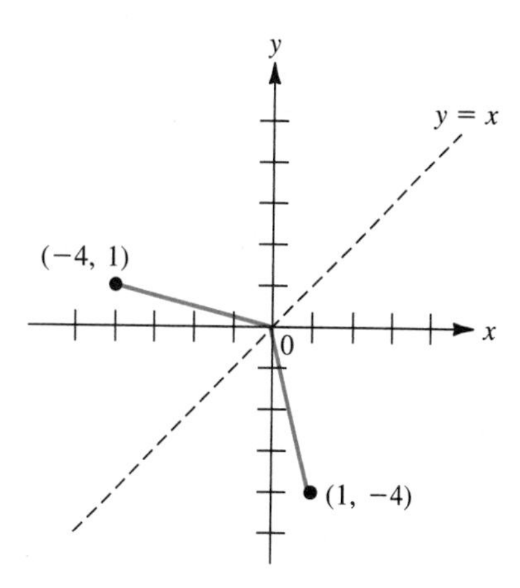


(b) (1, −4) is obtained by reflecting (−4, 1) about the line $y = x$

**Figure 5.7** Reflecting a vector in $\mathbb{R}^2$ about the line $y = x$

If $T\begin{pmatrix} x \\ y \end{pmatrix} = \begin{pmatrix} y \\ x \end{pmatrix}$, then $T\begin{pmatrix} 1 \\ 0 \end{pmatrix} = \begin{pmatrix} 0 \\ 1 \end{pmatrix}$ and $T\begin{pmatrix} 0 \\ 1 \end{pmatrix} = \begin{pmatrix} 1 \\ 0 \end{pmatrix}$, so the matrix representation of the linear transformation that reflects a vector in $\mathbb{R}^2$ about the line $y = x$ is $A = \begin{pmatrix} 0 & 1 \\ 1 & 0 \end{pmatrix}$.

**Shears.** A **shear along the $x$-axis** is a transformation that takes a vector $\begin{pmatrix} x \\ y \end{pmatrix}$ into a new vector $\begin{pmatrix} x + cy \\ y \end{pmatrix}$, where $c$ is a constant that may be positive or negative. Two shears along the $x$-axis are illustrated in Figure 5.8.

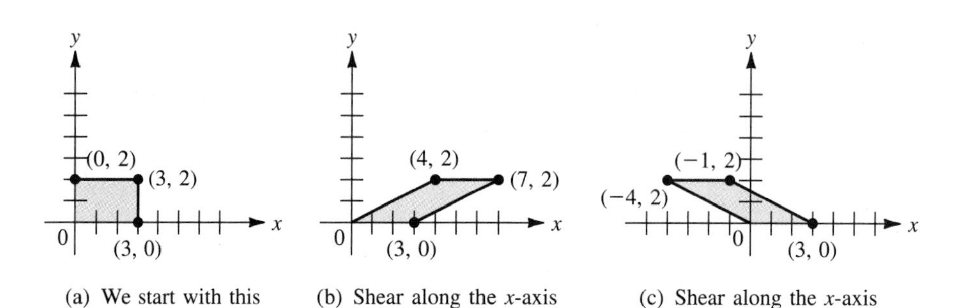


(a) We start with this rectangle

(b) Shear along the $x$-axis with $c = 2$

(c) Shear along the $x$-axis with $c = -2$

**Figure 5.8** Two shears along the $x$-axis

Let $T$ be a shear along the $x$-axis. Then

$$T\begin{pmatrix} 1 \\ 0 \end{pmatrix} = \begin{pmatrix} 1 \\ 0 \end{pmatrix} \quad \text{and} \quad T\begin{pmatrix} 0 \\ 1 \end{pmatrix} = \begin{pmatrix} 0 + c \cdot 1 \\ 1 \end{pmatrix} = \begin{pmatrix} c \\ 1 \end{pmatrix}$$

so the matrix representation of $T$ is $\begin{pmatrix} 1 & c \\ 0 & 1 \end{pmatrix}$. For example, in Figure 5.8$b$, $c = 2$, so

$$A_T = \begin{pmatrix} 1 & 2 \\ 0 & 1 \end{pmatrix} \quad \text{and} \quad A_T\begin{pmatrix} 3 \\ 2 \end{pmatrix} = \begin{pmatrix} 1 & 2 \\ 0 & 1 \end{pmatrix}\begin{pmatrix} 3 \\ 2 \end{pmatrix} = \begin{pmatrix} 7 \\ 2 \end{pmatrix}$$

In Figure 5.8$c$, $c = -2$, so $A_T = \begin{pmatrix} 1 & -2 \\ 0 & 1 \end{pmatrix}$,

$$A_T\begin{pmatrix} 3 \\ 2 \end{pmatrix} = \begin{pmatrix} 1 & -2 \\ 0 & 1 \end{pmatrix}\begin{pmatrix} 3 \\ 2 \end{pmatrix} = \begin{pmatrix} -1 \\ 2 \end{pmatrix}$$

and

$$A_T\begin{pmatrix} 0 \\ 2 \end{pmatrix} = \begin{pmatrix} 1 & -2 \\ 0 & 1 \end{pmatrix}\begin{pmatrix} 0 \\ 2 \end{pmatrix} = \begin{pmatrix} -4 \\ 2 \end{pmatrix}$$

Note that $A_T\begin{pmatrix} 3 \\ 0 \end{pmatrix} = \begin{pmatrix} 1 & -2 \\ 0 & 1 \end{pmatrix}\begin{pmatrix} 3 \\ 0 \end{pmatrix} = \begin{pmatrix} 3 \\ 0 \end{pmatrix}$. That is, vectors with $y$-coordinate zero are left unchanged by shears along the $x$-axis.

A **shear along the *y*-axis** is a transformation that takes a vector $\begin{pmatrix} x \\ y \end{pmatrix}$ into a new vector $\begin{pmatrix} x \\ y + cx \end{pmatrix}$, where $c$ is a constant that may be positive or negative. Two shears along the $y$-axis are illustrated in Figure 5.9.

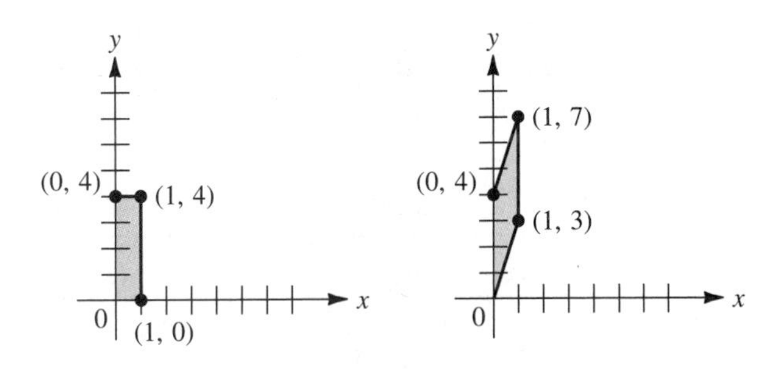


(a) We start with this rectangle

(b) Shear along the $y$-axis with $c = 3$

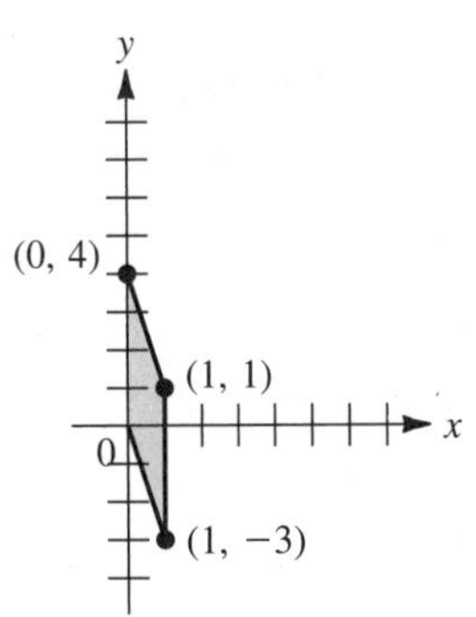


(c) Shear along the $y$-axis with $c = -3$

**Figure 5.9** Two shears along the $y$-axis

If $T$ is a shear along the $y$-axis, then

$$T\begin{pmatrix} 1 \\ 0 \end{pmatrix} = \begin{pmatrix} 1 \\ c \end{pmatrix} \quad \text{and} \quad T\begin{pmatrix} 0 \\ 1 \end{pmatrix} = \begin{pmatrix} 0 \\ 1 \end{pmatrix}$$

so $A_T = \begin{pmatrix} 1 & 0 \\ c & 1 \end{pmatrix}$. For example, in Figure 5.9*b* $c = 3$, so

$$A_T = \begin{pmatrix} 1 & 0 \\ 3 & 1 \end{pmatrix} \quad \text{and} \quad A_T\begin{pmatrix} 1 \\ 4 \end{pmatrix} = \begin{pmatrix} 1 & 0 \\ 3 & 1 \end{pmatrix}\begin{pmatrix} 1 \\ 4 \end{pmatrix} = \begin{pmatrix} 1 \\ 7 \end{pmatrix}$$

In Figure 5.9c $c = -3$, so

$$A_T\begin{pmatrix}1\\4\end{pmatrix} = \begin{pmatrix}1 & 0\\-3 & 1\end{pmatrix}\begin{pmatrix}1\\4\end{pmatrix} = \begin{pmatrix}1\\1\end{pmatrix} \quad \text{and} \quad A_T\begin{pmatrix}1\\0\end{pmatrix} = \begin{pmatrix}1 & 0\\-3 & 1\end{pmatrix}\begin{pmatrix}1\\0\end{pmatrix} = \begin{pmatrix}1\\-3\end{pmatrix}$$

Note that $A_T\begin{pmatrix}0\\4\end{pmatrix} = \begin{pmatrix}1 & 0\\-3 & 1\end{pmatrix}\begin{pmatrix}0\\4\end{pmatrix} = \begin{pmatrix}0\\4\end{pmatrix}$. That is, vectors with $x$-coordinate zero are left unchanged by shears along the $y$-axis.

In Table 5.1 we summarize these types of linear transformations.

**Table 5.1** Special Linear Transformations from $\mathbb{R}^2$ to $\mathbb{R}^2$

| **Transformation** | **Matrix Representation of Transformation: $A_T$** |
|---|---|
| Expansion along the $x$-axis | $\begin{pmatrix}c & 0\\0 & 1\end{pmatrix}$, $c > 1$ |
| Expansion along the $y$-axis | $\begin{pmatrix}1 & 0\\0 & c\end{pmatrix}$, $c > 1$ |
| Compression along the $x$-axis | $\begin{pmatrix}c & 0\\0 & 1\end{pmatrix}$, $0 < c < 1$ |
| Compression along the $y$-axis | $\begin{pmatrix}1 & 0\\0 & c\end{pmatrix}$, $0 < c < 1$ |
| Reflection about the line $y = x$ | $\begin{pmatrix}0 & 1\\1 & 0\end{pmatrix}$ |
| Reflection about the $x$-axis | $\begin{pmatrix}1 & 0\\0 & -1\end{pmatrix}$ |
| Reflection about the $y$-axis | $\begin{pmatrix}-1 & 0\\0 & 1\end{pmatrix}$ |
| Shear along the $x$-axis | $\begin{pmatrix}1 & c\\0 & 1\end{pmatrix}$ |
| Shear along the $y$-axis | $\begin{pmatrix}1 & 0\\c & 1\end{pmatrix}$ |

In Section 1.10 we discussed elementary matrices. Multiplication of a matrix by an elementary matrix has the effect of performing an elementary row operation on that matrix. Table 5.2 lists the elementary matrices in $\mathbb{R}^2$.

**Table 5.2** The Elementary Matrices in $\mathbb{R}^2$

| Elementary Row Operation | Elementary Matrix | Illustration |
|---|---|---|
| $R_1 \to cR_1$ | $\begin{pmatrix} c & 0 \\ 0 & 1 \end{pmatrix}$ | $\begin{pmatrix} c & 0 \\ 0 & 1 \end{pmatrix}\begin{pmatrix} x & y \\ z & w \end{pmatrix} = \begin{pmatrix} cx & cy \\ z & w \end{pmatrix}$ |
| $R_2 \to cR_2$ | $\begin{pmatrix} 1 & 0 \\ 0 & c \end{pmatrix}$ | $\begin{pmatrix} 1 & 0 \\ 0 & c \end{pmatrix}\begin{pmatrix} x & y \\ z & w \end{pmatrix} = \begin{pmatrix} x & y \\ cz & cw \end{pmatrix}$ |
| $R_1 \to R_1 + cR_2$ | $\begin{pmatrix} 1 & c \\ 0 & 1 \end{pmatrix}$ | $\begin{pmatrix} 1 & c \\ 0 & 1 \end{pmatrix}\begin{pmatrix} x & y \\ z & w \end{pmatrix} = \begin{pmatrix} x + cz & y + cw \\ z & w \end{pmatrix}$ |
| $R_2 \to R_2 + cR_1$ | $\begin{pmatrix} 1 & 0 \\ c & 1 \end{pmatrix}$ | $\begin{pmatrix} 1 & 0 \\ c & 1 \end{pmatrix}\begin{pmatrix} x & y \\ z & w \end{pmatrix} = \begin{pmatrix} x & y \\ z + cx & w + cy \end{pmatrix}$ |
| $R_1 \rightleftarrows R_2$ | $\begin{pmatrix} 0 & 1 \\ 1 & 0 \end{pmatrix}$ | $\begin{pmatrix} 0 & 1 \\ 1 & 0 \end{pmatrix}\begin{pmatrix} x & y \\ z & w \end{pmatrix} = \begin{pmatrix} z & w \\ x & y \end{pmatrix}$ |

**THEOREM 6** Every $2 \times 2$ elementary matrix $E$ is one of the following:

**i.** The matrix representation of an expansion along the $x$- or $y$-axis

**ii.** The matrix representation of a compression along the $x$- or $y$-axis

**iii.** The matrix representation of a reflection about the line $y = x$

**iv.** The matrix representation of a shear along the $x$- or $y$-axis

**v.** The matrix representation of a reflection about the $x$- or $y$-axis

**vi.** The product of the matrix representation of a reflection about the $x$- or $y$-axis and the matrix representation of an expansion or compression

**Proof** We refer to Tables 5.1 and 5.2.

*Case 1:* $E = \begin{pmatrix} c & 0 \\ 0 & 1 \end{pmatrix}$, $c > 0$ This is the matrix representation of an expansion along the $x$-axis if $c > 1$ or a compression along the $x$-axis if $0 < c < 1$.

*Case 2:* $E = \begin{pmatrix} c & 0 \\ 0 & 1 \end{pmatrix}$, $c < 0$

*Case 2a:* $c = -1$ Then $E = \begin{pmatrix} -1 & 0 \\ 0 & 1 \end{pmatrix}$, which is the matrix representation of a reflection about the $y$-axis.

*Case 2b:* $c < 0$, $c \neq -1$. Then $-c > 0$ and

$$E = \begin{pmatrix} c & 0 \\ 0 & 1 \end{pmatrix} = \begin{pmatrix} -1 & 0 \\ 0 & 1 \end{pmatrix}\begin{pmatrix} -c & 0 \\ 0 & 1 \end{pmatrix}$$

which is the product of the matrix representation of a reflection about the $y$-axis and the matrix representation of an expansion (if $-c > 1$) or compression (if $0 < -c < 1$) along the $x$-axis.

*Case 3:* $E = \begin{pmatrix} 1 & 0 \\ 0 & c \end{pmatrix}$, $c > 0$ Same as case 1 with the $y$-axis replacing the $x$-axis.

*Case 4:* $E = \begin{pmatrix} 1 & 0 \\ 0 & c \end{pmatrix}$, $c < 0$ Same as case 2 with the axes interchanged.

*Case 5:* $E = \begin{pmatrix} 1 & c \\ 0 & 1 \end{pmatrix}$ This is the matrix representation of a shear along the $x$-axis.

*Case 6:* $E = \begin{pmatrix} 1 & 0 \\ c & 1 \end{pmatrix}$ This is the matrix representation of a shear along the $y$-axis.

*Case 7:* $E = \begin{pmatrix} 0 & 1 \\ 1 & 0 \end{pmatrix}$ This is the matrix representation of a reflection about the line $y = x$.

In Theorem 1.10.3 on page 130 we showed that every invertible matrix can be written as the product of elementary matrices. In Theorem 6 we showed that every elementary matrix in $\mathbb{R}^2$ can be written as a product of matrix representations of expansions, compressions, shears, and reflections. Thus we have the following result.

**THEOREM 7** Let $T$: $\mathbb{R}^2 \to \mathbb{R}^2$ be a linear transformation such that its matrix representation is invertible. Then $T$ can be obtained as a succession of expansions, compressions, shears, and reflections.

*Note.* By the Summing Up Theorem on page 360, $A_T$ is invertible if and only if $\rho(A_T) = 2$. But by Theorem 4, $\rho(A_T) = \rho(T)$. This means that $A_T$ is invertible with respect to all choices of bases in $\mathbb{R}^2$ or it is not invertible with respect to any choices of bases.

**EXAMPLE 10** **Decomposing a Linear Transformation in $\mathbb{R}^2$ into a Succession of Expansions, Compressions, Shears, and Reflections** Consider the transformation $T$: $\mathbb{R}^2 \to \mathbb{R}^2$ with matrix representation $A_T = \begin{pmatrix} 1 & 2 \\ 3 & 4 \end{pmatrix}$. Using the technique of Section 1.10 (look at Example 3 on page 131), we can write $A_T$ as a product of three elementary matrices:

$$\begin{pmatrix} 1 & 2 \\ 3 & 4 \end{pmatrix} = \begin{pmatrix} 1 & 0 \\ 3 & 1 \end{pmatrix}\begin{pmatrix} 1 & 0 \\ 0 & -2 \end{pmatrix}\begin{pmatrix} 1 & 2 \\ 0 & 1 \end{pmatrix} \quad \textbf{(4)}$$

Now

$$\begin{pmatrix} 1 & 0 \\ 3 & 1 \end{pmatrix} \quad \text{represents a shear along the } y\text{-axis (with } c = 3)$$

$$\begin{pmatrix} 1 & 2 \\ 0 & 1 \end{pmatrix} \quad \text{represents a shear along the } x\text{-axis (with } c = 2)$$

$$\begin{pmatrix} 1 & 0 \\ 0 & -2 \end{pmatrix} = \begin{pmatrix} 1 & 0 \\ 0 & -1 \end{pmatrix}\begin{pmatrix} 1 & 0 \\ 0 & 2 \end{pmatrix}$$

represents an expansion along the $y$-axis (with $c = 2$) followed by a reflection about the $x$-axis

Thus to apply $T$ to a vector in $\mathbb{R}^2$, we

**i.** Shear along the $x$-axis with $c = 2$.
**ii.** Expand along the $y$-axis with $c = 2$.
**iii.** Reflect about the $x$-axis.
**iv.** Shear along the $y$-axis with $c = 3$.

Note that we do these operations in the reverse order in which we write the matrices in (4).

To illustrate this, suppose that $\mathbf{v} = \begin{pmatrix} 3 \\ -2 \end{pmatrix}$.

Then

$$T\mathbf{v} = A_T\mathbf{v} = \begin{pmatrix} 1 & 2 \\ 3 & 4 \end{pmatrix}\begin{pmatrix} 3 \\ -2 \end{pmatrix} = \begin{pmatrix} -1 \\ 1 \end{pmatrix}$$

Using the operations ($i$) to ($iv$), we have

$$\begin{pmatrix} 3 \\ -2 \end{pmatrix} \xrightarrow{\text{Shear}} \begin{pmatrix} 1 & 2 \\ 0 & 1 \end{pmatrix}\begin{pmatrix} 3 \\ -2 \end{pmatrix} = \begin{pmatrix} -1 \\ -2 \end{pmatrix} \xrightarrow{\text{Expansion}} \begin{pmatrix} 1 & 0 \\ 0 & 2 \end{pmatrix}\begin{pmatrix} -1 \\ -2 \end{pmatrix} = \begin{pmatrix} -1 \\ -4 \end{pmatrix}$$

$$\xrightarrow{\text{Reflection}} \begin{pmatrix} 1 & 0 \\ 0 & -1 \end{pmatrix}\begin{pmatrix} -1 \\ -4 \end{pmatrix} = \begin{pmatrix} -1 \\ 4 \end{pmatrix} \xrightarrow{\text{Shear}} \begin{pmatrix} 1 & 0 \\ 3 & 1 \end{pmatrix}\begin{pmatrix} -1 \\ 4 \end{pmatrix} = \begin{pmatrix} -1 \\ 1 \end{pmatrix}$$

We sketch these steps in Figure 5.10.

y
x
0
(3, −2)

(a) We start with this vector

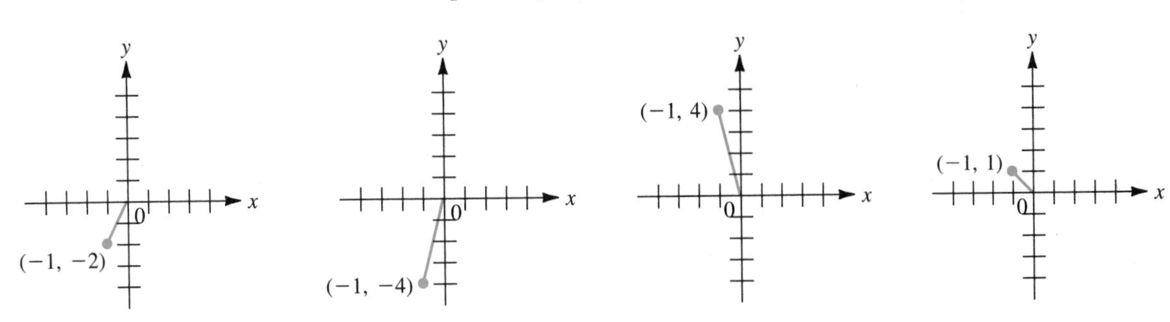


(b) Vector obtained by shearing along the $x$-axis with $c = 2$

(c) Vector obtained by expanding along the $y$-axis with $c = 2$

(d) Vector obtained by reflecting about the $x$-axis

(e) Vector obtained by shearing along the $y$-axis with $c = 3$

**Figure 5.10** Decomposing the linear transformation $T\begin{pmatrix} 3 \\ -2 \end{pmatrix} = \begin{pmatrix} 1 & 2 \\ 3 & 4 \end{pmatrix}\begin{pmatrix} 3 \\ -2 \end{pmatrix}$ into a succession of shears, expansions, and reflections

# PROBLEMS 5.3

## Self-Quiz

**I.** If $T$: $\mathbb{R}^3 \to \mathbb{R}^3$ is the linear transformation $T\begin{pmatrix} x \\ y \\ z \end{pmatrix} = \begin{pmatrix} z \\ -x \\ y \end{pmatrix}$, then $A_T =$

**a.** $\begin{pmatrix} 0 & -1 & 0 \\ 0 & 0 & 1 \\ 1 & 0 & 0 \end{pmatrix}$ **b.** $\begin{pmatrix} 0 & 0 & 1 \\ -1 & 0 & 0 \\ 0 & 1 & 0 \end{pmatrix}$ **c.** $\begin{pmatrix} 1 & 0 & 0 \\ 0 & -1 & 0 \\ 0 & 0 & 1 \end{pmatrix}$ **d.** $\begin{pmatrix} 0 & 0 & 1 \\ 0 & 1 & 0 \\ -1 & 0 & 0 \end{pmatrix}$

**II.** ________ represent(s) an expansion along the $y$-axis.

**a.** $\begin{pmatrix} 2 & 0 \\ 0 & 1 \end{pmatrix}$ **b.** $\begin{pmatrix} \frac{1}{2} & 0 \\ 0 & 1 \end{pmatrix}$ **c.** $\begin{pmatrix} 1 & 0 \\ 0 & 2 \end{pmatrix}$ **d.** $\begin{pmatrix} 1 & 0 \\ 0 & \frac{1}{2} \end{pmatrix}$

**III.** ________ represent(s) a shear along the $x$-axis.

**a.** $\begin{pmatrix} -1 & 0 \\ 0 & 1 \end{pmatrix}$ **b.** $\begin{pmatrix} 1 & 0 \\ 0 & -1 \end{pmatrix}$ **c.** $\begin{pmatrix} 1 & 3 \\ 0 & 1 \end{pmatrix}$

**d.** $\begin{pmatrix} 1 & \frac{1}{3} \\ 0 & 1 \end{pmatrix}$ **e.** $\begin{pmatrix} 1 & 0 \\ 3 & 1 \end{pmatrix}$ **f.** $\begin{pmatrix} 1 & 0 \\ \frac{1}{3} & 1 \end{pmatrix}$

In Problems 1–30 find the matrix representation $A_T$ of the linear transformation $T$, ker $T$, range $T$, $\nu(T)$, and $\rho(T)$. Unless otherwise stated, assume that $B_1$ and $B_2$ are standard bases.

**1.** $T$: $\mathbb{R}^2 \to \mathbb{R}^2$; $T\begin{pmatrix} x \\ y \end{pmatrix} = \begin{pmatrix} x - 2y \\ -x + y \end{pmatrix}$

**2.** $T$: $\mathbb{R}^2 \to \mathbb{R}^3$; $T\begin{pmatrix} x \\ y \end{pmatrix} = \begin{pmatrix} x + y \\ x - y \\ 2x + 3y \end{pmatrix}$

**3.** $T$: $\mathbb{R}^3 \to \mathbb{R}^2$; $T\begin{pmatrix} x \\ y \\ z \end{pmatrix} = \begin{pmatrix} x - y + z \\ -2x + 2y - 2z \end{pmatrix}$

**4.** $T$: $\mathbb{R}^2 \to \mathbb{R}^2$; $T\begin{pmatrix} x \\ y \end{pmatrix} = \begin{pmatrix} ax + by \\ cx + dy \end{pmatrix}$

**5.** $T$: $\mathbb{R}^3 \to \mathbb{R}^3$; $T\begin{pmatrix} x \\ y \\ z \end{pmatrix} = \begin{pmatrix} x - y + 2z \\ 3x + y + 4z \\ 5x - y + 8z \end{pmatrix}$

**6.** $T$: $\mathbb{R}^3 \to \mathbb{R}^3$; $T\begin{pmatrix} x \\ y \\ z \end{pmatrix} = \begin{pmatrix} -x + 2y + z \\ 2x - 4y - 2z \\ -3x + 6y + 3z \end{pmatrix}$

---

**Answers to Self-Quiz**

**I.** b **II.** c **III.** c, d

**7.** $T: \mathbb{R}^4 \to \mathbb{R}^3$; $T\begin{pmatrix} x \\ y \\ z \\ w \end{pmatrix} = \begin{pmatrix} x - y + 2z + 3w \\ y + 4z + 3w \\ x + 6z + 6w \end{pmatrix}$

**8.** $T: \mathbb{R}^4 \to \mathbb{R}^4$; $T\begin{pmatrix} x \\ y \\ z \\ w \end{pmatrix} = \begin{pmatrix} x - y + 2z + w \\ -x + z + 2w \\ x - 2y + 5z + 4w \\ 2x - y + z - w \end{pmatrix}$

**9.** $T: \mathbb{R}^2 \to \mathbb{R}^2$; $T\begin{pmatrix} x \\ y \end{pmatrix} = \begin{pmatrix} x - 2y \\ 2x + y \end{pmatrix}$; $B_1 = B_2 = \left\{\begin{pmatrix} 1 \\ -2 \end{pmatrix}, \begin{pmatrix} 3 \\ 2 \end{pmatrix}\right\}$

**10.** $T: \mathbb{R}^2 \to \mathbb{R}^2$; $T\begin{pmatrix} x \\ y \end{pmatrix} = \begin{pmatrix} 4x - y \\ 3x + 2y \end{pmatrix}$; $B_1 = B_2 = \left\{\begin{pmatrix} -1 \\ 1 \end{pmatrix}, \begin{pmatrix} 4 \\ 3 \end{pmatrix}\right\}$

**11.** $T: \mathbb{R}^3 \to \mathbb{R}^2$; $T\begin{pmatrix} x \\ y \\ z \end{pmatrix} = \begin{pmatrix} 2x + y + z \\ y - 3z \end{pmatrix}$;

$$B_1 = \left\{\begin{pmatrix} 1 \\ 0 \\ 1 \end{pmatrix}, \begin{pmatrix} 1 \\ 1 \\ 0 \end{pmatrix}, \begin{pmatrix} 1 \\ 1 \\ 1 \end{pmatrix}\right\};\ B_2 = \left\{\begin{pmatrix} 1 \\ -1 \end{pmatrix}, \begin{pmatrix} 2 \\ 3 \end{pmatrix}\right\}$$

**12.** $T: \mathbb{R}^2 \to \mathbb{R}^3$; $T\begin{pmatrix} x \\ y \end{pmatrix} = \begin{pmatrix} x - y \\ 2x + y \\ y \end{pmatrix}$; $B_1 = \left\{\begin{pmatrix} 2 \\ 1 \end{pmatrix}, \begin{pmatrix} 1 \\ 2 \end{pmatrix}\right\}$; $B_2 = \left\{\begin{pmatrix} 1 \\ -1 \\ 0 \end{pmatrix}, \begin{pmatrix} 0 \\ 2 \\ 0 \end{pmatrix}, \begin{pmatrix} 0 \\ 2 \\ 5 \end{pmatrix}\right\}$

**13.** $T: P_2 \to P_3$; $T(a_0 + a_1x + a_2x^2) = a_1 - a_1x + a_0x^3$

**14.** $T: \mathbb{R} \to P_3$; $T(a) = a + ax + ax^2 + ax^3$

**15.** $T: P_3 \to \mathbb{R}$; $T(a_0 + a_1x + a_2x^2 + a_3x^3) = a_2$

**16.** $T: P_3 \to P_1$; $T(a_0 + a_1x + a_2x^2 + a_3x^3) = (a_1 + a_3)x - a_2$

**17.** $T: P_3 \to P_2$; $T(a_0 + a_1x + a_2x^2 + a_3x^3) = (a_0 - a_1 + 2a_2 + 3a_3) + (a_1 + 4a_2 + 3a_3)x + (a_0 + 6a_2 + 5a_3)x^2$

**18.** $T: M_{22} \to M_{22}$; $T\begin{pmatrix} a & b \\ c & d \end{pmatrix} = \begin{pmatrix} a - b + 2c + d & -a + 2c + 2d \\ a - 2b + 5c + 4d & 2a - b + c - d \end{pmatrix}$

**19.** $T: M_{22} \to M_{22}$; $T\begin{pmatrix} a & b \\ c & d \end{pmatrix} = \begin{pmatrix} a + b + c + d & a + b + c \\ a + b & a \end{pmatrix}$

**20.** $T: P_2 \to P_3$; $T[p(x)] = xp(x)$; $B_1 = \{1, x, x^2\}$; $B_2 = \{1, (1 + x), (1 + x)^2, (1 + x)^3\}$

Calculus **21.** $D: P_4 \to P_3$; $Dp(x) = p'(x)$

Calculus **22.** $T: P_4 \to P_4$; $Tp(x) = xp'(x) - p(x)$

*Calculus **23.** $D: P_n \to P_{n-1}$; $Dp(x) = p'(x)$

Calculus **24.** $D: P_4 \to P_2$; $Dp(x) = p''(x)$

*Calculus **25.** $T: P_4 \to P_4$; $Tp(x) = p''(x) + xp'(x) + 2p(x)$

*Calculus **26.** $D: P_n \to P_{n-k}$; $Dp(x) = p^{(k)}(x)$

*Calculus **27.** $T: P_n \to P_n$; $Tp(x) = x^np^{(n)}(x) + x^{n-1}p^{(n-1)}(x) + \cdots + xp'(x) + p(x)$

Calculus **28.** $J: P_n \to \mathbb{R}$; $Jp = \int_0^1 p(x)\,dx$

**29.** $T: \mathbb{R}^3 \to P_2$; $T\begin{pmatrix} a \\ b \\ c \end{pmatrix} = a + bx + cx^2$

**30.** $T\colon P_3 \to \mathbb{R}^3$; $T(a_0 + a_1x + a_2x^2 + a_3x^3) = \begin{pmatrix} a_3 - a_2 \\ a_1 + a_3 \\ a_2 - a_1 \end{pmatrix}$

**31.** Let $T\colon M_{mn} \to M_{nm}$ be given by $TA = A^t$. Find $A_T$ with respect to the standard bases in $M_{mn}$ and $M_{nm}$.

***32.** Let $T\colon \mathbb{C}^2 \to \mathbb{C}^2$ be given by $T\begin{pmatrix} x \\ y \end{pmatrix} = \begin{pmatrix} x + iy \\ (1 + i)y - x \end{pmatrix}$. Find $A_T$.

Calculus **33.** Let $V = \text{span}\,\{1, \sin x, \cos x\}$. Find $A_D$, where $D\colon V \to V$ is defined by $Df(x) = f'(x)$. Find range $D$ and ker $D$.

Calculus **34.** Answer the questions of Problems 33 given $V = \text{span}\,\{e^x, xe^x, x^2e^x\}$.

**35.** Let $T\colon \mathbb{C}^2 \to \mathbb{C}^2$ be given by $T\mathbf{x} = \text{proj}_H\,\mathbf{x}$, where $H = \text{span}\,\{(1/\sqrt{2})(1, i)\}$. Find $A_T$.

**36.** Prove Theorem 2. **37.** Prove Theorem 4.

In Problems 38–45 describe in words the linear transformation $T\colon \mathbb{R}^2 \to \mathbb{R}^2$ with the given matrix representation $A_T$.

**38.** $A_T = \begin{pmatrix} 4 & 0 \\ 0 & 1 \end{pmatrix}$ **39.** $A_T = \begin{pmatrix} 1 & 0 \\ 0 & \frac{1}{4} \end{pmatrix}$ **40.** $A_T = \begin{pmatrix} 1 & 0 \\ 0 & -1 \end{pmatrix}$

**41.** $A_T = \begin{pmatrix} 1 & 2 \\ 0 & 1 \end{pmatrix}$ **42.** $A_T = \begin{pmatrix} 1 & -3 \\ 0 & 1 \end{pmatrix}$ **43.** $A_T = \begin{pmatrix} 1 & 0 \\ \frac{1}{2} & 1 \end{pmatrix}$

**44.** $A_T = \begin{pmatrix} 1 & 0 \\ -5 & 1 \end{pmatrix}$ **45.** $A_T = \begin{pmatrix} 0 & 1 \\ 1 & 0 \end{pmatrix}$

In Problems 46–55 write the $2 \times 2$ matrix representation of the given linear transformation and draw a sketch of the region obtained when the transformation is applied to the given rectangle.

**46.** Expansion along the $y$-axis with $c = 2$

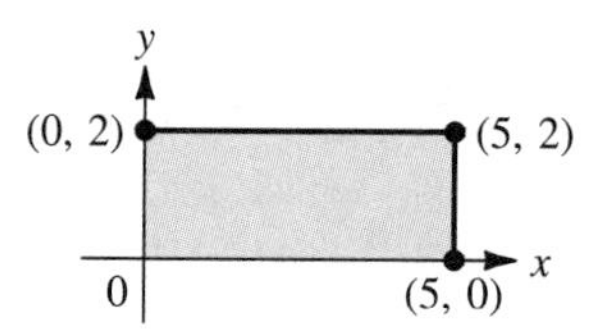


**47.** Compression along the $x$-axis with $c = \frac{1}{4}$

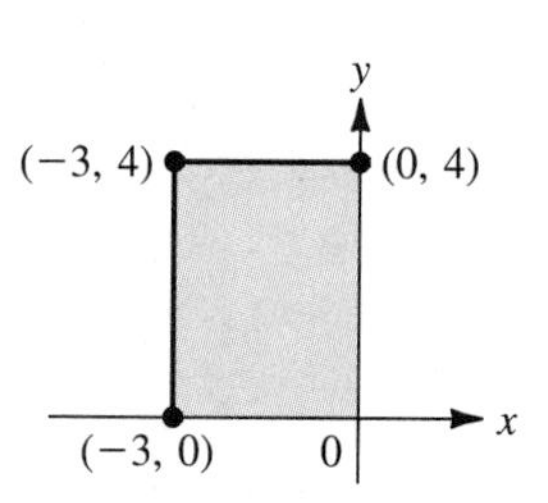


**48.** Shear along the $x$-axis with $c = -2$

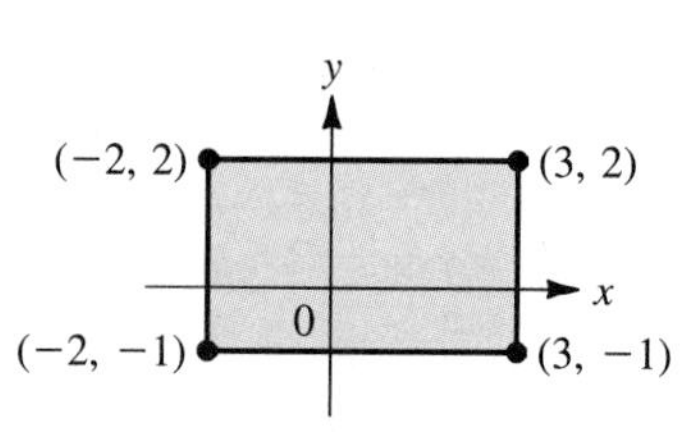

**49.** Shear along the $y$-axis with $c = 3$

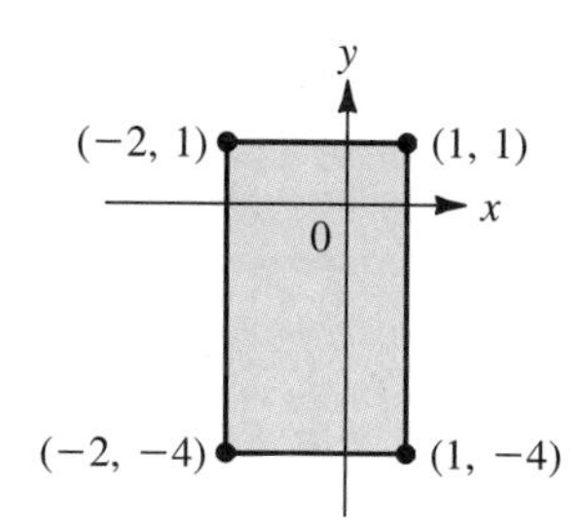


**50.** Shear along the $y$-axis with $c = -\frac{1}{2}$

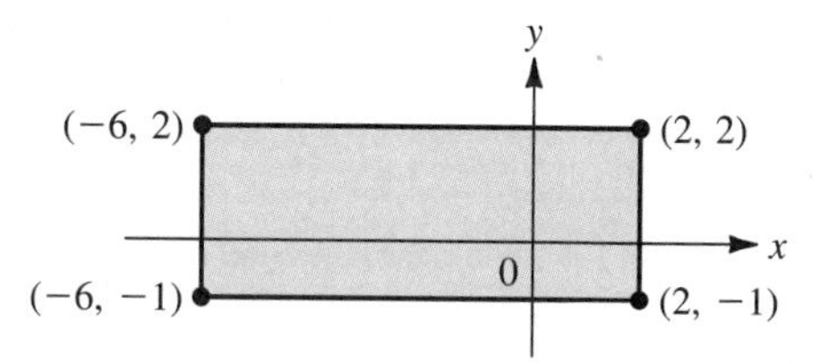


**51.** Shear along the $x$-axis with $c = \frac{1}{5}$

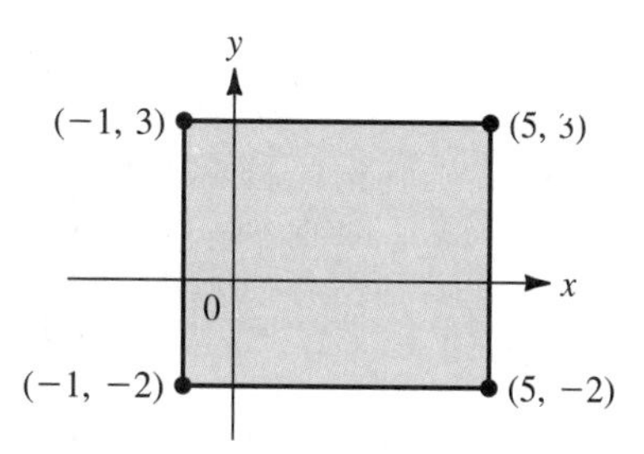


**52.** Reflection about the $x$-axis

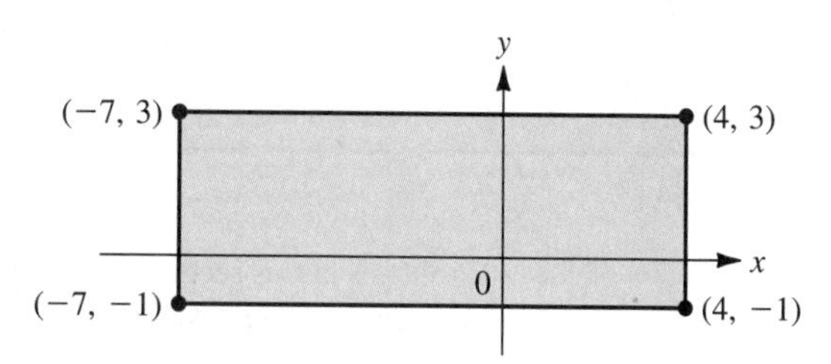


**53.** Reflection about the $y$-axis

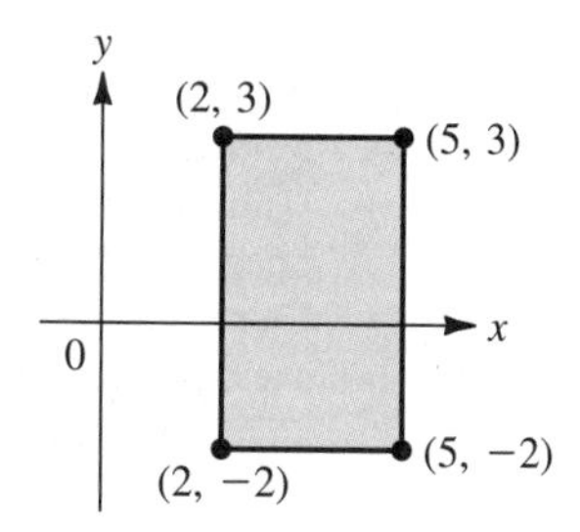

**54.** Reflection about the line $y = x$

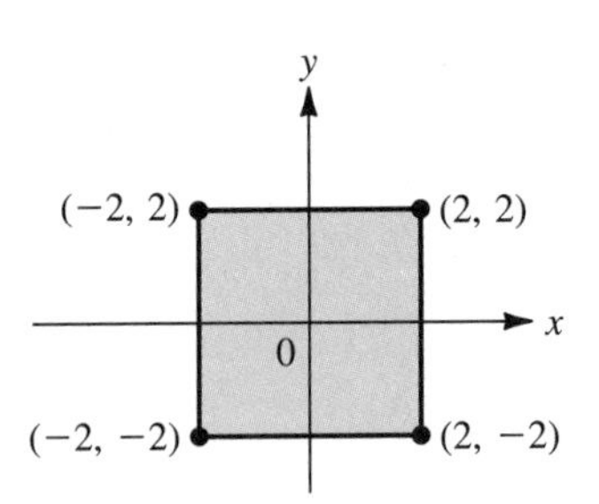


**55.** Reflection about the line $y = x$

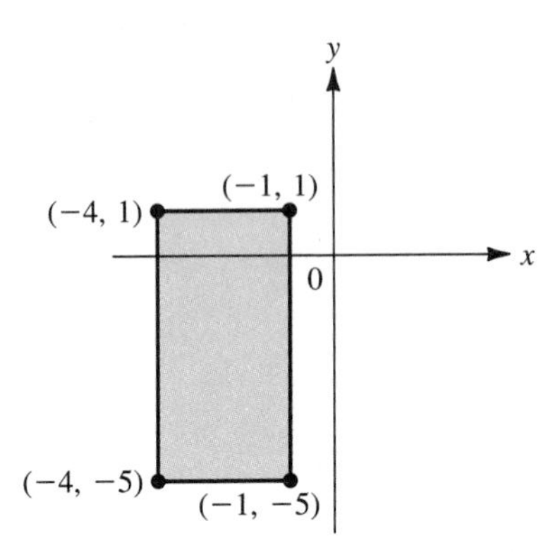


In Problems 56–63 write each linear transformation with given transformation matrix $A_T$ as a succession of expansions, compressions, reflections, and shears.

**56.** $A_T = \begin{pmatrix} 2 & -1 \\ 5 & 0 \end{pmatrix}$ **57.** $A_T = \begin{pmatrix} 3 & 2 \\ -1 & 4 \end{pmatrix}$ **58.** $A_T = \begin{pmatrix} 0 & -2 \\ 3 & -5 \end{pmatrix}$

**59.** $A_T = \begin{pmatrix} 3 & 6 \\ 4 & 2 \end{pmatrix}$ **60.** $A_T = \begin{pmatrix} 0 & 3 \\ 1 & -2 \end{pmatrix}$ **61.** $A_T = \begin{pmatrix} 0 & -2 \\ 5 & 7 \end{pmatrix}$

**62.** $A_T = \begin{pmatrix} 3 & 7 \\ -4 & -8 \end{pmatrix}$ **63.** $A_T = \begin{pmatrix} -1 & 10 \\ 6 & 2 \end{pmatrix}$

## MATLAB 5.3

M Problems in this section that refer to the use of the m-file *grafics* (on the accompanying disk) assume that you have worked the MATLAB problems in Section 5.1.

1. Consider the rectangle in Figure 5.8(a) in the text. Create a points and lines matrix for it.
   a. Let $T$ be the transformation that expands along the $y$-axis by a factor of 3 and compresses along the $x$-axis by a factor of $\frac{1}{2}$. Find its matrix representation and, on the same set of axes, plot the original rectangle and its transformed image using the m-file *grafics.*
   b. Using the appropriate matrix representations and the m-file *grafics,* reproduce the images of the shear transformations in Figures 5.8(b) and 5.8(c).
   c. Using the appropriate matrix representation and the m-file *grafics,* on the same set of axes, plot the original rectangle and the image after applying a shear transformation along the $y$-axis with $c = -2$.

2. The matrix representation of a composition of linear transformations is the product of the matrix representations of the individual transformations *in the appropriate order.* If $T:\mathbb{R}^2 \to \mathbb{R}^2$ with matrix representation $A$ and $S:\mathbb{R}^2 \to \mathbb{R}^2$ with matrix representation $B$, then $T(S(\mathbf{x})) = AB\mathbf{x}$.
   a. *(Paper and pencil)* Find the matrix $R$ that represents rotation around the origin counterclockwise by $\pi/2$ and the matrix $E$ that represents expansion along the $x$-axis by a factor of 2.
   b. Enter the points and lines matrices for the figure given in MATLAB Problem 1(a) in Section 5.1. Using the m-file *grafics,* on the same set of axes, plot the figure, the image of the figure after first rotating and then expanding, and the image of the figure after first expanding and then rotating. Use a different color (and point symbol) for each of the three plots. You will need a **hold on** statement after each call to *grafics.* You will need to adjust the parameter $M$ in the call to *grafics* until all three figures fit nicely on the graphics screen. Do not save this graph. Finding the appropriate $M$ is what is important.

      Using the $M$ that you found, on the same set of axes, plot the figure and the image after first rotating and then expanding. Label this plot, being sure to tell which images are plotted. (Use the on-line help to explore the commands **title, xlabel,** and **ylabel.**) Repeat for the figure and the image after first expanding and then rotating.

      Write a description comparing the two plots. Explain at least one feature of the geometry of the plots that would enable you to tell which kind of transformation was performed first.

3. **Projections** Let $\mathbf{v}$ be a vector in $\mathbb{R}^n$ with length 1. Let $T:\mathbb{R}^n \to \mathbb{R}^n$ be given by

$$T(\mathbf{x}) = \text{proj}_{\mathbf{v}}\mathbf{x} = (\mathbf{v} \cdot \mathbf{x})\mathbf{v}$$

   a. *(Paper and pencil)* Show that $T$ is linear. Show that the matrix representation, $P$, of $T$ (with respect to the standard basis), is given by

$$P = (v_1\mathbf{v} \quad v_2\mathbf{v} \quad \cdots \quad v_n\mathbf{v})$$

      Here $v_i$ refers to the $i$th component of $\mathbf{v}$. Recall that we are assuming that $\mathbf{v}$ has length 1.
   b. Suppose $\mathbf{v}$ is the vector of length 1 in $\mathbb{R}^2$ given by $\mathbf{v} = (1 \quad 0)^t$.
      i. Using the m-file *grafics,* find the matrix $P$ that represents projection onto $\mathbf{v}$. Enter the points and lines matrices from MATLAB Problem 1(a) in Section 5.1. On the same set of axes, plot the original figure and the image of the figure after applying the transformation given by $P$. Use different colors and/or symbols. For each key point in the original figure, identify the point that is its image after applying the transformation. Do the same for two of the line segments of the original figure.
      ii. *(Paper and pencil)* Use $P$ to find a basis for the kernel and range of the transformation. Describe how the geometry of projection onto $\mathbf{v}$ explains these results.
   c. Repeat the instructions of part (b) for the vector $\mathbf{v}$ of length 1 in the direction of $\mathbf{w} = (1 \quad 1)^t$. (To find $\mathbf{v}$, divide $\mathbf{w}$ by its length.)
   d. Repeat the instructions of part (b) for the vector $\mathbf{v}$ of length 1 in the direction of $\mathbf{w} = (-1 \quad 1)^t$.
   e. Repeat parts (b) through (d) for a figure of your own creation.

4. **Reflections** Let $\mathbf{v}$ be a vector in $\mathbb{R}^2$ of length 1. The transformation that reflects a given vector $\mathbf{x}$ in $\mathbb{R}^2$ across the line determined by $\mathbf{v}$ is a linear transformation. Therefore, it has a matrix representation. We will call the matrix representation $F$.

   a. *(Paper and pencil)* Explain why $2\text{proj}_{\mathbf{v}}\mathbf{x} = \mathbf{x} + F\mathbf{x}$, using the diagram below. Hence argue that $F = 2P - I$, where $P$ is the matrix representation of projection onto $\mathbf{v}$ and $I$ is the $2 \times 2$ identity matrix.

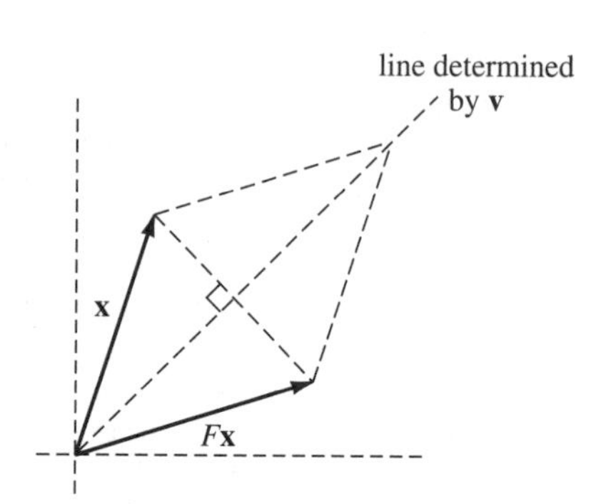


   b. Find the matrix $F$, as in the preceding discussion, representing the transformation of reflection across the $x$-axis. Here $\mathbf{v} = (1 \quad 0)^t$.

      Using the points and lines matrix from MATLAB Problem 1(a) in Section 5.1 and the m-file *grafics,* on the same set of axes, plot the original figure and the image of the figure after applying the given reflection. For each key point in the original figure, identify the point that is its image after applying the transformation. Do the same for two of the line segments of the original figure. Verify that the images are the given reflection of the original line segments.

   c. Repeat the instructions of part (b) for reflection across the $y = -x$ line. Here the vector $\mathbf{v}$ is the vector of length 1 in the direction of $\mathbf{w} = (-1 \quad 1)^t$.

   d. Repeat parts (b) and (c) for a figure of your own creation.

**PROJECT PROBLEM**

5. Create a design or picture by using one or two original figures and applying various transformations to them. Use *grafics* and the **hold on** feature. (You will need to enter the command **hold on** after each call to *grafics* since the m-file includes a final **hold off** command.)

   If you plot a transformed figure that you decide you do not want, you can "erase" it by replotting it using the invisible color option **'i'** in the call to *grafics.* One problem, though, is that this might erase parts of lines of other figures that you want to keep. If so, just replot the ones that you want to keep that were affected.

   If you want to translate a figure $a$ units in the $x$ direction and $b$ units in the $y$ direction and you have $n$ points, use the points matrix given by **newpts = pts + [a*ones(1,n); b*ones(1,n)],** where *pts* is the original points matrix for the figure.

6. Let $T:\mathbb{R}^4 \to \mathbb{R}^4$ be a linear transformation defined by

$$T\begin{pmatrix} 1 \\ 0 \\ 3 \\ -1 \end{pmatrix} = \begin{pmatrix} 3 \\ -1 \\ 7 \\ 2 \end{pmatrix}, \quad T\begin{pmatrix} 2 \\ -1 \\ 4 \\ 3 \end{pmatrix} = \begin{pmatrix} 2 \\ 0 \\ 6 \\ -2 \end{pmatrix}$$

$$T\begin{pmatrix}3\\2\\0\\-2\end{pmatrix}=\begin{pmatrix}1\\-1\\1\\4\end{pmatrix},\qquad T\begin{pmatrix}4\\2\\1\\1\end{pmatrix}=\begin{pmatrix}5\\1\\17\\-10\end{pmatrix}$$

a. Verify that the set $\{\mathbf{v}_1, \mathbf{v}_2, \mathbf{v}_3, \mathbf{v}_4\}$ given below is a basis for $\mathbb{R}^4$ and hence that $T$ is well defined.

$$\left\{\begin{pmatrix}1\\0\\3\\-1\end{pmatrix},\begin{pmatrix}2\\-1\\4\\3\end{pmatrix},\begin{pmatrix}3\\2\\0\\-2\end{pmatrix},\begin{pmatrix}4\\2\\1\\1\end{pmatrix}\right\}$$

b. Find the matrix representation, $C$, of $T$ with respect to the standard bases. Recall that you need to find $T(\mathbf{e}_i)$ for $i = 1, \ldots, 4$ and that $T(\mathbf{e}_i)$ is a linear combination of $\{T(\mathbf{v}_1), \ldots, T(\mathbf{v}_4)\}$, where the coefficients of the linear combinations are the coordinates of $\mathbf{e}_i$ with respect to the basis $\{\mathbf{v}_1, \mathbf{v}_2, \mathbf{v}_3, \mathbf{v}_4\}$.

c. Let $A$ be the matrix $[\mathbf{v}_1 \quad \mathbf{v}_2 \quad \mathbf{v}_3 \quad \mathbf{v}_4]$ and let $B$ be the matrix whose columns are the right-hand sides of the equalities in the definition of $T$; that is,

$$B=\begin{pmatrix}3&2&1&5\\-1&0&-1&1\\7&6&1&17\\2&-2&4&-10\end{pmatrix}.$$

Verify that the matrix representation, $C$, of the transformation $T$ satisfies $C = BA^{-1}$. Explain why this is true using the ideas of coordinates and transition matrices.

d. Using $C$, find a basis for the kernel and range of $T$.

7. Let $T:\mathbb{R}^2 \to \mathbb{R}^2$ be the transformation defined by first rotating *clockwise* around the origin by $\pi/4$, then expanding along the $x$-axis by a factor of 2 and expanding along the $y$-axis by a factor of 3, and then rotating *counterclockwise* around the origin by $\pi/4$.

a. Find the matrix representation of $T$ with respect to the standard basis.

b. Find the matrix representation of $T$ with respect to the basis

$$B=\left\{\begin{pmatrix}1\\1\end{pmatrix},\begin{pmatrix}-1\\1\end{pmatrix}\right\}$$

c. Explain how the geometry of $T$ can be described solely in terms of expansions in certain directions.

## 5.4 ISOMORPHISMS

In this section we introduce some important terminology and then prove a theorem which says that all $n$-dimensional vector spaces are "essentially" the same.

**DEFINITION 1** **One-to-One Transformation** Let $T: V \to W$ be a linear transformation. Then $T$ is **one-to-one,** written 1–1, if

$$T\mathbf{v}_1 = T\mathbf{v}_2 \quad \text{implies that} \quad \mathbf{v}_1 = \mathbf{v}_2 \tag{1}$$

That is, $T$ is 1–1 if and only if every vector $\mathbf{w}$ in the range of $T$ is the image of exactly one vector in $V$.

*Note.* A 1–1 transformation is also called an **injection.**

**THEOREM 1** Let $T: V \to W$ be a linear transformation. Then $T$ is 1–1 if and only if $\ker T = \{\mathbf{0}\}$.

**Proof** Suppose $\ker T = \{\mathbf{0}\}$ and $T\mathbf{v}_1 = T\mathbf{v}_2$. Then $T\mathbf{v}_1 - T\mathbf{v}_2 = T(\mathbf{v}_1 - \mathbf{v}_2) = \mathbf{0}$, which means that $(\mathbf{v}_1 - \mathbf{v}_2) \in \ker T = \{\mathbf{0}\}$. Thus $\mathbf{v}_1 - \mathbf{v}_2 = \mathbf{0}$, so $\mathbf{v}_1 = \mathbf{v}_2$, which shows that $T$ is 1–1. We now prove that if $T$ is 1–1, then $\ker T = \{\mathbf{0}\}$. Suppose that $T$ is 1–1 and $\mathbf{v} \in \ker T$. Then $T\mathbf{v} = \mathbf{0}$. But $T\mathbf{0} = \mathbf{0}$ also. Thus since $T$ is 1–1, $\mathbf{v} = \mathbf{0}$. This completes the proof. ■

**EXAMPLE 1** **A 1–1 Transformation from $\mathbb{R}^2$ to $\mathbb{R}^2$** Let $T: \mathbb{R}^2 \to \mathbb{R}^2$ be defined by $T\begin{pmatrix} x \\ y \end{pmatrix} = \begin{pmatrix} x - y \\ 2x + y \end{pmatrix}$. We easily find $A_T = \begin{pmatrix} 1 & -1 \\ 2 & 1 \end{pmatrix}$ and $\rho(A_T) = 2$; hence $\nu(A_T) = 0$ and $N_{A_T} = \ker T = \{\mathbf{0}\}$. Thus $T$ is 1–1. ■

**EXAMPLE 2** **A Transformation from $\mathbb{R}^2$ to $\mathbb{R}^2$ That Is Not 1–1** Let $T: \mathbb{R}^2 \to \mathbb{R}^2$ be defined by $T\begin{pmatrix} x \\ y \end{pmatrix} = \begin{pmatrix} x - y \\ 2x - 2y \end{pmatrix}$. Then $A_T = \begin{pmatrix} 1 & -1 \\ 2 & -2 \end{pmatrix}$, $\rho(A_T) = 1$, and $\nu(A_T) = 1$; hence $\nu(T) = 1$ and $T$ is not 1–1. Note, for example, that $T\begin{pmatrix} 1 \\ 1 \end{pmatrix} = \mathbf{0} = T\begin{pmatrix} 0 \\ 0 \end{pmatrix}$. ■

**DEFINITION 2** **Onto Transformation** Let $T: V \to W$ be a linear transformation. Then $T$ is said to be **onto** $W$ or simply **onto,** if for every $\mathbf{w} \in W$ there is at least one $\mathbf{v} \in V$ such that $T\mathbf{v} = \mathbf{w}$. That is, *$T$ is onto $W$ if and only if range $T = W$.*

*Note.* An onto transformation is also called a **surjection.**

**EXAMPLE 3** **Determining Whether a Transformation Is Onto** In Example 1, $\rho(A_T) = 2$; hence range $T = \mathbb{R}^2$ and $T$ is onto. In Example 2, $\rho(A_T) = 1$ and range $T = \text{span}\left\{\begin{pmatrix} 1 \\ 2 \end{pmatrix}\right\} \neq \mathbb{R}^2$; hence $T$ is not onto. ■

**THEOREM 2** Let $T$: $V \to W$ be a linear transformation and suppose that $\dim V = \dim W = n$.

**i.** If $T$ is 1–1, then $T$ is onto.

**ii.** If $T$ is onto, then $T$ is 1–1.

**Proof** Let $A_T$ be a matrix representation of $T$. Then if $T$ is 1–1, $\ker T = \{\mathbf{0}\}$ and $\nu(A_T) = 0$, which means that $\rho(T) = \rho(A_T) = n - 0 = n$ so that range $T = W$. If $T$ is onto, then $\rho(A_T) = n$ so that $\nu(T) = \nu(A_T) = 0$ and $T$ is 1–1. ■

**THEOREM 3** Let $T$: $V \to W$ be a linear transformation. Suppose that $\dim V = n$ and $\dim W = m$. Then

**i.** If $n > m$, $T$ is not 1–1.

**ii.** If $m > n$, $T$ is not onto.

**Proof**

**i.** Let $\{\mathbf{v}_1, \mathbf{v}_2, \ldots, \mathbf{v}_n\}$ be a basis for $V$. Let $\mathbf{w}_i = T\mathbf{v}_i$ for $i = 1, 2, \ldots, n$ and look at the set $S = \{\mathbf{w}_1, \mathbf{w}_2, \ldots, \mathbf{w}_n\}$. Since $m = \dim W < n$, the set $S$ is linearly dependent. Thus there exist scalars not all zero such that $c_1\mathbf{w}_1 + c_2\mathbf{w}_2 + \cdots + c_n\mathbf{w}_n = \mathbf{0}$. Let $\mathbf{v} = c_1\mathbf{v}_1 + c_2\mathbf{v}_2 + \cdots + c_n\mathbf{v}_n$. Since the $\mathbf{v}_i$'s are linearly independent and since not all the $c_i$'s are zero, we see that $\mathbf{v} \neq \mathbf{0}$. But $T\mathbf{v} = T(c_1\mathbf{v}_1 + c_2\mathbf{v}_2 + \cdots + c_n\mathbf{v}_n) = c_1T\mathbf{v}_1 + c_2T\mathbf{v}_2 + \cdots + c_nT\mathbf{v}_n = c_1\mathbf{w}_1 + c_2\mathbf{w}_2 + \cdots + c_n\mathbf{w}_n = \mathbf{0}$. Thus $\mathbf{v} \in \ker T$ and $\ker T \neq \{\mathbf{0}\}$.

**ii.** If $\mathbf{v} \in V$, then $\mathbf{v} = a_1\mathbf{v}_1 + a_2\mathbf{v}_2 + \cdots + a_n\mathbf{v}_n$ for some scalars $a_1, a_2, \ldots, a_n$ and $T\mathbf{v} = a_1T\mathbf{v}_1 + a_2T\mathbf{v}_2 + \cdots + a_nT\mathbf{v}_n = a_1\mathbf{w}_1 + a_2\mathbf{w}_2 + \cdots + a_n\mathbf{w}_n$. Thus $\{\mathbf{w}_1, \mathbf{w}_2, \ldots, \mathbf{w}_n\} = \{T\mathbf{v}_1, T\mathbf{v}_2, \ldots, T\mathbf{v}_n\}$ spans range $T$. Then from Problem 4.6.29 on page 346, $\rho(T) = \dim \text{range } T \leq n$. Since $m > n$, this shows that range $T \neq W$. Thus $T$ is not onto. ■

**EXAMPLE 4** **A Transformation from $\mathbb{R}^3 \to \mathbb{R}^2$ Is Not 1–1** Let $T$: $\mathbb{R}^3 \to \mathbb{R}^2$ be given by $T\begin{pmatrix} x \\ y \\ z \end{pmatrix} = \begin{pmatrix} 1 & 2 & 3 \\ 4 & 5 & 6 \end{pmatrix}\begin{pmatrix} x \\ y \\ z \end{pmatrix}$. Here $n = 3$ and $m = 2$, so $T$ is not 1–1. To see this, observe that

$$T\begin{pmatrix} -1 \\ 2 \\ 0 \end{pmatrix} = \begin{pmatrix} 1 & 2 & 3 \\ 4 & 5 & 6 \end{pmatrix}\begin{pmatrix} -1 \\ 2 \\ 0 \end{pmatrix} = \begin{pmatrix} 3 \\ 6 \end{pmatrix} \quad \text{and} \quad T\begin{pmatrix} 2 \\ -4 \\ 3 \end{pmatrix} = \begin{pmatrix} 1 & 2 & 3 \\ 4 & 5 & 6 \end{pmatrix}\begin{pmatrix} 2 \\ -4 \\ 3 \end{pmatrix} = \begin{pmatrix} 3 \\ 6 \end{pmatrix}$$

That is, two different vectors in $\mathbb{R}^3$ have the same image in $\mathbb{R}^2$. ■

**EXAMPLE 5** **A Transformation from $\mathbb{R}^2$ to $\mathbb{R}^3$ Is Not Onto** Let $T$: $\mathbb{R}^2 \to \mathbb{R}^3$ be given by $T\begin{pmatrix} x \\ y \end{pmatrix} = \begin{pmatrix} 1 & 2 \\ 3 & 4 \\ 5 & 6 \end{pmatrix}\begin{pmatrix} x \\ y \end{pmatrix}$. Here $n = 2$ and $m = 3$, so $T$ is not onto. To show this, we must find a vector in $\mathbb{R}^3$ that is not in the range of $T$. One such vector is $\begin{pmatrix} 0 \\ 0 \\ 1 \end{pmatrix}$. That is, there is no vector $\mathbf{x} = \begin{pmatrix} x \\ y \end{pmatrix}$ in $\mathbb{R}^2$ such that $T\mathbf{x} = \begin{pmatrix} 0 \\ 0 \\ 1 \end{pmatrix}$. We prove this by assuming that $T\begin{pmatrix} x \\ y \end{pmatrix} = \begin{pmatrix} 0 \\ 0 \\ 1 \end{pmatrix}$. That is,

$$\begin{pmatrix} 1 & 2 \\ 3 & 4 \\ 5 & 6 \end{pmatrix}\begin{pmatrix} x \\ y \end{pmatrix} = \begin{pmatrix} 0 \\ 0 \\ 1 \end{pmatrix} \quad \text{or} \quad \begin{pmatrix} x + 2y \\ 3x + 4y \\ 5x + 6y \end{pmatrix} = \begin{pmatrix} 0 \\ 0 \\ 1 \end{pmatrix}$$

Row-reducing, we have

$$\left(\begin{array}{cc|c} 1 & 2 & 0 \\ 3 & 4 & 0 \\ 5 & 6 & 1 \end{array}\right) \longrightarrow \left(\begin{array}{cc|c} 1 & 2 & 0 \\ 0 & -2 & 0 \\ 0 & -4 & 1 \end{array}\right) \longrightarrow \left(\begin{array}{cc|c} 1 & 2 & 0 \\ 0 & -2 & 0 \\ 0 & 0 & 1 \end{array}\right)$$

The last line reads $0 \cdot x + 0 \cdot y = 1$, so the system is inconsistent and $\begin{pmatrix} 0 \\ 0 \\ 1 \end{pmatrix}$ is not in the range of $T$.

**DEFINITION 3** **Isomorphism** Let $T$: $V \to W$ be a linear transformation. Then $T$ is an **isomorphism** if $T$ is 1–1 and onto.

*Note.* A transformation that is both 1–1 and onto is called a **bijection.**

**DEFINITION 4** **Isomorphic Vector Spaces** The vector spaces $V$ and $W$ are said to be **isomorphic** if there exists an isomorphism $T$ from $V$ onto $W$. In this case we write $V \cong W$.

*Remark.* The word "isomorphism" comes from the Greek *isomorphos* meaning "of equal form" (*iso* = equal; *morphos* = form). After a few examples we shall see how closely related are the "forms" of isomorphic vector spaces.

Let $T$: $\mathbb{R}^n \to \mathbb{R}^n$ and let $A_T$ be the matrix representation of $T$. Now $T$ is 1–1 if and only if $\ker T = \{\mathbf{0}\}$, which is true if and only if $\nu(A_T) = 0$ if and only if $\det A_T \neq 0$. Thus we can extend our Summing Up Theorem (last seen on page 360) in another direction.

**THEOREM 4** **Summing Up Theorem—View 7** Let $A$ be an $n \times n$ matrix. Then the following 11 statements are equivalent; that is, each one implies the other 10 (so that if one is true, all are true):

**i.** $A$ is invertible.

**ii.** The only solution to the homogeneous system $A\mathbf{x} = \mathbf{0}$ is the trivial solution ($\mathbf{x} = \mathbf{0}$).

**iii.** The system $A\mathbf{x} = \mathbf{b}$ has a unique solution for every $n$-vector $\mathbf{b}$.

**iv.** $A$ is row equivalent to the $n \times n$ identity matrix $I_n$.

**v.** $A$ can be written as the product of elementary matrices.

**vi.** The row echelon form of $A$ has $n$ pivots.

**vii.** The rows (and columns) of $A$ are linearly independent.

**viii.** $\det A \neq 0$.

**ix.** $\nu(A) = 0$.

**x.** $\rho(A) = n$.

**xi.** The linear transformation $T$ from $\mathbb{R}^n$ to $\mathbb{R}^n$ defined by $T\mathbf{x} = A\mathbf{x}$ is an isomorphism.

We now look at some examples of isomorphisms between other pairs of vector spaces.

**EXAMPLE 6** **An Isomorphism Between $\mathbb{R}^3$ and $P_2$** Let $T$: $\mathbb{R}^3 \to P_2$ be defined by $T\begin{pmatrix} a \\ b \\ c \end{pmatrix} = a + bx + cx^2$. It is easy to verify that $T$ is linear. Suppose that $T\begin{pmatrix} a \\ b \\ c \end{pmatrix} = \mathbf{0} = 0 + 0x + 0x^2$. Then $a = b = c = 0$. That is, $\ker T = \{\mathbf{0}\}$ and $T$ is 1–1. If $p(x) = a_0 + a_1x + a_2x^2$, then $p(x) = T\begin{pmatrix} a_0 \\ a_1 \\ a_2 \end{pmatrix}$. This means that range $T = P_2$ and $T$ is onto. Thus $\mathbb{R}^3 \cong P_2$.

*Note.* $\dim \mathbb{R}^3 = \dim P_2 = 3$. Thus by Theorem 2, once we know that $T$ is 1–1, we also know that it is onto. We verified that it was onto; but it was unnecessary to do so.

**EXAMPLE 7** Calculus **An Isomorphism Between Two Infinite Dimensional Vector Spaces** Let $V = \{f \in C^1[0, 1]\colon f(0) = 0\}$ and $W = C[0, 1]$. Let $D$: $V \to W$ be given by $Df = f'$. Suppose that $Df = Dg$. Then $f' = g'$ or $(f - g)' = 0$ and $f(x) - g(x) = c$, a constant. But $f(0) = g(0) = 0$, so $c = 0$ and $f = g$. Thus $D$ is 1–1. Let $g \in C[0, 1]$ and let $f(x) = \int_0^x g(t)\,dt$. Then from the fundamental theorem of calculus, $f \in C^1[0, 1]$ and $f'(x) = g(x)$ for every $x \in [0, 1]$. Moreover, since $\int_0^0 g(t)\,dt = 0$, we have $f(0) = 0$. Thus for every $g$ in $W$ there is an $f \in V$ such that $Df = g$. Hence $D$ is onto and we have shown that $V \cong W$.

The following theorem illustrates the similarity of two isomorphic vector spaces.

**THEOREM 5** Let $T\colon V \to W$ be an isomorphism:

**i.** If $\mathbf{v}_1, \mathbf{v}_2, \ldots, \mathbf{v}_n$ span $V$, then $T\mathbf{v}_1, T\mathbf{v}_2, \ldots, T\mathbf{v}_n$ span $W$.

**ii.** If $\mathbf{v}_1, \mathbf{v}_2, \ldots, \mathbf{v}_n$ are linearly independent in $V$, then $T\mathbf{v}_1, T\mathbf{v}_2, \ldots, T\mathbf{v}_n$ are linearly independent in $W$.

**iii.** If $\{\mathbf{v}_1, \mathbf{v}_2, \ldots, \mathbf{v}_n\}$ is a basis in $V$, then $\{T\mathbf{v}_1, T\mathbf{v}_2, \ldots, T\mathbf{v}_n\}$ is a basis in $W$.

**iv.** If $V$ is finite dimensional, then $W$ is finite dimensional and $\dim V = \dim W$.

**Proof**

**i.** Let $\mathbf{w} \in W$. Then since $T$ is onto, there is a $\mathbf{v} \in V$ such that $T\mathbf{v} = \mathbf{w}$. Since the $\mathbf{v}_i$'s span $V$, we can write $\mathbf{v} = a_1\mathbf{v}_1 + a_2\mathbf{v}_2 + \cdots + a_n\mathbf{v}_n$ so that $\mathbf{w} = T\mathbf{v} = a_1T\mathbf{v}_1 + a_2T\mathbf{v}_2 + \cdots + a_nT\mathbf{v}_n$, and this shows that $\{T\mathbf{v}_1, T\mathbf{v}_2, \ldots, T\mathbf{v}_n\}$ spans $W$.

**ii.** Suppose $c_1T\mathbf{v}_1 + c_2T\mathbf{v}_2 + \cdots + c_nT\mathbf{v}_n = \mathbf{0}$. Then $T(c_1\mathbf{v}_1 + c_2\mathbf{v}_2 + \cdots + c_n\mathbf{v}_n) = \mathbf{0}$. Thus since $T$ is 1–1, $c_1\mathbf{v}_1 + c_2\mathbf{v}_2 + \cdots + c_n\mathbf{v}_n = \mathbf{0}$, which implies that $c_1 = c_2 = \cdots = c_n = 0$ since the $\mathbf{v}_i$'s are independent.

**iii.** This follows from parts (*i*) and (*ii*).

**iv.** This follows from part (*iii*).

In general, it is difficult to show that two infinite dimensional vector spaces are isomorphic. For finite dimensional spaces, however, it is remarkably easy. Theorem 3 shows that if $\dim V \neq \dim W$, then $V$ and $W$ are not isomorphic. The next theorem shows that if $\dim V = \dim W$, and if $V$ and $W$ are real vector spaces, then $V$ and $W$ are isomorphic. That is,

> Two real finite dimensional spaces of the same dimension are isomorphic.

**THEOREM 6** Let $V$ and $W$ be two real† finite dimensional vector spaces with $\dim V = \dim W$. Then $V \cong W$.

**Proof** Let $\{\mathbf{v}_1, \mathbf{v}_2, \ldots, \mathbf{v}_n\}$ be a basis for $V$ and let $\{\mathbf{w}_1, \mathbf{w}_2, \ldots, \mathbf{w}_n\}$ be a basis for $W$. Define the linear transformation $T$ by

$$T\mathbf{v}_i = \mathbf{w}_i \qquad \text{for } i = 1, 2, \ldots, n \tag{2}$$

By Theorem 5.2.2 on page 478 there is exactly one linear transformation that satisfies equation (2). Suppose $\mathbf{v} \in V$ and $T\mathbf{v} = \mathbf{0}$. Then if $\mathbf{v} = c_1\mathbf{v}_1 + c_2\mathbf{v}_2 + \cdots + c_n\mathbf{v}_n$, we have $T\mathbf{v} = c_1T\mathbf{v}_1 + \cdots + c_nT\mathbf{v}_n = c_1\mathbf{w}_1 + c_2\mathbf{w}_2 + \cdots + c_n\mathbf{w}_n = \mathbf{0}$. But since $\mathbf{w}_1, \mathbf{w}_2, \ldots, \mathbf{w}_n$ are linearly independent, $c_1 = c_2 = \cdots = c_n = 0$. Thus $\mathbf{v} = \mathbf{0}$ and $T$ is 1–1. Since $V$ and $W$ are finite dimensional and $\dim V = \dim W$, $T$ is onto by Theorem 2 and the proof is complete. ■

This last result is one of the central results of linear algebra. It says that if you know one real $n$-dimensional vector space, you know all real vector spaces of dimension $n$. That is, if one associates all isomorphic vector spaces, then $\mathbb{R}^n$ is the only $n$-dimensional vector space over the reals.

# PROBLEMS 5.4

## Self-Quiz

### True-False

**I.** A linear transformation from $\mathbb{R}^n \to \mathbb{R}^m$ with $n \neq m$ cannot be both 1–1 and onto.

**II.** If $\dim V = 5$ and $\dim W = 7$, it may be possible to find an isomorphism $T$ from $V$ onto $W$.

**III.** If $T$ is 1–1, then $\ker T = \{\mathbf{0}\}$.

**IV.** If $T$ is an isomorphism from a vector space $V$ into $\mathbb{R}^6$, then $\rho(T) = 6$.

**V.** If $A_T$ is the transformation matrix of an isomorphism from $\mathbb{R}^8$ into $\mathbb{R}^8$, then $\det A_T \neq 0$.

### Answers to Self-Quiz

**I.** True **II.** False **III.** True **IV.** True **V.** True

† We need the word "real" here because it is important that the sets of scalars in $V$ and $W$ be the same. Otherwise the condition $T(\alpha\mathbf{v}) = \alpha T\mathbf{v}$ might not hold because $\mathbf{v} \in V$, $T\mathbf{v} \in W$, and either $\alpha\mathbf{v}$ or $\alpha T\mathbf{v}$ might not be defined. Theorem 6 is true if the word "real" is omitted and, instead, we impose the conditions that $V$ and $W$ be defined with the same set of scalars (like $\mathbb{C}$, for example).

**1.** Show that $T$: $M_{mn} \to M_{nm}$ defined by $TA = A^t$ is an isomorphism.

**2.** Show that $T$: $\mathbb{R}^n \to \mathbb{R}^n$ is an isomorphism if and only if $A_T$ is invertible.

* **3.** Let $V$ and $W$ be $n$-dimensional real vector spaces and let $B_1$ and $B_2$ be bases for $V$ and $W$, respectively. Let $A_T$ be the transformation matrix relative to the bases $B_1$ and $B_2$. Show that $T$: $V \to W$ is an isomorphism if and only if $\det A_T \neq 0$.

**4.** Find an isomorphism between $D_n$, the $n \times n$ diagonal matrices with real entries, and $\mathbb{R}^n$. [*Hint:* Look first at the case $n = 2$.]

**5.** For what value of $m$ is the set of $n \times n$ symmetric matrices isomorphic to $\mathbb{R}^m$?

**6.** Show that the set of $n \times n$ symmetric matrices is isomorphic to the set of $n \times n$ upper triangular matrices.

**7.** Let $V = P_4$ and $W = \{p \in P_5\colon p(0) = 0\}$. Show that $V \cong W$.

Calculus **8.** Define $T$: $P_n \to P_n$ by $Tp = p + p'$. Show that $T$ is an isomorphism.

**9.** Find a condition on the numbers $m$, $n$, $p$, $q$ such that $M_{mn} \cong M_{pq}$.

**10.** Show that $D_n \cong P_{n-1}$.

**11.** Prove that any two finite dimensional complex vector spaces $V$ and $W$ with $\dim V = \dim W$ are isomorphic.

**12.** Define $T$: $C[0, 1] \to C[3, 4]$ by $Tf(x) = f(x - 3)$. Show that $T$ is an isomorphism.

**13.** Let $B$ be an invertible $n \times n$ matrix. Show that $T$: $M_{nn} \to M_{nn}$ defined by $TA = AB$ is an isomorphism.

Calculus **14.** Show that the transformation $Tp(x) = xp'(x)$ is not an isomorphism from $P_n$ into $P_n$.

**15.** Let $H$ be a subspace of the finite dimensional inner product space $V$. Show that $T$: $V \to H$ defined by $T\mathbf{v} = \text{proj}_H \mathbf{v}$ is onto. Under what circumstances will it be 1–1?

**16.** Show that if $T$: $V \to W$ is an isomorphism, then there exists an isomorphism $S$: $W \to V$ such that $S(T\mathbf{v}) = \mathbf{v}$. Here $S$ is called the **inverse transformation** of $T$ and is denoted by $T^{-1}$.

**17.** Show that if $T$: $\mathbb{R}^n \to \mathbb{R}^n$ is defined by $T\mathbf{x} = A\mathbf{x}$ and if $T$ is an isomorphism, then $A$ is invertible and the inverse transformation $T^{-1}$ is given by $T^{-1}\mathbf{x} = A^{-1}\mathbf{x}$.

**18.** Find $T^{-1}$ for the isomorphism of Problem 7.

***19.** Consider the space $C = \{z = a + ib$, where $a$ and $b$ are real numbers and $i^2 = -1\}$. Show that if the scalars are taken to be the reals, then $C \cong \mathbb{R}^2$.

***20.** Consider the space $\mathbb{C}^n_{\mathbb{R}} = \{(\mathbf{c}_1, \mathbf{c}_2, \ldots, \mathbf{c}_n)\colon \mathbf{c}_i \in C$ and the scalars are the reals$\}$. Show that $\mathbb{C}^n_{\mathbb{R}} \cong \mathbb{R}^{2n}$. [*Hint:* See Problem 19.]

# MATLAB 5.4

1. Let $T$:$\mathbb{R}^4 \to \mathbb{R}^4$ be the transformation defined by $T(\mathbf{v}_i) = \mathbf{w}_i$ for $i = 1, \ldots, 4$, where

$$\{\mathbf{v}_1, \mathbf{v}_2, \mathbf{v}_3, \mathbf{v}_4\} = \left\{\begin{pmatrix}1\\0\\0\\0\end{pmatrix}, \begin{pmatrix}2\\1\\0\\0\end{pmatrix}, \begin{pmatrix}-2\\1\\2\\0\end{pmatrix}, \begin{pmatrix}3\\4\\2\\1\end{pmatrix}\right\}$$

$$\{\mathbf{w}_1, \mathbf{w}_2, \mathbf{w}_3, \mathbf{w}_4\} = \left\{ \begin{pmatrix} 1 \\ 2 \\ 1 \\ 0 \end{pmatrix}, \begin{pmatrix} 2 \\ 5 \\ 3 \\ 2 \end{pmatrix}, \begin{pmatrix} -1 \\ -1 \\ -1 \\ 2 \end{pmatrix}, \begin{pmatrix} 0 \\ 3 \\ 7 \\ 7 \end{pmatrix} \right\}$$

a. Verify that the set $\{\mathbf{v}_1, \mathbf{v}_2, \mathbf{v}_3, \mathbf{v}_4\}$ is a basis for $\mathbb{R}^4$, and therefore that $T$ is well-defined.

b. Verify that the set $\{\mathbf{w}_1, \mathbf{w}_2, \mathbf{w}_3, \mathbf{w}_4\}$ is a basis for $\mathbb{R}^4$. Why can you conclude that $T$ is an isomorphism?

c. Find the matrix representation, $A$, for $T$ with respect to the standard basis. (See MATLAB Problem 6 in Section 5.3.) Use this matrix representation to find a basis for the kernel and range of $T$, and thus verify that $T$ is an isomorphism. Verify that $A$ is invertible.

d. Suppose $S:\mathbb{R}^4 \to \mathbb{R}^4$ is the transformation defined by $S(\mathbf{w}_i) = \mathbf{v}_i$ for $i = 1, \ldots, 4$. Find the matrix representation, $B$, for $S$ and verify that $B = A^{-1}$.

## 5.5 ISOMETRIES

In this section we describe a special kind of linear transformation between vector spaces. We begin with a very useful result.

**THEOREM 1** Let $A$ be an $m \times n$ matrix with real entries.† Then for any vectors $\mathbf{x} \in \mathbb{R}^n$ and $\mathbf{y} \in \mathbb{R}^m$:

$$(A\mathbf{x}) \cdot \mathbf{y} = \mathbf{x} \cdot (A^t\mathbf{y}) \tag{1}$$

**Proof**

Equation (6) on p. 124; Theorem 1(*ii*) on p. 122; Associative law for matrix multiplication

$$A\mathbf{x} \cdot \mathbf{y} = (A\mathbf{x})^t\mathbf{y} = (\mathbf{x}^tA^t)\mathbf{y} = \mathbf{x}^t(A^t\mathbf{y})$$

Equation (6) on p. 124

$$= \mathbf{x} \cdot (A^t\mathbf{y})$$

■

Recall from Section 4.9, page 399, that a matrix $Q$ with real entries is **orthogonal** if $Q$ is invertible and $Q^{-1} = Q^t$. In Theorem 4.9.3 on page 399 we proved that $Q$ is orthogonal if and only if the columns of $Q$ form an orthonormal basis for $\mathbb{R}^n$. Now

† This result can easily be extended to matrices with complex components. See Problem 21.

let $Q$ be an $n \times n$ orthogonal matrix and let $T$: $\mathbb{R}^n \rightarrow \mathbb{R}^n$ be the linear transformation defined by $T\mathbf{x} = Q\mathbf{x}$. Then, using equation (1), we compute

$$(T\mathbf{x} \cdot T\mathbf{y}) = Q\mathbf{x} \cdot Q\mathbf{y} = \mathbf{x} \cdot (Q^tQ\mathbf{y}) = \mathbf{x} \cdot (I\mathbf{y}) = \mathbf{x} \cdot \mathbf{y}$$

In particular, if $\mathbf{x} = \mathbf{y}$, we see that $T\mathbf{x} \cdot T\mathbf{x} = \mathbf{x} \cdot \mathbf{x}$ or

$$|T\mathbf{x}| = |\mathbf{x}|$$

for every $\mathbf{x}$ in $\mathbb{R}^n$.

**DEFINITION 1** **Isometry** A linear transformation $T$: $\mathbb{R}^n \rightarrow \mathbb{R}^n$ is called an **isometry** if for every $\mathbf{x}$ in $\mathbb{R}^n$,

$$\boxed{|T\mathbf{x}| = |\mathbf{x}|} \qquad (2)$$

Because of equation (2) we can say: An isometry in $\mathbb{R}^n$ is a linear transformation that preserves length in $\mathbb{R}^n$. Note that (2) implies that

$$\boxed{|T\mathbf{x} - T\mathbf{y}| = |\mathbf{x} - \mathbf{y}|} \qquad (3)$$

[since $T\mathbf{x} - T\mathbf{y} = T(\mathbf{x} - \mathbf{y})$].

**THEOREM 2** Let $T$ be an isometry from $\mathbb{R}^n \rightarrow \mathbb{R}^n$ and suppose that $\mathbf{x}$ and $\mathbf{y}$ are in $\mathbb{R}^n$. Then

$$\boxed{T\mathbf{x} \cdot T\mathbf{y} = \mathbf{x} \cdot \mathbf{y}} \qquad (4)$$

That is, an isometry in $\mathbb{R}^n$ preserves the scalar product.

**Proof**

$$|T\mathbf{x} - T\mathbf{y}|^2 = (T\mathbf{x} - T\mathbf{y}) \cdot (T\mathbf{x} - T\mathbf{y}) = |T\mathbf{x}|^2 - 2T\mathbf{x} \cdot T\mathbf{y} + |T\mathbf{y}|^2 \qquad (5)$$

$$|\mathbf{x} - \mathbf{y}|^2 = (\mathbf{x} - \mathbf{y}) \cdot (\mathbf{x} - \mathbf{y}) = |\mathbf{x}|^2 - 2\mathbf{x} \cdot \mathbf{y} + |\mathbf{y}|^2 \qquad (6)$$

Since $|T\mathbf{x} - T\mathbf{y}|^2 = |\mathbf{x} - \mathbf{y}|^2$, $|T\mathbf{x}|^2 = |\mathbf{x}|^2$ and $|T\mathbf{y}|^2 = |\mathbf{y}|^2$, equations (5) and (6) show that

$$-2T\mathbf{x} \cdot T\mathbf{y} = -2\mathbf{x} \cdot \mathbf{y} \quad \text{or} \quad T\mathbf{x} \cdot T\mathbf{y} = \mathbf{x} \cdot \mathbf{y}$$ ■

In the derivation of equation (2) we showed that if the matrix representation of $T$ is an orthogonal matrix, then $T$ is an isometry. Conversely, suppose that $T$ is an isometry. If $A$ is the matrix representation of $T$, then for any $\mathbf{x}$ and $\mathbf{y}$ in $\mathbb{R}^n$

$$\mathbf{x}\cdot\mathbf{y} \overset{\text{from (4)}}{=} T\mathbf{x}\cdot T\mathbf{y} = A\mathbf{x}\cdot A\mathbf{y} \overset{\text{from (1)}}{=} \mathbf{x}\cdot A^tA\mathbf{y}$$

$$\mathbf{x}\cdot\mathbf{y} - \mathbf{x}\cdot A^tA\mathbf{y} = 0 \quad \text{or} \quad \mathbf{x}\cdot(\mathbf{y} - A^tA\mathbf{y}) = 0$$

Thus (see page 402)

$$\mathbf{y} - A^tA\mathbf{y} \in (\mathbb{R}^n)^\perp = \{\mathbf{0}\}$$

We see that for every $\mathbf{y} \in \mathbb{R}^n$

$$\mathbf{y} = A^tA\mathbf{y} \tag{7}$$

This implies that $A^tA = I$, so $A$ is orthogonal.

We have proved the following theorem.

**THEOREM 3** A linear transformation $T$: $\mathbb{R}^n \to \mathbb{R}^n$ is an isometry if and only if the matrix representation of $T$ is orthogonal.

## Isometries of $\mathbb{R}^2$

Let $T$ be an isometry from $\mathbb{R}^2$ to $\mathbb{R}^2$. Let

$$\mathbf{u}_1 = T\begin{pmatrix}1\\0\end{pmatrix} \quad \text{and} \quad \mathbf{u}_2 = T\begin{pmatrix}0\\1\end{pmatrix}$$

Then $\mathbf{u}_1$ and $\mathbf{u}_2$ are unit vectors [because of (2)] and

$$\mathbf{u}_1\cdot\mathbf{u}_2 \overset{\text{from (4)}}{=} \begin{pmatrix}1\\0\end{pmatrix}\cdot\begin{pmatrix}0\\1\end{pmatrix} = 0$$

Thus $\mathbf{u}_1$ and $\mathbf{u}_2$ are orthogonal. From equation (3.1.7) on page 235 there exists a number $\theta$, with $0 \le \theta < 2\pi$ such that

$$\mathbf{u}_1 = \begin{pmatrix}\cos\theta\\ \sin\theta\end{pmatrix}$$

Since $\mathbf{u}_1$ and $\mathbf{u}_2$ are orthogonal,

$$\text{Direction of } \mathbf{u}_2 = \text{direction of } \mathbf{u}_1 \pm \frac{\pi}{2}$$

In the first case

$$\mathbf{u}_2 = \begin{pmatrix} \cos\left(\theta + \dfrac{\pi}{2}\right) \\ \sin\left(\theta + \dfrac{\pi}{2}\right) \end{pmatrix} = \begin{pmatrix} -\sin\theta \\ \cos\theta \end{pmatrix}$$

In the second case

$$\mathbf{u}_2 = \begin{pmatrix} \cos\left(\theta - \dfrac{\pi}{2}\right) \\ \sin\left(\theta - \dfrac{\pi}{2}\right) \end{pmatrix} = \begin{pmatrix} \sin\theta \\ -\cos\theta \end{pmatrix}$$

So the matrix representation of $T$ is either

$$Q_1 = \begin{pmatrix} \cos\theta & -\sin\theta \\ \sin\theta & \cos\theta \end{pmatrix} \quad \text{or} \quad Q_2 = \begin{pmatrix} \cos\theta & \sin\theta \\ \sin\theta & -\cos\theta \end{pmatrix}$$

From Example 5.1.8 on page 469, we see that $Q_1$ is the matrix representation of a rotation transformation (counterclockwise through an angle of $\theta$). Now as is easily verified,

$$\begin{pmatrix} \cos\theta & \sin\theta \\ \sin\theta & -\cos\theta \end{pmatrix} = \begin{pmatrix} \cos\theta & -\sin\theta \\ \sin\theta & \cos\theta \end{pmatrix} \begin{pmatrix} 1 & 0 \\ 0 & -1 \end{pmatrix}$$

But the transformation $T\colon \mathbb{R}^2 \to \mathbb{R}^2$ given by

$$T\begin{pmatrix} x \\ y \end{pmatrix} = \begin{pmatrix} 1 & 0 \\ 0 & -1 \end{pmatrix}\begin{pmatrix} x \\ y \end{pmatrix} = \begin{pmatrix} x \\ -y \end{pmatrix}$$

is a reflection of $\begin{pmatrix} x \\ y \end{pmatrix}$ about the $x$-axis (see Example 5.1.1. on page 465). Thus we have the following theorem.

**THEOREM 4** Let $T\colon \mathbb{R}^2 \to \mathbb{R}^2$ be an isometry. Then $T$ is either

**i.** a rotation transformation

or

**ii.** a reflection about the $x$-axis followed by a rotation transformation. ■

Isometries have some interesting properties.

**THEOREM 5** Let $T$: $\mathbb{R}^n \to \mathbb{R}^n$ be an isometry. Then

**i.** If $\mathbf{u}_1, \mathbf{u}_2, \ldots, \mathbf{u}_n$ is an orthogonal set, then $T\mathbf{u}_1, T\mathbf{u}_2, \ldots, T\mathbf{u}_n$ is an orthogonal set.

**ii.** $T$ is an isomorphism.

**Proof** **i.** If $i \neq j$ and $\mathbf{u}_i \cdot \mathbf{u}_j = 0$, then $(T\mathbf{u}_i) \cdot (T\mathbf{u}_j) = \mathbf{u}_i \cdot \mathbf{u}_j = 0$, which proves $(i)$.

**ii.** Let $\mathbf{u}_1, \mathbf{u}_2, \ldots, \mathbf{u}_n$ be an orthonormal basis for $\mathbb{R}^n$. Then by part $(i)$ and the fact that $|T\mathbf{u}_i| = |\mathbf{u}_i| = 1$, we find that $T\mathbf{u}_1, T\mathbf{u}_2, \ldots, T\mathbf{u}_n$ is an orthonormal set in $\mathbb{R}^n$. By Theorem 4.9.1 on page 394 these vectors are linearly independent and hence form a basis for $\mathbb{R}^n$. Thus range $T = \mathbb{R}^n$, which proves that ker $T = \{\mathbf{0}\}$ [since $\nu(T) + \rho(T) = n$]. ■

We conclude this section by outlining how we can extend the notion of isometry to an arbitrary inner product space. Recall from page 441 that in an inner product space $V$

$$\|\mathbf{v}\| = (\mathbf{v}, \mathbf{v})^{1/2}$$

(Recall that in order to avoid confusion, we use double bars to denote a norm.)

**DEFINITION 2** **Isometry** Let $V$ and $W$ be real (or complex) inner product spaces and let $T$: $V \to W$ be a linear transformation. Then $T$ is an **isometry** if for every $\mathbf{v} \in V$

$$\|\mathbf{v}\|_V = \|T\mathbf{v}\|_W \tag{8}$$

Two facts follow immediately: First, since $T(\mathbf{v}_1 - \mathbf{v}_2) = T\mathbf{v}_1 - T\mathbf{v}_2$, we have for every $\mathbf{v}_1$ and $\mathbf{v}_2$ in $V$

$$\|T\mathbf{v}_1 - T\mathbf{v}_2\|_W = \|\mathbf{v}_1 - \mathbf{v}_2\|_V$$

**THEOREM 6** Let $T$: $V \to W$ be an isometry. Then for every $\mathbf{v}_1$ and $\mathbf{v}_2$ in $V$

$$(T\mathbf{v}_1, T\mathbf{v}_2) = (\mathbf{v}_1, \mathbf{v}_2) \tag{9}$$

That is, an isometry preserves inner products.

The proof of Theorem 6 is identical to the proof of Theorem 2 with inner products in $V$ and $W$ replacing the scalar product in $\mathbb{R}^n$. ■

**DEFINITION 3** **Isometrically Isomorphic Vector Spaces** Two vector spaces $V$ and $W$ with the same set of scalars are said to be **isometrically isomorphic** if there exists a linear transformation $T\colon V \to W$ that is both an isometry and an isomorphism.

**THEOREM 7** Any two $n$-dimensional real inner product spaces are isometrically isomorphic.

**Proof** Let $\{\mathbf{u}_1, \mathbf{u}_2, \ldots, \mathbf{u}_n\}$ and $\{\mathbf{w}_1, \mathbf{w}_2, \ldots, \mathbf{w}_n\}$ be orthonormal bases for $V$ and $W$, respectively. Let $T\colon V \to W$ be the linear transformation defined by $T\mathbf{u}_i = \mathbf{w}_i$, $i = 1, 2, \ldots, n$. If we can show that $T$ is an isometry, then we shall be done since reasoning as in the proof of Theorem 5 shows us that $T$ is also an isomorphism. Let $\mathbf{x}$ and $\mathbf{y}$ be in $V$. Then there exist sets of real numbers $c_1, c_2, \ldots, c_n$ and $d_1, d_2, \ldots, d_n$ such that $\mathbf{x} = c_1\mathbf{u}_1 + c_2\mathbf{u}_2 + \cdots + c_n\mathbf{u}_n$ and $\mathbf{y} = d_1\mathbf{u}_1 + d_2\mathbf{u}_2 + \cdots + d_n\mathbf{u}_n$. Since the $\mathbf{u}_i$'s are orthonormal, $(\mathbf{x}, \mathbf{y}) = ((c_1\mathbf{u}_1 + c_2\mathbf{u}_2 + \cdots + c_n\mathbf{u}_n), (d_1\mathbf{u}_1 + d_2\mathbf{u}_2 + \cdots + d_n\mathbf{u}_n)) = c_1d_1 + c_2d_2 + \cdots + c_nd_n$. Similarly, since $T\mathbf{x} = c_1T\mathbf{u}_1 + c_2T\mathbf{u}_2 + \cdots + c_nT\mathbf{u}_n = c_1\mathbf{w}_1 + c_2\mathbf{w}_2 + \cdots + c_n\mathbf{w}_n$, we obtain $(T\mathbf{x}, T\mathbf{y}) = ((c_1\mathbf{w}_1 + c_2\mathbf{w}_2 + \cdots + c_n\mathbf{w}_n), (d_1\mathbf{w}_1 + d_2\mathbf{w}_2 + \cdots + d_n\mathbf{w}_n)) = c_1d_1 + c_2d_2 + \cdots + c_nd_n$ because the $\mathbf{w}_i$'s are orthonormal. This completes the proof. ∎

**EXAMPLE 1** **An Isometry Between $\mathbb{R}^3$ and $P_2[0, 1]$** We illustrate this theorem by showing that $\mathbb{R}^3$ and $P_2[0, 1]$ are isometrically isomorphic. In $\mathbb{R}^3$ we use the standard basis $\left\{\begin{pmatrix}1\\0\\0\end{pmatrix}, \begin{pmatrix}0\\1\\0\end{pmatrix}, \begin{pmatrix}0\\0\\1\end{pmatrix}\right\}$. In $P_2$ we use the orthonormal basis $\{1, \sqrt{3}(2x - 1), \sqrt{5}(6x^2 - 6x + 1)\}$. (See Example 4.11.8 on page 442.) Let $\mathbf{x} = \begin{pmatrix}a_1\\b_1\\c_1\end{pmatrix}$ and $\mathbf{y} = \begin{pmatrix}a_2\\b_2\\c_2\end{pmatrix}$ be in $\mathbb{R}^3$. Then $(\mathbf{x}, \mathbf{y}) = \mathbf{x} \cdot \mathbf{y} = a_1a_2 + b_1b_2 + c_1c_2$. Recall that in $P_2[0, 1]$ we defined $(p, q) = \int_0^1 p(x)q(x)\,dx$. We define $T\begin{pmatrix}1\\0\\0\end{pmatrix} = 1$, $T\begin{pmatrix}0\\1\\0\end{pmatrix} = \sqrt{3}(2x - 1)$, and $T\begin{pmatrix}0\\0\\1\end{pmatrix} = \sqrt{5}(6x^2 - 6x + 1)$; hence

$$T\begin{pmatrix}a\\b\\c\end{pmatrix} = a + b\sqrt{3}(2x - 1) + c\sqrt{5}(6x^2 - 6x + 1)$$

and

$$
\begin{aligned}
(T\mathbf{x}, T\mathbf{y}) &= \int_0^1 [a_1 + b_1\sqrt{3}(2x - 1) + c_1\sqrt{5}(6x^2 - 6x + 1)] \\
&\quad \times [a_2 + b_2\sqrt{3}(2x - 1) + c_2\sqrt{5}(6x^2 - 6x + 1)]\, dx \\
&= a_1a_2 \int_0^1 dx + \int_0^1 b_1b_2 3(2x - 1)^2\, dx + \int_0^1 c_1c_2[5(6x^2 - 6x + 1)^2]\, dx \\
&\quad + (a_1b_2 + a_2b_1) \int_0^1 \sqrt{3}(2x - 1)\, dx \\
&\quad + (a_1c_2 + a_2c_1) \int_0^1 \sqrt{5}(6x^2 - 6x + 1)\, dx \\
&\quad + (b_1c_2 + b_2c_1) \int_0^1 [\sqrt{3}(2x - 1)][\sqrt{5}(6x^2 - 6x + 1)]\, dx \\
&= a_1a_2 + b_1b_2 + c_1c_2
\end{aligned}
$$

Here we saved time by using the fact that $\{1, \sqrt{3}(2x - 1), \sqrt{5}(6x^2 - 6x + 1)\}$ is an orthonormal set. Thus $T\colon \mathbb{R}^3 \to P_2[0, 1]$ is an isometry.

## PROBLEMS 5.5

### Self-Quiz

#### *True-False*

**I.** The linear transformation $T\colon \mathbb{R}^n \to \mathbb{R}^n$ is an isometry if $\|T\mathbf{x}\| = \|\mathbf{x}\|$ for every $\mathbf{x}$ in $\mathbb{R}$.

**II.** The linear transformation $T\colon \mathbb{R}^n \to \mathbb{R}^n$ is an isometry if the columns of its matrix representation are pairwise orthogonal.

**III.** The linear transformation $T\colon \mathbb{R}^n \to \mathbb{R}^n$ is an isometry if the columns of its matrix representation are pairwise orthogonal and each column has norm 1.

**IV.** If $T\colon \mathbb{R}^2 \to \mathbb{R}^2$ is an isometry, then $T\begin{pmatrix} 3 \\ -2 \end{pmatrix}$ is orthogonal to $T\begin{pmatrix} 2 \\ 3 \end{pmatrix}$.

**V.** If $T\colon \mathbb{R}^n \to \mathbb{R}^n$ is an isomorphism, then $T$ is an isometry.

**VI.** If $T\colon \mathbb{R}^n \to \mathbb{R}^n$ is an isometry, then $T$ is an isomorphism.

---

**Answers to Self-Quiz**

**I.** True **II.** False **III.** True **IV.** True **V.** False **VI.** True

**1.** Show that for any real number $\theta$, the transformation $T$: $\mathbb{R}^3 \to \mathbb{R}^3$ defined by $T\mathbf{x} = A\mathbf{x}$, where

$$A = \begin{pmatrix} \sin\theta & \cos\theta & 0 \\ \cos\theta & -\sin\theta & 0 \\ 0 & 0 & 1 \end{pmatrix}$$

is an isometry.

**2.** Do the same for the transformation $T$, where

$$A = \begin{pmatrix} \cos\theta & 0 & -\sin\theta \\ 0 & 1 & 0 \\ \sin\theta & 0 & \cos\theta \end{pmatrix}$$

**3.** Let $A$ and $B$ be orthogonal $n \times n$ matrices. Show that $T$: $\mathbb{R}^n \to \mathbb{R}^n$ defined by $T\mathbf{x} = AB\mathbf{x}$ is an isometry.

**4.** Find $A_T$ if $T$ is the transformation from $\mathbb{R}^3 \to \mathbb{R}^3$ defined by

$$T\begin{pmatrix} 2/3 \\ 1/3 \\ -2/3 \end{pmatrix} = \begin{pmatrix} 1/\sqrt{2} \\ 1/\sqrt{2} \\ 0 \end{pmatrix} \qquad T\begin{pmatrix} 1/3 \\ 2/3 \\ 2/3 \end{pmatrix} = \begin{pmatrix} -1/\sqrt{6} \\ 1/\sqrt{6} \\ 2/\sqrt{6} \end{pmatrix} \qquad T\begin{pmatrix} 2/3 \\ -2/3 \\ 1/3 \end{pmatrix} = \begin{pmatrix} 1/\sqrt{3} \\ -1/\sqrt{3} \\ 1/\sqrt{3} \end{pmatrix}$$

Show that $A_T$ is orthogonal.

**5.** Prove Theorem 6.

**6.** Let $T$: $\mathbb{R}^2 \to \mathbb{R}^2$ be an isometry. Show that $T$ preserves angles. That is, (angle between $\mathbf{x}$ and $\mathbf{y}$) = (angle between $T\mathbf{x}$ and $T\mathbf{y}$).

**7.** Give an example of a linear transformation from $\mathbb{R}^2$ onto $\mathbb{R}^2$ that preserves angles and is *not* an isometry.

**8.** For $\mathbf{x}, \mathbf{y} \in \mathbb{R}^n$ and $\mathbf{x}$ and $\mathbf{y} \neq \mathbf{0}$, define: (angle between $\mathbf{x}$ and $\mathbf{y}$) = $\measuredangle(\mathbf{x}, \mathbf{y}) = \cos^{-1}[(\mathbf{x} \cdot \mathbf{y})/|\mathbf{x}|\,|\mathbf{y}|]$. Show that if $T$: $\mathbb{R}^n \to \mathbb{R}^n$ is an isometry, then $T$ preserves angles.

**9.** Let $T$: $\mathbb{R}^n \to \mathbb{R}^n$ be an isometry and let $T\mathbf{x} = A\mathbf{x}$. Show that $S\mathbf{x} = A^{-1}\mathbf{x}$ is an isometry.

In Problems 10–14 find an isometry between the given pair of spaces.

Calculus **10.** $P_1[-1, 1]$, $\mathbb{R}^2$

*Calculus **11.** $P_3[-1, 1]$, $\mathbb{R}^4$

***12.** $M_{22}$, $\mathbb{R}^4$

*Calculus **13.** $M_{22}$, $P_3[-1, 1]$

**14.** $D_n$ and $\mathbb{R}^n$ ($D_n$ = set of diagonal $n \times n$ matrices)

**15.** Let $A$ be an $n \times n$ matrix with complex components. Then the **conjugate transpose** of $A$, denoted by $A^*$, is defined by $(A^*)_{ij} = \overline{a_{ji}}$. Compute $A^*$ if $A = \begin{pmatrix} 1+i & -4+2i \\ 3 & 6-3i \end{pmatrix}$.

**16.** The $n \times n$ complex matrix $A$ is called **Hermitian**† if $A^* = A$. Show that the matrix $A = \begin{pmatrix} 4 & 3-2i \\ 3+2i & 6 \end{pmatrix}$ is Hermitian.

**17.** Show that if $A$ is Hermitian, then the diagonal components of $A$ are real.

† Named after the French mathematician Charles Hermite (1822–1901).

**18.** The $n \times n$ complex matrix $A$ is called **unitary** if $A^* = A^{-1}$. Show that the matrix

$$A = \begin{pmatrix} \frac{1+i}{2} & \frac{3-2i}{\sqrt{26}} \\ \frac{1+i}{2} & \frac{-3+2i}{\sqrt{26}} \end{pmatrix}$$

is unitary.

**19.** Show that $A$ is unitary if and only if the columns of $A$ form an orthonormal basis in $\mathbb{C}^n$.

**20.** Show that if $A$ is unitary, then $|\det A| = 1$.

**21.** Let $A$ be an $n \times n$ matrix with complex components. In $\mathbb{C}^n$, if $\mathbf{x} = (c_1, c_2, \ldots, c_n)$ and $y = (d_1, d_2, \ldots, d_n)$, define the inner product $(\mathbf{x}, \mathbf{y}) = c_1\overline{d_1} + c_2\overline{d_2} + \cdots + c_n\overline{d_n}$. (See Example 4.11.2.) Prove that $(A\mathbf{x}, \mathbf{y}) = (\mathbf{x}, A^*\mathbf{y})$.

***22.** Show that any two complex inner product spaces of the same (finite) dimension are isometrically isomorphic.

## MATLAB 5.5

1. a. *(Paper and pencil)* By considering the *definition* of isometry, explain, using geometry, why rotation about the origin and reflection across a line determined by a vector of length 1 in $\mathbb{R}^2$ are isometries.

   b. Choose three values for an angle $\theta$ and verify for each that the matrix representation (with respect to the standard basis) of rotation counterclockwise by $\theta$ is an orthogonal matrix.

      Choose three random vectors $\mathbf{v}$ of length 1. For each, verify that the matrix representation (with respect to the standard basis) of reflection across $\mathbf{v}$ is an orthogonal matrix. Refer to MATLAB Problem 4 in Section 5.3 for a discussion of reflection.

   c. *(Paper and pencil)* Prove in general that the matrix representation of a rotation is an orthogonal matrix and that the matrix representation of a reflection is an orthogonal matrix.

   d. The theory of isometries from $\mathbb{R}^2$ to $\mathbb{R}^2$ implies that a reflection across a vector $\mathbf{v}$ of length 1 should be a reflection across the $x$-axis followed by a rotation. A vector of length 1 can be represented as $(\cos(\alpha)\ \sin(\alpha))^t$. Generate a random vector $\mathbf{w}$ and divide by its length to produce a vector $\mathbf{v}$ of length 1. Find $\alpha$ by alpha = **atan(v(2)/v(1)).** (If the first component of $\mathbf{v}$ is zero, then $\alpha = \pi/2$.) Find the matrix representation $F$ of reflection across $\mathbf{v}$ and verify that $F = RX$, where $R$ is the matrix representation for rotation counterclockwise by $\theta = 2\alpha$ and $X$ is the matrix representation for reflection across the $x$-axis. Repeat for two more choices of $\mathbf{w}$.

   e. *(Paper and pencil)* Prove the result from part (d). *Hint.* Find a general expression for $F$ in terms of $\alpha$ and use trigonometric identities.

2. Work Problem 4 in this section. Also, verify that the given transformation $T$ maps an orthonormal basis onto an orthonormal basis. Would this always be true for an isometry? Why or why not?

# SUMMARY

- ***Linear transformation***

Let $V$ and $W$ be vector spaces. A **linear transformation** $T$ from $V$ into $W$ is a function that assigns to each vector $\mathbf{v} \in V$ a unique vector $T\mathbf{v} \in W$ and that satisfies, for each $\mathbf{u}$ and $\mathbf{v}$ in $V$ and each scalar $\alpha$, (p. 467)

$$T(\mathbf{u} + \mathbf{v}) = T\mathbf{u} + T\mathbf{v}$$

and

$$T(\alpha\mathbf{v}) = \alpha T\mathbf{v}$$

- ***Basic properties of linear transformations***

Let $T$: $V \to W$ be a linear transformation. Then for all vectors $\mathbf{u}$, $\mathbf{v}$, $\mathbf{v}_1$, $\mathbf{v}_2$, . . . , $\mathbf{v}_n$ in $V$ and all scalars $\alpha_1, \alpha_2, \ldots, \alpha_n$ (p. 478)

**i.** $T(\mathbf{0}) = \mathbf{0}$

**ii.** $T(\mathbf{u} - \mathbf{v}) = T\mathbf{u} - T\mathbf{v}$

**iii.** $T(\alpha_1\mathbf{v}_1 + \alpha_2\mathbf{v}_2 + \cdots + \alpha_n\mathbf{v}_n) = \alpha_1 T\mathbf{v}_1 + \alpha_2 T\mathbf{v}_2 + \cdots + \alpha_n T\mathbf{v}_n$

- ***Kernel and range of a linear transformation***

Let $V$ and $W$ be vector spaces and let $T$: $V \to W$ be a linear transformation. Then the **kernel** of $T$, denoted by ker $T$, is given by (pp. 481, 482)

$$\ker T = \{\mathbf{v} \in V\colon\ T\mathbf{v} = \mathbf{0}\}$$

The **range** of $T$, denoted by range $T$, is given by

$$\text{Range } T = \{\mathbf{w} \in W\colon\ \mathbf{w} = T\mathbf{v} \text{ for some } \mathbf{v} \in V\}$$

Ker $T$ is a subspace of $V$ and range $T$ is a subspace of $W$.

- ***Nullity and rank of a linear transformation***

If $T$ is a linear transformation from $V$ to $W$, then (p. 482)

$$\textbf{Nullity of } T = \nu(T) = \dim \ker T$$
$$\textbf{Rank of } T = \rho(T) = \dim \text{range } T$$

- ***Transformation matrix***

Let $T$: $\mathbb{R}^n \to \mathbb{R}^m$ be a linear transformation. Then there is a unique $m \times n$ matrix $A_T$ such that (pp. 486, 487)

$$T\mathbf{x} = A_T\mathbf{x} \quad \text{for every } \mathbf{x} \in \mathbb{R}^n$$

The matrix $A_T$ is called the **transformation matrix** of $T$.

- Let $A_T$ be the transformation matrix corresponding to the linear transformation $T$. Then (p. 487)

**i.** range $T = R_{A_T} = C_{A_T}$

**ii.** $\rho(T) = \rho(A_T)$

**iii.** $\ker T = N_{A_T}$

**iv.** $\nu(T) = \nu(A_T)$

- ***Matrix representation of a linear transformation***

Let $V$ be a real $n$-dimensional vector space, $W$ be a real $m$-dimensional vector space, and $T: V \to W$ be a linear transformation. Let $B_1 = \{\mathbf{v}_1, \mathbf{v}_2, \ldots, \mathbf{v}_n\}$ be a basis for $V$ and let $B_2 = \{\mathbf{w}_1, \mathbf{w}_2, \ldots, \mathbf{w}_m\}$ be a basis for $W$. Then there is a unique $m \times n$ matrix $A_T$ such that (p. 490)

$$(T\mathbf{x})_{B_2} = A_T(\mathbf{x})_{B_1}$$

$A_T$ is called the **matrix representation** of $T$ with respect to the bases $B_1$ and $B_2$.

- Let $V$ and $W$ be finite dimensional vector spaces with $\dim V = n$. Let $T: V \to W$ be a linear transformation and let $A_T$ be a matrix representation of $T$. Then (p. 492)

  **i.** $\rho(T) = \rho(A_T)$

  **ii.** $\nu(T) = \nu(A_T)$

  **iii.** $\nu(T) + \rho(T) = n$

- ***One-to-one transformation***

Let $T: V \to W$ be a linear transformation. Then $T$ is **one-to-one,** written 1–1, if $T\mathbf{v}_1 = T\mathbf{v}_2$ implies that $\mathbf{v}_1 = \mathbf{v}_2$. That is, $T$ is 1–1 if every vector $\mathbf{w}$ in the range of $T$ is the image of exactly one vector in $V$. (p. 512)

- Let $T: V \to W$ be a linear transformation. Then $T$ is 1–1 if and only if $\ker T = \{\mathbf{0}\}$. (p. 512)

- ***Onto transformation***

Let $T: V \to W$ be a linear transformation. Then $T$ is said to be **onto** $W$ or simply **onto,** if for every $\mathbf{w} \in W$ there is at least one $\mathbf{v} \in V$ such that $T\mathbf{v} = \mathbf{w}$. That is, *$T$ is onto $W$ if and only if range $T = W$.* (p. 512)

- Let $T: V \to W$ be a linear transformation and suppose that $\dim V = \dim W = n$: (p. 513)

  **i.** If $T$ is 1–1, then $T$ is onto.

  **ii.** If $T$ is onto, then $T$ is 1–1.

- Let $T: V \to W$ be a linear transformation. Suppose that $\dim V = n$ and $\dim W = m$. Then (p. 513)

  **i.** If $n > m$, $T$ is not 1–1.

  **ii.** If $m > n$, $T$ is not onto.

- ***Isomorphism***

Let $T: V \to W$ be a linear transformation. Then $T$ is an **isomorphism** if $T$ is 1–1 and onto. (p. 514)

- ***Isomorphic vector spaces***

The vector spaces $V$ and $W$ are said to be **isomorphic** if there exists an isomorphism $T$ from $V$ onto $W$. In this case we write $V \cong W$. (p. 514)

- Any two real finite dimensional vector spaces of the same dimension are isomorphic. (p. 517)

- ***Summing Up Theorem***

Let $A$ be an $n \times n$ matrix. Then the following 11 statements are equivalent: (p. 515)

  **i.** $A$ is invertible.

  **ii.** The only solution to the homogeneous system $A\mathbf{x} = \mathbf{0}$ is the trivial solution ($\mathbf{x} = \mathbf{0}$).

**iii.** The system $A\mathbf{x} = \mathbf{b}$ has a unique solution for every $n$-vector $\mathbf{b}$.

**iv.** $A$ is row equivalent to the $n \times n$ identity matrix $I_n$.

**v.** $A$ can be written as the product of elementary matrices.

**vi.** The row echelon form of $A$ has $n$ pivots.

**vii.** The rows (and columns) of $A$ are linearly independent.

**viii.** $\det A \neq 0$.

**ix.** $\nu(A) = 0$.

**x.** $\rho(A) = n$.

**xi.** The linear transformation $T$ from $\mathbb{R}^n$ to $\mathbb{R}^n$ defined by $T\mathbf{x} = A\mathbf{x}$ is an isomorphism.

- Let $T\colon V \to W$ be an isomorphism: (p. 516)

**i.** If $\mathbf{v}_1, \mathbf{v}_2, \ldots, \mathbf{v}_n$ span $V$, then $T\mathbf{v}_1, T\mathbf{v}_2, \ldots, T\mathbf{v}_n$ span $W$.

**ii.** If $\mathbf{v}_1, \mathbf{v}_2, \ldots, \mathbf{v}_n$ are linearly independent in $V$, then $T\mathbf{v}_1, T\mathbf{v}_2, \ldots, T\mathbf{v}_n$ are linearly independent in $W$.

**iii.** If $\{\mathbf{v}_1, \mathbf{v}_2, \ldots, \mathbf{v}_n\}$ is a basis in $V$, then $\{T\mathbf{v}_1, T\mathbf{v}_2, \ldots, T\mathbf{v}_n\}$ is a basis in $W$.

**iv.** If $V$ is finite dimensional, then $W$ is finite dimensional and $\dim V = \dim W$.

- ***Isometry***

A linear transformation $T\colon \mathbb{R}^n \to \mathbb{R}^n$ is called an **isometry** if for every $\mathbf{x}$ in $\mathbb{R}^n$ (p. 520)

$$|T\mathbf{x}| = |\mathbf{x}|$$

- If $T$ is an isometry from $\mathbb{R}^n \to \mathbb{R}^n$, then for every $\mathbf{x}$ and $\mathbf{y}$ in $\mathbb{R}^n$, (p. 520)

$$|T\mathbf{x} - T\mathbf{y}| = |\mathbf{x} - \mathbf{y}| \quad \text{and} \quad T\mathbf{x} \cdot T\mathbf{y} = \mathbf{x} \cdot \mathbf{y}$$

- Let $T\colon \mathbb{R}^n \to \mathbb{R}^n$ be an isometry. Then: (p. 523)

**i.** If $\mathbf{u}_1, \mathbf{u}_2, \ldots, \mathbf{u}_n$ is an orthogonal set, then $T\mathbf{u}_1, T\mathbf{u}_2, \ldots, T\mathbf{u}_n$ is an orthogonal set.

**ii.** $T$ is an isomorphism.

- A linear transformation $T\colon \mathbb{R}^n \to \mathbb{R}^n$ is an isometry if and only if the matrix representation of $T$ is orthogonal. (p. 521)

- ***Isometry***

Let $V$ and $W$ be real (or complex) inner product spaces and let $T\colon V \to W$ be a linear transformation. Then $T$ is an **isometry** if for every $\mathbf{v} \in V$ (p. 523)

$$\|\mathbf{v}\|_V = \|T\mathbf{v}\|_W$$

- ***Isometrically isomorphic vector spaces***

Two vector spaces $V$ and $W$ are said to be **isometrically isomorphic** if there exists a linear transformation $T\colon V \to W$ that is both an isometry and an isomorphism. (p. 524)

- Any two $n$-dimensional real inner product spaces are isometrically isomorphic. (p. 524)

# REVIEW EXERCISES

In Exercises 1–6 determine whether the given transformation from $V$ to $W$ is linear.

**1.** $T\colon \mathbb{R}^2 \to \mathbb{R}^2$; $T(x, y) = (0, -y)$

**2.** $T\colon \mathbb{R}^3 \to \mathbb{R}^3$; $T(x, y, z) = (1, y, z)$

**3.** $T\colon \mathbb{R}^2 \to \mathbb{R}$; $T(x, y) = x/y$

**4.** $T\colon P_1 \to P_2$; $(Tp)(x) = xp(x)$

**5.** $T\colon P_2 \to P_2$; $(Tp)(x) = 1 + p(x)$

**6.** $T\colon C[0, 1] \to C[0, 1]$; $Tf(x) = f(1)$

In Exercises 7–12 find the kernel, range, rank, and nullity of the given linear transformation.

**7.** $T\colon \mathbb{R}^2 \to \mathbb{R}^2$; $T\begin{pmatrix} x \\ y \end{pmatrix} = \begin{pmatrix} 2 & -1 \\ 4 & 7 \end{pmatrix}\begin{pmatrix} x \\ y \end{pmatrix}$

**8.** $T\colon \mathbb{R}^3 \to \mathbb{R}^3$; $T\begin{pmatrix} x \\ y \\ z \end{pmatrix} = \begin{pmatrix} 1 & 2 & -1 \\ 2 & 4 & 3 \\ 1 & 2 & -6 \end{pmatrix}\begin{pmatrix} x \\ y \\ z \end{pmatrix}$

**9.** $T\colon \mathbb{R}^3 \to \mathbb{R}^2$; $T\begin{pmatrix} x \\ y \\ z \end{pmatrix} = \begin{pmatrix} y \\ -x \end{pmatrix}$

**10.** $T\colon P_2 \to P_4$; $Tp(x) = x^2p(x)$

**11.** $T\colon M_{22} \to M_{22}$; $T(A) = AB$, where $B = \begin{pmatrix} 1 & 1 \\ -1 & 1 \end{pmatrix}$

**12.** $T\colon C[0, 1] \to \mathbb{R}$; $Tf = f(1)$

In Exercises 13–18 find the matrix representation of the given linear transformation and find the kernel, range, nullity, and rank of the transformation.

**13.** $T\colon \mathbb{R}^2 \to \mathbb{R}^2$; $T(x, y) = (0, -y)$

**14.** $T\colon \mathbb{R}^3 \to \mathbb{R}^2$; $T(x, y, z) = (y, z)$

**15.** $T\colon \mathbb{R}^4 \to \mathbb{R}^2$; $T(x, y, z, w) = (x - 2z, 2y + 3w)$

**16.** $T\colon P_3 \to P_4$; $(Tp)(x) = xp(x)$

**17.** $T\colon M_{22} \to M_{22}$; $TA = AB$, where $B = \begin{pmatrix} -1 & 0 \\ 1 & 2 \end{pmatrix}$

**18.** $T\colon \mathbb{R}^2 \to \mathbb{R}^2$; $T(x, y) = (x - y, 2x + 3y)$; $B_1 = \left\{\begin{pmatrix} 1 \\ 1 \end{pmatrix}, \begin{pmatrix} 1 \\ 2 \end{pmatrix}\right\}$; $B_2 = \left\{\begin{pmatrix} -1 \\ 3 \end{pmatrix}, \begin{pmatrix} 4 \\ 1 \end{pmatrix}\right\}$

In Exercises 19–22 describe in words the linear transformation $T\colon \mathbb{R}^2 \to \mathbb{R}^2$ with the given matrix representation $A_T$.

**19.** $A_T = \begin{pmatrix} 3 & 0 \\ 0 & 1 \end{pmatrix}$

**20.** $A_T = \begin{pmatrix} 1 & 0 \\ 0 & \frac{1}{3} \end{pmatrix}$

**21.** $A_T = \begin{pmatrix} 1 & 0 \\ -2 & 1 \end{pmatrix}$

**22.** $A_T = \begin{pmatrix} 1 & -5 \\ 0 & 1 \end{pmatrix}$

In Exercises 23–26 write the $2 \times 2$ matrix representation of the given linear transformation and draw a sketch of the region obtained when the transformation is applied to the given rectangle.

**23.** Expansion along the $x$-axis with $c = 3$

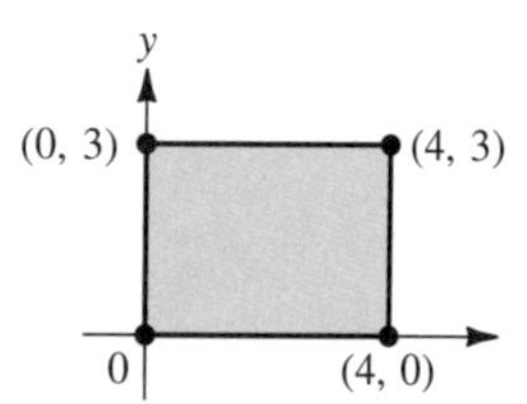

**24.** Compression along the $y$-axis with $c = \frac{1}{3}$

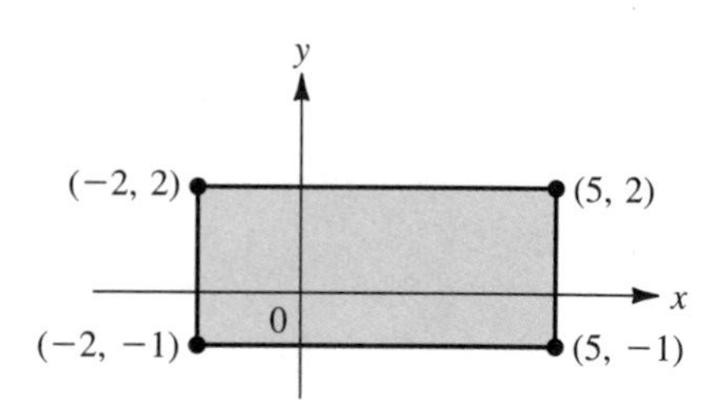


**25.** Reflection about the line $y = x$

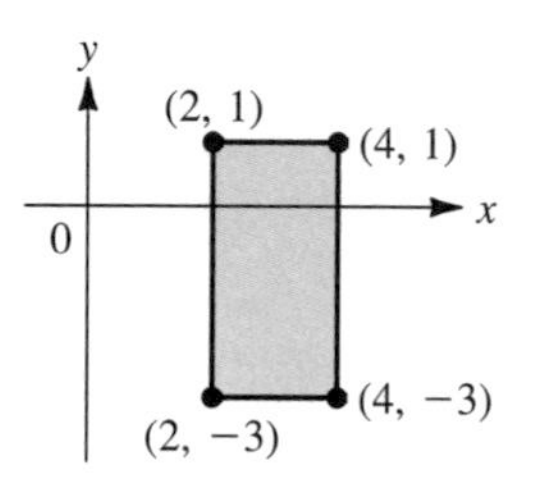


**26.** Shear along the $x$-axis with $c = -3$

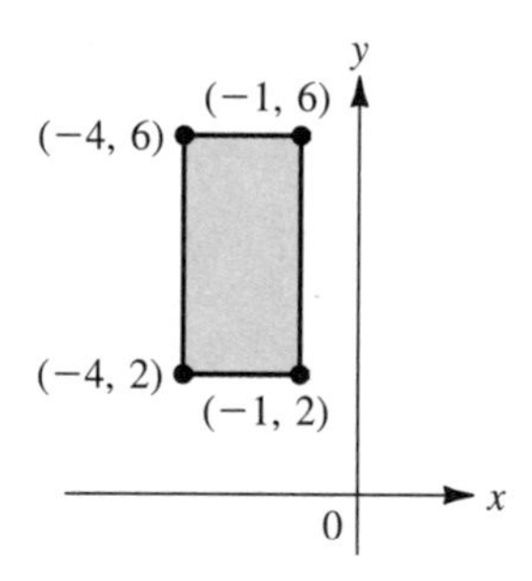


In Exercises 27–30 write each linear transformation matrix $A_T$ as a succession of expansions, compressions, reflections, and shears.

**27.** $A_T = \begin{pmatrix} 1 & 3 \\ -2 & 2 \end{pmatrix}$ **28.** $A_T = \begin{pmatrix} 0 & 5 \\ -3 & 2 \end{pmatrix}$ **29.** $A_T = \begin{pmatrix} -6 & 4 \\ 1 & 3 \end{pmatrix}$ **30.** $A_T = \begin{pmatrix} 2 & 1 \\ 1 & 5 \end{pmatrix}$

**31.** Find an isomorphism $T\colon P_2 \to \mathbb{R}^3$.

Calculus **32.** Find an isometry $T\colon \mathbb{R}^2 \to P_1[-1, 1]$.

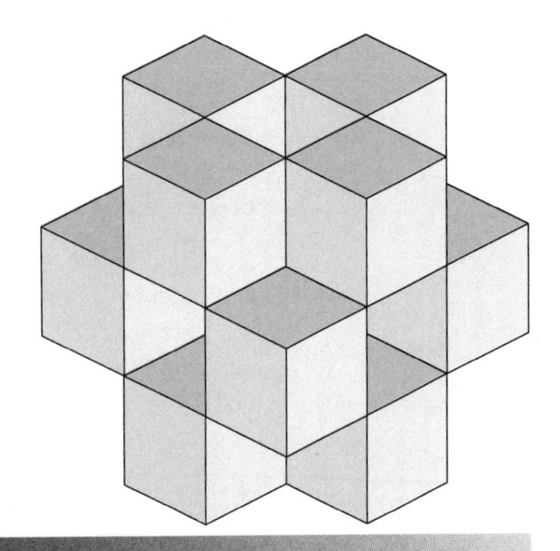

# 6

# Eigenvalues, Eigenvectors, and Canonical Forms

## 6.1 EIGENVALUES AND EIGENVECTORS

Let $T$: $V \to V$ be a linear transformation. In a great variety of applications (one of which is given in the next section) it is useful to find a vector $\mathbf{v}$ in $V$ such that $T\mathbf{v}$ and $\mathbf{v}$ are parallel. That is, we seek a vector $\mathbf{v}$ and a scalar $\lambda$ such that

$$T\mathbf{v} = \lambda\mathbf{v} \tag{1}$$

If $\mathbf{v} \neq \mathbf{0}$ and $\lambda$ satisfy (1), then $\lambda$ is called an *eigenvalue* of $T$ and $\mathbf{v}$ is called an *eigenvector* of $T$ corresponding to the eigenvalue $\lambda$. The purpose of this chapter is to investigate properties of eigenvalues and eigenvectors. If $V$ is finite dimensional, then $T$ can be represented by a matrix $A_T$. For that reason we shall discuss eigenvalues and eigenvectors of $n \times n$ matrices.

**DEFINITION 1** **Eigenvalue and Eigenvector** Let $A$ be an $n \times n$ matrix with real† components. The number $\lambda$ (real or complex) is called an **eigenvalue** of $A$ if there is a *nonzero* vector $\mathbf{v}$ in $\mathbb{C}^n$ such that

$$\boxed{A\mathbf{v} = \lambda\mathbf{v}} \tag{2}$$

The vector $\mathbf{v} \neq \mathbf{0}$ is called an **eigenvector** *of $A$ corresponding to the eigenvalue $\lambda$.*

† This definition is also valid if $A$ has complex components; but as the matrices we shall be dealing with will, for the most part, have real components, the definition is sufficient for our purposes.

*Note.* The word "eigen" is the German word for "own" or "proper." Eigenvalues are also called **proper values** or **characteristic values** and eigenvectors are called **proper vectors** or **characteristic vectors.**

*Remark.* As we shall see (for instance, in Example 6) a matrix with real components can have complex eigenvalues and eigenvectors. That is why, in the definition, we have asserted that $\mathbf{v} \in \mathbb{C}^n$. We shall not be using many facts about complex numbers in this book. For a discussion of those few facts we do need, see Appendix 2.

**EXAMPLE 1** **Eigenvalues and Eigenvectors of a 2 × 2 Matrix** Let $A = \begin{pmatrix} 10 & -18 \\ 6 & -11 \end{pmatrix}$. Then $A\begin{pmatrix} 2 \\ 1 \end{pmatrix} = \begin{pmatrix} 10 & -18 \\ 6 & -11 \end{pmatrix}\begin{pmatrix} 2 \\ 1 \end{pmatrix} = \begin{pmatrix} 2 \\ 1 \end{pmatrix}$. Thus $\lambda_1 = 1$ is an eigenvalue of $A$ with corresponding eigenvector $\mathbf{v}_1 = \begin{pmatrix} 2 \\ 1 \end{pmatrix}$. Similarly, $A\begin{pmatrix} 3 \\ 2 \end{pmatrix} = \begin{pmatrix} 10 & -18 \\ 6 & -11 \end{pmatrix}\begin{pmatrix} 3 \\ 2 \end{pmatrix} = \begin{pmatrix} -6 \\ -4 \end{pmatrix} = -2\begin{pmatrix} 3 \\ 2 \end{pmatrix}$ so that $\lambda_2 = -2$ is an eigenvalue of $A$ with corresponding eigenvector $\mathbf{v}_2 = \begin{pmatrix} 3 \\ 2 \end{pmatrix}$. As we soon shall see, these are the only eigenvalues of $A$.

**EXAMPLE 2** **Eigenvalues and Eigenvectors of the Identity Matrix** Let $A = I$. Then for any $\mathbf{v} \in \mathbb{C}^n$, $A\mathbf{v} = I\mathbf{v} = \mathbf{v}$. Thus 1 is the only eigenvalue of $A$ and every $\mathbf{v} \neq \mathbf{0} \in \mathbb{C}^n$ is an eigenvector of $I$.

We shall compute the eigenvalues and eigenvectors of many matrices in this section. But first we need to prove some techniques that will simplify our computations.

Suppose that $\lambda$ is an eigenvalue of $A$. Then there exists a nonzero vector $\mathbf{v} = \begin{pmatrix} x_1 \\ x_2 \\ \vdots \\ x_n \end{pmatrix} \neq \mathbf{0}$ such that $A\mathbf{v} = \lambda\mathbf{v} = \lambda I\mathbf{v}$. Rewriting this, we have

$$(A - \lambda I)\mathbf{v} = \mathbf{0} \tag{3}$$

If $A$ is an $n \times n$ matrix, equation (3) corresponds to a homogeneous system of $n$ equations in the unknowns $x_1, x_2, \ldots, x_n$. Since by assumption the system has nontrivial solutions, we conclude that $\det(A - \lambda I) = 0$. Conversely, if $\det(A - \lambda I) = 0$, then equation (3) has nontrivial solutions and $\lambda$ is an eigenvalue of $A$. On the other hand, if $\det(A - \lambda I) \neq 0$, then (3) has only the solution $\mathbf{v} = \mathbf{0}$ so that $\lambda$ is *not* an eigenvalue of $A$. Summing up these facts, we have the following theorem.

**THEOREM 1** Let $A$ be an $n \times n$ matrix. Then $\lambda$ is an eigenvalue of $A$ if and only if

$$p(\lambda) = \det(A - \lambda I) = 0 \tag{4}$$

■

**DEFINITION 2** **Characteristic Equation and Polynomial** Equation (4) is called the **characteristic equation** of $A$; $p(\lambda)$ is called the **characteristic polynomial** of $A$.

As will become apparent in the examples, $p(\lambda)$ is a polynomial of degree $n$ in $\lambda$. For example, if $A = \begin{pmatrix} a & b \\ c & d \end{pmatrix}$, then $A - \lambda I = \begin{pmatrix} a & b \\ c & d \end{pmatrix} - \begin{pmatrix} \lambda & 0 \\ 0 & \lambda \end{pmatrix} = \begin{pmatrix} a-\lambda & b \\ c & d-\lambda \end{pmatrix}$ and $p(\lambda) = \det(A - \lambda I) = (a-\lambda)(d-\lambda) - bc = \lambda^2 - (a+d)\lambda + (ad - bc)$.

According to the **fundamental theorem of algebra,** any polynomial of degree $n$ with real or complex coefficients has exactly $n$ roots (counting multiplicities). By this we mean, for example, that the polynomial $(\lambda - 1)^5$ has five roots, all equal to the number 1. Since any eigenvalue of $A$ is a root of the characteristic equation of $A$, we conclude that

*Counting multiplicities, every $n \times n$ matrix has exactly $n$ eigenvalues.*

**THEOREM 2** Let $\lambda$ be an eigenvalue of the $n \times n$ matrix $A$ and let $E_\lambda = \{\mathbf{v}\colon A\mathbf{v} = \lambda\mathbf{v}\}$. Then $E_\lambda$ is a subspace of $\mathbb{C}^n$.

**Proof** If $A\mathbf{v} = \lambda\mathbf{v}$, then $(A - \lambda I)\mathbf{v} = \mathbf{0}$. Thus $E_\lambda$ is the null space of the matrix $A - \lambda I$, which by Example 4.6.10 on page 343 is a subspace† of $\mathbb{C}^n$. ■

**DEFINITION 3** **Eigenspace** Let $\lambda$ be an eigenvalue of $A$. The subspace $E_\lambda$ is called the **eigenspace**‡ of $A$ corresponding to the eigenvalue $\lambda$.

† In Example 4.6.10 on page 343 we saw that $N_A$ is a subspace of $\mathbb{R}^n$ if $A$ is a real matrix. The extension of this result to $\mathbb{C}^n$ presents no difficulties.

‡ Note that $\mathbf{0} \in E_\lambda$ since $E_\lambda$ is a subspace. However, $\mathbf{0}$ is *not* an eigenvector.

We now prove another useful result.

**THEOREM 3** Let $A$ be an $n \times n$ matrix and let $\lambda_1, \lambda_2, \ldots, \lambda_m$ be distinct eigenvalues of $A$ (that is, $\lambda_i \neq \lambda_j$ if $i \neq j$) with corresponding eigenvectors $\mathbf{v}_1, \mathbf{v}_2, \ldots, \mathbf{v}_m$. Then $\mathbf{v}_1, \mathbf{v}_2, \ldots, \mathbf{v}_m$ are linearly independent. That is: *Eigenvectors corresponding to distinct eigenvalues are linearly independent.*

**Proof** We prove this by mathematical induction. We start with $m = 2$. Suppose that

$$c_1\mathbf{v}_1 + c_2\mathbf{v}_2 = \mathbf{0} \tag{5}$$

Then multiplying both sides of (5) by $A$, we have

$$\mathbf{0} = A(c_1\mathbf{v}_1 + c_2\mathbf{v}_2) = c_1A\mathbf{v}_1 + c_2A\mathbf{v}_2$$

or (since $A\mathbf{v}_i = \lambda_i\mathbf{v}_i$ for $i = 1, 2$)

$$c_1\lambda_1\mathbf{v}_1 + c_2\lambda_2\mathbf{v}_2 = \mathbf{0} \tag{6}$$

We then multiply (5) by $\lambda_1$ and subtract it from (6) to obtain

$$(c_1\lambda_1\mathbf{v}_1 + c_2\lambda_2\mathbf{v}_2) - (c_1\lambda_1\mathbf{v}_1 + c_2\lambda_1\mathbf{v}_2) = \mathbf{0}$$

or

$$c_2(\lambda_2 - \lambda_1)\mathbf{v}_2 = \mathbf{0}$$

Since $\mathbf{v}_2 \neq \mathbf{0}$ (by the definition of an eigenvector) and since $\lambda_2 \neq \lambda_1$, we conclude that $c_2 = 0$. Then inserting $c_2 = 0$ in (5), we see that $c_1 = 0$, which proves the theorem in the case $m = 2$. Now suppose that the theorem is true for $m = k$. That is, we assume that any $k$ eigenvectors corresponding to distinct eigenvalues are linearly independent. We prove the theorem for $m = k + 1$. So we assume that

$$c_1\mathbf{v}_1 + c_2\mathbf{v}_2 + \cdots + c_k\mathbf{v}_k + c_{k+1}\mathbf{v}_{k+1} = \mathbf{0} \tag{7}$$

Then multiplying both sides of (7) by $A$ and using the fact that $A\mathbf{v}_i = \lambda_i\mathbf{v}_i$, we obtain

$$c_1\lambda_1\mathbf{v}_1 + c_2\lambda_2\mathbf{v}_2 + \cdots + c_k\lambda_k\mathbf{v}_k + c_{k+1}\lambda_{k+1}\mathbf{v}_{k+1} = \mathbf{0} \tag{8}$$

We multiply both sides of (7) by $\lambda_{k+1}$ and subtract it from (8):

$$c_1(\lambda_1 - \lambda_{k+1})\mathbf{v}_1 + c_2(\lambda_2 - \lambda_{k+1})\mathbf{v}_2 + \cdots + c_k(\lambda_k - \lambda_{k+1})\mathbf{v}_k = \mathbf{0}$$

But by the induction assumption, $\mathbf{v}_1, \mathbf{v}_2, \ldots, \mathbf{v}_k$ are linearly independent. Thus $c_1(\lambda_1 - \lambda_{k+1}) = c_2(\lambda_2 - \lambda_{k+1}) = \cdots = c_k(\lambda_k - \lambda_{k+1}) = 0$; and since $\lambda_i \neq \lambda_{k+1}$ for $i = 1, 2, \ldots, k$, we conclude that $c_1 = c_2 = \cdots = c_k = 0$. But from (7) this means that $c_{k+1} = 0$. Thus the theorem is true for $m = k + 1$ and the proof is complete. ■

If

$$A = \begin{pmatrix} a_{11} & a_{12} & \cdots & a_{1n} \\ a_{21} & a_{22} & \cdots & a_{2n} \\ \vdots & \vdots & & \vdots \\ a_{n1} & a_{n2} & \cdots & a_{nn} \end{pmatrix}$$

then

$$p(\lambda) = \det(A - \lambda I) = \begin{vmatrix} a_{11} - \lambda & a_{12} & \cdots & a_{1n} \\ a_{21} & a_{22} - \lambda & \cdots & a_{2n} \\ \vdots & \vdots & & \vdots \\ a_{n1} & a_{n2} & \cdots & a_{nn} - \lambda \end{vmatrix}$$

and $p(\lambda) = 0$ can be written in the form

$$p(\lambda) = (-1)^n[\lambda^n + b_{n-1}\lambda^{n-1} + \cdots + b_1\lambda + b_0] = 0 \qquad \textbf{(9)}$$

Equation (9) has $n$ roots, some of which may be repeated. If $\lambda_1, \lambda_2, \ldots, \lambda_m$ are the distinct roots of (9) with multiplicities $r_1, r_2, \ldots, r_m$, respectively, then (9) may be factored to obtain

$$(-1)^n p(\lambda) = (\lambda - \lambda_1)^{r_1}(\lambda - \lambda_2)^{r_2} \cdots (\lambda - \lambda_m)^{r_m} = 0 \qquad \textbf{(10)}$$

Algebraic multiplicity

The numbers $r_1, r_2, \ldots, r_m$ are called the **algebraic multiplicities** of the eigenvalues $\lambda_1, \lambda_2, \ldots, \lambda_m$, respectively.

We now calculate eigenvalues and corresponding eigenspaces. We do this by a three-step procedure:

**Procedure for Computing Eigenvalues and Eigenvectors**

**i.** Find $p(\lambda) = \det(A - \lambda I)$.

**ii.** Find the roots $\lambda_1, \lambda_2, \ldots, \lambda_m$ of $p(\lambda) = 0$.

**iii.** Corresponding to each eigenvalue $\lambda_i$, solve the homogeneous system $(A - \lambda_i I)\mathbf{v} = \mathbf{0}$.

*Remark 1.* Step (*ii*) is usually the most difficult one.

*Remark 2.* A relatively easy way to find eigenvalues and eigenvectors of $2 \times 2$ matrices is suggested in Problems 35 and 36.

EXAMPLE 3 **Computing Eigenvalues and Eigenvectors** Let $A = \begin{pmatrix} 4 & 2 \\ 3 & 3 \end{pmatrix}$. Then $\det(A - \lambda I) = \begin{vmatrix} 4-\lambda & 2 \\ 3 & 3-\lambda \end{vmatrix} = (4-\lambda)(3-\lambda) - 6 = \lambda^2 - 7\lambda + 6 = (\lambda - 1)(\lambda - 6) = 0$. Thus the eigenvalues of $A$ are $\lambda_1 = 1$ and $\lambda_2 = 6$. For $\lambda_1 = 1$, we solve $(A - I)\mathbf{v} = \mathbf{0}$ or $\begin{pmatrix} 3 & 2 \\ 3 & 2 \end{pmatrix}\begin{pmatrix} x_1 \\ x_2 \end{pmatrix} = \begin{pmatrix} 0 \\ 0 \end{pmatrix}$. Clearly, any eigenvector corresponding to $\lambda_1 = 1$ satisfies $3x_1 + 2x_2 = 0$. One such eigenvector is $\mathbf{v}_1 = \begin{pmatrix} 2 \\ -3 \end{pmatrix}$. Thus $E_1 = \text{span}\left\{\begin{pmatrix} 2 \\ -3 \end{pmatrix}\right\}$. Similarly, the equation $(A - 6I)\mathbf{v} = \mathbf{0}$ means that $\begin{pmatrix} -2 & 2 \\ 3 & -3 \end{pmatrix}\begin{pmatrix} x_1 \\ x_2 \end{pmatrix} = \begin{pmatrix} 0 \\ 0 \end{pmatrix}$ or $x_1 = x_2$. Thus $\mathbf{v}_2 = \begin{pmatrix} 1 \\ 1 \end{pmatrix}$ is an eigenvector corresponding to $\lambda_2 = 6$ and $E_6 = \text{span}\left\{\begin{pmatrix} 1 \\ 1 \end{pmatrix}\right\}$. Note that $\mathbf{v}_1$ and $\mathbf{v}_2$ are linearly independent since one is not a multiple of the other.

*Note.* It makes no difference if we set $\lambda_1 = 1$ and $\lambda_2 = 6$ or $\lambda_1 = 6$ and $\lambda_2 = 1$. The results are the same.

EXAMPLE 4 **A 3 × 3 Matrix with Distinct Eigenvalues** Let $A = \begin{pmatrix} 1 & -1 & 4 \\ 3 & 2 & -1 \\ 2 & 1 & -1 \end{pmatrix}$. Then

$$\det(A - \lambda I) = \begin{vmatrix} 1-\lambda & -1 & 4 \\ 3 & 2-\lambda & -1 \\ 2 & 1 & -1-\lambda \end{vmatrix}$$
$$= -(\lambda^3 - 2\lambda^2 - 5\lambda + 6) = -(\lambda - 1)(\lambda + 2)(\lambda - 3)$$

Thus the eigenvalues of $A$ are $\lambda_1 = 1$, $\lambda_2 = -2$, and $\lambda_3 = 3$. Corresponding to $\lambda_1 = 1$ we have

$$(A - I)\mathbf{v} = \begin{pmatrix} 0 & -1 & 4 \\ 3 & 1 & -1 \\ 2 & 1 & -2 \end{pmatrix}\begin{pmatrix} x_1 \\ x_2 \\ x_3 \end{pmatrix} = \begin{pmatrix} 0 \\ 0 \\ 0 \end{pmatrix}$$

Solving by row reduction, we obtain, successively,

$$\left(\begin{array}{ccc|c} 0 & -1 & 4 & 0 \\ 3 & 1 & -1 & 0 \\ 2 & 1 & -2 & 0 \end{array}\right) \longrightarrow \left(\begin{array}{ccc|c} 0 & -1 & 4 & 0 \\ 3 & 0 & 3 & 0 \\ 2 & 0 & 2 & 0 \end{array}\right)$$
$$\longrightarrow \left(\begin{array}{ccc|c} 0 & -1 & 4 & 0 \\ 1 & 0 & 1 & 0 \\ 2 & 0 & 2 & 0 \end{array}\right) \longrightarrow \left(\begin{array}{ccc|c} 0 & -1 & 4 & 0 \\ 1 & 0 & 1 & 0 \\ 0 & 0 & 0 & 0 \end{array}\right)$$

Thus $x_1 = -x_3$, $x_2 = 4x_3$, an eigenvector is $\mathbf{v}_1 = \begin{pmatrix} -1 \\ 4 \\ 1 \end{pmatrix}$, and $E_1 = \text{span}\left\{\begin{pmatrix} -1 \\ 4 \\ 1 \end{pmatrix}\right\}$.

For $\lambda_2 = -2$ we have $[A - (-2I)]\mathbf{v} = (A + 2I)\mathbf{v} = \mathbf{0}$ or

$$\begin{pmatrix} 3 & -1 & 4 \\ 3 & 4 & -1 \\ 2 & 1 & 1 \end{pmatrix}\begin{pmatrix} x_1 \\ x_2 \\ x_3 \end{pmatrix} = \begin{pmatrix} 0 \\ 0 \\ 0 \end{pmatrix}$$

This leads to

$$\left(\begin{array}{ccc|c} 3 & -1 & 4 & 0 \\ 3 & 4 & -1 & 0 \\ 2 & 1 & 1 & 0 \end{array}\right) \longrightarrow \left(\begin{array}{ccc|c} 3 & -1 & 4 & 0 \\ 15 & 0 & 15 & 0 \\ 5 & 0 & 5 & 0 \end{array}\right)$$

$$\longrightarrow \left(\begin{array}{ccc|c} 3 & -1 & 4 & 0 \\ 1 & 0 & 1 & 0 \\ 5 & 0 & 5 & 0 \end{array}\right) \longrightarrow \left(\begin{array}{ccc|c} -1 & -1 & 0 & 0 \\ 1 & 0 & 1 & 0 \\ 0 & 0 & 0 & 0 \end{array}\right)$$

Thus $x_2 = -x_1$, $x_3 = -x_1$, and an eigenvector is $\mathbf{v}_2 = \begin{pmatrix} 1 \\ -1 \\ -1 \end{pmatrix}$. Then $E_{-2} = \text{span}\left\{\begin{pmatrix} 1 \\ -1 \\ -1 \end{pmatrix}\right\}$. Finally, for $\lambda_3 = 3$ we have

$$(A - 3I)\mathbf{v} = \begin{pmatrix} -2 & -1 & 4 \\ 3 & -1 & -1 \\ 2 & 1 & -4 \end{pmatrix}\begin{pmatrix} x_1 \\ x_2 \\ x_3 \end{pmatrix} = \begin{pmatrix} 0 \\ 0 \\ 0 \end{pmatrix}$$

and

$$\left(\begin{array}{ccc|c} -2 & -1 & 4 & 0 \\ 3 & -1 & -1 & 0 \\ 2 & 1 & -4 & 0 \end{array}\right) \longrightarrow \left(\begin{array}{ccc|c} -2 & -1 & 4 & 0 \\ 5 & 0 & -5 & 0 \\ 0 & 0 & 0 & 0 \end{array}\right)$$

$$\longrightarrow \left(\begin{array}{ccc|c} -2 & -1 & 4 & 0 \\ 1 & 0 & -1 & 0 \\ 0 & 0 & 0 & 0 \end{array}\right) \longrightarrow \left(\begin{array}{ccc|c} 2 & -1 & 0 & 0 \\ 1 & 0 & -1 & 0 \\ 0 & 0 & 0 & 0 \end{array}\right)$$

Hence $x_3 = x_1$, $x_2 = 2x_1$, and $\mathbf{v}_3 = \begin{pmatrix} 1 \\ 2 \\ 1 \end{pmatrix}$ so that $E_3 = \text{span}\left\{\begin{pmatrix} 1 \\ 2 \\ 1 \end{pmatrix}\right\}$. ■

*Remark.* In this and every other example there is always an infinite number of choices for each eigenvector. We arbitrarily choose a simple one by setting one or more of the $x_i$'s equal to a convenient number. Here we have set one of the $x_i$'s equal to 1. Another common choice is to scale the eigenvector so that it is a unit vector.

**EXAMPLE 5** **A 2 × 2 Matrix with One of Its Eigenvalues Equal to Zero** Let $A = \begin{pmatrix} 2 & -1 \\ -4 & 2 \end{pmatrix}$. Then $\det(A - \lambda I) = \begin{vmatrix} 2-\lambda & -1 \\ -4 & 2-\lambda \end{vmatrix} = \lambda^2 - 4\lambda = \lambda(\lambda - 4)$. Thus the eigenvalues are $\lambda_1 = 0$ and $\lambda_2 = 4$. The eigenspace corresponding to zero is simply the null space of $A$. We calculate $\begin{pmatrix} 2 & -1 \\ -4 & 2 \end{pmatrix}\begin{pmatrix} x_1 \\ x_2 \end{pmatrix} = \begin{pmatrix} 0 \\ 0 \end{pmatrix}$, so $2x_1 = x_2$ and an eigenvector is $\mathbf{v}_1 = \begin{pmatrix} 1 \\ 2 \end{pmatrix}$. Thus $E_0 = \text{span}\left\{\begin{pmatrix} 1 \\ 2 \end{pmatrix}\right\}$. Corresponding to $\lambda_2 = 4$ we have $\begin{pmatrix} -2 & -1 \\ -4 & -2 \end{pmatrix}\begin{pmatrix} x_1 \\ x_2 \end{pmatrix} = \begin{pmatrix} 0 \\ 0 \end{pmatrix}$, so $E_4 = \text{span}\left\{\begin{pmatrix} 1 \\ -2 \end{pmatrix}\right\}$.

**EXAMPLE 6** **A 2 × 2 Matrix with Complex Conjugate Eigenvalues** Let $A = \begin{pmatrix} 3 & -5 \\ 1 & -1 \end{pmatrix}$. Then $\det(A - \lambda I) = \begin{vmatrix} 3-\lambda & -5 \\ 1 & -1-\lambda \end{vmatrix} = \lambda^2 - 2\lambda + 2 = 0$ and

$$\lambda = \frac{-(-2) \pm \sqrt{4 - 4(1)(2)}}{2} = \frac{2 \pm \sqrt{-4}}{2} = \frac{2 \pm 2i}{2} = 1 \pm i$$

Thus $\lambda_1 = 1 + i$ and $\lambda_2 = 1 - i$. We compute

$$[A - (1+i)I]\mathbf{v} = \begin{pmatrix} 2-i & -5 \\ 1 & -2-i \end{pmatrix}^{\dagger}\begin{pmatrix} x_1 \\ x_2 \end{pmatrix} = \begin{pmatrix} 0 \\ 0 \end{pmatrix}$$

and we obtain $(2 - i)x_1 - 5x_2 = 0$ and $x_1 + (-2 - i)x_2 = 0$. Thus $x_1 = (2 + i)x_2$, which yields the eigenvector $\mathbf{v}_1 = \begin{pmatrix} 2+i \\ 1 \end{pmatrix}$ and $E_{1+i} = \text{span}\left\{\begin{pmatrix} 2+i \\ 1 \end{pmatrix}\right\}$. Similarly, $[A - (1-i)I]\mathbf{v} = \begin{pmatrix} 2+i & -5 \\ 1 & -2+i \end{pmatrix}\begin{pmatrix} x_1 \\ x_2 \end{pmatrix} = \begin{pmatrix} 0 \\ 0 \end{pmatrix}$ or $x_1 + (-2 + i)x_2 = 0$, which yields $x_1 = (2 - i)x_2$, $\mathbf{v}_2 = \begin{pmatrix} 2-i \\ 1 \end{pmatrix}$, and $E_{1-i} = \text{span}\left\{\begin{pmatrix} 2-i \\ 1 \end{pmatrix}\right\}$.

† Note that the columns of this matrix are linearly dependent because $\begin{pmatrix} -5 \\ -2-i \end{pmatrix} = (-2 - i)\begin{pmatrix} 2-i \\ 1 \end{pmatrix}$.

*Remark 1.* This example illustrates that a real matrix may have complex eigenvalues and eigenvectors. Some texts define eigenvalues of real matrices to be *real* roots of the characteristic equation. With this definition the matrix of the last example has *no* eigenvalues. This might make the computations simpler, but it also greatly reduces the usefulness of the theory of eigenvalues and eigenvectors. We shall see a significant illustration of the use of complex eigenvalues in Section 6.7.

*Remark 2.* Note that $\lambda_2 = 1 - i$ is the complex conjugate of $\lambda_1 = 1 + i$. Also, the components of $\mathbf{v}_2$ are complex conjugates of the components of $\mathbf{v}_1$. This is no coincidence. In Problem 33 you are asked to prove that

> The eigenvalues of a *real* matrix occur in complex conjugate pairs
> and
> corresponding eigenvectors are complex conjugates of one another.

Before giving further examples, we prove a theorem that, in some special cases, simplifies the computation of eigenvalues.

**THEOREM 4** The eigenvalues of a triangular matrix are the diagonal components of the matrix.

**Proof** If $A = \begin{pmatrix} a_{11} & a_{12} & \cdots & a_{1n} \\ 0 & a_{22} & \cdots & a_{2n} \\ \vdots & \vdots & \ddots & \vdots \\ 0 & 0 & \cdots & a_{nn} \end{pmatrix}$, then $A - \lambda I = \begin{pmatrix} a_{11} - \lambda & a_{12} & \cdots & a_{1n} \\ 0 & a_{22} - \lambda & \cdots & a_{2n} \\ \vdots & \vdots & \ddots & \vdots \\ 0 & 0 & \cdots & a_{nn} - \lambda \end{pmatrix}$ and since the determinant of a triangular matrix is equal to the product of its diagonal components (see page 178), we see that $\det(A - \lambda I) = (a_{11} - \lambda)(a_{22} - \lambda)\cdots(a_{nn} - \lambda)$ with zeros $a_{11}, a_{22}, \ldots, a_{nn}$. The proof for a lower triangular matrix is virtually identical. ■

**EXAMPLE 7** **The Eigenvalues of a Triangular Matrix** Let $A = \begin{pmatrix} 2 & 5 & 6 \\ 0 & -3 & 2 \\ 0 & 0 & 5 \end{pmatrix}$. Then

$$\det(A - \lambda I) = \begin{vmatrix} 2 - \lambda & 5 & 6 \\ 0 & -3 - \lambda & 2 \\ 0 & 0 & 5 - \lambda \end{vmatrix} = (2 - \lambda)(-3 - \lambda)(5 - \lambda) \text{ with zeros}$$

(and eigenvalues) 2, −3, and 5. ■

We now give more examples of the computation of eigenvalues and eigenvectors for matrices that are not triangular.

**EXAMPLE 8** **A 2 × 2 Matrix with One Eigenvalue and Two Linearly Independent Eigenvectors** Let $A = \begin{pmatrix} 4 & 0 \\ 0 & 4 \end{pmatrix}$. Then $\det(A - \lambda I) = \begin{vmatrix} 4-\lambda & 0 \\ 0 & 4-\lambda \end{vmatrix} = (\lambda - 4)^2 = 0$; hence $\lambda = 4$ is an eigenvalue of algebraic multiplicity 2. Since $A = 4I$, we know that $A\mathbf{v} = 4\mathbf{v}$ for every vector $\mathbf{v} \in \mathbb{R}^2$ so that $E_4 = \mathbb{R}^2 = \text{span}\left\{\begin{pmatrix} 1 \\ 0 \end{pmatrix}, \begin{pmatrix} 0 \\ 1 \end{pmatrix}\right\}$.

**EXAMPLE 9** **A 2 × 2 Matrix with One Eigenvalue and Only One Linearly Independent Eigenvector** Let $A = \begin{pmatrix} 4 & 1 \\ 0 & 4 \end{pmatrix}$. Then $\det(A - \lambda I) = \begin{vmatrix} 4-\lambda & 1 \\ 0 & 4-\lambda \end{vmatrix} = (\lambda - 4)^2 = 0$; thus $\lambda = 4$ is again an eigenvalue of algebraic multiplicity 2. But this time we have $(A - 4I)\mathbf{v} = \begin{pmatrix} 0 & 1 \\ 0 & 0 \end{pmatrix}\begin{pmatrix} x_1 \\ x_2 \end{pmatrix} = \begin{pmatrix} x_2 \\ 0 \end{pmatrix}$. Thus $x_2 = 0$, $\mathbf{v}_1 = \begin{pmatrix} 1 \\ 0 \end{pmatrix}$ is an eigenvector, and $E_4 = \text{span}\left\{\begin{pmatrix} 1 \\ 0 \end{pmatrix}\right\}$.

**EXAMPLE 10** **A 3 × 3 Matrix with Two Distinct Eigenvalues and Three Linearly Independent Eigenvectors** Let $A = \begin{pmatrix} 3 & 2 & 4 \\ 2 & 0 & 2 \\ 4 & 2 & 3 \end{pmatrix}$. Then $\det(A - \lambda I) =$

$$\begin{vmatrix} 3-\lambda & 2 & 4 \\ 2 & -\lambda & 2 \\ 4 & 2 & 3-\lambda \end{vmatrix} = -\lambda^3 + 6\lambda^2 + 15\lambda + 8\dagger = -(\lambda + 1)^2(\lambda - 8) = 0 \quad \text{so}$$

that the eigenvalues are $\lambda_1 = 8$ and $\lambda_2 = -1$ (with algebraic multiplicity 2). For $\lambda_1 = 8$, we obtain

$$(A - 8I)\mathbf{v} = \begin{pmatrix} -5 & 2 & 4 \\ 2 & -8 & 2 \\ 4 & 2 & -5 \end{pmatrix}\begin{pmatrix} x_1 \\ x_2 \\ x_3 \end{pmatrix} = \begin{pmatrix} 0 \\ 0 \\ 0 \end{pmatrix}$$

or, row-reducing, we have

$$\left(\begin{array}{ccc|c} -5 & 2 & 4 & 0 \\ 2 & -8 & 2 & 0 \\ 4 & 2 & -5 & 0 \end{array}\right) \longrightarrow \left(\begin{array}{ccc|c} -5 & 2 & 4 & 0 \\ -18 & 0 & 18 & 0 \\ 9 & 0 & -9 & 0 \end{array}\right)$$

$$\longrightarrow \left(\begin{array}{ccc|c} -5 & 2 & 4 & 0 \\ -1 & 0 & 1 & 0 \\ 9 & 0 & -9 & 0 \end{array}\right) \longrightarrow \left(\begin{array}{ccc|c} 0 & 2 & -1 & 0 \\ -1 & 0 & 1 & 0 \\ 0 & 0 & 0 & 0 \end{array}\right)$$

† This computation is not obvious. We have left out the algebraic details in the calculation of a 3 × 3 determinant. We will do this from now on.

Hence $x_3 = 2x_2$ and $x_1 = x_3$, we obtain the eigenvector $\mathbf{v}_1 = \begin{pmatrix} 2 \\ 1 \\ 2 \end{pmatrix}$, and $E_8 = \text{span}\left\{\begin{pmatrix} 2 \\ 1 \\ 2 \end{pmatrix}\right\}$. For $\lambda_2 = -1$ we have $(A + I)\mathbf{v} = \begin{pmatrix} 4 & 2 & 4 \\ 2 & 1 & 2 \\ 4 & 2 & 4 \end{pmatrix}\begin{pmatrix} x_1 \\ x_2 \\ x_3 \end{pmatrix} = \begin{pmatrix} 0 \\ 0 \\ 0 \end{pmatrix}$, which gives us the single equation $2x_1 + x_2 + 2x_3 = 0$ or $x_2 = -2x_1 - 2x_3$. If $x_1 = 1$ and $x_3 = 0$, we obtain $\mathbf{v}_2 = \begin{pmatrix} 1 \\ -2 \\ 0 \end{pmatrix}$. If $x_1 = 0$ and $x_3 = 1$, we obtain $\mathbf{v}_3 = \begin{pmatrix} 0 \\ -2 \\ 1 \end{pmatrix}$. Thus $E_{-1} = \text{span}\left\{\begin{pmatrix} 1 \\ -2 \\ 0 \end{pmatrix}, \begin{pmatrix} 0 \\ -2 \\ 1 \end{pmatrix}\right\}$. There are other convenient choices for eigenvectors. For example, $\mathbf{v} = \begin{pmatrix} 1 \\ 0 \\ -1 \end{pmatrix}$ is in $E_{-1}$ since $\mathbf{v} = \mathbf{v}_2 - \mathbf{v}_3$.

EXAMPLE 11 **A 3 × 3 Matrix with One Eigenvalue and Only One Linearly Independent Eigenvector** Let $A = \begin{pmatrix} -5 & -5 & -9 \\ 8 & 9 & 18 \\ -2 & -3 & -7 \end{pmatrix}$. Then $\det(A - \lambda I) =$

$$\begin{vmatrix} -5-\lambda & -5 & -9 \\ 8 & 9-\lambda & 18 \\ -2 & -3 & -7-\lambda \end{vmatrix} = -\lambda^3 - 3\lambda^2 - 3\lambda - 1 = -(\lambda + 1)^3 = 0.$$

Thus $\lambda = -1$ is an eigenvalue of algebraic multiplicity 3. To compute $E_{-1}$, we set $(A + I)\mathbf{v} = \begin{pmatrix} -4 & -5 & -9 \\ 8 & 10 & 18 \\ -2 & -3 & -6 \end{pmatrix}\begin{pmatrix} x_1 \\ x_2 \\ x_3 \end{pmatrix} = \begin{pmatrix} 0 \\ 0 \\ 0 \end{pmatrix}$ and row-reduce to obtain, successively,

$$\left(\begin{array}{ccc|c} -4 & -5 & -9 & 0 \\ 8 & 10 & 18 & 0 \\ -2 & -3 & -6 & 0 \end{array}\right) \longrightarrow \left(\begin{array}{ccc|c} 0 & 1 & 3 & 0 \\ 0 & -2 & -6 & 0 \\ -2 & -3 & -6 & 0 \end{array}\right)$$

$$\longrightarrow \left(\begin{array}{ccc|c} 0 & 1 & 3 & 0 \\ 0 & 0 & 0 & 0 \\ -2 & 0 & 3 & 0 \end{array}\right)$$

This yields $x_2 = -3x_3$ and $2x_1 = 3x_3$. Setting $x_3 = 2$, we obtain only one linearly independent eigenvector: $\mathbf{v}_1 = \begin{pmatrix} 3 \\ -6 \\ 2 \end{pmatrix}$. Thus $E_{-1} = \text{span}\left\{\begin{pmatrix} 3 \\ -6 \\ 2 \end{pmatrix}\right\}$.

**EXAMPLE 12** **A 3 × 3 Matrix with One Eigenvalue and Two Linearly Independent Eigenvectors** Let $A = \begin{pmatrix} -1 & -3 & -9 \\ 0 & 5 & 18 \\ 0 & -2 & -7 \end{pmatrix}$. Then $\det(A - \lambda I) = \begin{vmatrix} -1-\lambda & -3 & -9 \\ 0 & 5-\lambda & 18 \\ 0 & -2 & -7-\lambda \end{vmatrix} = -(\lambda+1)^3 = 0$. Thus, as in Example 10, $\lambda = -1$ is an eigenvalue of algebraic multiplicity 3. To find $E_{-1}$, we compute $(A + I)\mathbf{v} = \begin{pmatrix} 0 & -3 & -9 \\ 0 & 6 & 18 \\ 0 & -2 & -6 \end{pmatrix}\begin{pmatrix} x_1 \\ x_2 \\ x_3 \end{pmatrix} = \begin{pmatrix} 0 \\ 0 \\ 0 \end{pmatrix}$. Thus $-2x_2 - 6x_3 = 0$ or $x_2 = -3x_3$, and $x_1$ is arbitrary. Setting $x_1 = 0, x_3 = 1$, we obtain $\mathbf{v}_1 = \begin{pmatrix} 0 \\ -3 \\ 1 \end{pmatrix}$. Setting $x_1 = 1, x_3 = 1$ yields $\mathbf{v}_2 = \begin{pmatrix} 1 \\ -3 \\ 1 \end{pmatrix}$. Thus $E_{-1} = \text{span}\left\{\begin{pmatrix} 0 \\ -3 \\ 1 \end{pmatrix}, \begin{pmatrix} 1 \\ -3 \\ 1 \end{pmatrix}\right\}$.

In each of the last five examples we found an eigenvalue with an algebraic multiplicity of 2 or more. But, as we saw in Examples 9, 11, and 12, the number of linearly independent eigenvectors is not necessarily equal to the algebraic multiplicity of the eigenvalue (as was the case in Examples 8 and 10). This observation leads to the following definition.

**DEFINITION 4** **Geometric Multiplicity** Let $\lambda$ be an eigenvalue of the matrix $A$. Then the **geometric multiplicity** of $\lambda$ is the dimension of the eigenspace corresponding to $\lambda$ (which is the nullity of the matrix $A - \lambda I$). That is,

$$\text{Geometric multiplicity of } \lambda = \dim E_\lambda = \nu(A - \lambda I)$$

In Examples 8 and 10 we saw that for the eigenvalues of algebraic multiplicity 2, the geometric multiplicities were also 2. In Example 9 the geometric multiplicity of $\lambda = 4$ was 1 while the algebraic multiplicity was 2. In Example 11 the algebraic multiplicity was 3 and the geometric multiplicity was 1. In Example 12 the algebraic multiplicity was 3 and the geometric multiplicity was 2. These examples illustrate the fact that if the algebraic multiplicity of $\lambda$ is greater than 1, then we cannot predict the geometric multiplicity of $\lambda$ without additional information.

If $A$ is a $2 \times 2$ matrix and $\lambda$ is an eigenvalue with algebraic multiplicity 2, then the geometric multiplicity of $\lambda$ is $\leq 2$ since there can be at most two linearly independent vectors in a two-dimensional space. Let $A$ be a $3 \times 3$ matrix having two eigenvalues $\lambda_1$ and $\lambda_2$ with algebraic multiplicities 1 and 2, respectively. Then the geometric multiplicity of $\lambda_2$ is $\leq 2$ because otherwise we would have at least four linearly independent vectors in a three-dimensional space. In fact, the geometric multiplicity of an eigenvalue is always less than or equal to its algebraic multiplicity. The proof of the following theorem is not difficult if additional facts about determinants are proved. Since this would take us too far afield, we omit the proof.†

**THEOREM 5** Let $\lambda$ be an eigenvalue of $A$. Then

$$\text{Geometric multiplicity of } \lambda \leq \text{algebraic multiplicity of } \lambda$$

*Note.* The geometric multiplicity of an eigenvalue is never zero. This follows from Definition 1, which states that if $\lambda$ is an eigenvalue, then there exists a *nonzero* eigenvector corresponding to $\lambda$.

In the rest of this chapter an important problem for us will be to determine whether a given $n \times n$ matrix does or does not have $n$ linearly independent eigenvectors. From what we have already discussed in this section, the following theorem is apparent.

**THEOREM 6** Let $A$ be an $n \times n$ matrix. Then $A$ has $n$ linearly independent eigenvectors if and only if the geometric multiplicity of every eigenvalue is equal to its algebraic multiplicity. In particular, $A$ has $n$ linearly independent eigenvectors if all the eigenvalues are distinct (since then the algebraic multiplicity of every eigenvalue is 1). ■

In Example 5 we saw a matrix for which zero was an eigenvalue. In fact, from Theorem 1 it is evident that zero is an eigenvalue of $A$ if and only if $\det A = \det (A - 0I) = 0$. This enables us to extend, for the last time, our Summing Up Theorem (see Theorem 5.4.4, page 515).

---

†For a proof see Theorem 11.2.6 in C. R. Wylie's book *Advanced Engineering Mathematics* (New York: McGraw-Hill, Inc., 1975).

**THEOREM 7** **Summing Up Theorem—View 8** Let $A$ be an $n \times n$ matrix. Then the following 12 statements are equivalent; that is, each one implies the other 11 (so that if one is true, all are true):

**i.** $A$ is invertible.

**ii.** The only solution to the homogeneous system $A\mathbf{x} = \mathbf{0}$ is the trivial solution ($\mathbf{x} = \mathbf{0}$).

**iii.** The system $A\mathbf{x} = \mathbf{b}$ has a unique solution for every $n$-vector $\mathbf{b}$.

**iv.** $A$ is row equivalent to the $n \times n$ identity matrix $I_n$.

**v.** $A$ can be written as the product of elementary matrices.

**vi.** The row echelon form of $A$ has $n$ pivots.

**vii.** The rows (and columns) of $A$ are linearly independent.

**viii.** $\det A \neq 0$.

**ix.** $\nu(A) = 0$.

**x.** $\rho(A) = n$.

**xi.** The linear transformation $T$ from $\mathbb{R}^n$ to $\mathbb{R}^n$ defined by $T\mathbf{x} = A\mathbf{x}$ is an isomorphism.

**xii.** Zero is *not* an eigenvalue of $A$.

## PROBLEMS 6.1

### Self-Quiz

#### *True-False*

**I.** The eigenvalues of a triangular matrix are the numbers on the diagonal of the matrix.

**II.** If the real $3 \times 3$ matrix $A$ has three distinct eigenvalues, then the eigenvectors corresponding to those eigenvalues constitute a basis for $\mathbb{R}^3$.

**III.** If the $3 \times 3$ matrix $A$ has two distinct eigenvalues, then $A$ has at most two linearly independent eigenvectors.

**IV.** If $A$ has real entries, then $A$ can have exactly one complex eigenvalue (i.e., an eigenvalue $a + ib$ with $b \neq 0$).

**V.** If $\det A = 0$, then 0 is an eigenvalue of $A$.

#### *Multiple Choice*

**VI.** 1 is an eigenvalue for the $3 \times 3$ identity matrix. Its geometric multiplicity is ______.

**a.** 1 **b.** 2 **c.** 3

**VII.** 1 is the only eigenvalue of $A = \begin{pmatrix} 1 & 2 & 0 \\ 0 & 1 & 0 \\ 0 & 0 & 1 \end{pmatrix}$. Its geometric multiplicity is ______.

**a.** 1 **b.** 2 **c.** 3

In Problems 1–20 calculate the eigenvalues and eigenspaces of the given matrix. If the algebraic multiplicity of an eigenvalue is greater than 1, calculate its geometric multiplicity.

**1.** $\begin{pmatrix} -2 & -2 \\ -5 & 1 \end{pmatrix}$

**2.** $\begin{pmatrix} -12 & 7 \\ -7 & 2 \end{pmatrix}$

**3.** $\begin{pmatrix} 2 & -1 \\ 5 & -2 \end{pmatrix}$

**4.** $\begin{pmatrix} -3 & 0 \\ 0 & -3 \end{pmatrix}$

**5.** $\begin{pmatrix} -3 & 2 \\ 0 & -3 \end{pmatrix}$

**6.** $\begin{pmatrix} 3 & 2 \\ -5 & 1 \end{pmatrix}$

**7.** $\begin{pmatrix} 1 & -1 & 0 \\ -1 & 2 & -1 \\ 0 & -1 & 1 \end{pmatrix}$

**8.** $\begin{pmatrix} 1 & 1 & -2 \\ -1 & 2 & 1 \\ 0 & 1 & -1 \end{pmatrix}$

**9.** $\begin{pmatrix} 5 & 4 & 2 \\ 4 & 5 & 2 \\ 2 & 2 & 2 \end{pmatrix}$

**10.** $\begin{pmatrix} 1 & 2 & 2 \\ 0 & 2 & 1 \\ -1 & 2 & 2 \end{pmatrix}$

**11.** $\begin{pmatrix} 0 & 1 & 0 \\ 0 & 0 & 1 \\ 1 & -3 & 3 \end{pmatrix}$

**12.** $\begin{pmatrix} -3 & -7 & -5 \\ 2 & 4 & 3 \\ 1 & 2 & 2 \end{pmatrix}$

**13.** $\begin{pmatrix} 1 & -1 & -1 \\ 1 & -1 & 0 \\ 1 & 0 & -1 \end{pmatrix}$

**14.** $\begin{pmatrix} 7 & -2 & -4 \\ 3 & 0 & -2 \\ 6 & -2 & -3 \end{pmatrix}$

**15.** $\begin{pmatrix} 4 & 6 & 6 \\ 1 & 3 & 2 \\ -1 & -5 & -2 \end{pmatrix}$

**16.** $\begin{pmatrix} 4 & 1 & 0 & 1 \\ 2 & 3 & 0 & 1 \\ -2 & 1 & 2 & -3 \\ 2 & -1 & 0 & 5 \end{pmatrix}$

**17.** $\begin{pmatrix} a & 0 & 0 & 0 \\ 0 & a & 0 & 0 \\ 0 & 0 & a & 0 \\ 0 & 0 & 0 & a \end{pmatrix}$

**18.** $\begin{pmatrix} a & b & 0 & 0 \\ 0 & a & 0 & 0 \\ 0 & 0 & a & 0 \\ 0 & 0 & 0 & a \end{pmatrix}$; $b \neq 0$

**19.** $\begin{pmatrix} a & b & 0 & 0 \\ 0 & a & c & 0 \\ 0 & 0 & a & 0 \\ 0 & 0 & 0 & a \end{pmatrix}$; $bc \neq 0$

**20.** $\begin{pmatrix} a & b & 0 & 0 \\ 0 & a & c & 0 \\ 0 & 0 & a & d \\ 0 & 0 & 0 & a \end{pmatrix}$; $bcd \neq 0$

**21.** Show that for any real numbers $a$ and $b$, the matrix $A = \begin{pmatrix} a & b \\ -b & a \end{pmatrix}$ has the eigenvectors $\begin{pmatrix} 1 \\ i \end{pmatrix}$ and $\begin{pmatrix} 1 \\ -i \end{pmatrix}$.

---

### Answers to Self-Quiz

**I.** True **II.** True **III.** False **IV.** False **V.** True **VI.** c **VII.** b

In Problems 22–28 assume that the matrix $A$ has the eigenvalues $\lambda_1, \lambda_2, \ldots, \lambda_k$.

**22.** Show that the eigenvalues of $A^t$ are $\lambda_1, \lambda_2, \ldots, \lambda_k$.

**23.** Show that the eigenvalues of $\alpha A$ are $\alpha\lambda_1, \alpha\lambda_2, \ldots, \alpha\lambda_k$.

**24.** Show that $A^{-1}$ exists if and only if $\lambda_1\lambda_2\cdots\lambda_k \neq 0$.

***25.** If $A^{-1}$ exists, show that the eigenvalues of $A^{-1}$ are $1/\lambda_1, 1/\lambda_2, \ldots, 1/\lambda_k$.

**26.** Show that the matrix $A - \alpha I$ has the eigenvalues $\lambda_1 - \alpha, \lambda_2 - \alpha, \ldots, \lambda_k - \alpha$.

***27.** Show that the eigenvalues of $A^2$ are $\lambda_1^2, \lambda_2^2, \ldots, \lambda_k^2$.

***28.** Show that the eigenvalues of $A^m$ are $\lambda_1^m, \lambda_2^m, \ldots, \lambda_k^m$ for $m = 1, 2, 3, \ldots$.

**29.** Let $\lambda$ be an eigenvalue of $A$ with corresponding eigenvector $\mathbf{v}$. Let $p(\lambda) = a_0 + a_1\lambda + a_2\lambda^2 + \cdots + a_n\lambda^n$. Define the matrix $p(A)$ by $p(A) = a_0I + a_1A + a_2A^2 + \cdots + a_nA^n$. Show that $p(A)\mathbf{v} = p(\lambda)\mathbf{v}$.

**30.** Using the result of Problem 29, show that if $\lambda_1, \lambda_2, \ldots, \lambda_k$ are eigenvalues of $A$, then $p(\lambda_1), p(\lambda_2), \ldots, p(\lambda_k)$ are eigenvalues of $p(A)$.

**31.** Show that if $A$ is a diagonal matrix, then the eigenvalues of $A$ are the diagonal components of $A$.

**32.** Let $A_1 = \begin{pmatrix} 2 & 0 & 0 & 0 \\ 0 & 2 & 0 & 0 \\ 0 & 0 & 2 & 0 \\ 0 & 0 & 0 & 2 \end{pmatrix}$, $A_2 = \begin{pmatrix} 2 & 1 & 0 & 0 \\ 0 & 2 & 0 & 0 \\ 0 & 0 & 2 & 0 \\ 0 & 0 & 0 & 2 \end{pmatrix}$, $A_3 = \begin{pmatrix} 2 & 1 & 0 & 0 \\ 0 & 2 & 1 & 0 \\ 0 & 0 & 2 & 0 \\ 0 & 0 & 0 & 2 \end{pmatrix}$, and $A_4 = \begin{pmatrix} 2 & 1 & 0 & 0 \\ 0 & 2 & 1 & 0 \\ 0 & 0 & 2 & 1 \\ 0 & 0 & 0 & 2 \end{pmatrix}$. Show that for each matrix $\lambda = 2$ is an eigenvalue of algebraic multiplicity 4. In each case compute the geometric multiplicity of $\lambda = 2$.

***33.** Let $A$ be a real $n \times n$ matrix. Show that if $\lambda_1$ is a complex eigenvalue of $A$ with eigenvector $\mathbf{v}_1$, then $\overline{\lambda}_1$ is an eigenvalue of $A$ with eigenvector $\overline{\mathbf{v}}_1$.

***34.** A **probability matrix** is an $n \times n$ matrix having two properties:

**i.** $a_{ij} \geq 0$ for every $i$ and $j$.
**ii.** The sum of the components in every column is 1.

Prove that 1 is an eigenvalue of every probability matrix.

***35.** Let $A = \begin{pmatrix} a & b \\ c & d \end{pmatrix}$ be a $2 \times 2$ matrix. Suppose that $b \neq 0$. Let $m$ be a root (real or complex) of the equation

$$bm^2 + (a - d)m - c = 0$$

show that $a + bm$ is an eigenvalue of $A$ with corresponding eigenvector $\mathbf{v} = \begin{pmatrix} 1 \\ m \end{pmatrix}$. This gives us an easy way to compute eigenvalues and eigenvectors of $2 \times 2$ matrices. [This procedure appeared in the paper "A Simple Algorithm for Finding Eigenvalues and Eigenvectors for $2 \times 2$ Matrices" by Tyre A. Newton in The *American Mathematical Monthly,* 97(1), January 1990, 57–60.]

**36.** Let $A = \begin{pmatrix} a & 0 \\ c & d \end{pmatrix}$ be a $2 \times 2$ matrix. Show that $d$ is an eigenvalue of $A$ with corresponding eigenvector $\begin{pmatrix} 0 \\ 1 \end{pmatrix}$.

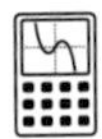

## CALCULATOR BOX

### TI-85

Eigenvalues and eigenvectors can be obtained directly on the TI-85. Suppose a square matrix $A$ is entered. Then the following key sequence will yield the eigenvalues of $A$:

2nd MATRX F3 ⟨MATH⟩ F4 ⟨eigV1⟩ ALPHA A ENTER

For example, if $A_1 = \begin{pmatrix} 1 & -1 & 4 \\ 3 & 2 & -1 \\ 2 & 1 & -1 \end{pmatrix}$ as in Example 4, the result is $\{3 \quad -2 \quad 1\}$. If $A_2 = \begin{pmatrix} 3 & -5 \\ 1 & -1 \end{pmatrix}$ as in Example 6, then $\{(1, 1)\ (1, -1)\}$ is displayed. This indicates that the eigenvalues are the complex numbers $1 + i$ and $1 - i$.

To obtain a set of eigenvectors, press the F5 ⟨eigVc⟩ key instead. For $A_1$ above the result is

```
[[-.442325868465  -.714920352984  -.527046276694]
 [-.884651736929   .714920352984  2.10818510678 ]
 [-.442325868465   .714920352984   .527046276695]]
```

Note that each column is an eigenvector. The third column is a multiple of $\begin{pmatrix} -1 \\ 4 \\ 1 \end{pmatrix}$. The second column is a multiple of $\begin{pmatrix} 1 \\ -1 \\ -1 \end{pmatrix}$ and the first column is a multiple of $\begin{pmatrix} 1 \\ 2 \\ 1 \end{pmatrix}$. For the matrix $A_2$ the result is

```
[[(-1, 2)  (-1, -2)]
 [ (0, 1)   (0, -1) ]]
```

Now the columns correspond to the eigenvectors $\begin{pmatrix} -1 + 2i \\ i \end{pmatrix}$ and $\begin{pmatrix} -1 - 2i \\ -i \end{pmatrix}$, respectively. Note that $-i\begin{pmatrix} -1 + 2i \\ i \end{pmatrix} = \begin{pmatrix} 2 + i \\ 1 \end{pmatrix}$, an eigenvector found in Example 6.

The geometric multiplicity of an eigenvalue may be found quickly by counting the linearly independent eigenvectors corresponding to that eigenvalue.

In Problems 37–40 find, on a calculator, the eigenvalues and a set of corresponding eigenvectors for each matrix.

**37.** $\begin{pmatrix} 4 & 1 & 2 & 6 \\ -1 & 3 & 4 & 2 \\ 5 & 2 & 0 & 6 \\ 3 & 8 & 1 & 5 \end{pmatrix}$

**38.** $\begin{pmatrix} 102 & -11 & 56 \\ 38 & -49 & 75 \\ 83 & 123 & -67 \end{pmatrix}$

**39.** $\begin{pmatrix} -0.031 & 0.082 & 0.095 \\ -0.046 & 0.067 & -0.081 \\ 0.055 & -0.077 & 0.038 \end{pmatrix}$

**40.** $\begin{pmatrix} 13 & 16 & 12 & 14 & 18 \\ 26 & 21 & 19 & 27 & 16 \\ 31 & 29 & 37 & 41 & 56 \\ 51 & 38 & 29 & 46 & 33 \\ 61 & 41 & 29 & 38 & 50 \end{pmatrix}$

In Problems 41–45 there is an eigenvalue of algebraic multiplicity 6. Determine its geometric multiplicity. Note that a number like $4E - 13 = 4 \times 10^{-13}$ is, effectively, equal to zero.

**41.** $\begin{pmatrix} 6 & 0 & 0 & 0 & 0 & 0 \\ 0 & 6 & 0 & 0 & 0 & 0 \\ 0 & 0 & 6 & 0 & 0 & 0 \\ 0 & 0 & 0 & 6 & 0 & 0 \\ 0 & 0 & 0 & 0 & 6 & 0 \\ 0 & 0 & 0 & 0 & 0 & 6 \end{pmatrix}$

**42.** $\begin{pmatrix} 6 & 1 & 0 & 0 & 0 & 0 \\ 0 & 6 & 0 & 0 & 0 & 0 \\ 0 & 0 & 6 & 0 & 0 & 0 \\ 0 & 0 & 0 & 6 & 0 & 0 \\ 0 & 0 & 0 & 0 & 6 & 0 \\ 0 & 0 & 0 & 0 & 0 & 6 \end{pmatrix}$

**43.** $\begin{pmatrix} 6 & 1 & 0 & 0 & 0 & 0 \\ 0 & 6 & 1 & 0 & 0 & 0 \\ 0 & 0 & 6 & 1 & 0 & 0 \\ 0 & 0 & 0 & 6 & 0 & 0 \\ 0 & 0 & 0 & 0 & 6 & 0 \\ 0 & 0 & 0 & 0 & 0 & 6 \end{pmatrix}$

**44.** $\begin{pmatrix} 6 & 1 & 0 & 0 & 0 & 0 \\ 0 & 6 & 1 & 0 & 0 & 0 \\ 0 & 0 & 6 & 1 & 0 & 0 \\ 0 & 0 & 0 & 6 & 1 & 0 \\ 0 & 0 & 0 & 0 & 6 & 0 \\ 0 & 0 & 0 & 0 & 0 & 6 \end{pmatrix}$

**45.** $\begin{pmatrix} 6 & 1 & 0 & 0 & 0 & 0 \\ 0 & 6 & 1 & 0 & 0 & 0 \\ 0 & 0 & 6 & 1 & 0 & 0 \\ 0 & 0 & 0 & 6 & 1 & 0 \\ 0 & 0 & 0 & 0 & 6 & 1 \\ 0 & 0 & 0 & 0 & 0 & 6 \end{pmatrix}$

## MATLAB 6.1

1. Consider the matrix below.

$$A = \begin{pmatrix} 38 & -95 & 55 \\ 35 & -92 & 55 \\ 35 & -95 & 58 \end{pmatrix}$$

a. Verify that $\mathbf{x} = (1 \quad 1 \quad 1)^t$ is an eigenvector for $A$ with eigenvalue $\lambda = -2$, that $\mathbf{y} = (3 \quad 4 \quad 5)^t$ is an eigenvector for $A$ with eigenvalue $\mu = 3$, and that $\mathbf{z} = (4 \quad 9 \quad 13)^t$ is an eigenvector for $A$ with eigenvalue $\mu = 3$. (*Note.* The best way to show $\mathbf{w}$ is an eigenvector for $A$ with eigenvalue $c$ is to show that $(A - cI)\mathbf{w} = \mathbf{0}$.)

b. Choose a random value for the scalar $a$. Verify that $a\mathbf{x}$ is an eigenvector for $A$ with eigenvalue $\lambda = -2$. Verify that $a\mathbf{y}$ and $a\mathbf{z}$ are eigenvectors for $A$ with eigenvalue $\mu = 3$. Repeat for three more choices of $a$.

c. Choose random values for scalars $a$ and $b$. Verify that $\mathbf{w} = a\mathbf{y} + b\mathbf{z}$ is an eigenvector for $A$ with eigenvalue $\mu = 3$. Repeat for three more choices of $a$ and $b$.

d. *(Paper and pencil)* What property of eigenvalues and eigenvectors is illustrated by parts (b) and (c)?

2. Consider the matrix below.

$$A = \begin{pmatrix} 1 & 1 & .5 & -1 \\ -2 & 1 & -1 & 0 \\ 0 & 2 & 0 & 2 \\ 2 & 1 & -1.5 & 2 \end{pmatrix}$$

a. Verify that $\mathbf{x} = (1 \quad i \quad 0 \quad -i)^t$ and $\mathbf{v} = (0 \quad i \quad 2 \quad 1 + i)^t$ are eigenvectors for $A$ with eigenvalue $\lambda = 1 + 2i$ and that $\mathbf{y} = (1 \quad -i \quad 0 \quad i)^t$ and $\mathbf{z} = (0 \quad -i \quad 2 \quad 1 - i)^t$ are eigenvectors for $A$ with eigenvalue $\mu = 1 - 2i$. (To find the transpose of a complex matrix A use **A.'**)

b. Choose a random *complex* value for the scalar $a$. (For example, choose **a = 5*(2*rand(1)−1)+i*3*rand(1).**) Verify that $a\mathbf{x}$ and $a\mathbf{y}$ are eigenvectors for $A$ with eigenvalue $\lambda = 1 + 2i$. Verify that $a\mathbf{y}$ and $a\mathbf{z}$ are eigenvectors for $A$ with eigenvalue $\mu = 1 - 2i$. Repeat for three more choices of $a$.

c. Choose random *complex* values for scalars $a$ and $b$. Verify that $\mathbf{u} = a\mathbf{x} + b\mathbf{v}$ is an eigenvector for $A$ with eigenvalue $\lambda = 2 + i$. Verify that $\mathbf{w} = a\mathbf{y} + b\mathbf{z}$ is an eigenvector for $A$ with eigenvalue $\mu = 2 - i$. Repeat for three more choices of $a$ and $b$.

d. *(Paper and pencil)* What property of eigenvalues and eigenvectors is illustrated by parts (b) and (c)?

3. Follow the directions below for each of the matrices $A$ in Problems 1, 6, 8, and 13 in this section of the text.

a. Find the characteristic polynomial *by hand* and check by finding **c = (−1)^n*poly(A).** (Here $n$ is the size of the matrix.) Enter **help poly** to help interpret the result of **poly** and explain why the $(-1)^n$ factor was included above.

b. Find the eigenvalues by finding the roots of the characteristic polynomial *by hand.* Check by finding **r = roots(c).**

c. For each eigenvalue $\lambda$ found, solve $(A - \lambda I)\mathbf{x} = \mathbf{0}$ *by hand* and check by using **rref(A − r(k)*eye(n))** for $k = 1, \ldots, n$, where **r** is the vector containing the eigenvalues and $n$ is the size of the matrix.

d. Verify that there are $n$ distinct eigenvalues (where $n$ is the size of the matrix) and there is a set of $n$ eigenvectors which is linearly independent.

e. Enter **[V,D] = eig(A).** For $k = 1, \ldots n$, verify that

**(A−D(k,k)*eye(n))*V(:,k) = 0**

Write a statement interpreting this in the language of eigenvalues and eigenvectors.

The routine **eig** finds eigenvectors of length 1. Since each eigenvalue has geometric and algebraic multiplicity 1, the vectors found in part (c), normalized to length 1, should match the columns of $V$ up to a possible multiple by a complex number of modulus 1 (usually 1, $-1$, $i$, or $-i$). Verify this.

4. Computations of eigenvalues (and associated eigenvectors) are sensitive to round-off error, especially when the eigenvalue has algebraic multiplicity greater than 1.
   a. *(Paper and pencil)* For the matrix below, compute the eigenvalues and eigenvectors by hand. Verify that $\lambda = 2$ is an eigenvalue with algebraic multiplicity 2 (and geometric multiplicity 1).

$$A = \begin{pmatrix} 1 & 2 & 2 \\ 0 & 2 & 1 \\ -1 & 2 & 2 \end{pmatrix}$$

   b. Find **c = poly(A)** and compare with your hand computations. Enter **format long.** Find **r = roots(c).** What do you notice about the eigenvalues? Try finding the eigenvectors by **rref(A−r(k)*eye(3))** for $k = 1$, 2, and 3. How successful are you?
   c. The **eig** routine is more numerically stable than using **roots.** (It uses a different process from the theoretical one described in this section.) However, it cannot avoid the basic fact about multiple roots and round-off error discussed in part (e). Still using **format long,** find **[V,D] = eig(A).** Compare the eigenvalues in $D$ with the true eigenvalues and with the eigenvalues computed in part (b). Argue why the computations from **eig** are somewhat closer to the true values.
   d. For $k = 1$, 2, and 3, verify that **(A−D(k,k)*eye(3))*V(:,k)** is close to zero. How could this lead one to say that even though there are inaccuracies, the computations are not too bad in some sense?

      Slight perturbations in the computation of the eigenvectors will lead to the eigenvectors being linearly independent: Find **rref(V).** By examining $V$, do you see any evidence that the eigenvectors associated with the eigenvalues close to $\lambda = 2$ are "nearly" dependent?
   e. *(Paper and pencil)* This part gives a rough explanation of the problems associated with numerical approximations of multiple roots (in this context, roots of the characteristic polynomial with algebraic multiplicity greater than 1). Below is a sketch of the characteristic polynomial $y = -(\lambda - 2)^2(\lambda - 1)$.

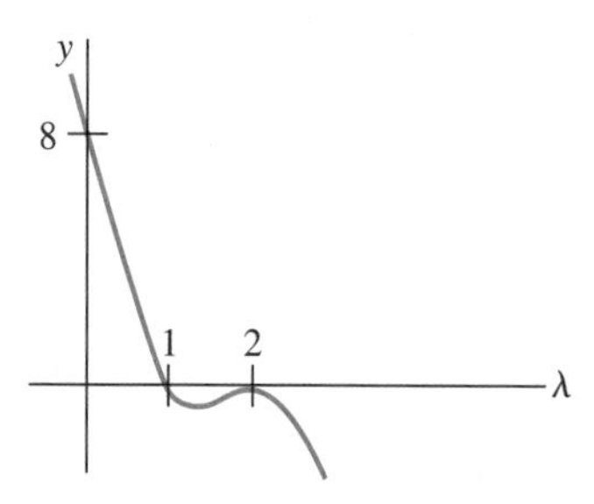


      Round-off error perturbs the values slightly. Suppose the perturbation is such that the graph is slightly shifted downward. Resketch the graph and explain why there is no longer a root of the function at $\lambda = 2$ and, in fact, two complex roots have been created where there was one real root. Suppose the graph is slightly shifted upward. Resketch the graph and explain what happens to the multiple root at $\lambda = 2$. Describe how these effects were noticed in the computations in previous parts of this problem.

5. a. For the matrices $A$ in Problems 6, 8, 12, 13, and 16 in this section of the text, find **poly(A)–poly(A′).** Regarding small numbers as zero (there is always round-off

error), formulate a conjecture about the characteristic polynomials of $A$ and $A^t$. What does this imply about the eigenvalues?

b. *(Paper and pencil)* Prove your conjecture.

6. a. Generate a random noninvertible matrix $A$. [Start with a random $A$ and change $A$ by replacing some of the columns (or rows) with linear combinations of some other columns (or rows).] Find **d = eig(A).** (If you give **eig** only one output variable, it returns a single vector containing the eigenvalues.) Repeat for three more noninvertible matrices. What do the eigenvalue sets of these matrices have in common? Explain why this should be so.

   b. i. For the matrices $A$ in Problems 1, 2, and 8 in the text and for the matrix below, find **d = eig(A)** and **e = eig(inv(A)).**

$$A = \begin{pmatrix} 3 & 9.5 & -2 & -10.5 \\ -10 & -42.5 & 10 & 44.5 \\ 6 & 23.5 & -5 & -24.5 \\ -10 & -43 & 10 & 45 \end{pmatrix}$$

   ii. Ignoring the order in which they appear in the vectors **d** and **e**, conjecture a relationship between the eigenvalues for $A$ and $A^{-1}$. Explain the evidence for your conjecture. Complete the following: If $\lambda$ is an eigenvalue for $A$, then _ is an eigenvalue for $A^{-1}$.

   iii. Test your conjecture on the matrices in Problems 3, 6, and 13 in this section of the text.

   c. For each of the matrices considered in part (b), compare the reduced row echelon forms of $A - \lambda I$ and $A^{-1} - \mu I$, where $\lambda$ is an eigenvalue for $A$ and $\mu$ is the corresponding eigenvalue for $A^{-1}$ discovered in part (b). Explain what this comparison tells you about the corresponding eigenvectors.

   d. *(Paper and pencil)* Formulate a conjecture about the relationship between the eigenvalues and eigenvectors of $A$ and the eigenvalues and eigenvectors of $A^{-1}$ and prove the conjecture. [*Hint.* Consider $AA^{-1}\mathbf{v}$, where $\mathbf{v}$ is an eigenvector for $A$.]

7. Follow the directions of MATLAB Problem 6, parts (b) through (d), in this section, except replace $A^{-1}$ with $A^2$ and **inv(A)** with **A*A**.

8. For each of the matrices $A$ in Problems 6, 7, 8, and 13 in this section of the text and for $A$, a random $4 \times 4$ matrix, generate a random invertible matrix $C$ of the same size as $A$ and form $B = CAC^{-1}$. Ignoring the order in which the values appear (and considering small numbers to be zero), compare the eigenvalues of $A$, **eig(A)**, with the eigenvalues of $B$, **eig(B)**. Describe any conclusions you can make from these comparisons.

9. We have seen that the eigenvalues of a random real $n \times n$ matrix can be any real or complex number as long as the complex numbers occur in complex conjugate pairs. We will examine some special categories of real matrices to see if special classes have special restrictions on the possible type of eigenvalues. (Because of round-off considerations, consider small numbers to be zero.)

   a. Generate a random real *symmetric* $n \times n$ matrix for some value of $n$. (Let $B$ be a random $n \times n$ matrix. Let **A = triu(B)+triu(B)′**.) Find **eig(A)**. Repeat for four more choices of symmetric matrices $A$. (Use more than one value of $n$.) Conjecture a property of eigenvalues of symmetric matrices.

   b. A special class of real symmetric matrices are the matrices $C$ formed by $C = AA^t$ for any matrix $A$. Generate five such matrices. (Do not use matrices all of the same size.)

Find **eig(C)** for each of these. Conjecture a property of eigenvalues of matrices of the form $AA^t$.

**10.** We have seen that if a matrix has distinct eigenvalues, then the eigenvectors are linearly independent. A nice class of linearly independent vectors is the class of orthogonal vectors. Generate a random real *symmetric* matrix $A$ as in MATLAB Problem 9. Find **[V,D] = eig(A)** and verify that the eigenvalues are distinct and the eigenvectors are orthogonal. Repeat for four more choices of $A$. (Use different sizes.)

**11. Graph Theory** For a graph of vertices and edges (like those below), we define the **adjacency matrix** $A$ for the graph as

$$a_{ij} = \begin{cases} 1 & \text{if } i \text{ and } j \text{ are connected by an edge} \\ 0 & \text{otherwise} \end{cases}$$

We use the convention that $a_{ii} = 0$.

We define the **chromatic number** of the graph to be the minimum number of colors needed to color the vertices of the graph in such a way that no two adjacent vertices are assigned the same color. Vertices are **adjacent** if they are connected by an edge.

The adjacency matrix for a graph is symmetric. (Why?) Therefore the eigenvalues will be real-valued (see Section 6.3 or MATLAB Problem 9 above), and hence can be ordered from the largest to the smallest. (Here we order as we would numbers on the real line; we are *not* just ordering the magnitudes.) Let $\lambda_1$ denote the largest eigenvalue and let $\lambda_n$ denote the smallest eigenvalue. It will turn out that $\lambda_1$ is positive and that $\lambda_n$ is negative.

Assume that the graph is **connected;** that is, there is a path from each vertex to any other vertex, possibly routing through other vertices. Let $\chi$ denote the chromatic number. Then it can be shown that

$$1 - \frac{\lambda_1}{\lambda_n} \leq \chi \leq 1 + \lambda_1$$

Using this theorem, find bounds on the chromatic number for the connected graphs below. Check the result by redrawing the graphs and coloring the vertices appropriately. For parts (a) through (c), based on looking at the graph, try to give some argument why you could not color the vertices with fewer colors than the theorem indicates. (*Note.* Recall that the chromatic number is an integer so that you are looking for integers that lie between the bounds given by the theorem.)

**a.**

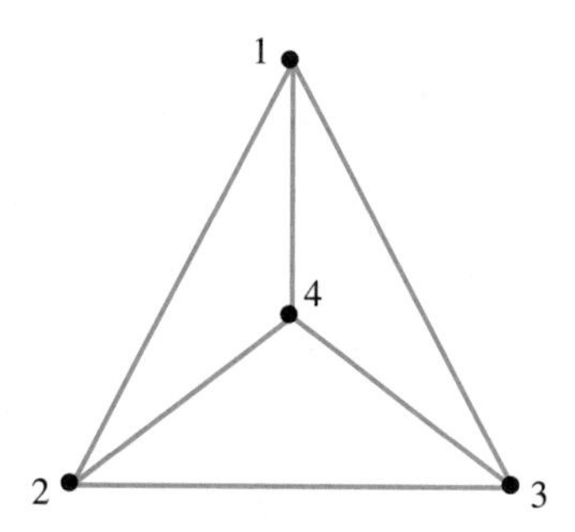

**b.**

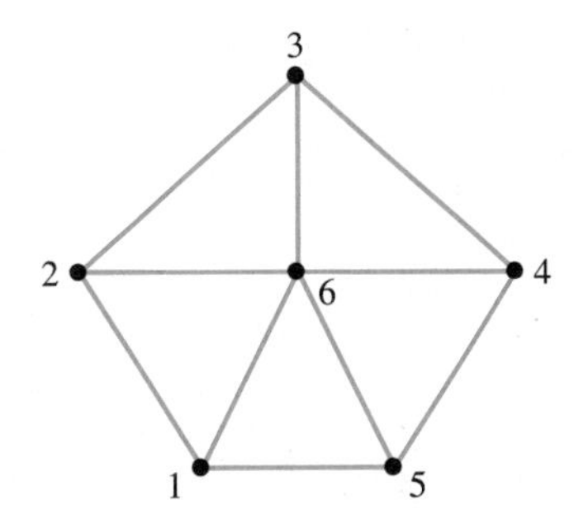


**c.**

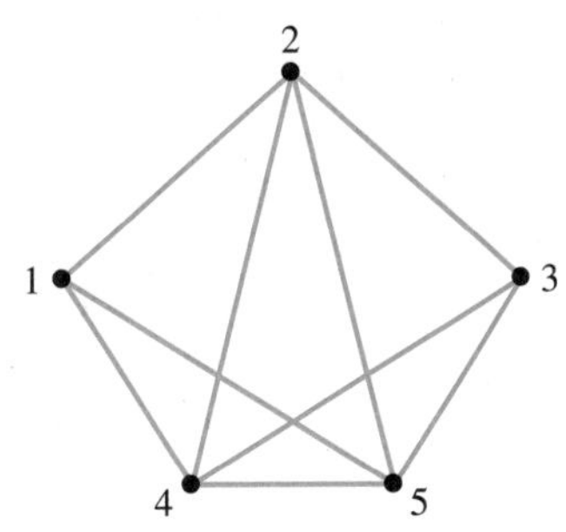


**d.**

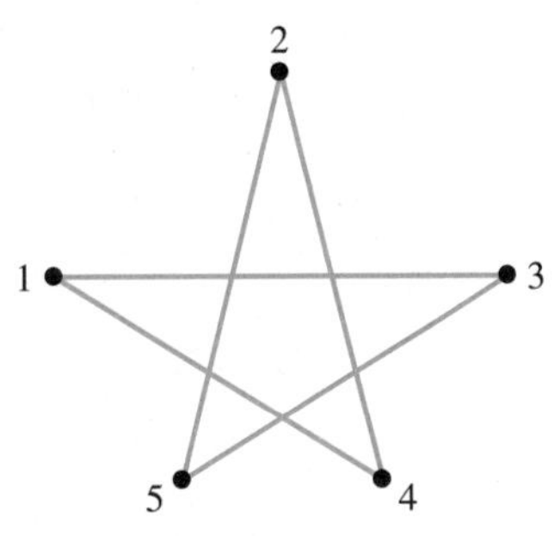


**e.**

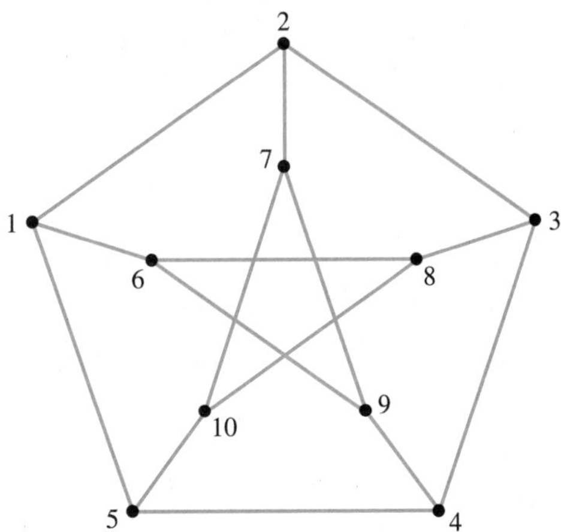


**f.** Draw your own graph and follow the directions above.

12. **Geology** One of the most important properties of deformed rock is its internal strain. One measurement of strain is based on the mechanical twinning of calcite. The initial shape of the calcite is known from the crystallography of calcite, and the deformed shape can be measured. Measurements are made from thin sections of rock samples containing calcite. Certain numbers are computed, representing measures of deformation (strain) with respect to a coordinate system determined by the thin section, and are placed in a $3 \times 3$ matrix. The eigenvectors of this matrix represent the directions of the principal strain axes. The associated eigenvalues represent the magnitudes of the strains in the direction of the principal axes with positive eigenvalues indicating extension and negative eigenvalues representing compression.
   a. For each of the matrices of deformation measurements below, find the direction (unit vector) of the principal axis of maximum extension and the direction (unit vector) of the principal axis of maximum compression:

$$A = \begin{pmatrix} -.01969633 & .01057339 & -.005030409 \\ .01057339 & .008020058 & -.006818069 \\ -.005030409 & -.006818069 & .01158627 \end{pmatrix}$$

$$A = \begin{pmatrix} -.01470626 & .01001909 & -.004158314 \\ .01001901 & .007722046 & -.004482362 \\ -.004158314 & -.004482362 & .006984212 \end{pmatrix}$$

   b. For each matrix from part (a), find the angle that the principal axis of maximum compressive strain makes with the $x$-axis (in thin section coordinates). [*Note.* The $x$-axis is represented by the vector $(1 \quad 0 \quad 0)^t$. Recall that the cosine of the angle between vectors $\mathbf{v}$ and $\mathbf{w}$ is found by $\mathbf{v} \cdot \mathbf{w}/|\mathbf{v}|\,|\mathbf{w}|$. Use the MATLAB function **acos** and multiply by $180/\pi$ to convert to degrees.]
   c. In a fold formation, the twinning strain is related to the total strain of the folding. The adequacy of a fold model to explain the structure of the folding can be tested using strain data. One model is that of *simple shear parallel to bedding.* (Here the bedding is parallel to the $x$-axis in the thin section coordinates.) For the sites from which the data in part (a) were obtained, the simple shear parallel to bedding model predicts that the acute angle between the *lines* determined by bedding (the $x$-axis) and the principal axis of maximum compressive strain is rather high, nearly 45° in many locations. Using the results from part (b), argue why this model is inadequate to explain the folding structure for the fold from which the data were obtained.

   *Acknowledgment.* The data and the interpretations that form the basis of this problem were derived from work by Dr. Richard Groshong, University of Alabama.

## 6.2 A MODEL OF POPULATION GROWTH (Optional)

In this section we show how the theory of eigenvalues and eigenvectors can be used to analyze a model of the growth of a bird population.† We begin by discussing a simple model of population growth. We assume that a certain species grows at a constant rate; that is, the population of the species after one time period (which

† The material in this section is based on a paper by D. Cooke: "A $2 \times 2$ Matrix Model of Population Growth," *Mathematical Gazette* 61(416): 120–123.

could be an hour, a week, a month, a year, etc.) is a constant multiple of the population in the previous time period. One way this could happen, for example, is that each generation is distinct and each organism produces $r$ offspring and then dies. If $p_n$ denotes the population after the $n$th time period, we would have

$$p_n = rp_{n-1}$$

For example, this model might describe a bacteria population, where, at a given time, an organism splits into two separate organisms. Then $r = 2$. Let $p_0$ denote the initial population. Then $p_1 = rp_0$, $p_2 = rp_1 = r(rp_0) = r^2p_0$, $p_3 = rp_2 = r(r^2p_0) = r^3p_0$, and so on, so that

$$p_n = r^n p_0 \tag{1}$$

From this model we see that the population increases without bound if $r > 1$ and decreases to zero if $r < 1$. If $r = 1$ the population remains at the constant value $p_0$.

This model is, evidently, very simplistic. One obvious objection is that the number of offspring produced depends, in many cases, on the ages of the adults. For example, in a human population the average female adult over 50 would certainly produce fewer children than the average 21-year-old female. To deal with this difficulty, we introduce a model that allows for age groupings with different fertility rates.

We look at a model of population growth for a species of birds. In this bird population we assume that the number of female birds equals the number of males. Let $p_{j,n-1}$ denote the population of juvenile (immature) females in the $(n-1)$st year and let $p_{a,n-1}$ denote the number of adult females in the $(n-1)$st year. Some of the juvenile birds will die during the year. We assume that a certain proportion $\alpha$ of the juvenile birds survive to become adults in the spring of the $n$th year. Each surviving female bird produces eggs later in the spring, which hatch to produce, on the average, $k$ juvenile female birds in the following spring. Adults also die, and the proportion of adults that survive from one spring to the next is $\beta$.

This constant survival rate of birds is not just a simplistic assumption. It appears to be the case with most of the natural bird populations that have been studied. This means that the adult survival rate of many bird species is independent of age. Perhaps few birds in the wild survive long enough to exhibit the effects of old age. Moreover, in many species the number of offspring seems to be uninfluenced by the age of the mother.

In the notation introduced above $p_{j,n}$ and $p_{a,n}$ represent, respectively, the populations of juvenile and adult females in the $n$th year. Putting together all the information given, we arrive at the following $2 \times 2$ system:

$$\begin{aligned} p_{j,n} &= \qquad\qquad kp_{a,n-1} \\ p_{a,n} &= \alpha p_{j,n-1} + \beta p_{a,n-1} \end{aligned} \tag{2}$$

or

$$\mathbf{p}_n = A\mathbf{p}_{n-1} \tag{3}$$

where $\mathbf{p}_n = \begin{pmatrix} p_{j,n} \\ p_{a,n} \end{pmatrix}$ and $A = \begin{pmatrix} 0 & k \\ \alpha & \beta \end{pmatrix}$. It is clear from (3) that $\mathbf{p}_1 = A\mathbf{p}_0$, $\mathbf{p}_2 = A\mathbf{p}_1 = A(A\mathbf{p}_0) = A^2\mathbf{p}_0, \ldots$, and so on. Hence

$$\boxed{\mathbf{p}_n = A^n\mathbf{p}_0} \tag{4}$$

where $\mathbf{p}_0$ is the vector of initial populations of juvenile and adult females.

Equation (4) is like equation (1), but now we are able to distinguish between the survival rates of juvenile and adult birds.

**EXAMPLE 1** **An Illustration of the Model Carried Through 20 Generations** Let $A = \begin{pmatrix} 0 & 2 \\ 0.3 & 0.5 \end{pmatrix}$. This means that each adult female produces two female offspring, and since the number of males is assumed equal to the number of females, at least four eggs—and probably many more since losses among fledglings are likely to be high. From the model it is apparent that $\alpha$ and $\beta$ lie in the interval $[0, 1]$. Since juvenile birds are not so likely as adults to survive, we must have $\alpha < \beta$.

In Table 6.1 we assume that, initially, there are 10 female (and 10 male) adults and no juveniles. The computations were done on a computer, but the work would not be too onerous if done on a hand calculator. For example, $\mathbf{p}_1 = \begin{pmatrix} 0 & 2 \\ 0.3 & 0.5 \end{pmatrix}\begin{pmatrix} 0 \\ 10 \end{pmatrix} = \begin{pmatrix} 20 \\ 5 \end{pmatrix}$ so that $p_{j,1} = 20$, $p_{a,1} = 5$, the total female population after 1 year is 25, and the ratio of juvenile to adult females is 4 to 1. In the second

**Table 6.1**

| Year $n$ | No. of juveniles $p_{j,n}$ | No. of adults $p_{a,n}$ | Total female population $T_n$ in $n$th year | $p_{j,n}/p_{a,n}$† | $T_n/T_{n-1}$† |
|---|---|---|---|---|---|
| 0 | 0 | 10 | 10 | 0 | — |
| 1 | 20 | 5 | 25 | 4.00 | 2.50 |
| 2 | 10 | 8 | 18 | 1.18 | 0.74 |
| 3 | 17 | 7 | 24 | 2.34 | 1.31 |
| 4 | 14 | 8 | 22 | 1.66 | 0.96 |
| 5 | 17 | 8 | 25 | 2.00 | 1.13 |
| 10 | 22 | 12 | 34 | 1.87 | 1.06 |
| 11 | 24 | 12 | 36 | 1.88 | 1.07 |
| 12 | 25 | 13 | 38 | 1.88 | 1.06 |
| 20 | 42 | 22 | 64 | 1.88 | 1.06 |

†The figures in these columns were obtained before the numbers in the previous columns were rounded. Thus, for example, in year 2, $p_{j,2}/p_{a,2} = 10/8.5 \approx 1.176470588 \approx 1.18$.

year $\mathbf{p}_2 = \begin{pmatrix} 0 & 2 \\ 0.3 & 0.5 \end{pmatrix}\begin{pmatrix} 20 \\ 5 \end{pmatrix} = \begin{pmatrix} 10 \\ 8.5 \end{pmatrix}$, which we round down to $\begin{pmatrix} 10 \\ 8 \end{pmatrix}$ since we cannot have $8\frac{1}{2}$ adult birds. Table 6.1 tabulates the ratios $p_{j,n}/p_{a,n}$ and the ratios $T_n/T_{n-1}$ of the total number of females in successive years.

In Table 6.1 it seems as if the ratio $p_{j,n}/p_{a,n}$ is approaching the constant 1.88 while the total population seems to be increasing at a constant rate of 6 percent a year. Let us see if we can determine why this is the case.

First, we return to the general case [equation (4)]. Suppose that $A$ has the real distinct eigenvalues $\lambda_1$ and $\lambda_2$ with corresponding eigenvectors $\mathbf{v}_1$ and $\mathbf{v}_2$. Since $\mathbf{v}_1$ and $\mathbf{v}_2$ are linearly independent, they form a basis for $\mathbb{R}^2$ and we can write

$$\mathbf{p}_0 = a_1\mathbf{v}_1 + a_2\mathbf{v}_2 \tag{5}$$

for some real numbers $a_1$ and $a_2$. Then (4) becomes

$$\mathbf{p}_n = A^n(a_1\mathbf{v}_1 + a_2\mathbf{v}_2) \tag{6}$$

But $A\mathbf{v}_1 = \lambda_1\mathbf{v}_1$ and $A^2\mathbf{v}_1 = A(A\mathbf{v}_1) = A(\lambda_1\mathbf{v}_1) = \lambda_1 A\mathbf{v}_1 = \lambda_1(\lambda_1\mathbf{v}_1) = \lambda_1^2\mathbf{v}_1$. Thus we can see that $A^n\mathbf{v}_1 = \lambda_1^n\mathbf{v}_1$, $A^n\mathbf{v}_2 = \lambda_2^n\mathbf{v}_2$, and from (6)

$$\mathbf{p}_n = a_1\lambda_1^n\mathbf{v}_1 + a_2\lambda_2^n\mathbf{v}_2 \tag{7}$$

The characteristic equation of $A$ is $\begin{vmatrix} -\lambda & k \\ \alpha & \beta - \lambda \end{vmatrix} = \lambda^2 - \beta\lambda - k\alpha = 0$ or $\lambda = (\beta \pm \sqrt{\beta^2 + 4\alpha k})/2$. By assumption, $k > 0$, $0 < \alpha < 1$, and $0 < \beta < 1$. Hence $4\alpha k > 0$ and $\beta^2 + 4\alpha k > 0$, which means that the eigenvalues are, indeed, real and distinct and that one eigenvalue, $\lambda_1$, is positive; the other, $\lambda_2$, is negative; and $|\lambda_1| > |\lambda_2|$. We can write (7) as

$$\mathbf{p}_n = \lambda_1^n\left[a_1\mathbf{v}_1 + \left(\frac{\lambda_2}{\lambda_1}\right)^n a_2\mathbf{v}_2\right] \tag{8}$$

Since $|\lambda_2/\lambda_1| < 1$, it is apparent that $(\lambda_2/\lambda_1)^n$ gets very small as $n$ gets large. Thus for $n$ large

$$\mathbf{p}_n \approx a_1\lambda_1^n\mathbf{v}_1 \tag{9}$$

This means that, in the long run, the age distribution stabilizes and is proportional to $\mathbf{v}_1$. Each age group will change by a factor of $\lambda_1$ each year. Thus, in the long run, equation (4) acts just like equation (1). In the short term—that is, before "stability" is reached—the numbers oscillate. The magnitude of this oscillation depends on the magnitude of $\lambda_2/\lambda_1$ (which is negative, thus explaining the oscillation).

EXAMPLE 1 (continued)

**The Eigenvalues and Eigenvectors of $A$ Determine the Behavior in Future Generations** For $A = \begin{pmatrix} 0 & 2 \\ 0.3 & 0.5 \end{pmatrix}$, we have $\lambda^2 - 0.5\lambda - 0.6 = 0$ or $\lambda = (0.5 \pm \sqrt{0.25 + 2.4})/2 = (0.5 \pm \sqrt{2.65})/2$ so that $\lambda_1 \approx 1.06$ and $\lambda_2 \approx$

$-0.56$. This explains the 6 percent increase in population noted in the last column of Table 6.1. Corresponding to the eigenvalue $\lambda_1 = 1.06$, we compute $(A - 1.06I)\mathbf{v}_1 = \begin{pmatrix} -1.06 & 2 \\ 0.3 & -0.56 \end{pmatrix}\begin{pmatrix} x_1 \\ x_2 \end{pmatrix} = \begin{pmatrix} 0 \\ 0 \end{pmatrix}$ or $1.06x_1 = 2x_2$ so that $\mathbf{v}_1 = \begin{pmatrix} 1 \\ 0.53 \end{pmatrix}$ is an eigenvector. Similarly, $(A + 0.56)\mathbf{v}_2 = \begin{pmatrix} 0.56 & 2 \\ 0.3 & 1.06 \end{pmatrix}\begin{pmatrix} x_1 \\ x_2 \end{pmatrix} = \begin{pmatrix} 0 \\ 0 \end{pmatrix}$ so that $0.56x_1 + 2x_2 = 0$ and $\mathbf{v}_2 = \begin{pmatrix} 1 \\ -0.28 \end{pmatrix}$ is a second eigenvector. Note that in $\mathbf{v}_1$ we have $1/0.53 \approx 1.88$. This explains the ratio $p_{j,n}/p_{a,n}$ in the fifth column of the table.

***Remark.*** In the preceding computations precision was lost because we rounded to only two decimal places of accuracy. Much greater accuracy is obtained by using a hand calculator or computer. For example, using a hand calculator, we easily calculate $\lambda_1 = 1.06394103$, $\lambda_2 = -0.5639410298$, $\mathbf{v}_1 = \begin{pmatrix} 1 \\ 0.531970515 \end{pmatrix}$, $\mathbf{v}_2 = \begin{pmatrix} 1 \\ -0.2819705149 \end{pmatrix}$, and the ratio of $p_{j,n}$ to $p_{a,n}$ is seen to be $1/0.5319710515 \approx 1.879801537$.

It is remarkable just how much information is available from a simple computation of eigenvalues. It is of great interest to know whether a population will ultimately increase or decrease. It will increase if $\lambda_1 > 1$, and the condition for that is $(\beta + \sqrt{\beta^2 + 4\alpha k})/2 > 1$ or $\sqrt{\beta^2 + 4\alpha k} > 2 - \beta$ or $\beta^2 + 4\alpha k > (2 - \beta)^2 = 4 - 4\beta + \beta^2$. This leads to $4\alpha k > 4 - 4\beta$ or

$$k > \frac{1 - \beta}{\alpha} \tag{10}$$

In Example 1 we had $\beta = 0.5$, $\alpha = 0.3$; thus (10) is satisfied if $k > 0.5/0.3 \approx 1.67$.

Before we close this section we indicate two limitations of this model:

**i.** Birth and death rates often change from year to year and are particularly dependent on the weather. This model assumes a constant environment.

**ii.** Ecologists have found that for many species birth and death rates vary with the size of the population. In particular, a population cannot grow when it reaches a certain size due to the effects of limited food resources and overcrowding. It is obvious that a population cannot grow indefinitely at a constant rate. Otherwise that population would overrun the earth.

## PROBLEMS 6.2

In Problems 1–3 find the numbers of juvenile and adult female birds after 1, 2, 5, 10, 19, and 20 years. Then find the long-term ratios of $p_{j,n}$ to $p_{a,n}$ and $T_n$ to $T_{n-1}$. [*Hint:* Use equations (7) and (9) and a calculator and round to three decimals.]

**1.** $\mathbf{p}_0 = \begin{pmatrix} 0 \\ 12 \end{pmatrix}$; $k = 3$, $\alpha = 0.4$, $\beta = 0.6$

**2.** $\mathbf{p}_0 = \begin{pmatrix} 0 \\ 15 \end{pmatrix}$; $k = 1$, $\alpha = 0.3$, $\beta = 0.4$

**3.** $\mathbf{p}_0 = \begin{pmatrix} 0 \\ 20 \end{pmatrix}$; $k = 4$, $\alpha = 0.7$, $\beta = 0.8$

**4.** Show that if $\alpha = \beta$ and $\alpha > \frac{1}{2}$, then the bird population will always increase in the long run if at least one female offspring on the average is produced by each female adult.

**5.** Show that, in the long run, the ratio $p_{j,n}/p_{a,n}$ approaches the limiting value $k/\lambda_1$.

**6.** Suppose we divide the adult birds into two age groups: those 1–5 years old and those more than 5 years old. Assume that the survival rate for birds in the first group is $\beta$, whereas in the second group it is $\gamma$ (and $\beta > \gamma$). Assume that the birds in the first group are equally divided as to age. (That is, if there are 100 birds in the group, then 20 are 1 year old, 20 are 2 years old, and so on.) Formulate a $3 \times 3$ matrix model for this situation.

## MATLAB 6.2

1. Consider the bird population given by

$$A = \begin{pmatrix} 0 & 3 \\ .4 & .6 \end{pmatrix} \quad \text{and} \quad \mathbf{p}_0 = \begin{pmatrix} 0 \\ 12 \end{pmatrix}$$

a. Find the numbers of juvenile and adult female birds after 2, 5, 10, and 20 years.

b. Find the numbers after 21 years and compute $p_{j,n}/p_{a,n}$ and $T_n/T_{n-1}$ for $n = 21$. [*Hint.* Use the MATLAB command **sum** to find $T_n$.] Repeat for $n = 22, 23, 24$, and 25. What do you conjecture for $\lim_{n\to\infty} p_{j,n}/p_{a,n}$ and $\lim_{n\to\infty} T_n/T_{n-1}$?

c. Find **[V,D] = eig(A)**. Verify that the eigenvalue of the largest magnitude is positive with algebraic multiplicity 1, that there exists an associated eigenvector whose components are all positive, and that the other eigenvalue is strictly less in magnitude. Compare this largest eigenvalue with $\lim_{n\to\infty} T_n/T_{n-1}$. Explain how these numbers indicate that the population is growing.

Let $\mathbf{w}$ be the eigenvector associated with this largest eigenvalue. Compare $w_1/w_2$ with $\lim_{n\to\infty} p_{j,n}/p_{a,n}$ and with $k/\lambda$, where $k = 3$ and $\lambda$ is the eigenvalue of the largest magnitude. Write a conclusion about these comparisons.

2. Consider the bird population given by

$$A = \begin{pmatrix} 0 & 3 \\ .3 & .15 \end{pmatrix} \quad \text{and} \quad \mathbf{p}_0 = \begin{pmatrix} 0 \\ 12 \end{pmatrix}$$

a. Compute **[V,D] = eig(A)** and use this information to find $\lim_{n\to\infty} p_{j,n}/p_{a,n}$ and $\lim_{n\to\infty} T_n/T_{n-1}$. Explain what properties of $V$ and $D$ justify your procedure.
b. Show that the ratios $p_{j,n}/p_{a,n}$ and $T_n/T_{n-1}$ have not stabilized yet after 25 years. Compute the ratios for $n = 46$ through 50 and show that they have stabilized after 50 years.
c. *(Paper and pencil)* Verify that for this population the second eigenvalue (the one smaller in magnitude) is closer to the eigenvalue of the largest magnitude than is true for the population in MATLAB Problem 1 in this section. Describe how that explains why the population ratios take longer to stabilize.

3. Suppose the information below represents a population of female deer:

$$A = \begin{pmatrix} 0 & 1 \\ .6 & .8 \end{pmatrix} \quad \text{and} \quad \mathbf{p}_0 = \begin{pmatrix} 100 \\ 200 \end{pmatrix}$$

a. Show that in the long run the population will grow by a factor of approximately 1.27. Justify your procedure.
b. *(Paper and pencil)* Farmers and other people in the area do not want the population to grow. They can control the population by "harvesting" it (allowing hunting). If $h$ is the proportion of the population harvested each time period, argue why the matrix modeling this would be

$$A = \begin{pmatrix} 0 & 1 \\ .6 & .8 - h \end{pmatrix}$$

c. Show that $h = .6$ is too large a harvest; that is, the deer population will die out. (The people in the area do not want the deer to die out.) Give two arguments for this: by looking at $A^n\mathbf{p}_0$ for $n$ getting larger and by examining the eigenvalues.
d. It is possible to choose $h$ so that the population neither grows nor dies out. Experiment with various values of $h$: Examine $A^n\mathbf{p}_0$ for $n$ getting larger and examine the eigenvalues for $A$. What can be said about the eigenvalues of $A$ when the desired $h$ is found?
c. *(Paper and pencil)* Explain the results observed in part (d) in terms of the theory presented in this section.

4. Consider a population of (female) birds grouped into three age classes: juvenile, 1 year old, and 2 years old or older. Suppose the matrix $A$ below is a model for the population growth and $\mathbf{p}_0$ is the initial population vector, where the first row represents juveniles; the second row, the 1 to 5 years age group; and the third row, the over 5 years age group.

$$A = \begin{pmatrix} 0 & 2 & 1 \\ .6 & 0 & 0 \\ 0 & .6 & .4 \end{pmatrix} \quad \text{and} \quad \mathbf{p}_0 = \begin{pmatrix} 0 \\ 50 \\ 50 \end{pmatrix}$$

a. *(Paper and pencil)* Explain what each entry in the matrix $A$ represents.
b. Find how many (female) birds of each age group are in the population after 30 years. For $n = 31$ through 35, using the MATLAB command **sum**, find $T_n/T_{n-1}$ and $\mathbf{w}_n = \mathbf{v}_n/\mathbf{sum}(\mathbf{v}_n)$, where $\mathbf{v}_n = A^n\mathbf{p}_0$. Explain how $\mathbf{w}_n$ gives you the proportion of each age group in the total population after $n$ years.

What appears to be the $\lim_{n\to\infty} T_n/T_{n-1}$? What is the interpretation of this limit? What appears to be the $\lim_{n\to\infty} \mathbf{w}_n$? What is the interpretation of this limit?
c. Find **[V,D] = eig(A)**. Verify that there is a positive eigenvalue of the largest magnitude and with multiplicity 1 (and the other eigenvalues are strictly smaller in magnitude) and that this "largest" eigenvalue has an associated eigenvector whose components are all positive. Find **zz = z/sum(z)**, where **z** is the eigenvector associated with

the largest eigenvalue. Compare the eigenvalue with the projected limit of $T_n/T_{n-1}$ from part (b) and compare **zz** with the limit of $\mathbf{w}_n$. Write a description of conclusions that you can make from this comparison.

d. *(Paper and pencil)* By extending the theory presented in this section, give an argument to explain your observations from the previous parts of this problem.

5. a. Rework MATLAB Problem 14, parts (a) through (c), in Section 1.6. By construction, the matrix $P$ in this problem is **stochastic;** that is, the entries in each column of $P$ sum to 1.

b. Find **[V,D] = eig(P).** Verify that there is a positive eigenvalue of the largest magnitude and with multiplicity 1 (and the other eigenvalues are strictly smaller in magnitude) and that this "largest" eigenvalue has an associated eigenvector whose components are all positive. What is this largest eigenvalue? How does this explain the behavior observed in part (a), that is, the fact that $P^n\mathbf{x}$ appears to converge to a fixed vector **y**?

Find 3000**z/sum(z)**, where **z** is the eigenvector associated with the largest eigenvalue. How does it compare with the limit vector **y**? What interpretation does **y** have?

c. Using eigenvalues and eigenvectors as in part (b), find the long-term car distribution for MATLAB Problem 14(g) in Section 1.6. Justify your procedure. Verify your findings by computing $P^n\mathbf{x}$ for $n$ getting large, where $P$ is the stochastic matrix modeling the problem and **x** is some initial car distribution vector whose components sum to 1000.

d. *(Paper and pencil)* Suppose $P$ is a $3 \times 3$ stochastic matrix; that is, the entries in each *column* of $P$ sum to 1. Argue why

$$P^t\begin{pmatrix}1\\1\\1\end{pmatrix} = \begin{pmatrix}1\\1\\1\end{pmatrix}$$

What does this tell you about the eigenvalues of $P^t$? What, in turn, does this tell you about the eigenvalues of $P$? What relevance does this have for the previous parts of this problem?

6. **Graph Theory** Refer to MATLAB Problem 11 in Section 6.1 for the definition of the **adjacency matrix** of a graph and other related definitions. For connected graphs, the adjacency matrix has the properties that all the eigenvalues are real, that there is a positive eigenvalue of largest magnitude, $\lambda_1$, with algebraic multiplicity 1, that there exists an associated eigenvector whose components are all positive, and that the other eigenvalues are strictly less in magnitude. Thus we would have that, for a given vector **x**, $A^n\mathbf{x} \approx \lambda_1^n a_1\mathbf{u}_1$ for large $n$, where $\mathbf{u}_1$ is the eigenvector associated with $\lambda_1$. (Here $a_1$ is the coordinate of **x** with respect to the basis of eigenvectors containing $\mathbf{u}_1$ as the first basis vector.)

a. *(Paper and pencil)* Explain why we can then conclude that the ratio of a component of $A^n\mathbf{x}$ to the sum of the components is approximately equal to the ratio of the corresponding component of $\mathbf{u}_1$ to the sum of its components.

b. *(Paper and pencil)* $(A^n)_{ij}$ has the interpretation of being the number of paths of length $n$ connecting vertex $i$ with vertex $j$. (Refer to Section 1.12. For example, a path of length 2 connecting $i$ with $j$ would consist of an edge connecting $i$ with some vertex $k$ and then an edge connecting $k$ with $j$.) If **x** is the vector with each component equal to 1, explain why the $i^{\text{th}}$ component of $A^n\mathbf{x}$ represents the total number of paths of length $n$ connecting vertex $i$ with all other vertices. Explain how we can thus conclude that the ratios of the components of $A^n\mathbf{x}$ to the sum of the components gives some indication of the relative *"importance"* of the vertices in the graph. Explain why and how we could thus use the ratios of the components of $\mathbf{u}_1$ to the sum of the components as an indication of the *"importance"* of each vertex in the graph. (A more sophisticated argument for the use of the eigenvector corresponding to the largest eigenvalue is known by the name **Gold's index.**)

c. For each of the graphs below, verify that the adjacency matrix has the properties stated in the discussion before part (a) and discuss the relative *"importance"* of the vertices in the graph. For the graphs in (i) through (iii), using your intuition from looking at the graph, argue why your results make sense. [*Note.* For ease in entering the adjacency matrix, refer to the discussion in MATLAB Problem 2 in Section 1.5.]

**i.** The graph in MATLAB Problem 11(a) in Section 6.1.

**ii.** The graph in MATLAB Problem 11(b) in Section 6.1.

**iii.** The graph in MATLAB Problem 11(c) in Section 6.1.

**iv.** Suppose we consider the graph below as representing airline routes between cities. A company wishes to choose a city in which to locate their head office. After analysis of the graph, write a report to the head of the company containing your recommendation (and justifications).

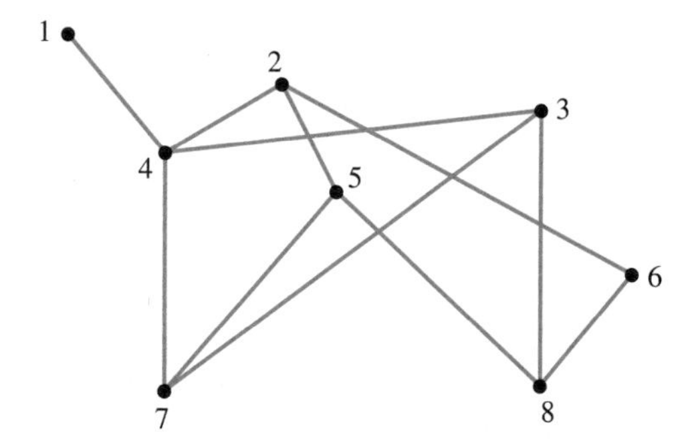


**PROJECT PROBLEM**

**v.** From a map of your state, create a graph whose vertices are key cities and whose edges are main roads connecting the cities. Determine the relative *"importance"* of each city. Justify and explain your procedure.

## 6.3 SIMILAR MATRICES AND DIAGONALIZATION

In this section we describe an interesting and useful relationship that can hold between two matrices.

**DEFINITION 1** **Similar Matrices** Two $n \times n$ matrices $A$ and $B$ are said to be **similar** if there exists an invertible $n \times n$ matrix $C$ such that

$$B = C^{-1}AC \tag{1}$$

Similarity transformation

The function defined by (1) that takes the matrix $A$ into the matrix $B$ is called a **similarity transformation.** We can write this linear transformation as

$$T(A) = C^{-1}AC$$

*Note.* $C^{-1}(A_1 + A_2)C = C^{-1}A_1C + C^{-1}A_2C$ and $C^{-1}(\alpha A)C = \alpha C^{-1}AC$ so that the function defined by (1) is, in fact, a linear transformation. This explains the use of the word "transformation" above.

The purpose of this section is to show that (1) similar matrices have several important properties in common and (2) most matrices are similar to diagonal matrices. (See the remark on page 569.)

*Note.* Suppose that $B = C^{-1}AC$. Then multiplying on the left by $C$, we obtain $CB = CC^{-1}AC$, or

$$CB = AC \tag{2}$$

Equation (2) is often taken as an alternative definition of similarity:

**Alternative Definition of Similarity**

$A$ and $B$ are similar if and only if there exists an invertible matrix $C$ such that

$$CB = AC$$

**EXAMPLE 1** **Two Similar Matrices** Let $A = \begin{pmatrix} 2 & 1 \\ 0 & -1 \end{pmatrix}$, $B = \begin{pmatrix} 4 & -2 \\ 5 & -3 \end{pmatrix}$, and $C = \begin{pmatrix} 2 & -1 \\ -1 & 1 \end{pmatrix}$. Then $CB = \begin{pmatrix} 2 & -1 \\ -1 & 1 \end{pmatrix}\begin{pmatrix} 4 & -2 \\ 5 & -3 \end{pmatrix} = \begin{pmatrix} 3 & -1 \\ 1 & -1 \end{pmatrix}$ and $AC = \begin{pmatrix} 2 & 1 \\ 0 & -1 \end{pmatrix}\begin{pmatrix} 2 & -1 \\ -1 & 1 \end{pmatrix} = \begin{pmatrix} 3 & -1 \\ 1 & -1 \end{pmatrix}$. Thus $CB = AC$. Since $\det C = 1 \neq 0$, $C$ is invertible. This shows, by equation (2), that $A$ and $B$ are similar.

**EXAMPLE 2** **A Matrix That Is Similar to a Diagonal Matrix** Let $D = \begin{pmatrix} 1 & 0 & 0 \\ 0 & -1 & 0 \\ 0 & 0 & 2 \end{pmatrix}$, $A = \begin{pmatrix} -6 & -3 & -25 \\ 2 & 1 & 8 \\ 2 & 2 & 7 \end{pmatrix}$, and $C = \begin{pmatrix} 2 & 4 & 3 \\ 0 & 1 & -1 \\ 3 & 5 & 7 \end{pmatrix}$. $C$ is invertible because $\det C = 3 \neq 0$. We then compute:

$$CA = \begin{pmatrix} 2 & 4 & 3 \\ 0 & 1 & -1 \\ 3 & 5 & 7 \end{pmatrix}\begin{pmatrix} -6 & -3 & -25 \\ 2 & 1 & 8 \\ 2 & 2 & 7 \end{pmatrix} = \begin{pmatrix} 2 & 4 & 3 \\ 0 & -1 & 1 \\ 6 & 10 & 14 \end{pmatrix}$$

$$DC = \begin{pmatrix} 1 & 0 & 0 \\ 0 & -1 & 0 \\ 0 & 0 & 2 \end{pmatrix} \begin{pmatrix} 2 & 4 & 3 \\ 0 & 1 & -1 \\ 3 & 5 & 7 \end{pmatrix} = \begin{pmatrix} 2 & 4 & 3 \\ 0 & -1 & 1 \\ 6 & 10 & 14 \end{pmatrix}$$

Thus $CA = DC$ and $A = C^{-1}DC$, so $A$ and $D$ are similar.

*Note.* In Examples 1 and 2 it was not necessary to compute $C^{-1}$. It was only necessary to know that $C$ was nonsingular.

**THEOREM 1** If $A$ and $B$ are similar $n \times n$ matrices, then $A$ and $B$ have the same characteristic polynomial, and therefore have the same eigenvalues.

**Proof** Since $A$ and $B$ are similar, $B = C^{-1}AC$ and

$$\begin{aligned} \det(B - \lambda I) &= \det(C^{-1}AC - \lambda I) = \det[C^{-1}AC - C^{-1}(\lambda I)C] \\ &= \det[C^{-1}(A - \lambda I)C] = \det(C^{-1}) \det(A - \lambda I) \det(C) \\ &= \det(C^{-1}) \det(C) \det(A - \lambda I) = \det(C^{-1}C) \det(A - \lambda I) \\ &= \det I \det(A - \lambda I) = \det(A - \lambda I) \end{aligned}$$

This means that $A$ and $B$ have the same characteristic equation, and since eigenvalues are roots of the characteristic equation, they have the same eigenvalues.

**EXAMPLE 3** **Eigenvalues of Similar Matrices Are the Same** In Example 2 it is obvious that the eigenvalues of $D = \begin{pmatrix} 1 & 0 & 0 \\ 0 & -1 & 0 \\ 0 & 0 & 2 \end{pmatrix}$ are 1, −1, and 2. Thus these are the eigenvalues of $A = \begin{pmatrix} -6 & -3 & -25 \\ 2 & 1 & 8 \\ 2 & 2 & 7 \end{pmatrix}$. Check this by verifying that $\det(A - I) = \det(A + I) = \det(A - 2I) = 0$.

In a variety of applications it is quite useful to "diagonalize" a matrix $A$—that is, to find a diagonal matrix similar to $A$.

**DEFINITION 2** **Diagonalizable Matrix** An $n \times n$ matrix $A$ is **diagonalizable** if there is a diagonal matrix $D$ such that $A$ is similar to $D$.

*Remark.* If $D$ is a diagonal matrix, then its eigenvalues are its diagonal components (see page 541). If $A$ is similar to $D$, then $A$ and $D$ have the same eigenvalues (by Theorem 1). Putting these two facts together, we observe that if $A$ is diagonalizable, then $A$ is similar to a diagonal matrix whose diagonal components are the eigenvalues of $A$.

The next theorem tells us when a matrix is diagonalizable.

**THEOREM 2** An $n \times n$ matrix $A$ is diagonalizable if and only if it has $n$ linearly independent eigenvectors. In that case the diagonal matrix $D$ similar to $A$ is given by

$$D = \begin{pmatrix} \lambda_1 & 0 & 0 & \cdots & 0 \\ 0 & \lambda_2 & 0 & \cdots & 0 \\ 0 & 0 & \lambda_3 & \cdots & 0 \\ \vdots & \vdots & \vdots & & \vdots \\ 0 & 0 & 0 & \cdots & \lambda_n \end{pmatrix}$$

where $\lambda_1, \lambda_2, \ldots, \lambda_n$ are the eigenvalues of $A$. If $C$ is a matrix whose columns are linearly independent eigenvectors of $A$, then

$$\boxed{D = C^{-1}AC} \qquad \textbf{(3)}$$

**Proof** We first assume that $A$ has $n$ linearly independent eigenvectors $\mathbf{v}_1, \mathbf{v}_2, \ldots, \mathbf{v}_n$ corresponding to the (not necessarily distinct) eigenvalues $\lambda_1, \lambda_2, \ldots, \lambda_n$.

Let

$$\mathbf{v}_1 = \begin{pmatrix} c_{11} \\ c_{21} \\ \vdots \\ c_{n1} \end{pmatrix}, \mathbf{v}_2 = \begin{pmatrix} c_{12} \\ c_{22} \\ \vdots \\ c_{n2} \end{pmatrix}, \ldots, \mathbf{v}_n = \begin{pmatrix} c_{1n} \\ c_{2n} \\ \vdots \\ c_{nn} \end{pmatrix}$$

and let

$$C = \begin{pmatrix} c_{11} & c_{12} & \cdots & c_{1n} \\ c_{21} & c_{22} & \cdots & c_{2n} \\ \vdots & \vdots & & \vdots \\ c_{n1} & c_{n2} & \cdots & c_{nn} \end{pmatrix}$$

Then $C$ is invertible since its columns are linearly independent. Now

$$AC = \begin{pmatrix} a_{11} & a_{12} & \cdots & a_{1n} \\ a_{21} & a_{22} & \cdots & a_{2n} \\ \vdots & \vdots & & \vdots \\ a_{n1} & a_{n2} & \cdots & a_{nn} \end{pmatrix} \begin{pmatrix} c_{11} & c_{12} & \cdots & c_{1n} \\ c_{21} & c_{22} & \cdots & c_{2n} \\ \vdots & \vdots & & \vdots \\ c_{n1} & c_{n2} & \cdots & c_{nn} \end{pmatrix}$$

and we see that the $i$th column of $AC$ is $A\begin{pmatrix} c_{1i} \\ c_{2i} \\ \vdots \\ c_{ni} \end{pmatrix} = A\mathbf{v}_i = \lambda_i \mathbf{v}_i$. Thus $AC$ is the matrix whose $i$th column is $\lambda_i \mathbf{v}_i$ and

$$AC = \begin{pmatrix} \lambda_1 c_{11} & \lambda_2 c_{12} & \cdots & \lambda_n c_{1n} \\ \lambda_1 c_{21} & \lambda_2 c_{22} & \cdots & \lambda_n c_{2n} \\ \vdots & \vdots & & \vdots \\ \lambda_1 c_{n1} & \lambda_2 c_{n2} & \cdots & \lambda_n c_{nn} \end{pmatrix}$$

But

$$CD = \begin{pmatrix} c_{11} & c_{12} & \cdots & c_{1n} \\ c_{21} & c_{22} & \cdots & c_{2n} \\ \vdots & \vdots & & \vdots \\ c_{n1} & c_{n2} & \cdots & c_{nn} \end{pmatrix} \begin{pmatrix} \lambda_1 & 0 & \cdots & 0 \\ 0 & \lambda_2 & \cdots & 0 \\ \vdots & \vdots & & \vdots \\ 0 & 0 & \cdots & \lambda_n \end{pmatrix}$$

$$= \begin{pmatrix} \lambda_1 c_{11} & \lambda_2 c_{12} & \cdots & \lambda_n c_{1n} \\ \lambda_1 c_{21} & \lambda_2 c_{22} & \cdots & \lambda_n c_{2n} \\ \vdots & \vdots & & \vdots \\ \lambda_1 c_{n1} & \lambda_2 c_{n2} & \cdots & \lambda_n c_{nn} \end{pmatrix}$$

Thus

$$AC = CD \tag{4}$$

and since $C$ is invertible, we can multiply both sides of (4) on the left by $C^{-1}$ to obtain

$$D = C^{-1}AC \tag{5}$$

This proves that if $A$ has $n$ linearly independent eigenvectors, then $A$ is diagonalizable. Conversely, suppose that $A$ is diagonalizable. That is, suppose that (5) holds for some invertible matrix $C$. Let $\mathbf{v}_1, \mathbf{v}_2, \ldots, \mathbf{v}_n$ be the columns of $C$. Then $AC = CD$, and, reversing the arguments above, we immediately see that $A\mathbf{v}_i = \lambda_i \mathbf{v}_i$ for $i = 1, 2, \ldots, n$. Thus $\mathbf{v}_1, \mathbf{v}_2, \ldots, \mathbf{v}_n$ are eigenvectors of $A$ and are linearly independent because $C$ is invertible. ■

***Notation.*** To indicate that $D$ is a diagonal matrix with diagonal components $\lambda_1, \lambda_2, \ldots, \lambda_n$, we write $D = \text{diag}\,(\lambda_1, \lambda_2, \ldots, \lambda_n)$.

Theorem 2 has a useful corollary that follows immediately from Theorem 6.1.3 on page 536.

**COROLLARY** If the $n \times n$ matrix $A$ has $n$ distinct eigenvalues, then $A$ is diagonalizable.

***Remark.*** If the real coefficients of a polynomial of degree $n$ are picked at random, then, with probability 1, the polynomial will have $n$ distinct roots. It is not difficult to see, intuitively, why this is so. If $n = 2$, for example, then the equation $\lambda^2 + a\lambda + b = 0$ has equal roots if and only if $a^2 = 4b$—a highly unlikely event if $a$ and $b$ are chosen at random. We can, of course, write down polynomials having roots of algebraic multiplicity greater than 1, but these polynomials are exceptional. Thus without attempting to be mathematically precise, it is fair to say that *most* polynomials have distinct roots. Hence *most* matrices have distinct eigenvalues, and as we stated at the beginning of the section, *most* matrices are diagonalizable.

**EXAMPLE 4** **Diagonalizing a 2 × 2 Matrix** Let $A = \begin{pmatrix} 4 & 2 \\ 3 & 3 \end{pmatrix}$. In Example 6.1.3 on page 538 we found the two linearly independent eigenvectors $\mathbf{v}_1 = \begin{pmatrix} 2 \\ -3 \end{pmatrix}$ and $\mathbf{v}_2 = \begin{pmatrix} 1 \\ 1 \end{pmatrix}$. Then setting $C = \begin{pmatrix} 2 & 1 \\ -3 & 1 \end{pmatrix}$, we find that

$$C^{-1}AC = \frac{1}{5}\begin{pmatrix} 1 & -1 \\ 3 & 2 \end{pmatrix}\begin{pmatrix} 4 & 2 \\ 3 & 3 \end{pmatrix}\begin{pmatrix} 2 & 1 \\ -3 & 1 \end{pmatrix}$$

$$= \frac{1}{5}\begin{pmatrix} 1 & -1 \\ 3 & 2 \end{pmatrix}\begin{pmatrix} 2 & 6 \\ -3 & 6 \end{pmatrix} = \frac{1}{5}\begin{pmatrix} 5 & 0 \\ 0 & 30 \end{pmatrix} = \begin{pmatrix} 1 & 0 \\ 0 & 6 \end{pmatrix}$$

which is the matrix whose diagonal components are the eigenvalues of $A$.

**EXAMPLE 5** **Diagonalizing a 3 × 3 Matrix with Three Distinct Eigenvalues** Let $A = \begin{pmatrix} 1 & -1 & 4 \\ 3 & 2 & -1 \\ 2 & 1 & -1 \end{pmatrix}$. In Example 6.1.4 on page 538 we computed the three linearly independent eigenvectors $\mathbf{v}_1 = \begin{pmatrix} -1 \\ 4 \\ 1 \end{pmatrix}$, $\mathbf{v}_2 = \begin{pmatrix} 1 \\ -1 \\ -1 \end{pmatrix}$, and $\mathbf{v}_3 = \begin{pmatrix} 1 \\ 2 \\ 1 \end{pmatrix}$. Then

$$C = \begin{pmatrix} -1 & 1 & 1 \\ 4 & -1 & 2 \\ 1 & -1 & 1 \end{pmatrix}$$ and

$$C^{-1}AC = -\frac{1}{6}\begin{pmatrix} 1 & -2 & 3 \\ -2 & -2 & 6 \\ -3 & 0 & -3 \end{pmatrix}\begin{pmatrix} 1 & -1 & 4 \\ 3 & 2 & -1 \\ 2 & 1 & -1 \end{pmatrix}\begin{pmatrix} -1 & 1 & 1 \\ 4 & -1 & 2 \\ 1 & -1 & 1 \end{pmatrix}$$

$$= -\frac{1}{6}\begin{pmatrix} 1 & -2 & 3 \\ -2 & -2 & 6 \\ -3 & 0 & -3 \end{pmatrix}\begin{pmatrix} -1 & -2 & 3 \\ 4 & 2 & 6 \\ 1 & 2 & 3 \end{pmatrix}$$

$$= -\frac{1}{6}\begin{pmatrix} -6 & 0 & 0 \\ 0 & 12 & 0 \\ 0 & 0 & -18 \end{pmatrix} = \begin{pmatrix} 1 & 0 & 0 \\ 0 & -2 & 0 \\ 0 & 0 & 3 \end{pmatrix}$$

with eigenvalues 1, $-2$, and 3.

*Remark.* Since there are an infinite number of ways to choose an eigenvector, there are an infinite number of ways to choose the diagonalizing matrix $C$. The only advice is to choose the eigenvectors and matrix $C$ that are, arithmetically, the easiest to work with. This usually means that you should insert as many 0's and 1's as possible.

**EXAMPLE 6** **Diagonalizing a 3 × 3 Matrix with Two Distinct Eigenvalues and Three Linearly Independent Eigenvectors** Let $A = \begin{pmatrix} 3 & 2 & 4 \\ 2 & 0 & 2 \\ 4 & 2 & 3 \end{pmatrix}$. Then from Example 6.1.10 on page 542 we have the three linearly independent eigenvectors $\mathbf{v}_1 = \begin{pmatrix} 2 \\ 1 \\ 2 \end{pmatrix}$, $\mathbf{v}_2 = \begin{pmatrix} 1 \\ -2 \\ 0 \end{pmatrix}$, and $\mathbf{v}_3 = \begin{pmatrix} 0 \\ -2 \\ 1 \end{pmatrix}$. Setting $C = \begin{pmatrix} 2 & 1 & 0 \\ 1 & -2 & -2 \\ 2 & 0 & 1 \end{pmatrix}$, we obtain

$$C^{-1}AC = -\frac{1}{9}\begin{pmatrix} -2 & -1 & -2 \\ -5 & 2 & 4 \\ 4 & 2 & -5 \end{pmatrix}\begin{pmatrix} 3 & 2 & 4 \\ 2 & 0 & 2 \\ 4 & 2 & 3 \end{pmatrix}\begin{pmatrix} 2 & 1 & 0 \\ 1 & -2 & -2 \\ 2 & 0 & 1 \end{pmatrix}$$

$$= -\frac{1}{9}\begin{pmatrix} -2 & -1 & -2 \\ -5 & 2 & 4 \\ 4 & 2 & -5 \end{pmatrix}\begin{pmatrix} 16 & -1 & 0 \\ 8 & 2 & 2 \\ 16 & 0 & -1 \end{pmatrix}$$

$$= -\frac{1}{9}\begin{pmatrix} -72 & 0 & 0 \\ 0 & 9 & 0 \\ 0 & 0 & 9 \end{pmatrix} = \begin{pmatrix} 8 & 0 & 0 \\ 0 & -1 & 0 \\ 0 & 0 & -1 \end{pmatrix}$$

This example illustrates that $A$ is diagonalizable even though its eigenvalues are not distinct.

**EXAMPLE 7** **A 2 × 2 Matrix with Only One Linearly Independent Eigenvector Cannot Be Diagonalized** Let $A = \begin{pmatrix} 4 & 1 \\ 0 & 4 \end{pmatrix}$. In Example 6.1.9 on page 542 we saw that $A$

did *not* have two linearly independent eigenvectors. Suppose that $A$ were diagonalizable (in contradiction to Theorem 2). Then $D = \begin{pmatrix} 4 & 0 \\ 0 & 4 \end{pmatrix}$ and there would be an invertible matrix $C$ such that $C^{-1}AC = D$. Multiplying this equation on the left by $C$ and on the right by $C^{-1}$, we find that $A = CDC^{-1} = C\begin{pmatrix} 4 & 0 \\ 0 & 4 \end{pmatrix}C^{-1} = C(4I)C^{-1} = 4CIC^{-1} = 4CC^{-1} = 4I = \begin{pmatrix} 4 & 0 \\ 0 & 4 \end{pmatrix} = D$. But $A \neq D$, so no such $C$ exists. ■

We have seen that many matrices are similar to diagonal matrices. However, two questions remain:

**i.** Is it possible to determine whether a given matrix is diagonalizable without computing eigenvalues and eigenvectors?

**ii.** What do we do if $A$ is not diagonalizable?

We shall find a partial answer to the first question in the next section and a complete answer to the second in Section 6.6. In Section 6.7 we shall see an important application of the diagonalizing procedure.

At the beginning of this chapter we defined eigenvectors and eigenvalues for a linear transformation $T\colon V \to V$, where $\dim V = n$. We stated then that as $T$ can be represented by an $n \times n$ matrix, we would limit our discussion to eigenvalues and eigenvectors of $n \times n$ matrices.

However, the linear transformation can be represented by many different $n \times n$ matrices—one for each basis chosen. Do these matrices have the same eigenvalues? The answer is yes, as proved in the next theorem.

**THEOREM 3** Let $V$ be a finite dimensional vector space with bases $B_1 = \{\mathbf{v}_1, \mathbf{v}_2, \ldots, \mathbf{v}_n\}$ and $B_2 = \{\mathbf{w}_1, \mathbf{w}_2, \ldots, \mathbf{w}_n\}$. Let $T\colon V \to V$ be a linear transformation. If $A_T$ is the matrix representation of $T$ with respect to the basis $B_1$ and if $C_T$ is the matrix representation of $T$ with respect to the basis $B_2$, then $A_T$ and $C_T$ are similar.

**Proof** $T$ is a linear transformation from $V$ to itself. From Theorem 5.3.3 on page 490 we have

$$(T\mathbf{x})_{B_1} = A_T(\mathbf{x})_{B_1} \tag{6}$$

and

$$(T\mathbf{x})_{B_2} = C_T(\mathbf{x})_{B_2} \tag{7}$$

Let $M$ denote the transition matrix from $B_1$ to $B_2$. Then by Theorem 4.8.1 on page 376

$$(\mathbf{x})_{B_2} = M(\mathbf{x})_{B_1} \tag{8}$$

for every $\mathbf{x}$ in $V$. Also,

$$(T\mathbf{x})_{B_2} = M(T\mathbf{x})_{B_1} \tag{9}$$

Inserting (8) and (9) in (7) yields

$$M(T\mathbf{x})_{B_1} = C_T M(\mathbf{x})_{B_1} \tag{10}$$

The matrix $M$ is invertible by the result of Theorem 4.8.2 on page 377. If we multiply both sides of (10) by $M^{-1}$ (which is the transition matrix from $B_2$ to $B_1$), we obtain

$$(T\mathbf{x})_{B_1} = M^{-1}C_T M(\mathbf{x})_{B_1} \tag{11}$$

Comparing (6) and (11), we have

$$A_T(\mathbf{x})_{B_1} = M^{-1}C_T M(\mathbf{x})_{B_1} \tag{12}$$

Since (12) holds for every $\mathbf{x} \in V$, we conclude that

$$A_T = M^{-1}C_T M$$

That is, $A_T$ and $C_T$ are similar. ■

## PROBLEMS 6.3

### Self-Quiz

#### *True-False*

**I.** If an $n \times n$ matrix has $n$ distinct eigenvalues, it can be diagonalized.

**II.** If the $5 \times 5$ matrix $A$ has three distinct eigenvalues, then $A$ cannot be similar to a diagonal matrix.

**III.** If $A$ is similar to the matrix $\begin{pmatrix} 1 & 2 & 5 \\ 0 & 2 & 4 \\ 0 & 0 & 3 \end{pmatrix}$, then its eigenvalues are 1, 2, and 3.

In Problems 1–15 determine whether the given matrix $A$ is diagonalizable. If it is, find a matrix $C$ such that $C^{-1}AC = D$. Verify that $AC = CD$ and that the nonzero entries of $D$ are the eigenvalues of $A$.

**1.** $\begin{pmatrix} -2 & -2 \\ -5 & 1 \end{pmatrix}$

**2.** $\begin{pmatrix} 3 & -1 \\ -2 & 4 \end{pmatrix}$

**3.** $\begin{pmatrix} 2 & -1 \\ 5 & -2 \end{pmatrix}$

**4.** $\begin{pmatrix} 3 & -5 \\ 1 & -1 \end{pmatrix}$

**5.** $\begin{pmatrix} 3 & 2 \\ -5 & 1 \end{pmatrix}$

**6.** $\begin{pmatrix} 1 & -1 & 0 \\ -1 & 2 & -1 \\ 0 & -1 & 1 \end{pmatrix}$

---

**Answers to Self-Quiz**

**I.** True **II.** False **III.** True

**7.** $\begin{pmatrix} 1 & 1 & -2 \\ -1 & 2 & 1 \\ 0 & 1 & -1 \end{pmatrix}$ **8.** $\begin{pmatrix} 2 & 1 & 0 \\ 0 & 0 & 1 \\ 0 & 0 & 0 \end{pmatrix}$ **9.** $\begin{pmatrix} 3 & 0 & 0 \\ 0 & 0 & 1 \\ 0 & 0 & 2 \end{pmatrix}$

**10.** $\begin{pmatrix} 3 & -1 & -1 \\ 1 & 1 & -1 \\ 1 & -1 & 1 \end{pmatrix}$ **11.** $\begin{pmatrix} 7 & -2 & -4 \\ 3 & 0 & -2 \\ 6 & -2 & -3 \end{pmatrix}$ **12.** $\begin{pmatrix} 4 & 6 & 6 \\ 1 & 3 & 2 \\ -1 & -5 & -2 \end{pmatrix}$

**13.** $\begin{pmatrix} -3 & -7 & -5 \\ 2 & 4 & 3 \\ 1 & 2 & 2 \end{pmatrix}$ **14.** $\begin{pmatrix} -2 & -2 & 0 & 0 \\ -5 & 1 & 0 & 0 \\ 0 & 0 & 2 & -1 \\ 0 & 0 & 5 & -2 \end{pmatrix}$ **15.** $\begin{pmatrix} 4 & 1 & 0 & 1 \\ 2 & 3 & 0 & 1 \\ -2 & 1 & 2 & -3 \\ 2 & -1 & 0 & 5 \end{pmatrix}$

**16.** Show that if $A$ is similar to $B$ and $B$ is similar to $C$, then $A$ is similar to $C$.

**17.** If $A$ is similar to $B$, show that $A^n$ is similar to $B^n$ for any positive integer $n$.

***18.** If $A$ is similar to $B$, show that $\rho(A) = \rho(B)$ and $\nu(A) = \nu(B)$. [*Hint:* First, prove that if $C$ is invertible, then $\nu(CA) = \nu(A)$ by showing that $\mathbf{x} \in N_A$ if and only if $\mathbf{x} \in N_{CA}$. Next prove that $\rho(AC) = \rho(A)$ by showing that $R_A = R_{AC}$. Conclude that $\rho(AC) = \rho(CA) = \rho(A)$. Finally, use the fact that $C^{-1}$ is invertible to show that $\rho(C^{-1}AC) = \rho(A)$.]

**19.** Let $D = \begin{pmatrix} 1 & 0 \\ 0 & -1 \end{pmatrix}$. Compute $D^{20}$.

**20.** If $A$ is similar to $B$, show that $\det A = \det B$.

**21.** Suppose that $C^{-1}AC = D$. Show that for any integer $n$, $A^n = CD^nC^{-1}$. This gives an easy way to compute powers of a diagonalizable matrix.

**22.** Let $A = \begin{pmatrix} 3 & -4 \\ 2 & -3 \end{pmatrix}$. Compute $A^{20}$. [*Hint:* Find a $C$ such that $A = CDC^{-1}$.]

***23.** Let $A$ be an $n \times n$ matrix whose characteristic equation is $(\lambda - c)^n = 0$. Show that $A$ is diagonalizable if and only if $A = cI$.

**24.** Use the result of Problem 21 and Example 6 to compute $A^{10}$, where $A = \begin{pmatrix} 3 & 2 & 4 \\ 2 & 0 & 2 \\ 4 & 2 & 3 \end{pmatrix}$.

***25.** Let $A$ and $B$ be real $n \times n$ matrices with distinct eigenvalues. Prove that $AB = BA$ if and only if $A$ and $B$ have the same eigenvectors.

**26.** If $A$ is diagonalizable, show that $\det A = \lambda_1\lambda_2 \cdots \lambda_n$, where $\lambda_1, \lambda_2, \ldots, \lambda_n$ are the eigenvalues of $A$.

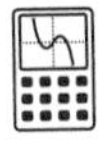

## CALCULATOR BOX

### TI-85

A matrix can easily be diagonalized on the TI-85. Start with an $n \times n$ matrix $A$ and find its eigenvalues and eigenvectors. If there are $n$ linearly independent eigenvectors (which must be the case if $A$ has $n$ distinct eigenvalues), then $A$ is diagonalizable. The TI-85 gives the eigenvectors as columns of a matrix. Store this matrix in memory C, say. Then the following key strokes will result in a diagonal matrix whose diagonal components are the eigenvalues of $A$.

ALPHA | C | 2nd | $x^{-1}$ | × | ALPHA | A | × | ALPHA | C | ENTER

$C^{-1} * A * C$ is displayed before ENTER is pressed.

Note, as on page 550, that a number like −2.2E − 13 is effectively the number zero. Note, too, that once the matrix $C$ is stored, you can determine whether its columns (eigenvectors) are linearly independent by computing either its determinant (which will be nonzero) or its rank (which will equal $n$).

In Problems 27–30 find a matrix $C$ such that $C^{-1}AC$ is a diagonal matrix.

**27.** $\begin{pmatrix} 4 & 1 & 2 & 6 \\ -1 & 3 & 4 & 2 \\ 5 & 2 & 0 & 6 \\ 3 & 8 & 1 & 5 \end{pmatrix}$

**28.** $\begin{pmatrix} 102 & -11 & 56 \\ 38 & -49 & 75 \\ 83 & 123 & -67 \end{pmatrix}$

**29.** $\begin{pmatrix} -0.031 & 0.082 & 0.095 \\ -0.046 & 0.067 & -0.081 \\ 0.055 & -0.077 & 0.038 \end{pmatrix}$

**30.** $\begin{pmatrix} 13 & 16 & 12 & 14 & 18 \\ 26 & 21 & 19 & 27 & 16 \\ 31 & 29 & 37 & 41 & 56 \\ 51 & 38 & 29 & 46 & 33 \\ 61 & 41 & 29 & 38 & 50 \end{pmatrix}$

## MATLAB 6.3

1. Rework MATLAB Problem 8 in Section 6.1.
2. Generate three random 4 × 4 matrices and three random 5 × 5 matrices. Find the eigenvalues and eigenvectors of each using **[V,D] = eig(A)**.
   a. How frequently are the eigenvalues distinct? Why do you think this is true?
   b. For the matrices for which $V$ is invertible, verify that $A = VDV^{-1}$.
3. a. For the matrix in MATLAB Problem 1 in Section 6.1, using the information given there (do not use **eig**), verify that the eigenvectors form a basis for $\mathbb{R}^3$ and find matrices $C$ and $D$, with $D$ diagonal, such that $A = CDC^{-1}$. Check your answer by verifying that $A = CDC^{-1}$.
   b. Follow the directions for part (a), except use the matrix and the information given in MATLAB Problem 2 in Section 6.1. [Here the eigenvectors will form a basis for $\mathbb{C}^4$.]
4. a. Consider the matrix $A$ given below.

$$A = \begin{pmatrix} 1 & -1 & 0 \\ -1 & 2 & -1 \\ 0 & -1 & 1 \end{pmatrix}$$

Form **d = eig(A)** and **dd = d.^20**. (Note the "." before the "^." It is important.) Form **E = diag(dd)**. Find **[V,D] = eig(A)**. Verify that $E = D^{20}$. Explain why this is true. Show that $A^{20} = VEV^{-1}$.

b. Repeat the directions of part (a) for the matrix

$$A = \begin{pmatrix} 3 & 9.5 & -2 & -10.5 \\ -10 & -42.5 & 10 & 44.5 \\ 6 & 23.5 & -5 & -24.5 \\ -10 & -43 & 10 & 45 \end{pmatrix}$$

c. *(Paper and pencil)* Work Problem 21 in the text.

5. **Geometry** An $n \times n$ matrix $A$ defines a linear transformation $T$ from $\mathbb{R}^n$ to $\mathbb{R}^n$ by $T(\mathbf{x}) = A\mathbf{x}$. We are interested in describing the geometry of such a linear transformation.

a. *(Paper and pencil)* If $\mathbf{x}$ is an eigenvector for $A$ with eigenvalue $\lambda$, then $A\mathbf{x} = \lambda\mathbf{x}$. If $\lambda > 0$, what is the geometric interpretation of the effect of the linear transformation on $\mathbf{x}$?

b. *(Paper and pencil)* Explain why and how the following statement is true: If $A$ is diagonalizable with positive eigenvalues, then the geometry of the linear transformation given by $A$ can be completely described in terms of expansions and compressions along vectors in a basis.

c. Verify that the matrix below is diagonalizable with positive eigenvalues. Describe the geometry [in the sense of part (b)] of the matrix below. Using this information, sketch the image (after applying the transformation determined by the matrix) of the rectangle with corners at $(1, 1)$, $(1, -1)$, $(-1, 1)$, and $(-1, -1)$. Describe your reasoning. (If you want a possibly nicer description of the eigenvectors than that given by **eig**, find the reduced row echelon form of $A - \lambda I$, where $\lambda$ is an eigenvalue.)

$$A = \begin{pmatrix} \frac{5}{2} & \frac{1}{2} \\ \frac{1}{2} & \frac{5}{2} \end{pmatrix}$$

d. For each matrix $A$ below, verify that $A$ is diagonalizable with positive eigenvalues. Write a description of the geometry as in part (b).

**i.** $A = \begin{pmatrix} 15 & -31 & 17 \\ 20.5 & -44 & 24.5 \\ 26.5 & -58 & 32.5 \end{pmatrix}$

**ii.** Let $B$ be a random $3 \times 3$ real matrix and let $A = B^tB$.

6. Consider the following matrices:

$$A = \begin{pmatrix} 22 & -10 \\ 50 & -23 \end{pmatrix}, \quad A = \begin{pmatrix} 8 & 3 \\ .5 & 5.5 \end{pmatrix}$$

$$A = \begin{pmatrix} 5 & -11 & 7 \\ -2 & 1 & 2 \\ -6 & 7 & 0 \end{pmatrix}, \quad A = \begin{pmatrix} 26 & -68 & 40 \\ 19 & -56 & 35 \\ 15 & -50 & 33 \end{pmatrix}$$

a. For each matrix $A$, find $\mathbf{e} = \mathbf{eig(A)}$ and $\mathbf{d} = \mathbf{det(A)}$. Explain why $A$ is diagonalizable. Conjecture a relationship between the eigenvalues of $A$ and the determinant of $A$.

b. Test your conjecture on the matrices given in MATLAB Problems 1 and 2 in Section 6.1.

c. *(Paper and pencil)* Complete the following statement with your conjecture and then prove it: If $A$ is diagonalizable, then $\det(A)$ is ____.

## 6.4 SYMMETRIC MATRICES AND ORTHOGONAL DIAGONALIZATION

In this section we shall see that real symmetric matrices† have a number of important properties. In particular, we show that any real symmetric matrix has $n$ linearly independent real eigenvectors, and therefore by Theorem 6.3.2 is diagonalizable. We begin by proving that the eigenvalues of a real symmetric matrix are real.

**THEOREM 1** Let $A$ be a real $n \times n$ symmetric matrix. Then the eigenvalues of $A$ are real.

**Proof‡** Let $\lambda$ be an eigenvalue of $A$ with eigenvector $\mathbf{v}$; that is, $A\mathbf{v} = \lambda\mathbf{v}$. Now $\mathbf{v}$ is a vector in $\mathbb{C}^n$, and an inner product in $\mathbb{C}^n$ (see Definition 4.11.1, page 439, and Example 4.11.2) satisfies

$$(\alpha\mathbf{x}, \mathbf{y}) = \alpha(\mathbf{x}, \mathbf{y}) \quad \text{and} \quad (\mathbf{x}, \alpha\mathbf{y}) = \overline{\alpha}(\mathbf{x}, \mathbf{y}) \tag{1}$$

Then

$$(A\mathbf{v}, \mathbf{v}) = (\lambda\mathbf{v}, \mathbf{v}) = \lambda(\mathbf{v}, \mathbf{v}) \tag{2}$$

Moreover, by Theorem 5.5.1 on page 519 and the fact that $A^t = A$

$$(A\mathbf{v}, \mathbf{v}) = (\mathbf{v}, A^t\mathbf{v}) = (\mathbf{v}, A\mathbf{v}) = (\mathbf{v}, \lambda\mathbf{v}) = \overline{\lambda}(\mathbf{v}, \mathbf{v}) \tag{3}$$

Thus equating (2) and (3), we have

$$\lambda(\mathbf{v}, \mathbf{v}) = \overline{\lambda}(\mathbf{v}, \mathbf{v}) \tag{4}$$

But $(\mathbf{v}, \mathbf{v}) = \|\mathbf{v}\|^2 \neq 0$, since $\mathbf{v}$ is an eigenvector. Thus we can divide both sides of (4) by $(\mathbf{v}, \mathbf{v})$ to obtain

$$\lambda = \overline{\lambda} \tag{5}$$

If $\lambda = a + ib$, then $\overline{\lambda} = a - ib$ and from (5) we have

$$a + ib = a - ib \tag{6}$$

which can hold only if $b = 0$. This shows that $\lambda = a$; hence $\lambda$ is real and the proof is complete. ■

We saw in Theorem 6.1.3 on page 536 that eigenvectors corresponding to distinct eigenvalues are linearly independent. For real symmetric matrices the result is stronger: *Eigenvectors of a real symmetric matrix corresponding to distinct eigenvalues are orthogonal.*

---

†Recall that $A$ is symmetric if and only if $A^t = A$.

‡This proof uses material in Sections 4.11 and 5.5 and should be omitted if those sections were not covered.

**THEOREM 2** Let $A$ be a real symmetric $n \times n$ matrix. If $\lambda_1$ and $\lambda_2$ are distinct eigenvalues with corresponding real eigenvectors $\mathbf{v}_1$ and $\mathbf{v}_2$, then $\mathbf{v}_1$ and $\mathbf{v}_2$ are orthogonal.

**Proof** We compute

$$A\mathbf{v}_1 \cdot \mathbf{v}_2 = \lambda_1 \mathbf{v}_1 \cdot \mathbf{v}_2 = \lambda_1(\mathbf{v}_1 \cdot \mathbf{v}_2) \tag{7}$$

and

$$A\mathbf{v}_1 \cdot \mathbf{v}_2 = \mathbf{v}_1 \cdot A^t\mathbf{v}_2 = \mathbf{v}_1 \cdot A\mathbf{v}_2 = \mathbf{v}_1 \cdot (\lambda_2\mathbf{v}_2) = \lambda_2(\mathbf{v}_1 \cdot \mathbf{v}_2) \tag{8}$$

Combining (7) and (8), we have $\lambda_1(\mathbf{v}_1 \cdot \mathbf{v}_2) = \lambda_2(\mathbf{v}_1 \cdot \mathbf{v}_2)$ and since $\lambda_1 \neq \lambda_2$, we conclude that $\mathbf{v}_1 \cdot \mathbf{v}_2 = 0$. This is what we wanted to show. ■

We now state the main result of this section. Its proof, which is difficult (and optional), is given at the end of this section.

**THEOREM 3** Let $A$ be a real symmetric $n \times n$ matrix. Then $A$ has $n$ real orthonormal eigenvectors. ■

*Remark.* It follows from this theorem that the geometric multiplicity of each eigenvalue of $A$ is equal to its algebraic multiplicity.

Theorem 3 tells us that if $A$ is symmetric, then $\mathbb{R}^n$ has a basis $B = \{\mathbf{u}_1, \mathbf{u}_2, \ldots, \mathbf{u}_n\}$ consisting of orthonormal eigenvectors of $A$. Let $Q$ be the matrix whose columns are $\mathbf{u}_1, \mathbf{u}_2, \ldots, \mathbf{u}_n$. Then by Theorem 4.9.3 on page 399, $Q$ is an orthogonal matrix. This leads to the following definition.

**DEFINITION 1** **Orthogonally Diagonalizable Matrix** An $n \times n$ matrix $A$ is said to be **orthogonally diagonalizable** if there exists an orthogonal matrix $Q$ such that

$$\boxed{Q^tAQ = D} \tag{9}$$

where $D = \text{diag}\,(\lambda_1, \lambda_2, \ldots, \lambda_n)$ and $\lambda_1, \lambda_2, \ldots, \lambda_n$ are the eigenvalues of $A$.

*Note.* Remember that $Q$ is orthogonal if $Q^t = Q^{-1}$; hence (9) could be written as $Q^{-1}AQ = D$.

**THEOREM 4** Let $A$ be a real $n \times n$ matrix. Then $A$ is orthogonally diagonalizable if and only if $A$ is symmetric.

**Proof** Let $A$ be symmetric. Then by Theorems 2 and 3, $A$ is orthogonally diagonalizable with $Q$ the matrix whose columns are the orthonormal eigenvectors given in Theorem 3. Conversely, suppose that $A$ is orthogonally diagonalizable. Then there exists an orthogonal matrix $Q$ such that $Q^tAQ = D$. Multiplying this equation on the left by $Q$ and on the right by $Q^t$ and using the fact that $Q^tQ = QQ^t = I$, we obtain

$$A = QDQ^t \tag{10}$$

Then $A^t = (QDQ^t)^t = (Q^t)^tD^tQ^t = QDQ^t = A$. Thus $A$ is symmetric and the theorem is proved. In the last series of equations we used the facts that $(AB)^t = B^tA^t$ [part (*ii*) of Theorem 1.9.1, page 122], $(A^t)^t = A$ [part (*i*) of Theorem 1.9.1], and $D^t = D$ for any diagonal matrix $D$. ∎

Before giving examples, we provide the following three-step procedure for finding the orthogonal matrix $Q$ that diagonalizes the symmetric matrix $A$.

**Procedure for Finding a Diagonalizing Matrix $Q$**

**i.** Find a basis for each eigenspace of $A$.

**ii.** Find an orthonormal basis for each eigenspace of $A$ by using the Gram-Schmidt or any other process.

**iii.** Write $Q$ as the matrix whose columns are the orthonormal eigenvectors obtained in step (*ii*).

**EXAMPLE 1** **Diagonalizing a 2 × 2 Symmetric Matrix Using an Orthogonal Matrix** Let $A = \begin{pmatrix} 1 & -2 \\ -2 & 3 \end{pmatrix}$. Then the characteristic equation of $A$ is $\det(A - \lambda I) = \begin{vmatrix} 1-\lambda & -2 \\ -2 & 3-\lambda \end{vmatrix} = \lambda^2 - 4\lambda - 1 = 0$, which has the roots $\lambda = (4 \pm \sqrt{20})/2 = (4 \pm 2\sqrt{5})/2 = 2 \pm \sqrt{5}$. For $\lambda_1 = 2 - \sqrt{5}$ we obtain $(A - \lambda I)\mathbf{v} = \begin{pmatrix} -1+\sqrt{5} & -2 \\ -2 & 1+\sqrt{5} \end{pmatrix}\begin{pmatrix} x_1 \\ x_2 \end{pmatrix} = \begin{pmatrix} 0 \\ 0 \end{pmatrix}$. An eigenvector is $\mathbf{v}_1 = \begin{pmatrix} 2 \\ -1+\sqrt{5} \end{pmatrix}$ and $|\mathbf{v}_1| = \sqrt{2^2 + (-1+\sqrt{5})^2} = \sqrt{10 - 2\sqrt{5}}$. Thus

$$\mathbf{u}_1 = \frac{1}{\sqrt{10 - 2\sqrt{5}}}\begin{pmatrix} 2 \\ -1+\sqrt{5} \end{pmatrix}$$

Next, for $\lambda_2 = 2 + \sqrt{5}$ we compute $(A - \lambda I)\mathbf{v} = \begin{pmatrix} -1-\sqrt{5} & -2 \\ -2 & 1-\sqrt{5} \end{pmatrix}\begin{pmatrix} x_1 \\ x_2 \end{pmatrix} =$

$\begin{pmatrix} 0 \\ 0 \end{pmatrix}$ and $\mathbf{v}_2 = \begin{pmatrix} 1-\sqrt{5} \\ 2 \end{pmatrix}$. Note that $\mathbf{v}_1 \cdot \mathbf{v}_2 = 0$ (which must be true according to Theorem 2). Then $|\mathbf{v}_2| = \sqrt{10-2\sqrt{5}}$ so that $\mathbf{u}_2 = \dfrac{1}{\sqrt{10-2\sqrt{5}}}\begin{pmatrix} 1-\sqrt{5} \\ 2 \end{pmatrix}$.

Finally,

$$Q = \frac{1}{\sqrt{10-2\sqrt{5}}}\begin{pmatrix} 2 & 1-\sqrt{5} \\ -1+\sqrt{5} & 2 \end{pmatrix}$$

$$Q^t = \frac{1}{\sqrt{10-2\sqrt{5}}}\begin{pmatrix} 2 & -1+\sqrt{5} \\ 1-\sqrt{5} & 2 \end{pmatrix}$$

and

$$\begin{aligned} Q^tAQ &= \frac{1}{10-2\sqrt{5}}\begin{pmatrix} 2 & -1+\sqrt{5} \\ 1-\sqrt{5} & 2 \end{pmatrix}\begin{pmatrix} 1 & -2 \\ -2 & 3 \end{pmatrix}\begin{pmatrix} 2 & 1-\sqrt{5} \\ -1+\sqrt{5} & 2 \end{pmatrix} \\ &= \frac{1}{10-2\sqrt{5}}\begin{pmatrix} 2 & -1+\sqrt{5} \\ 1-\sqrt{5} & 2 \end{pmatrix}\begin{pmatrix} 4-2\sqrt{5} & -3-\sqrt{5} \\ -7+3\sqrt{5} & 4+2\sqrt{5} \end{pmatrix} \\ &= \frac{1}{10-2\sqrt{5}}\begin{pmatrix} 30-14\sqrt{5} & 0 \\ 0 & 10+6\sqrt{5} \end{pmatrix} = \begin{pmatrix} 2-\sqrt{5} & 0 \\ 0 & 2+\sqrt{5} \end{pmatrix} \end{aligned}$$

**EXAMPLE 2** **Diagonalizing a 3 × 3 Symmetric Matrix Using an Orthogonal Matrix** Let $A = \begin{pmatrix} 5 & 4 & 2 \\ 4 & 5 & 2 \\ 2 & 2 & 2 \end{pmatrix}$. Then $A$ is symmetric and $\det(A - \lambda I) = \begin{vmatrix} 5-\lambda & 4 & 2 \\ 4 & 5-\lambda & 2 \\ 2 & 2 & 2-\lambda \end{vmatrix} = -(\lambda-1)^2(\lambda-10)$. Corresponding to $\lambda = 1$ we compute the linearly independent eigenvectors $\mathbf{v}_1 = \begin{pmatrix} -1 \\ 1 \\ 0 \end{pmatrix}$ and $\mathbf{v}_2 = \begin{pmatrix} -1 \\ 0 \\ 2 \end{pmatrix}$. Corresponding to $\lambda = 10$ we find that $\mathbf{v}_3 = \begin{pmatrix} 2 \\ 2 \\ 1 \end{pmatrix}$. To find $Q$, we apply the Gram-Schmidt process to $\{\mathbf{v}_1, \mathbf{v}_2\}$, a basis for $E_1$. Since $|\mathbf{v}_1| = \sqrt{2}$, we set $\mathbf{u}_1 = \begin{pmatrix} -1/\sqrt{2} \\ 1/\sqrt{2} \\ 0 \end{pmatrix}$. Next

$$\mathbf{v}_2' = \mathbf{v}_2 - (\mathbf{v}_2 \cdot \mathbf{u}_1)\mathbf{u}_1 = \begin{pmatrix} -1 \\ 0 \\ 2 \end{pmatrix} - \frac{1}{\sqrt{2}}\begin{pmatrix} -1/\sqrt{2} \\ 1/\sqrt{2} \\ 0 \end{pmatrix}$$

$$= \begin{pmatrix} -1 \\ 0 \\ 2 \end{pmatrix} - \begin{pmatrix} -1/2 \\ 1/2 \\ 0 \end{pmatrix} = \begin{pmatrix} -1/2 \\ -1/2 \\ 2 \end{pmatrix}$$

Then $|\mathbf{v}_2| = \sqrt{18/4} = 3\sqrt{2}/2$ and $\mathbf{u}_2 = \dfrac{2}{3\sqrt{2}}\begin{pmatrix} -1/2 \\ -1/2 \\ 2 \end{pmatrix} = \begin{pmatrix} -1/3\sqrt{2} \\ -1/3\sqrt{2} \\ 4/3\sqrt{2} \end{pmatrix}$. We check this by noting that $\mathbf{u}_1 \cdot \mathbf{u}_2 = 0$. Finally, we have $\mathbf{u}_3 = \mathbf{v}_3/|\mathbf{v}_3| = \frac{1}{3}\mathbf{v}_3 = \begin{pmatrix} 2/3 \\ 2/3 \\ 1/3 \end{pmatrix}$.

We can check this too by noting that $\mathbf{u}_1 \cdot \mathbf{u}_3 = 0$ and $\mathbf{u}_2 \cdot \mathbf{u}_3 = 0$. Thus

$$Q = \begin{pmatrix} -1/\sqrt{2} & -1/3\sqrt{2} & 2/3 \\ 1/\sqrt{2} & -1/3\sqrt{2} & 2/3 \\ 0 & 4/3\sqrt{2} & 1/3 \end{pmatrix}$$

and

$$Q^tAQ = \begin{pmatrix} -1/\sqrt{2} & 1/\sqrt{2} & 0 \\ -1/3\sqrt{2} & -1/3\sqrt{2} & 4/3\sqrt{2} \\ 2/3 & 2/3 & 1/3 \end{pmatrix}\begin{pmatrix} 5 & 4 & 2 \\ 4 & 5 & 2 \\ 2 & 2 & 2 \end{pmatrix}\begin{pmatrix} -1/\sqrt{2} & -1/3\sqrt{2} & 2/3 \\ 1/\sqrt{2} & -1/3\sqrt{2} & 2/3 \\ 0 & 4/3\sqrt{2} & 1/3 \end{pmatrix}$$

$$= \begin{pmatrix} -1/\sqrt{2} & 1/\sqrt{2} & 0 \\ -1/3\sqrt{2} & -1/3\sqrt{2} & 4/3\sqrt{2} \\ 2/3 & 2/3 & 1/3 \end{pmatrix}\begin{pmatrix} -1/\sqrt{2} & -1/3\sqrt{2} & 20/3 \\ 1/\sqrt{2} & -1/3\sqrt{2} & 20/3 \\ 0 & 4/3\sqrt{2} & 10/3 \end{pmatrix}$$

$$= \begin{pmatrix} 1 & 0 & 0 \\ 0 & 1 & 0 \\ 0 & 0 & 10 \end{pmatrix}$$

■

Conjugate transpose
Hermitian matrix
Unitary matrix

In this section we have proved results for real symmetric matrices. These results can be extended to complex matrices as well. If $A = (a_{ij})$ is a complex matrix, then the **conjugate transpose** of $A$, denoted by $A^*$, is defined by: the $ij$th element of $A^* = \overline{a_{ji}}$. The matrix $A$ is called **Hermitian**† if $A^* = A$. It turns out that Theorems 1, 2, and 3 are also true for Hermitian matrices. Moreover, if we define a **unitary** matrix to be a complex matrix $U$ with $U^* = U^{-1}$, then, using the proof of Theorem 4, we can show that a Hermitian matrix is unitarily diagonalizable. We leave all these facts as exercises (see Problems 15–17).

We conclude this section with a proof of Theorem 3.

**Proof of Theorem 3‡**

We prove that to every eigenvalue $\lambda$ of algebraic multiplicity $k$, there correspond $k$ orthonormal eigenvectors. This step, combined with Theorem 2, will prove

† See the footnote on page 526.

‡ If time permits.

the theorem. Let $\mathbf{u}_1$ be an eigenvector of $A$ corresponding to $\lambda_1$. We can assume that $|\mathbf{u}_1| = 1$. We can also assume that $\mathbf{u}_1$ is real because $\lambda_1$ is real and $\mathbf{u}_1 \in N_{A-\lambda_1 I}$, the null space of the real matrix $A - \lambda_1 I$. This null space is a subspace of $\mathbb{R}^n$ by Example 4.6.10 on page 343. Next we note that $\{\mathbf{u}_1\}$ can be expanded into a basis $\{\mathbf{u}_1, \mathbf{v}_2, \mathbf{v}_3, \ldots, \mathbf{v}_n\}$ for $\mathbb{R}^n$, and by the Gram-Schmidt process we can turn this basis into the orthonormal basis $\{\mathbf{u}_1, \mathbf{u}_2, \ldots, \mathbf{u}_n\}$. Let $Q$ be the orthogonal matrix whose columns are $\mathbf{u}_1, \mathbf{u}_2, \ldots, \mathbf{u}_n$. For convenience of notation we write $Q = (\mathbf{u}_1, \mathbf{u}_2, \ldots, \mathbf{u}_n)$. Now $Q$ is invertible and $Q^t = Q^{-1}$, so $A$ is similar to $Q^tAQ$, and by Theorem 6.3.1 on page 566, $Q^tAQ$ and $A$ have the same characteristic polynomial: $|Q^tAQ - \lambda I| = |A - \lambda I|$. Then

$$Q^t = \begin{pmatrix} \mathbf{u}_1^t \\ \mathbf{u}_2^t \\ \vdots \\ \mathbf{u}_n^t \end{pmatrix}$$

so that

$$Q^tAQ = \begin{pmatrix} \mathbf{u}_1^t \\ \mathbf{u}_2^t \\ \vdots \\ \mathbf{u}_n^t \end{pmatrix} A(\mathbf{u}_1, \mathbf{u}_2, \ldots, \mathbf{u}_n) = \begin{pmatrix} \mathbf{u}_1^t \\ \mathbf{u}_2^t \\ \vdots \\ \mathbf{u}_n^t \end{pmatrix} (A\mathbf{u}_1, A\mathbf{u}_2, \ldots, A\mathbf{u}_n)$$

$$= \begin{pmatrix} \mathbf{u}_1^t \\ \mathbf{u}_2^t \\ \vdots \\ \mathbf{u}_n^t \end{pmatrix} (\lambda_1\mathbf{u}_1, A\mathbf{u}_2, \ldots, A\mathbf{u}_n) = \begin{pmatrix} \lambda_1 & \mathbf{u}_1^tA\mathbf{u}_2 & \cdots & \mathbf{u}_1^tA\mathbf{u}_n \\ 0 & \mathbf{u}_2^tA\mathbf{u}_2 & \cdots & \mathbf{u}_2^tA\mathbf{u}_n \\ \vdots & \vdots & & \vdots \\ 0 & \mathbf{u}_n^tA\mathbf{u}_2 & \cdots & \mathbf{u}_n^tA\mathbf{u}_n \end{pmatrix}$$

The zeros appear because $\mathbf{u}_1^t\mathbf{u}_j = \mathbf{u}_1 \cdot \mathbf{u}_j = 0$ if $j \neq 1$. Now $[Q^tAQ]^t = Q^tA^t(Q^t)^t = Q^tAQ$. Thus $Q^tAQ$ is symmetric, which means that there must be zeros in the first row of $Q^tAQ$ to match the zeros in the first column. Thus

$$Q^tAQ = \begin{pmatrix} \lambda_1 & 0 & 0 & \cdots & 0 \\ 0 & q_{22} & q_{23} & \cdots & q_{2n} \\ 0 & q_{32} & q_{33} & \cdots & q_{3n} \\ \vdots & \vdots & \vdots & & \vdots \\ 0 & q_{n2} & q_{n3} & \cdots & q_{nn} \end{pmatrix}$$

and

$$|Q^tAQ - \lambda I| = \begin{vmatrix} \lambda_1 - \lambda & 0 & 0 & \cdots & 0 \\ 0 & q_{22} - \lambda & q_{23} & \cdots & q_{2n} \\ 0 & q_{32} & q_{33} - \lambda & \cdots & q_{3n} \\ \vdots & \vdots & \vdots & & \vdots \\ 0 & q_{n2} & q_{n3} & \cdots & q_{nn} - \lambda \end{vmatrix}$$

$$= (\lambda_1 - \lambda)\begin{vmatrix} q_{22}-\lambda & q_{23} & \cdots & q_{2n} \\ q_{32} & q_{33}-\lambda & \cdots & q_{3n} \\ \vdots & \vdots & & \vdots \\ q_{n2} & q_{n3} & \cdots & q_{nn}-\lambda \end{vmatrix} = (\lambda - \lambda_1)|M_{11}(\lambda)|$$

where $M_{11}(\lambda)$ is the 1, 1 minor of $Q^tAQ - \lambda I$. If $k = 1$, there is nothing to prove. If $k > 1$, then $|A - \lambda I|$ contains the factor $(\lambda - \lambda_1)^2$, and therefore $|Q^tAQ - \lambda I|$ also contains the factor $(\lambda - \lambda_1)^2$. Thus $|M_{11}(\lambda)|$ contains the factor $\lambda - \lambda_1$, which means that $|M_{11}(\lambda_1)| = 0$. This means that the last $n - 1$ columns of $Q^tAQ - \lambda_1 I$ are linearly dependent. Since the first column of $Q^tAQ - \lambda_1 I$ is the zero vector, this means that $Q^tAQ - \lambda_1 I$ contains at most $n - 2$ linearly independent columns. In other words, $\rho(Q^tAQ - \lambda_1 I) \le n - 2$. But $Q^tAQ - \lambda_1 I$ and $A - \lambda_1 I$ are similar; hence by Problem 6.3.18, $\rho(A - \lambda_1 I) \le n - 2$. Therefore $\nu(A - \lambda_1 I) \ge 2$, which means that $E_\lambda$ = kernel of $(A - \lambda_1 I)$ contains at least two linearly independent eigenvectors. If $k = 2$, we are done. If $k > 2$, then we take two orthonormal vectors $\mathbf{u}_1, \mathbf{u}_2$ in $E_\lambda$ and expand them into a new orthonormal basis $\{\mathbf{u}_1, \mathbf{u}_2, \ldots, \mathbf{u}_n\}$ for $\mathbb{R}^n$ and define $P = \{\mathbf{u}_1, \mathbf{u}_2, \ldots, \mathbf{u}_n\}$. Then, exactly as before, we show that

$$P^tAP - \lambda I = \begin{pmatrix} \lambda_1 - \lambda & 0 & 0 & 0 & \cdots & 0 \\ 0 & \lambda_1 - \lambda & 0 & 0 & \cdots & 0 \\ 0 & 0 & \beta_{33}-\lambda & \beta_{34} & \cdots & \beta_{3n} \\ 0 & 0 & \beta_{43} & \beta_{44}-\lambda & \cdots & \beta_{4n} \\ \vdots & \vdots & \vdots & \vdots & & \vdots \\ 0 & 0 & \beta_{n3} & \beta_{n4} & \cdots & \beta_{nn}-\lambda \end{pmatrix}$$

Since $k > 2$, we show, as before, that the determinant of the matrix in brackets is zero when $\lambda = \lambda_1$—which shows that $\rho(P^tAP - \lambda_1 I) \le n - 3$ so that $\nu(P^tAP - \lambda_1 I) = \nu(A - \lambda_1 I) \ge 3$. Then $\dim E_{\lambda_1} \ge 3$, and so on. We can clearly continue this process to show that $\dim E_{\lambda_1} = k$. Finally, in each $E_{\lambda_1}$ we can find an orthonormal basis. This completes the proof. ■

## PROBLEMS 6.4

### Self-Quiz

#### *True-False*

**I.** The eigenvalues of a real symmetric matrix are real.

**II.** The eigenvectors of a real symmetric matrix are real.

**III.** Every real symmetric matrix is similar to a diagonal matrix.

**IV.** If the real matrix $A$ can be diagonalized, then there is an orthogonal matrix $Q$ such that $Q^tAQ$ is diagonal.

**V.** If $A$ is real and symmetric, then there is an orthogonal matrix $Q$ such that $Q^tAQ$ is diagonal.

**VI.** A symmetric matrix is Hermitian.

**VII.** A Hermitian matrix is symmetric.

In Problems 1–8 find an orthogonal matrix $Q$ that diagonalizes the given symmetric matrix. Then verify that $Q^tAQ = D$, a diagonal matrix whose diagonal components are the eigenvalues of $A$.

**1.** $\begin{pmatrix} 3 & 4 \\ 4 & -3 \end{pmatrix}$ **2.** $\begin{pmatrix} 2 & 1 \\ 1 & 2 \end{pmatrix}$ **3.** $\begin{pmatrix} 1 & -1 \\ -1 & 1 \end{pmatrix}$

**4.** $\begin{pmatrix} 1 & -1 & -1 \\ -1 & 1 & -1 \\ -1 & -1 & 1 \end{pmatrix}$ **5.** $\begin{pmatrix} -1 & 2 & 2 \\ 2 & -1 & 2 \\ 2 & 2 & 1 \end{pmatrix}$ **6.** $\begin{pmatrix} 1 & -1 & 0 \\ -1 & 2 & -1 \\ 0 & -1 & 1 \end{pmatrix}$

**7.** $\begin{pmatrix} 3 & 2 & 2 \\ 2 & 2 & 0 \\ 2 & 0 & 4 \end{pmatrix}$ **8.** $\begin{pmatrix} 1 & -1 & 0 & 0 \\ -1 & 0 & 0 & 0 \\ 0 & 0 & 0 & 0 \\ 0 & 0 & 0 & 2 \end{pmatrix}$

**9.** Let $Q$ be a symmetric orthogonal matrix. Show that if $\lambda$ is an eigenvalue of $Q$, then $\lambda = \pm 1$.

**10.** $A$ is **orthogonally similar** to $B$ if there exists an orthogonal matrix $Q$ such that $B = Q^tAQ$. Suppose that $A$ is orthogonally similar to $B$ and that $B$ is orthogonally similar to $C$. Show that $A$ is orthogonally similar to $C$.

**11.** Show that if $Q = \begin{pmatrix} a & b \\ c & d \end{pmatrix}$ is orthogonal, then $b = \pm c$. [*Hint:* Write out the equations that result from the equation $Q^tQ = I$.]

**12.** Suppose that $A$ is a real symmetric matrix every one of whose eigenvalues is zero. Show that $A$ is the zero matrix.

**13.** Show that if a real $2 \times 2$ matrix $A$ has eigenvectors that are orthogonal, then $A$ is symmetric.

**14.** Let $A$ be a real skew-symmetric matrix ($A^t = -A$). Prove that every eigenvalue of $A$ is of the form $i\alpha$, where $\alpha$ is a real number. That is, prove that every eigenvalue of $A$ is an **imaginary** number.

***15.** Show that the eigenvalues of a complex $n \times n$ Hermitian matrix are real. [*Hint:* Use the fact that in $\mathbb{C}^n$, $(A\mathbf{x}, \mathbf{y}) = (\mathbf{x}, A^*\mathbf{y})$.]

***16.** If $A$ is an $n \times n$ Hermitian matrix, show that eigenvectors corresponding to different eigenvalues are orthogonal.

****17.** By repeating the proof of Theorem 3, except that $\overline{\mathbf{v}}_i^t$ replaces $\mathbf{v}_i^t$ where appropriate, show that any $n \times n$ Hermitian matrix has $n$ orthonormal eigenvectors.

**18.** Find a unitary matrix $U$ such that $U^*AU$ is diagonal, where $A = \begin{pmatrix} 1 & 1-i \\ 1+i & 0 \end{pmatrix}$.

---

## Answers to Self-Quiz

**I.** True **II.** True **III.** True **IV.** False **V.** True **VI.** False **VII.** False

**19.** Do the same for $A = \begin{pmatrix} 2 & 3-3i \\ 3+3i & 5 \end{pmatrix}$.

**20.** Prove that the determinant of a Hermitian matrix is real.

## MATLAB 6.4

1. a. *(Paper and pencil)* If $A$ is a *random* $n \times n$ symmetric matrix, then we expect $A$ to have distinct eigenvalues and that the associated eigenvectors will be orthogonal. Explain why we can thus say that we expect that there is an orthonormal basis for $\mathbb{R}^n$ consisting of eigenvectors of $A$.
   b. Generate five random symmetric matrices $A$ (not all the same size) by generating random real matrices $B$ and then forming **A = triu(B)+triu(B)′**. For each matrix $A$ generated, verify the expectation discussed in part (a). Verify that there is a matrix $Q$ and a diagonal matrix $D$ such that $A = QDQ^t$.

2. If $A$ is a complex-valued matrix, then $A^*$ can be found using MATLAB as **A′**. Generate a random $4 \times 4$ complex-valued matrix, $A$. (Use **A = B+i∗C**, where $B$ and $C$ are random real-valued matrices found using the **rand** command.) Generate the matrix **H = triu(A)+triu(A)′**.
   a. Verify that $H$ is hermitian. Find the eigenvalues for $H$. Even though $H$ is complex-valued, what do you notice about the eigenvalues?
   b. Repeat the directions of MATLAB Problem 1 in this section except change the word *symmetric* to *hermitian,* change $\mathbb{R}^n$ to $\mathbb{C}^n$, and change $Q^t$ to $Q^*$.

3. **Geometry** Suppose $A$ is a $2 \times 2$ symmetric real matrix. Then there exists a diagonal matrix $D$ and an orthogonal matrix $Q$ such that $A = QDQ^t$.
   a. *(Paper and pencil)* Since $Q$ is orthogonal, we have $\det(Q)$ is either $+1$ or $-1$. *Why?* We know that if $\det(Q) = -1$, multiplying a column of $Q$ by $-1$ will produce a new $Q$ that is still orthogonal but now has $\det(Q) = 1$. Why? Explain why the new $Q$ still contains an orthonormal basis of eigenvectors correctly corresponding with the eigenvalues in $D$ so that $A = QDQ^t$ for the new $Q$.
   b. *(Paper and pencil)* Using the facts that $Q$ is orthogonal, that $\det(Q) = 1$, and that a vector of length 1 can be written as $(\cos(\theta) \quad \sin(\theta))$ for some angle $\theta$, explain why we can write
   $$Q = \begin{pmatrix} \cos(\theta) & -\sin(\theta) \\ \sin(\theta) & \cos(\theta) \end{pmatrix}.$$
   Verify that $Q$ is thus a rotation matrix.
   c. *(Paper and pencil)* Combining the results of parts (a) and (b), we can conclude that a symmetric $2 \times 2$ real matrix $A$ can be diagonalized as $A = QDQ^t$, where $Q$ is a matrix representation for a rotation transformation. This allows us to give a description of the geometry of the linear transformation determined by $A$ in terms of rotations of the standard basis and expansions or compressions if the eigenvalues of $A$ are positive. Explain this description by first interpreting the action of $Q^t$, followed by the action of $D$, followed by the action of $Q$.
   d. For the following matrices, describe the geometry of $A$ as outlined in part (c). Use the description to draw a sketch of the image of the unit circle after applying the transformation determined by $A$. Adjust $Q$ if necessary so that $\det(Q) = 1$. [*Hint.* You will

need to use the adjusted $Q$ to find the angle $\theta$. Note that $Q(2, 1)/Q(1, 1) = \tan(\theta)$. Use the MATLAB command **atan**, adjust the answer by adding $\pi$ if the numbers in $Q$ indicate that the angle is in the second or third quadrant, and multiply by $180/\pi$.]

**i.** $A = \begin{pmatrix} \frac{7}{2} & \frac{1}{2} \\ \frac{1}{2} & \frac{7}{2} \end{pmatrix}$

**ii.** $A = \begin{pmatrix} 2.75 & -.433 \\ -.433 & 2.25 \end{pmatrix}$

## 6.5 QUADRATIC FORMS AND CONIC SECTIONS

In this section we use the material of Section 6.4 to discover information about the graphs of quadratic equations. Quadratic equations and quadratic forms, which are defined below, arise in a variety of ways. For example, we can use quadratic forms to obtain information about the conic sections in $\mathbb{R}^2$ (circles, parabolas, ellipses, hyperbolas) and extend this theory to describe certain surfaces, called *quadric surfaces,* in $\mathbb{R}^3$. These topics are discussed later in the section. Although we shall not discuss it in this text, quadratic forms arise in a number of applications ranging from a description of cost functions in economics to an analysis of the control of a rocket traveling in space.

**DEFINITION 1** **Quadratic Equation and Quadratic Form**

**i.** A **quadratic equation in two variables with no linear terms** is an equation of the form

$$ax^2 + bxy + cy^2 = d \tag{1}$$

where $|a| + |b| + |c| \neq 0$. That is, at least one of the numbers $a$, $b$, and $c$ is nonzero.

**ii.** A **quadratic form in two variables** is an expression of the form

$$F(x, y) = ax^2 + bxy + cy^2 \tag{2}$$

where $|a| + |b| + |c| \neq 0$.

Obviously, quadratic equations and quadratic forms are closely related. We begin our analysis of quadratic forms with a simple example.

Consider the quadratic form $F(x, y) = x^2 - 4xy + 3y^2$. Let $\mathbf{v} = \begin{pmatrix} x \\ y \end{pmatrix}$ and $A = \begin{pmatrix} 1 & -2 \\ -2 & 3 \end{pmatrix}$. Then

$$A\mathbf{v}\cdot\mathbf{v} = \begin{pmatrix} 1 & -2 \\ -2 & 3 \end{pmatrix}\begin{pmatrix} x \\ y \end{pmatrix}\cdot\begin{pmatrix} x \\ y \end{pmatrix} = \begin{pmatrix} x-2y \\ -2x+3y \end{pmatrix}\cdot\begin{pmatrix} x \\ y \end{pmatrix}$$
$$= (x^2 - 2xy) + (-2xy + 3y^2) = x^2 - 4xy + 3y^2 = F(x, y)$$

Thus we have "represented" the quadratic form $F(x, y)$ by the symmetric matrix $A$ in the sense that

$$\boxed{F(x, y) = A\mathbf{v}\cdot\mathbf{v}} \tag{3}$$

Conversely, if $A$ is a symmetric matrix, then equation (3) defines a quadratic form $F(x, y) = A\mathbf{v}\cdot\mathbf{v}$.

We can represent $F(x, y)$ by many matrices but only one symmetric matrix. To see this, let $A = \begin{pmatrix} 1 & a \\ b & 3 \end{pmatrix}$, where $a + b = -4$. Then $A\mathbf{v}\cdot\mathbf{v} = F(x, y)$. If $A = \begin{pmatrix} 1 & 3 \\ -7 & 3 \end{pmatrix}$, for example, then $A\mathbf{v} = \begin{pmatrix} x+3y \\ -7x+3y \end{pmatrix}$ and $A\mathbf{v}\cdot\mathbf{v} = x^2 - 4xy + 3y^2$. If, however, we insist that $A$ be symmetric, then we must have $a + b = -4$ and $a = b$. This pair of equations has the unique solution $a = b = -2$.

If $F(x, y) = ax^2 + bxy + cy^2$ is a quadratic form, let

$$\boxed{A = \begin{pmatrix} a & b/2 \\ b/2 & c \end{pmatrix}} \tag{4}$$

Then

$$A\mathbf{v}\cdot\mathbf{v} = \left[\begin{pmatrix} a & b/2 \\ b/2 & c \end{pmatrix}\begin{pmatrix} x \\ y \end{pmatrix}\right]\cdot\begin{pmatrix} x \\ y \end{pmatrix} = \begin{pmatrix} ax + (b/2)y \\ (b/2)x + cy \end{pmatrix}\cdot\begin{pmatrix} x \\ y \end{pmatrix}$$
$$= ax^2 + bxy + cy^2 = F(x, y)$$

Now let us return to the quadratic equation (1). Using (3), we can write (1) as

$$\boxed{A\mathbf{v}\cdot\mathbf{v} = d} \tag{5}$$

where $A$ is symmetric. By Theorem 6.4.4 on page 578, there is an orthogonal matrix $Q$ such that $Q^tAQ = D$, where $D = \text{diag}(\lambda_1, \lambda_2)$ and $\lambda_1$ and $\lambda_2$ are the eigenvalues of $A$. Then $A = QDQ^t$ (remember that $Q^t = Q^{-1}$) and (5) can be written

$$(QDQ^t\mathbf{v})\cdot\mathbf{v} = d \tag{6}$$

But from Theorem 5.5.1 on page 519, $A\mathbf{v}\cdot\mathbf{y} = \mathbf{v}\cdot A^t\mathbf{y}$. Thus

$$Q(DQ^t\mathbf{v}) \cdot \mathbf{v} = DQ^t\mathbf{v} \cdot Q^t\mathbf{v} \tag{7}$$

so that (6) reads

$$[DQ^t\mathbf{v}] \cdot Q^t\mathbf{v} = d \tag{8}$$

Let $\mathbf{v}' = Q^t\mathbf{v}$. Then $\mathbf{v}'$ is a 2-vector and (8) becomes

$$\boxed{D\mathbf{v}' \cdot \mathbf{v}' = d} \tag{9}$$

Let us look at (9) more closely. We can write $\mathbf{v}' = \begin{pmatrix} x' \\ y' \end{pmatrix}$. Since a diagonal matrix is symmetric, (9) defines a quadratic form $\overline{F}(x', y')$ in the variables $x'$ and $y'$. If $D = \begin{pmatrix} a' & 0 \\ 0 & c' \end{pmatrix}$, then $D\mathbf{v}' = \begin{pmatrix} a' & 0 \\ 0 & c' \end{pmatrix}\begin{pmatrix} x' \\ y' \end{pmatrix} = \begin{pmatrix} a'x' \\ c'y' \end{pmatrix}$ and

$$\overline{F}(x', y') = D\mathbf{v}' \cdot \mathbf{v}' = \begin{pmatrix} a'x' \\ c'y' \end{pmatrix} \cdot \begin{pmatrix} x' \\ y' \end{pmatrix} = a'x'^2 + c'y'^2$$

That is, *$\overline{F}(x', y')$ is a quadratic form with the $x'y'$ term missing.* Hence equation (9) is a quadratic equation in the new variables $x'$, $y'$ with the $x'y'$ term missing.

EXAMPLE 1 **Writing a Quadratic Form in New Variables $x'$ and $y'$ with the $x'y'$ Term Missing** Consider the quadratic equation $x^2 - 4xy + 3y^2 = 6$. Then as we have seen, the equation can be written in the form $A\mathbf{x} \cdot \mathbf{x} = 6$, where $A = \begin{pmatrix} 1 & -2 \\ -2 & 3 \end{pmatrix}$. In Example 6.4.1 on page 578 we saw that $A$ can be diagonalized to $D = \begin{pmatrix} 2 - \sqrt{5} & 0 \\ 0 & 2 + \sqrt{5} \end{pmatrix}$ by using the orthogonal matrix

$$Q = \frac{1}{\sqrt{10 - 2\sqrt{5}}}\begin{pmatrix} 2 & 1 - \sqrt{5} \\ -1 + \sqrt{5} & 2 \end{pmatrix}$$

Then

$$\mathbf{x}' = \begin{pmatrix} x' \\ y' \end{pmatrix} = Q^t\mathbf{x} = \frac{1}{\sqrt{10 - 2\sqrt{5}}}\begin{pmatrix} 2 & -1 + \sqrt{5} \\ 1 - \sqrt{5} & 2 \end{pmatrix}\begin{pmatrix} x \\ y \end{pmatrix}$$
$$= \frac{1}{\sqrt{10 - 2\sqrt{5}}}\begin{pmatrix} 2x + (-1 + \sqrt{5})y \\ (1 - \sqrt{5})x + 2y \end{pmatrix}$$

and in the new variables the equation can be written as

$$(2 - \sqrt{5})x'^2 + (2 + \sqrt{5})y'^2 = 6$$

Let us take another look at the matrix $Q$. Since $Q$ is real and orthogonal, $1 = \det QQ^{-1} = \det QQ^t = \det Q \det Q^t = \det Q \det Q = (\det Q)^2$. Thus $\det Q =$

$\pm 1$. If $\det Q = -1$, we can interchange the rows of $Q$ to make the determinant of this new $Q$ equal to 1. Then it can be shown (see Problem 36) that $Q = \begin{pmatrix} \cos\theta & -\sin\theta \\ \sin\theta & \cos\theta \end{pmatrix}$ for some number $\theta$ with $0 \le \theta < 2\pi$. But from Example 5.1.8 on page 469 this means that $Q$ is a rotation matrix. We have therefore proved the following theorem.

**THEOREM 1** **Principal Axes Theorem in $\mathbb{R}^2$** Let

$$ax^2 + bxy + cy^2 = d \tag{10}$$

be a quadratic equation in the variables $x$ and $y$. Then there exists a unique number $\theta$ in $[0, 2\pi)$ such that equation (10) can be written in the form

$$a'x'^2 + c'y'^2 = d \tag{11}$$

where $x'$, $y'$ are the axes obtained by rotating the $x$- and $y$-axes through an angle of $\theta$ in the counterclockwise direction. Moreover, the numbers $a'$ and $c'$ are the eigenvalues of the matrix $A = \begin{pmatrix} a & b/2 \\ b/2 & c \end{pmatrix}$. The $x'$- and $y'$-axes are called the **principal axes** of the graph of the quadratic equation (10).

Principal axes

We can use Theorem 1 to identify three important conic sections. Recall that the **standard equations** of a circle, ellipse, and hyperbola are

Circle: $$x^2 + y^2 = r^2 \tag{12}$$

Ellipse: $$\frac{x^2}{a^2} + \frac{y^2}{b^2} = 1 \tag{13}$$

Hyperbola: $$\frac{x^2}{a^2} - \frac{y^2}{b^2} = 1 \tag{14}$$

or

$$\frac{y^2}{a^2} - \frac{x^2}{b^2} = 1 \tag{15}$$

**EXAMPLE 2** **Identifying a Hyperbola** Identify the conic section whose equation is

$$x^2 - 4xy + 3y^2 = 6 \tag{16}$$

**Solution** In Example 1 we found that this can be written as $(2 - \sqrt{5})x'^2 + (2 + \sqrt{5})y'^2 = 6$ or

$$\frac{y'^2}{6/(2+\sqrt{5})} - \frac{x'^2}{6/(\sqrt{5}-2)} = 1$$

This is equation (15) with $a = \sqrt{6/(2+\sqrt{5})} \approx 1.19$ and $b = \sqrt{6/(\sqrt{5}-2)} \approx 5.04$. Since

$$Q = \frac{1}{\sqrt{10-2\sqrt{5}}}\begin{pmatrix} 2 & 1-\sqrt{5} \\ -1+\sqrt{5} & 2 \end{pmatrix}$$

and $\det Q = 1$, we have, using Problem 36 and the fact that 2 and $-1+\sqrt{5}$ are positive,

$$\cos\theta = \frac{2}{\sqrt{10-2\sqrt{5}}} \approx 0.85065$$

Thus $\theta$ is in the first quadrant, and using a calculator, we find that $\theta \approx 0.5536$ rad $\approx 31.7°$. Thus (16) is the equation of a standard hyperbola rotated through an angle of 31.7° (see Figure 6.1).

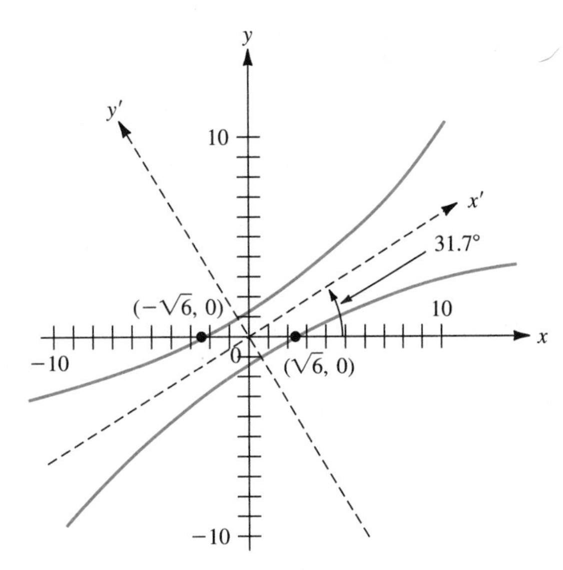


**Figure 6.1** The hyperbola $x^2 - 4xy + 3y^2 = 6$

**EXAMPLE 3** **An Ellipse** Identify the conic section whose equation is

$$5x^2 - 2xy + 5y^2 = 4 \tag{17}$$

**Solution** Here $A = \begin{pmatrix} 5 & -1 \\ -1 & 5 \end{pmatrix}$, the eigenvalues of $A$ are $\lambda_1 = 4$ and $\lambda_2 = 6$, and two orthonormal eigenvectors are $\mathbf{v}_1 = \begin{pmatrix} 1/\sqrt{2} \\ 1/\sqrt{2} \end{pmatrix}$ and $\mathbf{v}_2 = \begin{pmatrix} 1/\sqrt{2} \\ -1/\sqrt{2} \end{pmatrix}$. Then $Q =$

$\begin{pmatrix} 1/\sqrt{2} & 1/\sqrt{2} \\ 1/\sqrt{2} & -1/\sqrt{2} \end{pmatrix}$. Before continuing, we note that $\det Q = -1$. For $Q$ to be a rotation matrix we need $\det Q = 1$. This is easily accomplished by reversing the eigenvectors. Thus we set $\lambda_1 = 6$, $\lambda_2 = 4$, $\mathbf{v}_1 = \begin{pmatrix} 1/\sqrt{2} \\ -1/\sqrt{2} \end{pmatrix}$, $\mathbf{v}_2 = \begin{pmatrix} 1/\sqrt{2} \\ 1/\sqrt{2} \end{pmatrix}$, and $Q = \begin{pmatrix} 1/\sqrt{2} & 1/\sqrt{2} \\ -1/\sqrt{2} & 1/\sqrt{2} \end{pmatrix}$; now $\det Q = 1$. Then $D = \begin{pmatrix} 6 & 0 \\ 0 & 4 \end{pmatrix}$ and (17) can be written as $D\mathbf{v} \cdot \mathbf{v} = 4$ or

$$6x'^2 + 4y'^2 = 4 \tag{18}$$

where

$$\begin{pmatrix} x' \\ y' \end{pmatrix} = Q^t \begin{pmatrix} x \\ y \end{pmatrix} = \begin{pmatrix} 1/\sqrt{2} & -1/\sqrt{2} \\ 1/\sqrt{2} & 1/\sqrt{2} \end{pmatrix} \begin{pmatrix} x \\ y \end{pmatrix} = \begin{pmatrix} 1/\sqrt{2}x - 1/\sqrt{2}y \\ 1/\sqrt{2}x + 1/\sqrt{2}y \end{pmatrix}$$

Rewriting (18), we obtain $x'^2/(\frac{4}{6}) + y'^2/1 = 1$, which is equation (13) with $a = \sqrt{\frac{2}{3}}$ and $b = 1$. Moreover, since $1/\sqrt{2} > 0$ and $-1/\sqrt{2} < 0$, we have from Problem 36, $\theta = 2\pi - \cos^{-1}(1/\sqrt{2}) = 2\pi - \pi/4 = 7\pi/4 = 315°$. Thus (17) is the equation of a standard ellipse rotated through an angle of 315° (or 45° in the clockwise direction). (See Figure 6.2.)

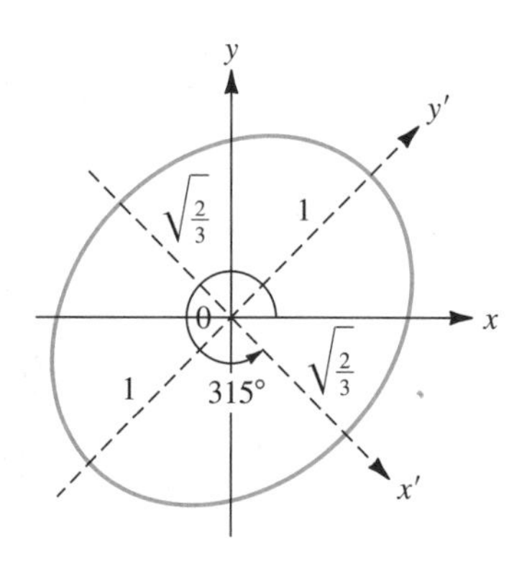


**Figure 6.2** The ellipse $5x^2 - 2xy + 5y^2 = 4$

**EXAMPLE 4** **A Degenerate Conic Section** Identify the conic section whose equation is

$$-5x^2 + 2xy - 5y^2 = 4 \tag{19}$$

**Solution** Referring to Example 3, equation (19) can be rewritten as

$$-6x'^2 - 4y'^2 = 4 \tag{20}$$

Since for any real numbers $x'$ and $y'$, $-6x'^2 - 4y'^2 \le 0$, we see that there are no real numbers $x$ and $y$ that satisfy (19). The conic section defined by (19) is called a **degenerate conic section.**

There is an easy way to identify the conic section defined by

$$ax^2 + bxy + cy^2 = d \tag{21}$$

If $A = \begin{pmatrix} a & b/2 \\ b/2 & c \end{pmatrix}$, then the characteristic equation of $A$ is

$$\lambda^2 - (a + c)\lambda + (ac - b^2/4) = 0 = (\lambda - \lambda_1)(\lambda - \lambda_2)$$

This means that $\lambda_1\lambda_2 = ac - b^2/4$. But equation (21) can, as we have seen, be rewritten as

$$\lambda_1 x'^2 + \lambda_2 y'^2 = d \tag{22}$$

If $\lambda_1$ and $\lambda_2$ have the same sign, then (21) defines an ellipse (or a circle) or a degenerate conic as in Examples 3 and 4. If $\lambda_1$ and $\lambda_2$ have opposite signs, then (21) is the equation of a hyperbola (as in Example 2). We can therefore prove the following.

**THEOREM 2** If $A = \begin{pmatrix} a & b/2 \\ b/2 & c \end{pmatrix}$, then the quadratic equation (21) is the equation of:

**i.** A hyperbola if $d \neq 0$ and $\det A < 0$.

**ii.** An ellipse, circle, or degenerate conic section if $d \neq 0$ and $\det A > 0$.

**iii.** A pair of straight lines or a degenerate conic section if $d \neq 0$ and $\det A = 0$.

**iv.** If $d = 0$, then (21) is the equation of two straight lines if $\det A \neq 0$ and the equation of a single line if $\det A = 0$.

**Proof** We have already shown why (*i*) and (*ii*) are true. To prove part (*iii*), suppose that $\det A = 0$. Then by our Summing Up Theorem (Theorem 6.1.7), $\lambda = 0$ is an eigenvalue of $A$ and equation (22) reads $\lambda_1 x'^2 = d$ or $\lambda_2 y'^2 = d$. If $\lambda_1 x'^2 = d$ and $d/\lambda_1 > 0$, then $x'_1 = \pm\sqrt{d/\lambda_1}$ is the equation of two straight lines in the $xy$-plane. If $d/\lambda_1 < 0$, then we have $x'^2 < 0$ (which is impossible) and we obtain a degenerate conic. The same facts hold if $\lambda_2 y'^2 = d$. Part (*iv*) is left as an exercise (see Problem 37). ■

*Note.* In Example 2 we had $\det A = ac - b^2/4 = -1$. In Examples 3 and 4 we had $\det A = 24$.

The methods described above can be used to analyze quadratic equations in more than two variables. We give one example below.

**EXAMPLE 5** **An Ellipsoid** Consider the quadratic equation

$$5x^2 + 8xy + 5y^2 + 4xz + 4yz + 2z^2 = 100 \tag{23}$$

If $A = \begin{pmatrix} 5 & 4 & 2 \\ 4 & 5 & 2 \\ 2 & 2 & 2 \end{pmatrix}$ and $\mathbf{v} = \begin{pmatrix} x \\ y \\ z \end{pmatrix}$, then (23) can be written in the form

$$A\mathbf{v} \cdot \mathbf{v} = 100 \tag{24}$$

From Example 6.4.2 on page 579, $Q^tAQ = D = \begin{pmatrix} 1 & 0 & 0 \\ 0 & 1 & 0 \\ 0 & 0 & 10 \end{pmatrix}$, where

$$Q = \begin{pmatrix} -1/\sqrt{2} & -1/3\sqrt{2} & 2/3 \\ 1/\sqrt{2} & -1/3\sqrt{2} & 2/3 \\ 0 & 4/3\sqrt{2} & 1/3 \end{pmatrix}$$

Let

$$\mathbf{v}' = \begin{pmatrix} x' \\ y' \\ z' \end{pmatrix} = Q^t\mathbf{v} = \begin{pmatrix} -1/\sqrt{2} & 1/\sqrt{2} & 0 \\ -1/3\sqrt{2} & -1/3\sqrt{2} & 4/3\sqrt{2} \\ 2/3 & 2/3 & 1/3 \end{pmatrix} \begin{pmatrix} x \\ y \\ z \end{pmatrix}$$
$$= \begin{pmatrix} (-1/\sqrt{2})x + (1/\sqrt{2})y \\ -(1/3\sqrt{2})x - (1/3\sqrt{2})y + (4/3\sqrt{2})z \\ (2/3)x + (2/3)y + (1/3)z \end{pmatrix}$$

Then as before, $A = QDQ^t$ and $A\mathbf{v} \cdot \mathbf{v} = QDQ^t\mathbf{v} \cdot \mathbf{v} = DQ^t\mathbf{v} \cdot Q^t\mathbf{v} = D\mathbf{v}' \cdot \mathbf{v}'$. Thus (24) can be written in the new variables $x'$, $y'$, $z'$ as $D\mathbf{v}' \cdot \mathbf{v}' = 100$ or

$$x'^2 + y'^2 + 10z'^2 = 100 \tag{25}$$

In $\mathbb{R}^3$ the surface defined by (25) is called an **ellipsoid** (see Figure 6.3).

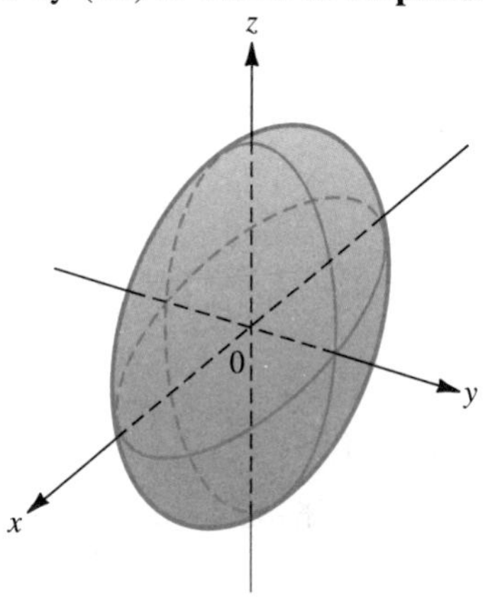


**Figure 6.3** The ellipsoid $5x^2 + 8xy + 5y^2 + 4xz + 4yz + 2z^2 = 100$, which can be written in new variables as $x'^2 + y'^2 + 10z'^2 = 100$

There is a great variety of three-dimensional surfaces of the form $A\mathbf{v} \cdot \mathbf{v} = d$, where $\mathbf{v} \in \mathbb{R}^2$. Such surfaces are called **quadric surfaces.**

We close this section by noting that quadratic forms can be defined in any number of variables.

**DEFINITION 2** **Quadratic Form** Let $\mathbf{v} = \begin{pmatrix} x_1 \\ x_2 \\ \vdots \\ x_n \end{pmatrix}$ and let $A$ be a symmetric $n \times n$ matrix. Then a **quadratic form** in $x_1, x_2, \ldots, x_n$ is an expression of the form

$$F(x_1, x_2, \ldots, x_n) = A\mathbf{v} \cdot \mathbf{v} \tag{26}$$

**EXAMPLE 6** **A Quadratic Form in Four Variables** Let

$$A = \begin{pmatrix} 2 & 1 & 2 & -2 \\ 1 & -4 & 6 & 5 \\ 2 & 6 & 7 & -1 \\ -2 & 5 & -1 & 3 \end{pmatrix} \quad \text{and} \quad \mathbf{v} = \begin{pmatrix} x_1 \\ x_2 \\ x_3 \\ x_4 \end{pmatrix}$$

Then

$$\begin{aligned} A\mathbf{v} \cdot \mathbf{v} &= \left[\begin{pmatrix} 2 & 1 & 2 & -2 \\ 1 & -4 & 6 & 5 \\ 2 & 6 & 7 & -1 \\ -2 & 5 & -1 & 3 \end{pmatrix}\begin{pmatrix} x_1 \\ x_2 \\ x_3 \\ x_4 \end{pmatrix}\right] \cdot \begin{pmatrix} x_1 \\ x_2 \\ x_3 \\ x_4 \end{pmatrix} \\ &= \begin{pmatrix} 2x_1 + x_2 + 2x_3 - 2x_4 \\ x_1 - 4x_2 + 6x_3 + 5x_4 \\ 2x_1 + 6x_2 + 7x_3 - x_4 \\ -2x_1 + 5x_2 - x_3 + 3x_4 \end{pmatrix} \cdot \begin{pmatrix} x_1 \\ x_2 \\ x_3 \\ x_4 \end{pmatrix} \\ &= 2x_1^2 + 2x_1x_2 - 4x_2^2 + 4x_1x_3 + 12x_2x_3 \\ &\quad + 7x_3^2 - 4x_1x_4 + 10x_2x_4 - 2x_3x_4 + 3x_4^2 \end{aligned}$$

(after simplification)

**EXAMPLE 7** **Finding a Symmetric Matrix That Corresponds to a Quadratic Form in Four Variables** Find the symmetric matrix $A$ corresponding to the quadratic form

$$5x_1^2 - 3x_1x_2 + 4x_2^2 + 8x_1x_3 - 9x_2x_3 + 2x_3^2 - x_1x_4 + 7x_2x_4 + 6x_3x_4 + 9x_4^2$$

**Solution** If $A = (a_{ij})$, then by looking at the earlier examples in this section, we see that $a_{ii}$ is the coefficient of the $x_i^2$ term and $a_{ij} + a_{ji}$ is the coefficient of the $x_i x_j$ term. Since $A$ is symmetric, $a_{ij} = a_{ji}$; hence $a_{ij} = a_{ji} = \frac{1}{2} \cdot$ (coefficient of $x_i x_j$ term). Putting this all together, we obtain

$$A = \begin{pmatrix} 5 & -\frac{3}{2} & 4 & -\frac{1}{2} \\ -\frac{3}{2} & 4 & -\frac{9}{2} & \frac{7}{2} \\ 4 & -\frac{9}{2} & 2 & 3 \\ -\frac{1}{2} & \frac{7}{2} & 3 & 9 \end{pmatrix}$$

# PROBLEMS 6.5

## Self-Quiz

**I.** If $A$ is a real symmetric matrix with positive eigenvalues, then $A\mathbf{v} \cdot \mathbf{v} = d > 0$ is the equation of
**a.** a parabola **b.** an ellipse **c.** a hyperbola
**d.** two straight lines **e.** none of the above

**II.** If $A$ is a real symmetric matrix with one positive and one negative eigenvalue, then $A\mathbf{v} \cdot \mathbf{v} = d > 0$ is the equation of
**a.** a parabola **b.** an ellipse **c.** a hyperbola
**d.** two straight lines **e.** none of the above

**III.** If $A$ is a real symmetric matrix with one positive eigenvalue and one eigenvalue equal to zero, then $A\mathbf{v} \cdot \mathbf{v} = d > 0$ is the equation of
**a.** a parabola **b.** an ellipse **c.** a hyperbola
**d.** two straight lines **e.** none of the above

**IV.** If $A$ is a real symmetric matrix with two negative eigenvalues, then $A\mathbf{v} \cdot \mathbf{v} = d > 0$ is the equation of
**a.** a parabola **b.** an ellipse **c.** a hyperbola
**d.** two straight lines **e.** none of the above

In Problems 1–13 write the quadratic equation in the form $A\mathbf{v} \cdot \mathbf{v} = d$ (where $A$ is a symmetric matrix) and eliminate the $xy$-term by rotating the axes through an angle of $\theta$. Write the equation in terms of the new variables and identify the conic section obtained.

**1.** $3x^2 - 2xy - 5 = 0$ **2.** $4x^2 + 4xy + y^2 = 9$ **3.** $4x^2 + 4xy - y^2 = 9$

**4.** $xy = 1$ **5.** $xy = a;\ a > 0$ **6.** $4x^2 + 2xy + 3y^2 + 2 = 0$

**7.** $xy = a;\ a < 0$ **8.** $x^2 + 4xy + 4y^2 - 6 = 0$ **9.** $-x^2 + 2xy - y^2 = 0$

**10.** $2x^2 + xy + y^2 = 4$ **11.** $3x^2 - 6xy + 5y^2 = 36$ **12.** $x^2 - 3xy + 4y^2 = 1$

**13.** $6x^2 + 5xy - 6y^2 + 7 = 0$

**14.** What are the possible forms of the graph of $ax^2 + bxy + cy^2 = 0$?

## Answers to Self-Quiz

**I.** b **II.** c **III.** d **IV.** e

In Problems 15–18 write the quadratic form in new variables $x'$, $y'$, and $z'$ so that no cross-product terms ($xy$, $xz$, $yz$) are present.

**15.** $x^2 - 2xy + y^2 - 2xz - 2yz + z^2$

**16.** $-x^2 + 4xy - y^2 + 4xz + 4yz + z^2$

**17.** $3x^2 + 4xy + 2y^2 + 4xz + 4z^2$

**18.** $x^2 - 2xy + 2y^2 - 2yz + z^2$

In Problems 19–21 find a symmetric matrix $A$ such that the quadratic form can be written in the form $A\mathbf{x} \cdot \mathbf{x}$.

**19.** $x_1^2 + 2x_1x_2 + x_2^2 + 4x_1x_3 + 6x_2x_3 + 3x_3^2 + 7x_1x_4 - 2x_2x_4 + x_4^2$

**20.** $x_1^2 - x_2^2 + x_1x_3 - x_2x_4 + x_4^2$

**21.** $3x_1^2 - 7x_1x_2 - 2x_2^2 + x_1x_3 - x_2x_3 + 3x_3^2 - 2x_1x_4 + x_2x_4 - 4x_3x_4 - 6x_4^2 + 3x_1x_5 - 5x_3x_5 + x_4x_5 - x_5^2$

**22.** Suppose that for some nonzero value of $d$, the graph of $ax^2 + bxy + cy^2 = d$ is a hyperbola. Show that the graph is a hyperbola for any other nonzero value of $d$.

**23.** Show that if $a \neq c$, the $xy$-term in quadratic equation (1) will be eliminated by rotation through an angle $\theta$ if $\theta$ is given by $\cot 2\theta = (a - c)/b$.

**24.** Show that if $a = c$ in Problem 23, then the $xy$-term will be eliminated by a rotation through an angle of either $\pi/4$ or $-\pi/4$.

***25.** Suppose that a rotation converts $ax^2 + bxy + cy^2$ into $a'(x')^2 + b'(x'y') + c'(y')^2$. Show that:

**a.** $a + c = a' + c'$ **b.** $b^2 - 4ac = b'^2 - 4a'c'$

***26.** A quadratic form $F(\mathbf{x}) = F(x_1, x_2, \ldots, x_n)$ is said to be **positive definite** if $F(\mathbf{x}) \geq 0$ for every $\mathbf{x} \in \mathbb{R}^n$ and $F(\mathbf{x}) = 0$ if and only if $\mathbf{x} = \mathbf{0}$. Show that $F$ is positive definite if and only if the symmetric matrix $A$ associated with $F$ has positive eigenvalues.

**27.** A quadratic form $F(\mathbf{x})$ is said to be **positive semidefinite** if $F(\mathbf{x}) \geq 0$ for every $\mathbf{x} \in \mathbb{R}^n$. Show that $F$ is positive semidefinite if and only if the eigenvalues of the symmetric matrix associated with $F$ are all nonnegative.

The definitions of **negative definite** and **negative semidefinite** are the definitions in Problems 26 and 27 with $\leq 0$ replacing $\geq 0$. A quadratic form is **indefinite** if it is none of the above. In Problems 28–35 determine whether the given quadratic form is positive definite, positive semidefinite, negative definite, negative semidefinite, or indefinite.

**28.** $3x^2 + 2y^2$ **29.** $-3x^2 - 3y^2$ **30.** $3x^2 - 2y^2$ **31.** $x^2 + 2xy + 2y^2$

**32.** $x^2 - 2xy + 2y^2$ **33.** $x^2 - 4xy + 3y^2$ **34.** $-x^2 + 4xy - 3y^2$ **35.** $-x^2 + 2xy - 2y^2$

***36.** Let $Q = \begin{pmatrix} a & b \\ c & d \end{pmatrix}$ be a real orthogonal matrix with $\det Q = 1$. Define the number $\theta \in [0, 2\pi)$:

**a.** If $a \geq 0$ and $c > 0$, then $\theta = \cos^{-1} a$ $(0 < \theta \leq \pi/2)$.

**b.** If $a \geq 0$ and $c < 0$, then $\theta = 2\pi - \cos^{-1} a$ $(3\pi/2 \leq \theta < 2\pi)$.

**c.** If $a \leq 0$ and $c > 0$, then $\theta = \cos^{-1} a$ $(\pi/2 \leq \theta < \pi)$.

**d.** If $a \leq 0$ and $c < 0$, then $\theta = 2\pi - \cos^{-1} a$ $(\pi < \theta \leq 3\pi/2)$.

**e.** If $a = 1$ and $c = 0$, then $\theta = 0$.

**f.** If $a = -1$ and $c = 0$, then $\theta = \pi$.

(Here $\cos^{-1} x \in [0, \pi]$ for $x \in [-1, 1]$.) With $\theta$ chosen as above, show that

$$Q = \begin{pmatrix} \cos\theta & -\sin\theta \\ \sin\theta & \cos\theta \end{pmatrix}$$

**37.** Prove, using formula (22), that equation (21) is the equation of two straight lines in the $xy$-plane when $d = 0$ and $\det A \neq 0$. If $\det A = d = 0$, show that equation (21) is the equation of a single line.

**38.** Let $A$ be the symmetric matrix representation of quadratic equation (1) with $d \neq 0$. Let $\lambda_1$ and $\lambda_2$ be the eigenvalues of $A$. Show that (1) is the equation of **(a)** a hyperbola if $\lambda_1\lambda_2 < 0$ and **(b)** a circle, ellipse, or degenerate conic section if $\lambda_1\lambda_2 > 0$.

## MATLAB 6.5

For each quadratic equation given in the problems below:

a. Find a symmetric matrix $A$ such that the equation can be written as $A\mathbf{v} \cdot \mathbf{v} = d$.
b. Find the eigenvalues and eigenvectors of $A$ by forming **[Q,D] = eig(A)**.
c. If $\det(Q) = -1$, adjust $Q$ appropriately so that $\det(Q) = 1$. [Refer to the discussion in Examples 2 and 3 in this section or see the discussion in MATLAB Problem 3(a) in Section 6.4.] Using the adjusted $Q$, find the angle $\theta$ of rotation. (Recall that the MATLAB command **acos** finds the inverse cosine and the MATLAB command **atan** finds the inverse tangent of an angle. You may want to convert radian measure into degrees by multiplying by $180/\pi$. The variable **pi** is built into MATLAB with the value $\pi$.)
d. Rewrite the equation in the form $a'x'^2 + b'y'^2 = d$ and identify the type of conic section described by the equation. Verify the result of Theorem 2.
e. *(Paper and pencil)* Using the angle $\theta$ of rotation and the rewritten equation from part (d), sketch the conic section described by the original equation. On the sketch, indicate the part of the geometry of the sketch that is obtained from knowing the eigenvalues.

1. Work Problem 10.
2. Work Problem 8.
3. Work Problem 4.
4. Work Problem 12.

## 6.6 JORDAN CANONICAL FORM

As we have seen, $n \times n$ matrices with $n$ linearly independent eigenvectors can be brought into an especially nice form by a similarity transformation. Fortunately, as "most" polynomials have distinct roots, "most" matrices will have distinct eigenvalues. As we shall see in Section 6.7, however, matrices that are not diagonalizable (that is, that do not have $n$ linearly independent eigenvectors) do arise in applications. In this case it is still possible to show that the matrix is similar to another, simpler matrix, but the new matrix is not diagonal and the transforming matrix $C$ is harder to obtain.

To discuss this case fully, we define the matrix $N_k$ to be the $k \times k$ matrix

$$N_k = \begin{pmatrix} 0 & 1 & 0 & \cdots & 0 \\ 0 & 0 & 1 & \cdots & 0 \\ \vdots & \vdots & \vdots & & \vdots \\ 0 & 0 & 0 & \cdots & 1 \\ 0 & 0 & 0 & \cdots & 0 \end{pmatrix} \tag{1}$$

Note that $N_k$ is the matrix with 1's above the main diagonal and 0's everywhere else.

Jordan block matrix

For a given scalar $\lambda$ we next define a $k \times k$ **Jordan† block matrix** $B(\lambda)$ by

$$B(\lambda) = \lambda I + N_k = \begin{pmatrix} \lambda & 1 & 0 & \cdots & 0 & 0 \\ 0 & \lambda & 1 & \cdots & 0 & 0 \\ \vdots & \vdots & \vdots & & \vdots & \vdots \\ 0 & 0 & & \cdots & \lambda & 1 \\ 0 & 0 & & \cdots & 0 & \lambda \end{pmatrix} \tag{2}$$

That is, $B(\lambda)$ is the $k \times k$ matrix with the scalar $\lambda$ on the diagonal, 1's above the diagonal, and 0's everywhere else.

***Note.*** We can (and often will) have a $1 \times 1$ Jordan block matrix. Such a matrix takes the form $B(\lambda) = (\lambda)$.

Jordan matrix

Finally, a **Jordan matrix** $J$ has the form

$$J = \begin{pmatrix} B_1(\lambda_1) & 0 & \cdots & 0 \\ 0 & B_2(\lambda_2) & \cdots & 0 \\ \vdots & \vdots & & \vdots \\ 0 & 0 & \cdots & B_r(\lambda_r) \end{pmatrix}$$

where each $B_j(\lambda_j)$ is a Jordan block matrix. Thus *a Jordan matrix is a matrix with Jordan block matrices down the diagonal and zeros everywhere else.*

**EXAMPLE 1** **Three Jordan Matrices** The following are examples of Jordan matrices. The Jordan blocks are outlined by the dotted lines:

**i.** $\begin{pmatrix} 2 & 1 & 0 \\ 0 & 2 & 0 \\ 0 & 0 & 4 \end{pmatrix}$ **ii.** $\begin{pmatrix} -3 & 0 & 0 & 0 & 0 \\ 0 & -3 & 1 & 0 & 0 \\ 0 & 0 & -3 & 1 & 0 \\ 0 & 0 & 0 & -3 & 0 \\ 0 & 0 & 0 & 0 & 7 \end{pmatrix}$

†Named for the French mathematician Camille Jordan (1838–1922). The results in this section first appeared in Jordan's brilliant *Traité des substitutions et des équations algebriques* (Treatise on substitutions and algebraic equations), which was published in 1870.

**iii.** $\begin{pmatrix} 4 & 1 & 0 & 0 & 0 & 0 & 0 \\ 0 & 4 & 0 & 0 & 0 & 0 & 0 \\ 0 & 0 & 3 & 1 & 0 & 0 & 0 \\ 0 & 0 & 0 & 3 & 1 & 0 & 0 \\ 0 & 0 & 0 & 0 & 3 & 0 & 0 \\ 0 & 0 & 0 & 0 & 0 & 5 & 1 \\ 0 & 0 & 0 & 0 & 0 & 0 & 5 \end{pmatrix}$

**EXAMPLE 2** **The 2 × 2 Jordan Matrices** The only $2 \times 2$ Jordan matrices are $\begin{pmatrix} \lambda_1 & 0 \\ 0 & \lambda_2 \end{pmatrix}$ and $\begin{pmatrix} \lambda & 1 \\ 0 & \lambda \end{pmatrix}$. In the first matrix the numbers $\lambda_1$ and $\lambda_2$ could be equal.

**EXAMPLE 3** **The 3 × 3 Jordan Matrices** The only $3 \times 3$ Jordan matrices are

$$\begin{pmatrix} \lambda_1 & 0 & 0 \\ 0 & \lambda_2 & 0 \\ 0 & 0 & \lambda_3 \end{pmatrix} \quad \begin{pmatrix} \lambda_1 & 0 & 0 \\ 0 & \lambda_2 & 1 \\ 0 & 0 & \lambda_2 \end{pmatrix} \quad \begin{pmatrix} \lambda_1 & 1 & 0 \\ 0 & \lambda_1 & 0 \\ 0 & 0 & \lambda_2 \end{pmatrix} \quad \begin{pmatrix} \lambda_1 & 1 & 0 \\ 0 & \lambda_1 & 1 \\ 0 & 0 & \lambda_1 \end{pmatrix}$$

where $\lambda_1$, $\lambda_2$, and $\lambda_3$ are not necessarily distinct.

The following result is one of the most important theorems in matrix theory. Although its proof is beyond the scope of this book,† we shall prove this theorem in the $2 \times 2$ case (see Theorem 3) and suggest a proof for the $3 \times 3$ case in Problem 19.

**THEOREM 1** Let $A$ be an $n \times n$ real or complex matrix. Then there exists an invertible $n \times n$ complex matrix $C$ such that

$$\boxed{C^{-1}AC = J} \tag{3}$$

where $J$ is a Jordan matrix whose diagonal elements are the eigenvalues of $A$. Moreover, the Jordan matrix $J$ is unique except for the order in which the Jordan blocks appear.

*Note 1.* The matrix $C$ in Theorem 1 need not be unique.

---

†For a proof see G. Birkhoff and S. MacLane, *A Survey of Modern Algebra,* 3rd ed. (New York: Macmillan, 1965), p. 311.

*Note 2.* By the last sentence of the theorem we mean, for example, that if $A$ is similar to

$$J_1 = \begin{pmatrix} 2 & 1 & 0 & 0 & 0 & 0 \\ 0 & 2 & 0 & 0 & 0 & 0 \\ 0 & 0 & 3 & 1 & 0 & 0 \\ 0 & 0 & 0 & 3 & 1 & 0 \\ 0 & 0 & 0 & 0 & 3 & 0 \\ 0 & 0 & 0 & 0 & 0 & 4 \end{pmatrix}$$

then $A$ is also similar to

$$J_2 = \begin{pmatrix} 3 & 1 & 0 & 0 & 0 & 0 \\ 0 & 3 & 1 & 0 & 0 & 0 \\ 0 & 0 & 3 & 0 & 0 & 0 \\ 0 & 0 & 0 & 4 & 0 & 0 \\ 0 & 0 & 0 & 0 & 2 & 1 \\ 0 & 0 & 0 & 0 & 0 & 2 \end{pmatrix} \quad \text{and} \quad J_3 = \begin{pmatrix} 4 & 0 & 0 & 0 & 0 & 0 \\ 0 & 2 & 1 & 0 & 0 & 0 \\ 0 & 0 & 2 & 0 & 0 & 0 \\ 0 & 0 & 0 & 3 & 1 & 0 \\ 0 & 0 & 0 & 0 & 3 & 1 \\ 0 & 0 & 0 & 0 & 0 & 3 \end{pmatrix}$$

and three other Jordan matrices. That is, the actual Jordan blocks remain the same but we can change the order in which they are written.

**DEFINITION 1** **Jordan Canonical Form** The matrix $J$ in Theorem 1 is called the **Jordan canonical form** of $A$.

*Remark.* If $A$ is diagonalizable, then $J = D = \text{diag}(\lambda_1, \lambda_2, \ldots, \lambda_n)$, where $\lambda_1, \lambda_2, \ldots, \lambda_n$ are the (not necessarily distinct) eigenvalues of $A$. Each diagonal component is a $1 \times 1$ Jordan block matrix.

We shall now see how to compute the Jordan canonical form of any $2 \times 2$ matrix. If $A$ has two linearly independent eigenvectors, we already know what to do. Therefore the only case of interest occurs when $A$ has a single eigenvalue $\lambda$ of algebraic multiplicity 2 and geometric multiplicity 1. That is, we assume that, corresponding to $\lambda$, $A$ has the single independent eigenvector $\mathbf{v}_1$. That is: *Any vector that is not a multiple of* $\mathbf{v}_1$ *is not an eigenvector.*

**THEOREM 2** Let the $2 \times 2$ matrix $A$ have an eigenvalue $\lambda$ of algebraic multiplicity 2 and geometric multiplicity 1. Let $\mathbf{v}_1$ be an eigenvector corresponding to $\lambda$. Then there exists a vector $\mathbf{v}_2$ that satisfies the equation

$$\boxed{(A - \lambda I)\mathbf{v}_2 = \mathbf{v}_1} \tag{4}$$

**Proof** Let $\mathbf{x} \in \mathbb{C}^2$ be a fixed vector that is *not* a multiple of $\mathbf{v}_1$ so that $\mathbf{x}$ is not an eigenvector of $A$. We first show that

$$\mathbf{w} = (A - \lambda I)\mathbf{x} \tag{5}$$

is an eigenvector of $A$. That is, we shall show that $\mathbf{w} = c\mathbf{v}_1$ for some constant $c$. Since $\mathbf{w} \in \mathbb{C}^2$ and $\mathbf{v}_1$ and $\mathbf{x}$ are linearly independent, there exist constants $c_1$ and $c_2$ such that

$$\mathbf{w} = c_1\mathbf{v}_1 + c_2\mathbf{x} \tag{6}$$

To show that $\mathbf{w}$ is an eigenvector of $A$, we must show that $c_2 = 0$. From (5) and (6) we find that

$$(A - \lambda I)\mathbf{x} = c_1\mathbf{v}_1 + c_2\mathbf{x} \tag{7}$$

Let $B = A - (\lambda + c_2)I$. Then from (7)

$$B\mathbf{x} = [A - (\lambda + c_2)I]\mathbf{x} = c_1\mathbf{v}_1 \tag{8}$$

If we assume that $c_2 \neq 0$, then $\lambda + c_2 \neq \lambda$ and $\lambda + c_2$ is not an eigenvalue of $A$ (since $\lambda$ is the only eigenvalue of $A$). Thus $\det B = \det [A - (\lambda + c_2)I] \neq 0$, which means that $B$ is invertible. Hence (8) can be written as

$$\mathbf{x} = B^{-1}c_1\mathbf{v}_1 = c_1B^{-1}\mathbf{v}_1 \tag{9}$$

Then multiplying both sides of (9) by $\lambda$, we have

$$\lambda\mathbf{x} = \lambda c_1B^{-1}\mathbf{v}_1 = c_1B^{-1}\lambda\mathbf{v}_1 = c_1B^{-1}A\mathbf{v}_1 \tag{10}$$

But $B = A - (\lambda + c_2)I$, so

$$A = B + (\lambda + c_2)I \tag{11}$$

Inserting (11) into (10), we have

$$\begin{aligned} \lambda\mathbf{x} &= c_1B^{-1}[B + (\lambda + c_2)I]\mathbf{v}_1 \\ &= c_1[I + (\lambda + c_2)B^{-1}]\mathbf{v}_1 \\ &= c_1\mathbf{v}_1 + (\lambda + c_2)c_1B^{-1}\mathbf{v}_1 \end{aligned} \tag{12}$$

But using (9), $c_1B^{-1}\mathbf{v}_1 = \mathbf{x}$ so that (12) becomes

$$\lambda\mathbf{x} = c_1\mathbf{v}_1 + (\lambda + c_2)\mathbf{x} = c_1\mathbf{v}_1 + c_2\mathbf{x} + \lambda\mathbf{x}$$

or

$$\mathbf{0} = c_1\mathbf{v}_1 + c_2\mathbf{x} \tag{13}$$

But $\mathbf{v}_1$ and $\mathbf{x}$ are linearly independent, so $c_1 = c_2 = 0$. This contradicts the assumption that $c_2 \neq 0$. Thus $c_2 = 0$, and by (6), $\mathbf{w}$ is a multiple of $\mathbf{v}_1$ so that $\mathbf{w} = c_1\mathbf{v}_1$ is an eigenvector of $A$. Moreover, $\mathbf{w} \neq \mathbf{0}$ since if $\mathbf{w} = \mathbf{0}$, then (5) tells us that $\mathbf{x}$ is an eigenvector of $A$. Therefore $c_1 \neq 0$. Let

$$\mathbf{v}_2 = \frac{1}{c_1}\mathbf{x} \tag{14}$$

Then $(A - \lambda I)\mathbf{v}_2 = (1/c_1)(A - \lambda I)\mathbf{x} = (1/c_1)\mathbf{w} = \mathbf{v}_1$. This proves the theorem. ■

**DEFINITION 2** **Generalized Eigenvector** Let $A$ be a $2 \times 2$ matrix with the single eigenvalue $\lambda$ having geometric multiplicity 1. Let $\mathbf{v}_1$ be an eigenvector of $A$. Then the vector $\mathbf{v}_2$ defined by $(A - \lambda I)\mathbf{v}_2 = \mathbf{v}_1$ is called a **generalized eigenvector** of $A$ corresponding to the eigenvalue $\lambda$.

**EXAMPLE 4** **Finding a Generalized Eigenvector** Let $A = \begin{pmatrix} 3 & -2 \\ 8 & -5 \end{pmatrix}$. The characteristic equation of $A$ is $\lambda^2 + 2\lambda + 1 = (\lambda + 1)^2 = 0$, so $\lambda = -1$ is an eigenvalue of algebraic multiplicity 2. Then

$$(A - \lambda I)\mathbf{v} = (A + I)\mathbf{v} = \begin{pmatrix} 4 & -2 \\ 8 & -4 \end{pmatrix}\begin{pmatrix} x_1 \\ x_2 \end{pmatrix} = \begin{pmatrix} 0 \\ 0 \end{pmatrix}$$

This yields the eigenvector $\mathbf{v}_1 = \begin{pmatrix} 1 \\ 2 \end{pmatrix}$. There is no other linearly independent eigenvector. To find a generalized eigenvector $\mathbf{v}_2$, we compute $(A + I)\mathbf{v}_2 = \mathbf{v}_1$ or $\begin{pmatrix} 4 & -2 \\ 8 & -4 \end{pmatrix}\begin{pmatrix} x_1 \\ x_2 \end{pmatrix} = \begin{pmatrix} 1 \\ 2 \end{pmatrix}$, which yields the system

$$\begin{aligned} 4x_1 - 2x_2 &= 1 \\ 8x_1 - 4x_2 &= 2 \end{aligned}$$

The second equation is double the first, so $x_2$ can be chosen arbitrarily and $x_1 = (1 + 2x_2)/4$. Therefore a possible choice for $\mathbf{v}_2$ is $\mathbf{v}_2 = \begin{pmatrix} \frac{1}{4} \\ 0 \end{pmatrix}$. ■

The reason for finding generalized eigenvectors is given in the following theorem.

**THEOREM 3** Let $A$, $\lambda$, $\mathbf{v}_1$, and $\mathbf{v}_2$ be as in Theorem 2 and let $C$ be the matrix whose columns are $\mathbf{v}_1$ and $\mathbf{v}_2$. Then $C^{-1}AC = J$, where $J = \begin{pmatrix} \lambda & 1 \\ 0 & \lambda \end{pmatrix}$ is the Jordan canonical form of $A$.

**Proof** Since $\mathbf{v}_1$ and $\mathbf{v}_2$ are linearly independent, we see that $C$ is invertible. Next note that $AC = A(\mathbf{v}_1, \mathbf{v}_2) = (A\mathbf{v}_1, A\mathbf{v}_2) = (\lambda\mathbf{v}_1, A\mathbf{v}_2)$. But from equation (4), $A\mathbf{v}_2 = \mathbf{v}_1 + \lambda\mathbf{v}_2$ so that $AC = (\lambda\mathbf{v}_1, \mathbf{v}_1 + \lambda\mathbf{v}_2)$. But $CJ = (\mathbf{v}_1, \mathbf{v}_2)\begin{pmatrix} \lambda & 1 \\ 0 & \lambda \end{pmatrix} = (\lambda\mathbf{v}_1, \mathbf{v}_1 + \lambda\mathbf{v}_2)$. Thus $AC = CJ$, which means that $C^{-1}AC = J$ and the theorem is proved. ■

**EXAMPLE 5** **Finding the Jordan Canonical Form of a 2 × 2 Matrix** In Example 4, $\mathbf{v}_1 = \begin{pmatrix} 1 \\ 2 \end{pmatrix}$ and $\mathbf{v}_2 = \begin{pmatrix} \frac{1}{4} \\ 0 \end{pmatrix}$. Then $C = \begin{pmatrix} 1 & \frac{1}{4} \\ 2 & 0 \end{pmatrix}$, $C^{-1} = -2\begin{pmatrix} 0 & -\frac{1}{4} \\ -2 & 1 \end{pmatrix} = \begin{pmatrix} 0 & \frac{1}{2} \\ 4 & -2 \end{pmatrix}$, and

$$\begin{aligned} C^{-1}AC &= \begin{pmatrix} 0 & \frac{1}{2} \\ 4 & -2 \end{pmatrix}\begin{pmatrix} 3 & -2 \\ 8 & -5 \end{pmatrix}\begin{pmatrix} 1 & \frac{1}{4} \\ 2 & 0 \end{pmatrix} \\ &= \begin{pmatrix} 0 & \frac{1}{2} \\ 4 & -2 \end{pmatrix}\begin{pmatrix} -1 & \frac{3}{4} \\ -2 & 2 \end{pmatrix} = \begin{pmatrix} -1 & 1 \\ 0 & -1 \end{pmatrix} = J \end{aligned}$$ ■

The method described above can be generalized to obtain the Jordan canonical form of every matrix. We shall not do this, but one generalization is suggested in Problem 19. Although we shall not prove this fact, it is always possible to determine the number of 1's above the diagonal in the Jordan canonical form of an $n \times n$ matrix $A$. Let $\lambda_i$ be an eigenvalue of $A$ with algebraic multiplicity $r_i$ and geometric multiplicity $s_i$. If $\lambda_1, \lambda_2, \ldots, \lambda_k$ are the eigenvalues of $A$, then

The number of 1's above the diagonal of the Jordan canonical form of $A$

$$\begin{aligned} &= (r_1 - s_1) + (r_2 - s_2) + \cdots + (r_k - s_k) \\ &= \sum_{i=1}^{k} r_i - \sum_{i=1}^{k} s_i = n - \sum_{i=1}^{k} s_i \end{aligned} \qquad \textbf{(15)}$$

If we know the characteristic equation of a matrix $A$, then we can determine the possible Jordan canonical forms of $A$.

**EXAMPLE 6** **Determining the Possible Jordan Canonical Forms of a 4 × 4 Matrix with Given Characteristic Equation** If the characteristic polynomial of $A$ is $(\lambda - 2)^3(\lambda + 3)$, then the possible Jordan canonical forms for $A$ are

$$J = \begin{pmatrix} 2 & 0 & 0 & 0 \\ 0 & 2 & 0 & 0 \\ 0 & 0 & 2 & 0 \\ 0 & 0 & 0 & -3 \end{pmatrix}, \quad \begin{pmatrix} 2 & 1 & 0 & 0 \\ 0 & 2 & 0 & 0 \\ 0 & 0 & 2 & 0 \\ 0 & 0 & 0 & -3 \end{pmatrix}, \quad \begin{pmatrix} 2 & 1 & 0 & 0 \\ 0 & 2 & 1 & 0 \\ 0 & 0 & 2 & 0 \\ 0 & 0 & 0 & -3 \end{pmatrix}$$

or any matrix obtained by rearranging the Jordan blocks in $J$. The first matrix corresponds to a geometric multiplicity of 3 (for $\lambda = 2$); the second corresponds to a geometric multiplicity of 2; and the third corresponds to a geometric multiplicity of 1.

# PROBLEMS 6.6

## Self-Quiz

**I.** Which of the following is not a Jordan matrix?

**a.** $\begin{pmatrix} 3 & 1 & 0 \\ 0 & 3 & 0 \\ 0 & 0 & 4 \end{pmatrix}$ **b.** $\begin{pmatrix} 3 & 1 & 0 \\ 0 & 4 & 0 \\ 0 & 0 & 5 \end{pmatrix}$ **c.** $\begin{pmatrix} 3 & 1 & 0 \\ 0 & 3 & 1 \\ 0 & 0 & 3 \end{pmatrix}$ **d.** $\begin{pmatrix} 3 & 0 & 0 \\ 0 & 4 & 0 \\ 0 & 0 & 5 \end{pmatrix}$

### *True-False*

**II.** Every matrix is similar to a Jordan matrix.

**III.** Suppose that $A$ is a $2 \times 2$ matrix having the eigenvalue 2 and corresponding eigenvector $\mathbf{v}_1$. Then there is a vector $\mathbf{v}_2$ that satisfies the equation $(A - 2I)\mathbf{v}_2 = \mathbf{v}_1$.

**IV.** Suppose that $A$ is a $2 \times 2$ matrix whose characteristic polynomial is $(\lambda - 2)^2$ such that the geometric multiplicity of 2 is 1. Then if $\mathbf{v}_1$ is an eigenvector of $A$, there is a vector $\mathbf{v}_2$ that satisfies the equation $(A - 2I)\mathbf{v}_2 = \mathbf{v}_1$.

In Problems 1–14 determine whether the given matrix is a Jordan matrix.

**1.** $\begin{pmatrix} 1 & 1 \\ 0 & -6 \end{pmatrix}$ **2.** $\begin{pmatrix} 1 & 0 \\ 0 & 0 \end{pmatrix}$ **3.** $\begin{pmatrix} 1 & 2 \\ 0 & 1 \end{pmatrix}$ **4.** $\begin{pmatrix} 1 & 0 & 0 \\ 0 & 3 & 1 \\ 0 & 0 & 3 \end{pmatrix}$

**5.** $\begin{pmatrix} 3 & 1 & 0 \\ 0 & 3 & 1 \\ 0 & 0 & 3 \end{pmatrix}$ **6.** $\begin{pmatrix} 3 & 1 & 0 \\ 0 & 3 & 1 \\ 0 & 0 & 2 \end{pmatrix}$ **7.** $\begin{pmatrix} 1 & 0 & 0 \\ 0 & 3 & 1 \\ 0 & 0 & 4 \end{pmatrix}$ **8.** $\begin{pmatrix} 1 & 1 & 0 \\ 0 & 3 & 1 \\ 0 & 0 & 3 \end{pmatrix}$

**9.** $\begin{pmatrix} 1 & 1 & 0 \\ 0 & 1 & 1 \\ 0 & 0 & 1 \end{pmatrix}$ **10.** $\begin{pmatrix} 1 & 0 & 0 & 0 & 0 \\ 0 & 2 & 1 & 0 & 0 \\ 0 & 0 & 2 & 1 & 0 \\ 0 & 0 & 0 & 2 & 0 \\ 0 & 0 & 0 & 0 & 2 \end{pmatrix}$ **11.** $\begin{pmatrix} 1 & 0 & 0 & 0 & 0 \\ 0 & 1 & 2 & 0 & 0 \\ 0 & 0 & 1 & 2 & 0 \\ 0 & 0 & 0 & 1 & 0 \\ 0 & 0 & 0 & 0 & 1 \end{pmatrix}$

## Answers to Self-Quiz

**I.** b **II.** True **III.** False **IV.** True

**12.** $\begin{pmatrix} 2 & 0 & 0 & 0 & 0 \\ 0 & 3 & 1 & 0 & 0 \\ 0 & 0 & 3 & 0 & 0 \\ 0 & 0 & 0 & 5 & 1 \\ 0 & 0 & 0 & 0 & 5 \end{pmatrix}$ **13.** $\begin{pmatrix} a & 0 & 0 & 0 & 0 \\ 0 & b & 0 & 0 & 0 \\ 0 & 0 & c & 0 & 0 \\ 0 & 0 & 0 & d & 0 \\ 0 & 0 & 0 & 0 & e \end{pmatrix}$ **14.** $\begin{pmatrix} a & 1 & 0 & 0 & 0 \\ 0 & a & 0 & 0 & 0 \\ 0 & 0 & c & 1 & 0 \\ 0 & 0 & 0 & c & 1 \\ 0 & 0 & 0 & 0 & c \end{pmatrix}$

In Problems 15–18 find an invertible matrix $C$ that transforms the $2 \times 2$ matrix to its Jordan canonical form.

**15.** $\begin{pmatrix} 6 & 1 \\ 0 & 6 \end{pmatrix}$ **16.** $\begin{pmatrix} -12 & 7 \\ -7 & 2 \end{pmatrix}$ **17.** $\begin{pmatrix} -10 & -7 \\ 7 & 4 \end{pmatrix}$ **18.** $\begin{pmatrix} 4 & -1 \\ 1 & 2 \end{pmatrix}$

***19.** Let $A$ be a $3 \times 3$ matrix. Assume that $\lambda$ is an eigenvalue of $A$ with algebraic multiplicity 3 and geometric multiplicity 1 and let $\mathbf{v}_1$ be the corresponding eigenvector.

**a.** Show that there is a solution, $\mathbf{v}_2$, to the system $(A - \lambda I)\mathbf{v}_2 = \mathbf{v}_1$ such that $\mathbf{v}_1$ and $\mathbf{v}_2$ are linearly independent.

**b.** With $\mathbf{v}_2$ defined by part (*a*), show that there is a solution, $\mathbf{v}_3$, to the system $(A - \lambda I)\mathbf{v}_3 = \mathbf{v}_2$ such that $\mathbf{v}_1$, $\mathbf{v}_2$, and $\mathbf{v}_3$ are linearly independent.

**c.** Show that if $C$ is a matrix whose columns are $\mathbf{v}_1$, $\mathbf{v}_2$, and $\mathbf{v}_3$, then

$$C^{-1}AC = \begin{pmatrix} \lambda & 1 & 0 \\ 0 & \lambda & 1 \\ 0 & 0 & \lambda \end{pmatrix}.$$

**20.** Apply the procedure described in Problem 19 to reduce the matrix $A = \begin{pmatrix} -2 & 1 & 0 \\ -2 & 1 & -1 \\ -1 & 1 & -2 \end{pmatrix}$ by a similarity transformation to its Jordan canonical form.

**21.** Do the same for $A = \begin{pmatrix} -1 & -2 & -1 \\ -1 & -1 & -1 \\ 2 & 3 & 2 \end{pmatrix}$.

**22.** Do the same for $A = \begin{pmatrix} -1 & -18 & -7 \\ 1 & -13 & -4 \\ -1 & 25 & 8 \end{pmatrix}$.

**23.** An $n \times n$ matrix $A$ is **nilpotent** if there is an integer $k$ such that $A^k = 0$. If $k$ is the smallest such integer, then $k$ is called the **index of nilpotency** of $A$. Prove that if $k$ is the index of nilpotency of $A$ and if $m \geq k$, then $A^m = 0$.

***24.** Let $N_k$ be the matrix defined by equation (1). Prove that $N_k$ is nilpotent with index of nilpotency $k$.

**25.** Write down all possible $4 \times 4$ Jordan matrices.

In Problems 26–31 the characteristic polynomial of a matrix $A$ is given. Write the possible Jordan canonical forms for $A$.

**26.** $(\lambda + 1)^2(\lambda - 2)^2$ **27.** $(\lambda - 3)^3(\lambda + 4)$ **28.** $(\lambda - 3)^4$

**29.** $(\lambda - 4)^3(\lambda + 3)^2$ **30.** $(\lambda - 6)(\lambda + 7)^4$ **31.** $(\lambda + 7)^5$

**32.** Using the Jordan canonical form, show that for any $n \times n$ matrix $A$, $\det A = \lambda_1\lambda_2 \cdots \lambda_n$, where $\lambda_1, \lambda_2, \ldots, \lambda_n$ are the eigenvalues of $A$.

## MATLAB 6.6

1. **a.** Let $A = CJC^{-1}$, where $C$ and $J$ are given below.

$$J = \begin{pmatrix} 2 & 0 & 0 & 0 \\ 0 & 2 & 0 & 0 \\ 0 & 0 & 3 & 1 \\ 0 & 0 & 0 & 3 \end{pmatrix}, \qquad C = \begin{pmatrix} 1 & 2 & 2 & -1 \\ 1 & 3 & 5 & 3 \\ 2 & 4 & 3 & 0 \\ 1 & 3 & 3 & 6 \end{pmatrix}$$

   **i.** Verify that columns 1 and 2 of $C$ are eigenvectors for $A$ with eigenvalue $\lambda = 2$. (Use the matrix $A - 2I$.)
   
   **ii.** Verify that column 3 of $C$ is an eigenvector for $A$ with eigenvalue $\mu = 3$. (Use the matrix $A - 3I$.) Verify that column 4 of $C$ is *not* an eigenvector for $A$ with eigenvalue $\mu = 3$ but that $(A - 3I)$ times column 4 is an eigenvector; that is, verify that $(A - 3I)^2$ (column 4) $= 0$. Column 4 of $C$ is called a **generalized eigenvector** for $A$ with eigenvalue $\mu = 3$.
   
   **iii.** Repeat for another choice of an invertible $4 \times 4$ matrix $C$. (Use the same $J$.)
   
   **iv.** *(Paper and pencil)* Explain why we can say that $\lambda = 2$ is an eigenvalue for $A$ with algebraic multiplicity 2 and geometric multiplicity 2 and that $\mu = 3$ is an eigenvalue for $A$ with algebraic multiplicity 2 and geometric multiplicity 1.

   **b.** For $J$ below and the matrix $C$ given in part (a), form $A = CJC^{-1}$.

$$J = \begin{pmatrix} 3 & 1 & 0 & 0 \\ 0 & 3 & 1 & 0 \\ 0 & 0 & 3 & 0 \\ 0 & 0 & 0 & 3 \end{pmatrix}$$

   For $k = 1, \ldots, 4$, let $\mathbf{c}_k$ denote the $k$th column of $C$.
   
   **i.** Verify that $(A - 3I)\mathbf{c}_1 = 0$, $(A - 3I)^2\mathbf{c}_2 = 0$, $(A - 3I)^3\mathbf{c}_3 = 0$, and $(A - 3I)\mathbf{c}_4 = 0$. Which of the columns of $C$ are eigenvectors for $A$? Which of the columns of $C$ are generalized eigenvectors for $A$?
   
   **ii.** Repeat for another choice of an invertible $4 \times 4$ matrix $C$.
   
   **iii.** *(Paper and pencil)* Explain why we can say that $\lambda = 3$ is an eigenvalue for $A$ with algebraic multiplicity 4 and geometric multiplicity 2.

   **c.** Form $A = CJC^{-1}$, where $C$ is the matrix given in part (a) and $J$ is given below.

$$J = \begin{pmatrix} 2 & 1 & 0 & 0 \\ 0 & 2 & 0 & 0 \\ 0 & 0 & 3 & 1 \\ 0 & 0 & 0 & 3 \end{pmatrix}$$

   **i.** Based on the patterns observed in parts (a) and (b), determine which columns of $C$ are eigenvectors for $A$ and which are generalized eigenvectors for $A$. Verify your answers by showing that the appropriate products are zero.
   
   **ii.** Repeat for another choice of $C$.
   
   **iii.** *(Paper and pencil)* What can you say about the algebraic and geometric multiplicities of the eigenvalues of $A$? Justify your answer.

2. Generate a $5 \times 5$ invertible matrix $C$. Form a matrix $A$ such that $\lambda = 2$ is an eigenvalue for $A$ with algebraic multiplicity 2 and geometric multiplicity 1 with columns 1 and 2 of

$C$ being the eigenvectors or generalized eigenvectors associated with $\lambda = 2$; $\mu = 4$ is an eigenvalue for $A$ with algebraic multiplicity 3 and geometric multiplicity 1 with columns 3 through 5 of $A$ being the eigenvectors or generalized eigenvectors associated with $\mu = 4$. Explain your procedure. Check your final answer for $A$ by showing that appropriate products are zero.

Calculus

## 6.7 AN IMPORTANT APPLICATION: MATRIX DIFFERENTIAL EQUATIONS

Let $x = f(t)$ represent some physical quantity such as the volume of a substance, the population of a certain species, the mass of a decaying radioactive substance, or the number of dollars invested in bonds. Then the rate of growth of $f(t)$ is given by its derivative $f'(t) = dx/dt$. If $f(t)$ is growing at a constant rate, then $dx/dt = k$ and $x = kt + C$; that is, $x = f(t)$ is a straight-line function.

It is often more interesting and more appropriate to consider the **relative rate of growth** defined by

$$\text{Relative rate of growth} = \frac{\text{actual size of growth}}{\text{size of } f(t)} = \frac{f'(t)}{f(t)} = \frac{x'(t)}{x(t)} \tag{1}$$

If the relative rate of growth is constant, then we have

$$\frac{x'(t)}{x(t)} = a \tag{2}$$

or

$$x'(t) = ax(t) \tag{3}$$

Differential equation

Equation (3) is called a **differential equation** because it is an equation involving a derivative. It is not difficult to prove that the only solutions to (3) are of the form

$$x(t) = ce^{at} \tag{4}$$

Initial value

where $c$ is an arbitrary constant. If, however, $x(t)$ represents some physical quantity, then it is the usual practice to specify an **initial value** $x_0 = x(0)$ of the quantity. Then substituting $t = 0$ in (4), we have $x_0 = x(0) = ce^{a \cdot 0} = c$ or

$$x(t) = x_0 e^{at} \tag{5}$$

The function $x(t)$ given by (5) is the unique solution to (3) satisfying the initial condition $x(0) = x_0$.

Equation (3) arises in a number of interesting applications. Some of these are undoubtedly given in your calculus text—in the chapter introducing the exponential function. In this section we consider a generalization of equation (3).

In the model discussed above we seek one unknown function. It often occurs that there are several functions linked by several differential equations. Examples are given later in the section. Consider the following system of $n$ differential equations in $n$ unknown functions:

$$\begin{aligned} x_1'(t) &= a_{11}x_1(t) + a_{12}x_2(t) + \cdots + a_{1n}x_n(t) \\ x_2'(t) &= a_{21}x_1(t) + a_{22}x_2(t) + \cdots + a_{2n}x_n(t) \\ &\vdots \\ x_n'(t) &= a_{n1}x_1(t) + a_{n2}x_2(t) + \cdots + a_{nn}x_n(t) \end{aligned} \tag{6}$$

where the $a_{ij}$'s are real numbers. System (6) is called an $n \times n$ **first-order system of linear differential equations.** The term "first order" means that only first derivatives occur in the system.

Now let

$$\mathbf{x}(t) = \begin{pmatrix} x_1(t) \\ x_2(t) \\ \vdots \\ x_n(t) \end{pmatrix}$$

Vector function

Here $\mathbf{x}(t)$ is called a **vector function.** We define

$$\mathbf{x}'(t) = \begin{pmatrix} x_1'(t) \\ x_2'(t) \\ \vdots \\ x_n'(t) \end{pmatrix}$$

Then if we define the $n \times n$ matrix

$$A = \begin{pmatrix} a_{11} & a_{12} & \cdots & a_{1n} \\ a_{21} & a_{22} & \cdots & a_{2n} \\ \vdots & \vdots & & \vdots \\ a_{n1} & a_{n2} & \cdots & a_{nn} \end{pmatrix}$$

system (6) can be written as

$$\boxed{\mathbf{x}'(t) = A\mathbf{x}(t)} \tag{7}$$

Note that equation (7) is almost identical to equation (3). The only difference is that now we have a vector function and a matrix whereas before we had a "scalar" function and a number ($1 \times 1$ matrix).

To solve equation (7), we might guess that a solution to (7) would have the form $e^{At}$. But what does $e^{At}$ mean? We shall answer that question in a moment. First, let us recall the series expansion of the function $e^t$:

$$e^t = 1 + t + \frac{t^2}{2!} + \frac{t^3}{3!} + \frac{t^4}{4!} + \cdots \tag{8}$$

This series converges for every real number $t$. Then for any real number $a$

$$e^{at} = 1 + at + \frac{(at)^2}{2!} + \frac{(at)^3}{3!} + \frac{(at)^4}{4!} + \cdots \tag{9}$$

**DEFINITION 1** **The Matrix $e^A$** Let $A$ be an $n \times n$ matrix with real (or complex) entries. Then $e^A$ is an $n \times n$ matrix defined by

$$e^A = I + A + \frac{A^2}{2!} + \frac{A^3}{3!} + \frac{A^4}{4!} + \cdots = \sum_{k=0}^{\infty} \frac{A^k}{k!} \tag{10}$$

Norm of a matrix

*Remark.* It is not difficult to prove that the series of matrices in equation (10) converges for every matrix $A$, but to do so would take us too far afield. We can, however, give an indication of why it is so. We first define $|A|_i$ to be the sum of the absolute values of the components in the $i$th row of $A$. We then define the **norm**† of $A$, denoted by $|A|$, by

$$|A| = \max_{1 \le i \le n} |A|_i \tag{11}$$

It can be shown that

$$|AB| \le |A|\,|B| \tag{12}$$

and

$$|A + B| \le |A| + |B| \tag{13}$$

Then using (12) and (13) in (10), we obtain

$$|e^A| \le 1 + |A| + \frac{|A|^2}{2!} + \frac{|A|^3}{3!} + \frac{|A|^4}{4!} + \cdots = e^{|A|}$$

Since $|A|$ is a real number, $e^{|A|}$ is finite. This shows that the series in (10) converges for any matrix $A$.

We shall now see the usefulness of the series in equation (10).

† This is called the **max-row sum norm** of $A$.

**THEOREM 1** For any constant vector $\mathbf{c}$, $\mathbf{x}(t) = e^{At}\mathbf{c}$ is a solution of (7). Moreover, the solution of (7) given by $\mathbf{x}(t) = e^{At}\mathbf{x}_0$ satisfies $\mathbf{x}(0) = \mathbf{x}_0$.

**Proof** We compute, using (10):

$$\mathbf{x}(t) = e^{At}\mathbf{c} = \left[I + At + A^2\frac{t^2}{2!} + A^3\frac{t^3}{3!} + \cdots\right]\mathbf{c} \tag{14}$$

But since $A$ is a constant matrix, we have

$$\begin{aligned} \frac{d}{dt}A^k\frac{t^k}{k!} &= \frac{d}{dt}\frac{t^k}{k!}A^k = \frac{kt^{k-1}}{k!}A^k \\ &= \frac{A^k t^{k-1}}{(k-1)!} = A\left[A^{k-1}\frac{t^{k-1}}{(k-1)!}\right] \end{aligned} \tag{15}$$

Then combining (14) and (15), we obtain (since $\mathbf{c}$ is a constant vector)

$$\mathbf{x}'(t) = \frac{d}{dt}e^{At}\mathbf{c} = A\left[I + At + A^2\frac{t^2}{2!} + A^3\frac{t^3}{3!} + \cdots\right]\mathbf{c} = Ae^{At}\mathbf{c} = A\mathbf{x}(t)$$

Finally, since $e^{A\cdot 0} = e^0 = I$, we have

$$\mathbf{x}(0) = e^{A\cdot 0}\mathbf{x}_0 = I\mathbf{x}_0 = \mathbf{x}_0$$

■

**DEFINITION 2** **Principal Matrix Solution** The matrix $e^{At}$ is called the **principal matrix solution** of the system $\mathbf{x}' = A\mathbf{x}$.

A major (and obvious) problem remains: How do we compute $e^{At}$ in a practical way? We begin with two examples.

**EXAMPLE 1** **Computing $e^{At}$ When $A$ Is a Diagonal Matrix** Let $A = \begin{pmatrix} 1 & 0 & 0 \\ 0 & 2 & 0 \\ 0 & 0 & 3 \end{pmatrix}$. Then

$$A^2 = \begin{pmatrix} 1 & 0 & 0 \\ 0 & 2^2 & 0 \\ 0 & 0 & 3^2 \end{pmatrix},\ A^3 = \begin{pmatrix} 1 & 0 & 0 \\ 0 & 2^3 & 0 \\ 0 & 0 & 3^3 \end{pmatrix},\ \ldots,\ A^m = \begin{pmatrix} 1 & 0 & 0 \\ 0 & 2^m & 0 \\ 0 & 0 & 3^m \end{pmatrix}$$

and

$$e^{At} = I + At + \frac{A^2t^2}{2!} + \frac{A^3t^3}{3!} + \cdots = \begin{pmatrix} 1 & 0 & 0 \\ 0 & 1 & 0 \\ 0 & 0 & 1 \end{pmatrix} + \begin{pmatrix} t & 0 & 0 \\ 0 & 2t & 0 \\ 0 & 0 & 3t \end{pmatrix}$$

$$
+ \begin{pmatrix} \frac{t^2}{2!} & 0 & 0 \\ 0 & \frac{2^2t^2}{2!} & 0 \\ 0 & 0 & \frac{3^2t^2}{2!} \end{pmatrix} + \begin{pmatrix} \frac{t^3}{3!} & 0 & 0 \\ 0 & \frac{2^3t^3}{3!} & 0 \\ 0 & 0 & \frac{3^3t^3}{3!} \end{pmatrix} + \cdots
$$

$$
= \begin{pmatrix} 1 + t + \frac{t^2}{2!} + \frac{t^3}{3!} + \cdots & 0 & 0 \\ 0 & 1 + (2t) + \frac{(2t)^2}{2!} + \frac{(2t)^3}{3!} + \cdots & 0 \\ 0 & 0 & 1 + (3t) + \frac{(3t)^2}{2!} + \frac{(3t)^3}{3!} + \cdots \end{pmatrix}
$$

$$
= \begin{pmatrix} e^t & 0 & 0 \\ 0 & e^{2t} & 0 \\ 0 & 0 & e^{3t} \end{pmatrix}
$$

EXAMPLE 2 **Computing $e^{At}$ When $A$ Is a $2 \times 2$ Matrix That Is Not Diagonalizable** Let $A = \begin{pmatrix} a & 1 \\ 0 & a \end{pmatrix}$. Then, as is easily verified,

$$
A^2 = \begin{pmatrix} a^2 & 2a \\ 0 & a^2 \end{pmatrix}, \; A^3 = \begin{pmatrix} a^3 & 3a^2 \\ 0 & a^3 \end{pmatrix}, \ldots, A^m = \begin{pmatrix} a^m & ma^{m-1} \\ 0 & a^m \end{pmatrix}, \ldots
$$

so that

$$
e^{At} = \begin{pmatrix} \displaystyle\sum_{m=0}^{\infty} \frac{(at)^m}{m!} & \displaystyle\sum_{m=1}^{\infty} \frac{ma^{m-1}t^m}{m!} \\ 0 & \displaystyle\sum_{m=0}^{\infty} \frac{(at)^m}{m!} \end{pmatrix}
$$

Now

$$
\sum_{m=1}^{\infty} \frac{ma^{m-1}t^m}{m!} = \sum_{m=1}^{\infty} \frac{a^{m-1}t^m}{(m-1)!} = t + at^2 + \frac{a^2t^3}{2!} + \frac{a^3t^4}{3!} + \cdots
$$

$$
= t\left(1 + at + \frac{a^2t^2}{2!} + \frac{a^3t^3}{3!} + \cdots\right) = te^{at}
$$

Thus

$$e^{At} = \begin{pmatrix} e^{at} & te^{at} \\ 0 & e^{at} \end{pmatrix}$$

As Example 1 illustrates, it is easy to calculate $e^{At}$ if $A$ is a diagonal matrix. Example 1 shows that if $D = \text{diag}(\lambda_1, \lambda_2, \ldots, \lambda_n)$, then

$$e^{Dt} = \text{diag}(e^{\lambda_1 t}, e^{\lambda_2 t}, \ldots, e^{\lambda_n t})$$

In Example 2 we calculated $e^{At}$ for a matrix $A$ in Jordan canonical form. It turns out that this is really all we need to be able to do, as the next theorem suggests.

**THEOREM 2** Let $J$ be the Jordan canonical form of a matrix $A$ and let $J = C^{-1}AC$. Then $A = CJC^{-1}$ and

$$\boxed{e^{At} = Ce^{Jt}C^{-1}} \tag{16}$$

**Proof** We first note that

$$\begin{aligned} A^n = (CJC^{-1})^n &= \overbrace{(CJC^{-1})(CJC^{-1})\cdots(CJC^{-1})}^{n \text{ times}} \\ &= CJ(C^{-1}C)J(C^{-1}C)J(C^{-1}C)\cdots(C^{-1}C)JC^{-1} \\ &= CJ^nC^{-1} \end{aligned}$$

It then follows that

$$(At)^n = C(Jt)^nC^{-1} \tag{17}$$

Thus

$$\begin{aligned} e^{At} &= I + (At) + \frac{(At)^2}{2!} + \cdots = CIC^{-1} + C(Jt)C^{-1} + C\frac{(Jt)^2}{2!}C^{-1} + \cdots \\ &= C\left[I + (Jt) + \frac{(Jt)^2}{2!} + \cdots\right]C^{-1} = Ce^{Jt}C^{-1} \end{aligned}$$

Theorem 2 tells us that to calculate $e^{At}$ we really need only to calculate $e^{Jt}$. When $J$ is diagonal (as is most often the case), then we know how to calculate $e^{Jt}$. If $A$ is a $2 \times 2$ matrix that is not diagonalizable, then $J = \begin{pmatrix} \lambda & 1 \\ 0 & \lambda \end{pmatrix}$ and $e^{Jt} = \begin{pmatrix} e^{\lambda t} & te^{\lambda t} \\ 0 & e^{\lambda t} \end{pmatrix}$ as we calculated in Example 2. In fact, it is not difficult to calculate $e^{Jt}$ where $J$ is any Jordan matrix. It is first necessary to compute $e^{Bt}$ for a Jordan block matrix $B$. A method for doing this is given in Problems 20–22.

We now apply our computations to a simple biological model of population growth. Suppose that in an ecosystem there are two interacting species $S_1$ and $S_2$. We denote the populations of the species at time $t$ by $x_1(t)$ and $x_2(t)$. One system governing the relative growth of the two species is

$$\begin{aligned} x_1'(t) &= ax_1(t) + bx_2(t) \\ x_2'(t) &= cx_1(t) + dx_2(t) \end{aligned} \tag{18}$$

We can interpret the constants $a$, $b$, $c$, and $d$ as follows: If the species are competing, then it is reasonable to have $b < 0$ and $c < 0$. This is true because increases in the population of one species will slow the growth of the other. A second model is a *predator-prey* relationship. If $S_1$ is the prey and $S_2$ is the predator ($S_2$ eats $S_1$), then it is reasonable to have $b < 0$ and $c > 0$ since an increase in the predator species will cause a decrease in the prey species, while an increase in the prey species will cause an increase in the predator species (since it will have more food). Finally, in a *symbiotic* relationship (each species lives off the other), we would likely have $b > 0$ and $c > 0$. Of course, the constants $a$, $b$, $c$, and $d$ depend on a wide variety of factors including available food, time of year, climate, limits due to overcrowding, other competing species, and so on. We shall analyze four different models by using the material in this section. We assume that $t$ is measured in years.

**EXAMPLE 3** **A Competitive Model** Consider the system

$$\begin{aligned} x_1'(t) &= \phantom{-}3x_1(t) - x_2(t) \\ x_2'(t) &= -2x_1(t) + 2x_2(t) \end{aligned}$$

Here an increase in the population of one species causes a decline in the growth rate of another. Suppose that the initial populations are $x_1(0) = 90$ and $x_2(0) = 150$. Find the populations of both species for $t > 0$.

**Solution** We have $A = \begin{pmatrix} 3 & -1 \\ -2 & 2 \end{pmatrix}$. The eigenvalues of $A$ are $\lambda_1 = 1$ and $\lambda_2 = 4$ with corresponding eigenvectors $\mathbf{v}_1 = \begin{pmatrix} 1 \\ 2 \end{pmatrix}$ and $\mathbf{v}_2 = \begin{pmatrix} 1 \\ -1 \end{pmatrix}$. Then

$$C = \begin{pmatrix} 1 & 1 \\ 2 & -1 \end{pmatrix} \quad C^{-1} = -\frac{1}{3}\begin{pmatrix} -1 & -1 \\ -2 & 1 \end{pmatrix} \quad J = D = \begin{pmatrix} 1 & 0 \\ 0 & 4 \end{pmatrix} \quad e^{Jt} = \begin{pmatrix} e^t & 0 \\ 0 & e^{4t} \end{pmatrix}$$

$$\begin{aligned} e^{At} = Ce^{Jt}C^{-1} &= -\frac{1}{3}\begin{pmatrix} 1 & 1 \\ 2 & -1 \end{pmatrix}\begin{pmatrix} e^t & 0 \\ 0 & e^{4t} \end{pmatrix}\begin{pmatrix} -1 & -1 \\ -2 & 1 \end{pmatrix} \\ &= -\frac{1}{3}\begin{pmatrix} 1 & 1 \\ 2 & -1 \end{pmatrix}\begin{pmatrix} -e^t & -e^t \\ -2e^{4t} & e^{4t} \end{pmatrix} \\ &= -\frac{1}{3}\begin{pmatrix} -e^t - 2e^{4t} & -e^t + e^{4t} \\ -2e^t + 2e^{4t} & -2e^t - e^{4t} \end{pmatrix} \end{aligned}$$

Finally, the solution to the system is given by

$$\mathbf{x}(t) = \begin{pmatrix} x_1(t) \\ x_2(t) \end{pmatrix} = e^{At}\mathbf{x}_0 = -\frac{1}{3}\begin{pmatrix} -e^t - 2e^{4t} & -e^t + e^{4t} \\ -2e^t + 2e^{4t} & -2e^t - e^{4t} \end{pmatrix}\begin{pmatrix} 90 \\ 150 \end{pmatrix}$$

$$= -\frac{1}{3}\begin{pmatrix} -240e^t - 30e^{4t} \\ -480e^t + 30e^{4t} \end{pmatrix} = \begin{pmatrix} 80e^t + 10e^{4t} \\ 160e^t - 10e^{4t} \end{pmatrix}$$

For example, after 6 months ($t = \frac{1}{2}$ year), $x_1(t) = 80e^{1/2} + 10e^2 \approx 206$ individuals, whereas $x_2(t) = 160e^{1/2} - 10e^2 \approx 190$ individuals. More significantly, $160e^t - 10e^{4t} = 0$ when $16e^t = e^{4t}$ or $16 = e^{3t}$ or $3t = \ln 16$ and $t = (\ln 16)/3 \approx 2.77/3 \approx 0.92$ years $\approx 11$ months. Thus the second species will be eliminated after only 11 months even though it started with a larger population. In Problems 10 and 11 you are asked to show that neither population will be eliminated if $x_2(0) = 2x_1(0)$ and that the first population will be eliminated if $x_2(0) > 2x_1(0)$. Thus, as was well known to Darwin, survival in this very simple model depends on the relative sizes of the competing species when competition begins.

**EXAMPLE 4** **A Predator-Prey Model** We consider the following system in which species 1 is the prey and species 2 is the predator:

$$\begin{aligned} x_1'(t) &= 2x_1(t) - x_2(t) \\ x_2'(t) &= x_1(t) + 4x_2(t) \end{aligned}$$

Find the populations of the two species for $t > 0$ if the initial populations are $x_1(0) = 500$ and $x_2(0) = 100$.

**Solution** Here $A = \begin{pmatrix} 2 & -1 \\ 1 & 4 \end{pmatrix}$ and the only eigenvalue is $\lambda = 3$ with the single eigenvector $\begin{pmatrix} 1 \\ -1 \end{pmatrix}$. One solution to the equation $(A - 3I)\mathbf{v}_2 = \mathbf{v}_1$ (see Theorem 6.6.2 on page 600 is $\mathbf{v}_2 = \begin{pmatrix} 1 \\ -2 \end{pmatrix}$. Then

$$C = \begin{pmatrix} 1 & 1 \\ -1 & -2 \end{pmatrix} \qquad C^{-1} = \begin{pmatrix} 2 & 1 \\ -1 & -1 \end{pmatrix} \qquad J = \begin{pmatrix} 3 & 1 \\ 0 & 3 \end{pmatrix}$$

$$e^{Jt} = \begin{pmatrix} e^{3t} & te^{3t} \\ 0 & e^{3t} \end{pmatrix} = e^{3t}\begin{pmatrix} 1 & t \\ 0 & 1 \end{pmatrix} \qquad \text{(from Example 2)}$$

and

$$e^{At} = Ce^{Jt}C^{-1} = e^{3t}\begin{pmatrix} 1 & 1 \\ -1 & -2 \end{pmatrix}\begin{pmatrix} 1 & t \\ 0 & 1 \end{pmatrix}\begin{pmatrix} 2 & 1 \\ -1 & -1 \end{pmatrix}$$

$$= e^{3t}\begin{pmatrix} 1 & 1 \\ -1 & -2 \end{pmatrix}\begin{pmatrix} 2-t & 1-t \\ -1 & -1 \end{pmatrix} = e^{3t}\begin{pmatrix} 1-t & -t \\ t & 1+t \end{pmatrix}$$

Thus the solution to the system is

$$\mathbf{x}(t) = \begin{pmatrix} x_1(t) \\ x_2(t) \end{pmatrix} = e^{At}\mathbf{x}_0 = e^{3t}\begin{pmatrix} 1-t & -t \\ t & 1+t \end{pmatrix}\begin{pmatrix} 500 \\ 100 \end{pmatrix} = e^{3t}\begin{pmatrix} 500-600t \\ 100+600t \end{pmatrix}$$

It is apparent that the prey species will be eliminated after $\frac{5}{6}$ year = 10 months—even though it started with a population five times as great as the predator species. In fact, it is easy to show (see Problem 12) that no matter how great the initial advantage of the prey species, the prey species will be eliminated in less than 1 year.

**EXAMPLE 5** **Another Predator-Prey Model** Consider the predator-prey model governed by the system

$$\begin{aligned} x_1'(t) &= x_1(t) + x_2(t) \\ x_2'(t) &= -x_1(t) + x_2(t) \end{aligned}$$

If the initial populations are $x_1(0) = x_2(0) = 1000$, determine the populations of the two species for $t > 0$.

**Solution** Here $A = \begin{pmatrix} 1 & 1 \\ -1 & 1 \end{pmatrix}$ with characteristic equation $\lambda^2 - 2\lambda + 2 = 0$, complex roots $\lambda_1 = 1 + i$ and $\lambda_2 = 1 - i$, and eigenvectors $\mathbf{v}_1 = \begin{pmatrix} 1 \\ i \end{pmatrix}$ and $\mathbf{v}_2 = \begin{pmatrix} 1 \\ -i \end{pmatrix}$.† Then

$$C = \begin{pmatrix} 1 & 1 \\ i & -i \end{pmatrix}, \quad C^{-1} = -\frac{1}{2i}\begin{pmatrix} -i & -1 \\ -i & 1 \end{pmatrix} = \frac{1}{2}\begin{pmatrix} 1 & -i \\ 1 & i \end{pmatrix},$$

$$J = D = \begin{pmatrix} 1+i & 0 \\ 0 & 1-i \end{pmatrix}$$

and

$$e^{Jt} = \begin{pmatrix} e^{(1+i)t} & 0 \\ 0 & e^{(1-i)t} \end{pmatrix}$$

Now by Euler's formula (see Appendix 2), $e^{it} = \cos t + i \sin t$. Thus

$$e^{(1+i)t} = e^t e^{it} = e^t(\cos t + i \sin t)$$

Similarly,

$$e^{(1-i)t} = e^t e^{-it} = e^t(\cos t - i \sin t)$$

† Note that $\lambda_2 = \overline{\lambda_1}$ and $\mathbf{v}_2 = \overline{\mathbf{v}_1}$. This should be no surprise, because according to the result of Problem 6.1.33 on page 548, eigenvalues of real matrices occur in complex conjugate pairs and their corresponding eigenvectors are complex conjugates.

Thus

$$e^{Jt} = e^t\begin{pmatrix} \cos t + i \sin t & 0 \\ 0 & \cos t - i \sin t \end{pmatrix}$$

and

$$e^{At} = Ce^{Jt}C^{-1} = \frac{e^t}{2}\begin{pmatrix} 1 & 1 \\ i & -i \end{pmatrix}\begin{pmatrix} \cos t + i \sin t & 0 \\ 0 & \cos t - i \sin t \end{pmatrix}\begin{pmatrix} 1 & -i \\ 1 & i \end{pmatrix}$$

$$= \frac{e^t}{2}\begin{pmatrix} 1 & 1 \\ i & -i \end{pmatrix}\begin{pmatrix} \cos t + i \sin t & -i \cos t + \sin t \\ \cos t - i \sin t & i \cos t + \sin t \end{pmatrix}$$

$$= \frac{e^t}{2}\begin{pmatrix} 2 \cos t & 2 \sin t \\ -2 \sin t & 2 \cos t \end{pmatrix} = e^t\begin{pmatrix} \cos t & \sin t \\ -\sin t & \cos t \end{pmatrix}$$

Finally,

$$\mathbf{x}(t) = e^{At}\mathbf{x}(0) = e^t\begin{pmatrix} \cos t & \sin t \\ -\sin t & \cos t \end{pmatrix}\begin{pmatrix} 1000 \\ 1000 \end{pmatrix} = \begin{pmatrix} 1000e^t(\cos t + \sin t) \\ 1000e^t(\cos t - \sin t) \end{pmatrix}$$

The prey species is eliminated when $1000e^t(\cos t - \sin t) = 0$ or when $\sin t = \cos t$. The first positive solution of this last equation is $t = \pi/4 \approx 0.7854$ year $\approx 9.4$ months.

**EXAMPLE 6** **A Model of Species Cooperation (Symbiosis)** Consider the symbiotic model governed by the system

$$\begin{aligned} x_1'(t) &= -\tfrac{1}{2}x_1(t) + x_2(t) \\ x_2'(t) &= \tfrac{1}{4}x_1(t) - \tfrac{1}{2}x_2(t) \end{aligned}$$

Note that in this model the population of each species increases proportionally to the population of the other and decreases proportionally to its own population. Suppose that $x_1(0) = 200$ and $x_2(0) = 500$. Determine the population of each species for $t > 0$.

**Solution** Here $A = \begin{pmatrix} -\frac{1}{2} & 1 \\ \frac{1}{4} & -\frac{1}{2} \end{pmatrix}$ with eigenvalues $\lambda_1 = 0$ and $\lambda_2 = -1$ and corresponding eigenvectors $\mathbf{v}_1 = \begin{pmatrix} 2 \\ 1 \end{pmatrix}$ and $\mathbf{v}_2 = \begin{pmatrix} 2 \\ -1 \end{pmatrix}$. Then

$$C = \begin{pmatrix} 2 & 2 \\ 1 & -1 \end{pmatrix}, \quad C^{-1} = -\frac{1}{4}\begin{pmatrix} -1 & -2 \\ -1 & 2 \end{pmatrix}, \quad J = D = \begin{pmatrix} 0 & 0 \\ 0 & -1 \end{pmatrix}$$

and

$$e^{Jt} = \begin{pmatrix} e^{0t} & 0 \\ 0 & e^{-t} \end{pmatrix} = \begin{pmatrix} 1 & 0 \\ 0 & e^{-t} \end{pmatrix}$$

Thus

$$
\begin{aligned}
e^{At} &= -\frac{1}{4}\begin{pmatrix} 2 & 2 \\ 1 & -1 \end{pmatrix}\begin{pmatrix} 1 & 0 \\ 0 & e^{-t} \end{pmatrix}\begin{pmatrix} -1 & -2 \\ -1 & 2 \end{pmatrix} \\
&= -\frac{1}{4}\begin{pmatrix} 2 & 2 \\ 1 & -1 \end{pmatrix}\begin{pmatrix} -1 & -2 \\ -e^{-t} & 2e^{-t} \end{pmatrix} \\
&= -\frac{1}{4}\begin{pmatrix} -2 - 2e^{-t} & -4 + 4e^{-t} \\ -1 + e^{-t} & -2 - 2e^{-t} \end{pmatrix}
\end{aligned}
$$

and

$$
\begin{aligned}
\mathbf{x}(t) = e^{At}\mathbf{x}(0) &= -\frac{1}{4}\begin{pmatrix} -2 - 2e^{-t} & -4 + 4e^{-t} \\ -1 + e^{-t} & -2 - 2e^{-t} \end{pmatrix}\begin{pmatrix} 200 \\ 500 \end{pmatrix} \\
&= -\frac{1}{4}\begin{pmatrix} -2400 + 1600e^{-t} \\ -1200 - 800e^{-t} \end{pmatrix} \\
&= \begin{pmatrix} 600 - 400e^{-t} \\ 300 + 200e^{-t} \end{pmatrix}
\end{aligned}
$$

Note that $e^{-t} \to 0$ as $t \to \infty$. This means that as time goes on, the two cooperating species approach the **equilibrium** populations 600 and 300, respectively. Neither population is eliminated.

## PROBLEMS 6.7

### Self-Quiz

**I.** If $C^{-1}AC = D$, then $e^{At} =$ ________.

**a.** $e^{Dt}$ **b.** $C^{-1}e^{Dt}C$ **c.** $Ce^{Dt}C^{-1}$ **d.** $e^{Ct}e^{Dt}e^{C^{-1}t}$

**II.** If $D = \begin{pmatrix} 3 & 0 \\ 0 & -4 \end{pmatrix}$, then $e^{Dt} =$ ________.

**a.** $\begin{pmatrix} e^{3t} & 0 \\ 0 & e^{-4t} \end{pmatrix}$ **b.** $\begin{pmatrix} e^{-3t} & 0 \\ 0 & e^{4t} \end{pmatrix}$ **c.** $\begin{pmatrix} e^{\frac{1}{3}t} & 0 \\ 0 & e^{-\frac{1}{4}t} \end{pmatrix}$ **d.** $\begin{pmatrix} e^{-4t} & 0 \\ 0 & e^{3t} \end{pmatrix}$

**III.** If $J = \begin{pmatrix} 2 & 1 \\ 0 & 2 \end{pmatrix}$, then $e^{Jt} =$ ________.

**a.** $\begin{pmatrix} e^{2t} & 0 \\ 0 & e^{2t} \end{pmatrix}$ **b.** $\begin{pmatrix} e^{2t} & e^{t} \\ 0 & e^{2t} \end{pmatrix}$ **c.** $\begin{pmatrix} e^{2t} & te^{2t} \\ 0 & e^{2t} \end{pmatrix}$ **d.** $\begin{pmatrix} e^{2t} & te^{2t} \\ te^{2t} & e^{2t} \end{pmatrix}$

**IV.** Suppose that

$$
\begin{aligned}
x' &= ax + by, & x(0) &= x_0 \\
y' &= cx + dy, & y(0) &= y_0
\end{aligned}
$$

$A = \begin{pmatrix} a & b \\ c & d \end{pmatrix}$ and $A$ is similar to a diagonal matrix $D$. Then there exists an invertible matrix $C$ such that $\begin{pmatrix} x(t) \\ y(t) \end{pmatrix} =$ ______.

**a.** $C^{-1}e^{Dt}C\begin{pmatrix} x_0 \\ y_0 \end{pmatrix}$ **b.** $Ce^{Dt}C^{-1}\begin{pmatrix} x_0 \\ y_0 \end{pmatrix}$ **c.** $e^{Dt}\begin{pmatrix} x_0 \\ y_0 \end{pmatrix}$

In Problems 1–9 find the principal matrix solution $e^{At}$ of the system $\mathbf{x}'(t) = A\mathbf{x}(t)$.

**1.** $A = \begin{pmatrix} -2 & -2 \\ -5 & 1 \end{pmatrix}$ **2.** $A = \begin{pmatrix} 3 & -1 \\ -2 & 4 \end{pmatrix}$ **3.** $A = \begin{pmatrix} 2 & -1 \\ 5 & -2 \end{pmatrix}$

**4.** $A = \begin{pmatrix} 3 & -5 \\ 1 & -1 \end{pmatrix}$ **5.** $A = \begin{pmatrix} -10 & -7 \\ 7 & 4 \end{pmatrix}$ **6.** $A = \begin{pmatrix} -2 & 1 \\ 5 & 2 \end{pmatrix}$

**7.** $A = \begin{pmatrix} -12 & 7 \\ -7 & 2 \end{pmatrix}$ **8.** $A = \begin{pmatrix} 1 & 1 & -2 \\ -1 & 2 & 1 \\ 0 & 1 & -1 \end{pmatrix}$ **9.** $A = \begin{pmatrix} 4 & 6 & 6 \\ 1 & 3 & 2 \\ -1 & -5 & -2 \end{pmatrix}$

**10.** In Example 3 show that if the initial vector $\mathbf{x}(0) = \begin{pmatrix} a \\ 2a \end{pmatrix}$, where $a$ is a constant, then both populations grow at a rate proportional to $e^t$.

**11.** In Example 3 show that if $x_2(0) > 2x_1(0)$, then the first population will be eliminated.

**12.** In Example 4 show that the first population will become extinct in $\alpha$ years, where $\alpha = x_1(0)/[x_1(0) + x_2(0)]$.

***13.** In a water desalinization plant there are two tanks of water. Suppose that tank 1 contains 1000 liters of brine in which 1000 kg of salt is dissolved and tank 2 contains 1000 liters of pure water. Suppose that water flows into tank 1 at the rate of 20 liters per minute and the mixture flows from tank 1 into tank 2 at a rate of 30 liters per minute. From tank 2, 10 liters is pumped back to tank 1 (establishing **feedback**) while 20 liters is flushed away. Find the amount of salt in both tanks at all times $t$. [*Hint:* Write the information as a $2 \times 2$ system and let $x_1(t)$ and $x_2(t)$ denote the amount of salt in each tank.]

**14.** A community of $n$ individuals is exposed to an infectious disease.† At any given time $t$, the community is divided into three groups: group 1 with population $x_1(t)$ is the susceptible group; group 2 with a population of $x_2(t)$ is the group of infected individuals in circulation; and group 3, population $x_3(t)$, consists of those who are isolated, dead, or immune. It is reasonable to assume that initially $x_2(t)$ and $x_3(t)$ will be small compared to

---

### Answers to Self-Quiz

**I.** c **II.** a **III.** c **IV.** b

---

† For a discussion of this model, see N. Bailey, "The Total Size of a General Stochastic Epidemic," *Biometrika* 40 (1953): 177–185.

$x_1(t)$. Let $\alpha$ and $\beta$ be positive constants denoting the rates at which susceptibles become infected and infected individuals join group 3, respectively. Then a reasonable model for the spread of the disease is given by the system

$$\begin{aligned} x_1'(t) &= -\alpha x_1(0)x_2 \\ x_2'(t) &= \alpha x_1(0)x_2 - \beta x_2 \\ x_3'(t) &= \beta x_2 \end{aligned}$$

**a.** Write this system in the form $\mathbf{x}' = A\mathbf{x}$ and find the solution in terms of $x_1(0)$, $x_2(0)$, and $x_3(0)$. Note that $x_1(0) + x_2(0) + x_3(0) = n$.

**b.** Show that if $\alpha x(0) < \beta$, then the disease will not produce an epidemic.

**c.** What will happen if $\alpha x(0) > \beta$?

**15.** Consider the **second-order differential equation** $x''(t) + ax'(t) + bx(t) = 0$.

**a.** Letting $x_1(t) = x(t)$ and $x_2(t) = x'(t)$, write the preceding equation as a first-order system in the form of equation (7), where $A$ is a $2 \times 2$ matrix.

**b.** Show that the characteristic equation of $A$ is $\lambda^2 + a\lambda + b = 0$.

In Problems 16–19 use the result of Problem 15 to solve the given equation.

**16.** $x'' + 5x' + 6x = 0$; $x(0) = 1$, $x'(0) = 0$

**17.** $x'' + 6x' + 9x = 0$; $x(0) = 1$, $x'(0) = 2$

**18.** $x'' + 4x = 0$; $x(0) = 0$, $x'(0) = 1$

**19.** $x'' - 3x' - 10x = 0$; $x(0) = 3$, $x'(0) = 2$

**20.** Let $N_3 = \begin{pmatrix} 0 & 1 & 0 \\ 0 & 0 & 1 \\ 0 & 0 & 0 \end{pmatrix}$. Show that $N_3^3 = 0$, the zero matrix.

**21.** Show that $e^{N_3 t} = \begin{pmatrix} 1 & t & t^2/2 \\ 0 & 1 & t \\ 0 & 0 & 1 \end{pmatrix}$. [*Hint:* Write down the series for $e^{N_3 t}$ and use the result of Problem 20.]

**22.** Let $J = \begin{pmatrix} \lambda & 1 & 0 \\ 0 & \lambda & 1 \\ 0 & 0 & \lambda \end{pmatrix}$. Show that $e^{Jt} = e^{\lambda t}\begin{pmatrix} 1 & t & t^2/2 \\ 0 & 1 & t \\ 0 & 0 & 1 \end{pmatrix}$. [*Hint:* $Jt = \lambda It + N_3 t$. Use the fact that $e^{A+B} = e^A e^B$ if $AB = BA$.]

**23.** Using the result of Problem 22, compute $e^{At}$, where $A = \begin{pmatrix} -2 & 1 & 0 \\ -2 & 1 & -1 \\ -1 & 1 & -2 \end{pmatrix}$. [*Hint:* See Problem 6.6.20 on page 604.]

**24.** Compute $e^{At}$, where $A = \begin{pmatrix} -1 & -18 & -7 \\ 1 & -13 & -4 \\ -1 & 25 & 8 \end{pmatrix}$.

**25.** Compute $e^{Jt}$, where $J = \begin{pmatrix} \lambda & 1 & 0 & 0 \\ 0 & \lambda & 1 & 0 \\ 0 & 0 & \lambda & 1 \\ 0 & 0 & 0 & \lambda \end{pmatrix}$.

**26.** Compute $e^{At}$, where $A = \begin{pmatrix} 2 & 1 & 0 & 0 \\ 0 & 2 & 0 & 0 \\ 0 & 0 & 3 & 1 \\ 0 & 0 & 0 & 3 \end{pmatrix}$.

**27.** Compute $e^{At}$, where $A = \begin{pmatrix} -4 & 1 & 0 & 0 \\ 0 & -4 & 1 & 0 \\ 0 & 0 & -4 & 0 \\ 0 & 0 & 0 & 3 \end{pmatrix}$.

## 6.8 A DIFFERENT PERSPECTIVE: THE THEOREMS OF CAYLEY-HAMILTON AND GERSHGORIN

There are many interesting results concerning the eigenvalues of a matrix. In this section we discuss two of them. The first says that any matrix satisfies its own characteristic equation. The second shows how to locate, crudely, the eigenvalues of any matrix with practically no computation.

Let $p(x) = x^n + a_{n-1}x^{n-1} + \cdots + a_1x + a_0$ be a polynomial and let $A$ be a square matrix. Then powers of $A$ are defined and we define

$$p(A) = A^n + a_{n-1}A^{n-1} + \cdots + a_1A + a_0I \tag{1}$$

EXAMPLE 1 **Evaluating $p(A)$** Let $A = \begin{pmatrix} -1 & 4 \\ 3 & 7 \end{pmatrix}$ and $p(x) = x^2 - 5x + 3$. Then

$$p(A) = A^2 - 5A + 3I = \begin{pmatrix} 13 & 24 \\ 18 & 61 \end{pmatrix} + \begin{pmatrix} 5 & -20 \\ -15 & -35 \end{pmatrix} + \begin{pmatrix} 3 & 0 \\ 0 & 3 \end{pmatrix} = \begin{pmatrix} 21 & 4 \\ 3 & 29 \end{pmatrix}$$

Expression (1) is a polynomial with scalar coefficients defined for a matrix variable. We can also define a polynomial with *square matrix* coefficients by

$$Q(\lambda) = B_0 + B_1\lambda + B_2\lambda^2 + \cdots + B_n\lambda^n \tag{2}$$

If $A$ is a matrix of the same size as $B$, then we define

$$Q(A) = B_0 + B_1A + B_2A^2 + \cdots + B_nA^n \tag{3}$$

We must be careful in (3) since matrices do not commute under multiplication.

**THEOREM 1** If $P(\lambda)$ and $Q(\lambda)$ are polynomials in the scalar variable $\lambda$ with square matrix coefficients and if $P(\lambda) = Q(\lambda)(A - \lambda I)$, then $P(A) = 0$.

**Proof** If $Q(\lambda)$ is given by equation (2), then

$$P(\lambda) = (B_0 + B_1\lambda + B_2\lambda^2 + \cdots + B_n\lambda^n)(A - \lambda I)$$

$$= B_0A + B_1A\lambda + B_2A\lambda^2 + \cdots + B_nA\lambda^n$$
$$- B_0\lambda - B_1\lambda^2 - B_2\lambda^3 - \cdots - B_n\lambda^{n+1} \quad \textbf{(4)}$$

Then substituting $A$ for $\lambda$ in (4), we obtain

$$P(A) = B_0A + B_1A^2 + B_2A^3 + \cdots + B_nA^{n+1}$$
$$- B_0A - B_1A^2 - B_2A^3 - \cdots - B_nA^{n+1} = 0$$

*Note.* We cannot prove this theorem by substituting $\lambda = A$ to obtain $P(A) = Q(A)(A - A) = 0$. This is because it is possible to find polynomials $P(\lambda)$ and $Q(\lambda)$ with matrix coefficients such that $F(\lambda) = P(\lambda)Q(\lambda)$ but $F(A) \neq P(A)Q(A)$. (See Problem 17.)

We can now state the first main theorem.

**THEOREM 2** **The Cayley-Hamilton Theorem**† Every square matrix satisfies its own characteristic equation. That is, if $p(\lambda) = 0$ is the characteristic equation of $A$, then $p(A) = 0$.

**Proof** We have

$$p(\lambda) = \det(A - \lambda I) = \begin{vmatrix} a_{11} - \lambda & a_{12} & \cdots & a_{1n} \\ a_{21} & a_{22} - \lambda & \cdots & a_{2n} \\ \vdots & \vdots & & \vdots \\ a_{n1} & a_{n2} & \cdots & a_{nn} - \lambda \end{vmatrix}$$

Clearly, any cofactor of $(A - \lambda I)$ is a polynomial in $\lambda$. Thus the adjoint of $A - \lambda I$ (see Definition 2.4.1, page 211) is an $n \times n$ matrix each of whose components is a polynomial in $\lambda$. That is,

$$\text{adj}(A - \lambda I) = \begin{pmatrix} p_{11}(\lambda) & p_{12}(\lambda) & \cdots & p_{1n}(\lambda) \\ p_{21}(\lambda) & p_{22}(\lambda) & \cdots & p_{2n}(\lambda) \\ \vdots & \vdots & & \vdots \\ p_{n1}(\lambda) & p_{22}(\lambda) & \cdots & p_{nn}(\lambda) \end{pmatrix}$$

This means that we can think of adj $(A - \lambda I)$ as a polynomial, $Q(\lambda)$, in $\lambda$ with $n \times n$ matrix coefficients. To see this, look at the following:

$$\begin{pmatrix} -\lambda^2 - 2\lambda + 1 & 2\lambda^2 - 7\lambda - 4 \\ 4\lambda^2 + 5\lambda - 2 & -3\lambda^2 - \lambda + 3 \end{pmatrix} = \begin{pmatrix} -1 & 2 \\ 4 & -3 \end{pmatrix}\lambda^2 + \begin{pmatrix} -2 & -7 \\ 5 & -1 \end{pmatrix}\lambda + \begin{pmatrix} 1 & -4 \\ -2 & 3 \end{pmatrix}$$

† Named after Sir William Rowan Hamilton and Arthur Cayley (1821–1895) (see pages 54 and 76). Cayley published the first discussion of this famous theorem in 1858. Independently, Hamilton discovered (but did not prove) the result in his work on quaternions.

Now from Theorem 2.4.2 on page 213

$$\det(A - \lambda I)I = [\text{adj}(A - \lambda I)][A - \lambda I] = Q(\lambda)(A - \lambda I) \qquad \textbf{(5)}$$

But $\det(A - \lambda I)I = p(\lambda)I$. If

$$p(\lambda) = \lambda^n + a_{n-1}\lambda^{n-1} + \cdots + a_1\lambda + a_0$$

then we define

$$P(\lambda) = p(\lambda)I = \lambda^n I + a_{n-1}\lambda^{n-1}I + \cdots + a_1\lambda I + a_0 I$$

Thus from (5) we have $P(\lambda) = Q(\lambda)(A - \lambda I)$. Finally, from Theorem 1, $P(A) = 0$. This completes the proof.

**EXAMPLE 2** **Illustration of the Cayley-Hamilton Theorem** Let $A = \begin{pmatrix} 1 & -1 & 4 \\ 3 & 2 & -1 \\ 2 & 1 & -1 \end{pmatrix}$. In Example 6.1.4 on page 538 we computed the characteristic equation $\lambda^3 - 2\lambda^2 - 5\lambda + 6 = 0$. Now we compute

$$A^2 = \begin{pmatrix} 6 & 1 & 1 \\ 7 & 0 & 11 \\ 3 & -1 & 8 \end{pmatrix}, \qquad A^3 = \begin{pmatrix} 11 & -3 & 22 \\ 29 & 4 & 17 \\ 16 & 3 & 5 \end{pmatrix}$$

and

$$A^3 - 2A^2 - 5A + 6I = \begin{pmatrix} 11 & -3 & 22 \\ 29 & 4 & 17 \\ 16 & 3 & 5 \end{pmatrix} + \begin{pmatrix} -12 & -2 & -2 \\ -14 & 0 & -22 \\ -6 & 2 & -16 \end{pmatrix}$$
$$+ \begin{pmatrix} -5 & 5 & -20 \\ -15 & -10 & 5 \\ -10 & -5 & 5 \end{pmatrix} + \begin{pmatrix} 6 & 0 & 0 \\ 0 & 6 & 0 \\ 0 & 0 & 6 \end{pmatrix}$$
$$= \begin{pmatrix} 0 & 0 & 0 \\ 0 & 0 & 0 \\ 0 & 0 & 0 \end{pmatrix}$$

In some situations the Cayley-Hamilton theorem is useful in calculating the inverse of a matrix. If $A^{-1}$ exists and $p(A) = 0$, then $A^{-1}p(A) = 0$. To illustrate, if $p(\lambda) = \lambda^n + a_{n-1}\lambda^{n-1} + \cdots + a_1\lambda + a_0$, then

$$p(A) = A^n + a_{n-1}A^{n-1} + \cdots + a_1 A + a_0 I = 0$$

and

$$A^{-1}p(A) = A^{n-1} + a_{n-1}A^{n-2} + \cdots + a_2 A + a_1 I + a_0 A^{-1} = 0$$

Thus

$$A^{-1} = \frac{1}{a_0}(-A^{n-1} - a_{n-1}A^{n-2} - \cdots - a_2A - a_1I) \tag{6}$$

Note that $a_0 \neq 0$ because $a_0 = \det A$ (why?) and we assumed that $A$ was invertible.

**EXAMPLE 3** **Using the Cayley-Hamilton Theorem to Compute $A^{-1}$**

Let $A = \begin{pmatrix} 1 & -1 & 4 \\ 3 & 2 & -1 \\ 2 & 1 & -1 \end{pmatrix}$. Then $p(\lambda) = \lambda^3 - 2\lambda^2 - 5\lambda + 6$. Here $n = 3$, $a_2 = -2$, $a_1 = -5$, $a_0 = 6$, and

$$A^{-1} = \frac{1}{6}(-A^2 + 2A + 5I)$$

$$= \frac{1}{6}\left[\begin{pmatrix} -6 & -1 & -1 \\ -7 & 0 & -11 \\ -3 & 1 & -8 \end{pmatrix} + \begin{pmatrix} 2 & -2 & 8 \\ 6 & 4 & -2 \\ 4 & 2 & -2 \end{pmatrix} + \begin{pmatrix} 5 & 0 & 0 \\ 0 & 5 & 0 \\ 0 & 0 & 5 \end{pmatrix}\right]$$

$$= \frac{1}{6}\begin{pmatrix} 1 & -3 & 7 \\ -1 & 9 & -13 \\ 1 & 3 & -5 \end{pmatrix}$$

Note that we computed $A^{-1}$ with a single division and with only one calculation of a determinant (in order to find $p(\lambda) = \det(A - \lambda I)$). This method is sometimes very efficient on a computer.

## Gershgorin's Circle Theorem

We now turn to the second important result of this section. Let $A$ be an $n \times n$ real or complex matrix. We write, as usual,

$$A = \begin{pmatrix} a_{11} & a_{12} & \cdots & a_{1n} \\ a_{21} & a_{22} & \cdots & a_{2n} \\ \vdots & \vdots & & \vdots \\ a_{n1} & a_{n2} & \cdots & a_{nn} \end{pmatrix}$$

Define the number

$$r_1 = |a_{12}| + |a_{13}| + \cdots + |a_{1n}| = \sum_{j=2}^{n} |a_{1j}| \tag{7}$$

Similarly, define

$$r_i = |a_{i1}| + |a_{i2}| + \cdots + |a_{i,i-1}| + |a_{i,i+1}| + \cdots + |a_{i,n}| = \sum_{\substack{j=1 \\ j \neq i}}^{n} |a_{ij}| \tag{8}$$

That is, $r_i$ is the sum of the absolute values of the numbers on the $i$th row of $A$ that are not on the main diagonal of $A$. Let

$$D_i = \{z \in \mathbb{C}: |z - a_{ii}| \leq r_i\} \tag{9}$$

Here $D_i$ is a disk in the complex plane centered at $a_{ii}$ with radius $r_i$ (see Figure 6.4).

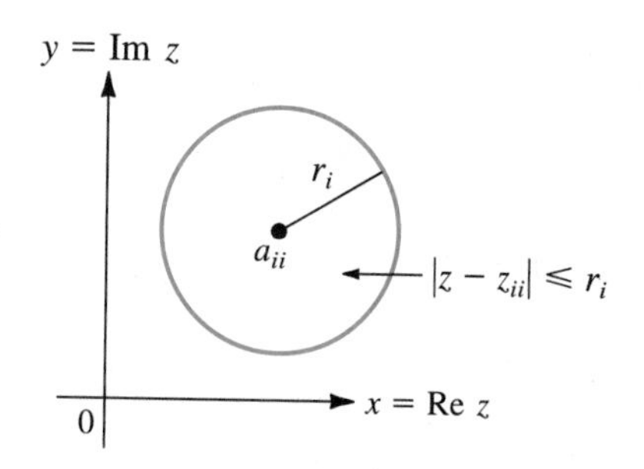


**Figure 6.4** A circle of radius $r_i$ centered at $a_{ii}$

The disk $D_i$ consists of all points in the complex plane on and inside the circle $C_i = \{z \in \mathbb{C}: |z - a_{ii}| = r_i\}$. The circles $C_i$, $i = 1, 2, \ldots, n$, are called **Gershgorin circles.**

**THEOREM 3** **Gershgorin's Circle Theorem**† Let $A$ be an $n \times n$ matrix and let $D_i$ be defined by equation (9). Then each eigenvalue of $A$ is contained in at least one of the $D_i$'s. That is, if the eigenvalues of $A$ are $\lambda_1, \lambda_2, \ldots, \lambda_k$, then

$$\{\lambda_1, \lambda_2, \ldots, \lambda_k\} \subset \bigcup_{i=1}^{n} D_i \tag{10}$$

† The Russian mathematician S. Gershgorin published this result in 1931.

**Proof** Let $\lambda$ be an eigenvalue of $A$ with eigenvector $\mathbf{v} = \begin{pmatrix} x_1 \\ x_2 \\ \vdots \\ x_n \end{pmatrix}$. Let $m = \max\{|x_1|, |x_2|, \ldots, |x_n|\}$. Then $(1/m)\mathbf{v} = \begin{pmatrix} y_1 \\ y_2 \\ \vdots \\ y_n \end{pmatrix}$ is an eigenvector of $A$ corresponding to $\lambda$ and $\max\{|y_1|, |y_2|, \ldots, |y_n|\} = 1$. Let $y_i$ be the component of $\mathbf{y}$ with $|y_i| = 1$. Now $A\mathbf{y} = \lambda\mathbf{y}$. The $i$th component of the $n$-vector $A\mathbf{y}$ is $a_{i1}y_1 + a_{i2}y_2 + \cdots + a_{in}y_n$. The $i$th component of $\lambda\mathbf{y}$ is $\lambda y_i$. Thus

$$a_{i1}y_1 + a_{i2}y_2 + \cdots + a_{in}y_n = \lambda y_i,$$

which we write as

$$\sum_{j=1}^{n} a_{ij}y_j = \lambda y_i \tag{11}$$

By subtracting $a_{ii}y_i$ from both sides, equation (11) can be rewritten as

$$\sum_{\substack{j=1 \\ j\neq i}}^{n} a_{ij}y_j = \lambda y_i - a_{ii}y_i = (\lambda - a_{ii})y_i \tag{12}$$

Next, taking the absolute value of both sides of (12) and using the triangle inequality ($|a + b| \le |a| + |b|$), we obtain

$$|(a_{ii} - \lambda)y_i| = \left| -\sum_{\substack{j=1 \\ j\neq i}}^{n} a_{ij}y_j \right| \le \sum_{\substack{j=1 \\ j\neq i}}^{n} |a_{ij}|\,|y_j| \tag{13}$$

We divide both sides of (13) by $|y_i|$ (which is equal to 1) to obtain

$$|a_{ii} - \lambda| \le \sum_{\substack{j=1 \\ j\neq i}}^{n} |a_{ij}| \frac{|y_j|}{|y_i|} \le \sum_{\substack{j=1 \\ j\neq 1}}^{n} |a_{ij}| = r_i \tag{14}$$

The last step followed the fact that $|y_j| \le |y_i|$ (by the way we chose $y_i$). But this proves the theorem since (14) shows that $\lambda \in D_i$. ∎

**EXAMPLE 4** **Using Gershgorin's Theorem** Let $A = \begin{pmatrix} 1 & -1 & 4 \\ 3 & 2 & -1 \\ 2 & 1 & -1 \end{pmatrix}$. Then $a_{11} = 1$, $a_{22} = 2$, $a_{33} = -1$, $r_1 = |-1| + |4| = 5$, $r_2 = |3| + |-1| = 4$, and $r_3 = |2| + |1| = 3$. Thus the eigenvalues of $A$ lie within the boundaries of the three circles drawn in

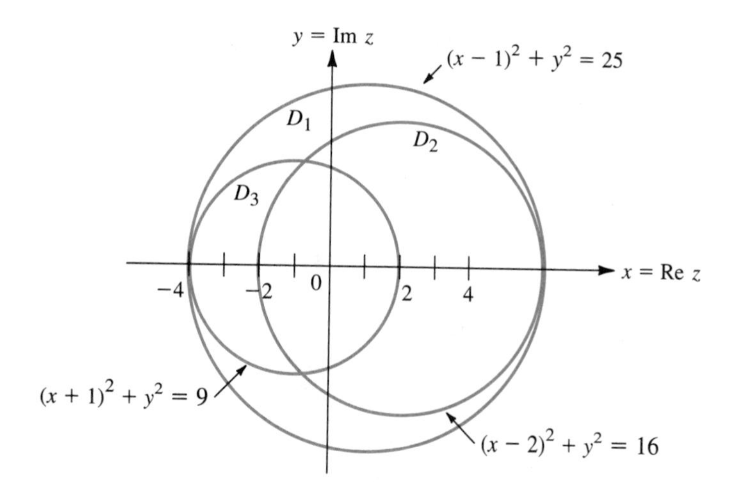


**Figure 6.5** All the eigenvalues of $A$ lie within these three circles

Figure 6.5. We can verify this since we know by Example 6.1.4 on page 538 that the eigenvalues of $A$ are 1, $-2$, and 3, which lie within the three circles. Note that the Gershgorin circles can intersect one another.

**EXAMPLE 5** **Using Gershgorin's Theorem** Find bounds on the eigenvalues of the matrix

$$A = \begin{pmatrix} 3 & 0 & -1 & -\frac{1}{4} & \frac{1}{4} \\ 0 & 5 & \frac{1}{2} & 0 & 1 \\ -\frac{1}{4} & 0 & 6 & \frac{1}{4} & \frac{1}{2} \\ 0 & -1 & \frac{1}{2} & -3 & \frac{1}{4} \\ \frac{1}{6} & -\frac{1}{6} & \frac{1}{3} & \frac{1}{3} & 4 \end{pmatrix}$$

**Solution** Here $a_{11} = 3$, $a_{22} = 5$, $a_{33} = 6$, $a_{44} = -3$, $a_{55} = 4$, $r_1 = \frac{3}{2}$, $r_2 = \frac{3}{2}$, $r_3 = 1$, $r_4 = \frac{7}{4}$, and $r_5 = 1$. The Gershgorin circles are drawn in Figure 6.6. It is clear from Theorem 3 and Figure 6.6 that if $\lambda$ is an eigenvalue of $A$, then $|\lambda| \le 7$ and $\text{Re } \lambda \ge -\frac{19}{4}$.

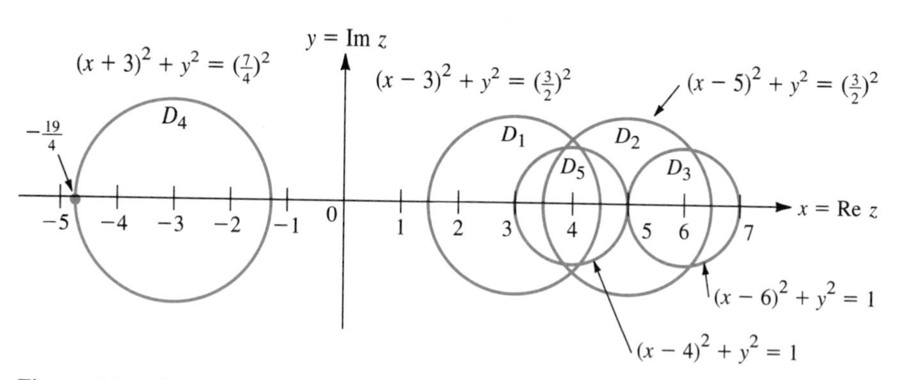


**Figure 6.6** All the eigenvalues of $A$ lie within these five circles

Note the power of the Gershgorin theorem to find the approximate location of eigenvalues after doing very little work.

## PROBLEMS 6.8

### Self-Quiz

**I.** Which equation is satisfied by $A = \begin{pmatrix} 1 & 3 \\ 0 & 2 \end{pmatrix}$?

**a.** $A^2 - 3A + 2I = 0$ **b.** $A^2 - 2A = 0$
**c.** $A^2 + 2A - 3I = 0$ **d.** $A^2 + 3A + 2I = 0$

**II.** According to Gershgorin's theorem, the eigenvalues of $\begin{pmatrix} 2 & -1 & 4 \\ 3 & 2 & 5 \\ 3 & 4 & 2 \end{pmatrix}$ lie in three circles centered at (2, 0) whose largest radius is ______.

**a.** 7 **b.** 8 **c.** $\sqrt{34}$ **d.** 10

In Problems 1–9: **(a)** Find the characteristic equation $p(\lambda) = 0$ of the given matrix; **(b)** verify that $p(A) = 0$; **(c)** use part (*b*) to compute $A^{-1}$.

**1.** $\begin{pmatrix} -2 & -2 \\ -5 & 1 \end{pmatrix}$ **2.** $\begin{pmatrix} 2 & -1 \\ 5 & -2 \end{pmatrix}$ **3.** $\begin{pmatrix} 1 & -1 & 0 \\ -1 & 2 & -1 \\ 0 & -1 & 1 \end{pmatrix}$

**4.** $\begin{pmatrix} 1 & 2 & 2 \\ 0 & 2 & 1 \\ -1 & 2 & 2 \end{pmatrix}$ **5.** $\begin{pmatrix} 0 & 1 & 0 \\ 0 & 0 & 1 \\ 1 & -3 & 3 \end{pmatrix}$ **6.** $\begin{pmatrix} -3 & -7 & -5 \\ 2 & 4 & 3 \\ 1 & 2 & 2 \end{pmatrix}$

**7.** $\begin{pmatrix} 2 & -1 & 3 \\ 4 & 1 & 6 \\ 1 & 5 & 3 \end{pmatrix}$ **8.** $\begin{pmatrix} 1 & 0 & 1 & 0 \\ 2 & -1 & 0 & 2 \\ -1 & 0 & 0 & 1 \\ 4 & 1 & -1 & 0 \end{pmatrix}$ **9.** $\begin{pmatrix} a & b & 0 & 0 \\ 0 & a & c & 0 \\ 0 & 0 & a & d \\ 0 & 0 & 0 & a \end{pmatrix}$; $bcd \neq 0$

In Problems 10–14 draw the Gershgorin circles for the given matrix $A$ and find a bound for $|\lambda|$ if $\lambda$ is an eigenvalue of $A$.

**10.** $\begin{pmatrix} 2 & 1 & 0 \\ \frac{1}{2} & 5 & \frac{1}{2} \\ 1 & 0 & 6 \end{pmatrix}$ **11.** $\begin{pmatrix} 3 & -\frac{1}{2} & -\frac{1}{3} & 0 \\ 0 & 6 & 1 & 0 \\ \frac{1}{3} & -\frac{1}{3} & 5 & \frac{1}{3} \\ -\frac{1}{2} & \frac{1}{4} & -\frac{1}{4} & 4 \end{pmatrix}$ **12.** $\begin{pmatrix} 1 & 3 & -1 & 4 \\ 2 & 5 & 0 & -7 \\ 3 & -1 & 6 & 1 \\ 0 & 2 & 3 & 4 \end{pmatrix}$

---

**Answers to Self-Quiz**

**I.** a **II.** b

**13.** $\begin{pmatrix} -7 & \frac{1}{5} & -\frac{1}{5} & \frac{2}{5} \\ -\frac{1}{10} & -10 & \frac{1}{10} & \frac{3}{10} \\ -\frac{1}{4} & \frac{1}{4} & 5 & \frac{1}{4} \\ 0 & -1 & 0 & 4 \end{pmatrix}$

**14.** $\begin{pmatrix} 3 & 0 & -\frac{1}{3} & \frac{1}{3} & 0 & \frac{1}{3} \\ \frac{1}{2} & 5 & -\frac{1}{2} & 0 & 1 & 0 \\ \frac{1}{10} & -\frac{1}{5} & 4 & \frac{3}{5} & -\frac{1}{5} & \frac{1}{10} \\ -1 & 0 & 0 & -3 & 0 & 0 \\ \frac{1}{2} & 0 & -\frac{1}{2} & 0 & 2 & \frac{1}{2} \\ -\frac{1}{4} & \frac{1}{4} & \frac{1}{4} & 0 & -\frac{1}{4} & 0 \end{pmatrix}$

**15.** Let $A = \begin{pmatrix} 2 & \frac{1}{2} & -\frac{1}{3} & \frac{1}{4} \\ \frac{1}{2} & 3 & \frac{1}{2} & 1 \\ -\frac{1}{3} & \frac{1}{2} & 5 & 2 \\ \frac{1}{4} & 1 & 2 & 4 \end{pmatrix}$. Prove that the eigenvalues of $A$ are positive real numbers.

**16.** Let $A = \begin{pmatrix} -4 & 1 & 1 & 1 \\ 1 & -6 & 2 & 1 \\ 1 & 2 & -5 & 1 \\ 1 & 1 & 1 & -4 \end{pmatrix}$. Prove that the eigenvalues of $A$ are negative and real.

**17.** Let $P(\lambda) = B_0 + B_1\lambda$ and $Q(\lambda) = C_0 + C_1\lambda$, where $B_0$, $B_1$, $C_0$, and $C_1$ are $n \times n$ matrices.

**a.** Compute $F(\lambda) = P(\lambda)Q(\lambda)$.

**b.** Let $A$ be an $n \times n$ matrix. Show that $F(A) = P(A)Q(A)$ if and only if $A$ commutes with both $C_0$ and $C_1$.

**18.** Let the $n \times n$ matrix $A$ have the eigenvalues $\lambda_1, \lambda_2, \ldots, \lambda_n$ and let $r(A) = \max_{1 \le i \le n} |\lambda_i|$. If $\|A\|$ is the max-row sum norm defined in Section 6.7, show that $r(A) \le \|A\|$.

**19.** The $n \times n$ matrix $A$ is said to be **strictly diagonally dominant** if $|a_{ii}| > r_i$ for $i = 1, 2, \ldots, n$, where $r_i$ is defined by equation (8). Show that if $A$ is a strictly diagonally dominant matrix, then $\det A \neq 0$.

## MATLAB 6.8

1. For the matrices in Problems 1 and 13 in the text in Section 6.1, find the characteristic polynomial *by hand.* Use MATLAB and the coefficients of the characteristic polynomial (found by hand) to verify the Cayley-Hamilton theorem for these matrices and to find the inverses of the matrices. Check your answers for the inverses.

2. a. For a random $4 \times 4$ matrix $A$, find **c = poly(A)**. Enter **help polyvalm** and then use **polyvalm** to illustrate the Cayley-Hamilton theorem.
   b. Use the Cayley-Hamilton theorem to find $A^{-1}$ and check your answer.
   c. Repeat parts (a) and (b) for a random $4 \times 4$ complex-valued matrix.

3. Let $A$ be a random $2 \times 2$ matrix. Consider the following MATLAB coding:

```
r1 = sum(abs(A(1,:)))−abs(A(1,1))
r2 = sum(abs(A(2,:)))−abs(A(2,2))
a1 = real(A(1,1)), b1 = imag(A(1,1))
a2 = real(A(2,2)), b2 = imag(A(2,2))
```

So far we have found the center and radius of each Gershgorin circle.

```
xx = -r1: 2*r1/100: r1;
 x = xx + a1;
 z = real(sqrt(r1*r1-xx.*xx));
 y = z + b1; yy = -z+b1;
x1 = [x fliplr(x)];
y1 = [y yy];
```

We have created vectors **x1** and **y1** containing the $x$-values and $y$-values for the circle (top and bottom) of radius **r1** around $A(1, 1)$. (Note the "**.**" before the "*****" in **xx.*xx** in the computation of **z**. The **real** command is used to ensure that round-off does not create values with small imaginary parts for **z**. You will want to use the ";" at the end of statements to suppress the display of the 100+ values.)

**Repeat the last set of coding except replace all the 1's with 2's.**

```
axis('square')
plot(x1, y1, 'b', x2, y2, 'g')
hold on
e = eig(A);
plot(real(e), imag(e), 'w*')
hold off
```

This coding plots the two Gershgorin circles (one in blue and one in green), finds the eigenvalues, and plots the eigenvalues as points (with the symbol "$*$" in white). The colors and the symbol can be changed. If you are using MATLAB 4.0, enter the **axis('square')** command *after* the plot statements. (*Note.* Because of the use of possibly different scales on the $x$- and $y$-axes, the circles may appear slightly elliptical.)

a. Enter a $2 \times 2$ real-valued matrix and enter the coding above. Explain your observations from the graph in light of Theorem 3.
b. Repeat part (a) for a $2 \times 2$ complex-valued matrix.
c. Repeat part (a) for a $3 \times 3$ complex-valued matrix. You will need to add some coding; that is, create **r3**, **a3**, **b3**, **x3**, and **y3** and modify the first plot statement.

# SUMMARY

- ***Eigenvalue and eigenvector***

  Let $A$ be an $n \times n$ matrix with real components. The number $\lambda$ (real or complex) is called an **eigenvalue** of $A$ if there is a *nonzero* vector $\mathbf{v}$ in $\mathbb{C}^n$ such that (p. 533)

  $$A\mathbf{v} = \lambda\mathbf{v}$$

  The vector $\mathbf{v} \neq \mathbf{0}$ is called an **eigenvector** of $A$ corresponding to the eigenvalue $\lambda$.
- Let $A$ be an $n \times n$ matrix. Then $\lambda$ is an eigenvalue of $A$ if and only if (p. 535)

  $$p(\lambda) = \det(A - \lambda I) = 0$$

The equation $p(\lambda) = 0$ is called the **characteristic equation** of $A$; $p(\lambda)$ is called the **characteristic polynomial** of $A$.

- Counting multiplicities, every $n \times n$ matrix has exactly $n$ eigenvalues. (p. 535)
- Eigenvectors corresponding to different eigenvalues are linearly independent. (p. 536)
- ***Algebraic multiplicity***

  If $p(\lambda) = (\lambda - \lambda_1)^{r_1}(\lambda - \lambda_2)^{r_2}\cdots(\lambda - \lambda_m)^{r_m}$, then $r_i$ is the **algebraic multiplicity** of $\lambda_i$. (p. 537)
- The eigenvalues of a real matrix occur in complex conjugate pairs. (p. 541)
- ***Eigenspace***

  If $\lambda$ is an eigenvalue of the $n \times n$ matrix $A$, then $E_\lambda = \{\mathbf{v}\colon A\mathbf{v} = \lambda\mathbf{v}\}$ is a subspace of $\mathbb{R}^n$ called the **eigenspace** of $A$ corresponding to $\lambda$. It is denoted by $E_\lambda$. (p. 535)
- ***Geometric multiplicity***

  The **geometric multiplicity** of an eigenvalue $\lambda$ of the matrix $A$ is equal to $\dim E_\lambda = \nu(A - \lambda I)$. (p. 544)
- For any eigenvalue $\lambda$ geometric multiplicity $\leq$ algebraic multiplicity. (p. 545)
- Let $A$ be an $n \times n$ matrix. Then $A$ has $n$ linearly independent eigenvectors if and only if the geometric multiplicity of every eigenvalue is equal to its algebraic multiplicity. In particular, $A$ has $n$ linearly independent eigenvectors if all the eigenvalues are distinct (since then the algebraic multiplicity of every eigenvalue is 1). (p. 545)
- ***Summing Up Theorem***

  Let $A$ be an $n \times n$ matrix. Then the following 12 statements are equivalent; that is, each one implies the other 11 (so that if one is true, all are true): (p. 546)

  **i.** $A$ is invertible.

  **ii.** The only solution to the homogeneous system $A\mathbf{x} = \mathbf{0}$ is the trivial solution ($\mathbf{x} = \mathbf{0}$).

  **iii.** The system $A\mathbf{x} = \mathbf{b}$ has a unique solution for every $n$-vector $\mathbf{b}$.

  **iv.** $A$ is row equivalent to the $n \times n$ identity matrix $I_n$.

  **v.** $A$ can be written as the product of elementary matrices.

  **vi.** The row echelon form of $A$ has $n$ pivots.

  **vii.** The rows (and columns) of $A$ are linearly independent.

  **viii.** $\det A \neq 0$.

  **ix.** $\nu(A) = 0$.

  **x.** $\rho(A) = n$.

  **xi.** The linear transformation $T$ from $\mathbb{R}^n$ to $\mathbb{R}^n$ defined by $T\mathbf{x} = A\mathbf{x}$ is an isomorphism.

  **xii.** Zero is *not* an eigenvalue of $A$.
- ***Similar matrices***

  Two $n \times n$ matrices $A$ and $B$ are said to be **similar** if there exists an invertible $n \times n$ matrix $C$ such that (pp. 564, 565)

$$B = C^{-1}AC$$

The function defined above which takes the matrix $A$ into the matrix $B$ is called a **similarity transformation.**

- $A$ and $B$ are similar if there exist an invertible matrix $C$ such that $CB = AC$. (p. 565)
- Similar matrices have the same eigenvalues. (p. 566)
- ***Diagonalizable matrix***

  An $n \times n$ matrix $A$ is **diagonalizable** if there is a diagonal matrix $D$ such that $A$ is similar to $D$. (p. 566)
- An $n \times n$ matrix $A$ is diagonalizable if and only if it has $n$ linearly independent eigenvectors. In that case the diagonal matrix $D$ similar to $A$ is given by (p. 567)

$$D = \begin{pmatrix} \lambda_1 & 0 & 0 & \cdots & 0 \\ 0 & \lambda_2 & 0 & \cdots & 0 \\ 0 & 0 & \lambda_3 & \cdots & 0 \\ \vdots & \vdots & \vdots & & \vdots \\ 0 & 0 & 0 & \cdots & \lambda_n \end{pmatrix}$$

where $\lambda_1, \lambda_2, \ldots, \lambda_n$ are the eigenvalues of $A$. If $C$ is a matrix whose columns are linearly independent eigenvectors of $A$, then

$$D = C^{-1}AC$$

- If the $n \times n$ matrix $A$ has $n$ distinct eigenvalues, then $A$ is diagonalizable. (p. 569)
- The eigenvalues of a real symmetric matrix are real. (p. 576)
- Eigenvectors of a real symmetric matrix corresponding to distinct eigenvalues are orthogonal. (p. 577)
- A real symmetric $n \times n$ matrix has $n$ real orthonormal eigenvectors. (p. 577)
- ***Orthogonally diagonalizable matrix***

  An $n \times n$ matrix $A$ is said to be **orthogonally diagonalizable** if there exists an orthogonal matrix $Q$ such that (p. 577)

$$Q^tAQ = D$$

where $D = \text{diag}(\lambda_1, \lambda_2, \ldots, \lambda_n)$ and $\lambda_1, \lambda_2, \ldots, \lambda_n$ are the eigenvalues of $A$.

- **Procedure for finding an orthogonal diagonalizing matrix $Q$ for a real symmetric matrix $A$:** (p. 578)

  **i.** Find a basis for each eigenspace of $A$.

  **ii.** Find an orthonormal basis for each eigenspace of $A$ by using the Gram-Schmidt process.

  **iii.** Write $Q$ as the matrix whose columns are the orthonormal eigenvectors obtained in step (*ii*).

- The **conjugate transpose** of an $m \times n$ matrix $A = (a_{ij})$, denoted by $A^*$, is the $n \times m$ matrix whose $ij$th component is $\overline{a_{ji}}$. (p. 580)
- A complex $n \times n$ matrix $A$ is **Hermitian** if $A^* = A$. (p. 580)
- A complex $n \times n$ matrix $U$ is **unitary** if $U^* = U^{-1}$. (p. 580)
- ***Quadratic equation and quadratic form***

  A **quadratic equation in two variables with no linear term** is an equation of the form (p. 585)

$$ax^2 + bxy + cy^2 = d$$

where $|a| + |b| + |c| \neq 0$ and $a$, $b$, $c$ are real numbers.

A **quadratic form in two variables** is an expression of the form

$$F(x, y) = ax^2 + bxy + cy^2 = d$$

where $|a| + |b| + |c| \neq 0$ and $a$, $b$, $c$ are real numbers.

- A quadratic form can be written as (p. 586)

$$F(x, y) = A\mathbf{v} \cdot \mathbf{v}$$

where $A = \begin{pmatrix} a & b/2 \\ b/2 & c \end{pmatrix}$ is a symmetric matrix.

- If the eigenvalues of $A$ are $a'$ and $c'$, then the quadratic form can be written as (p. 587)

$$\overline{F}(x', y') = a'x'^2 + c'y'^2$$

where $\begin{pmatrix} x' \\ y' \end{pmatrix} = Q^t \begin{pmatrix} x \\ y \end{pmatrix}$ and $Q$ is the orthogonal matrix that diagonalizes $A$.

- ***Principal axes theorem in*** $\mathbb{R}^2$

Let

$$ax^2 + bxy + cy^2 = d \qquad (*) \qquad \text{(p. 588)}$$

be a quadratic equation in the variables $x$ and $y$. Then there exists a unique number $\theta$ in $[0, 2\pi)$ such that equation $(*)$ can be written in the form

$$a'x'^2 + c'y'^2 = d$$

where $x'$, $y'$ are the axes obtained by rotating the $x$- and $y$-axes through an angle of $\theta$ in the counterclockwise direction. Moreover, the numbers $a'$ and $c'$ are the eigenvalues of the matrix $A = \begin{pmatrix} a & b/2 \\ b/2 & c \end{pmatrix}$. The $x'$- and $y'$-axes are called the **principal axes** of the graph of the quadratic equation.

- If $A = \begin{pmatrix} a & b/2 \\ b/2 & c \end{pmatrix}$, then the quadratic equation $(*)$ is the equation of: (p. 591)

  **i.** A hyperbola if $d \neq 0$ and $\det A < 0$.

  **ii.** An ellipse, circle, or degenerate conic section if $d \neq 0$ and $\det A > 0$.

  **iii.** A pair of straight lines or a degenerate conic section if $d \neq 0$ and $\det A = 0$.

  **iv.** If $d = 0$, then $(*)$ is the equation of two straight lines if $\det A \neq 0$ and the equation of a single line if $\det A = 0$.

- ***Quadratic form in*** $\mathbb{R}^n$

Let $\mathbf{v} = \begin{pmatrix} x_1 \\ x_2 \\ \vdots \\ x_n \end{pmatrix}$ and let $A$ be a symmetric $n \times n$ matrix. Then a **quadratic form** in $x_1, x_2, \ldots, x_n$ is an expression of the form (p. 593)

$$F(x_1, x_2, \ldots, x_n) = A\mathbf{v} \cdot \mathbf{v}$$

- The matrix $N_k$ is the $k \times k$ matrix (p. 597)

$$N_k = \begin{pmatrix} 0 & 1 & 0 & \cdots & 0 \\ 0 & 0 & 1 & \cdots & 0 \\ \vdots & \vdots & \vdots & & \vdots \\ 0 & 0 & 0 & \cdots & 1 \\ 0 & 0 & 0 & \cdots & 0 \end{pmatrix}$$

- The $k \times k$ **Jordan block matrix** $B(\lambda)$ is given by (p. 597)

$$B(\lambda) = \lambda I + N_k = \begin{pmatrix} \lambda & 1 & 0 & \cdots & 0 & 0 \\ 0 & \lambda & 1 & \cdots & 0 & 0 \\ \vdots & \vdots & \vdots & & \vdots & \vdots \\ 0 & 0 & & \cdots & \lambda & 1 \\ 0 & 0 & & \cdots & 0 & \lambda \end{pmatrix}$$

- A **Jordan matrix** $J$ has the form (p. 597)

$$J = \begin{pmatrix} B_1(\lambda_1) & 0 & \cdots & 0 \\ 0 & B_2(\lambda_2) & \cdots & 0 \\ \vdots & \vdots & & \vdots \\ 0 & 0 & \cdots & B_r(\lambda_r) \end{pmatrix}$$

where each $B_j(\lambda_j)$ is a Jordan block matrix.

- ***Jordan canonical form***

Let $A$ be an $n \times n$ matrix. Then there exists an invertible $n \times n$ matrix $C$ such that (pp. 598, 599)

$$C^{-1}AC = J$$

where $J$ is a Jordan matrix whose diagonal elements are the eigenvalues of $A$. Moreover, $J$ is unique except for the order in which the Jordan blocks appear.

The matrix $J$ is called the **Jordan canonical form** of $A$.

- Suppose $A$ is a $2 \times 2$ matrix with one eigenvalue $\lambda$ of geometric multiplicity 1. Then the Jordan canonical form of $A$ is (pp. 600, 601, 602)

$$J = \begin{pmatrix} \lambda & 1 \\ 0 & \lambda \end{pmatrix}$$

The matrix $C$ has columns $\mathbf{v}_1$ and $\mathbf{v}_2$, where $\mathbf{v}_1$ is an eigenvector and $\mathbf{v}_2$ is a **generalized eigenvector** of $A$. That is, $\mathbf{v}_2$ satisfies

$$(A - \lambda I)\mathbf{v}_2 = \mathbf{v}_1$$

- Let $A$ be an $n \times n$ matrix. Then $e^A$ is defined by (p. 608)

$$e^A = I + A + \frac{A^2}{2!} + \frac{A^3}{3!} + \cdots = \sum_{k=0}^{\infty} \frac{A^k}{k!}$$

- The **principal matrix solution** to the vector differential equation $\mathbf{x}'(t) = A\mathbf{x}(t)$ is $e^{At}$. (p. 609)

- The unique solution to the differential equation $\mathbf{x}'(t) = A\mathbf{x}(t)$ that satisfies $\mathbf{x}(0) = \mathbf{x}_0$ is $\mathbf{x}(t) = e^{At}\mathbf{x}_0$. (p. 609)
- If $J$ is the Jordan canonical form of the matrix $A$ and if $J = C^{-1}AC$, then (p. 611)

$$e^{At} = Ce^{Jt}C^{-1}$$

- ***The Cayley-Hamilton theorem***

Every square matrix satisfies its own characteristic equation. That is, if $p(\lambda) = 0$ is the characteristic equation of $A$, then $p(A) = 0$. (p. 620)

- ***Gershgorin circles***

Let

$$A = \begin{pmatrix} a_{11} & a_{12} & \cdots & a_{1n} \\ a_{21} & a_{22} & \cdots & a_{2n} \\ \vdots & \vdots & & \vdots \\ a_{n1} & a_{n2} & \cdots & a_{nn} \end{pmatrix}$$

and define the numbers (pp. 622, 623)

$$r_1 = |a_{12}| + |a_{13}| + \cdots + |a_{1n}| = \sum_{j=2}^{n} |a_{1j}|$$

$$r_i = |a_{i1}| + |a_{i2}| + \cdots + |a_{i,i-1}| + |a_{i,i+1}| + \cdots + |a_{i,n}|$$

$$= \sum_{\substack{j=1 \\ j\neq i}}^{n} |a_{ij}|$$

The **Gershgorin circles** are the circles that bound the disks

$$D_i = \{z \in \mathbb{C}\colon |z - a_{ii}| \leq r_i\} \qquad (**)$$

- ***Gershgorin circle theorem***

Let $A$ be an $n \times n$ matrix and let $D_i$ be defined by equation (* *). Then each eigenvalue of $A$ is contained in at least one of the $D_i$'s. That is, if the eigenvalues of $A$ are $\lambda_1, \lambda_2, \ldots, \lambda_k$, then (p. 623)

$$\{\lambda_1, \lambda_2, \ldots, \lambda_k\} \subset \bigcup_{i=1}^{n} D_i$$

# REVIEW EXERCISES

In Exercises 1–6 calculate the eigenvalues and eigenspaces of the given matrix.

**1.** $\begin{pmatrix} -8 & 12 \\ -6 & 10 \end{pmatrix}$ **2.** $\begin{pmatrix} 2 & 5 \\ 0 & 2 \end{pmatrix}$ **3.** $\begin{pmatrix} 1 & 0 & 0 \\ 3 & 7 & 0 \\ -2 & 4 & -5 \end{pmatrix}$

**4.** $\begin{pmatrix} 1 & -1 & 0 \\ 1 & 2 & 1 \\ -2 & 1 & -1 \end{pmatrix}$ **5.** $\begin{pmatrix} 5 & -2 & 0 & 0 \\ 4 & -1 & 0 & 0 \\ 0 & 0 & 3 & -1 \\ 0 & 0 & 2 & 3 \end{pmatrix}$ **6.** $\begin{pmatrix} -2 & 1 & 0 \\ 0 & -2 & 1 \\ 0 & 0 & -2 \end{pmatrix}$

In Exercises 7–15 determine whether the given matrix $A$ is diagonalizable. If it is, find a matrix $C$ such that $C^{-1}AC = D$. If $A$ is symmetric, find an orthogonal matrix $Q$ such that $Q^tAQ = D$.

**7.** $\begin{pmatrix} -18 & -15 \\ 20 & 17 \end{pmatrix}$ **8.** $\begin{pmatrix} \frac{17}{2} & \frac{9}{2} \\ -15 & -8 \end{pmatrix}$ **9.** $\begin{pmatrix} 1 & 1 & 1 \\ -1 & -1 & 0 \\ -1 & 0 & -1 \end{pmatrix}$

**10.** $\begin{pmatrix} 4 & 2 & 0 \\ 2 & 4 & 0 \\ 0 & 0 & -3 \end{pmatrix}$ **11.** $\begin{pmatrix} -3 & 2 & 1 \\ -7 & 4 & 2 \\ -5 & 3 & 2 \end{pmatrix}$ **12.** $\begin{pmatrix} 8 & 0 & 12 \\ 0 & -2 & 0 \\ 12 & 0 & -2 \end{pmatrix}$

**13.** $\begin{pmatrix} 2 & 2 & 0 \\ 2 & 2 & 0 \\ 0 & 0 & -3 \end{pmatrix}$ **14.** $\begin{pmatrix} 4 & 2 & -2 & 2 \\ 1 & 3 & 1 & -1 \\ 0 & 0 & 2 & 0 \\ 1 & 1 & -3 & 5 \end{pmatrix}$ **15.** $\begin{pmatrix} 3 & 4 & -4 & 0 \\ 0 & -1 & 0 & 0 \\ 0 & 0 & -1 & 0 \\ 0 & -4 & 4 & 3 \end{pmatrix}$

In Exercises 16–20 identify the conic section and write it in new variables with the $xy$ term absent.

**16.** $xy = -4$ **17.** $4x^2 + 2xy + 2y^2 = 8$ **18.** $4x^2 - 3xy + y^2 = 1$

**19.** $3y^2 - 2xy - 5 = 0$ **20.** $x^2 - 4xy + 4y^2 + 1 = 0$

**21.** Write the quadratic form $2x^2 + 4xy + 2y^2 - 3z^2$ in new variables $x'$, $y'$, and $z'$ so that no cross-product terms are present.

In Exercises 22–24 find a matrix $C$ such that $C^{-1}AC = J$, the Jordan canonical form of the matrix.

**22.** $\begin{pmatrix} -9 & 4 \\ -25 & 11 \end{pmatrix}$ **23.** $\begin{pmatrix} -4 & 4 \\ -1 & 0 \end{pmatrix}$ **24.** $\begin{pmatrix} 0 & -18 & -7 \\ 1 & -12 & -4 \\ -1 & 25 & 9 \end{pmatrix}$

In Exercises 25–27 compute $e^{At}$.

**25.** $A = \begin{pmatrix} -3 & 4 \\ -2 & 3 \end{pmatrix}$ **26.** $A = \begin{pmatrix} -4 & 4 \\ -1 & 0 \end{pmatrix}$ **27.** $A = \begin{pmatrix} -3 & -4 \\ 2 & 1 \end{pmatrix}$

**28.** Using the Cayley-Hamilton theorem, compute the inverse of

$$A = \begin{pmatrix} 2 & 3 & 1 \\ -1 & 1 & 0 \\ -2 & -1 & 4 \end{pmatrix}.$$

**29.** Use the Gershgorin circle theorem to find a bound on the eigenvalues of

$$A = \begin{pmatrix} 3 & \frac{1}{2} & -\frac{1}{2} & 0 \\ 0 & 4 & \frac{1}{3} & -\frac{1}{3} \\ 1 & 0 & 2 & -1 \\ \frac{1}{2} & -\frac{1}{2} & 1 & -3 \end{pmatrix}$$

# Appendix 1 Mathematical Induction

**Mathematical induction** is the name given to a fundamental principle of logic that can be used to prove a certain type of mathematical statement. Typically, we use mathematical induction to prove that a certain statement or equation holds for every positive integer. For example, we may need to prove that $2^n > n$ for all integers $n \geq 1$. To do this, we proceed in two steps:

> Step 1. We prove that the statement is true for some integer $N$ (usually $N = 1$).
>
> Step 2. We *assume* that the statement is true for an integer $k$ greater than or equal to the $N$ in Step 1 and then *prove* that it is true for the integer $k + 1$.

If we can complete these two steps, then we shall have demonstrated the validity of the statement for *all* positive integers greater than or equal to $N$. To convince you of this fact, we reason as follows: Since the statement is true for $N$ [by step (1)], it is true for the integer $N + 1$ [by step (2)]. Then it is also true for the integer $(N + 1) + 1 = N + 2$ [again by step (2)], and so on. We now demonstrate the procedure with some examples.

**EXAMPLE 1** Prove that $2^n > n$ for every integer $n \geq 1$.

**Solution** **Step 1.** If $n = 1$, then $2^1 = 2 > 1$, so the result is true for $n = 1$.

**Step 2.** Assume that $2^k > k$. Then

$$2^{k+1} = 2 \cdot 2^k = 2^k + 2^k > k + k > k + 1$$

Thus if the result is true for $n = k$, it is also true for $n = k + 1$.

This completes the proof by mathematical induction.

**EXAMPLE 2** Prove that the sum of the first $n$ positive integers is equal to $n(n + 1)/2$.

**Solution** We are asked to show that

$$1 + 2 + 3 + \cdots + n = \frac{n(n+1)}{2} \tag{1}$$

You may first wish to try a few examples to illustrate that formula (1) really works. (This, of course, does not prove the statement but it may help to persuade you of its truth.) For example,

$$1 + 2 + 3 + 4 + 5 + 6 + 7 + 8 + 9 + 10 = \frac{10(11)}{2} = 55$$

That is, formula (1) is true for $N = 10$.

**Step 1.** If $n = 1$, then the sum of the first 1 integer is 1. But $(1)(1 + 1)/2 = 1$, so equation (1) holds in the case $n = 1$.

**Step 2.** Assume that (1) is true for $n = k$; that is,

$$1 + 2 + 3 + \cdots + k = \frac{k(k+1)}{2}$$

We must now show that it is true for $n = k + 1$. That is, we must show that

$$1 + 2 + 3 + \cdots + k + (k + 1) = \frac{(k+1)(k+2)}{2}$$

But

$$\begin{aligned}
1 + 2 + 3 + \cdots + k + (k + 1) &= \overbrace{(1 + 2 + 3 + \cdots + k)}^{= k(k+1)/2 \text{ by assumption}} + (k + 1) \\
&= \frac{k(k+1)}{2} + (k + 1) \\
&= \frac{k(k+1) + 2(k+1)}{2} \\
&= \frac{(k+1)(k+2)}{2}
\end{aligned}$$

and the proof is complete.

## Where the Difficulty Lies

Mathematical induction is sometimes difficult at first sight because of Step 2. Step 1 is usually easy to carry out. In Example 1, for instance, we inserted the value $n = 1$ on both sides of equation (1) and verified that $1 = 1(1 + 1)/2$. Step 2 was much more difficult. Let us look at it again.

Induction hypothesis

We *assumed* that equation (1) was valid for $n = k$. We did not prove it. That assumption is called the **induction hypothesis.** We then used the induction hypothesis to show that equation (1) holds for $n = k + 1$. Perhaps this will be clearer if we look at a particular value for $k$, say, $k = 10$. Then we have

**Assumption**

$$1 + 2 + 3 + 4 + 5 + 6 + 7 + 8 + 9 + 10 = \frac{10(10 + 1)}{2} = \frac{10(11)}{2} = 55 \qquad \textbf{(2)}$$

**To prove**

$$1 + 2 + 3 + 4 + 5 + 6 + 7 + 8 + 9 + 10 + 11 = \frac{11(11 + 1)}{2} = \frac{11(12)}{2} = 66 \qquad \textbf{(3)}$$

**The actual proof**

$(1 + 2 + 3 + 4 + 5 + 6 + 7 + 8 + 9 + 10) + 11$

By the induction hypothesis (2)
↓

$$= \frac{10(11)}{2} + 11 = \frac{10(11)}{2} + \frac{2(11)}{2}$$

$$= \frac{11(10 + 2)}{2} = \frac{11(12)}{2}$$

which is equation (3). Thus *if* (2) is true, then (3) is true.

The beauty of the method of mathematical induction is that we do not have to prove each case separately, as we did in this illustration. Rather, we prove it for a first case, *assume* it for a general case, and then prove it for the general case plus 1. Two steps take care of an infinite number of cases. It's really quite a remarkable idea.

**EXAMPLE 3** Prove that the sum of the squares of the first $n$ positive integers is $n(n + 1)(2n + 1)/6$.

**Solution** We must prove that

$$1^2 + 2^2 + 3^2 + \cdots + n^2 = \frac{n(n + 1)(2n + 1)}{6} \qquad \textbf{(4)}$$

**Step 1.** Since $\dfrac{1(1+1)(2\cdot 1+1)}{6} = 1 = 1^2$, equation (4) is valid for $n = 1$.

**Step 2.** Suppose that equation (4) is true for $n = k$; that is

Induction hypothesis
$$1^2 + 2^2 + 3^2 + \cdots + k^2 = \frac{k(k+1)(2k+1)}{6}$$

Then to prove that (4) is true for $n = k + 1$, we have

$$\begin{aligned}
1^2 + 2^2 + 3^2 + \cdots + k^2 + (k+1)^2 &= (1^2 + 2^2 + 3^2 \cdots + k^2) + (k+1)^2 \\
&\overset{\text{Induction hypothesis}}{=} \frac{k(k+1)(2k+1)}{6} + (k+1)^2 \\
&= \frac{k(k+1)(2k+1) + 6(k+1)^2}{6} \\
&= \frac{k+1}{6}[k(2k+1) + 6(k+1)] \\
&= \frac{k+1}{6}[2k^2 + 7k + 6] \\
&= \frac{k+1}{6}[(k+2)(2k+3)] \\
&= \frac{(k+1)(k+2)[2(k+1)+1]}{6}
\end{aligned}$$

which is equation (4) for $n = k + 1$, and the proof is complete. To illustrate the formula, note that

$$\begin{aligned}
1^2 + 2^2 + 3^2 + 4^2 + 5^2 + 6^2 + 7^2 &= \frac{7(7+1)(2\cdot 7+1)}{6} \\
&= \frac{7\cdot 8\cdot 15}{6} = 140
\end{aligned}$$

**EXAMPLE 4** Use mathematical induction to prove the formula for the sum of a geometric progression:

$$1 + a + a^2 + \cdots + a^n = \frac{1 - a^{n+1}}{1 - a}, \qquad a \neq 1 \tag{5}$$

**Solution** **Step 1.** If $n = 0$ (the first integer in this case), then

$$\frac{1 - a^{0+1}}{1 - a} = \frac{1 - a}{1 - a} = 1 = a^0$$

Thus equation (5) holds for $n = 0$. (We use $n = 0$ instead of $n = 1$ since $a^0 = 1$ is the first term.)

**Step 2.** Assume that (5) holds for $n = k$; that is,

Induction hypothesis $$1 + a + a^2 + \cdots + a^k = \frac{1 - a^{k+1}}{1 - a}$$

Then

$$1 + a + a^2 + \cdots + a^k + a^{k+1} = (1 + a + a^2 + \cdots + a^k) + a^{k+1}$$

Induction hypothesis ↓

$$= \frac{1 - a^{k+1}}{1 - a} + a^{k+1}$$

$$= \frac{1 - a^{k+1} + (1 - a)a^{k+1}}{1 - a} = \frac{1 - a^{k+2}}{1 - a}$$

so that equation (5) also holds for $n = k + 1$, and the proof is complete.

**EXAMPLE 5** Use mathematical induction to prove that $2n + n^3$ is divisible by 3 for every positive integer $n$.

**Solution** **Step 1.** If $n = 1$, then $2n + n^3 = 2 \cdot 1 + 1^3 = 2 + 1 = 3$, which is divisible by 3. Thus the statement $2n + n^3$ is divisible by $n$ is true for $n = 1$.

**Step 2.** Assume that $2k + k^3$ is divisible by 3. Induction hypothesis

This means that $\dfrac{2k + k^3}{3} = m$, an integer. Then expanding $(k + 1)^3$, we obtain

$$\begin{aligned} 2(k + 1) + (k + 1)^3 &= 2k + 2 + (k^3 + 3k^2 + 3k + 1) \\ &= k^3 + 2k + 3k^2 + 3k + 3 \\ &= k^3 + 2k + 3(k^2 + k + 1) \end{aligned}$$

Then

$$\begin{aligned} \frac{2(k + 1) + (k + 1)^3}{3} &= \frac{k^3 + 2k}{3} + \frac{3(k^2 + k + 1)}{3} \\ &= m + k^2 + k + 1 = \text{an integer} \end{aligned}$$

Thus $2(k + 1) + (k + 1)^3$ is divisible by 3. This shows that the statement is true for $n = k + 1$.

**EXAMPLE 6** Let $A_1, A_2, \ldots, A_k$ be $k$ invertible $n \times n$ matrices. Show that

$$(A_1 A_2 \cdots A_m)^{-1} = A_m^{-1} A_{m-1}^{-1} \cdots A_2^{-1} A_1^{-1} \qquad \textbf{(6)}$$

For $m = 2$ we have $(A_1A_2)^{-1} = A_2^{-1}A_1^{-1}$ by Theorem 1.8.3. Thus equation (6) holds for $m = 2$. We assume it is true for $m = k$ and prove it for $m = k + 1$. Let $B = A_1A_2 \cdots A_k$. Then

$$(A_1A_2 \cdots A_kA_{k+1})^{-1} = (BA_{k+1})^{-1} = A_{k+1}^{-1}B^{-1} \tag{7}$$

But by the induction assumption

$$B^{-1} = (A_1A_2 \cdots A_k)^{-1} = A_k^{-1}A_{k-1}^{-1} \cdots A_2^{-1}A_1^{-1} \tag{8}$$

Substituting (8) into (7) completes the proof. ■

## Focus on . . .

### Mathematical Induction

The first mathematician to give a formal proof by the explicit use of mathematical induction was the Italian clergyman Franciscus Maurolicus (1494–1575), who was the abbot of Messina in Sicily, and is considered the greatest geometer of the sixteenth century. In his *Arithmetic,* published in 1575, Maurolicus used mathematical induction to prove, among other things, that for every positive integer $n$

$$1 + 3 + 5 + \cdots + (2n - 1) = n^2$$

You are asked to prove this in Problem 4.

The induction proofs of Maurolicus were given in a sketchy style that is difficult to follow. A clearer exposition of the method was given by the French mathematician Blaise Pascal (1623–1662). In his *Traité du Triangle Arithmétique,* published in 1662, Pascal proved a formula for the sum of binomial coefficients. He used his formula to develop what is today called the *Pascal triangle.*

Although the method of mathematical induction was used formally in 1575, the term *mathematical induction* was not used until 1838. In that year, one of the originators of set theory, Augustus de Morgan (1806–1871), published an article in the *Penny Cyclopedia* (London) entitled "Induction (Mathematics)." At the end of that article, he used the term we use today. However, the term did not enjoy widespread use until the 20th century.

## PROBLEMS A1

In Problems 1–20 use mathematical induction to prove that the given formula holds for all $n = 1, 2, \ldots$ unless some other set of values is specified.

**1.** $2 + 4 + 6 + \cdots + 2n = n(n + 1)$

**2.** $1 + 4 + 7 + \cdots + (3n - 2) = \dfrac{n(3n - 1)}{2}$

**3.** $2 + 5 + 8 + \cdots + (3n - 1) = \dfrac{n(3n + 1)}{2}$

**4.** $1 + 3 + 5 + \cdots + (2n - 1) = n^2$

**5.** $\left(\dfrac{1}{2}\right)^n < \dfrac{1}{n}$

**6.** $2^n < n!$ for $n = 4, 5, 6, \ldots,$ where

$$n! = 1 \cdot 2 \cdot 3 \cdots (n - 1) \cdot n$$

**7.** $1 + 2 + 4 + 8 + \cdots + 2^n = 2^{n+1} - 1$

**8.** $1 + 3 + 9 + 27 + \cdots + 3^n = \dfrac{3^{n+1} - 1}{2}$

**9.** $1 + \dfrac{1}{2} + \dfrac{1}{4} + \cdots + \dfrac{1}{2^n} = 2 - \dfrac{1}{2^n}$

**10.** $1 - \dfrac{1}{3} + \dfrac{1}{9} - \cdots + \left(-\dfrac{1}{3}\right)^n = \dfrac{3}{4}\left[1 - \left(-\dfrac{1}{3}\right)^{n+1}\right]$

**11.** $1^3 + 2^3 + 3^3 + \cdots + n^3 = \dfrac{n^2(n + 1)^2}{4}$

**12.** $1 \cdot 2 + 2 \cdot 3 + 3 \cdot 4 + \cdots + n(n + 1) = \dfrac{n(n + 1)(n + 2)}{3}$

**13.** $1 \cdot 2 + 3 \cdot 4 + 5 \cdot 6 + \cdots + (2n - 1)(2n) = \dfrac{n(n + 1)(4n - 1)}{3}$

**14.** $\dfrac{1}{2^2 - 1} + \dfrac{1}{3^2 - 1} + \dfrac{1}{4^2 - 1} + \cdots + \dfrac{1}{(n + 1)^2 - 1} = \dfrac{3}{4} - \dfrac{1}{2(n + 1)} - \dfrac{1}{2(n + 2)}$

**15.** $n + n^2$ is even.

**16.** $n < \dfrac{n^2 - n}{12} + 2$ if $n > 10$.

**17.** $n(n^2 + 5)$ is divisible by 6.

***18.** $3n^5 + 5n^3 + 7n$ is divisible by 15.

***19.** $x^n - 1$ is divisible by $x - 1$.

***20.** $x^n - y^n$ is divisible by $x - y$.

***21.** Give a formal proof that $(ab)^n = a^n b^n$ for every positive integer $n$.

**22.** Assuming that every polynomial has at least one complex root, prove that a polynomial of degree $n$ has exactly $n$ roots (counting multiplicities).

**23.** Given that $\det AB = \det A \det B$ for all $n \times n$ matrices $A$ and $B$, prove that $\det A_1 A_2 \cdots A_m = \det A_1 \det A_2 \cdots \det A_m$, where $A_1, \ldots, A_m$ are $n \times n$ matrices.

**24.** If $A_1, A_2, \ldots, A_k$ are $m \times n$ matrices, show that $(A_1 + A_2 + \cdots + A_k)^t = A_1^t + A_2^t + \cdots + A_k^t$. You may assume that $(A + B)^t = A^t + B^t$.

**25.** Prove that there are exactly $2^n$ subsets of a set containing $n$ elements.

**26.** Prove that if $2k - 1$ is an even integer for some integer $k$, then $2(k + 1) - 1 = 2k + 2 - 1 = 2k + 1$ is also an even integer. What, if anything, can you conclude by the proof?

**27.** What is wrong with the following proof that each horse in a set of $n$ horses has the same color as every other horse in the set?

**Step 1.** It is true for $n = 1$ since there is only one horse in the set and it obviously has the same color as itself.

**Step 2.** Suppose it is true for $n = k$. That is, each horse in a set containing $k$ horses is the same color as every other horse in the set. Let $h_1, h_2, \ldots, h_k, h_{k+1}$ denote the $k + 1$ horses in a set S. Let $S_1 = \{h_1, h_2, \ldots, h_k\}$ and $S_2 = \{h_2, h_3, \ldots, h_k, h_{k+1}\}$. Then both $S_1$ and $S_2$ contain $k$ horses, so the horses in each set are of the same color. We write $h_i = h_j$ to indicate that horse $i$ has the same color as horse $j$. Then we have

$$h_1 = h_2 = h_3 = \cdots = h_k$$

and

$$h_2 = h_3 = h_4 = \cdots = h_k = h_{k+1}$$

This means that

$$h_1 = h_2 = h_3 = \cdots = h_k = h_{k+1}$$

so all the horses in S have the same color. This proves the statement in the case $n = k + 1$, so the statement is true for all $n$.

# Appendix 2 Complex Numbers

In Chapter 6 we encountered the problem of finding the roots of the polynomial

$$\lambda^2 + b\lambda + c = 0 \tag{1}$$

To find the roots, we use the quadratic formula to obtain

$$\lambda = \frac{-b \pm \sqrt{b^2 - 4c}}{2} \tag{2}$$

If $b^2 - 4c > 0$, there are two real roots. If $b^2 - 4c = 0$, we obtain the single root (of multiplicity 2) $\lambda = -b/2$. To deal with the case $b^2 - 4c < 0$, we introduce the **imaginary unit:**†

$$\boxed{i = \sqrt{-1}} \tag{3}$$

† You should not be troubled by the term "imaginary." It's just a name. The British mathematician Alfred North Whitehead, in the chapter on imaginary numbers in his *Introduction to Mathematics,* wrote:

> At this point it may be useful to observe that a certain type of intellect is always worrying itself and others by discussion as to the applicability of technical terms. Are the incommensurable numbers properly called numbers? Are the positive and negative numbers really numbers? Are the imaginary numbers imaginary, and are they numbers?—are types of such futile questions. Now, it cannot be too clearly understood that, in science, technical terms are names arbitrarily assigned, like Christian names to children. There can be no question of the names being right or wrong. They may be judicious or injudicious; for they can sometimes be so arranged as to be easy to remember, or so as to suggest relevant and important ideas. But the essential principle involved was quite clearly enunciated in Wonderland to Alice by Humpty Dumpty, when he told her, apropos of his use of words, "I pay them extra and make them mean what I like." So we will not bother as to whether imaginary numbers are imaginary, or as to whether they are numbers, but will take the phrase as the arbitrary name of a certain mathematical idea, which we will now endeavour to make plain.

so $i^2 = -1$. Then for $b^2 - 4c < 0$

$$\sqrt{b^2 - 4c} = \sqrt{(4c - b^2)(-1)} = \sqrt{4c - b^2}\,i$$

and the two roots of (1) are given by

$$\lambda_1 = -\frac{b}{2} + \frac{\sqrt{4c - b^2}}{2}i \quad \text{and} \quad \lambda_2 = -\frac{b}{2} - \frac{\sqrt{4c - b^2}}{2}i$$

**EXAMPLE 1** Find the roots of the quadratic equation $\lambda^2 + 2\lambda + 5 = 0$.

**Solution** We have $b = 2$, $c = 5$, and $b^2 - 4c = -16$. Thus $\sqrt{b^2 - 4c} = \sqrt{-16} = \sqrt{16}\sqrt{-1} = 4i$ and the roots are

$$\lambda_1 = \frac{-2 + 4i}{2} = -1 + 2i \quad \text{and} \quad \lambda_2 = -1 - 2i$$

**DEFINITION 1** A **complex number** is an expression of the form

$$\boxed{z = \alpha + i\beta} \tag{4}$$

where $\alpha$ and $\beta$ are real numbers. $\alpha$ is called the **real part** of $z$ and is denoted by Re $z$. $\beta$ is called the **imaginary part** of $z$ and is denoted by Im $z$. Representation (4) is sometimes called the **Cartesian form** of the complex number $z$.

***Remark.*** If $\beta = 0$ in equation (4), then $z = \alpha$ is a real number. In this context we can regard the set of real numbers as a subset of the set of complex numbers.

**EXAMPLE 2** In Example 1, Re $\lambda_1 = -1$ and Im $\lambda_1 = 2$.

We can add and multiply complex numbers by using standard rules of algebra.

**EXAMPLE 3** Let $z = 2 + 3i$ and $w = 5 - 4i$. Calculate **(i)** $z + w$, **(ii)** $3w - 5z$, and **(iii)** $zw$.

**Solution**

**i.** $z + w = (2 + 3i) + (5 - 4i) = (2 + 5) + (3 - 4)i = 7 - i$.

**ii.** $3w = 3(5 - 4i) = 15 - 12i$; $5z = 10 + 15i$; *and* $3w - 5z = (15 - 12i) - (10 + 15i) = (15 - 10) + i(-12 - 15) = 5 - 27i$.

**iii.** $zw = (2 + 3i)(5 - 4i) = (2)(5) + 2(-4i) + (3i)(5) + (3i)(-4i) = 10 - 8i + 15i - 12i^2 = 10 + 7i + 12 = 22 + 7i$. Here we used the fact that $i^2 = -1$.

We can plot a complex number $z$ in the $xy$-plane by plotting Re $z$ along the $x$-axis and Im $z$ along the $y$-axis. Thus each complex number can be thought of as a point in the $xy$-plane. With this representation the $xy$-plane is called the **complex plane.** Some representative points are plotted in Figure A.1.

Complex plane

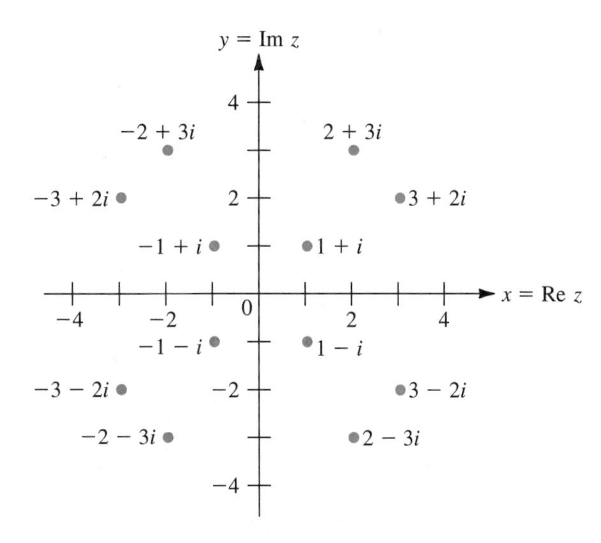


**Figure A.1** Twelve points in the complex plane

Conjugate

If $z = \alpha + i\beta$, then we define the **conjugate** of $z$, denoted by $\bar{z}$, by

$$\boxed{\bar{z} = \alpha - i\beta} \tag{5}$$

Figure A.2 depicts a representative value of $z$ and $\bar{z}$.

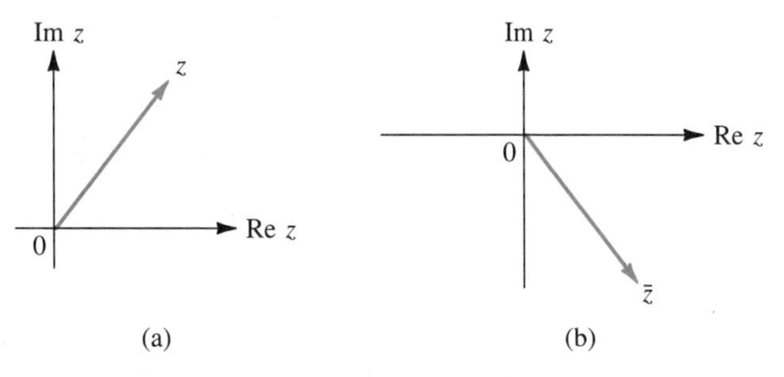


**Figure A.2** $\bar{z}$ is obtained by reflecting $z$ about the real axis

**EXAMPLE 4** Compute the conjugate of **(i)** $1 + i$, **(ii)** $3 - 4i$, **(iii)** $-7 + 5i$, and **(iv)** $-3$.

**Solution** **(i)** $\overline{1 + i} = 1 - i$; **(ii)** $\overline{3 - 4i} = 3 + 4i$; **(iii)** $\overline{-7 + 5i} = -7 - 5i$; **(iv)** $\overline{-3} = -3$.

It is not difficult to show (see Problem 35) that

$$\bar{z} = z \qquad \text{if and only if } z \text{ is real} \tag{6}$$

Imaginary number

If $z = \beta i$ with $\beta$ real, then $z$ is said to be **imaginary.** We can then show (see Problem 36) that

$$\bar{z} = -z \qquad \text{if and only if } z \text{ is imaginary} \tag{7}$$

Let $p_n(x) = a_0 + a_1x + a_2x^2 + \cdots + a_nx^n$ be a polynomial with real coefficients. Then it can be shown (see Problem 41) that the complex roots of the equation $p_n(x) = 0$ occur in complex conjugate pairs. That is, if $z$ is a root of $p_n(x) = 0$, then so is $\bar{z}$. We saw this fact illustrated in Example 1 in the case $n = 2$.

Magnitude

For $z = \alpha + i\beta$ we define the **magnitude**† of $z$, denoted by $|z|$, by

$$\text{Magnitude of } z = |z| = \sqrt{\alpha^2 + \beta^2} \tag{8}$$

Argument

and we define the **argument** of $z$, denoted by arg $z$, as the angle $\theta$ between the line $0z$ and the positive $x$-axis. It is conventional to take

$$-\pi < \arg z \leq \pi$$

From Figure A.3 we see that $r = |z|$ is the distance from $z$ to the origin. If $\alpha > 0$, then

$$\theta = \arg z = \tan^{-1} \frac{\beta}{\alpha}$$

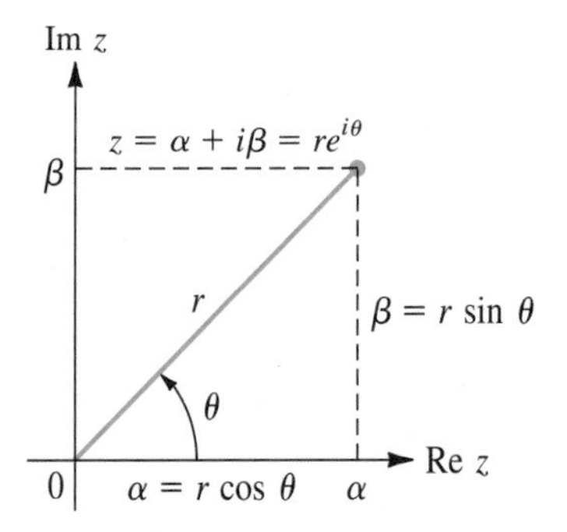


**Figure A.3** If $z = \alpha + i\beta$, then $\alpha = r\cos\theta$ and $\beta = r\sin\theta$

†The magnitude of a complex number is often referred to as its **modulus.**

where we observe the convention that $\tan^{-1} x$ takes values in the interval $\left(-\frac{\pi}{2}, \frac{\pi}{2}\right)$. If $\alpha = 0$ and $\beta > 0$, then $\theta = \arg z = \frac{\pi}{2}$. If $\alpha = 0$ and $\beta < 0$, then $\theta = \arg z = -\frac{\pi}{2}$. If $\alpha < 0$ and $\beta > 0$, then $\theta$ is in the second quadrant and is given by

$$\theta = \arg z = \pi - \tan^{-1}\left|\frac{\beta}{\alpha}\right|$$

Finally, if $\alpha < 0$ and $\beta < 0$, then $\theta$ is the third quadrant and

$$\theta = \arg z = -\pi + \tan^{-1}\frac{\beta}{\alpha}$$

In sum, we have

**Argument of $z$**

Let $z = \alpha + \beta i$. Then

$$\arg z = \tan\frac{\beta}{\alpha} \quad \text{if } \alpha > 0$$

$$\arg z = \frac{\pi}{2} \quad \text{if } \alpha = 0 \text{ and } \beta > 0$$

$$\arg z = -\frac{\pi}{2} \quad \text{if } \alpha = 0 \text{ and } \beta < 0 \tag{9}$$

$$\arg z = \pi - \tan^{-1}\left|\frac{\beta}{\alpha}\right| \quad \text{if } \alpha < 0 \text{ and } \beta > 0$$

$$\arg z = -\pi + \tan^{-1}\frac{\beta}{\alpha} \quad \text{if } \alpha < 0 \text{ and } \beta < 0 \tag{10}$$

arg 0 is not defined.

From Figure A.4 we see that

$$|\bar{z}| = |z| \tag{11}$$

and

$$\arg \bar{z} = -\arg z \tag{12}$$

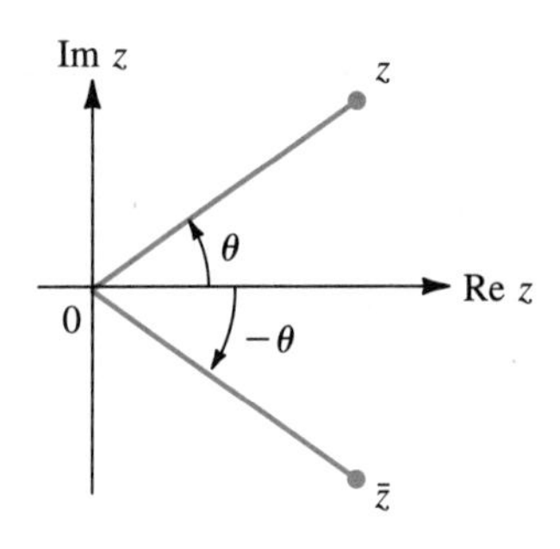


**Figure A.4** $\arg \bar{z} = -\arg z$

We can use $|z|$ and $\arg z$ to describe what is often a more convenient way to represent complex numbers.† From Figure A.3 it is evident that if $z = \alpha + i\beta$, $r = |z|$, and $\theta = \arg z$, then

$$\boxed{\alpha = r\cos\theta \quad \text{and} \quad \beta = r\sin\theta} \tag{13}$$

We shall see at the end of this appendix that

$$\boxed{e^{i\theta} = \cos\theta + i\sin\theta} \tag{14}$$

Since $\cos(-\theta) = \cos\theta$ and $\sin(-\theta) = -\sin\theta$, we also have

$$\boxed{e^{-i\theta} = \cos(-\theta) + i\sin(-\theta) = \cos\theta - i\sin\theta} \tag{14'}$$

Formula (14) is called **Euler's formula.**‡ Using Euler's formula and equation (13), we have

$$z = \alpha + i\beta = r\cos\theta + ir\sin\theta = r(\cos\theta + i\sin\theta)$$

or

$$\boxed{z = re^{i\theta}} \tag{15}$$

Polar form

Representation (15) is called the **polar form** of the complex number $z$.

---

†Those of you who have studied polar coordinates will find this representation very familiar.

‡Named for the great Swiss mathematician Leonhard Euler (1707–1783).

**EXAMPLE 5** Determine the polar forms of the following complex numbers: **(i)** 1, **(ii)** $-1$, **(iii)** $i$, **(iv)** $1 + i$, **(v)** $-1 - \sqrt{3}i$, and **(vi)** $-2 + 7i$.

**Solution** The six points are plotted in Figure A.5.

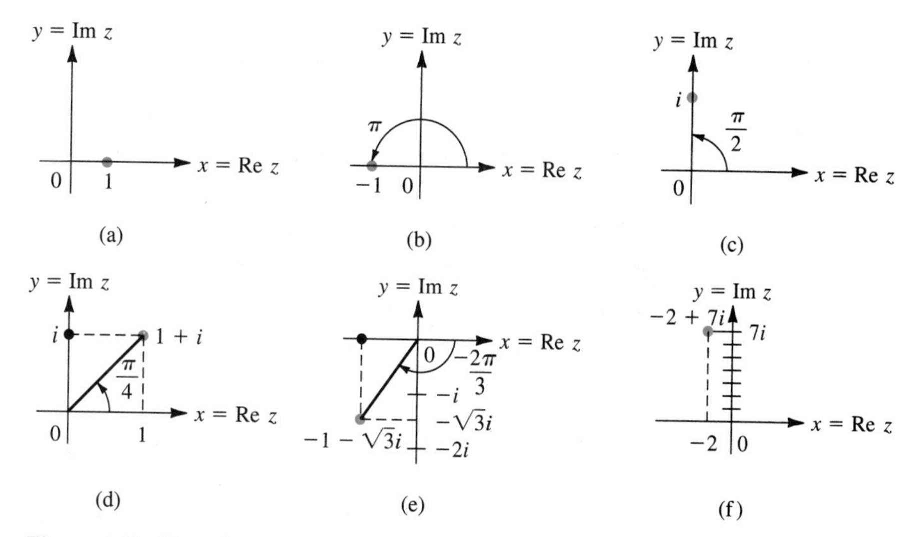


**Figure A.5** Six points in the complex plane

**i.** From Figure A.5*a* it is clear that arg $1 = 0$. Since Re $1 = 1$, we see that, in polar form, $1 = 1e^{i0} = 1e^0 = 1$.

**ii.** Since $\arg(-1) = \pi$ (Figure A.5*b*) and $|-1| = 1$, we have

$$-1 = 1e^{\pi i} = e^{i\pi}$$

**iii.** From Figure A.5*c* we see that $\arg i = \pi/2$. Since $|i| = \sqrt{0^2 + 1^2} = 1$, it follows that

$$i = e^{i\pi/2}$$

**iv.** $\arg(1 + i) = \tan^{-1}(1/1) = (\pi/4)$ and $|1 + i| = \sqrt{1^2 + 1^2} = \sqrt{2}$ so that

$$1 + i = \sqrt{2}e^{i\pi/4}$$

**v.** Here $\tan^{-1}(\beta/\alpha) = \tan^{-1}\sqrt{3} = \pi/3$. However, arg $z$ is in the third quadrant, so $\theta = \pi/3 - \pi = -2\pi/3$. Also, $|-1 - \sqrt{3}| = \sqrt{1^2 + (\sqrt{3})^2} = \sqrt{1 + 3} = 2$ so that

$$-1 - \sqrt{3} = 2e^{-2\pi i/3}$$

**vi.** To compute this we need a calculator. We find that, in radians,

$$\arg z = \tan^{-1}(-\tfrac{7}{2}) = \tan^{-1}(-3.5) \approx -1.2925$$

But $\tan^{-1} x$ is defined as a number in the interval $(-\pi/2, \pi/2)$. Since, from Figure A.5*f*, $\theta$ is in the second quadrant, we see that $\arg z = \pi - \tan^{-1}(3.5) \approx 1.8491$. Next, we see that

$$|-2 + 7i| = \sqrt{(-2)^2 + 7^2} = \sqrt{53}$$

Hence

$$-2 + 7i \approx \sqrt{53}e^{1.8491i}$$

**EXAMPLE 6** Convert the following complex numbers from polar to Cartesian form: **(i)** $2e^{i\pi/3}$; **(ii)** $4e^{3\pi i/2}$.

**Solution**

**i.** $e^{i\pi/3} = \cos \pi/3 + i \sin \pi/3 = \frac{1}{2} + (\sqrt{3}/2)i$. Thus $2e^{i\pi/3} = 1 + \sqrt{3}i$.

**ii.** $e^{3\pi i/2} = \cos 3\pi/2 + i \sin 3\pi/2 = 0 + i(-1) = -i$. Thus $4e^{3\pi i/2} = -4i$.

If $\theta = \arg z$, then by equation (12), $\arg \bar{z} = -\theta$. Thus since $|\bar{z}| = |z|$;

$$\text{If } z = re^{i\theta}, \quad \text{then } \bar{z} = re^{-i\theta} \tag{16}$$

Suppose we write a complex number in its polar form $z = re^{i\theta}$. Then

$$z^n = (re^{i\theta})^n = r^n(e^{i\theta})^n = r^n e^{in\theta} = r^n(\cos n\theta + i \sin n\theta) \tag{17}$$

Formula (17) is useful for a variety of computations. In particular, when $r = |z| = 1$, we obtain the **De Moivre formula.**†

**De Moivre Formula**

$$(\cos \theta + i \sin \theta)^n = \cos n\theta + i \sin n\theta \tag{18}$$

**EXAMPLE 7** Compute $(1 + i)^5$.

**Solution** In Example 5(**iv**) we showed that $1 + i = \sqrt{2}e^{\pi i/4}$. Then

$$(1 + i)^5 = (\sqrt{2}e^{\pi i/4})^5 = (\sqrt{2})^5 e^{5\pi i/4} = 4\sqrt{2}\left(\cos \frac{5\pi}{4} + i \sin \frac{5\pi}{4}\right)$$

$$= 4\sqrt{2}\left(-\frac{1}{\sqrt{2}} - \frac{1}{\sqrt{2}}i\right) = -4 - 4i$$

† Abraham De Moivre (1667–1754) was a French mathematician who was well known for his work in probability theory, infinite series, and trigonometry. He was so highly regarded that Newton often told those who came to him with questions on mathematics, "Go to M. De Moivre; he knows these things better than I do."

This can be checked by direct calculation. If the direct calculation seems no more difficult, then try to compute $(1 + i)^{20}$ directly. Proceeding as above, we obtain

$$\begin{aligned}(1 + i)^{20} &= (\sqrt{2})^{20}e^{20\pi i/4} = 2^{10}(\cos 5\pi + i \sin 5\pi) \\ &= 2^{10}(-1 + 0) = -1024\end{aligned}$$

**Proof of Euler's Formula**

We shall show that

$$e^{i\theta} = \cos\theta + i\sin\theta \qquad \textbf{(19)}$$

by using power series. If these are unfamiliar to you, skip the proof. We have

$$e^x = 1 + x + \frac{x^2}{2!} + \frac{x^3}{3!} + \cdots \qquad \textbf{(20)}$$

$$\sin x = x - \frac{x^3}{3!} + \frac{x^5}{5!} - \cdots \qquad \textbf{(21)}$$

$$\cos x = 1 - \frac{x^2}{2!} + \frac{x^4}{4!} - \cdots \qquad \textbf{(22)}$$

Although we do not prove this here, these three series converge for every complex number $x$. Then

$$e^{i\theta} = 1 + (i\theta) + \frac{(i\theta)^2}{2!} + \frac{(i\theta)^3}{3!} + \frac{(i\theta)^4}{4!} + \frac{(i\theta)^5}{5!} + \cdots \qquad \textbf{(23)}$$

Now $i^2 = -1$, $i^3 = -i$, $i^4 = 1$, $i^5 = i$, and so on. Thus (23) can be written

$$\begin{aligned}e^{i\theta} &= 1 + i\theta - \frac{\theta^2}{2!} - \frac{i\theta^3}{3!} + \frac{\theta^4}{4!} + \frac{i\theta^5}{5!} - \cdots \\ &= \left(1 - \frac{\theta^2}{2!} + \frac{\theta^4}{4!} - \cdots\right) + i\left(\theta - \frac{\theta^3}{3!} + \frac{\theta^5}{5!} - \cdots\right) \\ &= \cos\theta + i\sin\theta\end{aligned}$$

This completes the proof.

## PROBLEMS A2

In Problems 1–5 perform the indicated operation.

**1.** $(2 - 3i) + (7 - 4i)$

**2.** $3(4 + i) - 5(-3 + 6i)$

**3.** $(1 + i)(1 - i)$

**4.** $(2 - 3i)(4 + 7i)$

**5.** $(-3 + 2i)(7 + 3i)$

In Problems 6–15 convert the complex number to its polar form.

**6.** $5i$

**7.** $5 + 5i$

**8.** $-2 - 2i$

**9.** $3 - 3i$

**10.** $2 + 2\sqrt{3}i$

**11.** $3\sqrt{3} + 3i$

**12.** $1 - \sqrt{3}i$

**13.** $4\sqrt{3} - 4i$

**14.** $-6\sqrt{3} - 6i$

**15.** $-1 - \sqrt{3}i$

In Problems 16–25 convert from polar to Cartesian form.

**16.** $e^{3\pi i}$ **17.** $2e^{-7\pi i}$ **18.** $\frac{1}{2}e^{3\pi i/4}$ **19.** $\frac{1}{2}e^{-3\pi i/4}$

**20.** $6e^{\pi i/6}$ **21.** $4e^{5\pi i/6}$ **22.** $4e^{-5\pi i/6}$ **23.** $3e^{-2\pi i/3}$

**24.** $\sqrt{3}e^{23\pi i/4}$ **25.** $e^{i}$

In Problems 26–34 compute the conjugate of the given number.

**26.** $3 - 4i$ **27.** $4 + 6i$ **28.** $-3 + 8i$

**29.** $-7i$ **30.** $16$ **31.** $2e^{\pi i/7}$

**32.** $4e^{3\pi i/5}$ **33.** $3e^{-4\pi i/11}$ **34.** $e^{0.012i}$

**35.** Show that $z = \alpha + i\beta$ is real if and only if $z = \bar{z}$. [*Hint:* If $z = \bar{z}$, show that $\beta = 0$.]

**36.** Show that $z = \alpha + i\beta$ is imaginary if and only if $z = -\bar{z}$. [*Hint:* If $z = -\bar{z}$, show that $\alpha = 0$.]

**37.** For any complex number $z$, show that $z\bar{z} = |z|^2$.

**38.** Show that the circle of radius 1 centered at the origin (the *unit circle*) is the set of points in the complex plane that satisfy $|z| = 1$.

**39.** For any complex number $z_0$ and positive real number $a$, describe $\{z: |z - z_0| = a\}$.

**40.** Describe $\{z: |z - z_0| \le a\}$, where $z_0$ and $a$ are as in Problem 39.

***41.** Let $p(\lambda) = \lambda^n + a_{n-1}\lambda^{n-1} + a_{n-2}\lambda^{n-2} + \cdots + a_1\lambda + a_0$ with $a_0, a_1, \ldots, a_{n-1}$ real numbers. Show that if $p(z) = 0$, then $p(\bar{z}) = 0$. That is: *The roots of polynomials with real coefficients occur in complex conjugate pairs.* [*Hint:* $0 = \bar{0}$; compute $\overline{p(z)}$.]

**42.** Derive expressions for $\cos 4\theta$ and $\sin 4\theta$ by comparing the De Moivre formula and the expansion of $(\cos\theta + i\sin\theta)^4$.

**43.** Prove De Moivre's formula by mathematical induction. [*Hint:* Recall the trigonometric identities $\cos(x + y) = \cos x \cos y - \sin x \sin y$ and $\sin(x + y) = \sin x \cos y + \cos x \sin y$.]

# Appendix 3 The Error in Numerical Computations and Computational Complexity

In every chapter of this book we have performed numerical computations. We have, among other things, solved linear equations, multiplied and inverted matrices, found bases, and computed eigenvalues and eigenvectors. With few exceptions, the examples involved $2 \times 2$ and $3 \times 3$ matrices—not because most applications have only two or three variables but because the computations would have been too tedious otherwise.

With the recent and widespread use of calculators and computers, the situation has been altered. The remarkable strides made in the last few years in the theory of numerical methods for solving certain computational problems have made it possible to perform, quickly and accurately, the calculations mentioned in the first paragraph on high-order matrices.

The use of the computer presents new difficulties, however. Computers do not store numbers such as $\frac{2}{3}$, $7\frac{3}{7}$, $\sqrt{2}$, and $\pi$. Rather, every digital computer uses what is called *floating-point arithmetic*. In this system every number is represented in the form

$$x = \pm 0.d_1 d_2 \cdots d_k \times 10^n \tag{1}$$

where $d_1, d_2, \ldots, d_k$ are single-digit positive integers and $n$ is an integer. Any number written in this form is called a *floating-point number*. In equation (1) the number $\pm 0.d_1 d_2 \cdots d_k$ is called the *mantissa* and the number $n$ is called the *exponent*. The number $k$ is called the *number of significant digits* in the expression.

Different computers have different capabilities in the range of numbers expressible in the form of equation (1). Digits are usually represented in binary rather than decimal form. One popular computer, for example, carries 28 binary digits. Since

$2^{28} = 268{,}435{,}456$, we can use the 28 binary digits to represent any eight-digit number. Hence $k = 8$.

EXAMPLE 1 **The Floating-Point Form of Four Numbers** The following numbers are expressed in floating-point form:

**i.** $\frac{1}{4} = 0.25$

**ii.** $2378 = 0.2378 \times 10^4$

**iii.** $-0.000816 = -0.816 \times 10^{-3}$

**iv.** $83.27 = 0.8327 \times 10^2$

If the number of significant digits were unlimited, then we would have no problem. Almost every time numbers are introduced into the computer, however, errors begin to accumulate. This can happen in one of two ways:

**i. Truncation.** All significant digits after $k$ digits are simply "cut off." For example, if truncation is used, $\frac{2}{3} = 0.666666\ldots$ is stored (with $k = 8$) as $\frac{2}{3} = 0.66666666 \times 10^0$.

**ii. Rounding.** If $d_{k+1} \geq 5$, then 1 is added to $d_k$ and the resulting number is truncated. Otherwise the number is simply truncated. For example, with rounding (and $k = 8$), $\frac{2}{3}$ is stored as $\frac{2}{3} = 0.66666667 \times 10^0$.

EXAMPLE 2 **Illustration of Truncation and Rounding** We can illustrate how some numbers are stored with truncation and rounding by using eight significant digits:

| Number | Truncated Number | Rounded Number |
|---|---|---|
| $\frac{8}{3}$ | $0.26666666 \times 10^1$ | $0.26666667 \times 10^1$ |
| $\pi$ | $0.31415926 \times 10^1$ | $0.31415927 \times 10^1$ |
| $-\frac{1}{57}$ | $-0.17543859 \times 10^{-1}$ | $-0.17543860 \times 10^{-1}$ |

Individual round-off or truncation errors do not seem very significant. When thousands of computational steps are involved, however, the **accumulated** round-off error can be devastating. Thus in discussing any numerical scheme, it is necessary to know not only whether you will get the right answer, theoretically, but also how badly the round-off errors will accumulate. To keep track of things, we define two types of error. If $x$ is the actual value of a number and if $x^*$ is the number that appears in the computer, then the **absolute error** $\epsilon_a$ is defined by

Absolute error

$$\epsilon_a = |x^* - x| \tag{2}$$

Relative error

More interesting in most situations is the **relative error** $\epsilon_r$, defined by

$$\boxed{\epsilon_r = \left|\frac{x^* - x}{x}\right|} \tag{3}$$

**EXAMPLE 3** **Illustration of Relative Error** Let $x = 2$ and $x^* = 2.1$. Then $\epsilon_a = 0.1$ and $\epsilon_r = 0.1/2 = 0.05$. If $x_1 = 2000$ and $x_1^* = 2000.1$, then again $\epsilon_a = 0.1$. But now $\epsilon_r = 0.1/2000 = 0.00005$. Most people would agree that the 0.1 error in the first case is more significant than the 0.1 error in the second.

Much of numerical analysis is concerned with questions of **convergence** and **stability.** If $x$ is the exact solution to a problem and our computational method gives us approximating values $x_n$, then the method converges if, theoretically, $x_n$ approaches $x$ as $n$ gets large. If, moreover, it can be shown that the round-off errors will not accumulate in such a way as to make the answer very inaccurate, then the method is **stable.**

It is easy to give an example of a procedure in which round-off error can be quite large. Suppose we wish to compute $y = 1/(x - 0.66666665)$. For $x = \frac{2}{3}$, if the computer truncates, then $x = 0.66666666$ and $y = 1/0.00000001 = 10^8 = 10 \times 10^7$. If the computer rounds, then $x = 0.66666667$ and $y = 1/0.00000002 = 5 \times 10^7$. The difference here is enormous. The exact solution is $1/(\frac{2}{3} - \frac{66666665}{100{,}000{,}000}) = 1/(\frac{200{,}000{,}000}{300{,}000{,}000} - \frac{199{,}999{,}995}{300{,}000{,}000}) = 1/\frac{5}{300{,}000{,}000} = \frac{300{,}000{,}000}{5} = 60{,}000{,}000 = 6 \times 10^7$.

*Note.* We will not worry further about stability here. However, the people who design mathematical software do worry a great deal about it. You should know that the numerical analysts who design this software choose algorithms (or develop new ones) that tend to minimize the adverse consequences. In particular, MATLAB uses very high quality code. Today no well-informed amateur writes his or her own numerical software. We get subroutines from the pros.

## Computational Complexity

In solving problems on a computer, two questions naturally arise:

How accurate are my answers?

How much time will it take?

(After all, you pay by the hour on a computer.)

We discussed the first question in the first part of this section. To answer the second one, we must estimate the number of steps required to carry out a certain computation. The **computational complexity** of a problem is a measure of the number of arithmetic operations needed to solve the problem and the time needed to carry out all the needed operations.

There are two basic arithmetic operations carried out on a computer:

| Operation | Average Time (in microseconds)† |
|---|---|
| Addition or subtraction | $\frac{1}{2}$ microsecond |
| Multiplication or division | 2 microseconds |

† 1 microsecond = 1 millionth of a second = $10^{-6}$ second.

Thus in order to estimate the time needed to solve a problem on a computer, it is first necessary to count the number of additions, subtractions, multiplications, and divisions involved in solving the problem.

Counting the number of operations needed to solve a problem is often very difficult. We illustrate how it can be done in the case of Gauss-Jordan elimination. To simplify matters, we treat addition and subtraction as the same operation and multiplication and division as the same (although, in fact, each division takes three times as long as a multiplication; the average time for both is 2 microseconds).

**EXAMPLE 4** **Counting Additions and Multiplications in Gauss-Jordan Elimination** Let $A$ be an invertible $n \times n$ matrix. Estimate the number of additions and multiplications needed to solve the system $A\mathbf{x} = \mathbf{b}$ by Gauss-Jordan elimination.

**Solution** We begin, as in Section 1.3, by writing the system in the augmented-matrix form

$$\left(\begin{array}{cccc|c} a_{11} & a_{12} & \cdots & a_{1n} & b_1 \\ a_{21} & a_{22} & \cdots & a_{2n} & b_2 \\ \vdots & \vdots & & \vdots & \vdots \\ a_{n1} & a_{n2} & \cdots & a_{nn} & b_n \end{array}\right)$$

Since $A$ is invertible by assumption, its reduced row echelon form is the $n \times n$ identity matrix. We assume that in the row reduction no rows are permuted (interchanged) since such an interchange does not involve any additions or multiplications. Moreover, keeping track of row numbers is a bookkeeping task that requires considerably less time than an addition.

To keep track of which numbers are being computed during a given step, we write the augmented matrix with $C$'s and $L$'s. A $C$ denotes a number just computed. An $L$ denotes a number to be left alone.

**Step 1.** Multiply each number in the first row by $1/a_{11}$ to obtain

$$\overbrace{\left(\begin{array}{cccccc|c} 1 & C & C & \cdots & C & C & C \\ L & L & L & \cdots & L & L & L \\ \vdots & \vdots & \vdots & & \vdots & \vdots & \vdots \\ L & L & L & \cdots & L & L & L \end{array}\right)}^{n+1 \text{ columns}}$$

**Total for Step 1**

$n$ multiplications

$\left(\dfrac{a_{11}}{a_{11}} = 1\right.$ requires no computation, the 1 is simply inserted in the 1,1 position.)

no additions

**Step 2.** Multiply row 1 by $-a_{i1}$ and add it to the $i$th row for $i = 2, 3, \ldots, n$:

$$\left(\begin{array}{cccccc|c} 1 & L & L & \cdots & L & L & L \\ 0 & C & C & \cdots & C & C & C \\ 0 & C & C & \cdots & C & C & C \\ \vdots & \vdots & \vdots & & \vdots & \vdots & \vdots \\ 0 & C & C & \cdots & C & C & C \end{array}\right)$$

Let us count the operations.

To obtain the new second row:

The 0 in the 2,1 position requires no work. We know that the number in the 2,1 position will be 0, so we simply place it there. There are $(n + 1) - 1 = n$ numbers in the second row that must be changed. For example, if we denote the new $a_{22}$ by $a'_{22}$, then

$$a'_{22} = a_{22} - a_{21}a_{12}$$

This requires 1 multiplication and 1 addition. Since there are $n$ numbers to be changed in the second row, there are $n$ multiplications and $n$ additions needed in the second row. The same is true in each of the $n - 1$ rows 2 through $n$. Thus

**Total for Step 2**

$(n - 1)n$ multiplications

$(n - 1)n$ additions

*Notation* From now on $a'_{ij}$ will now denote the latest entry in the $i$th row and $j$th column.

**Step 3.** Multiply everything in the second row by $1/a'_{22}$:

$$\left(\begin{array}{cccccc|c} 1 & L & L & \cdots & L & L & L \\ 0 & 1 & C & \cdots & C & C & C \\ 0 & L & L & \cdots & L & L & L \\ \vdots & \vdots & \vdots & & \vdots & \vdots & \vdots \\ 0 & L & L & \cdots & L & L & L \end{array}\right)$$

**Total for Step 3**

$n - 1$ multiplications. (As before, the 1 in the 2,2 position is simply placed there.)

no additions

**Step 4.** Multiply row 2 by $-a'_{i2}$ and add it to the $i$th row, for $i = 1, 3, 4, \ldots, n$:

$$\left(\begin{array}{cccccc|c} 1 & 0 & C & \cdots & C & C & C \\ 0 & 1 & L & \cdots & L & L & L \\ 0 & 0 & C & \cdots & C & C & C \\ \vdots & \vdots & \vdots & & \vdots & \vdots & \vdots \\ 0 & C & C & \cdots & C & C & C \end{array}\right)$$

**Total for Step 4**

$(n - 1)(n - 1)$ multiplications

$(n - 1)(n - 1)$ additions

As in Step 2, each change requires 1 multiplication and 1 addition. But now the first two components in each row require no computation; that is, $(n + 1) - 2 = n - 1$

numbers in each row are computed. As before, computations are done in $n-1$ rows. This explains the numbers above.

You should now see the pattern. In step 5 we shall have $n-2$ multiplications (divide each component in the third row, besides the first three, by $a'_{33}$). In step 6 there will be $n-2$ multiplications and $n-2$ additions needed in each of $n-1$ rows for a total of $(n-1)(n-2)$ multiplications and $(n-1)(n-2)$ additions. We continue in this manner until there are four steps to go. Here is how the augmented matrix will appear:

$$\left(\begin{array}{cccccc|c} 1 & 0 & 0 & \cdots & a'_{1,n-1} & a'_{1n} & b'_1 \\ 0 & 1 & 0 & \cdots & a'_{2,n-1} & a'_{2n} & b'_2 \\ 0 & 0 & 1 & \cdots & a'_{3,n-1} & a'_{3n} & b'_3 \\ \vdots & \vdots & \vdots & & \vdots & \vdots & \vdots \\ 0 & 0 & 0 & \cdots & a'_{n-1,n-1} & a'_{n-1,n} & b'_{n-1} \\ 0 & 0 & 0 & \cdots & a'_{n,n-1} & a'_{nn} & b'_n \end{array}\right)$$

**Last Step Minus 3.** Divide the $(n-1)$st row by $a'_{n-1,n-1}$:

$$\left(\begin{array}{cccccc|c} 1 & 0 & 0 & \cdots & L & L & L \\ 0 & 1 & 0 & \cdots & L & L & L \\ 0 & 0 & 1 & \cdots & L & L & L \\ \vdots & \vdots & \vdots & & \vdots & \vdots & \vdots \\ 0 & 0 & 0 & \cdots & 1 & C & C \\ 0 & 0 & 0 & \cdots & L & L & L \end{array}\right)$$

2 multiplications
no additions

**Last Step Minus 2.** Multiply the $(n-1)$st row by $-a'_{i,n-1}$ and add it to the $i$th row, for $i = 1, 2, \ldots, n-2, n$

$$\left(\begin{array}{cccccc|c} 1 & 0 & 0 & \cdots & 0 & C & C \\ 0 & 1 & 0 & \cdots & 0 & C & C \\ 0 & 0 & 1 & \cdots & 0 & C & C \\ \vdots & \vdots & \vdots & & \vdots & \vdots & \vdots \\ 0 & 0 & 0 & \cdots & 1 & L & L \\ 0 & 0 & 0 & \cdots & 0 & C & C \end{array}\right)$$

$2(n-1)$ multiplications
$2(n-1)$ additions

**Last Step Minus 1.** Divide the $n$th row by $a'_{nn}$:

$$\left(\begin{array}{cccccc|c} 1 & 0 & 0 & \cdots & 0 & L & L \\ 0 & 1 & 0 & \cdots & 0 & L & L \\ 0 & 0 & 1 & \cdots & 0 & L & L \\ \vdots & \vdots & \vdots & & \vdots & \vdots & \vdots \\ 0 & 0 & 0 & \cdots & 1 & L & L \\ 0 & 0 & 0 & \cdots & 0 & 1 & C \end{array}\right)$$

1 multiplication
no additions

**Last Step.** Multiply the $n$th row by $-a'_{in}$ and add it to the $i$th row, for $i = 1, 2, \ldots, n-1$:

$$\left(\begin{array}{cccccc|c} 1 & 0 & 0 & \cdots & 0 & 0 & C \\ 0 & 1 & 0 & \cdots & 0 & 0 & C \\ 0 & 0 & 1 & \cdots & 0 & 0 & C \\ \vdots & \vdots & \vdots & & \vdots & \vdots & \vdots \\ 0 & 0 & 0 & \cdots & 1 & 0 & C \\ 0 & 0 & 0 & \cdots & 0 & 1 & L \end{array}\right) \qquad \begin{array}{l} 1(n-1) \text{ multiplications} \\ 1(n-1) \text{ additions} \end{array}$$

We now find the totals:

For the odd-numbered steps there are

$$n + (n-1) + (n-2) + \cdots + 3 + 2 + 1 \text{ multiplications}$$

and

$$\text{no additions}$$

For the even-numbered steps there are

$$(n-1)[n + (n-1) + (n-2) + \cdots + 3 + 2 + 1] \text{ multiplications}$$

and

$$(n-1)[n + (n-1) + (n-2) + \cdots + 3 + 2 + 1] \text{ additions}$$

In Example 2 in Appendix 1 (page A-2) we prove that

$$1 + 2 + 3 + \cdots + n = \frac{n(n+1)}{2} \tag{4}$$

Thus the total number of multiplications is

$$\underbrace{\frac{n(n+1)}{2}}_{\text{From odd-numbered steps}} + \underbrace{(n-1)\left[\frac{n(n+1)}{2}\right]}_{\text{From even-numbered steps}}$$

$$= \left[\frac{n(n+1)}{2}\right][1 + (n-1)] = n^2\left(\frac{n+1}{2}\right) = \frac{n^3}{2} + \frac{n^2}{2}$$

and the total number of additions is $(n-1)\left[\frac{n(n+1)}{2}\right] = \frac{n^3 - n}{2} = \frac{n^3}{2} - \frac{n}{2}$ ∎

## A Modification of Gauss-Jordan Elimination

There is a more efficient way to row-reduce $A$ to the identity matrix: First, reduce $A$ to its row echelon form to obtain the matrix

$$\left(\begin{array}{cccccc|c} 1 & a'_{12} & a'_{13} & \cdots & a'_{1,n-1} & a'_{1n} & b'_1 \\ 0 & 1 & a'_{23} & \cdots & a'_{2,n-1} & a'_{2n} & b'_2 \\ \vdots & \vdots & \vdots & & \vdots & \vdots & \vdots \\ 0 & 0 & 0 & \cdots & 1 & a'_{n-1,n} & b'_{n-1} \\ 0 & 0 & 0 & \cdots & 0 & 1 & b'_n \end{array}\right)$$

The next step is to make zero all entries in the $n$th column above the 1 in the $n,n$ position. This results in

$$\left(\begin{array}{cccccc|c} 1 & L & L & \cdots & L & 0 & C \\ 0 & 1 & L & \cdots & L & 0 & C \\ \vdots & \vdots & \vdots & & \vdots & \vdots & \vdots \\ 0 & 0 & 0 & \cdots & 1 & 0 & C \\ 0 & 0 & 0 & \cdots & 0 & 1 & L \end{array}\right)$$

Finally, working from right to left, make all the remaining entries above the diagonal equal to 0. In Problem 22 you are asked to show that with this modification, the number of multiplications is $\frac{1}{3}n^3 + n^2 - \frac{1}{3}n$ and the number of additions is $\frac{1}{3}n^3 + \frac{1}{2}n^2 - \frac{5}{6}n$.

For $n$ large

$$\frac{n^3}{2} + \frac{n^2}{2} \approx \frac{n^3}{2}$$

For example, when $n = 10{,}000$,

$$\frac{n^3}{2} + \frac{n^2}{2} = 500{,}050{,}000{,}000 = 5.0005 \times 10^{11}$$

and

$$\frac{n^3}{2} = 500{,}000{,}000{,}000 = 5 \times 10^{11}$$

Similarly, for $n$ large

$$\frac{1}{3}n^3 + n^2 - \frac{1}{3}n \approx \frac{n^3}{3}$$

Since $\frac{n^3}{3}$ is less than $\frac{n^3}{2}$, we see that the modification described above is more efficient when $n$ is large. (In fact, it is better when $n \geq 3$.)

In Table A.1 we give the number of additions and multiplications required for several of the processes discussed in Chapters 1 and 2.

You are asked to derive these formulas in Problems 22–25.

# PROBLEMS A3

In Problems 1–13 convert the number to a floating-point number with eight decimal places of accuracy. Either truncate (T) or round off (R) as indicated.

**1.** $\frac{1}{3}$ (T) **2.** $\frac{7}{8}$ **3.** $-0.000035$ **4.** $\frac{7}{9}$ (R)

**5.** $\frac{7}{9}$ (T) **6.** $\frac{33}{7}$ (T) **7.** $\frac{85}{11}$ (R) **8.** $-18\frac{5}{6}$ (T)

**9.** $-18\frac{5}{6}$ (R) **10.** 237,059,628 (T) **11.** 237,059,628 (R)

**12.** $-23.7 \times 10^{15}$ **13.** $8374.2 \times 10^{-24}$

**Table A.1** Number of Arithmetic Operations for an Invertible $n \times n$ Matrix $A$

| Technique | Number of Multiplications | Approximate Number of Multiplications for $n$ large | Number of Additions | Approximate Number of Additions for $n$ large |
|---|---|---|---|---|
| 1. Solve $A\mathbf{x} = \mathbf{b}$ by Gauss-Jordan elimination | $\frac{n^3}{2} + \frac{n^2}{2}$ | $\frac{n^3}{2}$ | $\frac{n^3}{2} - \frac{n}{2}$ | $\frac{n^3}{2}$ |
| 2. Solve $A\mathbf{x} = \mathbf{b}$ by modified Gauss-Jordan elimination | $\frac{n^3}{3} + n^2 - \frac{n}{3}$ | $\frac{n^3}{3}$ | $\frac{n^3}{3} + \frac{n^2}{2} - \frac{5n}{6}$ | $\frac{n^3}{3}$ |
| 3. Solve $A\mathbf{x} = \mathbf{b}$ by Gaussian elimination with back substitution | $\frac{n^3}{3} + n^2 - \frac{n}{3}$ | $\frac{n^3}{3}$ | $\frac{n^3}{3} + \frac{n^2}{2} - \frac{5n}{6}$ | $\frac{n^3}{3}$ |
| 4. Find $A^{-1}$ by Gauss-Jordan elimination | $n^3$ | $n^3$ | $n^3 - 2n^2 + n$ | $n^3$ |
| 5. Compute $\det A$ by reducing $A$ to a triangular matrix and multiplying the diagonal components | $\frac{n^3}{3} + \frac{2n}{3} - 1$ | $\frac{n^3}{3}$ | $\frac{n^3}{3} - \frac{n^2}{2} + \frac{n}{6}$ | $\frac{n^3}{3}$ |

In Problems 14–21 the number $x$ and an approximation $x^*$ are given. Find the absolute and relative errors $\epsilon_a$ and $\epsilon_r$.

**14.** $x = 5$; $x^* = 0.49 \times 10^1$

**15.** $x = 500$; $x^* = 0.4999 \times 10^3$

**16.** $x = 3720$; $x^* = 0.3704 \times 10^4$

**17.** $x = \frac{1}{8}$; $x^* = 0.12 \times 10^0$

**18.** $x = \frac{1}{800}$; $x^* = 0.12 \times 10^{-2}$

**19.** $x = -5\frac{5}{6}$; $x^* = -0.583 \times 10^1$

**20.** $x = 0.70465$; $x^* = 0.70466 \times 10^0$

**21.** $x = 70465$; $x^* = 0.70466 \times 10^5$

**22.** Derive the formulas in row 2 of Table A.1. [*Hint:* You will need the following formula, which is proved in Example 3 in Appendix 1:

$$1^2 + 2^2 + 3^2 + \cdots + n^2 = \frac{n(n+1)(2n+1)}{6}]$$

**23.** Derive the formulas in row 3 of Table A.1.

**24.** Derive the formulas in row 4 of Table A.1.

***25.** Derive the formulas in row 5 of Table A.1.

**26.** How many seconds would it take, on average, to solve $A\mathbf{x} = \mathbf{b}$ on a computer using Gauss-Jordan elimination if $A$ is a $20 \times 20$ matrix?

**27.** Answer Problem 26 if the modification described in the text is used.

**28.** How many seconds would it take, on average, to invert a $50 \times 50$ matrix? a $200 \times 200$ matrix? a $10{,}000 \times 10{,}000$ matrix?

**29.** Derive a formula for the number of multiplications and additions required to compute the product $AB$ where $A$ is an $m \times n$ and $B$ is an $n \times q$ matrix.

# Appendix 4 Gaussian Elimination with Pivoting

It is not difficult to program a computer to solve a system of linear equations by the Gaussian or Gauss-Jordan elimination method used throughout this text. There is, however, a variation of the method that was designed to reduce the accumulated round-off error in solving an $n \times n$ system of equations. This method, or a variation of it, is used in many software systems. Once you understand this simple modification of Gaussian elimination, you will understand why, for example, LU-decompositions or row echelon forms found on a calculator or in MATLAB are sometimes different from those you compute by hand.

In Chapter 1 we found that any matrix can be reduced to row echelon form by Gaussian elimination. There is a computational problem with this method, however. If we divide by a small number that has been rounded, the result could contain a significant round-off error. For example, $1/0.00074 \approx 1351$ while $1/0.0007 \approx 1429$. To avoid this problem, we use a method called **Gaussian elimination with partial pivoting.** The idea is always to divide by the largest (in absolute value) component in a column, thereby avoiding, so far as is possible, the type of error just illustrated. We describe the method with a simple example.

EXAMPLE 1 **Solving a System by Gaussian Elimination with Partial Pivoting** Solve the following system by Gaussian elimination with partial pivoting:

$$\begin{aligned} x_1 - x_2 + x_3 &= 1 \\ -3x_1 + 2x_2 - 3x_3 &= -6 \\ 2x_1 - 5x_2 + 4x_3 &= 5 \end{aligned}$$

**Solution** **Step 1.** Write the system in augmented matrix form. From the first column with nonzero components (called the **pivot column**), select the component with the *largest absolute value.* This component is called the **pivot:**

$$\text{pivot} \rightarrow \left(\begin{array}{rrr|r} 1 & -1 & 1 & 1 \\ \boxed{-3} & 2 & -3 & -6 \\ 2 & -5 & 4 & 5 \end{array}\right)$$

**Step 2.** Rearrange the rows to move the pivot to the top:

$$\left(\begin{array}{rrr|r} \boxed{-3} & 2 & -3 & -6 \\ 1 & -1 & 1 & 1 \\ 2 & -5 & 4 & 5 \end{array}\right)$$

(first and second rows were interchanged)

**Step 3.** Divide the first row by the pivot:

$$\left(\begin{array}{rrr|r} 1 & -\frac{2}{3} & 1 & 2 \\ 1 & -1 & 1 & 1 \\ 2 & -5 & 4 & 5 \end{array}\right)$$

(first row divided by $-3$)

**Step 4.** Add multiples of the first row to the other rows to make all the other components in the pivot column equal to zero:

$$\left(\begin{array}{rrr|r} 1 & -\frac{2}{3} & 1 & 2 \\ 0 & -\frac{1}{3} & 0 & -1 \\ 0 & -\frac{11}{3} & 2 & 1 \end{array}\right)$$

(first row multiplied by $-1$ and $-2$ and added to the second and third rows)

**Step 5.** Cover the first row and perform steps 1–4 on the resulting *submatrix:*

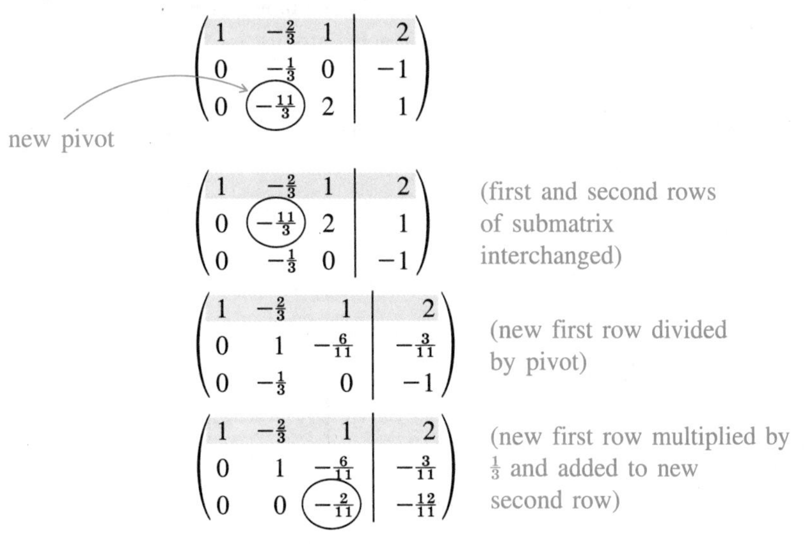


$$\left(\begin{array}{rrr|r} 1 & -\frac{2}{3} & 1 & 2 \\ 0 & -\frac{1}{3} & 0 & -1 \\ 0 & \boxed{-\frac{11}{3}} & 2 & 1 \end{array}\right)$$

$$\left(\begin{array}{rrr|r} 1 & -\frac{2}{3} & 1 & 2 \\ 0 & \boxed{-\frac{11}{3}} & 2 & 1 \\ 0 & -\frac{1}{3} & 0 & -1 \end{array}\right)$$

(first and second rows of submatrix interchanged)

$$\left(\begin{array}{rrr|r} 1 & -\frac{2}{3} & 1 & 2 \\ 0 & 1 & -\frac{6}{11} & -\frac{3}{11} \\ 0 & -\frac{1}{3} & 0 & -1 \end{array}\right)$$

(new first row divided by pivot)

$$\left(\begin{array}{rrr|r} 1 & -\frac{2}{3} & 1 & 2 \\ 0 & 1 & -\frac{6}{11} & -\frac{3}{11} \\ 0 & 0 & \boxed{-\frac{2}{11}} & -\frac{12}{11} \end{array}\right)$$

(new first row multiplied by $\frac{1}{3}$ and added to new second row)

**Step 6.** Continue in this manner until the matrix is in row echelon form:

$$\left(\begin{array}{ccc|c} 1 & -\frac{2}{3} & 1 & 2 \\ 0 & 1 & -\frac{6}{11} & -\frac{3}{11} \\ 0 & 0 & \boxed{-\frac{2}{11}} & -\frac{12}{11} \end{array}\right)$$

new pivot

$$\left(\begin{array}{ccc|c} 1 & -\frac{2}{3} & 1 & 2 \\ 0 & 1 & -\frac{6}{11} & -\frac{3}{11} \\ 0 & 0 & 1 & 6 \end{array}\right)$$

(divided new first row by pivot)

**Step 7.** Use **back substitution** to find the solution (if any) to the system. Evidently, we have $x_3 = 6$. Then $x_2 - \frac{6}{11}x_3 = -\frac{3}{11}$ or

$$x_2 = -\tfrac{3}{11} + \tfrac{6}{11}x_3 = -\tfrac{3}{11} + \tfrac{6}{11}(6) = 3$$

Finally, $x_1 - \frac{2}{3}x_2 + x_3 = 2$ or

$$x_1 = 2 + \tfrac{2}{3}x_2 - x_3 = 2 + \tfrac{2}{3}(3) - 6 = -2$$

The unique solution is given by the vector $(-2, 3, 6)$.

*Remark.* **Complete pivoting** involves finding the component in $A$ with the largest absolute value, not just the component in the first nonzero column. The problem with this method is that it usually involves the relabeling of variables when the columns are interchanged to bring the pivot to the first column. In most problems complete pivoting is not much more accurate than partial pivoting—at least not enough to justify the extra work involved. For this reason the partial pivoting method described above is more popular.

We now examine the method of partial pivoting applied to a computationally more difficult system. Calculations were done on a hand calculator and were rounded to six significant digits.

**EXAMPLE 2** **Solving a System by Gaussian Elimination with Partial Pivoting** Solve the system

$$\begin{aligned} 2x_1 - 3.5x_2 + x_3 &= 22.35 \\ -5x_1 + 3x_2 + 3.3x_3 &= -9.08 \\ 12x_1 + 7.8x_2 + 4.6x_3 &= 21.38 \end{aligned}$$

**Solution** Using the steps outlined above, we obtain, successively,

$$\left(\begin{array}{ccc|c} 2 & -3.5 & 1 & 22.35 \\ -5 & 3 & 3.3 & -9.08 \\ \boxed{12} & 7.8 & 4.6 & 21.38 \end{array}\right) \xrightarrow{R_1 \rightleftarrows R_3} \left(\begin{array}{ccc|c} \boxed{12} & 7.8 & 4.6 & 21.38 \\ -5 & 3 & 3.3 & -9.08 \\ 2 & -3.5 & 1 & 22.35 \end{array}\right)$$

pivot

$$\xrightarrow{R_1 \to \frac{1}{12}R_1} \left(\begin{array}{ccc|c} 1 & 0.65 & 0.383333 & 1.78167 \\ -5 & 3 & 3.3 & -9.08 \\ 2 & -3.5 & 1 & 22.35 \end{array}\right)$$

$$\xrightarrow[R_3 \to R_3 - 2R_1]{R_2 \to R_2 + 5R_1} \left(\begin{array}{ccc|c} 1 & 0.65 & 0.383333 & 1.78167 \\ 0 & \boxed{6.25} & 5.21667 & -0.17165 \\ 0 & -4.8 & 0.233334 & 18.7867 \end{array}\right)$$

new pivot

$$\xrightarrow{R_2 \to \frac{1}{6.25}R_2} \left(\begin{array}{ccc|c} 1 & 0.65 & 0.383333 & 1.78167 \\ 0 & 1 & 0.834667 & -0.027464 \\ 0 & -4.8 & 0.233334 & 18.7867 \end{array}\right)$$

$$\xrightarrow{R_3 \to R_3 + 4.8R_2} \left(\begin{array}{ccc|c} 1 & 0.65 & 0.383333 & 1.78167 \\ 0 & 1 & 0.834667 & -0.027464 \\ 0 & 0 & \boxed{4.23974} & 18.6549 \end{array}\right)$$

new pivot

$$\xrightarrow{R_3 \to \frac{1}{4.23974}R_3} \left(\begin{array}{ccc|c} 1 & 0.65 & 0.383333 & 1.78167 \\ 0 & 1 & 0.834667 & -0.027464 \\ 0 & 0 & 1 & 4.40001 \end{array}\right)$$

The matrix is now in row echelon form. Using back substitution, we obtain

$$x_3 \approx 4.40001$$

$$x_2 \approx -0.027464 - 0.834667x_3 = -0.027464 - (0.834667)(4.40001)$$
$$= -3.70001$$

$$x_1 \approx 1.78167 - (0.65)(x_2) - (0.383333)x_3 = 1.78167 - (0.65)(-3.70001)$$
$$- (0.383333)(4.40001) = 2.50001$$

The exact solution is $x_1 = 2.5$, $x_2 = -3.7$, and $x_3 = 4.4$. Our answers are very accurate indeed. ■

***Remark.*** Example 2 illustrates the fact that it is tedious to use this method without a calculator—especially if several significant digits of accuracy are required.

The next example shows how pivoting can significantly reduce the errors. Here we round to only three significant digits, thereby introducing greater round-off errors.

**EXAMPLE 3** **Partial Pivoting Can Make a Significant Difference** Consider the system

$$\begin{aligned} 0.0002x_1 - 0.00031x_2 + 0.0017x_3 &= 0.00609 \\ 5x_1 - 7x_2 + 6x_3 &= 7 \\ 8x_1 + 6x_2 + 3x_3 &= 2 \end{aligned}$$

The exact solution is $x_1 = -2$, $x_2 = 1$, $x_3 = 4$. Let us first solve the system by Gaussian elimination without pivoting, rounding to three significant figures.

$$\left(\begin{array}{ccc|c} 0.0002 & -0.00031 & 0.0017 & 0.00609 \\ 5 & -7 & 6 & 7 \\ 8 & 6 & 3 & 2 \end{array}\right) \xrightarrow{R_1 \to \frac{1}{0.0002}R_1} \left(\begin{array}{ccc|c} 1 & -1.55 & 8.5 & 30.5 \\ 5 & -7 & 6 & 7 \\ 8 & 6 & 3 & 2 \end{array}\right)$$

$$\xrightarrow[R_3 \to R_3 - 8R_1]{R_2 \to R_2 - 5R_1} \left(\begin{array}{ccc|c} 1 & -1.55 & 8.5 & 30.5 \\ 0 & 0.75 & -36.5 & -146 \\ 0 & 18.4 & -65 & -242 \end{array}\right) \xrightarrow{R_2 \to \frac{1}{0.75}R_2} \left(\begin{array}{ccc|c} 1 & -1.55 & 8.5 & 30.5 \\ 0 & 1 & -48.7 & -195 \\ 0 & 18.4 & -65 & -242 \end{array}\right)$$

$$\xrightarrow{R_3 \to R_3 - 18.4R_1} \left(\begin{array}{ccc|c} 1 & -1.55 & 8.5 & 30.5 \\ 0 & 1 & -48.7 & -195 \\ 0 & 0 & 831 & 3350 \end{array}\right) \xrightarrow{R_3 \to \frac{1}{831}R_3} \left(\begin{array}{ccc|c} 1 & -1.55 & 8.5 & 30.5 \\ 0 & 1 & -48.7 & -195 \\ 0 & 0 & 1 & 4.03 \end{array}\right)$$

This yields

$$\begin{aligned} x_3 &\approx 4.03 \\ x_2 &\approx -195 + (48.7)(4.03) = 1.26 \\ x_1 &\approx 30.5 + (1.55)(1.26) - 8.5(4.03) = -1.8 \end{aligned}$$

Here the errors are significant. The relative errors, given as percentages, are

$$x_1\colon\ \epsilon_r = \left|\frac{-0.2}{2}\right| = 10\%$$

$$x_2\colon\ \epsilon_r = \left|\frac{0.26}{1}\right| = 26\%$$

$$x_3\colon\ \epsilon_r = \left|\frac{0.03}{4}\right| = 0.75\%$$

Let us now repeat the procedure *with* pivoting. We obtain (with the pivots circled)

$$\left(\begin{array}{ccc|c} 0.0002 & -0.00031 & 0.0017 & 0.00609 \\ 5 & -7 & 6 & 7 \\ \textcircled{8} & 6 & 3 & 2 \end{array}\right)$$

$$\xrightarrow{R_1 \rightleftarrows R_3} \left(\begin{array}{ccc|c} \textcircled{8} & 6 & 3 & 2 \\ 5 & -7 & 6 & 7 \\ 0.0002 & -0.00031 & 0.0017 & 0.00609 \end{array}\right)$$

$$\xrightarrow{R_1 \to \frac{1}{8}R_1} \left(\begin{array}{ccc|c} 1 & 0.75 & 0.375 & 0.25 \\ 5 & -7 & 6 & 7 \\ 0.0002 & -0.00031 & 0.0017 & 0.00609 \end{array}\right)$$

$$\xrightarrow[R_3 \to R_3 - 0.0002R_1]{R_2 \to R_2 - 5R_1} \left(\begin{array}{ccc|c} 1 & 0.75 & 0.375 & 0.25 \\ 0 & \textcircled{-10.8} & 4.13 & 5.75 \\ 0 & -0.00046 & 0.00163 & 0.00604 \end{array}\right)$$

$$\xrightarrow{R_2 \to -\frac{1}{10.8}R_2} \left(\begin{array}{ccc|c} 1 & 0.75 & 0.375 & 0.25 \\ 0 & 1 & -0.382 & -0.532 \\ 0 & -0.00046 & 0.00163 & 0.00604 \end{array}\right)$$

$$\xrightarrow{R_3 \to R_3 + 0.00046R_2} \left(\begin{array}{ccc|c} 1 & 0.75 & 0.375 & 0.25 \\ 0 & 1 & -0.382 & -0.532 \\ 0 & 0 & \boxed{0.00145} & 0.0058 \end{array}\right)$$

$$\xrightarrow{R_3 \to \frac{1}{0.00145}R_3} \left(\begin{array}{ccc|c} 1 & 0.75 & 0.375 & 0.25 \\ 0 & 1 & -0.382 & -0.532 \\ 0 & 0 & 1 & 4.00 \end{array}\right)$$

Hence

$$x_3 = 4.00$$
$$x_2 = -0.532 + (0.382)(4.00) = 0.996$$
$$x_1 = 0.25 - 0.75(0.996) - (0.375)(4.00) = -2.00$$

Thus with pivoting and three-significant-digit rounding, $x_1$ and $x_3$ are obtained exactly and $x_2$ is obtained with the relative error of $0.004/1 = 0.4\%$.

Before leaving this section, we note that there are some matrices for which a small change in the entries can lead to a large change in the solution. Such matrices are called **ill-conditioned.**

EXAMPLE 4 **An Ill-Conditioned System** Consider the system

$$x_1 + \quad x_2 = 1$$
$$x_1 + 1.005x_2 = 0$$

The exact solution is easily seen to be $x_1 = 201$, $x_2 = -200$. If the coefficients are rounded to three significant digits, we obtain the system

$$x_1 + \quad x_2 = 1$$
$$x_1 + 1.01x_2 = 0$$

with the exact solution $x_1 = 101$, $x_2 = -100$. Changing one entry in the matrix of coefficients by $0.005/1.005 \approx 0.5$ percent caused a change of about 50 percent in the final solution!

There are techniques for recognizing and dealing with ill-conditioned matrices. One of these, the MATLAB function cond(A), gives a measure of the sensitivity of the solution of a system of linear equations to changes in the data.

# PROBLEMS A4

In Problems 1–4 solve the given system by Gaussian elimination with partial pivoting. Use a hand calculator and round to six significant digits at every step.

**1.**
$$\begin{aligned} 2x_1 - x_2 + x_3 &= 0.3 \\ -4x_1 + 3x_2 - 2x_3 &= -1.4 \\ 3x_1 - 8x_2 + 3x_3 &= 0.1 \end{aligned}$$

**2.**
$$\begin{aligned} 4.7x_1 + 1.81x_2 + 2.6x_3 &= -5.047 \\ -3.4x_1 - 0.25x_2 + 1.1x_3 &= 11.495 \\ 12.3x_1 + 0.06x_2 + 0.77x_3 &= 7.9684 \end{aligned}$$

**3.**
$$\begin{aligned} -7.4x_1 + 3.61x_2 + 8.04x_3 &= 25.1499 \\ 12.16x_1 - 2.7x_2 - 0.891x_3 &= 3.2157 \\ -4.12x_1 + 6.63x_2 - 4.38x_3 &= -36.1383 \end{aligned}$$

**4.**
$$\begin{aligned} 4.1x_1 - 0.7x_2 + 8.3x_3 + 3.9x_4 &= -4.22 \\ 2.6x_1 + 8.1x_2 + 0.64x_3 - 0.8x_4 &= 37.452 \\ -5.3x_1 - 0.2x_2 + 7.4x_3 - 0.55x_4 &= -25.73 \\ 0.8x_1 - 1.3x_2 + 3.6x_3 + 1.6x_4 &= -7.7 \end{aligned}$$

In Problems 5 and 6 solve the system by Gaussian elimination with and without pivoting by rounding to three significant figures. Then solve the system exactly and compute the relative errors of all six computed values.

**5.**
$$\begin{aligned} 0.1x_1 + 0.05x_2 + 0.2x_3 &= 1.3 \\ 12x_1 + 25x_2 - 3x_3 &= 10 \\ -7x_1 + 8x_2 + 15x_3 &= 2 \end{aligned}$$

**6.**
$$\begin{aligned} 0.02x_1 + 0.03x_2 - 0.04x_3 &= -0.04 \\ 16x_1 + 2x_2 + 4x_3 &= 0 \\ 50x_1 + 10x_2 + 8x_3 &= 6 \end{aligned}$$

**7.** Show that the system

$$\begin{aligned} x_1 + x_2 &= 50 \\ x_1 + 1.026x_2 &= 20 \end{aligned}$$

is ill-conditioned if rounding is done to three significant figures. What is the approximate relative error in each answer induced by rounding?

**8.** Do the same for the system

$$\begin{aligned} -0.0001x_1 + x_2 &= 2 \\ -x_1 + x_2 &= 3 \end{aligned}$$

# Appendix 5
# Using MATLAB

MATLAB, a computational and visual software for performing matrix analysis and other linear algebra activities is a product of:

The MathWorks, Inc.
24 Prime Park Way
Natick, MA 01760-1520
phone: (508) 653-1415
fax: (508) 653-2997
E-mail: info@mathworks.com

The MATLAB Problems in this text are designed to introduce the user to MATLAB commands as required by each problem set. The comments that follow focus on supporting issues.

## Elementary Linear Algebra Toolbox (M-file disk)

Several MATLAB Problems in the text have corresponding M-files (small programs) written to enable fuller exploration of certain concepts. The M-files are described in the problems. MATLAB 3.5 versions (including the Student Edition) of the files, or MATLAB 4.0 versions of the files, are available on a disk (either PC or Mac format) called the *Elementary Linear Algebra Toolbox,* from The MathWorks, free of charge, by sending the card provided in this text, or by contacting The MathWorks.

The disk contains a README file that contains a brief description of the M-files, information on setting the search path so that MATLAB will access these files, and contact information for questions or problems concerning use of the M-files. These M-files are User Contributed Routines that have been contributed to The MathWorks and are being redistributed by The MathWorks, upon request, on an "as is" basis. A User Contributed Routine is not a product of The MathWorks, Inc., who assumes no responsibility for errors that may exist in these files.

## MATLAB Primer

In addition to reading the manuals that accompany the software, it would be helpful to obtain a copy of the *MATLAB Primer* by Kermit Sigmon of the University of Florida. It is a general guide intended to serve as an introduction to MATLAB. An excellent feature of the *Primer* is the lists of MATLAB commands categorized according to the basic function of the command. MATLAB has an excellent on-line help feature for those who know the name of a certain command. (Enter **help,** followed by the name of the command, and a description of the use and outcome of the command is displayed.) The combination of the on-line help with the list of commands in the *Primer* is a powerful tool for learning MATLAB.

The plain TEX source of the latest version of the *MATLAB Primer* is available via anonymous ftp from

math.ufl.edu    directory:pub/matlab    file:primer.tex

This location also has a PostScript file *primer.ps* for the *Primer,* as well as a Spanish version *primersp.tex.* See the README file there for more detailed information. You are advised to download anew the latest version of each term, because minor improvements and corrections may have been made, and perhaps a new edition will have appeared. If you are unable to obtain a suitable copy from these sources you may contact the *MATLAB Primer* author: Kermit Sigmon, Department of Mathematics, University of Florida, Gainesville, FL 32611, sigmon@math.ufl.edu.

The *MATLAB Primer* is intended to be easily distributed via a local copy center. If an instructor wishes to have copies made for a class, Dr. Sigmon will give copyright permission.

## Obtaining a Record of Work and Results

The user will frequently wish to save a record of the work performed, both the commands and the results from MATLAB. The **diary** command can store this information in conjunction with any editor program. Before entering any commands to be recorded, enter the command **diary** followed by a filename, which should start with a letter and be no more than eight characters. Any text that appears on the command screen after this command will be saved to the file. You must enter the command **diary off** (at the end of the work to be recorded) in order to record the last portion of the work. If a diary command is used again with the same file, the new work will be appended to the old work. Once the work has been recorded, the file can be read, edited, and printed using the editor.

MATLAB 3.5 facilitates use of an editor without having to exit the program. To call the editor from MATLAB, begin the call with !. An exit from the editor will be a return to the command screen for MATLAB. In MATLAB 4.0, the editor screen and the MATLAB command screen will appear in separate windows. More information on the use of **diary** or ! can be obtained from the on-line help in MATLAB or from the *MATLAB Primer* mentioned above.

Some care should be exercised in the choice of an editor to avoid certain frustrations. Some editors place a Ctrl-Z character at the end of a file that has been edited (changes made and saved). In this situation, even if a **diary** command continues to save new work to the file after it has been edited, the editor will *not* be able to access the new work, so in effect it is lost.

## Graphics Considerations

Graphics commands are introduced in several MATLAB Problems. Here are some things you should know about them.

When working with MATLAB 3.5, including the student edition, hit any key to return to the command screen after viewing a graphics screen. Enter the command **shg** to return to a current graphics screen from the command screen. When working with MATLAB 4.0, the command screen and graphics screen will be in separate windows that can be toggled between each other easily or arranged to be viewed simultaneously.

After completing any problem or distinct part of a problem involving graphics, you should clear the graphics screen and release any frozen features (after saving or printing the graph if desired). In MATLAB 3.5, enter **clg** and **hold off.** In MATLAB 4.0, enter **clf.** Some of these instructions appear in the text problems.

There are several considerations for obtaining *graphics output.* If you are using MATLAB on a system that has access to extended memory or if you are using MATLAB 4.0, enter **help print** for directions. If your system does not have access to extended memory, enter **help meta** for information about saving a graphics screen to a file. Then, read the MATLAB manual for information about using the program **gpp.** If you are using the Student Edition of MATLAB, the only way to obtain a graphics printout is to perform a screen dump (Shift-PrtScr keys). Before using MATLAB you must load the graphics package for your operating system. *Warning:* Output from a screen dump does not necessarily preserve the aspect ratio of the screen; thus right angles may not appear as right angles and equal lengths may not appear equal.

## MATLAB 4.0 Considerations

Some MATLAB text problems include short blocks of MATLAB code for the user to enter. The code as written is compatible with MATLAB 3.5, including the Student

Edition. Efforts have been made to include a description of the modifications needed to make the code compatible with MATLAB 4.0. In the event that any description is not included, here are general guidelines for such modifications:

1. In problems involving setting axes using the **axis** command, MATLAB 3.5 requires the axes to be set before plotting, whereas in MATLAB 4.0 the axes should be set *after* plotting.
2. If a problem involves the use of **hold on,** in MATLAB 4.0 use the commands **hold on** and **axis(axis)** (or re-enter any **axis** commands).
3. If a problem involves the use of **clg,** in MATLAB 4.0 use **clf.**
4. If a problem involves the use of the **rat** command, in MATLAB 4.0 use **format rat.** Use the on-line help for a description of the complete use of **format rat.**

## Special Variable Names

The variable **i** is built into MATLAB to represent the complex number $i$, and the variable **pi** is built into MATLAB to represent the number $\pi$ as long as these variables have not been used for another purpose. One is unlikely to use **pi** inadvertently, but quite likely to use **i**. The variable **eps** is used globally in many MATLAB routines and should *not* be used in any other way.

# Answers to Odd-Numbered Problems

## Chapter 1

### Problems 1.2, page 5

**1.** $x = -\frac{13}{5}$, $y = -\frac{11}{5}$; $a_{11}a_{22} - a_{12}a_{21} = -10$

**3.** no solutions; $a_{11}a_{22} - a_{12}a_{21} = 0$

**5.** $x = \frac{11}{2}$, $y = -30$; $a_{11}a_{22} - a_{12}a_{21} = -2$

**7.** infinite number of solutions; $y = \frac{2}{3}x$, where $x$ is arbitrary; $a_{11}a_{22} - a_{12}a_{21} = 0$

**9.** $x = -1$, $y = 2$; $a_{11}a_{22} - a_{12}a_{21} = -1$

**11.** $a_{11}a_{22} - a_{12}a_{21} = a^2 - b^2$; if $a^2 - b^2 \neq 0$ (i.e., if $a \neq \pm b$), then $x = y = c/(a + b)$. If $a^2 - b^2 = 0$, then $a = \pm b$. If $a \neq 0$ and $a = b$, then there is an infinite number of solutions given by $y = c/a - x$. If $a \neq 0$ and $a = -b$, then there are no solutions (unless $c = 0$ in which case $x = y$ is a solution).

**13.** $a_{11}a_{22} - a_{12}a_{21} = -2ab$; so unique solution if both $a$ and $b$ are nonzero

**15.** $a = b = 0$ and $c \neq 0$ or $d \neq 0$

**17.** no point of intersection

**19.** The lines are coincident. Any point of the form $(x, (4x - 10)/6)$ is a point of intersection.

**21.** $(\frac{67}{45}, \frac{2}{15})$ **23.** $\sqrt{13}/13$

**25.** $\sqrt{61}/5$ **27.** $\sqrt{5}$

**29.** Since the slope of the given line $L$ is $-\dfrac{a}{b}$, the slope of $L_\perp$ is $\dfrac{b}{a}$. The equation of a line $L_\perp$ perpendicular to $L$ and passing through $(x_1, y_1)$ is given by $\dfrac{y - y_1}{x - x_1} = \dfrac{b}{a}$, or $bx - ay = bx_1 - ay_1$. The unique point of intersection of $L$ and $L_\perp$ is found to be $(x_0, y_0) = \left(\dfrac{ac - aby_1 + b^2x_1}{a^2 + b^2}, \dfrac{bc - abx_1 + a^2y_1}{a^2 + b^2}\right)$. Then $d$ is the distance between $(x_0, y_0)$ and $(x_1, y_1)$, and after some algebra, $d^2 = \dfrac{1}{(a^2 + b^2)^2} \times (a^2c^2 - 2a^2bcy_1 + a^2b^2y_1^2 - 2a^3cx_1 +$

$2a^3bx_1y_1 + a^4x_1^2 + c^2b^2 - 2ab^2cx_1 + a^2b^2x_1^2 - 2b^3cy_1 + 2ab^3x_1y_1 + b^4y_1^2) = \dfrac{a^2+b^2}{(a^2+b^2)^2}(c^2 - 2abcy_1 + b^2y_1^2 - 2acx_1 + 2abx_1y_1 + a^2x_1^2) = \dfrac{1}{(a^2+b^2)}(ax_1 + by_1 - c)^2$. Thus

$$d = \frac{|ax_1 + by_1 - c|}{\sqrt{a^2+b^2}}.$$

**31.** If $a_{11}a_{22} - a_{12}a_{21} = 0$, $-a_{11}a_{22} = -a_{12}a_{21}$ and $-\dfrac{a_{11}}{a_{12}} = -\dfrac{a_{21}}{a_{22}}$ if $a_{12}a_{22} \neq 0$. Thus the two lines are parallel since they have equal slopes. If $a_{12} = 0$ and $a_{22} = 0$, then the lines are parallel because they are both vertical.

**33.** The unique solution can be found to be $x = \dfrac{a_{22}b_1 - a_{12}b_2}{a_{11}a_{22} - a_{12}a_{21}}$ and $y = \dfrac{a_{11}b_2 - a_{21}b_1}{a_{11}a_{22} - a_{12}a_{21}}$.

**35.** Let $x$ = no. of cups; $y$ = no. of saucers; solutions are $(x, 240 - \frac{3}{2}x)$. [There is a finite number of solutions because $x$ and $y$ must be positive integers.]

**37.** 32 sodas, 128 milk shakes

## Problems 1.3, page 24

*Note:* Where there were an infinite number of solutions, we wrote the solutions with the last variable chosen arbitrarily. The solutions can be written in other ways as well.

**1.** $(2, -3, 1)$

**3.** $(3 + \frac{2}{9}x_3, \frac{8}{9}x_3, x_3)$, $x_3$ arbitrary

**5.** $(-9, 30, 14)$

**7.** no solution

**9.** $(-\frac{4}{5}x_3, \frac{9}{5}x_3, x_3)$, $x_3$ arbitrary

**11.** $(-1, \frac{5}{2} + \frac{1}{2}x_3, x_3)$, $x_3$ arbitrary

**13.** no solution

**15.** $(\frac{20}{13} - \frac{4}{13}x_4, -\frac{28}{13} + \frac{3}{13}x_4, -\frac{45}{13} + \frac{9}{13}x_4, x_4)$, $x_4$ arbitrary

**17.** $(18 - 4x_4, -\frac{15}{2} + 2x_4, -31 + 7x_4, x_4)$, $x_4$ arbitrary

**19.** no solution

**21.** row echelon form

**23.** reduced row echelon form

**25.** neither

**27.** reduced row echelon form

**29.** neither

**31.** row echelon form: $\begin{pmatrix} 1 & -6 \\ 0 & 1 \end{pmatrix}$; reduced row echelon form: $\begin{pmatrix} 1 & 0 \\ 0 & 1 \end{pmatrix}$

**33.** row echelon form:

$$\begin{pmatrix} 1 & -2 & 4 \\ 0 & 1 & -\frac{4}{11} \\ 0 & 0 & 1 \end{pmatrix}$$

reduced row echelon form:

$$\begin{pmatrix} 1 & 0 & 0 \\ 0 & 1 & 0 \\ 0 & 0 & 1 \end{pmatrix}$$

**35.** row echelon form:

$$\begin{pmatrix} 1 & -\frac{7}{2} \\ 0 & 1 \\ 0 & 0 \end{pmatrix}$$

reduced row echelon form:

$$\begin{pmatrix} 1 & 0 \\ 0 & 1 \\ 0 & 0 \end{pmatrix}$$

**37.** $x_1 = 30{,}000 - 5x_3$
$x_2 = x_3 - 5000$
$5000 \le x_3 \le 6000$; no

**39.** no unique solution (2 equations in 3 unknowns); if 200 shares of McDonald's, then 100 shares of Hilton and 300 shares of Delta

**41.** The row echelon form of the augmented matrix representing this system is

$$\left(\begin{array}{ccc|c} 1 & -\frac{1}{2} & \frac{3}{2} & a/2 \\ 0 & 1 & -\frac{19}{5} & \frac{2}{5}(b - \frac{3}{2}a) \\ 0 & 0 & 0 & -2a + 3b + c \end{array}\right)$$

which is inconsistent if $-2a + 3b + c \neq 0$ or $c \neq 2a - 3b$.

**43.** $a_{11}a_{22}a_{33} + a_{12}a_{23}a_{31} + a_{13}a_{32}a_{21} - a_{13}a_{22}a_{31} - a_{12}a_{21}a_{33} - a_{11}a_{32}a_{23} \neq 0$

**45.** $(-3, 5, 0, 2)$

**47.** $(-17.29018527, -0.2927858589, -12.91757558, 39.93531770)$

**49.** $\begin{pmatrix} 1 & 1 & -2.333 & -0.333 & 1.333 \\ 0 & 1 & -0.5 & -2 & 1 \\ 0 & 0 & 1 & -0.263 & -0.526 \\ 0 & 0 & 0 & 1 & 2 \end{pmatrix}$

**51.** $\begin{pmatrix} 1 & 0.887 & 0.37 & 0.623 & 2.562 \\ 0 & 1 & 0.086 & 0.653 & 24.665 \\ 0 & 0 & 1 & -0.307 & -25.187 \\ 0 & 0 & 0 & 1 & 39.935 \end{pmatrix}$

**53.** $\begin{pmatrix} 1 & -0.381 & 1.662 & 0.394 & -0.257 & 1.643 \\ 0 & 1 & -0.74 & -0.258 & -0.768 & -0.129 \\ 0 & 0 & 1 & 1.292 & -1.235 & 0.754 \\ 0 & 0 & 0 & 1 & -0.246 & -2.847 \\ 0 & 0 & 0 & 0 & 1 & 0.652 \end{pmatrix}$

**55.** $(1.275x_3 + 0.961, -0.403x_3 - 0.090, x_3)$, $x_3$ arbitrary

**57.** $(7.616x_4 - 11.870x_5 + 31.348, 4.876x_4 - 6.775x_5 + 11.043, -1.121x_4 + 3.072x_5 - 2.696, x_4, x_5)$, $x_4$, $x_5$ arbitrary

**59.** $(-11.870x_5 + 50.540, -6.775x_5 + 23.33, 3.072x_5 - 5.52, 2.52, x_5)$, $x_5$ arbitrary

## MATLAB Tutorial

**1.** **A = [2 2 3 4 5;−6 −1 2 0 7;1 2 −1 3 4]** or
```
A = [2    2    3    4    5;
    -6   -1    2    0    7;
     1    2   -1    3    4]
b = [-1;2;5]
```

**3.** **D = 2*(2*rand(3,4)−1)**

**5.** **K = B, K([1 4],:) = K([4 1],:)**

**7.** To write a comment statement, first enter %. The command gives you the submatrix of $B$ given by $\begin{pmatrix} b_{21} & b_{23} \\ b_{41} & b_{43} \end{pmatrix}$.

**9.** **C(2,:) = C(2,:)+3*C(1,:)**

**11.** The equivalent system of equations is

$$\begin{aligned} x_1 \qquad - .1915x_4 - 1.4681x_5 &= -1.1489 \\ x_2 \quad + 1.7447x_4 + 3.0426x_5 &= 2.4681 \\ x_3 + .2979x_4 + .6170x_5 &= -1.2128 \end{aligned}$$

## MATLAB 1.3

**1.** There is a unique solution since every column of the reduced row echelon form of the coefficient matrix has a pivot. If the *augmented* matrix has five columns, for example, and its reduced row echelon form is assigned to the variable **R**, then the solution will be **x = R(:,5).**

**3.** Here is the answer for (iv) as a sample. The reduced row echelon form is

$$\begin{pmatrix} 1 & 0 & .5 & 0 & 5 & 1 \\ 0 & 1 & 1 & 0 & 0 & 2 \\ 0 & 0 & 0 & 1 & -3 & -1 \\ 0 & 0 & 0 & 0 & 0 & 0 \\ 0 & 0 & 0 & 0 & 0 & 0 \end{pmatrix}.$$

The pivots are in the (1, 1), (2, 2) and (3, 4) positions. The equivalent system of the equations is

$$\begin{aligned} x_1 \quad + .5x_3 \quad + 5x_5 &= 1 \\ x_2 + x_3 \qquad &= 2 \\ x_4 - 3x_5 &= -1 \end{aligned}$$

Columns 3 and 5 do not have pivots so:

$$\begin{aligned} x_1 &= 1 - .5x_3 - 5x_5 \\ x_2 &= 2 - x_3 \\ x_4 &= -1 + 3x_5 \end{aligned}$$

**5.** Here is the coding for part (iii) after entering the matrix as **A**. There are variations that are possible.
To create zeros in column one below the (1, 1) position:
```
A(2,:) = A(2,:)-2*A(1,:)
A(3,:) = A(3,:)+3*A(1,:)
A(4,:) = A(4,:)-A(1,:)
```
The next pivot is in the (2, 3) position. To create zeros elsewhere in column 3 (there is a zero already in row 4) and to create a 1 in the pivot position:

**A(3,:) = A(3,:)−2*A(2,:)**
**A(1,:) = A(1,:)+2/3*A(2,:)**
**A(2,:) = 1/3*A(2,:)**

The next pivot is in the (3, 4) position. To create zeros elsewhere in column 4 (there are already zeros above the pivot) and to create a 1 in the pivot position:

**A(4,:) = A(4,:)+2*A(3,:)**
**A(3,:) = 1/2*A(3,:)**

This completes the reduction to

$$\begin{pmatrix} 1 & 2 & 0 & 0 & -3 & -12 \\ 0 & 0 & 1 & 0 & -2 & -5 \\ 0 & 0 & 0 & 1 & 1.5 & 8 \\ 0 & 0 & 0 & 0 & 0 & 0 \end{pmatrix}.$$

**7. a.** First system: $\begin{pmatrix} 2 \\ 3 \\ -1 \end{pmatrix}$. Second system: $\begin{pmatrix} -1 \\ 2 \\ 3 \end{pmatrix}$.

**b.** First system: Let $x_3$ be arbitrary. Then $x_1 = 2 - x_3$ and $x_2 = -1 + 2x_3$. Second system: Let $x_3$ be arbitrary. Then $x_1 = 1 - x_3$ and $x_2 = -1 + 2x_3$. Third system: no solution.

**c.** If a square system has a unique solution for one right-hand side, it will have a unique solution for any right-hand side. In explaining why, discuss why there is a pivot in every row and every column and what this implies (respectively) about existence and uniqueness of solutions. It is possible for a square system to have infinitely many solutions for one right-hand side and no solution for another right-hand side as illustrated by part (b).

**9. b.** The appropriate coefficient matrix is $\begin{pmatrix} .8 & -.1 & -.3 \\ -.15 & .75 & -.25 \\ -.1 & -.05 & 1 \end{pmatrix}$.

The solution is that industry 1 needs to produce \$537,197.63, industry 2 needs to produce \$466,453.67, and industry 3 needs to produce \$277,042.45.

**11. a.** Coefficient matrix:

$$\begin{pmatrix} 1 & 1 & 1 \\ 9 & 3 & 1 \\ 16 & 4 & 1 \end{pmatrix}$$

The polynomial is $-2.3333x^2 + 11.3333x - 10$.

**b.** Coefficient matrix:

$$\begin{pmatrix} 0 & 0 & 0 & 1 \\ 1 & 1 & 1 & 1 \\ 27 & 9 & 3 & 1 \\ 64 & 16 & 4 & 1 \end{pmatrix}$$

The polynomial is $-1.4167x^3 + 8.8333x^2 - 14.4167x + 5$.

## Problems 1.4, page 42

**1.** (0, 0) **3.** (0, 0, 0)

**5.** $(\frac{1}{6}x_3, \frac{5}{6}x_3, x_3)$, $x_3$ arbitrary

**7.** (0, 0)

**9.** $(-4x_4, 2x_4, 7x_4, x_4)$, $x_4$ arbitrary

**11.** (0, 0) **13.** (0, 0, 0)

**15.** $k = \frac{95}{11}$

**17.** $(1.6621x_3, 0.0023x_3, x_3)$, $x_3$ arbitrary

**19.** $(0.3305x_4 + 0.147x_5, 2x_4 - 3.25x_5, -0.7124x_4 + 0.1019x_5, x_4, x_5)$, $x_4$, $x_5$ arbitrary

## MATLAB 1.4

**3. a.** $x_1 = 6$ of $CO_2$, $x_2 = 6$ of $H_2O$, $x_3 = 1$ of $C_6H_{12}O_6$, and $x_4 = 6$ of $O_2$.

**b.** $x_1 = 15$ of $Pb(N_3)_2$, $x_2 = 44$ of $Cr(MnO_4)_2$, $x_3 = 22$ of $Cr_2O_3$, $x_4 = 88$ of $MnO_2$, $x_5 = 5$ of $Pb_3O_4$, and $x_6 = 90$ of NO.

## Problems 1.5, page 56

**1.** $\begin{pmatrix} 2 \\ -3 \\ 11 \end{pmatrix}$ **3.** $\begin{pmatrix} -4 \\ 0 \\ 4 \end{pmatrix}$

**5.** $\begin{pmatrix} -31 \\ 22 \\ -27 \end{pmatrix}$ **7.** $\begin{pmatrix} 0 \\ 0 \\ 0 \end{pmatrix}$

**9.** $\begin{pmatrix} -11 \\ 11 \\ -10 \end{pmatrix}$ **11.** $(1, 2, 5, 7)$

**13.** $(-8, 12, 4, 20)$

**15.** $(8, -5, 7, -1)$

**17.** $(7, 2, 4, 11)$

**19.** $(-11, 9, 18, 18)$

**21.** $\begin{pmatrix} 3 & 9 \\ 6 & 15 \\ -3 & 6 \end{pmatrix}$ **23.** $\begin{pmatrix} 2 & 2 \\ -2 & -1 \\ 6 & -1 \end{pmatrix}$

**25.** $\begin{pmatrix} 0 & 0 \\ 0 & 0 \\ 0 & 0 \end{pmatrix}$ **27.** $\begin{pmatrix} -2 & 4 \\ 7 & 15 \\ -15 & 10 \end{pmatrix}$

**29.** $\begin{pmatrix} 4 & 10 \\ 17 & 22 \\ -9 & 1 \end{pmatrix}$ **31.** $\begin{pmatrix} 0 & 6 \\ 5 & 14 \\ -9 & 9 \end{pmatrix}$

**33.** $\begin{pmatrix} 1 & -5 & 0 \\ -3 & 4 & -5 \\ -14 & 13 & -1 \end{pmatrix}$

**35.** $\begin{pmatrix} 1 & 1 & 5 \\ 9 & 5 & 10 \\ 7 & -7 & 3 \end{pmatrix}$

**37.** $\begin{pmatrix} -1 & -1 & -1 \\ -3 & -3 & -10 \\ -7 & 3 & 5 \end{pmatrix}$

**39.** $\begin{pmatrix} -1 & -1 & -5 \\ -9 & -5 & -10 \\ -7 & 7 & -3 \end{pmatrix}$

**41.** $0A = \overline{0}$ because $0a = 0$ for any scalar $a$;
$1A = A$ because $1a = a$ for any scalar $a$;
$\overline{0} + A = A$ because $0 + a = a$ for any scalar $a$.

**43.** follows because $\alpha(a + b) = \alpha a + \alpha b$ for scalars $a$ and $b$. Also, $(\alpha + \beta)a = \alpha a + \beta a$.

**45.** $\begin{pmatrix} 0 & 1 & 0 & 1 & 0 \\ 1 & 0 & 1 & 1 & 0 \\ 0 & 1 & 0 & 0 & 1 \\ 1 & 1 & 0 & 0 & 0 \\ 0 & 0 & 1 & 0 & 0 \end{pmatrix}$

## MATLAB 1.5

**1. a.** Here is a possible coding:

```
c = −A(2,1)/A(1,1),
    A(2,:) = A(2,:)+c*A(1,:)
c = −A(3,1)/A(1,1),
    A(3,:) = A(3,:)+c*A(1,:)
c = −A(4,1)/A(1,1),
    A(4,:) = A(4,:)+c*A(1,:)
```

Note that column 2 has no pivot. The next pivot is in the (2, 3) position.

```
c = −A(3,3)/A(2,3),
    A(3,:) = A(3,:)+c*A(2,:)
c = −A(4,3)/A(2,3),
    A(4,:) = A(4,:)+c*A(2,:)
```

The last row of commands was included to ensure that the (4, 3) position is truly a zero. The next pivot is in position (3, 4).

```
c = −A(4,4)/A(3,4),   A(4,:) =
  A(4,:)+c*A(3,:)
```

There are no more pivots. The row echelon form is:

$$\begin{pmatrix} 1 & 2 & -2 & 0 & 1 \\ 0 & 0 & 3 & 0 & -6 \\ 0 & 0 & 0 & 2 & 3 \\ 0 & 0 & 0 & 0 & 0 \end{pmatrix}.$$

**3. b.** $s(A + B) = sA + sB$

## Problems 1.6, page 78

**1.** $-14$ **3.** $1$ **5.** $ac + bd$

**7.** $51$ **9.** $\mathbf{a} = \mathbf{0}$ **11.** $4$

**13.** $28$

**15.** $\begin{pmatrix} 8 & 20 \\ -4 & 11 \end{pmatrix}$ **17.** $\begin{pmatrix} -3 & -3 \\ 1 & 3 \end{pmatrix}$

**19.** $\begin{pmatrix} 13 & 35 & 18 \\ 20 & 26 & 20 \end{pmatrix}$

**21.** $\begin{pmatrix} 19 & -17 & 34 \\ 8 & -12 & 20 \\ -8 & -11 & 7 \end{pmatrix}$

**23.** $\begin{pmatrix} 18 & 15 & 35 \\ 9 & 21 & 13 \\ 10 & 9 & 9 \end{pmatrix}$ **25.** $(7 \quad 16)$

**27.** $\begin{pmatrix} 3 & -2 & 1 \\ 4 & 0 & 6 \\ 5 & 1 & 9 \end{pmatrix}$ **29.** $\begin{pmatrix} a & b & c \\ d & e & f \\ g & h & j \end{pmatrix}$

**31.** If $D = a_{11}a_{22} - a_{12}a_{21}$, then

$$\begin{pmatrix} b_{11} & b_{12} \\ b_{21} & b_{22} \end{pmatrix} = \begin{pmatrix} a_{22}/D & -a_{12}/D \\ -a_{21}/D & a_{11}/D \end{pmatrix}$$

**33. a.** 3 in group 1, 4 in group 2, 5 in group 3

**b.** $\begin{pmatrix} 2 & 1 & 1 & 0 & 0 \\ 1 & 1 & 0 & 1 & 0 \\ 1 & 0 & 2 & 0 & 1 \end{pmatrix}$

**35.** orthogonal

**37.** orthogonal **39.** orthogonal

**41.** all $\alpha$ and $\beta$ that satisfy $5\alpha + 4\beta = 25$ ($\beta = (25 - 5\alpha)/4$, $\alpha$ arbitrary)

**43. a.** (2, 3, 5, 1)

**b.** $\begin{pmatrix} 1 \\ \frac{3}{2} \\ \frac{1}{2} \\ 2 \end{pmatrix}$ **c.** 11

**45. a.** $\begin{pmatrix} 80{,}000 & 45{,}000 & 40{,}000 \\ 50 & 20 & 10 \end{pmatrix}$

**b.** $\begin{pmatrix} 1 \\ 3 \\ 1 \end{pmatrix}$ **c.** money: 255,000; shares: 120

**47.** $\begin{pmatrix} 0 & -8 \\ 32 & 32 \end{pmatrix}$ **49.** $\begin{pmatrix} 11 & 38 \\ 57 & 106 \end{pmatrix}$

**51.** $A^2 = \begin{pmatrix} 0 & 0 & 1 & 0 & 0 \\ 0 & 0 & 0 & 1 & 0 \\ 0 & 0 & 0 & 0 & 1 \\ 0 & 0 & 0 & 0 & 0 \\ 0 & 0 & 0 & 0 & 0 \end{pmatrix}$

$A^3 = \begin{pmatrix} 0 & 0 & 0 & 1 & 0 \\ 0 & 0 & 0 & 0 & 1 \\ 0 & 0 & 0 & 0 & 0 \\ 0 & 0 & 0 & 0 & 0 \\ 0 & 0 & 0 & 0 & 0 \end{pmatrix}$

$A^4 = \begin{pmatrix} 0 & 0 & 0 & 0 & 1 \\ 0 & 0 & 0 & 0 & 0 \\ 0 & 0 & 0 & 0 & 0 \\ 0 & 0 & 0 & 0 & 0 \\ 0 & 0 & 0 & 0 & 0 \end{pmatrix}$

$A^5 = \begin{pmatrix} 0 & 0 & 0 & 0 & 0 \\ 0 & 0 & 0 & 0 & 0 \\ 0 & 0 & 0 & 0 & 0 \\ 0 & 0 & 0 & 0 & 0 \\ 0 & 0 & 0 & 0 & 0 \end{pmatrix}$

**53.** $PQ = \begin{pmatrix} \frac{11}{90} & \frac{41}{90} & \frac{19}{45} \\ \frac{11}{120} & \frac{71}{120} & \frac{19}{60} \\ \frac{1}{5} & \frac{1}{5} & \frac{3}{5} \end{pmatrix}$; all entries are nonnegative and $\frac{11}{90} + \frac{41}{90} + \frac{19}{45} = \frac{11}{120} + \frac{71}{120} + \frac{19}{60} = \frac{1}{5} + \frac{1}{5} + \frac{3}{5} = 1$.

**55.** Let $P = (p_{ij})$ and $Q = (q_{ij})$ be $k \times k$ probability matrices. Let $PQ = C = (c_{ij})$. The sum of the elements in the $m$th row of $PQ$ is

$c_{m1} + c_{m2} + c_{m3} + \cdots$
$+ c_{mk} = p_{m1}q_{11} + p_{m2}q_{21}$
$+ p_{m3}q_{31} + \cdots + p_{mk}q_{k1}$
$+ p_{m1}q_{12} + p_{m2}q_{22} + p_{m3}q_{32}$
$+ \cdots + p_{mk}q_{k2} + p_{m1}q_{13}$
$+ p_{m2}q_{23} + p_{m3}p_{33} + \cdots$
$+ p_{mk}q_{k3}$
$\vdots$
$+ p_{m1}q_{1k} + p_{m2}q_{2k} + p_{m3}q_{3k}$
$+ \cdots + p_{mk}q_{kk}$

(The elements in parentheses are those of a row of $Q$, whose sum is 1.)

$\downarrow$

$= p_{m1}(q_{11} + q_{12} + q_{13} + \cdots$
$+ q_{1k}) + p_{m2}(q_{21} + q_{22}$
$+ q_{23} + \cdots + q_{2k})$
$+ p_{m3}(q_{31} + q_{32} + q_{33} + \cdots$
$+ q_{3k}) + \cdots + p_{mk}(q_{k1} + q_{k2}$
$+ q_{k3} + \cdots + q_{kk})$
$= p_{m1}(1) + p_{m2}(1) + p_{m3}(1)$
$+ \cdots + p_{mk}(1) = 1$

**57.** **a.** player 2 > player 4 > player 1 > player 3

**b.** score = number of games won plus one-half the number of games that were won by each player that this given player beat

**59.** $A(B + C)$

$$= \begin{pmatrix} 1 & 2 & 4 \\ 3 & -1 & 0 \end{pmatrix} \begin{pmatrix} 1 & 9 \\ 2 & 11 \\ 10 & 1 \end{pmatrix}$$

$$= \begin{pmatrix} 45 & 35 \\ 1 & 16 \end{pmatrix}$$

$$AB + AC = \begin{pmatrix} 24 & 15 \\ 7 & 17 \end{pmatrix} + \begin{pmatrix} 21 & 20 \\ -6 & -1 \end{pmatrix} = \begin{pmatrix} 45 & 35 \\ 1 & 16 \end{pmatrix}$$

**61.** $\begin{pmatrix} 3 & 7 & 1 & 5 \\ 18 & 42 & 6 & 30 \\ 6 & 14 & 2 & 10 \end{pmatrix}$

**63.** $\left(\begin{array}{cc|cc} e & f & 0 & 0 \\ g & h & 0 & 0 \\ \hline 0 & 0 & a & b \\ 0 & 0 & c & d \end{array}\right)$

**65.** $AB = BA = \begin{pmatrix} I & 0 \\ C + D & I \end{pmatrix}$

Note that $D + C = C + D$

**67.** 36 **69.** 9840

**71.** $\frac{13}{3} + \frac{15}{4} + \frac{17}{5} = \frac{689}{60}$

**73.** $(1^2 + 2^2 + 3^2)(2^3 + 3^3 + 4^3) = 1386$

**75.** $\sum_{k=0}^{5} (-3)^k$ **77.** $\sum_{k=1}^{n} k^{1/k}$

**79.** $\sum_{k=0}^{9} \frac{(-1)^{k+1}}{a^k}$

**81.** $\sum_{k=2}^{7} k^2 \cdot 2k = \sum_{k=2}^{7} 2k^3$

**83.** $\sum_{i=1}^{3} \sum_{j=1}^{2} a_{ij}$ **85.** $\sum_{k=1}^{5} a_{3k} b_{k2}$

**87.** $\sum_{k=M}^{N} (a_k + b_k) = (a_M + b_M) + (a_{M+1} + b_{M+1}) + (a_{M+2} + b_{M+2}) + \cdots + (a_N + b_N)$
$= (a_M + a_{M+1} + a_{M+2} + \cdots + a_N) + (b_M + b_{M+1} + b_{M+2} + \cdots + b_N)$
$= \sum_{k=M}^{N} a_k + \sum_{k=M}^{N} b_k$

**89.** $\sum_{k=M}^{N} a_k = a_M + a_{M+1} + \cdots + a_m + a_{m+1} + \cdots + a_N = (a_M + a_{M+1} + \cdots + a_m) + (a_{m+1} + a_{m+2} + \cdots + a_N) = \sum_{k=M}^{m} a_k + \sum_{k=m+1}^{N} a_k$

**91.** $\begin{pmatrix} 34{,}192 & 38{,}621 \\ 50{,}408 & 44{,}115 \\ 62{,}661 & 71{,}731 \\ 59{,}190 & 55{,}046 \end{pmatrix}$

**93.** **a.** The numbers in each row of each matrix are positive and sum to 1.

**b.** $PQ =$

$$\begin{pmatrix} 0.31118 & 0.18444 & 0.14174 & 0.36264 \\ 0.32625 & 0.27585 & 0.08454 & 0.31336 \\ 0.17955 & 0.22651 & 0.19619 & 0.39775 \\ 0.30047 & 0.15251 & 0.33558 & 0.21144 \end{pmatrix}$$

is a probability matrix because the entries in each row are positive and sum to 1.

**95.** $A^n = \begin{pmatrix} a^n & u & v \\ 0 & b^n & w \\ 0 & 0 & c^n \end{pmatrix}$ where $u$, $v$, and $w$ are real numbers.

## MATLAB 1.6

**1.** $AB$ is defined; $BA$ is not defined and produces an error message.

**3.** You will find that $A(X + sZ) = B$.

**5.** **A = 10*(2*rand(5,6)−1).** The given expression will equal 0 showing that $A\mathbf{x}$ has the interpretation of a linear combination of the columns of $A$.

7. $(A + B)^2 = A^2 + 2AB + B^2$ only for those pairs $A$ and $B$ that commute. *Proof hint:* Multiply out $(A + B)^2$.

9. a. The product of upper triangular matrices is upper triangular. *Proof hint:* Suppose $T$ and $S$ are upper triangular. Use the fact that $(TS)_{ij}$ is a sum of entries of the form $t_{ik}s_{kj}$ and that $t_{ik} = 0$ for $i > k$ and $s_{kj} = 0$ for $k > j$ to show that $(TS)_{ij} = 0$ for $i > j$.

11. The pattern holds true: **AA*BB−K = 0.**

13. a. The desired indirect contact matrix $K$ is $XYZ$, where $X$ is the contact matrix from group 1 to group 2, $Y$ is the contact matrix from group 2 to group 3, and $Z$ is the contact matrix from group 3 to group 4. Below is an example of the coding to enter $Y$ easily:

```
Y = zeros(5,8);
Y(1,[1 3 5]) = [1 1 1]
Y(2,[3 4 7]) = [1 1 1]
Y(3,[1 5 6 8]) = [1 1 1 1]
Y(4,8) = 1
Y(5,[5 6 7]) = [1 1 1]
```

The indirect contact matrix between group 1 and group 4:

$$K = \begin{pmatrix} 3 & 1 & 1 & 2 & 2 & 1 & 1 & 3 & 3 & 2 \\ 4 & 3 & 1 & 4 & 2 & 2 & 2 & 2 & 3 & 2 \\ 5 & 3 & 1 & 4 & 1 & 3 & 1 & 3 & 3 & 2 \end{pmatrix}.$$

b. There are no zeros in $K$, so each person of group 1 has indirect contact with each person in group 4.

c. **[1 1 1]*K** will produce the column sums, that is, the total of the number of indirect contacts each member of group 4 has with group 1. $C1$ has 12 contacts, which is the largest number in the column sums. **K*ones(10,1)** will produce the row sums, that is, the total of the number of indirect contacts that each member of group 1 has with group 4. $A3$ is the most dangerous with 26 indirect contacts.

## Problems 1.7, page 94

1. $\begin{pmatrix} 2 & -1 \\ 4 & 5 \end{pmatrix}\begin{pmatrix} x_1 \\ x_2 \end{pmatrix} = \begin{pmatrix} 3 \\ 7 \end{pmatrix}$

3. $\begin{pmatrix} 3 & 6 & -7 \\ 2 & -1 & 3 \end{pmatrix}\begin{pmatrix} x_1 \\ x_2 \\ x_3 \end{pmatrix} = \begin{pmatrix} 0 \\ 1 \end{pmatrix}$

5. $\begin{pmatrix} 0 & 1 & -1 \\ 1 & 0 & 1 \\ 3 & 2 & 0 \end{pmatrix}\begin{pmatrix} x_1 \\ x_2 \\ x_3 \end{pmatrix} = \begin{pmatrix} 7 \\ 2 \\ -5 \end{pmatrix}$

7. $$\begin{aligned} x_1 + x_2 - x_3 &= 7 \\ 4x_1 - x_2 + 5x_3 &= 4 \\ 6x_1 + x_2 + 3x_3 &= 20 \end{aligned}$$

9. $$\begin{aligned} 2x_1 \qquad + x_3 &= 2 \\ -3x_1 + 4x_2 \qquad &= 3 \\ 5x_2 + 6x_3 &= 5 \end{aligned}$$

11. $$\begin{aligned} x_1 &= 2 \\ x_2 &= 3 \\ x_3 &= -5 \\ x_4 &= 6 \end{aligned}$$

13. $$\begin{aligned} 6x_1 + 2x_2 + x_3 &= 2 \\ -2x_1 + 3x_2 + x_3 &= 4 \\ 0x_1 + 0x_2 + 0x_3 &= 2 \end{aligned}$$

15. $$\begin{aligned} 7x_1 + 2x_2 &= 1 \\ 3x_1 + x_2 &= 2 \\ 6x_1 + 9x_2 &= 3 \end{aligned}$$

17. The simplest solution to the nonhomogeneous equation is obtained by setting $x_2 = 0$. Then the general solution is $(2, 0) + x_2(3, 1)$; $x_2$ arbitrary.

19. If $x_3 = 0$, one nonhomogeneous solution is $(2, 0, 0)$ and the general solution is $(2, 0, 0) + x_3(-\frac{1}{3}, -\frac{4}{3}, 1)$; $x_3$ arbitrary.

21. If $x_3 = x_4 = 0$, one nonhomogeneous solution is $(-1, 4, 0, 0)$ and the general solution is $(-1, 4, 0, 0) + x_3(-3, 4, 1, 0) + x_4(5, -7, 0, 1)$.

**23.** $(c_1y_1 + c_2y_2)''$
$+ a(x)(c_1y_1 + c_2y_2)'$
$+ b(x)(c_1y_1 + c_2y_2)$
$= c_1y_1'' + c_2y_2'' + a(x)c_1y_1'$
$+ a(x)c_2y_2' + b(x)c_1y_1$
$+ b(x)c_2y_2$
$= c_1(y_1'' + a(x)y_1' + b(x)y_1)$
$+ c_2(y_2'' + a(x)y_2' + b(x)y_2)$
$= c_1 \cdot 0 + c_2 \cdot 0 = 0$, since $y_1$ and $y_2$ solve (7).

## MATLAB 1.7

**1.** For part (b), $x_1 = -2x_3 + x_4 + 5$ and $x_2 = -x_3 - x_4 - 1$; an example of a solution, from choosing $x_3 = -1$ and $x_4 = -2$, is **x = [1;0;1;−2].** One should find that $A\mathbf{x} = \mathbf{y} = \mathbf{b}$, illustrating that if **x** is a solution to the system whose augmented matrix is $[A \quad \mathbf{b}]$, then $A\mathbf{x} = \mathbf{b}$ and **b** is a linear combination of the columns of $A$ where the coefficients in the linear combination are the components of **x**.

**3.** To verify that a vector **w** is a solution, show that $A\mathbf{w} = \mathbf{b}$.

**5. a.** Letting $x_3 = 1$ in the solution $x_1 = -x_3$, $x_2 = 2x_3$, $x_4 = 0$ yields that

$$\mathbf{0} = x_1(\text{col } 1) + x_2(\text{col } 2) + x_3(\text{col } 3) + x_4(\text{col } 4) = -(\text{col } 1) + 2(\text{col } 2) + (\text{col } 3),$$

which in turn yields col 3 = (col 1) − 2(col 2).

**b.** The solution is $x_1 = -2x_3 + x_4$ and $x_2 = -x_3 - x_4$. Letting $x_3 = 1$ and $x_4 = 0$ gives col 3 = 2(col 1) + (col 2). Letting $x_3 = 0$ and $x_4 = 1$ gives col 4 = −(col 1) + (col 2).

## Problems 1.8, page 112

**1.** $\begin{pmatrix} 2 & -1 \\ -3 & 2 \end{pmatrix}$ **3.** $\begin{pmatrix} 0 & 1 \\ 1 & 0 \end{pmatrix}$

**5.** not invertible

**7.** $\begin{pmatrix} \frac{1}{3} & -\frac{1}{3} & -\frac{1}{3} \\ 0 & \frac{1}{2} & 1 \\ 0 & 0 & -1 \end{pmatrix}$

**9.** not invertible

**11.** not invertible

**13.** $\begin{pmatrix} \frac{7}{3} & -\frac{1}{3} & -\frac{1}{3} & -\frac{2}{3} \\ \frac{4}{9} & -\frac{1}{9} & -\frac{4}{9} & \frac{1}{9} \\ -\frac{1}{9} & -\frac{2}{9} & \frac{1}{9} & \frac{2}{9} \\ -\frac{5}{3} & \frac{2}{3} & \frac{2}{3} & \frac{1}{3} \end{pmatrix}$

**15.** $\begin{pmatrix} 0 & 1 & 0 & 2 \\ 1 & -1 & -2 & 2 \\ 0 & 1 & 3 & -3 \\ -2 & 2 & 3 & -2 \end{pmatrix}$

**17.** $(A_1A_2\cdots A_m)^{-1} = A_m^{-1}A_{m-1}^{-1}\cdots A_2^{-1}A_1^{-1}$ since $(A_m^{-1}A_{m-1}^{-1}\cdots A_2^{-1}A_1^{-1})(A_1A_2\cdots A_{m-1}A_m) = (A_m^{-1}A_{m-1}^{-1}\cdots A_2^{-1})(A_1^{-1}A_1) \times A_2\cdots A_{m-1}A_m = (A_m^{-1}A_{m-1}^{-1}\cdots A_2^{-1}) \times (A_2\cdots A_{m-1}A_m) = \cdots = I$.

**19.** $A^{-1} = \dfrac{1}{a_{11}a_{22} - a_{21}a_{12}} \times \begin{pmatrix} a_{22} & -a_{12} \\ -a_{21} & a_{11} \end{pmatrix}$. If $A = \pm I$, then $A^{-1} = A$. If $a_{11} = -a_{22}$ and $a_{21}a_{12} = 1 - a_{11}^2$, then $a_{11}a_{22} - a_{21}a_{12} = -a_{11}^2 - (1 - a_{11}^2) = -1$. Thus

$$A^{-1} = \begin{pmatrix} -a_{22} & a_{12} \\ a_{21} & -a_{11} \end{pmatrix} = \begin{pmatrix} a_{11} & a_{12} \\ a_{21} & a_{22} \end{pmatrix} = A$$

**21.** The system $B\mathbf{x} = \mathbf{0}$ has an infinite number of solutions (by Theorem 1.4.1). But if $B\mathbf{x} = \mathbf{0}$, then $AB\mathbf{x} = \mathbf{0}$. Thus, from Theorem 6 [parts **(i)** and **(ii)**], $AB$ is not invertible.

**23.** $\begin{pmatrix} \sin\theta & \cos\theta & 0 \\ \cos\theta & -\sin\theta & 0 \\ 0 & 0 & 1 \end{pmatrix}$ is its own inverse (since $\sin^2\theta + \cos^2\theta = 1$).

**25.** If the $i$th diagonal component is 0, then in the row reduction of $A$ the

$i$th row is zero so that, by the statement in step 3(b) on page 103, $A$ is not invertible. Otherwise, if

$$A = \text{diag}(a_1, a_2, \ldots, a_n)$$

then

$$A^{-1} = \text{diag}\left(\frac{1}{a_1}, \frac{1}{a_2}, \ldots, \frac{1}{a_n}\right)$$

**27.** $\begin{pmatrix} \frac{1}{2} & -\frac{1}{6} & \frac{7}{30} \\ 0 & \frac{1}{3} & -\frac{4}{15} \\ 0 & 0 & \frac{1}{5} \end{pmatrix}$

**29.** We prove the result in the case $A$ is upper triangular. The proof in the lower triangular case is similar. Consider the homogeneous system

$$\begin{pmatrix} a_{11} & a_{12} & a_{13} & \cdots & a_{1,n-1} & a_{1n} \\ 0 & a_{22} & a_{23} & \cdots & a_{2,n-1} & a_{2n} \\ \vdots & \vdots & \vdots & & \vdots & \vdots \\ 0 & 0 & 0 & \cdots & a_{n-1,n-1} & a_{n-1,n} \\ 0 & 0 & 0 & \cdots & 0 & a_{nn} \end{pmatrix} \times \begin{pmatrix} x_1 \\ x_2 \\ \vdots \\ x_{n-1} \\ x_n \end{pmatrix} = \begin{pmatrix} 0 \\ 0 \\ \vdots \\ 0 \\ 0 \end{pmatrix}$$

Suppose that $a_{11}, a_{22}, \ldots, a_{nn}$ are all nonzero. The last equation in the homogeneous system is $a_{nn}x_n = 0$, and since $a_{nn} \neq 0$, $x_n = 0$. The next-to-the-last equation is

$$a_{n-1,n-1}x_{n-1} + a_{n-1,n}x_n = 0$$

and $a_{n-1,n-1} \neq 0$, $x_n = 0$ implies that $x_{n-1} = 0$. Similarly, we conclude that $x_1 = x_2 = \cdots = x_{n-1} = x_n = 0$, so the only solution to the homogeneous system is the trivial solution. By Theorem 6 [parts **(i)** and **(ii)**], $A$ is invertible. Conversely, suppose one of the diagonal components, say, $a_{11}$, is equal to 0. Then the homogeneous system $A\mathbf{x} = \mathbf{0}$ has the solution

$$\mathbf{x} = \begin{pmatrix} 1 \\ 0 \\ \vdots \\ 0 \end{pmatrix}$$

[If $a_{jj} = 0$ with $j \neq 1$, then choose $\mathbf{x}$ to be the vector with 1 in the $j$th position and 0 everywhere else.] Using Theorem 6 again, we conclude that $A$ is not invertible.

**31.** any nonzero multiple of (1, 2)

**33.** 3 chairs and 2 tables

**35.** 4 units of $A$ and 5 units of $B$

**37. a.** $A = \begin{pmatrix} 0.293 & 0 & 0 \\ 0.014 & 0.207 & 0.017 \\ 0.044 & 0.010 & 0.216 \end{pmatrix}$;

$$I - A = \begin{pmatrix} 0.707 & 0 & 0 \\ -0.014 & 0.793 & -0.017 \\ -0.044 & -0.010 & 0.784 \end{pmatrix}$$

**b.** $\begin{pmatrix} 18{,}689 \\ 22{,}598 \\ 3{,}615 \end{pmatrix}$

**39.** $\begin{pmatrix} 1 & \frac{1}{2} \\ 0 & 1 \end{pmatrix}$; yes

**41.** $\begin{pmatrix} 1 & \frac{2}{3} & \frac{1}{3} \\ 0 & 1 & 1 \\ 0 & 0 & 1 \end{pmatrix}$; yes

**43.** $\begin{pmatrix} 1 & -\frac{1}{2} & 2 \\ 0 & 1 & -14 \\ 0 & 0 & 0 \end{pmatrix}$; no

**45.** $\begin{pmatrix} 1 & 0 & 2 & 3 \\ 0 & 1 & 2 & 7 \\ 0 & 0 & 1 & \frac{10}{7} \\ 0 & 0 & 0 & 0 \end{pmatrix}$; no

**49.** $\begin{pmatrix} A_{11}^{-1} & 0 \\ -A_{22}^{-1}A_{21}A_{11}^{-1} & A_{22}^{-1} \end{pmatrix}$

In Problems 51–55 answers are given to four decimal places.

**51.** $\begin{pmatrix} -1.4075 & 0.4560 & 0.3034 \\ 0.6571 & -0.1670 & -0.1544 \\ 0.4399 & -0.2675 & -0.0323 \end{pmatrix}$

**53.** $\begin{pmatrix} 0.0398 & 0.0095 & 0.0352 & 0.0106 \\ -0.0037 & 0.0043 & 0.0004 & -0.0056 \\ 0.0183 & 0.0094 & 0.0085 & 0.0030 \\ 0.0194 & 0.0070 & 0.0255 & 0.0158 \end{pmatrix}$

**55.** The inverse of the given matrix is

$$\begin{pmatrix} 0.0433 & -0.1257 & -0.1964 & 0.1269 & 0.2034 \\ 0 & -0.0690 & -0.0671 & 0.0357 & 0.1133 \\ 0 & 0 & -0.0269 & 0.0191 & 0.0078 \\ 0 & 0 & 0 & 0.0110 & -0.0032 \\ 0 & 0 & 0 & 0 & 0.0213 \end{pmatrix}$$

### MATLAB 1.8

**1.** Let **S = R(:,[4 5 6]),** where **R** holds the reduced row echelon form. You should have that **A*S = S*A**, both equal to the identity, and **S** should match **inv(A).** We have that **S** = $\begin{pmatrix} 54 & -23 & -7 \\ -16 & 7 & 2 \\ -7 & 3 & 1 \end{pmatrix}$.

**3.** To show $A$ is not invertible, show that the reduced row echelon form is not equal to the identity. *Proof hint:* If $R_3 = 3R_1 + 5R_2$, what will be the end result of the following row operations: $R_3 \to R_3 - 3R_1$ followed by $R_3 \to R_3 - 5R_2$?

**5. a.** An upper triangular matrix is not invertible if a diagonal entry is zero. The diagonal entries of the inverse of an upper triangular matrix are the multiplicative inverses of the diagonal entries of the original matrix. *Proof hint:* Consider $[A \quad I]$. If you first perform row operations to create 1's in the pivot positions, what does that create in the part of the augmented matrix corresponding to $I$? Argue why these positions then do not change with the further row operations needed.

**b.** All such matrices are not invertible.

**c.** For vectors **x** with distinct entries, the associated matrix $V$ is invertible.

**7. c.** The entries in the inverse are large and become larger as $f$ becomes smaller, that is, as the matrix gets closer to being noninvertible.

**d.** The accuracy is getting worse as the matrix gets closer to being noninvertible as the computed solution and the exact solution agree to fewer and fewer digits.

**9.** Multiplying on the right in effect computes $(MA)A^{-1}$. The decoded message is, *''Are you having fun.''*

### Problems 1.9, page 124

**1.** $\begin{pmatrix} -1 & 6 \\ 4 & 5 \end{pmatrix}$ **3.** $\begin{pmatrix} 2 & -1 & 1 \\ 3 & 2 & 4 \end{pmatrix}$

**5.** $\begin{pmatrix} 1 & -1 & 1 \\ 2 & 0 & 5 \\ 3 & 4 & 5 \end{pmatrix}$ **7.** $\begin{pmatrix} 1 & 0 \\ 0 & 1 \\ 1 & 0 \\ 0 & 1 \end{pmatrix}$

**9.** $\begin{pmatrix} a & d & g \\ b & e & h \\ c & f & j \end{pmatrix}$

**11.** $[(A + B)^t]_{ij} = (A + B)_{ji} = a_{ji} + b_{ji} = (A^t)_{ij} + (B^t)_{ij}$. Thus the $ij$th component of $(A + B)^t$ equals the $ij$th component of $A^t$ plus the $ij$th component of $B^t$.

**13.** $(A + B)^t = A^t + B^t = A + B$

**15.** If $A$ is $m \times n$, then $A^t$ is $n \times m$ and $AA^t$ is $m \times m$. Also, $(AA^t)^t = (A^t)^t A^t = AA^t$.

**17.** If $A$ is upper triangular and $B = A^t$, then $b_{ij} = a_{ji} = 0$ if $j > i$. Thus $B$ is lower triangular.

**19.** $(A + B)^t = A^t + B^t$
$= -A - B = -(A + B)$

**21.** $(AB)^t = B^t A^t = (-B)(-A)$
$= (-1)^2 BA = BA$

**23.** $[\frac{1}{2}(A - A^t)]^t = \frac{1}{2}(A^t - (A^t)^t)$
$= \frac{1}{2}(A^t - A)$
$= -[\frac{1}{2}(A - A^t)]$

**25.** (ii) tells us that $a_{11}a_{21} + a_{12}a_{22} = 0$. So

$$AA^t = \begin{pmatrix} a_{11}^2 + a_{12}^2 & a_{11}a_{21} + a_{12}a_{22} \\ a_{11}a_{21} + a_{12}a_{22} & a_{21}^2 + a_{22}^2 \end{pmatrix} = \begin{pmatrix} 1 & 0 \\ 0 & 1 \end{pmatrix}$$

Then, from Theorem 1.8.7, we see that $A^t = A^{-1}$.

**27.** $\begin{pmatrix} 2 & -3 \\ -1 & 2 \end{pmatrix}$

**29.** $\begin{pmatrix} \frac{13}{8} & -\frac{15}{8} & \frac{5}{4} \\ -\frac{1}{2} & \frac{1}{2} & 0 \\ -\frac{1}{8} & \frac{3}{8} & -\frac{1}{4} \end{pmatrix}$

### MATLAB 1.9

**1.** $(AB)^t = B^tA^t$.

**3.** $B$ and $G$ are symmetric; that is, $b_{ij} = b_{ji}$ and $g_{ij} = g_{ji}$. $C$ is antisymmetric; that is, $c_{ij} = -c_{ji}$.

### Problems 1.10, page 134

**1.** yes, $R_1 \rightleftarrows R_2$

**3.** no [two operations are used: $R_1 \rightleftarrows R_2$ followed by $R_2 \to R_2 + R_1$]

**5.** no [two operations are used: $R_1 \to 3R_1$ and $R_2 \to 3R_2$]

**7.** no [two operations are used: $R_1 \rightleftarrows R_3$ followed by $R_1 \rightleftarrows R_2$]

**9.** yes, $R_2 \to R_2 + 2R_1$

**11.** no [two operations are used: $R_2 \to R_2 + R_1$ and $R_4 \to R_4 + R_3$]

**13.** $\begin{pmatrix} 1 & 0 & 0 \\ 0 & 4 & 0 \\ 0 & 0 & 1 \end{pmatrix}$

**15.** $\begin{pmatrix} 1 & -3 & 0 \\ 0 & 1 & 0 \\ 0 & 0 & 1 \end{pmatrix}$

**17.** $\begin{pmatrix} 0 & 0 & 1 \\ 0 & 1 & 0 \\ 1 & 0 & 0 \end{pmatrix}$ **19.** $\begin{pmatrix} 1 & 0 & 0 \\ 0 & 1 & 1 \\ 0 & 0 & 1 \end{pmatrix}$

**21.** $\begin{pmatrix} 1 & 0 \\ 0 & -2 \end{pmatrix}$ **23.** $\begin{pmatrix} 1 & 2 \\ 0 & 1 \end{pmatrix}$

**25.** $\begin{pmatrix} 0 & 0 & 1 \\ 0 & 1 & 0 \\ 1 & 0 & 0 \end{pmatrix}$ **27.** $\begin{pmatrix} -1 & 0 & 0 \\ 0 & 1 & 0 \\ 0 & 0 & 1 \end{pmatrix}$

**29.** $\begin{pmatrix} 1 & 0 & 0 \\ 0 & 1 & 0 \\ -5 & 0 & 1 \end{pmatrix}$

**31.** $\begin{pmatrix} 0 & 1 \\ 1 & 0 \end{pmatrix}$ **33.** $\begin{pmatrix} 1 & 0 \\ 0 & \frac{1}{4} \end{pmatrix}$

**35.** $\begin{pmatrix} 1 & 2 & 0 \\ 0 & 1 & 0 \\ 0 & 0 & 1 \end{pmatrix}$

**37.** $\begin{pmatrix} 1 & 0 & 0 \\ 0 & -2 & 0 \\ 0 & 0 & 1 \end{pmatrix}$

**39.** $\begin{pmatrix} 1 & 0 & 0 & -5 \\ 0 & 1 & 0 & 0 \\ 0 & 0 & 1 & 0 \\ 0 & 0 & 0 & 1 \end{pmatrix}$

**41.** $\begin{pmatrix} 2 & 0 \\ 0 & 1 \end{pmatrix}\begin{pmatrix} 1 & 0 \\ 3 & 1 \end{pmatrix}\begin{pmatrix} 1 & 0 \\ 0 & \frac{1}{2} \end{pmatrix}\begin{pmatrix} 1 & \frac{1}{2} \\ 0 & 1 \end{pmatrix}$

**43.** $\begin{pmatrix} 1 & 0 & 0 \\ 0 & 1 & 0 \\ 5 & 0 & 1 \end{pmatrix}\begin{pmatrix} 1 & 0 & 0 \\ 0 & 2 & 0 \\ 0 & 0 & 1 \end{pmatrix}\begin{pmatrix} 1 & 1 & 0 \\ 0 & 1 & 0 \\ 0 & 0 & 1 \end{pmatrix}$
$\times \begin{pmatrix} 1 & 0 & 0 \\ 0 & 1 & 0 \\ 0 & 0 & -4 \end{pmatrix}\begin{pmatrix} 1 & 0 & -\frac{1}{2} \\ 0 & 1 & 0 \\ 0 & 0 & 1 \end{pmatrix}\begin{pmatrix} 1 & 0 & 0 \\ 0 & 1 & \frac{3}{2} \\ 0 & 0 & 1 \end{pmatrix}$

**45.** $\begin{pmatrix} 0 & 0 & 1 \\ 0 & 1 & 0 \\ 1 & 0 & 0 \end{pmatrix}\begin{pmatrix} 1 & 0 & 0 \\ 0 & 1 & 0 \\ 0 & -1 & 1 \end{pmatrix}\begin{pmatrix} 1 & 0 & 0 \\ 0 & 1 & 0 \\ 0 & 0 & -1 \end{pmatrix}$
$\times \begin{pmatrix} 1 & 0 & 1 \\ 0 & 1 & 0 \\ 0 & 0 & 1 \end{pmatrix}\begin{pmatrix} 1 & 0 & 0 \\ 0 & 1 & -1 \\ 0 & 0 & 1 \end{pmatrix}$

**47.** $\begin{pmatrix} 2 & 0 & 0 & 0 \\ 0 & 1 & 0 & 0 \\ 0 & 0 & 1 & 0 \\ 0 & 0 & 0 & 1 \end{pmatrix}\begin{pmatrix} 1 & 0 & 0 & 0 \\ 0 & 3 & 0 & 0 \\ 0 & 0 & 1 & 0 \\ 0 & 0 & 0 & 1 \end{pmatrix}$
$\times \begin{pmatrix} 1 & 0 & 0 & 0 \\ 0 & 1 & 0 & 0 \\ 0 & 0 & -4 & 0 \\ 0 & 0 & 0 & 1 \end{pmatrix}\begin{pmatrix} 1 & 0 & 0 & 0 \\ 0 & 1 & 0 & 0 \\ 0 & 0 & 1 & 0 \\ 0 & 0 & 0 & 5 \end{pmatrix}$

**49.** $\begin{pmatrix} a & 0 \\ 0 & 1 \end{pmatrix}\begin{pmatrix} 1 & 0 \\ 0 & c \end{pmatrix}\begin{pmatrix} 1 & b/a \\ 0 & 1 \end{pmatrix}$; the first two matrices are elementary because $a \neq 0$ and $c \neq 0$

**51.** The $2 \times 2$ and $3 \times 3$ cases are the results of Problems 49 and 50. In the answer to Problem 1.8.29 we proved this result. Another proof can be given by showing, as in Problems 49 and 50, that $A$ can be written as the product of elementary matrices. The key step is to reduce $A$ to $I$ by noting that the only times we divide, we divide by the numbers on the diagonal, which are nonzero by assumption.

**53.** $A^t$ is upper triangular, so $(A^t)^{-1}$ is upper triangular by the result of Problem 52. But $(A^t)^{-1} = (A^{-1})^t$, so $(A^{-1})^t$ is upper triangular, which means that $A^{-1} = [(A^{-1})^t]^t$ is lower triangular.

**55.** Let $B = A_{ij}$ and $D = A_{ij}A$. Then the $kr$th component $d_{kr}$ of $D$ is given by

$$d_{kr} = \sum_{l=1}^{n} b_{kl}a_{lr} \qquad (*)$$

If $k \neq j$, the $k$th row of $B$ is the $k$th row of the identity, so $b_{kl} = 1$ if $l = k$ and 0 otherwise. Thus

$$d_{kr} = b_{kk}a_{kr} = a_{kr} \qquad \text{if } k \neq j$$

If $k = j$, then

$$b_{jl} = \begin{cases} 1, & \text{if } l = j \\ c, & \text{if } l = i \\ 0, & \text{otherwise} \end{cases}$$

and $(*)$ becomes

$$\begin{aligned} a_{jr} &= b_{jj}a_{jr} + b_{ji}a_{ir} \\ &= a_{jr} + ca_{jr} \end{aligned}$$

Thus each component in the $j$th row of $A_{ij}A$ is the sum of the corresponding component in the $j$th row of $A$ and $c$ times the corresponding component in the $i$th row of $A$.

**57.** $\begin{pmatrix} 1 & 0 \\ 2 & 1 \end{pmatrix}\begin{pmatrix} 1 & 2 \\ 0 & 0 \end{pmatrix}$

**59.** $\begin{pmatrix} 0 & 1 \\ 1 & 0 \end{pmatrix}\begin{pmatrix} 1 & 0 \\ 0 & 0 \end{pmatrix}$

**61.** $\begin{pmatrix} 1 & 0 & 0 \\ 0 & 1 & 0 \\ 1 & 0 & 1 \end{pmatrix}\begin{pmatrix} 1 & 0 & 0 \\ 0 & -3 & 0 \\ 0 & 0 & 1 \end{pmatrix} \times \begin{pmatrix} 1 & 0 & 0 \\ 0 & 1 & 0 \\ 0 & 3 & 1 \end{pmatrix}\begin{pmatrix} 1 & -3 & 3 \\ 0 & 1 & -\frac{1}{3} \\ 0 & 0 & 0 \end{pmatrix}$

## MATLAB 1.10

**1. a.** **i.** **F = eye(4); F(3,3) = 4**
**ii.** **F = eye(4); F(1,2) = −3**
**iii.** **F = eye(4); F([1 4],:) = F([4 1],:)**

**b.** **i.** The inverse is the identity except with $\frac{1}{4}$ in the $(3, 3)$ position.
**ii.** The inverse is the identity except with 3 in the $(1, 2)$ position.
**iii.** The inverse of $F$ is the same as $F$.

**3. a.** Below is a possible coding. Some steps may not be necessary for this particular matrix but are included to be complete.

```
U = A;
F1 = eye(3);
F1(2,1) = −U(2,1)/U(1,1),
U = F1*U
F2 = eye(3);
F2(3,1) = −U(3,1)/U(1,1),
U = F2*U
F3 = eye(3);
F3(3,2) = −U(3,2)/U(2,2),
U = F3*U
L = inv(F1)*inv(F2)*inv(F3)
```

$U = \begin{pmatrix} 1 & 2 & 3 \\ 0 & -1 & 4 \\ 0 & 0 & -1 \end{pmatrix}$ and $L = \begin{pmatrix} 1 & 0 & 0 \\ 1 & 1 & 0 \\ 2 & 0 & 1 \end{pmatrix}$

The matrix $L$ is lower triangular with 1's on the diagonal. The $(1, 2)$ entry of $L$ is the negative of the $(1, 2)$ entry in

$F1$, so it holds the negative of the first multiplier used and its position tells which entry was zeroed out. The (1, 3) entry of $L$ is the negative of the (1, 3) entry of $F2$ and the (3, 2) entry $L$ is the negative of the (3, 2) entry in $F3$.

**d.** $U = \begin{pmatrix} 6 & 2 & 7 & 3 \\ 0 & 7.3333 & -8.3333 & 0 \\ 0 & 0 & -1.5 & 3 \\ 0 & 0 & 0 & 26.8182 \end{pmatrix}$

and

$$L = \begin{pmatrix} 1 & 0 & 0 & 0 \\ 1.3333 & 1 & 0 & 0 \\ 1.6667 & .5 & 1 & 0 \\ .6667 & .9091 & -7.9394 & 1 \end{pmatrix}.$$

The coding is similar except that the elementary matrices are $4 \times 4$ and more steps are needed in the reduction.

**Problems 1.11,** page 152

**1.** $\begin{pmatrix} 1 & 0 \\ 3 & 1 \end{pmatrix}\begin{pmatrix} 1 & 2 \\ 0 & -2 \end{pmatrix}$

**3.** $\begin{pmatrix} 1 & 0 \\ -6 & 1 \end{pmatrix}\begin{pmatrix} -1 & 5 \\ 0 & 33 \end{pmatrix}$

**5.** $\begin{pmatrix} 1 & 0 & 0 \\ 2 & 1 & 0 \\ 1 & 0 & 1 \end{pmatrix}\begin{pmatrix} 2 & 1 & 7 \\ 0 & 1 & -9 \\ 0 & 0 & -1 \end{pmatrix}$

**7.** $\begin{pmatrix} 1 & 0 & 0 & 0 \\ 0 & 1 & 0 & 0 \\ 2 & 1 & 1 & 0 \\ 1 & 3 & 4 & 1 \end{pmatrix}\begin{pmatrix} 1 & 2 & -1 & 4 \\ 0 & -1 & 5 & 8 \\ 0 & 0 & -2 & -12 \\ 0 & 0 & 0 & 24 \end{pmatrix}$

**9.** $(8, -5)$

**11.** $(\frac{25}{33}, \frac{5}{33})$

**13.** $(-\frac{63}{2}, 34, 5)$

**15.** $(\frac{71}{12}, -\frac{17}{12}, -\frac{7}{4}, -\frac{11}{24})$

**17. a.** $L = \begin{pmatrix} 1 & 0 \\ 0 & 1 \end{pmatrix}$ $U = \begin{pmatrix} 1 & 4 \\ 0 & 2 \end{pmatrix}$

$P = \begin{pmatrix} 0 & 1 \\ 1 & 0 \end{pmatrix}$

**b.** $(-11, \frac{3}{2})$

**19. a.** $L = \begin{pmatrix} 1 & 0 & 0 \\ 0 & 1 & 0 \\ 0 & \frac{2}{3} & 1 \end{pmatrix}$

$U = \begin{pmatrix} 4 & 1 & 5 \\ 0 & 3 & 7 \\ 0 & 0 & -\frac{2}{3} \end{pmatrix}$

$P = \begin{pmatrix} 0 & 0 & 1 \\ 0 & 1 & 0 \\ 1 & 0 & 0 \end{pmatrix}$

**b.** $(-\frac{1}{2}, -\frac{7}{2}, \frac{3}{2})$

**21. a.** $L = \begin{pmatrix} 1 & 0 & 0 & 0 \\ 0 & 1 & 0 & 0 \\ 0 & \frac{1}{2} & 1 & 0 \\ \frac{1}{2} & -1 & \frac{5}{7} & 1 \end{pmatrix}$

$U = \begin{pmatrix} 2 & 0 & 3 & 1 \\ 0 & 4 & -1 & 5 \\ 0 & 0 & \frac{7}{2} & -\frac{3}{2} \\ 0 & 0 & 0 & \frac{81}{7} \end{pmatrix}$

$P = \begin{pmatrix} 0 & 0 & 1 & 0 \\ 0 & 1 & 0 & 0 \\ 1 & 0 & 0 & 0 \\ 0 & 0 & 0 & 1 \end{pmatrix}$

**b.** $(-\frac{73}{162}, \frac{4}{81}, \frac{53}{54}, -\frac{7}{162})$

**23. a.** $L = \begin{pmatrix} 1 & 0 & 0 & 0 \\ 0 & 1 & 0 & 0 \\ 0 & -\frac{1}{2} & 1 & 0 \\ -\frac{1}{2} & 0 & -\frac{1}{3} & 1 \end{pmatrix}$

$U = \begin{pmatrix} -2 & -4 & 5 & -10 \\ 0 & 4 & -3 & 2 \\ 0 & 0 & \frac{3}{2} & 2 \\ 0 & 0 & 0 & -\frac{7}{3} \end{pmatrix}$

$P = \begin{pmatrix} 0 & 0 & 0 & 1 \\ 0 & 1 & 0 & 0 \\ 1 & 0 & 0 & 0 \\ 0 & 0 & 1 & 0 \end{pmatrix}$

**b.** $(12, \frac{13}{2}, 7, -2)$

**25.** Let $C = AB$, where $A$ and $B$ are $n \times n$ upper triangular matrices. By the definition of an upper triangular matrix, $a_{ij} = 0$ if $i > j$. Similarly,

$b_{ij} = 0$ if $i > j$. Suppose $i > j$. Then

$$c_{ij} = \sum_{k=1}^{n} a_{ik}b_{kj}$$

If $k > j$, then $b_{kj} = 0$, so

$$c_{ij} = \sum_{k=1}^{j} a_{ik}b_{kj}$$

But for $1 \le k \le j$, $i > k$
so $a_{ik} = 0$, which means that $c_{ij} = 0$
for $i > j$ and $C$ is upper triangular.

**27.** The matrix can be factored as $LU$, where

$$U = \begin{pmatrix} 3 & -3 & 2 & 5 \\ 0 & 3 & -\frac{22}{3} & -\frac{10}{3} \\ 0 & 0 & 0 & 0 \\ 0 & 0 & 0 & 0 \end{pmatrix}$$

and

$$L = \begin{pmatrix} 1 & 0 & 0 & 0 \\ \frac{2}{3} & 1 & 0 & 0 \\ \frac{5}{3} & 1 & 1 & 0 \\ \frac{1}{3} & -1 & c & 1 \end{pmatrix}$$

where $c$ can be any real number.

**29.** $\begin{pmatrix} 1 & 0 & 0 \\ -2 & 1 & 0 \\ -1 & 1 & 1 \end{pmatrix}\begin{pmatrix} -1 & 2 & 3 \\ 0 & 5 & 13 \\ 0 & 0 & 0 \end{pmatrix}$

**31.** $\begin{pmatrix} 1 & 0 & 0 & 0 \\ \frac{3}{2} & 1 & 0 & 0 \\ \frac{1}{2} & 1 & 1 & 0 \\ 2 & 2 & c & 1 \end{pmatrix}\begin{pmatrix} 2 & -1 & 1 & 7 \\ 0 & \frac{7}{2} & -\frac{1}{2} & -\frac{9}{2} \\ 0 & 0 & 0 & 0 \\ 0 & 0 & 0 & 0 \end{pmatrix}$
where $c$ is any real number.

**33.** $\begin{pmatrix} 1 & 0 \\ -1 & 1 \end{pmatrix}\begin{pmatrix} 1 & 2 & 3 \\ 0 & 4 & 7 \end{pmatrix}$

**35.** $\begin{pmatrix} 1 & 0 \\ -\frac{2}{7} & 1 \end{pmatrix}\begin{pmatrix} 7 & 1 & 3 & 4 \\ 0 & \frac{37}{7} & \frac{48}{7} & \frac{64}{7} \end{pmatrix}$

**37.** $\begin{pmatrix} 1 & 0 & 0 & 0 & 0 \\ -\frac{2}{5} & 0 & 0 & 0 & 0 \\ \frac{1}{5} & \frac{29}{22} & 1 & 0 & 0 \\ -\frac{2}{5} & \frac{6}{11} & \frac{1}{7} & 1 & 0 \\ 1 & -\frac{10}{11} & -\frac{5}{21} & c & 1 \end{pmatrix}\begin{pmatrix} 5 & 1 & 3 \\ 0 & \frac{22}{5} & \frac{16}{5} \\ 0 & 0 & -\frac{42}{11} \\ 0 & 0 & 0 \\ 0 & 0 & 0 \end{pmatrix}$

**Note:** The TI-85 provides a $U$ with 1's down the diagonal; the $L$ provided does not have 1's down the diagonal. Therefore, in the answers to Problems 39, 41, and 43, we provide two answers: the first gives an $L$ with 1's on the diagonal; the second is the answer given on the TI-85.

**39.** $L = \begin{pmatrix} 1 & 0 & 0 \\ 0 & 1 & 0 \\ \frac{2}{3} & -\frac{5}{6} & 1 \end{pmatrix}$

$$L = \begin{pmatrix} 3 & 0 & 0 \\ 0 & 2 & 0 \\ 2 & -1.6667 & 8.5 \end{pmatrix}$$

$$U = \begin{pmatrix} 3 & 1 & 7 \\ 0 & 2 & 5 \\ 0 & 0 & \frac{17}{2} \end{pmatrix}$$

$$U = \begin{pmatrix} 1 & .3333 & 2.3333 \\ 0 & 1 & 2.5 \\ 0 & 0 & 1 \end{pmatrix}$$

$$P = \begin{pmatrix} 0 & 1 & 0 \\ 1 & 0 & 0 \\ 0 & 0 & 1 \end{pmatrix}$$

$$P = \begin{pmatrix} 0 & 1 & 0 \\ 1 & 0 & 0 \\ 0 & 0 & 1 \end{pmatrix}$$

**41.** $L =$

$$\begin{pmatrix} 1 & 0 & 0 & 0 \\ 0 & 1 & 0 & 0 \\ 0.3125 & -0.6518 & 1 & 0 \\ 0.125 & 0.0536 & -0.1945 & 1 \end{pmatrix}$$

$L =$

$$\begin{pmatrix} 16 & 0 & 0 & 0 \\ 0 & -7 & 0 & 0 \\ 5 & 4.5625 & 8.1696 & 0 \\ 2 & -.3750 & -1.5893 & 2.9760 \end{pmatrix}$$

$$U = \begin{pmatrix} 16 & -5 & 11 & 8 \\ 0 & -7 & 4 & 1 \\ 0 & 0 & 8.1696 & 0.1518 \\ 0 & 0 & 0 & 2.976 \end{pmatrix}$$

$U =$

$$\begin{pmatrix} 1 & -.3125 & .6875 & .5 \\ 0 & 1 & -.5714 & -.1429 \\ 0 & 0 & 1 & .0186 \\ 0 & 0 & 0 & 1 \end{pmatrix}$$

$$P = \begin{pmatrix} 0 & 0 & 0 & 1 \\ 1 & 0 & 0 & 0 \\ 0 & 1 & 0 & 0 \\ 0 & 0 & 1 & 0 \end{pmatrix}$$

$$P = \begin{pmatrix} 0 & 0 & 0 & 1 \\ 1 & 0 & 0 & 0 \\ 0 & 1 & 0 & 0 \\ 0 & 0 & 1 & 0 \end{pmatrix}$$

**43.** $L =$

$$\begin{pmatrix} 1 & 0 & 0 & 0 \\ 0.9121 & 1 & 0 & 0 \\ 0.5055 & -0.0725 & 1 & 0 \\ 0.2308 & 0.5336 & 0.3372 & 1 \end{pmatrix}$$

$$L = \begin{pmatrix} .91 & 0 & 0 & 0 \\ .83 & .5002 & 0 & 0 \\ .46 & -.0363 & .1892 & 0 \\ .21 & .2669 & .0638 & .0495 \end{pmatrix}$$

$U =$

$$\begin{pmatrix} 0.91 & 0.23 & 0.16 & -0.2 \\ 0 & 0.5002 & -0.8259 & 0.9524 \\ 0 & 0 & 0.1892 & -0.4199 \\ 0 & 0 & 0 & 0.0495 \end{pmatrix}$$

$$U = \begin{pmatrix} 1 & .2527 & .1758 & -.2198 \\ 0 & 1 & -1.6511 & 1.9040 \\ 0 & 0 & 1 & -2.2186 \\ 0 & 0 & 0 & 1 \end{pmatrix}$$

$$P = \begin{pmatrix} 0 & 1 & 0 & 0 \\ 0 & 0 & 0 & 1 \\ 0 & 0 & 1 & 0 \\ 1 & 0 & 0 & 0 \end{pmatrix}$$

$$P = \begin{pmatrix} 0 & 1 & 0 & 0 \\ 0 & 0 & 0 & 1 \\ 0 & 0 & 1 & 0 \\ 1 & 0 & 0 & 0 \end{pmatrix}$$

## MATLAB 1.11

**1.** The coding would be the same as in MATLAB Problem 3(a) in Section 1.10:

$$U = \begin{pmatrix} 8 & 2 & -4 & 6 \\ 0 & -1.5 & -3 & 1.5 \\ 0 & 0 & 0 & 6 \end{pmatrix} \quad \text{and}$$

$$L = \begin{pmatrix} 1 & 0 & 0 \\ 1.25 & 1 & 0 \\ .5 & -4 & 1 \end{pmatrix}$$

**3.** For the interpretation, refer to MATLAB Problem 3 in Section 1.10. The coding is very similar except that you will need six elementary matrices to complete the reduction and each elementary matrix is $4 \times 4$.

## Problems 1.12, page 163

**1.** $\begin{pmatrix} 0 & 1 & 1 & 0 \\ 0 & 0 & 0 & 0 \\ 1 & 1 & 0 & 0 \\ 1 & 0 & 0 & 0 \end{pmatrix}$ **3.** $\begin{pmatrix} 0 & 0 & 1 & 0 & 0 \\ 1 & 0 & 0 & 1 & 0 \\ 0 & 1 & 0 & 1 & 0 \\ 1 & 1 & 1 & 0 & 1 \\ 0 & 1 & 0 & 0 & 0 \end{pmatrix}$

**5.**

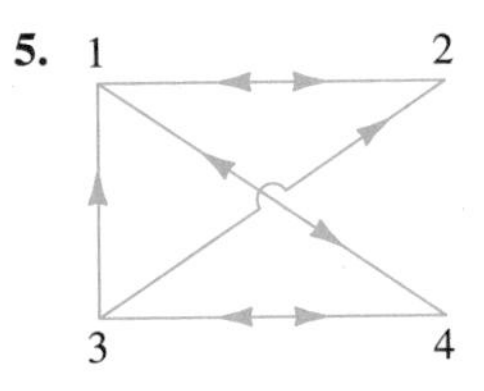


**7.**

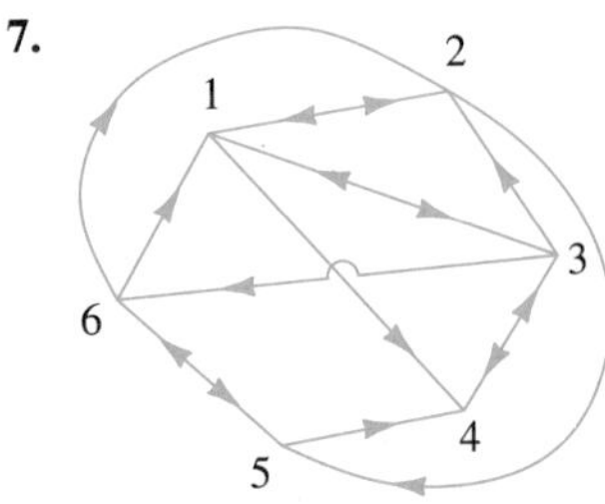


**9.** There is one 2-chain between 12, 14, 21, 22, 25, 32, 33, 34, 35, 41, 43, 51, 54; two 2-chains between 23, 31, 42, 44; one 3-chain between 12, 13, 14, 15, 21, 23, 35, 51, 52, 55; two 3-chains between 11, 31, 34, 45, 53; three 3-chains between 22, 24, 32, 33, 42, 43, 44; four 3-chains between 41; one 4-chain between 15, 51, 53; two 4-chains between 11, 14, 35; three 4-chains between 12, 13, 25, 45, 52, 54; four 4-chains between 22, 23, 24, 33; five 4-chains between 31;

six 4-chains between 21, 32, 34, 41, 44; seven 4-chains between 43; eight 4-chains between 42.

**11.** Let $B = A + A^2$. Then $(B)_{ij} = (A)_{ij} + (A^2)_{ij}$. But $(A)_{ij}$ is the number of one-step links between vertices $i$ and $j$, and $(A^2)_{ij}$ is the number of two-step links between vertices $i$ and $j$. So $(B)_{ij}$ is the total number of one-step or two-step links between vertices $i$ and $j$.

## Chapter 1—Review, page 168

**1.** $(\frac{1}{7}, \frac{10}{7})$ **3.** no solution

**5.** $(0, 0, 0)$ **7.** $(-\frac{1}{2}, 0, \frac{5}{2})$

**9.** $(\frac{1}{3}x_3, \frac{7}{3}x_3, x_3)$, $x_3$ arbitrary

**11.** no solution

**13.** $(0, 0, 0, 0)$

**15.** $\begin{pmatrix} -6 & 3 \\ 0 & 12 \\ 6 & 9 \end{pmatrix}$

**17.** $\begin{pmatrix} 16 & 2 & 3 \\ -20 & 10 & -1 \\ -36 & 8 & 16 \end{pmatrix}$

**19.** $\begin{pmatrix} 17 & 39 & 41 \\ 14 & 20 & 42 \end{pmatrix}$

**21.** $\begin{pmatrix} 9 & 10 \\ 30 & 32 \end{pmatrix}$

**23.** reduced row echelon form

**25.** neither

**27.** row echelon form:

$$\begin{pmatrix} 1 & 4 & -1 \\ 0 & 1 & \frac{5}{4} \end{pmatrix}$$

reduced row echelon form:

$$\begin{pmatrix} 1 & 0 & -6 \\ 0 & 1 & \frac{5}{4} \end{pmatrix}$$

**29.** $\begin{pmatrix} 1 & \frac{3}{2} \\ 0 & 1 \end{pmatrix}$; inverse is

$$\begin{pmatrix} \frac{4}{11} & -\frac{3}{11} \\ \frac{1}{11} & \frac{2}{11} \end{pmatrix}$$

**31.** $\begin{pmatrix} 1 & 2 & 0 \\ 0 & 1 & \frac{1}{3} \\ 0 & 0 & 1 \end{pmatrix}$; inverse is

$$\begin{pmatrix} -\frac{1}{4} & \frac{1}{4} & \frac{1}{4} \\ \frac{5}{8} & -\frac{1}{8} & -\frac{1}{8} \\ \frac{1}{8} & -\frac{5}{8} & \frac{3}{8} \end{pmatrix}$$

**33.** $\begin{pmatrix} 1 & 0 & 2 \\ 0 & 1 & 1 \\ 0 & 0 & 1 \end{pmatrix}$; inverse is

$$\begin{pmatrix} \frac{5}{6} & \frac{2}{3} & -2 \\ \frac{1}{3} & \frac{2}{3} & -1 \\ -\frac{1}{6} & -\frac{1}{3} & 1 \end{pmatrix}$$

**35.** $\begin{pmatrix} 1 & 2 & 0 \\ 2 & 1 & -1 \\ 3 & 1 & 1 \end{pmatrix}\begin{pmatrix} x_1 \\ x_2 \\ x_3 \end{pmatrix} = \begin{pmatrix} 3 \\ -1 \\ 7 \end{pmatrix}$; $A^{-1}$ is given in Exercise 31;

$$x_1 = \tfrac{3}{4}, \quad x_2 = \tfrac{9}{8}, \quad x_3 = \tfrac{29}{8}$$

**37.** $\begin{pmatrix} 2 & -1 \\ 3 & 0 \\ 1 & 2 \end{pmatrix}$; neither

**39.** $\begin{pmatrix} 2 & 3 & 1 \\ 3 & -6 & -5 \\ 1 & -5 & 9 \end{pmatrix}$; symmetric

**41.** $\begin{pmatrix} 1 & -1 & 4 & 6 \\ -1 & 2 & 5 & 7 \\ 4 & 5 & 3 & -8 \\ 6 & 7 & -8 & 9 \end{pmatrix}$; symmetric

**43.** $\begin{pmatrix} 1 & 0 & 0 \\ 0 & -2 & 0 \\ 0 & 0 & 1 \end{pmatrix}$

**45.** $\begin{pmatrix} 1 & 0 & 0 \\ 0 & 1 & 0 \\ -5 & 0 & 1 \end{pmatrix}$

**47.** $\begin{pmatrix} 1 & 0 & 0 \\ 0 & 1 & \frac{1}{5} \\ 0 & 0 & 1 \end{pmatrix}$

**49.** $\begin{pmatrix} 0 & 1 & 0 \\ 1 & 0 & 0 \\ 0 & 0 & 1 \end{pmatrix}$

**51.** $\begin{pmatrix} 2 & 0 \\ 0 & 1 \end{pmatrix}\begin{pmatrix} 1 & 0 \\ -1 & 1 \end{pmatrix} \times$
$\begin{pmatrix} 1 & 0 \\ 0 & \frac{1}{2} \end{pmatrix}\begin{pmatrix} 1 & -\frac{1}{2} \\ 0 & 1 \end{pmatrix}$

**53.** $\begin{pmatrix} 2 & 0 \\ 0 & 1 \end{pmatrix}\begin{pmatrix} 1 & 0 \\ -4 & 1 \end{pmatrix}\begin{pmatrix} 1 & -\frac{1}{2} \\ 0 & 0 \end{pmatrix}$

**55.** $\begin{pmatrix} 1 & 0 & 0 \\ 2 & 1 & 0 \\ 4 & -5 & 1 \end{pmatrix}\begin{pmatrix} 1 & -2 & 5 \\ 0 & -1 & -3 \\ 0 & 0 & -27 \end{pmatrix}$;
$(\frac{76}{27}, -\frac{7}{9}, -\frac{29}{27})$

**57.** $L = \begin{pmatrix} 1 & 0 & 0 \\ \frac{1}{3} & 1 & 0 \\ 0 & -\frac{3}{4} & 1 \end{pmatrix}$,
$U = \begin{pmatrix} 3 & 5 & 8 \\ 0 & \frac{4}{3} & -\frac{14}{3} \\ 0 & 0 & \frac{1}{2} \end{pmatrix}$
$P = \begin{pmatrix} 0 & 1 & 0 \\ 0 & 0 & 1 \\ 1 & 0 & 0 \end{pmatrix}$; $(-47, 19, \frac{11}{2})$

**59.** $\begin{pmatrix} 0 & 1 & 0 & 0 \\ 0 & 0 & 1 & 0 \\ 1 & 0 & 0 & 1 \\ 0 & 1 & 0 & 0 \end{pmatrix}$

**61.**

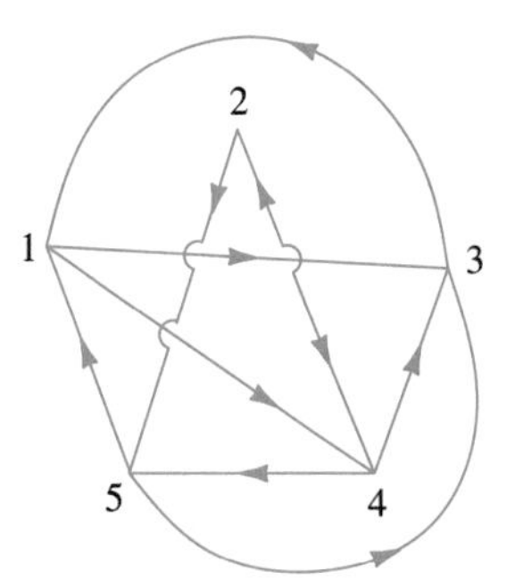


## Chapter 2

**Problems 2.1,** page 182

**1.** $-10$ **3.** 47 **5.** 4 **7.** 56

**9.** 274

**11.** Let $A = \begin{pmatrix} a_{11} & 0 & 0 & \cdots & 0 \\ 0 & a_{22} & 0 & \cdots & 0 \\ 0 & 0 & a_{33} & \cdots & 0 \\ \vdots & \vdots & \vdots & & \vdots \\ 0 & 0 & 0 & \cdots & a_{nn} \end{pmatrix}$

and

$B = \begin{pmatrix} b_{11} & 0 & 0 & \cdots & 0 \\ 0 & b_{22} & 0 & \cdots & 0 \\ 0 & 0 & b_{33} & \cdots & 0 \\ \vdots & \vdots & \vdots & & \vdots \\ 0 & 0 & 0 & \cdots & b_{nn} \end{pmatrix}$

Then $\det A = a_{11}a_{22}a_{33}\cdots a_{nn}$, $\det B = b_{11}b_{22}b_{33}\cdots b_{nn}$,

$AB =$
$$\begin{pmatrix} a_{11}b_{11} & 0 & 0 & \cdots & 0 \\ 0 & a_{22}b_{22} & 0 & \cdots & 0 \\ 0 & 0 & a_{33}b_{33} & \cdots & 0 \\ \vdots & \vdots & \vdots & & \vdots \\ 0 & 0 & 0 & \cdots & a_{nn}b_{nn} \end{pmatrix}$$

and

$$\begin{aligned} \det AB &= (a_{11}b_{11})(a_{22}b_{22}) \\ &\quad \times (a_{33}b_{33})\cdots(a_{nn}b_{nn}) \\ &= (a_{11}a_{22}a_{33}\cdots a_{nn}) \\ &\quad \times (b_{11}b_{22}b_{33}\cdots b_{nn}) \\ &= \det A \det B \end{aligned}$$

**13.** Almost any example will work. For instance, $\det \begin{pmatrix} 1 & 0 \\ 0 & 1 \end{pmatrix} = 1$, but

$\det\begin{pmatrix}1 & 0\\ 0 & 0\end{pmatrix} + \det\begin{pmatrix}0 & 0\\ 0 & 1\end{pmatrix} =$
$0 + 0 \neq 1.$

As another example, let $A = \begin{pmatrix}1 & 2\\ 3 & 4\end{pmatrix}$ and $B = \begin{pmatrix}5 & 6\\ 7 & 8\end{pmatrix}$; then $(A + B) = \begin{pmatrix}6 & 8\\ 10 & 12\end{pmatrix}$, $\det A = -2$, $\det B = -2$, and $\det(A + B) = -8 \neq \det A + \det B$.

**15.** Suppose that $A = (0, c)$. Then the situation is as depicted in the figure.

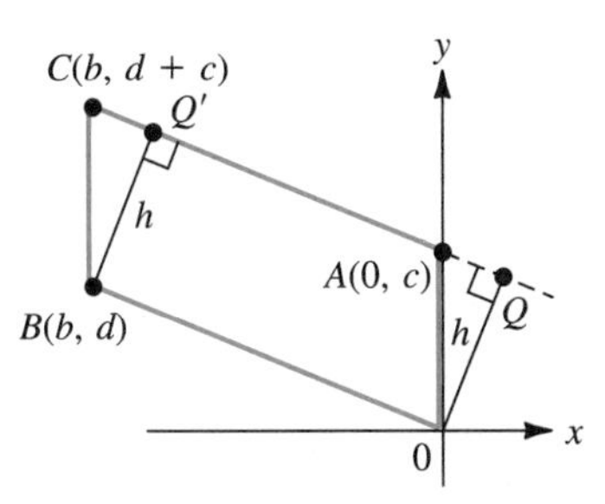


The base of the parallelogram is $\overline{AC} = \sqrt{b^2 + d^2}$. The slope of $AC$ is $\frac{d}{b}$, so the slope of $BQ' = -\frac{b}{d}$ and the equation of the line passing through 0 and $Q$ is $y = -\frac{b}{d}x$. The equation of line $AC$ is $y = c + \frac{d}{b}x$. $Q$ is the point of intersection of these two lines and is

$$\left(\frac{d(-bc)}{b^2 + d^2}, \frac{-b(-bc)}{b^2 + d^2}\right)$$

$$= \left(\frac{d \det A}{b^2 + d^2}, -\frac{b \det A}{b^2 + d^2}\right)$$

Then $h = \overline{0Q} = \frac{|\det A|}{\sqrt{b^2 + d^2}}$, so base $\times$ height $= \overline{AC} \times \overline{0Q} = |\det A|$.

A similar proof works in the case $A = (a, 0)$.

**17.** 40,954

**19.** $1.91524617423 \times 10^{14}$

## MATLAB 2.1

**1.** $A$ is invertible if $\det(A) \neq 0$ and is not invertible if $\det(A) = 0$. The matrices constructed in part (b)(ii) are never invertible and the determinants will be zero (consider very small numbers to be zero due to round-off).

**3.** $\det(A + B) \neq \det(A) + \det(B)$

**5.** $\det(A^{-1}) = 1/\det(A)$. *Proof hint:* Use $AA^{-1} = I$ and take the determinant of both sides.

**7. a.** $\det(M) = \det(A)\det(D)$
**b.** $\det(M) = \det(A)\det(D)\det(F)$.

## Problems 2.2, page 200

**1.** 28 **3.** 2 **5.** 32 **7.** −36

**9.** −260 **11.** −183 **13.** 24

**15.** −296 **17.** 138

**19.** *abcde* **21.** −8 **23.** 16

**25.** −16 **27.** −16

**29.** proof by induction: true for $n = 2$ since

$$\begin{vmatrix}1 + x_1 & x_2\\ x_1 & 1 + x_2\end{vmatrix}$$
$$= (1 + x_1)(1 + x_2) - x_1x_2$$
$$= 1 + x_1 + x_2$$

Assume true for $n = k$. That is,

$$\begin{vmatrix} 1+x_1 & x_2 & x_3 & \cdots & x_k \\ x_1 & 1+x_2 & x_3 & \cdots & x_k \\ x_1 & x_2 & 1+x_3 & \cdots & x_k \\ \vdots & \vdots & \vdots & & \vdots \\ x_1 & x_2 & x_3 & \cdots & 1+x_k \end{vmatrix}$$
$$= 1 + x_1 + x_2 + \cdots + x_k$$

Then, for $n = k + 1$,

$$\begin{vmatrix} 1+x_1 & x_2 & x_3 & \cdots & x_k & x_{k+1} \\ x_1 & 1+x_2 & x_3 & \cdots & x_k & x_{k+1} \\ x_1 & x_2 & 1+x_3 & \cdots & x_k & x_{k+1} \\ \vdots & \vdots & \vdots & & \vdots & \vdots \\ x_1 & x_2 & x_3 & \cdots & x_k & 1+x_{k+1} \end{vmatrix}$$

(using Property 3 in the first column)

$$= \begin{vmatrix} 1 & x_2 & x_3 & \cdots & x_k & x_{k+1} \\ 0 & 1+x_2 & x_3 & \cdots & x_k & x_{k+1} \\ 0 & x_2 & 1+x_3 & \cdots & x_k & x_{k+1} \\ \vdots & \vdots & \vdots & & \vdots & \vdots \\ 0 & x_2 & x_3 & \cdots & x_k & 1+x_{k+1} \end{vmatrix} \text{①}$$

$$+ \begin{vmatrix} x_1 & x_2 & x_3 & \cdots & x_k & x_{k+1} \\ x_1 & 1+x_2 & x_3 & \cdots & x_k & x_{k+1} \\ x_1 & x_2 & 1+x_3 & \cdots & x_k & x_{k+1} \\ \vdots & \vdots & \vdots & & \vdots & \vdots \\ x_1 & x_2 & x_3 & \cdots & x_k & 1+x_{k+1} \end{vmatrix} \text{②}$$

But, expanding det ① in its first column, we have

$$\det \text{①} = \begin{vmatrix} 1+x_2 & x_3 & \cdots & x_k & x_{k+1} \\ x_2 & 1+x_3 & \cdots & x_k & x_{k+1} \\ \vdots & \vdots & & \vdots & \vdots \\ x_2 & x_3 & \cdots & x_k & 1+x_{k+1} \end{vmatrix}$$
$$= 1 + x_2 + x_3 + \cdots + x_{k+1}$$

by the induction assumption (since ① is a $k \times k$ determinant). To evaluate det ②, subtract the first row from all other rows:

$$\det \text{②} = \begin{vmatrix} x_1 & x_2 & x_3 & \cdots & x_k & x_{k+1} \\ 0 & 1 & 0 & \cdots & 0 & 0 \\ 0 & 0 & 1 & \cdots & 0 & 0 \\ \vdots & \vdots & \vdots & & \vdots & \vdots \\ 0 & 0 & 0 & \cdots & 0 & 1 \end{vmatrix} = x_1$$

Adding det ① and det ② completes the proof.

**31.** If $n$ is odd, $\det A = -\det A$, so that $2 \det A = 0$ and $\det A = 0$.

**33.** $\dfrac{1}{2}\begin{vmatrix} 1 & x_1 & y_1 \\ 1 & x_2 & y_2 \\ 1 & x_3 & y_3 \end{vmatrix}$

$$= \frac{1}{2}\begin{vmatrix} x_2 - x_1 & x_3 - x_1 \\ y_2 - y_1 & y_3 - y_1 \end{vmatrix}$$

Look at the figures below.

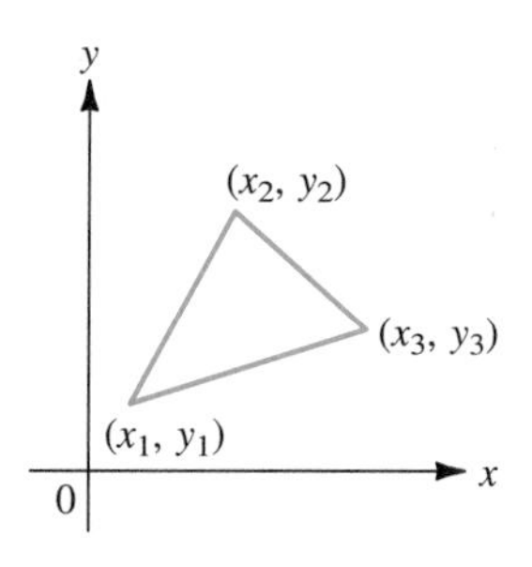


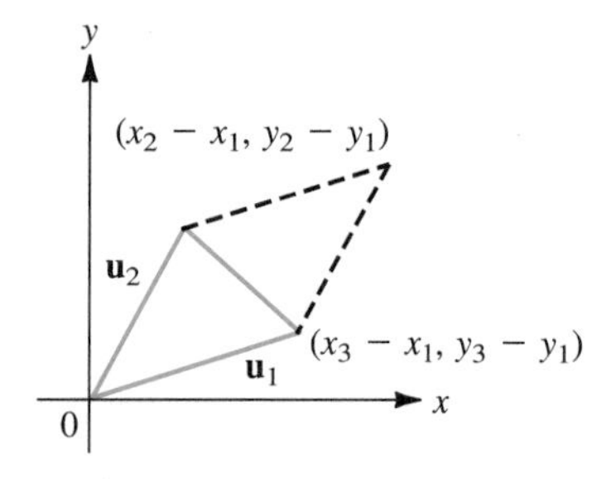


The area $A$ of the triangle is half the area of the parallelogram generated by the vectors $\mathbf{u}_1$ and $\mathbf{u}_2$, which, by the result of Problem 2.1.16, is given by

$$A = \pm\frac{1}{2}\begin{vmatrix} x_2 - x_1 & x_3 - x_1 \\ y_2 - y_1 & y_3 - y_1 \end{vmatrix}$$

**35.** $$D_3 = \begin{vmatrix} 1 & 1 & 1 \\ a_1 & a_2 & a_3 \\ a_1^2 & a_2^2 & a_3^2 \end{vmatrix}$$

$$= \begin{vmatrix} 1 & 0 & 0 \\ a_1 & a_2 - a_1 & a_3 - a_1 \\ a_1^2 & a_2^2 - a_1^2 & a_3^2 - a_1^2 \end{vmatrix}$$

$$= \begin{vmatrix} a_2 - a_1 & a_3 - a_1 \\ (a_2 + a_1)(a_2 - a_1) & (a_3 - a_1)(a_3 - a_1) \end{vmatrix}$$

$$= (a_2 - a_1)(a_3 - a_1) \times \begin{vmatrix} 1 & 1 \\ a_2 + a_1 & a_3 + a_1 \end{vmatrix}$$

$$= (a_2 - a_1)(a_3 - a_1) \times (a_3 - a_2)$$

**37. a.** $$D_n = \begin{vmatrix} 1 & 1 & \cdots & 1 \\ a_1 & a_2 & \cdots & a_n \\ a_1^2 & a_2^2 & \cdots & a_n^2 \\ \vdots & \vdots & & \vdots \\ a_1^{n-1} & a_2^{n-1} & \cdots & a_n^{n-1} \end{vmatrix}$$

**b.** We prove this by induction. The result is true for $n = 3$ by the result of Problem 35. We assume it true for $n = k$. Now

$$D_{k+1} = \begin{vmatrix} 1 & 1 & \cdots & 1 & 1 \\ a_1 & a_2 & \cdots & a_k & a_{k+1} \\ a_1^2 & a_2^2 & \cdots & a_k^2 & a_{k+1}^2 \\ \vdots & \vdots & & \vdots & \vdots \\ a_1^{k-1} & a_2^{k-1} & \cdots & a_k^{k-1} & a_{k+1}^{k-1} \\ a_1^k & a_2^k & \cdots & a_k^k & a_{k+1}^k \end{vmatrix}$$

We subtract the first column from each of the other $k$ columns:

$$D_{k+1} = \begin{vmatrix} 1 & 0 & \cdots & 0 & 0 \\ a_1 & a_2 - a_1 & \cdots & a_k - a_1 & a_{k+1} - a_1 \\ a_1^2 & a_2^2 - a_1^2 & \cdots & a_k^2 - a_1^2 & a_{k+1}^2 - a_1^2 \\ \vdots & \vdots & & \vdots & \vdots \\ a_1^{k-1} & a_2^{k-1} - a_1^{k-1} & \cdots & a_k^{k-1} - a_1^{k-1} & a_{k+1}^{k-1} - a_1^{k-1} \\ a_1^k & a_2^k - a_1^k & \cdots & a_k^k - a_1^k & a_{k+1}^k - a_1^k \end{vmatrix}$$

$$= \begin{vmatrix} a_2 - a_1 & a_3 - a_1 & \cdots & a_k - a_1 & a_{k+1} - a_1 \\ a_2^2 - a_1^2 & a_3^2 - a_1^2 & \cdots & a_k^2 - a_1^2 & a_{k+1}^2 - a_1^2 \\ \vdots & \vdots & & \vdots & \vdots \\ a_2^{k-1} - a_1^{k-1} & a_3^{k-1} - a_1^{k-1} & \cdots & a_k^{k-1} - a_1^{k-1} & a_{k+1}^{k-1} - a_1^{k-1} \\ a_2^k - a_1^k & a_3^k - a_1^k & \cdots & a_k^k - a_1^k & a_{k+1}^k - a_1^k \end{vmatrix}$$

Now $a_2^k - a_1^k = (a_2 - a_1) \times (a_2^{k-1} + a_2^{k-2}a_1 + a_2^{k-3}a_1^2 + \cdots + a_2^2a_1^{k-3} + a_2a_1^{k-2} + a_1^{k-1})$, and $a_2^{k-1} - a_1^{k-1} = (a_2 - a_1) \times (a_2^{k-2} + a_2^{k-3}a_1 + \cdots + a_2^2a_1^{k-4} + a_2a_1^{k-3} + a_1^{k-2})$. Note that if the terms in the second factor of the last expression are multiplied by $a_1$ and then subtracted from the second factor of $a_2^k - a_1^k$, only the term $a_2^{k-1}$ remains. Thus we **(i)** expand the last determinant obtained above in the first row, **(ii)** factor $a_{j-1} - a_1$ from the $j$th column for $1 \le j \le k$, and **(iii)** multiply the $l$th row by $a_1$ and subtract it from the $(l + 1)$st row, for $l = k - 1, k - 2, \ldots, 3, 2$ in succession. This yields $D_{k+1} = (a_2 - a_1)(a_3 - a_1)\cdots(a_{k+1} - a_1)$

$$\times \begin{vmatrix} 1 & 1 & \cdots & 1 & 1 \\ a_2 & a_3 & \cdots & a_k & a_{k+1} \\ a_2^2 & a_3^2 & \cdots & a_k^2 & a_{k+1}^2 \\ \vdots & \vdots & & \vdots & \vdots \\ a_2^{k-1} & a_3^{k-1} & \cdots & a_k^{k-1} & a_{k+1}^{k-1} \\ a_2^k & a_3^k & \cdots & a_k^k & a_{k+1}^k \end{vmatrix}$$

$$= \prod_{j=2}^{k+1} (a_j - a_1) \prod_{\substack{i=2 \\ j>i}}^{k+1} (a_j - a_i)$$

(from the induction assumption since the last determinant is $k \times k$)

$$= \prod_{\substack{i=1 \\ j>i}}^{k+1} (a_j - a_i)$$

This completes the proof.

**39.** **a.** $A^2 = \begin{pmatrix} 0 & 0 \\ 0 & 0 \end{pmatrix}$; $k = 2$

**b.** $A^3 = \begin{pmatrix} 0 & 0 & 0 \\ 0 & 0 & 0 \\ 0 & 0 & 0 \end{pmatrix}$; $k = 3$

**41.** $\det A^2 = \det A \det A = \det A$. If $\det A \neq 0$, then $\det A = 1$. The answer is 0 or 1.

**43.** Let $Q$ be an elementary permutation matrix so $Q$ is obtained by interchanging two rows, rows $i$ and $j$, of $I$. The $j$th row of $I$ has a 1 in the $j$th column, so the $i$th row of $Q$ has a 1 in the $j$th column. That is, $Q_{ij} = 1$. Similarly, $Q_{ji} = 1$. Thus $Q_{ij} = Q_{ji}$. The only other nonzero components of $Q$ are 1's on the diagonal and diagonal components stay put when the transpose is taken. Thus $Q^t = Q$. Now, if $P$ is a permutation matrix, then

$$P = P_n P_{n-1} \cdots P_2 P_1$$

where each $P_i$ is an elementary permutation matrix. Then, by Theorem 1:

$$\det P = \det P_n \det P_{n-1} \cdots \det P_2 \det P_1 = (-1)^n$$

by the result of Problem 42. Also, by Theorem 1.9.1 (ii),

$$P^t = P_1^t P_2^t \cdots P_{n-1}^t P_n^t = P_1 P_2 \cdots P_{n-1} P_n$$

Thus $P^t$ is a permutation matrix and, as above,

$$\det P^t = (-1)^n = \det P$$

## MATLAB 2.2

**1.** $\det(kA) = k^n \det(A)$, where $A$ is $n \times n$. *Proof hint*: $kA$ multiplies each of the $n$ rows of $A$ by $k$.

## Problems 2.3, page 210

**1.** $EB$ is the matrix obtained by permuting two rows of $B$. By Property 4, $\det EB = -\det B$. By Problem 2.2.42, $\det E = -1$. Thus $-\det B = \det E \det B$.

**3.** $EB$ is the matrix obtained by multiplying the $i$th row of $B$ by $c$. By Property 2, $\det EB = c \det B$. $E$ is the matrix obtained by multiplying the $i$th row of $I$ by $c$. Thus

$$\det E = c \det I = c \quad \text{and} \quad \det EB = c \det B = \det E \det B$$

## Problems 2.4, page 216

**1.** $\begin{pmatrix} \frac{1}{2} & -\frac{1}{2} \\ -\frac{1}{4} & \frac{3}{4} \end{pmatrix}$ **3.** $\begin{pmatrix} 0 & 1 \\ 1 & 0 \end{pmatrix}$

**5.** $\begin{pmatrix} \frac{1}{3} & -\frac{1}{4} & -\frac{1}{6} \\ 0 & \frac{1}{4} & \frac{1}{2} \\ 0 & \frac{1}{4} & -\frac{1}{2} \end{pmatrix}$

**7.** $\begin{pmatrix} 0 & 1 & -1 \\ 2 & -2 & -1 \\ -1 & 1 & 1 \end{pmatrix}$

**9.** not invertible

**11.** $\begin{pmatrix} \frac{7}{3} & -\frac{1}{3} & -\frac{1}{3} & -\frac{2}{3} \\ \frac{4}{9} & -\frac{1}{9} & -\frac{4}{9} & \frac{1}{9} \\ -\frac{1}{9} & -\frac{2}{9} & \frac{1}{9} & \frac{2}{9} \\ -\frac{5}{3} & \frac{2}{3} & \frac{2}{3} & \frac{1}{3} \end{pmatrix}$

**13.** Follows from the fact that $\det A^t = \det A$.

**15.** $A^{-1} = \begin{pmatrix} \frac{1}{14} & \frac{1}{14} & \frac{9}{28} \\ -\frac{5}{7} & \frac{2}{7} & -\frac{3}{14} \\ \frac{1}{14} & \frac{1}{14} & -\frac{5}{28} \end{pmatrix}$,

$\det A = -28, \quad \det A^{-1} = -\frac{1}{28}$

**17.** no inverse if $\alpha$ is any real number

**19.** Its determinant is $\cos^2\theta + \sin^2\theta = 1$; Its inverse is $\begin{pmatrix} \cos\theta & -\sin\theta \\ \sin\theta & \cos\theta \end{pmatrix}$

### MATLAB 2.4

**1.** For $n < m$, that is, more columns than rows, $\det(A^tA) = 0$ (or very small due to round-off), so $A^tA$ is not invertible. For $n > m$, that is, more rows than columns, $A^tA$ can be invertible.

**3.** We have that the reduced row echelon form of $A$ is the identity so that $A$ is invertible, although, by construction, it is close to being noninvertible. We have that $\det(A) = 6.55$, which is not close to zero.

### Problems 2.5, page 222

**1.** $x_1 = -5, x_2 = 3$

**3.** $x_1 = 2, x_2 = 5, x_3 = -3$

**5.** $x_1 = \frac{45}{13}, x_2 = -\frac{11}{13}, x_3 = \frac{23}{13}$

**7.** $x_1 = \frac{3}{2}, x_2 = \frac{3}{2}, x_3 = \frac{1}{2}$

**9.** $x_1 = \frac{21}{29}, x_2 = \frac{171}{29}, x_3 = -\frac{284}{29}, x_4 = -\frac{182}{29}$

### MATLAB 2.5

**1.** Here is a sample of the coding for part (a):

```
d = det(A); C = A; C(:,1) = b;
                    x1 = det(C)/d
C = A; C(:,2) = b; x2 = det(C)/d
```

Continue in this fashion and let **x = [x1;x2;x3;x4;x5].**

The flop count is greater for Cramer's rule than for the "\" command (*LU* decomposition). The difference in flop counts is even greater for the larger size matrix.

### Chapter 2—Review, page 225

**1.** $-4$ **3.** 24 **5.** 60 **7.** 34

**9.** $\begin{pmatrix} -\frac{1}{11} & \frac{4}{11} \\ \frac{2}{11} & \frac{3}{11} \end{pmatrix}$

**11.** not invertible

**13.** $\begin{pmatrix} \frac{1}{11} & \frac{1}{11} & 0 & \frac{3}{11} \\ \frac{9}{11} & -\frac{2}{11} & 0 & -\frac{6}{11} \\ \frac{3}{11} & \frac{3}{11} & 0 & -\frac{2}{11} \\ \frac{1}{22} & \frac{1}{22} & -\frac{1}{2} & \frac{3}{22} \end{pmatrix}$

**15.** $x_1 = \frac{11}{7}, x_2 = \frac{1}{7}$

**17.** $x_1 = \frac{1}{4}, x_2 = \frac{5}{4}, x_3 = -\frac{3}{4}$

## Chapter 3

### Problems 3.1, page 236

**1.** $|\mathbf{v}| = 4\sqrt{2}, \theta = \pi/4$

**3.** $|\mathbf{v}| = 4\sqrt{2}, \theta = 7\pi/4$

**5.** $|\mathbf{v}| = 2, \theta = \pi/6$

**7.** $|\mathbf{v}| = 2, \theta = 2\pi/3$

**9.** $|\mathbf{v}| = 2, \theta = 4\pi/3$

**11.** $|\mathbf{v}| = \sqrt{89}, \theta = \pi + \tan^{-1}(-\frac{8}{5}) \approx 2.13$ (in the second quadrant)

**13. a.** (6, 9)

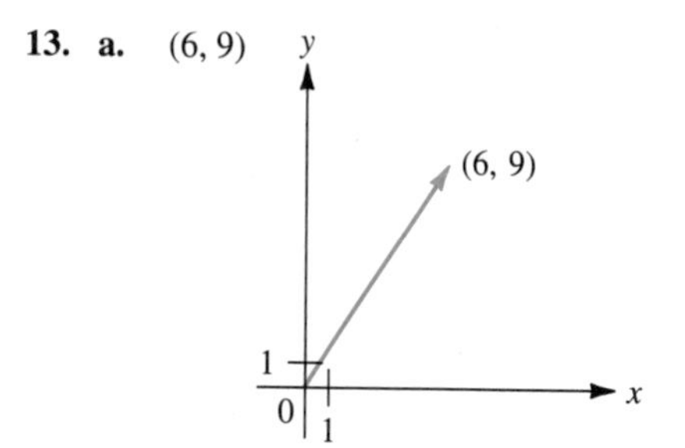


**b.** $(-3, 7)$

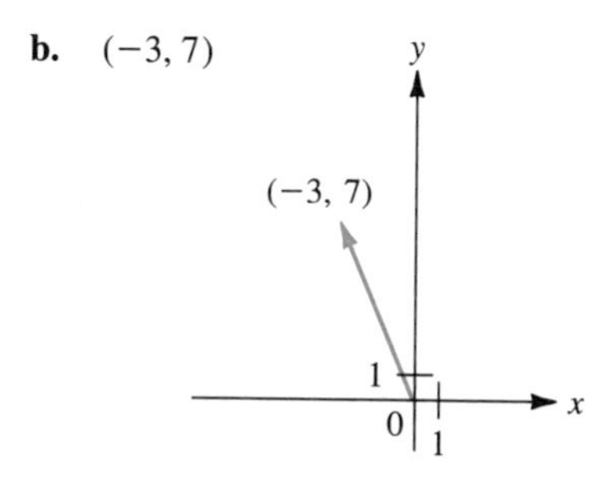

**c.** $(-7, 1)$

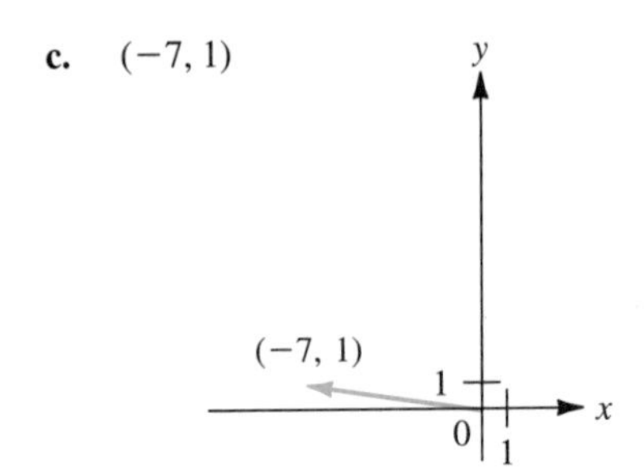


**d.** $(39, -22)$

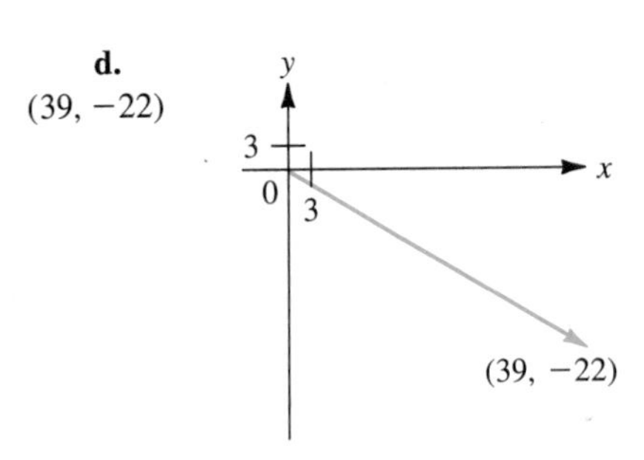


**15.** $|\mathbf{i}| = |(1, 0)| = \sqrt{1^2 + 0^2} = \sqrt{1} = 1;$
$|\mathbf{j}| = |(0, 1)| = \sqrt{0^2 + 1^2} = \sqrt{1} = 1$

**17.** $$|\mathbf{u}| = \sqrt{\left(\frac{a}{\sqrt{a^2+b^2}}\right)^2 + \left(\frac{b}{\sqrt{a^2+b^2}}\right)^2} = \sqrt{\frac{a^2}{a^2+b^2} + \frac{b^2}{a^2+b^2}} = 1$$

$$\text{Direction of } |\mathbf{u}| = \tan^{-1}\frac{\dfrac{b}{\sqrt{a^2+b^2}}}{\dfrac{a}{\sqrt{a^2+b^2}}} = \tan^{-1}\left(\frac{b}{a}\right) = \text{direction of } \mathbf{v}.$$

**19.** $(1/\sqrt{2})\mathbf{i} - (1/\sqrt{2})\mathbf{j}$

**21.** $(1/\sqrt{2})\mathbf{i} + (1/\sqrt{2})\mathbf{j}$ if $a > 0$; $-(1/\sqrt{2})\mathbf{i} - (1/\sqrt{2})\mathbf{j}$ if $a < 0$

**23.** $\sin\theta = -3/\sqrt{13}$, $\cos\theta = 2/\sqrt{13}$

**25.** $-(1/\sqrt{2})\mathbf{i} - (1/\sqrt{2})\mathbf{j}$

**27.** $\frac{3}{5}\mathbf{i} - \frac{4}{5}\mathbf{j}$

**29.** **a.** $(1/\sqrt{2})\mathbf{i} - (1/\sqrt{2})\mathbf{j}$
**b.** $(7/\sqrt{193})\mathbf{i} - (12/\sqrt{193})\mathbf{j}$
**c.** $-(2/\sqrt{53})\mathbf{i} + (7/\sqrt{53})\mathbf{j}$

**31.** $\overrightarrow{PQ}$ is a representation of $(c + a - c)\mathbf{i} + (d + b - d)\mathbf{j} = a\mathbf{i} + b\mathbf{j}$. Thus $\overrightarrow{PQ}$ and $(a, b)$ are representations of the same vector.

**33.** $4\mathbf{i} + 4\sqrt{3}\mathbf{j}$ **35.** $-3\mathbf{i} + 3\sqrt{3}\mathbf{j}$

**37.** **i.** Suppose $\mathbf{u} = \alpha\mathbf{v}$ where $\alpha > 0$. Then $|\mathbf{u} + \mathbf{v}| = |\alpha\mathbf{v} + \mathbf{v}| = |(\alpha + 1)\mathbf{v}| = |\alpha + 1|\,|\mathbf{v}| = (\alpha + 1)|\mathbf{v}|$ (since $\alpha + 1 > 0$) $= \alpha|\mathbf{v}| + |\mathbf{v}| = |\alpha\mathbf{v}| + |\mathbf{v}| = |\mathbf{u}| + |\mathbf{v}|$.

**ii.** Conversely, suppose $\mathbf{u} = (a, b)$, $\mathbf{v} = (c, d)$, and $|\mathbf{u} + \mathbf{v}| = |\mathbf{u}| + |\mathbf{v}|$. Then $\mathbf{u} + \mathbf{v} = (a + c, b + d)$, and $|\mathbf{u} + \mathbf{v}|^2 = (|\mathbf{u}| + |\mathbf{v}|)^2 = |\mathbf{u}|^2 + 2|\mathbf{u}|\,|\mathbf{v}| + |\mathbf{v}|^2$, which implies that

$$(a + c)^2 + (b + d)^2 = a^2 + b^2 + 2\sqrt{(a^2 + b^2)(c^2 + d^2)} + c^2 + d^2$$

and, after multiplying through and canceling like terms, we obtain $ac + bd = \sqrt{a^2c^2 + a^2d^2 + b^2c^2 + b^2d^2}$. Then squaring both sides and again canceling like terms, we have $2abcd = a^2d^2 + b^2c^2$, or $(ad - bc)^2 = a^2d^2 - 2abcd + b^2c^2 = 0$ so that $ad = bc$. If $d \neq 0$, then $a = \dfrac{b}{d}c$, $b = \dfrac{b}{d}d$, and $|\mathbf{u}| = \left|\dfrac{b}{d}\right|\,|\mathbf{v}|$. Then

$$\left|\frac{b + d}{d}\right|\,|\mathbf{v}| = \left|\frac{b + d}{d}(c, d)\right| = \left|\left(\frac{b}{d}c + c, \frac{b}{d}d + d\right)\right| = |(a + c, b + d)| = |\mathbf{u} + \mathbf{v}| =$$

$$|\mathbf{u}| + |\mathbf{v}| = \left|\frac{b}{d}\mathbf{v}\right| + |\mathbf{v}| = \left(\left|\frac{b}{d}\right| + 1\right)|\mathbf{v}| = \frac{|b| + |d|}{|d|}|\mathbf{v}|.$$

Thus $\left|\dfrac{b + d}{d}\right| = \dfrac{|b| + |d|}{|d|}$ so that $|b + d| = |b| + |d|$. Since $b$

and $d$ are real numbers, this implies that $b$ and $d$ have the same sign so that $\frac{b}{d}$ is positive. Thus if $\alpha = \left|\frac{b}{d}\right| = \frac{b}{d}$, then $\mathbf{u} = \alpha\mathbf{v}$. If $d = 0$, then $c \neq 0$ and $\mathbf{u} = \frac{a}{c}\mathbf{v}$ by the same reasoning as above where $\frac{a}{c} > 0$. Thus $\mathbf{u}$ is a positive scalar multiple of $\mathbf{v}$.

In the following answers $|\mathbf{v}|$ is given first and the direction is given in radians.

**39.** 2.99152034925; −0.952101203437

**41.** 2.99152034925; −2.18949145015

**43.** 114.738833879; −2.10075283072

**45.** 114.738833879; −1.04083982287

**47.** 0.086425871705; 1.40011222941

**49.** 0.086425871705; 1.74148042418

## MATLAB 3.1

**1. b.** An example of the coding (for Problem 40) after entering the vector as **v**:

```
mag = sqrt(v'*v),
direc = atan(v(2)/v(1))+pi
deg = 180/pi*direc
```

Note that for Problem 40 the vector **v** lies in the second quadrant, hence the need to add $\pi$ to the direction angle. *Problem 38:* magnitude = 2.9915, direction = .9521 radians = 54.55°; *Problem 40:* magnitude = 2.9915, direction = 2.1895 radians = 125.45°; *Problem 42:* magnitude = 114.7388, direction = 2.1008 radians = 120.36°; *Problem 44:* magnitude = 114.7388, direction = 1.0408 radians = 59.64°; *Problem 46:* magnitude = .0864, direction = −1.4001 radians = −80.22°; *Problem 48:* magnitude = .0864, direction = 4.5417 radians = 260.22°.

## Problems 3.2, page 247

**1.** 0; 0 **3.** 0; 0 **5.** 20; $\frac{20}{29}$

**7.** $-22$; $-22/5\sqrt{53}$

**9.** $\mathbf{u} \cdot \mathbf{v} = \alpha\beta - \beta\alpha = 0$

**11.** parallel

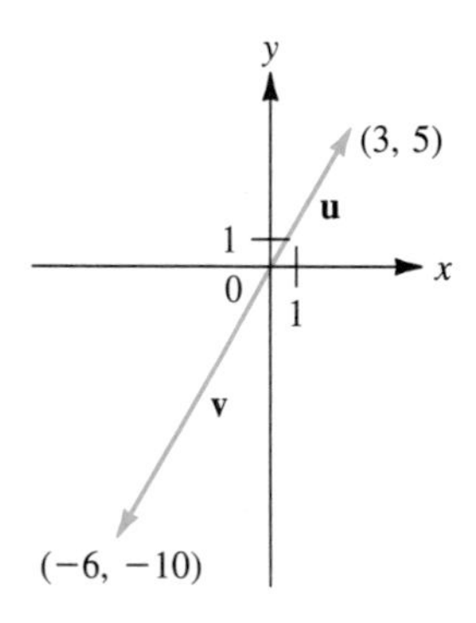


**13.** neither

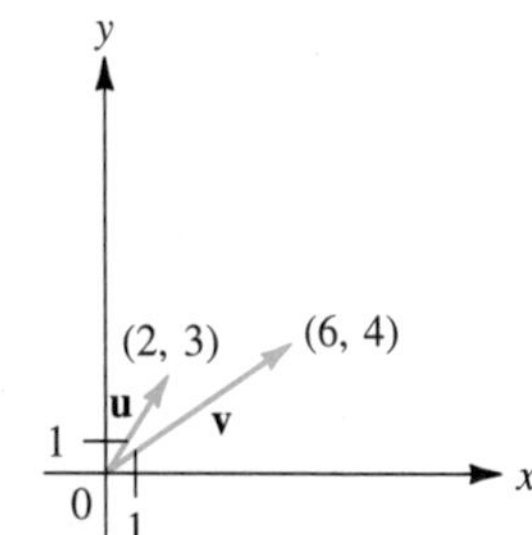


**15.** orthogonal

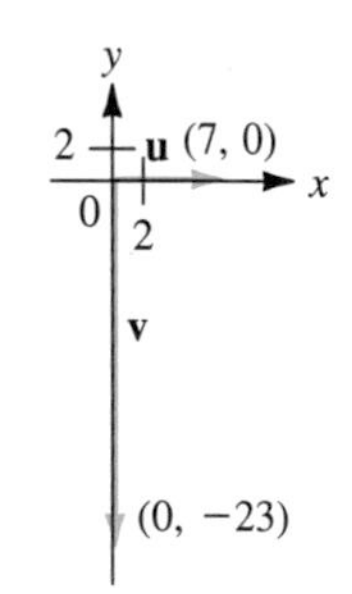


**17. a.** $-\frac{3}{4}$ **b.** $\frac{4}{3}$ **c.** $\frac{1}{7}$

**d.** $(-96 + \sqrt{7500})/78 \approx -0.12$

**19.** If **u** and **v** have opposite directions, then $\theta_{\mathbf{u}} = \theta_{\mathbf{v}} + \pi$. Thus $\cos\theta_{\mathbf{u}} = \cos(\theta_{\mathbf{v}} + \pi) = -\cos\theta_{\mathbf{v}}$. This implies that $\cos\theta_{\mathbf{u}} = \frac{3}{5} = -\frac{1}{\sqrt{1+\alpha^2}} = -\cos\theta_{\mathbf{v}}$, or $\sqrt{1+\alpha^2} = -\frac{5}{3} < 0$, which is impossible since $\sqrt{1+\alpha^2} = |\mathbf{v}| \geq 0$.

**21.** $\frac{3}{2}\mathbf{i} + \frac{3}{2}\mathbf{j}$ **23.** $\mathbf{0}$ **25.** $-\frac{2}{13}\mathbf{i} + \frac{3}{13}\mathbf{j}$

**27.** $[(\alpha+\beta)/2]\mathbf{i} + [(\alpha+\beta)/2]\mathbf{j}$

**29.** $[(\alpha-\beta)/2]\mathbf{i} + [(\alpha-\beta)/2]\mathbf{j}$

**31.** $a_1a_2 + b_1b_2 \geq 0$

**33.** $\text{Proj}_{\overrightarrow{PQ}}\,\overrightarrow{RS} = \frac{51}{25}\mathbf{i} + \frac{68}{25}\mathbf{j}$; $\text{Proj}_{\overrightarrow{RS}}\,\overrightarrow{PQ} = -\frac{17}{26}\mathbf{i} + \frac{85}{26}\mathbf{j}$

**35.** **i.** If $\mathbf{u} = \alpha\mathbf{v}$, then $\mathbf{u}\cdot\mathbf{v} = \alpha\mathbf{v}\cdot\mathbf{v} = \alpha|\mathbf{v}|^2$, $|\mathbf{u}| = |\alpha|\,|\mathbf{v}|$ so that $\cos\varphi = \frac{\alpha|\mathbf{v}|^2}{|\alpha|\,|\mathbf{v}|\,|\mathbf{v}|} = \frac{\alpha}{|\alpha|} = \pm 1$.

**ii.** Suppose that $\mathbf{u}$ and $\mathbf{v}$ are parallel. Then, if $\mathbf{u} = (a, b)$ and $\mathbf{v} = (c, d)$, we have $1 = \cos^2\varphi = \frac{|\mathbf{u}\cdot\mathbf{v}|^2}{|\mathbf{u}|^2|\mathbf{v}|^2} = \frac{(ac+bd)^2}{(a^2+b^2)(c^2+d^2)}$. Multiplying through and simplifying, we obtain $0 = a^2d^2 - 2abcd + b^2c^2$. Thus $ad = bc$. If $a \neq 0$, then $d = \left(\frac{c}{a}\right)b$ and $c = \left(\frac{c}{a}\right)a$ so that $\mathbf{v} = \left(\frac{c}{a}\right)\mathbf{u}$. If $a = 0$, then $b \neq 0$ and $\mathbf{v} = \frac{d}{b}\mathbf{u}$.

**37.** The line $ax + by + c = 0$ has slope $-\frac{a}{b}$. A vector parallel to the line is $\mathbf{u} = \mathbf{i} - \frac{a}{b}\mathbf{j}$, and

$$\mathbf{u}\cdot\mathbf{v} = 1\cdot a - \frac{a}{b}\cdot b = 0.$$

**39.** $52/5\sqrt{113} \approx 0.9783$; $61/\sqrt{34}\sqrt{113} \approx 0.9841$; $-27/5\sqrt{34} \approx -0.9261$

**41.** If either $a_1 = a_2 = 0$ or $b_1 = b_2 = 0$, both sides of the inequality are zero. If at least one of $a_1$ and $a_2 \neq 0$ and at least one of $b_1$ and $b_2 \neq 0$, let $\mathbf{u} = a_1\mathbf{i} + a_2\mathbf{j}$, $\mathbf{v} = b_1\mathbf{i} + b_2\mathbf{j}$. Then $\mathbf{u} \neq \mathbf{0}$, $\mathbf{v} \neq \mathbf{0}$, $|\mathbf{u}|\,|\mathbf{v}| \neq 0$, and $\left|\frac{\mathbf{u}\cdot\mathbf{v}}{|\mathbf{u}|\,|\mathbf{v}|}\right| = |\cos\varphi| \leq 1$. Thus $|a_1b_1 + a_2b_2| = |\mathbf{u}\cdot\mathbf{v}| \leq |\mathbf{u}|\,|\mathbf{v}| = \sqrt{a_1^2 + a_2^2}\sqrt{b_1^2 + b_2^2}$. Equality holds when $|\cos\varphi| = 1$, which is true if and only if $\mathbf{u}$ and $\mathbf{v}$ are parallel.

**43.** $\sqrt{5}$

**45.** Let $A = \begin{pmatrix} a_1 & b_1 \\ a_2 & b_2 \end{pmatrix}$. Then $A^t = \begin{pmatrix} a_1 & a_2 \\ b_1 & b_2 \end{pmatrix}$. Let $\mathbf{u} = (a_1, a_2)$ and $\mathbf{v} = (b_1, b_2)$. Then $\mathbf{u}\cdot\mathbf{v} = 0$, $|\mathbf{u}| = 1$ and $|\mathbf{v}| = 1$. $A^tA = \begin{pmatrix} a_1^2 + b_1^2 & a_1a_2 + b_1b_2 \\ a_2a_1 + b_2b_1 & a_2^2 + b_2^2 \end{pmatrix} = \begin{pmatrix} |\mathbf{u}|^2 & \mathbf{u}\cdot\mathbf{v} \\ \mathbf{u}\cdot\mathbf{v} & |\mathbf{v}|^2 \end{pmatrix} = \begin{pmatrix} 1 & 0 \\ 0 & 1 \end{pmatrix}$. Similarly, $AA^t = I$. Hence $A$ is invertible and $A^{-1} = A^t$.

**47.** $(-0.88449617907, 0.466547435114)$

**49.** $(-0.27075830549, -0.962647360152)$

**51.** $(-1.42889364286, -2.66927522328)$

**53.** $(-6164.36315451, 3523.92922513)$

## MATLAB 3.2

**1.** If $\mathbf{v} = a\mathbf{i} + b\mathbf{j}$, then enter **v = [a;b].** $\text{proj}_{\mathbf{v}}\mathbf{u}$ = **p = ((u′*v)/(v′*v))*v.** *Problem 22:* $\mathbf{p} = -2.5(\mathbf{i} + \mathbf{j})$; *Problem 24:* $\mathbf{p} = 2.5882\mathbf{i} + .6471\mathbf{j}$; *Problem 26:* $\mathbf{p} = .7692\mathbf{i} + 1.1538\mathbf{j}$.

## Problems 3.3, page 258

**1.** $\sqrt{40}$ **3.** 6

**5.** 3; $-1, 0, 0$

**7.** $\sqrt{5}$; $1/\sqrt{5}, 0, 2/\sqrt{5}$

**9.** $\sqrt{3}$; $1/\sqrt{3}, 1/\sqrt{3}, -1/\sqrt{3}$

**11.** $\sqrt{3}$; $1/\sqrt{3}, -1/\sqrt{3}, -1/\sqrt{3}$

**13.** $\sqrt{3}$; $-1/\sqrt{3}, -1/\sqrt{3}, 1/\sqrt{3}$

**15.** $\sqrt{78}$; $2/\sqrt{78}, 5/\sqrt{78}, -7/\sqrt{78}$

**17.** $\sqrt{29}$; $-2/\sqrt{29}, -3/\sqrt{29}, -4/\sqrt{29}$

**19.** $4\sqrt{3}\mathbf{i} + 4\sqrt{3}\mathbf{j} + 4\sqrt{3}\mathbf{k}$

**21.** $(1/\sqrt{26})\mathbf{i} - (3/\sqrt{26})\mathbf{j} + (4/\sqrt{26})\mathbf{k}$

**23.** $R = (-3, y, z)$, $y, z$ arbitrary; this set of points constitutes a plane parallel to the $yz$-plane.

**25.** $\left|\dfrac{\mathbf{u}\cdot\mathbf{v}}{|\mathbf{u}|\,|\mathbf{v}|}\right| = |\cos\varphi| \le 1$. Thus $|\mathbf{u}\cdot\mathbf{v}| \le |\mathbf{u}|\,|\mathbf{v}|$. Then

$$\begin{aligned}|\mathbf{u}+\mathbf{v}|^2 &= (\mathbf{u}+\mathbf{v})\cdot(\mathbf{u}+\mathbf{v})\\ &= |\mathbf{u}|^2 + 2\mathbf{u}\cdot\mathbf{v} + |\mathbf{v}|^2\\ &\le |\mathbf{u}|^2 + 2|\mathbf{u}|\,|\mathbf{v}| + |\mathbf{v}|^2\\ &= (|\mathbf{u}| + |\mathbf{v}|)^2\end{aligned}$$

**27.** $-6\mathbf{j} + 9\mathbf{k}$ **29.** $8\mathbf{i} - 14\mathbf{j} + 9\mathbf{k}$

**31.** $16\mathbf{i} + 29\mathbf{j} + 42\mathbf{k}$ **33.** $\sqrt{59}$

**35.** $\cos^{-1}(35/\sqrt{29}\sqrt{59})$
$\approx \cos^{-1}(0.8461)$
$\approx 0.5621$
$\approx 32.21°$

**37.** $\frac{25}{29}\mathbf{u} = \frac{50}{29}\mathbf{i} - \frac{75}{29}\mathbf{j} + \frac{100}{29}\mathbf{k}$

**39.** Since the line segments $PS$ and $SR$ are perpendicular (in Figure 3.26), the triangle $PSR$ is a right triangle and

$$\overline{PR}^2 = \overline{PS}^2 + \overline{SR}^2 \qquad \text{(i)}$$

But triangle $PRQ$ is also a right triangle so that

$$\overline{PQ}^2 = \overline{PR}^2 + \overline{RQ}^2 \qquad \text{(ii)}$$

So combining (i) and (ii), we get

$$\overline{PQ}^2 = \overline{PS}^2 + \overline{SR}^2 + \overline{RQ}^2 \qquad \text{(iii)}$$

Since the $x$- and $z$-coordinates of $P$ and $S$ are equal,

$$\overline{PS}^2 = (y_2 - y_1)^2 \qquad \text{(iv)}$$

Similarly,

$$\overline{RS}^2 = (x_2 - x_1)^2 \qquad \text{(v)}$$

and $$\overline{RQ}^2 = (z_2 - z_1)^2 \qquad \text{(vi)}$$

Thus, using (iv), (v), and (vi) in (iii) yields

$$\overline{PQ}^2 = (x_2 - x_1)^2 + (y_2 - y_1)^2 + (z_2 - z_1)^2$$

**41.** **i.** If $\mathbf{v} = \alpha\mathbf{u}$, then $\cos\varphi = \dfrac{\mathbf{u}\cdot\mathbf{v}}{|\mathbf{u}|\,|\mathbf{v}|} = \dfrac{\alpha|\mathbf{u}|^2}{|\alpha|\,|\mathbf{u}|^2} = \pm 1$. If $\mathbf{u}$ and $\mathbf{v}$ are parallel, then $\dfrac{\mathbf{u}}{|\mathbf{u}|} = \pm\dfrac{\mathbf{v}}{|\mathbf{v}|}$ so that

$$\mathbf{v} = \pm\frac{|\mathbf{v}|}{|\mathbf{u}|}\mathbf{u} = \alpha\mathbf{u}$$

**ii.** If $\mathbf{u}\cdot\mathbf{v} = 0$, then $\cos\varphi = 0$ and $\varphi = \dfrac{\pi}{2}$. If $\varphi = \dfrac{\pi}{2}$, then

$$\mathbf{u}\cdot\mathbf{v} = |\mathbf{u}|\,|\mathbf{v}|\cos\varphi = 0$$

In Problems 43–49 the magnitude is given first.

**43.** 0.707129874917; (0.327521164379, 0.590980546606, −0.737205453328)

**45.** 85.2279883606; (0.20298496225, 0.91988560927, 0.335570515627)

**47.** (−18.3995893751, −16.8662902605, 11.1711792634)

**49.** (57.4451474781, 271.495923758, 310.507180628)

## Problems 3.4, page 270

**1.** $-6\mathbf{i} - 3\mathbf{j}$ **3.** $-\mathbf{i} - \mathbf{j} + \mathbf{k}$

**5.** $12\mathbf{i} + 8\mathbf{j} - 21\mathbf{k}$

**7.** $(bc - ad)\mathbf{j}$

**9.** $-5\mathbf{i} - \mathbf{j} + 7\mathbf{k}$ **11.** $\mathbf{0}$

**13.** $42\mathbf{i} + 6\mathbf{j}$

**15.** $-9\mathbf{i} + 39\mathbf{j} + 61\mathbf{k}$

**17.** $-4\mathbf{i} + 8\mathbf{k}$ **19.** $\mathbf{0}$

**21.** $\pm[-(9/\sqrt{181})\mathbf{i} - (6/\sqrt{181})\mathbf{j} + (8/\sqrt{181})\mathbf{k}]$

**23.** $\sqrt{30}/\sqrt{6}\sqrt{29} \approx 0.415$

**25.** $5\sqrt{5}$ **27.** $\sqrt{523}$

**29.** $\sqrt{a^2b^2 + a^2c^2 + b^2c^2}$

**31.** Let $\mathbf{u} = a_1\mathbf{i} + b_1\mathbf{j} + c_1\mathbf{k}$ and $\mathbf{v} = a_2\mathbf{i} + b_2\mathbf{j} + c_2\mathbf{k}$. Then $\mathbf{u}\times\mathbf{v} = (b_1c_2 - c_1b_2)\mathbf{i} + (c_1a_2 - a_1c_2)\mathbf{j} +$

$(a_1b_2 - b_1a_2)\mathbf{k}$ so that $|\mathbf{u} \times \mathbf{v}|^2 = (b_1c_2 - c_1b_2)^2 + (c_1a_2 - a_1c_2)^2 + (a_1b_2 - b_1a_2)^2 = b_1^2c_2^2 - 2b_1c_2c_1b_2 + c_1^2b_2^2 + c_1^2a_2^2 - 2c_1a_2a_1c_2 + a_1^2c_2^2 + a_1^2b_2^2 - 2a_1b_2b_1a_2 + b_1^2a_2^2$. This is equal to $|\mathbf{u}|^2|\mathbf{v}|^2 - (\mathbf{u} \cdot \mathbf{v})^2 = (a_1^2 + b_1^2 + c_1^2) \cdot (a_2^2 + b_2^2 + c_2^2) - (a_1a_2 + b_1b_2 + c_1c_2)^2$.

**33.** Let $\mathbf{u} = a_1\mathbf{i} + b_1\mathbf{j} + c_1\mathbf{k}$, $\mathbf{v} = a_2\mathbf{i} + b_2\mathbf{j} + c_2\mathbf{k}$, and $\mathbf{w} = a_3\mathbf{i} + b_3\mathbf{j} + c_3\mathbf{k}$. Then

$$\begin{aligned}(\mathbf{u} \times \mathbf{v}) \cdot \mathbf{w} &= [(b_1c_2 - c_1b_2)\mathbf{i} \\ &\quad + (c_1a_2 - a_1c_2)\mathbf{j} \\ &\quad + (a_1b_2 - b_1a_2)\mathbf{k}] \\ &\quad \cdot [a_3\mathbf{i} + b_3\mathbf{j} + c_3\mathbf{k}] \\ &= b_1c_2a_3 - c_1b_2a_3 \\ &\quad + c_1a_2b_3 - a_1c_2b_3 \\ &\quad + a_1b_2c_3 - b_1a_2c_3\end{aligned}$$

and

$$\begin{aligned}\mathbf{u} \cdot (\mathbf{v} \times \mathbf{w}) &= [a_1\mathbf{i} + b_1\mathbf{j} + c_1\mathbf{k}] \\ &\quad \cdot [(b_2c_3 - c_2b_3)\mathbf{i} \\ &\quad + (c_2a_3 - a_2c_3)\mathbf{j} \\ &\quad + (a_2b_3 - b_2a_3)\mathbf{k}] \\ &= a_1b_2c_3 - a_1c_2b_3 \\ &\quad + b_1c_2a_3 - b_1a_2c_3 \\ &\quad + c_1a_2b_3 - c_1b_2a_3\end{aligned}$$

**35.** If $\mathbf{u}$ and $\mathbf{v}$ are parallel and neither is $\mathbf{0}$, then $\mathbf{v} = t\mathbf{u}$ for some constant $t$. Then if $\mathbf{u} = a\mathbf{i} + b\mathbf{j} + c\mathbf{k}$,

$$\mathbf{u} \times \mathbf{v} = \begin{vmatrix} \mathbf{i} & \mathbf{j} & \mathbf{k} \\ a & b & c \\ ta & tb & tc \end{vmatrix}$$

$= 0$ by Property 6 on page 196

Conversely, if $\mathbf{u} \times \mathbf{v} = \mathbf{0}$ and neither $\mathbf{u}$ nor $\mathbf{v}$ is $\mathbf{0}$, then by Theorem 3 $\sin\varphi = |\mathbf{u} \times \mathbf{v}|/|\mathbf{u}||\mathbf{v}| = 0$ so $\varphi = 0$ or $\pi$ and $\mathbf{u}$ and $\mathbf{v}$ are parallel.

**37.** 23

**39.** This problem will rely heavily on Property 3 of the determinants, which states that if the $i$th column (or row) of a determinant consists of a pair of elements, the determinant can be rewritten as a sum of two determinants whose columns (or rows) are identical, except for the $i$th column (or row). The first determinant contains one of the elements of the pair, while the other member of each pair of elements of the $i$th column (or row) appears in the second determinant. Also note that the volume generated by $\mathbf{u}$, $\mathbf{v}$, $\mathbf{w}$ is given by

$$\text{Volume} = \begin{vmatrix} u_1 & u_2 & u_3 \\ v_1 & v_2 & v_3 \\ w_1 & w_2 & w_3 \end{vmatrix}$$

Let

$$A = \begin{pmatrix} a_{11} & a_{12} & a_{13} \\ a_{21} & a_{22} & a_{23} \\ a_{31} & a_{32} & a_{33} \end{pmatrix}$$

$\mathbf{u}_1 = A\mathbf{u}, \quad \mathbf{v}_1 = A\mathbf{v}, \quad \mathbf{w}_1 = A\mathbf{w}$

Thus

$$\mathbf{u}_1 = \begin{pmatrix} a_{11} & a_{12} & a_{13} \\ a_{21} & a_{22} & a_{23} \\ a_{31} & a_{32} & a_{33} \end{pmatrix}\begin{pmatrix} u_1 \\ u_2 \\ u_3 \end{pmatrix} = \begin{pmatrix} a_{11}u_1 + a_{12}u_2 + a_{13}u_3 \\ a_{21}u_1 + a_{22}u_2 + a_{23}u_3 \\ a_{31}u_1 + a_{32}u_2 + a_{33}u_3 \end{pmatrix}$$

Similarly,

$$\mathbf{v}_1 = \begin{pmatrix} a_{11}v_1 + a_{12}v_2 + a_{13}v_3 \\ a_{21}v_1 + a_{22}v_2 + a_{23}v_3 \\ a_{31}v_1 + a_{32}v_2 + a_{33}v_3 \end{pmatrix}$$

and

$$\mathbf{w}_1 = \begin{pmatrix} a_{11}w_1 + a_{12}w_2 + a_{13}w_3 \\ a_{21}w_1 + a_{22}w_2 + a_{23}w_3 \\ a_{31}w_1 + a_{32}w_2 + a_{33}w_3 \end{pmatrix}$$

By Problem 36, the volume generated by $\mathbf{u}_1$, $\mathbf{v}_1$, $\mathbf{w}_1$ is

$$V = \begin{vmatrix} a_{11}u_1 + a_{12}u_2 + a_{13}u_3 & a_{21}u_1 + a_{22}u_2 + a_{23}u_3 & a_{31}u_1 + a_{32}u_2 + a_{33}u_3 \\ a_{11}v_1 + a_{12}v_2 + a_{13}v_3 & a_{21}v_1 + a_{22}v_2 + a_{23}v_3 & a_{31}v_1 + a_{32}v_2 + a_{33}v_3 \\ a_{11}w_1 + a_{12}w_2 + a_{13}w_3 & a_{21}w_1 + a_{22}w_2 + a_{23}w_3 & a_{31}w_1 + a_{32}w_2 + a_{33}w_3 \end{vmatrix}$$

By expansion, it can be verified that

$$V = \begin{vmatrix} a_{11} & a_{12} & a_{13} \\ a_{21} & a_{22} & a_{23} \\ a_{31} & a_{32} & a_{33} \end{vmatrix} \begin{vmatrix} u_1 & u_2 & u_3 \\ v_1 & v_2 & v_3 \\ w_1 & w_2 & w_3 \end{vmatrix}$$

= (det $A$)(volume generated by **u**, **v**, **w**), if det $A \geq 0$, otherwise use $-\det A$.

**41.** Let $\mathbf{u} = (u_1, u_2, u_3)$, $\mathbf{v} = (v_1, v_2, v_3)$, and $\mathbf{w} = (w_1, w_2, w_3)$. Then

$$\mathbf{v} \times \mathbf{w} = \begin{vmatrix} \mathbf{i} & \mathbf{j} & \mathbf{k} \\ v_1 & v_2 & v_3 \\ w_1 & w_2 & w_3 \end{vmatrix}$$

$= (v_2w_3 - v_3w_2, v_3w_1 - v_1w_3, v_1w_2 - v_2w_1)$;

$$\mathbf{u} \times (\mathbf{v} \times \mathbf{w}) = \begin{vmatrix} \mathbf{i} & \mathbf{j} & \mathbf{k} \\ u_1 & u_2 & u_3 \\ v_2w_3 - v_3w_2 & v_3w_1 - v_1w_3 & v_1w_2 - v_2w_1 \end{vmatrix}$$

$= (-u_3v_3w_1 + u_3v_1w_3 + u_2v_1w_2 - u_2v_2w_1, u_3v_2w_3 - u_3v_3w_2 - u_1v_1w_2 + u_1v_2w_1, u_1v_3w_1 - u_1v_1w_3 - u_2v_2w_3 + u_2v_3w_2)(*)$

$(\mathbf{u} \cdot \mathbf{w})\mathbf{v} = (u_1w_1 + u_2w_2 + u_3w_3) \times (v_1, v_2, v_3) = (u_1v_1w_1 + u_2v_1w_2 + u_3v_1w_3, u_1v_2w_1 + u_2v_2w_2 + u_3v_2w_3, u_1v_3w_1 + u_2v_3w_2 + u_3v_3w_3)$

$-(\mathbf{u} \cdot \mathbf{v})\mathbf{w} = -(u_1v_1 + u_2v_2 + u_3w_3)(w_1, w_2, w_3) = (-u_1v_1w_1 - u_2v_2w_1 - u_3v_3w_1, -u_1v_1w_2 - u_2v_2w_2 - u_3v_3w_2, -u_1v_1w_3 - u_2v_2w_3 - u_3v_3w_3)$

If we add the last two vectors, we obtain the vector (*).

**43.** $(0.294473, 0.676166, -0.547895)$

**45.** $(0.004852, 0.003952, -0.004704)$

### MATLAB 3.4

**1.** Let **c** denote the cross product. Check your answers by showing that **c'*u** and **c'*v** are both 0. *Problem 2:* $\mathbf{c} = -7\mathbf{i} - 3\mathbf{j} + 7\mathbf{k}$; *Problem 4:* $\mathbf{c} = 7\mathbf{i}$; *Problem 10:* $\mathbf{c} = -14\mathbf{i} - 3\mathbf{j} + 15\mathbf{k}$.

### Problems 3.5, page 281

In the answers to Problems 1–5 we assume that the first point is $P$ and the second point is $Q$. The vector equations are of the form $\overrightarrow{QR} = \overrightarrow{QP} + t\mathbf{v}$. Only **v** is given in the answers.

**1.** $\mathbf{v} = -\mathbf{i} + \mathbf{j} - 4\mathbf{k}$; $x = 2 - t$, $y = 1 + t$, $z = 3 - 4t$; $(x - 2)/(-1) = y - 1 = (z - 3)/(-4)$

**3.** $\mathbf{v} = -\mathbf{j} - 2\mathbf{k}$; $x = -4$, $y = 1 - t$, $z = 3 - 2t$; $x = -4$ and $z = 1 + 2y$

**5.** $\mathbf{v} = 2\mathbf{i} - 2\mathbf{k}$; $x = 1 + 2t$, $y = 2$, $z = 3 - 2t$; $y = 2$ and $x = 4 - z$

In Problems 7–11 **v** is already given.

**7.** $x = 2 + 2t$, $y = 2 - t$, $z = 1 - t$; $(x - 2)/2 = (y - 2)/(-1) = (z - 1)/(-1)$

**9.** $x = -1$, $y = -2 - 3t$, $z = 5 + 7t$; $x = -1$ and $7y + 3z = 1$

**11.** $x = a + dt$, $y = b + et$, $z = c$; $(x - a)/d = (y - b)/e$ and $z = c$

**13.** $\mathbf{v} = 3\mathbf{i} + 6\mathbf{j} + 2\mathbf{k}$; $x = 4 + 3t$, $y = 1 + 6t$, $z = -6 + 2t$; $(x - 4)/3 = (y - 1)/6 = (z + 6)/2$

**15.** The vector $\mathbf{v}_1 = a_1\mathbf{i} + b_1\mathbf{j} + c_1\mathbf{k}$ is parallel to $L_1$, while the vector $\mathbf{v}_2 = a_2\mathbf{i} + b_2\mathbf{j} + c_2\mathbf{k}$ is parallel to $L_2$. Thus $L_1 \perp L_2$ if $\mathbf{v}_1 \perp \mathbf{v}_2$ or $\mathbf{v}_1 \cdot \mathbf{v}_2 = 0$. But $\mathbf{v}_1 \cdot \mathbf{v}_2 = a_1a_2 + b_1b_2 + c_1c_2$.

**17.** $3\mathbf{i} + 6\mathbf{j} + 9\mathbf{k} = 3(\mathbf{i} + 2\mathbf{j} + 3\mathbf{k})$, so the direction vectors of the lines are parallel. Note that they are not coincident since, for example, the point $(1, -3, -3)$ is on $L_1$ but not on $L_2$.

**19.** If they had a point in common, we would have

$$2 - t = 1 + s$$
$$1 + t = -2s$$
$$-2t = 3 + 2s$$

The unique solution of the first two of these equations is $s = -2$, $t = 3$; but this pair does not satisfy the third equation.

**21.** **a.** $(\sqrt{186}/3)(t = \frac{1}{3})$
**b.** $\sqrt{1518}/11 = \sqrt{138/11}$, $(t = -\frac{4}{11})$
**c.** $\sqrt{30}/2$ $(t = -\frac{3}{2})$

**23.** $(x + 4)/26 = (y - 7)/1 = (z - 3)/37$

**25.** $(x - 4)/(-4) = (y - 6)/16 = z/24$

**27.** 3 **29.** $y = 0$ ($xz$-plane)

**31.** $x + y = 3$ **33.** $y + z = 5$

**35.** $-3x - 4y + z = 45$

**37.** $2x - 7y - 8z = -20$

**39.** $-12x - 21y + 22z = 63$

**41.** $2x + y = 7$ **43.** coincident

**45.** none of these

**47.** $(x, y, z) = (-1, -3, 0) + t(1, 2, 1)$

**49.** $(x, y, z) = (-11/4, 3/2, 0) + t(9, 16, 2)$

**51.** $13/\sqrt{69}$ **53.** $19/\sqrt{35}$

**55.** $\cos^{-1}(9/\sqrt{3}\sqrt{29}) = \cos^{-1}(0.9649) \approx 0.2657 \approx 15.23°$

**57.** $\cos^{-1}(20/\sqrt{294}\sqrt{6}) = \cos^{-1}(\frac{20}{42}) \approx 1.074 \approx 61.56°$

**59.** $\mathbf{n} = \mathbf{v} \times \mathbf{w}$ is orthogonal to $\mathbf{v}$ and $\mathbf{w}$. If $\mathbf{u} \cdot (\mathbf{v} \times \mathbf{w}) = 0$, then $\mathbf{u} \perp \mathbf{n}$, which means that $\mathbf{u}$ lies in the plane determined by $\mathbf{v}$ and $\mathbf{w}$.

**61.** coplanar; $29x - y + 11z = 0$

**63.** not coplanar; $\mathbf{u} \cdot (\mathbf{v} \times \mathbf{w}) = -9$.

## Chapter 3—Review, page 288

**1.** $|\mathbf{v}| = 3\sqrt{2}$, $\theta = \pi/4$

**3.** $|\mathbf{v}| = 4$, $\theta = 5\pi/3$

**5.** $|\mathbf{v}| = 12\sqrt{2}$, $\theta = 5\pi/4$

**7.** $2\mathbf{i} + 2\mathbf{j}$

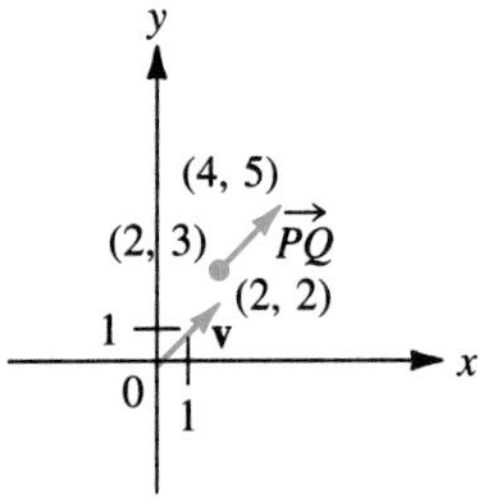

**9.** $4\mathbf{i} + 2\mathbf{j}$

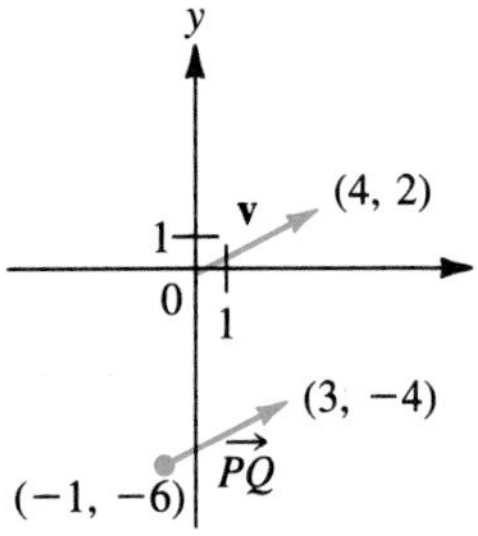

**11.** **a.** $(10, 5)$, **b.** $(5, -3)$, **c.** $(-31, 12)$

**13.** $(1/\sqrt{2})\mathbf{i} + (1/\sqrt{2})\mathbf{j}$

**15.** $(2/\sqrt{29})\mathbf{i} + (5/\sqrt{29})\mathbf{j}$ **17.** $\frac{3}{5}\mathbf{i} + \frac{4}{5}\mathbf{j}$

**19.** $(1/\sqrt{2})\mathbf{i} - (1/\sqrt{2})\mathbf{j}$ if $a > 0$ and $-(1/\sqrt{2})\mathbf{i} + (1/\sqrt{2})\mathbf{j}$ if $a < 0$

**21.** $-(5/\sqrt{29})\mathbf{i} - (2/\sqrt{29})\mathbf{j}$

**23.** $-(10/\sqrt{149})\mathbf{i} + (7/\sqrt{149})\mathbf{j}$

**25.** $\mathbf{j}$ **27.** $-\frac{7}{2}\sqrt{3}\mathbf{i} + \frac{7}{2}\mathbf{j}$ **29.** 0; 0

**31.** $-14$, $-14/\sqrt{5}\sqrt{41}$

**33.** neither

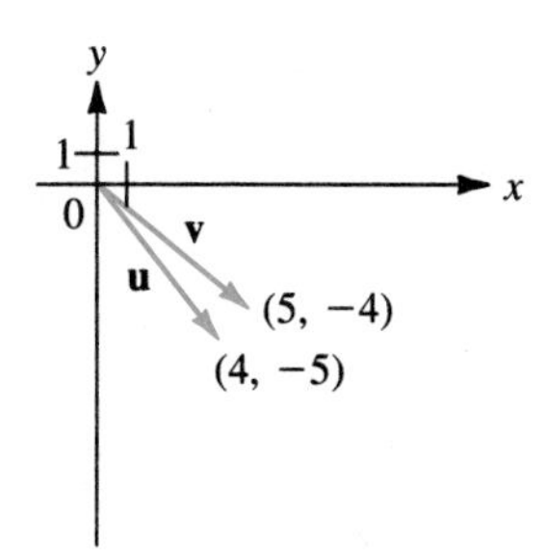


**35.** parallel

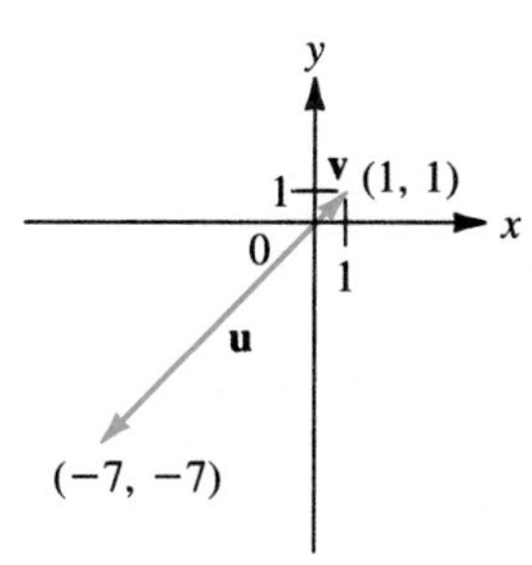


**37.** parallel

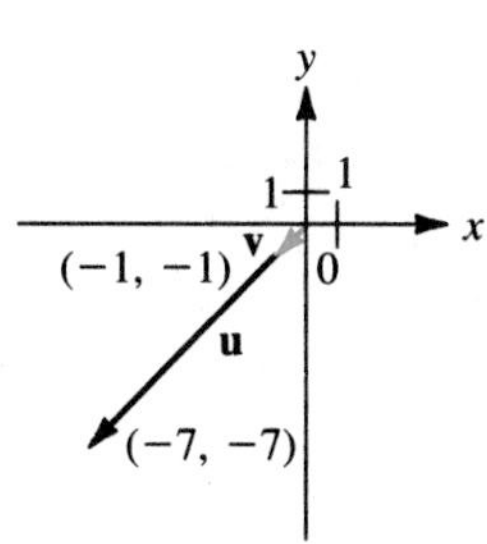


**39.** $7\mathbf{i} + 7\mathbf{j}$

**41.** $\frac{15}{13}\mathbf{i} + \frac{10}{13}\mathbf{j}$ **43.** $-\frac{3}{2}\mathbf{i} - \frac{7}{2}\mathbf{j}$

**45.** $\text{Proj}_{\overrightarrow{RS}}\, \overrightarrow{PQ} = -\frac{99}{25}\mathbf{i} + \frac{132}{25}\mathbf{j}$;
$\text{Proj}_{\overrightarrow{PQ}}\, \overrightarrow{RS} = -\frac{33}{82}\mathbf{i} - \frac{297}{82}\mathbf{j}$

**47.** $\sqrt{216}$

**49.** $\sqrt{130}$; $0, 3/\sqrt{130}, 11/\sqrt{130}$

**51.** $\sqrt{53}$; $-4/\sqrt{53}, 1/\sqrt{53}, 6/\sqrt{53}$

**53.** $(2/\sqrt{6})\mathbf{i} - (1/\sqrt{6})\mathbf{j} + (1/\sqrt{6})\mathbf{k}$

**55.** $\mathbf{i} - 14\mathbf{j} + 20\mathbf{k}$

**57.** $\frac{26}{21}\mathbf{i} - \frac{52}{21}\mathbf{j} + \frac{13}{21}\mathbf{k}$ **59.** 22

**61.** $\cos^{-1}(-9/\sqrt{798}) \approx 1.895 \approx 108.6°$

**63.** $-7\mathbf{i} - 7\mathbf{k}$

**65.** $-26\mathbf{i} - 8\mathbf{j} + 7\mathbf{k}$

**67.** $\sqrt{2065}$

**69.** $\overrightarrow{OR} = (-4\mathbf{i} + \mathbf{j}) + t(7\mathbf{i} - \mathbf{j} + 7\mathbf{k})$;
$x = -4 + 7t,\ y = 1 - t,\ z = 7t$;

$$(x + 4)/7 = (y - 1)/(-1) = z/7$$

**71.** $\overrightarrow{OR} = \mathbf{i} - 2\mathbf{j} - 3\mathbf{k} + t(5\mathbf{i} - 3\mathbf{j} + 2\mathbf{k})$;
$x = 1 + 5t,\ y = -2 - 3t,\ z = -3 + 2t$;

$$(x - 1)/5 = (y + 2)/(-3) = (z + 3)/2$$

**73.** $\sqrt{165}/3$ **75.** $x + z = -1$

**77.** $2x - 3y + 5z = 19$

**79.** $x = \frac{1}{2} - \frac{9}{2}t,\ y = \frac{7}{2} - \frac{11}{2}t,\ z = t$

**81.** $x = \frac{4}{3} - \frac{5}{2}t,\ y = -4 - \frac{7}{2}t,\ z = t$

**83.** $\cos^{-1}|-1/\sqrt{207}|$
$= \cos^{-1} 1/\sqrt{207}$
$\approx 1.501 \approx 86.01°$

## Chapter 4

### Problems 4.2, page 297

**1.** yes

**3.** no; (iv); also (vi) does not hold if $\alpha < 0$

**5.** yes **7.** yes

**9.** no (i), (iii), (iv), (vi) do not hold

**11.** yes **13.** yes

**15.** no; (i), (iii), (iv), (vi) do not hold

**17.** yes **19.** yes

**21.** Suppose that $\mathbf{0}$ and $\mathbf{0}'$ are additive identities. Then, by the definition of additive identity, $\mathbf{0} = \mathbf{0} + \mathbf{0}'$ and $\mathbf{0}' = \mathbf{0}' + \mathbf{0} = \mathbf{0} + \mathbf{0}'$. Thus $\mathbf{0} = \mathbf{0}'$.

**23.** For $\mathbf{x}, \mathbf{y}$ in $V$ define $\mathbf{z}$ by $\mathbf{z} = -\mathbf{x} + \mathbf{y}$. $\mathbf{z}$ exists since every $\mathbf{x}$ has an additive inverse $-\mathbf{x}$ and $V$ is closed under addition. Then $\mathbf{x} + \mathbf{z} = \mathbf{x} + (-\mathbf{x} + \mathbf{y}) = (\mathbf{x} - \mathbf{x}) + \mathbf{y} = \mathbf{0} + \mathbf{y} = \mathbf{y}$. Assume there exist $\mathbf{z}$ and $\mathbf{z}'$ such that $\mathbf{x} + \mathbf{z} = \mathbf{y}$ and $\mathbf{x} + \mathbf{z}' = \mathbf{y}$. Then $\mathbf{z} = -\mathbf{x} + \mathbf{y} = \mathbf{z}'$. So $\mathbf{z}$ is unique.

**25.** Let $y_1$ and $y_2$ be solutions to the equation. Then

$$y_1'' + a(x)y_1' + b(x)y_1(x) = 0$$

and

$y_2'' + a(x)y_2' + b(x)y_2(x) = 0$

Then

$$(y_1 + y_2)'' + a(x)(y_1 + y_2)' + b(x)(y_1 + y_2)$$
$$= [y_1'' + a(x)y_1' + b(x)y_1] + [y_2'' + a(x)y_2' + b(x)y_2]$$
$$= 0 + 0 = 0$$

so $y_1 + y_2$ is a solution. Similarly, $(\alpha y_1)'' + a(x)(\alpha y_1') + b(x)(\alpha y) = \alpha[y_1'' + a(x)y_1' + b(x)y_1] = \alpha \cdot 0 = 0$, so $\alpha y_1$ is also a solution. Thus the closure rules hold. Since $-y_1 = (-1)y$ is also a solution, we have the additive inverse. The other axioms follow easily.

**Problems 4.3,** page 303

**1.** no; because $\alpha(x, y) \notin H$ if $\alpha < 0$

**3.** yes **5.** yes **7.** yes

**9.** yes **11.** yes **13.** yes

**15.** no; the zero polynomial $\notin H$

**17.** no; the function $f(x) \equiv 0 \notin V$

**19.** yes

**21. a.** If $A_1, A_2 \in H_1$, then $(A_1 + A_2)_{11} = (A_1)_{11} + (A_2)_{11} = 0 + 0 = 0$, and $(\alpha A_1)_{11} = \alpha(A_1)_{11} = \alpha 0 = 0$ so that $H_1$ is a subspace. If $A_1, A_2 \in H_2$, then

$$A_1 = \begin{pmatrix} -b_1 & a_1 \\ a_1 & b_1 \end{pmatrix}$$

$$A_2 = \begin{pmatrix} -b_2 & a_2 \\ a_2 & b_2 \end{pmatrix}$$

$$A_1 + A_2 = \begin{pmatrix} -(b_1 + b_2) & (a_1 + a_2) \\ (a_1 + a_2) & (b_1 + b_2) \end{pmatrix} = \begin{pmatrix} -c & d \\ d & c \end{pmatrix} \in H_2. \text{ Also,}$$

$$\alpha A_1 = \begin{pmatrix} -\alpha b_1 & \alpha a_1 \\ \alpha a_1 & \alpha b_1 \end{pmatrix} = \begin{pmatrix} -c & d \\ d & c \end{pmatrix} \in H_2$$

and so $H_2$ is also a subspace.

**b.** $H = H_1 \cap H_2 = \left\{A \in M_{22}: A = \begin{pmatrix} 0 & a \\ a & 0 \end{pmatrix} \text{ for some scalar } a\right\}$. If $A_1, A_2 \in H$, then $A_1 = \begin{pmatrix} 0 & a_1 \\ a_1 & 0 \end{pmatrix}$, $A_2 = \begin{pmatrix} 0 & a_2 \\ a_2 & 0 \end{pmatrix}$, $A_1 + A_2 = \begin{pmatrix} 0 & a_1 + a_2 \\ a_1 + a_2 & 0 \end{pmatrix} = \begin{pmatrix} 0 & b \\ b & 0 \end{pmatrix} \in H$ and $\alpha A_1 = \begin{pmatrix} 0 & \alpha a_1 \\ \alpha a_1 & 0 \end{pmatrix} = \begin{pmatrix} 0 & c \\ c & 0 \end{pmatrix} \in H.$

**23.** If $\mathbf{x}_1, \mathbf{x}_2 \in H$, then $A(\mathbf{x}_1 + \mathbf{x}_2) = A\mathbf{x}_1 + A\mathbf{x}_2 = \mathbf{0} + \mathbf{0} = \mathbf{0}$, so $\mathbf{x}_1 + \mathbf{x}_2 \in H$. Also, $A(\alpha\mathbf{x}_1) = \alpha A\mathbf{x}_1 = \alpha\mathbf{0} = \mathbf{0}$ so $\alpha\mathbf{x}_1 \in H$ and $H$ is a subspace.

**25.** Let $\mathbf{u} = (x_1, y_1, z_1, w_1)$ and $\mathbf{v} = (x_2, y_2, z_2, w_2) \in H$. Then $\mathbf{u} + \mathbf{v} = (x_1 + x_2, y_1 + y_2, z_1 + z_2, w_1 + w_2)$ and $a(x_1 + x_2) + b(y_1 + y_2) + c(z_1 + z_2) + d(w_1 + w_2) = (ax_1 + by_1 + cz_1 + dw_1) + (ax_2 + by_2 + cz_2 + dw_2) = 0 + 0 = 0$, so $\mathbf{u} + \mathbf{v} \in H$. Similarly, $\alpha\mathbf{u} = (\alpha x_1, \alpha y_1, \alpha z_1, \alpha w_1)$ and $a(\alpha x_1) + b(\alpha y_1) + c(\alpha z_1) + d(\alpha w_1) = \alpha(ax_1 + by_1 + cz_1 + dw_1) = \alpha 0 = 0$, so $\alpha\mathbf{u} \in H$. Thus $H$ is a subspace.

**27.** Let $\mathbf{x}, \mathbf{y} \in H$. Then $\mathbf{x} = \mathbf{u}_1 + \mathbf{v}_1$ and $\mathbf{y} = \mathbf{u}_2 + \mathbf{v}_2$, where $\mathbf{u}_1, \mathbf{u}_2 \in H_1$ and $\mathbf{v}_1, \mathbf{v}_2 \in H_2$. Then $\mathbf{x} + \mathbf{y} = (\mathbf{u}_1 + \mathbf{v}_1) + (\mathbf{u}_2 + \mathbf{v}_2) = (\mathbf{u}_1 + \mathbf{u}_2) + (\mathbf{v}_1 + \mathbf{v}_2)$. Since $H_1$ and $H_2$ are subspaces, $\mathbf{u}_1 + \mathbf{u}_2 \in H_1$ and $\mathbf{v}_1 + \mathbf{v}_2 \in H_2$, so $\mathbf{x} + \mathbf{y} \in H$. Similarly, $\alpha\mathbf{x} = \alpha(\mathbf{u}_1 + \mathbf{v}_1) = \alpha\mathbf{u}_1 + \alpha\mathbf{v}_1$. But $\alpha\mathbf{u}_1 \in H_1$ and $\alpha\mathbf{v}_1 \in H_2$, so $\alpha\mathbf{x} \in H$ and $H$ is a subspace.

**29.** Let $\mathbf{v}_1 = \begin{pmatrix} x_1 \\ y_1 \end{pmatrix}$ and $\mathbf{v}_2 = \begin{pmatrix} x_2 \\ y_2 \end{pmatrix}$. $\mathbf{v}_1$ is not a multiple of $\mathbf{v}_2$ since the vectors are not collinear. Let $A = \begin{pmatrix} x_1 & x_2 \\ y_1 & y_2 \end{pmatrix}$. Then $\det A = x_1y_2 - x_2y_1$. If $\det A = 0$, then $x_1y_2 = x_2y_1$ or $x_1/x_2 = y_1/y_2$ (if $x_2 = 0$ or $y_2 = 0$, a similar conclusion can be drawn). Let $c = x_1/x_2 = y_1/y_2$. Then $x_1 = cx_2$ and $y_1 = cy_2$, so $\mathbf{v}_1 = c\mathbf{v}_2$ contradicting what was stated above. Thus $\det A \neq 0$. Let $\mathbf{v} = \begin{pmatrix} x \\ y \end{pmatrix}$ be any other vector in $\mathbb{R}^2$. We wish to find scalars $a$ and $b$ such that $\mathbf{v} = a\mathbf{v}_1 + b\mathbf{v}_2$, or

$$\begin{pmatrix} x \\ y \end{pmatrix} = a\begin{pmatrix} x_1 \\ y_1 \end{pmatrix} + b\begin{pmatrix} x_2 \\ y_2 \end{pmatrix} = \begin{pmatrix} ax_1 + bx_2 \\ ay_1 + by_2 \end{pmatrix}$$

or

$$\begin{pmatrix} x_1 & x_2 \\ y_1 & y_2 \end{pmatrix}\begin{pmatrix} a \\ b \end{pmatrix} = \begin{pmatrix} x \\ y \end{pmatrix}$$

or

$$A\begin{pmatrix} a \\ b \end{pmatrix} = \begin{pmatrix} x \\ y \end{pmatrix}$$

Since $\det A \neq 0$, this system has the unique solution $\begin{pmatrix} a \\ b \end{pmatrix} = A^{-1}\begin{pmatrix} x \\ y \end{pmatrix}$. Thus $\mathbf{v} \in H$, which shows that $\mathbb{R}^2 \subset H$. But since $H \subset \mathbb{R}^2$, we have

$$H = \mathbb{R}^2$$

### MATLAB 4.3

**1.** Check $\mathbf{S} - \mathbf{S}' = 0$ for verification. *Proof hint:* Use the definition to show that a matrix $W$ is symmetric; that is, show that $w_{ij} = w_{ji}$.

### Problems 4.4, page 310

**1.** yes

**3.** no; for example, $\begin{pmatrix} 1 \\ 2 \end{pmatrix} \notin$ span of the three vectors $\left[\text{each is a multiple of } \begin{pmatrix} 1 \\ 1 \end{pmatrix}\right]$.

**5.** yes **7.** yes

**9.** no; for example, $x \notin \text{span}\,\{1 - x, 3 - x^2\}$

**11.** yes **13.** yes

**15.** Let $p_i(x) = a_{0i} + a_{1i}x + \cdots + a_{ni}x^n$ for $i = 1, 2, \ldots, m$ and define the vector

$$\mathbf{a}_i = (a_{0i}, a_{1i}, \ldots, a_{ni})$$

Choose a vector $\mathbf{b} = (b_0, \ldots, b_n)$ such that $\mathbf{a}_i \cdot \mathbf{b} = 0$ for $i = 1, 2, \ldots, m$. If $m < n + 1$, this last homogeneous system of equations always has a nontrivial solution $\mathbf{b}$ (by Theorem 1.4.1). Suppose that $\mathbf{b}$ can be written as $\mathbf{b} = \alpha_1\mathbf{a}_1 + \alpha_2\mathbf{a}_2 + \cdots + \alpha_m\mathbf{a}_m$. Then $\mathbf{b} \cdot \mathbf{b} = \alpha_1\mathbf{a}_1 \cdot \mathbf{b} + \alpha_2\mathbf{a}_2 \cdot \mathbf{b} + \cdots + \alpha_m\mathbf{a}_m \cdot \mathbf{b} = 0$. But $\mathbf{b} \neq \mathbf{0}$. This contradiction shows that $\mathbf{b}$ cannot be written as a linear combination of the $\mathbf{a}_i$'s if $m < n + 1$. Let $q(x) = b_0 + b_1x + \cdots + b_nx^n$. Then $q(x)$ cannot be written as a linear combination of the $p_i$'s, and therefore the $p_i$'s do not span. We are left to conclude that $m \geq n + 1$.

**17.** If $p(x) \in P$, then $p(x) = a_0 + a_1x + \cdots + a_kx^k$ for some integer $k$. Thus $p(x)$ can be written as a linear combination of $1, x, x^2, \ldots, x^k$.

**19.** $\alpha\mathbf{v}_1 + \beta\mathbf{v}_2 = (\alpha + \beta c)\mathbf{v}_1$. Let $\mathbf{v}_1 = (x_1, y_1, z_1)$. Then any vector $\mathbf{v} = (x, y, z)$ in span $\{\mathbf{v}_1, \mathbf{v}_2\}$ can be written as $(x, y, z) = ((\alpha + \beta c)x_1, (\alpha + \beta c)y_1, (\alpha + \beta c)z_1)$, or

$$x = (\alpha + \beta c)x_1$$
$$y = (\alpha + \beta c)y_1$$
$$z = (\alpha + \beta c)z_1$$

which, from Section 3.5, is the equation of a line passing through $(0, 0, 0)$ with direction $(x_1, y_1, z_1)$.

**21.** Let $\mathbf{v} \in V$. Then there are scalars $\alpha_1, \alpha_2, \ldots, \alpha_n$ such that $\mathbf{v} = \alpha_1\mathbf{v}_1 + \alpha_2\mathbf{v}_2 + \cdots + \alpha_n\mathbf{v}_n = \alpha_1\mathbf{v}_1 + \alpha_2\mathbf{v}_2 + \cdots + \alpha_n\mathbf{v}_n + 0\mathbf{v}_{n+1}$. The last equation shows that $\mathbf{v}_1, \mathbf{v}_2, \ldots, \mathbf{v}_{n+1}$ span $V$.

**23.** Let $\mathbf{v}_i = (v_{i1}, v_{i2}, \ldots, v_{in})$ and $\mathbf{u}_i = (u_{i1}, u_{i2}, \ldots, u_{in})$. Also, let

$$\mathbf{w}_j = \begin{pmatrix} v_{1j} \\ v_{2j} \\ \vdots \\ v_{nj} \end{pmatrix}, \quad \mathbf{z}_j = \begin{pmatrix} u_{1j} \\ u_{2j} \\ \vdots \\ u_{nj} \end{pmatrix}$$

and $A = (a_{ij})$. Then writing out the components in the given inequalities, we have

$$(*) \quad v_{ij} = a_{i1}u_{1j} + a_{i2}u_{2j} + \cdots + a_{in}u_{nj}$$

or

$$\mathbf{w}_j = A\mathbf{z}_j; \text{ so } \mathbf{z}_j \overset{\substack{\text{given} \\ \det A \neq 0 \\ \downarrow}}{=} A^{-1}\mathbf{w}_j$$

Let $A^{-1} = B = (b_{ij})$. Then the expression for $\mathbf{z}_j$ can be written

$$u_{ij} = b_{i1}v_{1j} + b_{i2}v_{2j} + \cdots + b_{in}v_{nj}$$

We see that this is similar to the expression (*) with $u$'s and $v$'s interchanged. Thus

$$\mathbf{u}_i = \sum_{k=1}^{n} b_{ik}\mathbf{v}_k \quad \text{for } i = 1, 2, \ldots, n$$

so $\mathbf{u}_i \in \text{span}\,\{\mathbf{v}_1, \mathbf{v}_2, \ldots, \mathbf{v}_n\}$ for $i = 1, 2, \ldots, n$. Since it was given that $\mathbf{v}_i \in \text{span}\,\{\mathbf{u}_1, \mathbf{u}_2, \ldots, \mathbf{u}_n\}$, we conclude that the spans are equal.

## MATLAB 4.4

**3. a.** **i.** The system of equations is

$$\begin{aligned} 1 &= 1c_1 - 1c_2 + 3c_3 \\ -4 &= 1c_1 + 1c_2 + 0c_3 \end{aligned}$$

The solution is $c_3$ arbitrary and $c_1 = -1.5 - 1.5c_3$ and $c_2 = -2.5 + 1.5c_3$.

**ii.** The solution is $c_3$ arbitrary and $c_1 = -2 - 3.2857c_3$ and $c_2 = 1 + .8571c_3$.

**b.** For (i), $\mathbf{w} = -1.5\mathbf{v}_1 - 2.5\mathbf{v}_2$, $\mathbf{w} = -4\mathbf{v}_1 + \frac{5}{3}\mathbf{v}_3$, and $\mathbf{w} = -4\mathbf{v}_2 - \mathbf{v}_3$.

**5. a.** The reason is that the reduced row echelon form of $A$ has no rows of zeros.

**b.** The reduced row echelon form of $A$ has a row of zeros, so there will be some $\mathbf{w}$ for which the system whose augmented matrix is $[A \quad \mathbf{w}]$ will have no solution, and hence $\mathbf{w}$ would not be a linear combination of the columns of $A$. Experiment to find such a $\mathbf{w}$ by trial and error, choosing values for $\mathbf{w}$ and checking to see if there is a solution.

**7. a.** The reduced row echelon form has no rows of zeros (so solutions exist) and there is at least one column without a pivot implying that an arbitrary variable will be in the solution.

**b.** For the first $\mathbf{w}$ given we have $x_1 = 2 - x_4$, $x_2 = -1 + 2x_4$, $x_3 = 2 - x_4$, and $x_5 = 1$. Hence $\mathbf{w} = 2\mathbf{v}_1 - \mathbf{v}_2 + 2\mathbf{v}_3 + \mathbf{v}_5$. For the second $\mathbf{w}$ given we have $x_1 = -3 - x_4$, $x_2 = 6 + 2x_4$, $x_3 = 1 - x_4$, and $x_5 = 1$. Hence $\mathbf{w} = -3\mathbf{v}_1 + 6\mathbf{v}_2 + \mathbf{v}_3 + \mathbf{v}_5$.

**c.** The fourth vector was unnecessary, corresponding to the fact that $x_4$ was the natural arbitrary variable.

**d.** We have that $\mathbf{v}_4 = \mathbf{v}_1 - 2\mathbf{v}_2 + \mathbf{v}_3$. Show that the reduced row echelon form of the matrix whose columns are the vectors in the subset has no row of zeros and has a pivot in every column.

**e.** The unnecessary vectors are the third and fifth vectors. The first $\mathbf{w} = 2\mathbf{v}_1 + 4\mathbf{v}_2 - \mathbf{v}_4$ and the second $\mathbf{w} = -1\mathbf{v}_1 + 7\mathbf{v}_2 - 2\mathbf{v}_4$; $\mathbf{v}_3 = -\mathbf{v}_1 + 2\mathbf{v}_2$ and $\mathbf{v}_5 = 2\mathbf{v}_1 + \mathbf{v}_2 - 2\mathbf{v}_4$.

**9. b.** The constant term of $r$ is $2 \times$ (constant term of $p$) $- 3 \times$ (constant term of $q$) and this holds true for the coefficients of the $x$, $x^2$, and $x^3$ terms.

**c.** Expressed as $3 \times 1$ vectors, we have

$$p = \begin{pmatrix} -1 \\ 2 \\ 0 \end{pmatrix}$$

$$\text{set} = \left\{ \begin{pmatrix} -2 \\ 0 \\ -5 \end{pmatrix}, \begin{pmatrix} 8 \\ -9 \\ -6 \end{pmatrix}, \begin{pmatrix} 9 \\ -7 \\ -1 \end{pmatrix} \right\}$$

$p$ is a linear combination with $p = p_1 - p_2 + p_3$, where $p_i$ refers to the $i$th polynomial in the set. The set of polynomials spans all of $P_2$.

**d.** $p = 3p_1 + 2p_2 + p_3$; The set cannot span by Problem 15.

**e.** Yes, since the reduced row echelon form of the matrix, whose columns are the vectors representing the given polynomials, has no rows of zeros.

**Problems 4.5,** page 328

**1.** independent

**3.** dependent;

$$-2\begin{pmatrix} 2 \\ -1 \\ 4 \end{pmatrix} + \begin{pmatrix} 4 \\ -2 \\ 8 \end{pmatrix} = \begin{pmatrix} 0 \\ 0 \\ 0 \end{pmatrix}$$

**5.** dependent (from Theorem 2)

**7.** independent

**9.** independent

**11.** independent

**13.** independent

**15.** independent

**17.** dependent

**19.** independent

**21.** independent

**23.** $ad - bc = 0$ **25.** $\alpha = -\frac{13}{2}$

**27.** System (7) can be written as

(*)
$$c_1\begin{pmatrix} a_{11} \\ a_{21} \\ \vdots \\ a_{m1} \end{pmatrix} + c_2\begin{pmatrix} a_{12} \\ a_{22} \\ \vdots \\ a_{m2} \end{pmatrix} + \cdots + c_n\begin{pmatrix} a_{1n} \\ a_{2n} \\ \vdots \\ a_{mn} \end{pmatrix} = \begin{pmatrix} 0 \\ 0 \\ \vdots \\ 0 \end{pmatrix}$$

If (7) has even one nontrivial solution, then the columns of $A$ are linearly dependent. If the columns of $A$ are dependent, then there are number $c_1, c_2, \ldots, c_n$ not all zero such that (*) holds.

**29.** If $\mathbf{0} = c_1\mathbf{v}_1 + c_2\mathbf{v}_2 + \cdots + c_k\mathbf{v}_k$, then $\mathbf{0} = c_1\mathbf{v}_1 + c_2\mathbf{v}_2 + \cdots + c_k\mathbf{v}_k + 0\mathbf{v}_{k+1} + 0\mathbf{v}_{k+2} + \cdots + 0\mathbf{v}_n$. Since $\mathbf{v}_1, \mathbf{v}_2, \ldots, \mathbf{v}_n$ are independent, we have $c_1 = c_2 = \cdots = c_k = 0$.

**31.** If $c_1\mathbf{v}_1 + c_2\mathbf{v}_2 + c_3\mathbf{v}_3 = \mathbf{0}$, then $0 = \mathbf{0} \cdot \mathbf{v}_1 = (c_1\mathbf{v}_1 + c_2\mathbf{v}_2 + c_3\mathbf{v}_3) \cdot \mathbf{v}_1 = c_1(\mathbf{v}_1 \cdot \mathbf{v}_1) + c_2(\mathbf{v}_2 \cdot \mathbf{v}_1) + c_3(\mathbf{v}_3 \cdot \mathbf{v}_1) = c_1\mathbf{v}_1 \cdot \mathbf{v}_1 + c_2 0 + c_3 0 = c_1\mathbf{v}_1 \cdot \mathbf{v}_1$. Since $\mathbf{v}_1 \neq \mathbf{0}$, $\mathbf{v}_1 \cdot \mathbf{v}_1 \neq 0$, so we must have $c_1 = 0$. A similar computation shows that $c_2 = c_3 = 0$.

**33.** $x_2\begin{pmatrix} -1 \\ 1 \\ 0 \end{pmatrix} + x_3\begin{pmatrix} -1 \\ 0 \\ 1 \end{pmatrix}$

**35.** $x_3\begin{pmatrix} 13 \\ -6 \\ 1 \end{pmatrix}$

**37.** $x_2\begin{pmatrix}-2\\1\\0\\0\end{pmatrix} + x_3\begin{pmatrix}3\\0\\1\\0\end{pmatrix} + x_4\begin{pmatrix}-5\\0\\0\\1\end{pmatrix}$

**39.** Any nonzero $\mathbf{u}$ will lead to a similar result. For example, let $\mathbf{u} = (1, 0, 0)$. Then $\mathbf{x} = (0, 1, 0)$ and $\mathbf{y} = (0, 0, 1)$ are in $H$.
$\mathbf{w} = \mathbf{x} \times \mathbf{y} = (1, 0, 0) = \mathbf{u}$.

**e.** $H$ is a plane orthogonal to $\mathbf{u}$ and $\mathbf{w}$ is orthogonal to this plane and so it must be parallel to $\mathbf{u}$.

**41.** Let $p(x) = a_0 + a_1x + a_2x^2$ and $q(x) = b_0 + b_1x + b_2x^2$ be two polynomials in $P_2$. Let $r(x) = c_0 + c_1x + c_2x^2$ be a third polynomial in $P_2$. If $p$ and $q$ span $P_2$, then there are scalars $\alpha$ and $\beta$ such that $r = \alpha p + \beta q$; that is,

$$c_0 = \alpha a_0 + \beta b_0$$
$$c_1 = \alpha a_1 + \beta b_1$$
$$c_2 = \alpha a_2 + \beta b_2$$

This "overdetermined" system of three equations in two unknowns will have a solution if and only if the third equation is a linear combination of the first two. Since $c_0$, $c_1$, and $c_2$ are arbitrary, this will rarely be the case. Thus $p$ and $q$ cannot span $P_2$. To see this in another way, suppose that $\begin{pmatrix}\alpha\\\beta\end{pmatrix}$ is a solution to the system. Then $\begin{pmatrix}c_0\\c_1\end{pmatrix} = \begin{pmatrix}a_0 & b_0\\a_1 & b_1\end{pmatrix}\begin{pmatrix}\alpha\\\beta\end{pmatrix}$ or $\begin{pmatrix}\alpha\\\beta\end{pmatrix} = \begin{pmatrix}a_0 & b_0\\a_1 & b_1\end{pmatrix}^{-1}\begin{pmatrix}c_0\\c_1\end{pmatrix}$. But, also, $\begin{pmatrix}\alpha\\\beta\end{pmatrix} = \begin{pmatrix}a_1 & b_1\\a_2 & b_2\end{pmatrix}^{-1}\begin{pmatrix}c_1\\c_2\end{pmatrix}$. In general, these two expressions for $\begin{pmatrix}\alpha\\\beta\end{pmatrix}$ will not be equal, which means that in general the system does not have a solution.

**43.** Let $S = \{\mathbf{v}_1, \mathbf{v}_2, \ldots, \mathbf{v}_k\}$ be a subset of the linearly independent set $T = \{\mathbf{v}_1, \mathbf{v}_2, \ldots, \mathbf{v}_n\}$ with $k < n$. Suppose $S$ is dependent. Then there exist scalars not all zero with $c_1\mathbf{v}_1 + c_2\mathbf{v}_2 + \cdots + c_k\mathbf{v}_k = \mathbf{0}$. This shows that $T$ is dependent, a contradiction. Simply form a linear combination of the vectors in $T$, using $c_i$ whenever $\mathbf{v}_i \in S$ and is 0 otherwise.

**45.** Write the matrices as $A^{(1)}, A^{(2)}, \ldots, A^{(mn+1)}$. Suppose that $A^{(i)} = (a_{jk}^{(i)})$. Consider the system

$$\begin{aligned} a_{11}^{(1)}\alpha_1 + a_{11}^{(2)}\alpha_2 + \cdots + a_{11}^{(mn+1)}\alpha_{mn+1} &= 0\\ \vdots \qquad\qquad\qquad & \ \ \vdots\\ a_{mn}^{(1)}\alpha_1 + a_{mn}^{(2)}\alpha_2 + \cdots + a_{mn}^{(mn+1)}\alpha_{mn+1} &= 0 \end{aligned}$$

This is a homogeneous system with $mn$ equations and $mn + 1$ unknowns. Therefore it has a nonzero solution, and there are scalars $\alpha_1, \alpha_2, \ldots, \alpha_{mn+1}$ not all zero such that $\alpha_1A^{(1)} + \alpha_2A^{(2)} + \cdots + \alpha_{mn+1}A^{mn+1} = 0$ (the $m \times n$ zero matrix).

**47.** Assume that $1, x, \ldots, x^k$ are linearly independent. Suppose that $c_0 + c_1x + \cdots + c_kx^k + c_{k+1}x^{k+1} = 0$. If $c_{k+1} \neq 0$, then $x^{k+1} = -\frac{c_0}{c_{k+1}} - \frac{c_1}{c_{k+1}}x - \cdots - \frac{c_k}{c_{k+1}}x^k$, which is clearly impossible. Thus $c_{k+1} = 0$. But then $c_0 + c_1x + \cdots + c_kx^k = 0$, which implies that $c_0 = c_1 = \cdots = c_k = 0$ since $1, x, x^2, \ldots, x^k$ are linearly independent. Thus $1, x, x^2, \ldots, x^k, x^{k+1}$ are also linearly independent, and this completes the induction proof (see Appendix 1).

**49.** There are scalars $\alpha_1, \alpha_2, \ldots, \alpha_n$ not all zero such that $\alpha_1\mathbf{v}_1 + \alpha_2\mathbf{v}_2 + \cdots + \alpha_n\mathbf{v}_n = \mathbf{0}$. Let $k$ be the largest integer for which $\alpha_k \neq 0$ ($k$ may equal $n$). Then the equation reads $\alpha_1\mathbf{v}_1 + \alpha_2\mathbf{v}_2 + \cdots + \alpha_k\mathbf{v}_k = \mathbf{0}$ so that

$$\mathbf{v}_k = -\frac{\alpha_1}{\alpha_k}\mathbf{v}_1 - \frac{\alpha_2}{\alpha_k}\mathbf{v}_2 - \cdots - \frac{\alpha_{k-1}}{\alpha_k}\mathbf{v}_{k-1}$$

**51.** Suppose $f$ and $g$ are dependent. Then $g = cf$ and $g' = cf'$ for some constant $c$ and

$$w(f, g)(x) = \begin{vmatrix} f(x) & g(x) \\ f'(x) & g'(x) \end{vmatrix} = \begin{vmatrix} f(x) & cf(x) \\ f'(x) & cf'(x) \end{vmatrix} = cf(x)f'(x) - cf(x)f'(x) = 0$$

**53.** Suppose that $c_1(\mathbf{u} + \mathbf{v}) + c_2(\mathbf{u} + \mathbf{w}) + c_3(\mathbf{v} + \mathbf{w}) = \mathbf{0}$. Then $(c_1 + c_2)\mathbf{u} + (c_1 + c_3)\mathbf{v} + (c_2 + c_3)\mathbf{w} = \mathbf{0}$. Since $\mathbf{u}$, $\mathbf{v}$, and $\mathbf{w}$ are linearly independent,

$$\begin{aligned} c_1 + c_2 \qquad &= 0 \\ c_1 \qquad + c_3 &= 0 \\ c_2 + c_3 &= 0 \end{aligned}$$

The determinant of this homogeneous system is

$$\begin{vmatrix} 1 & 1 & 0 \\ 1 & 0 & 1 \\ 0 & 1 & 1 \end{vmatrix} = -2 \neq 0$$

so the only solution is $c_1 = c_2 = c_3 = 0$ and the three vectors are independent.

**55.** $\begin{vmatrix} 1 & 1 & 1 \\ a & b & c \\ a^2 & b^2 & c^2 \end{vmatrix} = (b - a)(c - a)(c - b)$, according to the result of Problem 2.2.35 on page 203.

**57.** $\begin{pmatrix} 2 \\ 1 \\ 2 \end{pmatrix}, \begin{pmatrix} -1 \\ 3 \\ 4 \end{pmatrix}, \begin{pmatrix} 1 \\ 2 \\ 2 \end{pmatrix}$. (There are many choices for the third vector.)

**59. a.** By the result of Problem 3.5.59 on page 285, $\mathbf{u} \cdot (\mathbf{v} \times \mathbf{w}) = 0$. From Theorem 2 part (vi) on page 263,

$$\mathbf{v} \cdot (\mathbf{v} \times \mathbf{w}) = \mathbf{w} \cdot (\mathbf{v} \times \mathbf{w}) = 0$$

Let $\mathbf{a} = \begin{pmatrix} a \\ b \\ c \end{pmatrix} = \mathbf{v} \times \mathbf{w}$. Then

$$\mathbf{a} \cdot \mathbf{u} = \mathbf{a} \cdot \mathbf{v} = \mathbf{a} \cdot \mathbf{w} = 0$$

and writing out the terms in each scalar product gives the desired result.

**b.** Think of the system in (a) as a homogeneous system of three equations in the three unknowns $a$, $b$, and $c$. Since the system has nontrivial solutions, its determinant is 0.

**c.** This follows from (a) since we have

$$a\mathbf{u} + b\mathbf{v} + c\mathbf{w} = \mathbf{0}$$

but $a$, $b$, and $c$ are not all equal to 0.

## MATLAB 4.5

**1.** The sets are independent for the even Problems 2 through 8 and are dependent for Problems 10 and 12.

**3.** The columns will always be dependent in a matrix that has more columns than rows. *Proof hint:* What can you say about the location of pivots in the reduced row echelon form of such a matrix?

**5. a.** $A\mathbf{z}$ is a linear combination of the columns of $A$.

**b.** First generate a random $\mathbf{z}$ of the right size and then let $\mathbf{w} = A\mathbf{z}$, where $A$ is the matrix whose columns are the vectors in the given set.

**c.** $\{\mathbf{v}_1, \ldots, \mathbf{v}_k, \mathbf{w}\}$ is linearly dependent.

**7. a.** Columns without pivots in the reduced row echelon form correspond to the columns that were created as linear combinations of other columns.

**c.** The columns of $A$ are dependent.

**e.** Some column(s) of $A$ is (are) a linear combination(s) of previous columns of $A$.

**f.** *Proof hint:* For part (c), rewrite the linear combination with $\mathbf{0}$ on one side of the equation. For part (e), suppose $a_1\mathbf{v}_1 + \cdots + a_k\mathbf{v}_k + \cdots + a_n\mathbf{v}_n = \mathbf{0}$, where $a_k \neq 0$. Use this to solve for $\mathbf{v}_k$.

**9. a.** The reduced row echelon form of the matrix, whose columns are the vectors in the given set, has a pivot in every column but a row of zeros.

**b.** The reduced row echelon form of the matrix, whose columns are the vectors in the given set, has no rows of zeros but has at least one column without a pivot.

**c.** No.

**11.** *Problems 13, 15, and 16:* The set is independent. *Problem 14:* The set is dependent and $3x + 5x^2 = -13(-x) + 5(x^2 - 2x)$. *Problem 17:* The set is dependent and $x^3 + 18x - 9 = 8.7273(2x) + 3.1818(x^3 - 3) + .5455(1 + x - 4x^3)$. (The exact coefficients are $\frac{96}{11}$, $\frac{35}{11}$, and $\frac{6}{11}$.)

**13.** Let $A$ be a random matrix of the desired size. Find **A(:)** and note that it creates the vector representation of $A$ as described in MATLAB Problem 10 in Section 4.4. Test for independence or dependence of the vectors. *Proof hint:* Matrices in $M_{mn}$ are represented by vectors with $mn$ components.

**Problems 4.6,** page 344

**1.** no; does not span

**3.** no; dependent

**5.** no; does not span

**7.** yes **9.** yes

**11.** $\left\{\begin{pmatrix}0\\1\\-1\end{pmatrix}, \begin{pmatrix}1\\0\\2\end{pmatrix}\right\}$ **13.** $\left\{\begin{pmatrix}2\\3\\4\end{pmatrix}\right\}$

**15.** Since $\dim \mathbb{R}^2 = 2$, a proper subspace $H$ must have dimension 1. Let $\{(x_0, y_0)\}$ be a basis for $H$. If $(x, y) \in H$, then $(x, y) = c(x_0, y_0)$ for some number $c$. This means that $x = cx_0$, $y = cy_0$ or $c = \dfrac{x}{x_0} = \dfrac{y}{y_0}$ and $y = \left(\dfrac{y_0}{x_0}\right)x$, which is the equation of a straight line through the origin with slope $\dfrac{y_0}{x_0}$ if $x_0 \neq 0$. If $x_0 = 0$, then the line is the $y$-axis.

**17.** Let $\{\mathbf{v}_1, \mathbf{v}_2, \ldots, \mathbf{v}_{n-1}\}$ be a basis for $H$. Let $\mathbf{v}_n$ be another vector in $H$. Then the vectors $\mathbf{v}_1, \mathbf{v}_2, \ldots, \mathbf{v}_{n-1}, \mathbf{v}_n$ are linearly dependent. Let $A$ be the matrix whose rows are $\mathbf{v}_1, \mathbf{v}_2, \ldots, \mathbf{v}_{n-1}, \mathbf{v}_n$. Then $\det A = 0$, and the equation $A\mathbf{a} = \mathbf{0}$ has a nontrivial solution

$$\mathbf{a} = \begin{pmatrix}a_1\\a_2\\\vdots\\a_n\end{pmatrix}$$

This means that $\mathbf{v}_i \cdot \mathbf{a} = \mathbf{0}$ for $i = 1, 2, \ldots, n$. In particular, if $\mathbf{v}_n = (x_1, x_2, \ldots, x_n)$, then $\mathbf{v}_n \cdot \mathbf{a} = \mathbf{0}$, or $a_1x_1 + a_2x_2 + \cdots + a_nx_n = 0$. Since $\mathbf{v}_n$ was an arbitrary vector in $H$, this proves the result.

**19.** $\left\{\begin{pmatrix}1\\1\end{pmatrix}\right\}$ **21.** $\left\{\begin{pmatrix}-2\\-3\\1\end{pmatrix}\right\}$

**23.** $\left\{\begin{pmatrix}3\\1\\0\end{pmatrix}, \begin{pmatrix}-2\\0\\1\end{pmatrix}\right\}$ **25.** $n$

**27.** $V$ has a basis $\{\mathbf{u}_1, \mathbf{u}_2, \ldots, \mathbf{u}_n\}$, so there exist scalars such that

$$\begin{aligned} \mathbf{v}_1 &= a_{11}\mathbf{u}_1 + a_{12}\mathbf{u}_2 + \cdots + a_{1n}\mathbf{u}_n \\ \mathbf{v}_2 &= a_{21}\mathbf{u}_1 + a_{22}\mathbf{u}_2 + \cdots + a_{2n}\mathbf{u}_n \\ &\vdots \\ \mathbf{v}_m &= a_{m1}\mathbf{u}_1 + a_{m2}\mathbf{u}_2 + \cdots + a_{mn}\mathbf{u}_n \end{aligned}$$

Let $\mathbf{a}_i = (a_{i1}, a_{i2}, \ldots, a_{in})$. The $\mathbf{a}_i$'s are $m$ linearly independent vectors in $\mathbb{R}^n$ (otherwise the $\mathbf{v}_i$'s would not be independent). Expand the $\mathbf{a}_i$'s into a basis $\mathbf{a}_1, \mathbf{a}_2, \ldots, \mathbf{a}_m, \mathbf{a}_{m+1}, \ldots, \mathbf{a}_n$ for $\mathbb{R}^n$ by adding $n - m$ linearly independent vectors to the set. Then if $\mathbf{a}_k = (a_{k1}, a_{k2}, \ldots, a_{kn})$ for $m < k \le n$, define $\mathbf{v}_k = a_{k1}\mathbf{u}_1 + a_{k2}\mathbf{u}_2 + \cdots + a_{kn}\mathbf{u}_n$ for $k = m + 1, m + 2, \ldots, n$. Since

$$\det\begin{pmatrix} a_{11} & a_{12} & \cdots & a_{1n} \\ a_{21} & a_{22} & \cdots & a_{2n} \\ \vdots & \vdots & & \vdots \\ a_{n1} & a_{n2} & \cdots & a_{nn} \end{pmatrix} \neq 0$$

the set $\{\mathbf{v}_1, \mathbf{v}_2, \ldots, \mathbf{v}_n\}$ forms a basis for $V$ since it consists of $n$ linearly independent vectors in $V$ with $\dim V = n$.

**29.** If the vectors are independent, then they form a basis and $\dim V = n$. If not, then by Problem 4.5.49, at least one of them can be written as a linear combination of the ones that precede it. Throw this vector out. Proceed in this manner until $m$ linearly independent vectors remain. These must still span $V$ by the manner in which they were chosen. Thus $\dim V = m < n$. In either event $\dim V \le n$.

**31. a.** See Problem 4.3.27.

**b.** Let $\{\mathbf{v}_1, \ldots, \mathbf{v}_n\}$ be a basis for $H$, and let $\{\mathbf{u}_1, \mathbf{u}_2, \ldots, \mathbf{u}_m\}$ be a basis for $K$. Clearly $B = \{\mathbf{v}_1, \mathbf{v}_2, \ldots, \mathbf{v}_n, \mathbf{u}_1, \mathbf{u}_2, \ldots, \mathbf{u}_m\}$ span $H + K$. Suppose that $\alpha_1\mathbf{v}_1 + \alpha_2\mathbf{v}_2 + \cdots + \alpha_n\mathbf{v}_n + \beta_1\mathbf{u}_1 + \beta_2\mathbf{u}_2 + \cdots + \beta_m\mathbf{u}_m = \mathbf{0}$, where not all of the coefficients are zero. Let $\mathbf{h} = \alpha_1\mathbf{v}_1 + \alpha_2\mathbf{v}_2 + \cdots + \alpha_n\mathbf{v}_n$ and $\mathbf{k} = \beta_1\mathbf{u}_1 + \beta_2\mathbf{u}_2 + \cdots + \beta_m\mathbf{u}_m$. Then by linear independence, neither $\mathbf{h}$ nor $\mathbf{k}$ is the zero vector. Also, $\mathbf{h} \in H$ and $\mathbf{k} \in K$. But then $\mathbf{h} + \mathbf{k} = \mathbf{0}$ or $\mathbf{h} = -\mathbf{k} \in K$. Thus $\mathbf{0} \neq \mathbf{h} \in H \cap K$, which contradicts the fact that $H \cap K = \{\mathbf{0}\}$. Hence all the $\alpha$'s and $\beta$'s are zero, which implies that the vectors in $B$ are linearly independent. Thus $B$ is a basis for $H + K$ and $\dim(H + K) = \dim H + \dim K$.

**33. i.** If dim span $\{\mathbf{v}_1, \mathbf{v}_2\} = 1$, then choose a basis $\{\mathbf{v}\}$ for span $\{\mathbf{v}_1, \mathbf{v}_2\}$. Then $\mathbf{v}_1 = \alpha\mathbf{v}$, and $\mathbf{v}_2 = \beta\mathbf{v}$. If $\mathbf{v}_1 = \mathbf{0}$, then $\mathbf{v}_1$ is a single point lying on $\mathbf{v}_2$. If $\mathbf{v}_1 \neq \mathbf{0}$, then $\alpha \neq 0$ so that $\mathbf{v}_2 = \beta\mathbf{v} = \dfrac{\beta}{\alpha}(\alpha\mathbf{v}) = \dfrac{\beta}{\alpha}\mathbf{v}_1$. In either case the vectors are collinear.

**ii.** If the vectors are collinear, then $\mathbf{v}_2 = c\mathbf{v}_1$ for some scalar $c$ so that $\mathbf{v}_1$ is a basis for span $\{\mathbf{v}_1, \mathbf{v}_2\}$ and dim span $\{\mathbf{v}_1, \mathbf{v}_2\} = 1$.

**35.** If they are not linearly independent, then, as in the answer to Problem 29, $\dim V < n$. Since $\dim V = n$, the vectors must be independent, and therefore constitute a basis for $V$.

**37.** $B_1 = \left\{\begin{pmatrix}1\\0\\1\\0\end{pmatrix}, \begin{pmatrix}0\\1\\0\\1\end{pmatrix}, \begin{pmatrix}1\\0\\0\\0\end{pmatrix}, \begin{pmatrix}0\\1\\0\\0\end{pmatrix}\right\}$

$B_2 = \left\{\begin{pmatrix}1\\0\\1\\0\end{pmatrix}, \begin{pmatrix}0\\1\\0\\1\end{pmatrix}, \begin{pmatrix}0\\0\\1\\0\end{pmatrix}, \begin{pmatrix}0\\0\\0\\1\end{pmatrix}\right\}$

There are infinitely many other choices.

MATLAB 4.6

1. The basis needs to span all of $\mathbb{R}^n$ and be independent. What properties of the reduced row echelon form of the matrix, whose columns are the vectors in the basis, reflect each of these basis properties?

3. **a.** The new set will not span all of $\mathbb{R}^n$.
   **b.** The new set will not be independent.
   **c.** *Proof hint:* Think about the properties of having or not having a row of zeros in reduced row echelon form and having or not having a pivot in every column.

5. **b.** $A$ is invertible if and only if the columns of $A$ form a basis.
   **c.** *Proof hint:* $A$ is invertible if and only if the reduced row echelon form of $A$ is the identity. How does this reflect the properties of the reduced row echelon form needed to conclude that the columns of $A$ form a basis for $\mathbb{R}^n$?

Problems 4.7, page 360

**1.** $\rho = 2,\ \nu = 0$ **3.** $\rho = 1,\ \nu = 2$

**5.** $\rho = 2,\ \nu = 1$ **7.** $\rho = 2,\ \nu = 2$

**9.** $\rho = 2,\ \nu = 0$ **11.** $\rho = 2,\ \nu = 2$

**13.** $\rho = 3,\ \nu = 1$ **15.** $\rho = 2,\ \nu = 1$

**17.** range basis $= \left\{\begin{pmatrix}1\\3\\5\end{pmatrix}, \begin{pmatrix}-1\\1\\-1\end{pmatrix}\right\}$; these are the first two columns of $A$.

null space basis $= \left\{\begin{pmatrix}-\frac{3}{2}\\ \frac{1}{2}\\ 1\end{pmatrix}\right\}$

**19.** range basis $= \left\{\begin{pmatrix}1\\0\\1\end{pmatrix}, \begin{pmatrix}-1\\1\\0\end{pmatrix}, \begin{pmatrix}3\\3\\5\end{pmatrix}\right\}$; these are the first three (linearly independent) columns of $A$ corresponding to pivots in the row echelon form of the matrix.

null space basis $= \left\{\begin{pmatrix}-6\\-4\\1\\0\end{pmatrix}\right\}$

**21.** range basis $= \left\{\begin{pmatrix}1\\-2\\2\\3\end{pmatrix}\right\}$

null space basis $=$

$$\left\{\begin{pmatrix}1\\1\\0\\0\end{pmatrix}, \begin{pmatrix}-2\\0\\1\\0\end{pmatrix}, \begin{pmatrix}-3\\0\\0\\1\end{pmatrix}\right\}$$

**23.** $\left\{\begin{pmatrix}1\\4\\-2\end{pmatrix}, \begin{pmatrix}0\\1\\-\frac{6}{7}\end{pmatrix}\right\}$

**25.** $\{(1, 0, 0, \frac{1}{2}), (0, 1, -1, \frac{3}{2}), (0, 0, 0, 1)\}$

**27.** no **29.** yes

**31.** If $\mathbf{c}_i$ denotes the $i$th column of $D$, then

$$\mathbf{c}_i = d_i\begin{pmatrix}0\\0\\\vdots\\1\\0\\\vdots\\0\end{pmatrix} \leftarrow i\text{th position.}$$

Thus the $\mathbf{c}_i$'s are linearly independent when $d_i \neq 0$, and the number of linearly independent columns is the rank.

**33.** $\rho(A^t) =$ dimension of column space of $A^t =$ dimension of row space of $A =$ dimension of column space of $A$ (by Theorem 4) $= \rho(A)$.

**35.** **i.** Let $H =$ range of $A$ and let $\{\mathbf{v}_1, \mathbf{v}_2, \ldots, \mathbf{v}_k\}$ be a basis for $H$. Since $B$ is invertible, $N_B = \{\mathbf{0}\}$, which means that

$\{B\mathbf{v}_1, B\mathbf{v}_2, \ldots, B\mathbf{v}_k\}$ is a linearly independent set in $\mathbb{R}^m$, and is therefore a basis for range $BA$. Then $\rho(BA) = k = \rho(A)$.

**ii.** Since $C$ is invertible, range of $C = \mathbb{R}^n$. Let $\mathbf{h} \in H$; then there is an $\mathbf{x} \in \mathbb{R}^n$ such that $A\mathbf{x} = \mathbf{h}$. Since range of $C = \mathbb{R}^n$, there is a $\mathbf{y} \in \mathbb{R}^n$ such that $C\mathbf{y} = \mathbf{x}$. Then $AC\mathbf{y} = \mathbf{h}$. Thus $H \subset$ range of $AC$. If $\mathbf{v} \in$ range of $AC$, there is a $\mathbf{u}$ in $\mathbb{R}^n$ such that $AC\mathbf{u} = \mathbf{v}$. But then $\mathbf{v} = A(C\mathbf{u})$ so that $\mathbf{v} \in$ range of $A = H$. Hence range of $AC \subset H$ so that range of $AC = H$ and $\rho(A) = \rho(AC)$.

**37.** Since $\rho(A) = 5$, the five rows of $A$ are linearly independent. Thus the five rows of $(A, \mathbf{b})$ are linearly independent and $\rho(A, \mathbf{b}) = 5$.

**39.** By Problem 35, $\rho(A) = \rho(AD) = \rho(C(AD)) = \rho(B)$.

**41. i.** If there is an $\mathbf{x} \neq \mathbf{0}$ such that $A\mathbf{x} = \mathbf{0}$, then $A(\alpha\mathbf{x}) = \alpha A\mathbf{x} = \mathbf{0}$ for every $\alpha \in \mathbb{R}$ so that $\nu(A) = \dim N_A \geq 1$, and $\rho(A) = n - \nu(A) \leq n - 1 < n$.

**ii.** If $\rho(A) < n$, then $\nu(A) = n - \rho(A) > 0$ so that there is an $\mathbf{x} \neq \mathbf{0}$ such that $A\mathbf{x} = \mathbf{0}$.

**43.** Suppose that $B$, the row echelon form of $A$, has $k$ pivots in its first $k$ rows. Since there are no other pivots, all the entries below the first $k$ rows are zero. Let $a_{1,m_1}, a_{2,m_2}, \ldots, a_{k,m_k}$ denote the pivots; let $\mathbf{r}_1, \mathbf{r}_2, \ldots, \mathbf{r}_k$ denote the first $k$ rows of $B$ and suppose that $c_1\mathbf{r}_1 + c_2\mathbf{r}_2 + \cdots + c_k\mathbf{r}_k = \mathbf{0}$. By the definition of a pivot, the $m_1$ component in the vector $\mathbf{0} = c_1\mathbf{r}_1 + \cdots + c_k\mathbf{r}_k$ is $c_1a_{1,m_1}$. Since $a_{1,m_1} \neq 0$, we conclude that $c_1 = 0$. The $m_2$ component of the vector is $c_1a_{1p} + c_2a_{2,m_2}$ for some $p > 1$. Since $c_1 = 0$ and $a_{2,m_2} \neq 0$, we conclude that $c_2 = 0$. Continuing in this manner, we see that $c_j = 0$ for $j = 1, 2, \ldots, k$ so the first $k$ rows of $B$ are linearly independent. Since all other rows in the row echelon form of $A$ are zero, we conclude that $\rho(A) = k$.

Now, suppose that $\rho(A) = k$. Let $B$ equal the row echelon form of $A$. As above, the first $k$ rows of $B$ are linearly independent and all entries below the first $k$ rows are zero. The first nonzero entry in each of the first $k$ rows of $B$ is a pivot, for if not, it would have been made zero by the row reduction of $A$ to its row echelon form. Thus $B$ has $k$ pivots.

**45.** range $A = \text{span}\left\{\begin{pmatrix}187\\-35\\257\end{pmatrix}, \begin{pmatrix}-46\\51\\-148\end{pmatrix}\right\}$; $\rho(A) = 2$; $\nu(A) = 3$

**47.** range $A =$

$$\text{span}\left\{\begin{pmatrix}.0284\\-0.5110\\-.0965\\.0795\\-.0110\end{pmatrix}, \begin{pmatrix}-.0311\\-.1216\\-.4270\\.0905\\-.3365\end{pmatrix}, \begin{pmatrix}-.0207\\-.1811\\-.5847\\.1604\\-.4243\end{pmatrix}, \begin{pmatrix}.0431\\.0904\\.3574\\-.4730\\.3101\end{pmatrix}\right\};$$

$\rho(A) = 4$, $\nu(A) = 1$

## MATLAB 4.7

**1. a.** The bases for the null spaces and their respective dimensions are given below.

*Problem 7:* $\left\{\begin{pmatrix}-6\\-4\\1\\0\end{pmatrix}, \begin{pmatrix}-6\\-3\\0\\1\end{pmatrix}\right\}$ Dimension = 2

*Problem 8:* $\left\{\begin{pmatrix}-6\\-4\\1\\0\end{pmatrix}\right\}$ Dimension = 1

*Problem 10:* $\left\{\begin{pmatrix}0\\0\\0\\0\end{pmatrix}\right\}$ Dimension = 0

*Problem 11:* $\left\{\begin{pmatrix}1\\3\\1\\0\end{pmatrix}, \begin{pmatrix}2\\3\\0\\1\end{pmatrix}\right\}$ Dimension = 2

*Problem 12:* $\left\{\begin{pmatrix}1\\1\\0\\0\end{pmatrix}, \begin{pmatrix}-2\\0\\1\\0\end{pmatrix}, \begin{pmatrix}-3\\0\\0\\1\end{pmatrix}\right\}$ Dimension = 3

*Problem 13:* $\left\{\begin{pmatrix}-1\\1\\-1.5\\1\end{pmatrix}\right\}$ Dimension = 1

*Problem (vii):* $\left\{\begin{pmatrix}-2\\-3\\1\\0\\0\end{pmatrix}, \begin{pmatrix}-1\\-1\\0\\-1\\1\end{pmatrix}\right\}$

Dimension = 2

**c.** In finding the vectors for the basis, the process involves writing the solution as a linear combination of the vectors with the coefficients of the linear combination being the arbitrary variables.

**d.** The dimension equals the number of arbitrary variables.

**3. a.** **ii.** The same number of vectors is in this basis for the null space as in the basis found from the reduced row echelon form.

**iii.** For example, **rref([B N])** will solve the system(s) whose coefficient matrix is $B$ with the columns of $N$ being the right-hand sides. For the first matrix in MATLAB Problem 2,

$$B = \begin{pmatrix}-2 & -1\\-3 & -1\\1 & 0\\0 & -1\\0 & 1\end{pmatrix}.$$

For the second matrix in MATLAB Problem 2, let **R = rref(A)**. Then **B = [[−R(:,4);1;0][−R(:,5);0;1]]**.

**b.** Let **R = rref(A)** and **B = [[2;1;0;0;0][−R(1,5);0;−R(2,5);−R(3,5);1]]**. You should notice that the entries in **A*N** are closer to zero, the true values.

**5. a.** Show that $A\mathbf{x} = \mathbf{b}$.

**b.** Use **N = null(A)**.

**c.** First generate a random $2 \times 1$ vector $\mathbf{z}$ (since the basis will contain two vectors). Then let $\mathbf{w} = N\mathbf{z}$ and show that $A(\mathbf{x} + \mathbf{w}) = \mathbf{b}$.

**7. a.** The nonzero rows of **rref(A′)** give a basis for the row space of $A^t$, so their transposes give a basis for the column space of $A$.

**b.** The desired matrix will be the transpose of the nonzero rows of the reduced row echelon form of $A^t$. To verify linear combinations, use the matrix just described as the coefficient matrix and use $A$ for the right-hand sides. Below are the matrices whose columns form a basis for the range of $A$.

*Problem 7:* $\begin{pmatrix}1 & 0\\0 & 1\\1 & 1\end{pmatrix}$

*Problem 11:* $\begin{pmatrix}1 & 0\\0 & 1\\2 & 1\\1 & -1\end{pmatrix}$

*Problem 12:* $\begin{pmatrix}1\\-2\\2\\3\end{pmatrix}$

*Problem 13:* $\begin{pmatrix}1 & 0 & 0\\0 & 1 & 0\\0 & 0 & 1\\1 & 1 & 1\end{pmatrix}$

**9. a.** Refer to Section 4.4.

**b.** For (i), the basis consists of columns 1 and 3 of $A$. For (ii), the basis consists of all columns of $A$. For (iii), the basis consists of columns 1, 2, and 4 of $A$. Be sure to use the columns of $A$, *not* the columns of the reduced row echelon form of $A$.

**c.** For Problem 7, a basis consists of columns 1 and 2 of $A$. For Problem 11, a basis consists of columns 1 and 2 of $A$. For Problem 12, a basis consists of the first column of $A$. For Problem 13, a basis consists of columns 1, 2, and 3 of $A$.

**11.** Let $O$ be the matrix whose columns form a basis obtained from using **orth** and let $B$ be the matrix whose columns form a basis from MATLAB Problem 7. They will have the same number of columns (that is, each basis contains the same number of vectors). To verify linear combinations, look at **rref([O B])** and **rref([B O])**.

**13. e.** Rank equals the number of pivots.

**f.** $A$ is invertible if and only if $\text{rank}(A) = n = \text{size of } A$.

**g.** For the $5 \times 5$ matrix, generate a random $5 \times 5$ matrix and check that it is invertible. If invertible, change three of the columns to be linear combinations of the other two columns.

**15.** For a solution to exist, the augmented matrix will have the same rank as $A$. *Proof hint:* Think about the location of pivots in the reduced row echelon form of the augmented matrix and in the reduced row echelon form of $A$.

**17. a.** The rank of $AB$ is equal to $k$.

**b.–d.** The final conjecture should be that $\text{rank}(AB) \le \min(\text{rank}(A), \text{rank}(B))$.

**Problems 4.8,** page 382

**1.** $\dfrac{x+y}{2}\begin{pmatrix}1\\1\end{pmatrix} + \dfrac{x-y}{2}\begin{pmatrix}1\\-1\end{pmatrix} = \begin{pmatrix}x\\y\end{pmatrix}$

**3.** $\dfrac{4x+3y}{41}\begin{pmatrix}5\\7\end{pmatrix} + \dfrac{7x-5y}{41}\begin{pmatrix}3\\-4\end{pmatrix} = \begin{pmatrix}x\\y\end{pmatrix}$

**5.** $\dfrac{dx-by}{ad-bc}\begin{pmatrix}a\\c\end{pmatrix} + \dfrac{-cx+ay}{ad-bc}\begin{pmatrix}b\\d\end{pmatrix} = \begin{pmatrix}x\\y\end{pmatrix}$

**7.** $(x-y)\begin{pmatrix}1\\0\\0\end{pmatrix} + (y-z)\begin{pmatrix}1\\1\\0\end{pmatrix} + z\begin{pmatrix}1\\1\\1\end{pmatrix} = \begin{pmatrix}x\\y\\z\end{pmatrix}$

**9.** $\dfrac{6x-11y+10z}{31}\begin{pmatrix}2\\1\\3\end{pmatrix} + \dfrac{2x+17y-7z}{31}\begin{pmatrix}-1\\4\\5\end{pmatrix} + \dfrac{7x+13y-9z}{31}\begin{pmatrix}3\\-2\\-4\end{pmatrix} = \begin{pmatrix}x\\y\\z\end{pmatrix}$

**11.** $a_0 + a_1x + a_2x^2 = (a_0 + a_1 + a_2)1 + a_1(x-1) + a_2(x^2-1)$

**13.** $a_0 + a_1x + a_2x_2 = \dfrac{(a_0+a_1+a_2)}{2}(x+1) + \dfrac{(a_1-a_0-a_2)}{2}(x-1) + a_2(x^2-1)$

**15.** $2(x^3 + x^2) - 5(x^2 + x) + 10(x + 1) - 16(1)$

**17.** $(\mathbf{x})_{B_2} = \begin{pmatrix} -\frac{1}{3} \\ 0 \end{pmatrix}$

**19.** $(\mathbf{x})_{B_2} = \begin{pmatrix} \frac{86}{33} \\ -\frac{20}{11} \\ \frac{7}{11} \end{pmatrix}$

**21.** independent

**23.** dependent

**25.** independent

**27.** independent

**29.** If they were linearly independent, they would span $P_n$. But $1 \in P_n$ and $1 \notin \text{span}\,\{p_1, p_2, \ldots, p_{n+1}\}$ since the constant term in each polynomial is 0.

**31.** If they were linearly independent, they would span $M_{mn}$. But the matrix $A = (a_{ij})$, where $a_{11} = 1$ and $a_{ij} = 0$ otherwise is not in the span of $A_1, A_2, \ldots, A_{mn}$ since a linear combination of matrices with a 0 in the 1, 1 position also has a 0 in the 1, 1 position.

**33.** $\begin{pmatrix} \cos\theta & -\sin\theta \\ \sin\theta & \cos\theta \end{pmatrix}\begin{pmatrix} 1 \\ 0 \end{pmatrix} = \begin{pmatrix} \cos\theta \\ \sin\theta \end{pmatrix}$; $\begin{pmatrix} \cos\theta & -\sin\theta \\ \sin\theta & \cos\theta \end{pmatrix}\begin{pmatrix} 0 \\ 1 \end{pmatrix} = \begin{pmatrix} -\sin\theta \\ \cos\theta \end{pmatrix}$.

$A^{-1}$ is obtained by rotating through an angle of $-\theta$. Thus

$$A^{-1} = \begin{pmatrix} \cos(-\theta) & -\sin(-\theta) \\ \sin(-\theta) & \cos(-\theta) \end{pmatrix} = \begin{pmatrix} \cos\theta & \sin\theta \\ -\sin\theta & \cos\theta \end{pmatrix}$$

Alternatively, since

$$A = \begin{pmatrix} \cos\theta & -\sin\theta \\ \sin\theta & \cos\theta \end{pmatrix},$$

$$A^{-1} = \begin{pmatrix} \cos\theta & \sin\theta \\ -\sin\theta & \cos\theta \end{pmatrix}$$

**35.** $A^{-1} =$

$$\begin{pmatrix} \cos(\pi/4) & \sin(\pi/4) \\ -\sin(\pi/4) & \cos(\pi/4) \end{pmatrix} = \begin{pmatrix} 1/\sqrt{2} & 1/\sqrt{2} \\ -1/\sqrt{2} & 1/\sqrt{2} \end{pmatrix}$$

(from Problem 33), so

$$A^{-1}\begin{pmatrix} 2 \\ -7 \end{pmatrix} = \begin{pmatrix} -5/\sqrt{2} \\ -9/\sqrt{2} \end{pmatrix}$$

**37.** Since $C$ is invertible, the columns of $C$ are linearly independent. That is, $\{\mathbf{c}_1, \mathbf{c}_2, \ldots, \mathbf{c}_n\}$ are $n$ linearly independent vectors in $V$, which are therefore a basis for $V$ since $\dim V = n$.

**39.** If $CA = I$, then $(\mathbf{x})_{B_1} = I(\mathbf{x})_{B_1} = CA(\mathbf{x})_{B_1}$. Conversely, suppose that $(\mathbf{x})_{B_1} = CA(\mathbf{x})_{B_1}$. Let $B_1 = \{\mathbf{v}_1, \mathbf{v}_2, \ldots, \mathbf{v}_n\}$. Then

$$(\mathbf{v}_1)_{B_1} = \begin{pmatrix} 1 \\ 0 \\ \vdots \\ 0 \end{pmatrix} = CA\begin{pmatrix} 1 \\ 0 \\ \vdots \\ 0 \end{pmatrix}$$

Let

$$CA = \begin{pmatrix} r_{11} & r_{12} & \cdots & r_{1n} \\ r_{21} & r_{22} & \cdots & r_{2n} \\ \vdots & \vdots & & \vdots \\ r_{n1} & r_{n2} & \cdots & r_{nn} \end{pmatrix}$$

Then $CA\begin{pmatrix} 1 \\ 0 \\ \vdots \\ 0 \end{pmatrix}$ = the first column of $CA = \begin{pmatrix} r_{11} \\ r_{22} \\ \vdots \\ r_{n1} \end{pmatrix} = \begin{pmatrix} 1 \\ 0 \\ \vdots \\ 0 \end{pmatrix}$. Similarly, the second column of $CA = \begin{pmatrix} 0 \\ 1 \\ 0 \\ \vdots \\ 0 \end{pmatrix}$ since $(\mathbf{v}_2)_{B_1} = \begin{pmatrix} 0 \\ 1 \\ 0 \\ \vdots \\ 0 \end{pmatrix}$. Continuing in this manner, we see that $CA = I$.

## MATLAB 4.8

**1. a.** **i.** $\mathbf{w} = 1.5\mathbf{v}_1 + .5\mathbf{v}_2$
**ii.** $\mathbf{w} = .5\mathbf{v}_1 + 3.5\mathbf{v}_2$

**3. a.** To find the coordinates of a vector **w** with respect to the basis $\{\mathbf{v}_1, \ldots, \mathbf{v}_4\}$, solve the system of equations to write **w** as a linear combination of the basis vectors; that is, solve the system $[\mathbf{v}_1\ \mathbf{v}_2\ \mathbf{v}_3\ \mathbf{v}_4 \mid \mathbf{w}]$.

**b.** $C = \begin{pmatrix} -84 & 45 & -5 & 21 \\ 19 & -10 & 1 & -5 \\ 21 & -11 & 1 & -5 \\ -4 & 2 & 0 & 1 \end{pmatrix}$

**c.** **i.** $(\mathbf{w})_B = (-105\quad 22\quad 26\quad -4)^t$.
**iii.** *Hint:* $(\mathbf{w})_B = C\mathbf{w}$ = linear combination of the columns of $C$, where the coefficients in the linear combination are the components of **w**.

**5. c.** $D = \begin{pmatrix} -7 & -13 & 19 \\ -6 & -11 & 18 \\ 1 & 2 & -2 \end{pmatrix}$

**d.** $(\mathbf{x})_B$ = solution to $[\mathbf{v}_1\quad \mathbf{v}_2\quad \mathbf{v}_3 \mid \mathbf{x}] = (-6\quad 2\quad -1)^t$. $(\mathbf{x})_C$ = solution to $[\mathbf{w}_1\quad \mathbf{w}_2\quad \mathbf{w}_3 \mid \mathbf{x}] = (-3\quad -4\quad 0)^t$.

**e.** $D = W^{-1}V$.

**f.** $D = \begin{pmatrix} -27 & -165 & 25 & -81 \\ 7 & 40 & -4 & 21 \\ 6 & 37 & -6 & 17 \\ -1 & -6 & 1 & -2 \end{pmatrix}$.

**7. a.** Let $B$ be the matrix whose columns are the basis vectors in the set $B$ and similarly for $C$. For example, the transition matrix $T$ from $B$ to $C$ will equal $C^{-1}B$. (Explain this.)

$T = \begin{pmatrix} -7 & -13 & 19 \\ -6 & -11 & 18 \\ 1 & 2 & -2 \end{pmatrix}$, $S = \begin{pmatrix} 0 & 0 & -1 \\ 0 & -1 & -9 \\ 2 & 6 & 80 \end{pmatrix}$,

$K = \begin{pmatrix} -1 & -2 & 2 \\ -3 & -7 & 0 \\ 30 & 68 & -14 \end{pmatrix}$

**b.** $K = ST$.

**9. a.** Use basic trigonometry.

**c.** **i.** Both $B$ and $C$ will have the form $\begin{pmatrix} \cos(\theta) & -\sin(\theta) \\ \sin(\theta) & \cos(\theta) \end{pmatrix}$ for some $\theta$. We have that $T = \begin{pmatrix} .2588 & .9659 \\ -.9659 & .2588 \end{pmatrix}$ and $S = \begin{pmatrix} .2588 & -.9659 \\ .9659 & .2588 \end{pmatrix}$.

**ii.** The $\frac{2\pi}{3}$ coordinates are $(3.0272\quad .2935)^t$ and the standard coordinates are $(-1.7678\quad 2.4749)^t$. For example, to find standard coordinates from the coordinates with respect to $B$, find $B\mathbf{x}$, where **x** contains the $B$-coordinates.

**iii.** The $\frac{\pi}{4}$ coordinates are $(1.8699\quad 1.5695)^t$ and the standard coordinates are $(.2124\quad 2.4321)^t$.

## Problems 4.9, page 407

**1.** $\begin{pmatrix} 1/\sqrt{2} \\ 1/\sqrt{2} \end{pmatrix}, \begin{pmatrix} -1/\sqrt{2} \\ 1/\sqrt{2} \end{pmatrix}$

**3.** **i.** If $a = b = 0$, $\{(1, 0), (0, 1)\}$
**ii.** If $a = 0$, $b \neq 0$, $\{(1, 0)\}$
**iii.** If $a \neq 0$, $b = 0$, $\{(0, 1)\}$
**iv.** If $a \neq 0$, $b \neq 0$, $\{(b/\sqrt{a^2 + b^2}, -a/\sqrt{a^2 + b^2})\}$

**5.** $\{(1/\sqrt{5}, 0, 2/\sqrt{5}), (2/\sqrt{30}, 5/\sqrt{30}, -1/\sqrt{30})\}$

**7.** $\{(2/\sqrt{29}, 3/\sqrt{29}, 4/\sqrt{29})\}$

**9.** $\{(1/\sqrt{5}, 0, 0, 2/\sqrt{5}), (2/\sqrt{30}, 5/\sqrt{30}, 0, -1/\sqrt{30}), (-2/\sqrt{10}, 1/\sqrt{10}, 2/\sqrt{10}, 1/\sqrt{10})\}$

**11.** $\{(a/\sqrt{a^2 + b^2 + c^2}, b/\sqrt{a^2 + b^2 + c^2}, c/\sqrt{a^2 + b^2 + c^2})\}$

**13.** $\{(-7/\sqrt{66}, -1/\sqrt{66}, 4/\sqrt{66})\}$

**15.** $Q^t = \begin{pmatrix} \frac{2}{3} & \frac{1}{3} & -\frac{2}{3} \\ \frac{1}{3} & \frac{2}{3} & \frac{2}{3} \\ \frac{2}{3} & -\frac{2}{3} & \frac{1}{3} \end{pmatrix}$ and $Q^tQ = I = QQ^t$

**17.** $PQ = \dfrac{1}{3\sqrt{2}} \times \begin{pmatrix} 1-\sqrt{8} & -1-\sqrt{8} \\ 1+\sqrt{8} & 1-\sqrt{8} \end{pmatrix}$

$(PQ)^t = \dfrac{1}{3\sqrt{2}} \times \begin{pmatrix} 1-\sqrt{8} & 1+\sqrt{8} \\ -1-\sqrt{8} & 1-\sqrt{8} \end{pmatrix}$

$(PQ)(PQ)^t = \dfrac{1}{18}\begin{pmatrix} 18 & 0 \\ 0 & 18 \end{pmatrix} = I$

**19.** $I = Q^{-1}Q = Q^tQ$. But $\det(Q^tQ) = \det Q^t \det Q = \det Q \det Q = (\det Q)^2$. Since

$$1 = \det I = \det Q^tQ = (\det Q)^2$$

we have

$$\det Q = \pm 1$$

**21.** If $\mathbf{v}_i = \mathbf{0}$, then $0\mathbf{v}_1 + 0\mathbf{v}_2 + \cdots + 0\mathbf{v}_{i-1} + \mathbf{v}_i + 0\mathbf{v}_{i+1} + \cdots + 0\mathbf{v}_n = \mathbf{0}$, which implies that the $\mathbf{v}_i$'s are linearly dependent. Thus $\mathbf{v}_i \neq \mathbf{0}$ for $i = 1, 2, \ldots, n$.

**23. a.** $\mathbf{0}$

**b.** $\dfrac{1}{\sqrt{a^2+b^2}}\begin{pmatrix} a \\ b \end{pmatrix}$

**c.** $\mathbf{v} = \begin{pmatrix} a \\ b \end{pmatrix} + \begin{pmatrix} 0 \\ 0 \end{pmatrix}$

**25. a.** $\dfrac{1}{49}\begin{pmatrix} -186 \\ 75 \\ 118 \end{pmatrix}$

**b.** $\dfrac{1}{7}\begin{pmatrix} 3 \\ -2 \\ 6 \end{pmatrix}$

**c.** $\mathbf{v} = \dfrac{1}{49}\begin{pmatrix} -186 \\ 75 \\ 118 \end{pmatrix} + \dfrac{13}{49}\begin{pmatrix} 3 \\ -2 \\ 6 \end{pmatrix}$

**27. a.** $\dfrac{1}{5}\begin{pmatrix} 1 \\ -3 \\ 4 \\ 17 \end{pmatrix}$

**b.** $\dfrac{1}{\sqrt{15}}\begin{pmatrix} 2 \\ -1 \\ 3 \\ -1 \end{pmatrix}$

**c.** $\dfrac{1}{5}\begin{pmatrix} 1 \\ -3 \\ 4 \\ 17 \end{pmatrix} + \dfrac{2}{5}\begin{pmatrix} 2 \\ -1 \\ 3 \\ -1 \end{pmatrix}$

**29.** $|\mathbf{u}_1 - \mathbf{u}_2|^2 = (\mathbf{u}_1 - \mathbf{u}_2) \cdot (\mathbf{u}_1 - \mathbf{u}_2) = \mathbf{u}_1 \cdot \mathbf{u}_1 - \mathbf{u}_2 \cdot \mathbf{u}_1 - \mathbf{u}_1 \cdot \mathbf{u}_2 + \mathbf{u}_2 \cdot \mathbf{u}_2 = 1 - 0 - 0 + 1 = 2$ since $\mathbf{u}_1$, $\mathbf{u}_2$ are orthonormal.

**31.** $a^2 + b^2 = 1$.

**33.** $|\mathbf{u} + \mathbf{v}|^2 = (|\mathbf{u}| + |\mathbf{v}|)^2$. This means that $(\mathbf{u} + \mathbf{v}) \cdot (\mathbf{u} + \mathbf{v}) = |\mathbf{u}|^2 + 2\mathbf{u} \cdot \mathbf{v} + |\mathbf{v}|^2 = (|\mathbf{u}| + |\mathbf{v}|)^2 = |\mathbf{u}|^2 + 2|\mathbf{u}|\,|\mathbf{v}| + |\mathbf{v}|^2$. Thus $\mathbf{u} \cdot \mathbf{v} = |\mathbf{u}|\,|\mathbf{v}|$, which can occur only if $\mathbf{u} = \lambda\mathbf{v}$; that is, $\mathbf{u}$ and $\mathbf{v}$ are linearly dependent.

**35.** We prove this by mathematical induction. If $k = 2$, this is the result of Problem 33. We assume it is true for $k = n$ and prove it for $k = n + 1$. Suppose that $|\mathbf{x}_1 + \mathbf{x}_2 + \cdots + \mathbf{x}_n + \mathbf{x}_{n+1}| = |\mathbf{x}_1| + |\mathbf{x}_2| + \cdots + |\mathbf{x}_n| + \mathbf{x}_{n+1}|$. (*) This implies that $|\mathbf{x}_1 + \mathbf{x}_2 + \cdots + \mathbf{x}_n| = |\mathbf{x}_1| + |\mathbf{x}_2| + \cdots + |\mathbf{x}_n|^{(\surd)}$, for if this is not true, then, by the triangle inequality,

$$|\mathbf{x}_1 + \mathbf{x}_2 + \cdots + \mathbf{x}_n| < |\mathbf{x}_1| + |\mathbf{x}_2| + \cdots + |\mathbf{x}_n|$$

But then

$$|\mathbf{x}_1 + \mathbf{x}_2 + \cdots + \mathbf{x}_n + \mathbf{x}_{n+1}| \le |\mathbf{x}_1 + \mathbf{x}_2 + \cdots + \mathbf{x}_n| + |\mathbf{x}_{n+1}| < |\mathbf{x}_1| + |\mathbf{x}_2| + \cdots + |\mathbf{x}_n| + |\mathbf{x}_{n+1}|$$

which contradicts (*). Thus, by the induction assumption, dim span $\{\mathbf{x}_1, \mathbf{x}_2, \ldots, \mathbf{x}_n\} = 1$. Let $\mathbf{u} = \mathbf{x}_1 + \mathbf{x}_2 + \cdots + \mathbf{x}_n$. By (*) and (√) $|\mathbf{u} + \mathbf{x}_{n+1}| = |\mathbf{u}| + |\mathbf{x}_{n+1}|$ so that by Problem 33, $\mathbf{x}_{n+1} = \lambda\mathbf{u}$ for some number $\lambda$. That is, $\mathbf{x}_{n+1} \in$ span $\{\mathbf{x}_1, \mathbf{x}_2, \ldots, \mathbf{x}_n\}$ so that dim span $\{\mathbf{x}_1, \mathbf{x}_2, \ldots, \mathbf{x}_n, \mathbf{x}_{n+1}\} = 1$ also. Thus the result is true for $k = n + 1$, and the proof is complete.

**37.** $(H^\perp)^\perp = \{\mathbf{v} \in \mathbb{R}^n;\ \mathbf{v} \cdot \mathbf{k} = 0$ for every $\mathbf{k} \in H^\perp\}$. Let $\mathbf{x} \in H$; then $\mathbf{x} \cdot \mathbf{k} = 0$ for every $\mathbf{k} \in H^\perp$ so that $\mathbf{x} \in (H^\perp)^\perp$, which shows that $H \subseteq (H^\perp)^\perp$. Conversely, if $\mathbf{v} \in (H^\perp)^\perp$, then $\mathbf{v} \cdot \mathbf{k} = 0$ for every $\mathbf{k} \in H^\perp$. But $\mathbf{v} = \mathbf{h}' + \mathbf{k}'$, where $\mathbf{h}' \in H$ and $\mathbf{k}' \in H^\perp$. Then $0 = \mathbf{v} \cdot \mathbf{k} = \mathbf{h}' \cdot \mathbf{k} + \mathbf{k}' \cdot \mathbf{k} = 0 + \mathbf{k}' \cdot \mathbf{k}$. Thus $\mathbf{k}' \cdot \mathbf{k} = 0$ for every $\mathbf{k} \in H$, which means, in particular, that $\mathbf{k}' \cdot \mathbf{k}' = 0$. Thus $\mathbf{k}' = \mathbf{0}$ and $\mathbf{v} = \mathbf{h}' \in H$. Thus $(H^\perp)^\perp \subset H$ and, together with $H \subset (H^\perp)^\perp$, shows that $(H^\perp)^\perp = H$.

**39.** Let $\mathbf{k} \in H_2^\perp$. Then $\mathbf{k} \cdot \mathbf{h} = 0$ for every $\mathbf{h} \in H_2$. Since $H_1 \subset H_2$, this shows that $\mathbf{k} \cdot \mathbf{h} = 0$ for every $\mathbf{h} \in H_1$. That is, $\mathbf{k} \in H_1^\perp$. Thus $H_2^\perp \subset H_1^\perp$.

## MATLAB 4.9

**1.** The coding for part (b) is

```
z1 = w1/norm(w1),
t2 = w2-(w2'*z1)*z1;
z2 = t2/norm(2),
t3 = w3-(w3'*z1)*z1-(w3'*z2)*z2;
z3 = t3/norm(t3)
```

Check orthogonality by finding $\mathbf{z}_1 \cdot \mathbf{z}_2$, $\mathbf{z}_1 \cdot \mathbf{z}_3$, and $\mathbf{z}_2 \cdot \mathbf{z}_3$. Check linear combinations by finding the reduced row echelon form of the matrix $[\mathbf{z}_1\ \ \mathbf{z}_2\ \ \mathbf{z}_3 \,|\, \mathbf{w}_1\ \ \mathbf{w}_2\ \ \mathbf{w}_3]$.

**3.** **a.–c.** $\mathbf{p}_1 = (1\ \ 2)^t$ and $\mathbf{p}_2 = (-4\ \ 2)^t$. Recall that a projection onto a vector drops a perpendicular to the line determined by the vector so that in the linear combination graph the parallelogram drawn will be a rectangle.

**d.** $\mathbf{p}_1 = (1.6\ \ 3.2)^t$ and $\mathbf{p}_2 = (2.4\ \ -1.2)^t$.

**f.** The projection onto $\mathbf{v}_2$ is the projection onto $H^\perp$.

**5.** **a.** $\{\mathbf{z}_1, \mathbf{z}_1\} = \left\{ \begin{pmatrix} -.2673 \\ .5345 \\ .8018 \end{pmatrix}, \begin{pmatrix} .8729 \\ -.2182 \\ .4364 \end{pmatrix} \right\}$

**b.** Because the dimension of $H^\perp$ is 1 and $\mathbf{n}$ is perpendicular to $H$.

**c.** Compare
**p = (z1'*w)*z1+(z2'*w)*z2**
with
**q = w−(n'*w)*n.**

**d.**

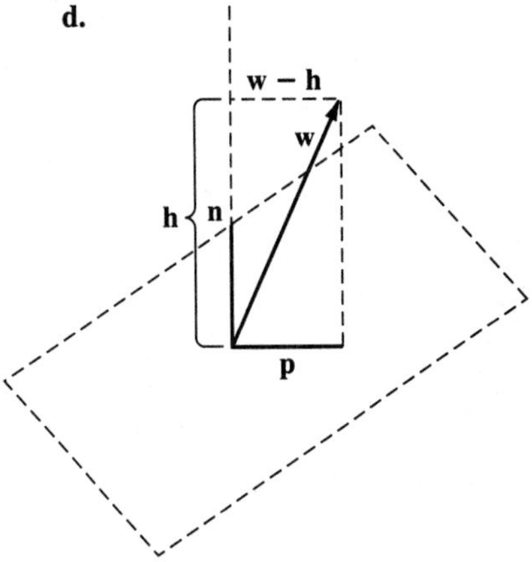

**7.** **a.** First compute **q = u1'*w*u1+u2'*w*u2 + u3'*w*u3+u4'*w*u4** and compare with **p = B*B'*w.**

**b.** Choose $\mathbf{x}$ to be a $4 \times 1$ vector. You will observe that $|\mathbf{w} - \mathbf{p}|$ is smaller than $|\mathbf{w} - \mathbf{h}|$ since $\mathbf{p}$

is the vector in $H$ that is the closest to $\mathbf{w}$.

**c.** $\mathbf{p} = BB^t\mathbf{w} = DD^t\mathbf{w}$.

**d.** *Proof hint:* Refer to part (a). Think of $\mathbf{w} \cdot \mathbf{u}_i$ as $\mathbf{u}_i' * \mathbf{w}$ and note that row $i$ of $B^t$ is $\mathbf{u}_i'$. Recall that a linear combination of vectors can be interpreted as multiplication by the matrix whose columns are the vectors.

**9. a.** *Hint:* Interpret $\mathbf{v} \cdot \mathbf{u}_i$ as $\mathbf{u}_i' * \mathbf{v}$ and use the fact that row $i$ of $B^t$ is $\mathbf{u}_i'$.

**b.** Use the facts that $\mathbf{u}_i$ and $\mathbf{w}$ have norm equal to 1.

**c.** **i.** Each angle is 45°. Be sure first to find $\mathbf{w}$ by entering the given vector and then dividing by its norm.

**ii.** The angle with $\mathbf{v}_1$ is 135° and the angle with $\mathbf{v}_2$ is 45°.

**d.** All angles are the same and equal to 54.74°.

**11.** *Proof hint:* Show that $(AB)^t(AB) = I$ using properties of $A$, $B$, and the transpose.

**13. a.** $\det(Q) = \pm 1$

**b.** *Proof hint:* Find the determinant of both sides of $QQ^t = I$.

**c.** $|\det(Q)| = 1$.

**15. b.** The standard basis is orthonormal and rotation preserves length and angle.

**c.** Products of orthogonal matrices are orthogonal.

**d.** $[\mathbf{v}_1 \ \ \mathbf{v}_2 \ \ \mathbf{v}_3]$ is not an orthogonal matrix.

**e.** $A = YRP =$

$$\begin{pmatrix} .9186 & -.2500 & .3062 \\ -.1768 & .4330 & .8839 \\ -.3536 & -.8660 & .3536 \end{pmatrix}$$

The matrix $B$ below contains all nine angles. The first column of $B$ contains the angles between the standard $x$-axis and the $x$-, $y$-, and $z$-axes of the spacecraft, respectively. Similarly, the second column contains the angles between the standard $y$-axis and the spacecraft coordinates, and the third column contains the angles between the standard $z$-axis and the spacecraft coordinates:

$$B = \begin{pmatrix} 23.28^\circ & 100.18^\circ & 110.70^\circ \\ 104.48^\circ & 64.34^\circ & 150.00^\circ \\ 72.17^\circ & 27.88^\circ & 69.30^\circ \end{pmatrix}$$

### Problems 4.10, page 427

**1.** $y = \frac{408}{126} - \frac{57}{126}x \approx 3.24 - 0.45x$

**3.** $y = \frac{162}{84} - \frac{10}{84}x \approx 1.93 - 0.12x$

**5.** $y = \frac{13,536}{5184} + \frac{10,800}{5184}x + \frac{1,584}{5,184}x^2 \approx 2.61 + 2.08x + 0.31x^2$

This is the equation of the parabola passing through the three points.

**7.** The argument here closely parallels the arguments given for linear and quadratic approximations.

**9.** This is a generalization of Problem 7.

**11.** $y \approx 108.71 + 4.906x - 0.00973x^2$

**13.** $y = 116.71766114 - 0.87933592x$

**15.** $y = -111.930576071 + 2.133265272x$

**17.** $y = 0.01357392x^2 - 3.39135882x + 217.42127707$

**19.** $y = 0.00004056437x^2 + 2.01798203x - 53.88518039$

### MATLAB 4.10

**1. b.** $\mathbf{u} = (2.9535 \ \ -1.1813)^t$; hence the line is $y = 2.9535 - 1.1813x$.

**c.** Use the MATLAB command **norm.** $|\mathbf{y}-A\mathbf{u}| = 4.066$ and $|\mathbf{y}-A\mathbf{w}| = 2.9712$. The sum of the squares of the difference in $y$-coordinates between the least squares line and the points is smaller than if any other line is used.

**d.** Recall that $\text{proj}_H\mathbf{y} = BB^t\mathbf{y}$.

**e.** $y = -.4722$.

**3.** $g \approx -30.6364$, $v_0 \approx 60.9470$, and the height above the ground is $\approx$ 10.8977.

**5.** **a.** The line fit is $y = -.1942 + 1.1921x$ with the least squares error norm equal to .4419. The quadratic fit is $y = -.0423 - .7078x - 5.7751x^2$, with the least squares error norm equal to .1171. The quadratic fit is apparently better: the norm is smaller and the *'s seem much closer to the quadratic. However, notice that one point could be considered an outlier.

**b.** The line fit is $y = 35.9357 - 83.4269x$, with the least squares error norm equal to 25.3326. The quadratic fit is $y = 41.5798 - 51.2577x - 59.5481x^2$ with the least squares error norm equal to 15.2469. The quadratic fit is slightly better: the norm is smaller and the *'s seem closer to the quadratic.

**7.** The line fit is $y = 40.8537 + .0066x$ and the quadratic fit is $y = -78 + .32x - .0002x^2$. The quadratic fit appears better, and by using this we can conclude that the product will be the strongest if the temperature is 800°.

**9.** **a.** **i.** Using **x = [0:10]′**, we have the fit $y = e^{1.7322+.2706x}$. The fit appears reasonable.

**ii.** Use $x = 15$. The projected population for 1950 is 327.1814 million.

**b.** **i.** The projected population for 1950 is larger than the actual population.

**ii.** An exponential model seems to be a reasonable fit. We have $y = e^{4.3999+.1287x}$. (Do not forget to use the logarithm of the population values for the **y** vector.)

**iii.** The growth of the exponential is mostly controlled by the coefficient of the $x$ term in the exponent. This coefficient for the second century is less than half the corresponding coefficient for the first century.

**iv.** $y = 294.89$ million.

**Problems 4.11,** page 447

**1.** **i.** $(A, A) = a_{11}^2 + a_{22}^2 + \cdots + a_{nn}^2 \geq 0$.

**ii.** $(A, A) = 0$ implies that $a_{ii}^2 = 0$ for $i = 1, 2, \ldots, n$ so that $A = 0$. If $A = 0$, then $(A, A) = 0$.

**iii.** $(A, B + C) = a_{11}(b_{11} + c_{11}) + \cdots + a_{nn}(b_{nn} + c_{nn}) = a_{11}b_{11} + a_{11}c_{11} + \cdots + a_{nn}b_{nn} + a_{nn}c_{nn} = (a_{11}b_{11} + \cdots + a_{nn}b_{nn}) + (a_{11}c_{11} + \cdots + a_{nn}c_{nn}) = (A, B) + (A, C)$

**iv.** Similarly, $(A + B, C) = (A, C) + (B, C)$

**v.** $(A, B) = (B, A) = \overline{(B, A)}$, since all components are real and $a_{ii}b_{ii} = b_{ii}a_{ii}$.

**vi.** $(\alpha A, B) = (\alpha a_{11})b_{11} + \cdots + (\alpha a_{nn})b_{nn} = \alpha[a_{11}b_{11} + \cdots + a_{nn}b_{nn}] = \alpha(A, B)$

**vii.** $(A, \alpha B) = \overline{(\alpha B, A)} = (\alpha B, A) = \alpha(B, A) = \alpha\overline{(A, B)} = \alpha(A, B)$

**3.** Let $E_i$ be the $n \times n$ matrix with a 1 in the $i, i$ position and 0 everywhere else. It is easy to see that $\{E_1, E_2, \ldots, E_n\}$ is an orthonormal basis for $D_n$.

**5.** $\left\{\left(\frac{1}{\sqrt{2}}, \frac{i}{\sqrt{2}}\right), \left(\frac{i}{\sqrt{2}}, \frac{1}{\sqrt{2}}\right)\right\}$

**7.** $\left\{\frac{1}{\sqrt{2}}, \sqrt{\frac{3}{2}}x, \sqrt{\frac{5}{8}}(3x^2 - 1)\right\}$

**9.** First note that if $A = (a_{ij})$ and $B^t = (b_{ji})$, then

$$(AB^t)_{ij} = \sum_{k=1}^{n} a_{ik}b_{jk}$$

so that

$$\text{tr}(AB^t) = \sum_{i=1}^{n}\sum_{j=1}^{n} a_{ij}b_{ij}$$

**i.** $(A, A) = \text{tr}(AA^t)$

$$= \sum_{i=1}^{n}\left(\sum_{j=1}^{n} a_{ij}^2\right) \geq 0$$

**ii.** $(A, A) = 0$ implies that $a_{ij}^2 = 0$ for every $i$ and $j$ so that $A = 0$. Conversely, if $A = 0$, then $A^t = 0$ and $AA^t = 0$ so that $\text{tr}(AA^t) = 0$.

**iii.** $(A, B + C) = \text{tr}[A(B + C)^t] + \text{tr}[A(B^t + C^t)] = \text{tr}(AB^t + AC^t) = \text{tr}(AB^t) + \text{tr}(AC^t) = (A, B) + (A, C)$

**iv.** Similarly, $(A + B, C) = (A, C) + (B, C)$

**v.** $(A, B) = \sum_{i=1}^{n}\sum_{j=1}^{n} a_{ij}b_{ij}$

$$= \text{tr}(BA^t) = (B, A)$$

**vi.** $(\alpha A, B) = \text{tr}(\alpha AB^t) = \alpha\text{tr}(AB^t) = \alpha(A, B)$

**vii.** $(A, \alpha B) = (\alpha B, A) = \alpha(B, A) = \alpha(A, B)$

**11.** $\begin{pmatrix} 1 & 0 \\ 0 & 0 \end{pmatrix}, \begin{pmatrix} 0 & 1 \\ 0 & 0 \end{pmatrix}, \begin{pmatrix} 0 & 0 \\ 1 & 0 \end{pmatrix}, \begin{pmatrix} 0 & 0 \\ 0 & 1 \end{pmatrix}$

**13. a. i.** $(p, p) = p(a)^2 + p(b)^2 + p(c)^2 \geq 0$

**ii.** $(p, p) = 0$ implies that $p(a) = p(b) = p(c) = 0$. But a quadratic can have at most two roots. Thus $p(x) = 0$ for all $x$. Conversely, if $p \equiv 0$, then $p(a) = p(b) = p(c) = 0$, so $(p, p) = 0$.

**iii.** $(p, q + r)$

$$\begin{aligned}
&= p(a)(q(a) + r(a)) \\
&\quad + p(b)(q(b) + r(b)) \\
&\quad + p(c)(q(c) + r(c)) \\
&= [p(a)q(a) + p(b)q(b) \\
&\quad + p(c)q(c)] \\
&\quad + [p(a)r(a) + p(b)r(b) \\
&\quad + p(c)r(c)] \\
&= (p, q) + (p, r)
\end{aligned}$$

**iv.** Similarly, $(p + q, r) = (p, r) + (q, r)$

**v.** $(p, q) = p(a)q(a) + p(b)q(b) + p(c)q(c)$

$$\begin{aligned}
&= q(a)p(a) + q(b)p(b) \\
&\quad + q(c)p(c) \\
&= (q, p)
\end{aligned}$$

**vi.** $(\alpha p, q) = [\alpha p(a)]q(a) + [\alpha p(b)]q(b) + [\alpha p(c)]q(c)$

$$\begin{aligned}
&= \alpha[p(a)q(a) \\
&\quad + p(b)q(b) \\
&\quad + p(c)q(c)] \\
&= \alpha(p, q)
\end{aligned}$$

**vii.** $(p, \alpha q) = (\alpha q, p) = \alpha(q, p) = \alpha(p, q)$

**b.** No, since (**ii**) is violated. For example, let $a = 1$, $b = -1$, and $p(x) = (x - 1)(x + 1) = x^2 - 1 \not\equiv 0$. Then $p(a) = p(b) = 0$ so that $(p, p) = 0$ even though $p \not\equiv 0$. In fact, for any polynomial $q$, we have $(p, q) = 0$.

**15.** $\sqrt{31}$

**17.** $0 \leq \left(\left(\frac{\mathbf{u}}{|\mathbf{u}|} - \frac{\mathbf{v}}{|\mathbf{v}|}\right), \left(\frac{\mathbf{u}}{|\mathbf{u}|} - \frac{\mathbf{v}}{|\mathbf{v}|}\right)\right)$

$$= \frac{(\mathbf{u}, \mathbf{u})}{|\mathbf{u}|^2} - \frac{(\mathbf{u}, \mathbf{v})}{|\mathbf{u}|\,|\mathbf{v}|} - \frac{(\mathbf{v}, \mathbf{u})}{|\mathbf{u}|\,|\mathbf{v}|} + \frac{(\mathbf{v}, \mathbf{v})}{|\mathbf{v}|^2}$$

$$= \frac{|\mathbf{u}|^2}{|\mathbf{u}|^2} - \left[\frac{(\mathbf{u}, \mathbf{v}) + \overline{(\mathbf{u}, \mathbf{v})}}{|\mathbf{u}|\,|\mathbf{v}|}\right] + \frac{|\mathbf{v}|^2}{|\mathbf{v}|^2}$$

Now if $z = a + bi$, then $z + \bar{z} = (a + bi) + (a - bi) = 2a = 2 \operatorname{Re} z$ (and $z - \bar{z} = 2bi = 2iImz$). Thus $(\mathbf{u}, \mathbf{v}) + \overline{(\mathbf{u}, \mathbf{v})} = 2 \operatorname{Re}(\mathbf{u}, \mathbf{v})$, and we have $2 - \dfrac{2 \operatorname{Re}(\mathbf{u}, \mathbf{v})}{|\mathbf{u}|\,|\mathbf{v}|} \geq 0$ or $\dfrac{\operatorname{Re}(\mathbf{u}, \mathbf{v})}{|\mathbf{u}|\,|\mathbf{v}|} \leq 1$. Let $\lambda$ be a real number. Then $0 \leq ((\lambda\mathbf{u} + (\mathbf{u}, \mathbf{v})\mathbf{v}), (\lambda\mathbf{u} + (\mathbf{u}, \mathbf{v})\mathbf{v})) = \lambda^2|\mathbf{u}|^2 + |(\mathbf{u}, \mathbf{v})|^2|\mathbf{v}|^2 + \lambda\overline{(\mathbf{u}, \mathbf{v})}(\mathbf{u}, \mathbf{v}) + \lambda(\mathbf{u}, \mathbf{v})(\mathbf{v}, \mathbf{u}) =$ (since $\lambda$ is real) $\lambda^2|\mathbf{u}|^2 + 2\lambda|(\mathbf{u}, \mathbf{v})|^2 + |(\mathbf{u}, \mathbf{v})|^2|\mathbf{v}|^2$. The last line is a quadratic equation in $\lambda$. If we have $a\lambda^2 + b\lambda + c \geq 0$, then the equation $a\lambda^2 + b\lambda + c = 0$ can have at most one real root, and therefore $b^2 - 4ac \leq 0$. Thus

$$4((|\mathbf{u}, \mathbf{v})|^2)^2 - 4|\mathbf{u}|^2|(\mathbf{u}, \mathbf{v})|^2|\mathbf{v}|^2 \leq 0$$

or $\quad |(\mathbf{u}, \mathbf{v})|^2 \leq |\mathbf{u}|^2|\mathbf{v}|^2$

and $\quad |(\mathbf{u}, \mathbf{v})| \leq |\mathbf{u}|\,|\mathbf{v}|$

**19.** $H^{\perp} = \operatorname{span}\{(-15x^2 + 16x - 3), (20x^3 - 30x^2 + 12x - 1)\}$

**21.** $1 + 2x + 3x^2 - x^3 = \dfrac{30x^2 + 52x + 19}{20} + \dfrac{(-20x^3 + 30x^2 - 12x + 1)}{20}$

**23.** $\dfrac{2}{\pi} + \sqrt{3}\left(\dfrac{2}{\pi} - \dfrac{8}{\pi^2}\right)\sqrt{3}(2x - 1) + \sqrt{5}\left(\dfrac{2}{\pi} + \dfrac{24}{\pi^2} - \dfrac{96}{\pi^3}\right) \times \sqrt{5}(6x^2 - 6x + 1) \approx -0.8346x^2 - 0.2091x + 1.0194$

**25.** $A^* = \begin{pmatrix} \dfrac{1}{\sqrt{2}} & \dfrac{1}{\sqrt{2}} \\ -\dfrac{1}{2} - \dfrac{i}{2} & \dfrac{1}{2} + \dfrac{i}{2} \end{pmatrix}$

Verify that $A^*A = I$.

**27.** We check the seven conditions on page 439.

**i.** $(f, f) = \int_a^b f\bar{f} = \int_a^b f_1^2 + f_2^2 \geq 0$ since $f_1^2 \geq 0$ and $f_2^2 \geq 0$

**ii.** follows from (i)

**iii.** $(f, g + h) = \int_a^b f\overline{(g + h)} = \int_a^b f\bar{g} + f\bar{h} = (f, g) + (f, h)$

**iv.** Similar to (iii)

**v.** $(f, g) = \int_a^b f\bar{g} = \int_a^b \overline{\bar{g}f}$ and $\overline{(g, f)} = \int_a^b \overline{g\bar{f}} = \int_a^b \bar{g}f$

**vi.** $(\alpha f, g) = \int_a^b \alpha f\bar{g} = \alpha\int_a^b f\bar{g}$

**vii.** $(f, \alpha g) = \int_a^b f(\overline{\alpha g}) = \int_a^b f\bar{\alpha}\bar{g} = \bar{\alpha}\int_a^b f\bar{g} = \bar{\alpha}(f, g)$

**29.** $\sqrt{\pi}$

## MATLAB 4.11

**1.** Refer to MATLAB Problem 1 in Section 4.9. You will need to compute a **t4** and a **z4**.

**3.** Part (a) works because $(\mathbf{w}, \mathbf{u}_i)$ equals $\mathbf{u}_i{}'*\mathbf{w}$ by definition. Refer to MATLAB Problem 7 in Section 4.9.

## Problems 4.12, page 457

**1.** Let $L$ be a linearly independent set in $V$. Let $S$ be the collection of all linearly independent subsets of $V$, partially ordered by inclusion such that every set in $S$ contains $L$. The proof then follows as in the proof of Theorem 2.

3. The result is true for $n = 2$. Assume it is true for $n = k$. Consider the $k + 1$ sets $A_1, A_2, \ldots, A_k, A_{k+1}$ in a chain. The first $k$ sets form a chain, and, by the induction assumption, one of them contains the other $k - 1$ sets. Call this set $A_i$. Then either $A_i \subseteq A_{k+1}$ or $A_{k+1} \subseteq A_i$. In either case we have found a set that contains the other $k$ sets and the result is true for $n = k + 1$. This completes the induction proof.

**Chapter 4—Review,** page 462

1. yes; dimension 2; basis $\{(1, 0, 1), (0, 1, 2)\}$
3. yes; dimension 3; basis $\{(1, 0, 0, -1), (0, 1, 0, -1), (0, 0, 1, -1)\}$
5. yes; dimension $[n(n + 1)]/2$; basis $\{(E_{ij}: j \geq i\}$, where $E_{ij}$ is the matrix with 1 in the $i, j$ position and 0 everywhere else
7. no; for example, $(x^5 - 2x) + (-x^5 + x^2) = x^2 - 2x$, which is not a polynomial of degree 5, so the set is not closed under addition.
9. no; for example, $\begin{pmatrix} 1 & 1 \\ 0 & 2 \\ 3 & 1 \end{pmatrix} + \begin{pmatrix} 2 & 1 \\ -1 & 2 \\ 1 & 0 \end{pmatrix} = \begin{pmatrix} 3 & 2 \\ -1 & 4 \\ 4 & 1 \end{pmatrix}$, which does not satisfy $a_{12} = 1$.
11. independent
13. dependent
15. independent
17. independent
19. independent
21. dimension 2; basis $\{(2, 0, 1), (0, 4, 3)\}$
23. dimension 3; basis $\{(1, 0, 3, 0), (0, 1, -1, 0), (0, 0, 1, 1)\}$
25. dimension 4; basis $\{D_1, D_2, D_3, D_4\}$, where $D_i$ is the matrix with a 1 in the $i, i$ position and 0 everywhere else
27. range $A = \text{span}\left\{\begin{pmatrix} 1 \\ -2 \end{pmatrix}\right\}$; $N_A = \text{span}\left\{\begin{pmatrix} 2 \\ 1 \end{pmatrix}\right\}$; $\rho(A) = \nu(A) = 1$
29. range $A = \mathbb{R}^3$; $N_A = \{\mathbf{0}\}$; $\rho(A) = 3$, $\nu(A) = 0$
31. range $A = \text{span}\left\{\begin{pmatrix} 2 \\ -1 \\ 4 \end{pmatrix}, \begin{pmatrix} 3 \\ 2 \\ 6 \end{pmatrix}\right\}$; $N_A = \{\mathbf{0}\}$; $\rho(T) = 2$, $\nu(T) = 0$
33. $\frac{3}{4}\begin{pmatrix} 1 \\ 2 \end{pmatrix} - \frac{5}{4}\begin{pmatrix} -1 \\ 2 \end{pmatrix} = \begin{pmatrix} 2 \\ -1 \end{pmatrix}$
35. $1(1 + x^2) + 0(1 + x) + 3(1) = 4 + x^2$
37. $\left\{\frac{1}{\sqrt{13}}\begin{pmatrix} 2 \\ 3 \end{pmatrix}, \frac{1}{\sqrt{13}}\begin{pmatrix} -3 \\ 2 \end{pmatrix}\right\}$
39. $\begin{pmatrix} 1/\sqrt{3} \\ 1/\sqrt{3} \\ 1/\sqrt{3} \end{pmatrix}$
41. **a.** $\begin{pmatrix} \frac{4}{3} \\ -\frac{1}{3} \\ \frac{5}{3} \end{pmatrix}$ **b.** $\begin{pmatrix} -1/\sqrt{3} \\ 1/\sqrt{3} \\ 1/\sqrt{3} \end{pmatrix}$
    **c.** $\begin{pmatrix} \frac{4}{3} \\ -\frac{1}{3} \\ \frac{5}{3} \end{pmatrix} + \begin{pmatrix} -\frac{7}{3} \\ \frac{7}{3} \\ \frac{7}{3} \end{pmatrix}$
43. **a.** $\begin{pmatrix} \frac{1}{2} \\ \frac{1}{2} \\ \frac{1}{2} \\ \frac{1}{2} \end{pmatrix}$
    **b.** $\left\{\begin{pmatrix} 1/\sqrt{2} \\ 0 \\ -1/\sqrt{2} \\ 0 \end{pmatrix}, \begin{pmatrix} 0 \\ 1/\sqrt{2} \\ 0 \\ -1/\sqrt{2} \end{pmatrix}\right\}$
    **c.** $\begin{pmatrix} \frac{1}{2} \\ \frac{1}{2} \\ \frac{1}{2} \\ \frac{1}{2} \end{pmatrix} + \begin{pmatrix} \frac{1}{2} \\ -\frac{1}{2} \\ -\frac{1}{2} \\ \frac{1}{2} \end{pmatrix}$

**45.** $\dfrac{e^2-1}{2} + 3(x-1) + (\frac{15}{4}e^2 - \frac{105}{4}) \times (x^2 - 2x + \frac{2}{3}) \approx 1.167 + 0.0821x + 1.459x^2$

**47.** $y = \frac{7}{6}x^2 + \frac{3}{2}x - \frac{8}{3}$

## Chapter 5

### Problems 5.1, page 472

**1.** linear **3.** linear

**5.** not linear, since

$$T\left(\alpha\begin{pmatrix}x\\y\\z\end{pmatrix}\right) = T\begin{pmatrix}\alpha x\\ \alpha y\\ \alpha z\end{pmatrix} = \begin{pmatrix}1\\ \alpha z\end{pmatrix}$$

while

$$\alpha T\begin{pmatrix}x\\y\\z\end{pmatrix} = \alpha\begin{pmatrix}1\\z\end{pmatrix} = \begin{pmatrix}\alpha\\ \alpha z\end{pmatrix}$$

**7.** linear

**9.** not linear, since

$$T\left(\alpha\begin{pmatrix}x\\y\end{pmatrix}\right) = T\begin{pmatrix}\alpha x\\ \alpha y\end{pmatrix} = (\alpha x)(\alpha y) = \alpha^2 xy$$

while $\alpha T\begin{pmatrix}x\\y\end{pmatrix} = \alpha xy$

**11.** linear

**13.** not linear, since

$$T\left(\alpha\begin{pmatrix}x\\y\\z\\w\end{pmatrix}\right) = \alpha^2 T\begin{pmatrix}x\\y\\z\\w\end{pmatrix} \neq \alpha T\begin{pmatrix}x\\y\\z\\w\end{pmatrix}$$

if $\alpha \neq 1$ or 0

**15.** not linear, since

$$\begin{aligned} T(A+B) &= (A+B)^t(A+B) \\ &= (A^t + B^t)(A+B) \\ &= A^tA + A^tB \\ &\quad + B^tA + B^tB \end{aligned}$$

But

$$\begin{aligned} T(A) + T(B) &= A^tA + B^tB \\ &\neq T(A+B) \end{aligned}$$

unless $A^tB + B^tA = 0$.

**17.** not linear, since $T(\alpha D) = (\alpha D)^2 = \alpha^2 D^2 \neq \alpha T(D) = \alpha D^2$ unless $\alpha = 1$ or 0.

**19.** linear **21.** linear

**23.** not linear, since $T(f+g) = (f+g)^2 \neq f^2 + g^2 = T(f) + T(g)$

**25.** linear **27.** linear

**29.** not linear, since

$$\begin{aligned} T(\alpha A) &= \det(\alpha A) \\ &= \alpha^n \det A \\ &\neq \alpha \det A \\ &= \alpha T(A) \end{aligned}$$

unless $\alpha = 0$ or 1. [$\det \alpha A = \alpha^n \det A$ by Problem 2.2.28.] Also, in general

$$\det(A+B) \neq \det A + \det B$$

**31.** **a.** $\begin{pmatrix}-14\\4\\26\end{pmatrix}$ **b.** $\begin{pmatrix}-31\\-6\\26\end{pmatrix}$

**33.** It rotates a vector counterclockwise around the $z$-axis through an angle of $\theta$ in a plane parallel to the $xy$-plane.

**35.** Suppose $\alpha < 0$. Then $T[(\alpha - \alpha)\mathbf{x}] = T(0\mathbf{x}) = 0T\mathbf{x} = \mathbf{0}$. Thus $T[(\alpha-\alpha)\mathbf{x}] = T(0\mathbf{x}) = 0T\mathbf{x} = \mathbf{0}$ and $T(\alpha\mathbf{x}) + T(-\alpha\mathbf{x}) = T((\alpha-\alpha)\mathbf{x}) = \mathbf{0}$. But $-\alpha > 0$ so that $T(-\alpha\mathbf{x}) = -\alpha T\mathbf{x}$. Therefore $T(\alpha\mathbf{x}) - \alpha T\mathbf{x} = \mathbf{0}$, or $T(\alpha\mathbf{x}) = \alpha T\mathbf{x}$ for $\alpha < 0$ as well.

**37.** $T(\mathbf{x}-\mathbf{y}) = T\mathbf{x} + T(-\mathbf{y}) = T\mathbf{x} + T[(-1)\mathbf{y}] = T\mathbf{x} + (-1)T\mathbf{y} = T\mathbf{x} - T\mathbf{y}$

**39.** $T(\mathbf{v}_1 + \mathbf{v}_2) = (\mathbf{v}_1 + \mathbf{v}_2, \mathbf{u}_0) = (\mathbf{v}_1, \mathbf{u}_0) + (\mathbf{v}_2, \mathbf{u}_0) = T\mathbf{v}_1 + T\mathbf{v}_2$ and $T(\alpha\mathbf{v}) = (\alpha\mathbf{v}, \mathbf{u}_0) = \alpha(\mathbf{v}, \mathbf{u}_0) = \alpha T\mathbf{v}$

**41.** $T(\mathbf{v}_1 + \mathbf{v}_2) = (\mathbf{v}_1 + \mathbf{v}_2, \mathbf{u}_1)\mathbf{u}_1 + (\mathbf{v}_1 + \mathbf{v}_2, \mathbf{u}_2)\mathbf{u}_2 + \cdots + (\mathbf{v}_1 + \mathbf{v}_2, \mathbf{u}_n)\mathbf{u}_n$
$= (\mathbf{v}_1, \mathbf{u}_1)\mathbf{u}_1 + (\mathbf{v}_2, \mathbf{u}_1)\mathbf{u}_1 + (\mathbf{v}_1, \mathbf{u}_2)\mathbf{u}_2 + (\mathbf{v}_2, \mathbf{u}_2)\mathbf{u}_2 + \cdots + (\mathbf{v}_1, \mathbf{u}_n)\mathbf{u}_n + (\mathbf{v}_2, \mathbf{u}_n)\mathbf{u}_n$
$= (\mathbf{v}_1, \mathbf{u}_1)\mathbf{u}_1 + (\mathbf{v}_1, \mathbf{u}_2)\mathbf{u}_2 + \cdots + (\mathbf{v}_1, \mathbf{u}_n)\mathbf{u}_n + (\mathbf{v}_2, \mathbf{u}_1)\mathbf{u}_1 + (\mathbf{v}_2, \mathbf{u}_2)\mathbf{u}_2 + \cdots + (\mathbf{v}_2, \mathbf{u}_n)\mathbf{u}_n$
$= T\mathbf{v}_1 + T\mathbf{v}_2$
$T(\alpha\mathbf{v}) = (\alpha\mathbf{v}, \mathbf{u}_1)\mathbf{u}_1 + (\alpha\mathbf{v}, \mathbf{u}_2)\mathbf{u}_2 + \cdots + (\alpha\mathbf{v}, \mathbf{u}_n)\mathbf{u}_n$
$= \alpha[(\mathbf{v}, \mathbf{u}_1)\mathbf{u}_1 + (\mathbf{v}, \mathbf{u}_2)\mathbf{u}_2 + \cdots + (\mathbf{v}, \mathbf{u}_n)\mathbf{u}_n] = \alpha T\mathbf{v}$

### MATLAB 5.1

**1.** The figure is a dog without a tail. The points are red *'s and we have $-20 \le x,y \le 20$.

**3. a.** Both the $x$ and $y$ scales of the dog are doubled. For example, the width of a leg is doubled and height is doubled.

**b.** The first matrix doubles the $x$ scale and leaves the $y$ scale the same. So, for example, the width of a leg is doubled but the height is the same. The second matrix doubles the $y$ scale and leaves the $x$ scale the same. So, for example, the width of a leg is the same but the height of the dog is doubled.

**c.** The matrix multiplies the $x$ scale by $r$ and the $y$ scale by $s$.

### Problems 5.2, page 484

**1.** kernel $= \{(0, y)\colon y \in \mathbb{R}\}$, that is, the $y$-axis; range $= \{(x, 0)\colon x \in \mathbb{R}\}$, that is, the $x$-axis; $\rho(T) = \nu(T) = 1$

**3.** kernel $= \{(x, -x)\colon x \in \mathbb{R}\}$—this is the line $x + y = 0$; range $= \mathbb{R}$; $\rho(T) = \nu(T) = 1$.

**5.** kernel $= \left\{\begin{pmatrix} 0 & 0 \\ 0 & 0 \end{pmatrix}\right\}$; range $= M_{22}$; $\rho(T) = 4$, $\nu(T) = 0$

**7.** kernel $= \{A\colon A^t = -A\} = \{A\colon A$ is skew-symmetric$\}$; range $= \{A\colon A$ is symmetric$\}$; $\rho(T) = (n^2 + n)/2$; $\nu(T) = (n^2 - n)/2$

**9.** kernel $= \{f \in C[0, 1]\colon f(\frac{1}{2}) = 0\}$; range $= \mathbb{R}$; $\rho(T) = 1$; the kernel is an infinite dimensional space so that $\nu(T) = \infty$. For example, the linearly independent functions $x - \frac{1}{2}$, $(x - \frac{1}{2})^2, (x - \frac{1}{2})^3, (x - \frac{1}{2})^4, \ldots, (x - \frac{1}{2})^n, \ldots$ all satisfy $f(\frac{1}{2}) = 0$.

**11.** If $\mathbf{v} \in V$, then $\mathbf{v} = c_1\mathbf{v}_1 + c_2\mathbf{v}_2 + \cdots + c_n\mathbf{v}_n$ so that $T\mathbf{v} = T(c_1\mathbf{v}_1 + c_2\mathbf{v}_2 + \cdots + c_n\mathbf{v}_n) = c_1T\mathbf{v}_1 + c_2T\mathbf{v}_2 + \cdots + c_nT\mathbf{v}_n) = c_1\mathbf{0} + c_2\mathbf{0} + \cdots + c_n\mathbf{0} = \mathbf{0}$. Thus $T\mathbf{v} = \mathbf{0}$ for every $\mathbf{v} \in V$, and is therefore the zero transformation.

**13.** The range of $T$ is a subspace of $\mathbb{R}^3$ and, by Example 4.6.9, the subspaces of $\mathbb{R}^3$ are $\{\mathbf{0}\}$, $\mathbb{R}^3$, and lines and planes passing through the origin.

**15.** $T\mathbf{x} = A\mathbf{x}$, where $A = \begin{pmatrix} 0 & a \\ b & c \end{pmatrix}$, $a$, $b$, $c$ real

**17.** $T\mathbf{x} = A\mathbf{x}$, where $A = \begin{pmatrix} 2 & -1 & 1 \\ 2 & -1 & 1 \\ 2 & -1 & 1 \end{pmatrix}$

**19. i.** If $A \in \ker T$, then $A - A^t = 0$, or $A = A^t$.

**ii.** If $A \in$ range of $T$, then there is a matrix $B$ such that $B - B^t = A$. Then $A^t = (B - B^t)^t = B^t - (B^t)^t = B^t - B = -A$ so that $A$ is skew-symmetric.

**21.** Let $T_{ij}(\mathbf{u}_i) = \mathbf{w}_j$ and $T_{ij}(\mathbf{u}_k) = \mathbf{0}$ if $k \ne i$. These form a basis for $L(V, W)$, so $\dim L(V, W) = nm$.

**23.** False. Let $S$ and $T$: $\mathbb{R}^2 \to \mathbb{R}^2$ be given by $S(\mathbf{x}) = A\mathbf{x}$ and $T(\mathbf{x}) = B\mathbf{x}$, where $A = \begin{pmatrix} 0 & 1 \\ 0 & 0 \end{pmatrix}$ and $B = \begin{pmatrix} 1 & 0 \\ 0 & 0 \end{pmatrix}$. Then $ST(\mathbf{x}) = AB\mathbf{x} = \begin{pmatrix} 0 & 0 \\ 0 & 0 \end{pmatrix}\mathbf{x} = \mathbf{0}$. However, $TS(\mathbf{x})$ is not the zero transformation because $BA = \begin{pmatrix} 0 & 1 \\ 0 & 0 \end{pmatrix} \neq$ the zero matrix.

**Problems 5.3, page 504**

**1.** $\begin{pmatrix} 1 & -2 \\ -1 & 1 \end{pmatrix}$; $\ker T = \{\mathbf{0}\}$; range $T = \mathbb{R}^2$; $\nu(T) = 0$, $\rho(T) = 2$

**3.** $\begin{pmatrix} 1 & -1 & 1 \\ -2 & 2 & -2 \end{pmatrix}$; range $T = \text{span}\left\{\begin{pmatrix} 1 \\ -2 \end{pmatrix}\right\}$; $\ker T = \text{span}\left\{\begin{pmatrix} 1 \\ 1 \\ 0 \end{pmatrix}, \begin{pmatrix} 0 \\ 1 \\ 1 \end{pmatrix}\right\}$; $\rho(T) = 1$, $\nu(T) = 2$

**5.** $\begin{pmatrix} 1 & -1 & 2 \\ 3 & 1 & 4 \\ 5 & -1 & 8 \end{pmatrix}$; range $T = \text{span}\left\{\begin{pmatrix} 1 \\ 3 \\ 5 \end{pmatrix}, \begin{pmatrix} -1 \\ 1 \\ -1 \end{pmatrix}\right\}$; $\ker T = \text{span}\left\{\begin{pmatrix} -3 \\ 1 \\ 2 \end{pmatrix}\right\}$; $\rho(T) = 2$, $\nu(T) = 1$

**7.** $\begin{pmatrix} 1 & -1 & 2 & 3 \\ 0 & 1 & 4 & 3 \\ 1 & 0 & 6 & 6 \end{pmatrix}$; range $T = \text{span}\left\{\begin{pmatrix} 1 \\ 0 \\ 1 \end{pmatrix}, \begin{pmatrix} -1 \\ 1 \\ 0 \end{pmatrix}\right\}$; $\ker T = \text{span}\left\{\begin{pmatrix} -6 \\ -4 \\ 1 \\ 0 \end{pmatrix}, \begin{pmatrix} -6 \\ -3 \\ 0 \\ 1 \end{pmatrix}\right\}$; $\rho(T) = 2$, $\nu(T) = 2$

**9.** $\begin{pmatrix} \frac{5}{4} & -\frac{13}{4} \\ \frac{5}{4} & \frac{3}{4} \end{pmatrix}$; range $T = \mathbb{R}^2$; $\ker T = \{\mathbf{0}\}$; $\rho(T) = 2$, $\nu(T) = 0$

**11.** $\begin{pmatrix} 3 & \frac{7}{5} & \frac{16}{5} \\ 0 & \frac{4}{5} & \frac{2}{5} \end{pmatrix}$; range $T = \mathbb{R}^2$; $(\ker T)_{B_1} = \text{span}\left\{\begin{pmatrix} 5 \\ 3 \\ -6 \end{pmatrix}\right\}$; $\rho(T) = 2$, $\nu(T) = 1$

**13.** $\begin{pmatrix} 0 & 1 & 0 \\ 0 & -1 & 0 \\ 0 & 0 & 0 \\ 1 & 0 & 0 \end{pmatrix}$; range $T = \text{span}\,\{1 - x, x^3\}$; $\ker T = \text{span}\,\{x^2\}$; $\rho(T) = 2$, $\nu(T) = 1$

**15.** $(0, 0, 1, 0)$; range $T = \mathbb{R}$; $\ker T = \text{span}\,\{1, x, x^3\}$; $\rho(T) = 1$, $\nu(T) = 3$

**17.** $\begin{pmatrix} 1 & -1 & 2 & 3 \\ 0 & 1 & 4 & 3 \\ 1 & 0 & 6 & 5 \end{pmatrix}$; range $T = \text{span}\,\{1 + x^2, -1 + x, 3 + 3x + 5x^2\} = P_2$; $\ker T = \text{span}\,\{x^2 - 4x - 6\}$; $\rho(T) = 3$, $\nu(T) = 1$

**19.** $\begin{pmatrix} 1 & 1 & 1 & 1 \\ 1 & 1 & 1 & 0 \\ 1 & 1 & 0 & 0 \\ 1 & 0 & 0 & 0 \end{pmatrix}$; range $T = M_{22}$; $\ker T = \left\{\begin{pmatrix} 0 & 0 \\ 0 & 0 \end{pmatrix}\right\}$; $\rho(T) = 4$, $\nu(T) = 0$

**21.** $\begin{pmatrix} 0 & 1 & 0 & 0 & 0 \\ 0 & 0 & 2 & 0 & 0 \\ 0 & 0 & 0 & 3 & 0 \\ 0 & 0 & 0 & 0 & 4 \end{pmatrix}$; range $D = P_3$; $\ker D = \mathbb{R}$; $\rho(D) = 4$, $\nu(D) = 1$

**23.** $\begin{pmatrix} 0 & 1 & 0 & 0 & \cdots & 0 \\ 0 & 0 & 2 & 0 & \cdots & 0 \\ 0 & 0 & 0 & 3 & \cdots & 0 \\ \vdots & \vdots & \vdots & \vdots & & \vdots \\ 0 & 0 & 0 & 0 & \cdots & n \end{pmatrix}$; range $D = P_{n-1}$; $\ker D = \mathbb{R}$; $\rho(D) = n$, $\nu(D) = 1$

**25.** $\begin{pmatrix} 2 & 0 & 2 & 0 & 0 \\ 0 & 3 & 0 & 6 & 0 \\ 0 & 0 & 4 & 0 & 12 \\ 0 & 0 & 0 & 5 & 0 \\ 0 & 0 & 0 & 0 & 6 \end{pmatrix}$; range $T = P_4$; ker $T = \{\mathbf{0}\}$; $\rho(T) = 5$, $\nu(T) = 0$

**27.** $A_T = \text{diag}\,(b_0, b_1, b_2, \ldots, b_n)$, where $b_j = \sum_{i=1}^{j+1} \frac{j!}{(j+1-i)!}$; range $T = P_n$; ker $T = \{\mathbf{0}\}$; $\rho(T) = n + 1$, $\nu(T) = 0$

**29.** $\begin{pmatrix} 1 & 0 & 0 \\ 0 & 1 & 0 \\ 0 & 0 & 1 \end{pmatrix}$; range $T = P_2$; ker $T = \{\mathbf{0}\}$, $\rho(T) = 3$, $\nu(T) = 0$

**31.** For example, in $M_{34}$,

$$A_T = \begin{pmatrix} 1 & 0 & 0 & 0 & 0 & 0 & 0 & 0 & 0 & 0 & 0 & 0 \\ 0 & 0 & 0 & 0 & 1 & 0 & 0 & 0 & 0 & 0 & 0 & 0 \\ 0 & 0 & 0 & 0 & 0 & 0 & 0 & 0 & 1 & 0 & 0 & 0 \\ 0 & 1 & 0 & 0 & 0 & 0 & 0 & 0 & 0 & 0 & 0 & 0 \\ 0 & 0 & 0 & 0 & 0 & 1 & 0 & 0 & 0 & 0 & 0 & 0 \\ 0 & 0 & 0 & 0 & 0 & 0 & 0 & 0 & 0 & 1 & 0 & 0 \\ 0 & 0 & 1 & 0 & 0 & 0 & 0 & 0 & 0 & 0 & 0 & 0 \\ 0 & 0 & 0 & 0 & 0 & 0 & 1 & 0 & 0 & 0 & 0 & 0 \\ 0 & 0 & 0 & 0 & 0 & 0 & 0 & 0 & 0 & 0 & 1 & 0 \\ 0 & 0 & 0 & 1 & 0 & 0 & 0 & 0 & 0 & 0 & 0 & 0 \\ 0 & 0 & 0 & 0 & 0 & 0 & 0 & 1 & 0 & 0 & 0 & 0 \\ 0 & 0 & 0 & 0 & 0 & 0 & 0 & 0 & 0 & 0 & 0 & 1 \end{pmatrix}$$

In general, $A_T = (a_{ij})$, where

$$a_{ij} = \begin{cases} 1, & \text{if } i = km + l, \\ & \text{and } j = (l-1)n + k + 1 \\ & \text{for } k = 1, 2, \ldots, n-1 \\ & \text{and } l = 1, 2, \ldots, m \\ 0, & \text{otherwise} \end{cases}$$

**33.** $\begin{pmatrix} 0 & 0 & 0 \\ 0 & 0 & -1 \\ 0 & 1 & 0 \end{pmatrix}$; range $D = \text{span}\,\{\sin x, \cos x\}$; ker $D = \mathbb{R}$; $\rho(D) = 2$, $\nu(D) = 1$

**35.** $\begin{pmatrix} \frac{1}{2} & -i/2 \\ i/2 & \frac{1}{2} \end{pmatrix}$

**37.** *Let* $B_1$ and $B_2$ be bases for $V$ and $W$, respectively. We have $(T\mathbf{v})_{B_2} = A_T(\mathbf{v})_{B_1}$ for every $\mathbf{v} \in V$. Then $\mathbf{v} \in \ker T$ if and only if $T\mathbf{v} = \mathbf{0}$ if and only if $A_T(\mathbf{v})_{B_1} = (\mathbf{0})_{B_2}$ if and only if $(\mathbf{v})_{B_1} \in \ker A_T$. Thus kernel of $T = N_{A_T}$ so that $\nu(T) = \nu(A_T)$. If $\mathbf{w} \in$ range $T$, then $T\mathbf{v} = \mathbf{w}$ for some $\mathbf{v} \in V$ so that $A_T(\mathbf{v})_{B_1} = (T\mathbf{v})_{B_2} = (\mathbf{w})_{B_2}$. This means that $(\mathbf{w})_{B_2} \in R_{A_T}$. Thus $R_{A_T} =$ range $T$ so that $\rho(T) = \rho(A_T)$. Since $\nu(A_T) + \rho(A_T) = n$ from Theorem 4.7.5, we see that $\nu(T) + \rho(T) = n$ also.

**39.** compression along the $y$-axis with $c = \frac{1}{4}$

**41.** shear along the $x$-axis with $c = 2$

**43.** shear along the $y$-axis with $c = \frac{1}{2}$

**45.** reflection about the line $y = x$

**47.** $\begin{pmatrix} \frac{1}{4} & 0 \\ 0 & 1 \end{pmatrix}$;

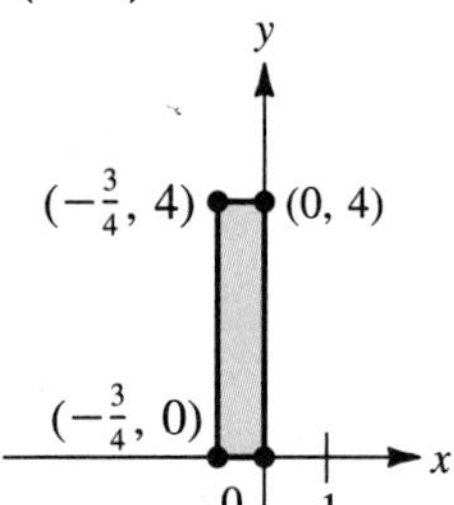


**49.** $\begin{pmatrix} 1 & 0 \\ 3 & 1 \end{pmatrix}$;

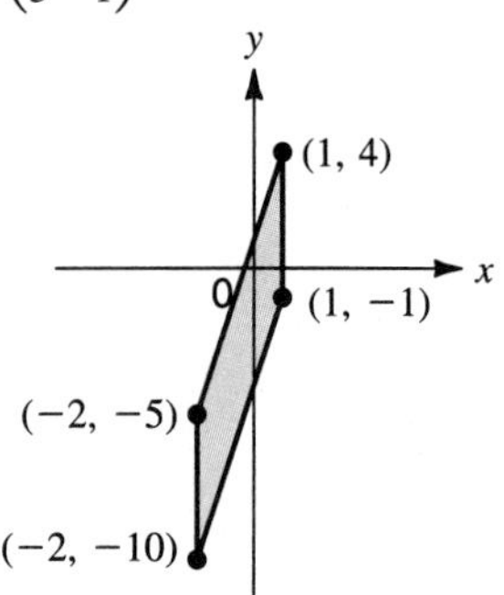

**51.** 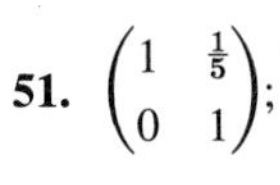 $\begin{pmatrix} 1 & \frac{1}{5} \\ 0 & 1 \end{pmatrix}$;

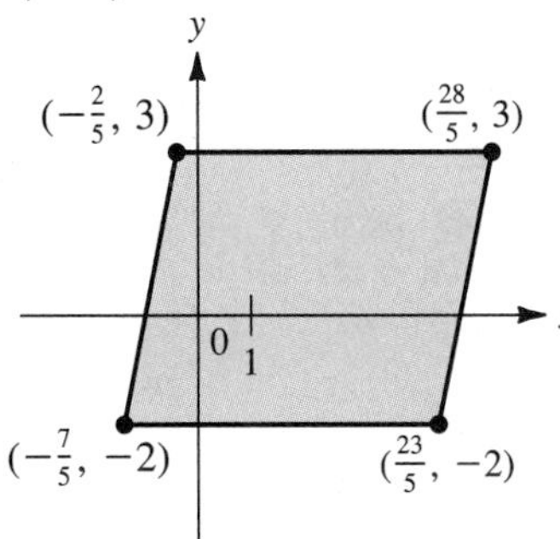


**53.** $\begin{pmatrix} -1 & 0 \\ 0 & 1 \end{pmatrix}$;

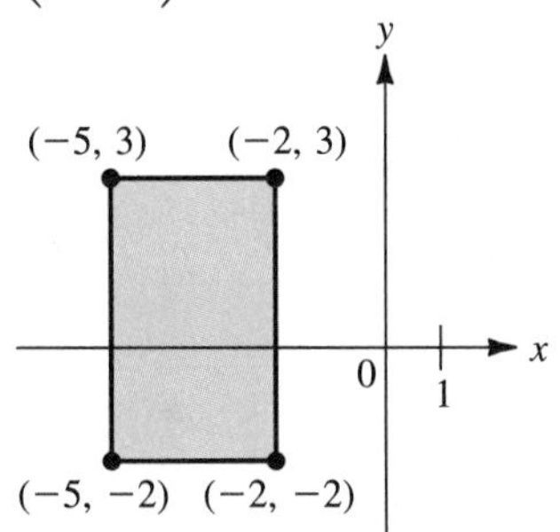


**55.** $\begin{pmatrix} 0 & 1 \\ 1 & 0 \end{pmatrix}$;

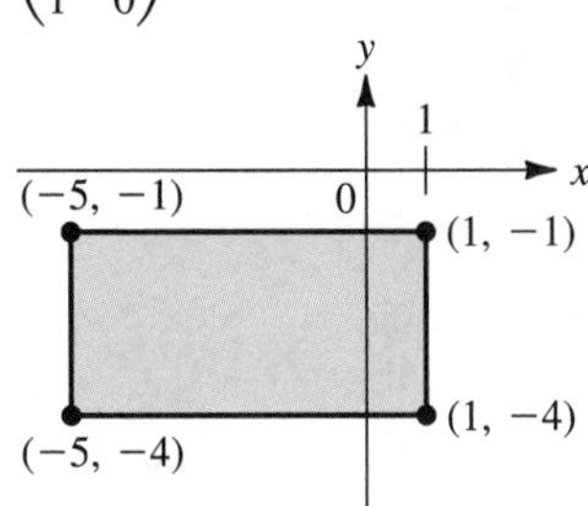


**57.** $\begin{pmatrix} 3 & 0 \\ 0 & 1 \end{pmatrix}\begin{pmatrix} 1 & 0 \\ -1 & 1 \end{pmatrix}\begin{pmatrix} 1 & 0 \\ 0 & \frac{14}{3} \end{pmatrix}\begin{pmatrix} 1 & \frac{2}{3} \\ 0 & 1 \end{pmatrix}$

**59.** $\begin{pmatrix} 3 & 0 \\ 0 & 1 \end{pmatrix}\begin{pmatrix} 1 & 0 \\ 4 & 1 \end{pmatrix}\begin{pmatrix} 1 & 0 \\ 0 & -1 \end{pmatrix}\begin{pmatrix} 1 & 0 \\ 0 & 6 \end{pmatrix}\begin{pmatrix} 1 & 2 \\ 0 & 1 \end{pmatrix}$

**61.** $\begin{pmatrix} 0 & 1 \\ 1 & 0 \end{pmatrix}\begin{pmatrix} 5 & 0 \\ 0 & 1 \end{pmatrix}\begin{pmatrix} 1 & 0 \\ 0 & -1 \end{pmatrix}\begin{pmatrix} 1 & 0 \\ 0 & 2 \end{pmatrix}\begin{pmatrix} 1 & \frac{7}{5} \\ 0 & 1 \end{pmatrix}$

**63.** $\begin{pmatrix} -1 & 0 \\ 0 & 1 \end{pmatrix}\begin{pmatrix} 1 & 0 \\ 6 & 1 \end{pmatrix}\begin{pmatrix} 1 & 0 \\ 0 & 62 \end{pmatrix}\begin{pmatrix} 1 & -10 \\ 0 & 1 \end{pmatrix}$

## MATLAB 5.3

**1. a.** Here is a possible coding:

```
pts = [0 3 3 0;0 0 2 2]
lns = [1 2 3 4;2 3 4 1]
A = [.5 0;0 3]
grafics(pts,lns,'b','*',10)
hold on
grafics(A*pts,lns,'g','o',10)
hold off
```

**b.** Use **A = [1 2;0 1]** for the shear in Figure 5.8(a) and use **A = [1 −2;0 1]** for the shear in Figure 5.8(b). Using $M = 7$ in the call to *grafics* will suffice.

**c.** Use **A = [1 0;−2 1].** Using $M = 6$ or 7 will suffice.

**3. a.** In showing that $T$ is linear, use the properties of dot product: $(\mathbf{v} \cdot a\mathbf{x}) = a(\mathbf{v} \cdot \mathbf{x})$ and $\mathbf{v} \cdot (\mathbf{x} + \mathbf{y}) = \mathbf{v} \cdot \mathbf{x} + \mathbf{v} \cdot \mathbf{y}$. To find the matrix representation, use the fact that $T(\mathbf{e}_i) = (\mathbf{v} \cdot \mathbf{e}_i)\mathbf{v} = v_i\mathbf{v}$.

**b.** **P = [1 0;0 0].** A basis for the range is **v** and a basis for the kernel is $\mathbf{w} = (0\ 1)^t$, a vector that is perpendicular to **v**. To project a vector onto **v**, one drops a perpendicular from the endpoint of the vector to the line determined by **v**. So, for example, if a vector is perpendicular to **v**, the projection is the zero vector. Every projection onto **v** is parallel to **v**, so **v** is clearly a basis for the range.

**c.** and **d.** Similar to (b). A basis for range will be **v** and a basis for kernel will be a vector perpendicular to **v**.

**7. a.** $A = \begin{pmatrix} 2.5 & -.5 \\ -.5 & 2.5 \end{pmatrix}$

Multiply the individual matrix representations together in the correct order: (rotate counterclockwise) (expansion) (rotate clockwise).

**b.** $A = \begin{pmatrix} 2 & 0 \\ 0 & 3 \end{pmatrix}$

**c.** $T$ expands by a factor of 2 in the direction of the first basis vector in $B$ and expands by a factor of 3 in the direction of the second basis vector in $B$.

## Problems 5.4, page 517

**1.** Since $(\alpha A)^t = \alpha A^t$ and $(A + B)^t = A^t + B^t$, $T$ is linear. $A^t = 0$ if and only if $A = 0$ so $\ker T = \{0\}$ and $T$ is $1 - 1$. For any matrix $A$, $(A^t)^t = A$, so $T$ is onto.

**3. i.** If $T$ is an isomorphism, then $T\mathbf{x} = A_T\mathbf{x} = \mathbf{0}$ if and only if $\mathbf{x} = \mathbf{0}$. Thus, by the Summing Up Theorem, $\det A_T \neq 0$.

**ii.** If $\det A_T \neq 0$, then $A_T\mathbf{x} = \mathbf{0}$ has only the trivial solution. Thus $T$ is $1 - 1$, and since $V$ and $W$ are finite dimensional, $T$ is also onto.

**5.** $m = [n(n + 1)]/2 = \dim \{A\text{: } A \text{ is } n \times n \text{ and symmetric}\}$.

**7.** Define $T\colon P_4 \to W$ by $Tp = xp$. $Tp = 0$ implies $p(x) = 0$; that is, $p$ is the zero polynomial. Thus $T$ is $1 - 1$, and since $\dim W = 5$, $T$ is also onto.

**9.** $mn = pq$

**11.** The proof of Theorem 6 proves the assertion with the understanding that the scalars $c_1, c_2, \ldots, c_n$ are complex numbers.

**13.** $T(A_1 + A_2) = (A_1 + A_2)B = A_1B + A_2B = TA_1 + TA_2$; $T(\alpha A) = (\alpha A)B = \alpha(AB) = \alpha TA$. Thus $T$ is linear. Suppose $TA = 0$. Then $AB = 0$. Since $B$ is invertible, we can multiply on the left by $B^{-1}$ to obtain $A = ABB^{-1} = 0B^{-1} = 0$ or $A = 0$. Thus $T$ is $1 - 1$, and since $\dim M_{nn} = n^2 < \infty$, $T$ is an isomorphism.

**15.** Choose $\mathbf{h} \in H$. Then $\text{Proj}_H\, \mathbf{h} = \mathbf{h}$ so that $T$ is onto. If $H = V$, then $T$ is also $1 - 1$.

**17.** Since $T$ is an isomorphism, $\ker T = \ker A = \{\mathbf{0}\}$ so that, by the Summing Up Theorem, $A$ is invertible. If $\mathbf{x} = T^{-1}\mathbf{y}$, then $T\mathbf{x} = A\mathbf{x} = \mathbf{y}$ so that $\mathbf{x} = A^{-1}\mathbf{y}$ since $A^{-1}$ exists. Thus $T^{-1}\mathbf{y} = A^{-1}\mathbf{y}$ for every $\mathbf{y} \in \mathbb{R}^n$.

**19.** For $z = a + ib \in \mathbb{C}$, define $Tz = (a, b) \in \mathbb{R}^2$. Then $T(z_1 + z_2) = T((a_1 + a_2) + i(b_1 + b_2)) = (a_1 + a_2, b_1 + b_2) = (a_1, b_1) + (a_2, b_2) = Tz_1 + Tz_2$. If $\alpha \in \mathbb{R}$, then $T(\alpha z) = T(\alpha(a + ib)) = T(\alpha a + i\alpha b) = (\alpha a, \alpha b) = \alpha(a, b) = \alpha Tz$. Thus $T$ is linear. Finally, if $T(z) = (0, 0)$, then clearly $z = a + ib = 0 + i0 = 0$. Thus $T$ is $1 - 1$ and because $\dim \mathbb{C}$ (over the reals) $= \dim \mathbb{R}^2 = 2$, $T$ is an isomorphism.

## MATLAB 5.4

**1. b.** Explain why $T$ is one to one and onto $\mathbb{R}^4$.

**c.** $A = WV^{-1}$, where column $\mathbf{i}$ of $W$ is $\mathbf{w}_i$ and column $i$ of $V$ is $\mathbf{v}_i$. (To see why, use the following: $T(\mathbf{e}) = a_1\mathbf{w}_1 + \cdots + a_4\mathbf{w}_4$, where the $a_i$ are the coordinates of $\mathbf{e}$ with respect to the basis in $V$, how one finds coordinates, and that a linear combination of vectors can be represented as multiplication by the matrix whose columns are the vectors.)

$$A = \begin{pmatrix} 1 & 0 & .5 & -4 \\ 2 & 1 & 1 & -9 \\ 1 & 1 & 0 & 0 \\ 0 & 2 & 0 & -1 \end{pmatrix},$$

$$A^{-1} = \begin{pmatrix} -2 & 1 & 1 & -1 \\ 2 & -1 & 0 & 1 \\ 38 & -18 & -2 & 10 \\ 4 & -2 & 0 & 1 \end{pmatrix}$$

The kernel will be 0, the range will be $\mathbb{R}^4$, and $A$ is invertible since the reduced row echelon form of $A$ is the identity.

**d.** The matrix for $S$ will be $VW^{-1}$, which is $A^{-1}$.

## Problems 5.5, page 525

**1.** $T\mathbf{x} \cdot T\mathbf{y}$

$$= \begin{pmatrix} x_1 \sin\theta + x_2 \cos\theta \\ x_1 \cos\theta - x_2 \sin\theta \\ x_3 \end{pmatrix} \cdot \begin{pmatrix} y_1 \sin\theta + y_2 \cos\theta \\ y_1 \cos\theta - y_2 \sin\theta \\ y_3 \end{pmatrix}$$

$$= x_1 y_1(\sin^2\theta + \cos^2\theta) + x_2 y_2(\sin^2\theta + \cos^2\theta) + x_3 y_3$$

$$= x_1 y_1 + x_2 y_2 + x_3 y_3$$

$$= \mathbf{x} \cdot \mathbf{y}$$

(all other terms in the scalar product drop out).

**3.** Using Theorem 1, $T\mathbf{x} \cdot T\mathbf{y} = (AB\mathbf{x}) \cdot (AB\mathbf{y}) = \mathbf{x} \cdot (AB)^t(AB\mathbf{y}) = \mathbf{x} \cdot (B^tA^t)(AB)\mathbf{y} = \mathbf{x} \cdot (B^{-1}A^{-1}AB)\mathbf{y} = \mathbf{x} \cdot \mathbf{y}$

**5.** Same proof as for Theorem 2 except replace $\mathbf{x} \cdot \mathbf{y}$ with $(\mathbf{x}, \mathbf{y})$ and $T\mathbf{x} \cdot T\mathbf{y}$ with $(T\mathbf{x}, T\mathbf{y})$.

**7.** $T\mathbf{x} = \alpha\mathbf{x}$, where $\alpha$ is a scalar and $\alpha \neq 0$ or 1.

**9.** $T\mathbf{x} \cdot T\mathbf{y} = \mathbf{x} \cdot \mathbf{y} = A\mathbf{x} \cdot A\mathbf{y}$ and $A^t = A^{-1}$ so that $A = (A^{-1})^t$. Then $\mathbf{x} \cdot \mathbf{y} = \mathbf{x} \cdot (I\mathbf{y}) = \mathbf{x} \cdot (A^{-1})^tA^{-1}\mathbf{y} = A^{-1}\mathbf{x} \cdot A^{-1}\mathbf{y} = S\mathbf{x} \cdot S\mathbf{y}$ so that $S\mathbf{x} = A^{-1}\mathbf{y}$ is an isometry.

**11.** $T(a_0 + a_1x + a_2x^2 + a_3x^3) = (a_0/\sqrt{2} - (\sqrt{5}/2\sqrt{2})a_2, \sqrt{(3/2)}a_1 - (3\sqrt{7}/2\sqrt{2})a_3, (3\sqrt{5}/2\sqrt{2})a_2, (5\sqrt{7}/2\sqrt{2})a_3)$

**13.** $T\begin{pmatrix} a & b \\ c & d \end{pmatrix} = (a/\sqrt{2} - (\sqrt{5}/2\sqrt{2})c, \sqrt{(3/2)}b - (3\sqrt{7}/2\sqrt{2})d, (3\sqrt{5}/2\sqrt{2})c, (5\sqrt{7}/2\sqrt{2})d)$

**15.** $A^* = \begin{pmatrix} 1-i & 3 \\ -4-2i & 6+3i \end{pmatrix}$

**17.** If $A$ is hermitian, then $A^* = A$. In particular, the diagonal components of $A$ do not move when we take the transpose so that $\overline{a_{ii}} = a_{ii}$, which means that $a_{ii}$ is real.

**19.** Let $A^* = B = (b_{ij})$ and let $\mathbf{c}_i$ be the $i$th column of $A$. Then $AB = I = (\delta_{ij})$ where

$$\delta_{ij} = \begin{cases} 1, & \text{if } i = j \\ 0, & \text{if } i \neq j \end{cases}$$

But $\delta_{ij} = \sum_{k=1}^{n} a_{ik}b_{kj} = \sum_{k=1}^{n} a_{ik}\overline{a_{jk}} = \mathbf{c}_i \cdot \mathbf{c}_j = \delta_{ij}$.

**21.** Since the $i$th component of $A\mathbf{x}$ is $\sum_{j=1}^{n} a_{ij}x_j$, we have $(A\mathbf{x}, \mathbf{y}) = \sum_{i=1}^{n}\sum_{j=1}^{n} a_{ij}x_j\overline{y_i}$. Similarly, if $A^* = B = (b_{ij})$, $(\mathbf{x}, A^*\mathbf{y}) = \sum_{i=1}^{n}\sum_{j=1}^{n} x_j\overline{b_{ji}y_i} = \sum_{j=1}^{n}\sum_{i=1}^{n} x_j\overline{b_{ji}y_i} = \sum_{j=1}^{n}\sum_{i=1}^{n} x_ja_{ij}\overline{y_i} = \sum_{i=1}^{n}\sum_{j=1}^{n} a_{ij}x_j\overline{y_i} = (A\mathbf{x}, \mathbf{y})$.

## MATLAB 5.5

**1. a.** Rotation and reflection preserve length.

**c.** Write out a general representation for each matrix and show that the matrix times its transpose equals the identity matrix. For reflection, use $F = 2P - I$, where you first show that $P = \begin{pmatrix} v_1^2 & v_1v_2 \\ v_1v_2 & v_2^2 \end{pmatrix}$. Use the fact that $v_1^2 + v_2^2 = 1$.

**d.** Some key points in the coding are

```
th = atan(v(2)/v(1))
```

**R = [cos(th) −sin(th);**
**sin(th) cos(th)];**
**F = 2*[v(1)*v**
**v(2)*v]−eye(2)**
**X = [1 0;0 −1]**

**e.** For reflection, let $F = 2P - I$, where $P$ is the same as in part (c) above, with $v_1 = \cos(\alpha)$ and $v_2 = \sin(\alpha)$, and simplify.

**Chapter 5—Review,** page 531

1. linear
3. not linear, since $T(\alpha(x, y)) = T(\alpha x, \alpha y) = \alpha x/\alpha y = x/y = T(x, y) \neq \alpha T(x, y)$ unless $\alpha = 1$.
5. not linear, since $T(p_1 + p_2) = 1 + p_1 + p_2$, but
$$Tp_1 + Tp_2 = (1 + p_1) + (1 + p_2) = 2 + p_1 + p_2.$$
7. $\ker T = \left\{\begin{pmatrix}0\\0\end{pmatrix}\right\}$; range $T = \mathbb{R}^2$; $\rho(T) = 2$; $\nu(T) = 0$
9. $\ker T = \text{span}\left\{\begin{pmatrix}0\\0\\1\end{pmatrix}\right\}$; range $T = \mathbb{R}^2$; $\rho(T) = 2$; $\nu(T) = 1$
11. $\ker T = \left\{\begin{pmatrix}0 & 0\\0 & 0\end{pmatrix}\right\}$; range $T = M_{22}$; $\rho(T) = 4$; $\nu(T) = 0$
13. $\begin{pmatrix}0 & 0\\0 & -1\end{pmatrix}$; range $T = \text{span}\left\{\begin{pmatrix}0\\1\end{pmatrix}\right\}$; $\ker T = \text{span}\left\{\begin{pmatrix}1\\0\end{pmatrix}\right\}$; $\rho(T) = \nu(T) = 1$
15. $\begin{pmatrix}1 & 0 & -2 & 0\\0 & 2 & 0 & 3\end{pmatrix}$; range $T = \mathbb{R}^2$; $\ker T = \text{span}\left\{\begin{pmatrix}2\\0\\1\\0\end{pmatrix}, \begin{pmatrix}0\\-3\\0\\2\end{pmatrix}\right\}$; $\rho(A) = \nu(A) = 2$
17. $\begin{pmatrix}-1 & 1 & 0 & 0\\0 & 2 & 0 & 0\\0 & 0 & -1 & 1\\0 & 0 & 0 & 2\end{pmatrix}$; range $T = M_{22}$; $\ker T = \{\mathbf{0}\}$; $\rho(T) = 4$, $\nu(T) = 0$
19. expansion along the $x$-axis with $c = 3$
21. shear along the $y$-axis with $c = -2$
23. $\begin{pmatrix}3 & 0\\0 & 1\end{pmatrix}$;

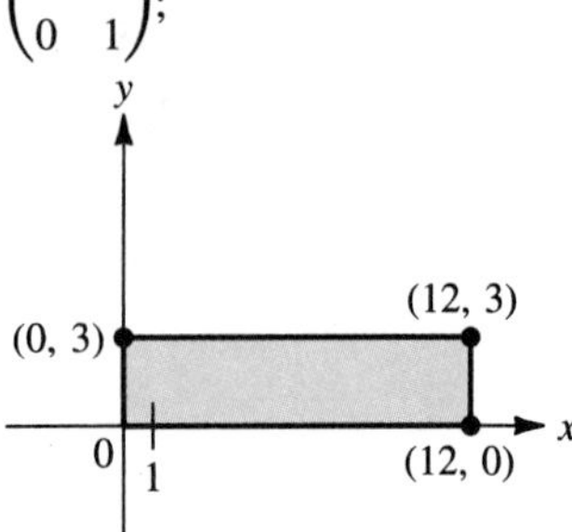


25. $\begin{pmatrix}0 & 1\\1 & 0\end{pmatrix}$;

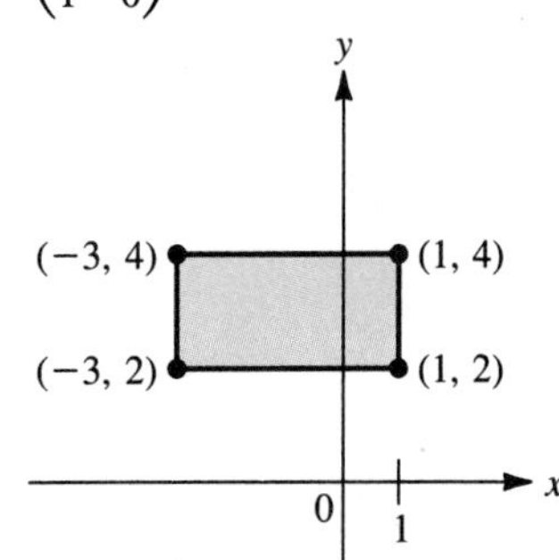


27. $\begin{pmatrix}1 & 0\\-2 & 1\end{pmatrix}\begin{pmatrix}1 & 0\\0 & 8\end{pmatrix}\begin{pmatrix}1 & 3\\0 & 1\end{pmatrix}$
29. $\begin{pmatrix}0 & 1\\1 & 0\end{pmatrix}\begin{pmatrix}1 & 0\\-6 & 1\end{pmatrix}\begin{pmatrix}1 & 0\\0 & 22\end{pmatrix}\begin{pmatrix}1 & 3\\0 & 1\end{pmatrix}$
31. $T(a_0 + a_1x + a_2x^2) = \begin{pmatrix}a_0\\a_1\\a_2\end{pmatrix}$

## Chapter 6

**Problems 6.1,** page 546

1. $-4, 3$; $E_{-4} = \text{span}\left\{\begin{pmatrix}1\\1\end{pmatrix}\right\}$; $E_3 = \text{span}\left\{\begin{pmatrix}2\\-5\end{pmatrix}\right\}$

**3.** $i, -i$; $E_i = \text{span}\left\{\begin{pmatrix} 2+i \\ 5 \end{pmatrix}\right\}$;

$E_{-i} = \text{span}\left\{\begin{pmatrix} 2-i \\ 5 \end{pmatrix}\right\}$

**5.** $-3, -3$; $E_{-3} = \text{span}\left\{\begin{pmatrix} 1 \\ 0 \end{pmatrix}\right\}$;

geom. mult. is 1

**7.** $0, 1, 3$; $E_0 = \text{span}\left\{\begin{pmatrix} 1 \\ 1 \\ 1 \end{pmatrix}\right\}$;

$E_1 = \text{span}\left\{\begin{pmatrix} -1 \\ 0 \\ 1 \end{pmatrix}\right\}$;

$E_3 = \text{span}\left\{\begin{pmatrix} 1 \\ -2 \\ 1 \end{pmatrix}\right\}$

**9.** $1, 1, 10$;

$E_1 = \text{span}\left\{\begin{pmatrix} 1 \\ 0 \\ -2 \end{pmatrix}, \begin{pmatrix} 0 \\ 1 \\ -2 \end{pmatrix}\right\}$;

$E_{10} = \text{span}\left\{\begin{pmatrix} 2 \\ 2 \\ 1 \end{pmatrix}\right\}$;

geom. mult. of 1 is 2

**11.** $1, 1, 1$; $E_1 = \text{span}\left\{\begin{pmatrix} 1 \\ 1 \\ 1 \end{pmatrix}\right\}$;

geom. mult. is 1 (alg. mult. is 3)

**13.** $-1, i, -i$;

$E_{-1} = \text{span}\left\{\begin{pmatrix} 0 \\ -1 \\ 1 \end{pmatrix}\right\}$;

$E_i = \text{span}\left\{\begin{pmatrix} 1+i \\ 1 \\ 1 \end{pmatrix}\right\}$;

$E_{-i} = \text{span}\left\{\begin{pmatrix} 1-i \\ 1 \\ 1 \end{pmatrix}\right\}$

**15.** $1, 2, 2$;

$E_1 = \text{span}\left\{\begin{pmatrix} 4 \\ 1 \\ -3 \end{pmatrix}\right\}$;

$E_2 = \text{span}\left\{\begin{pmatrix} 3 \\ 1 \\ -2 \end{pmatrix}\right\}$;

geom. mult. of 2 is 1

**17.** $a, a, a, a$; $E_a = \mathbb{R}^4$; geom. mult. of $a$ = alg. mult. of $a = 4$

**19.** $a, a, a, a$;

$E_a = \text{span}\left\{\begin{pmatrix} 1 \\ 0 \\ 0 \\ 0 \end{pmatrix}, \begin{pmatrix} 0 \\ 0 \\ 0 \\ 1 \end{pmatrix}\right\}$;

alg. mult. of $a = 4$; geom. mult. of $a = 2$.

**21.** Eigenvalues are $a \pm ib$. Then

$$[A - (a+ib)I]\left\{\begin{pmatrix} 1 \\ i \end{pmatrix}\right\}$$

$$= \begin{pmatrix} -ib & b \\ -b & -ib \end{pmatrix}\begin{pmatrix} 1 \\ i \end{pmatrix} = \begin{pmatrix} 0 \\ 0 \end{pmatrix}$$

Similarly,

$$[A - (a-ib)I]\begin{pmatrix} 1 \\ -i \end{pmatrix} = \begin{pmatrix} 0 \\ 0 \end{pmatrix}$$

**23.** Let $\beta_1, \beta_2, \ldots, \beta_m$ be the eigenvalues of $\alpha A$. Then, for each $i$, there is a vector $\mathbf{v}_i \neq \mathbf{0}$ such that $(\alpha A)\mathbf{v}_i = \beta_i \mathbf{v}_i$. Thus $(\alpha A - \beta_i I)\mathbf{v} = \mathbf{0}$, or $\alpha\left(A - \dfrac{\beta_i}{\alpha} I\right)\mathbf{v} = \mathbf{0}$, which implies that $\det\left(A - \dfrac{\beta_i}{\alpha} I\right) = 0$.

Thus $\dfrac{\beta_i}{\alpha}$ is an eigenvalue of $A$ and $\dfrac{\beta_i}{\alpha} = \lambda_j$ for some $j$ and $\beta_i = \alpha\lambda_j$. Thus each eigenvalue of $\alpha A$ is of the form $\alpha\lambda_j$. Conversely, if $\mu_i =$

$\alpha\lambda_i$, choose a vector $\mathbf{v}_i$ such that $A\mathbf{v}_i = \lambda_i\mathbf{v}_i$. Then $(\alpha A)\mathbf{v}_i = \alpha\lambda_i\mathbf{v}_i = \mu_i\mathbf{v}_i$ so that $\mu_i$ is an eigenvalue of $\alpha A$.

**25.** $\det(A - \lambda_i I) = 0$ and $\det A^{-1} \neq 0$ because $(A^{-1})^{-1} = A$ exists. Thus $0 = \det(A - \lambda_i I)\det A^{-1} = \det[(A - \lambda_i I)A^{-1}] = \det(I - \lambda_i A^{-1}) = \det \lambda_i \left(\frac{1}{\lambda_i} I - A^{-1}\right)$

$\lambda_i^n \det\left(\frac{1}{\lambda_i} I - A^{-1}\right)$. In the last step we used the fact that $\det \alpha A = \alpha^n \det A$ if $A$ is an $n \times n$ matrix (see Problem 2.2.28), and $\lambda_i \neq 0$ by Theorem 6, parts (**i**) and (**x**). Thus

$$\det\left(A^{-1} - \frac{1}{\lambda_i} I\right) = (-1)^n \det\left(\frac{1}{\lambda_i} I - A^{-1}\right) = 0$$

and $\frac{1}{\lambda_i}$ is an eigenvalue of $A^{-1}$. Conversely, if $\mu_i$ is an eigenvalue of $A^{-1}$, then since $(A^{-1})^{-1} = A$, $\frac{1}{\mu_i}$ is an eigenvalue of $A$ so that $\frac{1}{\mu_i} = \lambda_j$ for some $j$, or $\mu_i = \frac{1}{\lambda_j}$. Thus all eigenvalues of $A^{-1}$ are of the form $\frac{1}{\lambda_i}$. Alternatively, if $A\mathbf{v} = \lambda\mathbf{v}$, then $\mathbf{v} = A^{-1}\lambda\mathbf{v}$ so that $A^{-1}\mathbf{v} = \frac{1}{\lambda}\mathbf{v}$ and $\frac{1}{\lambda}$ is an eigenvalue of $A^{-1}$.

**27.** Since $\det(A - \lambda_i I) = 0$, we see that $\det(A^2 - \lambda_i^2 I) = \det(A - \lambda_i I)(A + \lambda_i I) = \det(A - \lambda_i I)\det(A + \lambda_i I) = 0$. Thus $\lambda_i^2$ is an eigenvalue of $A^2$. Conversely, if $\mu_i$ is an eigenvalue of $A^2$, then $0 = \det(A^2 - \mu_i I) = \det[(A - \sqrt{\mu_i} I)(A + \sqrt{\mu_i} I)]$ so that either $\det(A - \sqrt{\mu_i} I)$ or $\det(A + \sqrt{\mu_i} I) = 0$. In either case $\pm\sqrt{\mu_i}$ is an eigenvalue of $A$ so that $\mu_i = (\pm\sqrt{\mu_i})^2 = (\pm\lambda_j)^2 = \lambda_j^2$ for some $j$.

**29.** From Problem 28, $a_i A^i \mathbf{v} = a_i \lambda^i \mathbf{v}$ so that $p(A)\mathbf{v} = \sum_{i=0}^{n} (a_i A^i)\mathbf{v} = \sum_{i=1}^{n} (a_i A^i \mathbf{v}) = \sum_{i=1}^{n} a_i \lambda^i \mathbf{v} = \left(\sum a_i \lambda^i\right)\mathbf{v} = p(\lambda)\mathbf{v}$.

**31.** If $A$ is diagonal, then so is $A - \lambda I$ so that, by Theorem 2.1.1, $\det(A - \lambda I) = (a_{ii} - \lambda)(a_{22} - \lambda)\cdots(a_{nn} - \lambda) = 0$ when $\lambda = a_{ii}$ for some $i = 1, 2, \ldots, n$.

**33.** $A\mathbf{v} = \lambda\mathbf{v}$ where $\mathbf{v} \neq \mathbf{0}$. Then $\overline{A\mathbf{v}} = \overline{\lambda\mathbf{v}}$, which implies that $\overline{A}\overline{\mathbf{v}} = \overline{\lambda}\overline{\mathbf{v}}$. But if $A$ is real, then $\overline{A} = A$. Thus $A\overline{\mathbf{v}} = \overline{\lambda}\overline{\mathbf{v}}$ and $\mathbf{v} \neq \mathbf{0}$ so that $\overline{\lambda}$ is an eigenvalue of $A$ with eigenvector $\overline{\mathbf{v}}$. Here we have used the easily verified fact that $\overline{A\mathbf{v}} = \overline{A}\overline{\mathbf{v}}$.

**35.** Let $\lambda = a + bm$; then $A - \lambda I = \begin{pmatrix} -bm & b \\ c & d - a - bm \end{pmatrix}$ and $\det(A - \lambda I) = b^2m^2 + b(a - d)m - bc = b[bm^2 + (a - d)m - c] = 0$ so $\lambda$ is an eigenvalue of $A$. Similarly,

$$(A - \lambda I)\begin{pmatrix} 1 \\ m \end{pmatrix} = \begin{pmatrix} 0 \\ c - m(a - d) - bm^2 \end{pmatrix} = \begin{pmatrix} 0 \\ 0 \end{pmatrix},$$

so $\begin{pmatrix} 1 \\ m \end{pmatrix}$ is an eigenvector.

**37.** eigenvalues: 12.7089974455, 1.20237103639, $-0.955684240966 \pm 0.628878731063i$

corresponding eigenvectors:

$$\begin{pmatrix} -0.621427151885 \\ -0.317162132863 \\ -0.600741512354 \\ -0.648893720131 \end{pmatrix}, \begin{pmatrix} 0.716877837391 \\ -0.102986436421 \\ 0.460859947991 \\ -0.470715276803 \end{pmatrix},$$

$$\begin{pmatrix} 0.045437941761 \\ -0.489156858693 \\ 0.568554600861 \\ 0.189095737443 \end{pmatrix} \pm i \begin{pmatrix} -3.192511943 \\ -1.06860325542 \\ -1.47374850265 \\ 3.31095939015 \end{pmatrix}$$

Note that on the TI-85 $a + bi$ is displayed as $(a, b)$.

**39.** eigenvalues: $-0.070207416895$, $0.013167178735$, $0.13104023816$ corresponding eigenvectors:

$$\begin{pmatrix} -0.8690418183 \\ -0.052743032554 \\ 0.404187510937 \end{pmatrix}, \begin{pmatrix} -1.13177594421 \\ -0.765119595137 \\ 0.134236382997 \end{pmatrix}, \begin{pmatrix} 0.038878374501 \\ 1.21858619393 \\ -0.985517966726 \end{pmatrix}$$

**41.** 6 **43.** 3 **45.** 1

MATLAB 6.1

**1.** **a.–c.** For example, to show $a\mathbf{y} + b\mathbf{z}$ is an eigenvector with eigenvalue 3: **y = [3;4;5]; z = [4;9;13]; a = 3*rand(1); b = 4*(2*rand(1)−1); w = a*y+b*z; ans=(A−3*eye(3))*w;.** Check that **ans = 0.**

**d.** The eigenvectors for a given eigenvalue form a subspace.

**3.** **a.–c.** (**1.**) characteristic poly = $\lambda^2 + \lambda - 12$, eigenvalues are $\lambda = -4$ with eigenvector $(1\ 1)^t$ and $\lambda = 3$ with eigenvector $(-.4\ 1)^t$. (**6.**) characteristic poly = $\lambda^2 - 4\lambda + 13$, eigenvalues are $\lambda = 2 + 3i$ with eigenvector $(-.2 - .6i\ 1)^t$ and $\lambda = 2 - 3i$ with eigenvector $(-.2+.6i\ 1)^t$. (**8.**) characteristic poly = $-\lambda^3 + 2\lambda^2 + \lambda - 2$, eigenvalues are $\lambda = 2$ with eigenvector $(1\ 3\ 1)^t$, $\lambda = 1$ with eigenvector $(3\ 2\ 1)^t$, and $\lambda = -1$ with eigenvector $(1\ 0\ 1)^t$. (**13.**) characteristic poly = $-\lambda^3 - \lambda^2 - \lambda - 1$, eigenvalues are $\lambda = -1$ with eigenvector $(0\ -1\ 1)^t$, $\lambda = i$ with eigenvector $(1+i\ 1\ 1)^t$, and $\lambda = -i$ with eigenvector $(1 - i\ 1\ 1)^t$. Note that the $(-1)^n$ is needed since **poly** finds $\det(\lambda I - A)$ rather than $\det(A - \lambda I)$.

**e.** **D(k,k)** is an eigenvalue for $A$ with eigenvector **V(:,k).** To normalize a vector **x** to have length 1, find **x/norm(x).**

**5.** **a.** Characteristic polynomials of $A$ and $A^t$ are the same. Hence the eigenvalues will be the same.

**b.** *Proof hint:* You need to consider $\det(A^t - \lambda I)$. How is $A^t - \lambda I$ related to $(A - \lambda I)^t$? How is $\det(C^t)$ related to $\det(C)$?

**7.** Final conjecture: If **x** is an eigenvector for $A$ with eigenvalue $\lambda$, then **x** is an eigenvector for $A^2$ with eigenvalue $\lambda^2$. In part (c), compare the row echelon forms of $A - \lambda I$ and $A^2 - \lambda^2 I$ to see that the eigenvectors will be the same. *Proof hint:* Assume $A\mathbf{x} = \lambda\mathbf{x}$ and rewrite/simplify $A^2\mathbf{x} = A(A\mathbf{x})$.

**9.** Symmetric matrices will have real eigenvalues. Symmetric matrices of the form $AA^t$ will have nonnegative real eigenvalues.

**11.** **a.** $4 \le \chi \le 4$, so one needs four colors.

**b.** $3.13 \le \chi \le 4.44$, so one needs four colors.

**c.** $3.2 \le \chi \le 4.6$, so one needs four colors.

**d.** $2.23 \le \chi \le 3$, so one needs three colors.

**e.** $2.5 \le \chi \le 4$, so one needs three or four colors. You will need to experiment to see whether or not it can be done in three.

**Problems 6.2,** page 561

**1.**

| $n$ | $p_{j,n}$ | $p_{a,n}$ | $T_n$ | $p_{j,n}/p_{a,n}$ | $T_n/T_{n-1}$ |
|---|---|---|---|---|---|
| 0 | 0 | 12 | 12 | 0 | — |
| 1 | 36 | 7 | 43 | 5.14 | 3.58 |
| 2 | 21 | 19 | 40 | 1.11 | 0.930 |
| 5 | 104 | 45 | 149 | 2.31 | — |
| 10 | 600 | 291 | 891 | 2.06 | — |
| 19 | 16,090 | 7737 | 23827 | 2.08 | — |
| 20 | 23,170 | 11140 | 34310 | 2.08 | 1.44 |

Note that the eigenvalues are 1.44 and $-0.836$. The corresponding eigenvectors are $\begin{pmatrix} 2.09 \\ 1 \end{pmatrix}$ and $\begin{pmatrix} -3.57 \\ 1 \end{pmatrix}$.

**3.**

| $n$ | $p_{j,n}$ | $p_{a,n}$ | $T_n$ | $p_{j,n}/p_{a,n}$ | $T_n/T_{n-1}$ |
|---|---|---|---|---|---|
| 0 | 0 | 20 | 20 | 0 | — |
| 1 | 80 | 16 | 96 | 5 | 4.8 |
| 2 | 64 | 69 | 133 | 0.928 | 1.39 |
| 5 | 1092 | 498 | 1590 | 2.19 | — |
| 10 | 42,412 | 22,807 | 65,219 | 1.86 | — |
| 19 | $3.69 \times 10^7$ | $1.95 \times 10^7$ | $5.64 \times 10^7$ | 1.89 | — |
| 20 | $7.82 \times 10^7$ | $4.14 \times 10^7$ | $11.96 \times 10^7$ | 1.89 | 2.12 |

The eigenvalues are 2.12 and $-1.32$ with corresponding eigenvectors $\begin{pmatrix} 1.89 \\ 1 \end{pmatrix}$ and $\begin{pmatrix} -3.03 \\ 1 \end{pmatrix}$.

**5.** From equation (9), $p_n \approx a_1 \lambda_1^n \mathbf{v}_i$ for $n$ large. If $\mathbf{v}_1 = \begin{pmatrix} x \\ y \end{pmatrix}$, then

$$\frac{p_{j,n}}{p_{a,n}} \approx \frac{a_1 \lambda_1^n x}{a_1 \lambda_1^n y} = \frac{x}{y}; \text{ but}$$

$$\begin{pmatrix} -\lambda_1 & k \\ \alpha & \beta - \lambda_1 \end{pmatrix} \begin{pmatrix} x \\ y \end{pmatrix} = \begin{pmatrix} 0 \\ 0 \end{pmatrix} \text{ so that}$$

$-\lambda_1 x + ky = 0$ and $\dfrac{x}{y} = \dfrac{k}{\lambda_1}$. Thus

$$\frac{p_{j,n}}{p_{a,n}} \approx \frac{x}{y} = \frac{k}{\lambda_1} \text{ for } n \text{ large.}$$

**MATLAB 6.2**

**1. a.** After 2 years (after rounding down), there are 21 juveniles and 18 adults; after 5 years, 103 juveniles and 44 adults; after 10 years, 587 juveniles and 282 adults; and after 20 years, 21,965 juveniles and 10,513 adults.

**b.** $p_{j,n}/p_{a,n}$ is 2.0895 or 2.0894 for $n = 21$ through 25 and $T_n/T_{n-1}$ is 1.4358 for $n = 21$ through 25. These numbers would be the corresponding conjectured limits.

**c.** The eigenvalues are 1.4358 and $-.8358$. The largest eigenvalue equals the projected limit of $T_n/T_{n-1}$. The population is growing since $T_n \approx 1.4358 T_{n-1}$ and 1.4358 is larger than 1. The ratios $w_1/w_2$, limit of $p_{j,n}/p_{a,n}$ and $k/\lambda$ are all 2.0895. The ratio of juvenile to adult birds in the long run can be found by the ratio of the components of the eigenvector associated with the largest eigenvalue or it can be found by dividing the birth rate by the largest eigenvalue.

**3. a.** Matrix has largest eigenvalue = 1.2718, the other eigenvalue is strictly less in magnitude, and there is an eigenvector for the largest eigenvalue with all components positive. Under these conditions we have seen that $T_n/T_{n-1}$ approaches the largest eigenvalue.

**b.** *Hint:* The adult population in the following year will consist of new adults from the juvenile population + the adults that normally survive − the adults killed by hunting. The adults killed by hunting is $h \times$ (adult population).

**c.** The components of the vectors $A^n\mathbf{p}_0$ will get smaller and smaller, decreasing to zero. The largest eigenvalue is less than 1.

**d.** The $h$ that maintains a steady-state population in the long run is $h = .4$. For this $h$, the largest eigenvalue is 1.

**5. a.** The $i$th component of $P^n\mathbf{x}$ represents the number of households buying product $i$ after $n$ months. As $n$ gets larger, $P^n\mathbf{x}$ seems to get closer and closer to a fixed vector, $(900 \quad 500 \quad 1600)^t$, implying that the market share of each product stabilizes over time.

**b.** The largest eigenvalue is 1 and the other eigenvalues are strictly smaller in magnitude. Any initial starting vector will not be perpendicular to the eigenvector corresponding to the eigenvalue of 1 (you should explain why), so the extension of the theory says that $P^n\mathbf{x}$ will approach some fixed multiple of the eigenvector. The limit vector $\mathbf{y}$ is seen to equal $3000 \times$ (the eigenvector normalized so that the components sum to 1).

**c.** The largest eigenvalue is 1. Finding 1000**w/sum(w)**, where **w** is the eigenvector associated with the eigenvalue 1, will give you a vector in the direction of the eigenvector whose components add to 1000. This long-term distribution has approximately 333 cars at office 1, 238 cars at office 2, and 429 cars at office 3.

**d.** $P^t$ has an eigenvalue of 1, and hence so does $P$.

**Problems 6.3,** page 572

**1.** yes; $C = \begin{pmatrix} 1 & 2 \\ 1 & -5 \end{pmatrix}$,

$C^{-1}AC = \begin{pmatrix} -4 & 0 \\ 0 & 3 \end{pmatrix}$

**3.** yes; $C = \begin{pmatrix} 1 & 1 \\ 2-i & 2+i \end{pmatrix}$;

$C^{-1}AC = \begin{pmatrix} i & 0 \\ 0 & -i \end{pmatrix}$

**5.** yes;

$C = \begin{pmatrix} 2 & 2 \\ -1+3i & -1-3i \end{pmatrix}$;

$C^{-1}AC = \begin{pmatrix} 2+3i & 0 \\ 0 & 2-3i \end{pmatrix}$

**7.** yes; $C = \begin{pmatrix} 3 & 1 & 1 \\ 2 & 3 & 0 \\ 1 & 1 & 1 \end{pmatrix}$;

$C^{-1}AC = \begin{pmatrix} 1 & 0 & 0 \\ 0 & 2 & 0 \\ 0 & 0 & -1 \end{pmatrix}$

**9.** yes; $C = \begin{pmatrix} 0 & 0 & 1 \\ 1 & 1 & 0 \\ 0 & 2 & 0 \end{pmatrix}$;

$$C^{-1}AC = \begin{pmatrix} 0 & 0 & 0 \\ 0 & 2 & 0 \\ 0 & 0 & 3 \end{pmatrix}$$

**11.** $C = \begin{pmatrix} 1 & 0 & 2 \\ 3 & -2 & 1 \\ 0 & 1 & 2 \end{pmatrix}$;

$$C^{-1}AC = \begin{pmatrix} 1 & 0 & 0 \\ 0 & 1 & 0 \\ 0 & 0 & 2 \end{pmatrix}$$

**13.** no, since 1 is an eigenvalue of algebraic multiplicity 3 and geometric multiplicity 1

**15.** yes;

$$C = \begin{pmatrix} 0 & -1 & 1 & 1 \\ 0 & 1 & 1 & 1 \\ 1 & 0 & 1 & -1 \\ 0 & 1 & -1 & 1 \end{pmatrix};$$

$$C^{-1}AC = \begin{pmatrix} 2 & 0 & 0 & 0 \\ 0 & 2 & 0 & 0 \\ 0 & 0 & 4 & 0 \\ 0 & 0 & 0 & 6 \end{pmatrix}$$

**17.** $B = C^{-1}AC$ so $B^n = (C^{-1}AC)(C^{-1}AC)\cdots(C^{-1}AC) = C^{-1}A(CC^{-1})A(CC^{-1})\cdots AC = C^{-1}AIA\cdots IAC = C^{-1}AA\cdots AC = C^{-1}A^nC$.

**19.** $\begin{pmatrix} 1 & 0 \\ 0 & 1 \end{pmatrix}$

**21.** If $D = C^{-1}AC$, then, as in Problem 17, $D^n = (C^{-1}AC) \times (C^{-1}AC)\cdots(C^{-1}AC) = C^{-1}A^nC$ so that $A^n = CD^nC^{-1}$.

**23.** Clearly $A$ has $c$ as an eigenvalue of algebraic multiplicity $n$. Thus if $A$ is diagonalizable, there must be an invertible matrix $E$ such that $E^{-1}AE = \text{diag}(c, c, \ldots, c) = cI$ so that $A = E(cI)E^{-1} = cEIE^{-1} = cI$.

**25.** If $A$ and $B$ have distinct eigenvalues, then both have $n$ linearly independent eigenvectors, and we have $D_1 = C_1^{-1}AC_1$ and $D_2 = C_2^{-1}BC_2$.

**i.** If $A$ and $B$ have the same eigenvectors, then $C_1 = C_2 = C$ and $AB = (CD_1C^{-1}) \times (CD_2C^{-1}) = CD_1D_2C^{-1} = CD_2D_1C^{-1} = (CD_2C^{-1}) \times (CD_1C^{-1}) = BA$ (since diagonal matrices of the same order always commute).

**ii.** If $BA = AB$, let $\mathbf{x}$ be an eigenvector of $B$ corresponding to $\lambda$. Then $BA\mathbf{x} = AB\mathbf{x} = A(\lambda\mathbf{x}) = \lambda A\mathbf{x}$ so that $y = A\mathbf{x}$ is an eigenvector of $B$ corresponding to $\lambda$. Thus $A\mathbf{x}$ and $\mathbf{x}$ are linearly dependent so that there is a scalar $\mu$ with $A\mathbf{x} = \mu\mathbf{x}$. But this shows that $\mathbf{x}$ is also an eigenvector of $A$. Thus every eigenvector of $B$ is an eigenvector of $A$. A similar argument shows that every eigenvector of $A$ is an eigenvector of $B$.

Entries in Problems 27 and 29 are given to four decimal places.

**27.** $\begin{pmatrix} -.6214 & .7169 & .0454 - 3.1925i & .0454 + 3.1925i \\ -.3172 & -.1030 & -.4892 - 1.0686i & -.4892 + 1.0686i \\ -.6007 & .4609 & .5686 - 1.4737i & .5686 + 1.4737i \\ -.6489 & -.4707 & .1891 + 3.3110i & .1891 - 3.3110i \end{pmatrix}$

**29.** $\begin{pmatrix} -.8690 & -1.1318 & .0389 \\ -.0527 & -.7651 & 1.2186 \\ .4042 & .1342 & -.9855 \end{pmatrix}$

## MATLAB 6.3

**1.** This problem illustrates that $CAC^{-1}$ and $A$ have the same eigenvalues.

**3. a.** $D = \begin{pmatrix} -2 & 0 & 0 \\ 0 & 3 & 0 \\ 0 & 0 & 3 \end{pmatrix}$,

$$C = \begin{pmatrix} 1 & 3 & 4 \\ 1 & 4 & 9 \\ 1 & 5 & 13 \end{pmatrix}$$

**b.** $D =$

$$\begin{pmatrix} 2+i & 0 & 0 & 0 \\ 0 & 2+i & 0 & 0 \\ 0 & 0 & 2-i & 0 \\ 0 & 0 & 0 & 2-i \end{pmatrix},$$

$$C = \begin{pmatrix} 1 & 0 & 1 & 0 \\ i & i & -i & -i \\ 0 & 2 & 0 & 2 \\ -i & 1+i & i & 1-i \end{pmatrix}$$

**5.** $A\mathbf{x} = \lambda\mathbf{x}$ says that $A$ expands or compresses $\mathbf{x}$. If $A$ is diagonalizable, then $A$ expands or compresses each eigenvector by a factor given by the associated eigenvalue.

**c.** Expands the direction given by $(1 \quad -1)^t$ by a factor of 2 and expands the direction given by $(1 \quad 1)^t$ by a factor of 3. To sketch the image of the rectangle, take the diagonal running from the $(-1, -1)$ corner to the $(1, 1)$ corner and stretch by a factor of 3 in each direction; take the other diagonal and stretch by a factor of 2 in each direction.

**d.** **i.** Neither expands nor compresses in the direction of $(1 \quad 1 \quad 1)^t$, expands by a factor of 2 in the direction of $(3 \quad 4 \quad 5)^t$, and compresses by a factor of .5 in the direction of $(4 \quad 9 \quad 13)^t$.

**7.** $Q = \begin{pmatrix} -\frac{2}{3} & \frac{1}{3} & \frac{2}{3} \\ \frac{2}{3} & \frac{2}{3} & \frac{1}{3} \\ \frac{1}{3} & -\frac{2}{3} & \frac{2}{3} \end{pmatrix}$, $D = \begin{pmatrix} 0 & 0 & 0 \\ 0 & 3 & 0 \\ 0 & 0 & 6 \end{pmatrix}$

**9.** Let $\mathbf{u}$ be an eigenvector corresponding to $\lambda$ with $|\mathbf{u}| = 1$. Then $Q\mathbf{u} = \lambda\mathbf{u}$ and $1 = |\mathbf{u}| = |Q^{-1}Q\mathbf{u}| = |\lambda Q^{-1}\mathbf{u}| = |\lambda Q^t\mathbf{u}| = |\lambda Q\mathbf{u}| =$ (since $Q$ is symmetric) $|\lambda^2\mathbf{u}| = \lambda^2|\mathbf{u}| = \lambda^2$. Thus $\lambda^2 = 1$ and $\lambda = \pm 1$.

**11.** $1 = \det I = \det(Q^{-1}Q) = \det(Q^tQ) = (\det Q^t)(\det Q) = (\det Q)^2$ – since $\det A^t = \det A$ for any matrix $A$. Thus $\det Q = \pm 1$ and $\begin{pmatrix} a & c \\ b & d \end{pmatrix} = Q^t =$

$$Q^{-1} = \begin{pmatrix} \dfrac{d}{\det Q} & -\dfrac{b}{\det Q} \\ -\dfrac{c}{\det Q} & \dfrac{a}{\det Q} \end{pmatrix}.$$

If $\det Q = 1$, then $c = -b$. If $\det Q = -1$, then $c = b$.

**13.** If the $2 \times 2$ matrix $A$ has orthogonal eigenvectors, then $A$ is orthogonally diagonalizable, which means that $A$ is symmetric by Theorem 4.

**15.** Let $\lambda$ be an eigenvalue of $A$ with eigenvector $\mathbf{v}$ and suppose that $A^* = A$. Then $\lambda(\mathbf{v}, \mathbf{v}) = (\lambda\mathbf{v}, \mathbf{v}) = (A\mathbf{v}, \mathbf{v}) = (\mathbf{v}, A^*\mathbf{v}) = (\mathbf{v}, A\mathbf{v}) = (\mathbf{v}, \lambda\mathbf{v}) = \overline{\lambda}(\mathbf{v}, \mathbf{v})$. Since $\mathbf{v} \neq \mathbf{0}$, this means that $\lambda = \overline{\lambda}$ so that $\lambda$ is real.

**17.** Use Problem 16 after showing that to every eigenvalue of algebraic multiplicity $k$ there correspond $k$ orthonormal eigenvectors. Let $Q$ be obtained exactly as in the proof of Theorem 3. Recall that $(\mathbf{u}, \mathbf{v}) = u_1 \cdot \overline{v}_1 + \cdots + u_n \cdot \overline{v}_n$. $Q^t = \overline{Q}^{-1}$ and $A$ is similar to $Q^tA\overline{Q}$ or $\overline{Q}^tAQ$. $|Q^tA\overline{Q} - I| = |A - I|$; $\overline{Q}^tAQ = (\overline{Q}^tA)Q$

## Problems 6.4, page 582

**1.** $Q = \begin{pmatrix} 2/\sqrt{5} & 1/\sqrt{5} \\ 1/\sqrt{5} & -2/\sqrt{5} \end{pmatrix}$, $D = \begin{pmatrix} 5 & 0 \\ 0 & -5 \end{pmatrix}$

**3.** $Q = \begin{pmatrix} 1/\sqrt{2} & 1/\sqrt{2} \\ 1/\sqrt{2} & -1/\sqrt{2} \end{pmatrix}$, $D = \begin{pmatrix} 0 & 0 \\ 0 & 2 \end{pmatrix}$

**5.** $Q = \begin{pmatrix} 1/\sqrt{2} & \frac{1}{2} & \frac{1}{2} \\ -1/\sqrt{2} & \frac{1}{2} & \frac{1}{2} \\ 0 & 1/\sqrt{2} & -1/\sqrt{2} \end{pmatrix}$. $D = \begin{pmatrix} -3 & 0 & 0 \\ 0 & 1+2\sqrt{2} & 0 \\ 0 & 0 & 1-2\sqrt{2} \end{pmatrix}$

$$= \begin{pmatrix} \overline{\mathbf{u}}_1^t A \\ \overline{\mathbf{u}}_2^t A \\ \vdots \\ \overline{\mathbf{u}}_n^t A \end{pmatrix} (\mathbf{u}_1, \mathbf{u}_2, \ldots, \mathbf{u}_n)$$

$$= \begin{pmatrix} \overline{\mathbf{u}}_1^t \\ \overline{\mathbf{u}}_2^t \\ \vdots \\ \overline{\mathbf{u}}_n^t \end{pmatrix} (A^*\mathbf{u}_1, A^*\mathbf{u}_2, \ldots, A^*\mathbf{u}_n).$$

Now $(\overline{\mathbf{u}}_1^t, A^*\mathbf{u}_1) = (\overline{\mathbf{u}}_1^t, A\mathbf{u}_1) = (\overline{\mathbf{u}}_1^t, \lambda_1\mathbf{u}_1) = \overline{\lambda}_1(\overline{\mathbf{u}}_1^t, \mathbf{u}_1) = \overline{\lambda}_1 = \lambda_1$ (by Problem 15) and since $(\overline{\mathbf{u}}_1^t, \mathbf{u}_1) = \overline{\mathbf{u}}_1^t \cdot \overline{\mathbf{u}}_1 = 1 = \overline{\mathbf{u}_1^t \cdot \mathbf{u}_1}$. Then $\overline{Q}_t AQ =$

$$\begin{pmatrix} \lambda_1 & \overline{\mathbf{u}}_1^t A\mathbf{u}_2 & \cdots & \overline{\mathbf{u}}_1^t A\mathbf{u}_n \\ 0 & \overline{\mathbf{u}}_2^t A\mathbf{u}_2 & \cdots & \overline{\mathbf{u}}_2^t A\mathbf{u}_n \\ \vdots & \vdots & & \vdots \\ 0 & \overline{\mathbf{u}}_n^t A\mathbf{u}_2 & \cdots & \overline{\mathbf{u}}_n^t A\mathbf{u}_n \end{pmatrix}$$

$\overline{\mathbf{u}}_1^t A\mathbf{u}_j = A\overline{\mathbf{u}}_1^t \cdot \overline{\mathbf{u}}_j = \overline{A\mathbf{u}_1^t \cdot \mathbf{u}_j} = 0$ if $j \neq 1$. Now $(\overline{Q}^t AQ)^t = \overline{Q^t A^t (\overline{Q}^t)^t} = \overline{Q^t A^t \overline{Q}} = \overline{Q}^t \overline{A}^t Q = \overline{Q}^t AQ$, since $\overline{A}^t = A^* = A$. Thus $\overline{Q}^t AQ$ is hermitian, which means that the zeros in the first row of $\overline{Q}^t AQ$ must match the zeros in the first column. The rest of the proof follows, as in the proof of Theorem 3, with $Q^t$ replaced by $\overline{Q}^t$.

**19.** $U = \dfrac{1}{\sqrt{3}}\begin{pmatrix} -1+i & 1 \\ 1 & 1+i \end{pmatrix}$;

$U^*AU = \begin{pmatrix} -1 & 0 \\ 0 & 8 \end{pmatrix}$

## MATLAB 6.4

**1. a.** *Ingredients:* Because of the random choice, we expect distinct eigenvalues. Any multiple of an eigenvector is an eigenvector. Eigenvectors for distinct eigenvalues for a symmetric matrix are orthogonal.

**b.** The **eig** command will produce an orthonormal set of eigenvectors. Use this set for $Q$. Since $Q$ will then be orthogonal, $Q^t$ will equal $Q^{-1}$.

**3. a.** Basic facts to use: Multiplying a column or row of a matrix by $c$ multiplies the determinant by $c$. A multiple of an eigenvector is still an eigenvector for the same eigenvalue.

**b.** Basic facts to use: An orthogonal matrix has orthonormal columns; a vector perpendicular to $(a \quad b)^t$ is either $(b \quad -a)^t$ or $(-b \quad a)^t$ (or a multiple of these). Use the fact that the determinant is 1.

**c.** Since $Q^t = Q^{-1}$, one first rotates clockwise by an angle, then expands or compresses along the $x$- and $y$-axes as indicated by the diagonal matrix, and then rotates back, that is, counterclockwise, by the angle.

**d. i.** Rotate clockwise by $\theta = -45°$, expand by 3 along the $x$-axis and expand by 4 along the $y$-axis, and then rotate counterclockwise by $\theta$. See figures below. This has the same effect as expanding by 3 in the direction of $(1 \quad -1)^t$ and expanding by 4 in the direction of $(1 \quad 1)^t$.

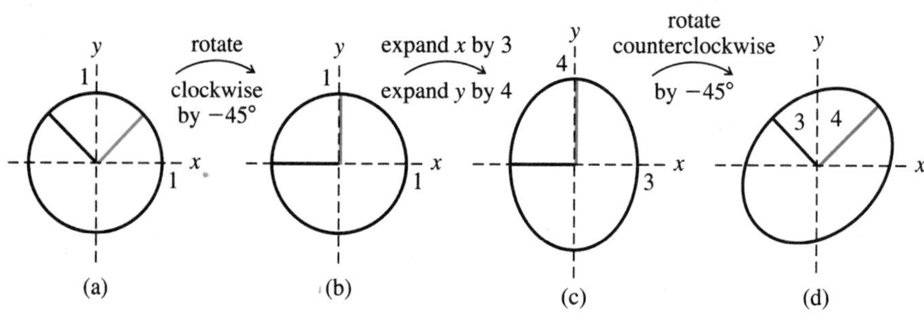

**ii.** Rotate clockwise by $\theta = 150°$, expand by 3 along the $x$-axis and expand by 2 along the $y$-axis, and then rotate counterclockwise by $\theta$. The image of the unit circle is sketched below.

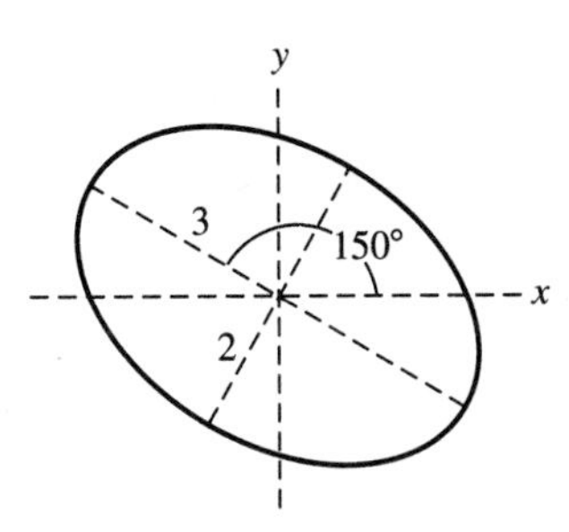


**Problems 6.5,** page 594

**1.** $\begin{pmatrix} 3 & -1 \\ -1 & 0 \end{pmatrix}\begin{pmatrix} x \\ y \end{pmatrix} \cdot \begin{pmatrix} x \\ y \end{pmatrix} = 5;$

$$Q = \begin{pmatrix} \dfrac{2}{\sqrt{26 - 6\sqrt{13}}} & \dfrac{2}{\sqrt{26 + 6\sqrt{13}}} \\ \dfrac{3 - \sqrt{13}}{\sqrt{26 - 6\sqrt{13}}} & \dfrac{3 + \sqrt{13}}{\sqrt{26 + 6\sqrt{13}}} \end{pmatrix} = \begin{pmatrix} 0.9571 & 0.2898 \\ -0.2898 & 0.9571 \end{pmatrix};$$

$$\frac{x'^2}{\left(\dfrac{10}{\sqrt{13} + 3}\right)} - \frac{y'^2}{\left(\dfrac{10}{\sqrt{13} - 3}\right)} = 1;$$

hyperbola; $\theta = 5.989 = 343°$

**3.** $\begin{pmatrix} 4 & 2 \\ 2 & -1 \end{pmatrix}\begin{pmatrix} x \\ y \end{pmatrix} \cdot \begin{pmatrix} x \\ y \end{pmatrix} = 9;$

$$Q = \begin{pmatrix} \dfrac{5 + \sqrt{41}}{\sqrt{82 + 10\sqrt{41}}} & \dfrac{5 - \sqrt{41}}{\sqrt{82 - 10\sqrt{41}}} \\ \dfrac{4}{\sqrt{82 + 10\sqrt{41}}} & \dfrac{4}{\sqrt{82 - 10\sqrt{41}}} \end{pmatrix} = \begin{pmatrix} 0.9436 & -0.3310 \\ 0.3310 & 0.9436 \end{pmatrix};$$

$$\frac{x'^2}{\left(\dfrac{18}{\sqrt{41} + 3}\right)} - \frac{y'^2}{\left(\dfrac{18}{\sqrt{41} - 3}\right)} = 1;$$

hyperbola; $\theta \approx 0.3374 \approx 19.33°$

**5.** $\begin{pmatrix} 0 & \frac{1}{2} \\ \frac{1}{2} & 0 \end{pmatrix}\begin{pmatrix} x \\ y \end{pmatrix} \cdot \begin{pmatrix} x \\ y \end{pmatrix} = a > 0;$

$$Q = \begin{pmatrix} 1/\sqrt{2} & 1/\sqrt{2} \\ -1/\sqrt{2} & 1/\sqrt{2} \end{pmatrix};$$

$\dfrac{x'^2}{2a} - \dfrac{y'^2}{2a} = 1$; hyperbola; $\theta = 7\pi/4 = 315°$.

**7.** Same as Problem 5 except that now we have a hyperbola with the roles of $x'$ and $y'$ reversed; since $a < 0$, we have

$$\frac{y'^2}{(-2a)} - \frac{x'^2}{(-2a)} = 1$$

**9.** $\begin{pmatrix} -1 & 1 \\ 1 & -1 \end{pmatrix}\begin{pmatrix} x \\ y \end{pmatrix} \cdot \begin{pmatrix} x \\ y \end{pmatrix} = 0;$

$$Q = \begin{pmatrix} 1/\sqrt{2} & -1/\sqrt{2} \\ 1/\sqrt{2} & 1/\sqrt{2} \end{pmatrix};$$

$y'^2 = 0$, which is the equation of a straight line through the origin; $\theta = \pi/4 = 45°$.

**11.** $\begin{pmatrix} 3 & -3 \\ -3 & 5 \end{pmatrix}\begin{pmatrix} x \\ y \end{pmatrix} \cdot \begin{pmatrix} x \\ y \end{pmatrix} = 36;$

$$Q = \begin{pmatrix} \dfrac{1 + \sqrt{10}}{\sqrt{20 + 2\sqrt{10}}} & \dfrac{1 - \sqrt{10}}{\sqrt{20 - 2\sqrt{10}}} \\ \dfrac{3}{\sqrt{20 + 2\sqrt{10}}} & \dfrac{3}{\sqrt{20 - 2\sqrt{10}}} \end{pmatrix} = \begin{pmatrix} 0.8112 & -0.5847 \\ 0.5847 & 0.8112 \end{pmatrix};$$

$$\frac{x'^2}{\left(\dfrac{36}{4 - \sqrt{10}}\right)} + \frac{y'^2}{\left(\dfrac{36}{4 + \sqrt{10}}\right)} = 1;$$

ellipse; $\theta \approx 0.6245 \approx 35.78°$

**13.** $\begin{pmatrix} 6 & \frac{5}{2} \\ \frac{5}{2} & -6 \end{pmatrix}\begin{pmatrix} x \\ y \end{pmatrix} \cdot \begin{pmatrix} x \\ y \end{pmatrix} = -7;$

$$Q = \begin{pmatrix} 5/\sqrt{26} & -1/\sqrt{26} \\ 1/\sqrt{26} & 5/\sqrt{26} \end{pmatrix};$$

$\dfrac{y'^2}{(14/13)} - \dfrac{x'^2}{(14/13)} = 1$; hyperbola; $\theta \approx 0.197 \approx 11.31°$

**15.** $\begin{pmatrix} 1 & -1 & -1 \\ -1 & 1 & -1 \\ -1 & -1 & 1 \end{pmatrix}\begin{pmatrix} x \\ y \\ z \end{pmatrix} \cdot \begin{pmatrix} x \\ y \\ z \end{pmatrix}$;

$Q = \begin{pmatrix} 1/\sqrt{3} & 1/\sqrt{2} & 1/\sqrt{6} \\ 1/\sqrt{3} & -1/\sqrt{2} & 1/\sqrt{6} \\ 1/\sqrt{3} & 0 & -2/\sqrt{6} \end{pmatrix}$;

$-x'^2 + 2y'^2 + 2z'^2$

**17.** $\begin{pmatrix} 3 & 2 & 2 \\ 2 & 2 & 0 \\ 2 & 0 & 4 \end{pmatrix}\begin{pmatrix} x \\ y \\ z \end{pmatrix} \cdot \begin{pmatrix} x \\ y \\ z \end{pmatrix}$;

$Q = \begin{pmatrix} -\frac{2}{3} & \frac{1}{3} & \frac{2}{3} \\ \frac{2}{3} & \frac{2}{3} & \frac{1}{3} \\ \frac{1}{3} & -\frac{2}{3} & \frac{2}{3} \end{pmatrix}$;

$3y'^2 + 6z'^2$

**19.** $\begin{pmatrix} 1 & 1 & 2 & \frac{7}{2} \\ 1 & 1 & 3 & -1 \\ 2 & 3 & 3 & 0 \\ \frac{7}{2} & -1 & 0 & 1 \end{pmatrix}$

**21.** $\begin{pmatrix} 3 & -\frac{7}{2} & \frac{1}{2} & -1 & \frac{3}{2} \\ -\frac{7}{2} & -2 & -\frac{1}{2} & \frac{1}{2} & 0 \\ \frac{1}{2} & -\frac{1}{2} & 3 & -2 & -\frac{5}{2} \\ -1 & \frac{1}{2} & -2 & -6 & \frac{1}{2} \\ \frac{3}{2} & 0 & -\frac{5}{2} & \frac{1}{2} & -1 \end{pmatrix}$

**23.** $\begin{pmatrix} \cos\theta & -\sin\theta \\ \sin\theta & \cos\theta \end{pmatrix}\begin{pmatrix} x \\ y \end{pmatrix} =$

$\begin{pmatrix} x\cos\theta & -y\sin\theta \\ x\sin\theta & +y\cos\theta \end{pmatrix} = \begin{pmatrix} x' \\ y' \end{pmatrix}$

Then the quadratic equation $ax'^2 + bx'y' + cy'^2$ becomes $a(x\cos\theta - y\sin\theta)^2 + b(x\cos\theta - y\sin\theta) \times (x\sin\theta + y\cos\theta) + c(x\sin\theta + y\cos\theta)^2$; the cross-product term is $-2axy(\sin\theta\cos\theta + bxy \times [\cos^2\theta - \sin^2\theta] + 2cxy \times \sin\theta\cos\theta = xy[-a\sin 2\theta + b\cos 2\theta + c\sin 2\theta] = 0$ so that $(c-a)\sin 2\theta + b\cos 2\theta = 0$ and $\dfrac{a-c}{b} = \dfrac{\cos 2\theta}{\sin 2\theta} = \cot 2\theta.$

**25.** Suppose that $ax^2 + bxy + cy^2$ is converted to $a'x'^2 + b'x'y' + c'y'^2$ by a rotation. Let

$A = \begin{pmatrix} a & \frac{b}{2} \\ \frac{b}{2} & c \end{pmatrix}$ and $A' = \begin{pmatrix} a' & \frac{b'}{2} \\ \frac{b'}{2} & c' \end{pmatrix}$

There is an orthogonal matrix $Q$ such that $A = QA'Q^t$ and $Q$ is also a rotation matrix. Thus $\det A = \det Q \det A' \det Q^t = \det QQ^t \det A' = \det A'$ since $QQ^t = I$. But $\det A = ac - \dfrac{b^2}{4}$ and $\det A' = a'c' - \dfrac{b'^2}{4}$. Finally, since $A$ and $A'$ are similar, they have the same eigenvalues. But the sum of the eigenvalues of $A$ is $a + c$, while the sum of the eigenvalues of $A'$ is $a' + c'$. Thus $a + c = a' + c'$.

**27.** Let $\lambda_1, \lambda_2, \ldots, \lambda_n$ be the eigenvalues of $A$. Then, removing the cross product terms, we have $F(\mathbf{x}) = F'(\mathbf{x}') = \lambda_1 \mathbf{x}_1'^2 + \lambda_2 \mathbf{x}_2'^2 + \cdots + \lambda_n \mathbf{x}_2'^2$ where $\mathbf{x}' = Q^t\mathbf{x}$. If $\lambda_i \geq 0$ for $i = 1, 2, \ldots, n$, then $F'(\mathbf{x}') \geq 0$. If $F'(\mathbf{x}') \geq 0$, then $\lambda_i \geq 0$ since, if not, there is a $\lambda_j$ with $\lambda_j < 0$. Let $\mathbf{x}^*$ be the vector with 0's in every position except the $j$th and a 1 in the $j$th position. Then $F(\mathbf{x}^*) = \lambda_j < 0$, which is a contradiction.

**29.** negative definite

**31.** positive definite

**33.** indefinite

**35.** negative definite

**37. i.** If $\det A \neq 0$, then neither $\lambda_1$ nor $\lambda_2$ is zero. Thus, with $d = 0$, equation (22) becomes $\lambda_1 x'^2 + \lambda_2 y'^2 = 0$. If now both $\lambda_1$ and $\lambda_2$ are positive or negative, then the equation is satisfied only when $x' = 0$ and $y' = 0$. These are the equations of two straight lines. If $\lambda_1$ and $\lambda_2$ have opposite signs, then the equations become $x' = \pm\sqrt{\dfrac{\lambda_2}{-\lambda_1}}y'$, which are again the equations of two straight lines. If $\det A = 0$, then one of $\lambda_1$ or $\lambda_2$ is zero, and the equation becomes $x' = 0$ or $y' = 0$, each of which is the equation of a single straight line.

## MATLAB 6.5

**1.** For Problem 10, $A = \begin{pmatrix} 2 & .5 \\ .5 & 1 \end{pmatrix}$. The rotation angle is $\theta = 202.5$ and the equation is $2.2071x'^2 + .7929y'^2 = 4$. It is an ellipse and $\det(A) > 0$. The sketch follows.

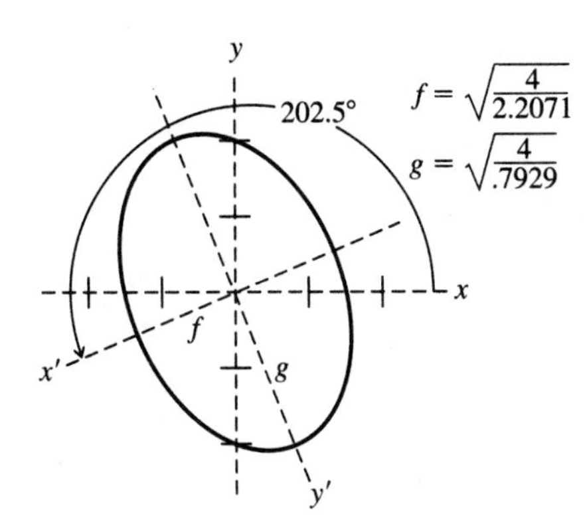


**3.** For Problem 4, $A = \begin{pmatrix} 0 & .5 \\ .5 & 0 \end{pmatrix}$. The rotation angle is $\theta = 315°$ and the equation is $-.5'^2 + .5y'^2 = 1$. It is a hyperbola and $\det(A) < 0$. The sketch follows.

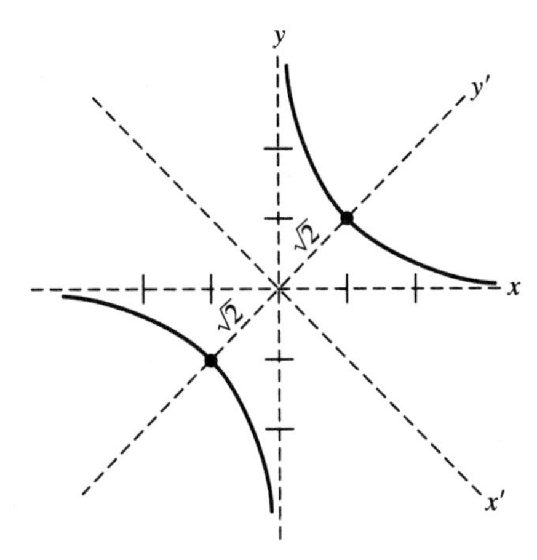


## Problems 6.6, page 603

**1.** no **3.** no **5.** yes **7.** no

**9.** yes **11.** no **13.** yes

**15.** $I$

**17.** $\begin{pmatrix} 1 & 0 \\ -1 & -\frac{1}{7} \end{pmatrix}$; $J = \begin{pmatrix} -3 & 1 \\ 0 & -3 \end{pmatrix}$

**19. a.** Let $\mathbf{x} \in \mathbb{C}^3$ be a fixed vector that is not an eigenvector. Since the geometric multiplicity of the eigenvalue $\lambda$ is one, $\mathbf{x}$ is not a multiple of $\mathbf{v}_1$, where $\mathbf{v}_1$ is an eigenvector. Then $\mathbf{x}$ and $\mathbf{v}_1$ are linearly independent. Let $\mathbf{w} = c_1\mathbf{v}_1 + c_2\mathbf{x}$. Assume that $\mathbf{w} = (A - \lambda I)\mathbf{x}$. Then $A\mathbf{x} - \lambda\mathbf{x} = c_1\mathbf{v}_1 + c_2\mathbf{x}$. Let $B = A - (c_2 + \lambda)I$ so that $B\mathbf{x} = c_1\mathbf{v}_1$. Assume that $c_2 \neq 0$. Then $\lambda + c_2$ is not an eigenvalue since $\lambda$ is the only eigenvalue. We have $\det B = \det[A - (\lambda + c_2)I] \neq 0$. Then $B^{-1}$ exists, $\mathbf{x} = c_1B^{-1}\mathbf{v}_1$, and $\lambda\mathbf{x} = c_1B^{-1}\lambda\mathbf{v}_1 = c_1B^{-1}A\mathbf{v}_1$. Since $A = B + (c_2 + \lambda)I$, $\lambda\mathbf{x} = c_1[B^{-1}B + (c_2 + \lambda)B^{-1}]\mathbf{v}_1 = c_1\mathbf{v}_1 + c_1(c_2 + \lambda)B^{-1}\mathbf{v}_1 = c_1\mathbf{v}_1 + (c_2 + \lambda)B^{-1}B\mathbf{x} = c_1\mathbf{v}_1 + c_2\mathbf{x} + \lambda\mathbf{x}$. Thus $c_1\mathbf{v}_1 + c_2\mathbf{x} = \mathbf{0}$ and $c_1 = c_2 = 0$ because $\mathbf{v}_1$ and $\mathbf{x}$ are linearly independent. This contradicts our previous assumption that $c_2 \neq 0$ and $\mathbf{w} = (A - \lambda I)\mathbf{x} = c_1\mathbf{v}_1$. Let $c_1\mathbf{x} = \mathbf{v}_2$; then $(A - \lambda I)\mathbf{v}_2 = \mathbf{v}_1$.

**b.** Let $\mathbf{y} \in \mathbb{C}^3$ with $\mathbf{y}$ not an eigenvector of $A$; $\mathbf{y}$ can be chosen linearly independent of $\mathbf{v}_2$ (it is already independent of $\mathbf{v}_1$) so that $\mathbf{z} = d_1\mathbf{v}_2 + d_2\mathbf{y}$ is not an eigenvector. Write $\mathbf{z}$ as $\mathbf{z} = (A - \lambda I)\mathbf{y}$. Then $A\mathbf{y} - \lambda\mathbf{y} = d_1\mathbf{v}_2 + d_2\mathbf{y}$. Let $D = A - (d_2 + \lambda)I$ so that $D\mathbf{y} = d_1\mathbf{v}_2$, since $A\mathbf{y} - (\lambda I)\mathbf{y} - d_2\mathbf{y} = d_1\mathbf{v}_2$. Assume that $d_2 \neq 0$; then $d_2 + \lambda$ is not an eigenvalue.

Clearly $\det D \neq 0$, $D^{-1}$ exists, $\mathbf{y} = d_1D^{-1}\mathbf{v}_2$, and $\lambda\mathbf{y} = d_1D^{-1}\lambda\mathbf{v}_2$. Then $\lambda\mathbf{v}_2 = \mathbf{v}_1 - A\mathbf{v}_2$; $\lambda\mathbf{y} = d_1D^{-1}(\mathbf{v}_1 - A\mathbf{v}_2) = d_1D^{-1}\mathbf{v}_1 - d_1D^{-1}A\mathbf{v}_2 . A = D + (d_2 + \lambda)I$. $\lambda\mathbf{y} = d_1D^{-1}\mathbf{v}_1 - d_1\mathbf{v}_2 - d_1d_2D^{-1}\mathbf{v}_2 - d_1D^{-1}\lambda\mathbf{v}_2 = d_1D^{-1}\mathbf{v}_1 - d_1D^{-1}A\mathbf{v}_2 - d_1\mathbf{v}_2 - d_1d_2D^{-1}\mathbf{v}_2 = d_1D^{-1}\lambda\mathbf{v}_2 - d_1(\mathbf{v}_2 + d_2D^{-1}\mathbf{v}_2) = \lambda\mathbf{y} - d_1(I + d_2D^{-1})\mathbf{v}_2$. So $\mathbf{0} = d_1(I + d_2D^{-1})\mathbf{v}_2$. $d_1 \neq 0$, otherwise $\lambda + d_2$ would be an eigenvalue and $\mathbf{y}$ an eigenvector. Then $(d_2DD^{-1} + D)\mathbf{v}_2 = D\mathbf{0} = \mathbf{0}$. $d_2\mathbf{v}_2 + [A - (d_2 + \lambda)I]\mathbf{v}_2 = \mathbf{0}$. $d_2\mathbf{v}_2 + (A - \lambda I)\mathbf{v}_2 - d_2\mathbf{v}_2 = \mathbf{0}$, or $(A - \lambda I)\mathbf{v}_2 = \mathbf{0}$, contrary to the result of part (a). Thus $d_2 = 0$ and $(A - I\lambda)\mathbf{y} = d_1\mathbf{v}_2$. Let $\mathbf{y} = d_1\mathbf{v}_3$; then $(A - I\lambda)\mathbf{v}_3 = \mathbf{v}_2$.

**c.** Let $C = (\mathbf{v}_1, \mathbf{v}_2, \mathbf{v}_3)$, where $\mathbf{v}_1$, $\mathbf{v}_2$, $\mathbf{v}_3$ are as above and linearly independent; then $C^{-1}$ exists. $AC = A(\mathbf{v}_1, \mathbf{v}_2, \mathbf{v}_3) = (A\mathbf{v}_1, A\mathbf{v}_2, A\mathbf{v}_3) = (\lambda\mathbf{v}_1, \mathbf{v}_1 + \lambda\mathbf{v}_2, \mathbf{v}_2 + \lambda\mathbf{v}_3)$;

$$CJ = (\mathbf{v}_1, \mathbf{v}_2, \mathbf{v}_3)\begin{pmatrix} \lambda & 1 & 0 \\ 0 & \lambda & 1 \\ 0 & 0 & \lambda \end{pmatrix} =$$

$(\lambda\mathbf{v}, \mathbf{v}_1 + \lambda\mathbf{v}_2, \mathbf{v}_2 + \lambda\mathbf{v}_3) = AC$; so $J = C^{-1}AC$.

**21.** $C = \begin{pmatrix} 1 & 1 & 0 \\ 0 & -1 & -2 \\ -1 & 0 & 3 \end{pmatrix}$; $J = \begin{pmatrix} 0 & 1 & 0 \\ 0 & 0 & 1 \\ 0 & 0 & 0 \end{pmatrix}$

**23.** If $m = k$, then $A^m = 0$ by definition of index of nilpotency. If $m > k$, then $A^m = A^{m-k}A^k = A^{m-k}0 = 0$.

**25.** $\begin{pmatrix} \lambda_1 & 0 & 0 & 0 \\ 0 & \lambda_2 & 0 & 0 \\ 0 & 0 & \lambda_3 & 0 \\ 0 & 0 & 0 & \lambda_4 \end{pmatrix}$ $\begin{pmatrix} \lambda_1 & 1 & 0 & 0 \\ 0 & \lambda_1 & 0 & 0 \\ 0 & 0 & \lambda_2 & 0 \\ 0 & 0 & 0 & \lambda_3 \end{pmatrix}$ $\begin{pmatrix} \lambda_1 & 1 & 0 & 0 \\ 0 & \lambda_1 & 1 & 0 \\ 0 & 0 & \lambda_1 & 0 \\ 0 & 0 & 0 & \lambda_2 \end{pmatrix}$ $\begin{pmatrix} \lambda_1 & 1 & 0 & 0 \\ 0 & \lambda_1 & 0 & 0 \\ 0 & 0 & \lambda_2 & 1 \\ 0 & 0 & 0 & \lambda_2 \end{pmatrix}$ $\begin{pmatrix} \lambda_1 & 1 & 0 & 0 \\ 0 & \lambda_1 & 1 & 0 \\ 0 & 0 & \lambda_1 & 1 \\ 0 & 0 & 0 & \lambda_1 \end{pmatrix}$

Here the $\lambda_i$'s are not necessarily distinct. Also, the blocks may be permuted on the diagonal.

**27.** $\begin{pmatrix} 3 & 0 & 0 & 0 \\ 0 & 3 & 0 & 0 \\ 0 & 0 & 3 & 0 \\ 0 & 0 & 0 & -4 \end{pmatrix}$ $\begin{pmatrix} 3 & 1 & 0 & 0 \\ 0 & 3 & 0 & 0 \\ 0 & 0 & 3 & 0 \\ 0 & 0 & 0 & -4 \end{pmatrix}$ $\begin{pmatrix} 3 & 1 & 0 & 0 \\ 0 & 3 & 1 & 0 \\ 0 & 0 & 3 & 0 \\ 0 & 0 & 0 & -4 \end{pmatrix}$

The Jordan blocks may be permuted along the diagonal.

**29.** $\begin{pmatrix} 4 & 0 & 0 & 0 & 0 \\ 0 & 4 & 0 & 0 & 0 \\ 0 & 0 & 4 & 0 & 0 \\ 0 & 0 & 0 & -3 & 0 \\ 0 & 0 & 0 & 0 & -3 \end{pmatrix}$ $\begin{pmatrix} 4 & 1 & 0 & 0 & 0 \\ 0 & 4 & 0 & 0 & 0 \\ 0 & 0 & 4 & 0 & 0 \\ 0 & 0 & 0 & -3 & 0 \\ 0 & 0 & 0 & 0 & -3 \end{pmatrix}$

$$\begin{pmatrix} 4 & 1 & 0 & 0 & 0 \\ 0 & 4 & 1 & 0 & 0 \\ 0 & 0 & 4 & 0 & 0 \\ 0 & 0 & 0 & -3 & 0 \\ 0 & 0 & 0 & 0 & -3 \end{pmatrix}$$

$$\begin{pmatrix} 4 & 1 & 0 & 0 & 0 \\ 0 & 4 & 1 & 0 & 0 \\ 0 & 0 & 4 & 0 & 0 \\ 0 & 0 & 0 & -3 & 1 \\ 0 & 0 & 0 & 0 & -3 \end{pmatrix}$$

$$\begin{pmatrix} 4 & 1 & 0 & 0 & 0 \\ 0 & 4 & 0 & 0 & 0 \\ 0 & 0 & 4 & 0 & 0 \\ 0 & 0 & 0 & -3 & 1 \\ 0 & 0 & 0 & 0 & -3 \end{pmatrix}$$

$$\begin{pmatrix} 4 & 0 & 0 & 0 & 0 \\ 0 & 4 & 0 & 0 & 0 \\ 0 & 0 & 4 & 0 & 0 \\ 0 & 0 & 0 & -3 & 1 \\ 0 & 0 & 0 & 0 & -3 \end{pmatrix}$$

The Jordan blocks may be permuted along the diagonal.

**31.**

$$\begin{pmatrix} -7 & 0 & 0 & 0 & 0 \\ 0 & -7 & 0 & 0 & 0 \\ 0 & 0 & -7 & 0 & 0 \\ 0 & 0 & 0 & -7 & 0 \\ 0 & 0 & 0 & 0 & -7 \end{pmatrix}$$

$$\begin{pmatrix} -7 & 1 & 0 & 0 & 0 \\ 0 & -7 & 0 & 0 & 0 \\ 0 & 0 & -7 & 0 & 0 \\ 0 & 0 & 0 & -7 & 0 \\ 0 & 0 & 0 & 0 & -7 \end{pmatrix}$$

$$\begin{pmatrix} -7 & 1 & 0 & 0 & 0 \\ 0 & -7 & 1 & 0 & 0 \\ 0 & 0 & -7 & 0 & 0 \\ 0 & 0 & 0 & -7 & 0 \\ 0 & 0 & 0 & 0 & -7 \end{pmatrix}$$

$$\begin{pmatrix} -7 & 1 & 0 & 0 & 0 \\ 0 & -7 & 1 & 0 & 0 \\ 0 & 0 & -7 & 1 & 0 \\ 0 & 0 & 0 & -7 & 0 \\ 0 & 0 & 0 & 0 & -7 \end{pmatrix}$$

$$\begin{pmatrix} -7 & 1 & 0 & 0 & 0 \\ 0 & -7 & 1 & 0 & 0 \\ 0 & 0 & -7 & 1 & 0 \\ 0 & 0 & 0 & -7 & 1 \\ 0 & 0 & 0 & 0 & -7 \end{pmatrix}$$

$$\begin{pmatrix} -7 & 1 & 0 & 0 & 0 \\ 0 & -7 & 0 & 0 & 0 \\ 0 & 0 & -7 & 1 & 0 \\ 0 & 0 & 0 & -7 & 0 \\ 0 & 0 & 0 & 0 & -7 \end{pmatrix}$$

$$\begin{pmatrix} -7 & 1 & 0 & 0 & 0 \\ 0 & -7 & 1 & 0 & 0 \\ 0 & 0 & -7 & 0 & 0 \\ 0 & 0 & 0 & -7 & 1 \\ 0 & 0 & 0 & 0 & -7 \end{pmatrix}$$

The Jordan blocks may be permuted along the diagonal.

### MATLAB 6.6

**1. a.** Show that $(A - 2I)(\text{col } 1) = 0$, $(A - 2I)(\text{col } 2) = 0$, and $(A - 3I)(\text{col } 3) = 0$. For part (iv), use the properties of similarity to conclude that $A$ and $J$ have the same eigenvalues with corresponding algebraic and geometric multiplicity. The algebraic and geometric multiplicities for eigenvalues of $J$ are easy to determine.

**c.** For $\lambda = 2$, column 1 is an eigenvector and column 2 is a generalized eigenvector; it has algebraic multiplicity 2 and geometric multiplicity 1. For $\lambda = 3$, column 3 is an eigenvector and column 4 is a generalized eigenvector; it has algebraic multiplicity 2 and geometric multiplicity 1.

### Problems 6.7, page 616

**1.** $\frac{1}{7}\begin{pmatrix} 5e^{-4t} + 2e^{3t} & 2e^{-4t} - 2e^{3t} \\ 5e^{-4t} - 5e^{3t} & 2e^{-4t} + 5e^{3t} \end{pmatrix}$

**3.** $\begin{pmatrix} 2\sin t + \cos t & -\sin t \\ 5\sin t & -2\sin t + \cos t \end{pmatrix}$

**5.** $e^{-3t}\begin{pmatrix} 1-7t & -7t \\ 7t & 1+7t \end{pmatrix}$

**7.** $e^{-5t}\begin{pmatrix} 1-7t & 7t \\ -7t & 1+7t \end{pmatrix}$

**9.** $\begin{pmatrix} 4e^t - 3e^{2t} + 6te^{2t} & -12e^t + 12e^{2t} - 6te^{2t} & 6te^{2t} \\ e^t - e^{2t} + 2te^{2t} & -3e^t + 4e^{2t} - 2te^{2t} & 2te^{2t} \\ -3e^t + 3e^{2t} - 4te^{2t} & 9e^t - 9e^{2t} + 4te^{2t} & -4te^{2t} + e^{2t} \end{pmatrix}$

**11.** $\mathbf{x}(t) = -\frac{1}{3}\begin{pmatrix} -e^t - 2e^{4t} & -e^t + e^{4t} \\ -2e^t + 2e^{4t} & -2e^t - e^{4t} \end{pmatrix} \times \begin{pmatrix} x_1(0) \\ x_2(0) \end{pmatrix}$, which leads to

$$x_1(t) = \tfrac{1}{3}[(x_1(0) + x_2(0)e^t + (2x_1(0) - x_2(0))e^{4t}] = \tfrac{1}{3}[(x_1(0) + x_2(0)) + (2x_1(0) - x_2(0))e^{3t}]e^t$$

If $2x_1(0) < x_2(0)$, then the first population will be extinct when $x_1(0) + x_2(0) = [x_2(0) - 2x_1(0)]e^{3t}$, or

$$t = \frac{1}{3}\ln\left(\frac{x_1(0) + x_2(0)}{x_2(0) - 2x_1(0)}\right)$$

**13.** $\begin{pmatrix} x_1 \\ x_2 \end{pmatrix}' = \frac{1}{1000}\begin{pmatrix} -30 & 10 \\ 30 & -30 \end{pmatrix}\begin{pmatrix} x_1 \\ x_2 \end{pmatrix}$; $\begin{pmatrix} x_1 \\ x_2 \end{pmatrix} = \begin{pmatrix} 500\,(e^{\alpha t} + e^{\beta t}) \\ 500\sqrt{3}\,(e^{\alpha t} - e^{\beta t}) \end{pmatrix}$ where $\alpha = -0.03 + \sqrt{0.0003} \approx -0.0127$ and $\beta = -0.03 - \sqrt{0.0003} \approx -0.0473$.

**15. a.** $\begin{pmatrix} x_1' \\ x_2' \end{pmatrix} = \begin{pmatrix} 0 & 1 \\ -b & -a \end{pmatrix}\begin{pmatrix} x_1 \\ x_2 \end{pmatrix}$

**b.** $\det\begin{pmatrix} -\lambda & 1 \\ -b & -a-\lambda \end{pmatrix} = \lambda^2 + a\lambda + b$ so that $p(\lambda) = \lambda^2 + a\lambda + b = 0$.

**17.** $(1 + 5t)e^{-3t}$ **19.** $\frac{8}{7}e^{5t} + \frac{13}{7}e^{-2t}$

**21.** By Problem 20, $N_3^k = 0$ for $k \geq 3$. Thus $e^{N_3t} = I + N_3t + N_3^2\frac{t^2}{2} =$

$$\begin{pmatrix} 1 & 0 & 0 \\ 0 & 1 & 0 \\ 0 & 0 & 1 \end{pmatrix} + \begin{pmatrix} 0 & t & 0 \\ 0 & 0 & t \\ 0 & 0 & 0 \end{pmatrix} + \begin{pmatrix} 0 & 0 & t^2/2 \\ 0 & 0 & 0 \\ 0 & 0 & 0 \end{pmatrix} = \begin{pmatrix} 1 & t & t^2/2 \\ 0 & 1 & t \\ 0 & 0 & 1 \end{pmatrix}.$$

**23.** From 6.6.20, $C = \begin{pmatrix} 1 & 1 & 0 \\ 1 & 2 & 1 \\ 0 & 1 & 0 \end{pmatrix}$ and $J = \begin{pmatrix} -1 & 1 & 0 \\ 0 & -1 & 1 \\ 0 & 0 & -1 \end{pmatrix}$ so that

$$e^{At} = Ce^{Jt}C^{-1} = \begin{pmatrix} 1 & 1 & 0 \\ 1 & 2 & 1 \\ 0 & 1 & 0 \end{pmatrix} e^{-t}\begin{pmatrix} 1 & t & t^2/2 \\ 0 & 1 & t \\ 0 & 0 & 1 \end{pmatrix}\begin{pmatrix} 1 & 0 & -1 \\ 0 & 0 & 1 \\ -1 & 1 & -1 \end{pmatrix}$$

$$= e^{-t}\begin{pmatrix} 1 - t - t^2/2 & t + t^2/2 & -t^2/2 \\ -2t - t^2/2 & 1 + 2t + t^2/2 & -t - t^2/2 \\ -t & t & 1-t \end{pmatrix}$$

**25.** $e^{\lambda t}\begin{pmatrix} 1 & t & t^2/2 & t^3/6 \\ 0 & 1 & t & t^2/2 \\ 0 & 0 & 1 & t \\ 0 & 0 & 0 & 1 \end{pmatrix}$

**27.** $\begin{pmatrix} e^{-4t} & te^{-4t} & (t^2/2)e^{-4t} & 0 \\ 0 & e^{-4t} & te^{-4t} & 0 \\ 0 & 0 & e^{-4t} & 0 \\ 0 & 0 & 0 & e^{3t} \end{pmatrix}$

## Problems 6.8, page 626

**1. a.** $p(\lambda) = \lambda^2 + \lambda - 12 = 0$;

**b.** $p(A) = A^2 + A - 12I = \begin{pmatrix} 14 & 2 \\ 5 & 11 \end{pmatrix} + \begin{pmatrix} -2 & -2 \\ -5 & 1 \end{pmatrix} + \begin{pmatrix} -12 & 0 \\ 0 & -12 \end{pmatrix} = \begin{pmatrix} 0 & 0 \\ 0 & 0 \end{pmatrix}$

**c.** $A^{-1} = \frac{1}{12}\begin{pmatrix} -1 & -2 \\ -5 & 2 \end{pmatrix}$

**3. a.** $p(\lambda) = -\lambda^3 + 4\lambda^2 - 3\lambda$;

**b.** $p(A) = -A^3 + 4A^2 - 3A$

$$= -\begin{pmatrix} 5 & -9 & 4 \\ -9 & 18 & -9 \\ 4 & -9 & 5 \end{pmatrix} + \begin{pmatrix} 8 & -12 & 4 \\ -12 & 24 & -12 \\ 4 & -12 & 8 \end{pmatrix} - \begin{pmatrix} 3 & -3 & 0 \\ -3 & 6 & -3 \\ 0 & -3 & 3 \end{pmatrix} = \begin{pmatrix} 0 & 0 & 0 \\ 0 & 0 & 0 \\ 0 & 0 & 0 \end{pmatrix}$$

**c.** $A^{-1}$ does not exist.

**5. a.** $p(\lambda) = -\lambda^3 + 3\lambda^2 - 3\lambda + 1 = 0$

**b.** $p(A) = -A^3 + 3A^2 - 3A + I$

$$= -\begin{pmatrix} 1 & -3 & 3 \\ 3 & -8 & 6 \\ 6 & -15 & 10 \end{pmatrix} + \begin{pmatrix} 0 & 0 & 3 \\ 3 & -9 & 9 \\ 9 & -24 & 18 \end{pmatrix} - \begin{pmatrix} 0 & 3 & 0 \\ 0 & 0 & 3 \\ 3 & -9 & 9 \end{pmatrix} + \begin{pmatrix} 1 & 0 & 0 \\ 0 & 1 & 0 \\ 0 & 0 & 1 \end{pmatrix} = \begin{pmatrix} 0 & 0 & 0 \\ 0 & 0 & 0 \\ 0 & 0 & 0 \end{pmatrix}$$

**c.** $A^{-1} = \begin{pmatrix} 3 & -3 & 1 \\ 1 & 0 & 0 \\ 0 & 1 & 0 \end{pmatrix}$

**7. a.** $p(\lambda) = -\lambda^3 + 6\lambda^2 + 18\lambda + 9 = 0$

**b.** $p(A) = -A^3 + 6A^2 + 18A + 9I$

$$= -\begin{pmatrix} 63 & 54 & 108 \\ 180 & 189 & 324 \\ 168 & 204 & 315 \end{pmatrix} + \begin{pmatrix} 18 & 72 & 54 \\ 108 & 162 & 216 \\ 150 & 114 & 252 \end{pmatrix} + \begin{pmatrix} 36 & -18 & 54 \\ 72 & -18 & 108 \\ 18 & 90 & 54 \end{pmatrix} + \begin{pmatrix} 9 & 0 & 0 \\ 0 & 9 & 0 \\ 0 & 0 & 9 \end{pmatrix} = \begin{pmatrix} 0 & 0 & 0 \\ 0 & 0 & 0 \\ 0 & 0 & 0 \end{pmatrix}$$

**c.** $A^{-1} = \frac{1}{9}\begin{pmatrix} -27 & 18 & -9 \\ -6 & 3 & 0 \\ 19 & -11 & 6 \end{pmatrix}$

**9. a.** $p(\lambda) = (a - \lambda)^4$

**b.** $p(A) = (aI - A)^4$

$$= \begin{pmatrix} 0 & -b & 0 & 0 \\ 0 & 0 & -c & 0 \\ 0 & 0 & 0 & -d \\ 0 & 0 & 0 & 0 \end{pmatrix}^4 = \begin{pmatrix} 0 & 0 & 0 & 0 \\ 0 & 0 & 0 & 0 \\ 0 & 0 & 0 & 0 \\ 0 & 0 & 0 & 0 \end{pmatrix}$$

**c.**

$$A^{-1} = \begin{pmatrix} 1/a & -b/a^2 & cb/a^3 & -bcd/a^4 \\ 0 & 1/a & -c/a^2 & cd/a^3 \\ 0 & 0 & 1/a & -d/a^2 \\ 0 & 0 & 0 & 1/a \end{pmatrix}$$

**11.** $|\lambda| \le 7$ and $\text{Re}\,\lambda \ge \frac{13}{6}$

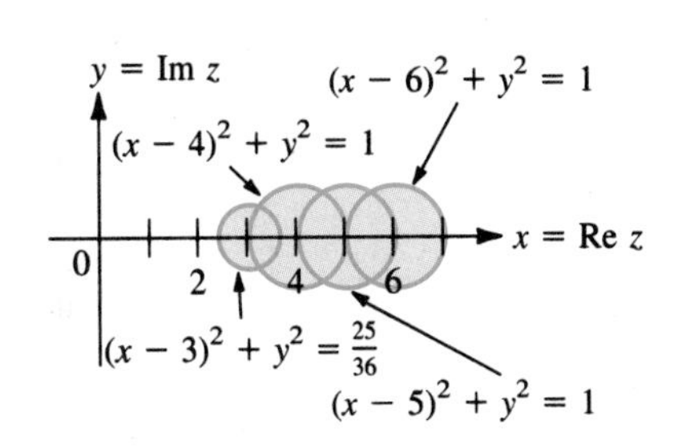

**13.** $|\lambda| \leq 10.5$ and $-10.5 \leq \text{Re } \lambda \leq 5.75$

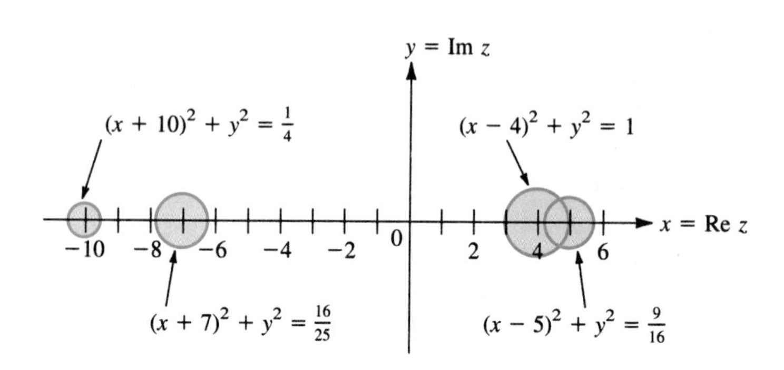


**15.** Since $A$ is symmetric, the eigenvalues of $A$ are real. Then, by Gershgorin's theorem, $\lambda = \text{Re } \lambda \geq 4 - (2 + 1 + \frac{1}{4}) = \frac{3}{4}$.

**17. a.** $F(\lambda) = B_0C_0 + B_0C_1\lambda + B_1C_0\lambda + B_1C_1\lambda^2$

**b.** $P(A)Q(A) = (B_0 + B_1A) \times (C_0 + C_1A) = B_0C_0 + B_0C_1A + B_1AC_0 + B_1AC_1A$
$F(A) = B_0C_0 + B_0C_1A + B_1C_0A + B_1C_1A^2$
$F(A) = P(A)Q(A)$ if and only if $C_0A = AC_0$ (in the third term) and $AC_1A = C_1A^2$ in the fourth term.

**19.** $\det A = \lambda_1\lambda_2 \cdots \lambda_n$. If $\det A = 0$, then $\lambda_i = 0$ for some $i$. But $|\lambda_i - a_{ii}| \leq r_i$ so that $|0 - a_{ii}| = |a_{ii}| \leq r_i$, which is impossible since $A$ is strictly diagonally dominant. Thus $\lambda_i \neq 0$ for $i = 1, 2, \ldots, n$ and $\det A \neq 0$.

### MATLAB 6.8

**1.** For Problem 1, $A^{-1} = \frac{1}{12}(I + A)$. For Problem 13, $A^{-1} = -I - A - A^2$.

**3.** You should observe that the white *'s are within the union of the circles.

### Chapter 6—Review, page 633

**1.** $4, -2$;

$$E_4 = \text{span}\left\{\begin{pmatrix}1\\1\end{pmatrix}\right\};$$

$$E_{-2} = \text{span}\left\{\begin{pmatrix}2\\1\end{pmatrix}\right\}$$

**3.** $1, 7, -5$;

$$E_1 = \text{span}\left\{\begin{pmatrix}-6\\3\\4\end{pmatrix}\right\};$$

$$E_7 = \text{span}\left\{\begin{pmatrix}0\\3\\1\end{pmatrix}\right\};$$

$$E_{-5} = \text{span}\left\{\begin{pmatrix}0\\0\\1\end{pmatrix}\right\}$$

**5.** $1, 3, 3 + \sqrt{2}i, 3 - \sqrt{2}i$;

$$E_1 = \text{span}\left\{\begin{pmatrix}1\\2\\0\\0\end{pmatrix}\right\};$$

$$E_3 = \text{span}\left\{\begin{pmatrix}1\\1\\0\\0\end{pmatrix}\right\};$$

$$E_{3+\sqrt{2}i} = \text{span}\left\{\begin{pmatrix}0\\0\\-1\\\sqrt{2}i\end{pmatrix}\right\};$$

$$E_{3-\sqrt{2}i} = \text{span}\left\{\begin{pmatrix}0\\0\\1\\\sqrt{2}i\end{pmatrix}\right\}$$

**7.** $C = \begin{pmatrix}-3 & 1\\4 & -1\end{pmatrix}$; $C^{-1}AC = \begin{pmatrix}2 & 0\\0 & -3\end{pmatrix}$

**9.** $C = \begin{pmatrix}0 & -1-i & -1+i\\1 & 1 & 1\\-1 & 1 & 1\end{pmatrix}$;

$$C^{-1}AC = \begin{pmatrix} -1 & 0 & 0 \\ 0 & i & 0 \\ 0 & 0 & -i \end{pmatrix}$$

**11.** not diagonalizable

**13.** $Q = \begin{pmatrix} 1/\sqrt{2} & 0 & 1/\sqrt{2} \\ 1/\sqrt{2} & 0 & -1/\sqrt{2} \\ 0 & 1 & 0 \end{pmatrix}$;

$$Q^tAQ = \begin{pmatrix} 4 & 0 & 0 \\ 0 & -3 & 0 \\ 0 & 0 & 0 \end{pmatrix}$$

**15.** $C = \begin{pmatrix} 1 & 1 & 1 & -1 \\ -1 & 0 & 0 & 0 \\ 0 & 1 & 0 & 0 \\ -1 & -1 & -1 & 2 \end{pmatrix}$;

$$C^{-1}AC = \begin{pmatrix} -1 & 0 & 0 & 0 \\ 0 & -1 & 0 & 0 \\ 0 & 0 & 3 & 0 \\ 0 & 0 & 0 & 3 \end{pmatrix}$$

**17.** $\dfrac{x'^2}{8/(3+\sqrt{2})} + \dfrac{y'^2}{8/(3-\sqrt{2})} = 1$: ellipse

**19.** $\dfrac{y'^2}{10/(\sqrt{13}+3)} - \dfrac{x'^2}{10/(\sqrt{13}-3)} = 1$: hyperbola

**21.** $4x'^2 - 3y'^2$

**23.** $C = \begin{pmatrix} 2 & -1 \\ 1 & 0 \end{pmatrix}$;

$$C^{-1}AC = \begin{pmatrix} -2 & 1 \\ 0 & -2 \end{pmatrix}$$

**25.** $\begin{pmatrix} -e^t + 2e^{-t} & 2e^t - 2e^{-t} \\ -e^t + e^{-t} & 2e^t - e^{-t} \end{pmatrix}$

**27.** $e^{-t}\begin{pmatrix} \cos 2t - \sin 2t & -2\sin 2t \\ \sin 2t & \cos 2t + \sin 2t \end{pmatrix}$

**29.** $|\lambda| \le 5$ and $-5 \le \text{Re}\,\lambda \le \frac{14}{3}$

## Appendix 1, page A-6

**1.** First, is it true for $n = 1$? $2 = 1(1 + 1)$, so it is. Now assume it is true for $n = k$. Then $2 + 4 + 6 + \cdots + 2k = k(k + 1)$. We must now show that it is true for $n = k + 1$; that is, we must show that $2 + 4 + 6 + \cdots + 2k + 2(k + 1) = (k + 1)[(k + 1) + 1]$. We know that $2 + 4 + 6 + \cdots + 2k + 2(k + 1) = k(k + 1) + 2(k + 1)$ (induction hypothesis) $= (k + 2)(k + 1) = (k + 1)[(k + 1) + 1]$

**3.** First, is it true for $n = 1$? $2 = \dfrac{1(3 \cdot 1 + 1)}{2}$, so it is. Now assume it is true for $n = k$. Then $2 + 5 + 8 + \cdots + (3k - 1) = \dfrac{k(3k + 1)}{2}$. We must now show it is true for $n = k + 1$; that is, we must show that

$$2 + 5 + 8 + \cdots + (3k - 1) + (3k + 2) = \frac{(k + 1)[3(k + 1) + 1]}{2} = \frac{(k + 1)(3k + 4)}{2} = \frac{3k^2 + 7k + 4}{2}$$

Add $3k + 2$ to both sides of the equation in the induction hypothesis and get

$$2 + 5 + 8 + \cdots + (3k - 1) + (3k + 2) = \frac{k(3k + 1)}{2} + (3k + 2) = \frac{3k^2 + k}{2} + \frac{6k + 4}{2} = \frac{3k^2 + 7k + 4}{2}$$

**5.** Is it true for $n = 1$? Yes, $\left(\frac{1}{2}\right)^1 = \frac{1}{2} < \frac{1}{1} = 1$. Now assume it is true for $n = k$. That is, $\left(\frac{1}{2}\right)^k < \frac{1}{k}$. Then

$\left(\frac{1}{2}\right)^{k+1} = \frac{1}{2}\left(\frac{1}{2}\right)^k < \frac{1}{2}\left(\frac{1}{k}\right) = \frac{1}{2k} < \frac{1}{k+1}$, since $2k > k + 1$ if $k > 1$.

**7.** Is it true for $n = 1$? Yes, since $1 + 2 = 2^2 - 1$. Now assume it is true for $n = k$; that is, $1 + 2 + 4 + \cdots + 2^k = 2^{k+1} - 1$. We must now prove that it is true for $n = k + 1$, or that $1 + 2 + 4 + \cdots + 2^k + 2^{k+1} = 2^{k+2} - 1$. Add $2^{k+1}$ to both sides of the induction hypothesis and obtain

$$\begin{aligned} 1 + 2 + 4 + \cdots + 2^k + 2^{k+1} &= 2^{k+1} - 1 + 2^{k+1} \\ &= 2 \cdot 2^{k+1} - 1 \\ &= 2^{k+2} - 1 \end{aligned}$$

**9.** Is it true for $n = 1$? Yes, since $1 + \frac{1}{2} = 2 - \frac{1}{2^1}$. Now assume it is true for $n = k$; that is,

$$1 + \frac{1}{2} + \frac{1}{4} + \cdots + \frac{1}{2^k} = 2 - \frac{1}{2^k}$$

We must now prove for $n = k + 1$; that is,

$$1 + \frac{1}{2} + \frac{1}{4} + \cdots + \frac{1}{2^k} + \frac{1}{2^{k+1}} = 2 - \frac{1}{2^{k+1}}$$

Add $\frac{1}{2^{k+1}}$ to both sides of the induction hypothesis and obtain

$$\begin{aligned} 1 + \frac{1}{2} + \frac{1}{4} + \cdots + \frac{1}{2^k} + \frac{1}{2^{k+1}} &= 2 - \frac{1}{2^k} + \frac{1}{2^{k+1}} \\ &= 2 - \frac{2}{2^{k+1}} + \frac{1}{2^{k+1}} \\ &= 2 - \frac{1}{2^{k+1}} \end{aligned}$$

**11.** Is it true for $n = 1$? Yes, $1^3 = \frac{1^2(1 + 1)^2}{4}$. Now assume it is true for $n = k$; that is,

$$1^3 + 2^3 + 3^3 + \cdots + k^3 = \frac{k^2(k + 1)^2}{4}$$

We must prove that it is true for $n = k + 1$; that is,

$$\begin{aligned} 1^3 + 2^3 + 3^3 + \cdots + k^3 + (k + 1)^3 &= \frac{(k + 1)^2[(k + 1) + 1]^2}{4} \\ &= \frac{k^4 + 6k^3 + 13k^2 + 12k + 4}{4} \end{aligned}$$

Add $(k + 1)^3$ to both sides of the induction hypothesis.

$$\begin{aligned} 1^3 + 2^3 + \cdots + k^3 + (k + 1)^3 &= \frac{k^2(k + 1)^2}{4} + (k + 1)^3 \\ &= \frac{k^4 + 2k^3 + k^2}{4} + k^3 + 3k^2 + 3k + 1 \\ &= \frac{k^4 + 2k^3 + k^2}{4} + \frac{4k^3 + 12k^2 + 12k + 4}{4} \\ &= \frac{k^4 + 6k^3 + 13k^2 + 12k + 4}{4} \end{aligned}$$

**13.** Is it true for $n = 1$? Yes, $1 \cdot 2 = \frac{1 \cdot (1 + 1) \cdot (4 - 1)}{3}$. Assume it is true for $n = k$; that is,

$$1 \cdot 2 + 3 \cdot 4 + \cdots + (2k - 1)(2k) = \frac{k(k + 1)(4k - 1)}{3}$$

Now prove for $n = k + 1$; that is,

$$\begin{aligned} 1 \cdot 2 + 3 \cdot 4 + (2k - 1)(2k) + (2k + 1)(2k + 2) &= \frac{(k + 1)(k + 2)(4k + 3)}{3} \\ &= \frac{4k^3 + 15k^2 + 17k + 6}{3} \end{aligned}$$

Add $[2(k+1)-1][2(k+1] = (2k+1)(2k+2)$ to both sides of the induction hypothesis. Obtain

$$1 \cdot 2 + 3 \cdot 4 + \cdots + (2k-1)(2k) + (2k+1)(2k+2)$$
$$= \frac{k(k+1)(4k-1)}{3} + (2k+1)(2k+2)$$
$$= \frac{4k^3+3k^2-k}{3} + 4k^2+6k+2$$
$$= \frac{4k^3+3k^2-k}{3} + \frac{12k^2+18k+6}{3}$$
$$= \frac{4k^3+15k^2+17k+6}{3}$$

Several of the following use the fact that if an integer $m$ divides evenly into an integer $a$ and divides evenly into an integer $b$, then $m$ divides evenly into $a + b$.

**15.** Is it true for $n = 1$? Yes, since $1^2 + 1 = 2$ is even. Assume that $k^2 + k$ is even. Now prove true for $k + 1$; that is, we now must prove that $(k+1)^2 + (k+1)$ is even. But

$$(k+1)^2 + (k+1) = k^2 + 2k + 1 + k + 1$$
$$= (k^2+k) + (2k+2)$$

Now 2 divides evenly into $k^2 + k$ by the induction hypothesis. It is evident that 2 divides evenly into $2k$ and that 2 divides evenly into 2. Therefore 2 divides evenly into $k^2 + k + 2k + 2$, meaning that the number is even.

**17.** Is it true for $n = 1$? Yes, because $1(1^2 + 5) = 6$ is divisible by 6. Now assume it is true for $k$; that is, $k(k^2 + 5)$ is divisible by 6. We now must prove that $(k+1)[(k+1)^2 + 5]$ is divisible by 6.

$$(k+1)[(k+1)^2+5]$$
$$= (k+1)(k^2+2k+6)$$
$$= (k+1)(k^2+5+2k+1)$$
$$= k(k^2+5) + (k^2+5) + k(2k+1) + (2k+1)$$
$$= k(k^2+5) + 3(k^2+k) + 6$$

Now $k(k^2 + 5)$ is divisible by 6 by the induction hypothesis; $3(k^2 + k)$ is clearly divisible by 3 and is even by Problem 15, so it is divisible by 6; and certainly 6 is divisible by 6, so the expression given is divisible by 6.

**19.** The problem is true if $n = 1$ since $x^1 - 1$ is divisible by $x - 1$. Now assume that $x^k - 1$ is divisible by $x - 1$. We have to prove that $x^{k+1} - 1$ is divisible by $x - 1$. Now

$$x^{k+1} - 1 = x^k x - 1 = x^k x - x + x - 1$$
$$= x(x^k - 1) + (x - 1).$$

The first term is divisible by $x - 1$ by the induction hypothesis, and the second term in the sum is divisible by $x - 1$, so the expression is divisible by $x - 1$.

**21.** If $n = 1$, $(ab)^1 = a^1b^1 = ab$, so it is true. Now assume that $n = k$; that is, $(ab)^k = a^kb^k$. We must prove for $k + 1$; that is, $(ab)^{k+1} = a^{k+1}b^{k+1}$. Now

$$(ab)^{k+1} = (ab)^k(ab) = a^kb^kab$$
$$= a^kab^kb \text{ (since multiplication is commutative)}$$
$$= a^{k+1}b^{k+1}$$

**23.** From Theorem 2.2.1, $\det A_1A_2 = \det A_1 \det A_2$, so the result holds for $n = 2$. Assume that it holds for $n = k$. Then

$$\det A_1A_2 \cdots A_kA_{k+1}$$
$$= \det A_1A_2 \cdots A_k \det A_{k+1} \text{ (using the result for } n = 2)$$
$$= (\det A_1 \det A_2 \cdots \det A_k) \det A_{k+1} \text{ (using the result for } n = k)$$
$$= \det A_1 \det A_2 \cdots \det A_k \det A_{k+1},$$

which is the result for $n = k + 1$.

**25.** $n = 1$; there are exactly two subsets of a set with one element: the set itself and the empty set. Now assume there are exactly $2^k$ subsets of a set with $k$ elements. Now consider a set $A$ with $k + 1$ elements. Remove one, call it $a_{k+1}$. The remaining elements form a set of $k$ elements. This set has $2^k$ subsets. Add $a_{k+1}$ to each of these $2^k$ subsets to obtain another $2^k$ subsets. In other words, $A$ has $2^k$ subsets containing the element $a_{k+1}$ and $2^k$ subsets not containing $a_{k+1}$, for a total of $2^k + 2^k = 2^{k+1}$ subsets.

**27.** It is not true for $n = 2$. In that case $S_1$ and $S_2$ are disjoint, and therefore you cannot say that $h_1 = h_2$.

### Appendix 2, page A-17

**1.** $9 - 7i$ **3.** 2 **5.** $-27 + 5i$

**7.** $5\sqrt{2}e^{i(\pi/4)}$

**9.** $3\sqrt{2}e^{i(7\pi/4)} = 3\sqrt{2}e^{-i(\pi/4)}$

**11.** $6e^{(\pi/6)i}$

**13.** $8e^{i(11\pi/6)} = 8e^{-i(\pi/6)}$

**15.** $2e^{i(4\pi/3)} = 2e^{-i(2\pi/3)}$

**17.** $-2$ **19.** $-\sqrt{2}/4 - i(\sqrt{2}/4)$

**21.** $-2\sqrt{3} + 2i$

**23.** $-\frac{3}{2} - \frac{3}{2}\sqrt{3}i$

**25.** $\cos 1 + i \sin 1 \approx 0.5403 + 0.8415i$

**27.** $4 - 6i$ **29.** $7i$

**31.** $2e^{-i(\pi/7)}$ **33.** $3e^{i(4\pi/11)}$

**35.** If $z = \bar{z}$, then $\alpha + i\beta = \alpha - i\beta$, or $i\beta = -i\beta$, which is possible if and only if $\beta = 0$ so that $z$ is real. If $z$ is real, then $z = \alpha = \bar{z}$.

**37.** $z\bar{z} = (\alpha + i\beta)(\alpha - i\beta) = \alpha^2 - (i^2\beta^2) = \alpha^2 + \beta^2 = |z|^2$

**39.** The locus of points on a circle in the complex plane centered at $z_0$ with radius $a$. If $z_0 = x_0 + iy_0$, then in $x$ and $y$ coordinates this is the circle whose equation is $(x - x_0)^2 + (y - y_0)^2 = a^2$.

**41.** Suppose that $p(z) = z^n + a_{n-1}z^{n-1} + \cdots + a_1z + a_0 = 0$. Then $\overline{z^n + a_{n-1}z^{n-1} + \cdots + a_1z + a_0} = \bar{0} = 0 = \overline{z^n} + \overline{a_{n-1}z^{n-1}} + \cdots + \overline{a_1z} + \overline{a_0} = \overline{z^n} + a_{n-1}\overline{z^{n-1}} + \cdots + a_1\bar{z} + a_0$ (since the $a_i$'s are real) $= \bar{z}^n + a_{n-1}\bar{z}^{n-1} + \cdots + a_1\bar{z} + a_0 = p(\bar{z}) = 0$. Here we have used the fact that for any integer $k$, $\overline{z^k} = \bar{z}^k$. This follows easily if we write $z$ in polar form. If $z = re^{i\theta}$, then $z^n = r^ne^{in\theta}$, $\overline{z^n} = r^ne^{-in\theta}$, $\bar{z} = re^{-i\theta}$, and $\bar{z}^n = r^ne^{-in\theta} = \overline{z^n}$.

**43.** Since $(\cos\theta + i\sin\theta)^1 = \cos 1 \cdot \theta + i \sin 1 \cdot \theta$, DeMoivre's formula holds for $n = 1$. Assume it holds for $n = k$. That is, $(\cos\theta + i\sin\theta)^k = \cos k\theta + i\sin k\theta$. Then $(\cos\theta + \sin\theta)^{k+1} = (\cos\theta + i\sin\theta)^k(\cos\theta + i\sin\theta) = (\cos k\theta + i\sin k\theta) \times (\cos\theta + i\sin\theta) = [\cos k\theta\cos\theta - \sin k\theta\sin\theta] + i[\sin k\theta\cos\theta + \cos k\theta\sin\theta] = \cos(k\theta + \theta) + i\sin(k\theta + \theta) = \cos(k+1)\theta + i\sin(k+1)\theta$, which is DeMoivre's formula for $n = k + 1$.

### Appendix 3, page A-26

**1.** $0.33333333 \times 10^0$

**3.** $-0.35 \times 10^{-4}$

**5.** $0.77777777 \times 10^0$

**7.** $0.77272727 \times 10^1$

**9.** $-0.18833333 \times 10^2$

**11.** $0.23705963 \times 10^9$

**13.** $0.83742 \times 10^{-20}$

**15.** $\epsilon_a = 0.1$, $\epsilon_r = 0.0002$

**17.** $\epsilon_a = 0.005$, $\epsilon_r = 0.04$

**19.** $\epsilon_a = 0.00333\ldots$, $\epsilon_r \approx 0.57143 \times 10^{-3}$

**21.** $\epsilon_a = 1$, $\epsilon_r \approx 0.1419144 \times 10^{-4}$

**23.** There are three different operations: (1) dividing row $i$ by $a_{ii}$; (2) multiplying row $i$ by $a_{ji}$, $j > i$, and subtracting it from row $j$; (3) doing back substitution. Operation (1) requires $\sum_{k=1}^{n} k = \frac{n(n+1)}{2}$ multiplications. Operation (2) requires

$$\sum_{k=1}^{n-1} k(k+1) = \sum_{k=1}^{n-1} k^2 + \sum_{k=1}^{n-1} k$$
$$= \frac{(n-1)n(2n-1)}{6} + \frac{(n-1)n}{2}$$
$$= \frac{n^3 - n}{3}$$

multiplications and additions. Operation (3) requires

$$\sum_{k=1}^{n-1} k = \frac{(n-1)n}{2} = \frac{n^2 - n}{2}$$

multiplications and additions. Adding these fractions together gives the desired results.

**25.** There are three different operations: (1) dividing row $i$ by $a_{ii}$; (2) multiplying row $i$ by $a_{ji}$, $j > i$ and subtracting it from row $j$; (3) keeping track of the $n$ diagonal elements and multiplying them together at the end. Operation (1) requires $\sum_{k=1}^{n-1} k = \frac{n(n-1)}{2}$ multiplications. Operation (2) requires $\sum_{k=1}^{n-1} k^2 = \frac{n(n-1)(2n-1)}{6}$ multiplications. Operation (3) requires $n - 1$ multiplications. The sum is $\frac{n-1}{6}[3n + n(2n-1) + 6] = \frac{1}{6}(n-1)(2n^2 + 2n + 6) = \frac{1}{6}(2n^3 + 4n - 6) = \frac{n^3}{3} + \frac{2}{3}n - 1$ multiplications. A similar computation yields the number of additions given in Table A.1.

**27.** 7545 microseconds $= 7.545 \times 10^{-3}$ seconds

**29.** $mqn$ multiplications and $mq(n-1)$ additions

## Appendix 4, page A-34

**1.** $x_1 = 1.6$, $x_2 = -0.800002$ (actual value is $-0.8$), $x_3 = -3.7$

**3.** $x_1 = -0.000001$, $x_2 = -2.61001$, $x_3 = 4.3$. Exact solution is $(0, -2.61, 4.3)$.

**5. a.** with pivoting: $x_1 = 5.99$, $x_2 = -2$, $x_3 = 3.99$

**b.** without pivoting: $x_1 = 6$, $x_2 = -2$, and $x_3 = 4$ (Yes, sometimes it's better to follow the simplest path. In Problem 6 pivoting gives much more accurate answers.) The relative errors with pivoting are $\frac{1}{600} = 0.0017$, 0, and $\frac{1}{400} = 0.0025$.

**7.** A solution with rounding to 3 significant figures is $x_1 = 1050$ and $x_2 = -1000$. The exact solution is $x_1 = \frac{15650}{13} \approx 1204$ and $x_2 = -\frac{15000}{13} \approx -1154$. The relative errors are $0.1279 \approx 13\%$ and $0.1334 \approx 13\%$

# Index

## Hardscrabble Books — Fiction of New England

Laurie Alberts, *Lost Daughters*

Laurie Alberts, *The Price of Land in Shelby*

Thomas Bailey Aldrich, *The Story of a Bad Boy*

Robert J. Begiebing, *The Adventures of Allegra Fullerton; Or, A Memoir of Startling and Amusing Episodes from Itinerant Life*

Anne Bernays, *Professor Romeo*

Chris Bohjalian, *Water Witches*

Dona Brown, ed., *A Tourist's New England: Travel Fiction, 1820–1920*

Joseph Bruchac, *The Waters Between: A Novel of the Dawn Land*

Joseph A. Citro, *The Gore*

Joseph A. Citro, *Guardian Angels*

Joseph A. Citro, *Shadow Child*

Sean Connolly, *A Great Place to Die*

Dorothy Canfield Fisher (Mark J. Madigan, ed.), *Seasoned Timber*

Dorothy Canfield Fisher, *Understood Betsy*

Joseph Freda, *Suburban Guerrillas*

Castle Freeman, Jr., *Judgment Hill*

Frank Gaspar, *Leaving Pico*

Ernest Hebert, *The Dogs of March*

Ernest Hebert, *Live Free or Die*

Ernest Hebert, *The Old American*

Sarah Orne Jewett (Sarah Way Sherman, ed.), *The Country of the Pointed Firs and Other Stories*

Lisa MacFarlane, ed., *This World Is Not Conclusion: Faith in Nineteenth-Century New England Fiction*

G. F. Michelsen, *Hard Bottom*

Anne Whitney Pierce, *Rain Line*

Kit Reed, *J. Eden*

Rowland E. Robinson (David Budbill, ed.), *Danvis Tales: Selected Stories*

Roxana Robinson, *Summer Light*

Rebecca Rule, *The Best Revenge: Short Stories*

R. D. Skillings, *How Many Die*

R. D. Skillings, *Where the Time Goes*

Lynn Stegner, *Pipers at the Gates of Dawn: A Triptych*

Theodore Weesner, *Novemberfest*

W. D. Wetherell, *The Wisest Man in America*

Edith Wharton (Barbara A. White, ed.), *Wharton's New England: Seven Stories and* Ethan Frome

Thomas Williams, *The Hair of Harold Roux*